W9-BIP-512

Instructor's Annotated Edition

Introductory Algebra

An Integrated Approach

Instructor's Annotated Edition

Introductory Algebra

An Integrated Approach

Preliminary Edition

Richard N. Aufmann
Palomar College

Joanne S. Lockwood
Plymouth State College

HOUGHTON MIFFLIN COMPANY Boston New York

Senior Sponsoring Editor: *Maureen O'Connor*
Senior Associate Editor: *Dawn Nuttall*
Senior Project Editor: *Chere Bemelmans*
Senior Production/Design Coordinator: *Carol Merrigan*
Senior Manufacturing Coordinator: *Marie Barnes*
Editorial Assistants: *Joy Park* and *John Brister*
Marketing Manager: *Ros Kane*

Cover designer: Harold Burch Designs, NYC.
Cover image: Tel Aviv, Israel. Superstock.
Chapter opening art by George McLean. Interior art by Network Graphics.

Printed in the U.S.A.

Library of Congress Number: 98-71977

ISBN Numbers:
Student Text: 0-395-88830-1
Instructor's Annotated Edition: 0-395-88864-6

23456789-WC-02 01 00 99

Contents

Preface

Introductory Algebra: An Integrated Approach is the second in a new series of three texts, the focus of which is to present mathematics as a cohesive subject and not one that is fragmented into many topics. Throughout each text there are themes of number sense, logic, geometry, statistics, probability, algebra, trigonometry, and discrete mathematics. The themes are woven throughout each text at increasingly more sophisticated levels, thereby providing students with an experience with each theme at a level that is appropriate for their particular course. The richness and diversity of these themes are demonstrated with applications taken from more than 100 disciplines.

We have paid special attention to the standards suggested by AMATYC and have made a serious attempt to incorporate those standards in each text. Problem solving, critical analysis, function concept, connecting mathematics to other disciplines through applications, multiple representations of concepts, and the appropriate use of technology are all integrated within each text. Our goal was to provide students with a variety of analytical tools that will make them more effective quantitative thinkers and problem solvers.

Instructional Features

Integrated Structure

The traditional approach to teaching mathematics has been to segment and divide mathematics into several courses such as algebra, geometry, statistics, and trigonometry. This approach makes it difficult for students to see mathematics as a unified subject of interacting themes.

By contrast, *Introductory Algebra: An Integrated Approach* is designed to reflect the fact that mathematics contains interrelated concepts. Each text in the series explores various themes including number sense, statistics, probability, algebra, logic, geometry, trigonometry, and discrete mathematics. Our approach is to weave these themes into each text at increasing levels of sophistication. There are significant advantages to this approach.

First, students can see that mathematics has a vast array of tools that can be used to solve meaningful problems. By integrating these themes in each text, students can explore multiple approaches to solving a problem. Modeling, analytic representation, and verbal representations of problems and their solutions are encouraged.

A second advantage of this approach applies to students who discontinue their formal math training after completing any one of these texts. Because many themes are integrated into each text, students will have acquired an understanding of principles that will enable them to select the ever increasing career and educational options for which these principles are a prerequisite.

Interactive Style

Introductory Algebra: An Integrated Approach uses an interactive style that encourages students to be active learners. Each example in the text is followed by a You-Try-It, which is a problem similar to the example, that the student should solve. To provide immediate feedback to the student, a complete worked-out solution to the You-Try-It is provided in the Solutions Section at the end of the book.

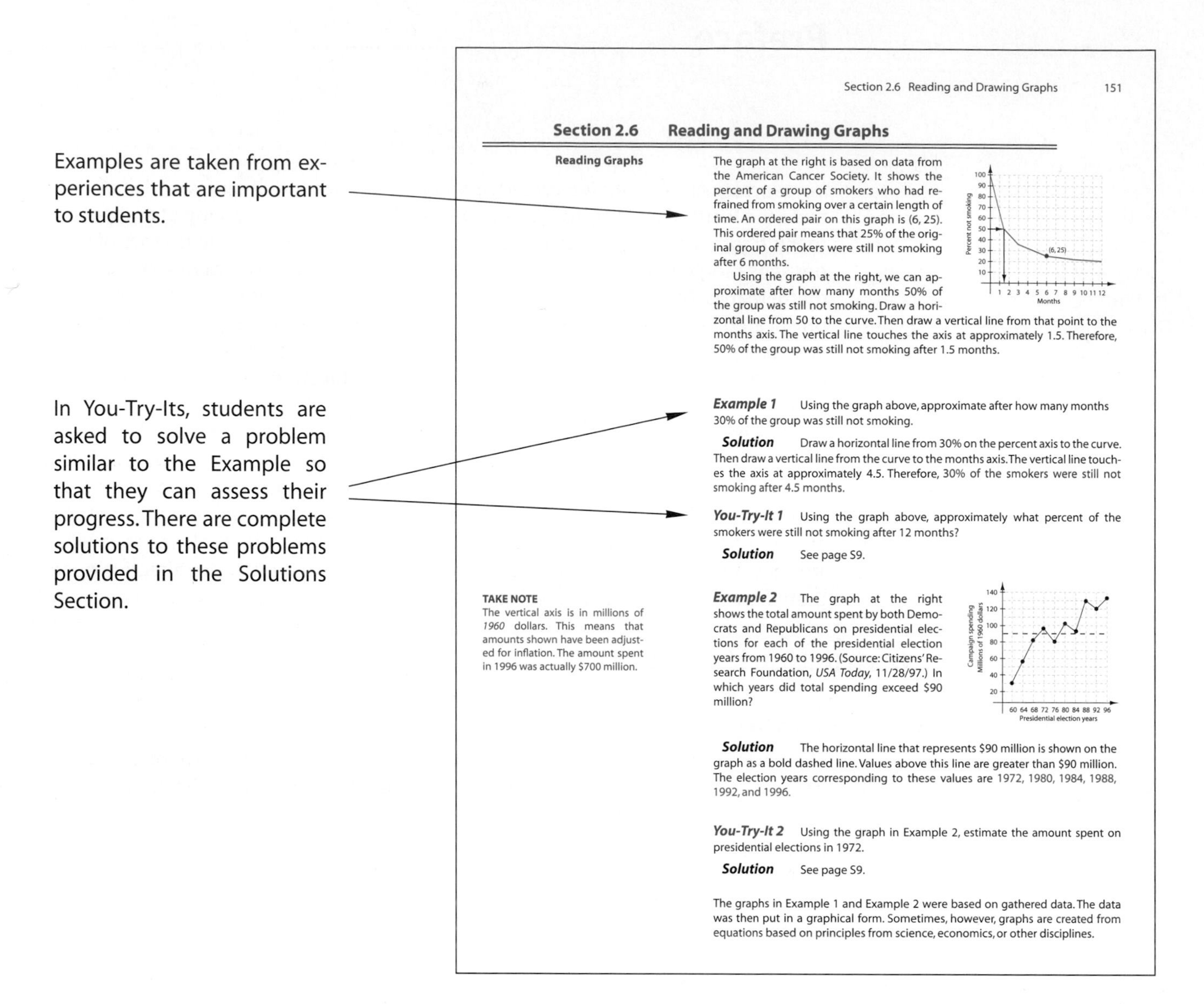

Section 2.6 Reading and Drawing Graphs 151

Section 2.6 Reading and Drawing Graphs

Reading Graphs

The graph at the right is based on data from the American Cancer Society. It shows the percent of a group of smokers who had refrained from smoking over a certain length of time. An ordered pair on this graph is (6, 25). This ordered pair means that 25% of the original group of smokers were still not smoking after 6 months.

Using the graph at the right, we can approximate after how many months 50% of the group was still not smoking. Draw a horizontal line from 50 to the curve. Then draw a vertical line from that point to the months axis. The vertical line touches the axis at approximately 1.5. Therefore, 50% of the group was still not smoking after 1.5 months.

Example 1 Using the graph above, approximate after how many months 30% of the group was still not smoking.

Solution Draw a horizontal line from 30% on the percent axis to the curve. Then draw a vertical line from the curve to the months axis. The vertical line touches the axis at approximately 4.5. Therefore, 30% of the smokers were still not smoking after 4.5 months.

You-Try-It 1 Using the graph above, approximately what percent of the smokers were still not smoking after 12 months?

Solution See page S9.

TAKE NOTE
The vertical axis is in millions of *1960* dollars. This means that amounts shown have been adjusted for inflation. The amount spent in 1996 was actually $700 million.

Example 2 The graph at the right shows the total amount spent by both Democrats and Republicans on presidential elections for each of the presidential election years from 1960 to 1996. (Source: Citizens' Research Foundation, *USA Today*, 11/28/97.) In which years did total spending exceed $90 million?

Solution The horizontal line that represents $90 million is shown on the graph as a bold dashed line. Values above this line are greater than $90 million. The election years corresponding to these values are 1972, 1980, 1984, 1988, 1992, and 1996.

You-Try-It 2 Using the graph in Example 2, estimate the amount spent on presidential elections in 1972.

Solution See page S9.

The graphs in Example 1 and Example 2 were based on gathered data. The data was then put in a graphical form. Sometimes, however, graphs are created from equations based on principles from science, economics, or other disciplines.

Another way we encourage students to interact with the text is by posing questions to students about what they are reading. To ensure that important points are not missed, the answers to these questions are given as footnotes on the same page as the question.

Multiple Representations of Concepts

A major focus of this text is to present multiple representations of concepts and to link concepts to applications. The page on the right is one illustration of how this is incorporated into this text. In Example 4, the student is given an application problem whose solution requires solving a radical equation. An algebraic solution is accompanied by a graphical check. Besides these two representations, the **Question** following the graph encourages students to provide a written or verbal interpretation of the graph. In other instances, we encourage students to give algebraic checks to graphical solutions (see page 294).

Equation solving skills are connected to applications to demonstrate how a concept is used. Students are shown an algebraic solution along with a graphical check. An organized problem-solving procedure accompanies application problems.

The **Question** feature is another way in which we ask students to be active learners. To ensure that important points are not missed, the answers to questions are given in footnotes on the same page as the question.

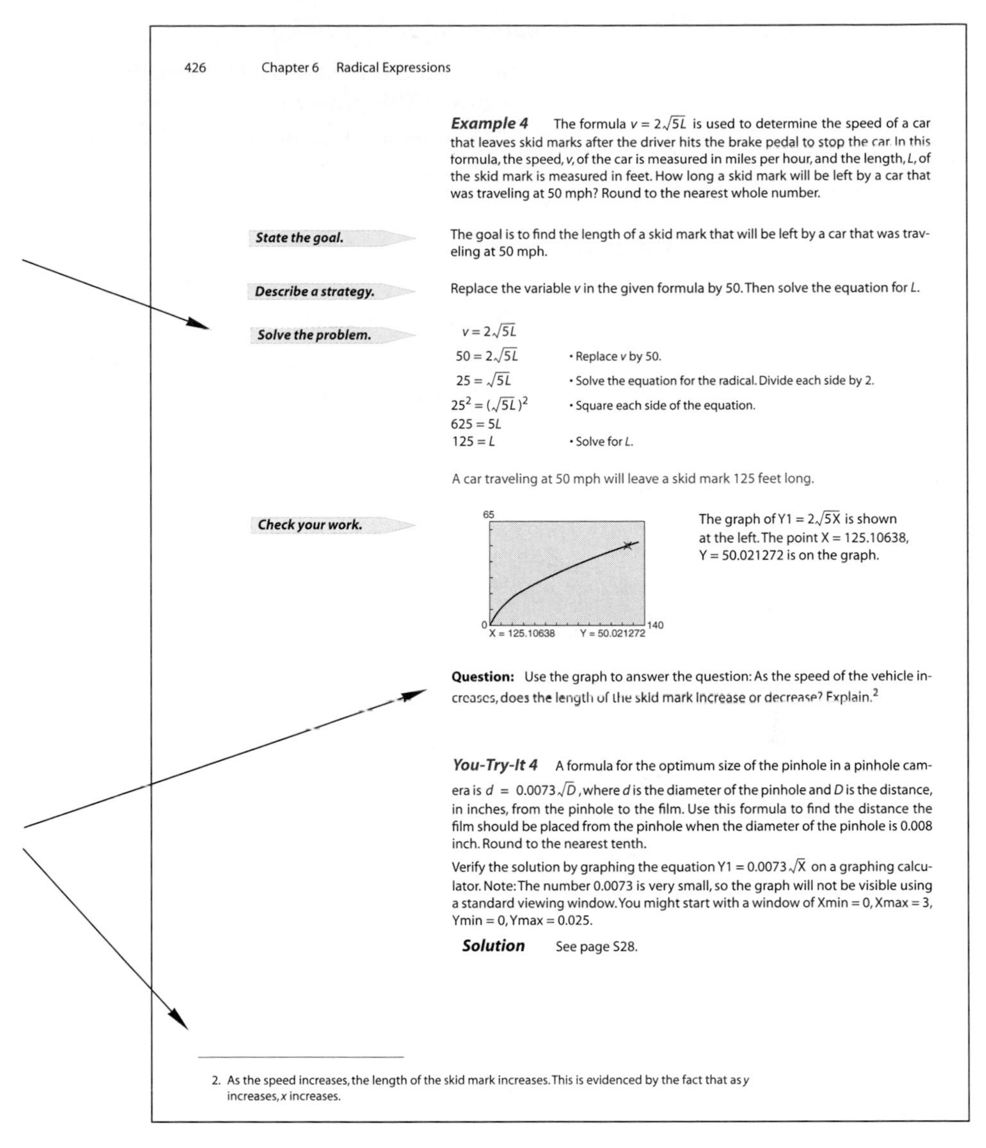

426 Chapter 6 Radical Expressions

Example 4 The formula $v = 2\sqrt{5L}$ is used to determine the speed of a car that leaves skid marks after the driver hits the brake pedal to stop the car. In this formula, the speed, v, of the car is measured in miles per hour, and the length, L, of the skid mark is measured in feet. How long a skid mark will be left by a car that was traveling at 50 mph? Round to the nearest whole number.

State the goal. The goal is to find the length of a skid mark that will be left by a car that was traveling at 50 mph.

Describe a strategy. Replace the variable v in the given formula by 50. Then solve the equation for L.

Solve the problem.

$v = 2\sqrt{5L}$

$50 = 2\sqrt{5L}$ • Replace v by 50.

$25 = \sqrt{5L}$ • Solve the equation for the radical. Divide each side by 2.

$25^2 = (\sqrt{5L})^2$ • Square each side of the equation.

$625 = 5L$

$125 = L$ • Solve for L.

A car traveling at 50 mph will leave a skid mark 125 feet long.

Check your work. The graph of $Y1 = 2\sqrt{5X}$ is shown at the left. The point X = 125.10638, Y = 50.021272 is on the graph.

Question: Use the graph to answer the question: As the speed of the vehicle increases, does the length of the skid mark increase or decrease? Explain.[2]

You-Try-It 4 A formula for the optimum size of the pinhole in a pinhole camera is $d = 0.0073\sqrt{D}$, where d is the diameter of the pinhole and D is the distance, in inches, from the pinhole to the film. Use this formula to find the distance the film should be placed from the pinhole when the diameter of the pinhole is 0.008 inch. Round to the nearest tenth.

Verify the solution by graphing the equation $Y1 = 0.0073\sqrt{X}$ on a graphing calculator. Note: The number 0.0073 is very small, so the graph will not be visible using a standard viewing window. You might start with a window of Xmin = 0, Xmax = 3, Ymin = 0, Ymax = 0.025.

Solution See page S28.

2. As the speed increases, the length of the skid mark increases. This is evidenced by the fact that as y increases, x increases.

Margin Notes

Point of Interest notes are interspersed throughout the text. These notes are interesting sidelights of the topic being discussed.

Take Note alerts students that a procedure may be particularly involved or reminds students that certain checks of their work should be performed.

Instructor Notes are printed only in the Instructor's Annotated Edition. These notes are suggestions for presenting a lesson or related material that can be used in class.

Suggested Activities are printed only in the Instructor's Annotated Edition. They are suggestions for group or class projects. They may also reference the Instructor's Resource Manual where longer and more involved Activities are described.

Chapter Summaries

At the end of each chapter is a Chapter Summary that includes the Definitions and Procedures that were covered in the chapter. These Chapter Summaries provide one focus for the student when preparing for a test.

Computer Tutor

This state-of-the-art Tutor is a networkable, interactive, algorithmically-driven software package. This powerful ancillary features full-color graphics, a glossary, extensive hints, animated solution steps, and a comprehensive class management system. The content is written by the authors and is in the same voice as the text.

Exercises

End-of-Section Exercises

These exercises are carefully developed to ensure that students can apply the concepts in the section to a variety of problem situations. We have tried to balance concept and practice to ensure that students have a mastery of both.

Explorations

The End-of-Section Exercises are followed by Explorations, which require that a student investigate a certain concept in depth or detail.

Chapter Review Exercises

Chapter Review Exercises are found at the end of each chapter. These exercises are selected to help the student integrate all of the topics presented in the chapter. The answers to all review exercises are given in the Answer Section.

Cumulative Review Exercises

Cumulative Review Exercises, which appear at the end of each chapter (beginning with Chapter 2), help the student maintain skills learned in previous chapters. The answers to all Cumulative Review Exercises are given in the Answer Section.

SUPPLEMENTS FOR THE INSTRUCTOR

Instructor's Annotated Edition

The Instructor's Annotated Edition is an exact replica of the student text, except that answers to the exercises are given in the text. Also, *Instructor Notes* in the margins offer suggestions for presenting the material in a lesson, and *Suggested Activity* notes offer additional group or class activities.

Instructor's Resource Manual with Test Bank

The Instructor's Resource Manual contains suggestions on course management, an Integrated Topics chart that shows where and how topics are integrated throughout the text, descriptions of those *Suggested Activities* that are too involved to fit in the margin of the text, and transparency blackline masters of selected art pieces from the text.

The Test Bank is a printout of all items in the Computerized Test Generator. Instructors who do not have access to a computer can use the Test Bank to select items to include on a test being prepared by hand. All items are free response and answers are provided at the back of the Test Bank.

Computerized Test Generator

The Computerized Test Generator's database contains more than 1000 test items. The Test Generator is designed to provide an unlimited number of tests for each chapter, cumulative chapter tests, and a final exam. The program also provides **on-line testing** and **gradebook** functions. It is available for Windows.

Instructor's Solutions Manual

The Instructor's Solutions Manual contains the complete worked-out solutions to all exercises in the text.

SUPPLEMENTS FOR THE STUDENT

Student Solutions Manual

The Student Solutions Manual contains the complete worked-out solutions to the odd-numbered exercises in the text.

Computer Tutor

The content of this interactive state-of-the-art tutorial software was written by the authors and is in the same voice as the text. The Tutor supports every topic in the text. Problems are algorithmically generated, solution steps are animated, lessons and problems are presented in a colorful, lively manner, and a management system tracks and reports student performance.

The Computer Tutor can be used in several ways: (1) to cover material the student missed because of absence; (2) to reinforce instruction on a concept that the student has not yet mastered; (3) to review material in preparation for exams.

This networkable Tutor is available for Windows and Macintosh systems. The Tutor is free to any school on adoption of this text; however, students can purchase a copy of the Tutor on floppy disk for home use.

ACKNOWLEDGMENTS

The authors would like to thank the following people who have reviewed this manuscript and provided many valuable suggestions.

Victor M. Cornell, *Mesa Community College,* AZ
Annalisa Ebanks, *Jefferson Community College,* KY
Michael A. Jones
Frank Pecchioni, *Jefferson Community College,* KY
James Ryan, *Madera Community College Center,* CA

The authors would like to give special thanks to the following people:

Lois Linnan, *Clarion University,* PA, for authoring the Test Bank as well as some of the Suggested Activities; and **Emily Keaton** for authoring the Instructor and Student Solutions Manuals.

Instructor's Annotated Edition

Introductory Algebra

An Integrated Approach

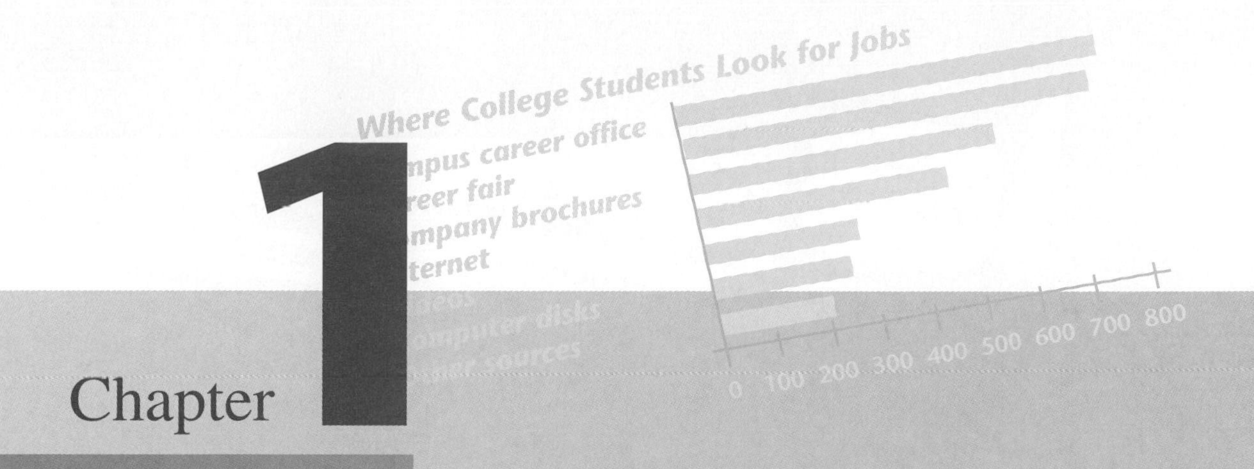

Chapter 1

Analyzing Real Number Data

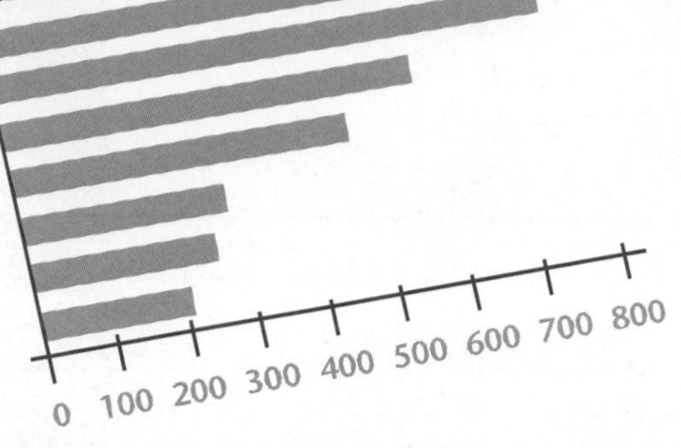

1
250
450
200
550
750

Section 1.1 Introduction to Problem Solving

Problem Solving

POINT OF INTEREST
George Polya taught at Stanford University from 1942 until his retirement. While at Stanford, he published *How to Solve It* in 1945. In this book, Polya outlines the problem-solving process discussed here. This process can be and is used to solve problems not only in mathematics but in many other disciplines as well.

An ewok was visiting an island on which there lived knights, who only make true statements, and knaves, who only make false statements. The ewok needed to find a knight to be a trusty guide. While walking along the shore, the ewok came upon three natives, named Arthur, Bernard, and Charles. The ewok first asked Arthur, "Are Bernard and Charles both knights?"

Arthur replied, "Yes."
The man then asked, "Is Bernard a knight?"
To his great surprise, Arthur answered, "No."

Who is a knight and who is a knave?[1]

Although the problem above is not strictly a math problem, solving a problem like this one requires some of the same problem-solving strategies that are used for math problems. One form of these strategies was stated by George Polya (1887 – 1985) as a four-step process.

1. Understand the problem.
2. Devise a strategy to solve the problem.
3. Execute the strategy and state the answer.
4. Review your solution.

State the goal.

Understand the Problem.
This part of problem solving is often overlooked. You must have a clear understanding of the problem.

- Read the problem carefully and try to determine the goal.
- Make sure you understand all the terms or words used in the problem.
- Make a list of known facts.
- Make a list of information that if known would help you solve the problem. Remember that it may be necessary to look up information you do not know in another book, an encyclopedia, the library, or perhaps on the Internet.

Describe a strategy.

Devise a Strategy to Solve the Problem.
Successful problem solvers use a variety of techniques when they attempt to solve a problem.

- Draw a diagram.
- Work backwards.
- Guess and check.
- Solve an easier problem.
- Look for a pattern.
- Make a table or chart.
- Write an equation.

Solve the problem.

Solve the Problem.

- Work carefully.
- Keep accurate and neat records of your attempts.
- When you have completed the solution, state the answer carefully.

Check your work.

Review the Solution.
Once you have found a solution, check the solution against the known facts and check for possible errors. Be sure the solution is consistent with the facts of the problem. Another important part of this review process is to ask if your solution can be used to solve other types of problems.

Here is an example of using this process.

1. This problem is based on a problem from Raymond Smullyan. The answer is at the end of this section.

Find the sum of the first 10,000 natural numbers.

State the goal.

Do you understand the meaning of all terms in the problem? For instance, a natural number is a number from the set {1, 2, 3, 4, 5, 6, 7, 8, 9, 10, 11, ...}, where the three dots indicate that there is no end to the natural numbers. What about the word *sum*? Do you know its meaning?[2] If you do, you will know that this problem is asking you to find

$$1 + 2 + 3 + \cdots + 9998 + 9999 + 10{,}000$$

Describe a strategy.

One strategy for this problem would be to use a calculator and just start adding the numbers. However, this plan may lead to mistakes because of all the numbers to enter and then add. Even if we could enter numbers accurately and quickly, say one number every two seconds, it would take over five hours to get the answer.

Instead, we will try to find a different strategy and try to solve an easier problem first. The idea is to see if solving the easier problem will lead to a strategy for solving the original problem.

Solve the problem.

Our easier problem will be to find the sum of the first 10 natural numbers. Note that when the natural numbers are paired, as shown below, each pair has the same sum.

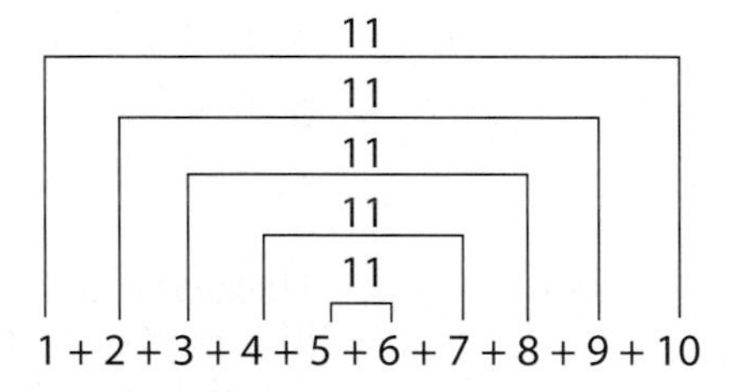

There are 5 pairs $\left(\frac{10}{2} = 5\right)$ whose sum is 11.

SUGGESTED ACTIVITY
Part of the review process is to ask whether the strategy can be used to solve other problems.
Ask students if the strategy will work for $47 + \cdots + 86$.
What about $28 + \cdots + 56$?
If not, is there a modification that will work?

Because there are 5 pairs whose sum is 11, the sum of the first 10 natural numbers is the product 11 times 5.

Sum of each pair
Number of pairs

$$1 + 2 + 3 + 4 + 5 + 6 + 7 + 8 + 9 + 10 = 11 \cdot 5 = 55$$

Using the easier problem as a model, a strategy we can use to find the sum of the first 10,000 natural numbers is to pair the numbers, find the sum of one pair, and then multiply that sum by the number of pairs.

Extending the pattern for the easier problem to the original problem, we have

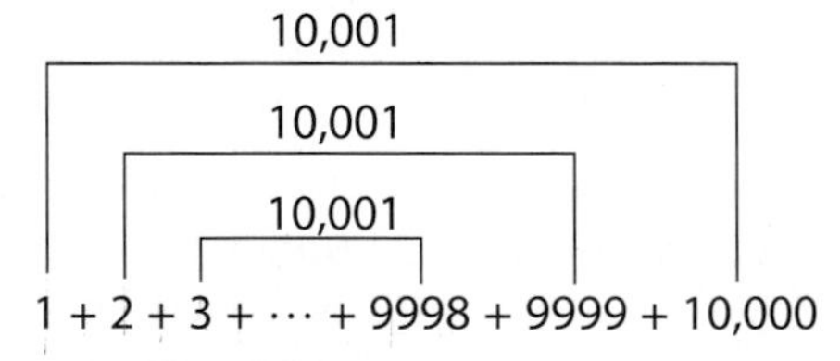

There are 5000 pairs $\left(\frac{10{,}000}{2} = 5000\right)$ whose sum is 10,001.

Thus,

$$1 + 2 + 3 + \cdots + 9998 + 9999 + 10{,}000 = 10{,}001 \cdot 5000 = 50{,}005{,}000$$

Check your work.

By repeating the calculations, you can verify that the solution is correct and is consistent with the problem we were given to solve.

2. To find a sum means to add all the numbers.

Example 1 What is the units digit of 7^{54}?

Solution

State the goal.

The goal is to find the units digit of 7^{54}. To achieve this goal, recall the place values of a number. For instance,

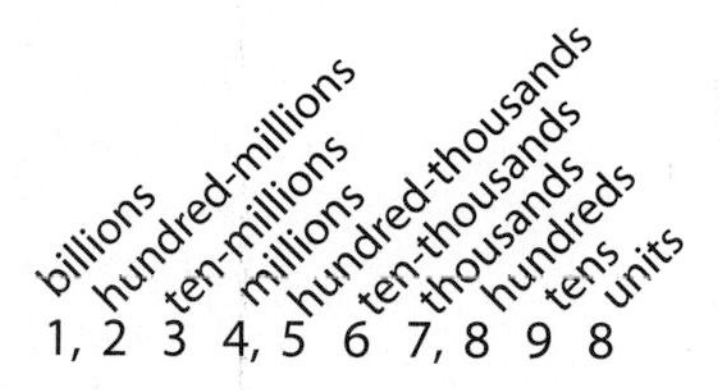

Therefore, the units digit is the rightmost digit. Also recall that

$$7^{54} = \overbrace{7 \cdot 7 \cdot 7 \cdot \cdots \cdot 7}^{54 \text{ times}}$$

For example, $7^1 = 7$, $7^2 = 7 \cdot 7 = 49$, and $7^3 = 7 \cdot 7 \cdot 7 = 343$. The expression 7^{54} is the fifty-fourth power of 7. The number 54 is called an *exponent.*

Describe a strategy.

One possible way to solve this problem that may occur to you is just to use a calculator to evaluate the expression. Trying this approach gives

$$7^{54} = 4.3181146\text{E}45$$

This answer is in *scientific notation* (studied more thoroughly later in the text) and means that the number is 46 digits long and only the first 10 digits are shown. Because our goal is to find the last digit, a calculator approach will not work.

Instead of a calculator approach, we will try to find a pattern for the units digit for small powers of 7.

Solve the problem.

Consider the table below:

$7^1 = \mathbf{7}$	$7^2 = 4\mathbf{9}$	$7^3 = 34\mathbf{3}$	$7^4 = 240\mathbf{1}$
$7^5 = 16{,}80\mathbf{7}$	$7^6 = 117{,}64\mathbf{9}$	$7^7 = 823{,}54\mathbf{3}$	$7^8 = 5{,}764{,}80\mathbf{1}$
$7^9 = 40{,}353{,}60\mathbf{7}$	$7^{10} = 282{,}475{,}24\mathbf{9}$	$7^{11} = 1{,}977{,}326{,}74\mathbf{3}$	

Note that the units digit repeats in the cycle 7, 9, 3, 1, 7, 9, 3, 1, 7 . . . . With this observation, we could solve the problem by writing out this repeating cycle of digits until we came to the 54th number. That would be the units digit of 7^{54}. However, a faster approach is to use division.

Recall that division separates objects into equal groups. In this case, the equal groups are the 4 digits in a cycle. Divide 4 into 54, the exponent.

$$\begin{array}{r} 13 \\ 4\overline{)54} \\ -4 \\ \hline 14 \\ -12 \\ \hline 2 \end{array}$$

← The remainder means that the units digit of 7^{54} is the second number in the cycle 7, 9, 3, 1.

Therefore, the units digit of 7^{54} is 9.

Observe that for 7^4 and 7^8, the remainder is zero when the exponent is divided by 4. Therefore, the units digit is 1 when the remainder is zero.

Check your work.

You could check this result by repeating 7, 9, 3, 1, 7, 9, 3, 1, . . . until the 54th number is reached.

You-Try-It 1 Determine the units digit of 5^{92}.

Solution See page S1.

Inductive Reasoning

Looking for patterns is one of the techniques used in *inductive reasoning.* Suppose, beginning in January, you save $25 each month. The total amount you have saved at the end of each month can be described in a list of numbers.

25	50	75	100	125	150	175	
Jan	Feb	Mar	Apr	May	June	July	...

The list of numbers that indicates your total savings is an *ordered* list of numbers called a **sequence.** Each of the numbers in a sequence is called a **term** of the sequence. The list is ordered because the position of a number in the list indicates the month in which that total amount has been saved. For example, the 7th term of the sequence (indicating July) is 175. This number means that a total of $175 has been saved by the end of the 7th month.

Question: What is the fourth term of the sequence?[3]

Now consider a person who has a different savings plan. The total amount saved by this person for the first seven months is given by the sequence

20, 35, 50, 65, 80, 95, 110, . . .

Question: Assuming this person continues to save in the same manner, what will be the total amount saved in the eighth month?[4]

The process you used to discover the next number in the above sequence is *inductive reasoning.* **Inductive reasoning** involves making generalizations from specific examples; in other words, we reach a conclusion by making observations about particular facts or cases. In the case of the above sequence, the person saved $15 per month after the first month.

Example 2 What are the next two letters of the sequence A, B, E, F, I, J, . . .?

Solution The pattern of this sequence is

A, B, C, D, E, F, G, H, I, J, . . .

To continue the pattern, we skip K, L. The next two letters of the sequence are M, N.

You-Try-It 2 What is the next term of the sequence

ban, ben, bin, bon, . . .

Solution See page S1.

TAKE NOTE
Recall that a proper fraction is one in which the numerator is greater than 0 but less than the denominator.

Example 3 Using a calculator, determine the decimal representation of several proper fractions that have a denominator of 99. For instance, you may use $\frac{8}{99}$, $\frac{23}{99}$, and $\frac{75}{99}$. Now use inductive reasoning to explain the pattern and use your reasoning to find the decimal representation of $\frac{53}{99}$ without a calculator.

Solution $\frac{8}{99} = 0.080808\ldots$; $\frac{23}{99} = 0.232323\ldots$; $\frac{75}{99} = 0.757575\ldots$.

Using inductive reasoning, the decimal representation of a proper fraction with a denominator of 99 is a repeating decimal where the digits of the numerator repeat. Using this reasoning, $\frac{53}{99} = 0.535353\ldots$.

SUGGESTED ACTIVITY
Remind students that one of the characteristics of a good problem solver is to try to extend a solution of a problem. Ask students whether the pattern for proper fractions works for improper fractions with a denominator of 99, for instance $\frac{137}{99}$.

---

3. 100
4. $125

You-Try-It 3 Using a calculator, find $11 \cdot 23$, $11 \cdot 36$, $11 \cdot 58$, and $11 \cdot 67$. Now use inductive reasoning to explain the pattern and use your reasoning to find the product of $11 \cdot 78$.

Solution See page S1.

Using inductive reasoning does not always guarantee that you will form a correct conclusion. No matter how many cases you examine that seem to support your conclusion, you cannot state your conclusion with absolute certainty unless you can show that the conclusion is true in *all* cases.

Consider the circles below. First two points, then three, then four, and then five points are chosen on a circle and all possible chords (line segments starting and ending on the circumference) are drawn between the points. Then the number of resulting regions is counted.

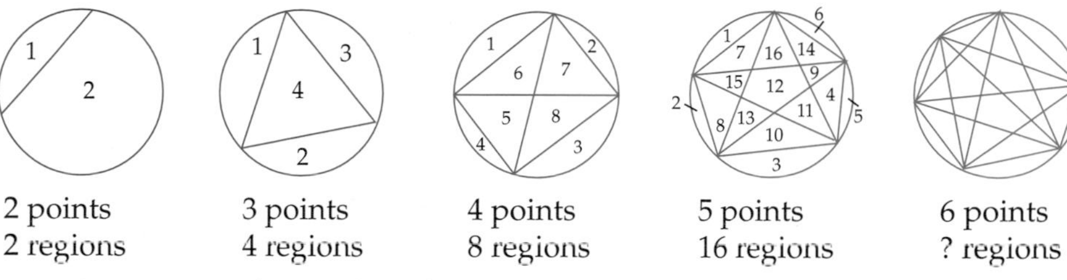

2 points 2 regions | 3 points 4 regions | 4 points 8 regions | 5 points 16 regions | 6 points ? regions

It appears as the number of points on the circle increases by one, the number of regions doubles from the previous case. However, by actually counting the regions for 6 points, you will find that there are 31 regions and *not* 32 as might be expected by inductive reasoning. This illustrates that showing something is true for a few cases does not make it true for all cases.

Deductive Reasoning

Sherlock Holmes, the fictional detective, was famous for using *deductive reasoning,* as illustrated in this excerpt from *The Adventure of the Priory School* by Arthur Conan Doyle.

> "This track, as you perceive, was made by a rider who was going from the direction of the school."
> "Or towards it?"
> "No, no, my dear Watson. The more deeply sunk impression is, of course, the hind wheel, upon which the weight rests. You perceive several places where it has passed across and obliterated the more shallow mark of the front one. It was undoubtedly heading away from the school."

Deductive reasoning involves drawing a conclusion which is based on given facts. Suppose that during the last week of your math class, your instructor tells you that if you receive an A on the final exam, you will earn an A in the course. When the final exam grades are posted, you learn that you received an A on the final exam. You can then assume that you will earn an A in the course.

Deductive reasoning is frequently used in reaching a conclusion from a sequence of known facts. As an example, consider the following: "If Amy studies computer programming, then she will get a well-paying job. If she gets a well-paying job, then she will be able to afford a new Corvette." From these statements, we can conclude that "If Amy studies computer programming, she will be able to afford a new Corvette." The diagram for this is shown below.

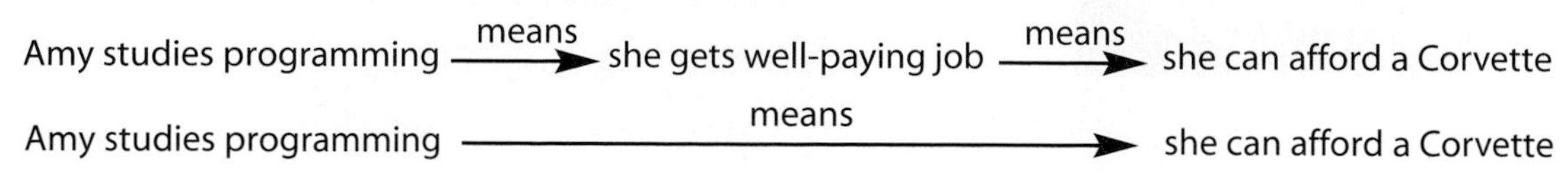

Example 4 below is another instance of reaching a conclusion from a sequence of known abstract facts.

Example 4 Given that ‡ ‡ = Σ Σ Σ and Σ Σ Σ = Δ, how many Δ's equal ‡ ‡ ‡ ‡?

Solution We are given that ‡ ‡ = Σ Σ Σ and Σ Σ Σ = Δ.

These are the same.

‡ ‡ = Σ Σ Σ   and   Σ Σ Σ = Δ

Therefore, these are equal.

‡ ‡ = Δ

Since 2 ‡'s equal 1 Δ, 4 ‡'s equal 2 Δ's. That is, ‡ ‡ ‡ ‡ = Δ Δ.

You-Try-It 4 Given that = ♣♣, and ♣♣♣♣ = ♦♦. Then ♦♦♦♦ equal how many 's?

Solution See page S1.

Example 5 For each of the problems below, determine whether inductive or deductive reasoning is being used.

a. Therese knows that any number that ends with 0, 2, 4, 6, or 8 is an even number. Since 35,716 ends in a 6, Therese concludes that 35,716 is an even number.

b. Kevin experiments with the following additions.

$$3 + 9 = 12,\ 11 + 15 = 26,\ 3 + 37 = 40, \text{ and } 27 + 55 = 82$$

One the basis of these examples, Kevin concludes that the sum of two odd numbers is an even number.

c. Winneta studies the sequence of numbers 1, 1, 2, 3, 5, 8, 13, 21, 34 and concludes that the next number in the sequence would be 55.

Solution

a. Theresa is drawing a conclusion based on known facts. She is using deductive reasoning.

b. Kevin is making a generalization based on some examples. He is using inductive reasoning.

c. Winneta has observed that, after the first two numbers, a number in the sequence is the sum of the two preceding numbers. She is using inductive reasoning because her observations are based on some examples.

You-Try-It 5 For each of the problems below, determine if inductive or deductive reasoning is being used.

a. Sara notices that in the sequence of numbers 3, 12, 48, 192, . . ., the next number in the sequence is 4 times the previous number. From this, Sara concludes that the sixth number in this sequence is 3072.

b. Shawn knows that a pentagon is a closed figure with five sides. He is told that the Pentagon building in Washington, D.C., is a pentagon. Shawn concludes that the Pentagon has five sides.

Solution See page S1.

Deductive reasoning, along with a chart, is used to solve problems like the following.

Four neighbors, Chris, Dana, Leslie, and Pat, each plant a different vegetable (beans, cucumbers, squash, or tomatoes) in their gardens. From the following statements, determine which vegetable each neighbor plants.

1. Pat's garden is bigger than the one that has tomatoes but smaller than the one that has cucumbers.
2. Chris, who planted the largest garden, didn't plant the beans.
3. The person who planted the beans has a garden the same size as Pat's.
4. Dana and the person who planted tomatoes also have flower gardens.

From statement 1, Pat did not plant tomatoes or cucumbers. In the chart below, write X1 for these conditions.

From statement 2, Chris didn't plant the beans. Chris planted the largest garden, and we know from statement 1 that the tomatoes were planted in a garden smaller than Pat's; therefore, Chris did not plant the tomatoes. Write X2 for these conditions.

From statement 3, Pat did not plant the beans. Write X3 for this condition. There are now X's for three of the four vegetables in Pat's row; therefore, Pat must have planted the squash. Place a √ in that box. Since Pat planted the squash, none of the other three people planted the squash. Write X3 for these conditions. There are now X's for three of the four vegetables in Chris's row; therefore, Chris must have planted the cucumbers. Place a √ in that box. Since Chris planted the cucumbers, neither Dana nor Leslie planted the cucumbers. Write X3 for these conditions.

From statement 4, Dana did not plant the tomatoes. Write X4 for this condition. Since there are 3 X's in the tomato column, Leslie must have planted the tomatoes. Place a √ in that box. Now Leslie cannot have planted the beans. Write X4 in that box. Since there are 3 X's in the beans column, Dana must have planted the beans. Place a √ in that box.

	Beans	Cucumbers	Squash	Tomatoes
Chris	X2	√	X3	X2
Dana	√	X3	X3	X4
Leslie	X4	X3	X3	√
Pat	X3	X1	√	X1

Chris planted the cucumbers, Dana the beans, Leslie the tomatoes, and Pat the squash.

Here is the answer to the problem posed at the beginning of this section. The only way Arthur's answers make sense is that he is lying. Therefore, Arthur is a knave. This means that Bernard is a knight (see the second question). From Arthur's answer to the first question and the fact that Bernard is a knight, we conclude that Charles must be a knave.

1.1 EXERCISES

Topics for Discussion

1. Discuss some of the techniques of good problem solvers.
Use Polya's suggestions as a guideline.

2. What is inductive reasoning?
Reasoning from specific cases to a general conclusion

3. What is deductive reasoning?
Drawing conclusions based on known facts

4. What is a sequence?
An ordered list

Problem Solving

5. How many times can 25 be subtracted from 1000?
Once

6. What is the units digit of 2^{521}?
2

7. What is the 96th digit in the decimal equivalent of $\frac{1}{7}$?

7

8. A perfect number is one for which the sum of the proper divisors is equal to the number. (The proper divisors are the ones that are less than the number and divide evenly into the number.) For instance, 496 is a perfect number. The proper divisors of 496 are 1, 2, 4, 8, 16, 31, 62, 124, and 248. The sum of the divisors is $1 + 2 + 4 + 8 + 16 + 31 + 62 + 124 + 248 = 496$. Find a perfect number between 20 and 30.
28

9. A certain type of bacteria doubles every minute. If one of these bacteria is placed in a jar at 1:00 P.M. and starts doubling, the jar will be full in one hour. At what time was the jar half-full?
1:59 P.M.

10. Suppose you have a balance scale and 8 coins. One of the coins is counterfeit and weighs slightly more than the other 7 coins. Explain how you can find the counterfeit coin in two weighings.
Complete solution in solutions manual. However, to start, place three coins on each pan of the balance scale.

11. Suppose you have a balance scale and 13 coins. One of the coins is counterfeit and weighs slightly more than the other 12 coins. Explain how you can find the counterfeit coin in three weighings.
Complete solution in solutions manual. However, to start, place four coins on each pan of the balance scale.

12. The number of ways 2 people, say Maria and Michael, can stand in a line are Maria then Michael or Michael then Maria. How many ways can 3 people stand in a line? How many ways can 4 people stand in a line? How many ways can 5 people stand in a line? Make a conjecture as to the number of ways 6 people can stand in a line.
6, 24, 120, 720

Inductive Reasoning

For Exercises 13 to 20, use inductive reasoning to predict the next term of the sequence.

13. 2, 4, 6, 8, 10, ?
12

14. 1, 3, 5, 7, 9, ?
11

15. 1, 4, 9, 16, 25, ?
36

16. 2, 4, 8, 16, 32, ?
64

17. 1, 3, 6, 10, 15, ?
21

18. 3, 7, 12, 18, 25, ?
33

19. a, z, b, y, c, x, ?
d

20. a, b, d, e, g, h, ?
j

21. Draw the next figure in the sequence: 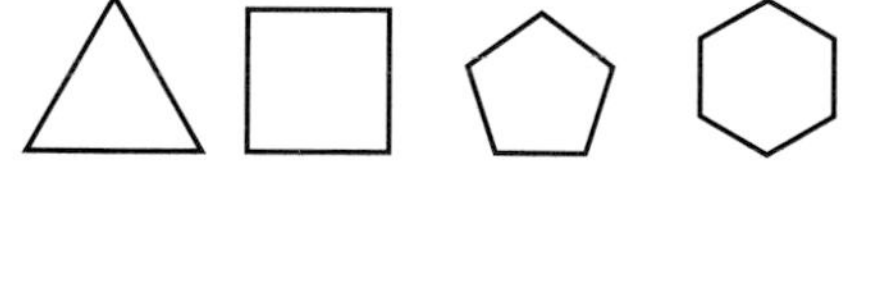

22. Draw the next figure in the sequence: 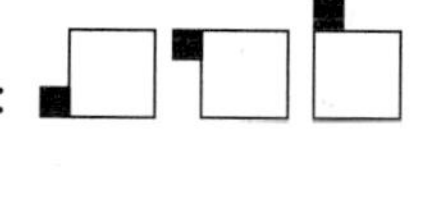

23. Draw the next figure in the sequence: |• ||• ||•• |||•• |||•••
||||•••

24. Draw the next figure in the sequence: 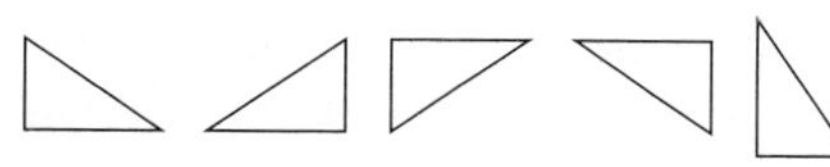

25. Take the number 7654 and reverse the digits to form the number 4567. Now subtract the smaller number from the larger one. (7654 - 4567 = 3087). Try this for other four-digit numbers whose digits are in descending order (largest to smallest). Make a conjecture from your observations.
The difference is always 3087.

Deductive Reasoning

26. The year 1998 was unusual in at least one respect: there were Friday the 13ths in two consecutive months. What were the months? (You should be able to do this problem without looking at a calendar for 1998.)
February and March

27. Given that ‡ ‡ = • • • and • • • = ¬, then ‡ ‡ ‡ ‡ = how many ¬'s?
2

28. Given that ÓÓÓ = ΩΩ and ¤ = ΩΩ, then ¤¤ = how many Ó's?
6

29. If ΔΔΔΔ = • • •, and • • = □□□□, and □ □ □ = + + + +, then ΔΔ = how many +'s?
4

30. If 🍎🍎🍎 = ♣♣♣♣♣♣, and ♣♣♣♣ = ◆ ◆, and ◆ ◆ ◆ ◆ = ∇∇∇∇∇, then 🍎🍎🍎🍎 = how many ∇'s?
5

31. If ❊❊❊❊ = ✈✈✈✈✈✈✈✈, and ✈✈ = ★★★★, and ★★ = ✂✂✂, then ✂✂✂✂✂✂ = how many ❊'s?
1

32. If ■■ = ✖✖✖, and ✖✖ = ☎☎☎☎, and ☎☎☎ = ☞☞☞☞, then ☞☞☞☞☞☞☞☞ = how many ■'s?
2

33. If ✪✪✪ = ●●◗, and ●● = ◆ ◆ ◆, and ◆◀ = ✎✎✎✎, then ✎✎✎✎✎✎✎✎✎✎ = how many ✪'s?
3

34. The Ontkeans, Kedrovas, McIvers, and Levinsons are neighbors. Each of the four families specializes in a different national cuisine (Chinese, French, Italian, or Mexican). From the following statements, determine which cuisine each family specializes in.

a. The Ontkeans invited the family that specializes in Chinese cuisine and the family that specializes in Mexican cuisine for dinner last night.

b. The McIvers live between the family that specializes in Italian cuisine and the Ontkeans. The Levinsons live between the Kedrovas and the family that specializes in Chinese cuisine.

c. The Kedrovas and the family that specializes in Italian cuisine both subscribe to the same culinary magazine.

Ontkeans—French; Kedrovas—Mexican; McIvers—Chinese; Levinsons—Italian

35. Anna, Kay, Megan, and Nicole decide to travel together during spring break, but they need to find a destination where each of them will be able to participate in her favorite sport (golf, horseback riding, sailing, or tennis). From the following statements, determine the favorite sport of each student.

a. Anna and the student whose favorite sport is sailing both like to swim, whereas Nicole and the student whose favorite sport is tennis would prefer to scuba dive.

b. Megan and the student whose favorite sport is sailing are roommates. Nicole and the student whose favorite sport is golf each live alone.

Anna—golf; Kay—sailing; Megan—tennis; Nicole—horseback riding

36. Chang, Nick, Pablo, and Saul each took a different form of transportation (bus, car, subway, or taxi) from the office to the airport. From the following statements, determine which form of transportation each took.

a. Chang spent more on transportation than did the fellow who took the bus, but less than the fellow who took the taxi.

b. Pablo, who did not travel by bus and who spent the least on transportation, arrived at the airport after Nick but before the fellow who took the subway.

c. Saul spent less on transportation than either Chang or Nick.

Chang—subway; Nick—taxi; Pablo—car; Saul—bus

37. Lomax, Parish, Thorpe, and Wong are neighbors. Each drives a different type of vehicle—a compact car, a sedan, a sports car, or a station wagon. From the following statements, determine which type of vehicle each of the neighbors drives.

a. Although the vehicle owned by Lomax has more mileage on it than does either the sedan or the sports car, it does not have the highest mileage of all four cars.

b. Wong and the owner of the sports car live on one side of the street, while Thorpe and the owner of the compact car live on the other side of the street.

c. Thorpe owns the vehicle with the most mileage on it.

Lomax—compact car; Parish—sports car; Thorpe—station wagon; Wong—sedan

Applying Concepts

38. Predict the next two terms of the sequence 1, 6, 12, 19, 27, 36, ...

46, 57

39. Determine the tens digit of 7^{123}.

4

40. Is 16,014,649,398 a perfect square? Explain your answer.

No. A perfect square must end with the digit 1, 4, 5, 6, or 9.

41. It is a fact that the fourth power of any number ends with a 1, 5, or 6. On the basis of this, can you conclude that 134,512,357,186 is the fourth power of some number?

No.

42. Dave and his three roommates, Joe, Tom, and Rick, each help pay their tuition by working part time on campus. Each has a different job; one works in the cafeteria, one works in the library, one works in the admissions office, and one works in the computer lab. Each roommate has a different major (business, education, history, or Spanish). From the following statements, determine each student's major and where he works on campus.

a. Rick, his roommate who works in the computer lab, and the Spanish major all work an even number of hours each week.

b. The education major works 15 hours per week, which is less than Joe works per week and more than the Spanish major, who works in the cafeteria.

c. Tom's hours plus those of the student who works in the library are less than the combined hours of the business major and the student working for admissions.

d. Dave and the student who works in the admissions office were roommates last year, while the student who works in the computer lab had a single room.

The table below may be helpful.

	Business	Education	History	Spanish	Cafeteria	Library	Admissions	Computer
Dave								
Joe								
Rick								
Tom								
Cafeteria								
Library								
Admissions								
Computer								

Dave is majoring in education and works in the library.
Joe's major is business, and he works in the computer lab.
Tom's major is Spanish; he works in the cafeteria.
Rick works for the admissions office, and his major is history.

Exploration

43. ***The RATS sequence*** As mentioned in the text, just giving the first few terms of a sequence does not make it possible to predict the next term with absolute certainty. We can use induction to guess at the next term, but that guess may not be correct. Consider the first five terms of the sequence 1, 2, 4, 8, 16. From this, you might expect that the next term is 32. However, the next two terms of this sequence are 77 and 145. This is a RATS sequence and is formed by **R**eversing the digits, then **A**dding the two numbers, **T**hen **S**orting the digits from smallest to largest. For instance, the next term after 145 is formed by

Reversing the digits

$145 + 541 = 686$ → 668

Adding the digits

Then **S**orting the digits from smallest to largest

Find the next two terms of the RATS sequence.

Section 1.2 Introduction to Logic and Sets

The Logical Operators *or* and *and*

> "I know what you're thinking about," said Tweedledum (*to Alice*): "but it isn't so, nohow."
> "Contrariwise," continued Tweedledee, "if it was so, it might be; and if it were so, it would be; but as it isn't, it ain't. That's logic."

POINT OF INTEREST

Lewis Carroll was the pen name of Charles Dodgson (1832–1898), a British mathematician and writer. His stories about Alice, invented to amuse the daughter of a friend, appear in *Alice's Adventures in Wonderland* (1865) and *Through the Looking-Glass* (1872). Queen Victoria of England was so amused by *Alice's Adventures in Wonderland* that she asked Dodgson to present her with the next book he wrote. That book was a math book and when he gave it to her, she was no longer amused.

This excerpt from Lewis Carroll's *Through the Looking-Glass* sums up Tweedledee's understanding of logic. But logic is a little more than that. It is the study of *statements* and their relationships to one another. A **statement** is a sentence that is either true or false. Here are some examples of statements.

- John Lennon wrote songs for the Beatles.
- Jackie Joyner-Kersee won an Olympic gold medal in the heptathlon.
- The moon is made of green cheese.
- There are exactly seven grains of sand on the beach.

Each of these is a statement because we can determine whether the sentence is true or false.

Question: Is the sentence "Join the Pepsi generation" a statement?[1]

Each statement has a *truth value* which is either true or false. The truth value of the statement "John Lennon wrote songs for the Beatles" is *true*. For the statement "The moon is made of green cheese," the truth value is *false*.

Question: What is the truth value of the statement "There are exactly seven grains of sand on the beach"?[2]

Addition and multiplication are examples of *mathematical operators*. They are used to combine numbers, as in $3 + 4 = 7$ or $5 \times 6 = 30$. In logic, the words *and, or,* and *not* are called *logical operators*. These words are used to combine words and sentences.

Consider the following sentence: "An unknown number begins with a 3 *and* ends with a 4." This sentence sets two conditions for an unknown number: the first number is a 3; the last number is a 4.

A possibility for the unknown number is 3794. This number begins with a 3 and ends with a 4, so it satisfies both conditions of the sentence. When statements are combined with *and,* each statement must be true for the combined statement to be true.

A number that is not possible for the unknown number is 34,512. Although this number begins with a 3, it does not end with a 4. Therefore, 34,512 is not a possibility for the unknown number.

Question: Is 533 a possibility for an unknown number that begins with a 5 and ends with a 3?[3]

In everyday language, the word *or* is used in two different ways. Consider the sentence "The tree is an oak tree *or* an elm tree."

In this case, either the tree is an oak or it is an elm. It cannot be both. This is called *exclusive or.* The conditions are such that they both cannot be true at the same time.

---

1. No. There is no way to determine whether the sentence is true or false.
2. False. The truth value is the word *true* or the word *false*, depending on whether the statement is true or false.
3. Yes. The number begins with a 5 and ends with a 3.

Now consider the sentence "To enter a community college in California, a person must be at least 18 years old or a high school graduate."

In this case, a person may be at least 18 years old *and* be a high school graduate. This is called *inclusive or*. For this use of *or*, both conditions may be true at the same time.

Generally in mathematics and logic, we assume that the word *or* is used in the *inclusive* sense unless it is clear from the problem or situation that the exclusive *or* is intended.

Question: Which form of *or* is used in the sentence "The car has exactly 2 doors or exactly 4 doors"?[4]

TAKE NOTE
We are assuming the use of the inclusive *or* for the discussion at the right. Remember, unless otherwise stated, the word *or* is used in the inclusive sense.

A sentence containing the word *or* is true if at least one of the statements that is being combined with *or* is true. To see how *or* differs from *and*, suppose an unknown number begins with a 5 *or* ends with a 6. Here are some possibilities for the unknown number.

5014	The number begins with a 5.
716	The number ends with a 6.
50,366	The number begins with a 5 and ends with a 6.

If, however, the unknown number had to begin with a 5 *and* end with a 6, neither the number 5014 nor the number 716 would satisfy the conditions. The word *and* requires that both conditions be satisfied. For the numbers given above, only 50,366 satisfies both conditions. It is a number that begins with a 5 *and* ends with a 6.

Sometimes it is convenient to use a letter to represent a number. For instance, if we want to symbolize the phrase "a number greater than 10," we could let x (or some other letter) represent the number and write "$x > 10$." A letter used in this way is called a **variable**. This idea is used in the next example.

Example 1 What is the truth value of $(x < 20)$ and $(x > 12)$ when $x = 15$?

Solution Because the inequalities are combined with *and*, the statement "$(x < 20)$ and $(x > 12)$" is true when both inequalities are true. Replace x by 15 in each inequality and determine whether the inequality is true or false.

$15 < 20$ True $\qquad 15 > 12$ True

Because the inequalities are combined with *and*, and each inequality is true, the statement $(x < 20)$ and $(x > 12)$ is true.

Question: If $x = 11$ in Example 1, what is the truth value of the statement?[5]

You-Try-It 1 What is the truth value of $(x > 12)$ or $(x < 6)$ when $x = 3$?

Solution See page S1.

INSTRUCTOR NOTE
Ask students to determine whether the definition of $\le$ uses the inclusive or exclusive *or*.

Two inequalities that are frequently used in mathematics are $\le$ (is less than *or* equal) and $\ge$ (is greater than *or* equal). Because the meaning of these symbols contains the word *or*, a statement containing one of these symbols is true when either condition is true. Here are some examples.

$3 \le 4$ is true because $3 < 4$. $\qquad 5 \ge 1$ is true because $5 > 1$.

$3 \le 3$ is true because $3 = 3$. $\qquad 1 \ge 1$ is true because $1 = 1$.

Question: If $x = 7$, is the statement $x \le 7$ true or false?[6]

---

4. Exclusive. The car cannot have both 2 doors and 4 doors.
5. $(11 < 20)$ is true but $(11 > 12)$ is false. Therefore, the truth value of the statement is false.
6. True, because replacing x by 7 results in the true statement $7 = 7$.

Logical operators and inequalities are such an important part of mathematics that these functions are built into graphing calculators. Typical graphing calculator screens are shown below.

TAKE NOTE
The Guidelines for Using Graphing Calculators appendix contains some suggested keystrokes for TEST and LOGIC operations for various calculators. The xor is used for exclusive *or*.

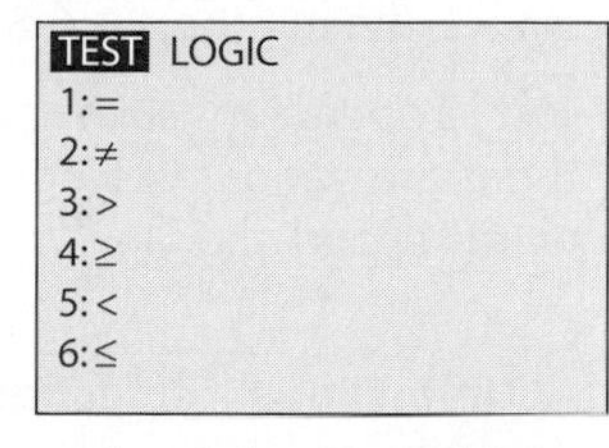

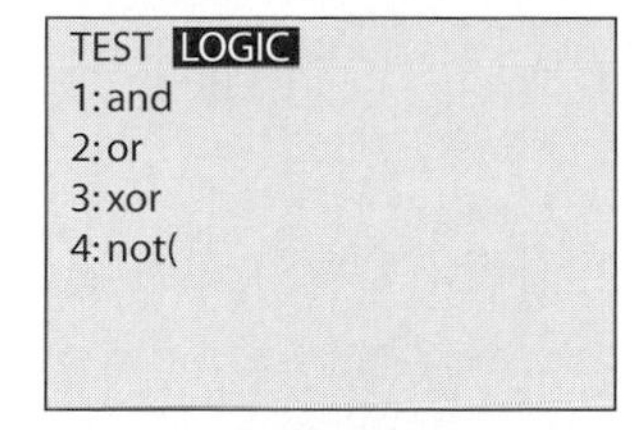

INSTRUCTOR NOTE
Have students check their solutions to You-Try-It 2 by using their calculators.

SUGGESTED ACTIVITY
See the Instructor's Resource Manual for an activity that will lead students to determine the Order of Operations Agreement for the logical operators *and*, *or*, and *not*.

As shown at the right, a graphing calculator can be used to check the answer to Example 1. Enter $(15 < 20)$ and $(15 > 12)$. The calculator places a 1 on the screen to mean that the statement is true. A 0 is placed on the screen if the answer is false. For instance, if we enter $(11 < 20)$ and $(11 > 12)$, a 0 is printed on the screen because the statement $(11 > 12)$ is false.

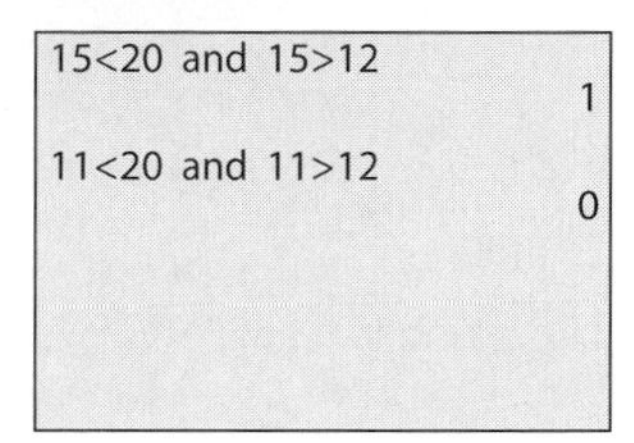

Example 2 Determine the truth value of the statement "$(3x - 4 \geq 5)$ or $(2x + 1 < 3)$" when $x = 3$.

Solution Because the inequalities are combined with *or*, the statement "$(3x - 4 \geq 5)$ or $(2x + 1 < 3)$" is true when either one of the inequalities is true. Replace x by 3 in each inequality and determine whether the inequality is true or false.

$3x - 4 \geq 5$	$2x + 1 < 3$
$3(3) - 4 \geq 5$	$2(3) + 1 < 3$
$9 - 4 \geq 5$	$6 + 1 < 3$
$5 \geq 5$ True	$7 < 3$ False

Because at least one of the inequalities is true and the inequalities are combined with *or*, the statement "$(3x - 4 \geq 5)$ or $(2x + 1 < 3)$" is true.

You-Try-It 2 Determine the truth value of the statement "$(7 \leq 3x + 1)$ and $(3x + 1 \leq 10)$" when $x = 3$.

Solution See page S1.

Introduction to Sets

It seems to be a human characteristic to group similar items. For instance, biologists place similar life forms into phyla and geologists divide the history of Earth into eras.

Mathematicians likewise place objects with similar properties in *sets*. A **set** is a collection of objects. The objects are called **elements** of the set. A set is denoted by placing braces around the elements of the set.

The numbers that we use to count things, such as the number of cars in a parking garage or the number of CDs in a music collection, are called the *natural numbers*. We can write the set of natural numbers as

$$\mathbf{N} = \{1, 2, 3, 4, 5, 6, 7, 8, 9, 10, 11, \ldots\},$$

where the letter N is used to represent the natural numbers. The three dots indicate the set continues on in a similar manner.

The symbol $\in$ is used to mean "is an element of." For instance, if $A = \{2, 4, 6\}$, then $2 \in A$, $4 \in A$, and $6 \in A$. However, $5 \notin A$ (5 is **not** an element of A).

A **prime number** is a natural number greater than 1 that is divisible (evenly) only by itself and 1. The first five prime numbers are 2, 3, 5, 7, and 11. A **composite number** is a natural number greater than 1 that is not a prime number. For instance, the numbers 4, 6, 8, 9, and 10 are the first five composite numbers.

Just as operations such as addition and multiplication are performed on numbers, there are various set operations. These operations are described by using the words *or* and *and*.

Union and Intersection of Two Sets

The **union** of two sets, written $A \cup B$, is the set of all elements that belong to either set *A* **or** set *B*.

The **intersection** of two sets, written $A \cap B$, is the set of all elements that are common to both set *A* **and** set *B*.

Example 3 Let $A = \{3, 6, 9, 12, 15, 18, 21, 24\}$ and $B = \{4, 8, 12, 16, 20, 24\}$. Find $A \cup B$ and $A \cap B$.

Solution

$A \cup B = \{3, 4, 6, 8, 9, 12, 15, 16, 18, 20, 21, 24\}$ • The union of two sets is the set of all elements that belong to *A* **or** *B*. Note that elements belonging to both sets are *not* repeated.

$A \cap B = \{12, 24\}$ • The intersection of two sets is the set of elements that are common to *A* **and** *B*.

You-Try-It 3 Let $A = \{1, 2, 3, 6, 7, 8\}$ and $B = \{3, 4, 5, 6, 7\}$. Find $A \cup B$ and $A \cap B$.

Solution See page S1.

It is possible to describe a set that has no elements. For instance, the set of people who are 500 years old has no elements. The set that contains no elements is called the **empty set**. The empty set is symbolized by $\varnothing$.

→ Let $X = \{a, b, c\}$ and $Y = \{x, y, z\}$. Find $X \cap Y$.

Because there are no elements that are common to sets *X* and *Y*, there are no elements in the intersection of the two sets. Therefore, $X \cap Y = \varnothing$.

When the intersection of two sets *A* and *B* is the empty set, the sets *A* and *B* are called **disjoint** sets.

Negations

The **negation** of a statement is the statement of opposite truth value. The negation of a true statement is a false statement. The negation of a false statement is a true statement.

Statement	**Negation**
John Lennon wrote songs for the Beatles. Truth value is *true*.	John Lennon did **not** write songs for the Beatles. Truth value is *false*.
Statement	**Negation**
The moon is made of green cheese. Truth value is *false*.	The moon is **not** made of green cheese. Truth value is *true*.

Example 4 Write the negation of the following sentences.

a. Tennis is a sport.
b. Louis Armstrong did not play the trumpet.

Solution

a. Tennis is not a sport.
b. Louis Armstrong did play the trumpet.

You-Try-It 4 Write the negation of the following sentences.

a. Smog is unhealthy.
b. Saturn does not have rings around it.

Solution See page S1.

The words **all, no** (or **none**), and **some** are called *quantifiers*. Writing the negation of a sentence that contains these words requires special attention.

Consider the sentence

"All pets are dogs." ← A false sentence

This sentence is not true because there are other pets, cats for example. Because the sentence is false, its negation must be true. You might be tempted to write the negation as "All pets are not dogs," but that sentence is not true either. There are dogs that are pets. The correct negation of "All pets are dogs" is

"Some pets are not dogs." ← A true sentence

Note the use of the word *some* in the negation.

Now consider the sentence

"Some computers are portable." ← A true sentence

Because that sentence is true, its negation must be false. Writing "Some computers are not portable" as the negation does not work because that sentence is also true. The correct negation of "Some computers are portable" is

"No computers are portable." ← A false sentence

The sentence "No flowers have red blooms" is false because there is at least one flower, roses for example, that has red blooms. Since the sentence is false, its negation must be true. The negation is "Some flowers have red blooms."

Here is a chart with statements and their negations.

Statement	Negation
All *A* are *B*.	Some *A* are not *B*.
No *A* are *B*.	Some *A* are *B*.
Some *A* are *B*.	No *A* are *B*.
Some *A* are not *B*.	All *A* are *B*.

Example 5 Write the negation of "Some trees are not pine trees."

Solution "All trees are pine trees."

You-Try-It 5 Write the negation of "All animals can fly."

Solution See page S1.

Example 6 Write the negation of "No physical exercises are healthy."

Solution "Some physical exercises are healthy."

You-Try-It 6 Write the negation of the sentence "Some houses are old."

Solution See page S1.

Counterexamples

An example that is given to show that a statement is not true is called a **counterexample**. For instance, suppose someone makes the statement "All colors are red." A counterexample to that statement would be to show someone the color blue or some other color.

If a statement is *always* true, there are no counterexamples. The statement "All even numbers are divisible by 2" is always true. It is not possible to give an example of an even number that is not divisible by 2. Therefore, it is not possible to give a counterexample to the statement.

A statement may be true in some cases but not in others. For instance, consider the statement "The square of any number is larger than the number." This statement is true for the number 7 because $7^2 = 49$ and $49 > 7$. However, there are other numbers for which the statement is not true. For example,

$$\left(\frac{1}{2}\right)^2 = \frac{1}{4} \quad \text{but} \quad \frac{1}{4} < \frac{1}{2} \qquad\qquad 1^2 = 1 \quad \text{but} \quad 1 = 1.$$

The numbers $\frac{1}{2}$ and 1 are counterexamples to the statement. Thus, the statement "The square of any number is larger than the number" is not always true.

→ Let x and y represent two positive numbers. Find a counterexample to the following statement: "If $x < y$, then $\frac{1}{x} < \frac{1}{y}$."

To find a counterexample, we use the guess and check problem-solving strategy mentioned earlier. Try $x = 2$ and $y = 3$. We know that $2 < 3$, but $\frac{1}{2} > \frac{1}{3}$.

Thus, we have found a counterexample.

Example 7 Find a counterexample to the statement "Only numbers that end in 5 are divisible by 5."

Solution To find a counterexample, you must find at least one number that is divisible by 5 but that does not end in 5. For instance, 20 is divisible by 5 but does not end in 5. The number 20 is a counterexample to the statement. There are other numbers we could have chosen.

You-Try-It 7 Find a counterexample to the statement "All prime numbers are odd numbers."

Solution See page S1.

TAKE NOTE
There are a number of conjectures in mathematics that have not been proved true but for which no one has been able to produce a counterexample. One such conjecture is the Goldbach conjecture. It states that any even number greater than 2 can be expressed as the sum of two prime numbers. For instance, $18 = 5 + 13$ and $26 = 7 + 19$. No one has been able to find an even number that is not the sum of two primes, but no one has been able to prove that all even numbers are the sum of two primes.

In mathematics, statements that are always true are called *theorems,* and mathematicians are always searching for theorems. Sometimes a conjecture by a mathematician appears to be a theorem—that is, the statement appears to be always true—but later on someone else finds a counterexample.

For instance, the mathematician Pierre de Fermat (1601–1665) conjectured that $2^{(2^n)} + 1$ was always a prime number for any natural number n. For instance, when $n = 3$, we have $2^{(2^3)} + 1 = 2^8 + 1 = 257$ and 257 is a prime number. However, in 1732 Leonhard Euler (1707–1783) showed that when $n = 5$, $2^{(2^5)} + 1 = 4{,}294{,}967{,}297$ and that $4{,}294{,}967{,}297 = 641 \cdot 6{,}700{,}417$. By the way, he did these calculations without a calculator! Therefore 4,294,967,297 is the product of two numbers (other than itself and 1) and hence is not a prime number. This counterexample showed that Fermat's conjecture was not a theorem.

1.2 EXERCISES

Topics for Discussion

1. What are logical operators? *And*, *or*, and *not* are the ones discussed here.
2. What is a statement in logic? A sentence that is either true or false.
3. What is the truth value of a statement?
   The truth value of a true statement is true, and the truth value of a false statement is false.
4. What is the relationship between the truth value of a statement and the truth value of its negation?
   The truth values are opposite.
5. Discuss the difference between the words *and* and *or*.
   Answers will vary.
6. What is a counterexample?
   An example that shows that a conjecture is not true.

The Logical Operators

For Exercises 7 through 14, determine which are statements.

7. Texas is a state in the United States.
   Yes
8. The temperature in Hawaii is less than 0°F.
   Yes
9. Whistle while you work.
   No
10. See the USA in your Chevrolet.
    No
11. What's in a name?
    No
12. Twenty-seven is bigger than four hundred.
    Yes
13. The sun is a star.
    Yes
14. Jump in the river.
    No

In Exercises 15 through 30, determine the truth value of the statement.

15. $17 < 40$
    True
16. $16 = 20$
    False
17. $5 > 5$
    False
18. $3 \leq 3$
    True
19. $(7 < 9)$ and $(12 > 3)$
    True
20. $(6 > 0)$ and $(5 < 4)$
    False
21. $(8 > 9)$ or $(5 < 15)$
    True
22. $(4 \leq 9)$ or $(5 \geq 2)$
    True
23. $(5 = 5)$ and $(7 \geq 9)$
    False
24. $(9 \geq 11)$ or $(2 = 2)$
    True
25. $(8 > 9)$ or $(5 < 15)$
    True
26. $(4 \leq 9)$ or $(5 \geq 2)$
    True

27. $(x < 9)$ and $(x > 1)$ when $x = 7$
True

28. $(x < 12)$ or $(x > 58)$ when $x = 30$
False

29. $(x < 12)$ or $(x \geq 22)$ when $x = 22$
True

30. $(x \geq 5)$ and $(x < 15)$ when $x = 4$
False

Introduction to Sets

31. Let $A = \{1, 2, 3, 5, 8, 13, 21, 34\}$. Determine which of the following are true and which are false.

a. $3 \in A$ True

b. $6 \in A$ False

c. $10 \notin A$ True

d. $13 \notin A$ False

Let $A = \{0, 1, 2, 3, 4, 5\}$, $B = \{2, 5, 7, 11\}$, $C = \{2, 4, 6, 8, 10\}$, and $D = \{1, 3, 5, 7, 9\}$. Find each of the following.

32. $A \cup B$ $\{0, 1, 2, 3, 4, 5, 7, 11\}$

33. $A \cap C$ $\{2, 4\}$

34. $C \cup D$ $\{1, 2, 3, 4, 5, 6, 7, 8, 9, 10\}$

35. $C \cap D$ $\varnothing$

36. $B \cap D$ $\{5, 7\}$

37. $B \cup D$ $\{1, 2, 3, 5, 7, 9, 11\}$

Negations

Write the negation for each of the following sentences.

38. The United States has 50 states.
The United States does not have 50 states.

39. A square does not have four sides.
A square has four sides.

40. Red is not a primary color.
Red is a primary color.

41. Water freezes at zero degrees Celsius.
Water does not freeze at zero degrees Celsius.

42. $17 \neq 19$
$17 = 19$

43. $41 = 63$
$41 \neq 63$

44. All trees have green leaves.
Some trees do not have green leaves.

45. Some prime numbers are odd numbers.
No prime numbers are odd numbers.

46. Some dogs are not friendly.
All dogs are friendly.

47. No even numbers are odd numbers.
Some even numbers are odd numbers.

48. All cars require fuel.
Some cars do not require fuel.

49. Some apples are green.
No apples are green.

50. Some diamonds are expensive.
No diamonds are expensive.

51. Some printers do not print in color.
All printers print in color.

52. No restaurants serve good food.
Some restaurants serve good food.

53. All music sounds nice.
Some music does not sound nice.

Counterexamples

Find a counterexample to each of the following.

54. All composite numbers are even numbers.
Answers will vary. One possible answer is 15.

55. If two prime numbers are multiplied, the result is always an odd number.
Answers will vary. One possible answer is $2 \cdot 5 = 10$.

56. When two numbers are multiplied together, the result is always bigger than either of the numbers.
Answers will vary. One possible answer is $0 \cdot 5 = 0$.

57. All fractions are less than 1.
Answers will vary. One possible answer is $\frac{7}{2}$.

58. If two numbers are added together, the result is smaller than when the same two numbers are multiplied together.
Answers will vary. One possible answer is $1 + 1 = 2$; $1 \cdot 1 = 1$.

59. If two prime numbers are added together, the result is not a prime number.
Answers will vary. One possible answer is $2 + 11 = 13$.

Applying Concepts

60. Give an example of a statement and an example that is not a statement.
Answers will vary. For instance, "The car has four wheels"; "Where's the beef?"

61. Suppose that $5 \in (A \cup B)$. Determine whether the following statements are always true, sometimes true, or never true.

a. $5 \in A$ Sometimes true

b. $5 \in B$ Sometimes true

c. $5 \in (A \cap B)$ Sometimes true

62. Suppose that $5 \in (A \cap B)$. Determine whether the following statements are always true, sometimes true, or never true.

a. $5 \in A$ Always true

b. $5 \in B$ Always true

c. $5 \in (A \cup B)$ Always true

63. Let A be a set. Complete the following to make a true statement.

a. $A \cup A = ?$ A

b. $A \cap A = ?$ A

c. $A \cup \varnothing = ?$ A

d. $A \cap \varnothing = ?$ $\varnothing$

64. Let A and B be two sets such that every element of A is an element of B.
a. Find $A \cup B$. B
b. Find $A \cap B$. A

65. Determine the truth value of each of the following.
a. $[(4 < 9) \text{ or } (5 > 15)] \text{ and } (1 > 0)$ True
b. $(3 \leq 9) \text{ and } [(17 > 20) \text{ or } (6 \leq 12)]$ True
c. $[(9 > 3) \text{ and } (25 > 7)] \text{ or } (0 > 12)$ True
d. $(11 \geq 13) \text{ or } [(6 > 10) \text{ and } (7 \leq 19)]$ False

66. Are there any natural numbers that can replace x in the statement "$(x < 10)$ and $(x \geq 5)$" so that the statement is true? If so, list the numbers.
Yes; 5, 6, 7, 8, 9

67. Are there any natural numbers that can replace x in the statement "$(x \geq 21)$ and $(x < 5)$" so that the statement is true? If so, list the numbers.
No

68. Find a counterexample to the statement: The sum of the proper divisors of a number equal the number. (The proper divisors of a number are less than the number but divide into the number evenly. For instance, the proper divisors of 6 are 1, 2, and 3.)
One possible answer is 8; $1 + 2 + 4 = 7 \neq 8$.

Exploration

69. ***Venn Diagrams*** A Venn diagram is a visual representation of operations on sets. For the Venn diagram at the right, the rectangle represents the set $U = \{1, 2, 3, 4, 5, 6, 7, 8\}$, sometimes called the universal set. One circle represents the set $A = \{2, 3, 4, 5\}$, and the other circle represents the set $B = \{4, 5, 6, 7\}$. From the Venn diagram, $A \cup B = \{2, 3, 4, 5, 6, 7\}$ and $A \cap B = \{4, 5\}$. The 1 and 8 shown outside of A or B mean that these two elements are in U but not in A or in B.

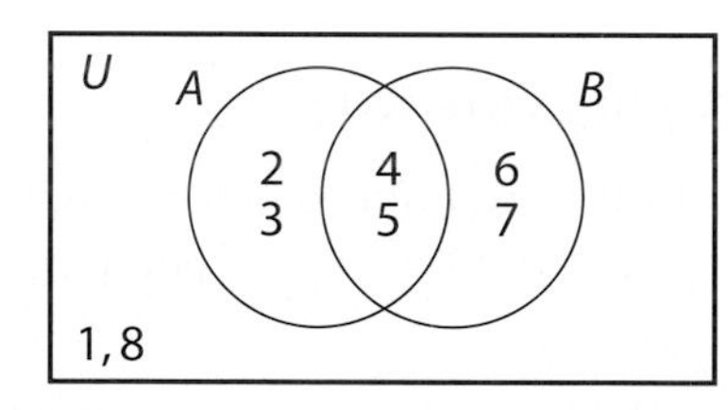

Three sets are shown in the second Venn diagram. This Venn diagram represents how sixty-five students belonging to a college art club are enrolled in three courses: English, math, and history.

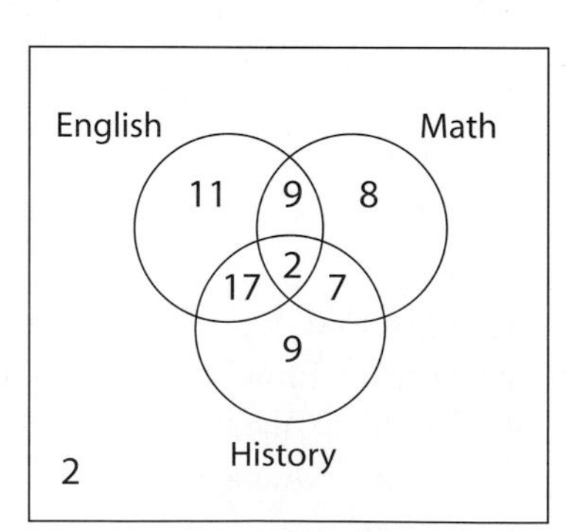

a. How many students were taking English, math, and history?
b. How many students were not taking any of these courses?
c. How many students were taking math but not English or history?
d. How many students were taking English and history but not math?
e. How many students were taking English or history but not math?
f. How many students were taking English and math but not history?
g. How many students were taking English or math but not history?
h. Notice that the only difference between part **d** and part **e** (and similarly between part **f** and part **g**) is the change from *and* to *or*. Does this affect the answer to the problem?

Section 1.3 Creating Tables and Graphs

Graphs and Frequency Tables

The **bar graph** below is based on a recent survey of college students who were asked where they look for jobs. For this bar graph, the bars are horizontal. Bar graphs may also be drawn with vertical bars.

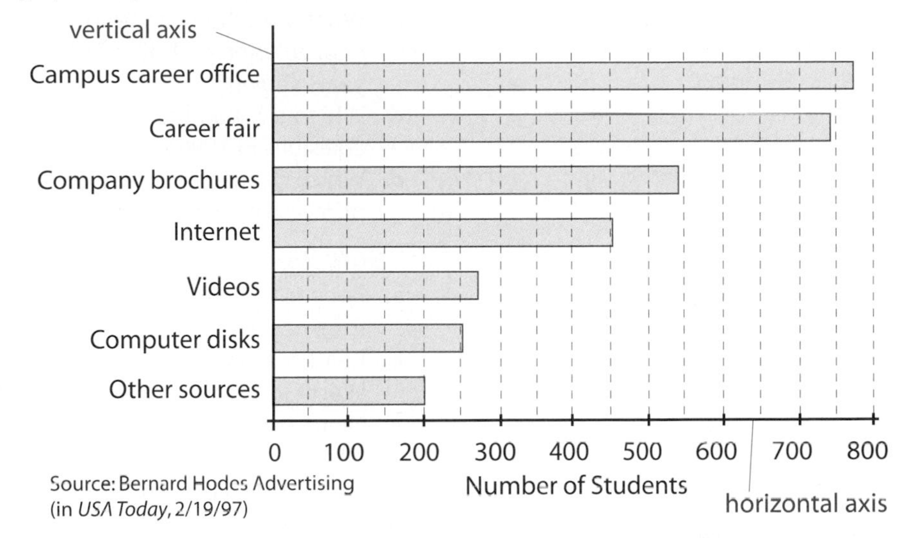

Source: Bernard Hodes Advertising (in *USA Today*, 2/19/97)

As shown on the graph, there is a **vertical axis** and a **horizontal axis.** The length of a bar indicates how many students used a particular method. For instance, since the career fair bar extends to about halfway between 700 and 800, approximately 750 students used a career fair as a means of looking for a job.

Question: According to this survey, approximately how many students used the Internet as a means of looking for a job?[1]

For bars that do not stop on one of the vertical dashed lines, it is more difficult to approximate the number of students. Looking at the graph, we can estimate that approximately 540 students used company brochures to look for a job. The actual number of students could be 542, 537, or some other number close to 540.

Question: Would 525 be a reasonable estimate for the number of students who used company brochures?[2]

A graph is a convenient method of displaying data to give the reader a general idea of the data. As an example, we can see that the number of students who used computer disks *is less than* the number who used videos. In mathematics, we use the symbol < to mean **is less than**.

The number of students who used a career fair *is greater than* the number of students who used the Internet. The symbol > is used to mean **is greater than**. The symbols < and > are called **inequality** symbols.

Question: Which symbol, > or <, must replace ?? in the following sentence to make the sentence true? "The number of students using the Internet ?? the number of students using computer disks."[3]

Some bar graphs use more than one bar so that quantities can be compared. The **double-bar** graph in the following example illustrates this.

---

1. 450 students
2. No. The bar for company brochures extends beyond the halfway point between 500 and 550. Therefore, the estimate must be more than 525.
3. The bar for the number of students using the Internet is longer than the bar for the number of students using computer disks. Therefore, replace the ?? by >.

Example 1 The double-bar graph at the right shows the number of near collisions on runways of airports in selected U.S. cities. Source: Federal Aviation Administration

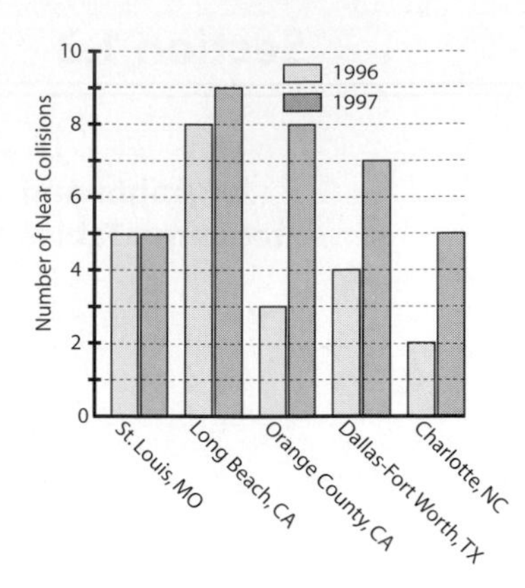

a. Did any airport report fewer near collisions in 1997 than in 1996?

b. What was the change in the number of near collisions at Dallas-Fort Worth for the years shown?

c. Which city had the same number of near collisions in 1996 as in 1997?

d. Which city had the greatest increase in the number of near collisions?

Solution

a. For each city, the bar for 1997 is equal to or taller than the bar for 1996. Therefore, no city reported fewer near collisions in 1997.

b. Read the graph to determine the number of near collisions for Dallas-Fort Worth in 1996 and 1997. Then subtract the numbers.

Near collisions for Dallas-Fort Worth in 1996: 4

Near collisions for Dallas-Fort Worth in 1997: 7 $7 - 4 = 3$

There were 3 more near collisions in 1997 than in 1996 at Dallas-Fort Worth.

c. Because the bars for St. Louis, MO, are the same height, there were the same number of near collisions in St. Louis for the two years shown.

d. By calculating the difference between the heights of the bars for each city, Orange County, CA, had the largest increase in the number of near collisions.

You-Try-It 1 The double-bar graph at the right is based on a survey of 200 men and 200 women. It shows the number of minutes they are willing to hold on the phone when they call the customer service line of a company.

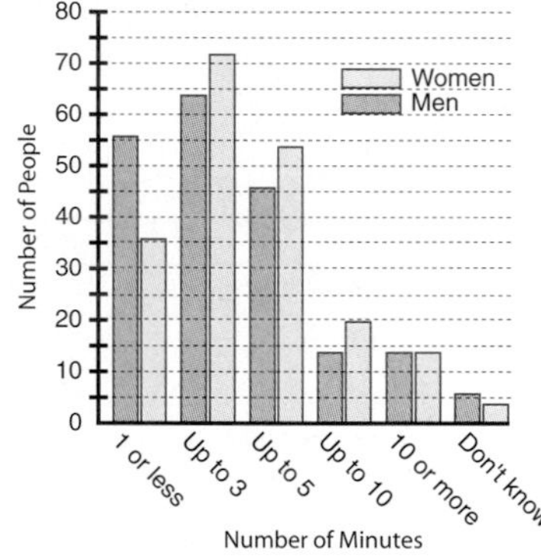

Source: Bruskin/Goldring for Inference

a. Were more men or more women willing to wait more than 1 minute but less than 3 minutes?

b. In which categories were men more willing to wait than women?

c. Were there more than 15 men or less than 15 men who were willing to wait more than 5 minutes but less than 10 minutes?

Solution See page S1.

There are many different types of graphs that can be used to display data. A **broken-line** graph uses line segments to connect points on the graph.

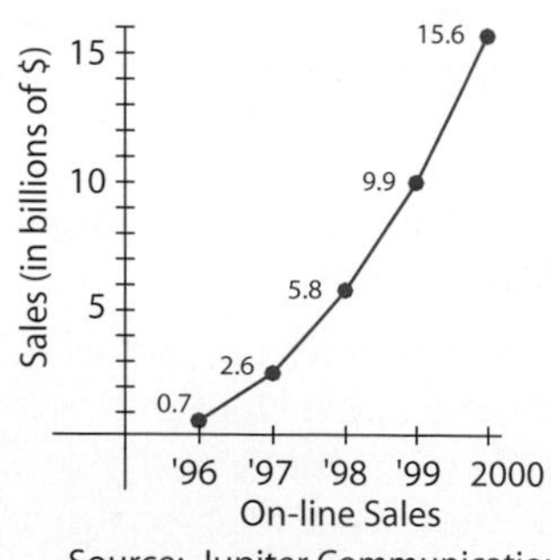

On-line Sales

Source: Jupiter Communications *USA Today*, 12/10/97

The broken-line graph at the left shows the total sales of all products sold on-line over the Internet and how sales are expected to increase through the year 2000. From the graph, we can estimate that in the year 2000, there will be $15.6 billion in sales on the Internet.

There are many other conclusions that can be drawn from this graph. For instance, we can estimate the expected increase in sales from 1999 to 2000.

Estimated sales in 1999: 9.9 billion

Estimated sales in 2000: 15.6 billion

Increase in sales $= 15.6 - 9.9 = 5.7$

The expected increase in sales is $5.7 billion.

INSTRUCTOR NOTE
A possible class/group discussion topic is the significance of the scale numbers 1200 and 1500 in the graph at the right. Some students may not understand that 1200 million is the same as 1.2 billion.

SUGGESTED ACTIVITY
Have several groups of students bring in a bar graph or a broken-line graph from a newspaper or magazine. Require that they write one conclusion from the data in the graph and present their finding to the class. Then ask the class to draw another conclusion from the data.

Two broken-line graphs are often shown in the same figure for comparison. The next example shows the change in video-game sales for a two-year period.

Example 2 The graph at the right shows the quarterly sales of video games in the United States for a two-year period.

a. Between which two quarters did sales decrease in 1997?
b. Between which two quarters did sales increase the most in 1996?
c. What was the increase in sales between the 3rd and 4th quarters for 1997?

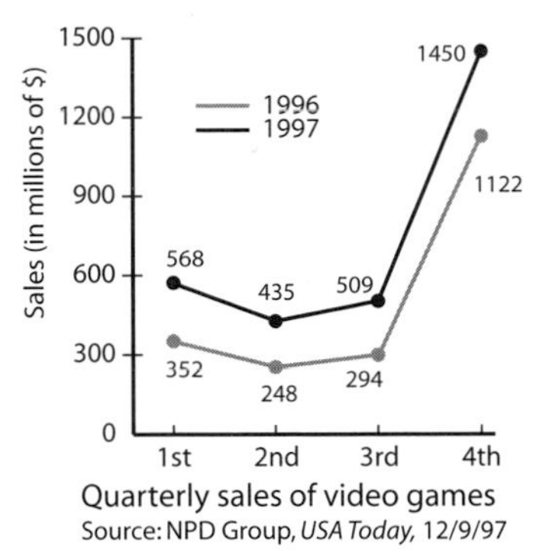

Quarterly sales of video games
Source: NPD Group, *USA Today*, 12/9/97

Solution

a. Sales decreased between the 1st and 2nd quarters of 1997.
b. The largest increase in sales occurred between the 3rd and 4th quarters.
c. To find the increase in sales between the 3rd and 4th quarters, read the graph to find the sales for each quarter. Then subtract the sales.
3rd quarter sales in 1997: 509 million
4th quarter sales in 1997: 1450 million
$1450 - 509 = 941$
Sales increased by $941 million between the 3rd and 4th quarters of 1997.

TAKE NOTE
The jagged lines in the graph at the right indicate that the axes do not begin at zero.

You-Try-It 2 The graph at the right shows the changes in life expectancy for women and men in the United States from 1940 to 1990.

a. In which ten-year period did life expectancy for men increase the least?
b. In which ten-year period did life expectancy for women increase the most?
c. What was the change in life expectancy for men between 1940 and 1990?

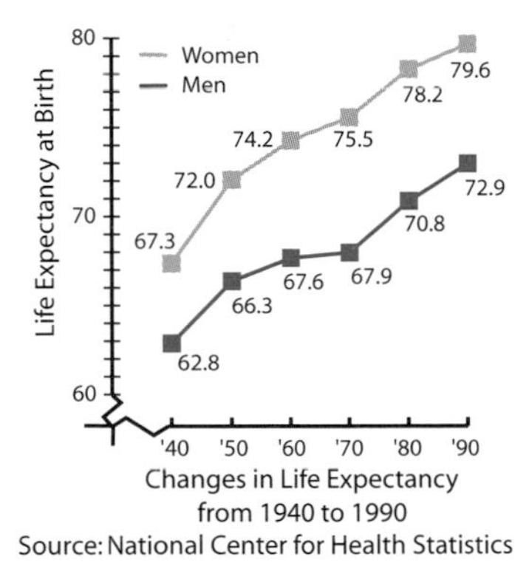

Changes in Life Expectancy from 1940 to 1990
Source: National Center for Health Statistics

Solution See page S1.

A **frequency table** is another method of organizing data. In a frequency table, data is combined into categories called **classes** and the number of data in each class, called the **frequency** of the class, is shown.

An example of a frequency table is shown below. This table, based on data from the *Information Please Almanac* (1996), shows the tuition for state residents at 60 selected public universities in the United States.

Tuition (class)	No. of Universities (frequency)
1500 – 1999	10
2000 – 2499	17
2500 – 2999	13
3000 – 3499	8
3500 – 3999	7
4000 – 4499	5

Each class has a **lower class limit** and an **upper class limit**. For instance, for the class 2000 – 2499, the lower class limit is 2000 and the upper class limit is 2499. Frequency tables are generally set up so that the difference between the successive lower class limits is the same number for each class. Note that for the table above, the difference between two successive lower class limits is 500. For instance, $2000 - 1500 = 500$ and $3000 - 2500 = 500$. The **class size** is the difference between two successive lower class limits.

By reading the frequency table, we can determine that 13 public universities in this survey had a tuition that was between $2500 and $2999 and that there were 5 public universities that had a tuition between $4000 and $4499.

Question: How many universities had an annual tuition between $3500 and $3999?[4]

One reason data is put into a frequency table is to be able to draw some conclusions from the data. For instance, the table below shows the annual salaries that were negotiated by 58 National Football League free agents (Source: *USA Today*, 4/7/97). By looking at the list, it is difficult to see any pattern or to draw any conclusions.

Annual Salaries for 58 NFL Free Agents (millions of dollars)
3.7, 1.8, 3.6, 2.79, 0.28, 2.5, 1.4, 3.3, 2.2, 2.3, 3.37, 0.44, 1.5, 0.6, 3.27, 3.4, 3.0, 2.6, 1.3, 0.55, 1.0, 3.86, 1.43, 3.8, 1.0, 0.37, 0.61, 4.0, 3.25, 0.67, 0.3, 0.33, 2.3, 1.25, 0.2, 0.95, 2.0, 2.42, 4.0, 1.07, 1.1, 1.5, 1.5, 1.5, 0.79, 0.33, 0.49, 0.35, 0.5, 0.88, 1.2, 0.75, 1.08, 0.7, 1.7, 0.44, 0.53, 2.17

Note that the data in this table is given in *millions of dollars*. For instance, an annual salary recorded as 0.44 means

$$0.44 \cdot 1{,}000{,}000 = 440{,}000$$

or $440,000.

Question: What is the dollar amount of an annual salary that is recorded in the table as 3.86?[5]

Example 3 Complete the frequency table below for the annual salaries for the NFL free agents.

Classes (Salary, in millions of $)	Frequency (Number of players with a salary in that class)
0.2 – 0.99	
1.0 – 1.79	
1.8 – 2.59	
2.6 – 3.39	
3.4 – 4.19	

Solution For each salary, place a tally mark, /, to the right of the class in which that salary belongs. For instance, for the first salary of 3.7, place a / to the right of the class 3.4 – 4.19 because 3.7 is greater than 3.4 and 3.7 is less than 4.19. The second number in the table is 1.8. Place a / to the right of the class 1.8 – 2.59 because 1.8 = 1.8. Do this for each salary. Count the number of tally marks in each class and record the frequency of that class. The result is shown below.

Classes (Salary, in millions of $)	Tally	Frequency (Number of players with a salary in that class)
0.2 – 0.99	~~////~~ ~~////~~ ~~////~~ ~~////~~ /	21
1.0 – 1.79	~~////~~ ~~////~~ ~~////~~	15
1.8 – 2.59	~~////~~ ///	8
2.6 – 3.39	~~////~~ //	7
3.4 – 4.19	~~////~~ //	7

4. 7 universities had a tuition between $3500 and $3999.
5. $3.86 \cdot 1{,}000{,}000$, which is $3,860,000.

TAKE NOTE
You-Try-It 3 illustrates that there is more than one way in which to make a frequency table. Generally, frequency tables contain from 5 to 12 classes, but there are no strict rules as to the number of classes.

You-Try-It 3 Complete the frequency table below (with different classes) for the annual salaries for the NFL free agents.

Classes (Salary, in millions of $)	Frequency (Number of players with a salary in that class)
0.2 – 0.69	
0.7 – 1.19	
1.2 – 1.69	
1.7 – 2.19	
2.2 – 2.69	
2.7 – 3.19	
3.2 – 3.69	
3.7 – 4.19	

Solution See page S2.

SUGGESTED ACTIVITY
Have each student determine the number of days to his or her next birthday. List these numbers on the board and then have students create a frequency table for the data. Ask students to calculate the relative frequencies for each class. Ask students to draw some conclusions from the data. As a suggestion, uses classes 0–62, 63–125, 126–188, 189–251, 252–314, 315–372.

Example 3 illustrates that once the data has been organized into a frequency table, it is easier to draw some conclusions about the data. For instance, 21 players negotiated an annual salary that was less than $1 million.

In a frequency table, the frequency column contains the actual number of instances of data for that class. For instance, in Example 3, 7 players negotiated annual salaries that were between $3.4 million and $4.19 million. A **relative frequency table** shows the *percent* of all the data that belongs to a class.

To calculate the relative frequency, divide the frequency of the class by the total number of all the data. For instance, in Example 3, 15 players of the 58 players in the survey negotiated annual salaries of between $1 million and $1.79 million. To find the relative frequency, divide 15 (the frequency of the class) by 58 (the number of players in the survey). Then change the decimal to a percent.

$$\frac{\text{frequency of a class}}{\text{total number of data}} = \frac{15}{58} \approx 0.259 = 25.9\%$$ Rounded to the nearest tenth of a percent.

The relative frequency table, rounded to the nearest tenth of a percent, for the free agent salaries is shown below.

Classes (Salary, in millions of $)	Relative Frequency (Percent of players with a salary in that class)
0.2 – 0.99	36.2%
1.0 – 1.79	25.9%
1.8 – 2.59	13.8%
2.6 – 3.39	12.1%
3.4 – 4.19	12.1%

Adding all the percents should result in 100%, which means all the data. However, due to rounding, the result may not be exactly 100%. For instance, in this relative frequency table, the percents add to 100.1%.

Histograms

TAKE NOTE

There are various methods that are used to draw a histogram. However, in most cases the bars are shown with no space between them, as in the histogram at the right. In this text, we follow the convention that the beginning of a class is shown on a horizontal axis. This means that 1.0 ($1,000,000) is not part of the first class but the beginning of the second class. This same approach is used for the remaining classes. For instance, 2.6 is the beginning of the fourth class and not part of the third class.

A **histogram** is a vertical bar graph of a frequency table. The height of a bar corresponds to the frequency of the class. The figure below is the histogram of the data in Example 3.

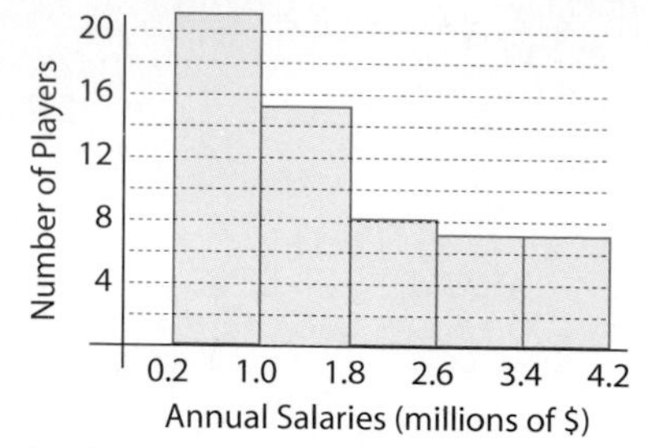

Looking at this graph, the bar whose height is 8 means that there were 8 players whose salary was between $1.8 million and $2.59 million.

Question: From the histogram, how many players had a salary between $1.0 million and $1.79 million?[6]

Example 4 Based on data from the U.S. Census Bureau, the number of people who spend less than one hour commuting to work is given in the histogram at the right. How many people commute between 10 and 30 minutes to work?

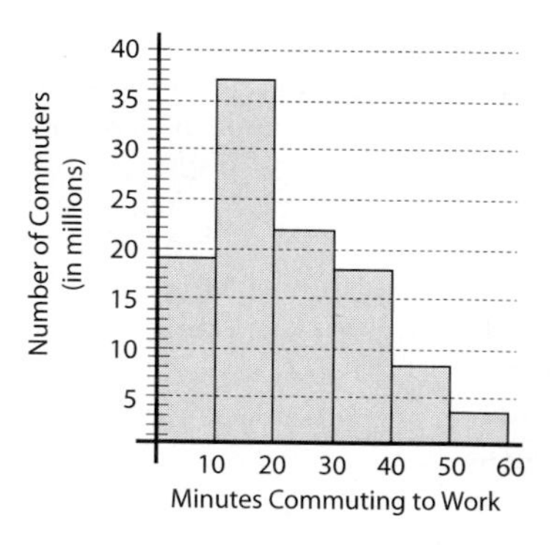

Solution From the histogram, determine the number of people who commute between 10 and 20 minutes (37 million) and the number who commute between 20 and 30 minutes (22 million). Then add the two numbers.

$$37 \text{ million} + 22 \text{ million} = 59 \text{ million}$$

There are 59 million people who commute between 10 and 30 minutes to work each day.

You-Try-It 4 The histogram at the right shows the number of states, including the District of Columbia, and the average annual income of state residents (Source: U.S. Bureau of the Census, 1995). In how many states is the average income less than $34,595?

Solution See page S2.

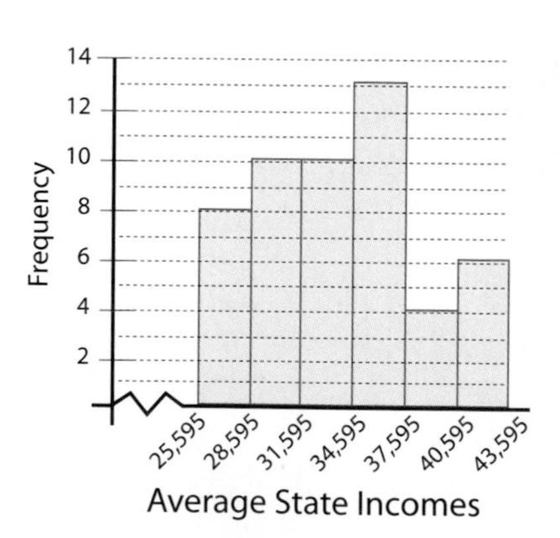

Question: For the histogram in Example 4, is it possible to determine the number of people who commute 15 minutes to work?[7]

---

6. Because the bar is halfway between 14 and 16, there were 15 players whose salary was between $1.0 million and $1.79 million.
7. No. We can only determine the number of people who commute between 10 and 20 minutes to work.

Stem-and-Leaf Plots

A way to display every number in a data set and still organize the data in intervals is to use a **stem-and-leaf plot.** Each piece of data is split into two parts, a **stem** and a **leaf**. The leaf is usually the rightmost digit of a number; the stem is the remaining digits. The stem and leaf for the numbers 26 and 349 are shown below.

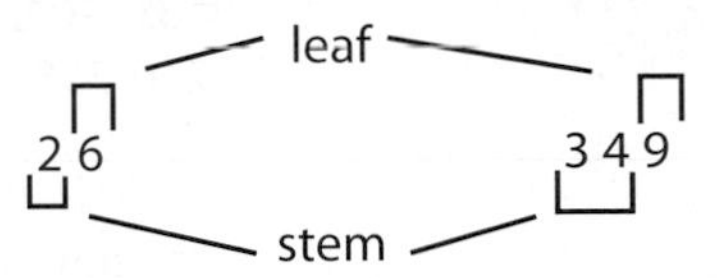

Data is placed into stem-and-leaf plots as a means of quickly showing the frequency of a particular event. The following stem-and-leaf plot shows the round-trip commuting distance for 29 employees of a photo-processing company.

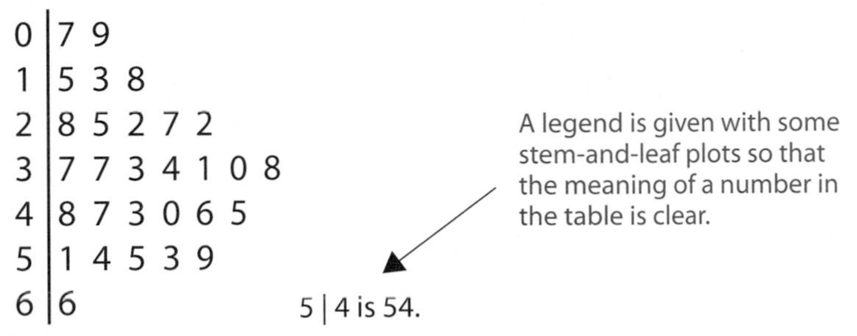

Stem	Leaves
0	7 9
1	5 3 8
2	8 5 2 7 2
3	7 7 3 4 1 0 8
4	8 7 3 0 6 5
5	1 4 5 3 9
6	6

A legend is given with some stem-and-leaf plots so that the meaning of a number in the table is clear.

5 | 4 is 54.

Looking at this display and the stem labeled 4, we can determine that because there are six numbers to the right of the bar, 6 employees have a round-trip commute between 40 and 49 miles to the photo-processing company. Compare this to looking at the raw data 28, 25, 37, 37, 33, 34, 31, 7, 9, 15, 51, 54, 55, 53, 59, 66, 13, 18, 22, 30, 38, 48, 47, 43, 27, 22, 40, 46, 45. From this list, it is difficult to see any pattern to the data. However, looking at the stem-and-leaf plot gives an idea of how far employees must commute to work.

Question: How many employees commute between 10 and 19 miles?[8]

The data from a stem-and-leaf plot can be turned into a histogram. Just think of rotating the stem-and-leaf plot.

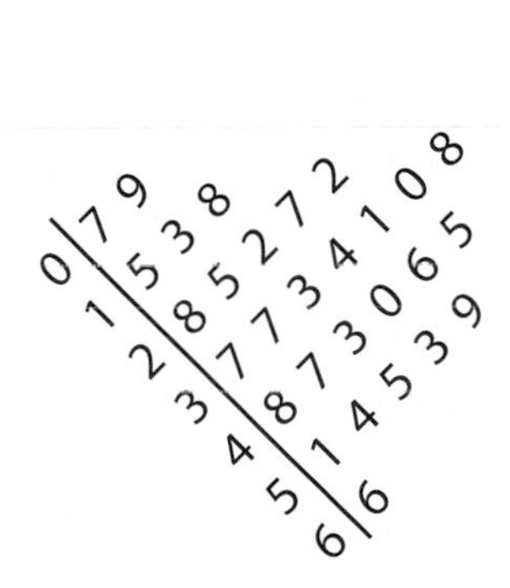

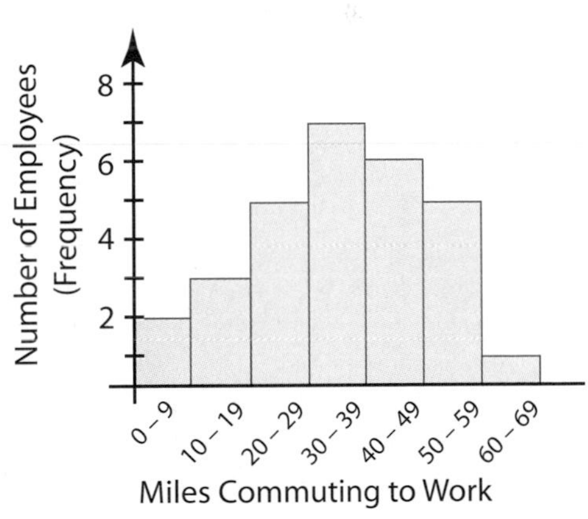

8. 3 employees commute between 10 and 19 miles.

Example 5 The data below are selected cholesterol levels of some women who participated in the Framingham Heart Study. Construct a stem-and-leaf plot for this data using the first two digits as the stem.

185, 204, 221, 213, 202, 217, 219, 209, 240, 223, 225, 204, 208, 232, 189, 220, 228, 192, 218, 212, 219, 205, 206, 245, 239, 227, 175, 216, 198, 209

Solution The smallest value in this set is 175 and the largest is 245. Therefore, stems from 17 (the first two digits of 175) to 24 (the first two digits of 245) are used. Draw a vertical line and list the stems to the left of the line. The beginning of the stem-and-leaf plot is shown below on the left.

Cholesterol Levels in Women

Stem	Leaves
17	
18	
19	
20	
21	
22	
23	
24	

Cholesterol Levels in Women

Stem	Leaves
17	5
18	5 9
19	2 8
20	4 2 9 4 8 5 6 9
21	3 7 9 8 2 9 6
22	1 3 5 0 8 7
23	2 9
24	0 5

To the right of the vertical line, list the leaves. For 185, the first number in the data, place a 5 to the right of 18. For 204, place a 4 to the right of 20. Continue until you have used all the data values. The result is shown above on the right.

You-Try-It 5 The table below gives the approximate annual rainfall total (in inches) for selected cities in the United States. Make a stem-and-leaf plot for this data. Use the first two digits of the number of inches of rainfall as the stem.

City	Rainfall	City	Rainfall
Little Rock, AR	104	Boston, MA	126
Norfolk, VA	115	Pittsburgh, PA	154
Des Moines, IA	107	Nashville, TN	119
Miami, FL	129	Houston, TX	104
Detroit, MI	135	Portland, ME	128
Mobile, AL	122	Philadelphia, PA	117
Spokane, WA	113	Milwaukee, WI	125
Hartford, CT	127	New York City, NY	142
Cleveland, OH	156	Washington, D.C	120
Chicago, IL	126	Kansas City, MO	104
Concord, NH	125	Atlanta, GA	115
Duluth, MN	134	Portland, OR	152

Solution See page S2.

1.3 EXERCISES

Topics for Discussion

1. What is a bar graph? Answers will vary.
2. What is a histogram? A bar graph of a frequency table.
3. How does a frequency table differ from a relative frequency table?
   A frequency table shows the actual number in each class; a relative frequency table shows the percent of the total in each class.
4. What is a stem-and-leaf plot?
   A method of organizing data into intervals and still displaying every number in the set of data.
5. How does a stem-and-leaf plot differ from a frequency table?
   A frequency table shows only the number of data in a class. A stem-and-leaf plot shows the value of every number in the data set.

Graphs and Frequency Tables

6. The bar graph at the right shows (at the beginning of the 1997 European indoor tennis season) the active women players with the most indoor singles titles. (Source: WTA Tour)
   a. Name the player who had the most wins. Graf
   b. Name the player who had the fewest wins. Seles
   c. How many victories did Novotna have? 10
   d. Graf had over 5 times more wins than which players? Huber, Seles

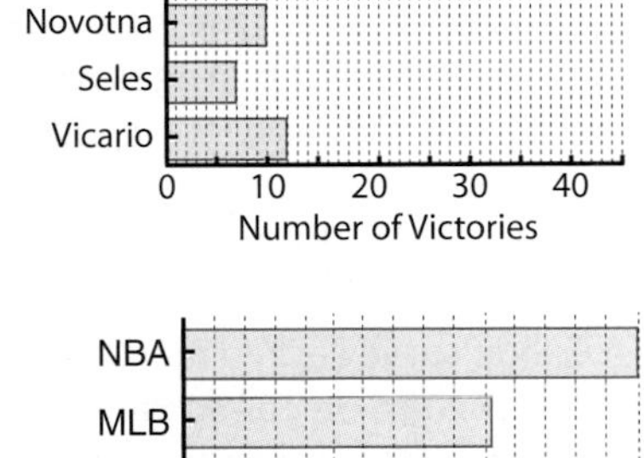

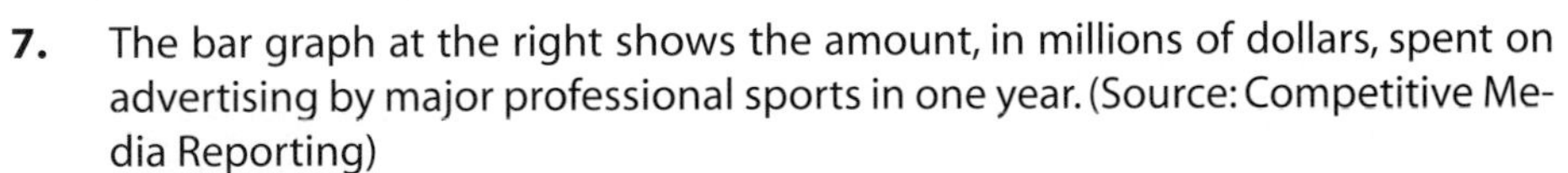

7. The bar graph at the right shows the amount, in millions of dollars, spent on advertising by major professional sports in one year. (Source: Competitive Media Reporting)
   a. How much was spent by the NBA (National Basketball Association)?
   b. Approximately how much was spent by MLB (Major League Baseball)?
   c. Did the National Hockey League (NHL) spend more or less than $3 million on advertising?
   d. Did the National Football League (NFL) spend approximately $7 million more than the NHL for advertising?
   a. $15 million; **b.** $10 million; **c.** less than; **d.** yes

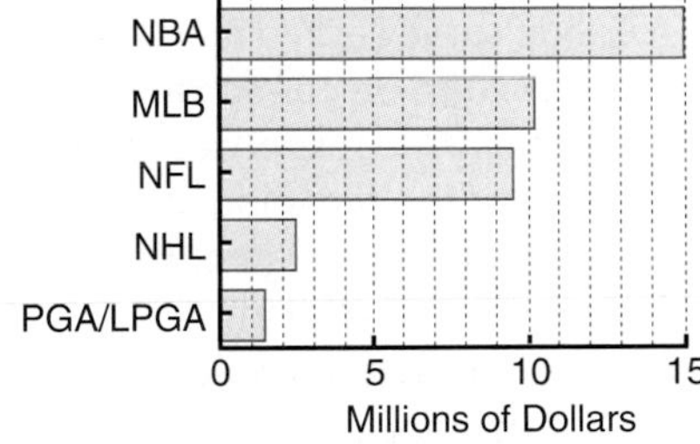

8. The vertical bar graph at the right shows the percent of people in each age group who have been designated drivers for people who have been drinking. (Source: Roper Starch Worldwide)
   a. In which age group did approximately 48% of the people act as a designated driver?
   b. In which age group, 30 – 44 or 45 – 50, did a larger percent of people act as a designated driver?
   c. Approximately what percent of the people 60 and older have acted as a designated driver?
   a. 30 – 44; **b.** 30 – 44; **c.** 16%

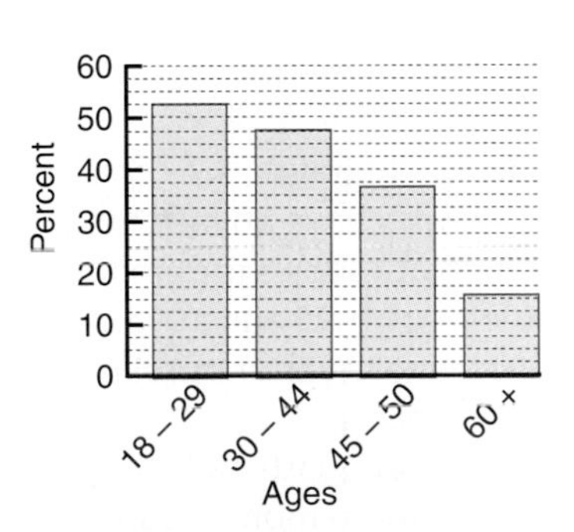

9. The graph at the right shows the states that had the greatest number of alternative fuel vehicles in the United States in 1997. (Source: Energy Information Administration)

a. Which state had the least number of alternative fuel vehicles?

b. Which two states had approximately the same number of alternative fuel vehicles?

c. Approximately how many alternative fuel vehicles were in Texas in 1997?

d. Approximately how many alternative fuel vehicles were in Ohio in 1997?

a. Michigan; **b.** Ohio and Illinois; **c.** 36,000; **d.** 20,000

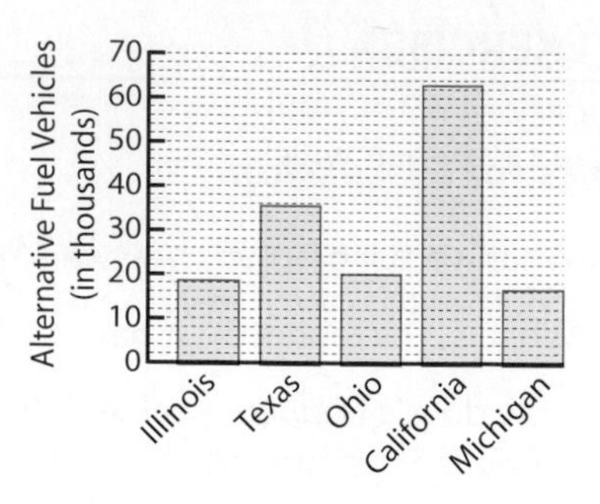

10. The broken-line graph at the right shows the average amount of money donated to charities during the Christmas and Hanukkah holiday season. (Source: Maritz AmeriPoll)

a. What age group donated approximately $58 during the holiday season?

b. What age group donated the most money during the holiday season?

c. What amount did the less-than-24 age group donate?

d. How much less did the 65-and-over age group give than the 55 – 64 age group?

a. 55 – 64; **b.** 35 – 44; **c.** $38; **d.** $6

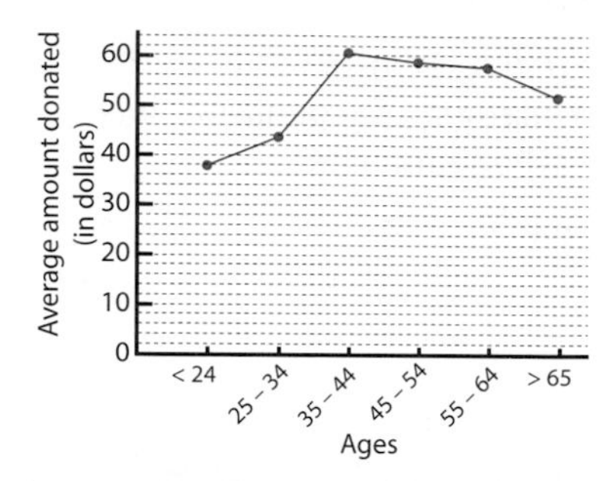

11. The broken-line graph at the right shows the percent of top executives in U.S. corporations by age group. (Source: Dun & Bradstreet)

a. What age group had the largest percent of top executives?

b. What age group had the smallest percent of top executives?

c. What percent of the executives were between 60 and 69?

d. Between what two ages were more than 50% of the top executives?

a. 50 – 59; **b.** 20 – 29; **c.** 18%; **d.** 40 – 59

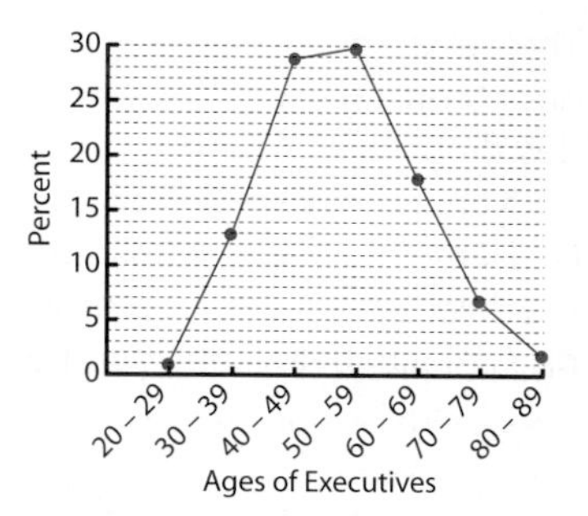

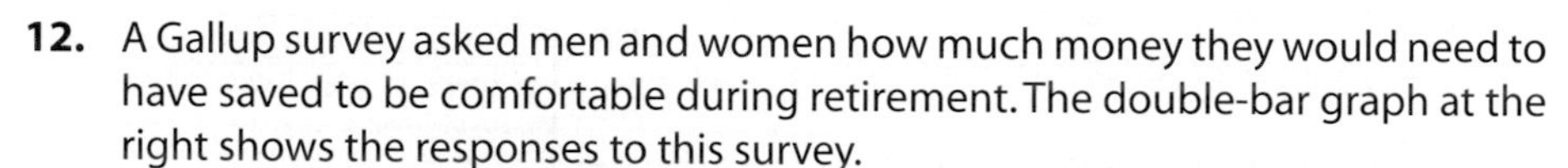

12. A Gallup survey asked men and women how much money they would need to have saved to be comfortable during retirement. The double-bar graph at the right shows the responses to this survey.

a. What is the meaning of the numbers in the category 200 – 499?

b. In what savings amount categories was the percent of women less than the percent of men?

c. In what savings amount category was the percent of women greater than the percent of men?

d. In what savings amount category was the percent of women equal to the percent of men?

a. $200,000 – $499,000; **b.** > $1,000,000 and $200,000 – $499,000; **c.** < $200,000; **d.** $500,000 – $1,000,000

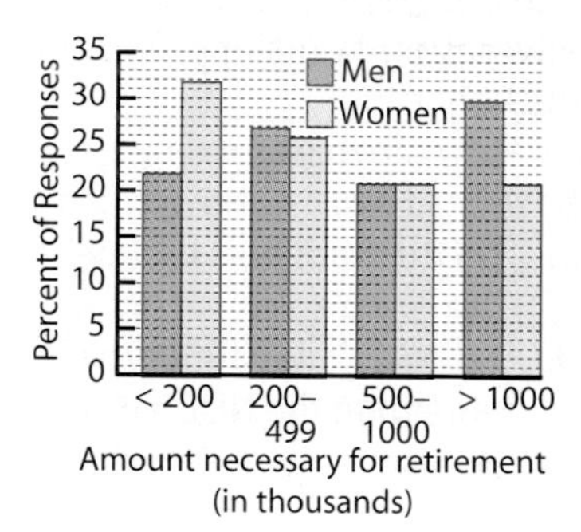

13. Venture capitalists are companies or individuals who loan money to entrepreneurs who have an idea for a new product or business. The double-bar graph at the right shows the amounts (in billions of dollars) that were loaned for each quarter of 1995 and 1996. (Source: Price Waterhouse)

a. In what quarter in 1996 was the most money loaned?

b. In what quarter in 1995 was the most money loaned?

c. In what quarter and year was the least money loaned?

d. Was more money loaned in 1995 or 1996?

a. 2nd; **b.** 4th; **c.** 1st quarter of 1995; **d.** 1996

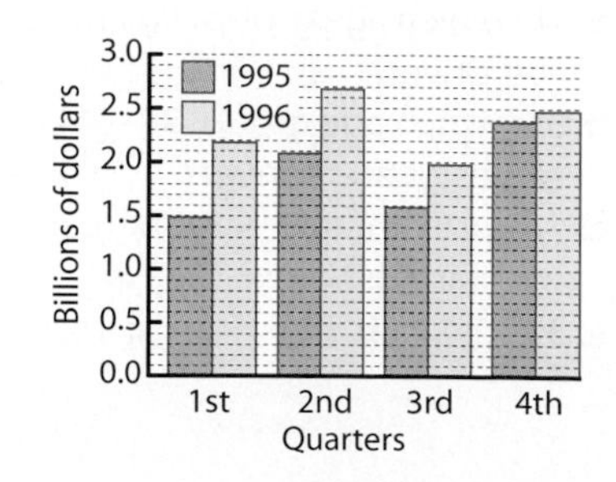

14. The numbers of laser printers and inkjet printers sold in the United States are shown in the double-line graph at the right. (Source: Dataquest)

a. Between which two years did inkjet printer sales first exceed laser printer sales?

b. What were the approximate sales of inkjet printers in 1994?

c. In which category of printer, inkjet or laser, are sales increasing more rapidly?

d. On the basis of the graphs, do you think that future laser printer sales will ever exceed inkjet sales? Why or why not?

a. 1992 and 1993; **b.** 6 million; **c.** inkjet; **d.** Answers will vary.

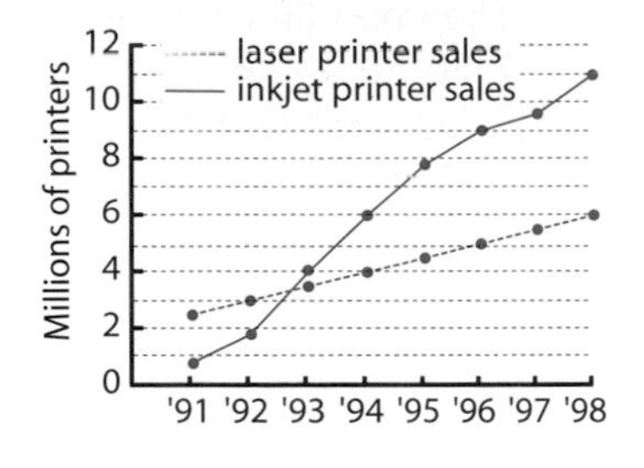

15. The U.S. Air Force expects to have 13 percent fewer pilots by the year 2002. The graph at the right shows the number of pilots that are needed and the number of pilots the Air Force projects it will have.

a. Is the number of pilots the Air Force is expected to have increasing or decreasing for the years shown?

b. Does the Air Force expect to need less than 14,000 pilots for the years shown?

c. In what year is the number of pilots needed by the Air Force approximately equal to the number the Air Force will have?

d. Is the difference between the number of pilots the Air Force needs and the number it expects to have increasing or decreasing for the years shown?

a. Decreasing; **b.** no; **c.** 1997; **d.** increasing

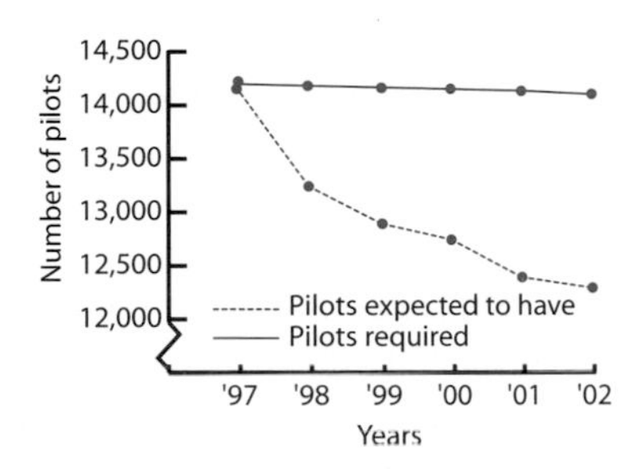

16. The total number of points scored by each team of the National Football League (NFL) for the 1997 season was: 348, 324, 237, 266, 299, 304, 327, 472, 394, 379, 375, 365, 339, 317, 307, 283, 422, 375, 369, 299, 263, 372, 265, 355, 354, 333, 326, 320, 225, 313. Complete the frequency table at the right for these data. (Source: NFL Website)

Classes Points scored	Frequency
425 – 474	1
375 – 424	5
325 – 374	10
275 – 324	9
225 – 276	5

17. The total number of touchdowns scored by each team of the National Football League (NFL) for the 1997 season was: 38, 50, 26, 37, 44, 24, 31, 32, 35, 35, 42, 38, 55, 43, 41, 36, 42, 46, 28, 32, 28, 29, 41, 43, 42, 27, 36, 40, 43, 41, 36. Complete the frequency table at the right for these data. (Source: NFL Website)

Classes No. of Touchdowns	Frequency
50 – 55	2
44 – 49	2
38 – 43	12
32 – 37	8
26 – 31	6
20 – 25	1

18. According to Campus Concepts, two-thirds of all college students have a job. The monthly amounts of money earned by surveyed students were: 213, 596, 498, 528, 109, 291, 354, 261, 489, 344, 415, 601, 584, 591, 402, 478, 483, 320, 390, 506, 549, 590, 85, 176, 298, 166, 161, 259, 566, 73, 574, 231, 174, 647, 399, 300, 272, 394, 187, 218. Complete the frequency table at the right for these data.

Classes Monthly Income	Frequency
0 – 99	2
100 – 199	6
200 – 299	8
300 – 399	7
400 – 499	6
500 +	11

19. Each state and the District of Columbia impose a tax on each package of cigarettes. The taxes (in cents) imposed as of January 1, 1998, were: 16.5, 100, 58, 31.5, 37, 20, 50, 24, 65, 33.9, 12, 80, 28, 58, 15.5, 36, 24, 3, 20, 74, 36, 76, 75, 48, 18, 17, 18, 34, 35, 37, 80, 21, 56, 5, 44, 24, 23, 68, 31, 71, 7, 33, 13, 41, 51.5, 44, 2.5, 82.5, 17, 29, 12. Complete the frequency table at the right for this data. (Source: Gannett News Service)

Classes Cigarette Tax	Frequency
90.5 – 105	1
75.5 – 90	4
60.5 – 75	5
45.5 – 60	6
30.5 – 45	13
15.5 – 30	15
0.5 – 15	7

20. Based on data from the U.S. Geological Service, the following earthquake magnitudes (on the Richter scale) were recorded for a region of southern California: 2.6, 3.3, 2.4, 4.4, 2.6, 2.7, 4.0, 2.7, 2.0, 2.0, 3.9, 2.2, 3.6, 2.5, 4.0, 3.3, 5.7, 2.9, 4.7, 3.4, 2.1, 2.4, 3.9, 2.6, 3.5, 5.2, 5.8, 4.7, 3.6, 2.4, 5.0, 2.0, 2.2, 3.1, 3.4. Complete the frequency table at the right for this data.

Classes Magnitude	Frequency
5.0 – 5.9	4
4.0 – 4.9	5
3.0 – 3.9	10
2.0 – 2.9	16
1.0 – 1.9	12

21. Make a relative frequency table for the data in Exercise 17. Round to the nearest tenth of a percent.

Classes No. of Touchdowns	Relative Frequency
50 – 55	6.5%
44 – 49	6.5%
38 – 43	38.7%
32 – 37	25.8%
26 – 31	19.4%
20 – 25	3.2%

22. Make a relative frequency table for the data in Exercise 18. Round to the nearest tenth of a percent.

Classes Monthly Income	Relative Frequency
0 – 99	5%
100 – 199	15%
200 – 299	20%
300 – 399	17.5%
400 – 499	15%
500 +	27.5%

23. Make a relative frequency table for the data in Exercise 19. Round to the nearest tenth of a percent.

Classes Cigarette Tax	Relative Frequency
90.5 – 105	2.0%
75.5 – 90	7.8%
60.5 – 75	9.8%
45.5 – 60	11.8%
30.5 – 45	25.5%
15.5 – 30	31.4%
0.5 – 15	13.7%

24. Make a relative frequency table for the data in Exercise 20. Round to the nearest tenth of a percent.

Classes Magnitude	Relative Frequency
5.0 – 5.9	8.5%
4.0 – 4.9	10.6%
3.0 – 3.9	21.3%
2.0 – 2.9	34.0%
1.0 – 1.9	25.5%

Histograms

25. The histogram at the right is based on data from the Toronto police department's web site and shows the approximate number of emergency 911 calls that can be expected during a 30-day period.

a. On how many days were the number of 911 calls between 2350 and 2500?
b. On how many days were the number of 911 calls less than 2050?
c. Is it possible to tell from the histogram the number of days on which there were 2350 calls?

a. 4; **b.** 8; **c.** no

26. The histogram at the right shows the number of home runs that were hit by all of the teams in Major League Baseball during the 1997 season. (Source: Major League Baseball)

a. How many teams hit 215 home runs or more?
b. How many teams hit fewer than 190 home runs?
c. Is it possible to determine how many teams hit 190 home runs?

a. 16; **b.** 5; **c.** no

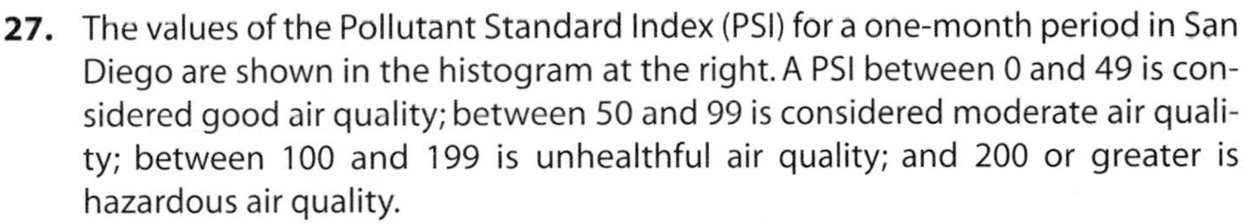

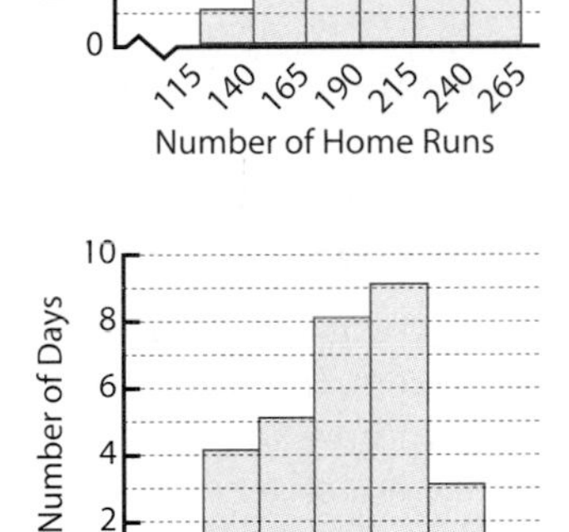

27. The values of the Pollutant Standard Index (PSI) for a one-month period in San Diego are shown in the histogram at the right. A PSI between 0 and 49 is considered good air quality; between 50 and 99 is considered moderate air quality; between 100 and 199 is unhealthful air quality; and 200 or greater is hazardous air quality.

a. How many days during the month was the PSI showing unhealthy air quality?
b. For 17 days the PSI was between what two numbers?
c. For the month shown, were there any days for which the PSI indicated good air quality?

a. 4; **b.** 80 and 100; **c.** no

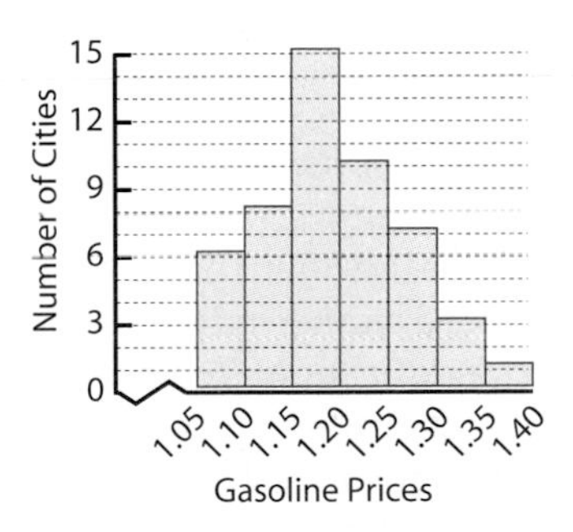

28. The results of a survey of the price of gasoline in 50 cities in the U.S. during March of 1997 are shown in the histogram at the right.

a. How many cities had gasoline prices between \$1.10 and \$1.30?
b. For the cities shown, were there more cities with gasoline prices that were less than \$1.30 or more cities with gasoline prices \$1.30 or more?
c. For March 1997, were there any cities whose gasoline prices were more than \$1.40? (Hint: Be careful with your answer.)

a. 40; **b.** less than \$1.30; **c.** Impossible to tell. The survey included only 50 cities in the U.S.

Stem-and-Leaf Plots

29. A survey was taken to determine the price, rounded to the nearest dollar, people pay for nonprescription sunglasses. The results of the survey are: 10, 60, 17, 53, 90, 75, 41, 61, 19, 77, 61, 92, 58, 38, 39, 91, 38, 82, 68, 82, 64, 63, 46, 47, 23, 69, 62, 59, 48, 28. Draw a stem-and-leaf plot for this data using the first digit as the stem and the second digit as the leaf.

1	0, 7, 9
2	3, 8
3	8, 9, 8
4	1, 6, 7, 8
5	3, 8, 9
6	0, 1, 1, 8, 4, 3, 9, 2
7	5, 7
8	2, 2
9	0, 2, 1

30. The scores for the winning Super Bowl team through the 1998 game are: 35, 33, 16, 23, 16, 24, 14, 24, 16, 21, 32, 27, 35, 31, 27, 38, 38, 46, 39, 42, 20, 55, 20, 37, 52, 30, 27, 35, 31. Draw a stem-and-leaf plot for this data using the first digit as the stem and the second digit as the leaf. (Source: *Sporting News* Internet site)

1	6, 6, 4, 6
2	3, 4, 4, 1, 7, 7, 0, 0, 7
3	5, 3, 2, 5, 1, 8, 8, 9, 7, 0, 5, 1
4	6, 2
5	5, 2

31. The average high temperatures in the 50 largest cities in the United States in July are: 94, 91, 91, 87, 86, 85, 85, 85, 84, 84, 83, 83, 83, 82, 82, 82, 81, 80, 80, 79, 79, 78, 78, 78, 78, 77, 77, 77, 77, 76, 76, 76, 75, 75, 75, 74, 74, 74, 74, 73, 73, 73, 73, 73, 73, 70, 70, 68, 65, 62. Draw a stem-and-leaf plot for this data using the first digit as the stem and the second digit as the leaf. (Source: Census Bureau web site.)

6	8, 5, 2
7	9, 9, 8, 8, 8, 8, 7, 7, 7, 7, 6, 6, 6, 5, 5, 5, 4, 4, 4, 4, 3, 3, 3, 3, 3, 3, 0, 0
8	7, 6, 5, 5, 5, 4, 4, 3, 3, 3, 2, 2, 2, 1, 0, 0
9	4, 1, 1

32. The following data represent average annual rainfall (in inches) for the 30 cities in the United States with the highest average annual rainfall: 44, 54, 56, 39, 41, 37, 36, 45, 49, 51, 64, 30, 62, 51, 49, 32, 36, 41, 40, 43, 45, 33, 32, 49, 40, 41, 36, 54, 31, 33. Draw a stem-and-leaf plot for this data using the first digit as the stem and the second digit as the leaf. (Source: Census Bureau web site.)

3	9, 7, 6, 0, 2, 6, 3, 2, 6, 1, 3
4	4, 1, 5, 9, 9, 1, 0, 3, 5, 9, 0, 1
5	4, 6, 1, 1, 4
6	4, 2

Applying Concepts

33. The graph at the right shows the states in which it is most difficult to pay the monthly rent. The graph shows the number of hours a month a person making the minimum wage of $5.15 per hour would have to work to pay the average monthly rent on a two-bedroom apartment. The graph also shows the national average. (Source: National Low Income Housing Coalition.)

a. In which two states shown is the monthly number of hours twice the national average or more?

b. What is the average monthly rent in Hawaii?

c. What is the average monthly rent in California?

a. Hawaii and New Jersey; **b.** $741.60 **c.** $597.40

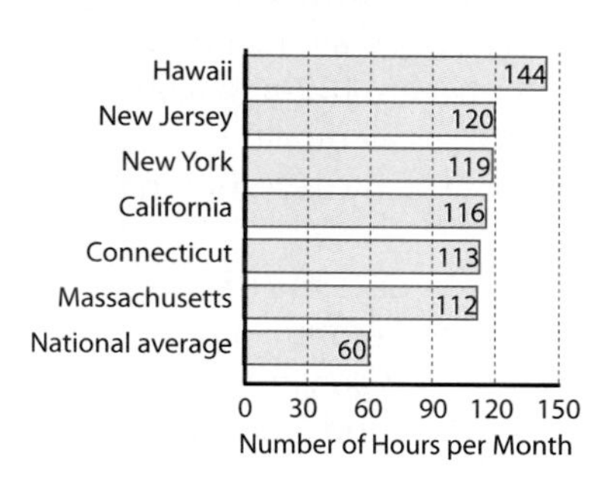

Exploration

34. ***Double Graphs*** Sometimes two graphs are drawn on the same coordinate grid but with vertical axes on each side of the graph. One such graph is shown at the right. The vertical axis on the left side shows the average hourly wage of a worker in the U.S. for March 1997 through February 1998. The axis on the right shows the average number of hours per week a worker spent working for the same time period. For instance, from the graph, the average hourly wage in March 1997 was approximately $12.15 per hour and the average number of hours worked was 34.8.

a. What was the approximate hourly wage for a worker in December?

b. What was the average number of hours per week a person worked in May?

c. Suppose Amanda earned the average hourly wage of $12.31 per hour in August. If she worked the average number of hours per week for that month, what were Amanda's earnings before any deductions?

d. Suppose Jorge earned the average hourly wage of $12.52 per hour in February. If he worked the average number of hours per week for that month, what were Jorge's earnings before any deductions?

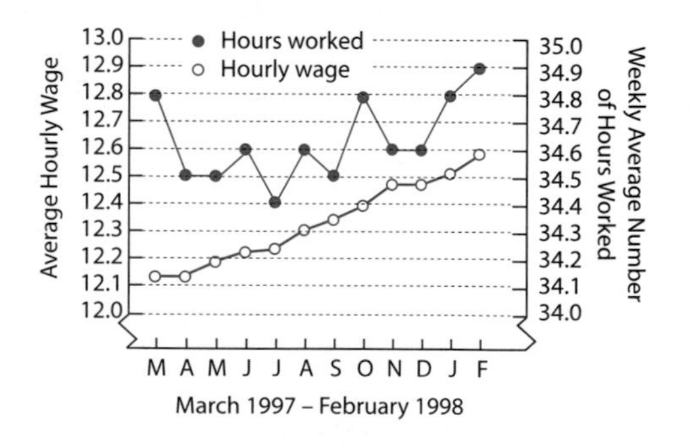

Section 1.4 Evaluating Numerical and Variable Expressions

Exponential Expressions

Genealogy is the study of the history of a family. A genealogist uses a family tree as a pictorial history of your ancestors. A sample of a family tree is shown below.

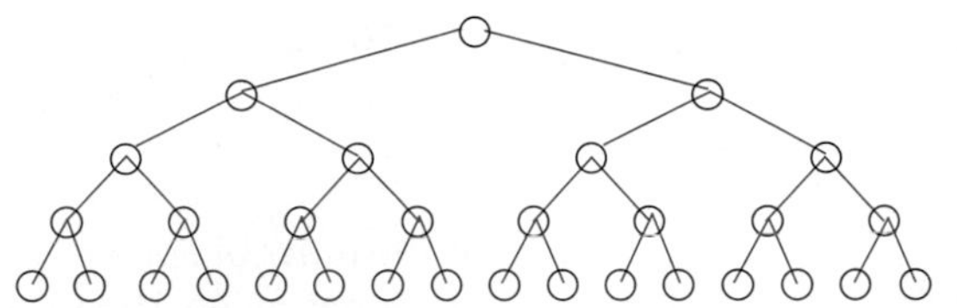

In each generation, you have twice the number of ancestors as in the previous generation.

$$2 \text{ Parents}$$
$$2 \cdot 2 = 4 \text{ Grandparents}$$
$$2 \cdot 2 \cdot 2 = 8 \text{ Great-grandparents}$$
$$2 \cdot 2 \cdot 2 \cdot 2 = 16 \text{ Great-great-grandparents}$$

POINT OF INTEREST

Rene Descartes (1596–1650) was the first mathematician to use exponential notation as it is used today.

Instead of showing the number of ancestors as a product, we could have written each product, called **expanded form**, in **exponential form.**

Expanded form		Exponential form	Read as
2	=	2^1	"2 to the first power" or just "two."
$2 \cdot 2$	=	2^2	"2 squared" or "2 to the second power."
$2 \cdot 2 \cdot 2$	=	2^3	"2 cubed" or "2 to the third power."
$2 \cdot 2 \cdot 2 \cdot 2$	=	2^4	"2 to the fourth power."

For each exponential expression, the **exponent** indicates the number of times the **base** is used in a product.

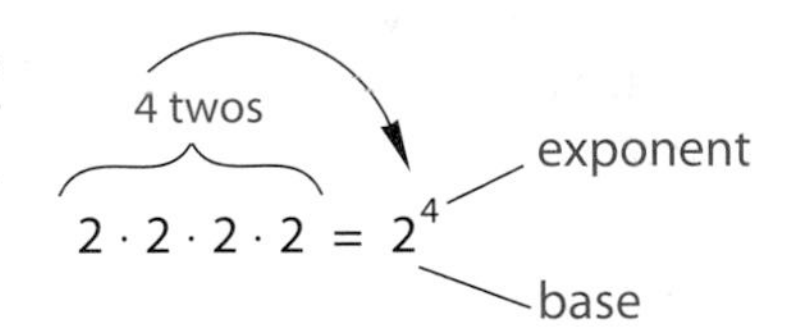

Question: Write the number of great-great-great grandparents you had in exponential form.[1]

Example 1 Evaluate: **a.** 5^3 **b.** $2^3 \cdot 3^2$

Solution Write the expressions in expanded form. Then multiply.

a. $5^3 = 5 \cdot 5 \cdot 5 = 125$

b. $2^3 \cdot 3^2 = (2 \cdot 2 \cdot 2) \cdot (3 \cdot 3) = 8 \cdot 9 = 72$

You-Try-It 1 Evaluate: **a.** 2^6 **b.** $3^3 \cdot 2^2$

Solution See page S2.

---

1. 2^5

Order of Operations Agreement

If a rock is dropped from a cliff, the distance the rock has fallen (in feet) after 2 seconds is given by the expression 16×2^2. There are two possible ways to evaluate this expression.

Evaluate the exponential expression. Then multiply.	Multiply. Then evaluate the exponential expression.
16×2^2	16×2^2
16×4	32^2
64	1024

Note that the *order* of applying the operations gives two different answers to how far the rock has fallen. The value of the expression on the left indicates that the rock has fallen 64 feet in 2 seconds. However, if we use the value of the expression on the right, the rock has fallen 1024 feet in 2 seconds. Scientists that have studied the motion of falling objects have determined that the rock will fall 64 feet in 2 seconds. Therefore, the correct evaluation of the expression is the one on the left. To ensure that everyone performs the order of mathematical operations in the same way, we use the following agreement.

POINT OF INTEREST

The cross × was first used as a symbol for multiplication in 1631 in a book titled *The Key to Mathematics*. Also in that year, another book, *Practice of the Analytical Art*, advocated the use of a dot to indicate multiplication.

SUGGESTED ACTIVITY

See the *Instructor's Resource Manual* for an activity that will allow students to discover some of the Order of Operations Agreement.

Order of Operations Agreement

1. Perform operations inside grouping symbols. These include parentheses, brackets, braces, and fraction bars.
2. Evaluate exponential expressions.
3. Perform multiplications and divisions as they occur from left to right.
4. Perform additions and subtractions as they occur from left to right.

We will seldom use the $\times$ sign for multiplication as we did in the expression 16×2^2. Instead, we will use a dot (·) or parentheses with no sign at all. For instance,

$$5 \cdot 6 = 30 \qquad 5(6) = 30 \qquad (5)(6) = 30 \qquad (6)5 = 30$$

Also, a fraction bar is frequently used to indicate division, as in $\frac{24}{3} = 8$, instead of writing $24 \div 3 = 8$.

Example 2 Evaluate: **a.** $\frac{8+4}{2} + 3(4^2)$ **b.** $2(7-3)^2 + 4 - \frac{12}{5-2}$

Solution Use the Order of Operations Agreement.

a. $\frac{8+4}{2} + 3(4^2)$

$= \frac{12}{2} + 3(4^2)$ • Simplify expressions within grouping symbols. Recall that a fraction bar is a grouping symbol.

$= \frac{12}{2} + 3(16)$ • Simplify exponential expressions.

$= 6 + 48$ • Multiply and divide from left to right as they occur.

$= 54$ • Add.

b. $2(7-3)^2 + 4 - \frac{12}{5-2}$

$= 2(4)^2 + 4 - \frac{12}{3}$ • Perform operations inside grouping symbols. Recall that a fraction bar is a grouping symbol.

$= 2(16) + 4 - \frac{12}{3}$ • Simplify exponential expressions.

$= 32 + 4 - 4$ • Multiply and divide from left to right as they occur.

$= 32$ • Add and subtract from left to right.

You-Try-It 2 Evaluate: **a.** $48 \div 2^3 - 2 \cdot 3$ **b.** $82 - 8(4 + 2 \cdot 3) + (4-1)^3$

Solution See page S2.

Question: How should parentheses be placed in $37 + 15 \div 5 + 8 \div 2$ so that when simplified, **(a)** the answer is 2; **(b)** the answer is 44?[2]

Evaluating Variable Expressions

Suppose that gasoline costs $1.30 per gallon. Then the amount you spend for gas for your car depends on how many gallons you purchase. Here are some examples.

A purchase of 5 gallons of gas costs 1.30(5) = $6.50.
A purchase of 7 gallons of gas costs 1.30(7) = $9.10.
A purchase of 11.3 gallons of gas costs 1.30(11.3) = $14.69.

If you decide to fill the tank, you may not know how many gallons of gas will be required. In that case you might use a letter like g to represent the number of gallons that will be purchased.

A purchase of g gallons of gas costs $1.30g$.

Recall that g is called a *variable*. A **variable** is a letter that represents a quantity that can change, or vary. The expression $1.30g$ (which means $1.30 \cdot g$) is called a **variable expression**. The number 1.30 is called the **coefficient** of the variable.

The expression $1.30g$ represents the cost to purchase g gallons of gas. As g (the number of gallons purchased) changes, the cost of the purchase changes.

Example 3 Using the variable expression $1.30g$, find the cost to purchase 12.5 gallons of gas.

Solution $1.30g$

$1.30(12.5)$ • Replace g by 12.5.

$= 16.25$ • Multiply.

The cost for 12.5 gallons of gas is $16.25.

You-Try-It 3 Using the variable expression $1.30g$, find the cost to purchase 9.7 gallons of gas.

Solution See page S2.

Replacing a variable in a variable expression, as we did in Example 3, by a given number and then simplifying the numerical expression is called **evaluating a variable expression**. The result (16.25) is called the **value of the variable expression**. The number substituted for g (12.5) is called the **value of the variable**.

---

2. Use the guess and check problem-solving strategy mentioned earlier.
(a) $(37 + 15) \div (5 + 8) \div 2 = 2$; **(b)** $37 + (15 \div 5) + (8 \div 2) = 44$

We could also use a variable to represent the total cost of the gasoline purchases discussed on the previous page. For instance, if we let T represent the total cost (in dollars) for a gasoline purchase, then $T = 1.30g$. This *equation* shows the relationship between T, the total cost of the purchase, and g, the number of gallons of gas purchased.

g	$1.30g$	T
0	1.30(0)	0
2	1.30(2)	2.60
4	1.30(4)	5.20
6	1.30(6)	7.80
8	1.30(8)	10.40
10	1.30(10)	13.00
12	1.30(12)	15.60

Using this equation, an **input/output table** can be made that shows how T changes as g changes. The input is g, the number of gallons purchased. The output is T, the cost of the purchase.

For the table above and at the right, we have chosen 0, 2, 4, 6, 8, 10, and 12 as the values of g. However, other values of g could have been used.

Question: What is the meaning of the number 10.40 in the table above?[3]

TAKE NOTE

Typical graphing calculator screens to create the table at the right are shown below.

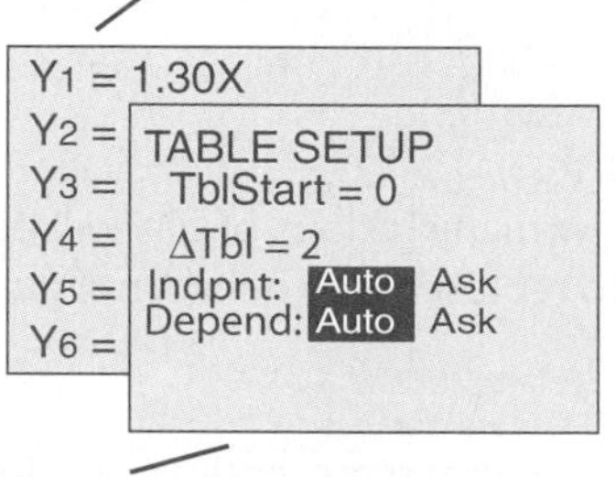

TblStart is the beginning value of X and ΔTbl is the difference between successive values of X.

See the appendix Guidelines for Using Graphing Calculators for additional information.

A graphing calculator can be used to create tables for equations such as $T = 1.30g$. This is accomplished by using the **Y = editor screen** as shown at the left. The output variable is designated by one of the calculator's Y variables. For this example we designate T as Y1. The input variable is usually designated by X. Thus the equation would appear as Y1 = 1.30X. The resulting table would look similar to the one at the right. The difference between any two successive X values is called the *change in X* or the *increment in X*. The symbol Δ is frequently used to represent the phrase "the change in." For the table at the right, the change in X is 2.

X	Y1	
0	0	
2	2.6	
4	5.2	
6	7.8	
8	10.4	
10	13	
12	15.6	
X= 0		

Besides getting a table of values as above, you can use your graphing calculator to find the value of the output variable for any value of the input variable. For example, once the total cost of a gasoline purchase equation is entered into your calculator, you can determine the value of T (or Y1) for any number of gallons of gasoline purchased. For instance, to find the total cost of purchasing 11.37 gallons of gasoline, store 11.37 in X. Then display the value of Y1. From the display at the right, the cost is \$14.78.

```
11.37 ->X
            11.37
Y1
           14.781
```

Example 4 If a rock is dropped off a cliff, the distance the rock falls is given by the equation $d = 16t^2$, where d is the distance (in feet) the rock has fallen and t is the time (in seconds) the rock has been falling. Use a graphing calculator to create an input/output table for this equation for increments of 0.5 second beginning with $t = 0$.

Solution The input variable is t, the number of seconds the rock was falling. The output variable is d, the distance the rock falls. Using the Y = editor, enter Y1 = 16X^2. Now adjust the table setting so that X begins with 0 and the increment for X (ΔTbl) is 0.5. The result should be a table similar to the one at the left.

X	Y1	
0	0	
.5	4	
1	16	
1.5	36	
2	64	
2.5	100	
3	144	
X= 0		

Question: What is the meaning of the number 64 in the table in Example 4?[4]

---

3. It costs \$10.40 for 8 gallons of gas.
4. The rock has fallen 64 feet in 2 seconds.

You-Try-It 4 Suppose that the average speed of an American Airlines flight from Los Angeles to Boston is 525 mph. Then the distance (in miles) the plane is from Boston is given by the equation $d = 2650 - 525t$, where t is the number of hours since the plane left Los Angeles. Create a table for this equation beginning with $t = 0$ and using increments of 1 hour.

Solution See page S2.

Example 5 The fuel economy of a car is given by the equation $M = -0.02v^2 + 1.6v + 3$, where M is the fuel economy in miles per gallon (mpg) and v is the speed of the car in miles per hour (mph). Create a table for this equation using ΔTbl = 5 and TblStart = 25. Use the table to answer the following questions.

a. What is the fuel economy when the speed of the car is 50 mph?

b. What is the speed of the car when the fuel economy is 35 mpg?

Solution Typical graphing calculator screens are shown at the right. Note that the fuel economy is given by Y1 and the speed is given by X.

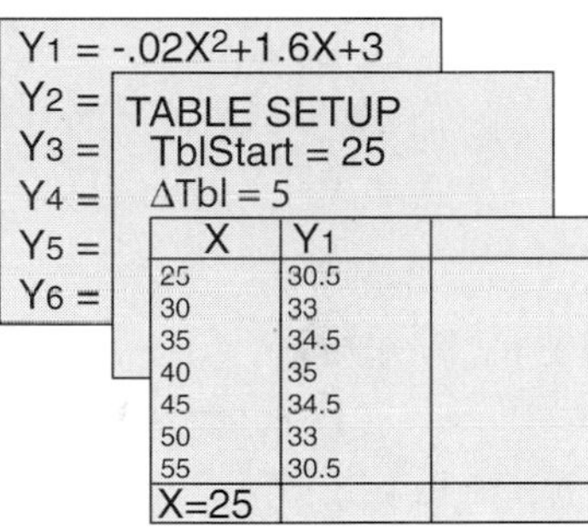

a. To find the fuel economy (Y1) when the speed (X) is 50 mph, read the output from the table when the input is 50. The fuel economy is 33 mpg.

b. To find the speed of the car when the fuel economy is 35 mpg, read the input from the table when the output is 35. The speed is 40 mph.

You-Try-It 5 The amount of garbage generated by each person living in the U.S. has been increasing. It is given by the equation $A = 0.05Y - 95$, where A is the pounds per person per day of garbage generated and Y is the year. Create a table for this equation using ΔTbl = 5 and TblStart = 1970. Use the table to answer the following questions.

a. What was the amount of garbage generated per person per day in 1990?

b. In what year was the amount of garbage generated per person per day 4.75 pounds?

Solution See page S2.

A **formula** is a special type of equation that states a rule about measurements. For instance, the formula for the area of a rectangle, $A = LW$, shows the relationship between the area, A, of the rectangle and its length, L, and width, W. In this formula, two variables, L and W, are written together. When two or more variables are written together in an expression, the operation is assumed to be multiplication. Thus $A = LW$ means $A = L \cdot W$.

W L

$A = LW$

In a similar manner,

xyz	means	$x \cdot y \cdot z$
$7rs$	means	$7 \cdot r \cdot s$
$\frac{2ab}{3}$	means	$\frac{2}{3} \cdot a \cdot b$

When evaluating variable expressions with more than one variable, replace each of the variables by its given value. Then use the Order of Operations Agreement to simplify the numerical expression.

For example, the area of a trapezoid is given by $A = \frac{1}{2}h(a+b)$, where h is the height and a and b are the lengths of the bases. To find the area of the trapezoid shown at the right, replace h, a, and b by their values and then simplify.

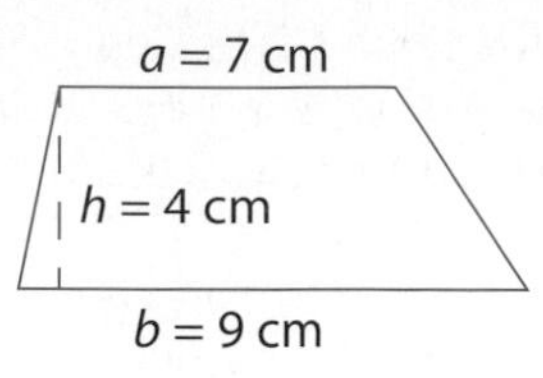

$A = \frac{1}{2}h(a+b)$

$= \frac{1}{2}(4)(7+9)$ • $h = 4, a = 7$, and $b = 9$.

$= \frac{1}{2}(4)(16) = 2(16)$ • Use the Order of Operations Agreement to simplify the numerical expression.

$= 32$

The area is 32 cm^2.

Here are some additional examples of evaluating variable expressions.

SUGGESTED ACTIVITY
Students can use their calculators to check the evaluation. The screen below shows one method.

```
3 → A
                 3
2 → B
                 2
5AB^3 + 2A²B² – 4
               188
```

Example 6 Evaluate the variable expression $5ab^3 + 2a^2b^2 - 4$ when $a = 3$ and $b = 2$.

Solution

$5ab^3 + 2a^2b^2 - 4$

$5(3)(2)^3 + 2(3)^2(2)^2 - 4$ • Replace a by 3 and b by 2.

$= 5(3)(8) + 2(9)(4) - 4$ • Use the Order of Operations Agreement to simplify the resulting numerical expression.

$= 120 + 72 - 4 = 188$

You-Try-It 6 What is the value of the variable expression $3xy^2 - 3x^2y$ when $x = 2$ and $y = 5$?

Solution See page S2.

In the preceding examples, an equation was given that showed the relationship between two quantities. Using the equation, a table was created. Sometimes, however, a relationship exists between two quantities, but there is no simple formula to show the relationship. A frequency table or relative frequency table is an example of using a table to show the relationship between two quantities.

The table at the left is based on data from the U.S. Census Bureau's Web site. It shows the relationship between age groups, A, in the U.S. and the percent, P, of the U.S. population in that age group. For instance, the age group between 20 and 29 makes up 14.83% of the U.S. population.

Age (A)	Percent (P)
0 – 9	15.05
10 – 19	14.80
20 – 29	14.83
30 – 39	12.97
40 – 49	15.47
50 – 59	10.45
60 – 69	7.69
70 – 79	6.13
80 – 89	2.75
90 – 99	0.53
100 +	0.02

Source: U.S. Bureau of the Census Web site, 9/97

Question: From the table at the left, what age group makes up the largest percent of the population?[5]

5. The age group 40 – 49 is the largest percent of the population.

1.4 EXERCISES

Topics for Discussion

1. Why must there be an Order of Operations Agreement? To ensure that everyone performs the order of mathematical operations in the same way.

2. What is the meaning of the phrase "evaluate the variable expression"? Answers will vary.

3. What is the difference between the meaning of "the value of the variable" and the meaning of "the value of the variable expression"? The first means the number substituted for the variable; the second means the result.

4. What is an input/output table? Answers will vary.

Exponential Expressions

Evaluate the exponential expressions.

5. 3^4 81 **6.** 2^5 32 **7.** 1^{20} 1 **8.** 0^{17} 0 **9.** 14^2 196 **10.** 9^3 729

Order of Operations Agreement

Evaluate.

11. $15 - 3 \cdot 4$ 3 **12.** $37 - 6 \cdot 5$ 7 **13.** $5^2 - 4(5)$ 5

14. $3^4 + \frac{15}{5}$ 84 **15.** $\frac{16(2+3)}{10}$ 8 **16.** $12 + 4 \cdot 2^3$ 44

17. $3^3 + 5(8-6)$ 37 **18.** $2^2(3^2) - 2 \cdot 3$ 30 **19.** $12 - (12-4) \div 4$ 10

20. $(8-2)^2 - 3 \cdot 4 + 6$ 30 **21.** $5(3+4)^2 - 6(7-5)^3$ 197 **22.** $2(8-5)^3 + 6(1+3)^3 - 9^2$ 357

Evaluating Variable Expressions

Evaluate the variable expressions when $a = 3, b = 2,$ and $c = 4$.

23. b^4 16 **24.** c^2 16 **25.** a^2b^3 72 **26.** $3b^4$ 48

27. $2a^2c$ 72 **28.** $4bc^2$ 128 **29.** $a^2 + b^2$ 13 **30.** $(a+b)^2$ 25

31. $5c - 3b$ 14 **32.** $9a + 3c$ 39 **33.** $5bc - 2ab$ 28 **34.** $6b^2 - 2c$ 16

35. $12a + 3b - 2c$ 34

36. $8 + 2a + 3bc$ 38

37. $8 - 2(c - b) + 3a^2$ 31

38. $(2a + b)^2 - 2(a^2 + b^2)$ 38

39. $a(4c - 2b) + 3b^2$ 48

40. $\dfrac{a^3 + b^3}{a + b}$ 7

Evaluate the formulas for Exercises 41 to 46.

41. The formula for the perimeter, P, of a rectangle is given by $P = 2L + 2W$, where L is the length and W is the width. Find the perimeter of the rectangle shown at the right. 220 centimeters

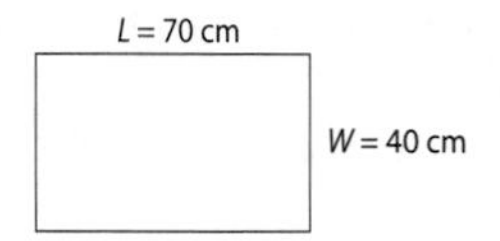

42. The formula for the perimeter, P, of a square is given by $P = 4s$, where s is the length of one of the equal sides. Find the perimeter of the square shown at the right. 20 feet

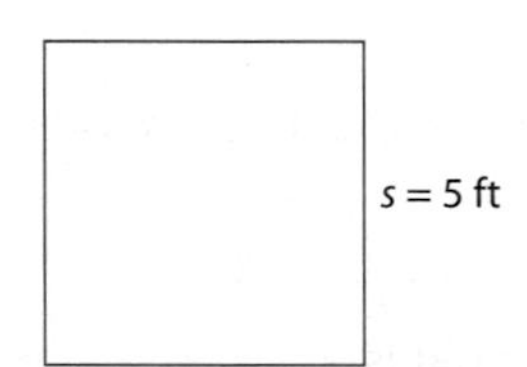

43. The formula for the area, A, of a rectangle is given by $A = LW$, where L is the length and W is the width. Find the area of the rectangle in Exercise 41.
2800 square centimeters

44. The formula for the area, A, of a square is given by $A = s^2$, where s is the length of one of the equal sides. Find the area of the square in Exercise 42.
25 square feet

45. The formula for the area, A, of a triangle is given by $A = \frac{1}{2}hb$, where h is the height and b is the base. Find the area of the triangle shown at the right.
3 square meters

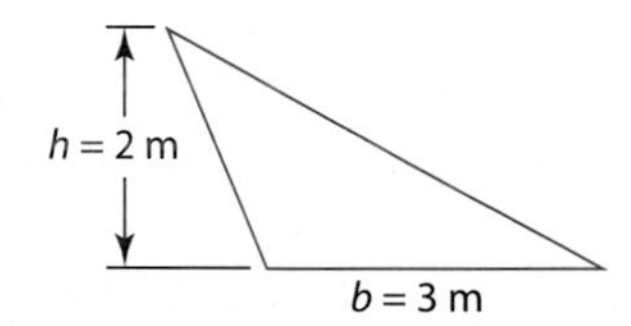

46. The formula for the area, A, of a trapezoid is given by $A = \frac{1}{2}h(b_1 + b_2)$, where h is the height, b_1 is one base, and b_2 is the other base. Find the area of the trapezoid shown at the right. 3900 square inches

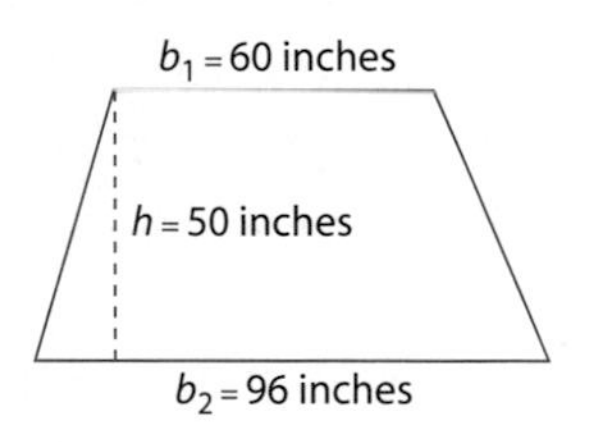

For Exercises 47 to 54, make an input/output table for the equation using the given information. Use the input/output table to answer the questions.

47. $y = 3x + 2$, ΔTbl = 1, TblStart = 0

Input, x	0	1	2	3	4	5
Output, $y = 3x + 2$	2	5	8	11	14	17

a. What is the value of y when $x = 3$? 11

b. What is the value of x when $y = 5$? 1

48. $y = 2x - 3$, ΔTbl = 1, TblStart = 2

Input, x	2	3	4	5	6	7
Output, $y = 2x - 3$	1	3	5	7	9	11

a. What is the value of y when $x = 6$? 9

b. What is the value of x when $y = 5$? 4

49. $y = x^2$, ΔTbl = 0.5, TblStart = 1

Input, x	1	1.5	2	2.5	3	3.5
Output, $y = x^2$	1	2.25	4	6.25	9	12.25

a. What is the value of y when $x = 2$? 4
b. What is the value of x when $y = 12.25$? 3.5

50. $y = x^2 - 1$, ΔTbl = 0.5, TblStart = 2

Input, x	2	2.5	3	3.5	4	4.5
Output, $y = x^2 - 1$	3	5.25	8	11.25	15	19.25

a. What is the value of y when $x = 4.5$? 19.25
b. What is the value of x when $y = 5.25$? 2.5

51. $y = 2x^2 + 1$, ΔTbl = 0.25, TblStart = 1.5

Input, x	1.5	1.75	2	2.25	2.5	2.75
Output, $y = 2x^2 + 1$	5.5	7.125	9	11.125	13.5	16.125

a. What is the value of y when $x = 1.75$? 7.125
b. What is the value of x when $y = 16.125$? 2.75

52. $y = 12 - x^2$, ΔTbl = 0.25, TblStart = 1

Input, x	1	1.25	1.5	1.75	2	2.25
Output, $y = 12 - x^2$	11	10.438	9.75	8.9375	8	6.9375

a. What is the value of y when $x = 1.5$? 9.75
b. What is the value of x when $y = 8.9375$? 1.75

53. $y = x^2 + 3x + 1$, ΔTbl = 1, TblStart = 0

Input, x	0	1	2	3	4	5
Output, $y = x^2 + 3x + 1$	1	5	11	19	29	41

a. What is the value of y when $x = 3$? 19
b. What is the value of x when $y = 29$? 4

54. $y = x^2 + 5x - 2$, ΔTbl = 1, TblStart = 1

Input, x	1	2	3	4	5	6
Output, $y = x^2 + 5x - 2$	4	12	22	34	48	64

a. What is the value of y when $x = 4$? 34
b. What is the value of x when $y = 48$? 5

Applying Concepts

55. In June, the temperature at various elevations of the Grand Canyon can be approximated by the equation $T = -0.005x + 113.25$, where T is the temperature in Fahrenheit and x is the elevation (distance above sea level) in feet. Use ΔTbl = 500 and TblStart = 2450.

a. According to this equation, what is the temperature at an elevation of 4450 feet (about halfway down the south rim of the Canyon)? $T = 91°$ F
b. At Inner Gorge, the bottom of the Canyon, the temperature is 101°F. What is the elevation of Inner Gorge? 2450 feet

56. The equation $F = \frac{9}{5}C + 32$ can be used to convert Celsius temperatures, C, to Fahrenheit temperatures, F. Use ΔTbl = 5 and TblStart = 0.

a. Find the Fahrenheit temperature when the Celsius temperature is 20°C. $F = 68°$ F
b. Find the Celsius temperature when the Fahrenheit temperature is 50°F. $C = 10°$ C

57. Old Faithful is a geyser in Yellowstone National Park that was so named because of its regular eruptions for the past 100 years. An equation that can predict the approximate time until the next eruption is given by $T = 12.4L + 32$, where T is the time (in minutes) to the next eruption and L is the length of time (in minutes) of the last eruption. Use ΔTbl = 0.5 and TblStart = 2.

a. According to this equation, how long will it be to the next eruption if the last eruption lasted 3.5 minutes? $T = 75.4$ minutes

b. If the time between two eruptions is 63 minutes, what was the length of time of the last eruption? $L = 2.5$ minutes

58. The altitude (in feet above sea level) of a hot-air balloon is given by the equation $H = 100t + 1250$, where H is the altitude of the balloon t seconds after it has been released. Use ΔTbl = 1 and TblStart = 0.

a. What is the altitude of the balloon 3 seconds after it is released? $H = 1550$ feet

b. How many seconds after it is released is the balloon at an altitude of 1750 feet? $t = 5$ seconds

59. The equation $h = -16t^2 + 64t + 5$ gives the height of a ball thrown straight up, where h is the height of the ball in feet and t is the time in seconds since the ball was released. Use ΔTbl = 0.5 and TblStart = 0.

a. What is the height of the ball 2 seconds after it is released? $h = 69$ feet

b. What are two times, once on the way up and once on the way down, that the ball is 65 feet above the ground? 1.5 s, 2.5 s

60. The distance d (in feet) of a platform diver above the water t seconds after the dive begins is given by the equation $d = 50 - 16t^2$. Use ΔTbl = 0.25 and TblStart = 0.

a. How many feet above the water is the diver 0.25 second after the dive begins? $d = 49$ feet

b. After how many seconds is the diver 25 feet above the water? $t = 1.25$ seconds

Explorations

61. The order in which the operations are performed for an expression such as a^{b^c} influence the final value of that expression.

a. Evaluate $(2^3)^2$ and $2^{(3^2)}$.

b. Is there a difference between the two expressions?

c. If the expression is written as 2^{3^2}, what should be its value?

d. What is the Order of Operations Agreement for a^{b^c}?

e. Does your calculator follow that agreement? (Note: Some calculators follow the Order of Operations Agreement for exponents and some do not.)

62. Consider the following variable expressions: $(a+b)^2$, a^2+b^2, $(a+b)^3$, a^3+b^3, $(a+b)^4$, and a^4+b^4.

a. By trying different values of a and b, is $(a+b)^2 = a^2+b^2$ always true?

b. By trying different values of a and b, is $(a+b)^3 = a^3+b^3$ always true?

c. By trying different values of a and b, is $(a+b)^4 = a^4+b^4$ always true?

d. Using inductive reasoning, make a conjecture as to whether $(a+b)^n = a^n+b^n$ is always true when n is a natural number.

Section 1.5 Translating Phrases into Variable Expressions

Translating Phrases into Variable Expressions

Creating a variable expression is an important goal in the application of mathematics. Many application problems are given in verbal or written form and must be translated to a mathematical expression. A partial list of the verbal phrases used to indicate the different mathematical operations follows.

Addition	added to	7 added to z	$z+7$
	more than	9 more than x	$x+9$
	the sum of	the sum of a and b	$a+b$
	the total of	the total of 5 and c	$5+c$
	increased by	9 increased by w	$9+w$
Subtraction	minus	x minus 3	$x-3$
	less than	7 less than y	$y-7$
	decreased by	a decreased by b	$a-b$
	the difference between	the difference between x and 5	$x-5$
Multiplication	times	10 times t	$10t$
	of	$\frac{1}{2}$ of p	$\frac{1}{2}p$
	the product of	the product of x and y	xy
	multiplied by	8 multiplied by r	$8r$
	twice	twice x	$2x$
Division	divided by	8 divided by z	$\frac{8}{z}$
	the quotient of	the quotient of x and 4	$\frac{x}{4}$
	the ratio of	the ratio of r to s	$\frac{r}{s}$
Power	the square of	the square of x	x^2
	the cube of	the cube of z	z^3

Translating a phrase that contains the word *sum, difference, product,* or *quotient* can sometimes cause a problem. In the examples at the right, note that the operation replaces the word *and.*

the *sum* of x *and* y (+)	$x+y$
the *difference between* x *and* y (−)	$x-y$
the *product* of x *and* y (·)	$x \cdot y$
the *quotient* of x *and* y (÷)	$\frac{x}{y}$

➤ Translate "the difference between three times a number and 7" into a variable expression.

Assign a variable, say n, to the unknown quantity. Identify words that indicate mathematical operations. Use the assigned variable to write the variable expression.

TAKE NOTE
Note that the minus sign replaces the word *and* for the phrase *difference between.*

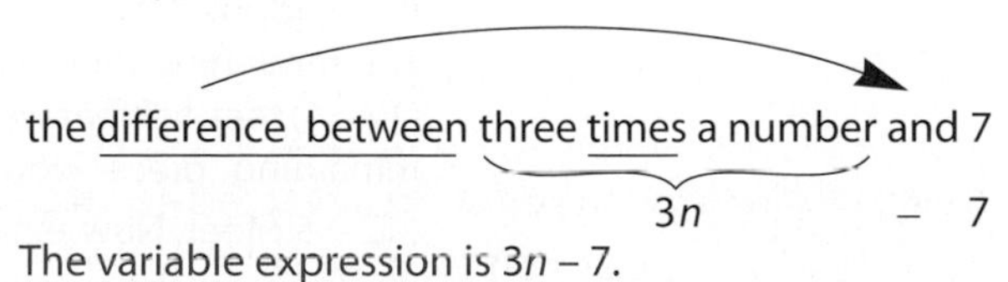

The variable expression is $3n-7$.

TAKE NOTE
Note the translation of *less than* at the right as $8n - 9$. It would be incorrect to write $9 - 8n$.

→ Translate "9 less than 8 times some number" into a variable expression.

Assign a variable to the unknown quantity. Identify words that indicate mathematical operations. Then write the variable expression.

9 less than 8 times some number: $8n - 9$

Example 1 Translate into a variable expression.

a. 8 less than 5 times some number

b. The product of 4 times a number and the sum of the number and 5

c. The quotient of the square of a number and 1 more than 3 times the number

Solution

a. The unknown number: n

8 less than 5 times some number — • Identify the words that indicate mathematical operations.

$5n - 8$ — • Use the operations to write the variable expression.

b. The unknown number: n

The product of 4 times a number and the sum of the number and 5 — • Identify the words that indicate mathematical operations.

4 times the number: $4n$
The sum of the number and 5: $n + 5$ — • Use the assigned variable to write a variable expression for any unknown quantity.

$4n(n + 5)$ — • Write the variable expression.

c. The unknown number: n

The quotient of the square of a number and 1 more than 3 times the number — • Identify the words that indicate mathematical operations.

The square of the number: n^2
1 more than 3 times the number: $3n + 1$ — • Use the assigned variable to write a variable expression for any unknown quantity.

$\frac{n^2}{3n + 1}$ — • Write the variable expression.

SUGGESTED ACTIVITY
Give students a mathematical expression and have them translate it to words. Here are some possibilities:

$5n, n + 7, \frac{x}{2y}, 3n + 4, 3(n + 4)$

You-Try-It 1 Translate into a variable expression.

a. 7 more than the product of a number and 12

b. The total of 18 and the quotient of a number and 9

c. The square of a number subtracted from the number increased by 9

Solution See page S2.

In application problems, it may be convenient to express the unknown quantities in terms of a single variable.

→ A wire for a guitar is 12 feet long and is cut into two pieces. Use one variable to express the lengths of the two pieces.

x | $12 - x$
12 ft

As an aid to writing the variable expressions, first suppose that 4 feet is cut from the wire. The remaining piece would be 8 feet, which is (12 – 4) feet. If 5 feet were cut from the wire, the remaining piece would be 7 feet, which is (12 – 5) feet. Now extend this idea and let x feet represent the length of one piece of the cut wire. Then the length of the remaining piece is $(12 - x)$ feet.

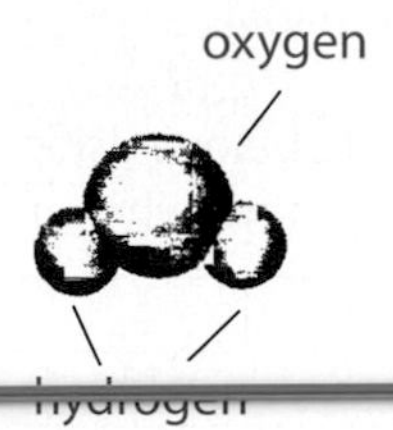

Example 2 The chemical formula for a molecule of water is H_2O. This formula means that there are 2 hydrogen atoms for each oxygen atom in one molecule of water. If x represents the number of atoms of oxygen in one liter of water, express the number of hydrogen atoms in one liter of water in terms of the number of oxygen atoms.

Solution In each molecule of water there are 2 hydrogen atoms for 1 oxygen atom. For instance, if there were 5 oxygen atoms, there would be 10 hydrogen atoms. If there were 100 oxygen atoms, there would be 200 hydrogen atoms. Using the same reasoning, if there are x oxygen atoms in one liter of water, there are $2x$ hydrogen atoms in one liter of water.

You-Try-It 2 A bread recipe requires four times as much wheat flour as rye flour. Express the amount of wheat flour needed for the recipe in terms of the amount of rye flour.

Solution See page S3.

If you are having difficulty writing a variable expression for a problem, first try using some specific numbers for the quantity that is changing. For instance, suppose the cost to rent a pair of skis is \$10 plus \$7 per day. Then the cost to rent the skis for 3 days is $10 + 7(3)$ or \$31. The cost to rent the skis for 8 days is $10 + 7(8)$ or \$66. Now replace the quantity that is changing with a variable. The variable expression is $10 + 7d$, where d is the number of days the skis are rented.

Example 3 The cost to rent a car is \$39.95 plus \$.15 per mile driven. Express the cost of renting the car in terms of the number of miles driven.

Solution Choose a variable to represent the number of miles driven.

Number of miles driven: m

39.95 plus 0.15 for each mile driven: $39.95 + 0.15m$

You-Try-It 3 A chef is paid \$640 per week plus \$32 for each hour of overtime worked. Express the chef's weekly pay in terms of the number of hours of overtime worked.

Solution See page S3.

In the next example, use inductive reasoning to find a pattern for the input/output table. Then write a variable expression based on the pattern you found.

Example 4 Complete the input/output table below. Using the variable expression you find for the last column, what is the output when n is 12?

Input	0	1	2	3	4	5	6	n
Output	0	2	4	6	8	10	?	?

Solution Using inductive reasoning, each output is twice the input. Let n represent the input. Then twice the input is the variable expression $2n$. This suggests that we complete the table as follows.

Input	0	1	2	3	4	5	6	n
Output	0	2	4	6	8	10	**12**	$\mathbf{2n}$

Use the variable expression $2n$ to find the output when n is 12.

$2n$

$2(12) = 24$ • Replace n by 12. Then simplify.

The output when n is 12 is 24.

You-Try-It 4 Complete the input/output table below. Using the variable expression you find for the last column, what is the output when n is 9?

Input	0	1	2	3	4	5	6	n
Output	1	3	5	7	9	11	?	?

Solution See page S3.

The next example is based on *Hooke's Law,* which states the relationship between the amount a spring is stretched and the weight placed on the spring.

Example 5 Complete the input/output table below, which shows the number of inches a spring is stretched as various weights are placed on the end of the spring. Using the variable expression you find for the last column, what is the output when a weight of 11 pounds is placed on the spring?

Input, weight (pounds)	1	2	3	4	5	6	w
Output, inches stretched	0.5	1	1.5	2	2.5	?	?

Solution Using inductive reasoning, it appears that the number of inches the spring is stretched is the weight placed on the spring divided by 2. For instance, using the input values 2, 3, and 5, we have

$$\frac{2}{2} = 1 \qquad \frac{3}{2} = 1.5 \qquad \frac{5}{2} = 2.5$$

This suggests that we complete the table as follows:

Input, weight (pounds)	1	2	3	4	5	6	w
Output, inches stretched	0.5	1	1.5	2	2.5	**3**	$\mathbf{\frac{w}{2}}$

TAKE NOTE
There is more than one way to write the answer to Example 5. For instance, you could write $0.5w$ or $\frac{1}{2} \cdot w$.

Use the variable expression $\frac{w}{2}$ to find the number of inches the spring is stretched when a weight of 11 pounds is placed on the spring.

$\frac{w}{2}$

$\frac{11}{2} = 5.5$ • Replace w by 11. Then simplify.

The spring will be stretched 5.5 inches.

You-Try-It 5 Complete the input/output table below. How many dots will be in pattern number 7?

Pattern number	1	2	3	4	n
Pattern	•	•• / ••	••• / ••• / •••	•••• / •••• / •••• / ••••	••···• / ••···• / ······ / ······ / ••···•
Number of dots	1	4	9	16	?

Solution See page S3.

1.5 EXERCISES

Topics for Discussion

1. In this section, a variable represented a number. Discuss whether a variable could represent some other quantity.
Answers will vary.

2. Give some examples of quantities that could be represented by a variable.
Answers will vary.

3. Can the value of a variable ever equal 0? If not, why not? If so, give an example of an actual situation where it makes sense for the value of a variable to be 0.
Answers will vary.

4. In the text, we stated that a variable was a letter that represented a quantity that can change. In the formula for the circumference of a circle, $C = \pi d$, is π a variable?
Answers will vary.

Translating Phrases into Variable Expressions

5. 4 divided by the difference between p and 6
$\frac{4}{p-6}$

6. the product of 7 and the total of r and 8
$7(r+8)$

7. the ratio of 8 more than d to d
$\frac{d+8}{d}$

8. the total of 9 times the cube of m and the square of m
$9m^3 + m^2$

9. three-eighths of the sum of t and 15
$\frac{3}{8}(t+15)$

10. the difference between the square of c and the total of c and 14
$c^2 - (c+14)$

11. a number less than 13
$13 - n$

12. forty more than a number
$n + 40$

13. three-sevenths of a number
$\frac{3}{7}n$

14. the square of the difference between a number and ninety
$(n-90)^2$

15. the quotient of twice a number and five

$\frac{2n}{5}$

16. the sum of four-ninths of a number and twenty

$\frac{4}{9}n + 20$

17. eight subtracted from the product of fifteen and a number

$15n - 8$

18. the product of a number and ten more than the number

$n(n + 10)$

19. six less than the total of the cube of a number and the number

$(n^3 + n) - 6$

20. fourteen added to the product of seven and a number

$7n + 14$

21. the quotient of three and the total of four and a number

$\frac{3}{4 + n}$

22. the quotient of twelve and the sum of a number and two

$\frac{12}{n + 2}$

23. eleven plus one-half of a number

$11 + \frac{1}{2}n$

24. the ratio of two and the sum of a number and one

$\frac{2}{n + 1}$

25. a number multiplied by the difference between twice the number and nine

$n(2n - 9)$

26. the difference between sixty and the quotient of a number and fifty

$60 - \frac{n}{50}$

27. the product of a number less than nine and the number

$(9 - n)n$

28. the sum of the square of a number and three times the number

$n^2 + 3n$

29. the quotient of seven more than twice a number and the number

$\frac{2n+7}{n}$

30. the sum of the cube of a number and three more than the number

$n^3 + (n+3)$

31. a number decreased by the difference between the cube of the number and ten

$n - (n^3 - 10)$

32. the square of a number decreased by one-fourth of the number

$n^2 - \frac{1}{4}n$

33. four less than seven times the square of a number

$7n^2 - 4$

34. the product of two and a number decreased by the quotient of seven and the number

$2n - \frac{7}{n}$

35. eighty decreased by the product of thirteen and a number

$80 - 13n$

36. the cube of a number decreased by the product of twelve and the number

$n^3 - 12n$

37. The cruising speed of a jet plane is twice the cruising speed of a propeller-driven plane. Express the cruising speed of the jet plane in terms of the cruising speed of the propeller-driven plane. $2s$

38. In football, the number of points awarded for a touchdown is three times the number of points awarded for a safety. Express the number of points awarded for a touchdown in terms of the number of points awarded for a safety.

$3s$

39. A mixture contains four times as many peanuts as cashews. Express the amount of peanuts in the mixture in terms of the amount of cashews.

$4c$

40. In a coin bank, there are ten more dimes than quarters. Express the number of dimes in the coin bank in terms of the number of quarters. $q + 10$

41. A 5-cent stamp in a stamp collection is 25 years older than an 8-cent stamp in the collection. Express the age of the 5-cent stamp in terms of the age of the 8-cent stamp.
$s + 25$

42. The length of a rectangle is five meters more than twice the width. Express the length of the rectangle in terms of the width. $2W + 5$

43. In a triangle, the measure of the smallest angle is three degrees less than one-half the measure of the largest angle. Express the measure of the smallest angle in terms of the measure of the largest angle.
$\frac{1}{2}L - 3$

44. The cost of renting a car for a day is \$24.50 plus 21 cents per mile driven. Express the cost of renting the car in terms of the number of miles driven. Use decimals for constants and coefficients. $24.50 + 0.21m$

45. First-class mail costs 32 cents for the first ounce and 23 cents for each additional ounce. Express the cost of mailing a package first class in terms of the weight of the package. Use decimals for constants and coefficients. $0.32 + 0.23(w - 1)$

46. The sum of two numbers is twenty-three. Use one variable to represent the two numbers. x and $23 - x$

47. A rope twelve feet long was cut into two pieces. Use one variable to express the lengths of the two pieces. S and $12 - S$

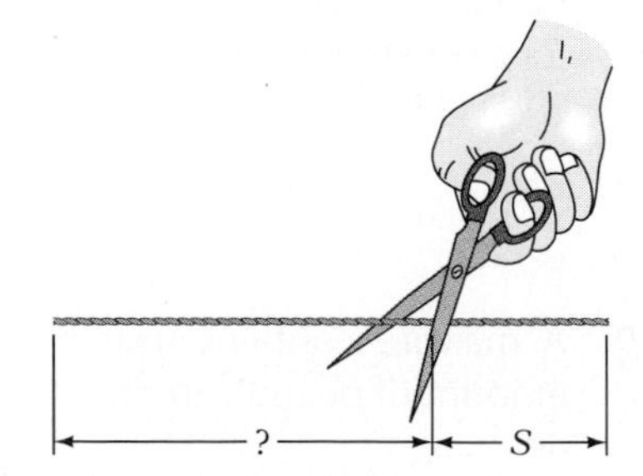

48. A coin purse contains thirty-five coins in nickels and dimes. Use one variable to express the number of nickels and the number of dimes in the coin purse.
n and $35 - n$

49. Twenty gallons of oil were poured into two containers of different sizes. Use one variable to express the amount of oil poured into each container.
x and $20 - x$

Applying Concepts

50. Two cars are traveling in opposite directions and at different rates. Two hours later the cars are two hundred miles apart. Express the distance traveled by the slower car in terms of the distance traveled by the faster car.
$200 - x$

51. An employee is paid $640 per week plus $24 for each hour of overtime worked. Express the employee's weekly pay in terms of the number of hours of overtime worked.
$640 + 24h$

52. An auto repair bill is $92 for parts and $25 for each hour of labor. Express the amount of the repair bill in terms of the number of hours of labor.
$92 + 25h$

53. A coin bank contains nickels and dimes. Using n for the number of nickels in the bank and d for the number of dimes in the bank, write an expression for the value, in pennies, of the coins in the bank. $5n + 10d$

54. A coin bank contains dimes and quarters. Using d for the number of dimes in the bank and q for the number of quarters in the bank, write an expression for the value, in dollars, of the coins in the bank. $0.10d + 0.25q$

55. Each molecule of octyl acetate, which gives air fresheners an orange scent, contains 10 carbon atoms, 20 hydrogen atoms, and 2 oxygen atoms. If x represents the number of atoms of oxygen in one gram of octyl acetate, express the number of carbon atoms in one gram of octyl acetate in terms of x.
$5x$ carbon atoms

56. Each molecule of glucose (sugar) contains 6 carbon atoms, 12 hydrogen atoms, and 6 oxygen atoms. If x represents the number of atoms of oxygen in a pound of sugar, express the number of hydrogen atoms in the pound of sugar in terms of x. $2x$ hydrogen atoms

57. A wire whose length is given as x inches is bent into a square. Express the length of a side of the square in terms of x.
$\frac{1}{4}x$

58. Complete each statement with the word *even* or *odd*.

a. If k is an odd natural number, then $k + 1$ is an **even** natural number.

b. If n is a natural number, then $2n$ is an **even** natural number.

c. If m and n are even natural numbers, then mn is an **even** natural number.

d. If m and n are odd natural numbers, then $m + n$ is an **even** natural number.

e. If m and n are odd natural numbers, then mn is an **odd** natural number.

f. If m is an even natural number and n is an odd natural number, then $m + n$ is an **odd** natural number.

Exploration

59. ***Translating Geometric Formulas*** Sometimes formulas are given verbally. Translate each of the following, replacing the word *is* with an equal sign. Use the letter P for perimeter and the letter A for area. The symbol for pi is π.

a. The perimeter of a rectangle is the sum of twice the length and twice the width.

b. The area of a rectangle is the product of the length and the width.

c. The area of a triangle is one-half of the base times the height.

d. The circumference of a circle is the product of pi and the diameter.

e. The area of a circle is the product of pi and the square of the radius.

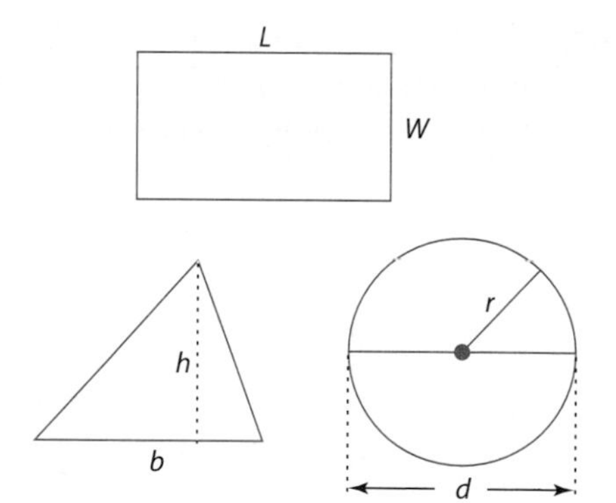

60. ***Card trick*** Ask a friend to draw a card from a regular deck of playing cards and not show you the card. You can determine the card from the following operations. Using the face value of the card (1 for an ace, 11 for a jack, 12 for a queen, 13 for a king), have your friend add 10 to the value, multiply the result by 8, subtract 40, divide by 2, and then subtract 20. To that result, ask your friend to add 0 if the card is a spade, 1 if it is a heart, 2 if it is a diamond, and 3 if it is a club. Now divide by 4. The quotient will be the value of your friend's card, and the remainder will be the suit. Can you explain how this trick works?

Section 1.6 Statistical Measures

Mean

TAKE NOTE
Some graphing calculators can calculate the mean and median for data entered into the calculator. See the appendix Guidelines for Using Graphing Calculators for assistance with some graphing calculators.

The average daily consumption of oil in the U.S. is 17 million barrels. The EPA estimates that a 1997 Eagle Talon averages 28 miles per gallon on the highway. The average rainfall for portions of Kauai is 350 inches per year. Each of these statements uses one number to describe an entire collection of numbers. Such a number is called an *average*.

In statistics, an average is called a *measure of central tendency*. Three of the most common averages, the *mean*, the *median*, and the *mode*, are discussed here. We begin with the mean.

The **mean** of a set of data is the sum of the measurements divided by the number of measurements. The symbol for the mean is $\bar{x}$.

Formula for the Mean

$$\bar{x} = \frac{\text{Sum of the data values}}{\text{Number of data values}}$$

Example 1 The scores for 15 selected games of the New York Liberty of the Women's National Basketball Association were 65, 50, 69, 62, 67, 76, 69, 71, 64, 61, 73, 78, 50, 69, and 67. Find the mean score for the 15 games.

Solution To find the mean, add the scores and then divide by the number of scores.

$$\bar{x} = \frac{65 + 50 + 69 + 62 + 66 + 76 + 69 + 71 + 64 + 61 + 73 + 78 + 50 + 69 + 67}{15}$$

$$= 66$$

The mean score for the 15 games was 66.

You-Try-It 1 An independent automotive lab tested 8 Corvettes to determine an average miles per gallon rating for these cars. The results for the 8 cars were 19, 16, 18, 14, 18, 15, 17, and 19. Find the mean miles per gallon for these cars.

Solution See page S3.

Median

TAKE NOTE
We will arrange data from smallest to largest when calculating the median.

Observe also that the calculation of the median depends on whether there are an even or odd number of data in the set.

The **median** of data is the number which separates the data into two equal parts when the numbers are arranged from smallest to largest (or largest to smallest). There are an equal number of values above the median and below the median.

To find the median of a set of numbers, first arrange the numbers from smallest to largest. If there is an *odd* number of data in the set, the median is the number in the middle. If a data set contains an *even* number of values, the median is the mean of the two middle numbers.

The result of arranging the points scored by the New York Liberty from smallest to largest is shown below.

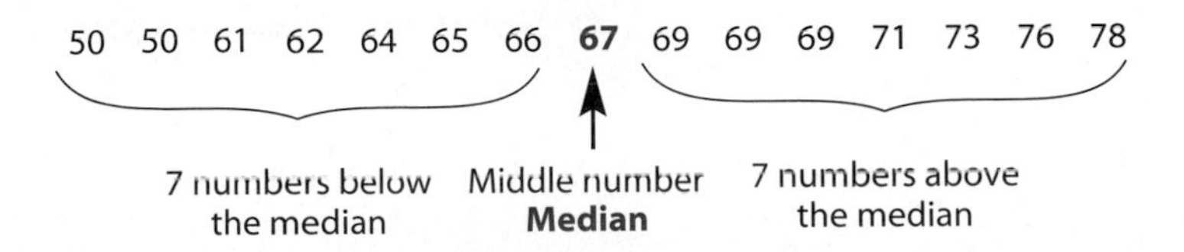

The median is 67. This indicates that in 7 games the New York Liberty scored below 67 points and in 7 games they scored above 67 points.

SUGGESTED ACTIVITY

Have students try to crease sets of numbers for which the mean is greater than the median, less than the median, and equal to the median.

Students can create their own sets (probably better) or create sets using a graphing calculator. The function seq(randInt(0, 50), A, 1, 25) will create of list of 25 integers between 0 and 50. The functions to create this list are under LIST and MATH on a TI-83 calculator.

As an example of calculating the median for a data set that contains an even number of data, suppose that the Director of Student Housing at a college surveyed 10 apartment owners to determine the various rents for a one-bedroom apartment. The monthly rents for the 10 apartments were $495, $550, $325, $425, $395, $575, $450, $375, $400, $515.

To find the median, first arrange the numbers from smallest to largest. The median is the mean of the two middle numbers.

325 375 395 400 **425 450** 495 515 550 575

Middle 2 numbers

$$\text{Median} = \frac{425 + 450}{2} = 437.5$$

The median rental price was $437.50.

Question: For the apartment data set, are there the same number of rents below the median as above the median?[1]

Example 2 A Lundberg survey in June of 1997 established the price for one gallon of regular unleaded gasoline for various cities. The result is shown in the following table.

City	Price per gallon
Los Angeles	$1.30
Honolulu	$1.59
Chicago	$1.33
Boston	$1.24
Washington	$1.20
Miami	$1.25
Dallas	$1.15
Denver	$1.18

Find the median price per gallon for these cities.

Solution First, arrange the data from smallest to largest.

1.15 1.18 1.20 1.24 1.25 1.30 1.33 1.59

Because there are an even number of data, the median is the mean of the middle two numbers.

$$\text{Median} = \frac{1.24 + 1.25}{2} = 1.245$$

The median price per gallon was $1.245.

You-Try-It 2 A survey of 16 students in a math class was conducted to determine the number of hours per week students spent working. The results were: 18, 21, 25, 8, 23, 21, 24, 22, 15, 22, 19, 18, 25, 28, 28, 17. Find the median number of hours worked by these students.

Solution See page S3.

1. Yes. The five values below the median are $325, $375, $395, $400, and $425. The five values above the median are $450, $495, $515, $550, and $575.

The mean is similar to a *balance point* for the data. For instance, suppose the ages of fifteen students in an anthropology class are 20, 18, 25, 27, 26, 24, 17, 19, 20, 18, 21, 22, 26, 25, and 19. Now consider a board on which one-pound weights are attached at points corresponding to a student's age as shown below.

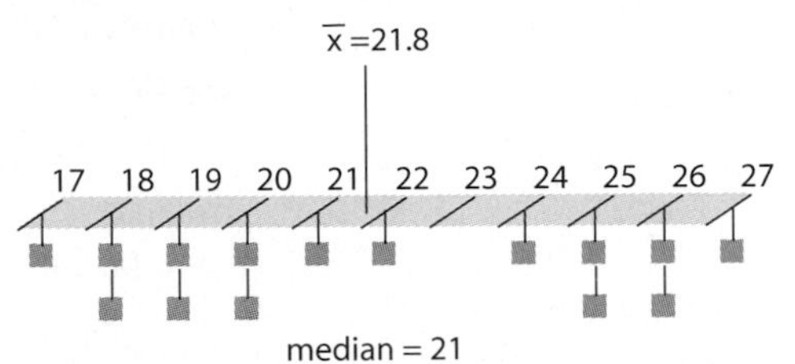

If a string were attached at the mean, 21.8, the board could be lifted and would balance. The median of this data is 21, the age for which there are equal numbers of students below that age and above that age.

Mode

The **mode** of a set of numbers is the value that occurs most frequently. If a set of numbers has no number occurring more than once, then the data has no mode.

TAKE NOTE
From the calculations of the mean, median, and mode for the New York Liberty, note that these three averages are not all the same.

The data for the points scored by the New York Liberty for 15 selected games is repeated below.

50 50 61 62 64 65 66 67 69 69 69 71 73 76 78

Because 69 occurs more than any other number, the mode for this data is 69.

Now consider the data for the apartment rents, which is repeated below.

325 375 395 400 425 450 495 515 550 575

Because no number occurs more than once, this set of data has no mode.

TAKE NOTE
Data that is obtained by measuring or counting is called *quantitative* data. Examples of quantitative data are 25 pounds, 98 people, 13 miles, and 0.5 milliliter.

Although the mode is used as just discussed, it is frequently used for *qualitative* data. Qualitative data results when people are asked questions such as "What is your favorite color?" or "Would you rate the food at a restaurant as good, fair, or poor?" In these cases, the mode is the category that receives the greatest number of responses. This is called the **modal response**.

Example 3 Students at a college were asked to rate the cafeteria in terms of providing opportunities to choose healthy selections. The results are shown in the table at the right. What was the modal response for this survey?

Cafeteria rating results	
Excellent	47
Good	76
Fair	93
Poor	32

Solution The modal response is the category that receives the largest number of responses. For this survey, the modal response was fair.

You-Try-It 3 A survey by the music industry asked people where they purchase music CDs. The results of the survey are shown at the right. What is the modal response for this survey?

CD Sales	
Record stores	499
Other stores	315
Record clubs	143
Mail order	290

Solution See page S3.

Box-and-Whiskers Plots

The purpose of calculating a mean or median is to obtain one number that describes some measurements. That one number alone, however, may not adequately represent the data. A **box-and-whiskers plot** is a graph that gives a more complete picture of the data. A box-and-whiskers plot shows five numbers: the smallest value, the *first quartile*, the median, the *third quartile*, and the greatest value.

The **first quartile**, symbolized by Q_1, is the number for which approximately one-quarter or 25% of the data lies *below* that number. Q_1 is the *median* of the lower half of the data values.

The **third quartile**, symbolized by Q_3, is the number for which about one-quarter of the data lies *above* that number. Q_3 is the median of the upper half of the data values.

To find Q_1 and Q_3 for the points scored by the New York Liberty, first find the median.

50 50 61 62 64 65 66 **67** 69 69 69 71 73 76 78

↑

Median

The first quartile, Q_1, is the median of the lower half of the data: $Q_1 = 62$. — 50 50 61 **62** 64 65 66

The third quartile, Q_3, is the median of the upper half of the data: $Q_3 = 71$. — 69 69 69 **71** 73 76 78

Because $Q_1 = 62$, approximately 25% of the New York Liberty's scores were below 62. The fact that $Q_3 = 71$ means that approximately 25% of their scores were above 71.

Because 25% of the data is below Q_1 and 25% of the data is above Q_3, that leaves the remaining 50% of the data between Q_1 and Q_3. The difference between these values is the *interquartile range.*

Interquartile Range

The interquartile range is the difference between the third quartile, Q_3, and the first quartile, Q_1.

$$\text{Interquartile range} = Q_3 - Q_1$$

Approximately 50 percent of the data in any distribution lies between Q_1 and Q_3.

TAKE NOTE

The interquartile range is a single number—9 for the basketball data at the right. It is *incorrect* to say that the interquartile range is between 62 and 71.

For the selected basketball scores for the New York Liberty, we have

$$\text{Interquartile range} = Q_3 - Q_1 = 71 - 62 = \mathbf{9}$$

The box-and-whiskers plot for the data for the New York Liberty is shown below.

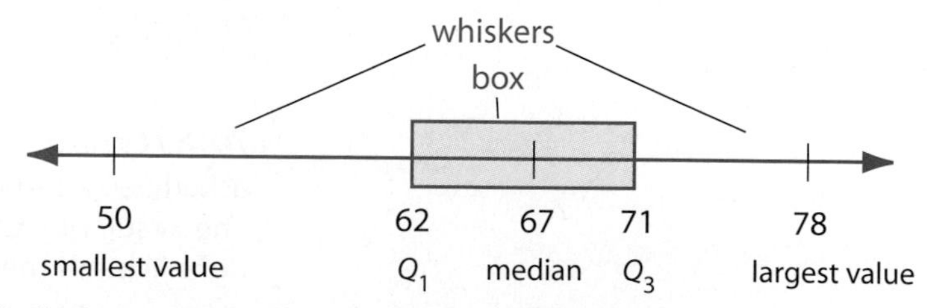

TAKE NOTE

The statistical features of some graphing calculators can calculate Q_1 and Q_3. There is also an option for drawing a box-and-whiskers plot for data entered into the calculator. See the appendix Guidelines for Using Graphing Calculators for assistance with some graphing calculators.

A box-and-whiskers plot shows the interquartile range as a box and five labeled values: the smallest number, 50; the first quartile Q_1, 62; the median, 67; the third quartile Q_3, 71; and the largest number, 78.

Question: For the box-and-whiskers plot above, 50% of the data lies between what two numbers?[2]

2. 50% of the data lies between 62 and 71, the values of Q_1 and Q_3.

Example 4 The grades of 12 students on a biology exam were as follows: 76, 75, 80, 66, 56, 75, 77, 81, 72, 89, 94, 96. Find the interquartile range and draw a box-and-whiskers plot.

Solution First find the median. Arrange the numbers from smallest to largest.

56 66 72 75 75 76 77 80 81 89 94 96

Because there is an even number of data, the median is the mean of the middle two numbers.

$$\text{Median} = \frac{76 + 77}{2} = 76.5$$

The first quartile, Q_1, is the median of the lower half of the data. Because there is an even number of values (6) below the median, Q_1 is the mean of 72 and 75. $Q_1 = 73.5$.

56 66 72 75 75 76

$$Q_1 = \frac{72 + 75}{2} = 73.5$$

The third quartile, Q_3, is the median of the upper half of the data. Because there is an even number of values (6) above the median, Q_3 is the mean of 81 and 89. $Q_3 = 85$.

77 80 81 89 94 96

$$Q_3 = \frac{81 + 89}{2} = 85$$

Interquartile range = 85 − 73.5 = 11.5. The box-and-whiskers plot is shown below.

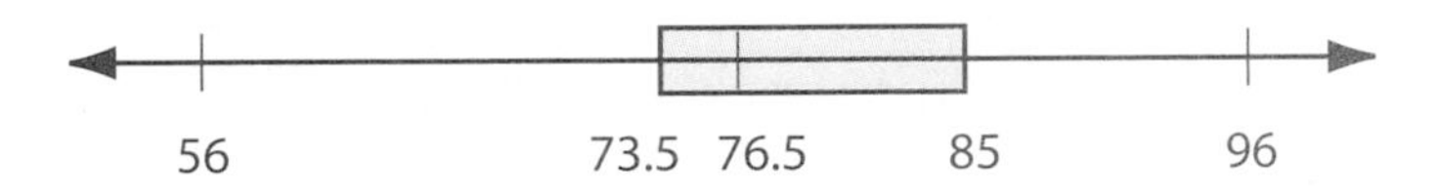

You-Try-It 4 Twelve newly designed batteries for portable CD players were tested by running the CD player continuously until the battery quit. The length of time (in minutes) each battery operated was: 87, 94, 78, 110, 124, 103, 99, 130, 100, 98, 136, 129. Find the interquartile range and draw a box-and-whiskers plot for these data.

Solution See page S3.

Question: Looking at the box-and-whiskers plot for Example 4, approximately 25% of the data is greater than what number?[3]

TAKE NOTE
In statistics, an *inference* is a conclusion based on data. It is similar to inductive reasoning, discussed in Section 1.1.

The values of Q_1, the median, and Q_3 are used to make inferences about the data. For instance, based on information from the Bureau of Labor Statistics, the values of these numbers for the incomes of physicians in the United States are Q_1 = \$108,000, median = \$156,000, and Q_3 = \$240,000. Here are some inferences we can make about this data.

- Because the median is \$156,000, 50% of the physicians in the U.S. earn over \$156,000 per year and 50% earn less than \$156,000.
- Because Q_1 is \$108,000, approximately 25% of the physicians earn less than \$108,000 per year.
- Approximately 50% of the physicians earn between \$108,000 and \$240,000 per year.

3. Approximately 25% of the data is greater than 85, which is Q_3.

Example 5 The values of Q_1, the median, and Q_3 for the hourly wages of a heating and air-conditioning technician in Los Angeles are Q_1 = \$15.85, median = \$22.86, and Q_3 = \$26.42. If there are approximately 5400 heating and air-conditioning technicians in Los Angeles, how many have an hourly wage between \$15.85 and \$26.42?

Solution Approximately 50% of the data in a distribution lies between Q_1 and Q_3. Therefore, 50% of the 5400 technicians have an hourly wage between \$15.85 and \$26.42. Let *N* represent the number of technicians that have an hourly wage between \$15.85 and \$26.42.

N = 50% of 5400

= 0.50(5400) • Recall that *of* translates as "multiply".

= 2700

Approximately 2700 technicians have an houry wage between \$15.85 and \$26.42.

You-Try-It 5 Recall the data for the annual salaries of physicians given on the previous page: Q_1 = \$108,000, median = \$156,000, Q_3 = \$240,000. In 1997, there were approximately 583,000 physicians in the U.S. How many physicians earned over \$240,000?

Solution See page S3.

The interquartile range is an example of what in statistics is called a *measure of dispersion*. The purpose of these is to measure how scattered the data is in a distribution. For instance, suppose the scores on two tests are represented by the box-and-whiskers plots given below.

For Test 1, the interquartile range is 10, which means that 50% of the scores are within 10 points of one another. For Test 2, the interquartile range is 20, so 50% of the scores are within 20 points of one another. The test scores for Test 2 are more scattered about the median than the ones for Test 1.

Another measure of dispersion is the *range*. The **range** is the difference between the largest data value and the smallest data value. For instance, for the New York Liberty basketball team scores, 50, 50, 61, 62, 64, 65, 67, 67, 69, 69, 69, 71, 73, 76, and 78, the range is 78 – 50 = 28.

SUGGESTED ACTIVITY

In the Instructor's Resource Manual, there is an activity that guides students through the use of a graphing calculator to calculate the descriptive statistics discussed in this section and to draw a box-and-whiskers plot.

Example 6 The distances (in miles) that 19 students commute to school are: 18, 13, 21, 14, 16, 10, 12, 16, 13, 17, 18, 13, 17, 14, 12, 10, 15, 18, 15. Find the interquartile range and the range for this data.

Solution To find the interquartile range, arrange the data from smallest to largest and find Q_1 and Q_3.

10, 10, 12, 12, **13**, 13, 13, 14, 14, **15**, 15, 16, 16, 17, **17**, 18, 18, 18, 21
(Q_1 = 13, median = 15, Q_3 = 17)

Interquartile range = 17 – 13 = 4 Range = 21 – 10 = 11

You-Try-It 6 The numbers of units 12 sophomores enrolled in were: 13, 16, 12, 13, 21, 15, 16, 17, 17, 17, 12, 15. Find the interquartile range and the range for this data.

Solution See page S3.

1.6 EXERCISES

Topics for Discussion

1. What is a measure of central tendency?
Answers will vary.

2. What is the meaning of Q_1 and Q_3?
25% of the data lies below Q_1. 25% of the data lies above Q_3.

3. What are the main features of a box-and-whiskers plot?
Smallest data value, Q_1, median, Q_3, largest data value

4. What is the significance of the data that lie between Q_1 and Q_3?
50% of the data lies between the two numbers.

5. Explain the difference between the range of data and the interquartile range.
Answers will vary.

6. How does the mean differ from the median?
Answers will vary.

Mean, Median, and Mode

7. Suppose that the scores for Amy Alcott's last 15 rounds of golf were: 70, 72, 70, 69, 69, 70, 70, 70, 71, 69, 69, 71, 68, 70, 70. Find the mean, median, and mode of her scores.
mean = 69.9; median = 70; mode = 70

8. Airline passengers were asked their opinions as to the value of a dinner served during a flight from Los Angeles to Dallas. The responses, in dollars, were: 7, 8, 4, 9, 8, 4, 6, 8, 11, 3, 5, 7, 11, 5, 5, 8, 5, 9, 4, 4. Find the mean, median, and mode of this data.
mean = 6.55; median = 6.5; mode = 4, 5, 8

9. The numbers of points scored by the Boston Celtics for 10 games were: 97, 101, 99, 98, 92, 91, 100, 98, 95, 101. Find the mean and median for this data.
mean = 97.2; median = 98

10. An airline schedule that indicates that a flight from Los Angeles to San Jose, California, takes 54 minutes bases that time on the average of the times it has taken previous flights to make that trip. Suppose that the times (in minutes) to fly between Boston and New York on a Delta Shuttle were: 66, 66, 70, 69, 69, 63, 72, 66, 62, 70, 71, 68, 65, 74. Find the mean and median of this data.
mean = 67.9; median = 68.5

11. The table below shows the verbal and math scores for 10 students who took the Scholastic Aptitude Test (SAT). Find the difference between the mean of the verbal scores and the mean of the math scores for these students. 14.3

Student	1	2	3	4	5	6	7	8	9	10
Verbal score	516	486	520	466	454	404	408	466	560	463
Math score	490	486	438	441	450	432	504	414	485	460

12. The table below shows the high and low temperatures for a 15-day period in Las Vegas, Nevada, in July. Find the difference between the mean of the high temperatures and the mean of the low temperatures. 15

Day	1	2	3	4	5	6	7	8	9	10	11	12	13	14	15
High temperature	96	91	95	94	98	99	96	96	96	91	96	94	90	105	99
Low temperature	74	76	74	77	76	93	73	84	74	81	94	81	85	87	82

13. The monthly amounts spent, in dollars, for long distance phone calls for a family for one year were: 46.96, 47.78, 51.77, 48.37, 51.49, 44.68, 51.90, 51.06, 47.85, 44.19, 58.83, 50.84. Find the mean and median monthly amount spent for long distance calls.
mean = 49.64; median = 49.605

14. The numbers of milliseconds a hard disk controller required to access selected information on a 2-gigabyte hard drive were: 8.1, 9.9, 9.5, 12.9, 11.7, 12.1, 8.1, 7.6, 10.3, 10.5, 160.6, 8.4. Find the mean and median times for the hard disk controller to access the information.
mean = 22.475; median = 10.1

15. According to a survey by First Market Research conducted for *PC World*, approximately one-third of executives responsible for company Internet policies monitor their employees' use of the Internet. The responses for these executives are shown at the right. What was the modal response? (USA Today, 3/26/98)
modal response = Download types

Time on the net	44
Data downloaded	36
Download types	46
E-mail content	17
Sites visited	73

16. Business travelers use increasingly more technology, such as laptops, voice mail, and cell phones, when on the road. The table at the right shows how 1000 travelers responded when asked how technology had affected travel stress. What was the modal response? (USA Today, 3/25/98)
modal response = Less stressful

More stressful	230
Less stressful	500
No difference	25
Don't know	2

17. The table at the right shows how 5000 people responded to the question "Did you pay the minimum due on one or more credit cards, or did you pay more?" What was the modal response? (USA Today, 10/2/97)
modal response = Paid more than the minimum on all cards

Paid more than the minimum on all cards	3500
Paid the minimum on one or more cards	900
Paid less than the minimum on one or more cards	450
No response	150

18. A survey of 725 students conducted by the housing department of a university asked students whether they were very satisfied, satisfied, unsatisfied, or very unsatisfied with dormitory rooms. The responses by these students are given at the right. What was the modal response?
modal response = satisfied

Very satisfied	103
Satisfied	312
Unsatisfied	226
Very unsatisfied	84

Box-and-Whiskers Plots

19. Internet kiosks have been set up in many airports to allow travelers to access the Internet while they are waiting for a plane. The following list of numbers is the number of minutes a certain kiosk was in use for a 15-day period: 222, 239, 204, 216, 139, 240, 234, 246, 176, 189, 202, 89, 233, 190, 233. Find Q_1 and Q_3 for these data.

$Q_1 = 189$; $Q_3 = 234$

20. The hourly wages of unionized electricians in 10 cites were as follows: 27.99, 27.10, 26.85, 27.63, 29.84, 20.98, 27.86, 26.14, 29.61, 27.67. Find Q_1 and Q_3 for these data.

$Q_1 = 26.85$; $Q_3 = 27.99$

21. The box-and-whiskers plot below is a graphical representation of the number of inches of rain that fell in a rainforest over a 20-year period. In what percent of the years was the rainfall less than 200 inches?

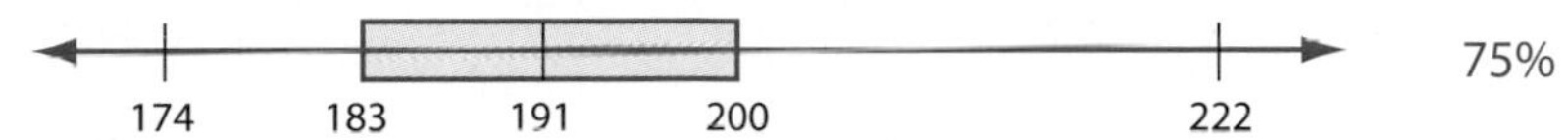

75%

22. The box-and-whiskers plot below is a graphical representation of the ages on inauguration day for United States presidents. What percent of the presidents were between the ages of 51 and 58 when inaugurated?

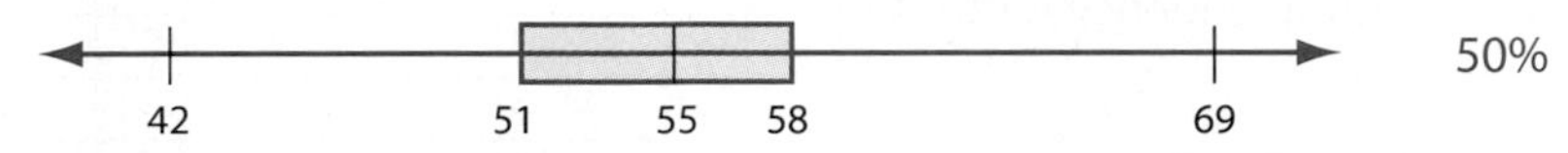

50%

23. The average baseball game ticket prices in 1998 for the teams of the National League were: 12.00, 14.55, 12.00, 11.35, 12.24, 14.63, 8.37, 13.00, 9.87, 9.67, 14.23, 17.48, 11.42, 9.81, 16.23, 11.02. Draw a box-and-whiskers plot for these data. What is the interquartile range? (USA Today, 1/22/98)

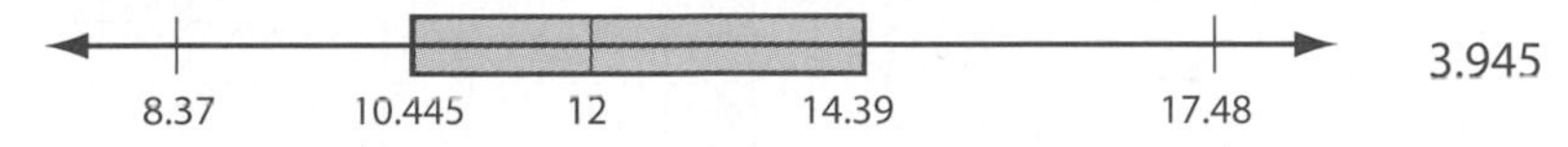

3.945

24. The costs for a lift ticket for several ski areas in the western United States are as follows: 23, 30, 30, 26, 28, 29, 28, 26, 30, 31, 32, 27, 28. Draw a box-and-whiskers plot for these data. What is the interquartile range?

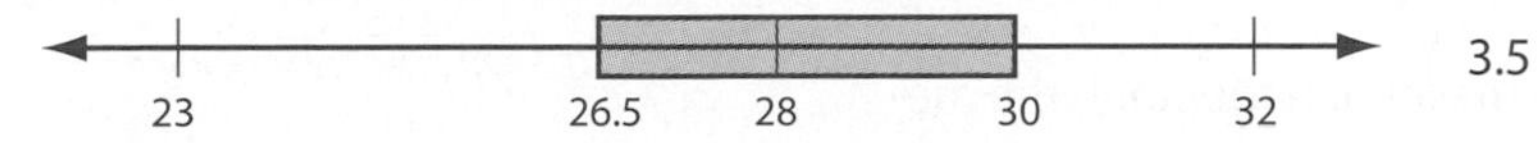

3.5

25. The semiannual premiums for car insurance for a 16-year-old eligible for a good student discount for various insurance companies were as follows: 881, 932, 949, 897, 930, 925, 982, 946, 795, 1002, 1007, 840. Draw a box-and-whiskers plot for these data. What is the interquartile range?

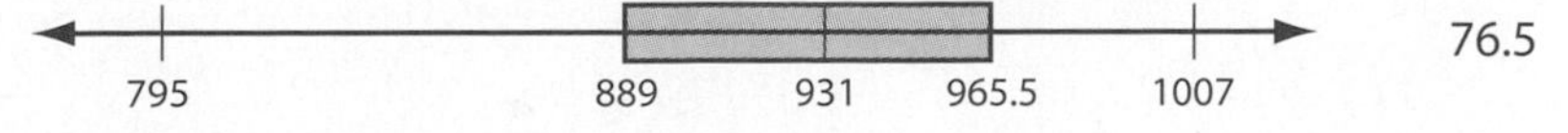

76.5

26. According to the Texas Transportation Institute, in 26 of 50 urban areas surveyed, traffic exceeded road capacity, thereby creating traffic jams. The annual delays for drivers in the 20 worst cities were: 63, 71, 65, 53, 35, 59, 57, 26, 75, 56, 40, 36, 61, 28, 35, 55, 38, 46, 28, 40. Draw a box-and-whiskers plot for these data. What is the interquartile range?

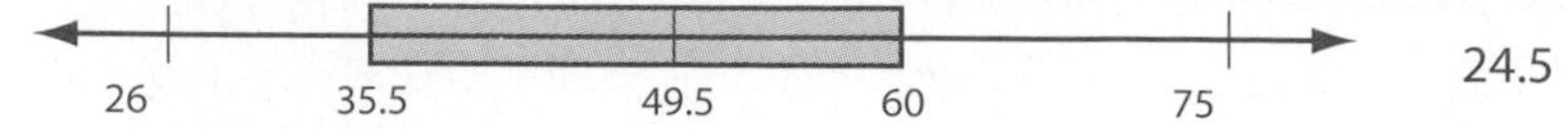

Applying Concepts

27. The average baseball game ticket prices in 1998 for teams in the American League were: 12.56, 10.50, 14.07, 15.80, 16.12, 19.98, 14.64, 10.40, 11.12, 10.33, 18.02, 15.15, 14.25, 16.00. Compare a box-and-whiskers plot for these data to the data in Exercise 23. On the basis of your comparison, is there evidence to suggest that it generally costs more to attend an American League baseball game than a National League baseball game? Yes

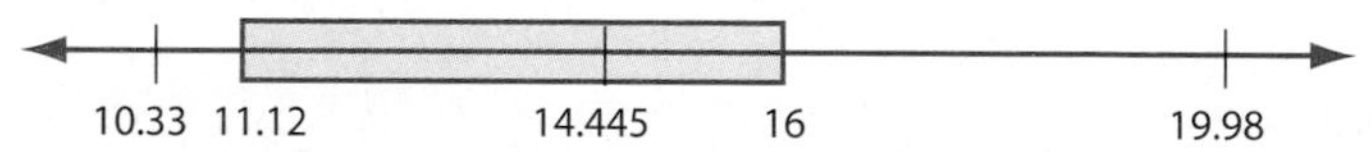

28. The data in the table at the right show the highest elevation in each of the 50 states.

a. Make a box-and-whiskers plot for these data.

b. Calculate the mean of the elevations.

c. Is the mean or the median a better measure of central tendency for these data? Why?

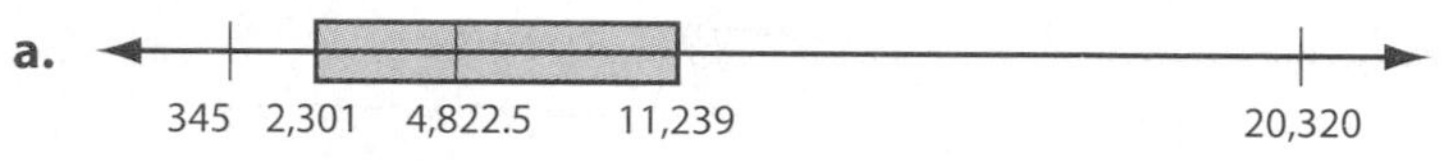

b. 6,228.08; **c.** median

Highest Elevations per State (feet)

AK	20,320	LA	535	OH	1,549
AL	2,405	MA	3,487	OK	4,973
AR	2,753	MD	3,360	OR	11,239
AZ	12,633	ME	5,267	PA	3,213
CA	14,494	MI	1,979	RI	812
CO	14,433	MN	2,301	SC	3,560
CT	2,380	MO	1,772	SD	7,242
DE	442	MS	806	TN	6,643
FL	345	MT	12,799	TX	8,749
GA	4,784	NC	6,684	UT	13,528
HI	13,796	ND	3,506	VA	5,729
IA	4,973	NE	5,426	VT	4,393
ID	12,662	NH	6,288	WA	14,410
IL	1,235	NJ	1,803	WI	1,951
IN	1,257	NM	13,161	WV	4,861
KS	4,039	NV	13,140	WY	13,804
KY	4,139	NY	5,344		

Explorations

29. ***Outliers*** An outlier is a data value whose distance from the nearer quartile is at least 1.5 times the interquartile range. Sometimes outliers are removed from a data set when analyzing data so that more accurate inferences can be drawn from the data. The following data is the thickness of the ozone layer, in Dobson units, for a 20-day period in Fresno, California: 427, 466, 372, 299, 293, 284, 298, 284, 314, 302, 286, 296, 306, 308, 320, 318, 344, 345, 354, 381.

a. Make a box-and-whiskers plot for these data.

b. What data value is an outlier?

c. Calculate the mean for this data. Is the mean smaller or larger than the median?

d. Which measure of central tendency, the mean or the median, better represents the average thickness of the ozone layer?

e. Remove the outlier from the data and recalculate the mean. Is the mean nearer the median? Why?

Section 1.7 Introduction to Probability

Sample Spaces and Events

A weather forecaster estimates that there is a 75% chance of rain. A state lottery director claims that there is a $\frac{1}{9}$ chance of winning a prize offered by the lottery. A health organization claims that there is a 0.15 chance that a person will get the flu.

Each of these statements involves uncertainty to some extent. The degree of uncertainty is called **probability**. For the statements above, the probability of rain is 75%; the probability of winning a prize in the lottery is $\frac{1}{9}$; and the probability of getting the flu is 0.15. Probabilities can be expressed as percents, fractions, or decimals.

A probability is determined from an **experiment**, which is any activity that has an observable outcome. Examples of experiments are:

- Tossing a coin and observing whether it lands heads or tails
- Interviewing voters to determine their preference for a political candidate
- Recording the percent change in the price of a stock

All the possible outcomes of an experiment is called the **sample space** of the experiment. The outcomes of an experiment are listed between braces and frequently designated by S.

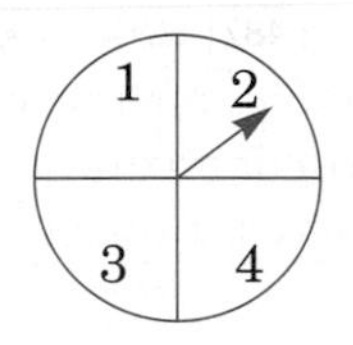

Example 1 List all the possible outcomes of the experiment.

a. One number is selected at random from the digits 0 through 9.
b. A fair coin is tossed once.
c. The spinner at the left is spun once. If it lands on a line, it is spun again.

Solution

a. The phrase "at random" means that each of the digits has an equal chance of being drawn. $S = \{0, 1, 2, 3, 4, 5, 6, 7, 8, 9\}$.
b. A fair coin is one for which heads and tails have an equal chance of being tossed. $S = \{H, T\}$, where H represents heads and T represents tails.
c. Assuming the spinner does not come to rest on a line, $S = \{1, 2, 3, 4\}$.

You-Try-It 1 List all the possible outcomes of the experiment of rolling a die once and recording the number of dots on the upward face.

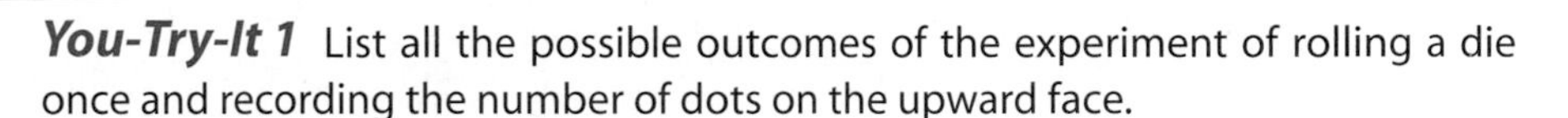

Solution See page S3.

An **event** is one or more outcomes of an experiment. Events are denoted by capital letters.

Consider the experiment of randomly choosing one number from the digits 0 through 9. Some possible events are:

- The number is even. $E = \{0, 2, 4, 6, 8\}$
- The number is a prime number. $P = \{2, 3, 5, 7\}$
- The number is less than 10. $T = \{0, 1, 2, 3, 4, 5, 6, 7, 8, 9\}$. Note that in this case, the event is the entire sample space.
- The number is greater than 20. This event is impossible for the given sample space. The impossible event is symbolized by $\varnothing$.

Question: For the experiment of choosing one of the digits from 0 through 9, list the elements of the event that an odd number is chosen.[1]

1. $E = \{1, 3, 5, 7, 9\}$

The Counting Principle

Suppose you are purchasing a car stereo system and have narrowed your choice to two different radios (Sony and Alpine), three different speakers (Bose, JVC, Alpine), and two different CD players (Kenwood, Kardon). One way to organize the choices you have is to use a *tree diagram*.

A **tree diagram** is a method of organizing the information and illustrating the answer. The tree diagram below illustrates all the possibilities for the options you have.

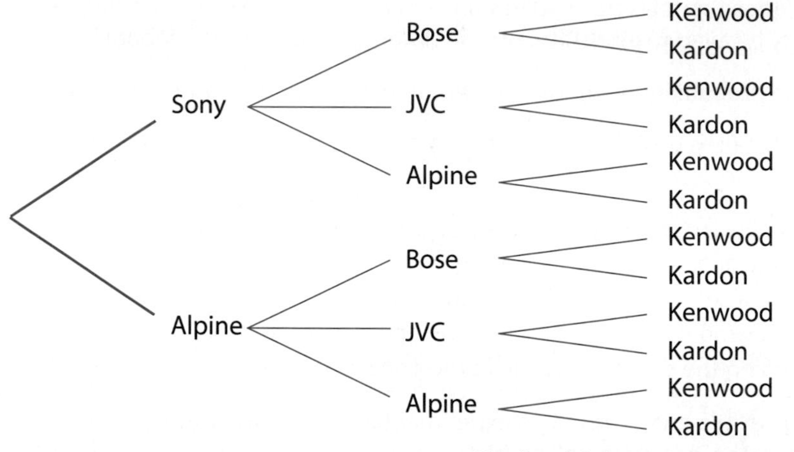

Sony radio, Bose speakers, Kenwood CD player
Sony radio, Bose speakers, Kardon CD player
Sony radio, JVC speakers, Kenwood CD player
Sony radio, JVC speakers, Kardon CD player
Sony radio, Alpine speakers, Kenwood CD player
Sony radio, Alpine speakers, Kardon CD player
Alpine radio, Bose speakers, Kenwood CD player
Alpine radio, Bose speakers, Kardon CD player
Alpine radio, JVC speakers, Kenwood CD player
Alpine radio, JVC speakers, Kardon CD player
Alpine radio, Alpine speakers, Kenwood CD player
Alpine radio, Alpine speakers, Kardon CD player

The "path" along Sony, Bose, Kenwood indicates that you would have selected the Sony radio, with Bose speakers, and a Kenwood CD player.

Another method of organizing your work is to make a table. For instance, in many board games (Monopoly and backgammon, for instance), two dice are rolled and the action taken by the player depends on the numbers that occur on the upward faces of the dice. Making a table similar to the one below shows all of the possible outcomes from the toss of two dice.

6 possible outcomes from white die

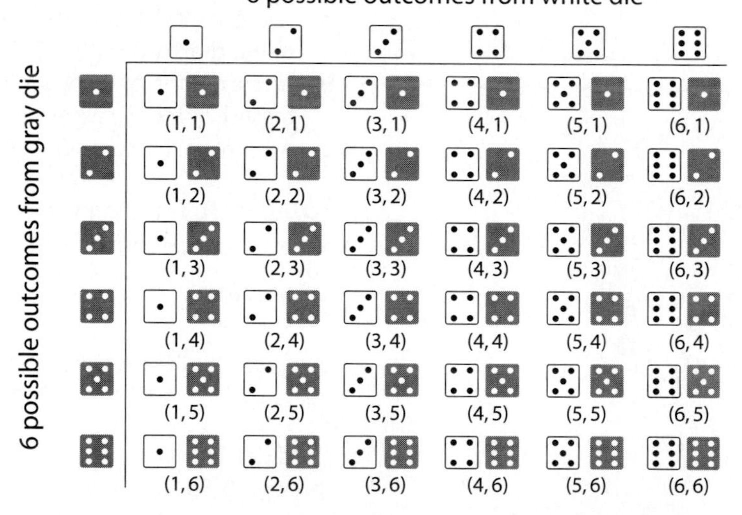

6 possible outcomes from gray die

	1	2	3	4	5	6
1	(1, 1)	(2, 1)	(3, 1)	(4, 1)	(5, 1)	(6, 1)
2	(1, 2)	(2, 2)	(3, 2)	(4, 2)	(5, 2)	(6, 2)
3	(1, 3)	(2, 3)	(3, 3)	(4, 3)	(5, 3)	(6, 3)
4	(1, 4)	(2, 4)	(3, 4)	(4, 4)	(5, 4)	(6, 4)
5	(1, 5)	(2, 5)	(3, 5)	(4, 5)	(5, 5)	(6, 5)
6	(1, 6)	(2, 6)	(3, 6)	(4, 6)	(5, 6)	(6, 6)

From the tree diagram, there are 12 different systems you could have chosen. Note that we obtain the same result by multiplying the number of choices for each option.

$$\begin{bmatrix}\text{Number of}\\ \text{radio choices}\end{bmatrix} \times \begin{bmatrix}\text{number of}\\ \text{speaker choices}\end{bmatrix} \times \begin{bmatrix}\text{number of}\\ \text{CD choices}\end{bmatrix} = \begin{bmatrix}\text{number of}\\ \text{possibilities}\end{bmatrix}$$

$$2 \times 3 \times 2 = 12$$

In a similar manner, counting the possible outcomes from rolling two dice, we find that there are 36 possible outcomes. This result can also be obtained by multiplying the number of choices for each option.

$$\begin{bmatrix}\text{Number of choices}\\ \text{for white die}\end{bmatrix} \times \begin{bmatrix}\text{number of choices}\\ \text{for gray die}\end{bmatrix} = \begin{bmatrix}\text{number of}\\ \text{possibilities}\end{bmatrix}$$

$$6 \times 6 = 36$$

This is expressed mathematically as the Counting Principle.

The Counting Principle

To find the number of possible ways in which a sequence of choices can be made, find the product of the number of choices available for each option.

TAKE NOTE
Note the difference between Example 2 and You-Try-It 2. Example 2 does not allow numbers such as 344 because the digit 4 is used more than once. In You-Try-It 2, it is permissible to use a digit more than once. Therefore, numbers such as 344 and 555 are included.

Example 2 Use the Counting Principle to determine how many three-digit numbers can be formed from the digits 2, 3, 4, 5, and 6, assuming that a digit cannot be used more than once.

Solution There are five possible choices for the first digit of the number. They are 2, 3, 4, 5, or 6. Because a digit cannot be used more than once, there are only four possible choices for the second digit and only three possible choices for the next digit. By the Counting Principle, the number of three-digit numbers is

$$5 \cdot 4 \cdot 3 = 60$$

There are 60 numbers that can be formed from the digits 2, 3, 4, 5, and 6, assuming no digit is used more than once.

You-Try-It 2 Use the Counting Principle to determine how many three-digit numbers can be formed from the digits 2, 3, 4, 5, and 6, assuming that a digit can be used more than once.

Solution See page S3.

Example 3 An experiment consists of tossing 4 coins, a penny, a dime, a nickel, and a quarter, and determining, for each coin, whether the upward face is a head or a tail. How many elements are in the sample space of this experiment?

Solution The sample space consists of all possible outcomes of this experiment. Rather than trying to list all the possible outcomes, we will use the Counting Principle to determine the number of outcomes of this experiment.

$$\begin{bmatrix}\text{Options for}\\ \text{first coin}\end{bmatrix} \times \begin{bmatrix}\text{options for}\\ \text{second coin}\end{bmatrix} \times \begin{bmatrix}\text{options for}\\ \text{third coin}\end{bmatrix} \times \begin{bmatrix}\text{options for}\\ \text{fourth coin}\end{bmatrix} = \begin{bmatrix}\text{total number}\\ \text{of outcomes}\end{bmatrix}$$

$$2 \times 2 \times 2 \times 2 = 16$$

There are 16 elements in the sample space.

You-Try-It 3 In California, a license plate number begins with a number, followed by 3 letters, and then followed by 3 more numbers. What is the maximum number of license plate numbers that can be issued in California?

Solution See page S3.

Probability of an Event

The probability of an event is defined in terms of the sample space and the event.

> ***Probability Formula***
> The probability of an event E, written $P(E)$, is the ratio of the number of elements in the event, $N(E)$, to the number of elements in the sample space, $N(S)$.
>
> $$P(E) = \frac{\text{number of elements in the event}}{\text{number of elements in the sample space}} = \frac{N(E)}{N(S)}$$

The outcomes of the experiment of tossing a fair coin are *equally likely*. Any one of the outcomes is just as likely as another. If a fair coin is tossed once, the probability of a head or a tail is $\frac{1}{2}$. Each event, heads or tails, is equally likely. The probability formula applies to experiments for which the outcomes are equally likely.

Not all experiments have equally likely outcomes. Consider an exhibition baseball game between a professional team and a college team. Although either team could win the game, the probability that the professional team will win a particular game is greater than that of the college team. The outcomes are not equally likely. For the experiments in this section, assume that the outcomes of an experiment are equally likely.

SUGGESTED ACTIVITY
Here is a game students can try to analyze.

A player chooses a number between 2 and 8 and then spins the spinner below twice. The player wins the game if the sum of the two spins is the number recorded by the player. What number should a player choose to have the best chance of winning the game?

Example 4 Using the table of outcomes from the roll of two dice, determine the probability of obtaining a roll for which the sum of the numbers on the upward faces is at least 10.

Solution From the table, there are 36 possible outcomes from the toss of two dice. Therefore, $N(S) = 36$. The event that the sum of the upward faces is at least 10 means that the sum is 10 or greater. That is, the sum is 10, 11, or 12. From the table, $E = \{(4,6),(6,4),(5,5),(5,6),(6,5),(6,6)\}$.

$$P(E) = \frac{\text{number of elements in the event}}{\text{number of elements in the sample space}} = \frac{6}{36} = \frac{1}{6}$$

The probability of obtaining a roll for which the sum of the numbers on the upward faces is at least 10 is $\frac{1}{6}$.

You-Try-It 4 Using the table of outcomes from the roll of two dice, determine the probability of obtaining a roll for which the sum of the numbers on the upward faces is exactly 7.

Solution See page S3.

When discussing experiments and events, it is convenient to refer to the *favorable outcomes* of an experiment. These are the outcomes of an experiment that confirm the particular event. For the experiment of tossing a fair die once, the sample space is $\{1, 2, 3, 4, 5, 6\}$. One possible event E would be rolling a number that is divisible by 3. The favorable outcomes of the experiment for E are 3 and 6. Thus $E = \{3, 6\}$.

Example 5 There are five choices, *a* through *e*, for each question on a multiple-choice test. By just guessing, what is the probability of choosing the correct answer, *d*, on a certain question?

Solution

The experiment is choosing one of the letters *a* through *e*.

$S = \{a, b, c, d, e\}$

There are 5 elements in the sample space, *S*.

The event is the correct answer. Since there is only one correct answer, there is just one favorable outcome that results in a correct answer.

$E = \{d\}$. There is 1 element in *E*.

Use the probability formula.

$$P(E) = \frac{N(E)}{N(S)} = \frac{1}{5}$$

The probability of guessing the correct answer is $\frac{1}{5}$.

You-Try-It 5 What is the probability that one card, randomly chosen from a deck of regular playing cards, is an ace?

Solution See page S4.

Example 6 A game show contestant has won a prize in one of six envelopes that are numbered from 1 to 6. The contestant rolls a fair die once and receives the prize that corresponds to the number of dots showing on the upward face of the die. The major prize, the keys to a new 4x4 truck, is contained in envelope number 5. What is the probability that the contestant will win the truck?

Solution

The experiment is rolling the die. The possible outcomes are any of the numbers 1 through 6.

$S = \{1, 2, 3, 4, 5, 6\}$

There are 6 elements in *S*.

The contestant wins the truck if the outcome from the toss is a 5.

$E = \{5\}$. There is 1 element in *E*.

Use the probability formula.

$$P(E) = \frac{N(E)}{N(S)} = \frac{1}{6}$$

The probability of winning the 4x4 truck is $\frac{1}{6}$.

You-Try-It 6 A dodecahedral die has 12 sides with the numbers 1 through 12 painted on the faces. If the die is tossed once, what is the probability that the number on the upward face will be divisible by 4?

Solution See page S4.

Calculating a probability consists of determining the number of possible outcomes of an experiment and the number of possible outcomes of the experiment that are favorable to some event.

→ Suppose a professor writes 3 true/false questions for a test. If the answers are randomly chosen to be true or false, what is the probability that the test will have 2 true and 1 false question?

The experiment *S* consists of choosing T or F for each of the 3 questions. The outcomes of the experiment are $S = \{TTT, TTF, TFT, TFF, FTT, FTF, FFT, FFF\}$. The event consists of 2 true questions and 1 false question. There are 3 favorable outcomes of the experiment with 2 T's and 1 F. $E = \{TTF, TFT, FTT\}$.

$$P(E) = \frac{N(E)}{N(S)} = \frac{3}{8}$$

The probability that there are 2 true questions and 1 false question is $\frac{3}{8}$.

TAKE NOTE

The tree diagram below shows the possible outcomes of the true/false test discussed at the right. The favorable events are in bold.

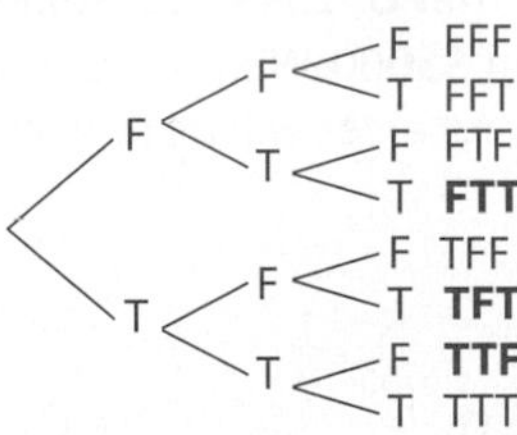

For the problem on the previous page, we could have determined the number of elements in the sample space by using the Counting Principle.

$$2 \times 2 \times 2 = 8$$

SUGGESTED ACTIVITY

See the *Instructor's Resource Manual* for an activity that uses a simple simulation to determine the number of game cards that are necessary for a player to win a prize.

Example 7 Lois is playing backgammon and wants to roll a 3 on one of the two dice she tosses. What is the probability of this event?

Solution Look at the table of outcomes from tossing two dice. By counting the entries in the table, the sample space consists of 36 possible outcomes. The elements that are favorable to E have 3 dots on one of the dice. These are $E = \{(3,1), (3,2), (3,3), (3,4), (3,5), (3,6), (1,3), (2,3), (4,3), (5,3), (6,3)\}$.
Therefore, there are 11 elements in E. Using the probability formula, we have

$$P(E) = \frac{11}{36}$$

The probability of a 3 on one of the two dice is $\frac{11}{36}$.

You-Try-It 7 Suppose that Erin is playing Monopoly and wants to land on Boardwalk, which is 6 spaces away. What is the probability that on a roll of two dice, the sum of the dots on the upward faces is 6?

Solution See page S4.

Empirical Probabilities

The probabilities that have been calculated so far are referred to as *mathematical* or *theoretical* probabilities. The calculations are based on the assumption that, for example, either side of a coin is equally likely or that any one of the six faces of a fair die is equally likely to be face up when the coin or die is tossed. Not all probabilities arise from such assumptions.

Empirical probabilities are based on observations of certain events. For instance, a weather forecast of a 75% chance of rain is an empirical probability. From historical records kept by the weather bureau, rain occurred 75% of the time that a similar weather pattern existed. It is theoretically impossible to predict the weather, and only observations of past weather patterns can be used to predict the future weather conditions.

Insurance rates are based on *empirical* probabilities. For instance, the premium for car insurance is based on the number of drivers insured by a company in various age groups and the number of times they are involved in traffic accidents. There is no theoretical way to predict whether a person will be involved in an accident.

Empirical Probability Formula

The empirical probability of an event E is the ratio of the number of observations of E to the total number of observations.

$$\text{Probability of } E = \frac{\text{number of observations of } E}{\text{total number of observations}}$$

Example 8 Records of an insurance company show that of 2549 claims filed by policy holders for theft, 927 were claims for more than \$5000. What is the empirical probability that a claim for theft that this company receives will be for more than \$5000?

Solution Let E be the event that a policy holder files a claim for more than \$5000. The empirical probability of E is the ratio of the number of claims for over \$5000 to the total number of claims.

$$P(E) = \frac{\text{number of observations of } E}{\text{total number of observations}}$$

$$= \frac{927}{2549} \approx 0.36$$

The empirical probability is 0.36.

You-Try-It 8 In a survey of 5892 adults, it was determined that 336 had less than a high school education. What is the probability that one adult chosen from this group has less than a high school education? Write the probability as a decimal to the nearest thousandth.

Solution See page S4.

Empirical probabilities can be used to test theoretical probabilities. For example, suppose you want to determine if a coin is fair. After tossing the coin 1000 times, you note that heads occurred 527 times. The empirical probability of heads is $\frac{527}{1000} = 0.527$. This differs from the theoretical value of 0.5, but is the difference enough to suggest that the coin is not fair? That is a more difficult question, but one that can be answered in a statistics course.

Odds of an Event

Sometimes the chances of an event occurring are given in terms of *odds*. This concept is closely related to probability.

Odds of an Event

The **odds in favor** of an event is the ratio of the number of favorable outcomes of an experiment to the number of unfavorable outcomes.

$$\text{Odds in favor} = \frac{\text{number of favorable outcomes}}{\text{number of unfavorable outcomes}}$$

The **odds against** an event is the ratio of the number of unfavorable outcomes of an experiment to the number of favorable outcomes.

$$\text{Odds against} = \frac{\text{number of unfavorable outcomes}}{\text{number of favorable outcomes}}$$

TAKE NOTE

Recall that the reciprocal of a fraction is formed by interchanging the numerator and denominator. For instance, the reciprocal of $\frac{2}{3}$ is $\frac{3}{2}$.

Note from the definition of the odds of an event that the fractions for the odds in favor and odds against are reciprocals of each other.

→ Find the odds in favor of a 4 when a single die is rolled once.

List the favorable outcomes and the unfavorable outcomes.

Favorable outcomes = $\{4\}$

Unfavorable outcomes = $\{1, 2, 3, 5, 6\}$

$$\text{The odds in favor of a 4} = \frac{\text{number of favorable outcomes}}{\text{number of unfavorable outcomes}} = \frac{1}{5}$$

Frequently the odds of an event are expressed as a ratio using the word TO. For the problem above, the odds in favor of a 4 are 1 TO 5.

Example 9 Calculate the odds against rolling a sum of 7 when two dice are rolled once.

Solution From the table on page 70, there are 30 unfavorable outcomes (all the pairs whose sum is not 7). There are 6 favorable outcomes (all the pairs whose sum is 7).

$$\text{Odds against a sum of 7} = \frac{\text{number of unfavorable outcomes}}{\text{number of favorable outcomes}} = \frac{30}{6} = \frac{5}{1}$$

The odds against a sum of 7 are 5 TO 1.

You-Try-It 9 Calculate the odds in favor of rolling a sum of 11 when two dice are rolled once.

Solution See page S4.

It is possible to compute the probability of an event from the odds in favor fraction. The probability of an event is the ratio of the numerator to the sum of the numerator and denominator.

→ According to the California Lottery Commission, the odds in favor of winning some prize are 1 TO 9. To calculate the probability of winning a prize:

Write the ratio 1 TO 9 as a fraction. $1 \text{ TO } 9 = \frac{1}{9}$

The probability of winning a prize is the ratio of the numerator to the sum of the numerator and denominator. $\text{Probability} = \frac{1}{1+9} = \frac{1}{10}$

The probability of winning a prize is $\frac{1}{10}$.

Example 10 For the 1997 Super Bowl game between the Denver Broncos and the Green Bay Packers, the odds against Denver winning the game were 11 TO 2. What was the probability of Denver's winning the game?

Solution In this problem, the odds *against* winning the game are given as 11 TO 2 or $\frac{11}{2}$. To calculate the probability of Denver's winning the game, we need the odds *in favor*, which is the reciprocal of $\frac{11}{2}$, or the odds in favor $= \frac{2}{11}$.

The probability of winning the game is the ratio of the numerator to the sum of the numerator and denominator.

$$\text{Probability of winning} = \frac{2}{2+11} = \frac{2}{13}$$

The probability of Denver's winning the game was $\frac{2}{13}$.

You-Try-It 10 Suppose the odds in favor of Martina Hingis's winning the New York Open Tennis tournament were given as 1 TO 9. What was the probability that she would win the tournament?

Solution See page S4.

1.7 EXERCISES

Topics for Discussion

1. The probability of an event lies between which two numbers?
Zero and one

2. What is a sample space? What is an event?
Outcomes of an experiment; subset of a sample space

3. What is an empirical probability? Give some examples.
Probabilities calculated from gathered data

4. How does a probability differ from odds?
$P(E) = \dfrac{n(E)}{n(S)}$; odds in favor $E = \dfrac{\text{number of favorable outcomes}}{\text{number of unfavorable outcomes}}$

5. Give an example of an event that has probability 0. Give an example of an event that has probability 1.
Answers will vary.

6. What does the phrase "equally likely" mean?
Each element of the sample space has the same probability of being selected.

Sample Spaces and Events

7. Two coins are tossed. List the outcomes of the sample space for this experiment.
{HH, HT, TH, TT}

8. A coin is tossed and then a die is rolled. List the outcomes of the sample space for this experiment.
{H1, H2, H3, H4, H5, H6, T1, T2, T3, T4, T5, T6}

9. One number is selected from the digits 2, 3, and 5. Now a second number is selected from the same three digits. List the outcomes of the sample space for this experiment.
{22, 23, 25, 32, 33, 35, 52, 53, 55}

10. One number is selected from the digits 2, 3, and 5. Now a second number is selected from the remaining two digits. List the outcomes of the sample space for this experiment.
{23, 25, 32, 35, 52, 53}

11. A sample space consists of every possible arrangement of the letters *x*, *y*, and *z*. List the elements of the sample space.
{*xyz*, *xzy*, *yxz*, *yzx*, *zxy*, *zyx*}

12. A product code on a washing machine begins with one of the letters A, B, or C, followed by an 8 or 9, followed by one of the letters Y or Z. List the elements of this sample space.
{A8Y, A9Y, A8Z, A9Z, B8Y, B9Y, B8Z, B9Z, C8Y, C9Y, C8Z, C9Z}

13. For fan appreciation day, each of the seat numbers in a baseball stadium is written on a piece of paper and placed in a box. One piece of paper is chosen, and the person occupying that seat wins a season ticket for next season's games. Does the event of choosing one piece of paper from the box satisfy the condition of being equally likely?
Yes

14. The spinner at the right is spun once. Are the probabilities of the pointer landing in the region for each of the numbers 1, 2, 3, and 4 equally likely?
No

15. A card is drawn from a regular deck of playing cards. List the elements of the event that the card is a king.
{king of hearts, king of spades, king of clubs, king of diamonds}

16. A ball is drawn from a hat that contains a red, a blue, a green, a white, and a black ball. List the elements of the event that the ball is not blue.
{red, green, white, black}

17. One day is randomly chosen from the days of the week. List the elements of the event that the day chosen is not a Saturday or Sunday.
{Monday, Tuesday, Wednesday, Thursday, Friday}

18. A two-digit number is formed by randomly choosing two digits from the digits 4, 5, 6, 7, and 8. List the elements of the event that the number formed is less than 38.
$\varnothing$

19. If two dice are rolled once, list the elements of the event that the sum of the numbers face up is greater than 10.
{(5, 6), (6, 5), (6, 6)}

20. If two dice are rolled once, list the elements of the event that the product of the numbers face up is less than 8.
{(1, 1), (1, 2), (1, 3), (1,4), (1, 5), (1, 6), (2,1), (2, 2), (2, 3), (3,1), (3, 2), (4, 1), (5, 1), (6, 1)}

The Counting Principle

21. A coin is tossed twice. Make a tree diagram that shows the possible outcomes.

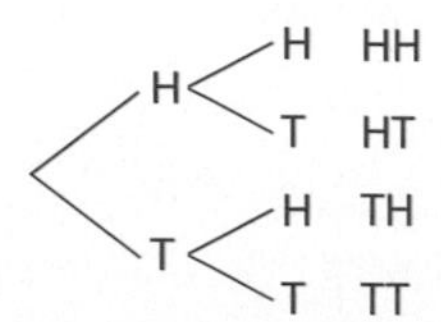

22. A luncheon special at a Mexican restaurant offers a main dish, a side dish, and a drink. The main dish consists of a taco, an enchilada, or a burrito. The side dish is either rice or beans, and the drink is either lemonade or ice tea. Make a tree diagram to show the different luncheon specials that are possible.

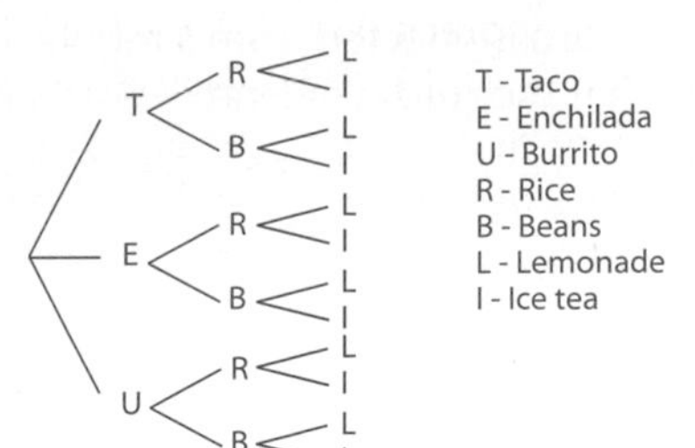

23. The owner of an electronic garage door opener can choose a security code for the opener by setting 8 switches to off or on. How many different security codes are possible?
256

24. The security code for a safe provided in a hotel room can be changed by the guest. If one security code is a sequence of 4 digits chosen from the digits 1 through 9, how many different security codes are possible?
6561

25. A combination lock is set by selecting 3 numbers from the numbers 1 through 36. How many different combinations can be created?
46,656

26. The telephone extensions for a corporation are four-digit numbers that must begin with a 4. How many different extensions can be created for this company?
1000

27. A serial number on a computer begins with 8, 9, or 0 (for 1998, 1999, and 2000); followed by one of the letters A, B, C, D, E, F, G, H, I, J, K, or L (where A corresponds to January and L corresponds to December); followed by a C, D, B, or L (indicating that it was manufactured in Chicago, Dallas, Boston, or Los Angeles). How many different serial numbers are possible?
144

Probability of an Event

28. A die is tossed once. What is the probability that the upward face is a 3?
$\frac{1}{6}$

29. Two coins are tossed. What is the probability that the upward faces are both tails?
$\frac{1}{4}$

30. One card is selected from a regular deck of playing cards. What is the probability that the card is a 7?
$\frac{1}{13}$

31. One card is selected from a regular deck of playing cards. What is the probability that the card is a heart?

$\frac{1}{4}$

32. Three coins are tossed. What is the probability that there are two heads and one tail?

$\frac{3}{8}$

33. Three coins are tossed. What is the probability that none of the coins shows a tail?

$\frac{1}{8}$

34. One card is randomly selected from a regular deck of playing cards. What is the probability that the card is not a spade?

$\frac{3}{4}$

35. If two dice are tossed, what is the probability that both dice have the same number on the upward faces?

$\frac{1}{6}$

36. If two dice are tossed, what is the probability that both dice have different numbers on the upward faces?

$\frac{5}{6}$

37. A charity raffle sold 25,000 raffle tickets for $1 each. If there is 1 grand prize of $5000, 2 second-place prizes of $1000 each, and 5 third-place prizes of $250 each, what is the probability that you will win one of the prizes if you purchase one ticket?

$\frac{1}{3125}$

38. A blindfolded lottery winner is offered the opportunity to reach into a large bowl containing 100 pieces of paper and select one piece of paper on which is written a prize. The bowl contains one prize of $1,000,000; 5 prizes of $100,000; 10 prizes of $10,000; and 50 prizes of $1000. The remaining pieces of paper contain a prize for $500. What is the probability that the lottery winner will choose a prize worth more than $10,000?

$\frac{3}{50}$

39. In American roulette, there are 18 numbers colored red, 18 numbers colored black, and 2 numbers (0 and 00) colored green on a wheel. If a roulette ball is spun around the wheel so that it has an equally likely chance of landing on any one of the numbers, what is the probability that it lands on green?

$\frac{1}{19}$

40. In European roulette, there are 18 numbers colored red, 18 numbers colored black, and 1 number (0) colored green on a wheel. If a roulette ball is spun around the wheel so that it has an equally likely chance of landing on any one of the numbers, what is the probability that it lands on green?

$\frac{1}{37}$

Odds of an Event

41. If a die is tossed once, what are the odds in favor of the upward face being a 4?
1 TO 5

42. If one card is randomly chosen from a regular deck of playing cards, what are the odds in favor of the card being a heart?
1 TO 3

43. One card is randomly selected from a bowl containing 3 red cards, 5 blue cards, and 7 green cards. What are the odds against the card being blue?
2 TO 1

44. If two dice are tossed, what are the odds against the sum of the upward faces being a 7?
5 TO 1

45. With each purchase at a fast food restaurant, a patron receives a "scratcher" card which may be good for a prize. The odds in favor of the prize being an order of french fries is 1 TO 15. What is the probability of the prize being an order of fries?

$\frac{1}{16}$

46. In a room of 25 people, the odds in favor of at least two people having the same birthday are approximately 1 TO 1. What is the probability of at least two people having a birthday on the same day?

$\frac{1}{2}$

47. The odds against a horse winning a race are given as 5 TO 2. What is the probability that the horse will win the race?

$\frac{2}{7}$

48. When two dice are tossed, the odds against a sum of eleven on the upward faces are 17 TO 1. What is the probability of rolling an eleven with the toss of two dice?

$\frac{1}{18}$

Empirical Probabilities

49. A survey of Americans who exercise regularly found that 650 exercise to stay healthier, 180 exercise to lose weight, 100 exercise to have fun, and 70 exercise to look good. If one person is selected from this survey, what is the probability the person exercises to stay healthy? (Marist Institute for Public Opinion)

$\frac{13}{20}$

50. Suppose a coin is tossed 100 times and a head turns up 90 times. Do you think the coin is fair? Why or why not? Answers will vary.

Applying Concepts

51. The scores on a psychology exam given to 250 students were analyzed and the following information obtained: mean = 74, median = 72, mode = 70. Suppose one student's test paper is selected at random from these tests.

a. Which test score, the mean, median, or mode, has the greatest probability of being selected?

b. Which probability is greater, that of the test score being greater than the mean or that of the test score being greater than the median?

a. mode; **b.** the probability of being greater than the median

52. If the probability of an event occurring is x, what is the probability of the event not occurring? $1 - x$

Exploration

53. ***Factorials*** Suppose that you are trying to form all possible 4-digit numbers from the digits 2, 3, 4, and 5 such that no digit is used more than once. There are 4 choices for the first number. Because a digit cannot repeat in a number, there are only 3 choices for the second number, 2 choices for the third number, and 1 choice for the last number. By the Counting Principle, there are $4 \cdot 3 \cdot 2 \cdot 1 = 24$ possible numbers that could be formed. The product $4 \cdot 3 \cdot 2 \cdot 1$ is frequently written as 4! and is read "4 factorial." The exclamation mark is read "factorial." As another example, $6! = 6 \cdot 5 \cdot 4 \cdot 3 \cdot 2 \cdot 1 = 720$.

a. Evaluate 5! and 10!. (Many calculators have a built-in factorial operation.)

b. Does $\frac{12!}{3!}$ equal 4!? If not, what is the value of $\frac{12!}{3!}$?

c. Does (5!)(3!) equal 15!? If not, what is the value of (5!)(3!)?

Chapter Summary

Definitions

A *sequence* is an ordered list of numbers. Each of the numbers in a sequence is called a *term*.

Inductive reasoning involves making generalizations from specific examples; in other words, we reach a conclusion by making observations about particular facts or cases. *Deductive reasoning* involves drawing a conclusion which is based on given facts. It is frequently used in reaching a conclusion from a sequence of known facts.

A *statement* is a sentence that is either true or false.

The *negation* of a statement is the statement of opposite truth value.

An example that is given to show that a statement is not true is called a *counter-example.*

The symbol $<$ means *is less than*; the symbol $>$ is used to mean *is greater than.* These symbols are called *inequality* symbols.

A *double-bar* graph uses more than one bar so that quantities can be compared. A *broken-line* graph uses line segments to connect points on the graph.

In a *frequency table,* organized data is combined into categories called *classes* and the number of data in each class, called the *frequency* of the class, is shown. Each class has a *lower class limit* and an *upper class limit.* The *class size* is the difference between two successive lower class limits. A *relative frequency table* shows the percent of all the data that belongs to a class.

A *histogram* is a vertical bar graph of a frequency table.

A way to display every number in a data set and still organize the data in intervals is to use a *stem-and-leaf plot.* Each piece of data is split into two parts, a *stem* and a *leaf.* The leaf is usually the rightmost digit of a number; the stem is the remaining digits. Data is placed into stem-and-leaf plots as a means of quickly showing the frequencies of a particular event.

Each product can be written out in *expanded form.* In *exponential form,* the *exponent* indicates the number of times the *base* is used in a product.

A *variable* is a letter that represents a quantity that can change, or vary. An expression using this letter representation is called a *variable expression.* The number multiplied by the variable is called the *coefficient.*

Replacing a variable in a variable expression by a given number and then simplifying the numerical expression is called *evaluating a variable expression.* The result is called the *value of the variable expression.* The number substituted for the variable is called the *value of the variable.*

An *input/output table* shows the relationship between two variables.

A *formula* is a special type of equation that states a rule about measurements.

The *mean* of a set of data is the sum of the measurements divided by the number of measurements. The *median* of data is the number which separates the data into two equal parts when the numbers are arranged from smallest to largest (or largest to smallest). The *mode* of a set of numbers is the value that occurs most frequently. If a set of numbers has no number occurring more than once, then the data has no mode. When the mode is used for qualitative data, it is the category that receives the greatest number of responses, called the *modal response*.

A *box-and-whiskers plot* is a graph that gives a more complete picture of the data. It shows five numbers: the smallest value, the first quartile, the median, the third quartile, and the greatest value. The *first quartile*, symbolized by Q_1, is the number for which approximately one-quarter or 25% of the data lies below that number. Q_1 is the median of the lower half of the data values. The *third quartile*, symbolized by Q_3, is the number for which about one-quarter of the data lies above that number. Q_3 is the median of the upper half of the data values. The *interquartile range* is the difference between the third quartile Q_3 and the first quartile Q_1.

The *range* is the difference between the largest data value and the smallest data value.

The degree of uncertainty of an outcome is called *probability*. A probability is determined from an *experiment*, which is any activity that has an observable outcome. All the possible outcomes of an experiment is called the *sample space*. An *event* is one or more outcomes of an experiment.

The *probability* of an event *E*, written *P*(*E*), is the ratio of the number of elements in the event to the number of elements in the sample space.

The *empirical probability* of an event *E* is the ratio of the number of observations of *E* to the total number of observations.

The *odds in favor* of an event is the ratio of the number of favorable outcomes of an experiment to the number of unfavorable outcomes. The *odds against* an event is the ratio of the number of unfavorable outcomes of an experiment to the number of favorable outcomes.

Procedures

The **four-step process in problem solving:**

1. Understand the problem.
2. Devise a strategy to solve the problem.
3. Execute the strategy and state the answer.
4. Review your solution.

Order of Operations Agreement:

1. Perform operations inside grouping symbols. These include parentheses, brackets, braces, and fraction bars.
2. Evaluate exponential expressions.
3. Perform multiplications and divisions as they occur from left to right.
4. Perform additions and subtractions as they occur from left to right.

Chapter Review Exercises

1. If $A = \{1, 3, 4, 6, 7\}$, $B = \{2, 4, 6, 9\}$, and $C = \{0, 5\}$, find:

a. $A \cup B$ **b.** $A \cap B$ **c.** $B \cap C$

a. $\{1, 2, 3, 4, 6, 7, 9\}$; **b.** $\{4, 6\}$; **c.** $\varnothing$

2. What is the next term in the sequence 1, 2, 6, 24, 120, . . .? 720

3. Write the negation: "All animals do not have feathers." Some animals have feathers.

4. Find a counterexample: "All numbers that end in 8 are divisible by 4." 18

5. Determine the truth value when $x = 4$: $(6 < 3x - 5)$ and $(2x + 1 < 10)$. True

6. Using the variable expression $1.25g$, where g is the number of gallons of gas, find the cost to purchase 10.8 gallons of gas. $13.50

7. Evaluate: $16 + 8 \div 2^2$ 18

8. Evaluate: $80 \div 2^4 - 2 \cdot 2$ 1

9. Evaluate: $4^2 + (11 - 2) \div 3$ 19

10. Evaluate: $(7 + 1)^2 - 3(4 - 4)^3$ 64

11. Translate into a variable expression: Four-sevenths of a number. $\frac{4}{7}n$

12. Translate into a variable expression: 4 less than 5 times a number. $5n - 4$

13. Translate into a variable expression: Twice a number divided by the difference between three times the number and seven. $\frac{2n}{3n - 7}$

14. Translate into a variable expression: The difference between 4 times the square of d and the sum of d and 8. $4d^2 - (d + 8)$

15. If $\varnothing\varnothing = \Delta\Delta\Delta$, and $\Delta\Delta = \infty\infty\infty\infty\infty$, and $\infty\infty\infty\infty = \Omega\Omega\Omega\Omega$, then $\Omega\Omega\Omega\Omega\Omega\Omega\Omega\Omega\Omega$ = how many $\varnothing$'s? 2

16. In a gumball machine, there are three times as many white gumballs as red ones. Express the amount of white gumballs in terms of the amount of red ones. $3r$

17. If three coins are tossed, what is the probability that none of the coins shows a head? $\frac{1}{8}$

18. If one card is randomly chosen from a regular deck of playing cards, what are the odds in favor of the card being a spade? 1 TO 3

19. If a piece of paper is folded in half, then in half again, and this process continues, how many layers of papers are there after 10 folds? What is the relationship of the number of folds to the number of layers? 1024; 2^n = number of folds.

20. Twenty-one students took their first math exam. Their scores were: 78, 64, 59, 87, 72, 61, 89, 94, 73, 67, 45, 74, 83, 78, 82, 74, 84, 79, 83, 92, 58. Make a stem-and-leaf plot for these data.

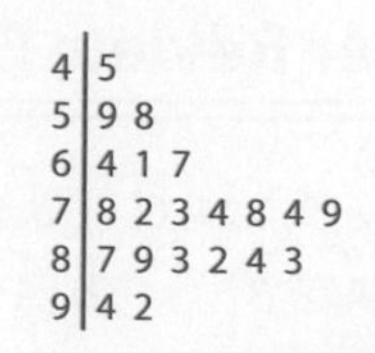

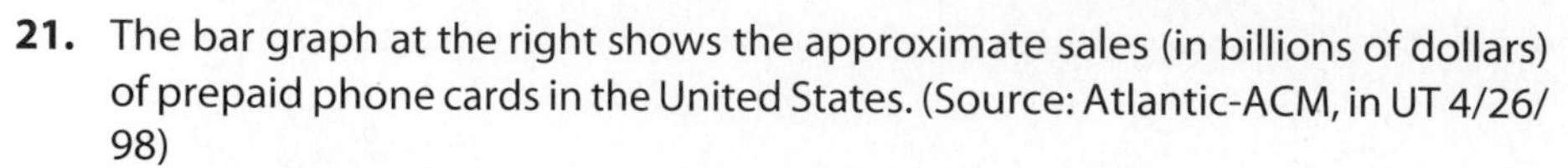

21. The bar graph at the right shows the approximate sales (in billions of dollars) of prepaid phone cards in the United States. (Source: Atlantic-ACM, in UT 4/26/98)

a. What were the approximate sales of phone cards in 1995?

b. For the years shown, between which two years did sales of phone cards increase the most?

c. Assuming the trend shown by the graph continues, would you expect 1998 sales of phone cards to be less than or more than that of 1997?

a. \$.7 billion; **b.** 1996 and 1997; **c.** more than

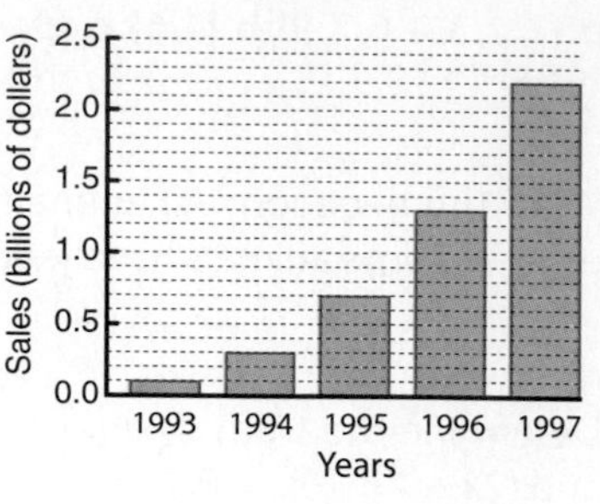

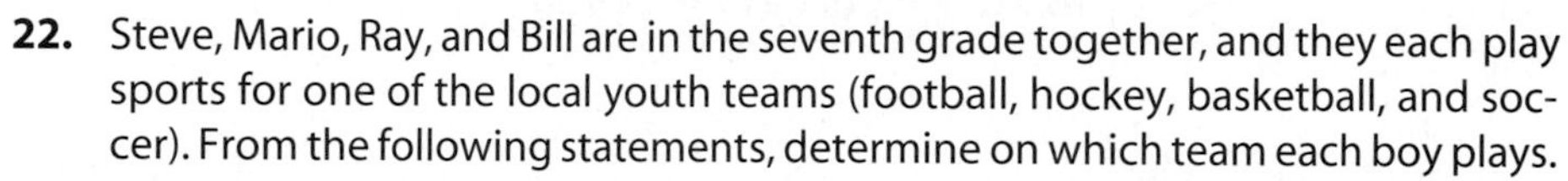

22. Steve, Mario, Ray, and Bill are in the seventh grade together, and they each play sports for one of the local youth teams (football, hockey, basketball, and soccer). From the following statements, determine on which team each boy plays.

a. Steve and the student who plays football both live in houses on Main Street, while Bill and the student who plays basketball live in an apartment complex on 3rd Avenue.

b. Ray and the student who plays football are partners in science class, while Bill and the student who plays soccer are in different classes.

Steve, soccer; Mario, football; Ray, basketball; Bill, hockey

23. Suppose that a group of golfers had a contest to see who could hit the ball the farthest, and their distances were 261, 226, 270, 247, 232, 263, 289, 260, 258, 238, 228, 272, 212, 230. Find the mean, median, and mode of this data.

mean = 249; median = 252.5; mode, none

24. Suppose a random survey was taken at a local college to see whether 700 students were very satisfied, satisfied, unsatisfied, or very unsatisfied with the food choices on campus. The responses by these students are given at the right. What was the modal response?

Very unsatisfied

Survey of campus food choices

Response	Count
Very satisfied	98
Satisfied	172
Unsatisfied	201
Very unsatisfied	229

25. A random sample of SAT scores in mathematics is as follows: 479, 486, 406, 453, 533, 413, 528, 538, 479, 578, 431, 525, 528, 624. Find Q_1 and Q_3 for these data.

$Q_1 = 453; Q_3 = 533$

26. The high school grade-point averages for 15 random college freshmen were: 4.00, 2.53, 3.45, 2.48, 2.69, 2.82, 2.33, 2.21, 3.25, 2.90, 2.75, 2.82, 3.51, 2.78, 2.46. Draw a box-and-whiskers plot for these data. What is the interquartile range?

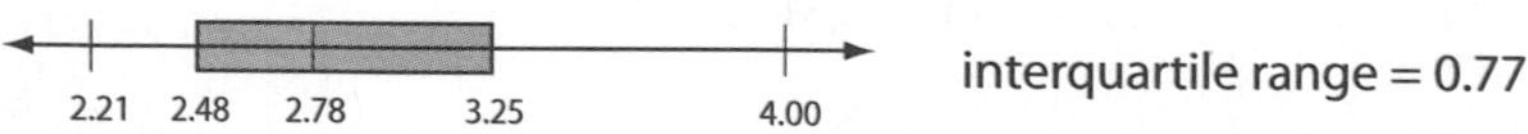

interquartile range = 0.77

27. A sample space consists of every possible arrangement of the letters a, b, and c. List the elements of the sample space.

$\{abc, acb, bac, bca, cab, cba\}$

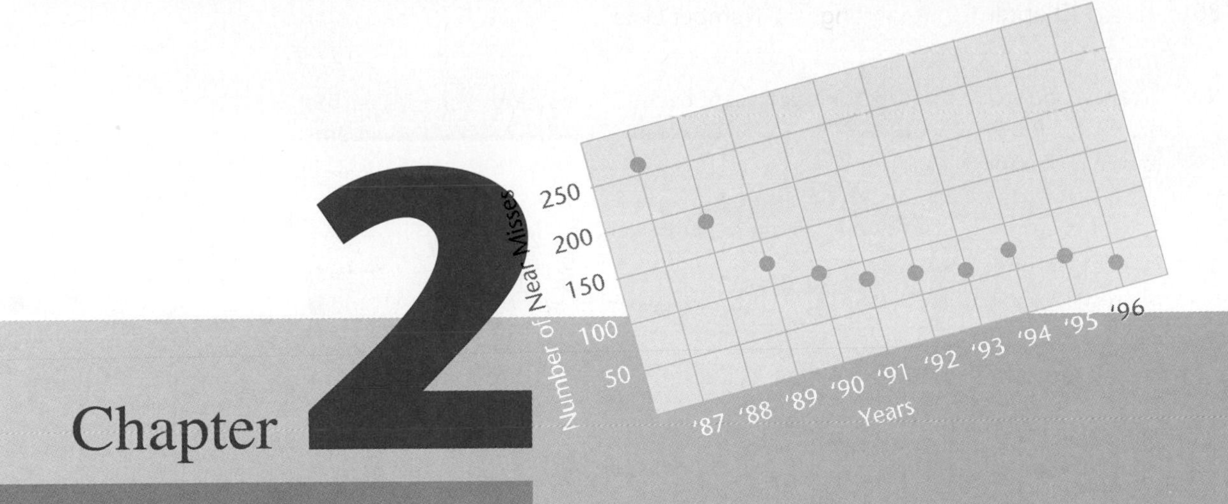

Chapter 2

Real Numbers and the Rectangular Coordinate System

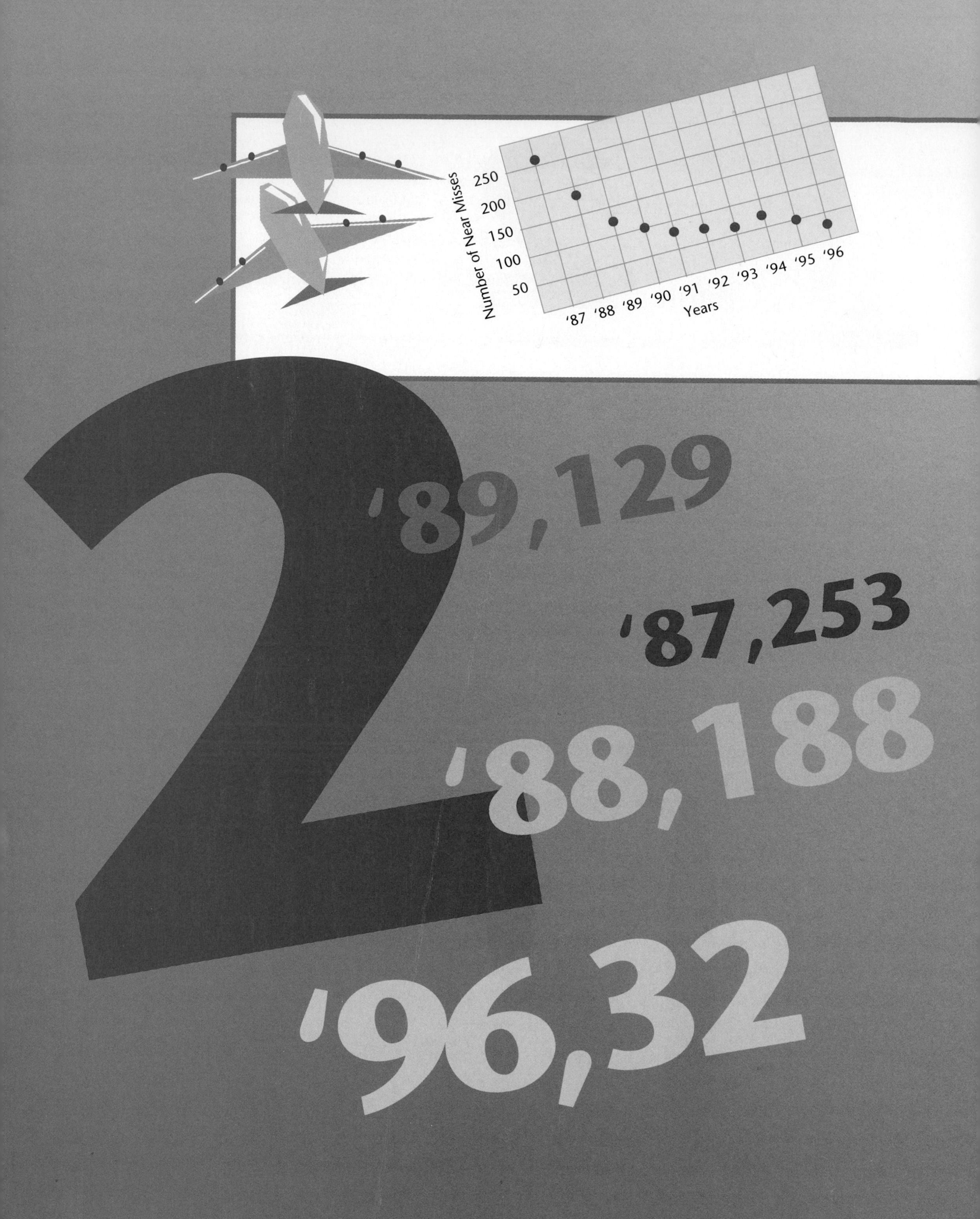

Number of Near Misses
250
200
150
100
50
'87 '88 '89 '90 '91 '92 '93 '94 '95 '96
Years
2
'89,129
'87,253
'88,188
'96,32

Section 2.1 Addition and Subtraction of Integers

Opposites and Absolute Value

On January 12, 1911, the temperature in Rapid City, South Dakota, dropped 62° in two hours. The temperature went from 49 degrees above zero to 13 degrees below zero. Numbers above zero are called **positive** numbers. Sometimes a positive number is written as +49, but generally the + sign is omitted.

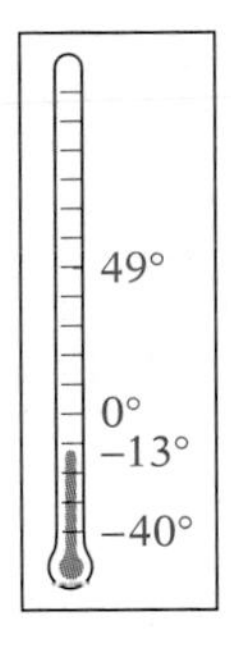

Numbers below zero are called **negative** numbers and are represented by placing a negative sign in front of the number. The number "negative 13" is written −13. The set of positive and negative whole numbers and zero is called the *integers*.

Integers = {…, −5, −4, −3, −2, −1, 0, 1, 2, 3, 4, 5, …}

A **number line** is used as a graphical representation of positive and negative numbers.

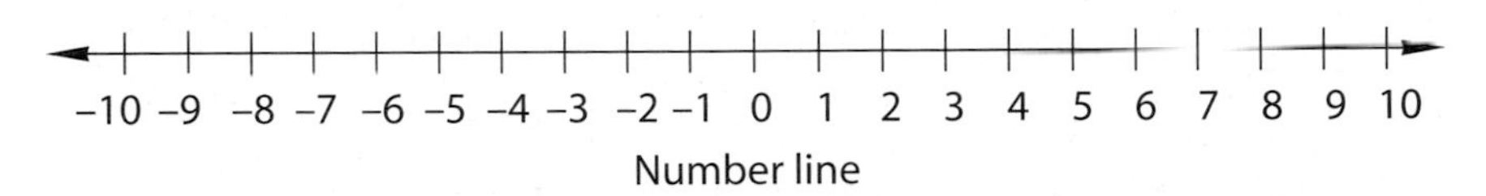

Number line

POINT OF INTEREST
The word *integer* comes directly from the Latin word *integer,* meaning "whole, complete, perfect, entire." In fact, *integer* and *entire* have the same word origin.

To **graph**, or **plot**, a number, place a dot at the location given by the number. For example, to graph the number 6, place a dot on the number line directly above 6. The number 6 is called the **coordinate** of the point (dot) that was placed at 6.

The graphs of −7, −1.5, $\frac{5}{3}$, and 6 are shown below.

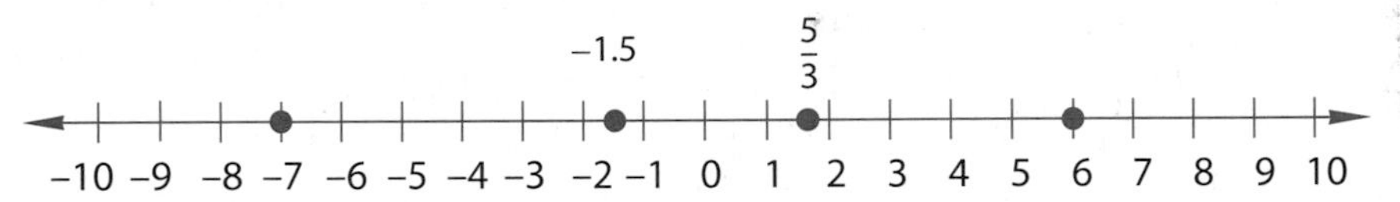

For the coordinates −1.5 and $\frac{5}{3}$, it is necessary to place the dot at the approximate location of these coordinates on the number line.

Two numbers that are the same distance from zero on the number line but are on opposite sides of zero are opposite numbers, or **opposites**. The opposite of a number is also called its **additive inverse**.

The opposite of 5 is −5.
The opposite of −5 is 5.

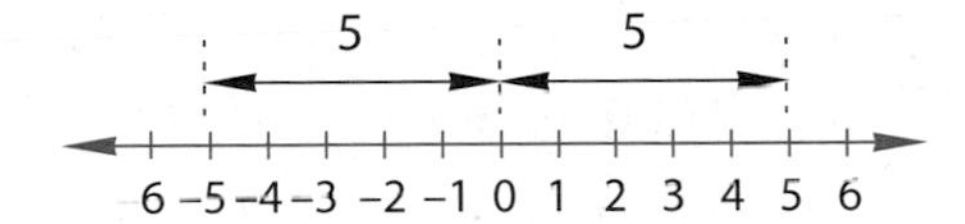

The negative sign can be used to indicate "the opposite of."

−(2) = −2. The opposite of 2 is negative 2.

−(−2) = 2. The opposite of −2 is 2.

Zero is neither a positive nor a negative number. The opposite of 0 is 0.

Question: What is the opposite of −12?[1]

On scientific calculators, the +/− key is used to obtain a negative number. For instance, 2 +/− will produce −2. On graphing calculators, the (−) key is used to obtain a negative number. For example, (−) 2 will produce −2.

1. The opposite of −12 is 12.

The absolute value of a number is its distance from zero on the number line. Therefore, the absolute value of a number is a positive number or zero. The symbol for absolute value is two vertical bars, | |.

POINT OF INTEREST

The definition of absolute value given in the box is written in what is called rhetorical style. That is, it is written without the use of variables. This is how all mathematics was written prior to the Renaissance. During that period, from the 14th to the 16th century, the idea of expressing a variable symbolically was developed. In terms of that symbolism, the definition of absolute value is

$$|x| = \begin{cases} x, & x > 0 \\ 0, & x = 0 \\ -x, & x < 0 \end{cases}$$

Absolute Value

The absolute value of a positive number is the number itself. The absolute value of zero is zero. The absolute value of a negative number is the opposite of the negative number.

The distance from 0 to 4 is 4. Therefore, the absolute value of 4 is 4.

$|4| = 4$

The distance from 0 to −4 is 4. Therefore, the absolute value of −4 is 4.

$|-4| = 4$

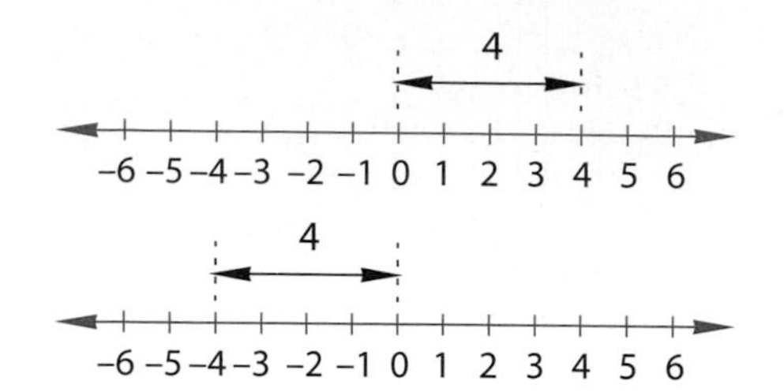

Question: What is the value of $|-36|$?[2]

TAKE NOTE

A graphing calculator uses the symbol ABS for absolute value. For Example 1a, enter ABS(−14). For Example 1b, enter −ABS(−23).

Example 1 Simplify each of the following.

a. $|-14|$ **b.** $-|-23|$ **c.** $-(|9| + |-20|)$

Solution

a. $|-14| = 14$

b. $-|-23| = -23$ • The negative sign *in front* of the absolute value sign means the opposite of $|-23|$.

c. $-(|9| + |-20|) = -(9 + 20) = -29$

You-Try-It 1 Simplify each of the following.

a. $|-42|$ **b.** $-|-31|$ **c.** $-(|-23| - |-17|)$

Solution See page S5.

Addition of Integers

A number can be represented by an arrow. A positive number is represented by an arrow pointing to the right, and a negative number is represented by an arrow pointing to the left. The magnitude (absolute value) of the number is represented by the length of the arrow.

POINT OF INTEREST

Using arrows to represent a number is done not only here but in engineering as well. Engineers call these arrows *vectors* and use them in many different situations.

Question: What number is represented by the arrow ?[3]

Addition is the process of finding the total of two numbers. The numbers being added are called **addends**. The total is called the **sum**. We can find rules for adding integers by using a number line and the arrow representation of numbers.

---

2. The absolute value of a negative number is a positive number. Therefore, $|-36| = 36$.
3. Because the arrow points to the left and is of length 5, the number is −5.

SUGGESTED ACTIVITY
There are a number of models of the addition of integers. Using arrows on the number line is just one of them. Another suggestion is to use a checking account. If there is a balance of $25 in a checking account, and a check is written for $30, the account will be overdrawn by $5 (−5).
Another model uses two colors of plastic chips, say blue for positive and red for negative, and the idea that a blue/ red pair is equal to zero. To add −8 + 3, place 8 red and 3 blue chips in a circle. Make as many blue/red pairs as possible and remove them from the region. There are 5 red chips remaining, or −5.

To add two integers, find the point on the number line corresponding to the first addend. Starting at that point, draw an arrow representing the second addend. The sum is the number directly below the tip of the arrow.

$3 + 5 = 8$

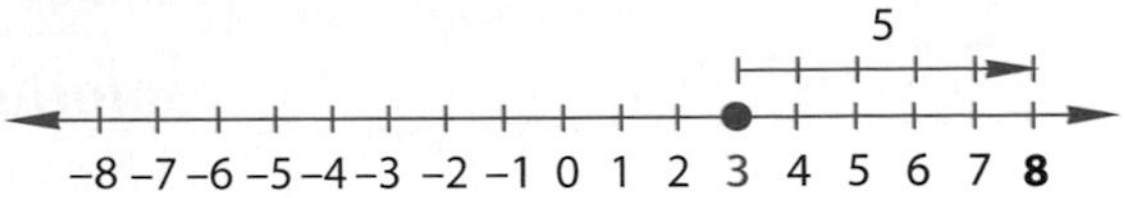

5
−8 −7 −6 −5 −4 −3 −2 −1 0 1 2 3 4 5 6 7 **8**

$-3 + (-5) = -8$

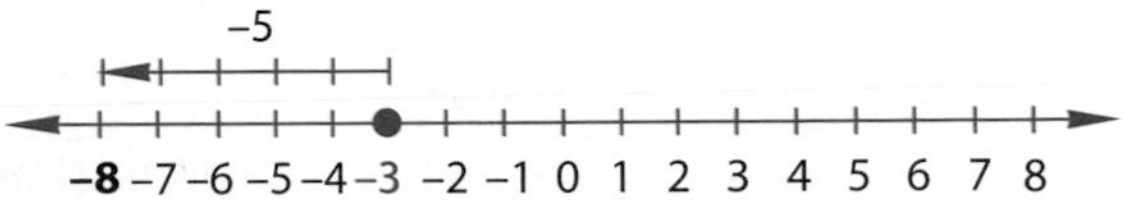

−5
−8 −7 −6 −5 −4 −3 −2 −1 0 1 2 3 4 5 6 7 8

$3 + (-5) = -2$

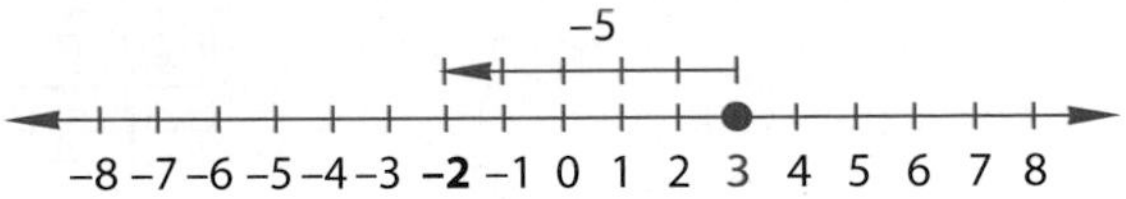

−5
−8 −7 −6 −5 −4 −3 **−2** −1 0 1 2 3 4 5 6 7 8

$-3 + 5 = 2$

5
−8 −7 −6 −5 −4 −3 −2 −1 0 1 **2** 3 4 5 6 7 8

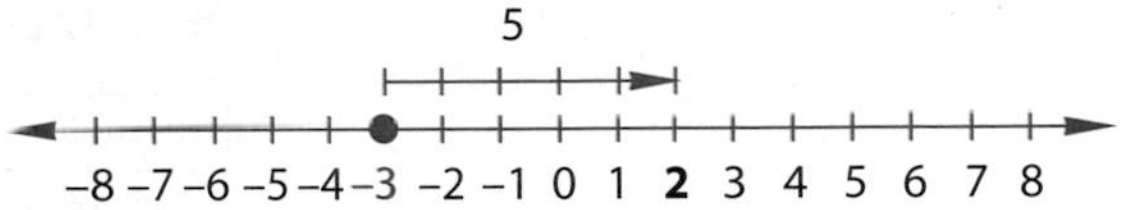

The arrow model for adding integers suggests the following rule.

Addition of Integers

Same Signs
To add two numbers with the same sign, add the absolute values of the numbers. Then attach the sign of the addends.

Different Signs
To add two numbers with different signs, find the absolute value of each number. Then subtract the smaller of these absolute values from the larger one. Attach the sign of the number with the larger absolute value.

Question: Which number has the larger absolute value, 37 or −62?[4]

Example 2 **a.** Simplify: $23 + (-51)$

b. Find the sum of −14 and −48.

Solution

a. $|23| = 23, |-51| = 51$ • Find the absolute value of each number.

$51 - 23 = 28$ • The signs are different. Subtract the smaller of these absolute values from the larger one.

$23 + (-51) = -28$ • Because $|-51| > |23|$, the answer is negative.

b. Recall that *sum* means to add.

$|-14| = 14, |-48| = 48$ • Find the absolute value of each number.

$14 + 48 = 62$ • The signs are the same. Add the absolute values of the two numbers.

$(-14) + (-48) = -62$ • Both numbers are negative. The answer is negative.

You-Try-It 2 **a.** Simplify: $(-28) + 71$

b. Find the total of −42 and 28.

Solution See page S5.

4. $|37| = 37$ and $|-62| = 62$. Therefore, $|-62| > |37|$.

Example 3 Complete the input/output table below.

Input, n	-3	-2	-1	0	1	2	3
Output, $(-4) + n$							

Solution Evaluate the variable expression $(-4) + n$ for each of the input values. Here is the calculation for the input values -2, 0, and 3.

$$(-4) + n \qquad (-4) + n \qquad (-4) + n$$
$$(-4) + (-2) = -6 \qquad (-4) + 0 = -4 \qquad (-4) + 3 = -1$$

After completing the remaining calculations, the table is

Input, n	-3	-2	-1	0	1	2	3
Output, $(-4) + n$	-7	-6	-5	-4	-3	-2	-1

You-Try-It 3 Complete the input/output table below.

Input, x	-3	-2	-1	0	1	2	3
Output, $x + (-6)$							

Solution See page S5.

Subtraction of Integers

POINT OF INTEREST
In mathematics manuscripts dating from the 1500s, an m was used to indicate minus. Some historians believe that, over time, the m went from
$m \rightarrow$ ~ $\rightarrow$ ~ $\rightarrow$ —

SUGGESTED ACTIVITY
A subtraction model based on blue and red chips similar to that for addition can be provided. Restrict the terms of the subtraction to, say, between -10 and 10, and start with 10 blue/red pairs in a circle. Because each blue/red pair is equal to zero, the circle contains 10 zeros. To model $-3 - (-7)$, place 3 more red chips in the circle and remove (subtract) any 7 red chips. Now pair as many blue and red chips as possible. There will be 4 blue chips left without a red chip. In other words, $-3 - (-7) = 4$.

The "–" sign is used in two different ways: to mean *subtract*, as in $9 - 4$ (9 **minus** 4), and to mean the *opposite*, as in -4 (the **opposite** of 4 or **negative** 4).

Look at the next four examples and be sure you understand the difference between *minus* (meaning subtract) and *negative* (meaning the opposite of).

a. $9 - 4$; positive 9 minus positive 4
b. $-9 - 4$; negative 9 minus positive 4
c. $9 - (-4)$; positive 9 minus negative 4
d. $-9 - (-4)$; negative 9 minus negative 4

Question: What is the meaning of $-6 - 4$ and $8 - (-7)$ in words?[5]

Subtraction is the process of finding the difference between two numbers. Look at the following two problems, $8 - 3$ and $8 + (-3)$.

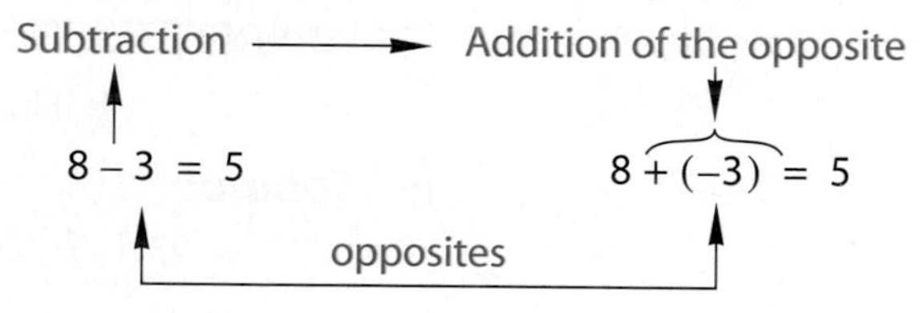

Subtracting Integers
To subtract an integer, add its opposite. In symbols,
$$a - b = a + (-b)$$

First number	–	second number	=	first number	+	opposite of the second number	
15	–	20	=	15	+	(–20)	= –5
15	–	(–20)	=	15	+	20	= 35
–15	–	20	=	–15	+	(–20)	= –35
–15	–	(–20)	=	–15	+	20	= 5

5. $-6 - 4$ means negative 6 minus positive 4; $8 - (-7)$ means positive 8 minus negative 7.

Example 4 **a.** Find the difference between –43 and 26.

b. Simplify: $2-(-24)-18-(-3)$

Solution

a. Recall that to find the difference between two numbers, subtract the numbers.

$-43-26 = -43+(-26) = -69$ • Add the opposite of the second number.

b. $2-(-24)-18-(-3) = 2+24+(-18)+3$ • Rewrite each subtraction as addition of the opposite.

$= 11$

You-Try-It 4 **a.** What is 72 less than 46?

b. Simplify: $-15-12-9-(-36)$

Solution See page S5.

Example 5 Complete the input/output table below.

Input, w	–3	–2	–1	0	1	2	3
Output, $1-w$							

TAKE NOTE
You can check your calculations for Example 5 by using the table feature of a graphing calculator. Some typical graphing calculator screens are shown below.

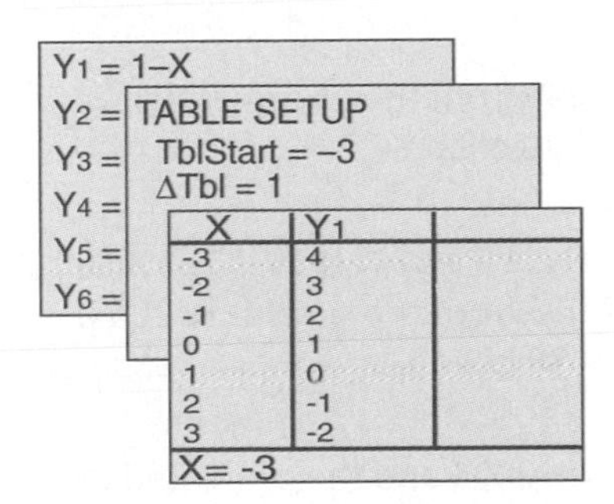

Solution Evaluate the variable expression $1-w$ for each of the input values. Here is the calculation for the input values –3, 1, and 3.

$1-w$ $\quad$ $1-w$ $\quad$ $1-w$

$1-(-3) = 1+3 = 4$ $\quad$ $1-1 = 1+(-1) = 0$ $\quad$ $1-3 = 1+(-3) = -2$

After completing the remaining calculations, the table is

Input, w	–3	–2	–1	0	1	2	3
Output, $1-w$	4	3	2	1	0	–1	–2

You-Try-It 5 Complete the input/output table below.

Input, z	–3	–2	–1	0	1	2	3
Output, $z-6$							

Solution See page S5.

Applications

Finding the distance between two points on a number line is one application of subtraction and of absolute value.

Distance Between Two Points on a Number Line
The distance between the points P and Q, denoted by $d(P, Q)$, whose coordinates are a and b is given by $d(P, Q) = |a-b|$.

TAKE NOTE
Note that, because of the absolute value, the order in which the coordinates are subtracted does not matter. For instance,
$d(Q, P) = |3-(-5)| = |8| = 8$
The absolute value is used to ensure that the distance between two points is always a positive number.

Example 6 Find the distance between the points labeled P and Q.

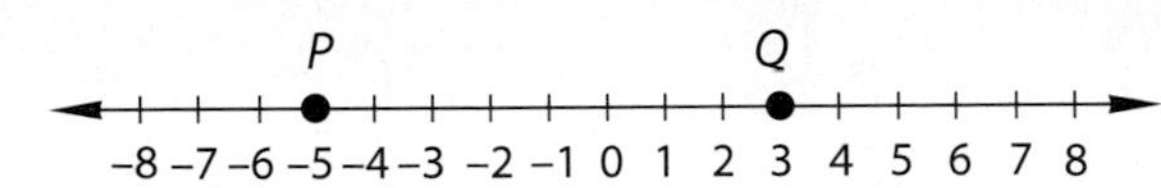

Solution The coordinate of P is –5 and the coordinate of Q is 3. The distance between P and Q is

$$d(P, Q) = |-5-3| = |-8| = 8$$

You-Try-It 6 Find the distance between the points labeled A and B.

A B

−8 −7 −6 −5 −4 −3 −2 −1 0 1 2 3 4 5 6 7 8

Solution See page S5.

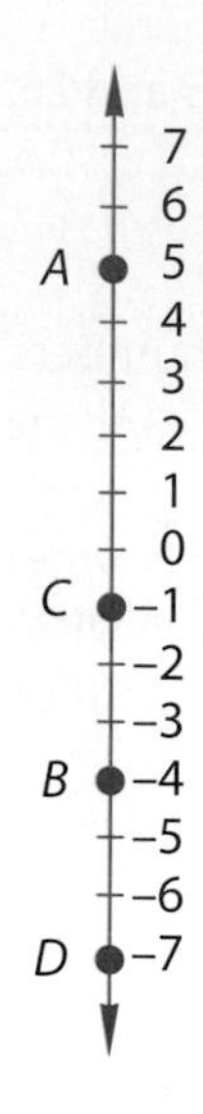

Calculating the distance between two points on a vertical line is similar to the calculation on a horizontal line. The distance between two points on a vertical line is the absolute value of the difference between the coordinates of the points. For instance, for the vertical number line at the left, the distance between A and B is

$$\begin{aligned} d(A, B) &= |5 - (-4)| \\ &= |5 + 4| = |9| \\ &= 9 \end{aligned}$$

Question: What is the distance between the points C and D on the number line at the left?[6]

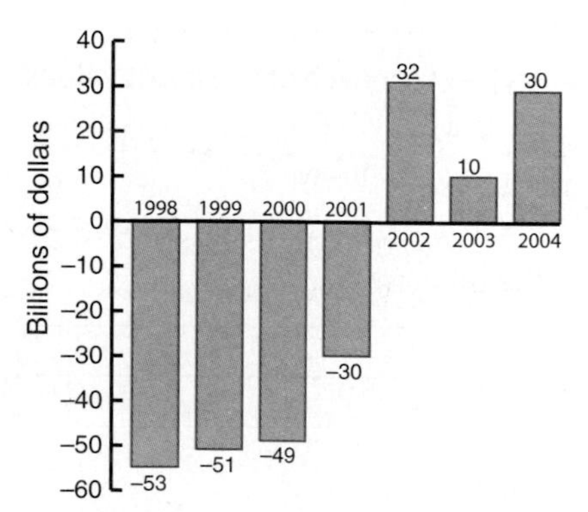

Estimated Federal Deficits and Surpluses
Source: Congressional Budget Office

Example 7 The graph at the left shows projected deficits (negative numbers) and projected surpluses (positive numbers) for the federal budget for a seven-year period.

a. What is the difference between the projected deficit in 2001 and the projected deficit in 1999?

b. What is the difference between the projected surplus in 2004 and the projected deficit in 1998?

Solution

a. To find the difference, subtract the deficit in 1999 from the one in 2001.
Deficit for 1999: −\$51 billion; deficit for 2001: −\$30 billion

$-30 - (-51) = -30 + 51 = 21$

The difference between the projected deficits in 2001 and in 1999 is \$21 billion.

b. To find the difference, subtract the deficit in 1998 from the surplus in 2004.
Deficit for 1998: −\$53 billion; surplus for 2004: \$30 billion

$30 - (-53) = 30 + 53 = 83$

The difference between the projected surplus in 2004 and the projected deficit in 1998 is \$83 billion.

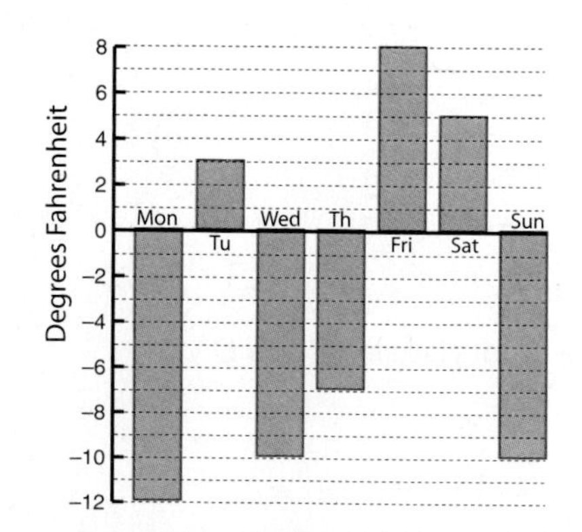

Temperatures for 1 week in Nome, Alaska

You-Try-It 7 The graph at the left shows the temperatures in degrees Fahrenheit for a one-week period in Nome, Alaska.

a. What was the difference between the temperature on Wednesday and the temperature on Monday?

b. What was the difference between the temperature on Friday and the temperature on Thursday?

Solution See page S5.

---

6. $d(C, D) = |-1 - (-7)| = |-1 + 7| = |6| = 6$

2.1 EXERCISES

Topics for Discussion

1. Is the absolute value of every number positive? No
2. Is subtraction a commutative operation? No
3. If two integers are added, is the sum greater than either of the two integers? No
4. If two integers are subtracted, is the difference smaller than either of the two integers? No
5. What number is its own opposite? Zero
6. Is the statement $|x| = x$ always a true statement? No

Opposites and Absolute Value

Find the opposite of the given number.

7. 25 −25 **8.** 42 −42 **9.** −34 34 **10.** −45 45

11. 0 0 **12.** −7 7 **13.** 12 −12 **14.** −3 3

Evaluate.

15. $-(-16)$ 16 **16.** $-(-30)$ 30 **17.** $-(49)$ −49 **18.** $-(32)$ −32

19. $|16|$ 16 **20.** $|25|$ 25 **21.** $|-32|$ 32 **22.** $|-21|$ 21

23. $-|86|$ −86 **24.** $-|40|$ −40 **25.** $-|-54|$ −54 **26.** $-|-27|$ −27

27. $|-23| + |19|$ 42 **28.** $|-14| + |-41|$ 55 **29.** $-(|-13| + |12|)$ −25 **30.** $-(|14| + |-12|)$ −26

31. $-(|12| + |-12|)$ −24 **32.** $-|14 + 7|$ −21 **33.** $-(|-61| + |-34|)$ −95 **34.** $-(|-9| + |-51|)$ −60

Addition of Integers

35. $-6 + 15$ 9 **36.** $12 + (-16)$ −4

37. $(-23) + (-17)$ −40 **38.** $30 + (-21)$ 9

39. $17 + (-3) + 29$ 43

40. $-3 + (-8) + 110$ 99

41. $13 + 62 + (-38)$ 37

42. $-32 + (-42) + (-18)$ −92

43. $13 + (-22) + 4 + (-5)$ −10

44. $22 + 10 + 2 + (-18)$ 16

45. Find the sum of −8 and 11. 3

46. What is the total of −12 and −5? −17

47. What number is −9 more than −11? −20

48. Find 17 added to −21. −4

49. Find the sum of −16, −8, and 14. −10

50. Find the total of 32, −61, 17, and −44. −56

51. Complete the input/output table below.

Input, x	−3	−2	−1	0	1	2	3
Output, $-2 + x$	−5	−4	−3	−2	−1	0	1

52. Complete the input/output table below.

Input, z	−3	−2	−1	0	1	2	3
Output, $z + (-4)$	−7	−6	−5	−4	−3	−2	−1

53. Complete the input/output table below.

Input, t	−3	−2	−1	0	1	2	3
Output, $t + 5$	2	3	4	5	6	7	8

54. Complete the input/output table below.

Input, p	−3	−2	−1	0	1	2	3
Output, $p + (-3)$	−6	−5	−4	−3	−2	−1	0

Subtraction of Integers

55. $-6 - 8$ −14

56. $2 - (-2)$ 4

57. $-12 - 16$ −28

58. $12 - (-7)$ 19

59. $-12 - (-3) - (-15)$ 6

60. $-6 - 19 - (-31)$ 6

61. $4 - 12 - (-8)$ 0

62. $-30 - (-65) - 29 - 6$ 0

63. $13 - 7 - (-15) - 9$ 12

64. $42 - (-82) - 65 - 7$ 52

65. $-16 - 47 - 63 - 12$ −138

66. $-47 - (-67) - 13 - 15$ −8

67. Find the difference between 7 and 14. −7

68. What is −3 subtracted from −6? −3

69. What number is −12 less than −6? 6

70. Find −15 minus 24. −39

71. Find 24 minus −24. 48

72. Find the difference between 32 and −27. 59

73. Complete the input/output table below.

Input, x	−3	−2	−1	0	1	2	3
Output, $x - 4$	−7	−6	−5	−4	−3	−2	−1

74. Complete the input/output table below.

Input, z	−3	−2	−1	0	1	2	3
Output, $z - 1$	−4	−3	−2	−1	0	1	2

75. Complete the input/output table below.

Input, t	−3	−2	−1	0	1	2	3
Output, $1 - t$	4	3	2	1	0	−1	−2

76. Complete the input/output table below.

Input, p	−3	−2	−1	0	1	2	3
Output, $-2 - p$	1	0	−1	−2	−3	−4	−5

Applications

77. The temperature at which mercury boils (called its boiling point) is 360°C. Mercury freezes at −39°C (the freezing point). Find the difference between the boiling point and the freezing point of mercury. 399°C

78. The temperature at which radon boils is −62°C. Radon freezes at −71°C. Find the difference between the temperature at which radon boils and the temperature at which it freezes. 9°C

79. On January 22, 1943, the temperature at Spearfish, South Dakota, rose from −4°F to 45°F in two minutes. How many degrees did the temperature rise during those two minutes? 49° F

80. In a 24-hour period in January 1916, the temperature in Browning, Montana, dropped from 44°F to −56°F. How many degrees did the temperature drop during that time? 100°F

81. Using the number line below, find the indicated distances.

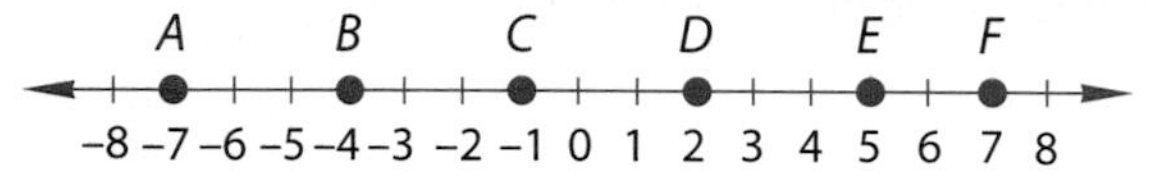

a. $d(A, C)$ 6 **b.** $d(B, F)$ 11 **c.** $d(C, E)$ 6

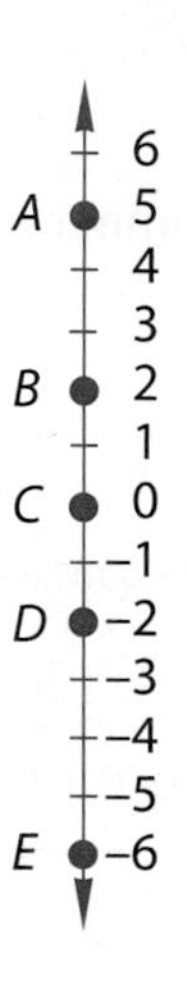

82. Using the vertical number line at the right, find the indicated distances.
a. $d(E, C)$ 6
b. $d(B, D)$ 4
c. $d(D, E)$ 4

83. The broken-line graph at the right shows the depth of a submarine for various times as it surfaces. When the submarine begins to surface ($t = 0$), the submarine is 100 meters below sea level.
a. What is the difference between the depth of the submarine 60 seconds after it began to surface and 20 seconds after it began to surface? 20 meters
b. How many meters did the submarine rise between 60 seconds and 120 seconds? 75 meters
c. Eight seconds after beginning to surface, is the submarine more than or less than 60 meters below sea level? The submarine is more than 60 meters below sea level.
d. Did the submarine rise more during the first 60 seconds or the last 60 seconds? The submarine rose more during the last 60 seconds.

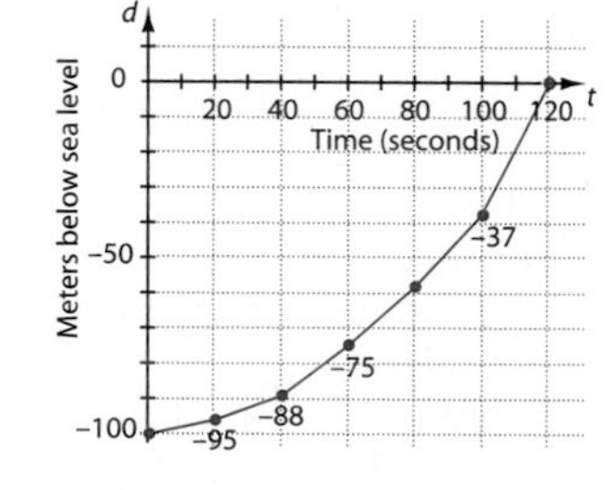

84. The broken-line graph at the right shows the height of a rocket that has been fired from a submarine. When the rocket is fired ($t = 0$), the rocket is 150 meters below sea level.
a. How far did the rocket travel in the first 0.5 second? (Note that on the horizontal axis each division represents 0.25 second.) 73 meters
b. How far did the rocket travel in the first 1.75 seconds of flight? 300 meters
c. How long after it was fired did it first pass sea level? 1 second

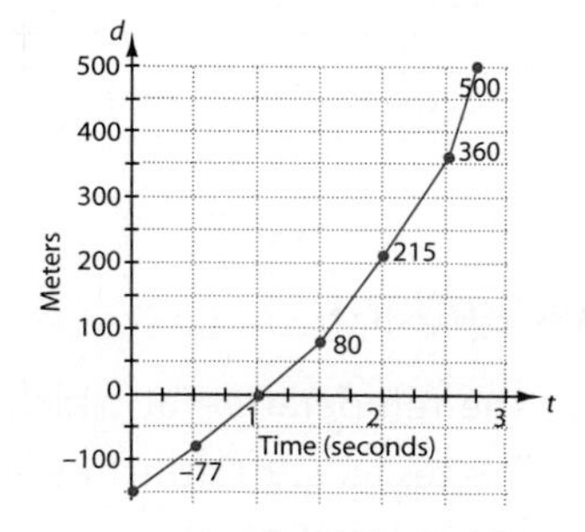

85. The bar graph at the right shows the net profit (rounded to the nearest million dollars) for Immune Response Corporation.

a. In which year did Immune Response have the least net profit? 1992

b. What was the difference between the net profit for 1997 and 1993 for Immune Response? 8 million dollars

c. Between which two years did the net profit increase the most? 1992 and 1993

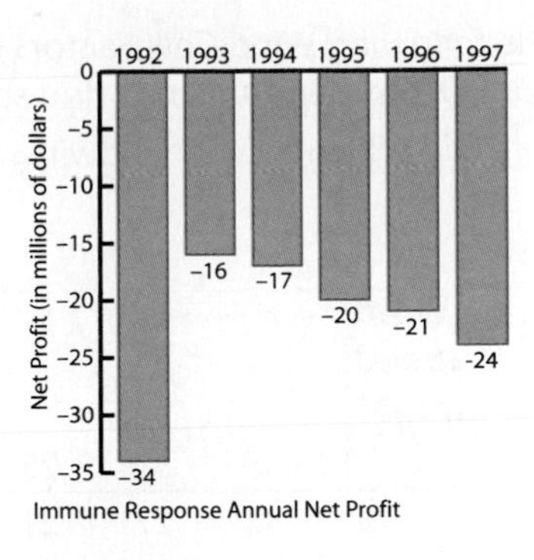

Immune Response Annual Net Profit

86. The bar graph at the right shows the net profit (rounded to the nearest million dollars) for Sunrise Medical Corporation.

a. In which year did Sunrise Medical have the least net profit? 1996

b. What was the difference between the net profit for 1996 and 1993 for Sunrise Medical? 59 million dollars

c. Between which two years did the net profit increase the most? 1996 and 1997

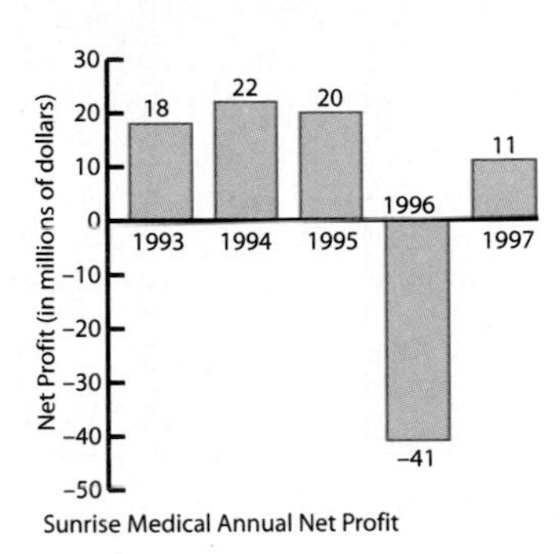

Sunrise Medical Annual Net Profit

Applying the Concepts

Evaluate the expression for $a = -4$ and $b = -7$.

87. $a + b$ -11

88. $-a + b$ -3

89. $-a + (-b)$ 11

90. $a + (-b)$ 3

91. $a - b$ 3

92. $-a - b$ 11

93. Use inductive reasoning to complete the table below.

Input	−3	−2	−1	0	1	2	n
Output	7	6	5	4	3	2	$4 + (-n)$

94. Use inductive reasoning to complete the table below.

Input	−3	−2	−1	0	1	2	n
Output	−5	−4	−3	−2	−1	0	$-2 + n$

95. Replace the ? with $\le$, $=$, or $\ge$ so that the statement $|x + y|$? $|x| + |y|$ is always true for any integers. $\le$

96. Replace the ? with $\le$, $=$, or $\ge$ so that the statement $||x| - |y||$? $|x| - |y|$ is always true for any integers. $\ge$

The following Wind Chill Factors table gives equivalent temperatures for combinations of temperature and wind speed. For example, the combination of a temperature of 15°F and a wind blowing at 10 mph has a cooling effect of –3°F.

Wind Chill Factors

Wind Speed (mph)	**Thermometer Reading (in degrees Fahrenheit)**																
	35	30	25	20	15	10	5	0	−5	−10	−15	−20	−25	−30	−35	−40	−45
5	33	27	21	19	12	7	0	−5	−10	−15	−21	−26	−31	−36	−42	−47	−52
10	22	16	10	3	−3	−9	−15	−22	−27	−34	−40	−46	−52	−58	−64	−71	−77
15	16	9	2	−5	−11	−18	−25	−31	−38	−45	−51	−58	−65	−72	−78	−85	−92
20	12	4	−3	−10	−17	−24	−31	−39	−46	−53	−60	−67	−74	−81	−88	−95	−103
25	8	1	−7	−15	−22	−29	−36	−44	−51	−59	−66	−74	−81	−88	−96	−103	−110
30	6	−2	−10	−18	−25	−33	−41	−49	−56	−64	−71	−79	−86	−93	−101	−109	−116
35	4	−4	−12	−20	−27	−35	−43	−52	−58	−67	−74	−82	−89	−97	−105	−113	−120
40	3	−5	−13	−21	−29	−37	−45	−53	−60	−69	−76	−84	−92	−100	−107	−115	−123
45	2	−6	−14	−22	−30	−38	−46	−54	−62	−70	−78	−85	−93	−102	−109	−117	−125

97. Use the Wind Chill Factors table to determine which of the conditions feels colder.

a. a temperature of 5°F with a 20 mph wind or a temperature of 10°F with a 15 mph wind 5°F with a 20 mph wind feels colder.

b. a temperature of –25°F with a 10 mph wind or a temperature of –15°F with a 20 mph wind –15°F with a 20 mph wind feels colder.

98. Use the Wind Chill Factors table to determine which of the conditions feels warmer.

a. a temperature of 25°F with a 25 mph wind or a temperature of 10°F with a 10 mph wind 25° F with a 25 mph wind feels warmer.

b. a temperature of –5°F with a 10 mph wind or a temperature of –15°F with a 5 mph wind –15° F with a 5 mph wind feels warmer.

Exploration—Mean Deviation, a Measure of Dispersion

99. In statistics, the range and interquartile range (discussed in Chapter 1) are referred to as **measures of dispersion.** Another measure of dispersion used in statistics is the **mean deviation.**

The number of hours of television that were watched in one week by 15 children between the ages of 7 and 8 are given below.

35, 24, 22, 25, 22, 26, 33, 17, 28, 26, 25, 25, 23, 24, 20

a. Find the mean of the data.

b. For each number in the data, find and record the absolute value of the difference between the number and the mean.

c. Find the mean of the numbers that were calculated in part b. This is the mean deviation of the data.

d. Find the mean deviation for the following low temperatures (in Fahrenheit) in Sitka, Alaska: –6, –8, 3, –3, 5, 0, –2, 6, –1, –4.

e. Add 5 to each number in part d. Does the value of the mean deviation change?

f. If the mean deviation of a data set is 0, what can be said about the data?

Section 2.2 Multiplication and Division of Integers

Multiplication and Division of Integers

SUGGESTED ACTIVITY
There is a suggested group activity in the Instructor's Resource Manual in which students are encouraged to discover the rules for multiplying integers by using a calculator.

One method we can use to develop rules for multiplication of integers is to look for a pattern. Consider multiplying a decreasing sequence of integers by 5.

Decreasing sequence of integers →

4	3	2	1	0	−1	−2	−3	−4
5(4)	5(3)	5(2)	5(1)	5(0)	5(−1)	5(−2)	5(−3)	5(−4)
20	15	10	5	0	?	?	?	?

Using inductive reasoning, it appears that each term of the sequence of products (20, 15, 10, 5, 0) is 5 less than the previous term. To continue this pattern, the question marks should be replaced by −5, −10, −15, and −20.

Decreasing sequence of integers →

4	3	2	1	0	−1	−2	−3	−4
5(4)	5(3)	5(2)	5(1)	5(0)	5(−1)	5(−2)	5(−3)	5(−4)
20	15	10	5	0	−5	−10	−15	−20

This suggests that *the product of a positive number and a negative number is a negative number.*

Now consider multiplying a decreasing sequence of integers by −5.

Decreasing sequence of integers →

4	3	2	1	0	−1	−2	−3	−4
−5(4)	−5(3)	−5(2)	−5(1)	−5(0)	−5(−1)	−5(−2)	−5(−3)	−5(−4)
−20	−15	−10	−5	0	?	?	?	?

Using inductive reasoning, it appears that each term of the sequence of products (−20, −15, −10, −5, 0) is 5 more than the previous term. To continue this pattern, the question marks should be replaced by 5, 10, 15, and 20.

Decreasing sequence of integers →

4	3	2	1	0	−1	−2	−3	−4
−5(4)	−5(3)	−5(2)	−5(1)	−5(0)	−5(−1)	−5(−2)	−5(−3)	−5(−4)
−20	−15	−10	−5	0	5	10	15	20

This suggests that *the product of two negative numbers is a positive number.* On the basis of these findings, we have the following rules for multiplying integers.

Multiplying Integers

Same signs
To multiply two numbers with the same sign, multiply the absolute values of the numbers. The product is positive.

Different signs
To multiply two numbers with different signs, multiply the absolute values of the numbers. The product is negative.

Example 1 **a.** Find the product of 6 and −13.

b. Simplify: $-5(4)(-2)(-7)$

Solution

a. Recall that to find a product means to multiply.

$6(-13) = -78$ • Multiply the absolute values. The product is negative.

b. $-5(4)(-2)(-7) = -20(-2)(-7)$ • Use the Order of Operations Agreement.

$= 40(-7) = -280$

You-Try-It 1 **a.** What is −12 times −32?

b. Simplify: $4(-3)(-7)(-6)$

Solution See page S5.

Example 2 Complete the following input-output table.

Input, n	−3	−2	−1	0	1	2	3
Output, $-4n$							

Solution Evaluate the variable expression $-4n$ for each of the input values. Here is the calculation for the input values −2, 0, and 3.

$-4n$ $-4n$ $-4n$

$-4(-2) = 8$ $-4(0) = 0$ $-4(3) = -12$

After completing the remaining calculations, the table is

Input, n	−3	−2	−1	0	1	2	3
Output, $-4n$	12	8	4	0	−4	−8	−12

You-Try-It 2 Complete the following input-output table.

Input, x	−3	−2	−1	0	1	2	3
Output, $5x$							

Solution See page S5.

Note the difference between the two exponential expressions below.

$-3^4 = -(3 \cdot 3 \cdot 3 \cdot 3) = -81$ • $-3^4 = -(3^4)$

$(-3)^4 = (-3)(-3)(-3)(-3) = 81$

Because of this difference, we recommend that you use parentheses around numbers when you substitute into a variable expression. The evaluation of $-x^3$ when $x = -2$ and then when $x = 2$ is shown below.

$-x^3$

$-(-2)^3$ • Replace x by −2.

$= -(-8)$

$= 8$

$-x^3$

$-(2)^3$ • Replace x by 2.

$= -(8)$

$= -8$

Question: If $a = -2$, what is the value of $-a^2$?[1]

1. $-a^2 = -(-2)^2 = -(-2)(-2) = -4$

Example 3 Evaluate $3x^2 - y^3$ when $x = -4$ and $y = -3$.

Solution

$$3x^2 - y^3$$
$$3(-4)^2 - (-3)^3 = 3(16) - (-27)$$
$$= 48 + 27 = 75$$

You-Try-It 3 Evaluate $b^2 - 4ac$ when $a = 5, b = -2,$ and $c = -4$.

Solution See page S5.

For every division problem there is a related multiplication problem. For instance,

Division		Related multiplication
$\frac{12}{4} = 3$	because	$3 \cdot 4 = 12$

Extending this to negative integers, we have

$$\frac{-12}{4} = -3 \quad \text{because} \quad -3 \cdot 4 = -12$$

$$\frac{-12}{-4} = 3 \quad \text{because} \quad 3 \cdot (-4) = -12$$

$$\frac{12}{-4} = -3 \quad \text{because} \quad -3 \cdot (-4) = 12$$

This suggests the following rules for dividing integers.

Dividing Integers

Same signs
To divide two numbers with the same sign, divide the absolute values of the numbers. The quotient is positive.

Different signs
To divide two numbers with different signs, divide the absolute values of the numbers. The quotient is negative.

Example 4 Simplify: **a.** the quotient of –72 and –9

b. $\frac{(-6)^2}{-12}$

Solution

a. Recall that to find a quotient means to divide.

$$\frac{-72}{-9} = 8$$

• Divide the absolute values. The quotient is positive.

b. $\frac{(-6)^2}{-12} = \frac{36}{-12} = -3$

• Use the Order of Operations Agreement.

You-Try-It 4 Simplify: **a.** 96 divided by –8

b. $\frac{-12^2}{-8}$

Solution See page S5.

Example 5 Complete the following input-output table.

Input, n	−6	−4	−2	0	2	4	6
Output, $\frac{n}{-2}$							

Solution Evaluate the variable expression $\frac{n}{-2}$ for each of the input values. Here is the calculation for the input values −2, 0, and 6.

$$\frac{n}{-2} \qquad \frac{n}{-2} \qquad \frac{n}{-2}$$

$$\frac{-2}{-2} = 1 \qquad \frac{0}{-2} = 0 \qquad \frac{6}{-2} = -3$$

After completing the remaining calculations, the table is

Input, n	−6	−4	−2	0	2	4	6
Output, $\frac{n}{-2}$	3	2	1	0	−1	−2	−3

You-Try-It 5 Complete the following input-output table.

Input, x	−4	−2	−1	1	2	4	8
Output, $\frac{4}{x}$							

Solution See page S6.

Example 6 Evaluate $\frac{x^2 - y^2}{z^2}$ for $x = -6, y = -9$, and $z = -3$.

Solution

$$\frac{x^2 - y^2}{z^2}$$

$$\frac{(-6)^2 - (-9)^2}{(-3)^2} = \frac{36 - 81}{9}$$

$$= \frac{-45}{9} = -5$$

• Replace x by −6, y by −9, and z by −3. Then use the Order of Operations Agreement to simplify the numerical expression.

You-Try-It 6 Evaluate $\frac{-2a}{b^3}$ for $a = 4$ and $b = -2$.

Solution See page S6.

INSTRUCTOR NOTE
To relate zero in division to a real situation, explain that $6 \div 3 = 2$ means that if \$6 is divided equally among 3 people, each person receives \$2. If \$0 is divided equally among 3 people, each person receives \$0 ($0 \div 3 = 0$). Now, if \$6 is divided equally among 0 people ($6 \div 0$), how many dollars does each person receive?

In a number of instances throughout this text, we will put *restrictions* on the value of a variable. For instance, for the variable expression $\frac{4}{b}$, we would state that $b \neq 0$ because division by zero is not allowed. To see this, consider dividing by zero and the related multiplication.

Division	Related multiplication
$\frac{4}{0} = ?$	$? \cdot 0 = 4$

Because the product of any number and zero is zero, it is not possible to find a replacement for the question mark. Therefore, division by zero is not allowed.

Question: For the expression $\frac{3}{x}$, what value can x *not* have?[2]

Note that $-\frac{12}{4} = -3$, $\frac{-12}{4} = -3$, and $\frac{12}{-4} = -3$. This suggests the following.

> If a and b are integers, and $b \neq 0$, then $-\frac{a}{b} = \frac{-a}{b} = \frac{a}{-b}$.

Question: What is the value of $-\frac{6}{-2}$?[3]

Here are some useful division principles involving zero and one.

> **Zero and One in Division**
>
> Division by zero is not defined.
>
> Zero divided by any number other than zero is zero. In symbols,
>
> $$\frac{0}{a} = 0, \quad a \neq 0$$
>
> Any number other than zero divided by itself is 1. In symbols,
>
> $$\frac{a}{a} = 1, \quad a \neq 0$$
>
> Any number divided by 1 is the number. In symbols,
>
> $$\frac{a}{1} = a$$

Applications

Example 7 The daily low temperatures, in degrees Celsius, for a one-week period were recorded as: -8°, -12°, -6°, -13°, 5°, -10°, 2°. What was the average temperature for the week?

Solution

State the goal.

You must determine the average temperature. Recall that to calculate an average, add the data values and then divide by the number of data.

Describe a strategy.

To find the average temperature, add the temperatures and divide by the number of temperatures.

Solve the problem.

$$\text{Average temperature} = \frac{-8 + (-12) + (-6) + (-13) + 5 + (-10) + 2}{7}$$

$$= \frac{-42}{7} = 6$$

The average temperature was -6°C.

Check your work.

Review your calculations to be sure they are accurate.

You-Try-It 7 To discourage guessing on a true/false test, a psychology professor awards 2 points for a correct answer, takes off 4 points for an incorrect answer, and takes off 2 points if an answer is left blank. What is the score for a student who answered 48 questions correctly, answered 14 questions incorrectly, and left 8 questions unanswered?

Solution See page S6.

---

2. Because division by zero is not allowed, x cannot equal zero.
3. By the Order of Operations Agreement, the fraction bar is a grouping symbol. Therefore, we can write $-\frac{6}{-2} = -\left(\frac{6}{-2}\right) = -(-3) = 3$.

2.2 EXERCISES

Topics for Discussion

1. If $-4x = 0$, what can be said about x? $x = 0$
2. Does $a \div b = b \div a$? No, unless $a = b$ and $a \neq 0, b \neq 0$.
3. If two integers are multiplied, is the product greater than either of the two integers? Sometimes
4. If the value of $-5y$ is a positive integer, is y a positive or a negative integer? Negative
5. Explain why division by zero is not allowed. The product of any number and zero is zero; it is not possible to find the other number.
6. Multiplication is frequently described as repeated addition. Can division be described as repeated subtraction? Yes

Multiplication and Division of Integers

Multiply.

7. $-6(14)$ -84

8. $-8 \cdot 12$ -96

9. $(-12)(-11)$ 132

10. $24(-14)$ -336

11. $12(0)$ 0

12. $0(-41)$ 0

13. $-5(-4)(-8)$ -160

14. $6(-7)(4)$ -168

15. $6(-5)(-3)$ 90

16. $-4(-3)(-7)$ -84

17. $8(-2)(-1)(-10)$ -160

18. $-4(-4)(-3)(-5)$ 240

Evaluate.

19. $(-4)^3$ -64

20. -4^3 -64

21. -5^4 -625

22. $(-5)^4$ 625

23. Find the product of -8 and 17. -136

24. What is 12 times -7? -84

25. What is 9 multiplied by -18? -162

26. Find the product of 10 and -11. -110

27. Find the product of -3, -5, and 6. 90

28. What is 21 times -4 times -3? 252

29. Complete the input/output table below.

Input, x	-6	-4	-2	0	1	3	5
Output, $-2x$	12	8	4	0	-2	-6	-10

30. Complete the input/output table below.

Input, z	−3	−2	−1	0	1	2	3
Output, $5z$	−15	−10	−5	0	5	10	15

31. Complete the input/output table below.

Input, t	−3	−2	−1	0	1	2	3
Output, t^2	9	4	1	0	1	4	9

32. Complete the input/output table below.

Input, p	−3	−2	−1	0	1	2	3
Output, $-p^2$	−9	−4	−1	0	−1	−4	−9

Evaluate the variable expression for $a = -3, b = 2,$ and $c = -4$.

33. $5a$ −15

34. $-4b$ −8

35. bc −8

36. $3ac$ 36

37. $a^2 - b^2$ 5

38. $c^3 - a^3$ −37

39. $2a^2b - 4c^2$ −28

40. $-3c^2 + 2a^2b$ −12

Divide.

41. $54 \div (-6)$ −9

42. $-56 \div 7$ −8

43. $-72 \div (-12)$ 6

44. $(-48) \div (-8)$ 6

45. $\frac{27}{-27}$ −1

46. $\frac{0}{-17}$ 0

47. $\frac{-30}{-15}$ 2

48. $\frac{-63}{9}$ −7

Evaluate.

49. $\frac{-3^2(4)}{-12}$ 3

50. $\frac{-5^2(6)}{10}$ −15

51. $\frac{3(-4)^2}{-8}$ −6

52. $\frac{-5(-3)^2}{-15}$ 3

53. Find the quotient of −12 and 4. −3

54. What is the value of the ratio of 64 to −8? −8

55. What is –84 divided by –6? 14

56. Find the quotient of 98 and –14. –7

57. What is the value of the ratio of –51 to –17? 3

58. What is –168 divided by –7? 24

59. Complete the input/output table below.

Input, x	–9	–6	–3	0	3	6	9
Output, $\frac{x}{-3}$	3	2	1	0	–1	–2	–3

60. Complete the input/output table below.

Input, z	–6	–4	–2	0	2	4	6
Output, $\frac{-z}{2}$	3	2	1	0	–1	–2	–3

61. Complete the input/output table below.

Input, t	–4	–2	–1	1	2	4
Output, $\frac{-8}{t}$	2	4	8	–8	–4	–2

62. Complete the input/output table below.

Input, p	–4	–3	–2	–1	1	2	3
Output, $\frac{-12}{p}$	3	4	6	12	–12	–6	–4

Evaluate the expression when $a = -4, b = 6$, and $c = -1$.

63. $-4ab$ 96

64. $-3a^2$ –48

65. abc^4 –24

66. $3(ac)^2$ 48

67. $\frac{-a^2}{4c}$ 4

68. $\frac{b-a^2}{b-a}$ –1

69. $\frac{-6c+3a^3}{b}$ –31

70. $\frac{2(2c^3-b^2)}{-a}$ –19

Applications

71. The daily low temperatures (in degrees Celsius) in Fargo, North Dakota, for a seven-day period were: $-4^\circ, 2^\circ, 7^\circ, -5^\circ, -4^\circ, -2^\circ$, and -1°. What was the average temperature for this seven-day period? The average temperature was -1°.

72. The daily low temperatures (in degrees Celsius) in Billings, Montana, for a five-day period were: $-8^\circ, -6^\circ, -5^\circ, 2^\circ$, and -3°. What was the average temperature for this five-day period? The average temperature was -4°.

73. The score on an aptitude test is the sum of 8 times the number of correct answers and -2 times the number of incorrect answers. Questions that were not answered are not counted in the score. What score does a person receive who answered 28 questions correctly, answered 5 questions incorrectly, and left 7 questions unanswered? The person's score is 214.

74. To discourage guessing, a professor scores a multiple-choice exam by awarding 6 points for a correct answer, -2 points for an incorrect answer, and -1 point for a question that is not answered. What score does a student receive who answered 37 questions correctly, 5 questions incorrectly, and left 8 questions unanswered? The student receives a score of 204.

Applying Concepts

75. If x and y are integers, use inductive reasoning to determine which of the symbols $\le, =$, or $\ge$ should replace ?? in $|xy|$?? $|x||y|$ to make a true statement. $=$

76. If x and y are integers and $y \neq 0$, use inductive reasoning to determine which of the symbols $\le$, $=$, or $\ge$ should replace ?? in $\left|\frac{x}{y}\right|$?? $\frac{|x|}{|y|}$ to make a true statement. $=$

77. If a and b are integers and $a < b$, which of the symbols $<$ or $>$ should replace ?? so that $-a$?? $-b$ is a true statement? $>$

78. If a and b are integers and $a > b$, which of the symbols $<$ or $>$ should replace ?? so that $-a$?? $-b$ is a true statement? $<$

79. If the value of $-\frac{-a}{-b}$ is a positive number, do a and b have the same sign or do they have opposite signs? Opposite signs

80. Evaluate $(-1)^1, (-1)^2, (-1)^3, (-1)^4, (-1)^5$, and $(-1)^6$. On the basis of your answers, determine the value of $(-1)^n$ when n is a positive even integer and when n is a positive odd integer. When n is a positive even integer, $(-1)^n = 1$; when n is a positive odd integer, $(-1)^n = -1$.

81. Is it possible that $x = -x$ for some value of x? Explain your answer.
Yes, when $x = 0$.

82. Is it possible that $|x| = -x$ for some value of x? Explain your answer.
Yes, when x is a negative number or zero.

Explorations—Finding a Pattern

83. In the following diagram, a series of triangles are formed with toothpicks.

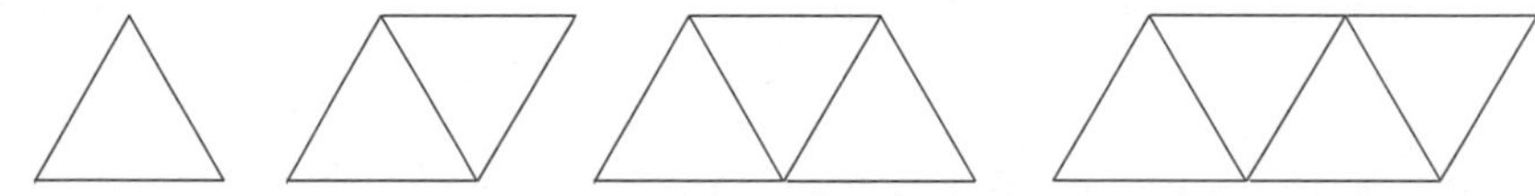

- **a.** Make an input/output table where the input is the number of triangles and the output is the number of toothpicks.
- **b.** Use inductive reasoning to determine the number of toothpicks needed to make 10 triangles.
- **c.** Write a variable expression for the number of toothpicks required for n triangles.

84. In the following diagram, a series of squares are formed with toothpicks.

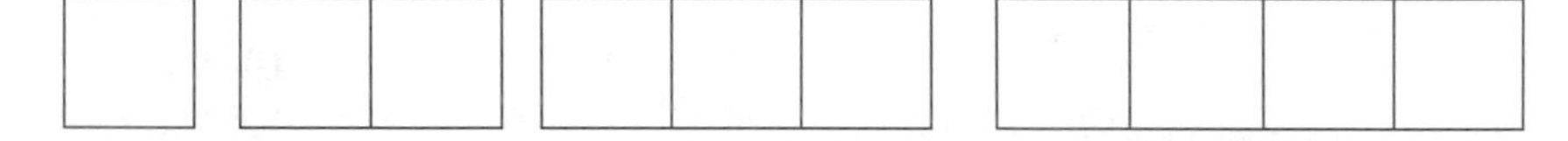

- **a.** Make an input/output table where the input is the number of squares and the output is the number of toothpicks.
- **b.** Use inductive reasoning to determine the number of toothpicks needed to make 10 squares.
- **c.** Write a variable expression for the number of toothpicks required for n squares.

85. In the following diagram, a series of pentagons are formed with toothpicks.

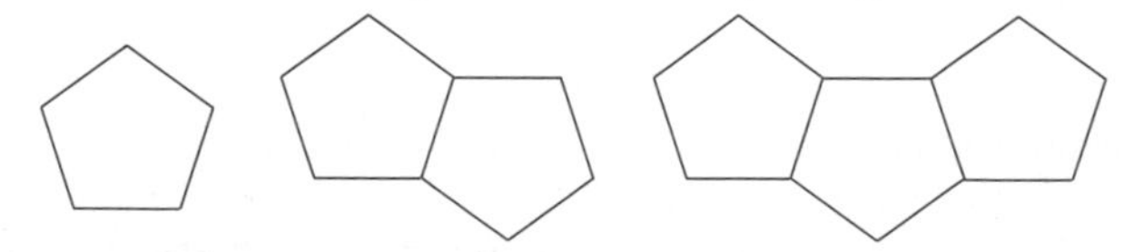

- **a.** Make an input/output table where the input is the number of pentagons and the output is the number of toothpicks.
- **b.** Use inductive reasoning to determine the number of toothpicks needed to make 10 pentagons.
- **c.** Write a variable expression for the number of toothpicks required for n pentagons.

Section 2.3 Introduction to Rectangular Coordinates

Graphing Points on a Rectangular Coordinate System

A cartographer (a person who makes a map) divides a map of a city into little squares as shown on a map of Washington, D.C., below. Telling a visitor to Washington that the White House is located in square A3 enables the visitor to locate the White House within a small area of the map.

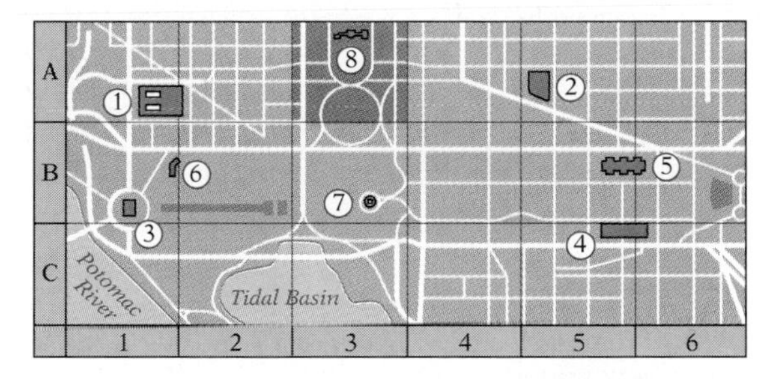

In mathematics we have a similar problem, that of locating a point in a plane. One way to solve the problem is to use a *rectangular coordinate system.*

A **rectangular coordinate system** is formed by two number lines, one horizontal and one vertical, that intersect at the zero point of each line. The point of intersection is called the **origin**. The two axes are called the **coordinate axes**, or simply the **axes**. Frequently, the horizontal axis is labeled the *x*-axis, and the vertical axis is labeled the *y*-axis. In this case, the axes form what is called the ***xy*-plane**.

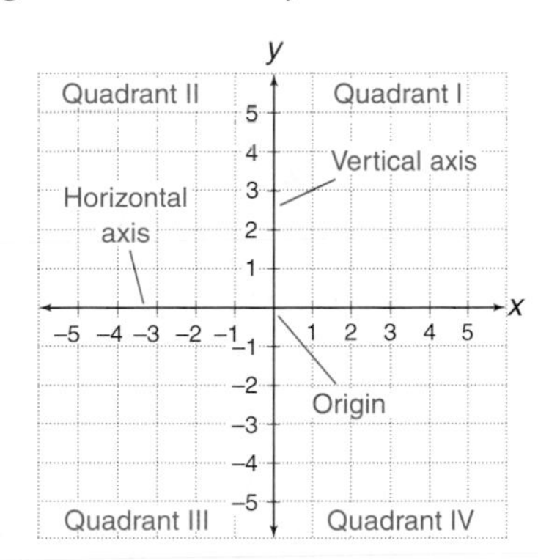

The two axes divide the plane into four regions called **quadrants**, which are numbered counterclockwise, using Roman numerals, from I to IV starting at the upper right.

Each point in the plane can be identified by a pair of numbers called an **ordered pair.** The first number of the ordered pair measures a horizontal change from the *y*-axis and is called the **abscissa**, or ***x*-coordinate**. The second number of the pair measures a vertical change from the *x*-axis and is called the **ordinate**, or ***y*-coordinate.** The ordered pair (x, y) associated with a point is also called the **coordinates** of the point.

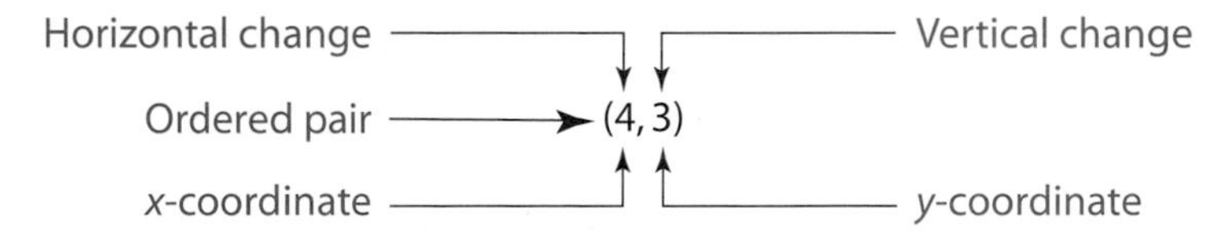

If the axes are labeled as other than x and y, then we refer to the ordered pair by the given labels. For instance, if the horizontal axis is labeled t and the vertical axis is labeled d, then the ordered pair is written as (t, d). In any case, we sometimes just refer to the first number in an ordered pair as the **first coordinate** of the ordered pair and the second number as the **second coordinate** of the ordered pair.

To **graph**, or **plot**, a point means to place a dot at the coordinates of the point. For example, to graph the ordered pair (4, 3), start at the origin. Move 4 units to the right and then 3 units up. Draw a dot. To graph (−3, −4), start at the origin. Move 3 units left and then 4 units down. Draw a dot.

The **graph of an ordered pair** is the dot drawn at the coordinates of the point in the plane. The graphs of the ordered pairs (4, 3) and (−3, −4) are shown at the right.

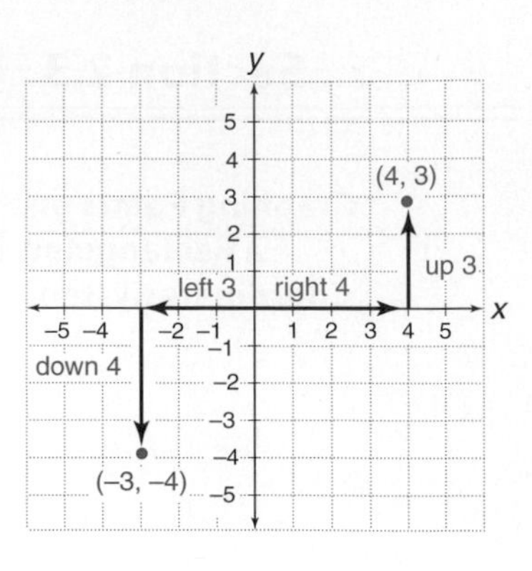

TAKE NOTE
This is *very* important. An *ordered pair* is a pair of coordinates and the *order* in which the coordinates appear is important.

The graphs of the points whose coordinates are (2, 3) and (3, 2) are shown at the right. Note that they are different points. The order in which the numbers in an ordered pair appear is important.

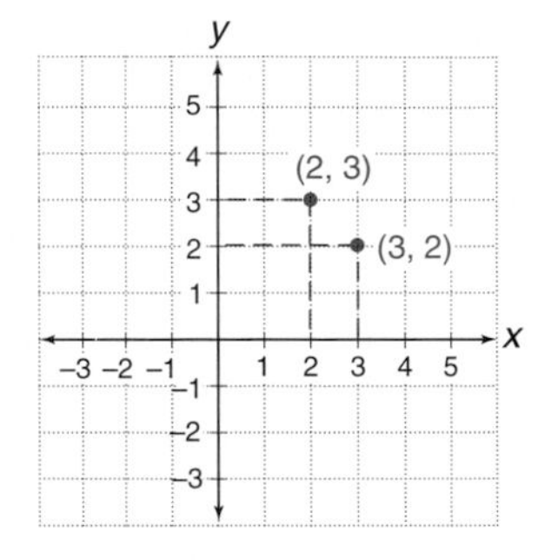

SUGGESTED ACTIVITY
See the *Instructor's Resource Manual* for a calculator activity that lets students explore a rectangular coordinate system.

Each point in the plane is associated with an ordered pair, and each ordered pair is associated with a point in the plane. Although only integers are labeled on the coordinate grid, any ordered pair can be graphed by approximating its location. The graphs of the ordered pairs (−3.5, 3.1) and $\left(\frac{1}{2}, -\frac{11}{3}\right)$ are shown at the right.

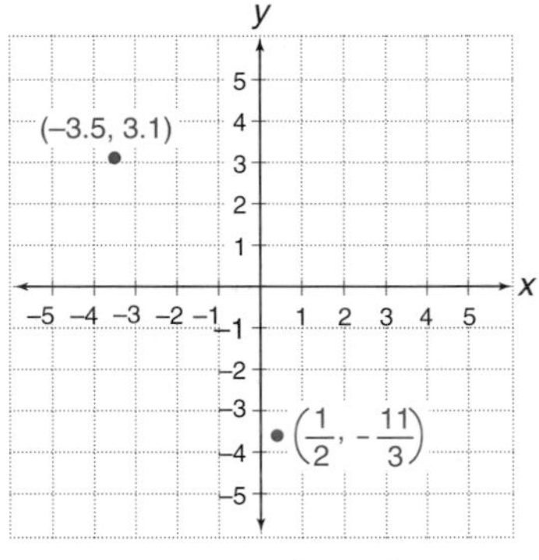

Just as we can state that two numbers are equal (for instance, $\frac{3}{4} = \frac{6}{8}$), two ordered pairs are equal when the corresponding coordinates are equal. That is,

$$(a, b) = (d, c) \text{ whenever } a = d \text{ and } b = c.$$

Example 1 Graph the ordered pairs (−2, −3), (0, 4), (−3, 0), and (4, −1).

Solution

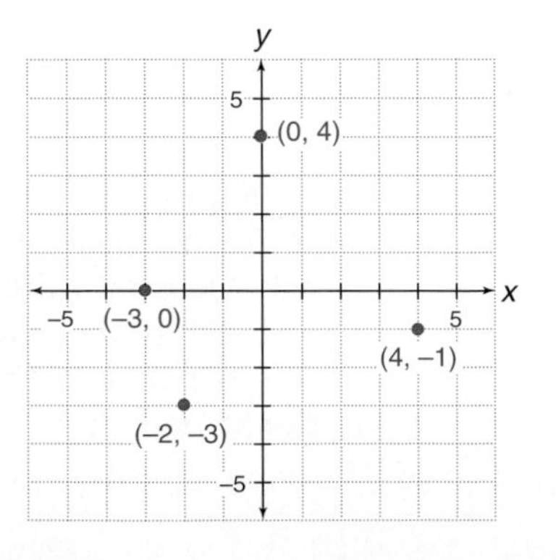

You-Try-It 1 Graph the ordered pairs (3, −4), (0, 0), (2, 3), and (−3, −2).

Solution See page S6.

Example 2 Find the coordinates of each of the points.

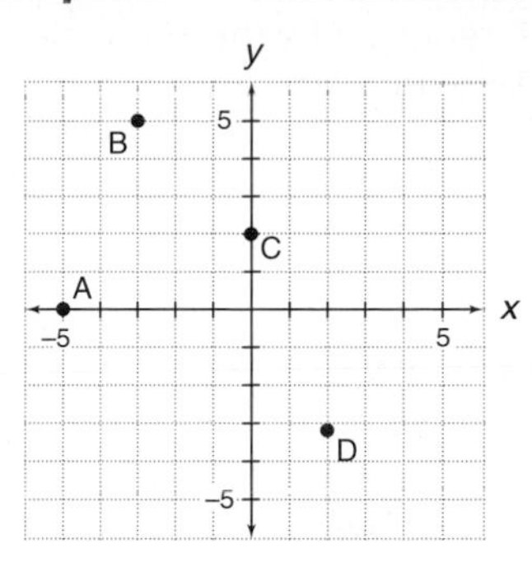

Solution $A(-5, 0), B(-3, 5), C(0, 2), D(2, -3)$

You-Try-It 2 Find the coordinates of each of the points.

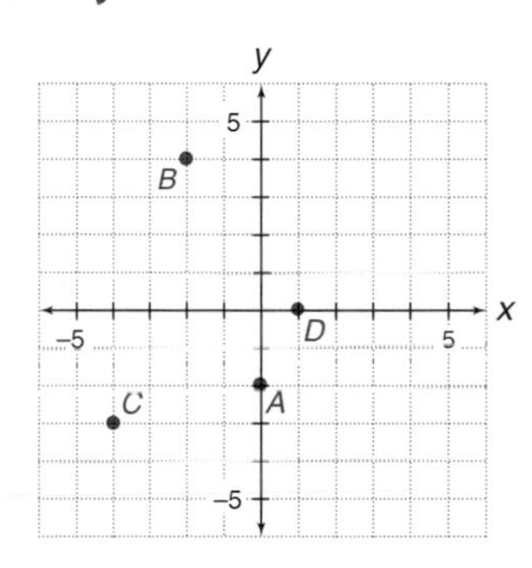

Solution See page S6.

Scatter Diagrams

One reason that coordinate grids are so important is that they are used to show relationships between different quantities. For instance, consider a recent report by the Federal Aviation Authority (FAA). The graph below, called a **scatter diagram**, shows how the number of reported incidences of "near misses" (approximately 500 feet between aircraft) has decreased over time.

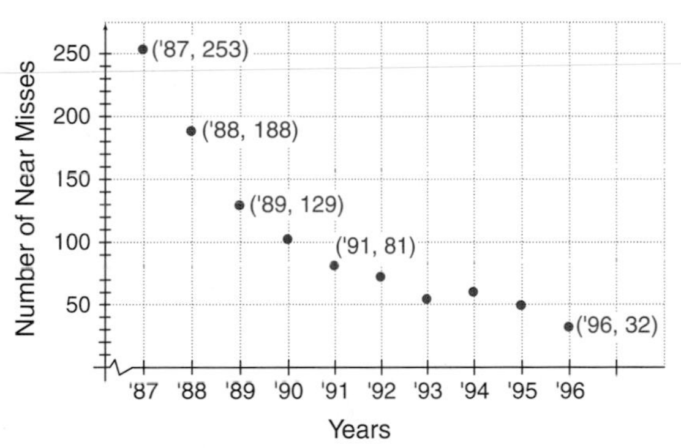

For this graph, the horizontal axis is the year and the vertical axis is the number of near misses. The coordinates of some of the ordered pairs are shown on the graph. The ordered pair ('91, 81) means that in 1991 there were 81 near misses reported. For those points whose coordinates are not given, we can only approximate the values. For instance, there were approximately 100 near misses in 1990.

Question: **a.** How many near misses were reported in 1996? **b.** Between which two years did the number of near misses increase?[1]

---

1. **a.** There were 32 near misses in 1996. **b.** Between 1993 and 1994.

Example 3 The number of critical near misses between aircraft (100 feet or less) reported by the FAA is shown in the table below. Draw a scatter diagram for these data on the given coordinate grid.

Year (x)	1990	1991	1992	1993	1994	1995	1996
Number of near misses (y)	18	16	9	11	8	10	3

Solution

The horizontal axis is the year and the vertical axis is the number of misses. Graph each of the ordered pairs ('90, 18), ('91, 16), ('92, 9), ('93, 11), ('94, 8), ('95, 10), and ('96, 3).

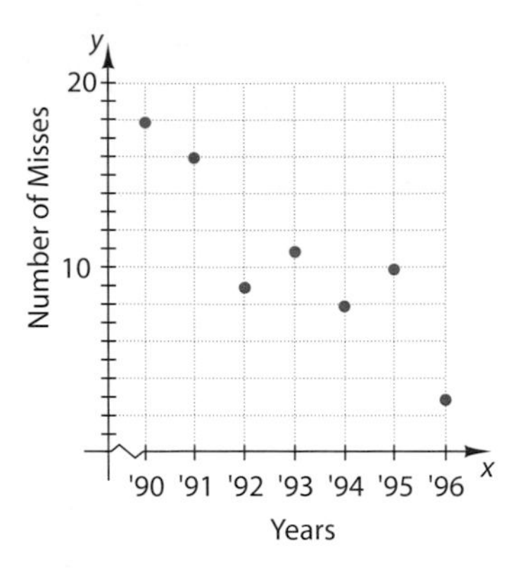

You-Try-It 3 The grams of sugar and the grams of fiber in a one-ounce serving of six breakfast cereals in 1997 are shown in the table below. Draw the scatter diagram for these data on the coordinate grid provided.

Cereal	Sugar (x)	Fiber (y)
Wheaties	4	3
Rice Krispies	3	0
Total	5	3
Life	6	2
Kix	3	1
Grape-Nuts	7	5

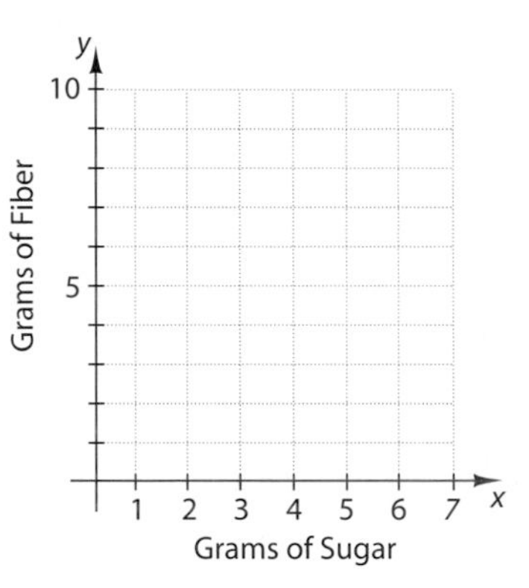

Solution See page S6.

Graphs of Input/Output Tables

One of the major reasons that coordinate systems are used is to give a graphical representation of an input/output table. Consider the input/output table below.

Input, x	−3	−2	−1	0	1	2	3
Output, y	6	4	2	0	−2	−4	−6

The input/output table gives rise to ordered pairs. The input is the x-coordinate of the ordered pair and the output is the y-coordinate of the ordered pair. For the table above, the ordered pairs are: (−3, 6), (−2, 4), (−1, 2), (0, 0), (1, −2), (2, −4), and (3, −6). The graphs of these ordered pairs are shown on the xy-coordinate system at the right.

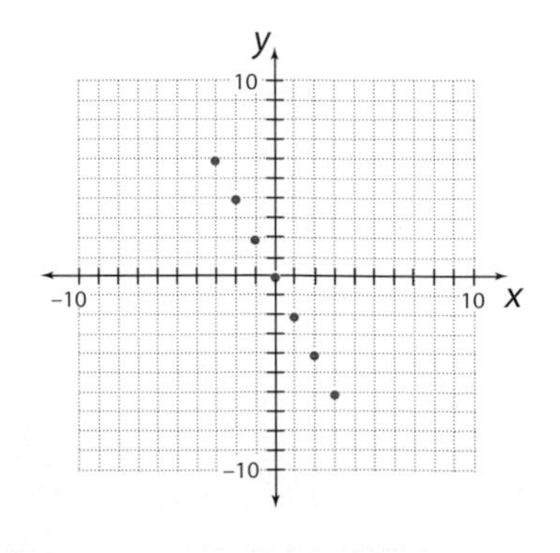

Example 4 Graph the ordered pairs that result from the following input/output table.

Input, x	−3	−2	−1	0	1	2	3
Output, $2x + 1$							

Solution First, complete the table. The result is shown below.

Input, x	−3	−2	−1	0	1	2	3
Output, $2x + 1$	−5	−3	−1	1	3	5	7

We will graph the ordered pairs on an *xy*-coordinate system. Therefore, the output, $2x + 1$, will be designated as *y*. The ordered pairs to graph are: (−3, −5), (−2, −3), (−1, −1), (0, 1), (1, 3), (2, 5), and (3, 7).

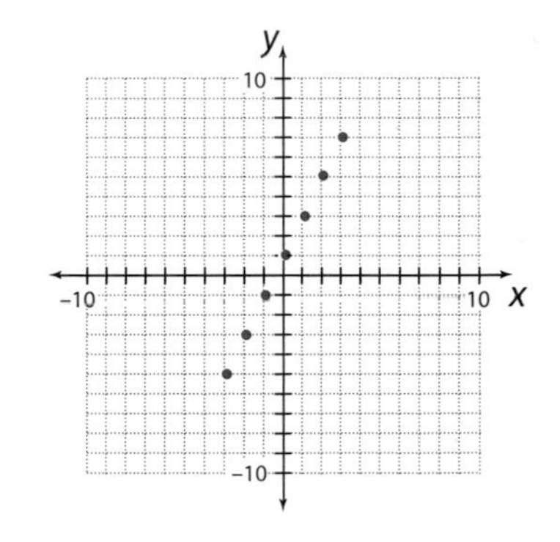

You-Try-It 4 Graph the ordered pairs that result from the following input/output table.

Input, x	−3	−2	−1	0	1	2	3
Output, $1 - 2x$							

Solution See page S6.

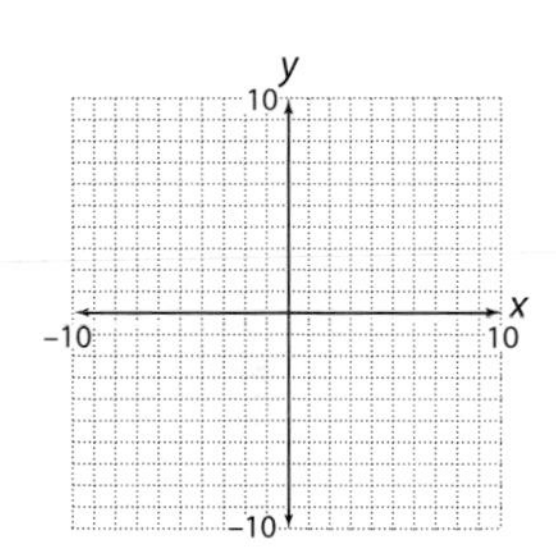

It is also possible to create a graph for an equation.

Example 5 Graph $y = 3 - x$ for $x = -3, -2, -1, 0, 1, 2,$ and 3.

Solution Create an input/output table for the equation where the inputs are the values of *x* and the outputs are the values of *y*, which, in this case, are $3 - x$. Then graph the ordered pairs. We will show the input/output table vertically rather than horizontally.

Input, x	Output, $3 - x = y$
−3	$3 - (-3) = 6$
−2	$3 - (-2) = 5$
−1	$3 - (-1) = 4$
0	$3 - 0 = 3$
1	$3 - 1 = 2$
2	$3 - 2 = 1$
3	$3 - 3 = 0$

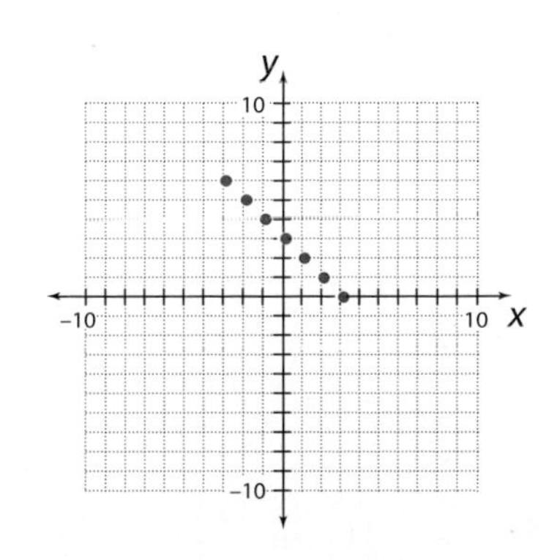

You-Try-It 5 Graph $y = 2x$ for $x = -4, -3, -2, 0, 1, 3, 5$.

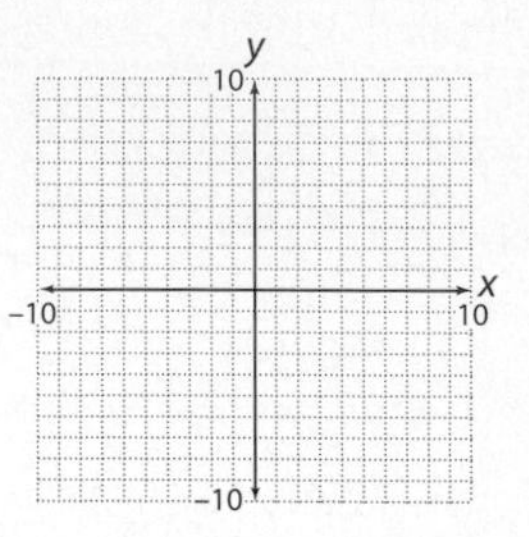

Solution See page S7.

Applications

Application problems that include graphs may require that the *scale* along the vertical axis or horizontal axis be adjusted to show the details of the application.

Example 6 As personal computers have become more powerful, the size of the monitor that is sold with the computer has increased. An equation that approximates the number of 15-inch monitors that will be sold each year for the next few years is $S = 3000T + 8000$, where S is the number of 15-inch monitors sold and T is the year. (For this particular equation, $T = 0$ corresponds to 1996. Therefore, $T = 1$ corresponds to 1997, $T = 2$ corresponds to 1998, and so on.) Graph this equation for the years 1996 through 2001.

Solution Create an input/output table for the equation where the inputs are the values of T and the outputs are the values of S, which, in this case, are $3000T + 8000$. Then graph the ordered pairs. Since the graph is to include the years 1996 through 2001, the input values of T are 0, 1, 2, 3, 4, and 5.

Input, T	Output, $S = 3000T + 8000$
0	8,000
1	11,000
2	14,000
3	17,000
4	20,000
5	23,000

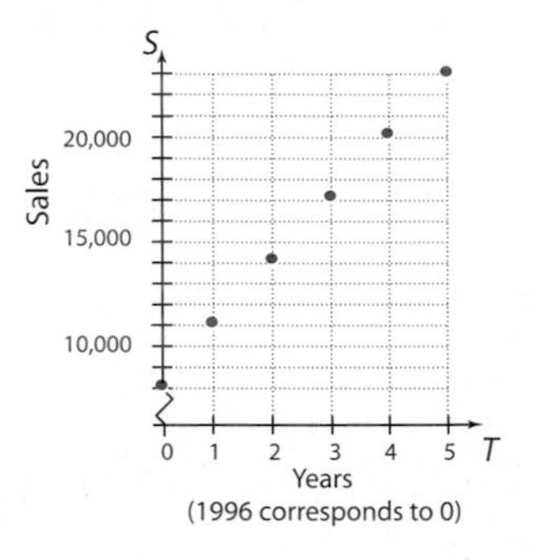

Note that the largest output value is 23,000. Therefore, the vertical axis must be scaled to include that number.

You-Try-It 6 Suppose that a new sports utility vehicle that originally cost $20,000 decreases in value by $2000 each year. The equation $V = 20{,}000 - 2000t$ gives the value V of the car after t years of ownership. Graph this equation for $t = 0$, 1, 2, 3, 4, and 5.

Solution See page S7.

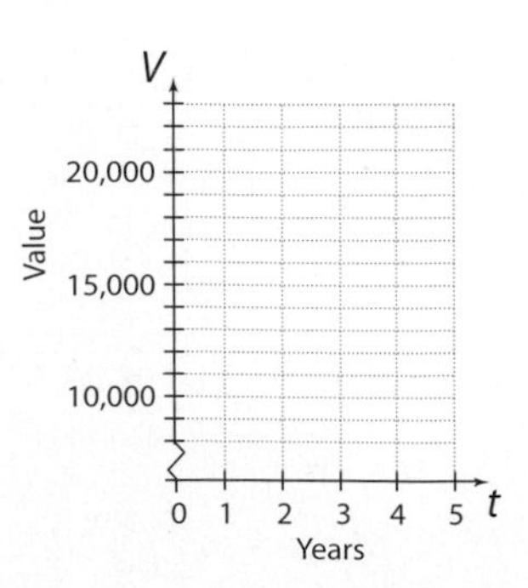

2.3 EXERCISES

Topics for Discussion

1. What is a rectangular coordinate system? Answers will vary.
2. Can a single point in a rectangular coordinate system be represented by two different ordered pairs? No
3. What is the x-coordinate of any point on the y-axis? $x = 0$
4. What is the y-coordinate of any point on the x-axis? $y = 0$
5. Describe the signs of the coordinates of points plotted in: **(a)** the first quadrant; **(b)** the second quadrant; **(c)** the third quadrant; **(d)** the fourth quadrant. **(a)** x and y both positive; **(b)** x negative, y positive; **(c)** x and y both negative; **(d)** x positive, y negative.

Graphing Points on a Rectangular Coordinate System

6. Graph the ordered pairs $(-2, 1)$, $(3, -5)$, $(-2, 4)$, and $(0, 3)$.

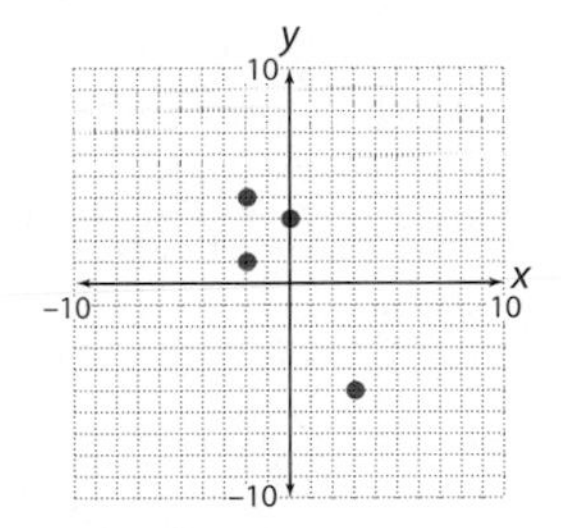

7. Graph the ordered pairs $(5, -1)$, $(-3, -3)$, $(-1, 0)$, and $(1, -1)$.

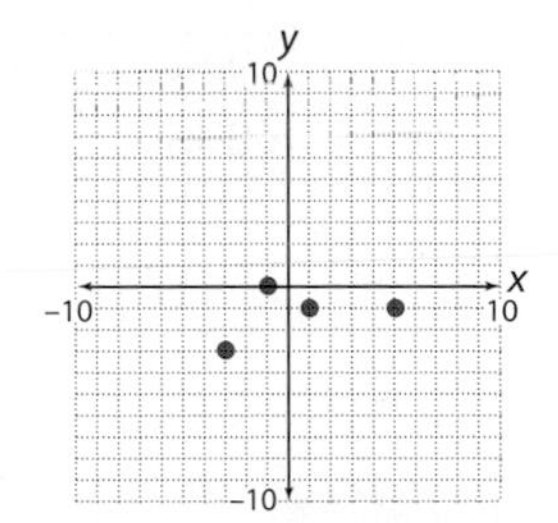

8. Graph the ordered pairs $(0, 0)$, $(0, -5)$, $(-3, 0)$, and $(0, 7)$.

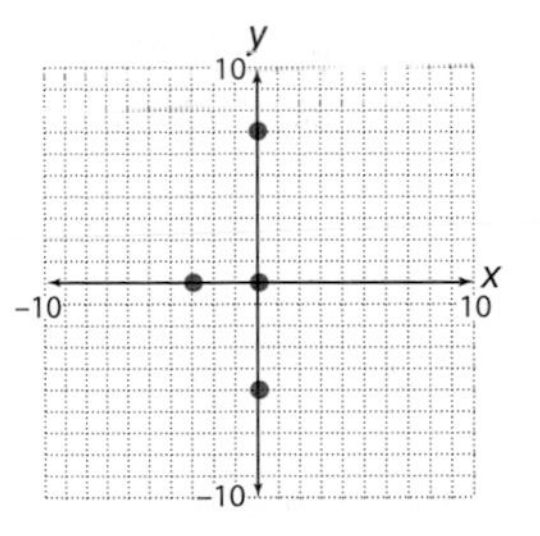

9. Graph the ordered pairs $(-4, 5)$, $(-3, 1)$, $(3, -4)$, and $(5, 0)$.

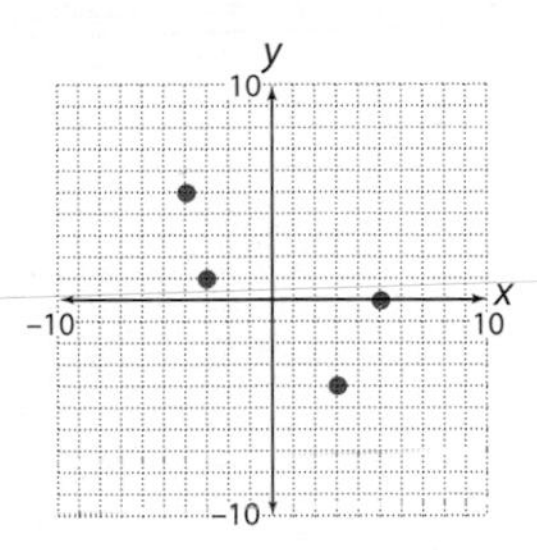

10. Graph the ordered pairs $(-1, 4)$, $(-2, -3)$, $(0, 7)$, and $(-8, 0)$.

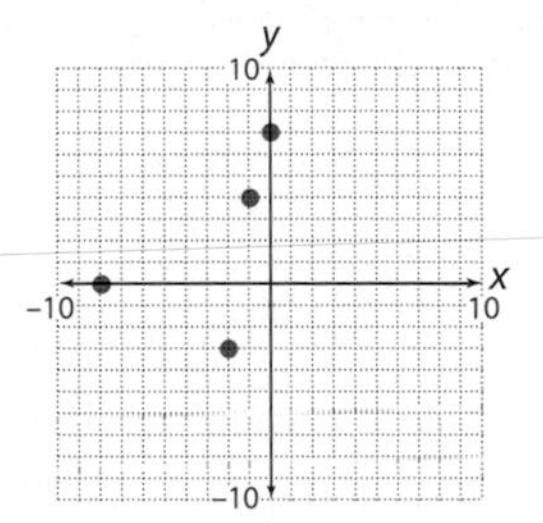

11. Graph the ordered pairs $(-6, 5)$, $(-5, -7)$, $(-4, 9)$, and $(0, -6)$.

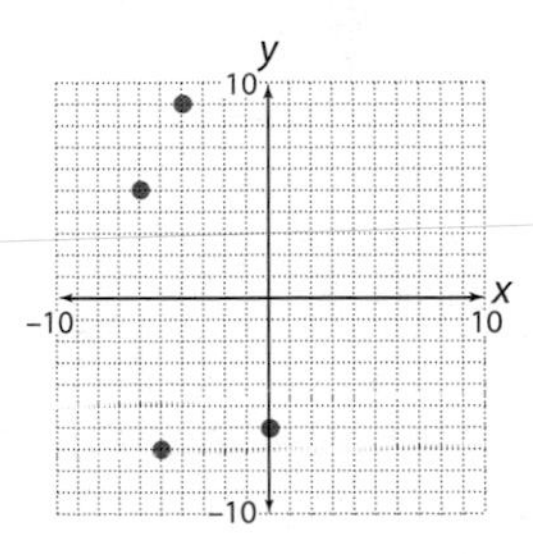

12. Find the coordinates of each of the points.

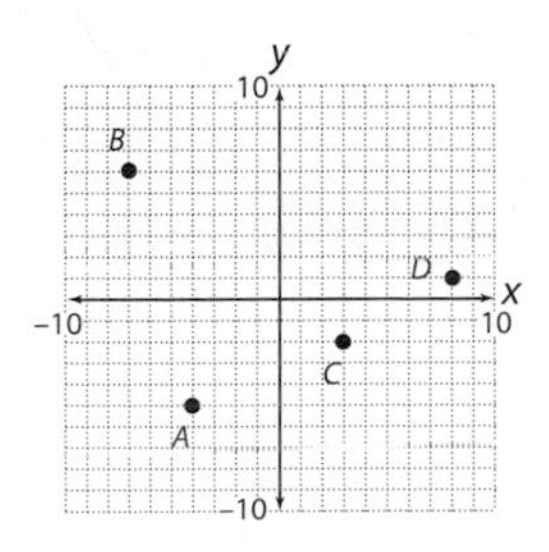

$A(-4, -5)$; $B(-7, 6)$; $C(3, -2)$; $D(8, 1)$

13. Find the coordinates of each of the points.

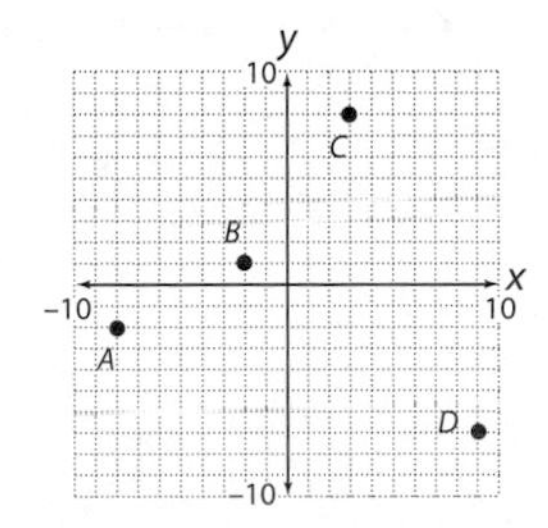

$A(-8, -2)$; $B(-2, 1)$; $C(3, 8)$; $D(9, -7)$

14. Find the coordinates of each of the points.

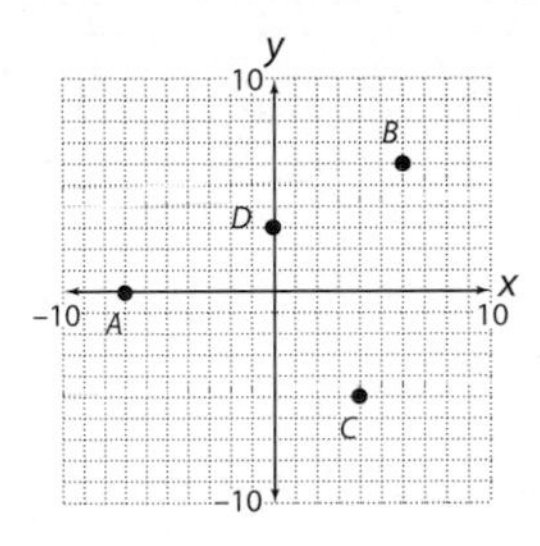

$A(-7, 0)$; $B(6, 6)$; $C(4, -5)$; $D(0, 3)$

15. Find the coordinates of each of the points.

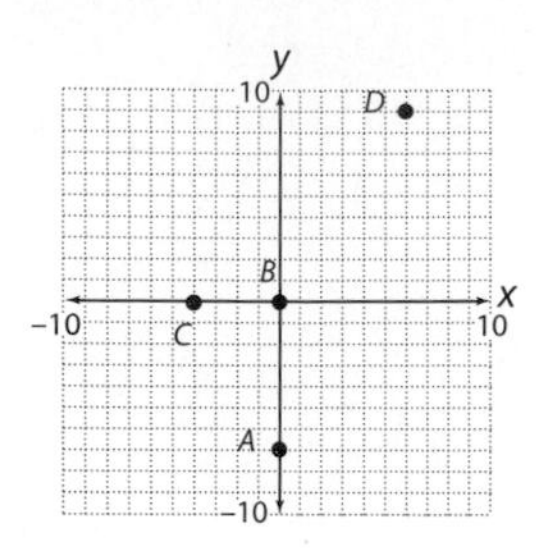

A (0, −7); *B* (0, 0); *C* (−4, 0); *D* (6, 9)

16. a. Name the abscissa of points *A* and *B*. **b.** Name the ordinate of points *C* and *D*.

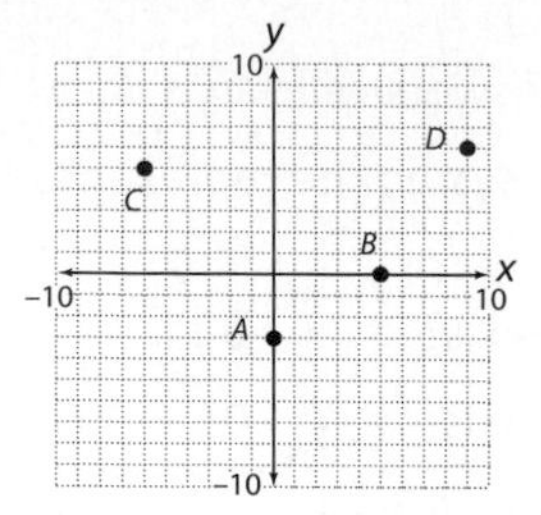

Abscissa: *A*, 0; *B*, 5; ordinate: *C*, 5; *D*, 6.

17. a. Name the abscissa of points *A* and *B*. **b.** Name the ordinate of points *C* and *D*.

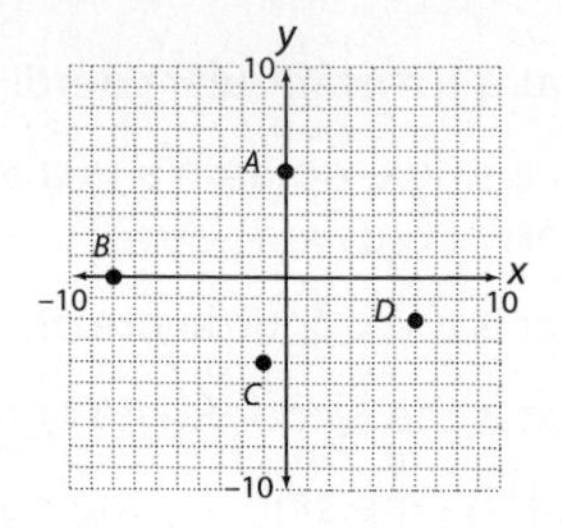

Abscissa: *A*, 0; *B*, −8; ordinate: *C*, −4 ; *D*, −2.

Scatter Diagrams

18. The table below shows the percent of all applicants to U.S. veterinary schools who are women. Make a scatter diagram for this table. Source: Association of American Veterinary Medical Colleges.

Year	1992	1993	1994	1995	1996	1997
Percent	66%	65%	67%	68%	68%	70%

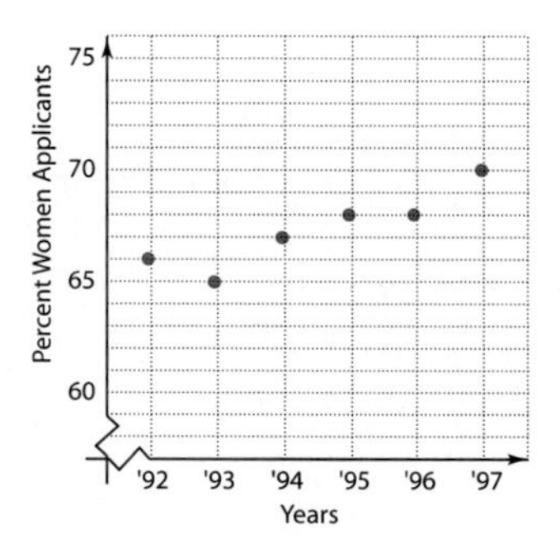

19. The table below shows how the average weight of All-Pro offensive linemen has changed from 1991 to 1996. Make a scatter diagram of this table. Source: Total Football in San Diego Union Tribune 1/24/98.

Year	1992	1993	1994	1995	1996	1997	1998
Weight, pounds	292	295	287	293	297	301	299

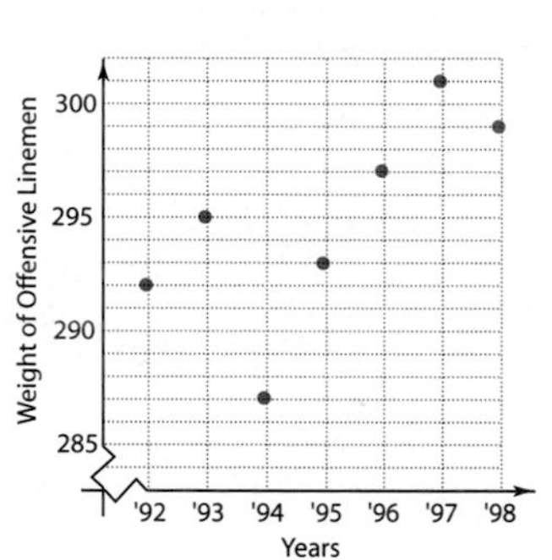

20. The table below shows the approximate annual sales (in millions of dollars) of digital cameras in the United States. Make a scatter diagram of this table. Source: International Data Corporation.

Year	1995	1996	1997	1998	1999
Sales, in millions of $	300	400	520	580	820

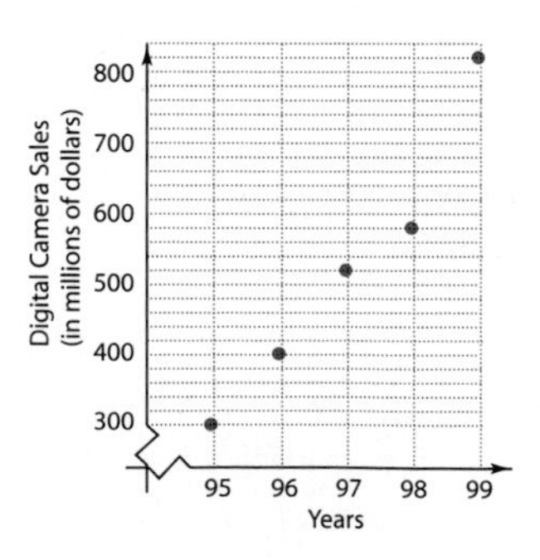

21. The table below shows the approximate annual federal funding for the arts (in millions of dollars). Make a scatter diagram of this table. Source: National Endowment for the Arts.

Year	1993	1994	1995	1996	1997	1998
Funding, in millions of $	170	165	160	100	100	100

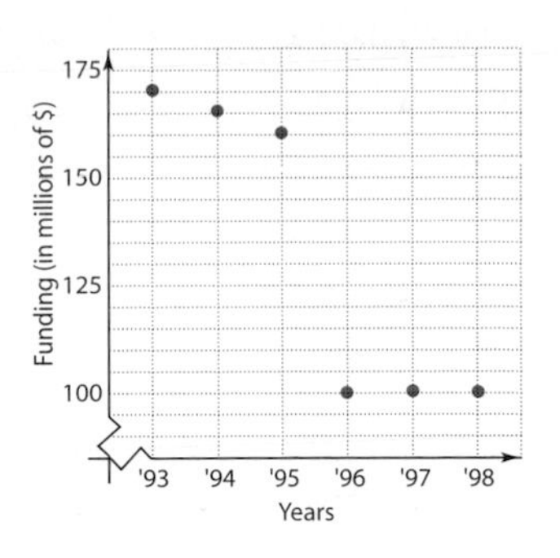

22. The scatter diagram at the right shows the approximate amount of charges (in billions of dollars) to credit cards for the December holiday season. Make an input/output table from the graph. Source: RAM Research (Internet).

Year	1992	1993	1994	1995	1996	1997
Charges, in billions of $	65	80	95	115	130	140

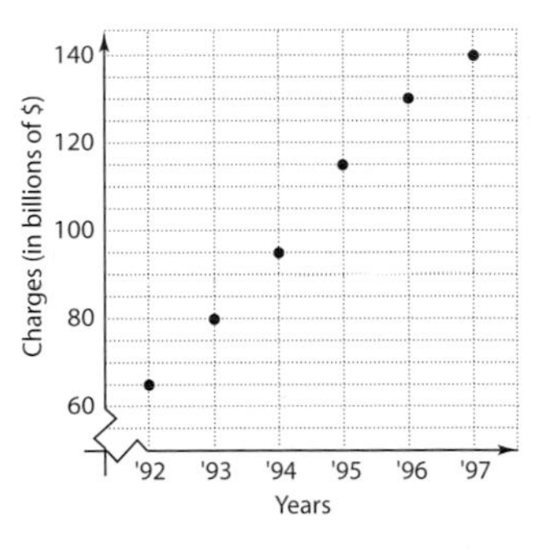

23. The scatter diagram at the right shows the percent of people who completely repay their credit card debt each month. Make an input/output table from the graph. Source: RAM Research.

Year	1992	1993	1994	1995	1996	1997
Percent who repay	31	32	33	34	36	39

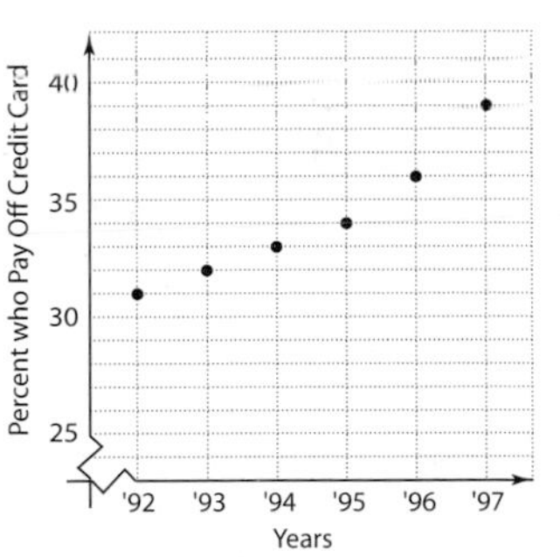

24. The scatter diagram at the right shows the number of games Cal Ripken (who holds the record for consecutive baseball games played) left early. Make an input/output table from the graph. Source: USA Today 9/25/97.

Year	1992	1993	1994	1995	1996	1997
Games left early	9	8	6	8	15	18

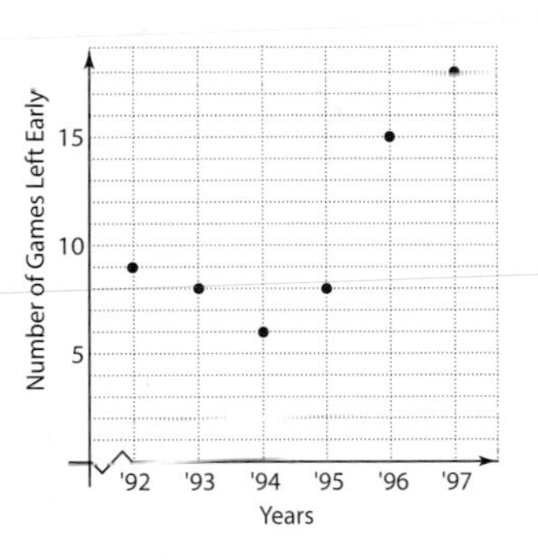

25. The scatter diagram at the right shows the average monthly cost (rounded to the nearest dollar) of cable TV service. Make an input/output table for this scatter diagram. Source: Paul Kagan Associates.

Year	1992	1993	1994	1995	1996	1997
Average monthly charge	18	19	21	23	24	26

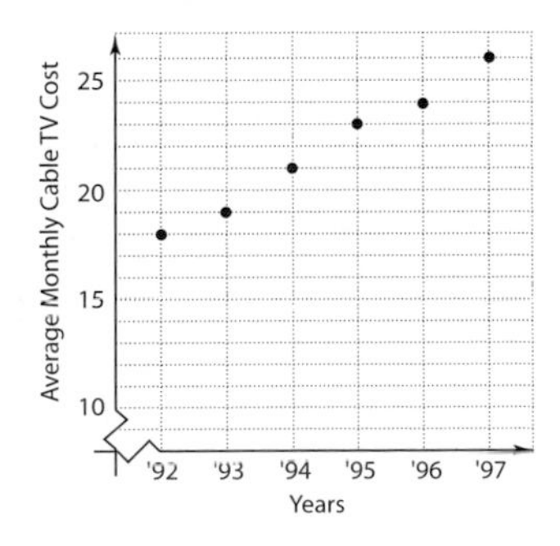

Graphs of Input/Output Tables

26. Graph $y = 2x + 1$ for $x = -3, -2, -1, 0, 1, 2,$ and 3.

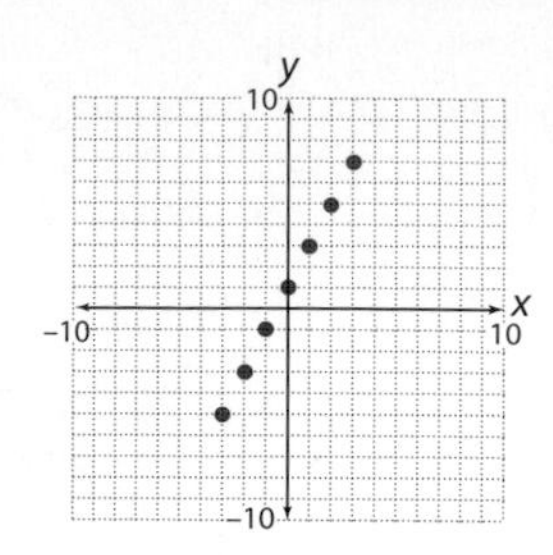

27. Graph $y = 2x - 3$ for $x = -3, -2, -1, 0, 1, 2,$ and 3.

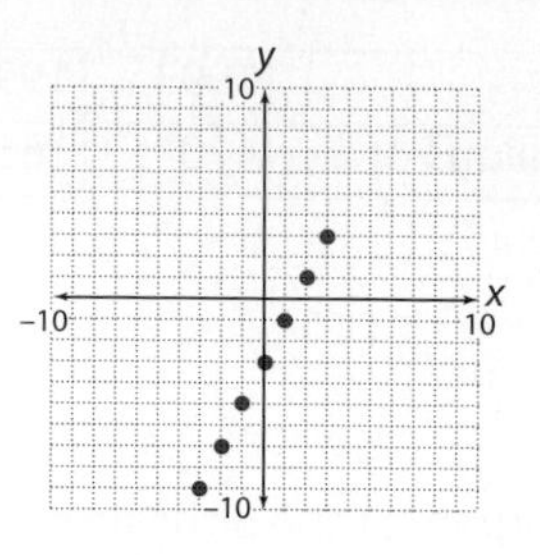

28. Graph $y = 1 + 3x$ for $x = -3, -2, -1, 0, 1, 2,$ and 3.

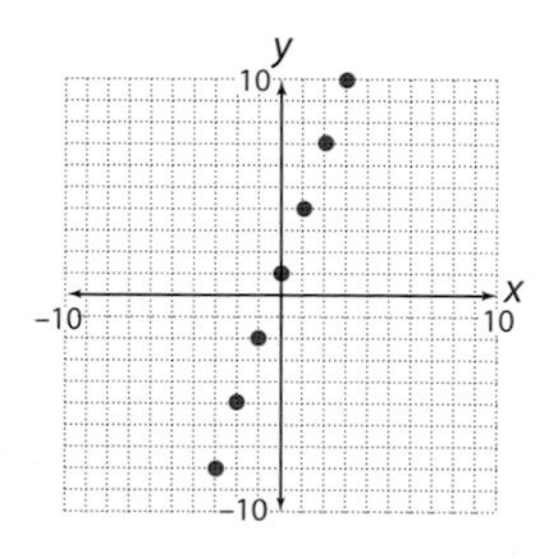

29. Graph $y = 3 - 2x$ for $x = -3, -2, -1, 0, 1, 2,$ and 3.

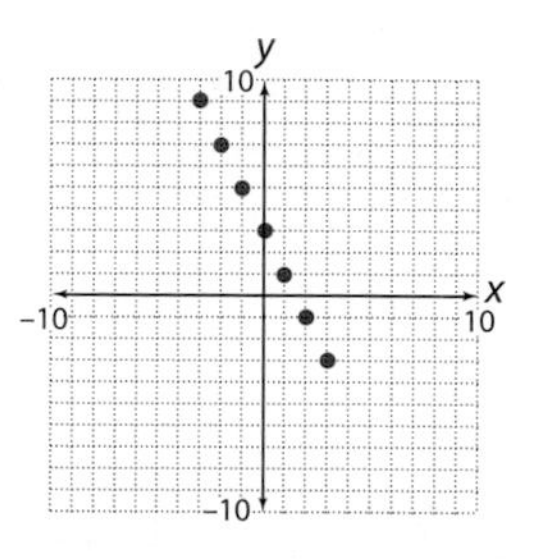

30. Graph $y = x^2$ for $x = -3, -2, -1, 0, 1, 2,$ and 3.

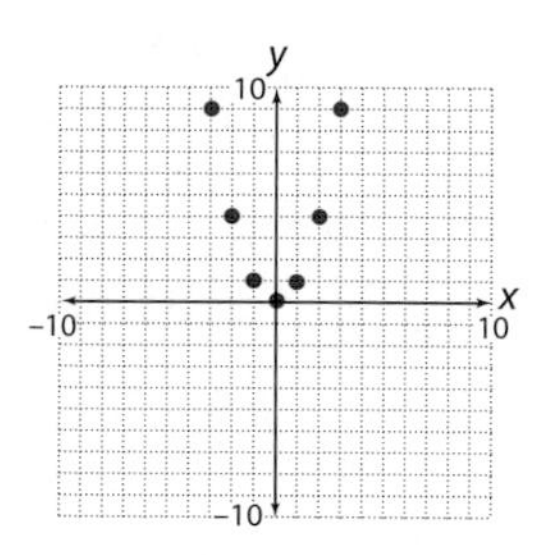

31. Graph $y = x^2 - 2$ for $x = -3, -2, -1, 0, 1, 2,$ and 3.

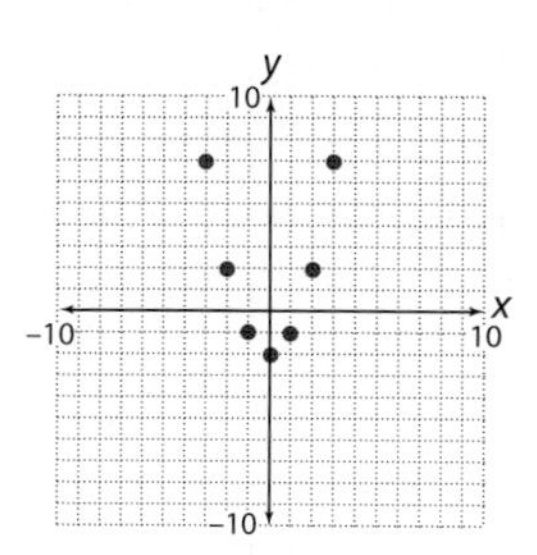

32. Graph $y = -x^2 + 1$ for $x = -3, -2, -1, 0, 1, 2,$ and 3.

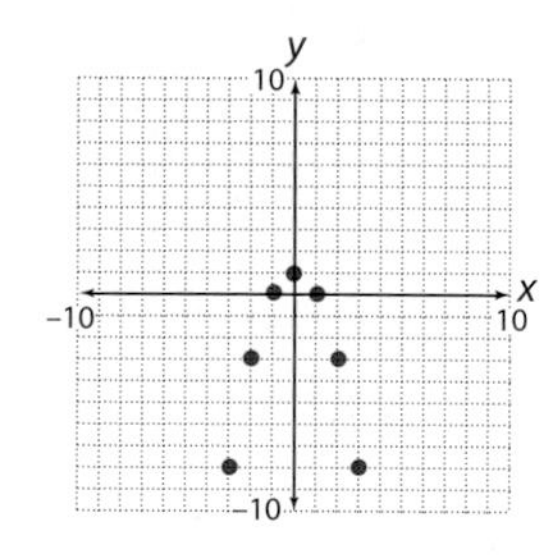

33. Graph $y = 2x^2 - 9$ for $x = -3, -2, -1, 0, 1, 2,$ and 3.

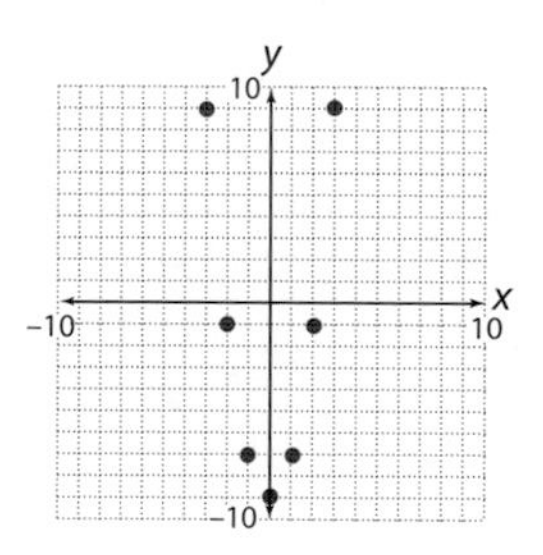

34. Graph $y = x^3$ for $x = -2, -1, 0, 1,$ and 2.

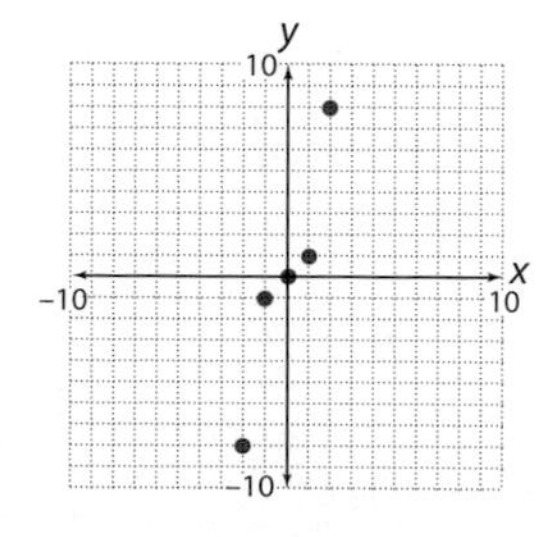

35. Graph $y = -x^3$ for $x = -2, -1, 0, 1,$ and 2.

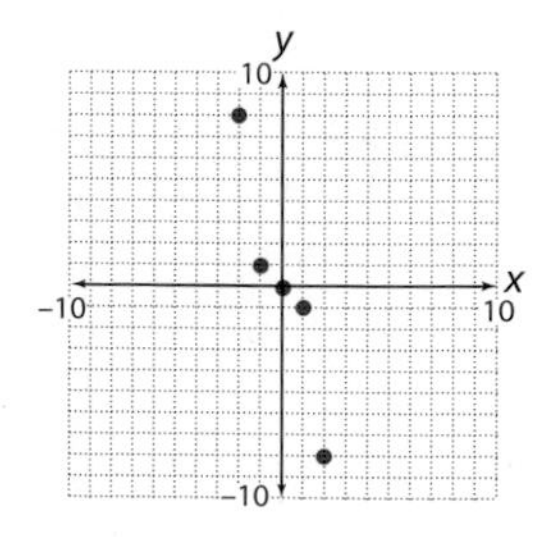

Applications

36. The equation $F = 1.8C + 32$ relates a Celsius temperature, C, to a Fahrenheit temperature, F. Graph this equation for $C = 0, 5, 10, 15$, and 20. What is the meaning of the ordered pair $(10, 50)$?
When the Celsius temperature is 10, the Fahrenheit temperature is 50.

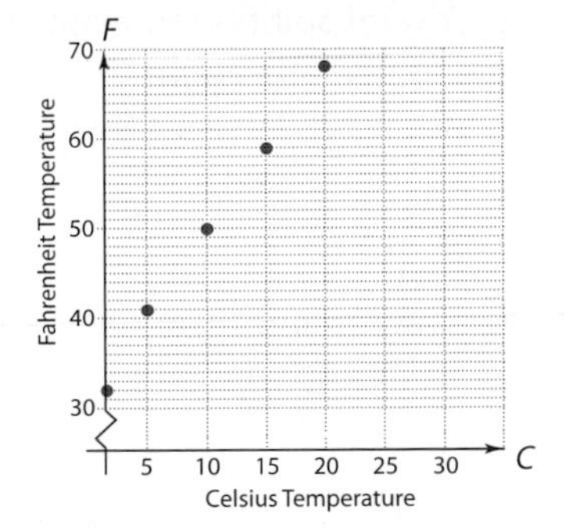

37. The equation $d = 900t$ relates the distance, d (in feet), a marathon runner has run t minutes after starting a race. Graph this equation for $t = 0, 1, 2, 3, 4$, and 5. What is the meaning of the ordered pair $(4, 3600)$?
After 4 minutes, the runner has traveled 3600 feet.

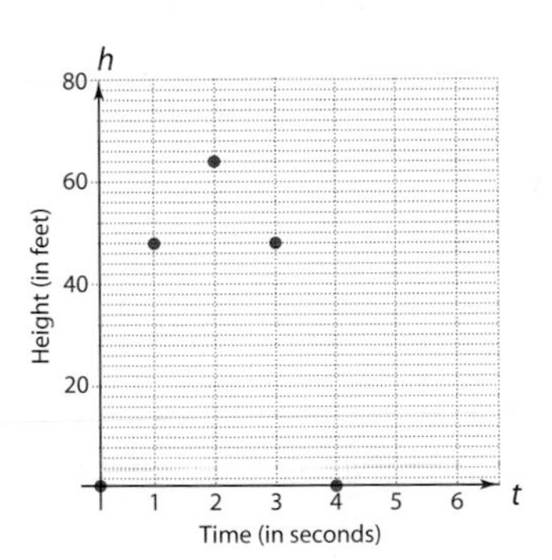

38. The equation $h = -16t^2 + 64t$ relates the height, h (in feet), of a rock above the ground t seconds after it has been thrown straight up. Graph this equation for $t = 0, 1, 2, 3$, and 4. What is the meaning of the ordered pair $(2, 64)$?
After 2 seconds, the height is 64 feet.

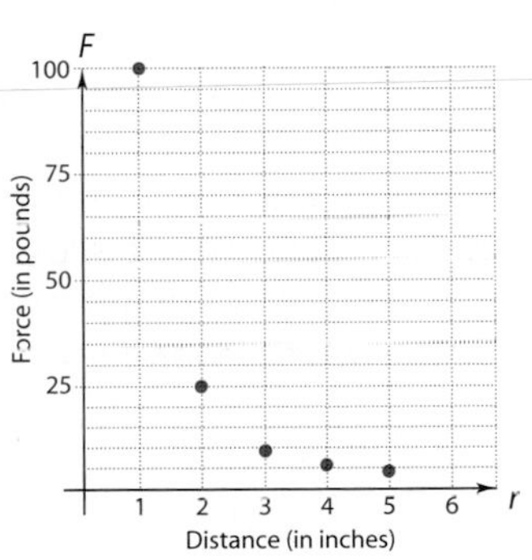

39. The equation $F = \dfrac{100}{r^2}$ relates the force, F (in pounds), between two magnets that are r inches apart. Graph this equation for $r = 1, 2, 3, 4$, and 5. What is the meaning of the ordered pair $(2, 25)$?
When the magnets are 2 inches apart, the force between them is 25 pounds.

F
100
Force (in pounds)
75
50
25
1 2 3 4 5 6
r
Distance (in inches)

40. The equation $h = -3t^2 + 24t$ approximates the height, h (in feet), of a rock above the surface of the moon t seconds after it has been thrown straight up by an astronaut. Graph this equation for $t = 0, 1, 2, 3, 4, 5$, and 6. What is the meaning of the ordered pair $(3, 45)$?
After 3 seconds, the height of the rock is 45 feet.

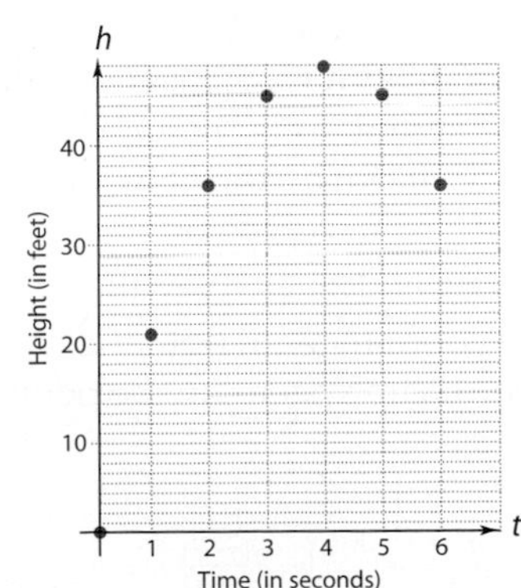

Applying Concepts

41. Plot several points with x-coordinate 3 on a rectangular coordinate system and draw a line through the points. Describe the line drawn through the points. The line is vertical passing through the point (3, 0).

42. Plot several points with y-coordinate −2 on a rectangular coordinate system and draw a line through the points. Describe the line drawn through the points. The line is horizontal passing through (0, −2).

43. If several points whose x- and y-coordinates are equal are plotted on a rectangular coordinate system, describe the line drawn through those points. The line is a diagonal that cuts quadrants I and III equally.

44. If several points whose y-coordinate is the opposite of the x-coordinate are plotted on a rectangular coordinate system, describe the line drawn through those points. The line is a diagonal that cuts quadrants II and IV equally.

45. If the points whose coordinates are (−4, 4), (4, 4), (4, −4), (−4, −4), and (−4, 4) are connected in the given order, what is the name of the geometric figure that is drawn? The figure is a square.

46. How many different lines can pass through the point whose coordinates are (2, 3)? How many different lines can pass through the points whose coordinates are (2, 3) and (4, 7)? Infinite; 1

47. Is it possible to draw one line that will **(a)** pass through 2 quadrants; **(b)** pass through 3 quadrants; **(c)** pass through all 4 quadrants of a rectangular coordinate system? (Remember, a line has infinite length.) a. Yes; b. yes; c. no

Exploration—Computer Screens

48. A computer screen has a coordinate system that is different from the xy-coordinate system we have discussed. In one mode, the origin of the computer coordinate system is the top left point of the screen, as shown at the right. Plot the points whose coordinates are (200, 400), (0, 100), and (100, 300) in this system.

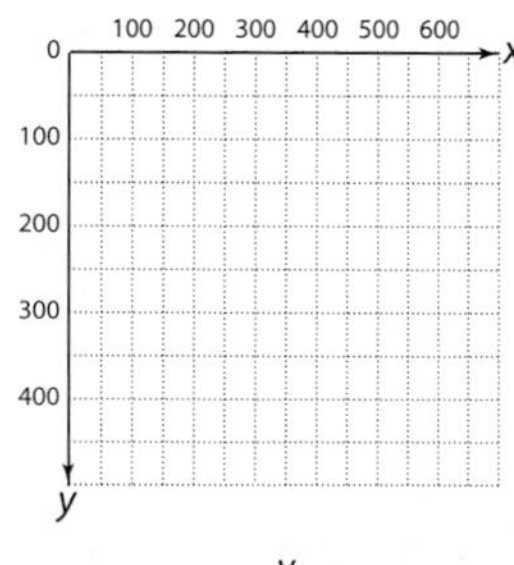

49. Sometimes it is inconvenient to think in terms of a computer screen's coordinate system as in the previous exercise, and so a rectangular coordinate system is drawn on top of the computer screen coordinate system. (See the graph at the right. We are using x_C and y_C as the labels for the computer coordinate system and x_R and y_R as the labels for the rectangular coordinate system.) Give the coordinates in both the computer coordinate system and the rectangular coordinate system for the points shown in the graph.

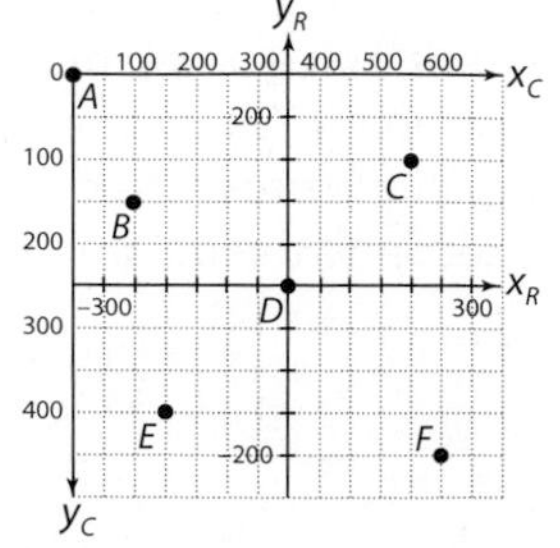

Section 2.4 Operations with Rational Numbers

Real Numbers

In the figures below, measurements have been given using *rational numbers.*

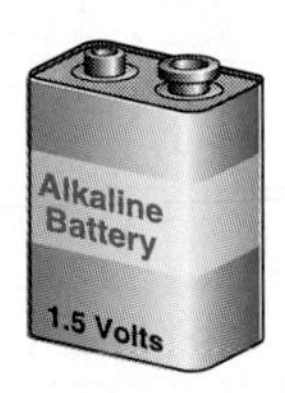

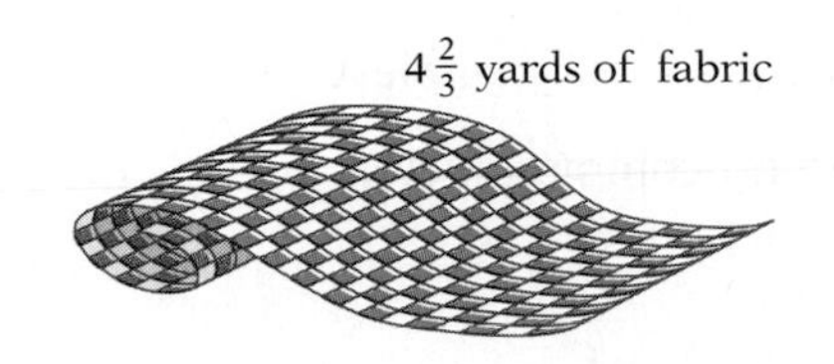

Rainfall Totals	
Yesterday	1.14 in.
Storm Total	2.37 in.
Year-to-date	6.48 in.
Normal	5.93 in.

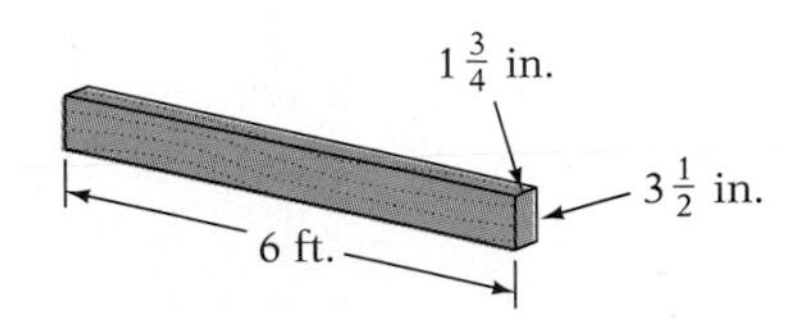

2 x 4 piece of lumber 6 ft long

> **Rational Number**
> A rational number is the quotient of two integers. In symbols, a rational number is one of the form $\frac{a}{b}$, where a and b are integers and $b \neq 0$.

POINT OF INTEREST
As early as A.D. 630 the Hindu mathematician Brahmagupta wrote a fraction as one number over another separated by a space. The Arab mathematician al Hassar (around A.D. 1050) was the first to show a fraction with a horizontal bar separating the numerator and denominator.

Examples of rational numbers are: $\frac{2}{3}, \frac{4}{-5}, \frac{-12}{7},$ and $\frac{5}{1}$.

Because any integer can be written as the quotient of the integer and 1, every integer is a rational number. For instance,

$$\frac{6}{1} = 6, \quad \frac{-5}{1} = -5, \quad \frac{0}{1} = 0$$

A number written in decimal notation is also a rational number.

seven-tenths: $0.7 = \frac{7}{10}$

negative fifty-three thousandths: $-0.053 = -\frac{53}{1000}$

SUGGESTED ACTIVITY
The *Instructor's Resource Manual* includes a procedure students can use to find the repeating digits for fractions such as $\frac{19}{23}$ whose repeating cycle is longer than the calculator display.

A rational number written as a fraction can be written in decimal notation by dividing the numerator of the fraction by the denominator.

$\frac{3}{4} = 0.75 \qquad \frac{13}{8} = 1.625 \qquad \frac{9}{16} = 0.5625$ terminating decimals

$\frac{1}{3} = 0.\overline{3} \qquad \frac{4}{7} = 0.\overline{571428} \qquad \frac{777}{9900} = 0.07\overline{84}$ repeating decimals

Each of the decimals in the first line is a **terminating** decimal because the decimal representation ends (or terminates). The decimals in the second line are **repeating** decimals and never end. The bar over the digits in a repeating decimal indicates that those digits repeat over and over. For instance,

$$\frac{1}{3} = 0.3333333\ldots \qquad \frac{4}{7} = 0.57142857\ldots \qquad \frac{777}{9900} = 0.07848484\ldots$$

Because the quotient of two integers is either a terminating or a repeating decimal, a rational number also can be described as a terminating decimal (such as 2.34) or a repeating decimal (such as $0.\overline{571428}$).

TAKE NOTE
If you use a calculator to find the decimal form of a fraction, be aware that the number of digits needed before you can determine the repeating digits may exceed the calculator's display. For instance,

$\frac{17}{23} = 0.\overline{7391304347826086956521}$

Question: Identify each of the following as a terminating decimal, a repeating decimal, or neither of those: -4.513427; $3.587\overline{2379}$; 2.13113111311113...[1]

1. terminating; repeating; neither

A fraction is in **simplest form** when the numerator and denominator do not contain a common factor greater than 1. The fractions $\frac{2}{3}$, $\frac{4}{6}$, and $\frac{8}{12}$ are **equivalent fractions** because they represent the same part of a whole. However, the fraction $\frac{2}{3}$ is in simplest form because there are no common factors greater than 1 in the numerator and denominator.

$\frac{2}{3}$ $\frac{4}{6}$ $\frac{8}{12}$

To write a fraction in simplest form, divide the numerator and denominator by the common factor. This is frequently shown by placing a slash through each common factor.

$$\frac{4}{6} = \frac{\overset{1}{\cancel{2}} \cdot 2}{\underset{1}{\cancel{2}} \cdot 3} = \frac{2}{3}$$ • Divide the numerator and denominator by the common factor, 2.

$$\frac{8}{12} = \frac{\overset{1}{\cancel{4}} \cdot 2}{\underset{1}{\cancel{4}} \cdot 3} = \frac{2}{3}$$ • Divide the numerator and denominator by the common factor, 4.

SUGGESTED ACTIVITY

For a somewhat different way of looking at equivalent fractions, ask students how they would divide up $\frac{3}{4}$ of a whole—say a pizza—among 6 people so that each gets the same amount. By dividing the pizza into eighths, there would be 6 eighths, and each person would receive $\frac{1}{8}$. Give other examples, such as dividing $\frac{2}{3}$ of a whole among 12 people.

Question: What is $\frac{12}{16}$ written in simplest form?[2]

A fraction can represent a part of a whole.

$\frac{3}{4}$ of a pizza remains

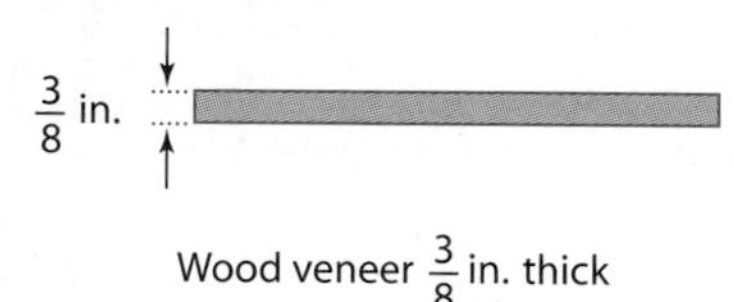

Wood veneer $\frac{3}{8}$ in. thick

A fraction also can represent part of a total. For instance, the table at the right, based on estimates from the National Foundation for Women Business Owners, shows the number of businesses owned by women in various sectors of the economy. (*San Diego Union-Tribune*, 5/5/97)

Services (medical, consulting, day care)	4,160,000
Transportation, communication, utilities	240,000
Finance, insurance, real estate	800,000
Manufacturing	720,000
Wholesale/retail	1,760,000
Other	320,000
Total	8,000,000

From the table,

$$\frac{\text{Number of businesses in finance, insurance, real estate}}{\text{Total number of businesses}} = \frac{800{,}000}{8{,}000{,}000} = \frac{1}{10}$$

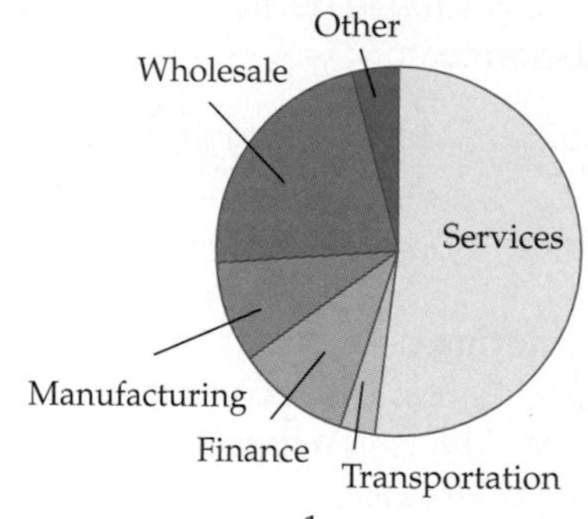

Finance is $\frac{1}{10}$ of the total.

This means that $\frac{1}{10}$ of all the businesses owned by women were in the area of finance, insurance, and real estate. The circle graph shown at the left is a convenient way to display how each fraction of a total contributes to the total.

2. $\frac{12}{16} = \frac{3 \cdot \overset{1}{\cancel{4}}}{4 \cdot \underset{1}{\cancel{4}}} = \frac{3}{4}$

TAKE NOTE
To find a common denominator, write the prime factorization of each number and determine the common factors.

$12 = 2 \cdot 2 \cdot 3$

$18 = 2 \cdot 3 \cdot 3$

Factors of 12

$\text{LCM} = 2 \cdot 2 \cdot 3 \cdot 3$

Factors of 18

The common denominator is the least common multiple (LCM) of the denominators.

The solution of the example below requires the comparison of two fractions to determine which of the two is larger. When comparing fractions, first rewrite each fraction with a common denominator. Here is an example.

→ Determine which is the larger fraction, $\frac{7}{12}$ or $\frac{11}{18}$.

A common denominator is the least common multiple (LCM) of 12 and 18, which is 36. Rewrite the fractions as equivalent fractions with denominator 36.

$\frac{7}{12} = \frac{7 \cdot 3}{12 \cdot 3} = \frac{21}{36}$ • Multiply the numerator and denominator by 3, which is $36 \div 12$.

$\frac{11}{18} = \frac{11 \cdot 2}{18 \cdot 2} = \frac{22}{36}$ • Multiply the numerator and denominator by 2, which is $36 \div 18$.

$22 > 21$ • Compare the numerators.

$\frac{11}{18} > \frac{7}{12}$ • The larger fraction has the larger numerator.

Example 1 In a recent survey people were asked whether they preferred nonfat, lowfat, or whole milk. The results of the survey are shown in the circle graph at the right. According to this survey, did more people prefer whole milk or lowfat milk?

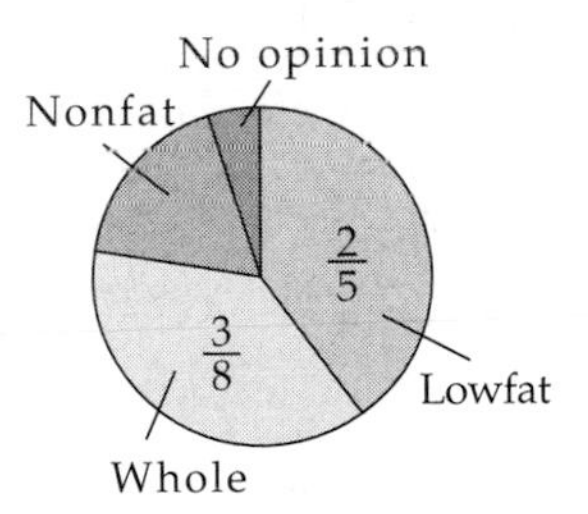

Solution

State the goal. Our goal is to determine which of the regions in the graph, whole milk or lowfat milk, is larger.

Describe a strategy. To answer this question, we could look at the graph and try to determine which of the regions appears larger. In this case, however, the two regions are so close to the same size that a different method of solving the problem is necessary.

One possibility is to compare the fraction $\frac{3}{8}$ to the fraction $\frac{2}{5}$.

Solve the problem. To compare fractions, rewrite the fractions as equivalent fractions with a common denominator. Use the least common multiple (LCM) of the denominators as the common denominator. The LCM of 5 and 8 is 40.

$$\frac{3}{8} = \frac{15}{40} \qquad \frac{2}{5} = \frac{16}{40}$$

Because $16 > 15$, $\frac{16}{40} > \frac{15}{40}$. Therefore, more people in this survey preferred lowfat milk than preferred whole milk.

Check your work. As a visual check, $\frac{15}{40}$ and $\frac{16}{40}$ are approximately equal, which is consistent with the graph. An alternative solution to this problem would be to compare the decimal equivalents of $\frac{3}{8}$ and $\frac{2}{5}$.

You-Try-It 1 The circle graph at the right is based on data provided by Interep Research and reported in *USA Today* (4/30/97). People were asked what type of radio station they listen to. According to this survey, do more people listen to adult contemporary radio stations or news/talk stations?

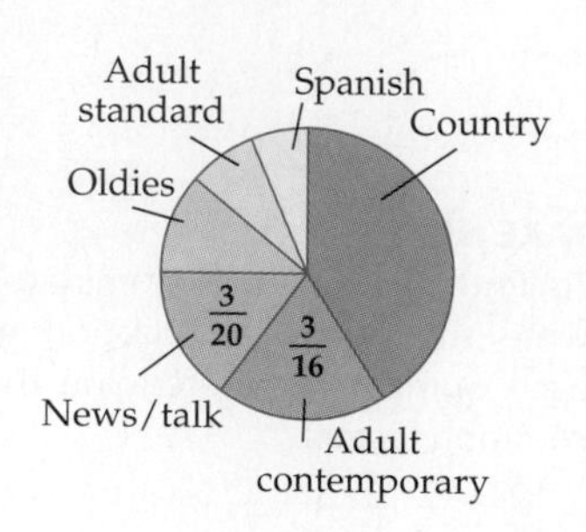

Solution See page S7.

Some numbers have neither terminating nor repeating decimal representations. These are the **irrational numbers**. For example, 2.45445444544445 . . . is an irrational number. Two other examples are $\sqrt{2}$ and π.

TAKE NOTE
The symbol $\approx$ is read "is approximately equal to."

$\sqrt{2} \approx 1.4142135...$ $\quad\quad$ $\pi \approx 3.1415926...$

The three dots mean that the digits continue on and on without ever repeating or terminating. Although we cannot write a decimal that is exactly equal to $\sqrt{2}$ or to π, we can approximate these numbers. Shown below is $\sqrt{2}$ rounded to the nearest thousandth and π rounded to the nearest hundredth.

$\sqrt{2} \approx 1.414$ $\quad\quad$ $\pi \approx 3.14$

The rational numbers and the irrational numbers taken together are called the **real numbers.**

Question: Which of the following numbers are rational and which are irrational: 0.101101110. . ., 2.31257, $\frac{-13}{-87}$, $-4.12\overline{731}$?[3]

Add and Subtract Rational Numbers

To add or subtract fractions, first rewrite each fraction as an equivalent fraction with a common denominator, using the least common multiple (LCM) of the denominators as the common denominator. Then add the numerators, and place the sum over the common denominator. Write the answer in simplest form.

The sign rules for adding integers apply to addition of rational numbers.

Example 2 Subtract: $-\frac{5}{6}-\left(-\frac{3}{10}\right)$

TAKE NOTE
Although we could have written the answer to Example 2 with the negative sign in the numerator, in this text we write negative fractions with the negative sign in front of the fraction.

Solution Find a common denominator. The LCM of 6 and 10 is 30. (The LCM of the denominators is sometimes called the least common denominator, LCD.)

$-\frac{5}{6}-\left(-\frac{3}{10}\right) = -\frac{5}{6}+\frac{3}{10}$ • Rewrite subtraction as addition of the opposite.

$= -\frac{25}{30}+\frac{9}{30}$ • Rewrite each fraction as an equivalent fraction using the LCM of the denominators as the common denominator.

$= \frac{-25+9}{30} = \frac{-16}{30}$ • Add the numerators, and place the sum over the common denominator.

$= -\frac{8}{15}$ • Write the answer in simplest form.

You-Try-It 2 Add: $-\frac{8}{9}+\frac{5}{6}$

Solution See page S7.

---

3. The first one is irrational; the last three are rational.

Example 3 Evaluate $x - y - z$ for $x = -\frac{1}{2}, y = \frac{2}{3}$, and $z = -\frac{3}{4}$.

Solution Replace x, y, and z with their values and then simplify.

$$x - y - z$$

$$-\frac{1}{2} - \frac{2}{3} - \left(-\frac{3}{4}\right) = -\frac{1}{2} + \left(-\frac{2}{3}\right) + \frac{3}{4}$$

• Rewrite subtraction as addition of the opposite.

$$= -\frac{6}{12} + \left(-\frac{8}{12}\right) + \frac{9}{12}$$

• The LCM of 2, 3, and 4 is 12.

$$= \frac{-6 + (-8) + 9}{12} = -\frac{5}{12}$$

You-Try-It 3 Evaluate $-a + b - c$ when $a = -\frac{5}{6}, b = \frac{1}{2}$, and $c = -\frac{2}{3}$.

Solution See page S7.

Rational numbers in decimal form are also added and subtracted by using the sign rules for integers.

Example 4 Add: $2.31 + (-4.2)$

Solution

$|2.31| = 2.31$

$|-4.2| = 4.2$

• The signs are different. Find the absolute value of each number.

$4.2 - 2.31 = 1.89$

• Subtract the smaller of these from the larger one.

$2.31 + (-4.2) = -1.89$

• Attach the sign of the number with the larger absolute value.

You-Try-It 4 Subtract: $-5.1 - (-10.39)$

Solution See page S7.

Multiply and Divide Rational Numbers

The sign rules for multiplying and dividing integers apply to multiplication and division of rational numbers. The product of two fractions is the product of the numerators divided by the product of the denominators.

TAKE NOTE
An alternative solution to Example 5 is to divide by the common factors of the numerators and denominators.

$$-\frac{5}{6} \cdot \frac{4}{15} = -\left(\frac{\overset{1}{\cancel{5}} \cdot \overset{2}{\cancel{4}}}{\underset{3}{\cancel{6}} \cdot \underset{3}{\cancel{15}}}\right) = -\frac{2}{9}$$

Example 5 Multiply: $-\frac{5}{6} \cdot \frac{4}{15}$

Solution

$$-\frac{5}{6} \cdot \frac{4}{15} = -\left(\frac{5 \cdot 4}{6 \cdot 15}\right)$$

• The signs are different. The product is negative. Multiply the numerators and multiply the denominators.

$$= -\frac{2}{9}$$

• Write the answer in simplest form.

You-Try-It 5 Multiply: $\left(-\frac{16}{9}\right)\left(-\frac{15}{28}\right)$

Solution See page S7.

To multiply decimals, multiply as with integers. The number of decimal places in the product is the sum of the decimal places of each factor.

→ Multiply: $-5.34 \cdot 0.0056$

$$\underbrace{-5.34}_{\text{2 places}} \cdot \underbrace{0.0056}_{\text{4 places}} = \underbrace{-0.029904}_{\text{6 places}}$$

• The signs are different. The product is negative.

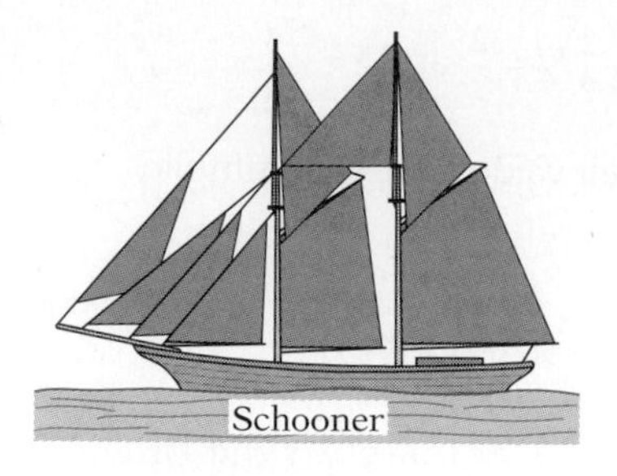

Example 6 A hobbyist is making a scale model of a schooner. If the hobbyist is using a scale of $\frac{3}{4}$ of an inch to represent an actual length of 1 foot, how tall should the mast in the model be if the mast in the schooner is $18\frac{1}{2}$ feet tall?

Solution Because $\frac{3}{4}$ of an inch in the model represents 1 foot in the schooner,

$\frac{3}{4}$ in. + $\frac{3}{4}$ in. = $1\frac{1}{2}$ in. represents 2 feet on the schooner.

$\frac{3}{4}$ in. + $\frac{3}{4}$ in. + $\frac{3}{4}$ in. = $2\frac{1}{4}$ in. represents 3 feet on the schooner.

To represent $18\frac{1}{2}$ feet, we must add $\frac{3}{4}$ in. $18\frac{1}{2}$ times. Since multiplication is the repeated addition of the same number, we have

$$\frac{3}{4} \cdot 18\frac{1}{2} = \frac{3}{4} \cdot \frac{37}{2} = \frac{111}{8} = 13\frac{7}{8}$$

The height of the mast in the model should be $13\frac{7}{8}$ inches.

You-Try-It 6 Computer drawing programs, such as Adobe Illustrator and Freehand, have a *Scale* function that allows an image to be redrawn either smaller or larger depending on the scale factor. A scale factor of $\frac{1}{2}$ will redraw the image one-half the current size. A scale factor of 2 will double the size of an image. Suppose a line displayed on the screen is $2\frac{2}{3}$ inches long. If the artist uses a scale factor of $\frac{3}{4}$ to redraw the line, what is the length of the new line?

Solution See page S8.

In the diagram below and on the left, a 6-foot piece of lumber is divided into pieces 2 feet long. The diagram on the right shows a 6-foot piece divided into pieces $\frac{2}{3}$ foot long. In each case, note that the number of pieces is found by division.

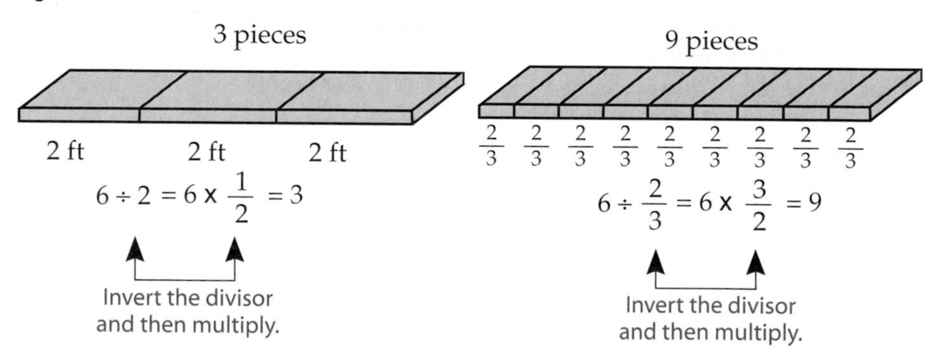

The model above suggests that to divide fractions, invert the divisor and then proceed as in multiplication of fractions.

Example 7 Divide: $-\frac{5}{8} \div \left(-\frac{7}{12}\right)$

Solution $-\frac{5}{8} \div \left(-\frac{7}{12}\right) = \frac{5}{8} \div \frac{7}{12}$ • The signs are the same. The quotient is positive.

$= \frac{5}{8} \cdot \frac{12}{7}$ • Invert the divisor and then multiply.

$= \frac{15}{14}$ • Write the answer in simplest form.

You-Try-It 7 Divide: $\frac{1}{8} \div \left(-\frac{5}{12}\right)$

Solution See page S8.

Dimensional Analysis

TAKE NOTE
See the Guidelines for Using Graphing Calculators appendix for a list of equivalent measurements.

In application problems, many numbers have *units of measurement* associated with the number. For instance,

9 **feet** 32 **pounds** $\frac{3}{4}$ **gallon** 23.8 **meters**

When a problem contains two different units, say feet and inches, it is usually necessary to convert one of the measurements so that both units are the same. This is accomplished by using **conversion factors.**

For instance, since 1 foot = 12 inches, we have the two conversion factors

$$\frac{1 \text{ foot}}{12 \text{ inches}} \quad \text{and} \quad \frac{12 \text{ inches}}{1 \text{ foot}}$$

We can show that each of these conversion factors is equal to 1 by replacing one of the measurements by its equivalent measurement. You can think of dividing the numerator and denominator by the common unit "inches" in one case and "foot" in the other case. It is important to remember that all conversion factors are equal to 1.

$$\frac{1 \text{ foot}}{12 \text{ inches}} = \frac{\overset{1}{\cancel{12 \text{ inches}}}}{\underset{1}{\cancel{12 \text{ inches}}}} = 1$$

$$\frac{12 \text{ inches}}{1 \text{ foot}} = \frac{\overset{1}{\cancel{1 \text{ foot}}}}{\underset{1}{\cancel{1 \text{ foot}}}} = 1$$

Example 8 Convert 5 pints to quarts.

TAKE NOTE
When converting from one unit to a second unit, use the conversion factor that has the second unit in the numerator.

Solution To convert from one unit to another, it is necessary to use the appropriate conversion factor. If you do not know the factor, you may need to look in a reference book. For pints and quarts, the conversion factors are $\frac{2 \text{ pints}}{1 \text{ quart}}$ and $\frac{1 \text{ quart}}{2 \text{ pints}}$. Because we are converting from pints to quarts, use the conversion factor that contains quart in the numerator. We use the abbreviations pt for pints and qt for quarts.

$$5 \text{ pt} = 5 \text{ pt} \cdot 1 = \frac{5 \text{ pt}}{1} \cdot \boxed{\frac{1 \text{ qt}}{2 \text{ pt}}} = \frac{5}{2} \text{ qt} = 2.5 \text{ quarts}$$

You-Try-It 8 Convert 29 feet to yards.

Solution See page S8.

TAKE NOTE
We are using the conversion factors $\frac{1 \text{ hour}}{3600 \text{ seconds}} = 1$ and $\frac{5280 \text{ feet}}{1 \text{ mile}} = 1$.

Conversions between units may require using more than one conversion factor.

→ Convert 60 miles per hour to feet per second.

$$\frac{60 \text{ miles}}{1 \text{ hour}} = \frac{60 \cancel{\text{ miles}}}{1 \cancel{\text{ hour}}} \cdot \frac{1 \cancel{\text{ hour}}}{3600 \text{ seconds}} \cdot \frac{5280 \text{ feet}}{1 \cancel{\text{ mile}}}$$
$$= \frac{60 \cdot 5280 \text{ feet}}{3600 \text{ seconds}} = 88 \text{ feet/second}$$

If the 6-foot board used earlier were a little longer, say $6\frac{1}{3}$ feet, and we tried dividing it into $\frac{2}{3}$-foot pieces, there would be a remainder at the end.

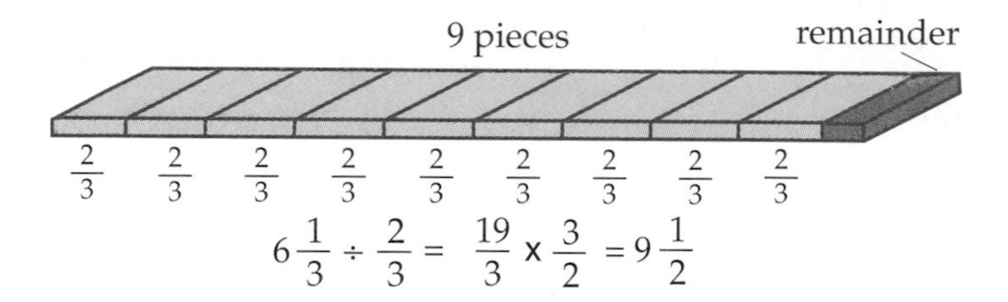

$$6\frac{1}{3} \div \frac{2}{3} = \frac{19}{3} \times \frac{3}{2} = 9\frac{1}{2}$$

The number $9\frac{1}{2}$ means 9 pieces, each $\frac{2}{3}$ foot long, and a remaining piece that is $\frac{1}{2}$ of a $\frac{2}{3}$ foot-long piece. The actual length of the remainder is

$$\frac{1}{2} \cdot \frac{2}{3} \text{ foot} = \frac{2}{6} \text{ foot} = \frac{1}{3} \text{ foot}$$ • Recall that *of* means multiply.

Example 9 A food warehouse purchases 100 gallons of orange juice for \$248. The juice is then repackaged in quart containers and sold for \$1.75 per quart. What is the markup on each quart of orange juice that is sold?

Solution

State the goal.

We must find the markup on each quart of orange juice.

Describe a strategy.

To find the markup, determine the cost per quart paid by the warehouse. This requires

a. changing 100 gallons to quarts;

b. finding the cost per quart by dividing 248 by the number of quarts, which is the answer to part **a**.

The markup is the selling price per quart minus the cost per quart.

Solve the problem.

$$100 \text{ gallons} = \frac{100 \cancel{\text{ gallons}}}{1} \cdot \frac{4 \text{ quarts}}{1 \cancel{\text{ gallon}}}$$ • Convert gallons to quarts.
$$= 400 \text{ quarts}$$

$$\text{Cost per quart} = 248 \div 400 = 0.62$$ • Find the cost per quart.

$$\text{Markup} = \text{selling price} - \text{cost}$$
$$= 1.75 - 0.62$$
$$= 1.13$$

The markup is \$1.13 per quart.

Check your work.

Review the solution, ensuring that all calculations are correct.

You-Try-It 9 A caterer has $7\frac{1}{2}$ gallons of iced tea for a party. How many $\frac{2}{3}$-cup servings are possible?

Solution See page S8.

93. Convert 44 feet per second to miles per hour.
30 miles per hour

94. Convert 66 feet per second to miles per hour.
45 miles per hour

95. Convert 12 quarts per minute to gallons per second.
$\frac{1}{20}$ gallon per second

96. Convert 15 miles per hour to feet per second.
22 feet per second

97. A candy store purchases 12 pounds of candy for \$6.50 and then repackages the candy in 4-ounce packages. The selling price for each 4-ounce package is \$.60. Find the profit earned on each 4-ounce package that is sold. Round to the nearest cent. The profit on each package is \$.46.

98. A mechanic purchases a 40-gallon container of oil for \$128 and charges customers \$1.50 per quart for an oil change. Find the profit the mechanic makes on one quart of oil. The profit on one quart of oil is \$.70.

99. A charity group is trying to raise money by recycling aluminum cans. The charity has found a recycler that will pay \$.50 per pound for the 10,000 cans collected. If four aluminum cans weigh 3 ounces, how much money will the recycler pay to the charity? The recycler will pay the charity \$234.38.

100. Each pleat in a drape requires 8 inches of fabric. How many pleats can be made in a drape material that is 4 yards long? 18 pleats can be made.

101. Studs for the support of a wall are required at each end and 16 inches apart. How many studs are required for a wall that is 13 feet long? 11 studs are required.

Applying Concepts

102. When two rational numbers are multiplied, it is possible for the product to be less than either factor, greater than either factor, or a number between the two factors. Give examples of each of these occurrences. Answers will vary.

103. If the same positive number is added to both the numerator and the denominator of $\frac{4}{7}$, is the new fraction greater than, less than, or equal to $\frac{4}{7}$? The new fraction is greater than $\frac{4}{7}$.

104. Suppose the numerator of a fraction is a fixed number, for instance, 5. How does the value of the fraction change as the denominator increases? As the denominator increases, the fraction decreases.

105. Given any two distinct rational numbers, is it always possible to find a rational number between the two given numbers? If so, explain how to find one.
Yes. Add the two numbers and then divide by 2.

106. Make an input/output table for $y = -\frac{2}{3}x$ for $x = -3, -2, -1, 0, 1, 2,$ and 3.

x	-3	-2	-1	0	1	2	3
y	2	$\frac{4}{3}$	$\frac{2}{3}$	0	$-\frac{2}{3}$	$-\frac{4}{3}$	-2

107. Make an input/output table for $y = 3x$ for $x = -\frac{3}{2}, -\frac{1}{2}, \frac{1}{2},$ and $\frac{3}{2}$.

x	$-\frac{3}{2}$	$-\frac{1}{2}$	$\frac{1}{2}$	$\frac{3}{2}$
y	$-\frac{9}{2}$	$-\frac{3}{2}$	$\frac{3}{2}$	$\frac{9}{2}$

Exploration—Moving Averages

108. One of the statistics a stock market analyst uses to evaluate a stock's performance is the **moving average** of a stock's price changes. This is the arithmetic mean of the changes in the value of a stock for a given number of days. To illustrate this procedure, we will calculate the five-day moving average of a stock's price changes. In actual practice, a stock market analyst may use 50 days, 200 days, or some other number of days.

The table below shows the dollar increase or decrease in the price of Intel Corporation stock from the previous closing price of a stock for a 10-day period.

Day 1	Day 2	Day 3	Day 4	Day 5	Day 6	Day 7	Day 8	Day 9	Day 10
–0.0625	–1.1875	0.8125	2.3125	3.625	1.25	2.375	–1.8125	1.625	–0.75

Calculate the five-day moving average for the change in the stock price by finding the average change in the stock price for days 1 through 5, days 2 through 6, days 3 through 7, and so on to days 6 through 10. The list of means you calculate is the five-day moving average for the change in the price of the stock.

109. Another statistic calculated by a stock market analyst is the moving average of the stock price, not the change in stock price as in Exercise 108. In this case, the arithmetic mean of the stock price is calculated over a certain period of time. The graph at the right shows the daily fluctuation in stock price, the 50-day moving average of the stock price, and the 200-day moving average of the stock price for Intel Corporation.

The stock price of Intel for a seven-day period is given in the table below.

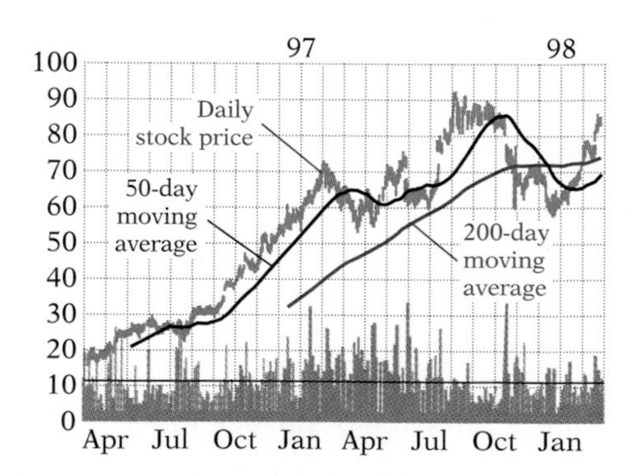

Day 1	Day 2	Day 3	Day 4	Day 5	Day 6	Day 7
85.0625	85.00	83.8125	84.625	86.9375	90.5625	91.8125

a. Calculate the moving average of the stock price for days 1 through 5, days 2 through 6, and days 3 through 7.

b. From the graph, is the 50-day moving average ever less than the 200-day moving average? Explain your answer.

c. From the graph, is the 200-day moving average always less than the closing price of the stock? Explain your answer.

d. Does the 200-day moving average fluctuate more or less than the 50-day moving average?

Section 2.5 Coordinate Geometry

Area of a Figure in the Coordinate Plane

Recall that the distance between two points on a *vertical* or *horizontal* line is the absolute value of the difference between the coordinates of the points. To find the vertical distance between two points in a plane, we use the y-coordinates of the two points. For the horizontal distance, we use the x-coordinates of the points. For the points in the graph at the right, we have

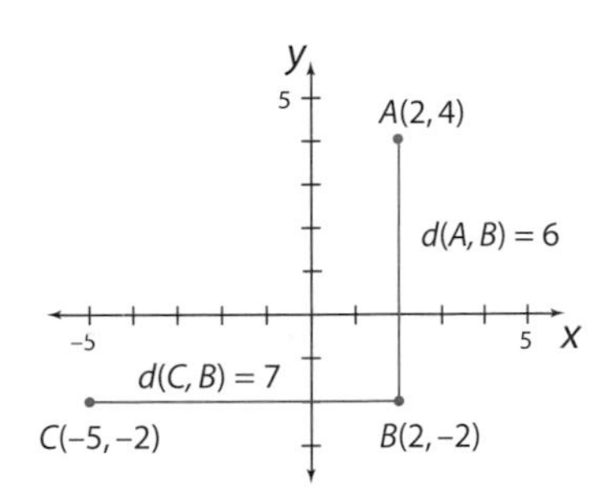

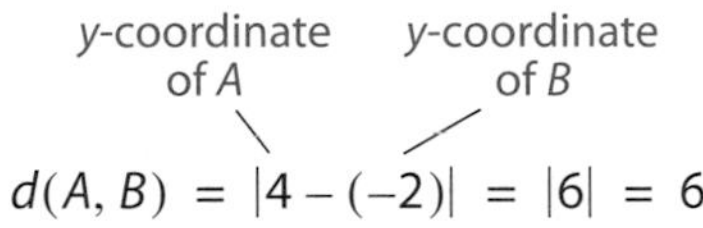

$$d(A, B) = |4 - (-2)| = |6| = 6$$

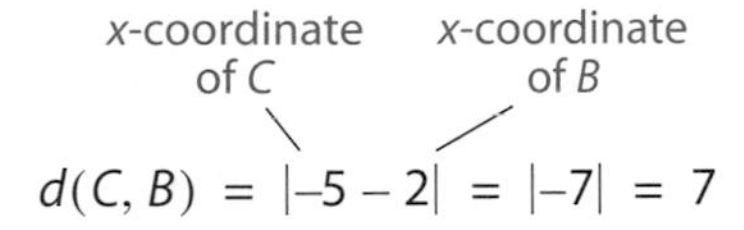

$$d(C, B) = |-5 - 2| = |-7| = 7$$

Question: What is the vertical and horizontal distance between the points $A(-2, 4)$ and $B(-6, -7)$?[1]

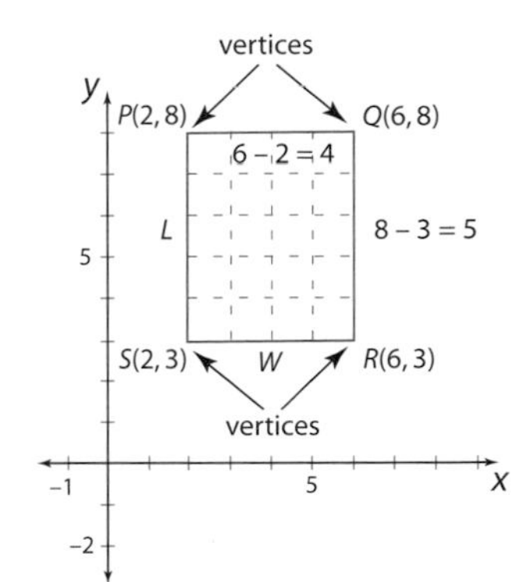

A rectangle has been drawn on the coordinate grid at the left. The **vertices** of the rectangle are the points P, Q, R, and S.

The length, L, of the rectangle is the vertical distance between points P and S or points Q and R. Points P and S are used here.

$$L = |8 - 3| = |5| = 5$$

The width, W, of the rectangle is the horizontal distance between points P and Q or points S and R. Points S and R are used here.

$$W = |2 - 6| = |-4| = 4$$

Once the length and width of the rectangle are determined, the area can be found. For the rectangle above, we have

$$A = LW$$
$$= 5(4) = 20$$ The area is 20 square units.

Because there are no specific units of measurement given on the coordinate grid (such as feet or meters), the area is stated in terms of square "units."

Example 1 Find the area of the parallelogram with vertices $P(1, 2)$, $Q(5, 2)$, $R(2, -3)$, and $S(-2, -3)$.

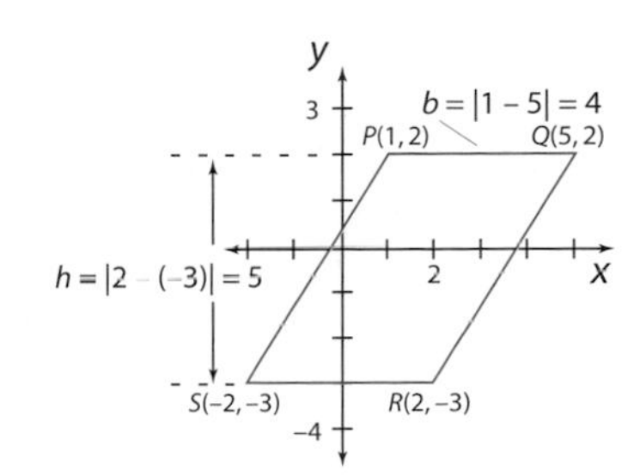

Solution Use the formula for the area of a parallelogram, $A = bh$. The base is the horizontal distance between P and Q or between S and R.

$$b = |1 - 5| = |-4| = 4$$ • We are using the x-coordinates of points P and Q.

The height is the vertical distance between P and S or between Q and R.

$$h = |2 - (-3)| = |5| = 5$$ • We are using the y-coordinates of points P and S.

Find the area.

$$A = bh$$
$$= 4(5) = 20$$ • $b = 4, h = 5$

The area is 20 square units.

1. Vertical distance = $|4 - (-7)| = |11| = 11$;
horizontal distance = $|-2 - (-6)| = |4| = 4$

You-Try-It 1 Find the area of the triangle with vertices $P(-4, 4)$, $Q(4, -2)$, and $R(-2, -2)$.

Solution See page S8.

Question: The two parallelograms at the right share the same base. Without actually calculating the area of each, are the areas equal?[2]

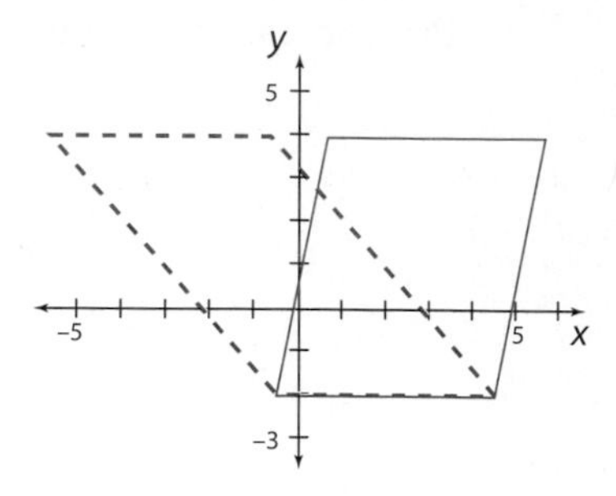

Translating Geometric Figures

Moving a geometric figure on the coordinate plane to a new location without changing its shape or turning it is called a **translation**. A **vertical translation** moves the figure up or down. A **horizontal translation** moves the figure left or right. Examples of vertical and horizontal translations are shown in the figures below.

SUGGESTED ACTIVITY
Have students find in a newspaper or a magazine some examples of patterns that include translated figures.

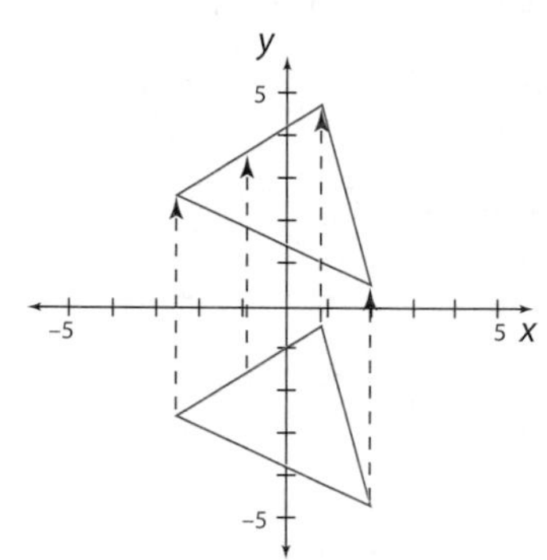

Vertical translation
Each point is moved
exactly the same distance.

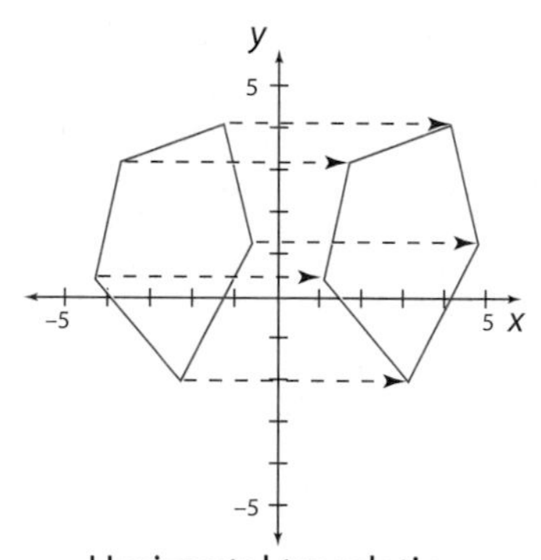

Horizontal translation
Each point is moved
exactly the same distance.

SUGGESTED ACTIVITY
Have students find in a newspaper or a magazine some examples of patterns with line symmetry and point symmetry.

For a vertical translation, the *y*-coordinate of each point is changed by exactly the same amount. For a horizontal translation, the *x*-coordinate is changed by exactly the same amount.

Example 2 Draw the rectangle with vertices $A(-2, -2)$, $B(1, 0)$, $C(-2, 5)$, and $D(-5, 3)$. Perform a horizontal translation by adding 3 to the *x*-coordinate of each vertex and then drawing the rectangle at the new location.

TAKE NOTE
It is common practice when translating points in the plane to use a prime (′) for the corresponding point. For instance, labeling points *A* and *A′* means that point *A* was translated to point *A′*. The same is true for points *B* and *B′*, *C* and *C′*, and *D* and *D′*.

Solution We will use a prime (′) to indicate the vertices of the rectangle that is formed by adding 3 to the *x*-coordinates of the vertices of the original rectangle.

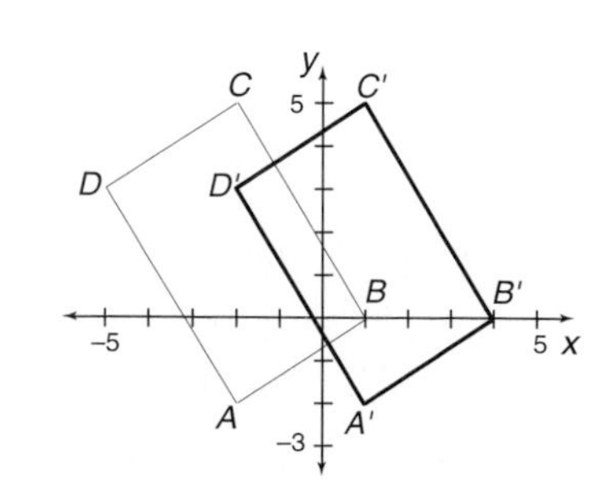

$A(-2, -2) \xrightarrow[\text{x-coordinate}]{\text{add 3 to the}} A'(1, -2)$

$B(1, 0) \xrightarrow[\text{x-coordinate}]{\text{add 3 to the}} B'(4, 0)$

$C(-2, 5) \xrightarrow[\text{x-coordinate}]{\text{add 3 to the}} C'(1, 5)$

$D(-5, 3) \xrightarrow[\text{x-coordinate}]{\text{add 3 to the}} D'(-2, 3)$

2. Yes because the base and height of each parallelogram is the same.

You-Try-It 2 Draw the triangle with vertices $A(-4, -1)$, $B(1, 3)$, and $C(-1, 5)$. Perform a vertical translation by subtracting 3 from the y-coordinate of each vertex and then drawing the triangle at the new location.

Solution See page S8.

POINT OF INTEREST
Translations of a pattern are used in wallpaper designs.

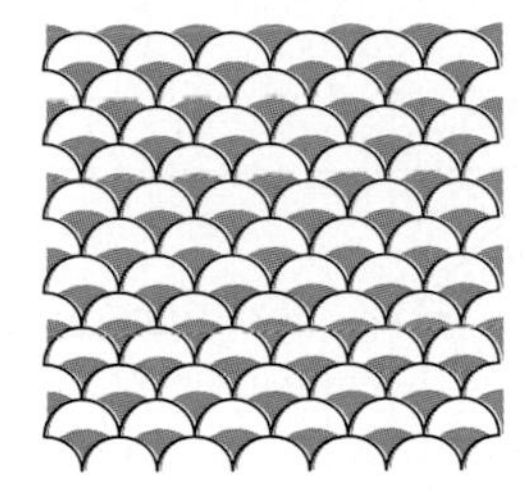

Translations are also used in the design of the background on Web pages. A small piece of the background is created and then a browser, such as Netscape or Internet Explorer, translates the pattern to every location on the computer screen.

Some translations are a combination of both a vertical and a horizontal translation. For the parallelogram with vertices A, B, C, and D, 3 was added to the x-coordinate and 4 was subtracted from the y-coordinate. The translated parallelogram has vertices A', B', C', and D'.

For the parallelogram $ABCD$, the base is

$b = |-2 - 2| = 4$ • Using the x-coordinates of points D and C.

and the height is

$h = |5 - (-1)| = 6$ • Using the y-coordinates of points A and D.

Therefore, the area is

$$A = bh$$
$$= 4 \cdot 6 = 24$$

The area is 24 square units.

For the parallelogram $A'B'C'D'$, the base is

$b = |1 - 5| = 4$ • Using the x-coordinates of points D' and C'.

The height is

$h = |1 - (-5)| = 6$ • Using the y-coordinates of points A' and D'.

Therefore, the area is

$$A = bh$$
$$= 4 \cdot 6 = 24$$

This shows that the area of the original parallelogram and that of the translated parallelogram are the same, which suggests that **a translation of a geometric figure does not change its area.**

Example 3 Translate the triangle with vertices at $A(-1, 2)$, $B(3, 4)$, and $C(4, -1)$ to the left 3 units and down 2 units. Draw the original triangle and the translated triangle.

Solution A translation of 3 units to the left means that 3 is subtracted from each x-coordinate and a translation of 2 units down means that 2 is subtracted from each y-coordinate.

$$A(-1, 2) \xrightarrow[\text{subtract 2 from 2}]{\text{subtract 3 from }-1} A'(-4, 0)$$

$$B(3, 4) \xrightarrow[\text{subtract 2 from 4}]{\text{subtract 3 from 3}} B'(0, 2)$$

$$C(4, -1) \xrightarrow[\text{subtract 2 from }-1]{\text{subtract 3 from 4}} C'(1, -3)$$

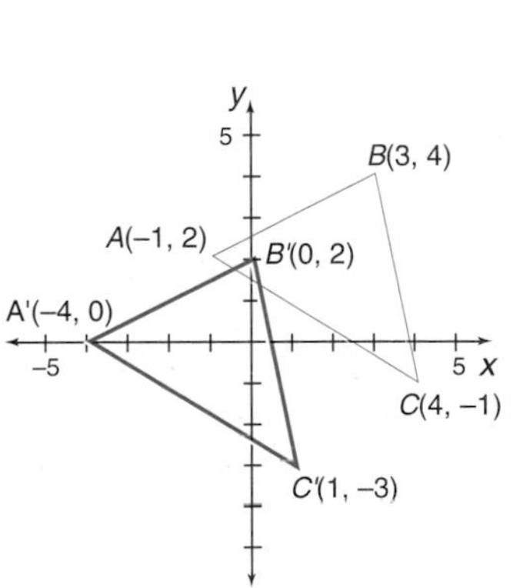

You-Try-It 3 Translate the square with vertices $A(-5, 2)$, $B(-2, 5)$, $C(1, 2)$, and $D(-2, -1)$ to the right 4 units and down 3 units.

Solution See page S8.

Stretching or Shrinking Geometric Figures

A **transformation** of a geometric figure is a change in either the position of the figure, like a translation, or the shape of the figure, like stretching it or shrinking it. Recall that translations occur by *adding* a number to either the x- or y-coordinates of all ordered pairs of the figure.

Stretching or **shrinking** a geometric figure is accomplished by *multiplying* the coordinates of the figure by a positive number. In some transformations, the x- and y-coordinates are multiplied by different numbers.

TAKE NOTE
Stretching and shrinking a figure is one way in which movie animators create "morphs". Morphing computer programs allow the animator to start with one figure and then, by performing a series of transformations, change the figure into something else.

Example 4 For the square with vertices $A(-4, -2)$, $B(-1, 1)$, $C(2, -2)$, and $D(-1, -5)$, multiply the x-coordinates by 3 and the y-coordinates by 2. Record the vertices of the new figure and then draw its graph.

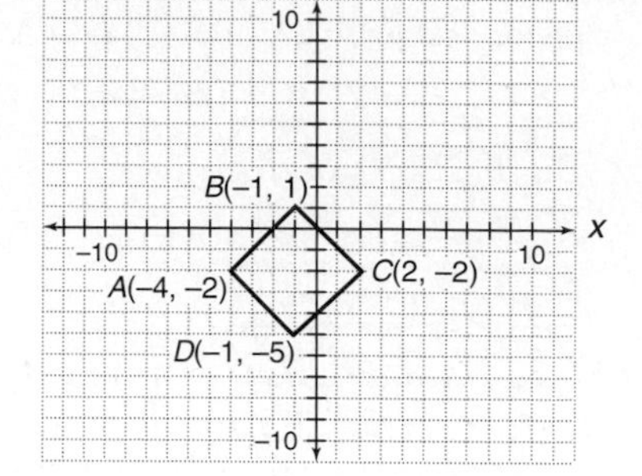

Solution We will use a prime (′) to indicate the vertices of the figure that is formed by multiplying the coordinates of the original square.

$$A(-4, -2) \xrightarrow[\text{multiply } -2 \text{ by } 2]{\text{multiply } -4 \text{ by } 3} A'(-12, -4)$$

$$B(-1, 1) \xrightarrow[\text{multiply } 1 \text{ by } 2]{\text{multiply } -1 \text{ by } 3} B'(-3, 2)$$

$$C(2, -2) \xrightarrow[\text{multiply } -2 \text{ by } 2]{\text{multiply } 2 \text{ by } 3} C'(6, -4)$$

$$D(-1, -5) \xrightarrow[\text{multiply } -5 \text{ by } 2]{\text{multiply } -1 \text{ by } 3} D'(-3, -10)$$

Graph the points A', B', C', and D'. Connect the points to draw the graph of the new figure.

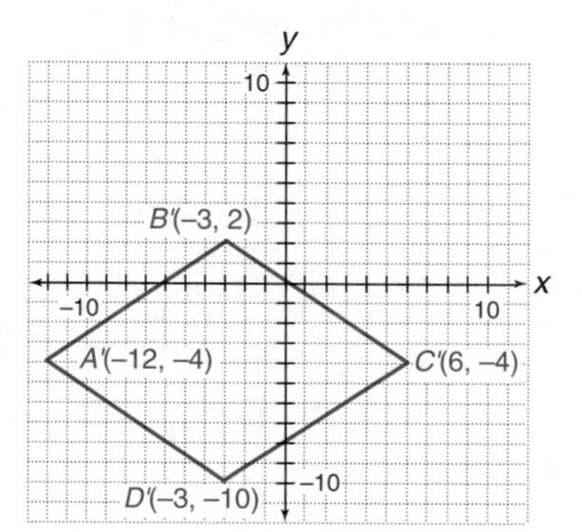

You-Try-It 4 Multiply the x-coordinates of the trapezoid with vertices $A(-6, -3)$, $B(-4, 4)$, $C(2, 4)$, and $D(4, -3)$ by $\frac{1}{2}$. Record the vertices of the new figure and then draw its graph.

Solution See page S9.

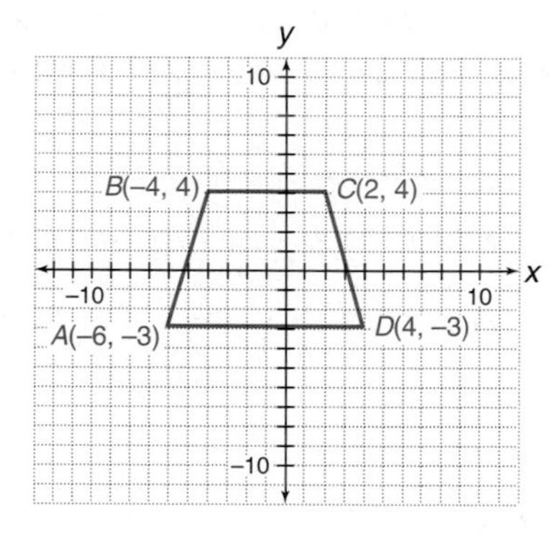

Example 4 shows that, unlike a translation, stretching or shrinking a figure may produce a different figure. For Example 4, a square was transformed into a quadrilateral with no parallel sides.

Question: After performing the transformation for You-Try-It 4, is the new figure still a trapezoid?[3]

3. Yes. See page S9 for the figure.

Examples of stretching (multiplying by a number greater than 1) or shrinking (multiplying by a number between 0 and 1) a geometric figure are shown below.

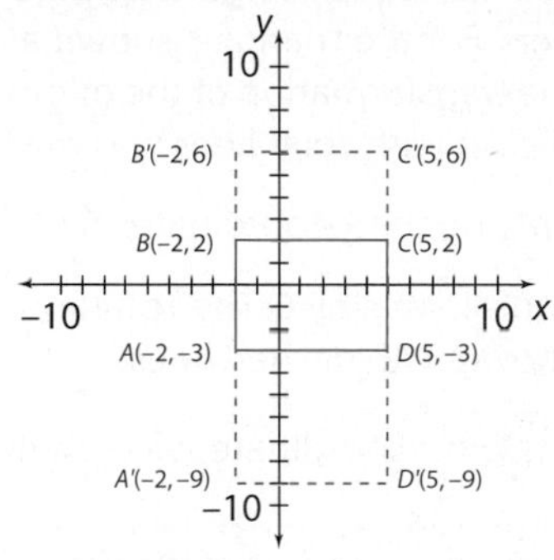

Figure 2.1
The *y*-coordinates of the original rectangle (solid lines) were multiplied by 3, *stretching* the rectangle vertically.

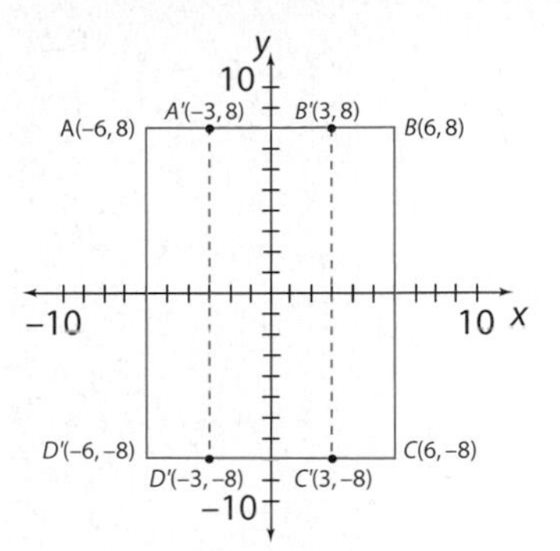

Figure 2.2
The *x*-coordinates of the original rectangle (solid lines) were multiplied by 1/2, *shrinking* the rectangle horizontally.

Because the stretched rectangle in Figure 2.1 (shown dashed) is larger than the original rectangle, the area of the stretched figure is larger than the area of the original one. Similarly, shrinking the rectangle in Figure 2.2 produced a smaller rectangle. Therefore its area is less than that of the original one. This suggests that **stretching or shrinking a geometric figure changes the area of the figure.**

The amount by which the area changes depends on the number that is used to multiply the coordinates. The calculations for the areas of the rectangles in Figure 2.1 are shown below.

TAKE NOTE
The calculation of distances between points on a vertical or horizontal line was shown in Section 2.1.

Original Rectangle

$$\begin{aligned} W &= d(A, D) = |-2-5| = 7 \\ L &= d(A, B) = |-3-2| = 5 \\ A &= LW \\ &= 7(5) \\ &= 35 \text{ square units} \end{aligned}$$

Stretched Rectangle

$$\begin{aligned} L &= d(A', B') = |-9-6| = 15 \\ W &= d(A', D') = |-2-5| = 7 \\ A &= LW \\ &= 15(7) \\ &= 105 \text{ square units} \end{aligned}$$

Note that the area of the stretched rectangle is 3 times the area of the original rectangle and that the *y*-coordinates were multiplied by 3.

$$\begin{aligned} \text{Area of stretched rectangle} &= 3(\text{area of original rectangle}) \\ 105 &= 3(35) \\ 105 &= 105 \end{aligned}$$

For the rectangle in Figure 2.2, we have a similar situation. The area of the rectangle after shrinking is $\frac{1}{2}$ the area of the original rectangle, and $\frac{1}{2}$ was the number by which the *x*-coordinates were multiplied.

$$\begin{aligned} \text{Area of rectangle after shrinking} &= \frac{1}{2}(\text{area of original rectangle}) \\ 96 &= \frac{1}{2}(192) \\ 96 &= 96 \end{aligned}$$

As discussed earlier in this section, it is possible to multiply both the x- and y-coordinates of a figure by different numbers. For the triangles shown at the right, the transformation of the original triangle shown with solid lines was made by multiplying the x-coordinates by $\frac{2}{3}$ and the y-coordinates by 6. The resulting triangle is shown with dashed lines.

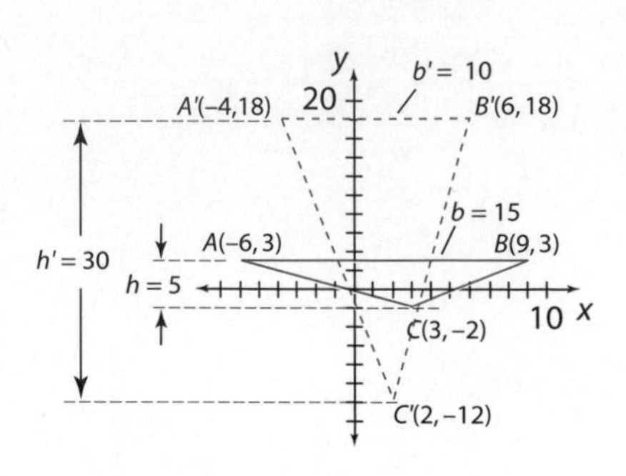

The x-coordinates were multiplied by $\frac{2}{3}$ and the y-coordinates were multiplied by 6. Thus, the area of the new triangle will be $\frac{2}{3} \cdot 6 = 4$ times the area of the original triangle.

Original Triangle	Transformed Triangle
$A = \frac{1}{2}bh$	$A = \frac{1}{2}b'h'$
$= \frac{1}{2}(15)(5)$	$= \frac{1}{2}(10)(30)$
$= \frac{75}{2}$ square units	$= 150$ square units

$$\text{Area of triangle after transformation} = 4(\text{area of original triangle})$$

$$150 = 4\left(\frac{75}{2}\right)$$

$$150 = 150$$

Reflections and Symmetry

The x-coordinates of points A, B, C, and D at the right have been multiplied by -1.

$A(2, 4)$ became $A'(-2, 4)$.
$B(-5, 1)$ became $B'(5, 1)$.
$C(3, -1)$ became $C'(-3, -1)$.
$D(-4, -3)$ became $D'(4, -3)$.

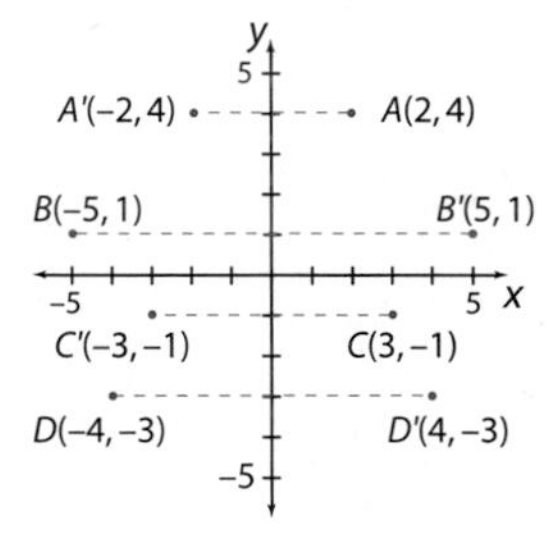

The x-coordinates of points A and A', B and B', C and C', and D and D' are opposites. Thus each pair of points is the same distance from the y-axis. Points that have opposite x-coordinates and the same y-coordinate are said to be **symmetric with respect to the *y*-axis.**

Question: **a.** Are the graphs of the ordered pairs (3, 4) and (−3, 6) symmetric with respect to the y-axis? **b.** Are the graphs of the ordered pairs (−2, −1) and (2, −1) symmetric with respect to the y-axis?[4]

Look at the graph at the right and note that the y-coordinates of points A and A', B and B', C and C', and D and D' are opposites. Thus each pair of points is the same distance from the x-axis. Points that have opposite y-coordinates and the same x-coordinate are said to be **symmetric with respect to the *x*-axis.**

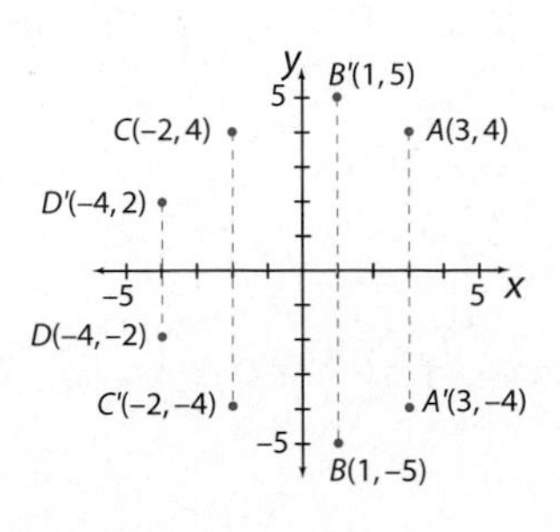

4. **a.** No. The x-coordinates are opposites but the y-coordinates are not equal. **b.** Yes. The x-coordinates are opposites and the y-coordinates are equal.

Question: **a.** Are the graphs of the ordered pairs (0, 4) and (0, –4) symmetric with respect to the *x*-axis? **b.** Are the graphs of the ordered pairs (–2, 3) and (2, 3) symmetric with respect to the *x*-axis?[5]

Multiplying the *x*-coordinate of each point of a geometric figure by –1 produces a geometric figure that is symmetric with respect to the *y*-axis to the original figure. The new figure is called a **reflection through the *y*-axis** of the original figure. (See Figure 2.3.) Multiplying the *y*-coordinate of each point of a geometric figure by –1 produces a geometric figure that is symmetric with respect to the *x*-axis to the original figure. The new figure is called a **reflection through the *x*-axis** of the original figure. Figure 2.4 illustrates a reflection about the *x*-axis.

POINT OF INTEREST

There are many instances of symmetry. Humans are symmetric with respect to a vertical line. Some letters of the alphabet are symmetric with respect to a horizontal line.

Multiplying each x-coordinate by –1 causes a *reflection through the y-axis* of the figure. The triangles *ABC* and *A'B'C'* are *symmetric* with respect to the y-axis.

Figure 2.3

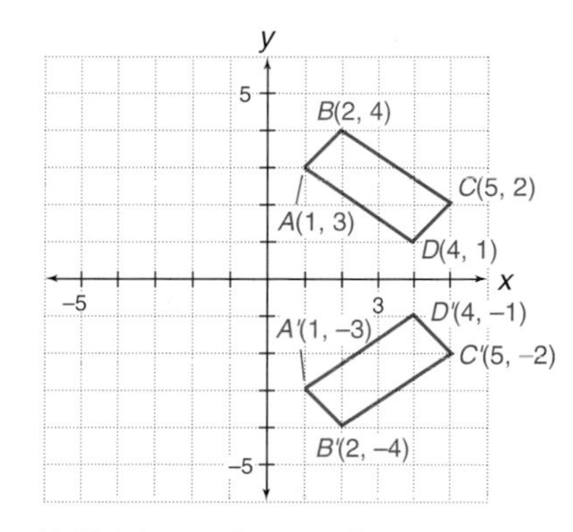

Multiplying each y-coordinate by –1 causes a *reflection through the x-axis* of the figure. The parallelograms *ABCD* and *A'B'C'D'* are *symmetric* with respect to the x-axis.

Figure 2.4

A convenient way to think of symmetry with respect to an axis is to think of folding the coordinate grid along that axis. The two figures are symmetric with respect to the axis if one figure exactly covers the other. For instance, if the coordinate grid in Figure 2.3 is folded along the *y*-axis, the two triangles will cover each other exactly. The triangles are symmetric with respect to the *y*-axis.

Similarly in Figure 2.4, if the coordinate grid is folded along the *x*-axis, the two parallelograms will cover each other exactly. The parallelograms are symmetric with respect to the *x*-axis.

POINT OF INTEREST

The Dutch artist M. C. Escher (1898 – 1972) used translations and reflections to create some unusual art. One of his creations is shown below.

Example 5 Using the diagram on the right, answer the following questions.

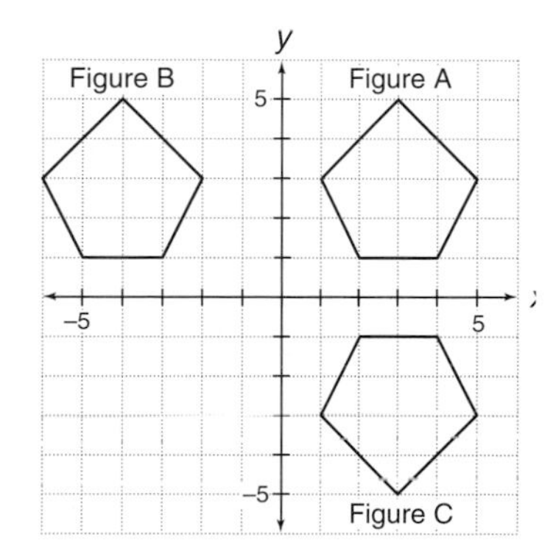

a. Are Figure A and Figure B symmetric with respect to the *y*-axis?

b. Are Figure A and Figure C symmetric with respect to the *x*-axis?

Solution

a. Because Figure A is closer to the *y*-axis than Figure B, if the coordinate grid were folded along the *y*-axis, Figure A would not lie on top of Figure B. Thus Figure A and Figure B are not symmetric with respect to the *y*-axis.

b. If the coordinate grid were folded along the *x*-axis, Figure A would lie on top of Figure C. Figure A and Figure C are symmetric with respect to the *x*-axis.

---

5. **a.** Yes. The *y*-coordinates are opposites and the *x*-coordinates are equal. **b.** No. The y-coordinates are not opposites.

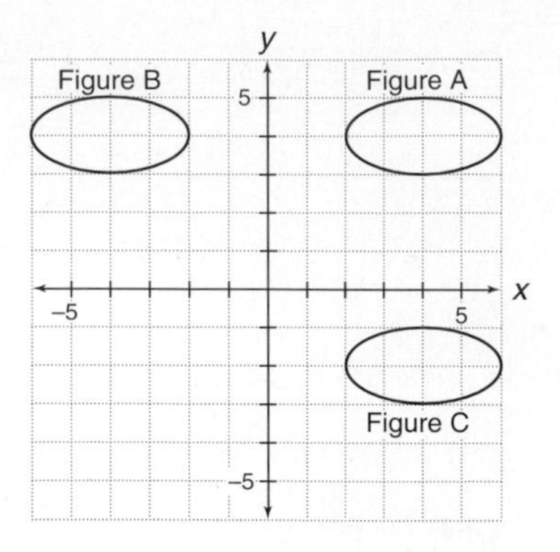

You-Try-It 5 Using the diagram on the right, answer the following questions.

a. Are Figure A and Figure B symmetric with respect to the *y*-axis?

b. Are Figure A and Figure C symmetric with respect to the *x*-axis?

Solution See page S9.

Example 5 and You-Try-It 5 show that symmetry with respect to a coordinate axis requires more than the figures looking the same. The two figures must cover one another when the coordinate grid is folded along an axis.

Symmetry is a very general concept and is used not only with geometric figures but with any type of graph.

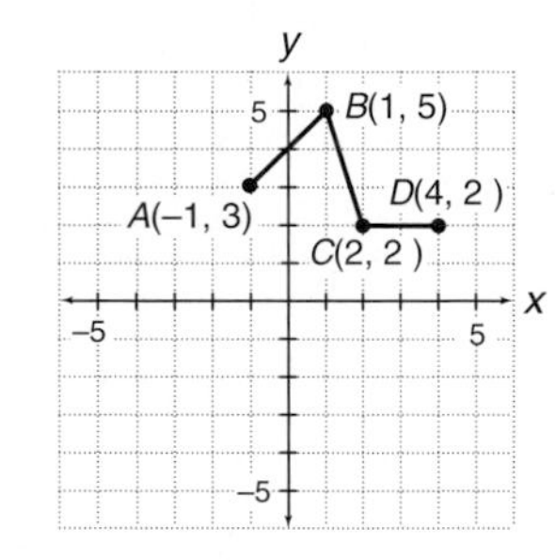

Example 6 Using the graph at the right, complete the following.

a. Draw the graph that is symmetric with respect to the *y*-axis to the given graph.

b. Draw the graph that is symmetric with respect to the *x*-axis to the given graph.

Solution

a. For each of the given ordered pairs, multiply the *x*-coordinate by −1. Graph the new ordered pairs and connect them so the resultant graph is symmetric with respect to the *y*-axis. See the graph below.

b. For each of the given ordered pairs, multiply the *y*-coordinate by −1. Graph the new ordered pairs and connect them so the resultant graph is symmetric with respect to the *x*-axis. See the graph below.

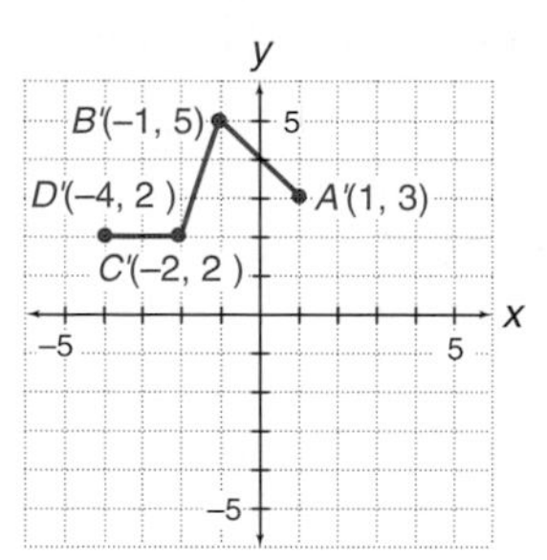

Graph for Example 6a.

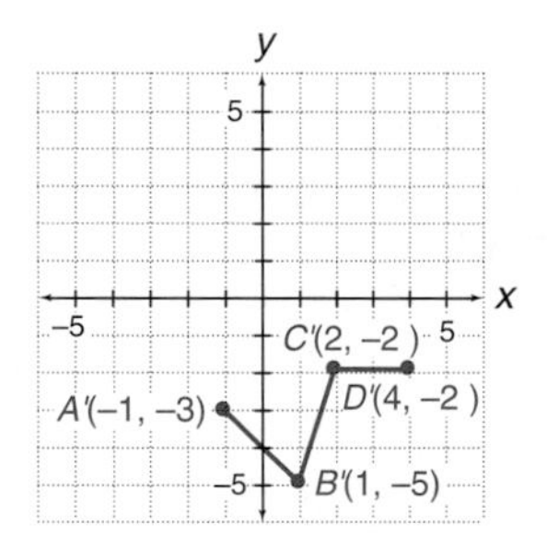

Graph for Example 6b.

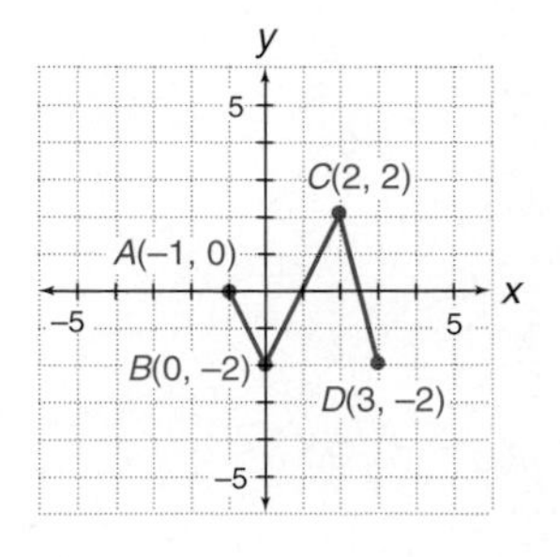

You-Try-It 6 Using the graph at the right, complete the following.

a. Draw the graph that is symmetric with respect to the *y*-axis to the given graph.

b. Draw the graph that is symmetric with respect to the *x*-axis to the given graph.

Solution See page S9.

2.5 EXERCISES

Topics for Discussion

1. What is a translation of a geometric figure?

1. A movement that does not change the shape of the figure or rotate the figure.

2. What is a transformation of a geometric figure? Is a transformation of a geometric figure a translation? Is a translation of a geometric figure a transformation?

2. Any change in the shape or position of a figure. A transformation is not necessarily a translation. A translation is a transformation.

3. Describe what it means to reflect a geometric figure through the x-axis. What is the relationship between the coordinates of the original figure and those of the reflected figure?

3. Draw the mirror image of the figure using the x-axis as the mirror. The x-coordinates of the two figures are equal. The y-coordinates are opposites.

4. Describe what it means to reflect a geometric figure through the y-axis. What is the relationship between the coordinates of the original figure and those of the reflected figure?

4. Draw the mirror image of the figure using the y-axis as the mirror. The y-coordinates of the two figures are equal. The x-coordinates are opposites.

5. Describe what it means for two graphs to be symmetric with respect to the x-axis.

5. The x-coordinates of the two figures are equal. The y-coordinates are opposites.

6. Describe what it means for two graphs to be symmetric with respect to the y-axis.

6. The y-coordinates of the two figures are equal. The x-coordinates are opposites.

Area of a Figure in the Coordinate Plane

7. Find the area of the rectangle. 48 square units

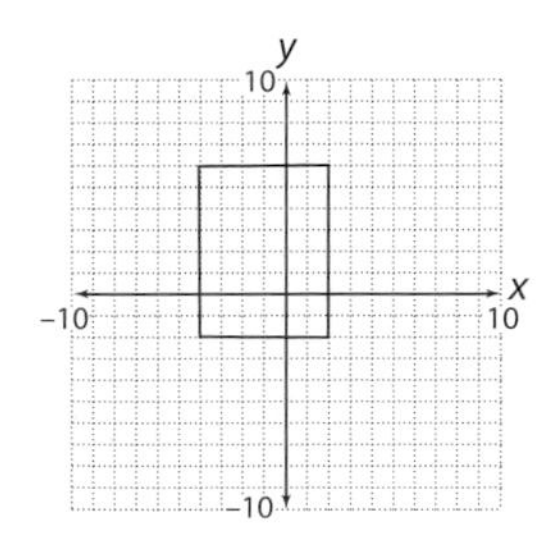

8. Find the area of the rectangle. 112 square units

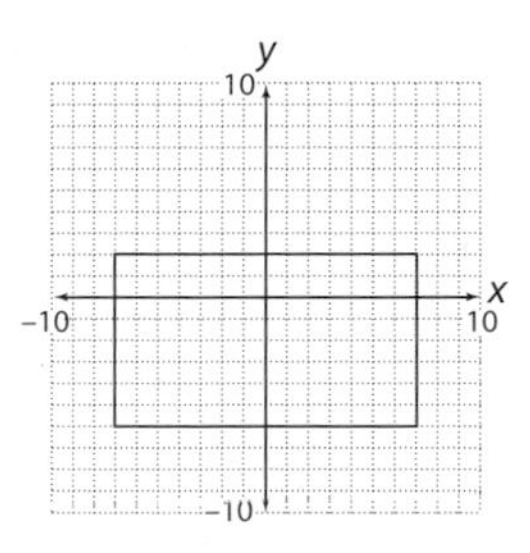

9. Find the area of the parallelogram. 54 square units

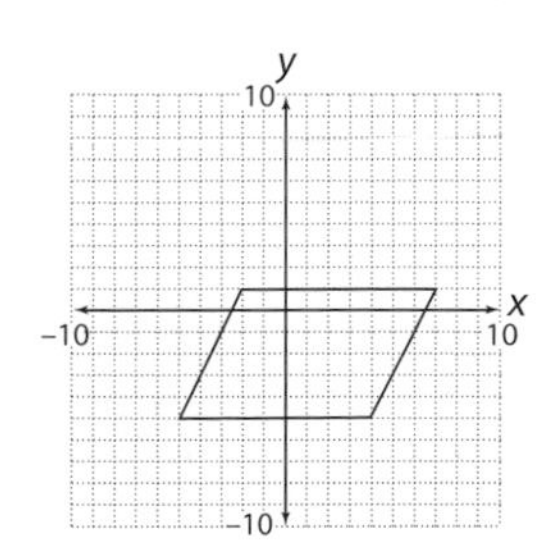

10. Find the area of the parallelogram. 70 square units

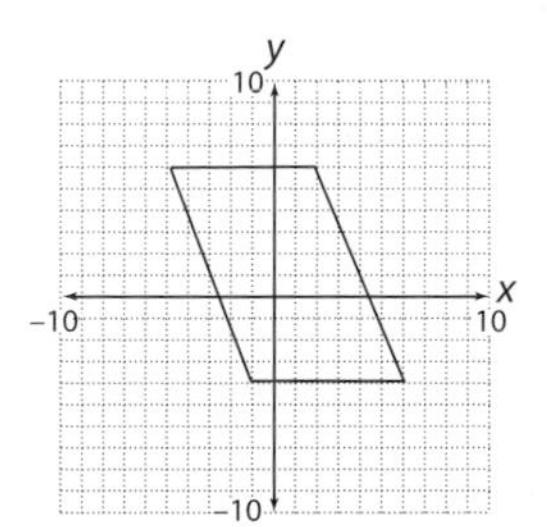

11. Find the area of the triangle. 15 square units

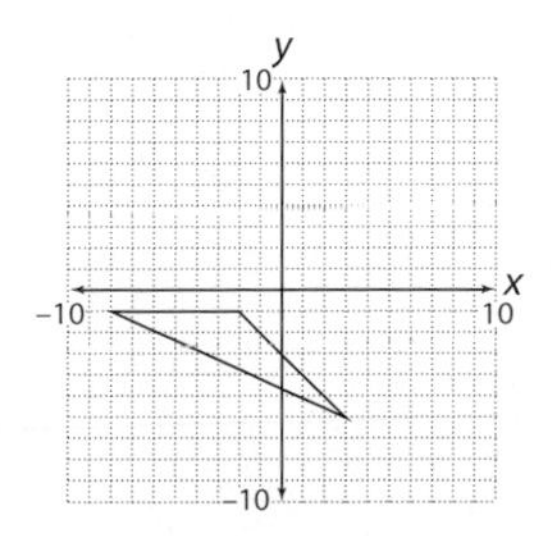

12. Find the area of the triangle. 66 square units

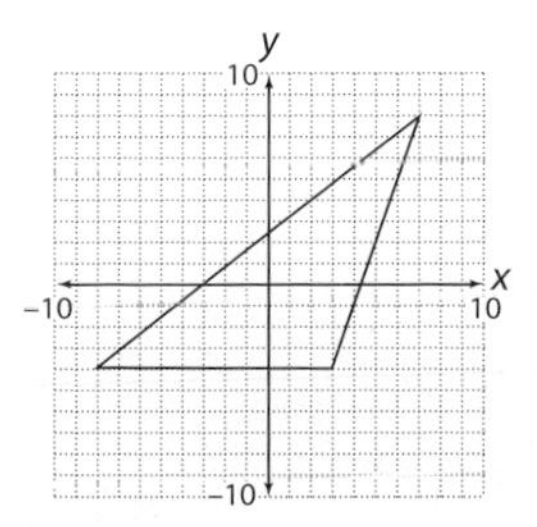

13. Find the area of the triangle. 45 square units

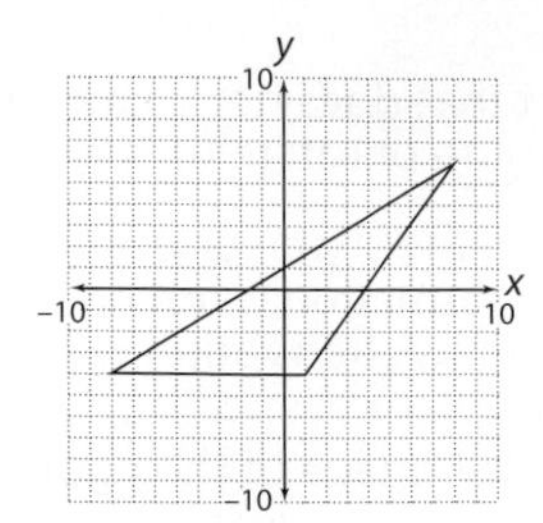

14. Find the area of the triangle. $67\frac{1}{2}$ square units

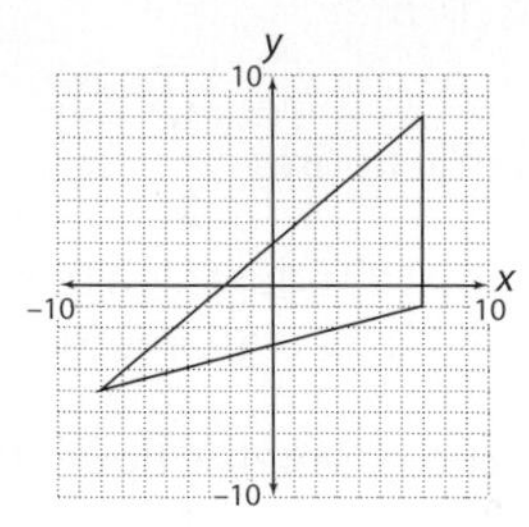

15. Find the area of the trapezoid. 99 square units

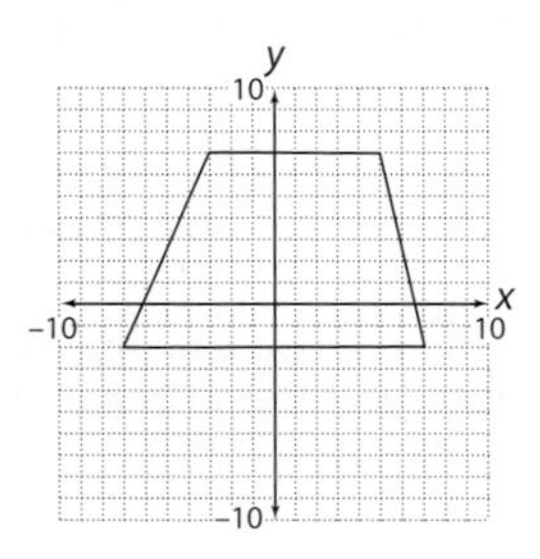

16. Find the area of the trapezoid. $59\frac{1}{2}$ square units

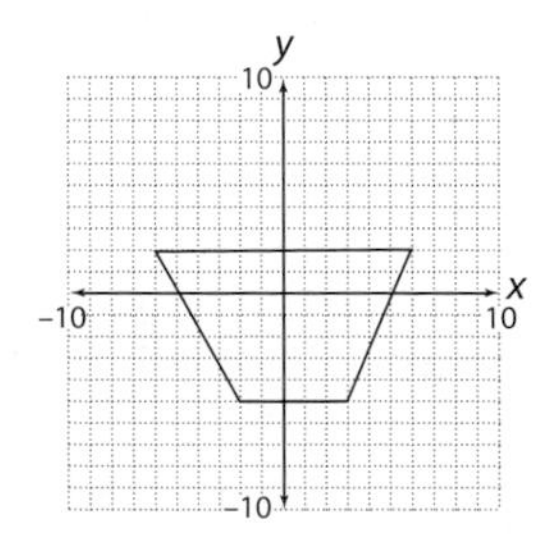

Translating Geometric Figures

17. Translate the rectangle 3 units to the right. What are the coordinates of the vertices of the new rectangle? (5, 6), (5, −2), (−1, −2), (−1, 6)

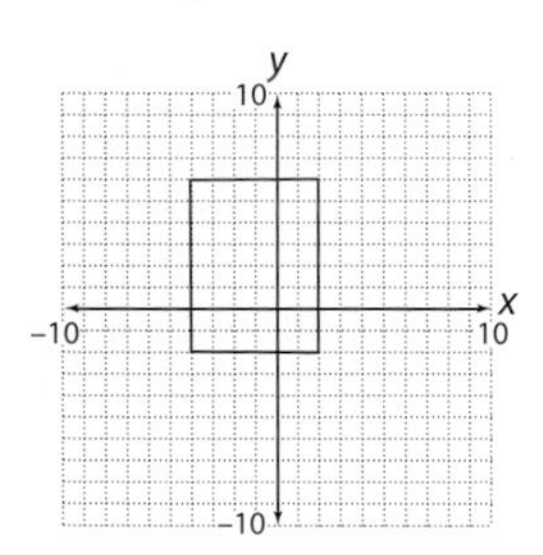

18. Translate the rectangle 2 units up. What are the coordinates of the vertices of the new rectangle?

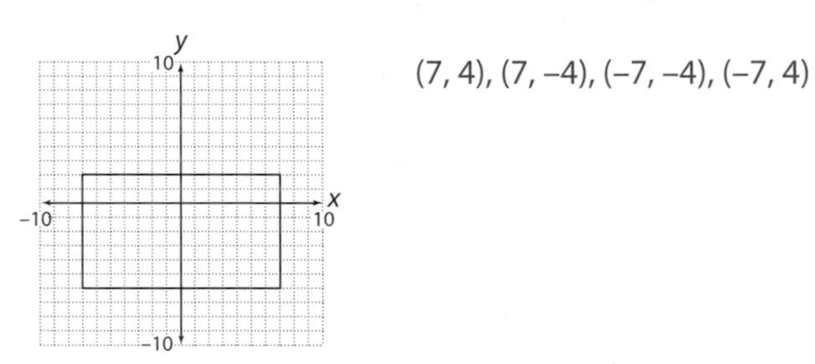

(7, 4), (7, −4), (−7, −4), (−7, 4)

19. Translate the parallelogram 2 units to the left and 3 units down. What are the coordinates of the vertices of the new parallelogram? (−4, −2), (5, −2), (−7, −8), (−2, −8)

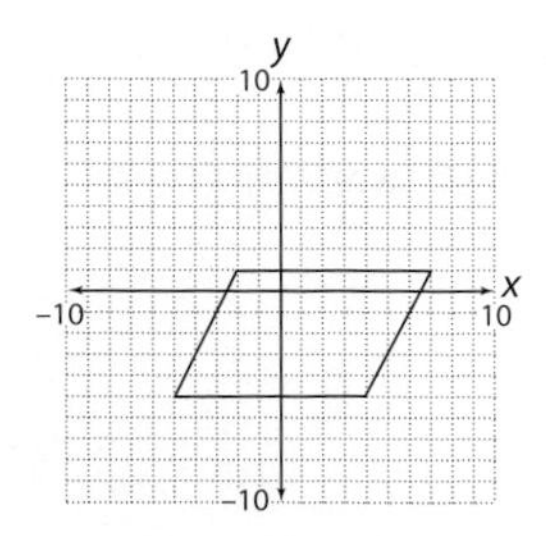

20. Translate the parallelogram 3 units to the right and 4 units up. What are the coordinates of the vertices of the new parallelogram? (−2, 10), (5, 10), (9, 0), (2, 0)

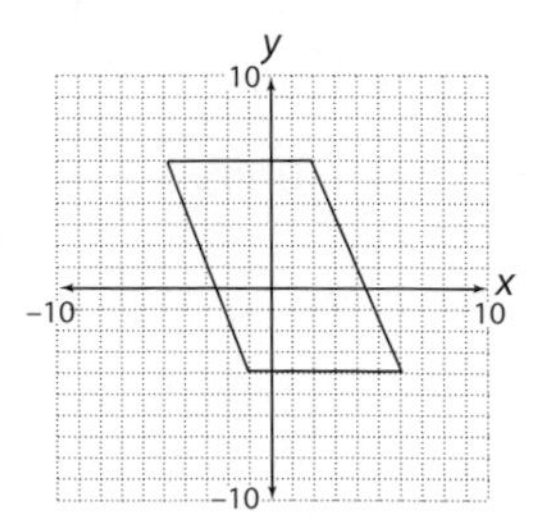

21. Translate the triangle 7 units to the right and 6 units up. What are the coordinates of the vertices of the new triangle? (−1, 5), (5, 5), (10, 0)

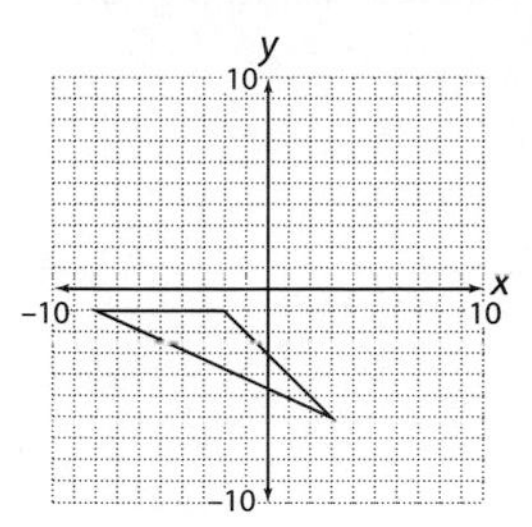

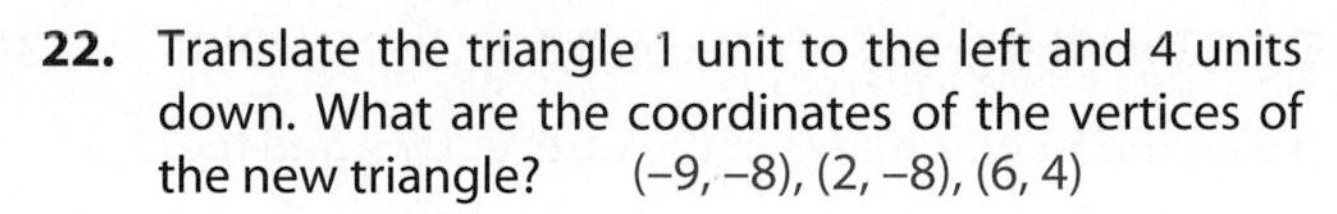

22. Translate the triangle 1 unit to the left and 4 units down. What are the coordinates of the vertices of the new triangle? (−9, −8), (2, −8), (6, 4)

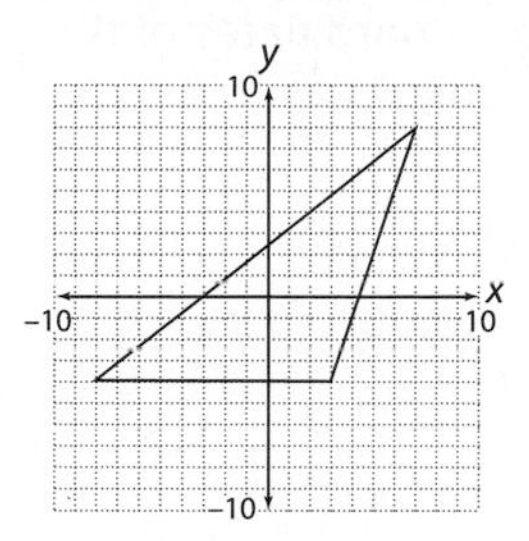

23. Translate the triangle 1 unit to the left and 3 units down. Calculate the area of each figure to verify that the translation did not change the area of the figure.

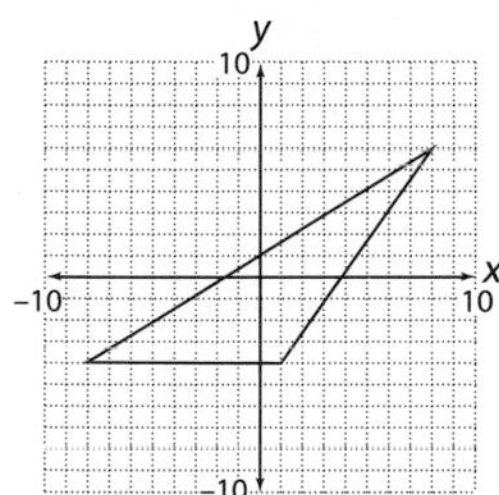

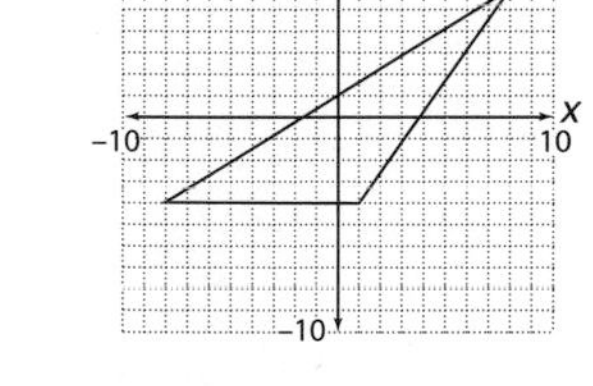

(−9, −7), (7, 3), (0, −7)
Area = 45 square units.

24. Translate the triangle 3 units to the right and 5 units down. Calculate the area of each figure to verify that the translation did not change the area of the figure.

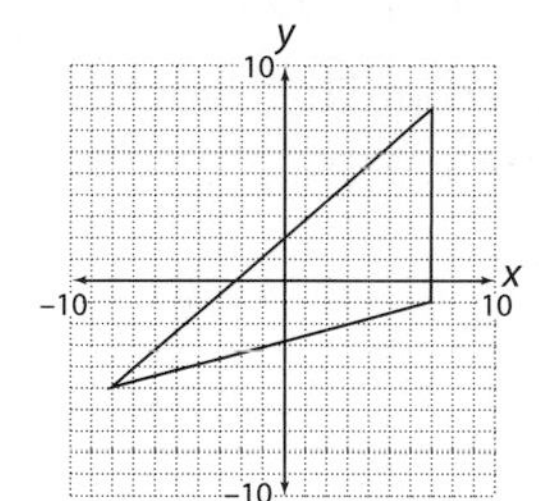

(−5, −10), (10, 3), (10, −6)
Area = 67.5 square units.

25. Translate the trapezoid 3 units to the right and 2 units up. Calculate the area of each figure to verify that the translation did not change the area of the figure.

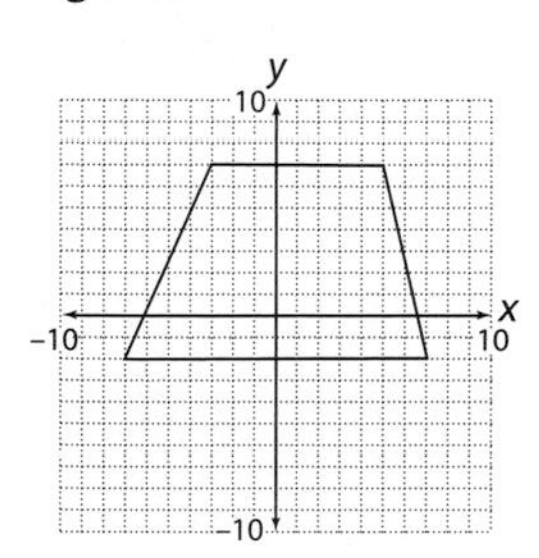

(−4, 0), (0, 9), (8, 9), (10, 0)
Area = 99 square units.

26. Translate the trapezoid 4 units to the left and 5 units down. Calculate the area of each figure to verify that the translation did not change the area of the figure.

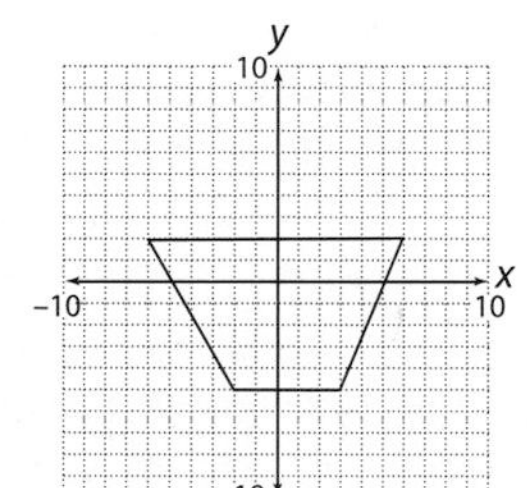

(−10, −3), (2, −3), (−1, −10), (−6, −10)
Area = 59.5 square units.

Stretching or Shrinking Geometric Figures

27. Transform the graph of the rectangle by multiplying the *x*-coordinates of the vertices by 2 and the *y*-coordinates by 3. What are the coordinates of the vertices of the new rectangle? (4, 6), (4, −3), (−6, −3), (−6, 6)

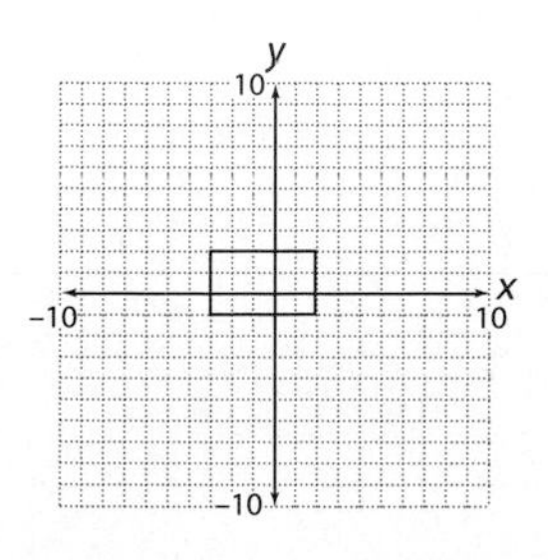

28. Transform the graph of the rectangle by multiplying the *x*-coordinates of the vertices by $\frac{1}{2}$ and the *y*-coordinates by $\frac{2}{3}$. What are the coordinates of the vertices of the new rectangle? (2, 2), (2, −4), (−3, −4), (−3, 2)

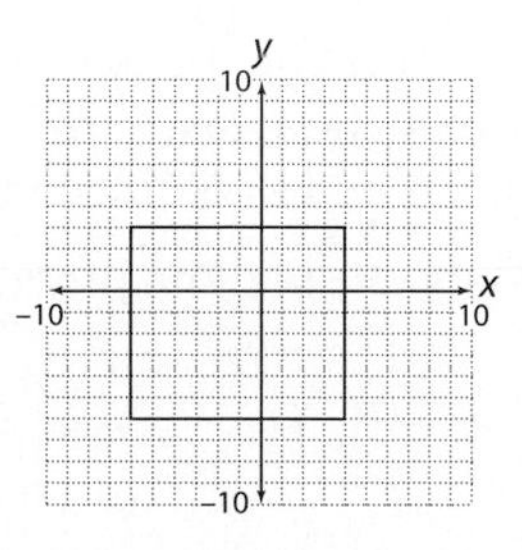

29. Transform the graph of the parallelogram by multiplying the x-coordinates of the vertices by $\frac{2}{3}$ and the y-coordinates by 3. What are the coordinates of the vertices of the new parallelogram?

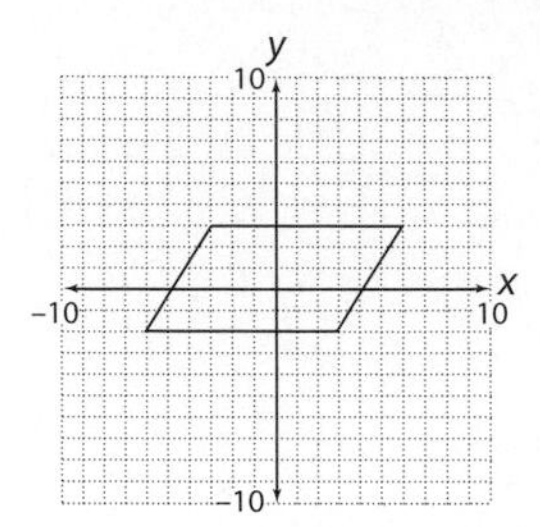

(4, 9), (2, –6), (–4, –6), (–2, 9)

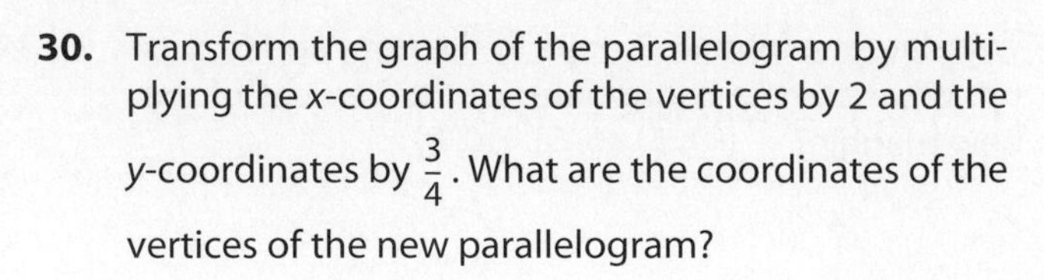

30. Transform the graph of the parallelogram by multiplying the x-coordinates of the vertices by 2 and the y-coordinates by $\frac{3}{4}$. What are the coordinates of the vertices of the new parallelogram?

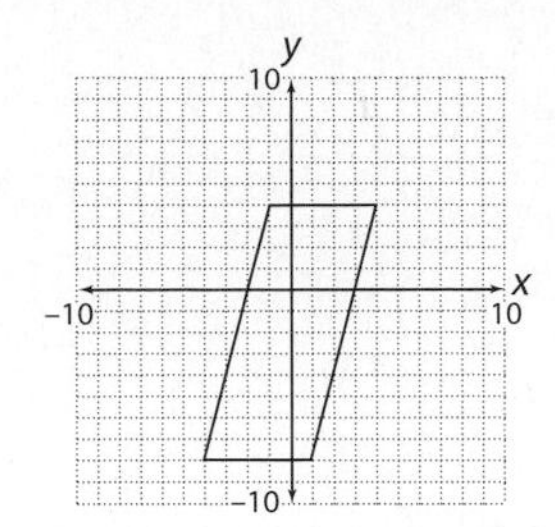

(8, 3), (2, –6), (–8, –6), (–2, 3)

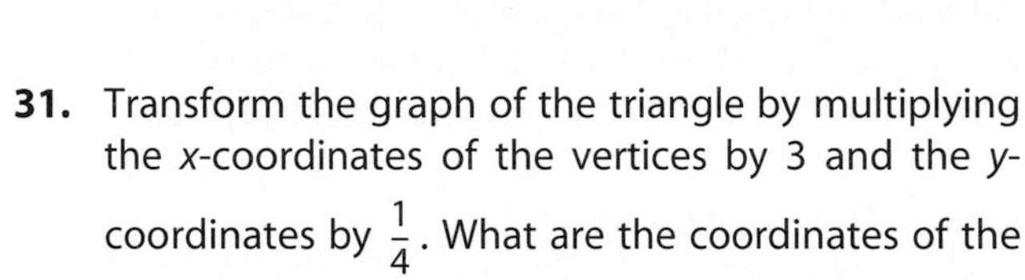

31. Transform the graph of the triangle by multiplying the x-coordinates of the vertices by 3 and the y-coordinates by $\frac{1}{4}$. What are the coordinates of the vertices of the new triangle?

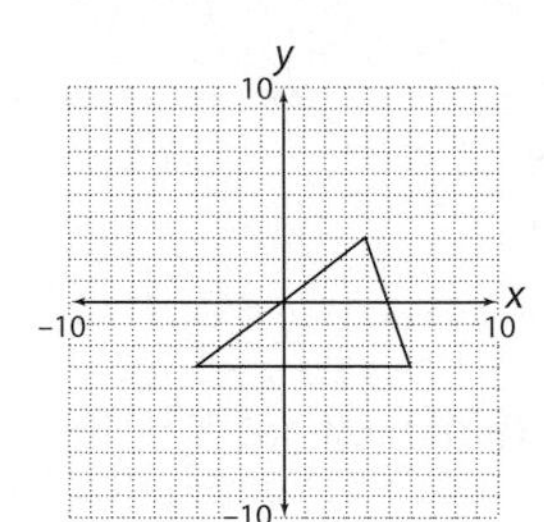

$(12, \frac{3}{4}), (18, -\frac{3}{4}), (-12, -\frac{3}{4})$

32. Transform the graph of the triangle by multiplying the x-coordinates of the vertices by $\frac{1}{2}$ and the y-coordinates by 2. What are the coordinates of the vertices of the new triangle?

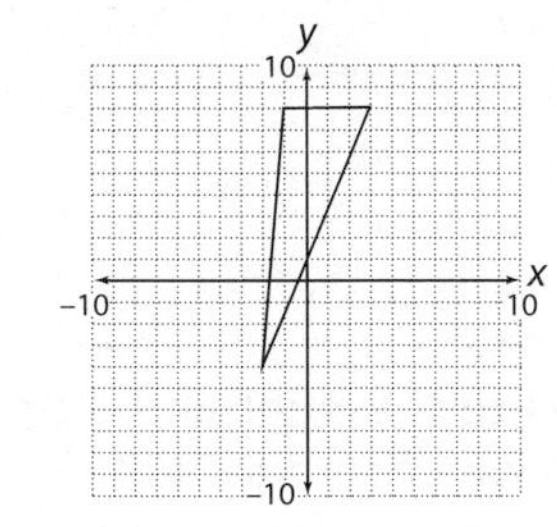

$(-\frac{1}{2}, 16), (\frac{3}{2}, 16), (-1, -8)$

33. Calculate the area of the original figure and the transformed figure for Exercise 27. What is the relationship between the areas? How is this relationship related to the factors used to multiply the x- and y-coordinates? The area of the transformed figure is 6 times the original area. The product of the factors is 6.

34. Calculate the area of the original figure and the transformed figure for Exercise 28. What is the relationship between the areas? How is this relationship related to the factors used to multiply the x- and y-coordinates? The area of the transformed figure is $\frac{1}{3}$ times the original area. The product of the factors is $\frac{1}{3}$.

35. Calculate the area of the original figure and the transformed figure for Exercise 29. What is the relationship between the areas? How is this relationship related to the factors used to multiply the x- and y-coordinates? The area of the transformed figure is 2 times the original area. The product of the factors is 2.

36. Calculate the area of the original figure and the transformed figure for Exercise 30. What is the relationship between the areas? How is this relationship related to the factors used to multiply the x- and y-coordinates? The area of the transformed figure is $1\frac{1}{2}$ times the original area. The product of the factors is $1\frac{1}{2}$.

Reflections and Symmetry

37. Draw the figure that is symmetric with respect to the x-axis.

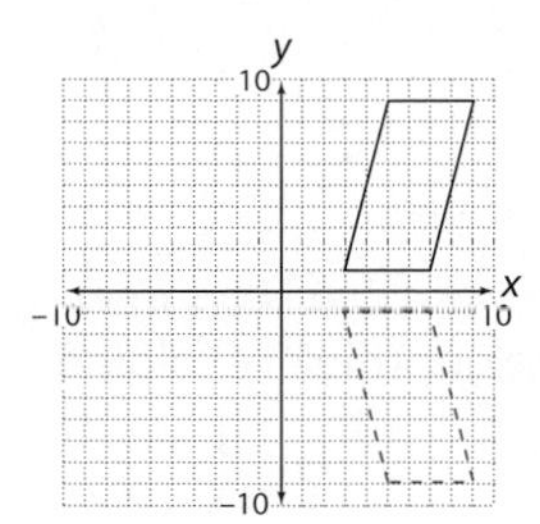

38. Draw the figure that is symmetric with respect to the y-axis.

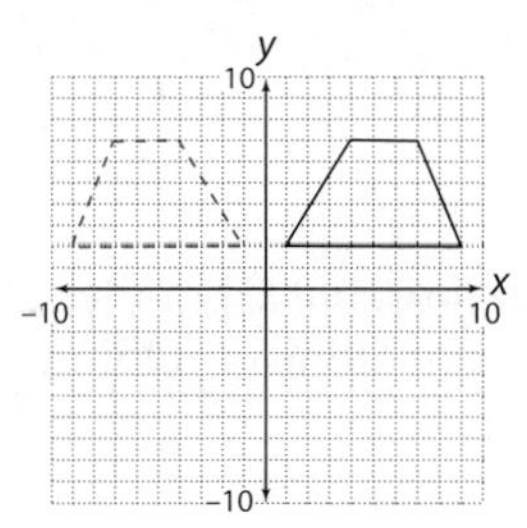

39. Draw the figure that is symmetric with respect to the y-axis.

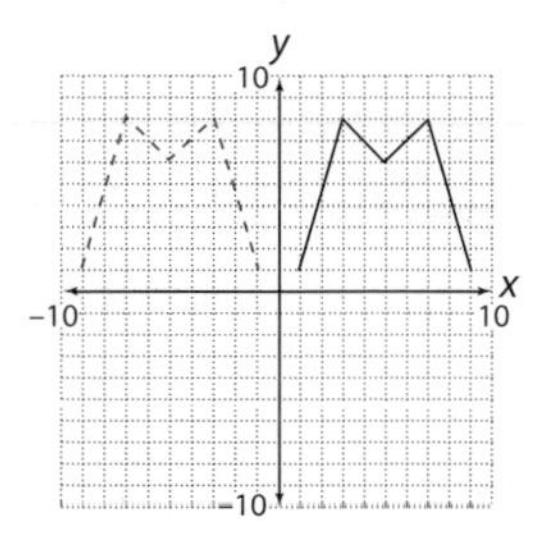

40. Draw the figure that is symmetric with respect to the x-axis.

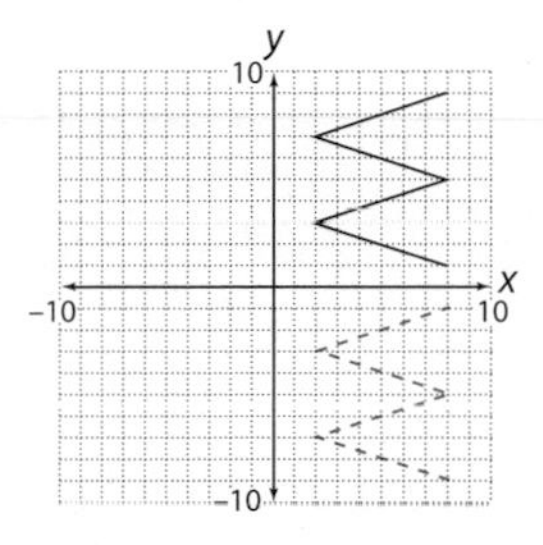

41. Draw the figure that is symmetric with respect to the x-axis.

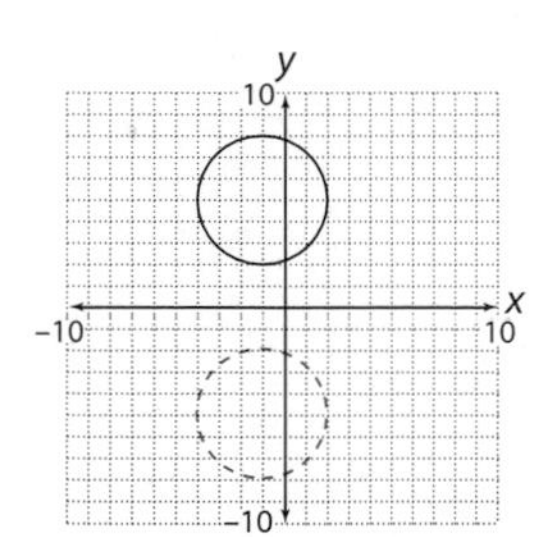

42. Draw the figure that is symmetric with respect to the y-axis.

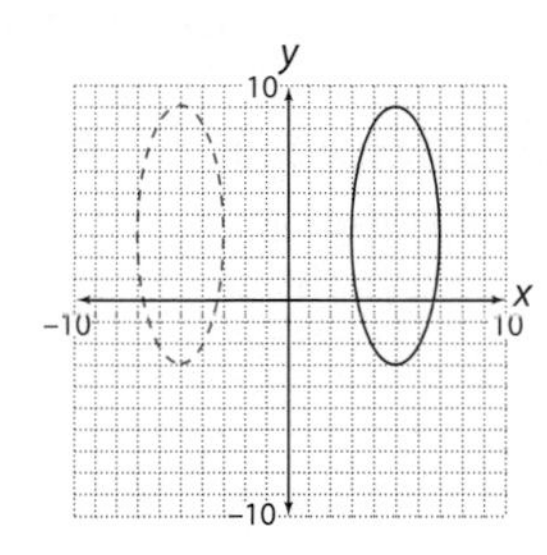

43. Draw the figure that is symmetric with respect to the y-axis.

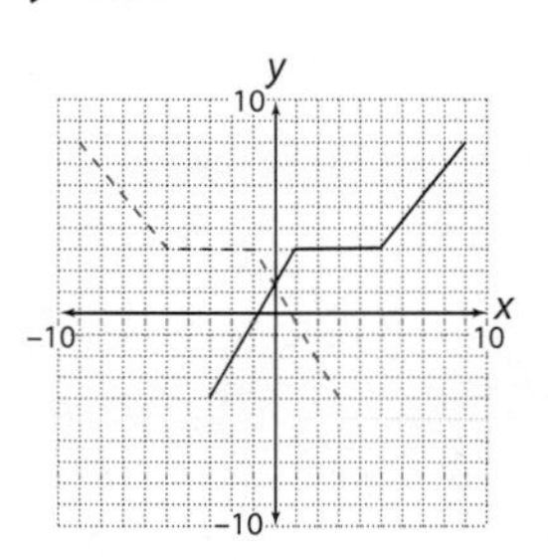

44. Draw the figure that is symmetric with respect to the x-axis.

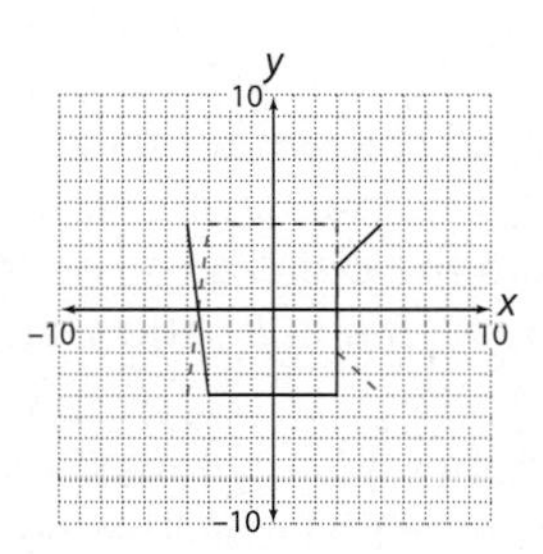

Applying Concepts

45. Suppose you were to perform a transformation on a figure by multiplying all of the x-coordinates by 0. What figure is the result of this transformation?
It is a line segment on the y-axis.

46. Suppose you were to perform a transformation on a figure by multiplying all of the y-coordinates by 0. What figure is the result of this transformation?
It is a line on the x-axis.

47. For the figure below, draw the figure that is symmetric to the dashed line. (Suggestion: Think of folding the paper along the dashed line.)

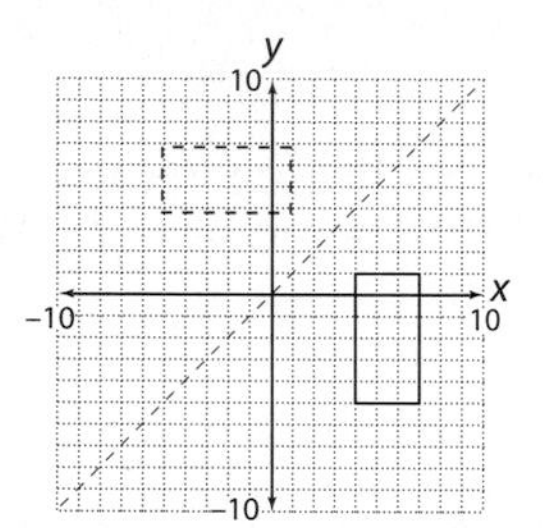

48. For the figure below, draw the figure that is symmetric to the dashed line. (Suggestion: Think of folding the paper along the dashed line.)

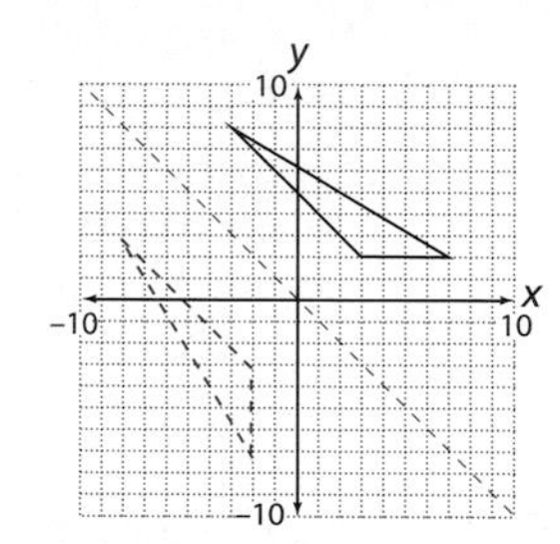

Explorations—Symmetry with Respect to a Point

49. Besides x-axis symmetry and y-axis symmetry, it is also possible to discuss **origin symmetry.** To draw a graph that is symmetric with respect to the origin, multiply both the x-coordinates and the y-coordinates of the original figure by -1 and then graph the new ordered pairs. The resulting graph is symmetric with respect to the origin.

a. For the figure below, draw the figure that is symmetric with respect to the origin.

b. For the figure below, draw the figure that is symmetric with respect to the origin.

50. Origin symmetry (see Exercise 49) is a special instance of a figure being symmetric with respect to a point. Two points are symmetric to a given point if they are the same distance from the point on a line through the point. In the figure at the right, points A and A', B and B', and C and C' are symmetric with respect to the point P.

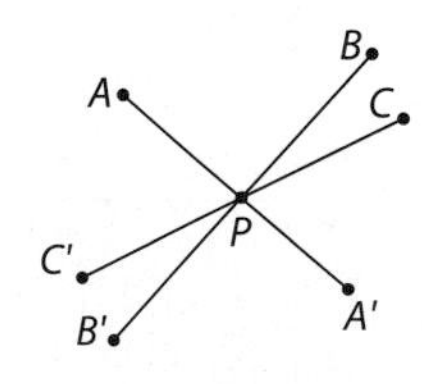

Some capital letters in the alphabet have symmetry with respect to a point in the middle of a box that surrounds the letter. The letter H has symmetry with respect to a point in the middle of the letter but the letter T does not.

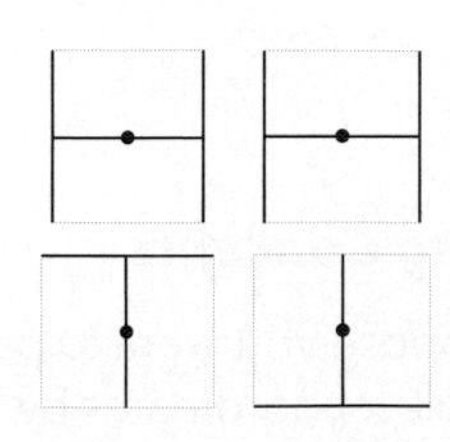

a. What capital letters in the English alphabet are symmetric with respect to a point in the middle of a box that surrounds the letter?

b. Are there any lowercase letters that are symmetric with respect to a point in the middle of a box that surrounds the letter?

Section 2.6 Reading and Drawing Graphs

Reading Graphs

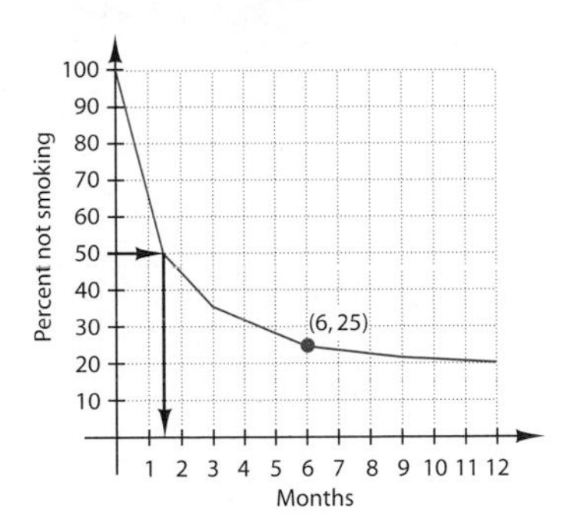

The graph at the right is based on data from the American Cancer Society. It shows the percent of a group of smokers who had refrained from smoking over a certain length of time. An ordered pair on this graph is (6, 25). This ordered pair means that 25% of the original group of smokers were still not smoking after 6 months.

Using the graph at the right, we can approximate after how many months 50% of the group was still not smoking. Draw a horizontal line from 50 to the curve. Then draw a vertical line from that point to the months axis. The vertical line touches the axis at approximately 1.5. Therefore, 50% of the group was still not smoking after 1.5 months.

Example 1 Using the graph above, approximate after how many months 30% of the group was still not smoking.

Solution Draw a horizontal line from 30% on the percent axis to the curve. Then draw a vertical line from the curve to the months axis. The vertical line touches the axis at approximately 4.5. Therefore, 30% of the smokers were still not smoking after 4.5 months.

You-Try-It 1 Using the graph above, approximately what percent of the smokers were still not smoking after 12 months?

Solution See page S9.

TAKE NOTE
The vertical axis is in millions of *1960* dollars. This means that amounts shown have been adjusted for inflation. The amount spent in 1996 was actually $700 million.

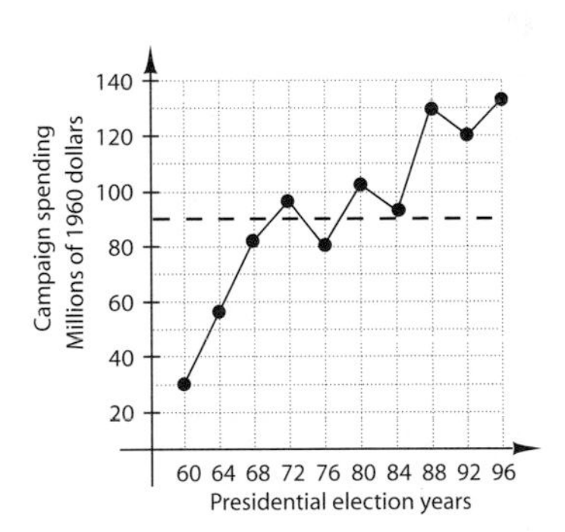

Example 2 The graph at the right shows the total amount spent by both Democrats and Republicans on presidential elections for each of the presidential election years from 1960 to 1996. (Source: Citizens' Research Foundation, *USA Today*, 11/28/97.) In which years did total spending exceed $90 million?

Solution The horizontal line that represents $90 million is shown on the graph as a bold dashed line. Values above this line are greater than $90 million. The election years corresponding to these values are 1972, 1980, 1984, 1988, 1992, and 1996.

You-Try-It 2 Using the graph in Example 2, estimate the amount spent on presidential elections in 1972.

Solution See page S9.

The graphs in Example 1 and Example 2 were based on gathered data. The data was then put in a graphical form. Sometimes, however, graphs are created from equations based on principles from science, economics, or other disciplines.

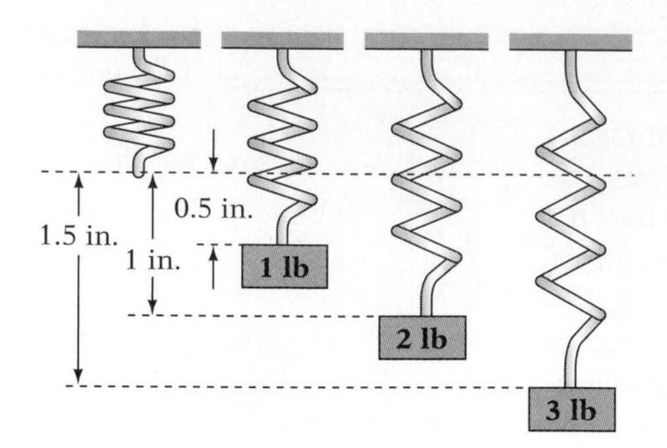

Here is an example from physics. Hooke's Law states that the distance (in inches) a spring will stretch depends on the weight (in pounds) that is placed on the end of the spring. The equation for the spring at the left is $d = 0.5w$.

The input/output table for the three weights shown at the left is given below. The graph of the table is shown at the right.

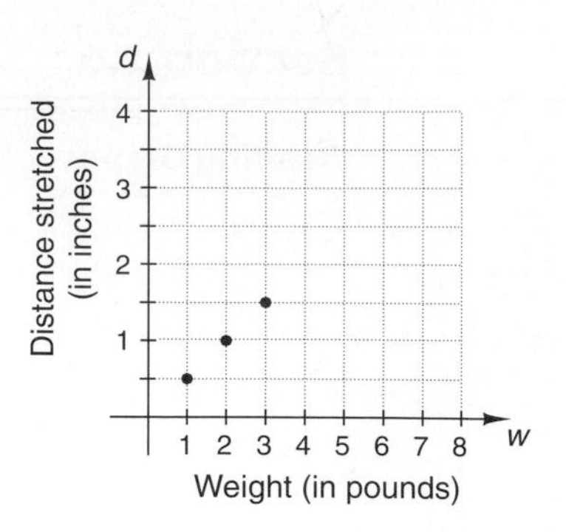

Input, w (weight)	1	2	3
Output, d (distance)	0.5	1	1.5

Using the Y= editor of a graphing calculator, we can enter the equation $d = 0.5w$ as $Y_1 = 0.5X$ and then create a table of values starting with 0 and using an increment of 0.5. A portion of the resulting input/output table and the graph of the table is shown below.

X	Y_1	
2	1	
2.5	1.25	
3	1.5	
3.5	1.75	
4	2	
4.5	2.25	
5	2.5	
X=2		

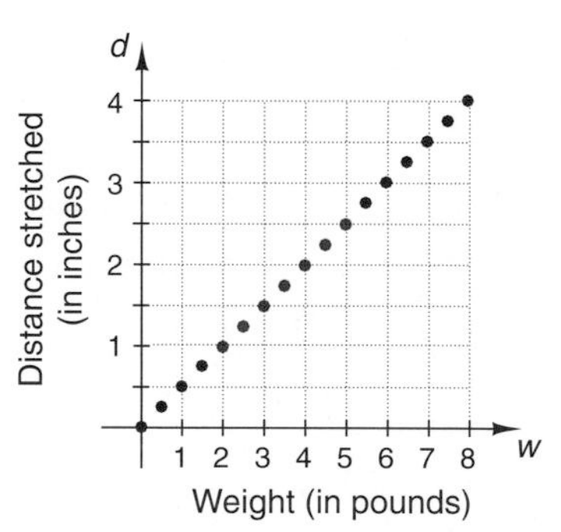

If the increment is changed to 0.25 from 0.5, the input/output table and graph will appear as shown below.

X	Y_1	
2	1	
2.25	1.125	
2.5	1.25	
2.75	1.375	
3	1.5	
3.25	1.625	
3.5	1.75	
X=2		

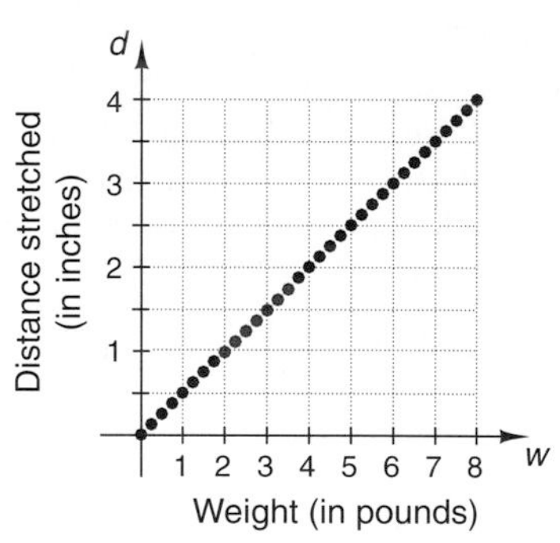

Note that as the increment changed to a smaller value, the number of points on the graph increased. If we continued to choose smaller and smaller increments, there would be so many points on the coordinate grid that the graph would appear as a line. This line is called the **graph of the equation** $d = 0.5w$. The graph is a visual representation of the equation.

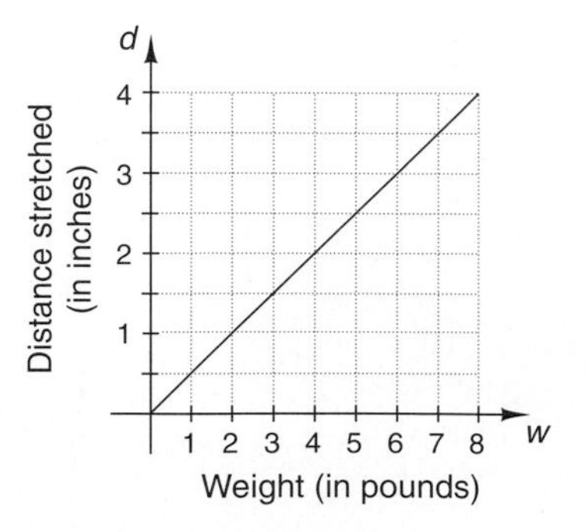

A graphing calculator can be used to draw the graph of an equation by entering an expression in the Y= editor window. The portion of the graph that is shown on the calculator's screen is called the *viewing window* or just the *window.* All graphing calculators have some built-in viewing windows. One of these windows is called the **standard window.** For this window, the graph of the equation is shown for points whose x-coordinates are between −10 and 10 and whose y-coordinates are between −10 and 10.

We will designate the minimum x-value for a window as Xmin and the maximum x-value as Xmax. For the standard viewing window, Xmin = −10 and Xmax = 10. Similarly, the minimum y-value is given as Ymin and the maximum y-value is given as Ymax.

Question: For the standard viewing window, what are Ymin and Ymax?[1]

There are two other entries on the viewing window: Xscl and Yscl. These represent the distance between the tic marks on the x- and y-axes, respectively. For instance, Xscl = 1 and Yscl = 1 means that there is 1 unit between tic marks on the x-axis and 1 unit between tic marks on the y-axis. (See Figure 2.5 below.) The values for Xscl and Yscl do not have to be equal.

Some typical screens that you might use to graph $y = 2x - 1$ in the standard viewing window are shown below. We encourage you to use your graphing calculator to produce this graph. Because each manufacturer engineers its calculators differently, your screens may not look exactly like the ones below.

TAKE NOTE
You can change the values of Xmin, Xmax, Ymin, and Ymax by selecting the *viewing window key* on your calculator. For the graph at the right, the values of Xmin, Xmax, Ymin, and Ymax are shown next to the coordinate axes. These numbers will not be part of your graph.

Y= editor screen

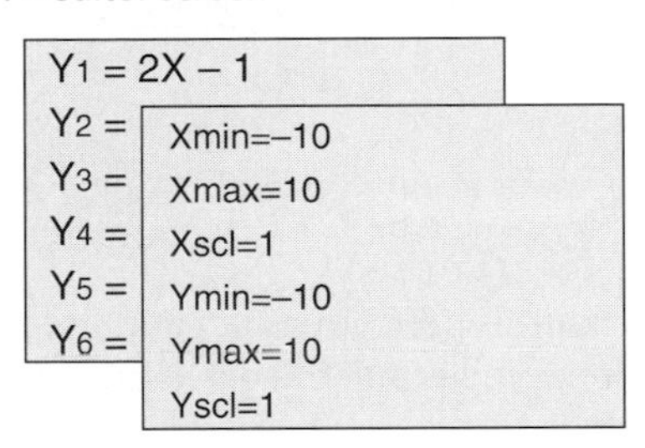

Viewing window screen

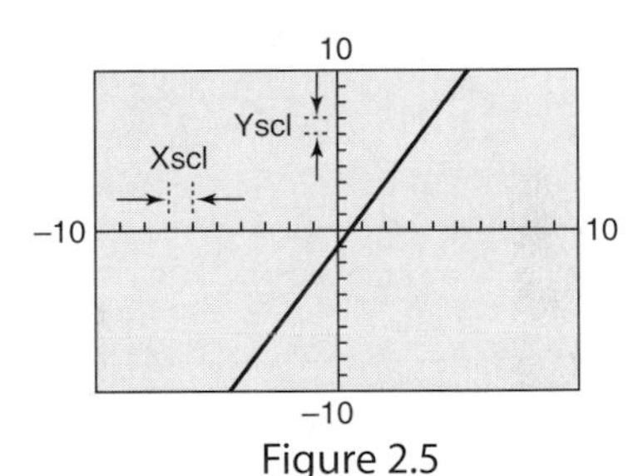

Figure 2.5

Once you have a graph of an equation in the viewing window, you can use the *TRACE* key to position a *cursor* on the curve. The equation of the graph is shown along with the coordinates of the ordered pair at the location of the cursor. By using the arrow keys, you can move the cursor along the graph. If you move the cursor outside the viewing window, the coordinates still show at the bottom of the screen but the cursor is no longer visible. For the graph at the right, the coordinates of the point under the cursor are (3.1914894, 5.3829787).

10
Y1=2X−1
cursor
−10
10
X=3.1914894
Y=5.3829787
−10

As you moved the cursor along the graph of $y = 2x - 1$, you may have noticed that many of the coordinates contain long decimal numbers. This is due to the way the calculator is designed and the fact that the standard viewing window is being used.

---

1. Ymin = −10; Ymax = 10.

Two other viewing windows that are built into most graphing calculators are the *decimal* window and the *integer* window. In these windows, the x-coordinates shown at the bottom of the screen are given in tenths (decimal window) or integers (integer window). The graph of $y = 2x - 1$ in the decimal window is shown at the right.

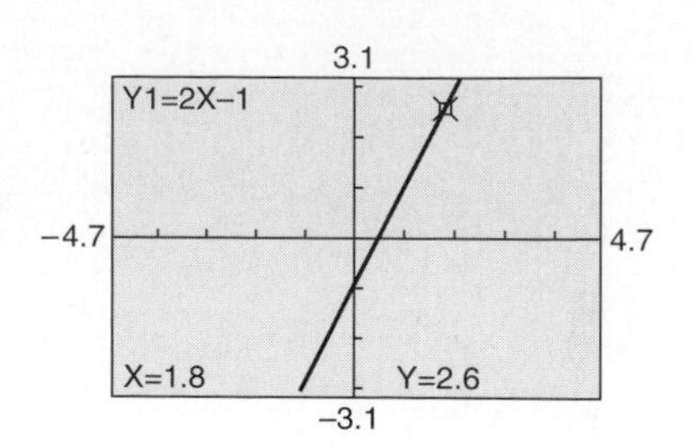

Example 3 Use a graphing calculator to produce the graph of $y = 0.2x^3$ for the decimal viewing window.

a. Trace along the graph to find the output value of $0.2x^3$ when the input is -1.7.

b. Trace along the curve to find the input for which the output value of $0.2x^3$ is 1.6.

Solution

a. Since $y = 0.2x^3$, the output value of $0.2x^3$ when the input is -1.7 will be the y-coordinate on the graph of $y = 0.2x^3$ when $x = -1.7$.

TAKE NOTE
Here are the Y= editor screen and the viewing window screen for Example 3.

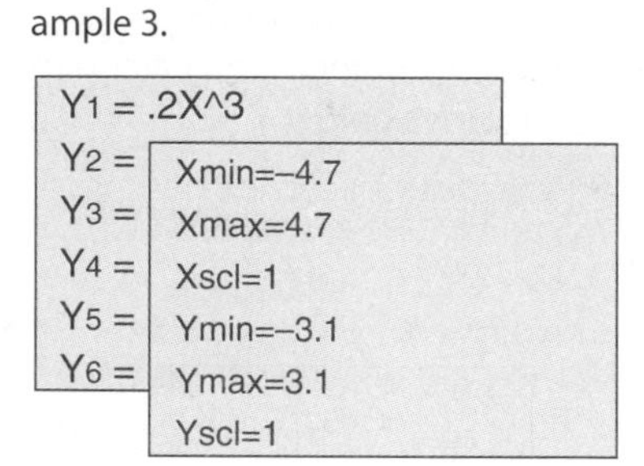

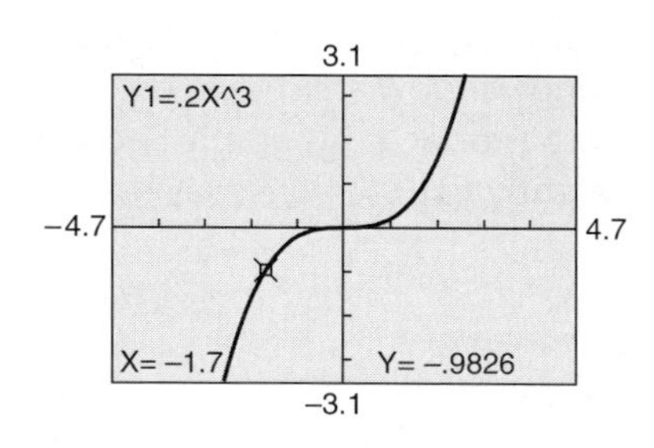

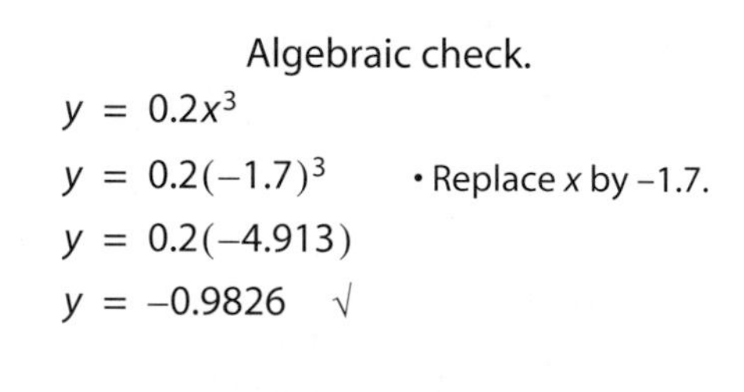
Algebraic check.

$$y = 0.2x^3$$
$$y = 0.2(-1.7)^3 \quad \text{• Replace } x \text{ by } -1.7.$$
$$y = 0.2(-4.913)$$
$$y = -0.9826 \quad \surd$$

The value of $0.2x^3$ when $x = -1.7$ is -0.9826. Note that this is confirmed by the algebraic check.

b. For this part, you must find the input value for a given output, 1.6. Trace along the graph until the given output value, Y, is shown at the bottom of the screen.

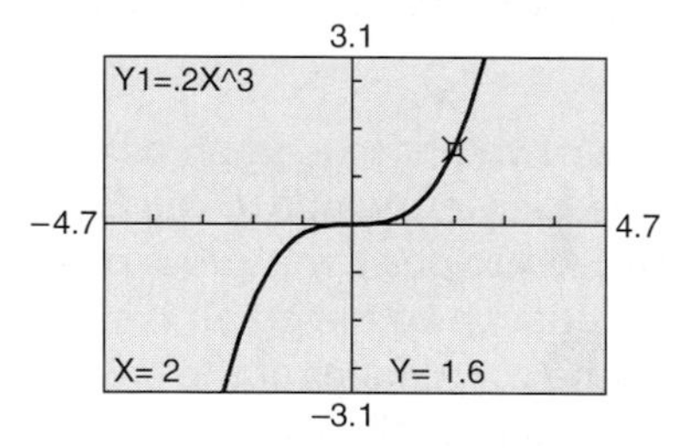

Algebraic check.

$$y = 0.2x^3$$
$$1.6 \stackrel{?}{=} 0.2(2)^3$$
$$1.6 \stackrel{?}{=} 0.2(8)$$
$$1.6 = 1.6 \quad \surd$$

From the graph above, the value of $0.2x^3$ is 1.6 when $x = 2$. This is confirmed by the algebraic check.

You-Try-It 3 Using a graphing calculator, produce the graph of $y = \frac{2x}{3} + 1$ for the decimal viewing window.

a. Trace along the graph to find the value of $\frac{2x}{3} + 1$ when $x = 2.1$.

b. Trace along the graph to find the value of x for which the value of $\frac{2x}{3} + 1$ is -0.8.

Solution See page S9.

TAKE NOTE
Check the Guidelines for Using Graphing Calculators appendix for some suggestions for the procedure shown at the right.

The algebraic check that was shown for Example 3a can be accomplished with a graphing calculator. With $0.2x^3$ entered as Y1 on the Y= editor screen, store −1.7 (the input value) in X. Place Y1 on the home screen and then hit ENTER. The output value is shown on the screen.

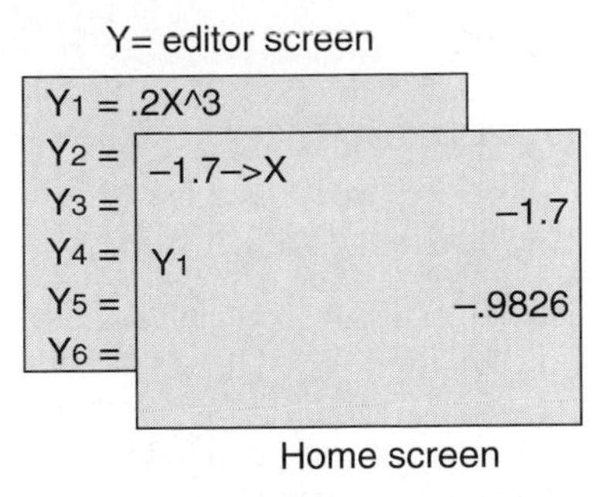

Question: For the screen at the right, which number is the input value and which number is the output value?[2]

```
2.5–>X
                2.5
Y1
              3.125
```

Calculating the output (*y*-value) for a given input (*x*-value) can be accomplished by the method shown above. However, finding the input for a given output frequently requires a graphing approach. Example 4 demonstrates such a method.

Example 4 Find the input value, *x*, for which the output of $2x^3$ is 11. Use a viewing window with Xmin = −1, Xmax = 4, Ymin = −5, Ymax = 15.

Solution The approach to this problem is similar to that of Example 1. A horizontal line is drawn from 11 (the output value) on the *y*-axis to the curve. Then we determine the *x*-coordinate (input value) where the line intersects the curve. Some typical screens that you might see are shown below. One *very important* point: for some calculators, the Ymin and Ymax values must be chosen so that the horizontal line is shown intersecting the graph.

The point at which the line intersects the curve can be found by using the INTERSECT feature of your calculator. See the Guidelines for Using Graphing Calculators appendix for some assistance.

TAKE NOTE
A calculator check of the answer to Example 4 is shown below. Because the input value, *x*, can only be approximated, the output value is also an approximation. The approximation, however, should be very close to 11.

```
1.7651742–>X
          1.7651742
Y1
         11.0000006
```

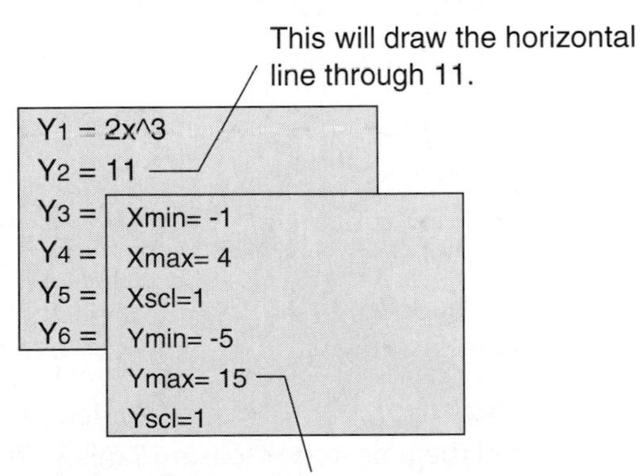

Choose this number to be greater than 11 so the point of intersection shows on the screen.

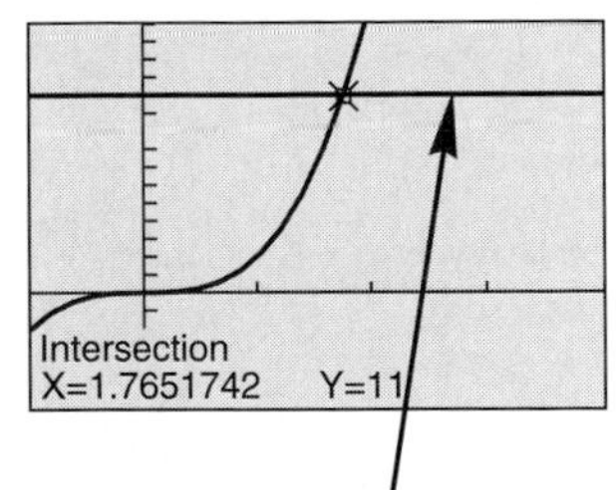

Unlike in Example 1, a graphing calculator will draw the horizontal line across the screen rather than just from the axis to the curve.

The input value is approximately −1.7651742

You-Try-It 4 Find the input value, *x*, for which the output of $x^3 + 2$ is −3. Use a viewing window with Xmin = −3, Xmax = 1, Ymin = −5, Ymax = 5.

Solution See page S9.

---

2. The input value is 2.5; the output value is 3.125.

Tracing along a curve to find a value of x for a given y value may produce more than one x-value. The graph of $y = 0.5x^2 - 2x - 5$ is shown below. Note that there are two values of x for which y is 5.5. They are −3 and 7.

SUGGESTED ACTIVITY
Ask students to create the graph at the right and verify the situation presented. Here are some additional questions you might ask students.
Are there other values of y that have more than one x-value?
Is there a y-value that has only one x-value?

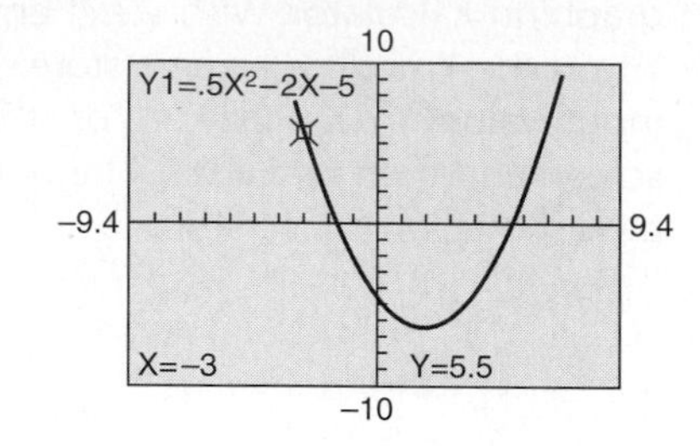

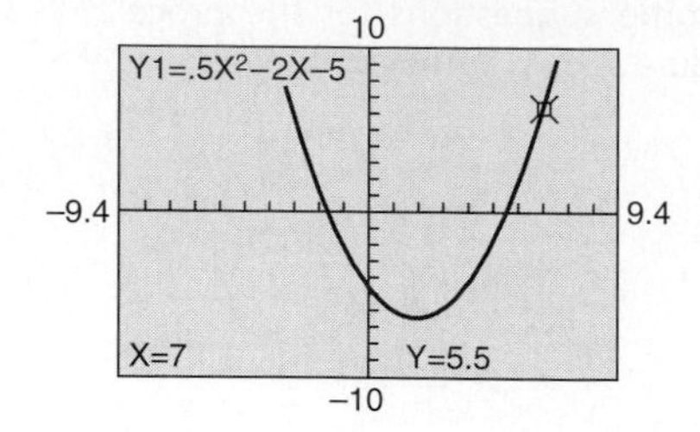

Applications to Uniform Motion

A train that travels constantly in a straight line at 50 mph is in *uniform motion.* **Uniform motion** means that the speed and direction of the object do not change.

The uniform motion equation is $d = rt$, where d is the distance traveled, r is the rate of travel, and t is the time spent traveling. This equation can be used to find, for instance, the distance a train whose speed is 50 mph will travel in 4 hours.

$$d = rt$$
$$= 50(4) \quad \text{• Replace } r \text{ by 50 (the rate) and } t \text{ by 4 (the time).}$$
$$= 200$$

The train will travel 200 miles in 4 hours.

SUGGESTED ACTIVITY
The *Instructor's Resource Manual* contains an activity that asks students to write an interpretation of four distance/time graphs.

Substituting 50 for r in the equation $d = rt$ produces the equation $d = 50t$. By graphing this equation with a graphing calculator (using Y1 for d and X for t) and tracing along the graph, we can determine the distance the train travels for many values of t. Two possible points are shown on the graph at the right. The ordered pair (4, 200) means that the train travels 200 miles in 4 hours.

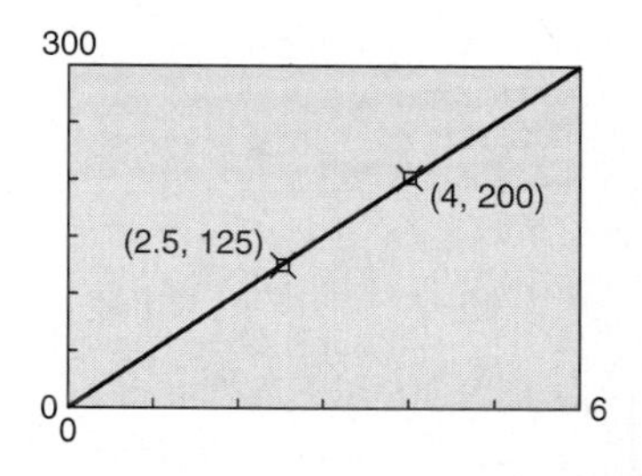

Question: What is the meaning of the ordered pair (2.5, 125) for the graph above?[3]

Example 5 The average running speed of Fatuma Roba, the winner in the women's division of the 1997 Boston Marathon, was approximately 10.75 miles per hour. The equation $d = 10.75t$ describes the distance, d (in miles), she had run t hours after she started the race. Graph this equation using a viewing window of Xmin = 0, Xmax = 3, Ymin = 0, Ymax = 27. To the nearest tenth of an hour, estimate the time it took Fatuma to run a distance of 15 miles.

Solution For this problem, we must find the time it took to run 15 miles. That is, we are given the output, 15 miles, and must determine the input. This is similar to Example 4.

Draw the graph of Y1 = 10.75X and Y2 = 15. Then use the INTERSECT feature to determine the coordinates of the point where the curves intersect. To the nearest tenth of an hour, it took Fatuma 1.4 hours to run 15 miles.

Y1 = 10.75X
Y2 = 15
Y3 =
Y4 =
Y5 =
Y6 =

Xmin= 0
Xmax= 3
Xscl=1
Ymin= 0
Ymax= 27
Yscl=1

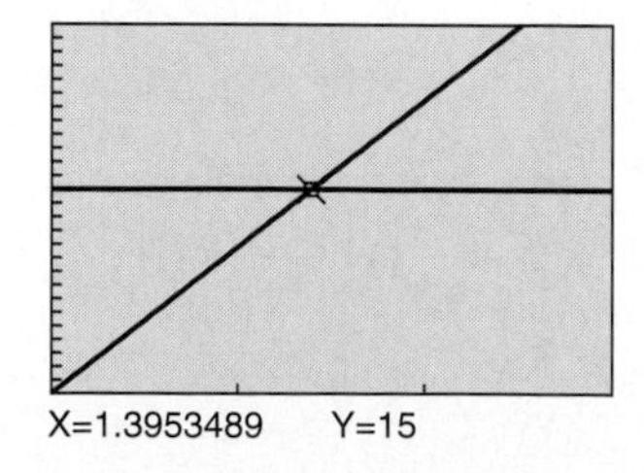

3. The train travels 125 miles in 2.5 hours.

You-Try-It 5 A ferry is traveling from Port Huron to Port Raven at an average speed of 12 miles per hour. The distance (in miles) the ferry is from Port Raven t hours after leaving Port Huron is given by $d = 46 - 12t$. Graph this equation and use your graph to estimate, to the nearest tenth of an hour, the time it takes until the ferry is 15 miles from Port Raven. Use the viewing window Xmin = 0, Xmax = 4, Ymin = 0, and Ymax = 50.

Solution See page S9.

Applications to Percent Concentration

The **percent concentration** of a liquid or alloy (a mixture of metals) is a measure of what percent of the quantity is pure. For instance, a 5% concentration of hydrogen peroxide means that 5% of the amount is hydrogen peroxide and the remaining 95% is something else.

Pure gold is called 24-karat gold. If you purchase a necklace that is 18-karat gold, the concentration of gold is $\frac{18}{24} = 0.75 = 75\%$. A gold ingot is 24-karat gold; the concentration of gold is 100%.

To form mixtures that are not pure, some other substance is added. As the substance is added, the percent concentration of the mixture will change. For instance, consider a chef making a chocolate sauce. Suppose the chef starts with 10 ounces of pure chocolate and begins adding sugar, 1 ounce at a time. Then the concentration of chocolate will change as more sugar is added.

TAKE NOTE
An input/output table for the concentration of chocolate is shown below.

Y1 = 10/(10+x)(100)
Y2 =
Y3 =
Y4 =
Y5 =
Y6 =

X	Y1
0	100
1	90.909
2	83.333
3	76.923
4	71.429
5	66.667
6	62.5

X=0

Chocolate	Sugar	Percent concentration of chocolate
10	0	$\frac{10}{10+0}(100) = 100$
10	1	$\frac{10}{10+1}(100) \approx 91$
10	2	$\frac{10}{10+2}(100) \approx 83$
10	3	$\frac{10}{10+3}(100) \approx 77$

Question: If the chef adds one more ounce of sugar, what is the percent concentration of the chocolate in the mixture now?[4]

An equation for the percent concentration of chocolate is $P = \frac{10}{10+x}(100)$, where P is the percent concentration of chocolate and x is the amount of sugar that has been added. A graph of this equation is shown at the right. Note that as more sugar is added (x increases), the percent concentration of chocolate decreases (P decreases).

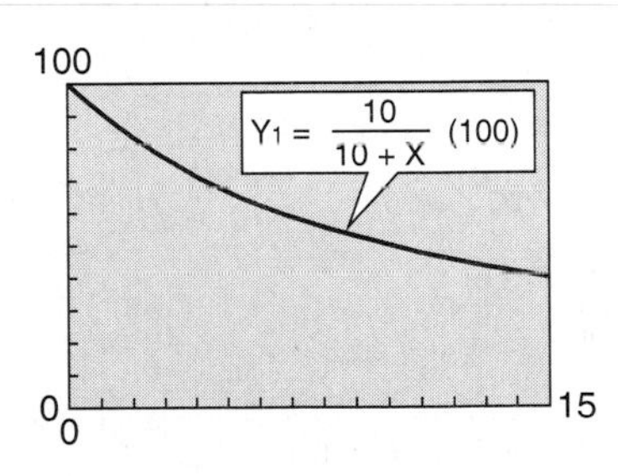

Question: Is it possible that the graph of the percent concentration of chocolate would ever intersect the x-axis?[5]

INSTRUCTOR NOTE
Have students respond to this question in class to ensure that all students understand the concept.

---

4. $\frac{10}{10+4} = \frac{10}{14} \approx 0.71 = 71\%$. The chocolate concentration is 71%.
5. No. That would mean that the concentration of chocolate would be less than zero, which is not possible.

The discussion so far has been on the percent concentration of chocolate. We could have also looked at the percent concentration of sugar.

Chocolate	Sugar	Percent concentration of sugar
10	0	$\frac{0}{10+0}(100) = 0$
10	1	$\frac{1}{10+1}(100) \approx 9$
10	2	$\frac{2}{10+2}(100) \approx 17$

Question: If the chef adds one more ounce of sugar, what is the percent concentration of sugar in the resulting mixture?[6]

An equation for the percent concentration of sugar is $P = \frac{100x}{10+x}$. A graph of that equation is shown at the right. As more sugar is added (x increases), the percent concentration of sugar increases (P increases).

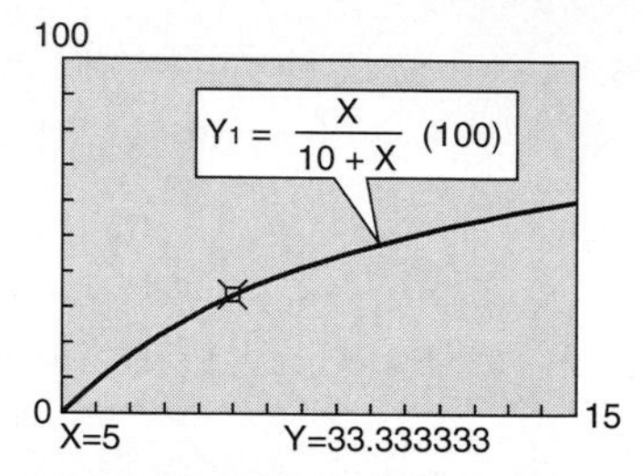

Question: Using the graph above, what is the percent concentration of the sugar for the point shown?[7]

Example 6 A saltwater solution is formed by adding salt to a container of water. The percent concentration of salt in the water is given by the equation $P = \frac{100x}{5+x}$, where P is the percent concentration of salt and x is the number of grams of salt that has been added to the solution. Use a graphing calculator to find the number of grams of salt, to the nearest hundredth, that must be added to the solution to produce a mixture that is 20% salt.

TAKE NOTE
Here is an algebraic check that the solution to Example 6 is correct.

$$P = \frac{100x}{5+x}$$
$$= \frac{100(1.25)}{5+1.25} \quad \text{• Replace } x \text{ by 1.25.}$$
$$= \frac{125}{6.25} = 20$$

Solution To find the number of grams, draw the graph of Y1 = 100X/(5 + X) and Y2 = 20. Then use the INTERSECT feature to determine the coordinates of the point where the curves intersect. 1.25 grams of salt are required to make a solution that has a 20% salt concentration.

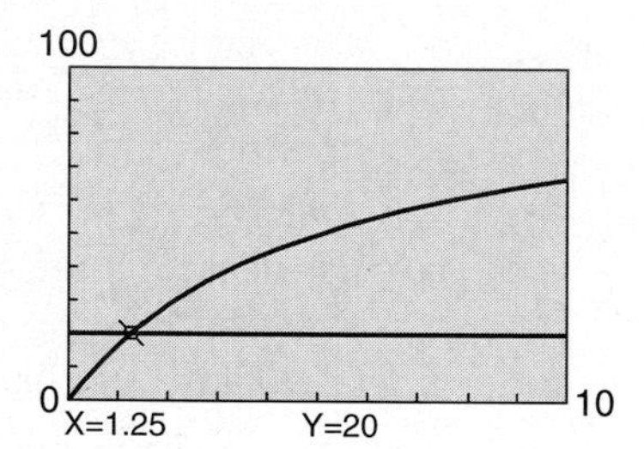

You-Try-It 6 A silver nitrate solution is formed by adding silver nitrate, $AgNO_3$, to water. The percent concentration of $AgNO_3$ in the water is given by the equation $P = \frac{100x}{10+x}$, where P is the percent concentration of $AgNO_3$ and x is the number of grams of $AgNO_3$ that has been added to the water. Use a graphing calculator to find the number of grams of $AgNO_3$, to the nearest hundredth, that must be added to the solution to produce a mixture that is 5% $AgNO_3$.

Solution See page S10.

---

6. $\frac{3}{10+3} = \frac{3}{13} \approx 0.23 = 23\%$. The sugar concentration is 23%.
7. The percent concentration is the output value (Y). Therefore, the concentration is 33.33%.

2.6 EXERCISES

Topics for Discussion

1. When a coordinate system is drawn, which axis represents the input variable and which axis represents the output variable? Input is x-axis, output is y-axis.
2. What is the graph of an equation? It is a visual representation of the equation.
3. What is the viewing window of a graphing calculator? Answers will vary.
4. When using the TRACE feature of a graphing calculator, what is the significance of the numbers on the bottom of the screen as you press the ▶ or ◀ key? They are the x- and y-coordinates of the points on the graph.
5. If two graphs intersect at a point, what can be said about the coordinates (x, y) at the point of intersection? The x-coordinates and the y-coordinates are equal.
6. If pure water is added to a saltwater solution, does the percent concentration of salt increase or decrease? The percent concentration decreases.

Reading Graphs

7. The graph at the right shows how many teams in the National Basketball Association (NBA) averaged over 100 points per game. Source: *USA Today,* 1/15/98.
   - **a.** How many teams averaged over 100 points in 1994? 20 teams
   - **b.** In what year did 20 teams average over 100 points per game? 1994
   - **c.** What is the difference between the number of teams that averaged over 100 points per game in 1998 and the number of teams that averaged over 100 points per game in 1990? 23 teams
   - **d.** Between which two years did the number of teams that averaged over 100 points per game not change? 1991 and 1992

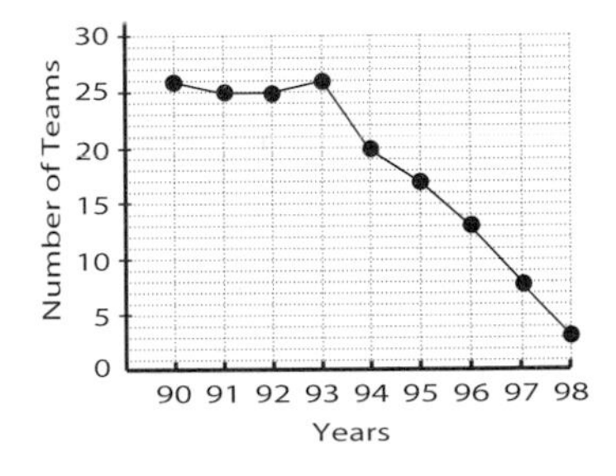

8. The graph at the right shows the sales of baseball cards in the United States for a 10-year period.
   - **a.** How many baseball cards were sold in 1993? 850 million cards
   - **b.** In which year were sales of cards the greatest? 1991
   - **c.** In which year were the least number of cards sold? 1988
   - **d.** By what amount did sales of cards decline between 1991 and 1997? 600 million
   - **e.** What was the average decline per year between 1991 and 1997? 100 million
   - **f.** In which years between 1991 and 1997 did the average decline exceed the actual decline? 1992 and 1993

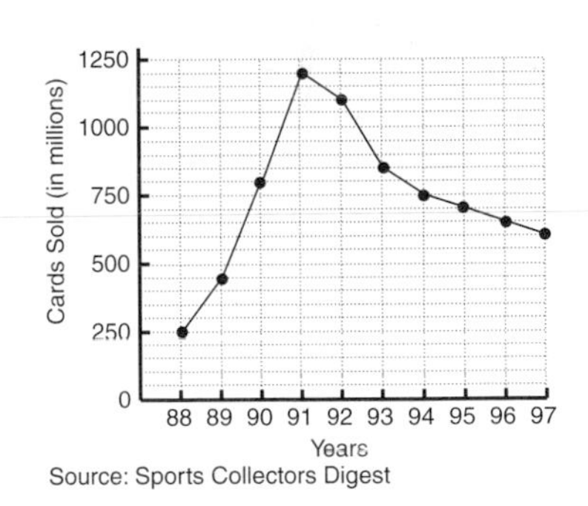

9. The graph at the right shows the amount of potassium nitrate, KNO_3, that can be dissolved in 100 grams of water at various temperatures.
   - **a.** How many grams of KNO_3 can be dissolved in 100 grams of water at a temperature of 50°C? 80 grams
   - **b.** What is the difference between the number of grams of KNO_3 that can be dissolved in 100 grams of water at 90°C and 70°C? 70 grams
   - **c.** Does the amount of KNO_3 that can be dissolved in 100 grams of water increase or decrease as temperature increases? The amount increases.
   - **d.** Approximate the number of grams of KNO_3 that can be dissolved in 100 grams of water at a temperature of 60°C. 105 grams

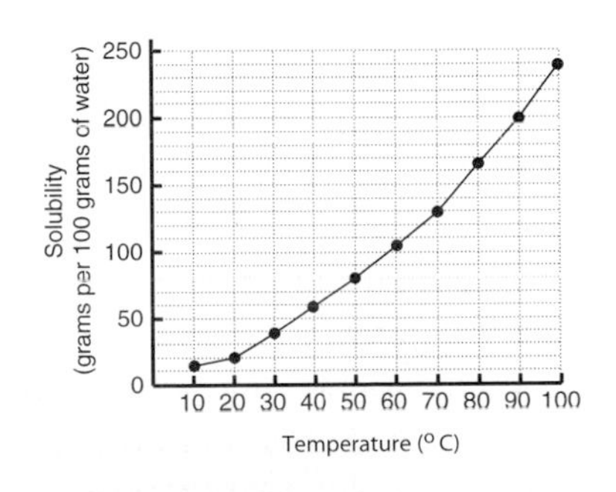

10. For some compounds, the amount of compound that can be dissolved in water decreases as the water becomes hotter. The graph at the right shows the amount of cerium selenate that can be dissolved in 100 grams of water for various temperatures.

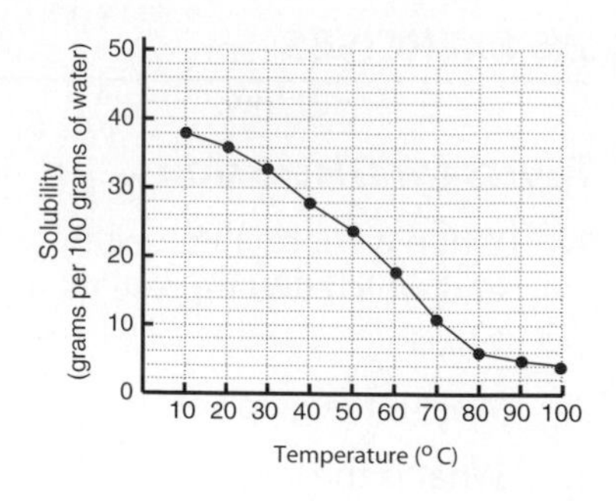

a. How many grams of cerium selenate can be dissolved in water that is 10°C? 38 grams

b. At what temperature will 24 grams of cerium selenate dissolve in 100 grams of water? 50°C

c. For the increments shown on the horizontal axis, between what two temperatures does the number of grams of cerium selenate that can be dissolved in 100 grams of water first drop below 30? 30°C and 40 °C

11. The thickness of the ozone layer at any one spot on Earth varies during the year and with the latitude. The graph at the right shows the thickness, in millimeters, of the ozone layer at a latitude of 40° north.

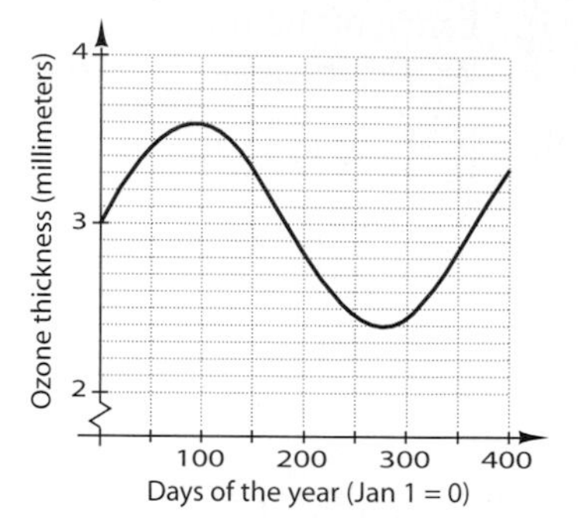

a. What was the thickness of the ozone layer on January 1? 3 millimeters

b. Is the ozone thickness at this latitude ever greater than 3.4 millimeters? Yes

c. Is the ozone thickness at this latitude ever less than 1.8 millimeters? No

d. The horizontal axis shows increments of 50. Between which two increments is the ozone layer thickest? Between 50 and 100

e. The horizontal axis shows increments of 50. Between which two increments is the ozone layer thinnest? Between 250 and 300

12. The risk of having an automobile accident increases as the amount of alcohol consumed by the driver increases. The graph at the right shows the increase in risk for various blood alcohol levels. For instance, the ordered pair (0.05, 2) means that a driver with a blood alcohol level of 0.05 (one-half of the maximum level for most states) is 2 times as likely to be in an accident as a driver who has not been drinking.

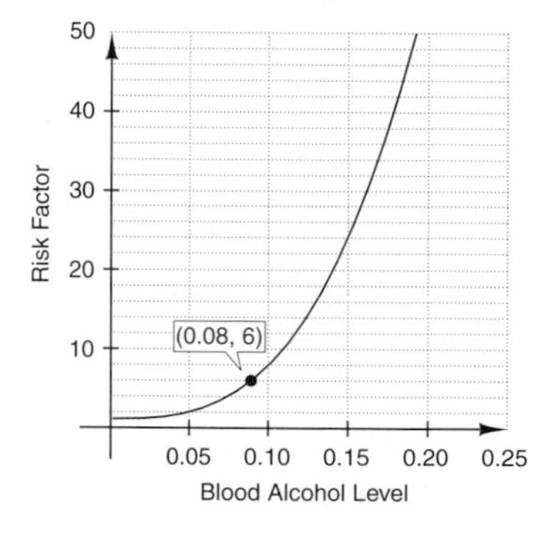

a. How many times as likely is a driver with a blood alcohol level of 0.10 to be in an accident than a driver who has not been drinking? 8 times

b. What is the blood alcohol level of a driver who is 24 times as likely to have an accident as a driver who has not been drinking? .15

c. Explain the meaning of the ordered pair (0.08, 6) shown on the graph. With a blood alcohol level of .08, a person is 6 times as likely to be in an accident.

13. Produce the graph of $y = 2x - 1$ on a graphing calculator using the decimal viewing window.

a. Trace along the graph to find the output value of $2x - 1$ when the input is 2. Algebraically check your result. 3

b. Trace along the graph to find the input value for which the output value of $2x - 1$ is 1. Algebraically check your result. 1

14. Produce the graph of $y = 1 - 0.5x$ on a graphing calculator using the decimal viewing window.

a. Trace along the graph to find the output value of $1 - 0.5x$ when the input is 4. Algebraically check your result. −1

b. Trace along the graph to find the input value for which the output value of $1 - 0.5x$ is −2. Algebraically check your result. 6

15. Produce the graph of $y = x^2$ on a graphing calculator using the decimal viewing window.

a. Trace along the graph to find the output value of x^2 when the input is −1.2. Algebraically check your result. 1.44

b. Trace along the graph to find the input value for which the output value of x^2 is 2.56. Algebraically check your result. 1.6 or −1.6

16. Produce the graph of $y = -x^2 + 1$ on a graphing calculator using the decimal viewing window.

a. Trace along the graph to find the output value of $-x^2 + 1$ when the input is -1.2. Algebraically check your result. -0.44

b. Trace along the graph to find the input value for which the output value of $-x^2 + 1$ is -1.25. Algebraically check your result. 1.5 or -1.5

17. Produce the graph of $y = x^3 + x - 1$ on a graphing calculator using the decimal viewing window.

a. Trace along the graph to find the output value of $(x^3 + x - 1)$ when the input is 1.9. Algebraically check your result. 7.759

b. Trace along the graph to find the input value for which the output value of $(x^3 + x - 1)$ is -7. Algebraically check your result. -1.7

18. Produce the graph of $y = x^3 + 2x - 1$ on a graphing calculator using the decimal viewing window.

a. Trace along the graph to find the output value of $(x^3 + 2x - 1)$ when the input is -1.3. Algebraically check your result. -5.797

b. Trace along the graph to find the input value for which the output value of $(x^3 + 2x - 1)$ is 2. Algebraically check your result. 1

Applications to Uniform Motion

19. The average rate of climb for the space shuttle *Endeavour* for its first 9 minutes of flight is approximately 8.4 miles per minute. The distance, d (in miles), *Endeavour* is above Earth t minutes after launch can be approximated by $d = 8.4t$. Graph this equation using a viewing window of Xmin = 0, Xmax = 9.4, Xscl = 1, Ymin = 0, Ymax = 80, and Yscl = 10.

a. What is the distance above Earth of *Endeavour* 5 minutes after lift-off? a. 42 miles

b. To the nearest tenth of a minute, how long does it take *Endeavour* to reach a distance of 5 miles above Earth? b. 0.6 minute

20. The distance, d (in feet), the Concorde is above Earth during the first 10 minutes after taking off from New York's Kennedy Airport can be approximated by $d = 2100t$, where t is minutes after takeoff. Graph this equation using a viewing window of Xmin = 0, Xmax = 9.4, Xscl = 1, Ymin = 0, Ymax = 25000, and Yscl = 2500.

a. What is the distance of the Concorde above Earth 8 minutes after taking off? Now use the TRACE feature of your calculator to trace to X = 8 and verify your evaluation of the expression. 16,800 feet

b. To the nearest tenth of a minute, how long does it take the Concorde to reach a distance of 12,000 feet above Earth? 5.7 minutes

21. The distance, d (in feet), the Concorde is above Earth during the last 30 minutes as it comes in for a landing at Orly airport in France can be approximated by $d = 45{,}000 - 1500t$, where t is the time in minutes before it lands. Graph this equation using a viewing window of Xmin = 0, Xmax = 37.6, Xscl = 5, Ymin = 0, Ymax = 45000, and Yscl = 5000.

a. What is the distance of the Concorde above Earth 15 minutes before landing? Now use the TRACE feature of your calculator to trace to X = 15 and verify your evaluation of the expression. 22,500 feet

b. To the nearest tenth of a minute, how long does it take before the Concorde is 5000 feet above the runway? 26.7 minutes

22. The distance, d (in feet), a hot-air balloon is above Earth during the last 20 minutes of its flight can be approximated by $d = 2500 - 125t$, where t is the time in minutes before it lands. Graph this equation using a viewing window of Xmin = 0, Xmax = 37.6, Xscl = 5, Ymin = 0, Ymax = 3000, and Yscl = 300.

a. What is the distance of the hot-air balloon above Earth 17 minutes before landing? Now use the TRACE feature of your calculator to trace to X = 17 and verify your evaluation of the expression. 375 feet

b. To the nearest tenth of a minute, how long does it take before the hot-air balloon is 500 feet above the landing area? 16 minutes

Applications to Percent Concentration

23. To make simple syrup, sugar is added to water. The percent concentration, P, of the sugar in water can be approximated by the equation $P = \frac{x}{25 + x}(100)$, where x is the number of ounces of sugar that are added to the water. Graph this equation using a viewing window of Xmin = 0, Xmax = 25, Xscl = 5, Ymin = 0, Ymax = 100, and Yscl = 10. To the nearest tenth, how many ounces of sugar must be added before the concentration is 35% sugar? 13.5 ounces

24. A cereal mixture is to contain oats and barley. The percent concentration, P, of the oats in the cereal can be approximated by the equation $P = \frac{x}{50 + x}(100)$, where x is the number of pounds of barley that are added to the oats. Graph this equation using a viewing window of Xmin = 0, Xmax = 250, Xscl = 25, Ymin = 0, Ymax = 100, and Yscl = 10. To the nearest tenth, how many pounds of barley must be added to produce a cereal that is 70% barley? 116.7 pounds

25. A fruit punch is made by adding water to a concentrate of fruit. The percent concentration, P, of the fruit concentrate in the water can be approximated by the equation $P = \frac{5000}{50 + x}$, where x is the number of grams of water that are added to the concentrate. Graph this equation using a viewing window of Xmin = 0, Xmax = 600, Xscl = 100, Ymin = 0, Ymax = 100, and Yscl = 10. To the nearest ten, how many grams of water were added to the fruit concentrate to produce a punch that is 10% fruit? 450 grams

26. A cleaning solution is made by adding water to vinegar. The percent concentration, P, of the vinegar in the water can be approximated by the equation $P = \dfrac{3200}{32 + x}$, where x is the number of grams of water that are added to the vinegar. Graph this equation using a viewing window of Xmin = 0, Xmax = 600, Xscl = 100, Ymin = 0, Ymax = 100, and Yscl = 10. To the nearest tenth, how many grams of water were added to the vinegar to produce a solution that is 5% vinegar? 608 grams

Applying Concepts

27. The graph at the right shows the numbers of U. S. households (in millions) owning certain technologies.

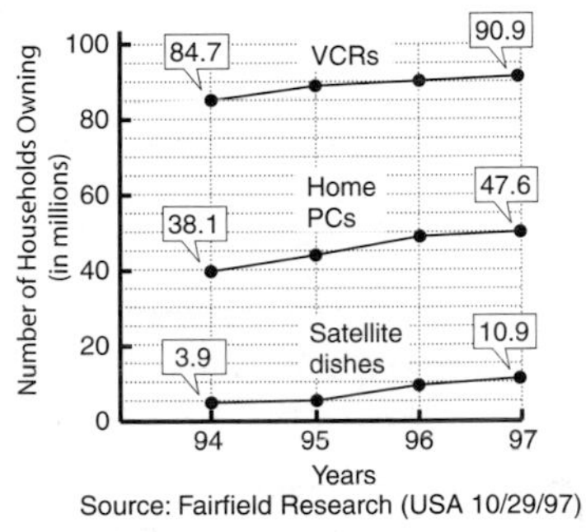

Source: Fairfield Research (USA 10/29/97)

a. For any of the technologies shown, did the number of households owning that technology decrease for the years shown? If so, which ones? No

b. What was the difference between the number of households that owned home PCs and the number that owned satellite dishes in 1994? 34.2 million households

c. Which of the technologies shown had the greatest increase in ownership for the years shown? Home PCs

d. Find the average increase in households owning home PCs for the years shown. The average increase was 3.167 million per year.

e. Assuming that the average increase in households owning home PCs remains the same for 1998 and 1999, approximate the number of households that will have home PCs in 1999. 53.934 million households

28. The graph at the right is a stacked-bar graph. It shows the amount (in millions of dollars) spent by the federal government for Medicare and Medicaid for the years shown. (Source: Healthcare Financing Administration, Department of Human Services, *USA Today*, 8/6/97.) The numbers in each bar show the amount spent for each program.

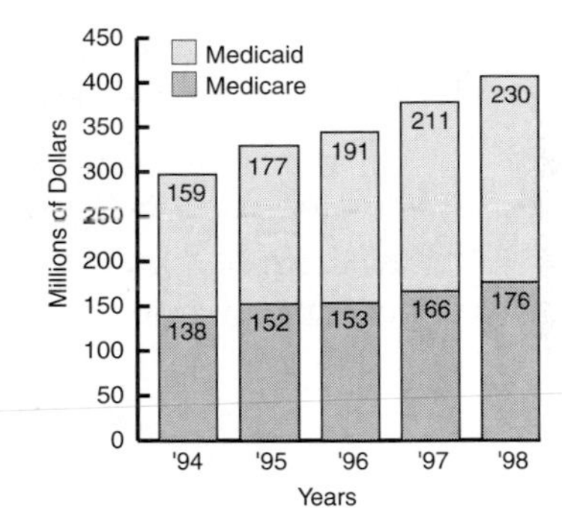

a. How much was spent for Medicaid in 1996? 191 million dollars

b. How much was spent on both programs in 1997? 377 million dollars

c. Are the amounts spent for each program increasing each year? Yes

d. For the years shown, which program increased by the greater amount? Medicaid

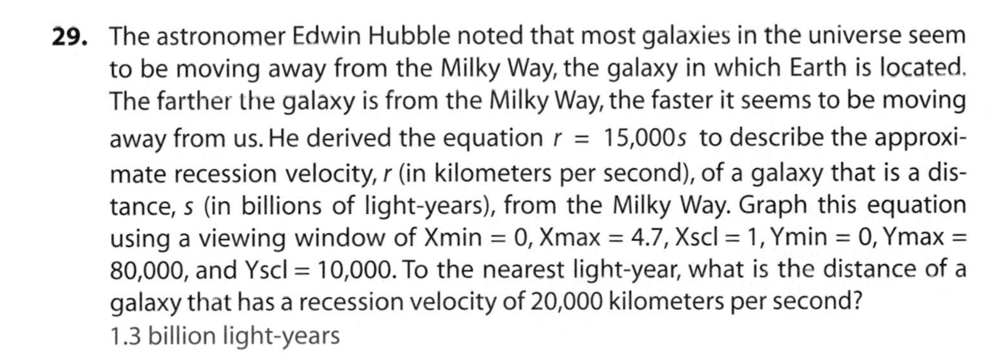

29. The astronomer Edwin Hubble noted that most galaxies in the universe seem to be moving away from the Milky Way, the galaxy in which Earth is located. The farther the galaxy is from the Milky Way, the faster it seems to be moving away from us. He derived the equation $r = 15{,}000s$ to describe the approximate recession velocity, r (in kilometers per second), of a galaxy that is a distance, s (in billions of light-years), from the Milky Way. Graph this equation using a viewing window of Xmin = 0, Xmax = 4.7, Xscl = 1, Ymin = 0, Ymax = 80,000, and Yscl = 10,000. To the nearest light-year, what is the distance of a galaxy that has a recession velocity of 20,000 kilometers per second?
1.3 billion light-years

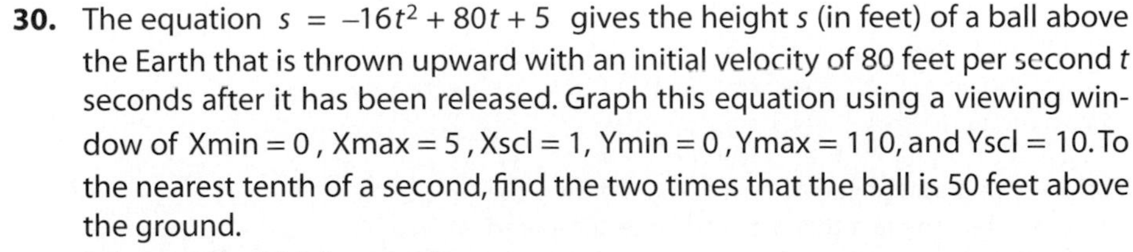

30. The equation $s = -16t^2 + 80t + 5$ gives the height s (in feet) of a ball above the Earth that is thrown upward with an initial velocity of 80 feet per second t seconds after it has been released. Graph this equation using a viewing window of Xmin = 0, Xmax = 5, Xscl = 1, Ymin = 0, Ymax = 110, and Yscl = 10. To the nearest tenth of a second, find the two times that the ball is 50 feet above the ground.
0.6 second and 4.4 seconds

31. Water is being added to 75 grams of salt. The percent concentration, S, of the salt is given by the equation $S = \frac{7500}{75 + x}$, where x is the number of grams of water added to the solution. The percent of the solution that is water, W, is given by $W = \frac{100x}{75 + x}$. Graph these two equations using Y1 for S and Y2 for W and a viewing window of Xmin = 0, Xmax = 200, Xscl = 20, Ymin = 0, Ymax = 100, and Yscl = 10.

a. Use the INTERSECT feature of your calculator to find the coordinates of the point where the two graphs intersect. (75, 50)

b. Explain why the ordered pair you found in part a is the point of intersection of the curves. Answers will vary.

Explorations—Graphing Calculator Coordinates

32. By understanding some of the characteristics of a graphing calculator screen, you can choose what are called *nice coordinates.* These coordinates are convenient when analyzing certain graphs.

a. Using a graphing calculator, graph $y = x + 1$ in the standard viewing window. Once you have drawn the graph, activate the TRACE feature. Record the current x-coordinate. Now press ▶. Record the new x-coordinate and then find the difference between the two values. If you press ▶ again, the x-coordinate should change by the same amount.

b. Now change the viewing window to those shown at the right and graph $y = x + 1$ again. Activate the TRACE feature and record the current x-coordinate. Now press ▶ several times. By what amount does the x-coordinate change each time ▶ is pressed?

TI - 82/83/85/86	Sharp EL-9600/Casio CFX-9850
Xmin= –4.7	Xmin= –6.3
Xmax= 4.7	Xmax= 6.3
Xscl=1	Xscl=1
Ymin= –10	Ymin= –10
Ymax= 10	Ymax= 10
Yscl=1	Yscl=1

c. The choices for Xmin and Xmax depend on the number of horizontal pixels[1] on the calculator screen. For the TI-82/83/85/86, there are 94 pixels; for the Sharp EL-9600 and Casio CFX-9850, there are 126 pixels. The change each time ▶ is pressed is called Δx and is calculated from the following formula.

TI-82/83/85/86

$$\frac{\text{Xmax} - \text{Xmin}}{94} = \Delta x$$

Sharp EL-9600/Casio CFX-9850

$$\frac{\text{Xmax} - \text{Xmin}}{126} = \Delta x$$

Show that the formula produces $\Delta x = 0.1$ for the values of Xmin and Xmax for the viewing window suggested in **b.**

d. The formula Xmax = Xmin + (No. of horizontal pixels)(Δx) can be used to create a viewing window on which the cursor will change by Δx each time ▶ is pressed. For the TI-82/83/85/86, the No. of horizontal pixels is 94. For the Sharp El-9600 and the Casio CFX-9850, the No. of horizontal pixels is 126. Suppose you wanted a viewing window for which Xmin = –2 and $\Delta x = 0.5$. What is the value of Xmax for your calculator?

e. Graph $y = x + 1$ for the viewing window in part **d.** and verify that the x-coordinate changes by 0.5 each time ▶ is pressed.

---

1. A pixel can be thought of as one square on a piece of graph paper. When a graph is drawn, the pixels (points) that belong to the graph are illuminated. The fact that a pixel is large compared to a point placed on graph paper contributes to the jagged nature of a calculator graph.

Chapter Summary

Definitions

The *integers* are . . . −4, −3, −2, −1, 0, 1, 2, 3, 4 . . . . The numbers . . . , −4, −3, −2, −1 are *negative integers*. The numbers 1, 2, 3, 4 . . . are *positive integers*. The number 0 is neither a negative nor a positive integer.

A *number line* is used as a graphical representation of positive and negative numbers.

To *graph*, or *plot*, a number, place a dot at the location given by the number. The graph of −2 and 5 is shown on the number line below.

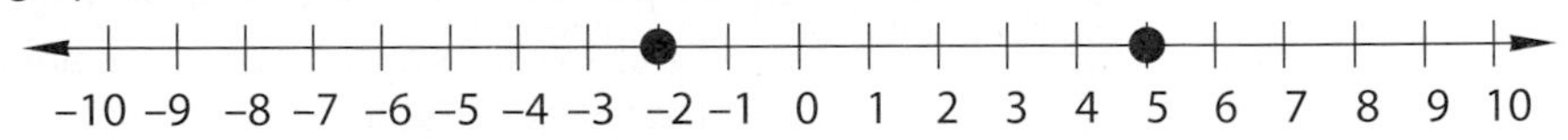

The numbers −2 and 5 are called the *coordinates* of the points (dots).

Two numbers that are the same distance from zero on the number line but are on opposite sides of zero are opposite numbers, or *opposites*. The opposite of a number is also called its *additive inverse*. The opposite of 7 is −7. The opposite of 8 is −8 .

The *absolute value* of a number is its distance from zero on the number line. Therefore, the absolute value of a number is a positive number or zero. The symbol for absolute value is two vertical bars, | |.

Addition is the process of finding the total of two numbers. The numbers being added are called *addends*. The total is called the *sum*. We can find rules for adding integers by using a number line and the arrow representation of numbers.

Subtraction is the process of finding the difference between two numbers.

The *distance* between the points *P* and *Q*, denoted by $d(P, Q)$, whose coordinates are *a* and *b* is given by $d(P, Q) = |a - b|$.

A *rectangular coordinate system* is formed by two number lines, one horizontal and one vertical, that intersect at the zero point of each line. The point of intersection is called the *origin*. The two axes are called the *coordinate axes*, or simply the *axes*. Frequently, the horizontal axis is labeled the *x*-axis, and the vertical axis is labeled the *y*-axis. In this case, the axes form what is called the *xy-plane*.

The two axes divide the plane into four regions called *quadrants*, which are numbered counterclockwise, using Roman numerals, from I to IV starting at the upper right.

Each point in the plane can be identified by a pair of numbers called an *ordered pair*. The first number of the ordered pair measures a horizontal change from the *y*-axis and is called the *abscissa*, or *x-coordinate*. The second number of the pair measures a vertical change from the *x*-axis and is called the *ordinate*, or *y-coordinate*. The ordered pair (x, y) associated with a point is also called the *coordinates* of the point.

Sometimes the first number in an ordered pair is called the *first coordinate* of the ordered pair and the second number is the *second coordinate* of the ordered pair.

The *graph of an ordered pair* is the dot drawn at the coordinates of the point in the plane.

A *scatter diagram* is a graph of the ordered pairs of related data.

A *rational number* is the quotient of two integers. In symbols, a rational number is one of the form $\frac{a}{b}$, where *a* and *b* are integers and $b \neq 0$.

A *terminating decimal* is a number whose decimal representation terminates. The number 0.375 is a terminating decimal. A *repeating decimal* is a number whose decimal representation repeats a block of digits. The number 0.3454545... is a repeating decimal. Numbers with neither terminating nor repeating decimal representations are called *irrational numbers.* The rational numbers and irrational numbers together are called *real numbers.*

A fraction is in *simplest form* when the numerator and denominator do not contain a common factor greater than 1. The fractions $\frac{2}{3}$, $\frac{4}{6}$, and $\frac{8}{12}$ are *equivalent fractions* because they represent the same part of a whole. However, the fraction $\frac{2}{3}$ is in simplest form because there are no common factors greater than 1 in the numerator and denominator.

Conversion factors are used to convert one unit of measurement to another unit of measurement.

Moving a geometric figure to a new location on the coordinate plane without changing its shape or turning it is called a *translation.* A *vertical translation* moves the figure up or down. A *horizontal translation* moves the figure left or right.

Stretching or *shrinking* a geometric figure is accomplished by multiplying the coordinates of the figure by a positive number.

A *transformation* of a geometric figure is a change in either the position of the figure, like a translation, or the shape of the figure, like stretching it or shrinking it.

Points that have opposite x-coordinates and the same y-coordinate are said to be *symmetric with respect to the y-axis.* Points that have opposite y-coordinates and the same x-coordinate are said to be *symmetric with respect to the x-axis.*

The *graph of an equation* is a visual representation of the equation.

Uniform motion means that the speed and direction of the object do not change.

The *percent concentration* of a substance is a measure of the percent of the substance that is pure.

Procedures

To add numbers with the same sign, add the absolute values of the numbers. Then attach the sign of the addends.

To add numbers with different signs, find the absolute value of each number. Then subtract the smaller of these absolute values from the larger one. Attach the sign of the number with the larger absolute value.

To subtract a number, add its opposite.

To multiply or divide numbers with different signs, multiply or divide the absolute values of the numbers. The sign is negative.

To multiply or divide numbers with the same sign, multiply or divide the absolute values of the numbers. The sign is positive.

To graphically find an input for a given output, draw the graph of the equation and the graph of the horizontal line representing the output. Use the INTERSECT feature of a graphing calculator to find the point at which the line and the graph of the equation meet. The x-coordinate of the intersection point is the required input value.

Chapter Review Exercises

1. Evaluate: $-(-14)$ 14

2. Evaluate: $-|-3|$ -3

3. Add: $-14 + 32$ 18

4. Add: $-26 + (-17) + 43$ 0

5. Subtract: $23 - (-51)$ 74

6. Subtract: $-15 - 42 - (-13)$ -44

7. Multiply: $23(-12)$ -276

8. Multiply: $-3(-17)$ 51

9. Divide: $-54 \div (-6)$ 9

10. Divide: $64 \div (-8)$ -8

11. What number is 12 more than -17? -5

12. Find the difference between -23 and -16. -7

13. Find the product of -14 and -4. 56

14. What is -81 divided by 3? -27

15. Write $\frac{5}{11}$ as a repeating decimal. $0.\overline{45}$

16. Simplify: $-\frac{2}{3} + \frac{1}{4} - \left(-\frac{1}{2}\right)$ $\frac{1}{12}$

17. Simplify: $-\frac{3}{4}\left(\frac{5}{9}\right) \div \left(-\frac{5}{8}\right)$ $\frac{2}{3}$

18. Simplify: $-\frac{(-2)^3}{6^2}$ $\frac{2}{9}$

19. Convert 2.5 gallons to pints. 20 pints

20. Convert 46 inches to feet. $3\frac{5}{6}$ feet

21. The graph at the right shows the losses for Amtrak in millions of dollars.
   a. In which year did Amtrak lose the most money? 1996
   b. What is the difference between the amount of money Amtrak lost in 1995 and the amount it lost in 1992? \$13 million

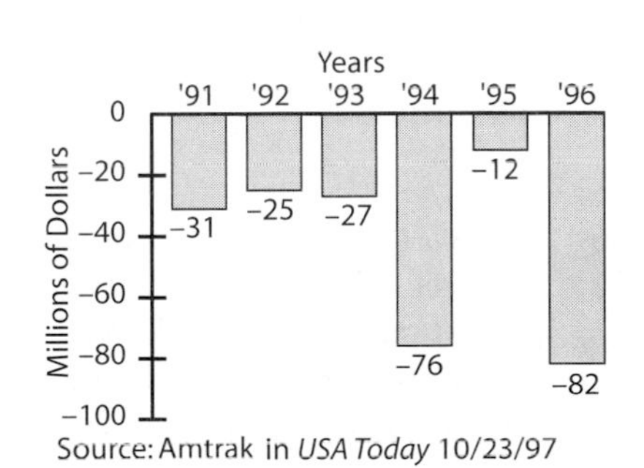

Source: Amtrak in *USA Today* 10/23/97

22. The changes (in dollars) in the price of a stock for a 5-day period were: $-1.25, 1, -0.625, -1.5, 0.875$. Find the average change in the price of the stock for the 5-day period. $-\$.3$

23. Complete the input/output table below and then graph the resulting ordered pairs.

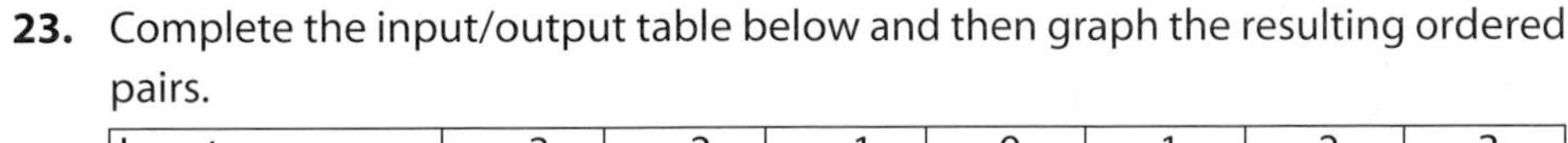

Input, x	-3	-2	-1	0	1	2	3
Output, $3 - 2x$	9	7	5	3	1	-1	-3

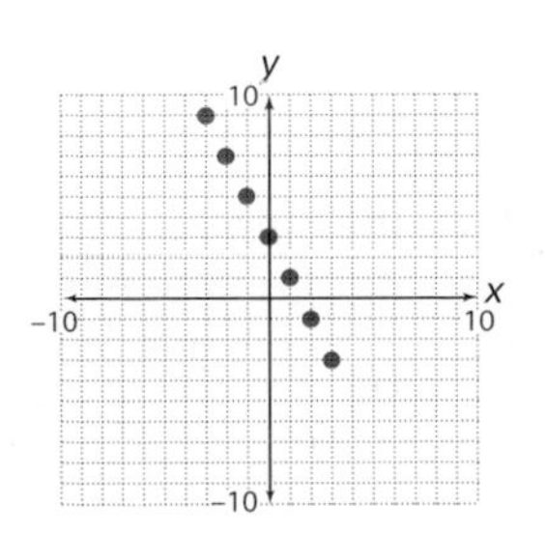

24. The table below shows how the number of wireless telephone subscribers has changed. Draw a scatter diagram of these data. (Source: Dataquest, Herschel Shosteck Associates)

Year, x	1992	1993	1994	1995	1996
Subscribers (in millions)	12	16	22	34	44

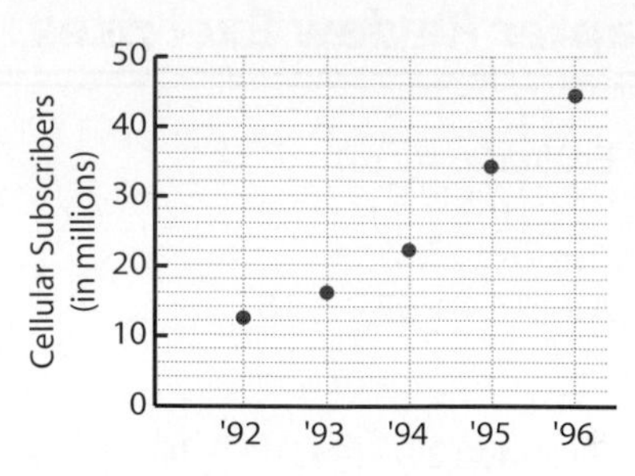

25. Find the area of the triangle.
54 square units

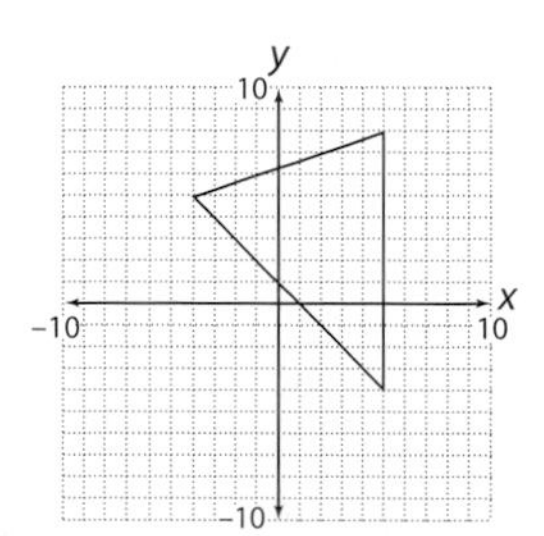

26. Translate the trapezoid 3 units down and 4 units to the right. What are the coordinates of the vertices of the translated trapezoid? (0, 4), (5, –6), (9, –2), (7, 2)

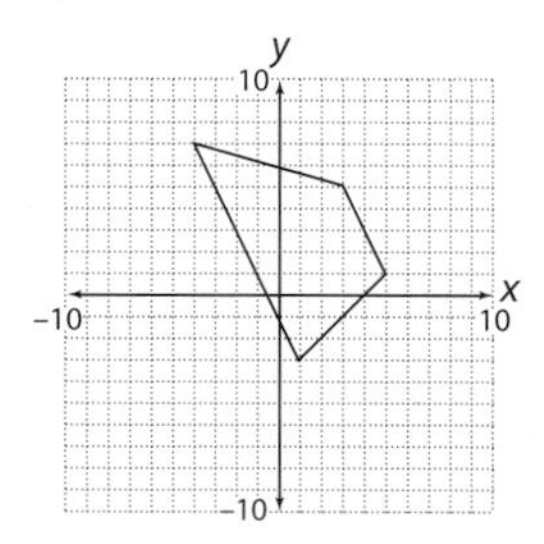

27. Draw the figure that is symmetric with respect to the x-axis.

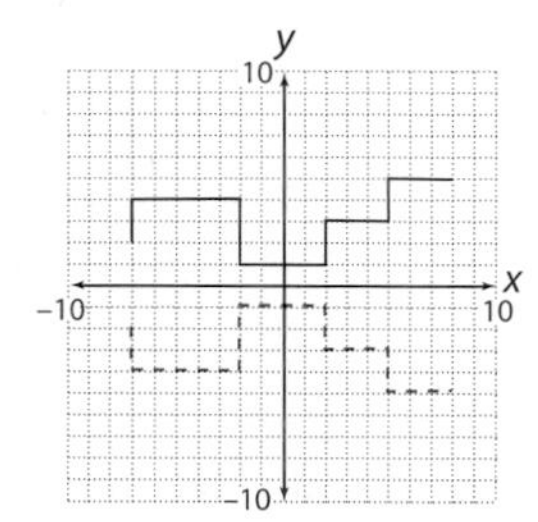

28. Draw the figure that is symmetric with respect to the y-axis.

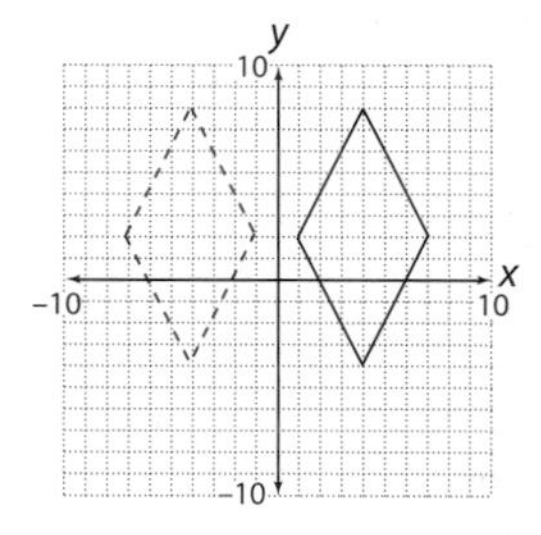

29. Use a graphing calculator to produce the graph of $y = 0.1x^3 - 2$ for the standard viewing window. Find the value of x, to the nearest hundredth, for which the output is 2. 3.42

30. The average rate of climb for a passenger jet during the first 10 minutes of flight is approximately 2000 feet per minute. The altitude, A (in feet), of the plane t minutes after takeoff is given by $A = 2000t$. Using a graphing calculator, find, to the nearest tenth of a minute, the time it takes for the plane to reach an altitude of 12,400 feet. 6.2 minutes

31. A salad dressing is made by adding olive oil to 2 ounces of vinegar. The percent concentration of the vinegar in the salad dressing is given by $P = \dfrac{200}{x + 2}$, where x is the number of ounces of olive oil that is being added. Using a graphing calculator, find the number of ounces of olive that must be added to have a salad dressing that is 25% vinegar. 6 ounces

Cumulative Review Exercises

1. Determine the truth value of the statement "$2x + 1 > 1$ or $6 - 2x < 3$" when $x = 2$. True

2. Evaluate the expression $4ab^2 + 3a^3$ when $a = -2$ and $b = 3$. -96

3. The heights, in inches, of 25 women in a college tennis program are: 68, 65, 61, 64, 65, 66, 61, 64, 62, 68, 61, 64, 61, 60, 67, 69, 64, 65, 66, 66, 63, 67, 65, 64, 66. Find the mean and median for this data.
Mean = 64.48; median = 65

4. Write the negation of the sentence "Some planes have jet engines."
No planes have jet engines.

5. For the north-south traffic lanes, a traffic light is red for 45 seconds, green for 1 minute, and yellow for 5 seconds. If a car comes to this traffic light from the north, what is the probability that the light is green? Write the answer as a fraction in simplest form.
$\frac{6}{11}$

6. The scores of 25 students on an archaeology exam were: 82, 70, 91, 78, 89, 63, 69, 80, 77, 67, 82, 93, 73, 79, 70, 62, 71, 90, 78, 76, 81, 85, 62, 62, 72. Draw a stem-and-leaf plot for these data.

Stem	Leaf
9	1, 3, 0
8	2, 9, 0, 2, 1, 5
7	0, 8, 7, 3, 9, 0, 1, 8, 6, 2
6	3, 9, 7, 2, 2, 2

7. If $x = 4$, what is the truth value of $(3x + 2 > 5)$ and $(2x - 1 \leq 7)$? True

8. Translate "the sum of the square of the number and four more than twice the number" into a variable expression. $x^2 + 2x + 4$

9. Complete the input/output table below. Using the variable expression you find for the last column, what is the output when n is 9? 28

Input	0	1	2	3	4	5	6	n
Output	1	4	7	10	13	16	**19**	**$3n + 1$**

10. Transform the figure below by multiplying the x-coordinates by 2 and the y-coordinates by $\frac{2}{3}$.

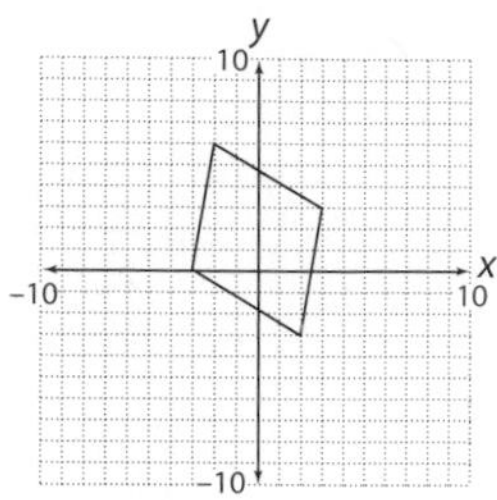
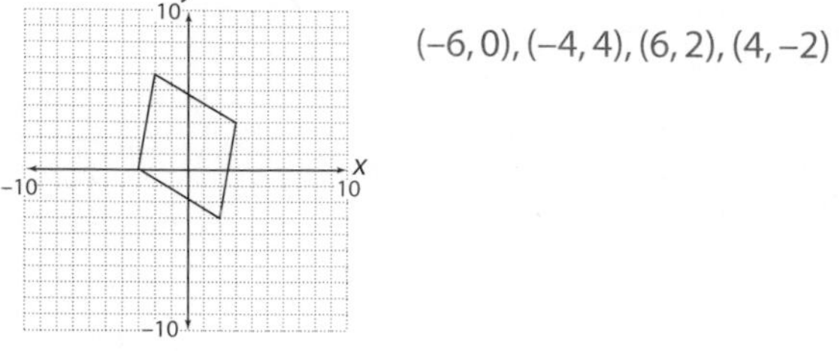

$(-6, 0), (-4, 4), (6, 2), (4, -2)$

11. Find the area of the triangle below.
50 square units

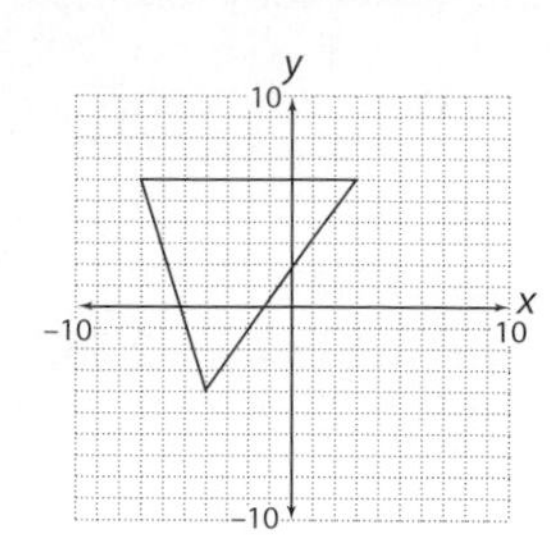

12. Draw the figure that is the reflection through the x-axis of the given figure.

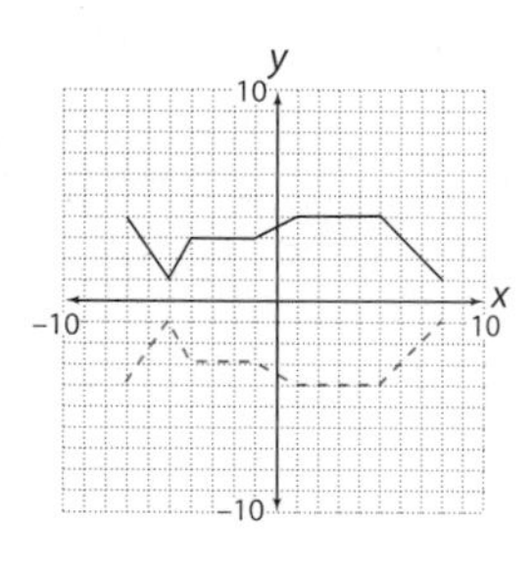

13. Translate the given figure 2 units down and 3 units right. What are the coordinates of the vertices of the translated figure?

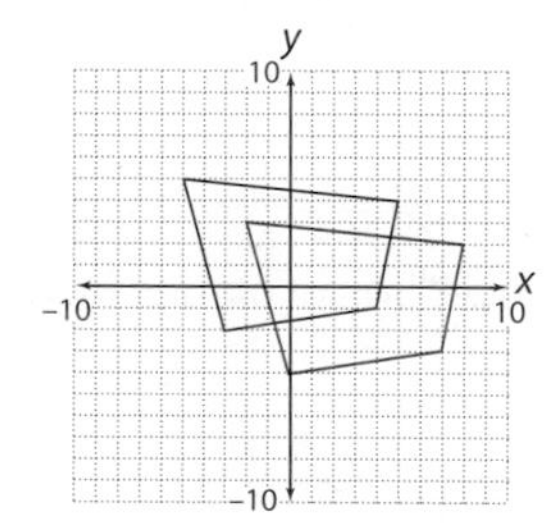

$(-2, 3), (8, 2), (0, -4), (7, -3)$

14. Complete the input/output table below. Graph the resulting ordered pairs.

Input, x	-2	-1	0	1	2	3	4
Output, $4 - 2x$	8	6	4	2	0	-2	-4

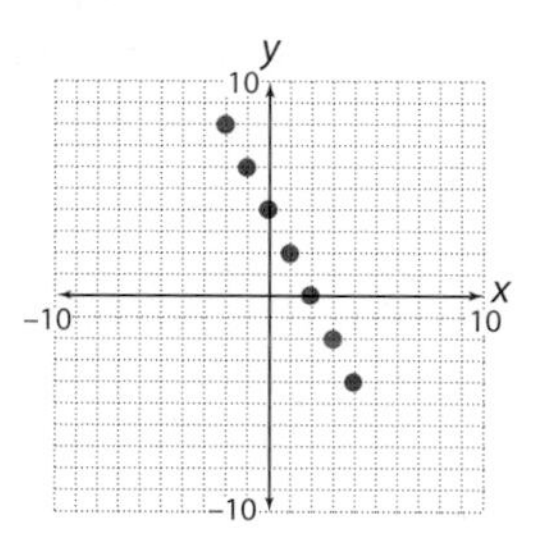

15. Use a graphing calculator to draw the graph of $y = -x^3 + 2x - 1$. Find the value of x, to the nearest hundredth, for which the output is 4. -2.09

16. A ferry leaves a dock and travels to an island that is 100 kilometers away. The distance, s (in kilometers), of the ferry from the island is given by the equation $s = 100 - 19t$, where t is the time, in hours, since the boat left the dock. Graph this equation using a viewing window of Xmin = 0, Xmax = 5, Xscl = 1, Ymin = 0, Ymax = 100, and Yscl = 10. Use the graph to estimate the time at which the ferry is 30 kilometers from the island. 3.68 hours

17. A garlic-flavored oil is being made by adding garlic to the oil. The percent concentration of the garlic is given by $P = \frac{100x}{50 + x}$, where x is the number of grams of garlic being added to the oil. How many grams of garlic must be added to produce a mixture that is 10% garlic? Use a viewing window of Xmin = 0, Xmax = 1000, Xscl = 100, Ymin = 0, Ymax = 100, and Yscl = 10. 5.6 grams

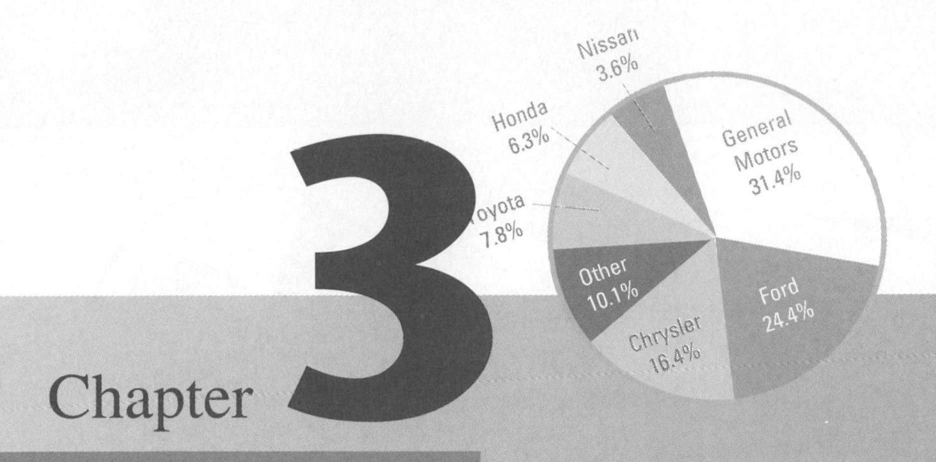

Chapter 3

First-Degree Equations and Inequalities

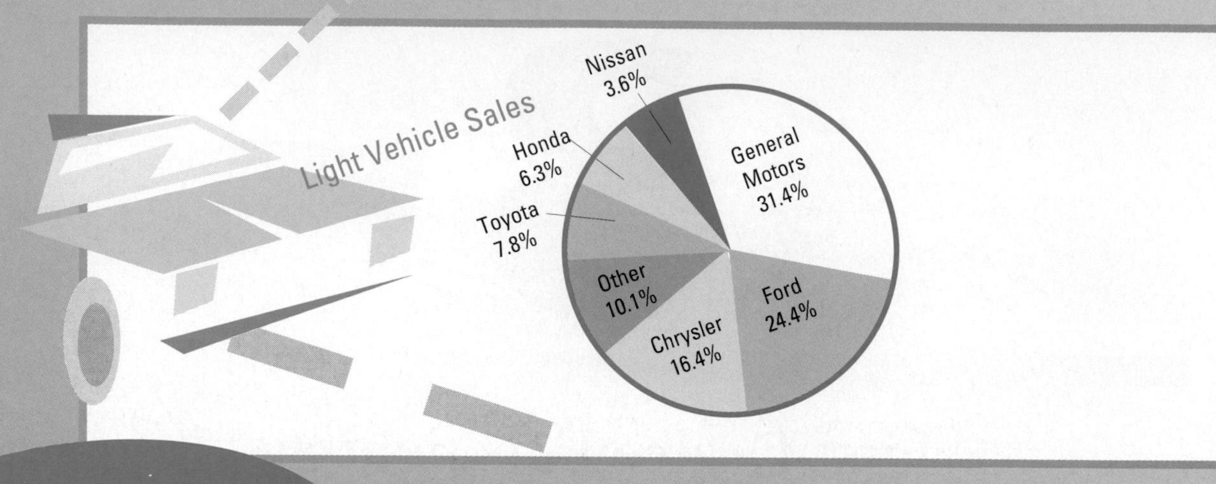
Light Vehicle Sales
Nissan
3.6%
Honda
6.3%
Toyota
7.8%
General
Motors
31.4%
Other
10.1%
Chrysler
16.4%
Ford
24.4%

3
16.4%
6.3%
10.1%
31.4%

Section 3.1 Simplifying Variable Expressions

Properties of Real Numbers

Earlier in the text, we *evaluated* a variable expression. That is, we replaced the variable by a number and then calculated a numerical result. In this section we look at *simplifying* a variable expression. This is accomplished by using the Properties of Real Numbers.

Note that numbers can be added in either order and the result is the same.

$$9 + (-12) = -3 \qquad \text{and} \qquad -12 + 9 = -3$$

This is the ***Commutative Property of Addition,*** which states that if a and b are any two numbers, then $a + b = b + a$.

When three numbers are added together, the numbers can be grouped in any order and the sum will be the same.

$$-6 + (3 + 7) = -6 + 10 = 4 \qquad \text{and} \qquad (-6 + 3) + 7 = -3 + 7 = 4$$

This is the ***Associative Property of Addition,*** which states that if a, b, and c are any three numbers, then $a + (b + c) = (a + b) + c$.

Two other properties of addition are also important. The first says that the sum of a number and its opposite is zero.

$$-8 + 8 = 0 \qquad \text{and} \qquad 8 + (-8) = 0$$

This is the ***Inverse Property of Addition,*** which states that $a + (-a) = 0$ and $-a + a = 0$. Recall that a and $-a$ are opposites or additive inverses of each other.

The second of these two other properties expresses the fact that the sum of a number and zero is the number.

$$-3 + 0 = -3 \qquad \text{and} \qquad 0 + (-3) = -3$$

The ***Addition Property of Zero*** states that if a is any number, then $a + 0 = a$ and $0 + a = a$.

Notice that numbers can be multiplied in either order and the result is the same.

$$9(-8) = -72 \qquad \text{and} \qquad (-8)9 = -72$$

This is the ***Commutative Property of Multiplication,*** which states that if a and b are any two numbers, then $ab = ba$.

When three numbers are multiplied together, the numbers can be grouped in any order and the product will be the same.

$$3(5 \cdot 2) = 3(10) = 30 \qquad \text{and} \qquad (3 \cdot 5)2 = 15 \cdot 2 = 30$$

This is the ***Associative Property of Multiplication,*** which states that if a, b, and c are any three numbers, then $a(bc) = (ab)c$.

Two other properties of multiplication are also important. The first says that the product of a number and its reciprocal is one.

$$\frac{1}{8} \cdot 8 = 1 \qquad \text{and} \qquad 8\left(\frac{1}{8}\right) = 1$$

This is the ***Inverse Property of Multiplication,*** which states that for $a \neq 0$, $a \cdot \frac{1}{a} = 1$ and $\frac{1}{a} \cdot a = 1$. The terms a and $\frac{1}{a}$ are **reciprocals.** They are also called **multiplicative inverses** of each other.

The second of these other two properties expresses the fact that the product of a number and one is the number.

$$9 \cdot 1 = 9 \qquad \text{and} \qquad 1 \cdot 9 = 9$$

The ***Multiplication Property of One*** states that if a is any number, then $a \cdot 1 = a$ and $1 \cdot a = a$.

TAKE NOTE
Here is a summary of the discussion at the right. If a coefficient is 1 or –1, it is usually not written. For instance, we write $1y$ as y and $-1xy$ as $-xy$.

Recall that the coefficient of a variable is the number that multiplies the variable. Note from the Multiplication Property of One that when the coefficient is 1, the 1 is not written. Thus we write x instead of $1x$ or $1 \cdot x$. A coefficient of –1 is treated in much the same way. For instance, we normally write $-1x$ and $-1 \cdot x$ as $-x$.

By the Order of Operations Agreement, the expression $6(4+7)$ is simplified by adding the numbers inside the parentheses first and then multiplying.

$$\begin{aligned} 6(4+7) &= 6(11) \\ &= 66 \end{aligned}$$

However, we can multiply each number inside the parentheses by the number outside the parentheses and achieve the same result.

$$\begin{aligned} 6(4+7) &= 6(4) + 6(7) \\ &= 24 + 42 \\ &= 66 \end{aligned}$$

This is an example of the **Distributive Property,** which states that if a, b, and c are any numbers, then $a(b + c) = ab + ac$.

Example 1 Complete the statement $4 + x = ?$ by using the Commutative Property of Addition.

Solution $4 + x = x + 4$ • The Commutative Property of Addition states that the order of addends can be interchanged.

You-Try-It 1 Complete the statement $4(3x) = ?$ by using the Associative Property of Multiplication.

Solution See page S11.

Example 2 Complete the statement by using the Distributive Property.

$4(6 + 9) = ?(6) + ?(9)$

Solution $4(6+9) = 4(6) + 4(9)$ • The Distributive Property states that each number inside the parentheses is multiplied by the number outside the parentheses.

You-Try-It 2 Complete the statement $12 + ? = 0$ by using the Inverse Property of Addition.

Solution See page S11.

Simplifying Variable Expressions

A variable expression is shown at the right. The expression can be rewritten by writing subtraction as addition of the opposite.

$3x^2 - 4xy + 5z - 2$

$3x^2 + (-4xy) + 5z + (-2)$

Note that the expression has 4 addends. The **terms** of a variable expression are the addends of the expression. The expression has 4 terms.

4 terms

$3x^2 - 4xy + 5z - 2$

Variable terms Constant term

The terms $3x^2$, $-4xy$, and $5z$ are **variable terms.** The term -2 is a **constant term** or simply a **constant**.

TAKE NOTE
The fact that $3x$ is positive and $-7x$ is negative comes from rewriting the expression as addition of the opposite. This step is rarely written but is always done mentally.

$3x^2 + 3x - 6 - 7x + 9$ equals

$3x^2 + 3x + (-6) + (-7x) + 9$.

Like terms of a variable expression are terms that have the same variable parts. The terms $3x$ and $-7x$ are like terms. Constant terms are also like terms. Thus -6 and 9 are like terms. The terms $3x^2$ and $3x$ are not like terms because $x^2 = x \cdot x$ and thus the variable parts are not the same.

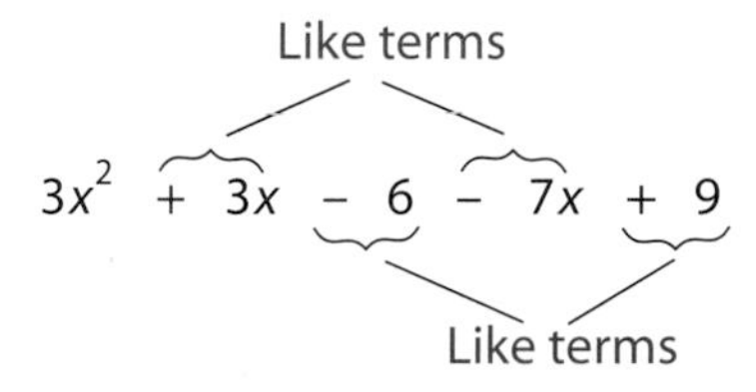

Terms such as $4xy$ and $-7yx$ are like terms because, by the Commutative Property of Multiplication, $xy = yx$. The same is true for terms such as $-4abc$ and $12bca$.

Question: Which of the following pairs of terms are like terms: **a.** $3a$ and $3b$; **b.** $7z^2$ and $7z^3$; **c.** $6ab$ and $3a$; **d.** $-4c^2$ and $6c^2$?[1]

TAKE NOTE
Combining like terms is an operation we do quite naturally, probably every day. For instance,
4 apples + 7 apples = 11 apples
\$5 + \$2 = \$7
6 pounds + 3 pounds = 9 pounds

Equally natural is the idea of not combining items that are not similar.
4 apples + 7 oranges
2 dogs + 5 cats

By using the Commutative Property of Multiplication, we can rewrite the Distributive Property as $ba + ca = (b + c)a$. This is sometimes called the *factoring form* of the Distributive Property. This form of the Distributive Property is used to **combine** like terms of a variable expression by adding their coefficients. For instance,

$$7x + 9x = (7 + 9)x \qquad \text{• Use the Distributive Property: } ba + ca = (b + c)a.$$
$$= 16x$$

Combining the like terms of a variable expression is called **simplifying the variable expression**.

Question: What is the result of simplifying $9y + 5y$?[2]

Because subtraction is defined as addition of the opposite, the Distributive Property also applies to subtraction. Thus, we can write $a(b - c) = ab - ac$ and $ac - bc = (a - b)c$. Here are the steps to combine the like terms of $8z - 12z$.

$$8z - 12z = (8 - 12)z \qquad \text{• Use the Distributive Property } ba - ca = (b - c)a.$$
$$= -4z \qquad \text{• Note: } 8 - 12 = 8 + (-12) = -4$$

Some variable expressions cannot be simplified. For instance, the variable expression $4a + 7b$ cannot be rewritten in a simpler form. The terms $4a$ and $7b$ do not have the same variable part. Therefore, the Distributive Property cannot be used. We say that $4a + 7b$ is in ***simplest form.*** As another example, $5x^2 + 8x$ is in simplest form because $5x^2$ and $8x$ are not like terms.

---

1. **a.** Not like terms; **b.** not like terms, $z^2 = z \cdot z$ and $z^3 = z \cdot z \cdot z$; **c.** not like terms; **d.** like terms.
2. $9y + 5y = (9 + 5)y = 14y$.

SUGGESTED ACTIVITY
The squares and rectangles used at the right are commonly referred to as *algebra tiles*. There is an activity in the Instructor's Resource Manual that describes ways these can be used to assist students with combining terms such as x^2, x, and constants.

Using squares and rectangles can help in understanding the concepts of adding like terms. Each ■ represents one x^2 and each ▬ represents one x. The figure below shows the simplification of $(2x^2 + 5x) + (3x^2 + 2x)$.

$$(2x^2 \quad + \quad 5x) \quad + \quad (3x^2 \quad + \quad 2x) \quad = \quad 5x^2 \quad + \quad 7x$$

The expression is simplified by combining the like terms, which, in this case, are the like geometric figures. Thus, $(2x^2 + 5x) + (3x^2 + 2x) = 5x^2 + 7x$.

Geometric models such as the one above are useful to visualize combining like terms. Generally, however, we use the Properties of Real Numbers to simplify a variable expression.

Example 3 Simplify: **a.** $2(-y)$ **b.** $-\frac{1}{3}(-3y)$

Solution

a. $2(-y) = 2(-1 \cdot y)$ •Recall: $-y = -1 \cdot y$.

$= [2(-1)]y$ •Use the Associative Property of Multiplication to regroup factors.

$= -2y$ • Multiply.

b. $-\frac{1}{3}(-3y) = \left[-\frac{1}{3}(-3)\right]y$ •Use the Associative Property of Multiplication to regroup factors.

$= 1y$ • Use the Inverse Property of Multiplication.

$= y$ • Use the Multiplication Property of One.

You-Try-It 3 Simplify: **a.** $-5(-3a)$ **b.** $\left(-\frac{1}{2}c\right)2$

Solution See page S11.

Example 4 Simplify: **a.** $5y + 3x - 5y$ **b.** $4x^2 + 5x - 6x^2 - 7x$

Solution

a. $5y + 3x - 5y = 3x + 5y - 5y$ • Use the Commutative Property of Addition to rearrange the terms.

$= 3x + (5y - 5y)$ • Use the Associative Property of Addition to group like terms.

$= 3x + 0$ • Use the Inverse Property of Addition.

$= 3x$ • Use the Addition Property of Zero.

b. $4x^2 + 5x - 6x^2 - 7x = 4x^2 - 6x^2 + 5x - 7x$ • Use the Commutative Property of Addition to rearrange the terms.

$= (4x^2 - 6x^2) + (5x - 7x)$ • Use the Associative Property of Addition to group like terms.

$= -2x^2 + (-2x)$ • Use the Distributive Property to combine like terms.

$= -2x^2 - 2x$ • Rewrite addition of the opposite as subtraction.

TAKE NOTE
As we did in the solution to Example 4b, it is customary to rewrite addition of the opposite as subtraction. For instance, we write $-2x^2 + (-2x)$ as $-2x^2 - 2x$.

You-Try-It 4 Simplify: **a.** $3a - 2b + 5a$ **b.** $2z^2 - 5z - 3z^2 + 6z$

Solution See page S11.

Question: Suppose you correctly simplify an expression and write the answer as $x + 7$ and another person writes the answer as $7 + x$. Are both answers correct?[3]

The Distributive Property also is used to remove parentheses from a variable expression. For instance,

$$\begin{aligned} 4(2x + 5z) &= 4(2x) + 4(5z) && \bullet \text{ Use the Distributive Property: } a(b + c) = ab + ac. \\ &= (4 \cdot 2)x + (4 \cdot 5)z && \bullet \text{ Use the Associative Property of Multiplication to regroup factors.} \\ &= 8x + 20z && \bullet \text{ Multiply.} \end{aligned}$$

When a negative number precedes the parentheses, be especially careful that all of the operations are performed correctly. Here are two examples.

→ Simplify: $-5(3x - 7) = -5(3x) - (-5)7$ • Use the Distributive Property.

$= -15x - (-35)$ • Multiply.

$= -15x + 35$ • Rewrite subtraction as addition of the opposite.

→ Simplify: $-3(-7a + 4) = -3(-7a) + (-3)4$ • Use the Distributive Property.

$= 21a + (-12)$ • Multiply.

$= 21a - 12$ • Rewrite addition of the opposite as subtraction.

The Distributive Property can be extended to expressions containing more than two terms. For instance,

$$\begin{aligned} 4(2x + 3y + 5z) &= 4(2x) + 4(3y) + 4(5z) \\ &= 8x + 12y + 20z \end{aligned}$$

Example 5 Simplify: **a.** $-3(2x + 4)$ **b.** $-(3z - 4)$ **c.** $(4a - 2c)5$ **d.** $6(3x - 4y + z)$

Solution

a. $-3(2x + 4) = -3(2x) + (-3)4$ • Use the Distributive Property.

$= -6x - 12$

b. $-(3z - 4) = -1(3z - 4)$ • Just as $-x = -1 \cdot x$, $-(3z - 4) = -1(3z - 4)$.

$= -1(3z) - (-1)(4)$ • Use the Distributive Property.

$= -3z + 4$

c. $(4a - 2c)5 = (4a)5 - (2c)5$ • Use the Distributive Property.

$= 20a - 10c$

d. $6(3x - 4y + z) = 6(3x) - 6(4y) + 6(z)$ • Use the Distributive Property.

$= 18x - 24y + 6z$

---

3. Yes; by the Commutative Property of Addition, $x + 7 = 7 + x$.

You-Try-It 5 Simplify: **a.** $-3(5y - 2)$ **b.** $-(6c + 5)$ **c.** $(3p - 7)(-3)$ **d.** $-2(4x + 2y - 6z)$

Solution See page S11.

TAKE NOTE
When we simplified $5 + 12x + 6$ (shown at the right), we wrote $12x + 11$. That is, the variable term was written first. Throughout the text we use this convention of writing variable terms first and then the constant term. If there is more than one variable term, we arrange the variable terms alphabetically. There is no mathematical reason to do this. It is just a convention that developed over time.

To simplify the expression $5 + 3(4x - 2)$, use the Distributive Property to remove the parentheses. Remember that $3(4x - 2)$ means $3 \cdot (4x - 2)$. Thus, by the Order of Operations Agreement, perform the multiplication before doing the addition.

$$5 + 3(4x + 2) = 5 + 3(4x) + 3(2)$$ • Use the Distributive Property.
$$= 5 + 12x + 6$$
$$= 12x + 11$$ • Add the like terms 5 and 6.

TAKE NOTE
The three steps in the dashed box at the right are usually done mentally. For the remainder of the text, we will show such a simplification as:

$$7 - 2[3y - 6] = 7 - 6y + 12$$
$$= -6y + 19$$

Here is another example, this time with subtraction. To simplify $7 - 2[3y - 6]$, use the Order of Operations Agreement to perform the multiplication before doing the subtraction.

$$7 - 2[3y - 6] = 7 + (-2)[3y + (-6)]$$ • Rewrite subtraction as addition of the opposite.
$$= 7 + (-2)(3y) + (-2)(-6)$$ • Use the Distributive Property.
$$= 7 + (-6y) + 12$$
$$= 7 - 6y + 12$$ • Rewrite addition of the opposite as subtraction.
$$= -6y + 19$$ • Add the like terms 7 and 12.

Example 6 Simplify: $3(2x - 4) - 5(3x + 2)$

Solution

$$3(2x - 4) - 5(3x + 2) = 6x - 12 - 15x - 10$$ • Use the Distributive Property to remove parentheses.
$$= -9x - 22$$ • Combine like terms.

You-Try-It 6 Simplify: $7(-3x - 4y) - 3(3x + y)$

Solution See page S11.

To simplify some expressions, the Distributive Property is used to remove parentheses and brackets so that like terms can be combined.

Example 7 Simplify: $3a - 2[7a - 2(2a + 1)]$

Solution

$$3a - 2[7a - 2(2a + 1)] = 3a - 2[7a - 4a - 2]$$ • Use the Distributive Property to remove parentheses.
$$= 3a - 2[3a - 2]$$ • Combine like terms inside the brackets.
$$= 3a - 6a + 4$$ • Use the Distributive Property to remove brackets.
$$= -3a + 4$$ • Combine like terms.

You-Try-It 7 Simplify: $2v - 3[5 - 3(3 + 2v)]$

Solution See page S11.

3.1 EXERCISES

Topics for Discussion

1. Explain the Commutative and Associative Property of Addition and Multiplication.
   Answers will vary.

2. Discuss the Distributive Property and how it is used to simplify a variable expression.
   Answers will vary.

3. Explain why a^2 and a are not like terms.
   Answers will vary.

4. When asked to simplify $3 + 5c$, Alexsis wrote the answer as $8c$. Discuss whether Alexsis's answer is correct or not.
   Because 3 and $5c$ are not like terms, the answer is not correct.

5. Does the expression $-x$ mean that the number represented by x is a negative number?
   No; $-x = -1 \cdot x$. If x represents a negative number, then $-x$ is a positive number; if x represents a positive number, then $-x$ is a negative number.

Properties of Real Numbers

Use the given property to complete the statement.

6. The Commutative Property of Multiplication
   $2 \cdot 5 = 5 \cdot ?$ 2

7. The Addition Property of Zero
   $? + x = x$ 0

8. The Commutative Property of Addition
   $9 + 17 = ? + 9$ 17

9. The Distributive Property
   $2(4 + 3) = 8 + ?$ 6

10. The Associative Property of Multiplication
    $4(5x) = (? \cdot 5)x$ 4

11. The Multiplication Property of One
    $? \cdot 1 = -4$ −4

12. The Associative Property of Addition
    $(4 + 5) + 6 = ? + (5 + 6)$ 4

13. The Inverse Property of Addition
    $8 + ? = 0$ −8

14. The Multiplication Property of Zero
$y \cdot ? = 0$ 0

15. The Inverse Property of Multiplication
$\left(-\frac{1}{5}\right)(-5) = ?$ 1

Identify the Property of Real Numbers that justifies the statement.

16. $1 \cdot a = a$
The Multiplication Property of One

17. $3(4x) = (3 \cdot 4)x$
The Associative Property of Multiplication

18. $0 + c = c$
The Addition Property of Zero

19. $z + (-z) = 0$
The Inverse Property of Addition

20. $\left(-\frac{2}{3}\right)\left(-\frac{3}{2}\right) = 1$
The Inverse Property of Multiplication

21. $3(4 + 7) = 12 + 21$
The Distributive Property

22. $2 + (4 + w) = (2 + 4) + w$
The Associative Property of Addition

23. $(-3 + 9)8 = -24 + 72$
The Distributive Property

24. $(3x)(4) = 4(3x)$
The Commutative Property of Multiplication

25. $(x + y) + z = z + (x + y)$
The Commutative Property of Addition

Simplifying Variable Expressions

Simplify each of the following. If the expression is already in simplest form, write "simplest form" as the answer.

26. $6x + 8x$ $14x$

27. $12y + 9y$ $21y$

28. $8b - 5b$ $3b$

29. $4y - 10y$ $-6y$

30. $2a + 7$ simplest form

31. $x + y$ simplest form

32. $-12a + 17a$ $5a$

33. $-12xy + 17xy$ $5xy$

34. $3x + 5x + 3x$ $11x$

35. $-5x^2 - 12x^2 + 3x^2$ $-14x^2$

36. $7x - 3y + 10x$ $17x - 3y$

37. $3x - 8y - 10x + 4x$ $-3x - 8y$

38. $5a + 6a - 2a$ $9a$

39. $-5x + 7x - 4x$ $-2x$

40. $2a - 5a + 3a$ 0

41. $12y^2 + 10y^2$ $22y^2$

42. $3x^2 - 15x^2$ $-12x^2$

43. $9z^2 - 9z^2$ 0

44. $\frac{3}{4}x - \frac{1}{4}x$ $\frac{1}{2}x$

45. $\frac{2}{5}y - \frac{3}{4}y$ $-\frac{7}{20}y$

46. $3x - 7 + 4x$ $7x - 7$

47. $4(3x)$ $12x$

48. $-2(-3y)$ $6y$

49. $(3a)(-2)$ $-6a$

50. $-5(3x^2)$ $-15x^2$

51. $\frac{1}{8}(8x)$ x

52. $\frac{12x}{5}\left(\frac{5}{12}\right)$ x

53. $\frac{1}{7}(14x)$ $2x$

54. $-\frac{5}{8}(24a^2)$ $-15a^2$

55. $(33y)\left(\frac{1}{11}\right)$ $3y$

56. $-(x + 2)$ $-x - 2$

57. $-2(a + 7)$ $-2a - 14$

58. $(5 - 3b)7$ $35 - 21b$

59. $3(5x^2 + 2x)$ $15x^2 + 6x$

60. $(-3x - 6)5$ $-15x - 30$

61. $-3(2y^2 - 7)$ $-6y^2 + 21$

62. $4(x^2 - 3x + 5)$
$4x^2 - 12x + 20$

63. $4(-3a^2 - 5a + 7)$
$-12a^2 - 20a + 28$

64. $5(2x^2 - 4xy - y^2)$
$10x^2 - 20xy - 5y^2$

65. $6a - (5a + 7)$ $a - 7$

66. $8 - (12 + 4y)$ $-4y - 4$

67. $6(2y - 7) - 3(3 - 2y)$ $18y - 51$

68. $2[x + 2(x + 7)]$ $6x + 28$

69. $-2[3x - (5x - 2)]$ $4x - 4$

70. $4a - 2[2b - (b - 2a)] + 3b$ b

Applying Concepts

Complete.

71. A number that has no reciprocal is 0.

72. Two numbers that are their own reciprocals are −1 and 1.

73. The additive inverse of $a - b$ is $-a + b$.

74. The multiplicative inverse of $-a$ is $-\frac{1}{a}$.

75. Determine whether the statement is true or false. If the statement is false, give an example that illustrates that it is false.

a. Division is a commutative operation. False; $8 \div 4 \neq 4 \div 8$

b. Division is an associative operation. False; $(8 \div 4) \div 2 \neq 8 \div (4 \div 2)$

c. Subtraction is an associative operation. False; $7 - (5 - 2) \neq (7 - 5) - 2$

d. Subtraction is a commutative operation. False; $7 - 4 \neq 4 - 7$

e. Addition is a commutative operation. True

Exploration

76. ***Binary operations*** Operations such as addition and multiplication are called *binary operations.* The word *binary* means "consisting of two parts." Addition consists of two parts called the addends; multiplication consists of two parts called factors. Other binary operations can be defined. For instance, define $\otimes$ as $a \otimes b = (a \cdot b) - (a + b)$. Then

$7 \otimes 4 = (7 \cdot 4) - (7 + 4) = 28 - 11 = 17$

a. Find $6 \otimes 8$.

b. Find $\frac{4}{3} \otimes \frac{3}{4}$.

c. Does $a \otimes 0 = 0$?

d. Does $a \otimes 1 = 1$?

e. Is $\otimes$ a commutative operation? Support your answer.

f. Is $\otimes$ an associative operation? Support your answer.

Section 3.2 Introduction to Equations

Solutions of Equations

In 1687, Isaac Newton (1642 – 1727) published *Philosophiae Naturalis Principia Mathematica* (Mathematical Principles of Natural Philosophy). An *equation* contained within this work was

$$F = \frac{GMm}{r^2}$$

POINT OF INTEREST
With the publication of Newton's work, Edmund Halley used the equation at the right to predict that a comet last seen in 1682 would appear again in 1758. Ever since, this comet has reappeared approximately every 76 years and has become known as Halley's Comet. Using the principles upon which this equation is based, astronomers have predicted that the comet Hale-Bopp, last seen in 1997, will reappear again around 6200.

Although it is not apparent by just looking at the equation, Newton was mathematically stating that there is a force of gravity between two objects (like Earth and a rock) that causes the objects to come together, say by dropping the rock.

An **equation** expresses the equality of two mathematical expressions. The expressions can be either numerical or variable expressions. Because an equation expresses the equality of two mathematical expressions, an equation *always* contains an equal sign. Besides the equation above, here are some other examples of equations.

$$x + 3 = 7 \qquad y = 2x \qquad 5 + 4z = z^2 - 5z$$

Question: Which of the following are equations and which are expressions: **a.** $p = 4$; **b.** $3x^2 + y$; **c.** $6x^2 + 4x - 8$; **d.** $\frac{1}{x+2} = \frac{3}{x-2}$?[1]

The equation $x + 3 = 7$ is true if the variable is replaced by **4**.

$x + 3 = 7$
$\mathbf{4} + 3 = 7$ A true equation

The equation $x + 3 = 7$ is false if the variable is replaced by **6**.

$x + 3 = 7$
$\mathbf{6} + 3 = 7$ A false equation

A **solution** of an equation is a number that, when substituted for the variable, results in a true equation. **4** is a solution of the equation $x + 3 = 7$. **6** is not a solution of the equation $x + 3 = 7$.

Example 1 Is −3 a solution of the equation $5 - 2x = 6x + 29$?

Solution

$5 - 2x$	$= 6x + 29$
$5 - 2(-3)$	$6(-3) + 29$
$5 + 6$	$-18 + 29$
11	$= 11$

- Replace the variable by the given number, −3.
- Evaluate the numerical expressions using the Order of Operations Agreement.
- Compare the results. If the results are equal, the given number is a solution. If the results are not equal, the given number is not a solution.

Yes, −3 is a solution of the equation.

You-Try-It 1 Is 4 a solution of the equation $3x = x^2 - 4$?

Solution See page S11.

1. **a.** Equation; **b.** expression; **c.** expression; **d.** equation.

TAKE NOTE
See the appendix Guidelines for Using Graphing Calculators.

A graphing calculator can be used to test whether a certain value of the variable is a solution of the equation. For You-Try-It 1, enter the left side of the equation $(3x)$ as Y1 and the right side of the equation $(x^2 - 4)$ as Y2. Store the proposed solution, 4, in X and then test to determine whether Y1 = Y2. Your screens will be similar to the ones below.

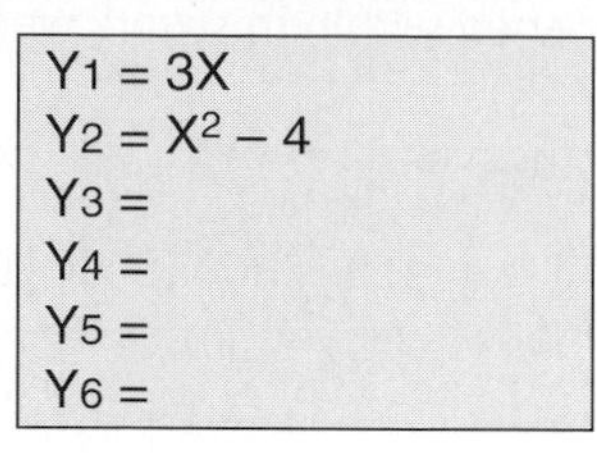

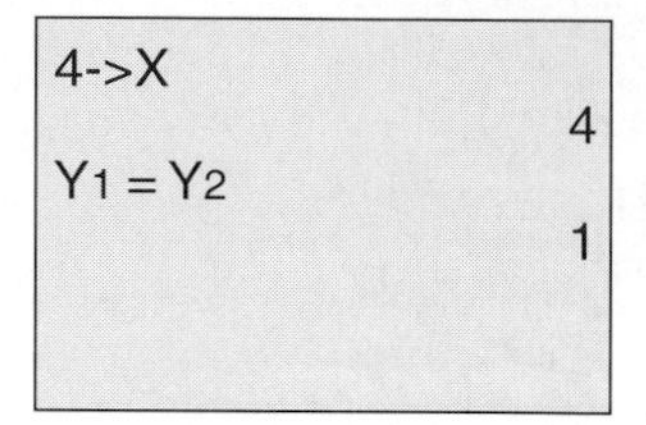

The 1 on the right-hand screen above indicates that the statement Y1 = Y2 is true when $x = 4$. A 0 (zero) would indicate the statement is false.

Solve Equations of the Form $x + a = b$

Read the three problems below.

a. What number plus three equals seven?

b. The distance that Julia drives from home to campus is seven miles. If Julia has driven three miles, how many additional miles must she drive to reach campus?

c. A chef has three quarts of milk. How many more quarts of milk does the chef need for a New England clam chowder recipe that requires seven quarts of milk?

SUGGESTED ACTIVITY
Have students make up another situation that requires solving the equation $x + 3 = 7$. Now have them make up a situation that requires solving $x - 3 = 7$.

Question: What do the three problems above have in common?[2]

You may have noticed that the answer to each of the problems above contained a 4. Although the problems are about different situations, they can be described by similar equations.

a. Let n represent the unknown number. The equation is

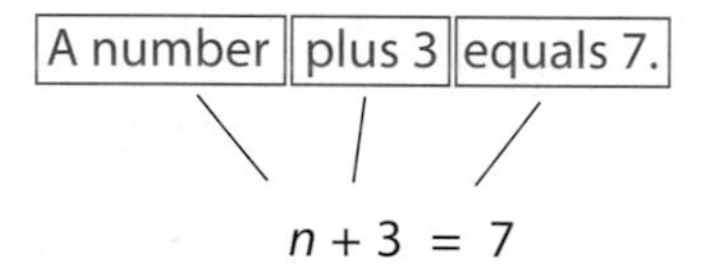

b. Let d represent the additional miles Julia must drive. The equation is

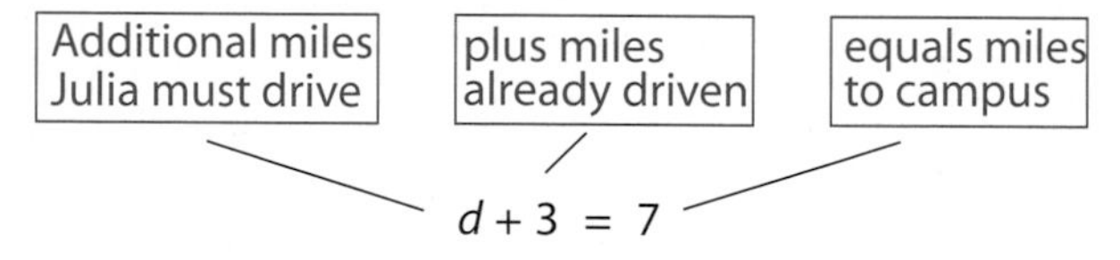

c. Let q represent the number of quarts needed to complete the recipe. The equation is

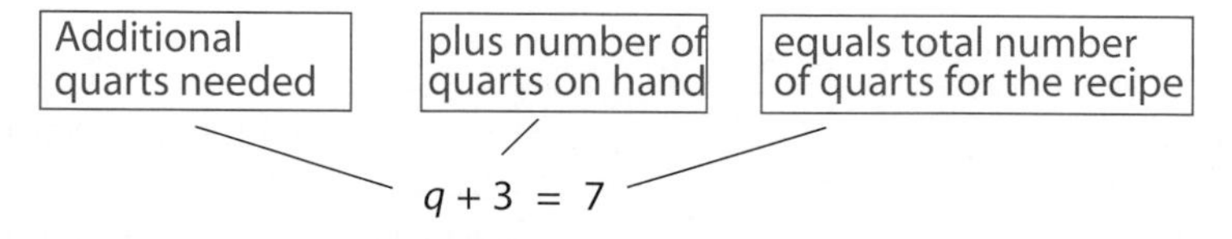

If you answered 4 to any one of these problems, you *solved an equation*. To **solve** an equation means to find a solution of the equation. The solution to each one of the equations is 4.

2. The answers are **4**, **4** miles, and **4** quarts. Each answer contains a **4**.

The simplest equation to solve is one of the form *variable* = *constant*. The solution is the constant. For instance, for the equation $x = 4$, 4 is the solution of the equation because 4 = 4 is a true equation.

$$\begin{aligned} x &= 4 \\ y &= -3 \\ b &= 7 \\ \text{Variable} &= \text{constant} \end{aligned}$$

Question: What is the solution of the equation $y = -3$?[3]

As shown at the right, the solution of the equation $x + 3 = 7$ is 4.

$$\begin{array}{c|c} x + 3 = 7 & \\ \hline 4 + 3 & 7 \\ 7 = 7 & \end{array}$$

Note that if 5 is added to each side of the equation, the solution of the new equation is still 4.

$$\begin{aligned} x + 3 &= 7 \\ x + 3 + 5 &= 7 + 5 \\ x + 8 &= 12 \end{aligned}$$

$$\begin{array}{c|c} x + 8 = 12 & \\ \hline 4 + 8 & 12 \\ 12 = 12 & \end{array}$$

Note that if −6 is added to each side of the equation, the solution of the new equation is still 4.

$$\begin{aligned} x + 3 &= 7 \\ x + 3 + (-6) &= 7 + (-6) \\ x - 3 &= 1 \end{aligned}$$

$$\begin{array}{c|c} x - 3 = 1 & \\ \hline 4 - 3 & 1 \\ 1 = 1 & \end{array}$$

This illustrates the Addition Property of Equations.

> **Addition Property of Equations**
> The same number can be added to each side of an equation without changing the solution of the equation. That is, the equations $a = b$ and $a + c = b + c$ have the same solutions.

This property is used in solving equations. Note the effect of adding, to each side of the equation $x + 3 = 7$, the *opposite of the constant term* 3. After each side of the equation is simplified, the equation is in the form *variable* = *constant*. The solution is the constant.

$$\begin{aligned} x + 3 &= 7 \\ x + 3 + (-3) &= 7 + (-3) \\ x + 0 &= 4 \\ x &= 4 \\ \text{Variable} &= \text{constant} \end{aligned}$$

The solution is 4.

In solving an equation, the goal is to rewrite the given equation in the form *variable* = *constant*. The Addition Property of Equations can be used to rewrite an equation in this form. The Addition Property of Equations is used to remove a term from one side of an equation by **adding the opposite of that term to each side of the equation**.

TAKE NOTE
You should always check your solution. The check for Example 2 is shown below.

$$\begin{array}{c|c} x - 4 = 9 & \\ \hline 13 - 4 & 9 \\ 9 = 9 & \end{array}$$

This is a true equation. The solution checks.

Example 2 Solve: $x - 4 = 9$

Solution

$x - 4 = 9$ • The goal is to rewrite the equation in the form variable = constant.

$x - 4 + 4 = 9 + 4$ • Add the opposite of the constant term −4 to each side of the equation. This step uses the Addition Property of Equations.

$x + 0 = 13$ • Simplify using the Inverse Property of Addition ($-4 + 4 = 0$).

$x = 13$ • Simplify using the Addition Property of Zero. The equation is in the form variable = constant.

The solution is 13.

---

3. The solution is −3, the constant.

You-Try-It 2 Solve: $y - 7 = -10$

Solution See page S11.

Because subtraction is defined in terms of addition, the Addition Property of Equations makes it possible to subtract the same number from each side of an equation without changing the solution of the equation.

➜ Solve: $z + \frac{1}{3} = -\frac{1}{2}$

The goal is to write the equation in the form *variable* = *constant*.

$$z + \frac{1}{3} = -\frac{1}{2}$$

Add the opposite of the constant term $\frac{1}{3}$ to each side of the equation. This is equivalent to subtracting $\frac{1}{3}$ from each side of the equation.

$$z + \frac{1}{3} - \frac{1}{3} = -\frac{1}{2} - \frac{1}{3}$$

$$z + 0 = -\frac{3}{6} - \frac{2}{6}$$

$$z = -\frac{5}{6}$$

The solution $-\frac{5}{6}$ checks.

The solution is $-\frac{5}{6}$.

For some equations, the goal may be to rewrite the equation in the form *constant* = *variable* rather than *variable* = *constant*. In each case, the solution is the constant.

Example 3 Solve: $\frac{3}{4} = \frac{2}{3} + a$

Solution

$$\frac{3}{4} = \frac{2}{3} + a$$

$$\frac{3}{4} - \frac{2}{3} = \frac{2}{3} - \frac{2}{3} + a$$

• Subtract $\frac{2}{3}$ from each side of the equation.

$$\frac{1}{12} = a$$

• Simplify each side.

Note: $\frac{3}{4} - \frac{2}{3} = \frac{9}{12} - \frac{8}{12} = \frac{1}{12}$

You-Try-It 3 Solve: $-7 = -3 + v$

Solution See page S11.

Solve Equations of the Form $ax = b$

SUGGESTED ACTIVITY
See the *Instructor's Resource Manual* for an activity that will have students create an equation of the form $ax = b$.

The solution of the equation shown at the right is 5.

$$\begin{array}{r|l} 2x & = 10 \\ \hline 2(5) & 10 \\ 10 & = 10 \end{array}$$

Note that if each side of the equation is multiplied by 4, the solution of the new equation is still 5.

$$2x = 10$$
$$4(2x) = 4(10)$$
$$8x = 40$$

$$\begin{array}{r|l} 8x & = 40 \\ \hline 8(5) & 40 \\ 40 & = 40 \end{array}$$

Note that if each side of the equation is multiplied by -3, the solution of the new equation is still 5.

$$2x = 10$$
$$-3(2x) = -3(10)$$
$$-6x = -30$$

$$\begin{array}{r|l} -6x & = -30 \\ \hline -6(5) & -30 \\ -30 & = -30 \end{array}$$

This illustrates the Multiplication Property of Equations.

> **Multiplication Property of Equations**
> Each side of an equation can be multiplied by the same nonzero number without changing the solution of the equation. That is, the equations $a = b$ and $ac = bc$, $c \neq 0$, have the same solutions.

This property is used in solving equations. Note the effect of multiplying each side of the equation $2x = 10$ by the reciprocal of the constant term 2. After each side of the equation is simplified, the equation is in the form *variable = constant*. The solution is the constant.

$$2x = 10$$
$$\frac{1}{2} \cdot 2x = \frac{1}{2} \cdot 10$$
$$1x = 5$$
$$x = 5$$

Variable = constant
The solution is 5.

In solving an equation, the goal is to rewrite the given equation in the form *variable = constant*. The Multiplication Property of Equations can be used to rewrite an equation in this form. The Multiplication Property of Equations is used to remove a coefficient from a variable term in an equation **by multiplying each side of the equation by the reciprocal of the coefficient**.

Question: Do you remember the meaning of coefficient?[4]

Example 4 Solve: $\frac{-2w}{3} = 8$

Solution

$-\frac{2}{3}w = 8$ • $\frac{-2w}{3} = -\frac{2}{3}w$

$-\frac{3}{2}\left(-\frac{2}{3}w\right) = -\frac{3}{2}(8)$ • Multiply each side of the equation by the reciprocal of the coefficient $-\frac{2}{3}$. This uses the Multiplication Property of Equations.

$1w = -12$ • Simplify using the Inverse Property of Multiplication $\left[-\frac{3}{2}\left(-\frac{2}{3}\right) = 1\right]$.

$w = -12$ • Simplify using the Multiplication Property of One. The equation is now in the form variable = constant.

The solution is −12. • Write the solution.

You-Try-It 4 Solve: $15 = \frac{3y}{5}$

Solution See page S11.

Because division is defined in terms of multiplication, the Multiplication Property of Equations makes it possible to divide each side of an equation by the same number without changing the solution of the equation.

---

4. A numerical coefficient is the number that multiplies a variable. The coefficient of $4x$ is 4. For the expression $\frac{3m}{4}$, the coefficient of m is $\frac{3}{4}$ because $\frac{3m}{4} = \frac{3}{4} \cdot \frac{m}{1} = \frac{3}{4}m$.

→ Solve: $18 = 6p$

The goal is to write the equation in the form *constant* = *variable.*

$18 = 6p$

Multiply each side of the equation by the reciprocal of 6, the coefficient of p. This is equivalent to dividing each side of the equation by 6.

$\frac{18}{6} = \frac{6p}{6}$

The solution 3 checks.

$3 = p$

The solution is 3.

When using the Multiplication Property of Equations to solve an equation, multiply each side of the equation by the reciprocal of the coefficient when the coefficient is a fraction. Divide each side of the equation by the coefficient when the coefficient is an integer or a decimal.

Example 5 Solve and check: $-6x = 8$

Solution

$-6x = 8$

$\frac{-6x}{-6} = \frac{8}{-6}$ • Divide each side of the equation by -6, the coefficient of x.

$x = -\frac{4}{3}$ • Simplify each side of the equation.

Check:

$$\begin{array}{c|c} -6x & 8 \\ \hline -6\left(-\frac{4}{3}\right) & 8 \end{array}$$

$8 = 8$ • This is a true equation. The solution checks.

The solution is $-\frac{4}{3}$.

You-Try-It 5 Solve and check: $9 = 27q$

Solution See page S12.

Before using one of the Properties of Equations, check to see if one or both sides of the equation can be simplified. In Example 6 below, like terms appear on the left side of the equation. The first step in solving this equation is to combine the like terms so that there is only one variable term on the left side of the equation.

Example 6 Solve: $6t - 11t = -15$

Solution

$6t - 11t = -15$

$-5t = -15$ • Combine like terms.

$\frac{-5t}{-5} = \frac{-15}{-5}$ • Divide each side of the equation by -5.

$t = 3$

The solution is 3.

You-Try-It 6 Solve: $18 = 2n + 6n$

Solution See page S12.

Solve the Basic Percent Equation

INSTRUCTOR NOTE
In application problems involving percent, the basic percent equation frequently results in an equation of the form $ax = b$.

The solution of a problem that involves a percent requires solving the basic percent equation shown below.

The Basic Percent Equation

$$\text{Percent} \cdot \text{Base} = \text{Amount}$$
$$P \cdot B = A$$

To translate a problem involving a percent into an equation, remember that the word *of* translates to "multiply" and the word *is* translates to "=". The base usually follows the word *of*.

SUGGESTED ACTIVITY
Students have difficulty finding the base and amount for percent equations. In the *Instructor's Resource Manual* (IRM), there are several sentences concerning percents that have no unknowns. Ask students to identify the amount and base for each sentence. Here are two instances (the IRM contains application examples that are more difficult):
1. 30 is 75% of 40.
2. 40% of 20 is 8.

➜ 30% of what number is 15?

Given: $P = 30\% = 0.30$; $A = 15$
Unknown: Base, B
Replace P and A by their values and then solve for B.

$$PB = A$$
$$(0.30)B = 15$$
$$\frac{0.30B}{0.30} = \frac{15}{0.30}$$
$$B = 50$$

The number is 50.

➜ What percent of 25 is 20?

Given: $B = 25$; $A = 20$
Unknown: Percent, P
Replace B and A by their values and then solve for P.

$$PB = A$$
$$P(25) = 20$$
$$25P = 20$$
$$\frac{25P}{25} = \frac{20}{25}$$
$$P = 0.80$$

Remember to write P as a percent. 20 is 80% of 25.

➜ Find $16\frac{2}{3}\%$ of 270.

TAKE NOTE
In most cases, the percent is written as a decimal before solving an equation. However, some percents are more easily written as a fraction.

$33\frac{1}{3}\% = \frac{1}{3}$; $66\frac{2}{3}\% = \frac{2}{3}$

$16\frac{2}{3}\% = \frac{1}{6}$; $83\frac{1}{3}\% = \frac{5}{6}$

Given: $P = 16\frac{2}{3}\% = \frac{1}{6}$; $B = 270$
Unknown: Amount, A
Replace P and B by their values and then solve for A.

$$PB = A$$
$$\frac{1}{6} \cdot 270 = A$$
$$45 = A$$

$16\frac{2}{3}\%$ of 270 is 45.

Example 7 12 is 15% of what number?

Solution

$$PB = A$$
$$0.15B = 12 \quad \text{• Write 15\% as 0.15. The base is unknown.}$$
$$\frac{0.15B}{0.15} = \frac{12}{0.15}$$
$$B = 80$$

The number is 80.

You-Try-It 7 27 is what percent of 60?

Solution See page S12.

Applications of Percents

Example 8 Based on data from Bausch and Lomb (*USA Today*, 11/21–23/97), 731 of the 850 computer executives responding to a survey reported eyestrain symptoms from viewing a computer monitor. What percent of the executives reported eyestrain?

Solution

State the goal. The goal is to find what percent of the executives reported eyestrain.

Describe a strategy. To find the percent, solve the basic percent equation for percent. The base is 850, the number of executives responding to the survey, and the amount is 731, the number of executives that reported eyestrain.

Solve the problem.

$PB = A$

$850P = 731$ • We have used the Commutative Property of Multiplication to write $P(850)$ as $850P$.

$\frac{850P}{850} = \frac{731}{850}$

$P = 0.86$

86% of the executives surveyed reported eyestrain.

Check your work. Note that $P = 0.86$ but the question asks for the percent of executives. Therefore, you must write the answer as a percent.

You-Try-It 8 Based on a national survey of workers, approximately 55% take a lunch break that is 15 minutes or less. If 1260 workers were surveyed, how many took a lunch break that was 15 minutes or less? (Source: *USA Today*, 1/21–23/97)

Solution See page S12.

A frequent application problem that involves percents is to find a percent increase or a percent decrease between two situations. For these problems, the base is always the number *before* the increase or decrease. The amount is the actual increase or decrease.

Example 9 The amount of snow in the Sierra Nevada increased from 125 inches in 1997 to 135 inches in 1998. What percent increase does this represent?

Solution

State the goal. The goal is to find the percent increase in snowfall between 1997 and 1998.

Describe a strategy. To find the percent increase, first find the amount of increase ($135 - 125 = 10$). Then solve the basic percent equation for percent. The base is 125, the amount of snow before the increase. The amount is 10, the increase in the amount of snow.

Solve the problem.

$PB = A$

$125P = 10$

$\frac{125P}{125} = \frac{10}{125}$

$P = 0.08$

There was an 8% increase in the amount of snowfall between the two years.

Check your work. Note that 8% of 125 is 10, the increase in the amount of snow.

You-Try-It 9 As a fog bank covered an airport, the distance a pilot could see decreased from 1000 feet to 550 feet. What percent decrease does this represent?

Solution See page S12.

3.2 EXERCISES

Topics for Discussion

1. How is the Addition Property of Equations used to solve an equation? Give an example of the use of this property.
Answers will vary.

2. How is the Multiplication Property of Equations used to solve an equation? Give an example of the use of this property.
Answers will vary.

3. When using the Multiplication Property of Equations, the number that multiplies each side of an equation must be nonzero. Explain why.
Answers will vary.

4. What is the basic percent equation?
Percent · Base = Amount

5. Explain the difference between an equation and an expression.
An equation has an equal sign; an expression does not have an equal sign.

Solutions of Equations

6. Is 4 a solution of $2x = 8$?
Yes

7. Is 1 a solution of $4 - 2m = 3$?
No

8. Is 0 a solution of $4a + 5 = 3a + 5$?
Yes

9. Is 3 a solution of $z^2 + 1 = 4 + 3z$?
No

10. Is 4 a solution of $x(x + 1) = x^2 + 5$?
No

11. Is $\frac{2}{5}$ a solution of $5m + 1 = 10m - 3$?
No

Solving Equations

Solve.

12. $x + 5 = 7$ 2

13. $2 + a = 8$ 6

14. $n - 5 = -2$ 3

15. $a - 3 = -5$ −2

16. $10 + m = 3$ −7

17. $b - 5 = -3$ 2

18. $4 = m - 11$ 15

19. $4 = -10 + b$ 14

20. $x - \frac{1}{2} = \frac{1}{2}$ 1

21. $m+\frac{1}{2}=-\frac{1}{4}$ $-\frac{3}{4}$

22. $-\frac{5}{6}=x-\frac{1}{4}$ $-\frac{7}{12}$

23. $\frac{5}{12}=n+\frac{3}{4}$ $-\frac{1}{3}$

24. $-1.926+t=-1.042$ 0.884

25. $2a=-14$ −7

26. $-6n=-30$ 5

27. $-56=7x$ −8

28. $35=-5x$ −7

29. $-12m=-144$ 12

30. $-\frac{b}{3}=6$ −18

31. $-\frac{4}{3}c=-8$ 6

32. $\frac{2n}{3}=2$ 3

33. $-6=-\frac{2}{3}y$ 9

34. $-\frac{2}{5}m=-\frac{6}{7}$ $\frac{15}{7}$

35. $10y-3y=21$ 3

Percent Equations

36. 12 is what percent of 50?
24%

37. 12% of what is 48?
400

38. What percent of 12 is 3?
25%

39. 12 is what percent of 6?
200%

40. Find 15.4% of 50.
7.7

41. $\frac{3}{4}$% of what is 3?
400

42. What percent of 125 is 50?
40%

43. $5\frac{1}{4}$% of what is 21?
400

44. What is 18.5% of 46?
8.51

45. 1 is 0.5% of what?
200

46. Find 125% of 16.
20

47. What is 250% of 12?
30

Applications

48. The circle graph at the right shows the market share of light vehicle sales for each manufacturer of cars. Assuming that there were 19,200,000 light vehicles sold by these companies through April 1998, answer the following questions. (Source: *New York Times*, 5/21/98, page D1)

a. How many cars did General Motors sell? Round to the nearest hundred thousand.

b. How many more cars did Toyota sell than Honda?

a. 6,000,000 cars; b. 288,000 cars

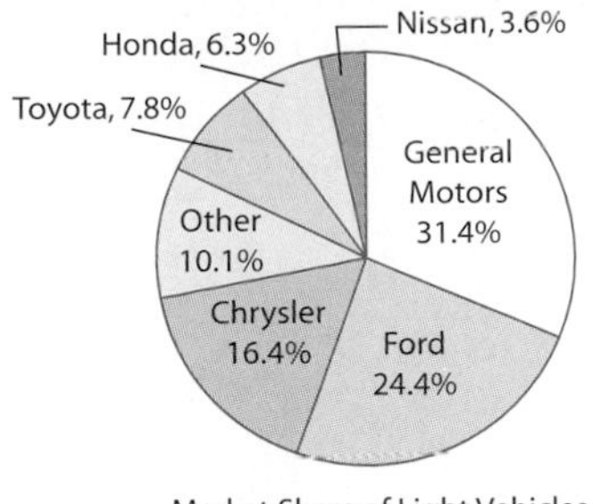

Market Share of Light Vehicles in U.S. through April 1998.
Source: From *The New York Times*, May 21, 1998, p. D1. Copyright©1998 by The New York Times. Reprinted by permission.

49. According to the National Association of Colleges and Employers, the average starting salary of computer science majors in 1997 was \$41,000. This was up from an average starting salary in 1996 of \$36,000. What was the percent increase in starting salary? Round to the nearest tenth of a percent.
13.9%

50. The graph at the right, based on data from the U.S. Department of Agriculture, shows the total world grain production from 1950 to 1997. What was the percent increase in world grain production from 1950 to 1997? Round to the nearest percent. 192%

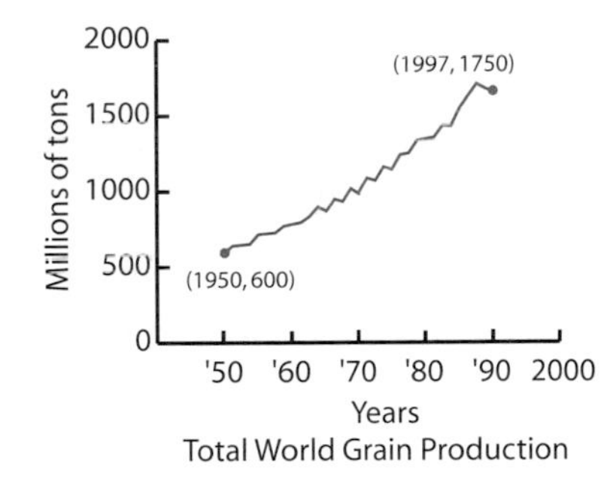

Total World Grain Production

51. In 1990 there were an average of 83 carbonated soft drinks consumed per person in the U.S. By 1997, the consumption had grown to 102 carbonated soft drinks per person. What percent increase, to the nearest tenth of a percent, does this represent? Source: NPD Group 12th Annual Report on Eating. 22.9%

52. The average balance for all credit cards of an individual in the U.S. declined from \$3362 in 1997 to \$2870 in 1998. What percent decrease does this represent? Round to the nearest tenth of a percent. Source: Marist Institute of Public Opinion. 14.6%

53. According to the Administrative Office of the United States Courts, the number of authorized wiretaps was 1094 in 1997. This was approximately a 6% increase from the previous year. How many authorized wiretaps were there in 1996? 1032 wiretaps

54. In 1997, the average annual health care benefit costs for workers were \$3924. This was a 3.8% increase from 1993. Find, to the nearest dollar, the average annual health care benefit costs in 1993. \$3780

55. To override a presidential veto, at least $66\frac{2}{3}$% of the Senate must vote to override the veto. There are 100 senators in the Senate. What is the minimum number of votes needed to override a veto? 67 votes

56. An airline knowingly overbooks certain flights by selling 18% more tickets than there are available seats. How many tickets would this airline sell for an airplane that has 150 seats? 177 tickets

Applying Concepts

Solve.

57. $\frac{2m + m}{5} = -9$ -15

58. $\frac{3y - 8y}{7} = 15$ -21

59. $\frac{1}{\frac{1}{x}} = 5$ 5

60. $\frac{1}{\frac{1}{x}} + 8 = 10$ 2

61. $\frac{4}{\frac{3}{y}} = 8$ 6

62. $\frac{5}{\frac{7}{a}} - \frac{3}{\frac{7}{a}} = 6$ 21

63. Solve for x: $x \div 28 = 1481$ remainder 25. 41,493

64. Your bill for dinner, including a 7.25% sales tax, was \$62.74. You want to leave a 15% tip on the cost of the dinner before the sales tax. Find the amount of the tip to the nearest dollar. \$9

65. Make up an equation of the form $x + a = b$ that has -2 as a solution.
One possible answer is $x + 9 = 7$.

66. Make up an equation of the form $ax = b$ that has $\frac{2}{3}$ as a solution.

One possible answer is $3x = 2$.

67. If a quantity increases by 100%, how many times its original value is the new value? 2

Exploration

68. ***Consumer Price Index (CPI)*** The CPI is a percent that is written without the percent sign. For instance, a CPI of 141.2 means 141.2% and a CPI of 78.3 means 78.3%. The CPI is used to compare the costs of items in the *base years*, between 1982 and 1984, and the current year. For instance, the CPI for tomatoes at the end of 1997 was 161.7. This means that \$10 worth of tomatoes in the base years cost \$16.17 at the end of 1997. The table at the right gives the CPI for various food products at the end of 1997. Use this table for the following exercises. Source: Bureau of Labor Statistics

Product	CPI
Orange juice	167.0
Oranges	58.3
Bananas	46.1
Eggs	117.2
Milk	161.4
Chicken	100.1
Ground chuck	181.1
White bread	88.4
Coffee	416.1

a. Of the items listed, which items cost less at the end of 1997 than they did in the base years?
b. Of the items listed, which item had the greatest increase in cost over the base years?
c. If a person paid \$1.29 per pound for ground chuck in the base years, what did a pound of ground chuck cost at the end of 1997?
d. If a person paid \$.98 for a dozen eggs in the base years, what did a dozen eggs cost at the end of 1997?
e. Of the items listed, which item cost approximately the same at the end of 1997 as it did in the base years?

Section 3.3 General Equations

Solving Equations Using the Addition and Multiplication Properties of Equations

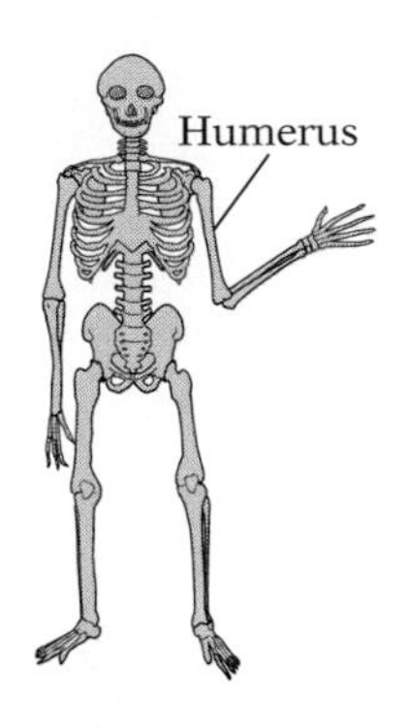

By measuring the bones of skeletal fossils, anthropologists have created the equation $H = 1.2L + 27.8$, which is used to approximate the height, H (in inches), of a primate based on the length, L (in inches), of its humerus (the bone extending from the shoulder to the elbow). For instance, if an anthropologist discovered the humerus of a primate that measured 28.5 inches, then the approximate height of the primate can be found by using the equation $H = 1.2L + 27.8$.

$$\begin{aligned} H &= 1.2L + 27.8 \\ &= 1.2(28.5) + 27.8 && \text{• Replace } L \text{ by 28.5.} \\ &= 62 \end{aligned}$$

The approximate height of the primate is 62 inches.

The input/output table at the right was produced by entering the equation $H = 1.2L + 27.8$ into a graphing calculator with Y1 as H and X as L. By looking at the table, we can determine that a primate whose height is 65 inches has a humerus that is 31 inches.

X	Y1	
28	61.4	
28.5	62	
29	62.6	
29.5	63.2	
30	63.8	
30.5	64.4	
31	65	
X=31		

Question: Using the table above, what is the length of the humerus for a primate whose height is 63.8 inches?[1]

By using the table, we can find the length of the humerus for heights given in the table. However, suppose we wanted to find the length of the humerus for a primate whose height is 63.5 inches. Because 63.5 is between 63.2 and 63.8, the approximate length of the humerus is between 29.5 inches and 30 inches.

A graph of the input/output table is shown at the right where the axes have been scaled to show the features of the graph. A line has been drawn through the points. Drawing a horizontal line from 63.5 to the line and then a vertical line to the L-axis is another method of approximating the solution of the equation. From the graph, it appears that the length of the humerus is approximately 29.8 inches.

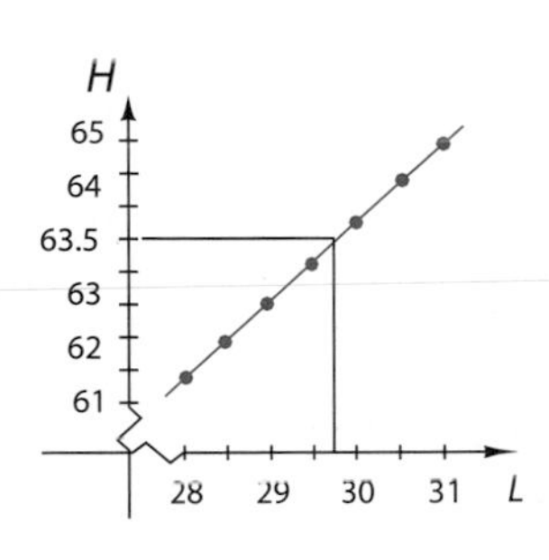

However, to find the exact length, we must solve the equation $H = 1.2L + 27.8$ for L when $H = 63.5$. This will require both the Addition and Multiplication Properties of Equations.

$$\begin{aligned} H &= 1.2L + 27.8 \\ 63.5 &= 1.2L + 27.8 && \text{• Replace } H \text{ by 63.5.} \\ 63.5 - 27.8 &= 1.2L + 27.8 - 27.8 && \text{• Subtract 27.8 from each side of the equation.} \\ 35.7 &= 1.2L \\ \frac{35.7}{1.2} &= \frac{1.2L}{1.2} && \text{• Divide each side of the equation by 1.2.} \\ 29.75 &= L \end{aligned}$$

The length of the humerus is 29.75 inches.

SUGGESTED ACTIVITY
A possible activity at this point is to ask students if the answer, 29.75, at the right is reasonable. Some students may reason that because 63.5 is halfway between 63.2 and 63.8, L should be halfway between 29.5 and 30.

1. The length is 30 inches, the value of X when Y1 is 63.8.

TAKE NOTE

Remember to check your solutions.

$2x - 3 = -5$	
$2(-1) - 3$	-5
$-2 - 3$	-5
-5	$= -5$

The solution checks.

Example 1 Solve: $2x - 3 = -5$

Solution

$$2x - 3 = -5$$

$$2x - 3 + 3 = -5 + 3 \quad \text{• Add 3 to each side of the equation.}$$

$$2x = -2 \quad \text{• Simplify.}$$

$$\frac{2x}{2} = \frac{-2}{2} \quad \text{• Divide each side of the equation by 2.}$$

$$x = -1 \quad \text{• Simplify.}$$

The solution is −1.

INSTRUCTOR NOTE

The activity below combines, along with the graph at the right, multiple representations of the concept of solution.

SUGGESTED ACTIVITY

Have students enter $Y1 = 2X - 3$ and $Y2 = -5$ into the Y= editor window, then create a table similar to the one below. Students can discuss how this table can be used as a check that the solution of $2x - 3 = -5$ is −1.

X	Y1	Y2
-3	-9	-5
-2	-7	-5
-1	-5	-5
0	-3	-5
1	-1	-5
2	1	-5
3	3	-5
X= -1		

Now have students use the logical operators as another method of verifying the solution. A sample screen for this operation is shown below.

```
-1-> X
                -1
Y1 = Y2
                 1
```

Ask students to discuss which method they believe is best and why.

As a final question, ask students if there are other methods that could be used to verify the solution of this equation. One possible student response might be the suggestion to store −1 in X and then display Y1. If the value of Y1 is −5, then −1 is a solution of the equation.

You-Try-It 1 Solve: $8 = 4x - 6$

Solution See page S12.

There are several different ways that you can visualize the solution of an equation such as the one in Example 1. We think of solving the equation $2x - 3 = -5$ as finding the input value, x, for which the output, y, is −5. Draw the graph of $Y_1 = 2X - 3$. Then draw a line from −5 on the y-axis to the graph by entering $Y_2 = -5$. The x-coordinate of the point of intersection is the solution of the equation. This is shown at the right.

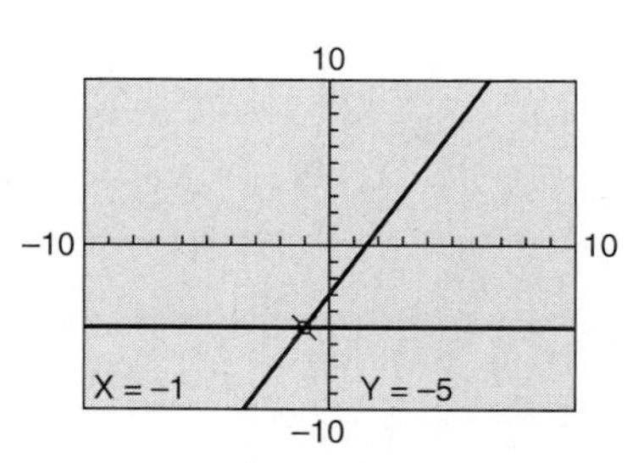

Example 2 Graphically solve $3 - 4x = 5$ and then check your solution algebraically.

Solution Using the Y= editor, enter $Y1 = 3 - 4x$ and $Y2 = 5$ and then graph the equations. Use the INTERSECT feature of the calculator to determine the point of intersection. Remember that for some calculators it is necessary to choose a viewing window that contains the point of intersection. The x-coordinate of the point of intersection is the solution of the equation.

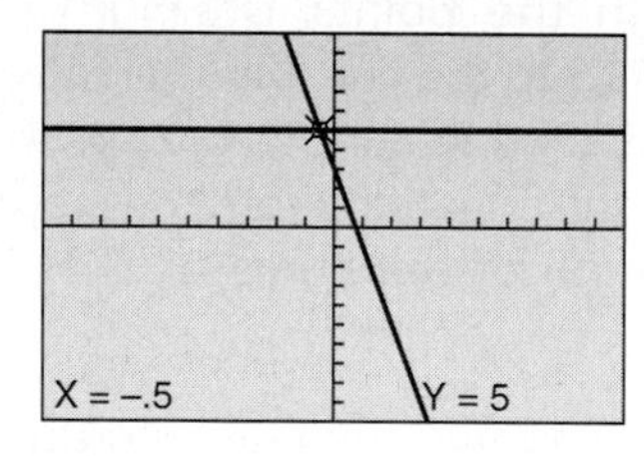

The solution is −0.5.

Algebraic check:

There are several ways to check the solution of this equation. Here is one possibility.

$3 - 4x = 5$	
$3 - 4(-0.5)$	5
$3 + 2$	5
5	$= 5$

The solution checks.

You-Try-It 2 Graphically solve $3 = 2x - 4$ and then check your solution algebraically.

Solution See page S12.

Some equations may have a variable expression on both the left and right sides of the equation. Example 3 on the next page is such an equation.

INSTRUCTOR NOTE
Ask students if the equation at the right could be solved by first subtracting 3 from each side of the equation. Complete the solution of the equation to show students that the answer is the same.

Example 3 Solve: $4x + 3 = 7x - 3$

Solution

$$4x + 3 = 7x - 3$$

$$4x - 7x + 3 = 7x - 7x - 3$$ • Subtract $7x$ from each side of the equation.

$$-3x + 3 = -3$$

$$-3x + 3 - 3 = -3 - 3$$ • Subtract 3 from each side of the equation.

$$-3x = -6$$

$$\frac{-3x}{-3} = \frac{-6}{-3}$$ • Divide each side of the equation by −3.

$$x = 2$$

The solution is 2.

You-Try-It 3 Solve: $2z - 5 = 7z + 8$

Solution See page S12.

It is possible to graphically check the solution to Example 3 by graphing the expression on each side of the equation and then determining the point of intersection by using the INTERSECT feature of your calculator. The x-coordinate of the point of intersection is the solution of the equation. Some typical graphing calculator screens for this check are shown below. Note that the viewing window has been chosen so that the point of intersection appears on the screen.

SUGGESTED ACTIVITY
Have students make a table for Y1 and Y2 starting with X = −2 and using an increment of 1. Using this table, ask students to explain how the table verifies that the solution of $4x + 3 = 7x - 3$ in Example 3 is 2. Now ask students how the table relates to the graph and the point of intersection.

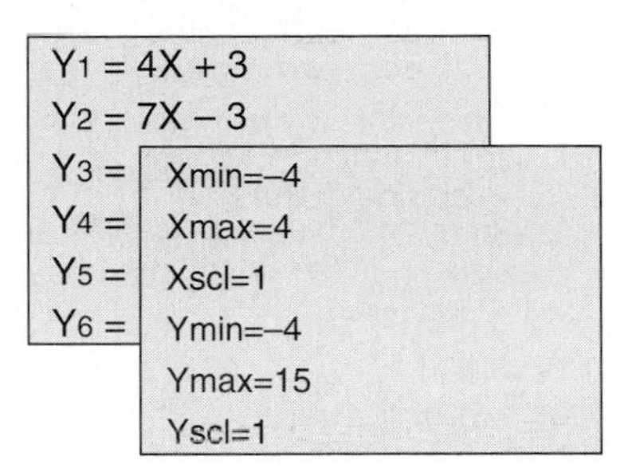

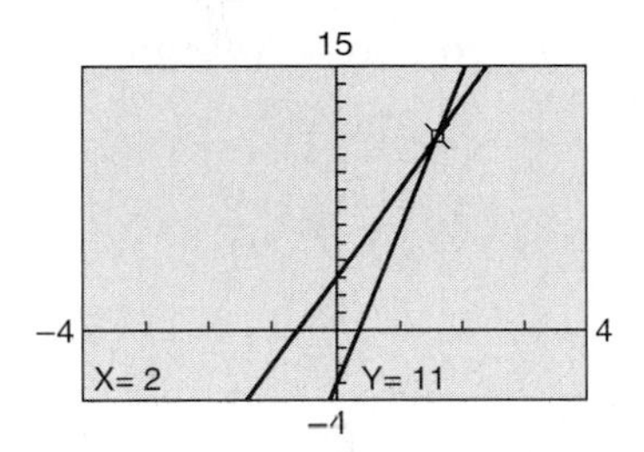

If an equation contains parentheses, use the Distributive Property to remove the parentheses.

TAKE NOTE
Remember to check a solution of an equation either algebraically or by using your graphing calculator. The check below uses the logical operators available within a graphing calculator.

Y1 = 2 − 3(3X − 1)
Y2 = 5(2 − 3X)
Y3 =
Y4 =
Y5 =
Y6 =
5/6 -> X
.8333333333
Y1 = Y2
1

Recall that the output of a 1 means that $\frac{5}{6}$ is a solution of the equation.

Example 4 Solve: $2 - 3(3x - 1) = 5(2 - 3x)$

Solution

$$2 - 3(3x - 1) = 5(2 - 3x)$$

$$2 - 9x + 3 = 10 - 15x$$ • Use the Distributive Property.

$$-9x + 5 = 10 - 15x$$ • Simplify.

$$-9x + 15x + 5 = 10 - 15x + 15x$$ • Add $15x$ to each side of the equation.

$$6x + 5 = 10$$

$$6x + 5 - 5 = 10 - 5$$ • Subtract 5 from each side of the equation.

$$6x = 5$$

$$\frac{6x}{6} = \frac{5}{6}$$ • Divide each side of the equation by 6.

$$x = \frac{5}{6}$$

The solution is $\frac{5}{6}$.

You-Try-It 4 Solve: $4x + 3(2x - 1) = 2(4x - 3)$

Solution See page S12.

To solve equations that contain fractions, it may be easier to first multiply each side of the equation by the least common multiple (LCM) of the denominators.

Example 5 Solve $\frac{6x}{5} + \frac{2}{3} = \frac{x}{2} + \frac{2}{5}$ and graphically check your solution.

Solution

The LCM of 2, 3, and 5 is 30. • Find the LCM of the denominators.

$$\frac{6x}{5} + \frac{2}{3} = \frac{x}{2} + \frac{2}{5}$$

$$30\left(\frac{6x}{5} + \frac{2}{3}\right) = 30\left(\frac{x}{2} + \frac{2}{5}\right)$$

• Multiply each side of the equation by the LCM (30).

$$30\left(\frac{6x}{5}\right) + 30\left(\frac{2}{3}\right) = 30\left(\frac{x}{2}\right) + 30\left(\frac{2}{5}\right)$$

• Use the Distributive Property.

$$36x + 20 = 15x + 12$$

$$36x - 15x + 20 = 15x - 15x + 12$$

• Subtract $15x$ from each side of the equation.

$$21x + 20 = 12$$

$$21x + 20 - 20 = 12 - 20$$

• Subtract 20 from each side of the equation.

$$21x = -8$$

$$\frac{21x}{21} = \frac{-8}{21}$$

• Divide each side of the equation by 21.

$$x = -\frac{8}{21}$$

The solution is $-\frac{8}{21}$.

Graphical check

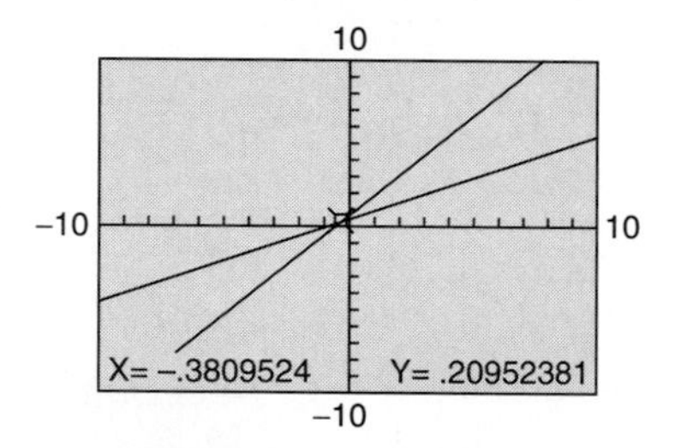

Note that the graphical solution shows the answer in decimal form. Converting $-\frac{8}{21}$ to a decimal, we have $-\frac{8}{21} \approx -0.38095238$.

You-Try-It 5 Solve $\frac{2x}{3} - \frac{1}{4} = \frac{3x}{4} + \frac{1}{3}$ and graphically check your solution.

Solution See page S13.

Translating Number Sentences into Equations

An equation states that two mathematical expressions are equal. Therefore, to translate a sentence into an equation requires recognizing the words or phrases that mean "equals." Some of these phrases are listed below.

equals
is
is equal to
amounts to
represents

} translates to =

Once the sentence is translated into an equation, the equation can be solved by rewriting the equation in the form *variable* = *constant*.

3.3 EXERCISES

Topics for Discussion

1. Explain what it means to check the solution of an equation.
   Answers will vary.

2. Explain the advantages and disadvantages of using a graphing calculator to find or check the solution of an equation.
   Answers will vary.

3. How is the Distributive Property used when solving equations?
   Answers will vary.

4. Explain what is wrong with the demonstration at the right, which suggests that $3 = 2$.
   Because the solution of the equation is zero, it is not possible to divide each side by x.

$$3x + 4 = 2x + 4$$
$$3x + 4 - 4 = 2x + 4 - 4 \quad \text{• Subtract 4 from each side.}$$
$$3x = 2x$$
$$\frac{3x}{x} = \frac{2x}{x} \quad \text{• Divide each side by } x.$$
$$3 = 2$$

5. How are the Addition and Multiplication Properties of Equations used to solve the equation $4x + 3 = 2x - 5$? Answers will vary.

Solving Equations

Solve.

6. $3x + 1 = 10$ 3

7. $5 = 4x + 9$ −1

8. $2 - x = 11$ −9

9. $-5d + 3 = -12$ 3

10. $-13 = -11y + 9$ 2

11. $-8x + 3 = -29$ 4

12. $6a + 5 = 9$ $\frac{2}{3}$

13. $9 - 4x = 6$ $\frac{3}{4}$

14. $8b - 3 = -9$ $-\frac{3}{4}$

15. $10 = -18x + 7$ $-\frac{1}{6}$

16. $9x + \frac{4}{5} = \frac{4}{5}$ 0

17. $7 = 9 - 5a$ $\frac{2}{5}$

18. $\frac{1}{3}m - 1 = 5$ 18

19. $-\frac{3}{8}b + 4 = 10$ −16

20. $\frac{3c}{7} - 1 = 8$ 21

21. $17 + \frac{5}{8}x = 7$ −16

22. $7 = \frac{2x}{5} + 4$ $\frac{15}{2}$

23. $5y + 9 + 2y = 23$ 2

24. $b - 8b + 1 = -6$ 1

25. $1.2x - 3.44 = 1.3$ 3.95

26. If $3x + 5 = -4$, evaluate $2x - 5$. -11

27. $6y + 2 = y + 17$ 3

28. $9a - 10 = 3a + 2$ 2

29. $7a - 5 = 2a - 20$ -3

30. $4y - 2 = -16 - 3y$ -2

31. $6d - 2 = 7d + 5$ -7

32. $10 - 4n = 16 - n$ -2

33. $2b - 10 = 7b$ -2

34. $-5a - 3 = 2a + 18$ -3

35. $8 - 4x = 18 - 5x$ 10

36. $6y - 1 = 2y + 2$ $\frac{3}{4}$

37. $8a - 2 = 4a - 5$ $-\frac{3}{4}$

38. If $5x = 3x - 8$, evaluate $4x + 2$. -14

39. $5x + 2(x + 1) = 23$ 3

40. $7a - (3a - 4) = 12$ 2

41. $10x + 1 = 2(3x + 5) - 1$ 2

42. $3y - 7 = 5(2y - 3) + 4$ $\frac{4}{7}$

43. $3[2 - 4(y - 1)] = 3(2y + 8)$ $-\frac{1}{3}$

44. If $4 - 3a = 7 - 2(2a + 5)$, evaluate $a^2 + 7a$. 0

Translating Number Sentences into Equations

Write an equation and solve.

45. The difference between a number and fifteen is seven. Find the number.
$n - 15 = 7; n = 22$

46. The sum of five and a number is three. Find the number.
$5 + n = 3; n = -2$

47. The product of seven and a number is negative twenty–one. Find the number.
$7n = -21; n = -3$

48. The quotient of a number and four is two. Find the number.
$\frac{n}{4} = 2; n = 8$

49. Four less than three times a number is five. Find the number.
$3n - 4 = 5; n = 3$

50. The difference between five and twice a number is one. Find the number.
$5 - 2n = 1; n = 2$

51. Four times the sum of twice a number and three is twelve. Find the number.
$4(2n + 3) = 12; n = 0$

52. Twenty-one is three times the difference between four times a number and five. Find the number. $21 = 3(4n - 5); n = 3$

53. Twelve is six times the difference between a number and three. Find the number. $12 = 6(n - 3); n = 5$

54. The difference between six times a number and four times the number is negative fourteen. Find the number. $6n - 4n = -14; n = -7$

55. Twenty-two is two less than six times a number. Find the number.
$22 = 6n - 2; n = 4$

56. Negative fifteen is three more than twice a number. Find the number.
$-15 = 2n + 3; n = -9$

57. Seven more than four times a number is three more than two times the number. Find the number. $4n + 7 = 2n + 3; n = -2$

58. The difference between three times a number and four is five times the number. Find the number. $3n - 4 = 5n; n = -2$

59. Eight less than five times a number is four more than eight times the number. Find the number. $5n - 8 = 8n + 4; n = -4$

60. The sum of a number and six is four less than six times the number. Find the number. $n + 6 = 6n - 4; n = 2$

61. Twice the difference between a number and twenty-five is three times the number. Find the number. $2(n - 25) = 3n; n = -50$

62. Four times a number is three times the difference between thirty-five and the number. Find the number. $4n = 3(35 - n); n = 15$

63. The sum of two numbers is twenty. Three times the smaller is equal to two times the larger. Find the two numbers. $3n = 2(20 - n)$; 8 and 12

64. The sum of two numbers is fifteen. One less than three times the smaller is equal to the larger. Find the two numbers. $3n - 1 = 15 - n$; 4 and 11

65. The sum of two numbers is eighteen. The total of three times the smaller and twice the larger is forty-four. Find the two numbers.
$3n + 2(18 - n) = 44$; 8 and 10

66. The sum of two numbers is two. The difference between eight and twice the smaller number is two less than four times the larger. Find the two numbers.
$8 - 2n = 4(2 - n) - 2$; -1 and 3

Applications

67. As a result of depreciation, the value of a car is now \$9600. This is three-fifths of its original value. Find the original value of the car. \$16,000

68. The operating speed of a personal computer is 100 megahertz. This is one-third the speed of a newer model. Find the speed of the newer personal computer. 300 megahertz

69. One measure of computer speed is mips (millions of instructions per second). One computer has a rating of 30 mips, which is two-thirds the speed of a second computer. Find the mips rating of the second computer. 45 mips

70. A university employs a total of 600 teaching assistants and research assistants. There are three times as many teaching assistants as research assistants. Find the number of research assistants employed by the university.
150 research assistants

71. A soil supplement contains iron, potassium, and a mulch. There is five times as much mulch as iron and twice as much potassium as iron. Find the amount of mulch in 24 pounds of the soil supplement. 15 pounds

72. A real estate agent sold two homes and received commissions totaling \$6000. The agent's commission on one home was one and one-half times the commission on the second home. Find the agent's commission on each home.
\$2400 and \$3600

73. The purchase price of a new big-screen TV, including finance charges, was \$3276. A down payment of \$450 was made. The remainder was paid in 24 equal monthly installments. Find the monthly payment. \$117.75

74. The purchase price of a new computer system, including finance charges, was \$6350. A down payment of \$350 was made. The remainder was paid in 24 equal monthly installments. Find the monthly payment. \$250

75. The cost to replace a water pump in a sports car was $600. This included $375 for the water pump and $45 per hour for labor. How many hours of labor were required to replace the water pump? 5 hours

76. The cost of electricity in a certain city is $.08 for each of the first 300 kWh (kilowatt-hours) and $.13 for each kWh over 300 kWh. Find the number of kilowatt-hours used by a family that receives a $51.95 electric bill. 515 kWh

77. An investor deposited $5000 into two accounts. Two times the smaller deposit is $1000 more than the larger deposit. Find the amount deposited into each account. $2000 and $3000

78. The length of a rectangular cement patio is 3 feet less than three times the width. The sum of the length and width of the patio is 21 feet. Find the length of the patio. 15 feet

79. Greek architects considered a rectangle whose length was approximately 1.6 times its width to be the most visually appealing. Find the length and width of a rectangle constructed in this manner if the sum of the length and width is 130 feet. Length: 80 feet; width: 50 feet

The Parthenon was the chief temple of the goddess Athena and was built on the acropolis in Athens around 440 B. C. The length is approximately 1.6 times the width.

80. A computer screen consists of tiny dots of light called pixels. In a certain graphics mode, there are 640 horizontal pixels. This is 40 more than three times the number of vertical pixels. Find the number of vertical pixels.
200 pixels

81. A wire 12 feet long is cut into two pieces. Each piece is bent into the shape of a square. The perimeter of the larger square is twice the perimeter of the smaller square. Find the perimeter of the larger square. 8 feet

82. Five thousand dollars is divided between two scholarships. Three times the smaller scholarship is equal to twice the larger. Find the amount of the larger scholarship. $3000

83. A carpenter is building a wood door frame. The height of the frame is 1 foot less than three times the width. What is the width of the largest door frame that can be constructed from a board 19 feet long? (Hint: A door frame consists of only three sides; there is no frame below a door.) 3 feet

84. A 20-foot board is cut into two pieces. Twice the length of the shorter piece is 4 feet longer than the length of the longer piece. Find the length of the shorter piece. 8 feet

Applying Concepts

85. The amount of liquid in a container triples every minute. The container becomes completely filled at 3:40 P.M. What fractional part of the container is filled at 3:39 P.M.? $\frac{1}{3}$

86. A cyclist traveling at a constant speed completes $\frac{3}{5}$ of a trip in $1\frac{1}{2}$ hours. In how many additional hours will the cyclist complete the entire trip? 1 hour

87. The charges for a long-distance telephone call are $1.21 for the first three minutes and $.42 for each additional minute or fraction of a minute. If the charges for a call were $6.25, how many minutes did the phone call last?
15 minutes

88. Four employees are paid at four consecutive levels on a wage scale. The difference between any two consecutive levels is $320 per month. The average of the four employees' monthly wages is $2880. What is the monthly wage of the highest-paid employee? $3360

89. A coin bank contains nickels, dimes, and quarters. There are 14 nickels in the bank, 16 of the coins are dimes, and $\frac{3}{5}$ of the coins are quarters. How many coins are in the coin bank? 75 coins

90. During one day at an office, one-half of the amount of money in the petty cash drawer was used in the morning, and one-third of the remaining money was used in the afternoon, leaving $5 in the petty cash drawer at the end of the day. How much money was in the petty cash drawer at the start of the day?
$15

Exploration

91. ***Business application*** Two people decide to open a business to recondition toner cartridges for copy machines. They rent a building for $7000 per year and estimate that building maintenance, taxes, and insurance will cost $6500 per year. Each person wants to make $12 per hour in the first year and will work 10 hours per day for 260 days of the year. Assume that it costs $28 to restore a cartridge and that they can sell the restored cartridge for $45.

a. How many cartridges must they restore and sell annually to break even, not including the hourly wage they wish to earn?

b. How many cartridges must they restore and sell annually to earn the hourly wage they desire?

c. Suppose the entrepreneurs are successful in their business and are restoring and selling 25 cartridges each day of the 260 days they are open. What would be their hourly wage for the year?

d. As the company becomes successful and is selling and restoring 25 cartridges each day of the 260 days they are open, the entrepreneurs decide to hire a part-time employee 4 hours per day and pay the employee $8 per hour. How many additional cartridges must be restored and sold each year to just cover the cost of the new employee? You can neglect employee costs such as social security, workers' compensation, and other costs.

Section 3.4 Applications to Geometry

Angles

A **ray** is part of a line that starts at a point, called the **endpoint** of the ray, but has no end. A ray is named by giving its endpoint and some other point on the ray. The ray below is named $\overrightarrow{PQ}$.

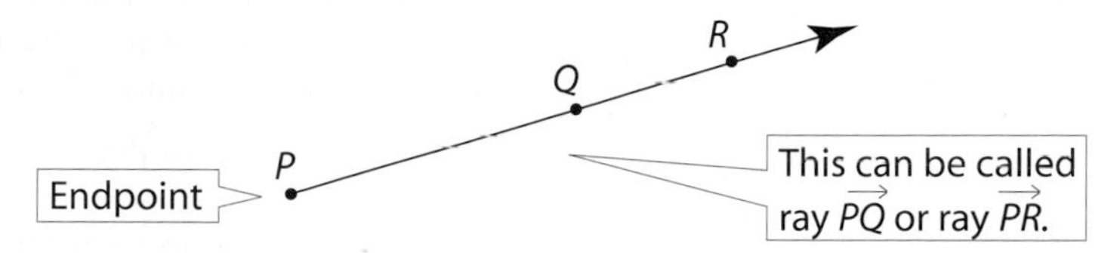

An **angle** is formed by two rays with a common endpoint. The endpoint is the **vertex** of the angle and the rays are the **sides** of the angle. An angle is named by giving its vertex or the vertex and a point on each ray, or by a letter inside the angle.

TAKE NOTE
Although Greek letters are frequently used to designate an angle, other letters, such as *x* or *y*, can be used.

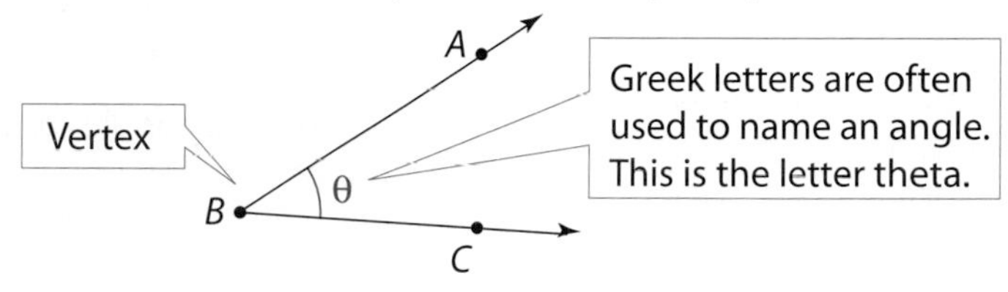

The symbol $\angle$ is used to denote an angle. Using this notation, the angle above can be named $\angle ABC$, $\angle B$, or $\angle\theta$. When three points are used to name an angle, the vertex is always given as the middle point.

A protractor can be used to measure an angle in **degrees** (°). The measure of $\angle ABC$ is written $m\angle ABC$; the measure of $\angle B$ is written $m\angle B$. In the figure below, a protractor was used to measure an angle of 60° and an angle of 137°.

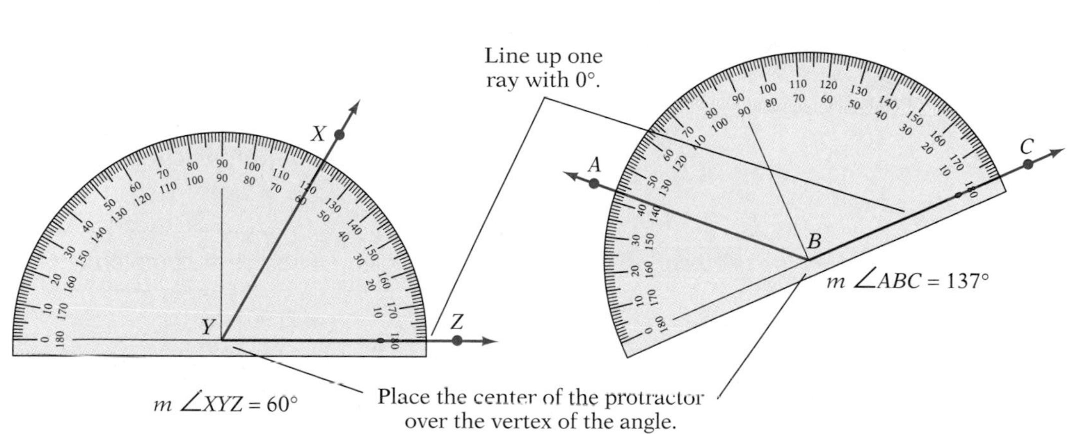

POINT OF INTEREST
When a space shuttle reenters the atmosphere of Earth, it must maintain a *reentry angle*. If the angle is too large, the shuttle will slow too quickly and may burn up. If the reentry angle is too small, the shuttle will bounce off the atmosphere.

A **right angle** has a measure of 90°. A right-angle symbol, ⌐, is frequently placed inside an angle to indicate a right angle.

An angle whose measure is between 0° and 90° is called an **acute** angle. An angle whose measure is between 90° and 180° is called an **obtuse** angle.

A **straight** angle measures 180°.

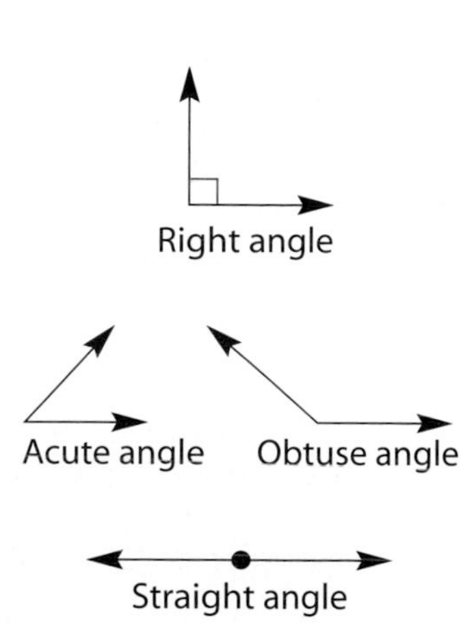

Two angles are **complements** of each other if the sum of the measures of the angles is 90°. The angles are called **complementary angles**.

$$63^\circ + 27^\circ = 90^\circ$$

In many instances, complementary angles of a right angle are discussed. For instance, for the figure at the right, $\angle ABC$ and $\angle CBD$ are complementary angles.

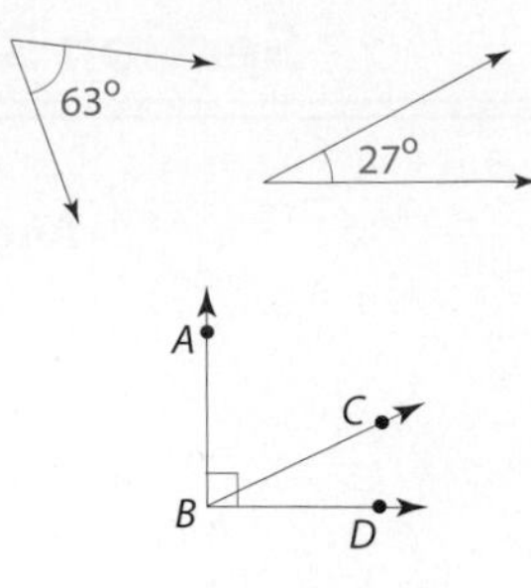

Complementary angles

Two angles are **supplements** of each other if the sum of the measures of the angles is 180°. The angles are called **supplementary angles.**

$$77^\circ + 103^\circ = 180^\circ$$

Supplementary angles of a straight angle are frequently discussed. In the figure at the right, $\angle RST$ and $\angle TSU$ are supplementary angles.

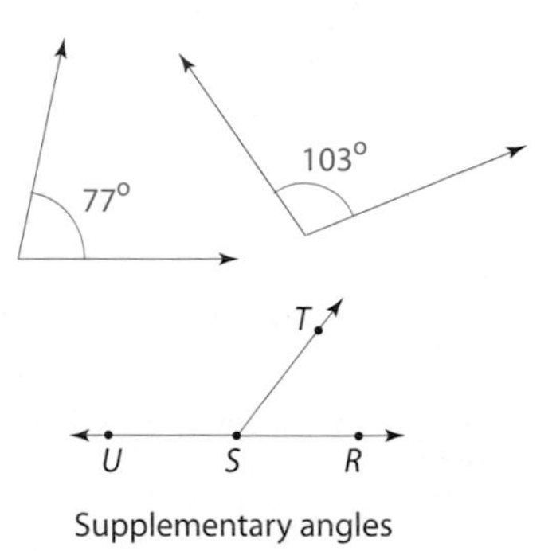

Supplementary angles

Example 1 One angle is twice its complement. Find the measure of each angle.

Solution

State the goal. The goal is to find two complementary angles such that one angle is twice the other.

Describe a strategy. Let x represent the measure of one angle. Because the angles are complements of each other, the measure of the complementary angle is $90^\circ - x$. Thus, we have:

Measure of one angle: x
Measure of the complement: $90 - x$

Note that $x + (90^\circ - x) = 90^\circ$, which shows that the angles are complements of each other.

Solve the problem.

One angle	=	twice its complement

$$x = 2(90 - x)$$

$$x = 180 - 2x \qquad \text{• Use the Distributive Property.}$$

$$3x = 180 \qquad \text{• Add } 2x \text{ to each side of the equation.}$$

$$x = 60 \qquad \text{• Divide each side of the equation by 3.}$$

One angle is 60°.

To find the measure of the complement, evaluate $90 - x$ when $x = 60$. The measure of the complement is 30°.

$$90 - x$$
$$90 - 60 = 30$$

Check your work. Note that $30^\circ + 60^\circ = 90^\circ$, so the two angles are complements. Also, 60° is twice 30°. This verifies that our solution is correct.

You-Try-It 1 One angle is 3° more than twice its supplement. Find the two angles.

Solution See page S13.

Intersecting Lines

POINT OF INTEREST
Many cities in the New World, unlike those in Europe, were designed using rectangular street grids. Washington, D.C., was planned that way, except that diagonal avenues were added to provide for quick troop movement in the event the city required defense. As an added precaution, monuments and statues were constructed at major intersections so that attackers would not have a straight shot down a boulevard.

Four angles are formed by the intersection of two lines. If each of the four angles is a right angle, then the two lines are **perpendicular**. Line p is perpendicular to line q. This is written $p \perp q$, where $\perp$ is read "is perpendicular to."

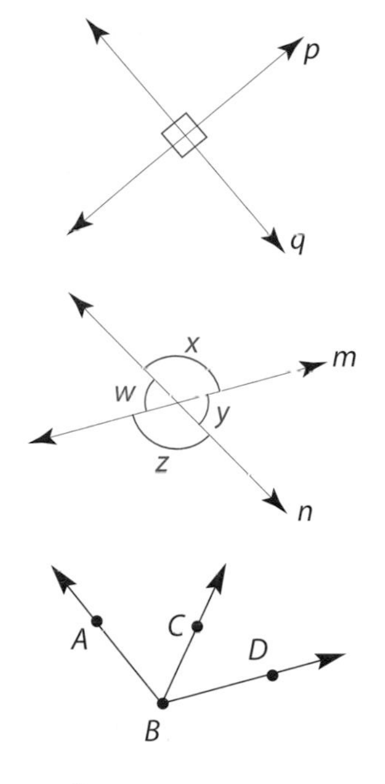

If the two lines are not perpendicular, then two of the angles are acute angles and two of the angles are obtuse angles. The two acute angles are always opposite each other and the two obtuse angles are always opposite each other. $\angle w$ and $\angle y$ are acute angles; $\angle x$ and $\angle z$ are obtuse angles.

Two angles that have the same vertex and share a common side are called **adjacent angles**. For the figure shown at the right, $\angle ABC$ and $\angle CBD$ are adjacent angles.

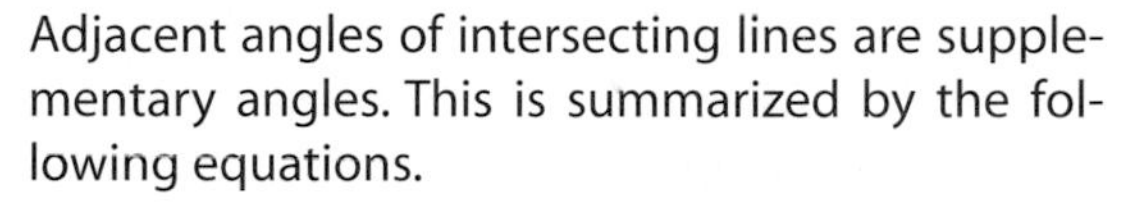

Adjacent angles of intersecting lines are supplementary angles. This is summarized by the following equations.

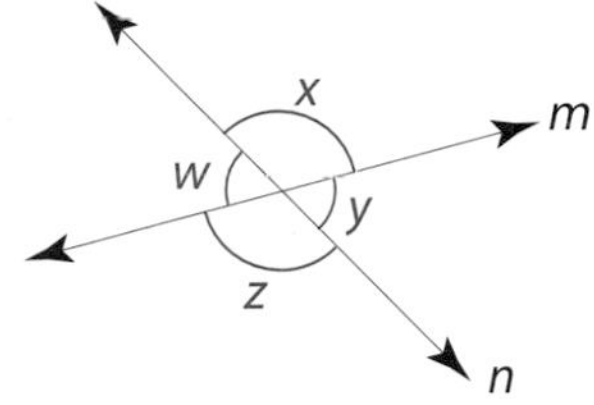

$$m\angle x + m\angle y = 180°$$
$$m\angle y + m\angle z = 180°$$
$$m\angle z + m\angle w = 180°$$
$$m\angle w + m\angle x = 180°$$

The angles that are on opposite sides of intersecting lines are called **vertical angles**. For the intersecting lines m and n above, $\angle x$ and $\angle z$ are vertical angles; $\angle w$ and $\angle y$ are also vertical angles. Vertical angles have the same measure. Thus,

$$m\angle x = m\angle z \qquad \text{and} \qquad m\angle w = m\angle y$$

Example 2 Find the value of x for the intersecting lines at the right.

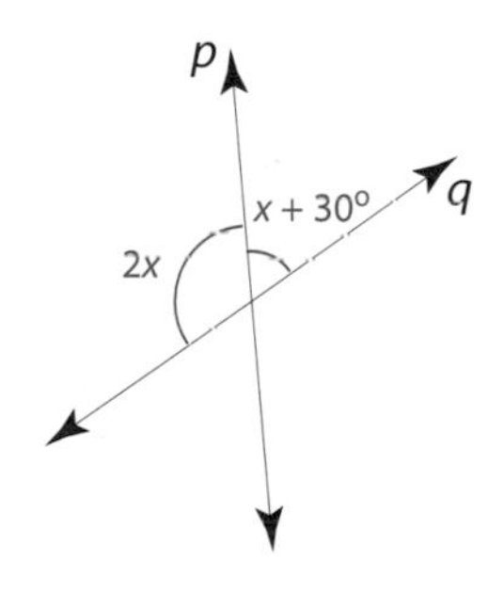

Solution

State the goal. The goal is to find the value of x.

Describe a strategy. The angles are adjacent angles of intersecting lines. Therefore, the angles are supplementary angles. This means that the sum of the measures of the two angles is 180°.

Solve the problem.

$$2x + (x + 30) = 180$$
$$3x + 30 = 180 \qquad \text{• Combine like terms.}$$
$$3x = 150 \qquad \text{• Subtract 30 from each side of the equation.}$$
$$x = 50 \qquad \text{• Divide each side of the equation by 3.}$$

The value of x is 50.

Check your work. Replacing x by 50 in the equation $2x + (x + 30) = 180$, we have $2(50) + (50 + 30) = 100 + 80 = 180$. The solution checks.

You-Try-It 2 The measures of two adjacent angles for a pair of intersecting lines are $2x$ and $3x$. Find the measure of the larger angle.

Solution See page S13.

A line that intersects two other lines at different points is called a **transversal.** If the lines cut by a transversal are parallel lines and the transversal is perpendicular to the parallel lines, all eight angles formed are right angles.

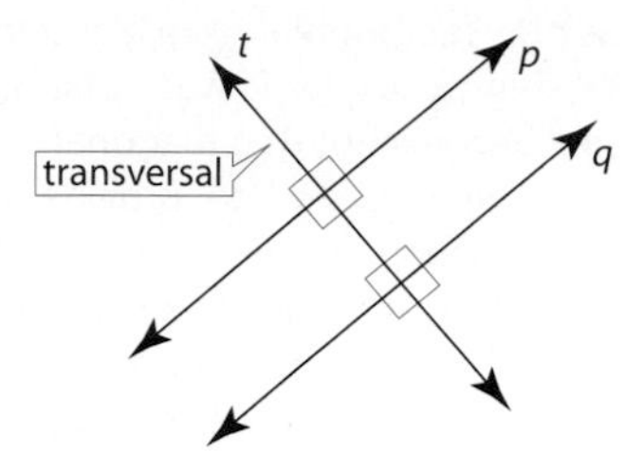

The symbol for parallel lines is $\|$. For the diagram at the right, $p \parallel q$.

If the lines cut by a transversal, t, are parallel lines and the transversal is not perpendicular to the parallel lines, all four acute angles have the same measure and all four obtuse angles have the same measure. For the figure at the right:

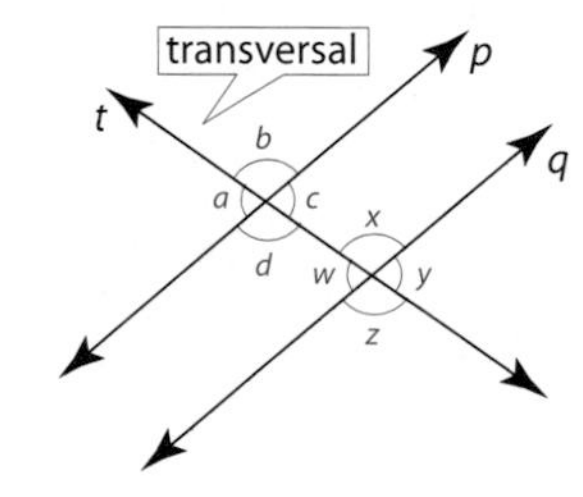

$$m\angle a = m\angle c = m\angle w = m\angle y$$
$$m\angle b = m\angle d = m\angle x = m\angle z$$

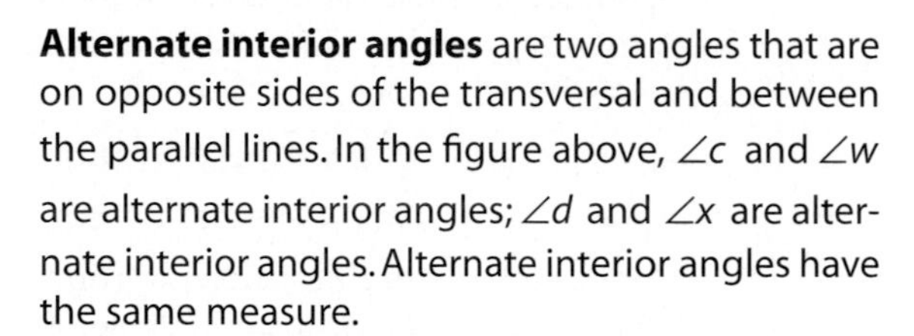

Alternate interior angles are two angles that are on opposite sides of the transversal and between the parallel lines. In the figure above, $\angle c$ and $\angle w$ are alternate interior angles; $\angle d$ and $\angle x$ are alternate interior angles. Alternate interior angles have the same measure.

Alternate interior angles have the same measure.

$$m\angle c = m\angle w$$
$$m\angle d = m\angle x$$

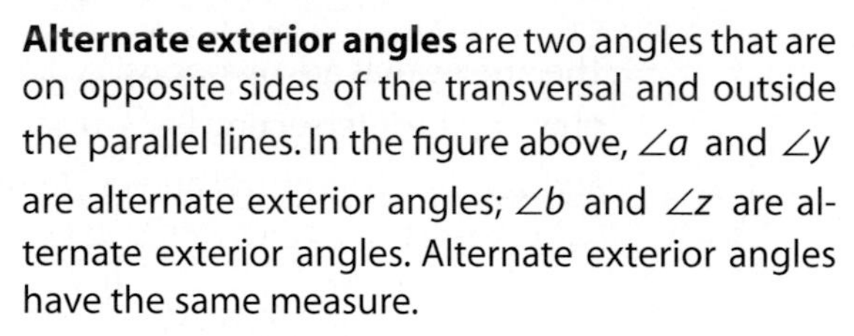

Alternate exterior angles are two angles that are on opposite sides of the transversal and outside the parallel lines. In the figure above, $\angle a$ and $\angle y$ are alternate exterior angles; $\angle b$ and $\angle z$ are alternate exterior angles. Alternate exterior angles have the same measure.

Alternate exterior angles have the same measure.

$$m\angle a = m\angle y$$
$$m\angle b = m\angle z$$

Corresponding angles are two angles that are on the same side of the transversal and are both acute angles or are both obtuse angles. For the figure above, the following pairs of angles are corresponding angles: $\angle a$ and $\angle w$, $\angle d$ and $\angle z$, $\angle b$ and $\angle x$, $\angle c$ and $\angle y$. Corresponding angles have the same measure.

Corresponding angles have the same measure.

$$m\angle a = m\angle w$$
$$m\angle d = m\angle z$$
$$m\angle b = m\angle x$$
$$m\angle c = m\angle y$$

Question: In the figure at the right, $p \parallel q$. Which of the angles a, b, c, and d have the same measure as $\angle m$? Which angles have the same measure as $\angle n$?[1]

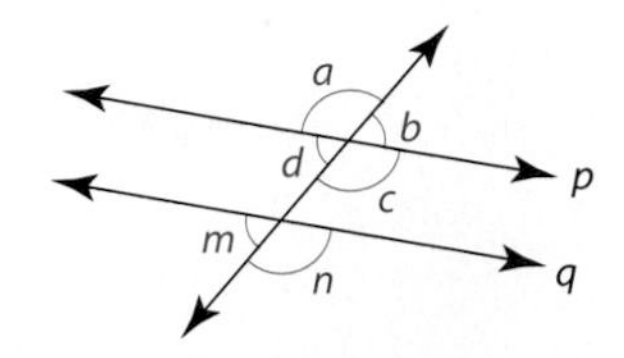

---

1. The angles that have the same measure as angle $\angle m$ are $\angle b$ and $\angle d$. The angles that have the same measure as angle $\angle n$ are $\angle a$ and $\angle c$.

Example 3 Given $p \parallel q$, find the value of x.

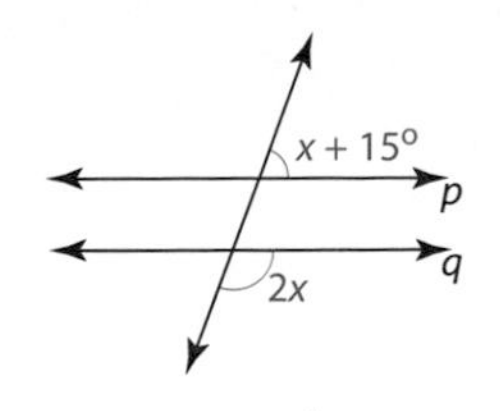

Solution

State the goal. To goal is to find the value of x.

Describe a strategy. Because alternate exterior angles are equal, we can label the angle above line p and adjacent to $\angle(x + 15°)$ as $2x$. Knowing that the sum of the measures of adjacent angles of intersecting lines is 180°, we have $m\angle 2x + m\angle(x + 15°) = 180°$.

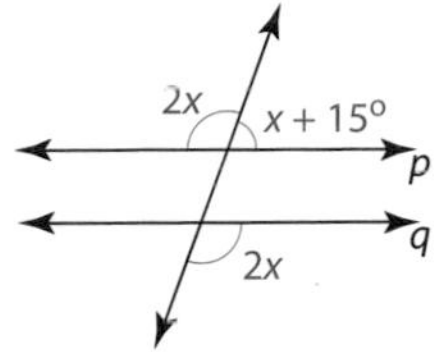

Solve the problem.

$$\begin{aligned} 2x + (x + 15°) &= 180° \\ 3x + 15° &= 180° \\ 3x &= 165° \\ x &= 55° \end{aligned}$$

The value of x is 55°.

Check your work. By replacing x by 55° in the equation $2x + (x + 15°) = 180°$, you can verify that the solution is correct.

You-Try-It 3 Given that $p \parallel q$, find the value of x.

Solution See page S13.

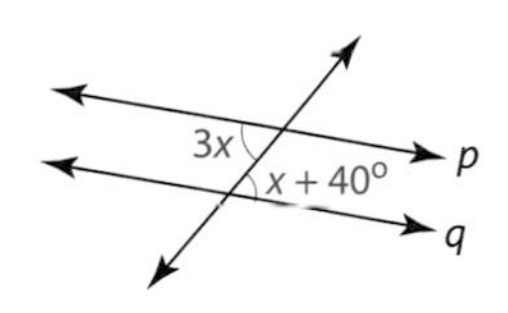

Angles of a Triangle

SUGGESTED ACTIVITY
Have students investigate the sum of the angles of a polygon by dividing the polygon into triangles. Have them create an input/output table in which to record their results. On the basis of their findings, have them determine a variable expression that will give the number of degrees in an n-sided polygon.

If the lines cut by a transversal are not parallel lines, the three lines will intersect at three points. In the figure at the right, the transversal t intersects lines p and q. The three lines intersect at points A, B, and C. These three points define three line segments $\overline{AB}$, $\overline{BC}$, and $\overline{AC}$. The plane figure formed by these line segments is a **triangle**.

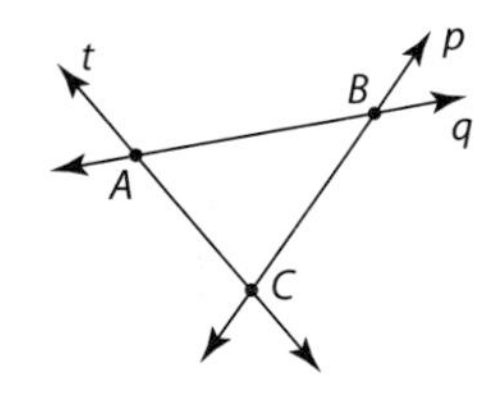

Each of the three points of intersection is the vertex of an angle of the triangle. The angles within the triangle are called **interior angles**. In the figure at the right, $\angle a$, $\angle b$, and $\angle c$ are interior angles. The sum of the measures of interior angles is 180°.

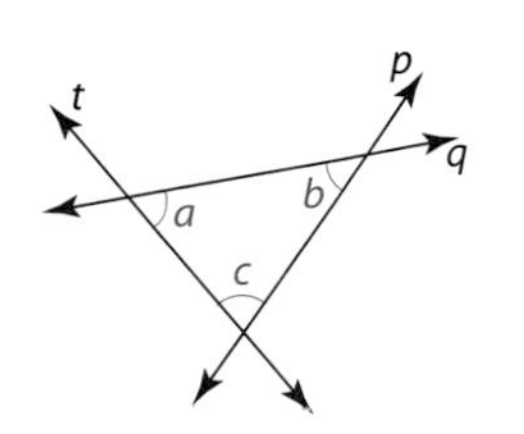

$m\angle a + m\angle b + m\angle c = 180°$

> ***The Sum of the Measures of the Interior Angles of a Triangle***
> The sum of the measures of the interior angles of a triangle is 180°.

As an example of this, suppose the measures of two angles of a triangle are 25° and 47°. If x is the measure of the third angle, then

$$\begin{aligned} x + 25° + 47° &= 180° && \text{• The sum of the measures of the angles is 180°.} \\ x + 72° &= 180° && \text{• Solve for } x. \\ x &= 108° \end{aligned}$$

The measure of the third angle is 108°.

An angle adjacent to an interior angle of a triangle is an **exterior angle** of the triangle. In the figure at the right, $\angle x$ and $\angle y$ are exterior angles for $\angle a$. The sum of the measures of an interior angle and an exterior angle of a triangle is 180°.

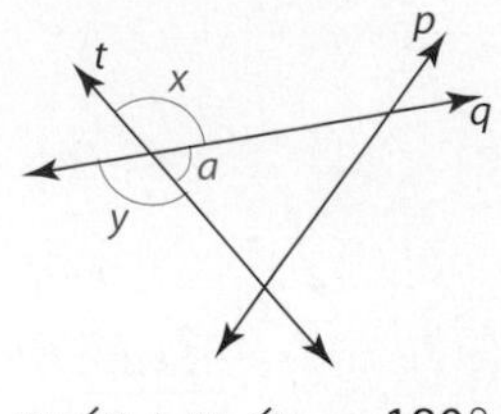

$$m\angle a + m\angle x = 180^\circ$$
$$m\angle a + m\angle y = 180^\circ$$

Example 4 Given that $m\angle c = 40^\circ$ and $m\angle d = 100^\circ$, find the measure of $\angle e$.

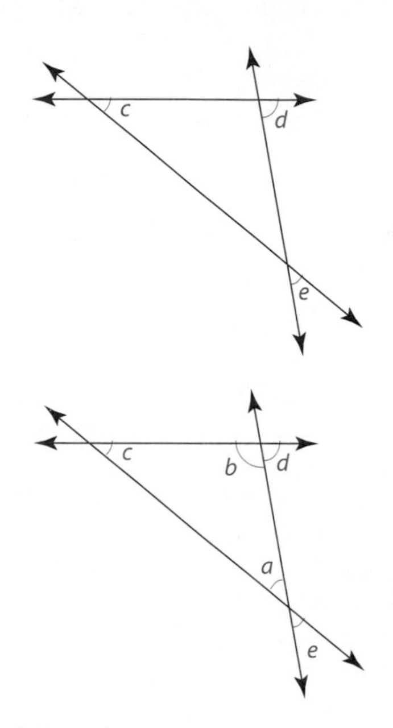

Solution

State the goal. The goal is to find $m\angle e$.

Describe a strategy. Because vertical angles have the same measure, we can find the solution to this problem by finding the measure of $\angle a$, the vertical angle for $\angle e$. The measure of this angle can be found by using two other facts: $\angle b$ and $\angle d$ are supplementary angles; and the sum of the measures of $\angle a$, $\angle b$, and $\angle c$ is 180°.

Solve the problem.

$$\begin{aligned} m\angle b + m\angle d &= 180^\circ \\ m\angle b + 100^\circ &= 180^\circ \\ m\angle b &= 80^\circ \end{aligned}$$

- $\angle b$ and $\angle d$ are supplementary angles.
- $m\angle d = 100^\circ$

The measure of $\angle b$ is 80°.

$$\begin{aligned} m\angle a + m\angle b + m\angle c &= 180^\circ \\ m\angle a + 80^\circ + 40^\circ &= 180^\circ \\ m\angle a + 120^\circ &= 180^\circ \\ m\angle a &= 60^\circ \end{aligned}$$

- The sum of the measures of the interior angles is 180°.

The measure of $\angle a$ is 60°.

Because $\angle a$ and $\angle e$ are vertical angles, $m\angle a = m\angle e$. Therefore $m\angle e = 60^\circ$.

Check your work. Be sure to check your work.

You-Try-It 4 Given that $m\angle y = 55^\circ$, find the measure of $\angle d$.

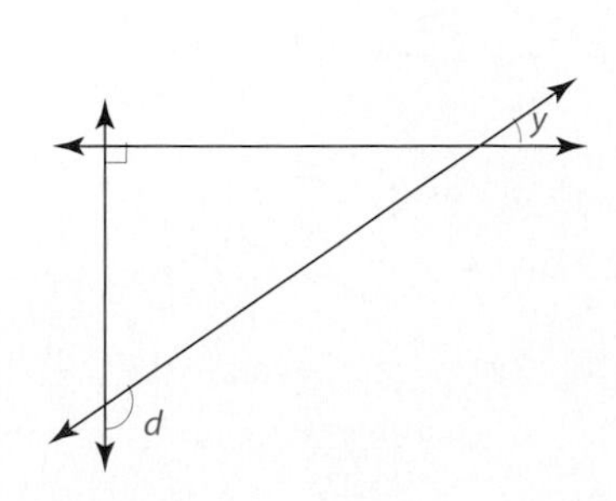

Solution See page S14.

3.4 EXERCISES

Topics for Discussion

1. What are vertical angles? If all four vertical angles are equal, what can be said about the lines?
The opposite angles of intersecting lines. The lines are perpendicular.

2. What are complementary angles? What are supplementary angles?
Two angles whose sum is 90°. Two angles whose sum is 180°.

3. If a transversal, t, cuts two lines p and q, and the alternate interior angles are not equal, what can be said about the lines p and q?
They are not parallel.

4. Suppose that $p \parallel q$ and t is a transversal that intersects both p and q. If $\angle a$ and $\angle b$ are alternate interior angles and $\angle x$ and $\angle y$ are alternate exterior angles, what is the sum of the measure of $\angle a$ and the measure of $\angle x$? 180°

Angles

Use a protractor to measure the angle. State whether the angle is acute, obtuse, or right.

5.

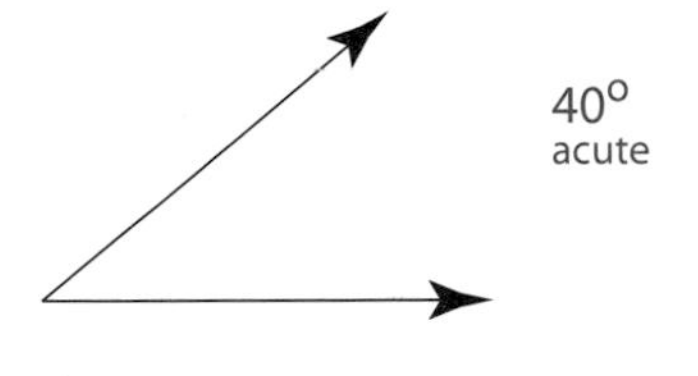

40° acute

6.

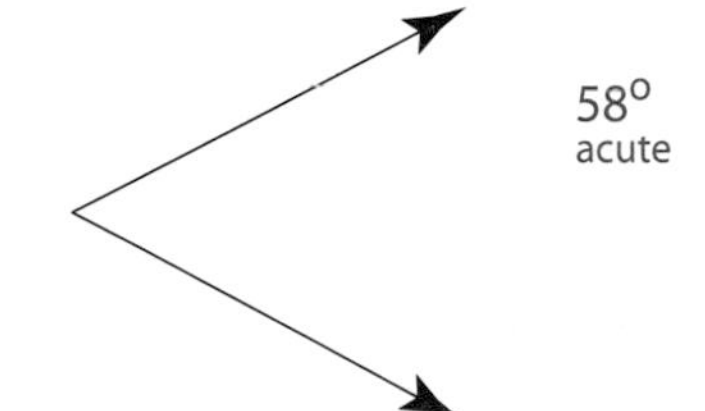

58° acute

7.

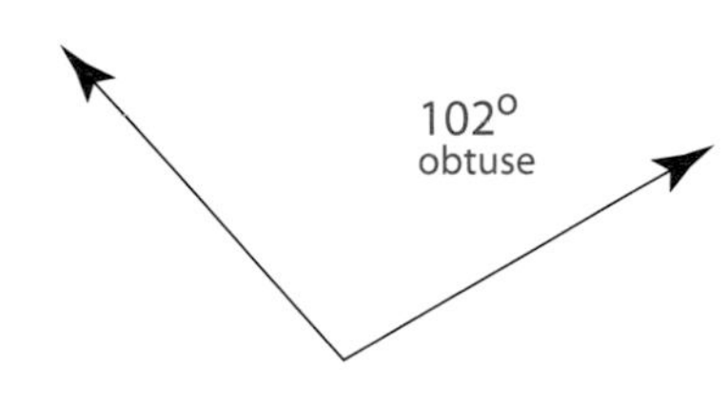

102° obtuse

8.

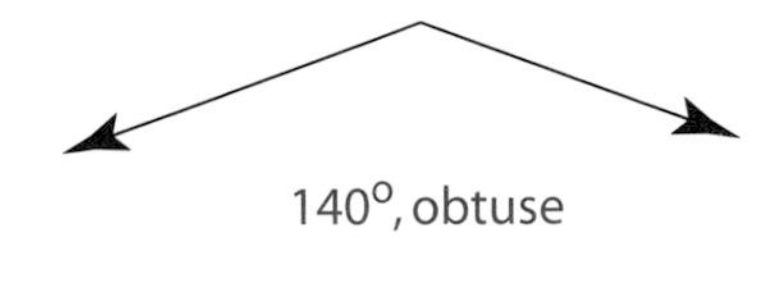

140°, obtuse

9.

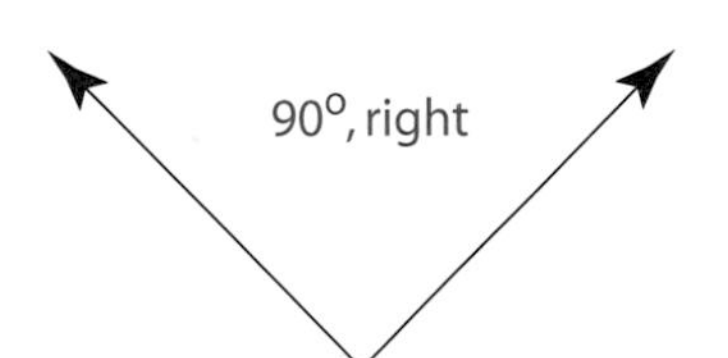

90°, right

10.

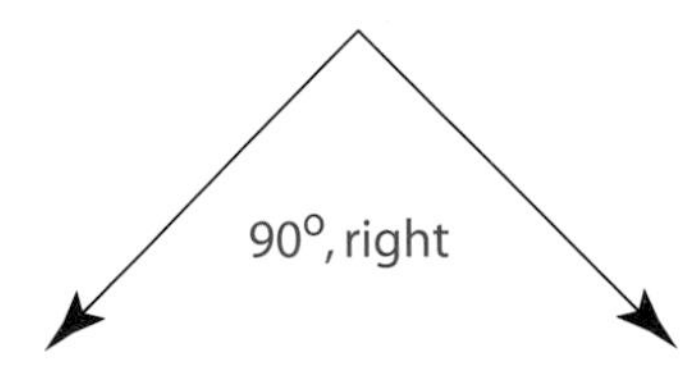

90°, right

Solve.

11. Find the complement of a 75° angle.
15°

12. Find the complement of a 12° angle.
78°

13. Find the supplement of a 143° angle.
37°

14. Find the supplement of a 63° angle.
117°

15. Find two complementary angles such that the larger angle is 6 degrees more than twice the smaller angle. 28° and 62°

16. Find two complementary angles such that the smaller angle is 15 degrees less than one-half the larger angle. 20° and 70°

17. Find two supplementary angles such that the larger angle is 12 degrees less than three times the smaller angle. 48° and 132°

18. Find two supplementary angles such that the smaller angle is 10 degrees less than two-thirds the larger angle. 66° and 114°

Intersecting Lines

Given that $\angle ABC$ is a right angle, find the value of x.

19.

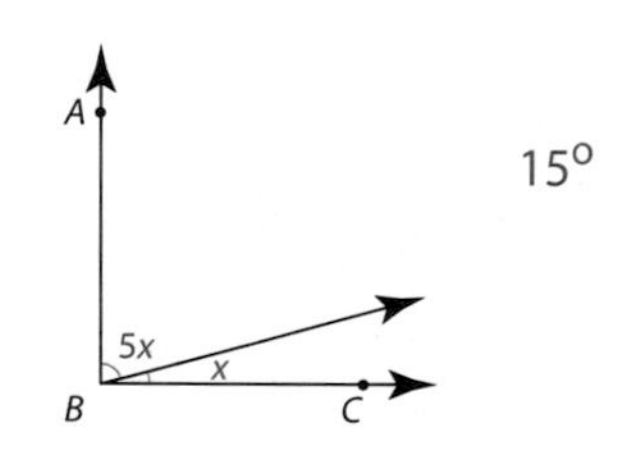

15°

20.

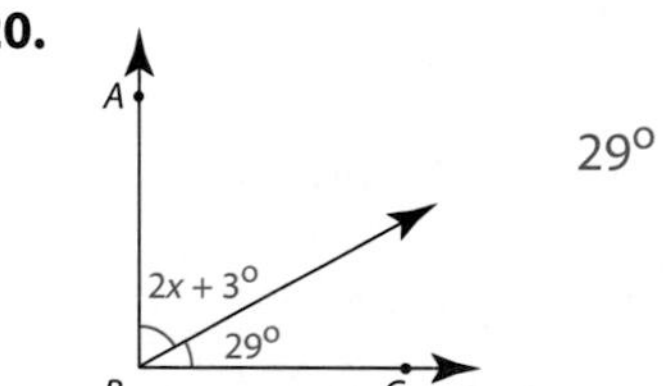

29°

21.

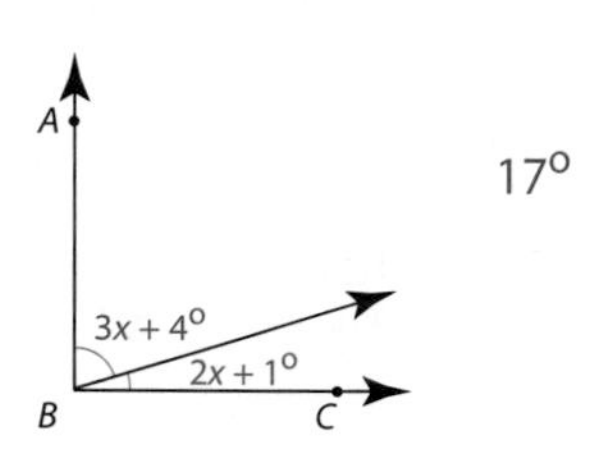

17°

22.

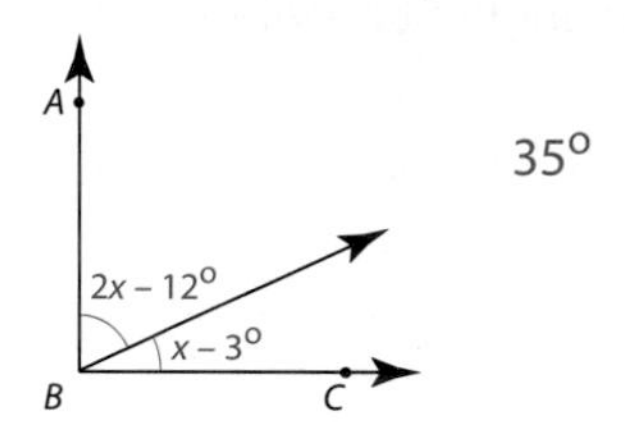

35°

Find the value of x.

23.

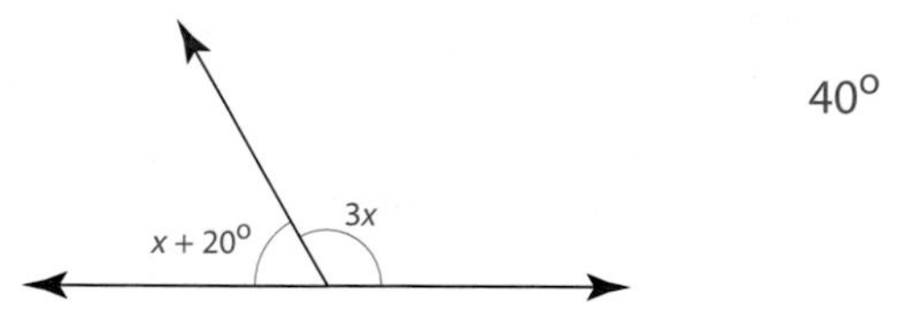

40°

24.

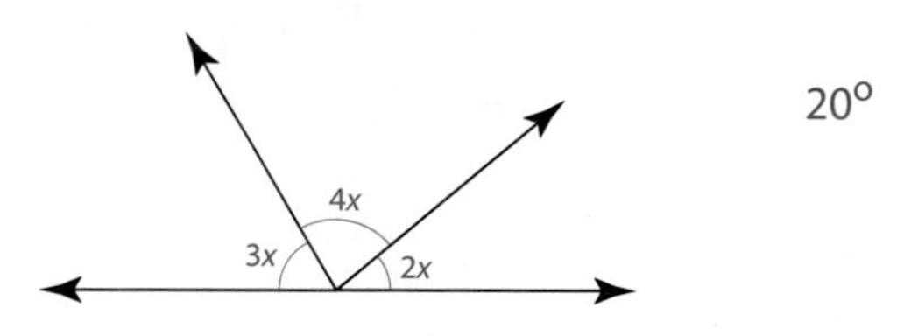

20°

25.

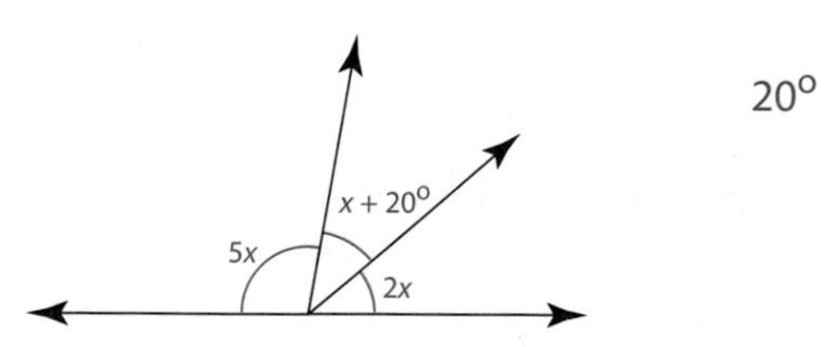

20°

26.

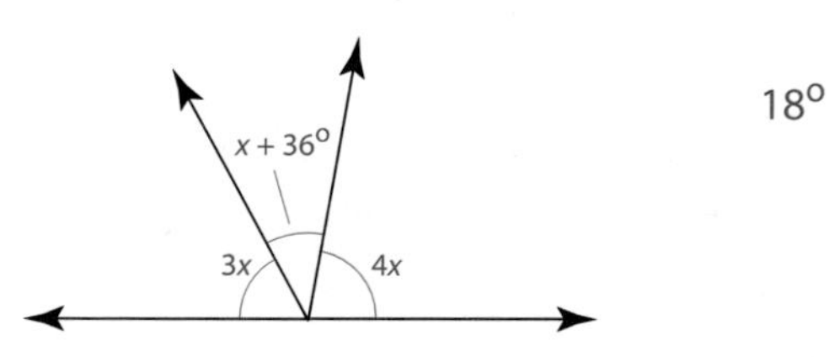

18°

Find the measure of $\angle b$.

27.

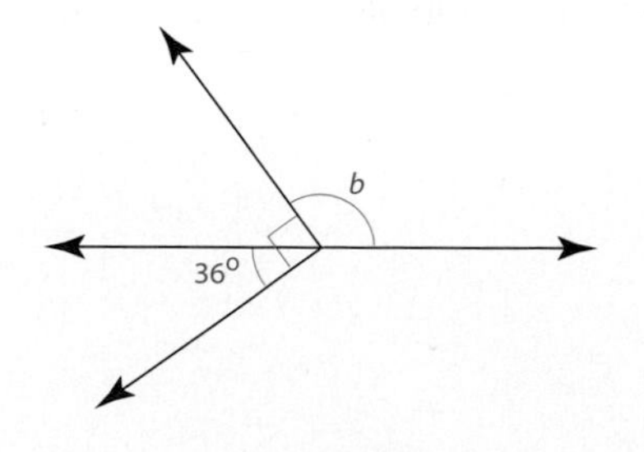

126°

28.

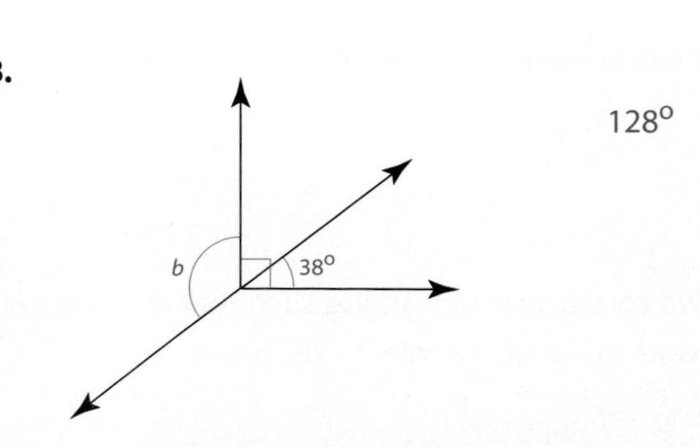

128°

Find x.

29.

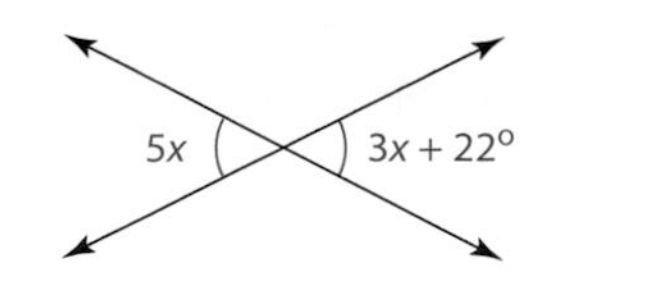

11°

30.

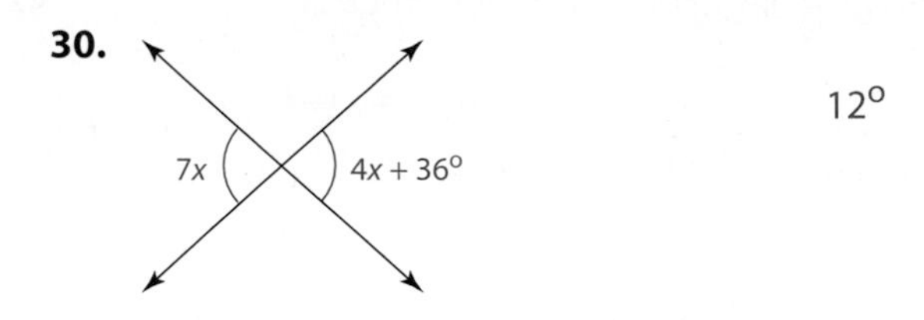

12°

Given that $p \parallel q$, find $m\angle a$ and $m\angle b$.

31.

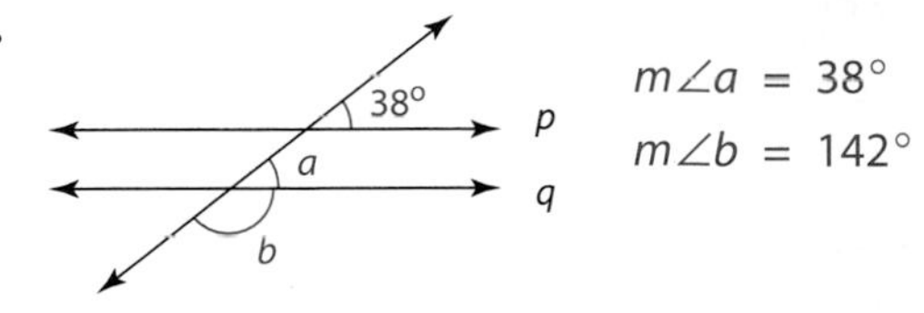

$m\angle a = 38°$

$m\angle b = 142°$

32.

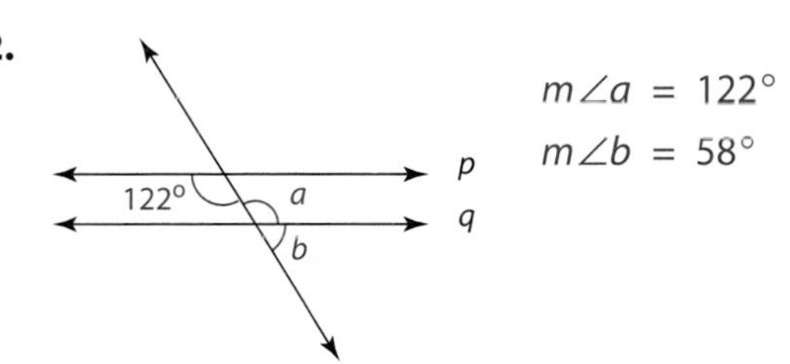

$m\angle a = 122°$

$m\angle b = 58°$

33.

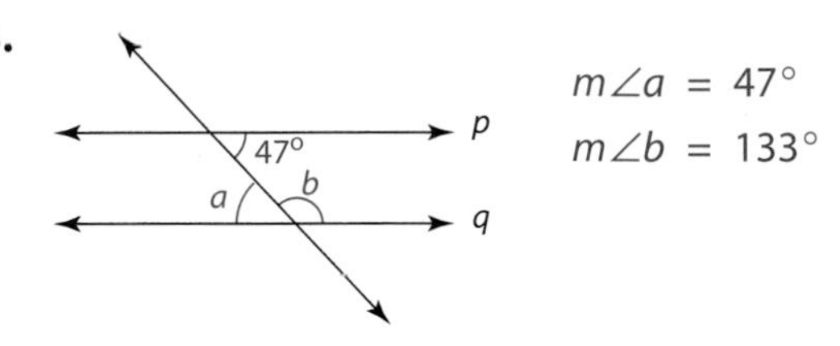

$m\angle a = 47°$

$m\angle b = 133°$

34.

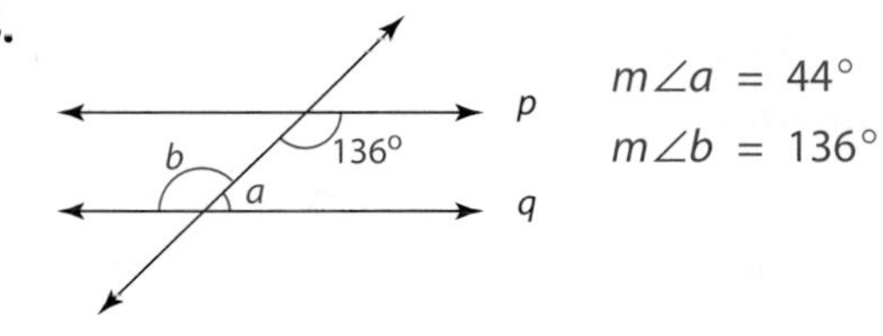

$m\angle a = 44°$

$m\angle b = 136°$

Given that $p \parallel q$, find x.

35.

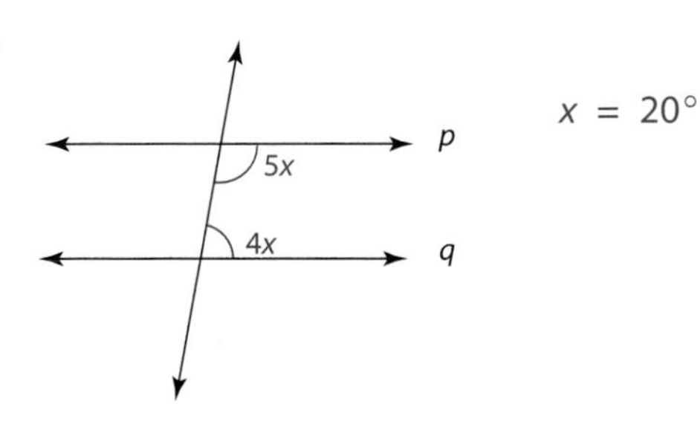

$x = 20°$

36.

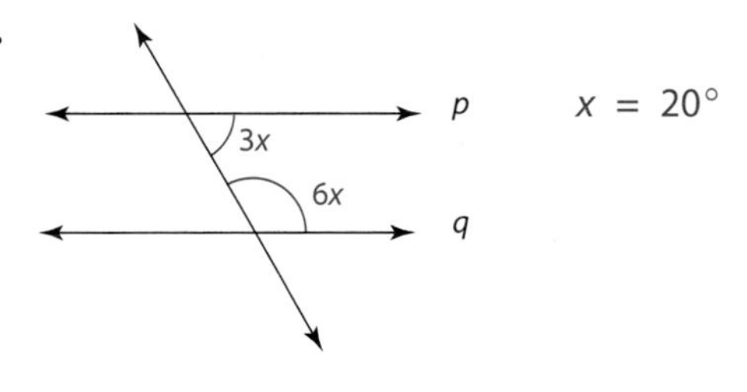

$x = 20°$

37.

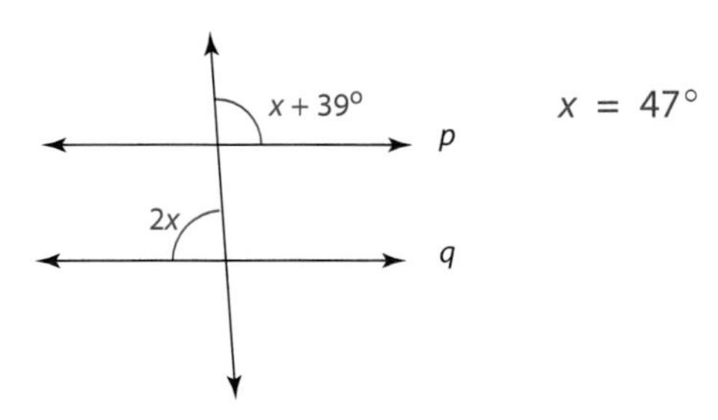

$x = 47°$

38.

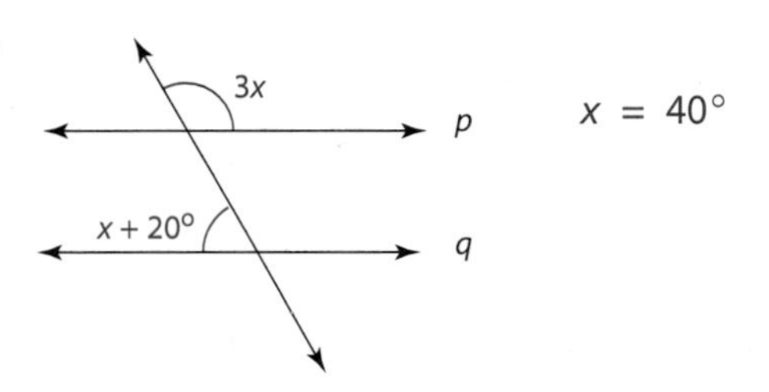

$x = 40°$

Angles of a Triangle

39. Given that $m\angle a = 95°$ and $m\angle b = 60°$, find $m\angle x$ and $m\angle y$.

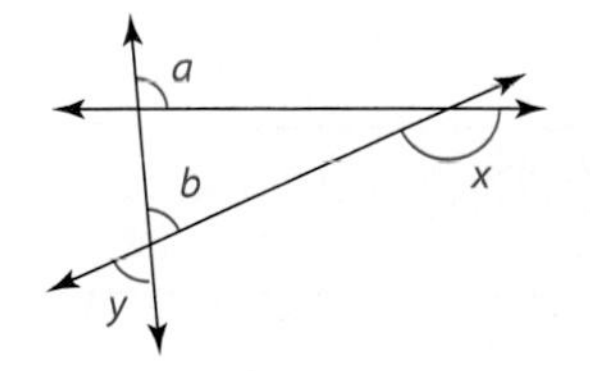

$m\angle x = 145°$

$m\angle y = 60°$

40. Given that $m\angle a = 35°$ and $m\angle b = 55°$, find $m\angle x$ and $m\angle y$.

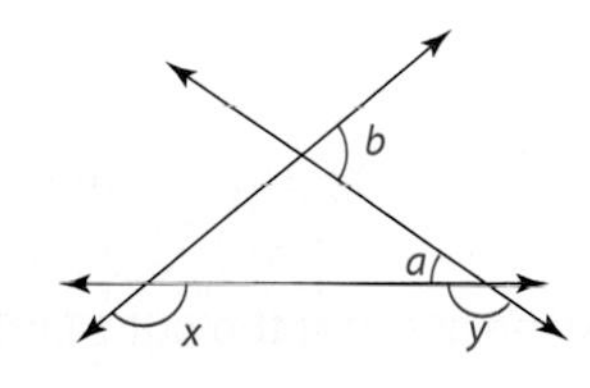

$m\angle x = 160°$

$m\angle y = 145°$

41. Given $m\angle y = 45°$, find $m\angle a$ and $m\angle b$.

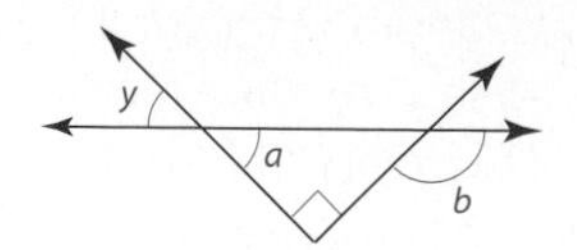

$m\angle a = 45°$

$m\angle b = 135°$

42. Given $m\angle y = 140°$, find $m\angle a$ and $m\angle b$.

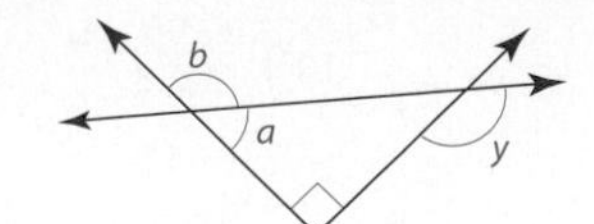

$m\angle a = 50°$

$m\angle b = 130°$

43. The measure of one of the acute angles of a right triangle is two degrees more than three times the measure of the other acute angle. Find the measure of each angle. 22° and 68°

44. The measure of the largest angle of a triangle is five times the measure of the smallest angle of the triangle. The measure of the third angle is three times the measure of the smallest angle. Find the measure of the largest angle. 100°

Applying Concepts

45. Suppose Earth is a perfect sphere with a belt wrapped around the equator. How much longer would the belt need to be so that it could be supported by poles that were 10 feet above the surface of Earth?
62.83 feet

46. Cut out a triangle and then tear off two of the angles, as shown in the figure at the right. Position the pieces you tore off so that $\angle a$ is adjacent to $\angle b$ and $\angle c$ is adjacent to $\angle b$. Describe what you observe. What does this demonstrate?
Answers will vary.

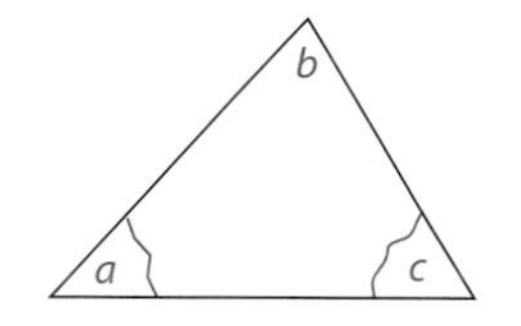

47. The measure of the supplement of the complement of $\angle a$ is 120°. What is the measure of $\angle a$? 30°

48. The measure of the complement of the supplement of $\angle a$ is 50°. What is the measure of $\angle a$? 140°

49. Determine whether the statement is always true, sometimes true, or never true.
a. Two lines that are parallel to a third line are parallel to each other.
b. A triangle contains two acute angles.
c. Vertical angles are complementary angles.
d. Adjacent angles are supplementary angles.
a. Always true; b. always true; c. sometimes true; d. sometimes true

Explorations

50. ***Properties of Triangles*** For the figure at the right, explain why $m\angle a + m\angle b = m\angle x$. Write a rule that describes the relationship between an exterior angle of a triangle and the opposite interior angles.

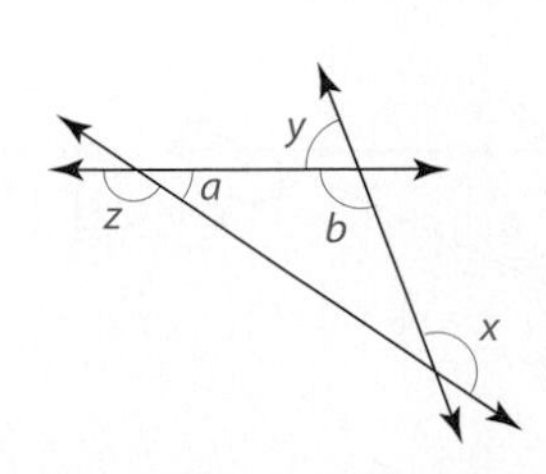

51. ***Properties of Triangles*** For the figure at the right, find $m\angle x + m\angle y + m\angle z$.

Section 3.5 Uniform Motion Problems

Uniform Motion Problems

Recall that uniform motion means that an object is moving in a straight line with constant speed. The distance, d, that an object will travel in a certain time, t, is given by the equation $d = rt$, where r is the speed of the object. For instance, if a marathon runner runs at a rate of 6 miles per hour, then the distance, in miles, traveled by the runner is given by the equation $d = 6t$, where t is the number of hours spent running. The table and graph below show the relationship between the distance run and the time running. The variable Y1 represents d, the distance, and X represents the time, t.

SUGGESTED ACTIVITY
The *Instructor's Resource Manual* contains an activity that uses a graphing calculator to model various uniform motion problems. These activities may help students "see" the relationships between the moving objects.

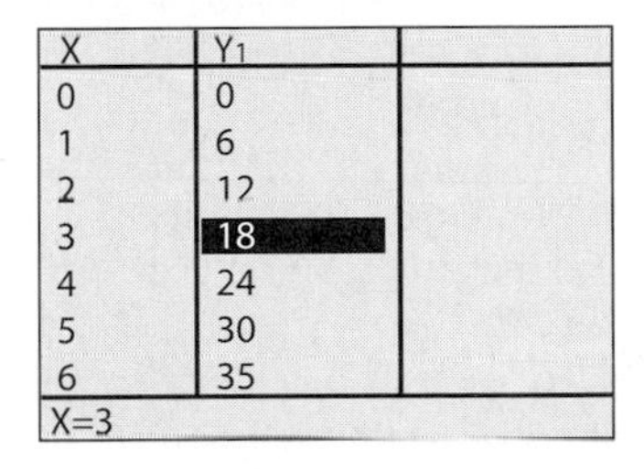

X	Y1	
0	0	
1	6	
2	12	
3	18	
4	24	
5	30	
6	35	
X=3		

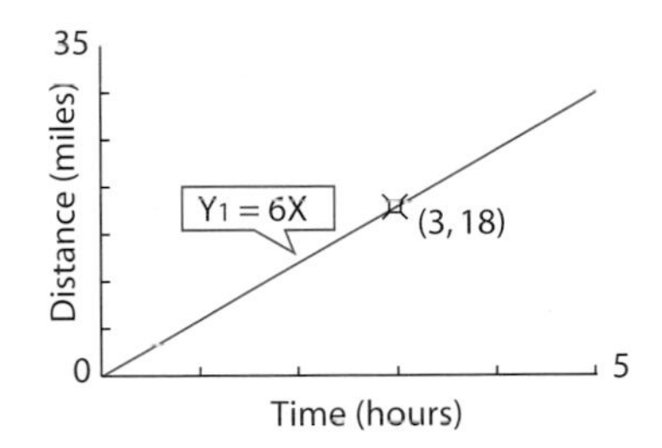

Question: How far has the runner traveled after 4 hours?[1]

The graph above is a **distance-time** graph and shows the relationship between the time of travel and the distance traveled. Time is on the horizontal axis and distance is on the vertical axis.

SUGGESTED ACTIVITY
Tell students that you are going to show them the graphs of two runners on the same coordinate axis. Then graph Y1 = 6X and Y2 = 4X without showing students the equations for the graphs. Ask students to identify which runner has the faster speed. You can approach the answer to the question by moving the cursor to X = 3 and showing students that Y1 is greater than Y2. Thus the distance traveled by Y1 is greater than the distance traveled by Y2. Consequently, the rate of Y1 is greater than the rate of Y2. This activity will help prepare students for the concept of slope in the next chapter.

Suppose a motorcycle leaves San Antonio, Texas, traveling at 40 miles per hour. Two hours later a car leaves from the same spot and travels the same road at 60 miles per hour. Let t represent the time the car has been traveling. Then the distance the car travels in t hours is given by $d = 60t$.

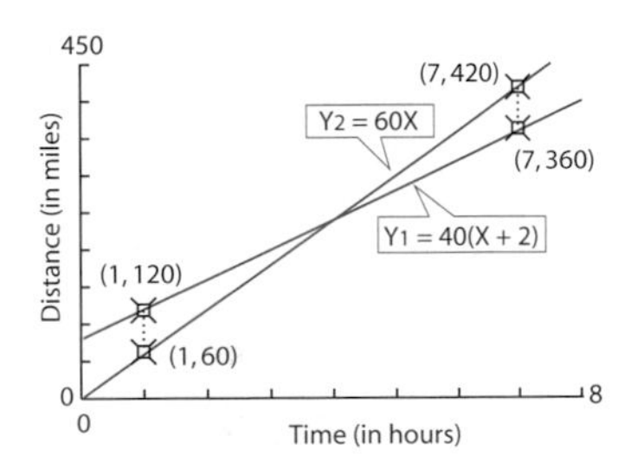

The motorcycle has been traveling 2 hours longer than the car. Therefore, the time of travel for the motorcycle is $t + 2$. The distance traveled by the motorcycle is $d = 40(t + 2)$. The graphs of the motorcycle and the car are shown above, where Y1 represents the motorcycle and Y2 represents the car. The variable X is used for t.

After the car has been traveling for 1 hour, the graph shows that the motorcycle has traveled 120 miles and the car has traveled 60 miles. The difference between the distances is $120 - 60 = 60$. The motorcycle is 60 miles ahead of the car. You can also see this by looking at the table of values for Y1 and Y2 at the right.

X	Y1	Y2
1	120	60
2	160	120
3	200	180
4	240	240
5	280	300
6	320	360
7	360	420
X = 4		

The graph and table show that the motorcycle is ahead of the car until the car has been traveling 4 hours. At that time, the car has caught up to the motorcycle. The car and the motorcycle have traveled the same distance. After 4 hours, the car is ahead of the motorcycle. At the end of 7 hours, the car is 60 miles ahead of the motorcycle.

1. From the table, the runner traveled 24 miles in 4 hours.

Another method that can be used to determine when the car will catch up to or overtake the motorcycle is to solve an equation.

Distance traveled by the car: $60t$
Distance traveled by the motorcycle: $40(t + 2)$

When the car overtakes the motorcycle, each has traveled the same distance.

Distance traveled by the car	equals	distance traveled by the motorcycle
$60t$	$=$	$40(t+2)$

$$60t = 40(t+2)$$
$$60t = 40t + 80 \quad \text{• Use the Distributive Property.}$$
$$20t = 80 \quad \text{• Subtract } 40t \text{ from each side of the equation.}$$
$$t = 4 \quad \text{• Divide each side of the equation by 20.}$$

The car overtakes the motorcycle in 4 hours.

Example 1 Two cars, one traveling 10 miles per hour faster than the second car, start at the same time from the same point and travel in opposite directions. In 3 hours, they are 300 miles apart. Find the rate of the second car.

Solution

State the goal. The goal is to find the rate of the second car.

Describe a strategy. Let r represent the unknown rate of the second car. Now represent the speed of the first car in terms of r. Since the first car is traveling 10 miles per hour faster than the second car, the rate of the first car is $r + 10$.

SUGGESTED ACTIVITY
Have students repeat this problem using r as the speed of the first car. When they complete the solution of their equation, it will be necessary for them to understand that they have the speed of the *first* car and must use that information to find the speed of the second car, which is the requested information.

Now represent the distance traveled by each car in 3 hours by using $d = rt$.

Distance traveled by first car: $3(r + 10)$
Distance traveled by second car: $3r$

To assist in writing an equation, draw a diagram showing how the distances are related. The cars are traveling in opposite directions, so show one car moving to the right and the other car moving to the left. The distance between the two cars after 3 hours is 300 miles.

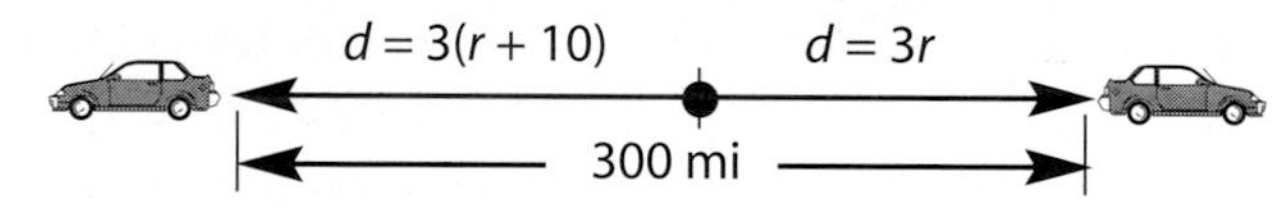

Note that the distance traveled by the first car plus the distance traveled by the second car is 300 miles. Translating the last sentence into an equation, we have $3(r + 10) + 3r = 300$.

Solve the problem.

$$3(r+10) + 3r = 300$$
$$3r + 30 + 3r = 300 \quad \text{• Use the Distributive Property.}$$
$$6r + 30 = 300 \quad \text{• Combine like terms.}$$
$$6r = 270 \quad \text{• Subtract 30 from each side of the equation.}$$
$$r = 45 \quad \text{• Divide each side of the equation by 6.}$$

Check your work. Replace r in the equation $3(r + 10) + 3r = 300$ and check that the left and right sides of the equation have the same value.

You-Try-It 1 Two trains, one traveling at twice the speed of the other, start at the same time from stations that are 306 miles apart and travel toward each other on parallel tracks. In 3 hours, the trains pass each other. Find the rate of each train.

Solution See page S14.

Example 2 On a survey mission, a pilot left an airport and flew out to a parcel of land and then back to the airport. The total trip took 5 hours. The rate of the plane out was 110 miles per hour. Due to head winds, the rate of the plane back was 90 miles per hour. How far from the airport was the parcel of land?

Solution

State the goal. The goal is to find the distance of the parcel of land from the airport.

Describe a strategy. If we can determine the time it took the pilot to fly to the parcel of land, then we can use the equation $d = rt$ to find the distance from the airport to the parcel of land. For instance, suppose it took 2 hours to fly to the parcel of land. Then the distance from the airport to the parcel would be:

INSTRUCTOR NOTE
This problem is particularly difficult for students because the choice of the variable is not the answer to the question. Relating this example to the Suggested Activity on the previous page may help these students.

rate of plane out → 110; time to parcel → 2

$$d = 110(2) = 220$$

220 ← distance to parcel

This suggests that we let t represent the time it takes the pilot to fly to the parcel of land. Because the total time of the trip was 5 hours, the time to return from the parcel of land is the total time for the trip (5 hours) minus the time out (t). Therefore, the time to return to the airport is $5 - t$.

Now write an expression for the distance traveled from the airport to the parcel of land and the distance from the parcel of land to the airport by using $d = rt$.

distance from the airport to the parcel: $110t$
distance from the parcel to the airport: $90(5 - t)$

To assist in writing an equation, draw a diagram showing the distances traveled to and from the parcel of land.

parcel ← $d = 110t$ — airport
parcel — $d = 90(5 - t)$ → airport

Note that the distance the plane travels to the parcel is the same as the distance the plane travels returning from the parcel. Translating the last sentence into an equation, we have $110t = 90(5 - t)$.

Solve the problem.

$110t = 90(5 - t)$
$110t = 450 - 90t$ • Use the Distributive Property.
$200t = 450$ • Add $90t$ to each side of the equation.
$t = 2.25$ • Divide each side of the equation by 200.

The time from the airport to the parcel was 2.25 hours. The distance to the parcel is $d = rt = 110(2.25) = 247.5$.

The distance to the parcel of land is 247.5 miles.

Check your work. It is important to note that the solution of the equation was 2.25 hours but that the answer to the question, how far is the parcel, required substituting into the equation $d = rt$. As a further check of your work, note that substituting 2.25 into $d = 90(5 - t) = 90(5 - 2.25) = 90(2.75) = 247.5$. This means that the distance back is equal to the distance out, as it should be.

You-Try-It 2 A bicycling club rides to the countryside at a speed of 16 miles per hour and returns over the same course at a rate of 12 miles per hour. How far does the club ride to the countryside if the time they spent riding was 7 hours?

Solution See page S14.

3.5 EXERCISES

Topics for Discussion

1. How are distance, rate, and time related?

   Answers will vary. For instance, $d = rt$; $t = \frac{d}{r}$; and $r = \frac{d}{t}$

2. Suppose a jogger starts on a 4-mile course. Eight minutes later a second jogger starts on the same course. If both joggers arrive at the finish line at the same time, which jogger is running faster?

   The second jogger

3. A Boeing 757 airplane leaves San Diego, California, and is flying to Dallas, Texas. One hour later, a Boeing 767 leaves San Diego taking the same route to Dallas. If t represents the time the Boeing 757 is in the air, how long has the Boeing 767 been in the air.

   $(t - 1)$ hours

4. If two objects started in the same place and are moving in opposite directions, how can the total distance between the two objects be expressed?

   The sum of the distances traveled by each object equals the total distance between the objects.

5. Two friends are standing 50 feet away and begin walking toward each other on a straight sidewalk. When they meet, what is the total distance covered by the two friends?

   50 feet

6. Suppose two planes are heading toward each other. One plane is traveling at 450 miles per hour and the other plane is traveling at 375 miles per hour. What is the rate at which the distance between the planes is changing?

   825 miles per hour

Uniform Motion Problems

7. Write the equation for the distance a car traveling at a constant rate of 45 miles per hour travels in t hours. Using a graphing calculator with viewing window Xmin = 0, Xmax = 6, Xscl = 1, Ymin = 0, Ymax = 300, Yscl = 50, graph this equation with Y1 as d and X as t.

   $d = 45t$

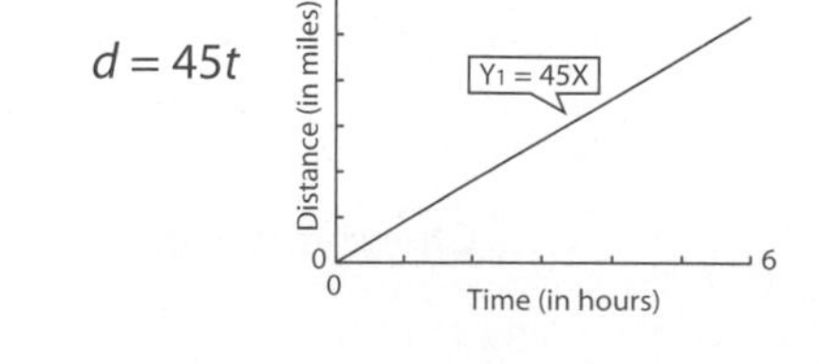

8. Write the equation for the distance a cyclist traveling at a constant rate of 12 miles per hour travels in t hours. Using a graphing calculator with viewing window Xmin = 0, Xmax = 4, Xscl = 1, Ymin = 0, Ymax = 50, Yscl = 10, graph this equation with Y1 as d and X as t.

   $d = 12t$

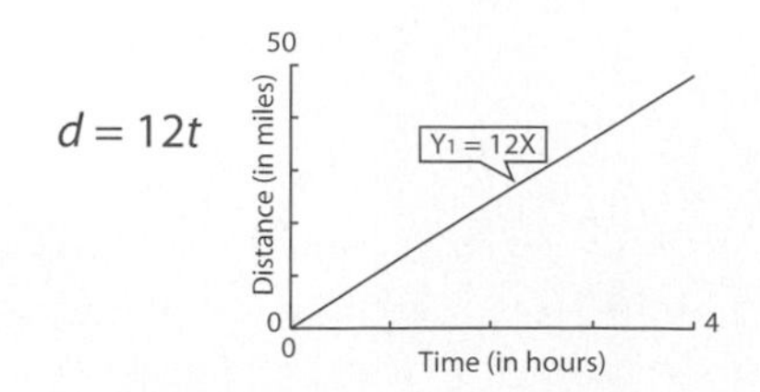

9. Jacob starts on a 16-mile hike on a path through a nature preserve. One hour later, Davadene starts on the same path walking in the same direction as Jacob. The graphs of the two hikers are shown at the right, where *t* is the time that Davadene has been walking.

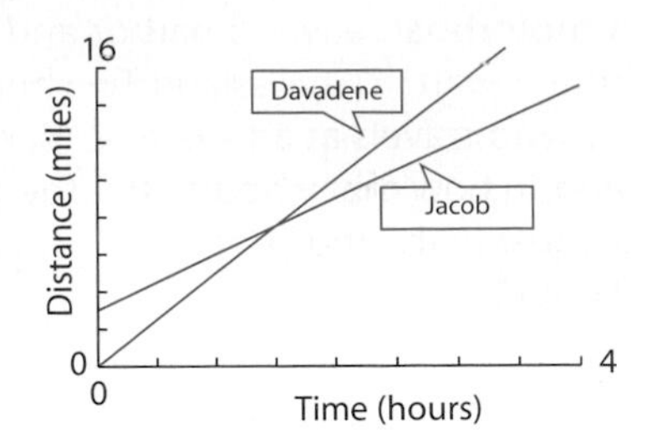

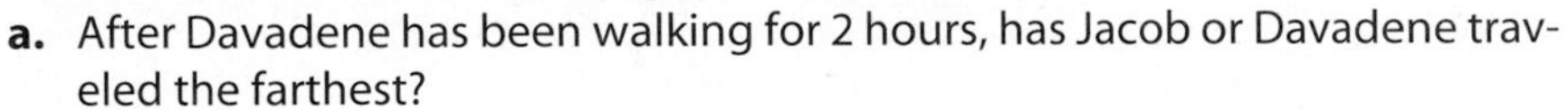

a. After Davadene has been walking for 2 hours, has Jacob or Davadene traveled the farthest?

b. Does Davadene ever pass Jacob on this hike? How can you tell?

a. Davadene; b. Yes. Her graph intersects Jacob's graph.

10. Imogene begins a 50-mile bicycle course. Two hours later, Alice starts on the same course riding in the same direction as Imogene. The graphs of the two cyclists are shown at the right, where *t* is the time that Alice has been riding.

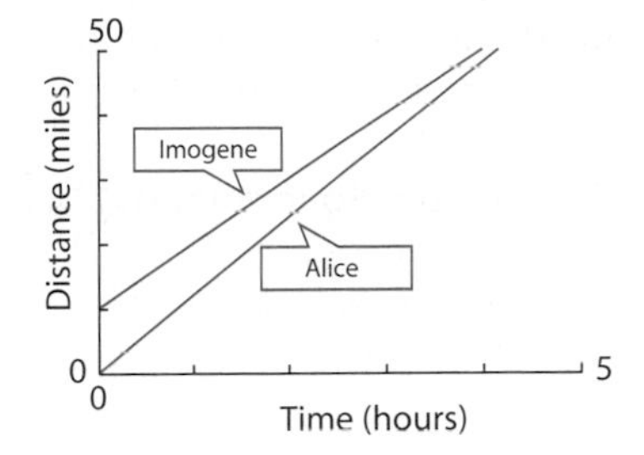

a. After Alice has been riding for 4 hours, has Imogene or Alice traveled the farthest?

b. Does Alice ever pass Imogene on this ride? How can you tell?

a. Imogene; b. No. Her graph does not intersect Imogene's graph.

11. Two small planes start from the same point and fly in opposite directions. The first plane is flying 25 miles per hour slower than the second plane. In 2 hours, the planes are 470 miles apart. Find the rate of each plane.

105 miles per hour; 130 miles per hour

12. Two cyclists start from the same point and ride in opposite directions. One cyclist rides twice as fast as the other. In 3 hours, they are 81 miles apart. Find the rate of each cyclist.

9 miles per hour; 18 miles per hour

13. Two planes leave an airport at 8 A.M., one flying north at 480 kilometers per hour and the other flying south at 520 kilometers per hour. At what time will they be 3000 kilometers apart?

11 A.M.

14. A long-distance runner started on a course, running at an average speed of 6 miles per hour. One-half hour later, a second runner began the same course at an average speed of 7 miles per hour. How long after the second runner started will the second runner overtake the first runner?

3 hours

15. A motorboat leaves a harbor and travels at an average speed of 9 miles per hour toward a small island. Two hours later a cabin cruiser leaves the same harbor and travels at an average speed of 18 miles per hour toward the same island. In how many hours after the cabin cruiser leaves will the cabin cruiser be alongside the motorboat?
2 hours

16. A 555-mile, 5-hour plane trip was flown at two speeds. For the first part of the trip, the average speed was 105 miles per hour. For the remainder of the trip, the average speed was 115 miles per hour. For how long did the plane fly at each speed?
105 miles per hour: 2 hours; 115 miles per hour: 3 hours

17. An executive drove from home at an average speed of 30 miles per hour to an airport where a helicopter was waiting. The executive boarded the helicopter and flew to the corporate offices at an average speed of 60 miles per hour. The entire distance was 150 miles. The entire trip took 3 hours. Find the distance from the airport to the corporate offices.
120 miles

18. After a sailboat had been on the water for 3 hours, a change in the wind direction reduced the average speed of the boat by 5 miles per hour. The entire distance sailed was 57 miles. The total time spent sailing was 6 hours. How far did the sailboat travel in the first 3 hours?
36 miles

19. A car and a bus set out at 3 P.M. from the same point headed in the same direction. The average speed of the car is twice the average speed of the bus. In 2 hours the car is 68 miles ahead of the bus. Find the rate of the car.
68 miles per hour

20. A passenger train leaves a train depot 2 hours after a freight train leaves the same depot. The freight train is traveling 20 miles per hour slower than the passenger train. Find the rate of each train if the passenger train overtakes the freight train in 3 hours.
Passenger train: 50 miles per hour; freight train: 30 miles per hour

21. As part of flight training, a student pilot was required to fly to an airport and then return. The average speed on the way to the airport was 100 miles per hour, and the average speed returning was 150 miles per hour. Find the distance between the two airports if the total flying time was 5 hours.
300 miles

22. A ship traveling east at 25 miles per hour is 10 miles from a harbor when another ship leaves the harbor traveling east at 35 miles per hour. How long does it take the second ship to catch up to the first ship? 1 hour

23. At 10 A.M. a plane leaves Boston, Massachusetts, for Seattle, Washington, a distance of 3000 miles. One hour later a plane leaves Seattle for Boston. Both planes are traveling at a speed of 500 miles per hour. How many hours after the plane leaves Seattle will the planes pass each other? 2.5 hours

24. At noon a train leaves Washington, D.C., headed for Charleston, South Carolina, a distance of 500 miles. The train travels at a speed of 60 miles per hour. At 1 P.M. a second train leaves Charleston headed for Washington, D.C., traveling at 50 miles per hour. How long after the train leaves Charleston will the two trains pass each other? 4 hours

25. Two cyclists start at the same time from opposite ends of a course that is 51 miles long. One cyclist is riding at a rate of 16 miles per hour, and the second cyclist is riding at a rate of 18 miles per hour. How long after they begin will they meet? 1.5 hours

26. A bus traveled on a level road for 2 hours at an average speed that was 20 miles per hour faster than its average speed on a winding road. The time spent on the winding road was 3 hours. Find the average speed on the winding road if the total trip was 210 miles. 34 miles per hour

27. A bus traveling at a rate of 60 miles per hour overtakes a car traveling at a rate of 45 miles per hour. If the car had a 1-hour head start, how far from the starting point does the bus overtake the car? 180 miles

28. A car traveling at 48 miles per hour overtakes a cyclist who, riding at 12 miles per hour, had a 3-hour head start. How far from the starting point does the car overtake the cyclist?
48 miles

Applying Concepts

29. Michele and Sarakit are racing box cars. The graph of each of their cars for a certain race is shown at the right. Who has the box car with the greater speed for the 10 seconds shown? Explain your answer.
Michele; her graph is above Sarakit's graph.

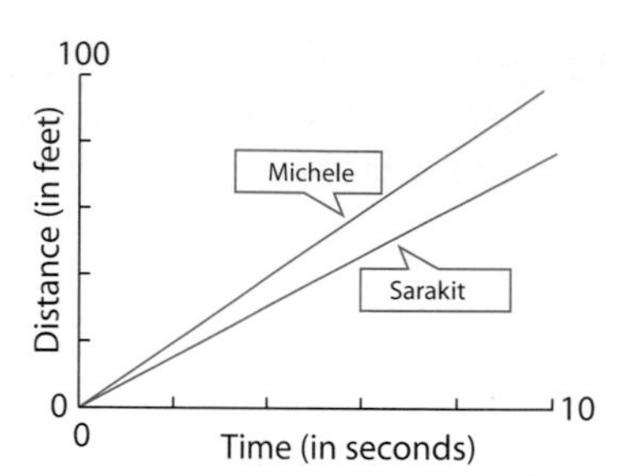

30. At 10 A.M., two campers left their campsite by canoe and paddled downstream at an average speed of 12 miles per hour. They then turned around and paddled back upstream at an average rate of 4 miles per hour. The total trip took 1 hour. At what time did the campers turn around downstream?
10:15 A.M.

31. Emily starts riding across a railroad bridge at point A. When $\frac{3}{8}$ of the way across the bridge, she notices a train approaching at 60 miles per hour. Emily can ride fast enough to return to A just as the train arrives at A or she can continue across the bridge and arrive at B at the same time the train does. How fast can Emily ride? 15 miles per hour

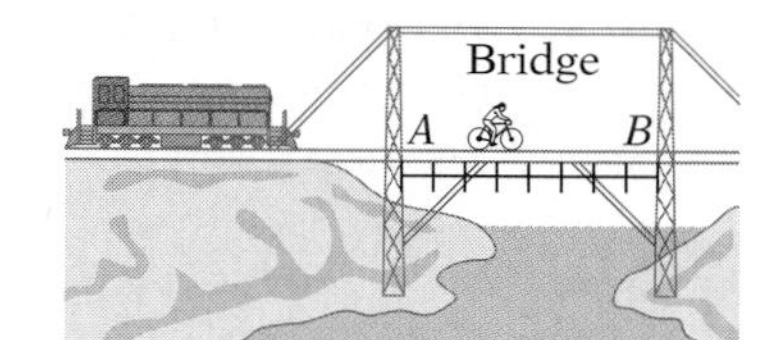

32. A car travels a 1-mile track at an average speed of 30 miles per hour. At what average speed must the car travel the next mile so that the average speed for the 2 miles is 60 miles per hour?
It is impossible to average 60 miles per hour.

Exploration

33. ***Uniform motion problems with a graphing calculator*** Another way that we can check the answers to Example 1 and Example 2 in this section is to use a graphing calculator.

a. For Example 1, enter $Y1 = 3(X + 10) + 3X$ and $Y2 = 300$. Graph these two equations. Then use the INTERSECT feature of your calculator to find the point of intersection of the two lines. The x-coordinate of the point of intersection is the solution of $3(r + 10) + 3r = 300$.

b. For Example 2, enter $Y1 = 110X$ and $Y2 = 90(5 - X)$. Graph these two equations. Then use the INTERSECT feature of your calculator to find the point of intersection of the two lines. The x-coordinate of the point of intersection is the solution of $110t = 90(5 - t)$.

c. Solve Exercises 11, 15, and 19 by using a graphing calculator.

Section 3.6 Percent Mixture Problems

The quantity of a substance in a solution can be given as a percent of the total solution. For instance, in a 15% sugar water solution, 15% of the total solution is sugar. The remaining 85% of the solution is water.

The equation $Q = Ar$ relates the quantity, Q, of a substance in the solution to the amount, A, of solution and the percent concentration, r, of the solution. For example, suppose there are 40 ounces of the sugar water solution mentioned above. Then the quantity of sugar in the solution is:

SUGGESTED ACTIVITY
See the *Instructor's Resource Manual* for an activity that will help students better understand percent mixture problems.

$Q = Ar$

$= 40(0.15)$ • The total amount, A, of solution is 40 ounces. The percent concentration, r, is 15% = 0.15.

$= 6$

There are 6 ounces of sugar in the solution. Because the total amount of solution is 40 ounces, there are 40 – 6 = 34 ounces of water.

SUGGESTED ACTIVITY
This is a mixture problem that can be solved by just thinking about the solution rather than trying to write an equation. Suppose you have one cup of cream and one cup of coffee. Take one tablespoon of cream and add it to the coffee and thoroughly stir the mixture. Now take one tablespoon of the coffee and add it to the cream. Is there more coffee in the cream or more cream in the coffee?

The solution of percent mixture problems is based on the fact that the sum of the quantities of some substance before mixing is equal to the quantity of the substance after mixing. For instance, suppose a 100-gram alloy (a mixture of two or more metals) of gold is 25% gold and a second 150-gram alloy of gold is 50% gold.

Quantity of gold in one alloy	Quantity of gold in second alloy
$Q_1 = 100(0.25) = 25$	$Q_2 = 150(0.50) = 75$

There are 25 grams of gold in the first alloy and 75 grams of gold in the second alloy. If the two alloys are now mixed together, the total quantity of gold would be 100 grams (25 + 75) and the total amount of alloy would be 250 grams (100 + 150). The percent of gold in the new alloy can be found from the equation $Q = Ar$.

$Q = Ar$

$100 = 250r$ • Quantity of gold in new alloy is 100 grams. Amount of new alloy is 250 grams.

$0.40 = r$ • Divide each side of the equation by 250.

The new alloy contains 40% gold.

Example 1 Six pounds of a 15% salt solution are mixed with 4 pounds of a 20% salt solution. What is the percent concentration of the new solution?

Solution

State the goal. The goal is to find the percent concentration of the new solution.

Describe a strategy. Find the quantity of salt in the first solution and the quantity of salt in the second solution. The quantity of salt in the new solution is the sum of these quantities. The amount of the new solution is 10 gallons (6 + 4). Now use the equation $Q = Ar$ to find the percent concentration for the new solution

Solve the problem.

Quantity of salt in first solution	$Q_1 = 6(0.15)$ $= 0.9$	Quantity of salt in second solution	$Q_2 = 4(0.20)$ $= 0.8$

There are 0.9 + 0.8 = 1.7 pounds of salt in the new solution.

$Q = Ar$

$1.7 = 10r$ • Quantity of salt in new solution is 1.7 pounds. Amount of new solution is 10 pounds.

$0.17 = r$ • Divide each side of the equation by 10.

The new solution has a 17% concentration of salt.

Check your work. Be sure to check your work.

You-Try-It 1 A chemist mixes 1.8 liters of a 9% acid solution with 1.2 liters of a 4% acid solution. What is the percent concentration of acid in the new solution?

Solution See page S14.

Many percent mixture problems require mixing two substances to produce a new substance that has a given percent concentration.

Example 2 A dermatologist has 0.5% hydrocortisone cream that must be mixed with 0.75% hydrocortisone cream to make 50 grams of cream that is 0.6% hydrocortisone. How many grams of each available cream must the dermatologist use?

State the goal.

Solution The goal is to find the number of grams of 0.5% hydrocortisone cream and the number of grams of 0.75% hydrocortisone cream that must be mixed to produce 50 grams of cream that is 0.6% hydrocortisone.

Describe a strategy.

Let x represent the number of grams of the 0.5% cream that is to be used. (We could have let x represent the number of grams of 0.75% cream.) Because the dermatologist needs to make 50 grams of cream and x grams are 0.5% hydrocortisone cream, the amount remaining for the 0.75% cream is $50 - x$.

Find the quantity of hydrocortisone in both the 0.5% cream and the 0.75% cream.

quantity of hydrocortisone in a cream: rA
quantity in 0.5% cream: $0.005x$
quantity in 0.75% cream: $0.0075(50 - x)$
quantity in 0.6% cream: $0.006(50)$

Write an equation using the fact that the sum of the quantity of hydrocortisone in the 0.5% cream and the quantity in the 0.75% cream equals the quantity in the 0.6% cream.

Solve the problem.

$$\begin{aligned} 0.005x + 0.0075(50 - x) &= 0.006(50) \\ 0.005x + 0.375 - 0.0075x &= 0.3 && \text{• Use the Distributive Property.} \\ -0.0025x + 0.375 &= 0.3 && \text{• Combine like terms.} \\ -0.0025x &= -0.075 && \text{• Subtract 0.375 from each side.} \\ x &= 30 && \text{• Divide each side by } -0.0025. \end{aligned}$$

Because x represents the number of grams of 0.5% hydrocortisone cream, the dermatologist must use 30 grams of the 0.5% cream. The quantity of 0.75% cream is $50 - x = 50 - 30 = 20$. The dermatologist must use 20 grams of the 0.75% cream.

Check your work.

One way to check is shown at the left. A second way to check is to calculate the percent concentration of the cream after mixing to ensure it is 0.6%.

Hydrocortisone in first cream $Q_1 = 30(0.005) = 0.15$ Hydrocortisone in second cream $Q_2 = 20(0.0075) = 0.15$

There is $0.15 + 0.15 = 0.3$ gram of hydrocortisone in the new mixture.

$$\begin{aligned} Q &= Ar \\ 0.3 &= 50r && \text{• Quantity of hydrocortisone in new mixture is 0.3 gram. Amount of new mixture is 50 grams.} \\ 0.006 &= r && \text{• Divide each side of the equation by 50.} \end{aligned}$$

The percent concentration is 0.6%. The solution checks.

TAKE NOTE
Here is one possible check.

$0.005x + 0.0075(50 - x)$	$= 0.006(50)$
$0.005(30) + 0.0075(50 - 30)$	0.30
$0.005(30) + 0.0075(20)$	0.30
$0.15 + 0.15$	0.30
0.30	$= 0.30$

The solution checks.

You-Try-It 2 How many pounds of coffee that is 30% mocha java beans must be mixed with 80 pounds of coffee that is 40% mocha java beans to produce a coffee blend that is 32% mocha java?

Solution See page S14.

Example 3 How many ounces of pure bran must be added to 50 ounces of a cereal that is 40% bran to produce a mixture that is 50% bran?

Solution

State the goal. The goal is to find the number of ounces of pure bran that must be added to a mixture to produce a new mixture that is 50% bran.

Describe a strategy. Let x represent the number of ounces of pure bran that must be added. Because the pure bran is being added to 50 ounces of existing cereal, the amount of the new mixture will be $(x + 50)$ ounces.

Find the quantity of bran in the bran that is being added, the bran in the 50 ounces of cereal, and the bran in the new mixture. Note that the bran that is being added is pure bran, which means its percent concentration is 100%. The decimal equivalent of 100% is 1.

quantity of bran: rA
quantity of bran in amount added: $1 \cdot x$
quantity of bran in 50 ounces of cereal: $0.40(50)$
quantity of bran in new mixture: $0.50(x + 50)$

Write an equation using the fact that the sum of the quantity of bran being added and the quantity in the original 50 ounces of cereal equals the quantity in the new mixture.

Solve the problem.

$$
\begin{aligned}
1 \cdot x + 0.40(50) &= 0.50(x + 50) \\
x + 20 &= 0.5x + 25 && \text{• Use the Distributive Property.} \\
0.5x + 20 &= 25 && \text{• Subtract } 0.5x \text{ from each side.} \\
0.5x &= 5 && \text{• Subtract 20 from each side.} \\
x &= 10 && \text{• Divide each side by 0.5.}
\end{aligned}
$$

It is necessary to add 10 ounces of pure bran to make a new mixture that is 50% bran.

Check your work. To check our answer, we will calculate the percent concentration of the bran after mixing it with the 50 ounces of cereal to ensure that the new mixture is 50% bran.

Quantity of bran being added $\quad Q_1 = 10 \cdot 1 = 10$

Quantity of bran in cereal $\quad Q_2 = 50(0.4) = 20$

There are $10 + 20 = 30$ ounces of bran in the new mixture. The amount of new mixture is the original amount of cereal plus the number of ounces of bran that was added: $50 + 10 = 60$.

$$
\begin{aligned}
Q &= Ar \\
30 &= 60r && \text{• Quantity of bran in new mixture is 30 ounces. Amount of new mixture is 60 ounces.} \\
0.5 &= r && \text{• Divide each side of the equation by 60.}
\end{aligned}
$$

The percent concentration is 50%. The solution checks.

You-Try-It 3 How many ounces of a 25% gold alloy must a jeweler mix with 8 ounces of a 30% gold alloy to produce a new alloy that is 27% gold?

Solution See page S14.

3.6 EXERCISES

Topics for Discussion

1. Suppose pure water is added to a saltwater solution.

a. Does the percent concentration of the salt increase, decrease, or remain the same? Decreases

b. Does the percent concentration of water increase, decrease, or remain the same? Increases

c. Does the amount of salt increase, decrease, or remain the same? Remains the same

2. Suppose a 20%-gold alloy is added to a gold alloy that is 50% gold.

a. Does the percent concentration of the gold increase, decrease, or remain the same? Decreases

b. Does the amount of gold increase, decrease, or remain the same? Increases

3. Suppose a solution that is 60% acid is added to a solution that is 30% acid.

a. Does the percent concentration of the acid increase, decrease, or remain the same? Increases

b. Does the amount of acid increase, decrease, or remain the same? Increases

4. Suppose pure sugar is added to a sugar water solution.

a. Does the percent concentration of the sugar increase, decrease, or remain the same? Increases

b. Does the percent concentration of water increase, decrease, or remain the same? Decreases

c. Does the amount of sugar increase, decrease, or remain the same? Increases

Percent Mixture Problems

5. Forty ounces of a 30% gold alloy are mixed with 60 ounces of a 20% gold alloy. Find the percent concentration of the resulting alloy.
24%

6. One hundred ounces of juice that is 50% tomato juice are added to 200 ounces of a vegetable juice that is 25% tomato juice. What is the percent concentration of the tomato juice in the resulting mixture?

$33\frac{1}{3}\%$

7. How many gallons of a 15% acid solution must be mixed with 5 gallons of a 20% acid solution to make a 16% acid solution?
20 gallons

8. How many pounds of a chicken feed that is 50% corn must be mixed with 400 pounds of a feed that is 80% corn to make a chicken feed that is 75% corn?
80 pounds

9. A rug is made by weaving 20 pounds of yarn that is 50% wool with a yarn that is 25% wool. How many pounds of the yarn that is 25% wool is used if the finished rug is 35% wool?
30 pounds

10. Five gallons of a dark green latex paint that is 20% yellow paint is combined with a lighter green latex paint that is 40% yellow paint. How many gallons of the lighter green paint must be used to create paint that is 25% yellow paint?
$1\frac{2}{3}$ gallons

11. How many gallons of a plant food that is 9% nitrogen must be combined with another plant food that is 25% nitrogen to make 10 gallons of a solution that is 15% nitrogen?
6.25 gallons

12. A chemist wants to make 50 milliliters of a 16% acid solution by mixing a 13% acid solution and an 18% acid solution. How many milliliters of each solution should the chemist use?
13% solution: 20 milliliters; 18% solution: 30 milliliters

13. Five grams of sugar are added to a 45-gram serving of a breakfast cereal that is 10% sugar. What is the percent concentration of sugar in the resulting mixture?
19%

14. Thirty ounces of pure silver are added to 50 ounces of a silver alloy that is 20% silver. What is the percent concentration of the resulting alloy?
50%

15. How many pounds of fertilizer that is 40% ammonium sulfate must be mixed with 80 pounds of fertilizer that is 30% ammonium sulfate to make a mixture that is 32% ammonium sulfate?
20 pounds

16. The manager of a garden shop mixes grass seed that is 60% rye grass with 70 pounds of grass seed that is 80% rye grass to make a mixture that is 74% rye grass. How much of the 60% rye grass is used?
30 pounds

17. A hair dye is made by blending a 7% hydrogen peroxide solution and a 4% hydrogen peroxide solution. How many milliliters of each are used to make a 300-milliliter solution that is 5% hydrogen peroxide?
7% solution: 100 milliliters; 4% solution: 200 milliliters

18. A tea that is 20% jasmine is blended with a tea that is 15% jasmine. How many pounds of each tea are used to make 5 pounds of tea that is 18% jasmine?
20% jasmine: 3 pounds; 15% jasmine: 2 pounds

19. How many ounces of pure chocolate must be added to 150 ounces of chocolate topping that is 50% chocolate to make a topping that is 75% chocolate?
150 ounces

20. How many ounces of pure silver must be added to 50 ounces of a silver alloy that is 20% silver to produce an alloy that is 50% silver?
30 ounces

21. A clothing manufacturer has some pure silk thread and some thread that is 85% silk. How many kilograms of each must be woven together to make 75 kilograms of cloth that is 96% silk?
Pure silk: 55 kilograms; 85% silk: 20 kilograms

Applying Concepts

22. How many ounces of pure water must be evaporated from 50 ounces of a 12% salt solution to produce a 15% salt solution?
10 ounces

23. A radiator contains 15 gallons of a 20% antifreeze solution. How many gallons must be drained from the radiator and replaced by pure antifreeze so that the radiator will contain 15 gallons of a 40% antifreeze solution?
3.75 gallons

24. How many grams of pure water must be added to 50 grams of pure acid to make a solution that is 40% acid?
75 grams

25. Make up a percent mixture problem similar to one in this section and then solve it.

Exploration

26. ***Changes in percent concentration*** Suppose you begin with a 30-milliliter solution that is 10% hydrochloric acid.

a. If 10 milliliters of a 30% solution of hydrochloric acid solution are added to this solution, what is the percent concentration of the new solution?

b. If x milliliters of a 30% hydrochloric acid solution are added to the original solution, what is the percent concentration of the new solution?

c. Letting Y_1 represent the percent concentration and X represent the number of milliliters of the 30% hydrochloric acid solution that is being added, graph the equation. Try using a viewing window of Xmin = 0, Xmax = 50, Ymin = 0, and Ymax = 0.3.

d. Using your graph, is the change in percent concentration from X = 10 to X = 20 the same as the change in percent concentration from X = 20 to X = 30?

e. If more and more of the 30% hydrochloric acid solution is added to the original mixture, will the percent concentration of the resulting solution ever exceed 30%? Explain your answer.

Section 3.7 Inequalities

Introduction to Inequalities

Recall that an equation contains an equal sign. An **inequality** contains the symbol $>$, $<$, $\leq$, or $\geq$. An inequality expresses the relative order of two mathematical expressions. Here are some examples of inequalities in one variable.

$$\left.\begin{array}{l} 4x \geq 12 \\ 2x + 7 \leq 9 \\ x^2 + 1 > 3x \end{array}\right\} \quad \text{Inequalities in one variable}$$

A **solution of an inequality in one variable** is a number that when substituted for the variable results in a true inequality. For the inequality $x < 6$ shown below, 5, 0, and -4 are solutions of the inequality because replacing the variable by these numbers results in a true inequality.

$$\begin{array}{lll} x < 6 & x < 6 & x < 6 \\ 5 < 6 \quad \text{True} & 0 < 6 \quad \text{True} & -4 < 6 \quad \text{True} \end{array}$$

The number 7 is not a solution of the inequality because $7 < 6$ is a false inequality.

Besides the numbers 5, 0, and -4, there are an infinite number of other solutions of the inequality $x < 6$. Any number less than 6 is a solution; for instance, -5.2, $\frac{5}{2}$, π, and 1 are also solutions of the inequality.

The set of all the solutions of an inequality is called the **solution set** of the inequality. Because there are an infinite number of elements in this set, it is not possible to use the roster method to list all the solutions. A second method of representing the elements of a set that is especially useful when writing infinite sets is **set-builder notation**. In set-builder notation, the set of real numbers less than 6 is written $\{x \mid x < 6, x \in \text{real numbers}\}$.

Note at the right how to read a set written in set-builder notation.

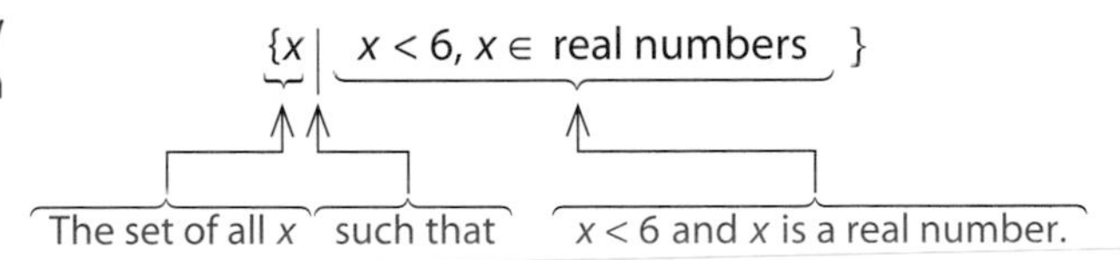

TAKE NOTE
Recall that the integers are the numbers $\ldots -3, -2, -1, 0, 1, 2, 3, \ldots$. The real numbers include the rational numbers (which include the integers) and the irrational numbers.

Example 1 Use set-builder notation to write the set of real numbers greater than or equal to -8.

Solution $\{x \mid x \geq -8, x \in \text{real numbers}\}$

You-Try-It 1 Use set-builder notation to write the set of integers less than or equal to 5.

Solution See page S15.

When using set-builder notation with real numbers, $x \in$ real numbers is frequently omitted. Using this convention, we would write $\{x \mid x < 6\}$ instead of $\{x \mid x < 6, x \in \text{real numbers}\}$. Similarly, the answer to Example 1 would be written $\{x \mid x \geq -8\}$. In this text, any time the type of number is not included in set-builder notation, we assume it is a real number.

The graph of the solution set of $x < 6$ is shown at the right. The parenthesis at 6 indicates that 6 is not part of the graph. This graph represents $\{x \mid x < 6\}$.

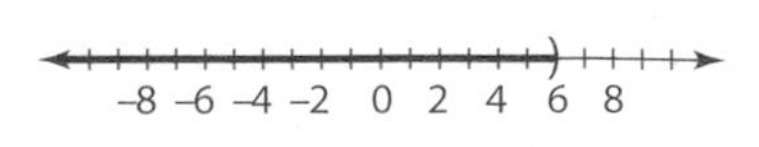

The graph of the solution set of $x \geq -8$ is shown at the right. The bracket at −8 indicates that −8 is part of the graph. This graph represents $\{x \mid x \geq -8\}$.

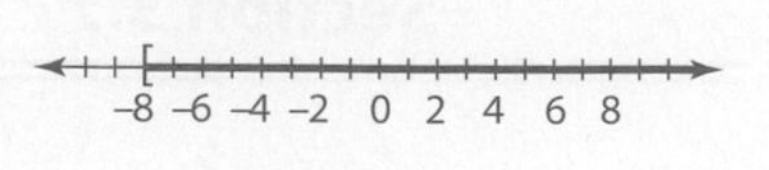

Example 2 Graph the solution set of $x \leq 3$.

Solution

–8 –6 –4 –2 0 2 4 6 8

You-Try-It 2 Graph the solution set of $x > -1$.

Solution See page S15.

Addition Property of Inequalities

In solving an inequality, the goal is to rewrite the given inequality in the form *variable* > *constant* or *variable* < *constant*. The Addition Property of Inequalities is used in order to rewrite an inequality in this form.

> **Addition Property of Inequalities**
> The same number can be added to each side of an inequality without changing the solution set of the inequality.
> If $a > b$, then $a + c > b + c$.
> If $a < b$, then $a + c < b + c$.

The Addition Property of Inequalities also holds true for an inequality containing the symbol ≤ or ≥.

The Addition Property of Inequalities is used when, in order to rewrite an inequality in the form *variable* > *constant* or *variable* < *constant*, a term must be removed from one side of the inequality. Add the opposite of the term to each side of the inequality.

Because subtraction is defined in terms of addition, the Addition Property of Inequalities makes it possible to subtract the same number from each side of an inequality without changing the solution set of the inequality.

Example 3 Solve and graph the solution set of $x + 5 > 3$.

Solution

$x + 5 > 3$

$x + 5 - 5 > 3 - 5$ • Subtract 5 from each side of the inequality.

$x > -2$

The solution set is $\{x \mid x > -2\}$. The graph is shown below.

–8 –6 –4 –2 0 2 4 6 8

You-Try-It 3 Solve and graph the solution set of $x - 4 \leq 1$.

Solution See page S15.

Multiplication Property of Inequalities

When multiplying or dividing an inequality by a number, the inequality symbol may be reversed depending on whether a positive or negative number is used. Look at the following two examples.

$3 < 5$

$2(3) < 2(5)$ • Multiply by positive 2. The inequality symbol remains the same.

$6 < 10$

$3 < 5$

$-2(3) > -2(5)$ • Multiply by negative 2. The inequality symbol is reversed.

$-6 > -10$

This is summarized in the Multiplication Property of Inequalities.

> **Multiplication Property of Inequalities**
>
> ***Rule 1***
>
> Each side of an inequality can be multiplied by the same **positive number** without changing the solution set of the inequality.
>
> If $a > b$ and $c > 0$, then $ac > bc$. If $a < b$ and $c > 0$, then $ac < bc$.
>
> ***Rule 2***
>
> If each side of an inequality is multiplied by the same **negative number** and the inequality symbol is reversed, then the solution set of the inequality is not changed.
>
> If $a > b$ and $c < 0$, then $ac < bc$. If $a < b$ and $c < 0$, then $ac > bc$.

TAKE NOTE
$c > 0$ means c is a positive number. Note that the inequality symbols are not changed.

TAKE NOTE
$c < 0$ means c is a negative number. Note that the inequality symbols are reversed.

The Multiplication Property of inequalities also holds for an inequality containing the symbol $\le$ or $\ge$.

In solving an inequality, the goal is to rewrite the given inequality in the form *variable* > *constant* or *constant* > *variable*. The Multiplication Property of Inequalities is used when, in order to rewrite an inequality in the form *variable* > *constant* or *constant* > *variable*, a coefficient must be removed from a variable term of the inequality.

To write the inequality at the right in the form *variable* $\le$ *constant*, multiply each side of the inequality by $-\frac{2}{3}$, the reciprocal of $-\frac{3}{2}$. Because each side of the inequality is multiplied by a negative number, the inequality symbol must be reversed.

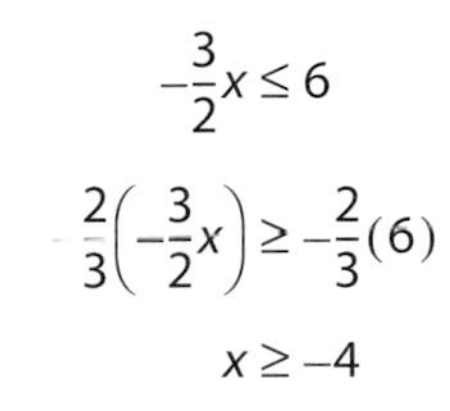

$$-\frac{3}{2}x \le 6$$

$$\frac{2}{3}\left(-\frac{3}{2}x\right) \ge -\frac{2}{3}(6)$$

$$x \ge -4$$

The graph of the solution set of the inequality is shown at the right.

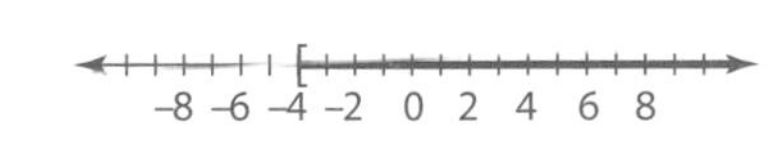

Recall that division is defined in terms of multiplication. Therefore, the Multiplication Property of Inequalities allows each side of an inequality to be divided by the same number. When each side of an inequality is divided by a positive number, the inequality symbol remains the same. When each side of an inequality is divided by a negative number, the inequality symbol must be reversed.

TAKE NOTE
Any time an inequality is multiplied or divided ***by*** a negative number, the inequality symbol must be reversed. Compare these two examples.

$$2x < -4$$
$$\frac{2x}{2} < \frac{-4}{2}$$
$$x < -2$$

Divide each side by *positive* 2. The inequality symbol *is not* reversed.

$$-2x < -4$$
$$\frac{-2x}{-2} > \frac{-4}{-2}$$
$$x > 2$$

Divide each side by *negative* 2. The inequality symbol *is* reversed.

Example 4 Solve and graph the solution set of $7x < -21$.

Solution

$$7x < -21$$

$$\frac{7x}{7} < \frac{-21}{7}$$

• Divide each side of the inequality by 7. Because 7 is a *positive* number, do not reverse the inequality symbol.

$$x < -3$$

The solution set is $\{x \mid x < -3\}$. The graph is shown below.

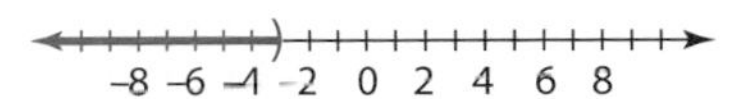

You-Try-It 4 Solve and graph the solution set of $-3x \ge 6$.

Solution See page S15.

General Inequalities

When solving an inequality, it is often necessary to apply both the Addition and the Multiplication Properties of Inequalities.

Example 5 Solve: $5x + 2 \le 8x - 7$

Solution

$$5x + 2 \le 8x - 7$$

$$5x - 8x + 2 \le 8x - 8x - 7$$
• Subtract $8x$ from each side of the inequality.

$$-3x + 2 \le -7$$

$$-3x + 2 - 2 \le -7 - 2$$
• Subtract 2 from each side of the inequality.

$$-3x \le -9$$

$$\frac{-3x}{-3} \ge \frac{-9}{-3}$$
• Divide each side of the inequality by −3. The inequality symbol must be reversed.

$$x \ge 3$$

The solution set is $\{x \mid x \ge 3\}$.

You-Try-It 5 Solve: $3x - 5 > x + 7$

Solution See page S15.

When an inequality contains parentheses, one of the steps in solving the inequality requires the use of the Distributive Property.

Example 6 Solve: $-2(x - 7) \le 3 - 4(2x - 3)$

Solution

$$-2(x - 7) \le 3 - 4(2x - 3)$$

$$-2x + 14 \le 3 - 8x + 12$$
• Use the Distributive Property.

$$-2x + 14 \le -8x + 15$$

$$-2x + 8x + 14 \le -8x + 8x + 15$$
• Add $8x$ to each side of the inequality.

$$6x + 14 \le 15$$

$$6x + 14 - 14 \le 15 - 14$$
• Subtract 14 from each side of the inequality.

$$6x \le 1$$

$$\frac{6x}{6} \le \frac{1}{6}$$
• Divide each side of the inequality by 6.

$$x \le \frac{1}{6}$$

The solution set is $\left\{x \mid x \le \frac{1}{6}\right\}$.

You-Try-It 6 Solve: $3(3 - 2x) \ge -5x - 2(3 - x)$

Solution See page S15.

Applications

SUGGESTED ACTIVITY

Have students translate the following sentences into inequalities.

1. x is at least 20.
2. The minimum value of y is 12.
3. w exceeds −3.
4. x is at most 7.
5. y is more than 15.
6. m is 17 or less.

Solving application problems requires recognition of the verbal phrases that translate into mathematical symbols. For instance, consider the sentence "To vote in an election for the president of the United States, a person must be *at least* 18 years old." This means the person must be 18 years old or older. If we let A represent a person's age, then this condition is represented mathematically as $A \ge 18$.

Now consider a plaque in an elevator that reads "The *maximum* allowable weight is 1050 pounds." This means that the total weight of all the occupants in the elevator must be 1050 pounds or less. If W represents the total weight of all the occupants, then this condition is expressed mathematically as $W \le 1050$.

There are additional words or phrases that translate into an inequality symbol. Here is a list of some of the phrases used to indicate each of the four inequality symbols.

$<$ is less than

$>$ { is greater than; is more than; exceeds }

$\leq$ { is less than or equal to; maximum; at most; or less }

$\geq$ { is greater than or equal to; minimum; at least; or more }

Example 7 A student must have an average of at least 90 on five tests in order to earn an A in an anthropology course. The student's first four test scores were: 88, 94, 82, and 92. What minimum score must the student receive on the last test to earn an A in the course?

Solution

State the goal. The goal is to find the minimum score on test 5 that allows the student to earn an A in the course.

Describe a strategy. Let x represent the minimum score on test 5 that is necessary to earn an A. Recall that the average score on 5 tests is calculated by adding the scores of the 5 tests and then dividing the sum by 5. Because the goal is to find the *minimum* score, we must write an inequality.

Solve the problem.

$$\frac{88+94+82+92+x}{5} \geq 90$$

$$\frac{356+x}{5} \geq 90$$

$$5\left(\frac{356+x}{5}\right) \geq 5(90)$$

$$356+x \geq 450$$

$$x \geq 94$$

The student must score 94 or more on test 5 to earn an A in the course.

Check your work. You can check this answer by calculating the average of the 5 tests assuming that the student scored 94 (the smallest value) and determining whether the average is 90.

$$\frac{88+94+82+92+94}{5} = \frac{450}{5} = 90$$

The solution checks.

You-Try-It 7 The base of a triangle is 8 inches and the height is $(3x-5)$ inches. Express as an integer the maximum height of the triangle when the area is less than 112 square inches.

Solution See page S15.

3.7 EXERCISES

Topics for Discussion

1. Does adding or subtracting the same term on each side of an inequality change the solution set of the inequality? No.

2. Does multiplying or dividing each side of an inequality by the same nonzero term change the solution set of the inequality? No, if the number is positive, or if the number is negative and the sign is reversed.

3. Explain how set-builder notation is used. Answers will vary.

4. How does the solution set of $x \le 4$ differ from the solution set of $x < 4$? How does the solution set of $x > -1$ differ from the solution set of $x \ge -1$? When the symbols $\le$ or $\ge$ are used, the constant term is included in the solution set.

Solving Inequalities

Solve and graph the solution set.

5. $x + 1 < 3$ $\{x \mid x < 2\}$; -8 -6 -4 -2 0 2 4 6 8

6. $x - 5 > -2$ $\{x \mid x > 3\}$; -8 -6 -4 -2 0 2 4 6 8

7. $5 + x \ge 4$ $\{x \mid x \ge -1\}$; -8 -6 -4 -2 0 2 4 6 8

8. $-2 + n \ge 0$ $\{n \mid n \ge 2\}$; -8 -6 -4 -2 0 2 4 6 8

9. $8x \le -24$ $\{x \mid x \le -3\}$; -8 -6 -4 -2 0 2 4 6 8

10. $-4x < 8$ $\{x \mid x > -2\}$ -8 -6 -4 -2 0 2 4 6 8

11. $3x > 0$ $\{x \mid x > 0\}$; -8 -6 -4 -2 0 2 4 6 8

12. $-2n \le -8$ $\{n \mid n \ge 4\}$; -8 -6 -4 -2 0 2 4 6 8

Solve.

13. $y - 3 \ge -12$ $\{y \mid y \ge -9\}$

14. $3x - 5 < 2x + 7$ $\{x \mid x < 12\}$

15. $8x - 7 \ge 7x - 2$ $\{x \mid x \ge 5\}$

16. $2x + 4 < x - 7$ $\{x \mid x < -11\}$

17. $4x - 8 \le 2 + 3x$ $\{x \mid x \le 10\}$

18. $6x + 4 \ge 5x - 2$ $\{x \mid x \ge -6\}$

19. $2x - 12 > x - 10$ $\{x \mid x > 2\}$

20. $d + \frac{1}{2} < \frac{1}{3}$ $\left\{d \mid d < -\frac{1}{6}\right\}$

21. $x + \frac{5}{8} \geq -\frac{2}{3}$ $\left\{x|x \geq -\frac{31}{24}\right\}$

22. $2x - \frac{1}{2} < x + \frac{3}{4}$ $\left\{x|x < \frac{5}{4}\right\}$

23. $x - 0.23 \leq 0.47$ $\{x \mid x \leq 0.70\}$

24. $1.2x < 0.2x - 7.3$ $\{x \mid x < -7.3\}$

25. $3x < 5$ $\left\{x|x < \frac{5}{3}\right\}$

26. $-8x \leq -40$ $\{x \mid x \geq 5\}$

27. $10x > -25$ $\left\{x|x > -\frac{5}{2}\right\}$

28. $-5x \geq \frac{10}{3}$ $\left\{x|x \leq -\frac{2}{3}\right\}$

29. $\frac{2}{3}x < -12$ $\{x \mid x < -18\}$

30. $-\frac{3}{8}x < 6$ $\{x \mid x > -16\}$

31. $\frac{2}{3}y \geq 4$ $\{y \mid y \geq 6\}$

32. $-\frac{2}{3}x \leq 4$ $\{x \mid x \geq -6\}$

33. $-\frac{2}{11}b \geq -6$ $\{b \mid b \leq 33\}$

34. $-\frac{2}{3}x \geq \frac{4}{7}$ $\left\{x|x \leq -\frac{6}{7}\right\}$

35. $-\frac{3}{4}y \geq -\frac{5}{8}$ $\left\{y|y \leq \frac{5}{6}\right\}$

36. $\frac{2}{3}x \leq \frac{9}{14}$ $\left\{x|x \leq \frac{27}{28}\right\}$

37. $-\frac{3}{5}y < \frac{9}{10}$ $\left\{y|y > -\frac{3}{2}\right\}$

38. $-0.27x < 0.135$ $\{x \mid x > -0.5\}$

39. $8.4y \geq -6.72$ $\{y \mid y \geq -0.8\}$

40. $1.5x \leq 6.30$ $\{x \mid x \leq 4.2\}$

41. $-3.9x \geq -19.5$ $\{x \mid x \leq 5\}$

42. $0.07 < -0.378$ $\{x \mid x < -5.4\}$

43. $4x - 8 < 2x$ $\{x \mid x < 4\}$

44. $2x - 8 > 4x$ $\{x \mid x < -4\}$

45. $8 - 3x \le 5x$ $\{x \mid x \ge 1\}$

46. $3x + 2 \ge 5x - 8$ $\{x \mid x \le 5\}$

47. $5x - 2 < 3x - 2$ $\{x \mid x < 0\}$

48. $0.1(180 + x) > x$ $\{x \mid x < 20\}$

49. $0.15x + 55 > 0.10x + 80$ $\{x \mid x > 500\}$

50. $2(3x - 1) > 3x + 4$ $\{x \mid x > 2\}$

51. $3(2x - 5) \ge 8x - 5$ $\{x \mid x \le -5\}$

52. $2(2y - 5) \le 3(5 - 2y)$ $\left\{y \mid y \le \frac{5}{2}\right\}$

53. $5(2 - x) > 3(2x - 5)$ $\left\{x \mid x < \frac{25}{11}\right\}$

54. $5(x - 2) > 9x - 3(2x - 4)$ $\{x \mid x > 11\}$

55. $4 - 3(3 - n) \le 3(2 - 5n)$ $\left\{n \mid n \le \frac{11}{18}\right\}$

56. $2x - 3(x - 4) \ge 4 - 2(x - 7)$ $\{x \mid x \ge 6\}$

57. $\frac{1}{2}(9x - 10) \le -\frac{1}{3}(12 - 6x)$ $\left\{x \mid x \le \frac{2}{5}\right\}$

58. $\frac{2}{3}(9t - 15) + 4 < 6 + \frac{3}{4}(4 - 12t)$ $\{t \mid t < 1\}$

59. $3[4(n - 2) - (1 - n)] > 5(n - 4)$ $\left\{n \mid n > \frac{7}{10}\right\}$

60. $2(m + 7) \le 4[3(m - 2) - 5(1 + m)]$ $\left\{m \mid m \le -\frac{29}{5}\right\}$

Applications

61. Three-fifths of a number is greater than two-thirds. Find the smallest integer that satisfies the inequality. 2

62. To avoid a tax penalty, at least 90% of a self-employed person's total annual income tax liability must be paid by January 15. What amount of income tax must be paid by January 15 by a person with an annual income tax liability of \$3500? \$3150 or more

63. A service organization will receive a bonus of $200 for collecting more than 1850 pounds of aluminum cans during its four collection drives. On the first three drives, the organization collected 505 pounds, 493 pounds, and 412 pounds. How many pounds of cans must the organization collect on the fourth drive to receive the bonus? More than 440 pounds

64. A student must have an average of at least 80 points on five tests to receive a B in a course. The student's grades on the first four tests were 75, 83, 86, and 78. What scores on the last test will enable this student to receive a B in the course? 78 or better

65. A sales representative for a stereo store has the option of a monthly salary of $2000 or a 35% commission on the selling price of each item sold by the representative. What dollar amounts in sales will make the commission more attractive than the monthly salary? More than $5714

66. The sales agent for a jewelry company is offered a flat monthly salary of $3200 or a salary of $1000 plus an 11% commission on the selling price of each item sold by the agent. If the agent chooses the $3200 salary, what dollar amount does the agent expect to sell in one month? $20,000 or less

67. A computer bulletin board service charges a flat fee of $10 per month or $4 per month plus $.10 for each minute the service is used. For how many minutes must a person use this service for the cost to exceed $10?
More than 60 minutes

68. For a product to be labeled orange juice, a state agency requires that at least 80% of the drink be real orange juice. How many ounces of artificial flavors can be added to 32 ounces of real orange juice if the drink is to be labeled orange juice? 8 or less ounces

69. A shuttle service taking skiers to a ski area charges $8 per person each way. Four skiers are debating whether to take the shuttle bus or rent a car for $45 plus $.25 per mile. Assuming that the skiers will share the cost of the car and that they want the least expensive method of transportation, how far away is the ski area if they choose to take the shuttle service? More than 76 miles

70. Company A rents cars for $25 per day and $.08 per mile driven. Company B rents cars for $16 per day and $.14 per mile driven. You want to rent a car for one day. Find the maximum number of miles you can drive a Company B car if it is to cost you less than a Company A car. 150 miles

Applying Concepts

Use the roster method to list the positive integers that are solutions of the inequalities.

71. $7 - 2b \le 15 - 5b$ $\{1, 2\}$

72. $13 - 8a \ge 2 - 6a$ $\{1, 2, 3, 4, 5\}$

73. $2(2c - 3) < 5(6 - c)$ $\{1, 2, 3\}$

74. $-6(2 - d) \ge 4(4d - 9)$ $\{1, 2\}$

Use the roster method to list the integers that are solutions of the intersection of the solution sets of the two inequalities.

75. $5x - 12 \le x + 8$ $\{3, 4, 5\}$
$3x - 4 \ge 2 + x$

76. $6x - 5 > 9x - 2$ $\{-4, -3, -2\}$
$5x - 6 < 8x + 9$

77. $4(x - 2) \le 3x + 5$ $\{10, 11, 12, 13\}$
$7(x - 3) \ge 5x - 1$

78. $3(x + 2) < 2(x + 4)$ $\{-1, 0, 1\}$
$4(x + 5) > 3(x + 6)$

Exploration

79. ***Measurements as approximations*** Recall the rules for rounding a number, which are given at the right. Given these rules, then some possible values of the number 2.7 that was rounded to the nearest tenth are: 2.73, 2.68, 2.65, and 2.749. If V represents the exact value of 2.7 before it was rounded, then the inequality $2.65 \le V < 2.75$ represents all possible values of 2.7 before it was rounded. This is read "V is greater than or equal to 2.65 and less than 2.75."

Now suppose a rectangle is measured as 3.4 meters by 4.8 meters, each measurement rounded to the nearest tenth of a meter. By using the smallest and largest possible values of each measurement, we can find the possible values of the area, A.

$$3.35(4.75) \le A < 3.45(4.85)$$
$$15.9125 \le A < 16.7325$$

The area is greater than or equal to 15.9125 square meters and less than 16.7325 square meters.

a. Suppose the length of a line is measured as 4.2 inches, rounded to the nearest tenth. Write an inequality that represents the possible lengths of the line.

b. The length of the side of a square was given as 6.4 centimeters, rounded to the nearest tenth of a centimeter. Write an inequality that represents the possible areas of the square.

c. The base of a triangle was measured as 5.43 meters and the height as 2.47 meters, each measurement rounded to the nearest hundredth. Write an inequality that represents the possible areas of the triangle.

d. A rectangle is measured as 3.0 meters by 4.0 meters, each measurement rounded to the nearest tenth of a meter. Write an inequality that represents the possible areas of the rectangle.

If the digit to the right of the given place-value is less than 5, that digit and all the digits to the right are dropped. For example, 2.73 rounded to the nearest tenth is 2.7.

If the digit to the right of the given place-value is greater than or equal to 5, increase the given place value by 1 and drop the remaining digits. For example, 2.65 rounded to the nearest tenth is 2.7.

Chapter Summary

Definitions

The *terms* of a variable expression are the addends of the expression. A number without a variable is a *constant term* or simply a *constant*. *Like terms* of a variable expression are the terms that have the same variable parts.

Combining the like terms of a variable expression is called *simplifying the variable expression*.

An *equation* expresses the equality of two mathematical expressions. The expressions can be either numerical or variable expressions.

A *solution* of an equation is a number that, when substituted for the variable, results in a true equation.

To *solve* an equation means to find a solution of the equation.

The Basic Percent Equation is: Percent · Base = Amount
$P \cdot B = A$

A *ray* is a part of a line that starts at a point, called the *endpoint* of the ray, but has no end.

A *protractor* can be used to measure an angle in *degrees* ($^\circ$). A *right angle* has a measure of 90°. An angle whose measure is between 0° and 90° is called an *acute* angle. An angle whose measure is between 90° and 180° is called an *obtuse* angle. A *straight* angle measures 180°.

Two angles are *complements* of each other if the sum of the measures of the angles is 90°. The angles are called *complementary angles*. Two angles are *supplements* of each other if the sum of the measures of the angles is 180°. The angles are called *supplementary angles*.

Four angles are formed by the intersection of two lines. If each of the four angles is a right angle, then the two lines are *perpendicular*.

Two angles that have the same vertex and share a common side are called *adjacent angles*. The angles that are on opposite sides of intersecting lines are called *vertical angles*.

A line that intersects two other lines at different points is called a *transversal*.

Alternate interior angles are two angles that are on opposite sides of the transversal and between the parallel lines. *Alternate exterior angles* are two angles that are on opposite sides of the transversal and outside the parallel lines.

If the lines cut by a transversal are not parallel lines, the three lines will intersect at three points. The plane figure formed by these line segments is a *triangle*. The angles within the triangle are called *interior angles*. An angle adjacent to an interior angle of a triangle is an *exterior angle* of the triangle.

The sum of the measures of the interior angles of a triangle is 180°.

A *distance-time* graph shows the relationship between the time of travel and the distance traveled. Time is on the horizontal axis; distance is on the vertical axis.

An expression that contains the symbol $>, <, \leq,$ or $\leq$ is called an *inequality.* The *solution set of an inequality* is a set of numbers, each element of which, when substituted for the variable, results in a true inequality.

Procedures

The ***Commutative Property of Addition*** states that if a and b are any two numbers, then $a + b = b + a$.

The ***Associative Property of Addition*** states that if $a, b,$ and c are any three numbers, then $a + (b + c) = (a + b) + c$.

The ***Inverse Property of Addition*** states that $a + (-a) = 0$ and $-a + a = 0$. Recall that a and $-a$ are opposites or additive inverses of each other.

The ***Addition Property of Zero*** states that if a is any number, then $a + 0 = a$ and $0 + a = a$.

The ***Commutative Property of Multiplication*** states that if a and b are any two numbers, then $ab = ba$.

The ***Associative Property of Multiplication*** states that if $a, b,$ and c are any three numbers, then $a(bc) = (ab)c$.

The ***Multiplication Property of Equations*** states that each side of an equation can be multiplied by the same nonzero number without changing the solution of the equation. That is, the equations $a = b$ and $ac = bc,\ c \neq 0$, have the same solutions.

The ***Inverse Property of Multiplication*** states that for $a \neq 0$, $a \cdot \frac{1}{a} = 1$ and $\frac{1}{a} \cdot a = 1$. The terms a and $\frac{1}{a}$ are *reciprocals.* They are also called *multiplicative inverses* of each other.

The ***Multiplication Property of One*** states that if a is any number, then $a \cdot 1 = a$ and $1 \cdot a = a$.

The ***Distributive Property*** states that if $a, b,$ and c are any numbers, then $a(b + c) = ab + ac$.

The ***Addition Property of Inequalities*** states that the same number can be added to each side of an inequality without changing the solution set of the inequality.

The ***Multiplication Properties of Inequalities*** state that each side of an inequality can be multiplied by the same *positive number* without changing the solution set of the inequality; and, if each side of an inequality is multiplied by the same *negative number* and the inequality symbol is reversed, then the solution set of the inequality is not changed.

The ***Addition Property of Equations*** is used to remove a term from one side of an equation by *adding the opposite of that term to each side of the equation.*

Chapter Review Exercises

1. Use the given property to complete the statement: The Associative Property of Addition. $(4 + 5) + 6 = ? + (5 + 6)$ 4

2. Identify the property that justifies the statement $(4 \cdot 3) \cdot 5 = 4 \cdot (3 \cdot 5)$.
Associative Property of Multiplication

3. Simplify: $-15xy + 3xy$ $-12xy$

4. Simplify: $\left(\frac{3x}{4}\right)\left(\frac{4}{3}\right)$ x

5. Multiply: $3(2x^2 + xy - 3y^2)$ $6x^2 + 3xy - 9y^2$

6. Simplify: $-5x - 2[2x - 4(x + 7)] - 6$ $-x + 50$

7. Solve: $5 + x = 12$ 7

8. Solve: $x - \frac{2}{5} = \frac{3}{5}$ 1

9. Solve: $-\frac{y}{2} = 5$ -10

10. Solve: $-15 = -\frac{3}{5}x$ 25

11. 45% of what is 9?   20

12. Find $16\frac{2}{3}$% of 30.   5

13. Solve: $8 = 7d - 1$ $\frac{9}{7}$

14. Solve: $3 = \frac{3a}{4} + 1$ $\frac{8}{3}$

15. Solve: $7x - 4 - 2x = 6$ 2

16. Solve: $4y - 8 = y - 8$ 0

17. Solve: $8b + 5 = 5b + 7$ $\frac{2}{3}$

18. Solve: $5x = 7x - 8$ 4

19. Solve: $6y + 2(2y + 3) = 16$ 1

20. Solve: $9 - 5x = 12 - (6x + 7)$ -4

21. Find the complement of a 68° angle. 22°

22. Find the supplement of a 123° angle. 57°

23. A telephone company estimates that the number N of phone calls made per day between two cities of population P_1 and P_2 that are d miles apart is given by the equation $N = \frac{2.51P_1P_2}{d^2}$.

a. Estimate the population P_2, given that P_1 is 48,000, the number of phone calls is 1,100,000, and the distance between the cities is 75 miles. Round to the nearest thousand. 51,000 people

b. Estimate the population P_1, given that P_2 is 125,000, the number of phone calls is 2,500,000, and the distance between the cities is 50 miles. Round to the nearest thousand. 20,000 people

24. Find two complementary angles such that the smaller angle is 10 degrees less than $\frac{1}{3}$ the larger angle. 15°, 75°

25. In a recent city election, 16,400 out of 80,000 registered voters voted. What percent of the registered voters voted in the election? 20.5%

26. An airline knowingly overbooks flights by selling 25% more tickets than there are seats available. How many tickets would this airline sell for a Boeing 757-200 that has 200 seats? 250 tickets

27. Two cars, one traveling 15 mph faster than the other car, start at the same time from the same point and travel in opposite directions. In 3 hours, they are 345 miles apart. Find the rate of each car. 50 mph, 65 mph

28. How many pounds of tea that cost \$4.20 per pound must be mixed with 12 pounds of tea that cost \$2.25 per pound to make a mixture that costs \$3.40 per pound? 17.25 pounds

29. A residential water bill is based on a flat fee of \$10 plus a charge of \$.75 for each 1000 gallons of water used. Find the number of gallons of water a family can use and have a monthly water bill that is less than \$55. Less than 60,000 gallons

Cumulative Review Exercises

1. What is the next term in the sequence 1, 2, 4, 7, 11, ...? 16

2. Find a counterexample: "All numbers that end in 6 are divisible by 3."
For example, 16

3. Evaluate: $3^4 + (12 - 7) \div 5$ 82

4. Evaluate: $(8 - 2)^2 - 4(2 + 1)^3$ −72

5. Translate into a variable expression: Three times a number divided by the sum of four times the number and three.

$\frac{3n}{4n + 3}$

6. If 4 coins are tossed, what is the probability that none of the coins shows a tail?

$\frac{1}{16}$

7. The bar graph at the right shows the approximate net income (in billions of dollars) of General Motors for the years 1990 through 1997.

a. For the years shown, in which year did General Motors have the greatest net income?

b. For the years shown, in which year did General Motors have the least net income?

c. For the years shown, what was the difference between the net income in 1997 and in 1990?

a. 1995; b. 1991; c. $8.7 billion

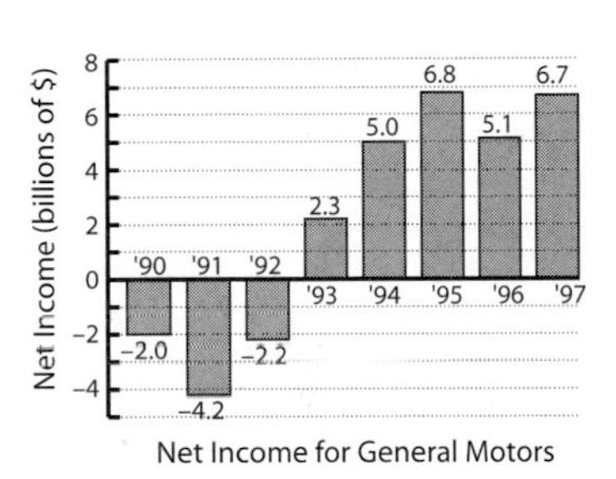

Net Income for General Motors

8. The temperatures for 11 consecutive days at a desert resort were 95°, 98°, 98°, 104°, 97°, 100°, 96°, 97°, 108°, 93°, and 104°. Find the mean and median of this data. Mean = 99.1°; median = 98°

9. A random sample of SAT scores in English is as follows: 479, 486, 453, 406, 413, 533, 528, 479, 538, 578, 431, 624, 528, 525. Find Q_1 and Q_3 for this data.

$Q_1 = 453$; $Q_3 = 533$

10. Evaluate: $-(-15)$ 15

11. Evaluate: $-|-17|$ −17

12. Add: $43 + (-12) + 7$ 38

13. Subtract: $7 - 13 - (-31)$ 25

14. Multiply: $-4(-16)$ 64

15. What number is 8 times −9? −72

16. What number is 9 less than −17? −26

17. Simplify: $\left(\frac{3}{5}\right)\left(-\frac{7}{12}\right) \div \frac{21}{(-2)^2}$ $-\frac{1}{15}$

18. Graph the ordered pairs from the following input/output table.

Input, x	−2	0	1	3	5
Output, y	4	−1	−3	−2	0

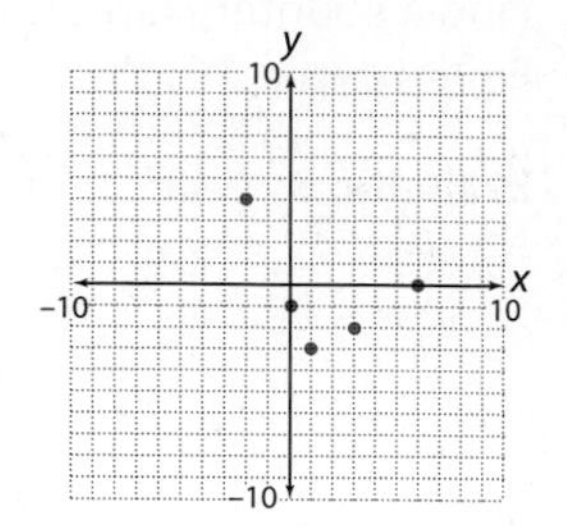

19. Identify the property that justifies the statement $4(3y) = (4 \cdot 3)y$.
Associative Property of Multiplication

20. Solve: $4 - 5v = 29$ −5

21. Solve: $5 - 3(2x + 1) = 8$ −1

22. Solve: $6x + 7 \geq 1$ $\{x \mid x \geq -1\}$

23. Solve: $2 - 5(2x + 3) < 8 - 3x$ $\{x \mid x > -3\}$

24. What percent of 84 is 6? $7\frac{1}{7}\%$

25. Find the complement of a 47° angle. 43°

26. The measure of one angle of a triangle is 15° more than the measure of the second angle. The third angle is 15° less than the measure of the second angle. Find the measure of each angle. 75°, 60°, and 45°

27. A bus traveling at a rate of 60 miles per hour overtakes a car traveling at a rate of 45 miles per hour. If the car had a one-hour head start, how far from the starting point does the bus overtake the car? 180 miles

28. An apple-flavored fruit drink is prepared that must contain at least 80% pure apple juice. How many liters of water can be mixed with 300 liters of pure apple juice and still meet the 80% requirement? 75 liters or less

29. What are the coordinates of the figure at the right after it has been translated 2 units to the right and 3 units down? $A'(-5,-6), B'(-1,3), C'(5,-9)$

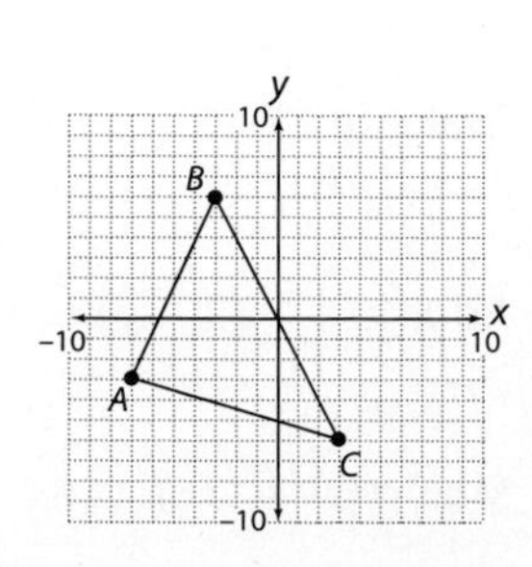

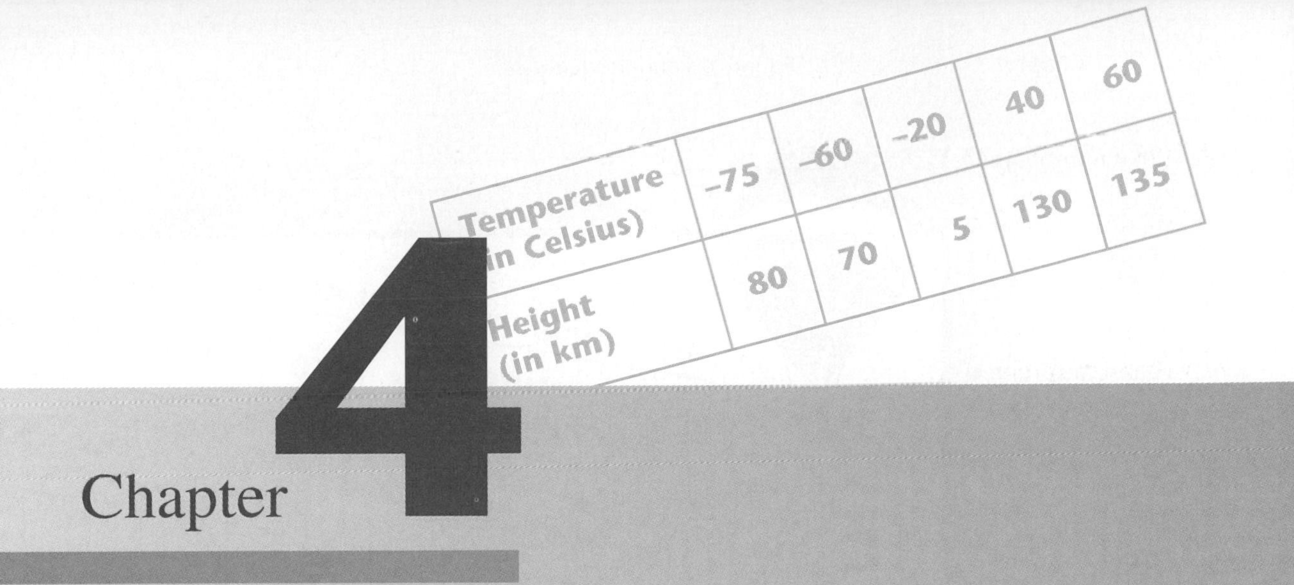

Chapter 4

Linear Equations in Two Variables and Systems of Linear Equations

4.1 Introduction to Functions

4.2 Equations of the Form $Ax + By = C$

4.3 Slopes of Straight Lines

4.4 Equations of Straight Lines

4.5 Solving Systems of Equations by Graphing or Substitution

4.6 Solving Systems of Equations by the Addition Method

Temperature vs. Height Above Earth

Temperature (in Celsius)	–75	–60	–20	40	60
Height (in km)	80	70	5	130	135

4

–75
80

–20
5

60
135

Section 4.1 Introduction to Functions

Evaluating Functions

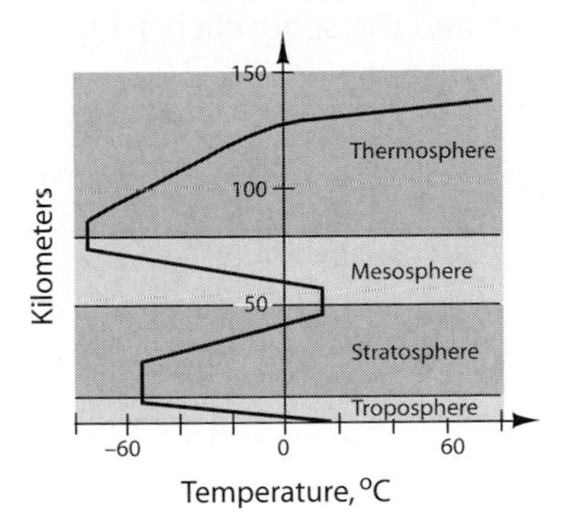

SUGGESTED ACTIVITY
Have students discuss the advantages of having a graph or having an equation as a description of a relationship instead of having just the input/output table or the ordered pairs. For instance, take away the graph and ask students to predict the height at which the temperature is 0°C or take away the equation and ask similar questions for the uniform motion problem.

Atmospheric scientists study the air above the surface of Earth. Smog, the ozone layer, and even the reason that radio waves can be transmitted across the ocean are influenced by the atmosphere. Four of the major atmospheric layers are shown at the left. The graph also shows a relationship between temperature and distance above the surface of Earth. Using the graph, it is possible to approximate some values of an input/output table.

Temperature (in Celsius)	−75	−60	−60	−20	−20	−20	−20	40	60
Height (in km)	80	70	100	5	40	60	120	130	135

The results of this input/output table can be written as the ordered pairs (−75, 80), (−60, 70), (−60, 100), (−20, 5), (−20, 40), (−20, 60), (−20, 120), (40, 130), and (60, 135).

The graph, the input/output table, and the ordered pairs are different ways of describing the relationship between temperature and height.

Now suppose that a car travels at a speed of 45 miles per hour for 4 hours. The distance the car has traveled at any time during those 4 hours depends on the time the car has been traveling. Using the equation $d = rt$, we can write the equation $d = 45t$ to express how d depends on t. The input/output table shows how distance depends on time for various values of t.

Time, t (in hours)	0.5	1	1.75	2	3	3.5	4
Distance, d (in miles)	22.5	45	78.75	90	135	157.5	180

The numbers in the input/output table can also be written as ordered pairs where the first component of the ordered pair is time traveled and the second component is distance. The ordered pairs are (0.5, 22.5), (1, 45), (1.75, 78.75), (2, 90), (3, 135), (3.5, 157.5), and (4, 180). These ordered pairs are graphed at the right with the horizontal axis as time and the vertical axis as distance.

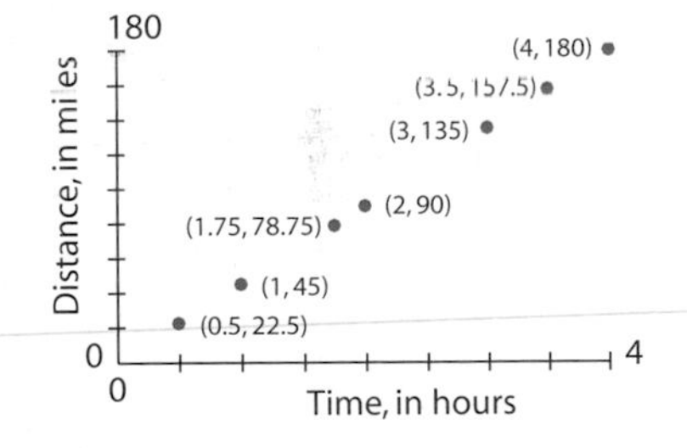

The ordered pairs from the input/output table above are only some of the possible ordered pairs. Other possibilities are (1.5, 67.5), (2.2, 99), and (3.6, 162). If all of the ordered pairs of the equation were drawn, the graph would appear as a line. The graph of the equation and the three additional ordered pairs are shown at the right. Note that the graphs of all the ordered pairs are on the same line.

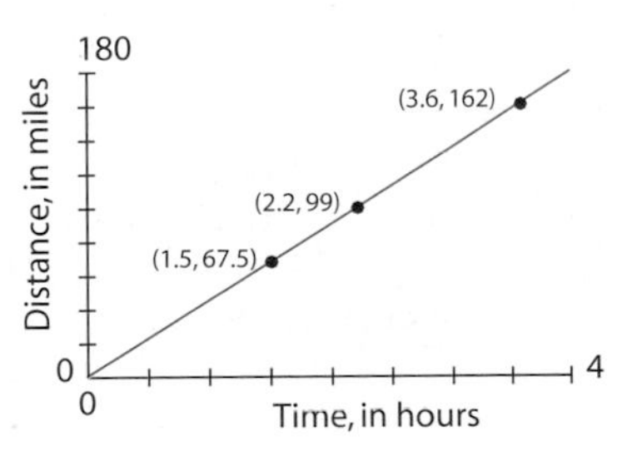

TAKE NOTE
Here are two advantages to having the equation, as in the time-distance relationship: (1) with the equation the graph can be drawn, and (2) with the equation exact values of d can be calculated for given values of t.

For the temperature-height relationship given above, we used the graph to determine ordered pairs that show specific instances of the relationship. For instance, the heights at which the temperature is −60°C are approximately 70 and 100 kilometers. For the time-distance relationship, we used an equation to determine ordered pairs and the graph. For instance, after 2 hours the car had traveled 90 miles.

In mathematics, a **relation** is a set of ordered pairs that describes how one quantity is related to another. The sets of ordered pairs of the temperature-height and time-distance examples on the previous page are relations.

The following table shows the number of hours that each of ten students spent in the foreign language lab one week prior to an exam and the score each achieved on the exam.

Hours	2	3	4	4	5	5	5	6	6	7
Score	70	75	70	80	75	80	90	95	85	90

This information can be written as the relation

$$\{(2, 70), (3, 75), (4, 70), (4, 80), (5, 75), (5, 80), (5, 90), (6, 95), (6, 85), (7, 90)\}$$

where the first coordinate of the ordered pair is the number of hours spent in the lab and the second coordinate is the score on the exam.

Question: What is the meaning of the ordered pair (5, 90)?[1]

The **domain** of a relation is the set of first components of the ordered pairs. The **range** of a relation is the set of second components of the ordered pairs. For the lab hours–exam score relation,

Domain = {2, 3, 4, 5, 6, 7}   Range = {70, 75, 80, 85, 90, 95}

TAKE NOTE
Recall that when listing the elements of a set, duplicate elements are listed only once.

The **graph of a relation** is the graph of the ordered pairs that belong to the relation. The graph of the relation above is shown at the right. The horizontal axis represents the domain of the relation (hours spent in the language lab); the vertical axis represents the range of the relation (score on the exam).

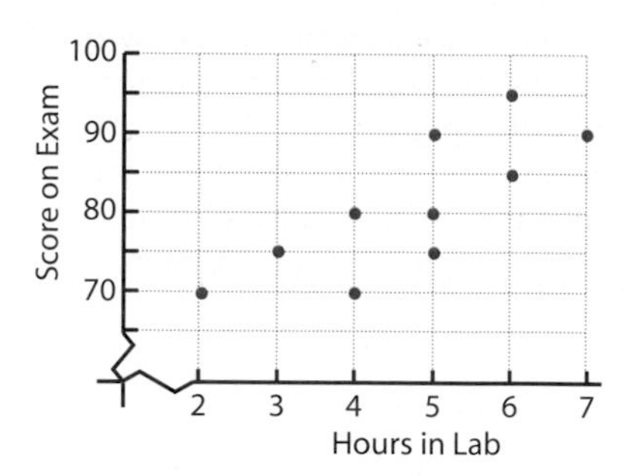

The temperature-height graph and time-distance graph on the previous page are also examples of graphs of a relation.

Looking at the time-distance graph on the preceding page, the domain of the relation is the set of numbers between 0 and 4 (the horizontal axis), and the range of the relation is the set of numbers between 0 and 180 (the vertical axis). The domain and range of this relation are both infinite sets of numbers. For instance, values of time could be 2.34 hours, 3.012 hours, or any other number between 0 and 4. Similarly, the distance traveled could be any number between 0 and 180.

TAKE NOTE
The idea of *function* is one of the most important concepts in math. It is a concept you will encounter in all your mathematics courses.

Although relations are important in mathematics, the concept of *function* is especially useful in applications. A **function** is a special type of relation in which no two ordered pairs have the same first coordinate and different second coordinates. The relation above is not a function because the ordered pairs (4, 70) and (4, 80) have the same first coordinate and different second coordinates. The ordered pairs (6, 85) and (6, 95) also have the same first coordinate and different second coordinates.

The temperature-height relation is also not a function because there are ordered pairs that have the same first coordinate but different second coordinates. For instance, (−20, 5) and (−20, 40) are ordered pairs of the relation, so there are at least two ordered pairs with the same first coordinate and different second coordinates.

The time-distance relation is a function. There are no two ordered pairs with the same first coordinate and different second coordinates.

1. A student who spent 5 hours in the lab scored 90 on the exam.

Using a table is another way of describing a relation. The table at the right describes a grading scale that defines a relationship between a test score and a letter grade. Some of the ordered pairs in this relation are (38, F), (73, C), and (94, A).

Score	Letter Grade
90 – 100	A
80 – 89	B
70 – 79	C
60 – 69	D
0 – 59	F

INSTRUCTOR NOTE
The concept of a function was beginning to form with the work of Fermat and Descartes. However, it was Euler who gave us the word and its first definition: "A function of a variable quantity is an analytic expression composed in any way whatsoever of the variable quantity and numbers or constant quantities." The sense of function in Euler's definition is that of "*y* is a function of *x*," but it does not encompass the more general notion of a function as a certain set of ordered pairs.

This relation defines a function because no two ordered pairs can have the same first coordinate and different second coordinates. For instance, it is not possible to have an average of 73 paired with any grade other than C. Both (73, C) and (73, A) cannot be ordered pairs belonging to the function, or two students with the same score would receive different grades. Note that (81, B) and (88, B) are ordered pairs of this function. Ordered pairs of a function may have different first coordinates paired with the same second coordinate.

The domain of this function is $\{0, 1, 2, 3, \ldots, 98, 99, 100\}$.

The range of this function is {A, B, C, D, F}.

Example 1 Find the domain and range of the relation

$$\{(-1, 2), (0, 2), (1, 3), (2, 4), (3, 5), (5, -2), (6, 0)\}$$

Is the relation a function? Graph the relation.

Solution

Domain: $\{-1, 0, 1, 2, 3, 5, 6\}$

• The domain of a relation is the set of first coordinates of the ordered pairs.

Range: $\{-2, 0, 2, 3, 4, 5\}$

• The range of a relation is the set of second coordinates of the ordered pairs.

Because no two ordered pairs have the same first coordinate, the relation is a function.

To graph the function, graph each of the ordered pairs of the function. The graph is shown at the right.

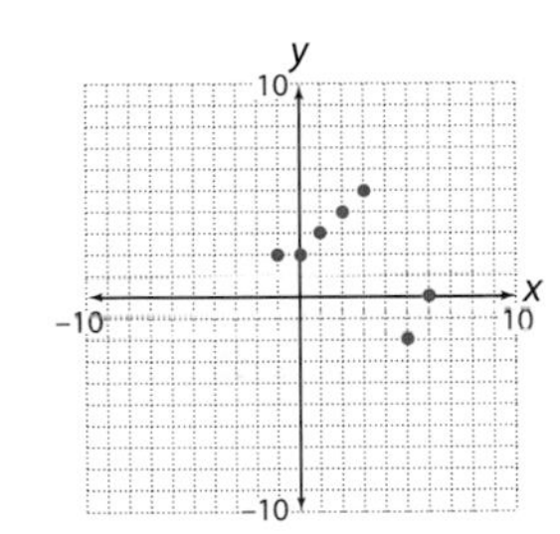

You-Try-It 1 Find the domain and range of the relation

$$\{(-3, 3), (-2, 2), (-1, 1), (0, 0), (1, 1), (2, 2), (3, 3)\}$$

Is the relation a function? Graph the relation.

Solution See page S16.

TAKE NOTE
The notation *A*(*s*) does not mean *A times s*. The letter *A* stands for the function, and *A*(*s*) is the value of the function at *s*.

Although a function can be described in terms of ordered pairs or in a table, functions are often described by an equation. For instance, we can write a function that describes the relationship between the area of a square and the length of one of its sides by the equation

$A(s) = s^2$ Read $A(s)$ as "*A* of *s*" or "the value of *A* at *s*."

INSTRUCTOR NOTE
One way to assist students with evaluating a function is to use open parentheses. For instance,

$$A(s) = s^2$$
$$A(\ ) = (\ )^2$$
$$\downarrow \quad \downarrow$$
$$A(4) = 4^2$$

The notation $A(s)$ is used to indicate that *A*, the area of a square, depends on *s*, the length of one of its sides. To find the area of a square with a side that is 4 feet long, we **evaluate the function**. This means we replace *s* by 4 and then simplify. The **value of the function** is 16 square feet.

$$A(s) = s^2$$
$$A(4) = 4^2$$
$$A(4) = 16$$

The ordered pair (4, 16) is one of the ordered pairs of the function. Note that the first coordinate is the length of a side of a square; the second coordinate is the area of the square for that length side.

Example 2 The area of a circle is given by $A(r) = \pi r^2$, where $A(r)$ is the area of a circle that has a radius that is r units in length. Find the area of the circle whose radius is 3 inches.

Solution Evaluate $A(r) = \pi r^2$ when $r = 3$ inches.

$A(r) = \pi r^2$

$A(3) = \pi(3)^2$ • Replace r by 3.

$A(3) = 9\pi$ • Simplify. This is an exact answer.

$A(3) \approx 9(3.14) = 28.26$ • This is an approximate answer using 3.14 for π.

The area is 9π square inches or approximately 28.26 square inches.

You-Try-It 2 Recall that a cube is a box with all three sides the same length. The volume of a cube is given by $V(x) = x^3$, where $V(x)$ is the volume of a cube whose side is x units in length. Find the volume of a cube whose sides measure 2.5 meters.

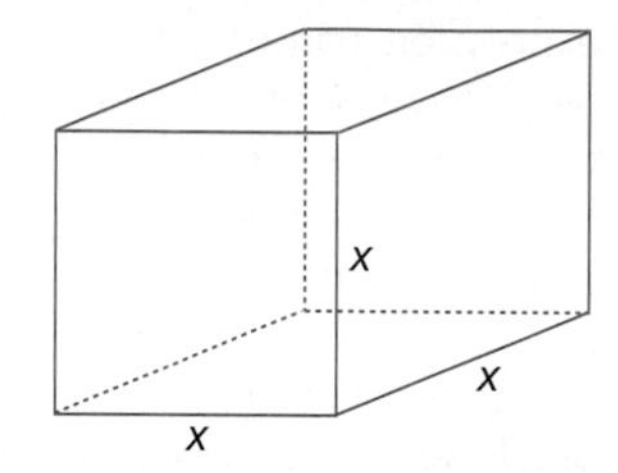

Solution See page S16.

SUGGESTED ACTIVITY
Students can use a graphing calculator to evaluate the function in Example 3 to check their work. This activity works particularly well with a TI-82 or higher calculator or a Sharp EL-9600 calculator.

Have students enter $-x^2 + 1$ into Y1. Now return to the home screen and select Y1 from the Y-Vars menu. Enter (–2) and press ENTER. Your screen should look like the one below.

By pressing 2nd ENTER and using the arrow keys, students can edit the expression Y1(–2), replacing –2 by any other value they wish to check.
Students using other calculators must store –2 in X/Θ/T and then display Y1.

When a function is described by an equation and the domain is specified, the range of the function can be found by evaluating the function at each point of the domain.

Example 3 Find the range of $f(x) = -x^2 + 1$ when the domain is $\{-2, -1, 0, 1, 2\}$.

Solution

$f(x) = -x^2 + 1$

$f(-2) = -(-2)^2 + 1 = -4 + 1 = -3$ • Replace x by each member of the domain.

$f(-1) = -(-1)^2 + 1 = -1 + 1 = 0$

$f(0) = -(0)^2 + 1 = 0 + 1 = 1$

$f(1) = -(1)^2 + 1 = -1 + 1 = 0$

$f(2) = -(2)^2 + 1 = -4 + 1 = -3$

When the domain is $\{-2, -1, 0, 1, 2\}$, the range of $f(x) = -x^2 + 1$ is $\{-3, 0, 1\}$.

You-Try-It 3 Find the range of $f(x) = 3x - 2$ when the domain is $\{-2, -1, 0, 1, 2\}$.

Solution See page S16.

Graphs of Linear Functions

Consider some of the possible values in an input/output table for $y = 2x - 1$.

Input, x	–4	–1	0	1	2	4	5
Output, $y = 2x - 1$	–9	–3	–1	1	3	7	9

The input and output values can be used to write ordered pairs (x, y), where the first coordinate is the input and the second coordinate is the output. For instance, the ordered pairs for the table are (–4, –9), (–1, –3), (0, –1), (1, 1), (2, 3), (4, 7) and (5, 9). Because the equation can be used to write a set of ordered pairs, the equation defines a relation.

It is not possible to substitute one value of x into the equation $y = 2x - 1$ and get two different values of y. For example, when $x = 2$, the only possible value of y is 3. Thus, there are no ordered pairs with the same first coordinate and different second coordinates, and the equation defines a function.

In the equation $y = 2x - 1$, the variable y is called the **dependent variable** because its value *depends* on the value of x. The variable x is called the **independent variable**. We choose a value for x and substitute that value into the equation to determine the value of y. We say that y is a function of x.

TAKE NOTE
When y is a function of x, then y and $f(x)$ are interchangeable.

When an equation defines y as a function of x, *functional notation* is frequently used to emphasize that the relation is a function. In this case, it is common to use the notation $f(x)$. Therefore, we can write the equation

$$y = 2x - 1$$

in functional notation as

$$f(x) = 2x - 1$$

The graph of a function is a graph of the ordered pairs (x, y) or $(x, f(x))$ of the function. The graph of $f(x) = 2x - 1$ is created by plotting the ordered pairs of the function. The graph in Figure A shows the ordered pairs $(-4, -9)$, $(-1, -3)$, $(0, -1)$, $(1, 1)$, $(2, 3)$, $(4, 7)$ and $(5, 9)$ from the input/output table on the previous page. Evaluating the function when x is not an integer produces more ordered pairs to graph, such as $\left(-\frac{7}{2}, -8\right)$, $\left(-\frac{5}{2}, -6\right)$, $\left(-\frac{4}{3}, -\frac{11}{3}\right)$, $\left(\frac{3}{4}, \frac{1}{2}\right)$, $\left(\frac{3}{2}, 2\right)$, and $(3.2, 5.4)$, as shown in Figure B. Evaluating the function for still other values of x would result in more and more ordered pairs to graph. The result would be so many dots that the graph would look like the straight line shown in Figure C, which is the graph of $f(x) = 2x - 1$.

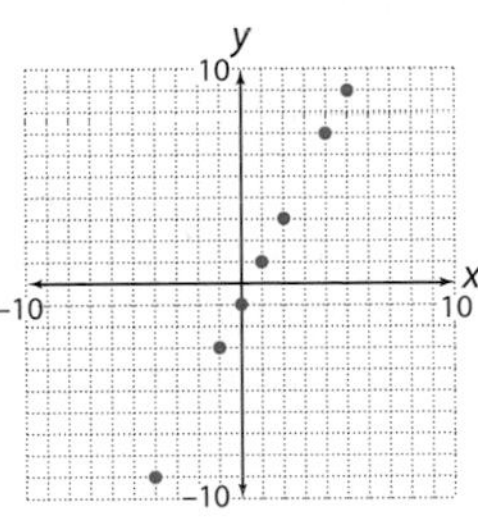

Figure A

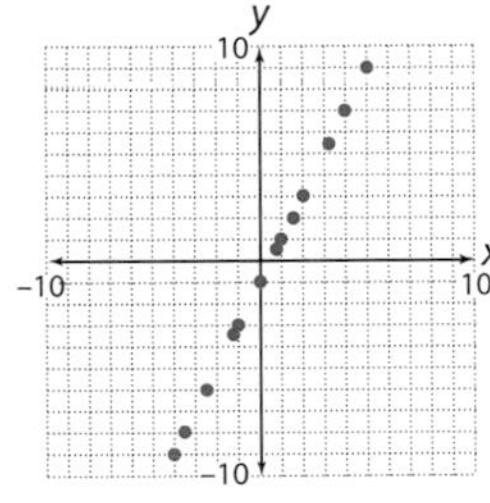

Figure B

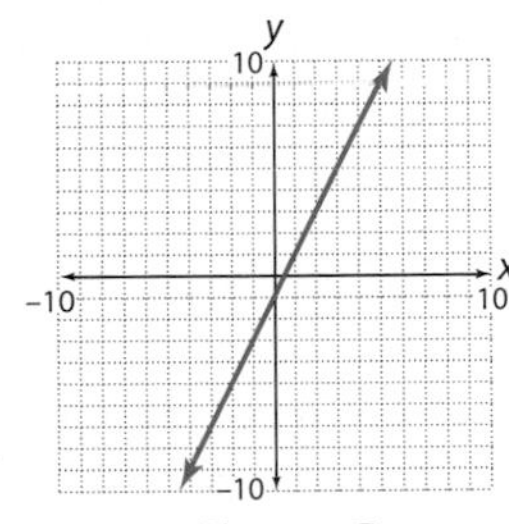

Figure C

TAKE NOTE
Note that the phrase <u>lin</u>ear function contains the word *line*. Every linear function has a graph that is a straight line.

The equation $f(x) = 2x - 1$ is an example of a *linear function*. The graph of a linear function is a straight line.

Linear Function

A linear function is one that can be represented by the equation $f(x) = mx + b$.

INSTRUCTOR NOTE
Students must make a connection between a function and its graph. A linear function has a characteristic graph that is different from the graph of a quadratic function, which is different from that of a cubic function, and so on.

Draw the graph of a parabola or some other nonlinear function and ask students if it represents a linear function.

Here are some examples of equations that represent linear functions.

$f(x) = 3x + 2$	• $m = 3, b = 2$.
$g(x) = -2x + 1$	• $m = -2, b = 1$.
$y = \frac{1}{2}x - 3$	• $m = \frac{1}{2}, b = -3$.
$y = \frac{2x}{3}$	• $\frac{2x}{3} = \frac{2}{3}x$. Therefore, $m = \frac{2}{3}, b = 0$.
$F(x) = 4$	• $m = 0, b = 4$.

The equation $f(x) = 2x^2 + 3$ is not a linear function because x is squared, $y = \frac{3}{x}$ is not a linear function because the variable is in the denominator, and $f(x) = |x|$ is not a linear function because the variable is inside the absolute value symbol. These equations are examples of *nonlinear functions*.

Because the graph of a linear function is a straight line, and a straight line is determined by two points, the graph of a linear function can be drawn by finding two of the ordered pairs of the function, plotting those points, and then drawing a line through the two points. However, you can ensure better accuracy by plotting at least three points.

Example 4 Graph: $f(x) = \frac{2}{3}x - 2$

Solution To draw the graph, evaluate the function for selected values of x. We will choose $x = -3$, $x = 0$, and $x = 6$. However, any values of x could have been chosen. Plot the ordered pairs and then draw a line through the points.

x	$y = f(x) = \frac{2}{3}x - 2$	y	(x, y)
-3	$f(-3) = \frac{2}{3}(-3) - 2$	-4	$(-3, -4)$
0	$f(0) = \frac{2}{3}(0) - 2$	-2	$(0, -2)$
6	$f(6) = \frac{2}{3}(6) - 2$	2	$(6, 2)$

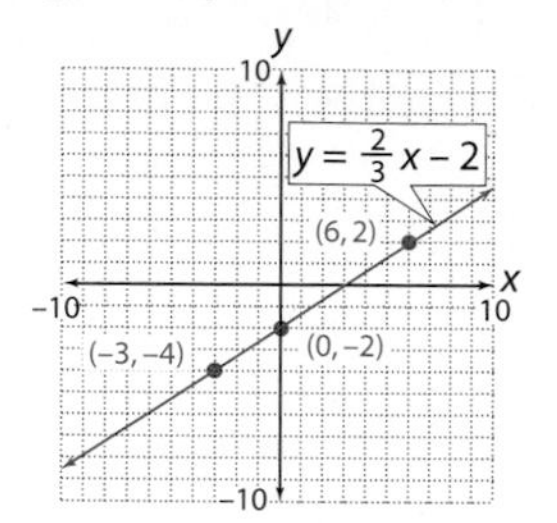

SUGGESTED ACTIVITY
Have students use a graphing calculator to check their graphs and check the points they plotted. They can enter Y1 as (2/3)x − 2 or as 2x/3 − 2. If they use the standard viewing window, their graphs will look something like the one below. The points on the graph can be entered on the home screen by using the Pt-On function. This will also reinforce evaluation of a function. Here are some typical screens. The last 2 in Pt-On places a square at the coordinates of the point.

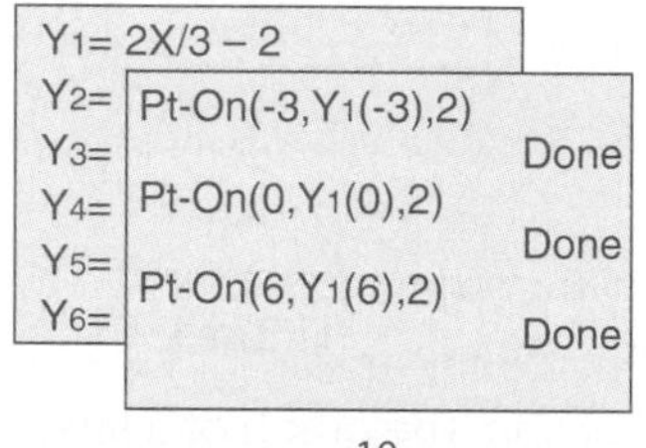

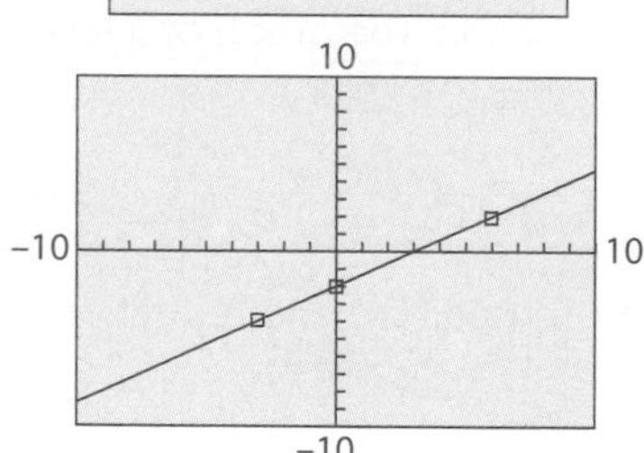

Students can verify the points by using TRACE. With the graph on the screen, press TRACE. Then enter the x-coordinate of the point to be checked and press ENTER. The cursor will move to that point and the coordinates will be shown on the screen below the graph.

You-Try-It 4 Graph: $y = 2 - \frac{x}{2}$

Solution See page S16.

The equation $F(x) = 4$ given on the previous page is an example of the *constant function.* No matter what value of x is chosen, the value of the function is 4. For instance,

$$F(-3) = 4 \qquad F(2) = 4 \qquad F(7) = 4$$

The graph of $F(x) = 4$ is shown at the right. It is a horizontal line passing through $(0, 4)$.

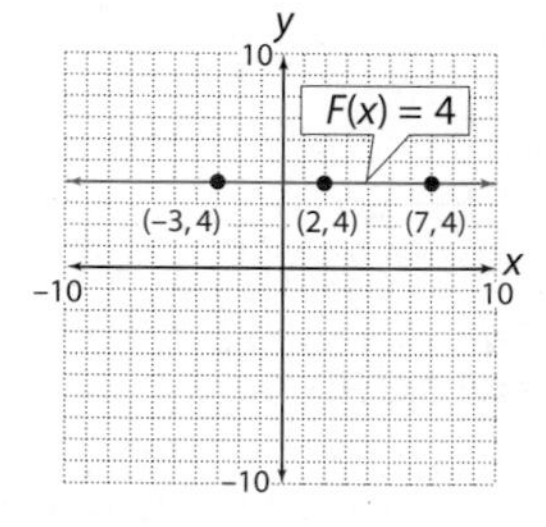

The **constant function** is written as $f(x) = c$, where c is some real number. No matter what value of x is chosen, the value of the constant function is c.

When using a graphing calculator to draw the graph of a function, you should choose a few values for x and evaluate the function for those values. Then verify that the ordered pairs $(x, f(x))$ are points on the graph.

Example 5 Graph $f(x) = -2x - 3$ by using a graphing calculator. Then find three ordered pairs of the function and ensure that those points are on the graph drawn by the calculator.

Solution Enter $f(x) = -2x - 3$ in the Y= editor as Y1 = −2X − 3. Then draw the graph using the standard viewing window. The graph is shown below. Choose three values of x. We will use −2, 0, and 3. However, you may choose any values you wish. Evaluate $f(x)$ for these values.

$$f(x) = -2x - 3$$
$$f(-2) = -2(-2) - 3 = 1$$
$$f(0) = -2(0) - 3 = -3$$
$$f(3) = -2(3) - 3 = -9$$

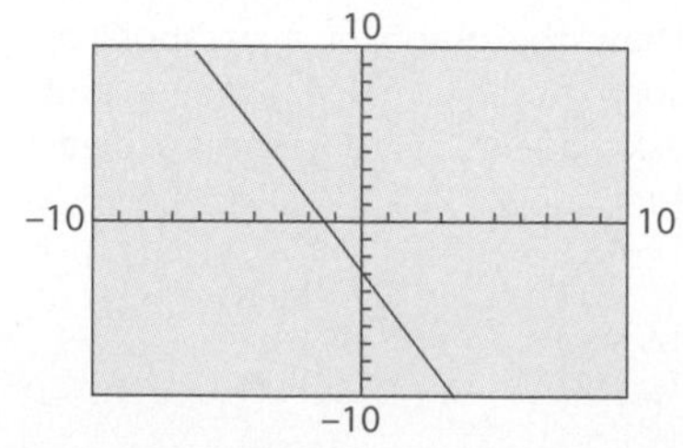

Now check the graph and be sure that $(-2, 1)$, $(0, -3)$, and $(3, -9)$ are points on the graph.

You-Try-It 5 Graph $g(x) = \frac{4x}{5} + 1$ by using a graphing calculator. Then find three ordered pairs of the function and ensure that those points are on your graph. Suggestion: When choosing values of *x*, use multiples of 5, such as −10, −5, 0, 5, or 10. This will lead to integer values for *y*, which will be easier to check.

Solution See page S16.

Consider $f(x) = -2x - 3$ from Example 5. No matter what value of *x* is selected, the value of $-2x - 3$ is a real number. For instance,

$$f(x) = -2x - 3$$

$$f(-5) = -2(-5) - 3 = 7 \qquad \text{• 7 is a real number.}$$

$$f\left(\frac{8}{5}\right) = -2\left(\frac{8}{5}\right) - 3 = -\frac{31}{5} \qquad \text{• } -\frac{31}{5} \text{ is a real number.}$$

$$f(\pi) = -2(\pi) - 3 \approx -9.28 \qquad \text{• } -9.28 \text{ is a real number.}$$

Thus the domain of the function is the real numbers. In general, the domain of any linear function is the real numbers. The numbers 7, $-\frac{31}{5}$, and −9.28 are in the range of the function. The range of a linear function with $m \neq 0$ is also the real numbers. If the linear function is the constant function ($m = 0$), then the range is the constant. For example, the range of $f(x) = 4$ is {4}.

When evaluating a function, the number in the range is calculated for a given value in the domain. It is also possible to give an element of the range of a function and find a number in the domain that corresponds to the selected range element.

TAKE NOTE
You can verify the solution to Example 6 by using a graphing calculator. Graph $Y_1 = -3x - 2$ and $Y_2 = 7$. Now use the INTERSECT feature to find the point of intersection.

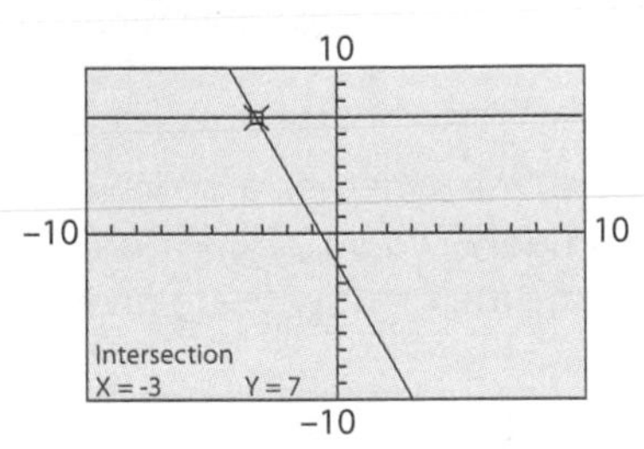

Example 6 Given that 7 is in the range of $f(x) = -3x - 2$, find the value of *x* in the domain so that $f(x) = 7$.

Solution Because $f(x)$ is a symbol for an element of the range, we must find *x* (in the domain) so that $f(x) = 7$.

$$f(x) = -3x - 2$$

$$7 = -3x - 2 \qquad \text{• Replace } f(x) \text{ by 7.}$$

$$9 = -3x \qquad \text{• Solve for } x.$$

$$-3 = x$$

The value of *x* is −3. This means that (−3, 7) is an ordered pair of the function.

You-Try-It 6 Given that 1 is in the range of $f(x) = 2x + 5$, find the value of *x* in the domain so that $f(x) = 1$.

Solution See page S16.

Applications of Linear Functions

Suppose that an architect charges a fee of \$500 plus \$2.75 per square foot to design a house. The equation that represents the architect's fee is given by $F = 2.75s + 500$, where *F* is the fee and *s* is the number of square feet in the house. This is a linear function with $m = 2.75$ and $b = 500$. The equation can be written in functional notation as $F(s) = 2.75s + 500$. Evaluating this function for a certain value of *s* gives the architect's fee to draw the plans for the house. For instance, when $s = 2500$, then

$$F(s) = 2.75s + 500$$

$$F(2500) = 2.75(2500) + 500$$

$$= 7375$$

Thus (2500, 7375) is an ordered pair of this function. The ordered pair means that the architect's fee is \$7375 to draw the plans for a 2500-square-foot house.

Some uniform motion problems are applications of linear functions. Consider a cyclist who is riding at 20 miles per hour. Then the distance traveled by the cyclist is given by $d = 20t$. This is a linear function with $m = 20$ and $b = 0$. The equation can be written in functional notation as $d(t) = 20t$.

Another application of linear equations is to scatter diagrams and a linear function that approximates the points in the scatter diagram. The linear function that approximates the points is called a **regression equation**. This equation is used to predict values that are not part of the original data. The **regression line** is the graph of the regression equation.

Example 7 A study was conducted to determine how the amount of rainfall affected air pollution as measured by the amount of particulate matter removed from the air. The results are recorded in the table below.

Rainfall, in centimeters	2	2.5	4	4.8	6	6.7	7.2	8.6
Particulates removed from air	142	145	120	119	112	99	102	88

TAKE NOTE
The units of measure for particulate matter are in micrograms per cubic meter. A microgram is one-millionth of a gram.

The regression equation for this data is $y = -8.5x + 160.29$, where x is the amount of rainfall in centimeters and y is the particulate matter removed from the air in micrograms per cubic meter. The scatter diagram and regression line are shown at the right. Use the regression equation to find how much particulate matter is removed from the air when there is 5 centimeters of rain. Round to the nearest whole number.

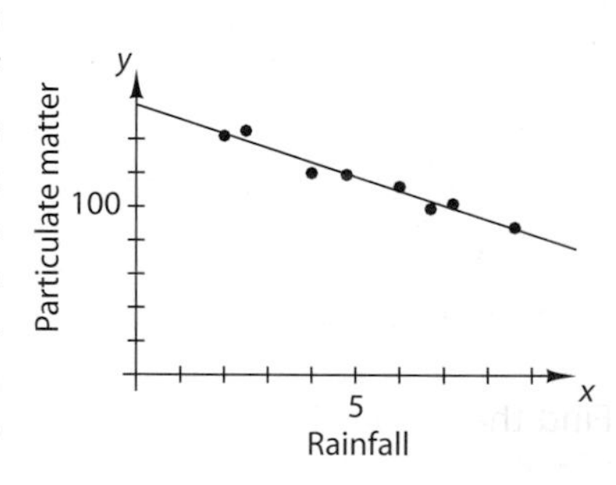

Solution Evaluate the regression equation when $x = 5$.

$$\begin{aligned} y &= -8.5x + 160.29 \\ &= -8.5(5) + 160.29 \\ &= 117.79 \end{aligned}$$

When the amount of rainfall is 5 centimeters, 118 micrograms per cubic meter of particulate matter are removed from the air.

You-Try-It 7 A study was conducted to determine whether a score on a placement test could predict how successful a student would be in a physics course. The results of the test and the course average for 7 students are given in the table below.

Placement test	20	25	28	30	34	36	37
Average in course	58	72	70	83	84	90	95

The regression equation for this data is $y = 2x + 18.7$, where x is the student's score on the placement test and y is the student's course average in the physics course. The scatter diagram and regression line are shown at the right. Use the regression equation to predict the average in the course for a student who scores 31 on the placement test. Round to the nearest whole number.

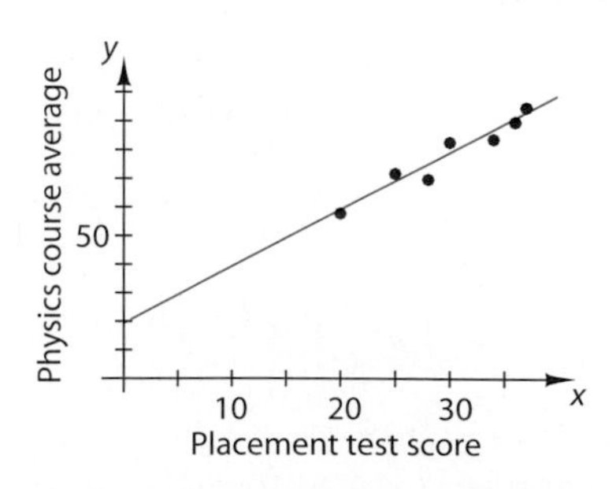

Solution See page S16.

4.1 EXERCISES

Topics for Discussion

1. Discuss the difference between a relation and a function.
   Answers will vary.

2. Discuss various ways a relation or function can be expressed.
   Answers will vary.

3. What is the meaning of the symbol $f(x)$?
   It is the value of the function at x.

4. What is a constant function?
   A function of the form $f(x) = c$.

5. Give an example of a relation that is not a function.
   Answers will vary. One possibility is the relation between age and hair color.

6. What is a linear function? Describe the shape of the graphs of all linear functions.
   A function of the form $f(x) = mx + b$. The graph of every linear function is a straight line.

Evaluating Functions

Find the domain and range of each relation. State whether or not the relation is a function.

7. $\{(0, 0), (2, 0), (4, 0), (6, 0)\}$
   D: $\{0, 2, 4, 6\}$; R: $\{0\}$; yes

8. $\{(-2, 2), (0, 2), (1, 2), (2, 2)\}$
   D: $\{-2, 0, 1, 2\}$; R: $\{2\}$; yes

9. $\{(2, 2), (2, 4), (2, 6), (2, 8)\}$
   D: $\{2\}$; R: $\{2, 4, 6, 8\}$; no

10. $\{(-4, 4), (-2, 2), (0, 0), (-2, -2)\}$
    D: $\{-4, -2, 0\}$; R: $\{-2, 0, 2, 4\}$; no

11. $\{(0, 0), (1, 1), (2, 2), (3, 3)\}$
    D: $\{0, 1, 2, 3\}$; R: $\{0, 1, 2, 3\}$; yes

12. $\{(0, 5), (1, 4), (2, 3), (3, 2), (4, 1), (5, 0)\}$
    D: $\{0, 1, 2, 3, 4, 5\}$; R: $\{0, 1, 2, 3, 4, 5\}$; yes

13. $\{(-2, -3), (2, 3), (-1, 2), (1, 2), (-3, 4), (3, 4)\}$
    D: $\{-3, -2, -1, 1, 2, 3\}$; R: $\{-3, 2, 3, 4\}$; yes

14. $\{(-1, 0), (0, -1), (1, 0), (2, 3), (3, 5)\}$
    D: $\{-1, 0, 1, 2, 3\}$; R: $\{-1, 0, 3, 5\}$; yes

Evaluate the function at the given value of x. Write an ordered pair that is an element of the function.

15. $f(x) = 4x$; $x = 10$ 40; (10, 40)

16. $f(x) = 8x$; $x = 11$ 88; (11, 88)

17. $f(x) = x - 5; x = -6$ $-11; (-6, -11)$

18. $f(x) = x + 7; x = -9$ $-2; (-9, -2)$

19. $f(x) = 3x^2; x = -2$ $12; (-2, 12)$

20. $f(x) = x^2 - 1; x = -8$ $63; (-8, 63)$

21. $f(x) = 5x + 1; x = \frac{1}{2}$ $\frac{7}{2}; \left(\frac{1}{2}, \frac{7}{2}\right)$

22. $f(x) = 2x - 6; x = \frac{3}{4}$ $-\frac{9}{2}; \left(\frac{3}{4}, -\frac{9}{2}\right)$

23. $f(x) = x^2 + 2x; x = 3$ $15; (3, 15)$

24. $f(x) = x^2 - 2x; x = -2$ $8; (-2, 8)$

25. $f(x) = \frac{1}{x}; x = 3$ $\frac{1}{3}; \left(3, \frac{1}{3}\right)$

26. $f(x) = \frac{2}{x + 1}; x = -3$ $-1; (-3, -1)$

Find the range of the function defined by the given equation. Write five ordered pairs that belong to the function.

27. $f(x) = 3x - 4$; domain $= \{-5, -3, -1, 1, 3\}$
R: $\{-19, -13, -7, -1, 5\}$;
$(-5, -19), (-3, -13), (-1, -7), (1, -1), (3, 5)$

28. $f(x) = 2x + 5$; domain $= \{-10, -5, 0, 5, 10\}$
R: $\{-15, -5, 5, 15, 25\}$;
$(-10, -15), (-5, -5), (0, 5), (5, 15), (10, 25)$

29. $f(x) = \frac{1}{2}x + 3$; domain $= \{-4, -2, 0, 2, 4\}$
R: $\{1, 2, 3, 4, 5\}$; $(-4, 1), (-2, 2), (0, 3), (2, 4), (4, 5)$

30. $f(x) = \frac{3}{4}x - 1$; domain $= \{-8, -4, 0, 4, 8\}$
R: $\{-7, -4, -1, 2, 5\}$; $(-8, -7), (-4, -4), (0, -1), (4, 2), (8, 5)$

31. $f(x) = x^2 + 6$; domain $= \{-3, -1, 0, 1, 3\}$
R: $\{6, 7, 15\}$; $(-3, 15), (-1, 7), (0, 6), (1, 7), (3, 15)$

32. $f(x) = 3x^2 + 6$; domain $= \{-2, -1, 0, 1, 2\}$
R: $\{6, 9, 18\}$; $(-2, 18), (-1, 9), (0, 6), (1, 9), (2, 18)$

Graphs of Linear Functions

Graph without using a graphing calculator.

33. $f(x) = 2x$

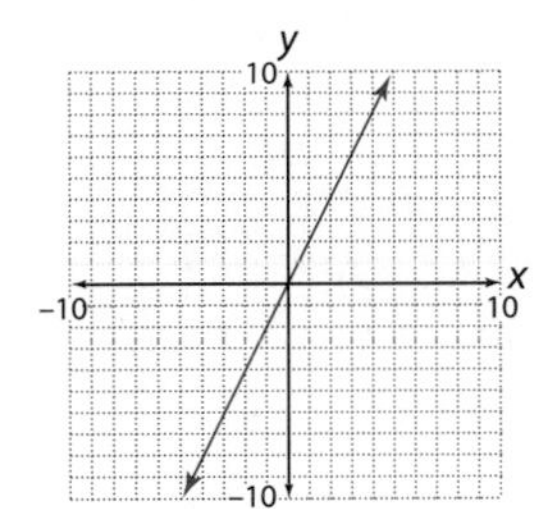

34. $f(x) = -3x$

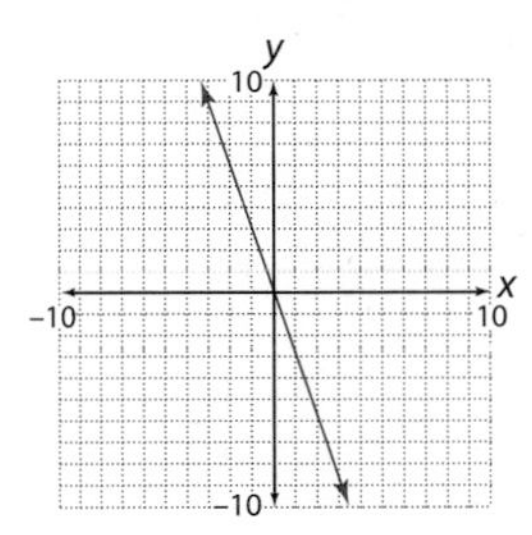

35. $f(x) = x + 2$

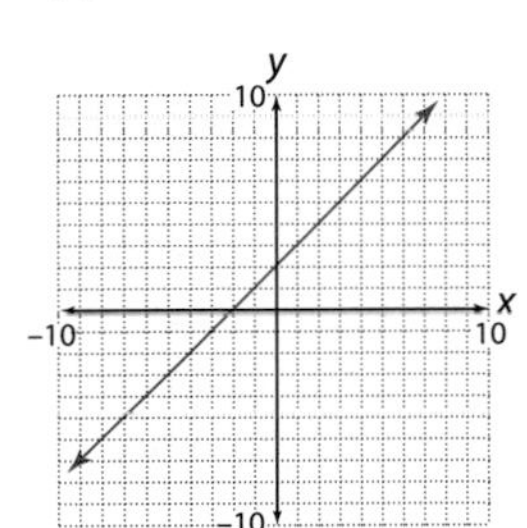

36. $f(x) = -x - 3$

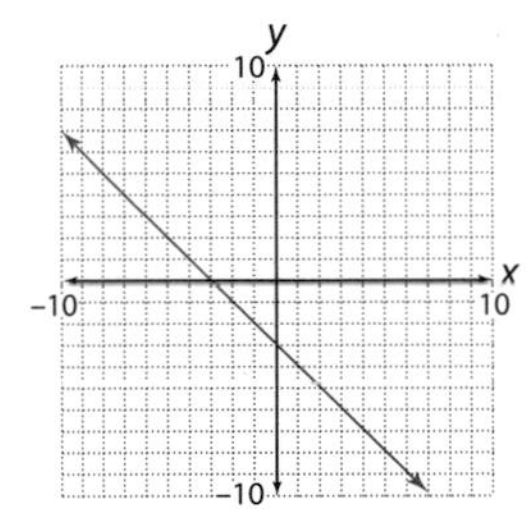

37. $f(x) = 2x - 1$

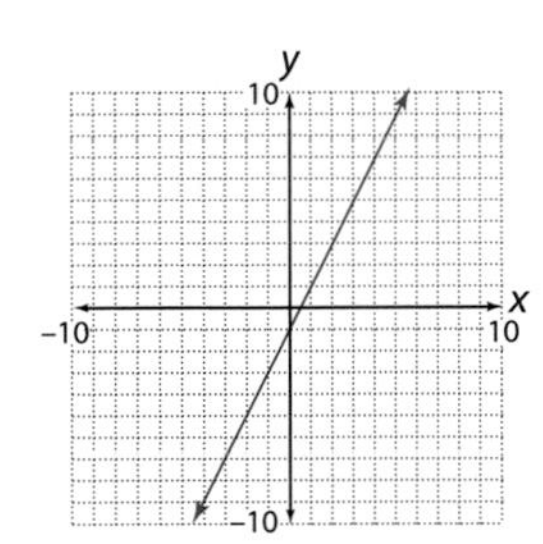

38. $f(x) = 3x + 4$

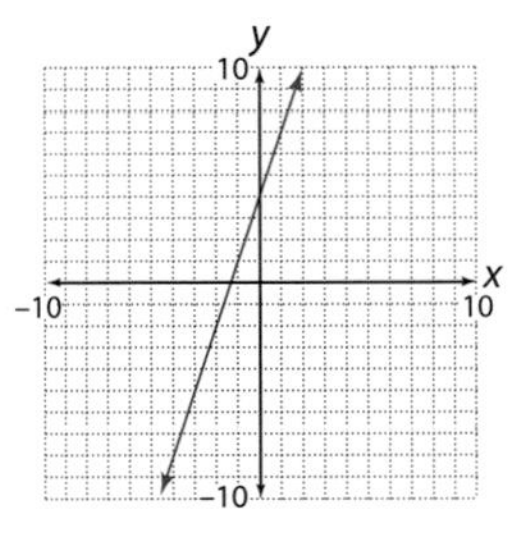

39. $f(x) = -2x + 3$

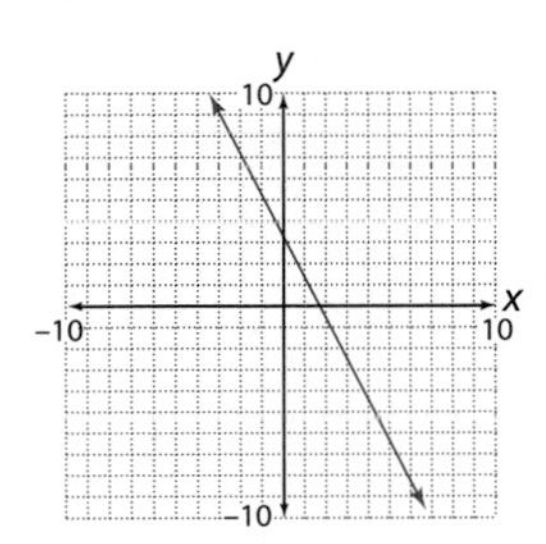

40. $f(x) = -3x - 2$

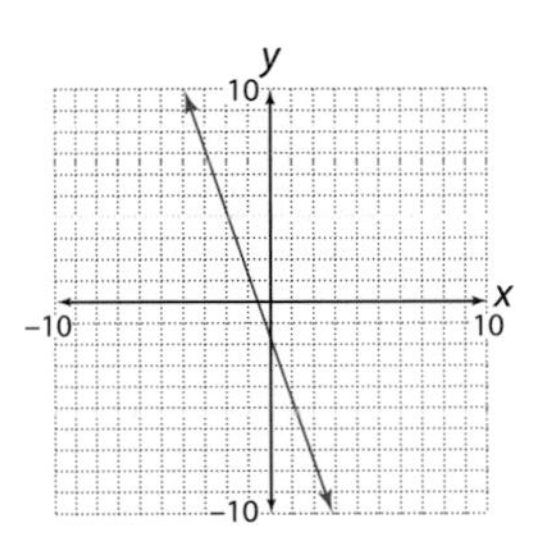

41. $f(x) = \frac{1}{3}x - 4$

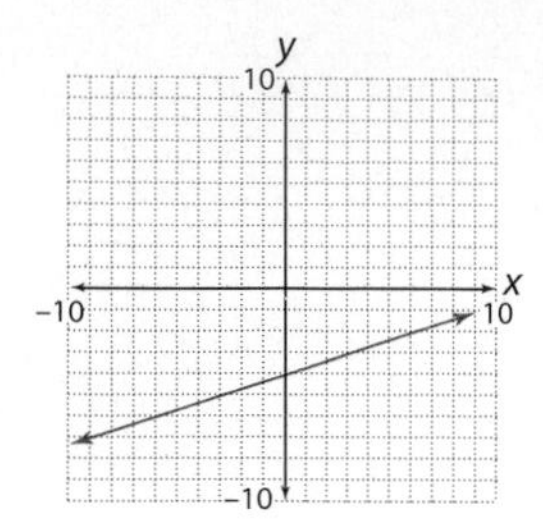

42. $f(x) = \frac{3}{2}x + 1$

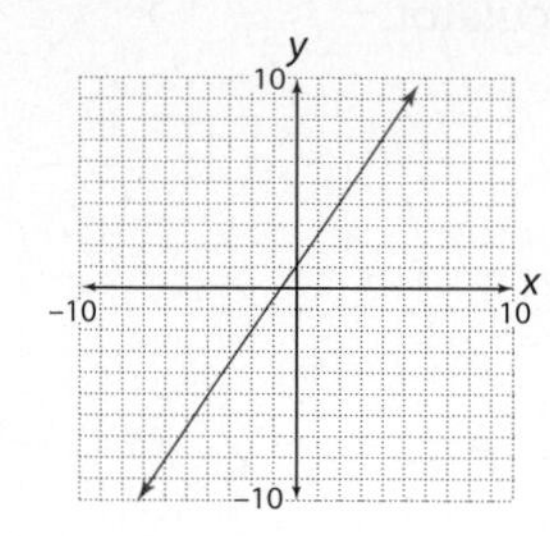

43. $f(x) = -\frac{2}{3}x + 1$

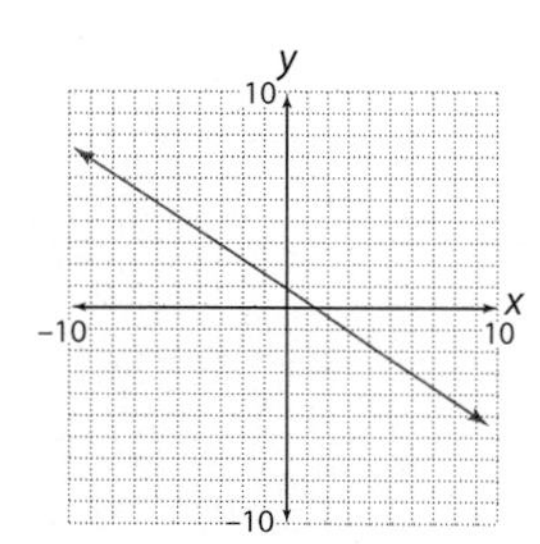

44. $f(x) = -\frac{3}{4}x + 4$

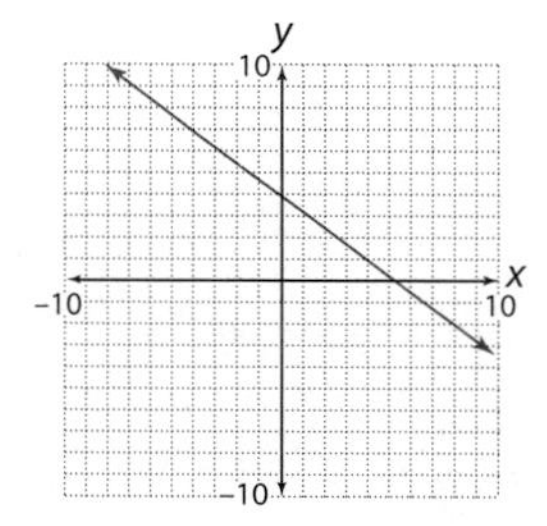

Graph using a graphing calculator and the standard viewing window.

45. $f(x) = 2x - 3$

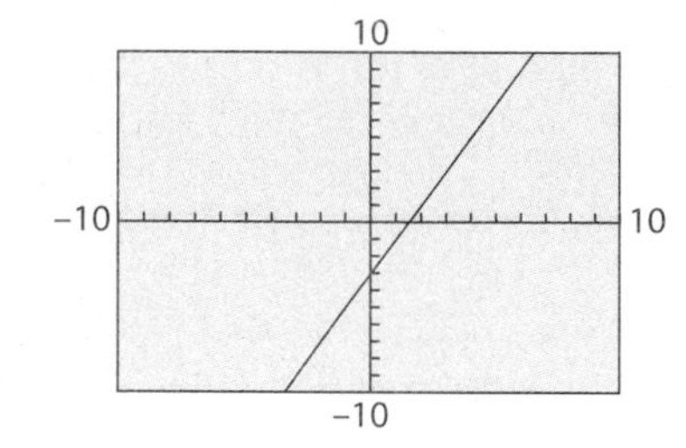

46. $f(x) = -3x + 4$

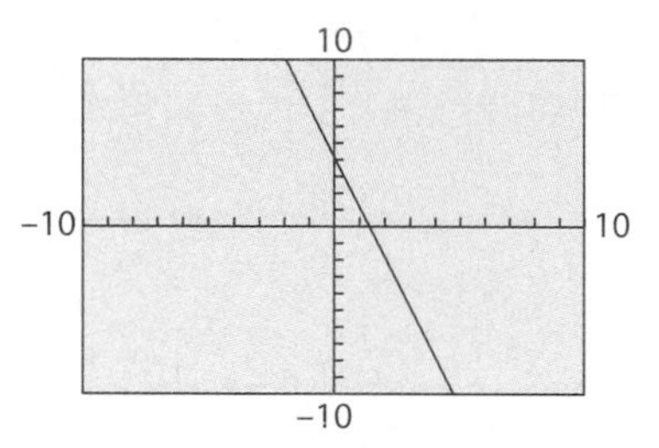

47. $f(x) = -\frac{4}{3}x + 5$

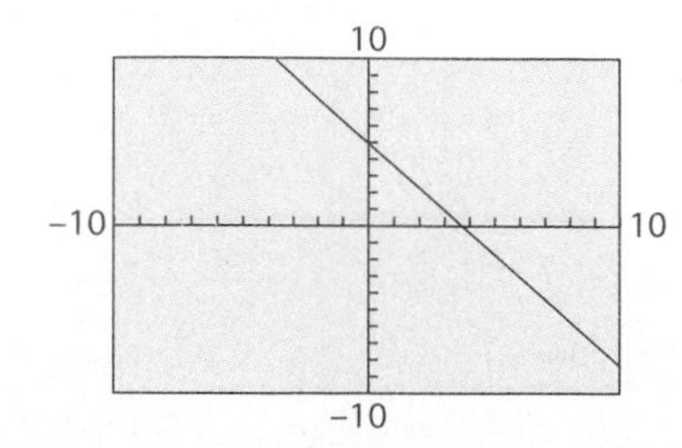

48. $f(x) = \frac{5}{2}x$

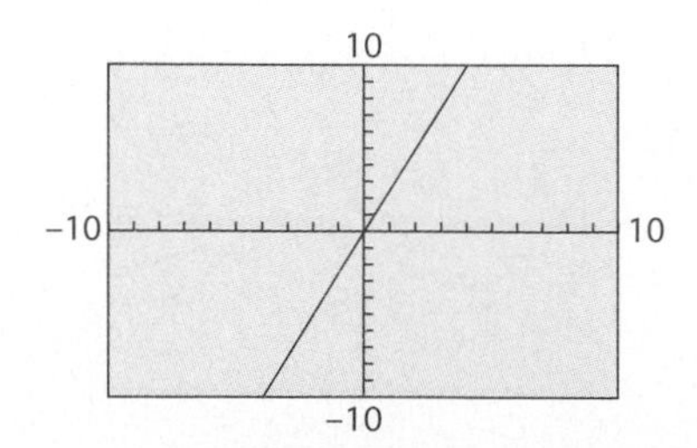

49. $f(x) = 0.3x + 2$

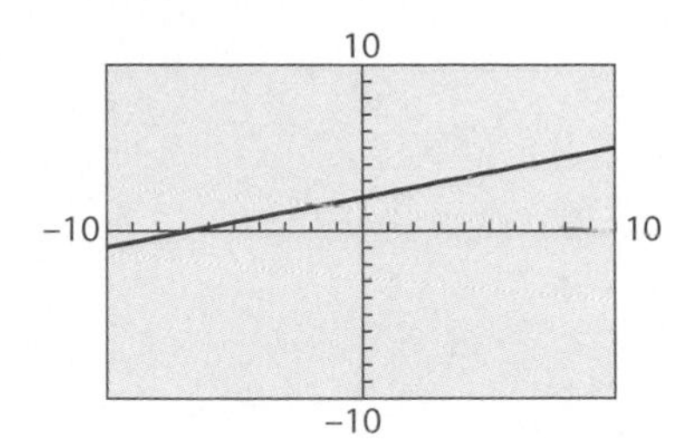

50. $f(x) = -1.8x - 5$

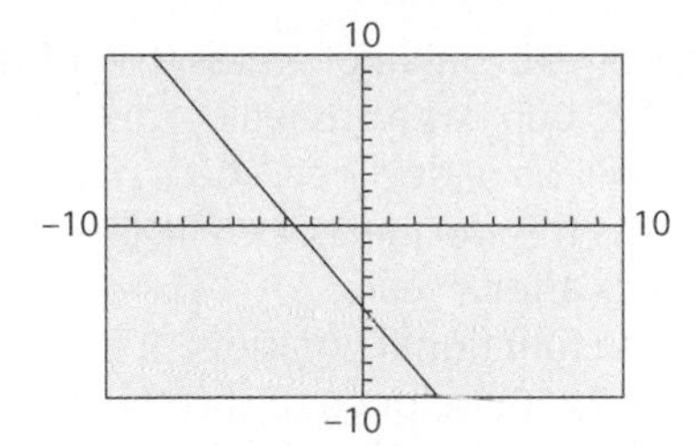

51. $f(x) = -2.5x + 1.4$

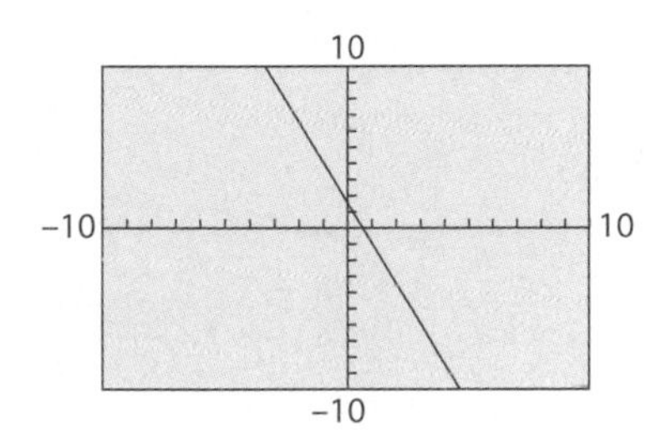

52. $f(x) = 3.2x - 4$

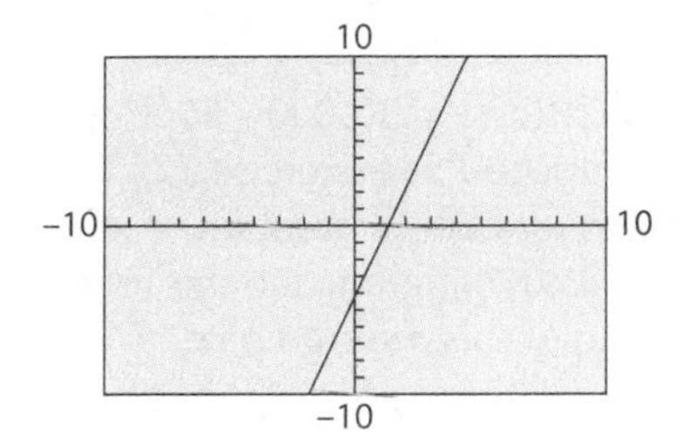

53. Given that 5 is in the range of $f(x) = 2x + 7$, find the value of x in the domain so that $f(x) = 5$.
-1

54. Given that -5 is in the range of $f(x) = 3x + 1$, find the value of x in the domain so that $f(x) = -5$.
-2

55. Given that 0 is in the range of $f(x) = 3x + 8$, find the value of x in the domain so that $f(x) = 0$.
$-\frac{8}{3}$

56. Given that 4 is in the range of $f(x) = 5x - 2$, find the value of x in the domain so that $f(x) = 4$.
$\frac{6}{5}$

Applications

57. Depreciation is the decline in value of an asset. For instance, a company that purchases a truck for \$20,000 has an asset worth \$20,000. In 5 years, however, the value of the truck will have declined, and it may be worth only \$4000. An equation that represents this decline is $V = 20{,}000 - 3200x$, where V is the value, in dollars, of the truck after x years.

a. Write the equation in functional notation.

b. The ordered pair (4, 7200) belongs to this function. Write a sentence that explains the meaning of this ordered pair.

a. $f(x) = 20{,}000 - 3200x$; b. The value of the truck after 4 years is \$7200.

58. A company uses the equation $V = 30{,}000 - 5000x$ to estimate the depreciated value, in dollars, of a computer. (See Exercise 57.)

a. Write the equation in functional notation.

b. The ordered pair (1, 25,000) belongs to this function. Write a sentence that explains the meaning of this ordered pair.

a. $f(x) = 30{,}000 - 5000x$; b. The value of the computer after 1 year is \$25,000.

59. A diagonal of a polygon is a line segment joining a vertex to a nonadjacent vertex. The 9 diagonals of a hexagon are shown at the right. Next to the hexagon there is a table that shows the number of diagonals for polygons from 3 to 11 sides.

a. Does this table represent a function?

b. How many diagonals does a polygon with 9 sides have?

c. How many sides does a polygon with 35 diagonals have?

d. How many diagonals does a polygon with 11 sides have?

e. How many sides does a polygon with 20 diagonals have?

f. Using inductive reasoning, how many diagonals will a polygon with 13 sides have?

a. Yes; b. 27; c. 10; d. 44; e. 8; f. 65

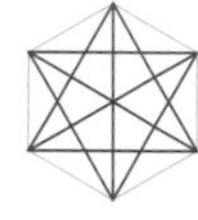

Number of sides, x	Number of diagonals, y
3	0
4	2
5	5
6	9
7	14
8	20
9	27
10	35
11	44

60. The table at the right shows the number of games that must be scheduled if each team of a league plays every other team twice.

a. Does this table represent a function?

b. How many games must be scheduled in a league with 5 teams?

c. How many teams are in a league for which 42 games are scheduled?

d. How many games must be scheduled in a league with 9 teams?

e. How many teams are in a league for which 90 games are scheduled?

f. Using inductive reasoning, how many games must be scheduled in a league with 12 teams?

a. Yes; b. 20; c. 7; d. 72; e. 10; f. 132

Number of teams, x	Number of games, y
2	2
3	6
4	12
5	20
6	30
7	42
8	56
9	72
10	90

61. According to Einstein's Theory of Relativity, as the speed of an object increases, the mass of the object also increases. The increase in mass is very small until the speed of the object approaches the speed of light, 186,000 miles per second. The vertical axis for the graph at the right shows the mass of an object. The horizontal axis gives the speed of the object as a percent of the speed of light. For instance, when $x = 20$, the object is traveling at 20% the speed of light.

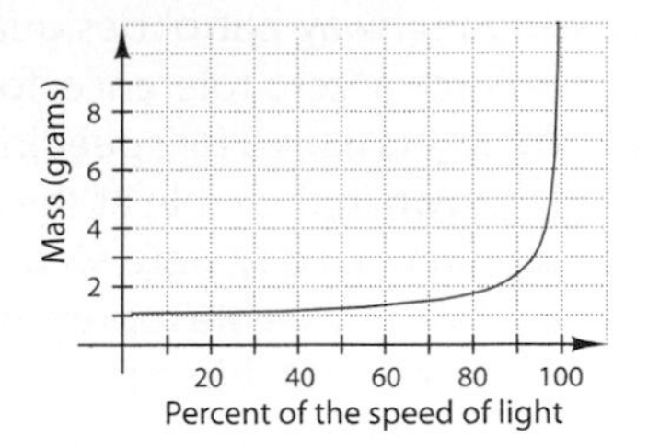

a. Does this graph represent a function?
b. Estimate the mass of the object when $x = 50$.
c. Approximate the percent of the speed of light required so that the mass of the object is 2 grams.
d. Approximate the percent of the speed of light required so that the mass of the object is 3 times its original mass.
a. Yes; b. 1.2 grams; c. 86%; d. 95%

62. If you have ever snow skied or hiked in the mountains, you know that breathing is a little more difficult because there is less oxygen in the atmosphere as altitude increases. The graph at the right shows the amount of oxygen available at various altitudes as a percent of the amount available at sea level.

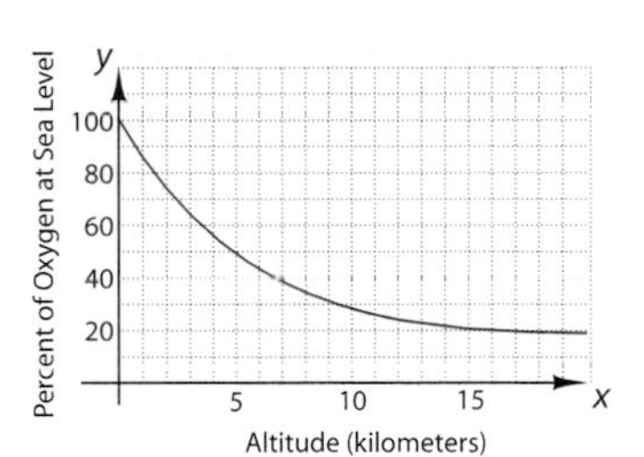

a. Does this graph represent a function?
b. Estimate the percent of oxygen available at an altitude of 9 kilometers.
c. At approximately what altitude is the percent of oxygen 50% of that at sea level?
a. Yes; b. 30%; c. 5 kilometers

Applying Concepts

63. A function f is defined by $\{(-4,-6),(-2,-2),(0,2),(2,6),(4,10)\}$. Find $f(2)$.
6

64. A function f is defined by $\{(0,9),(1,8),(2,7),(3,6),(4,5)\}$. Find $f(1)$.
8

65. The table at the right shows the amount of a monthly payment at various interest rates for each \$1000 borrowed to purchase a car. The term of the loan is 5 years.

Interest rate, x	Monthly payment per \$1000, y
8%	20.28
8.25%	20.40
8.50%	20.52
8.75%	20.64
9%	20.76
9.25%	20.88
9.5%	21.00
9.75%	21.12
10%	21.24

a. Does this table represent a function?
b. What is the monthly payment to borrow \$1000 at an interest rate of 9.25% for 5 years?
c. What is the monthly payment to borrow \$1000 at an interest rate of 8.5% for 5 years?
d. Suppose you purchase a car and wish to finance \$11,000 at an interest rate of 8.75% for 5 years. What is your monthly payment?
e. Suppose you purchase a car and wish to finance \$15,455 at an interest rate of 9.5% for 5 years. What is your monthly payment?
a. Yes; b. \$20.88; c. \$20.52; d. \$227.04; e. \$324.56

66. The highway patrol designated a certain stretch of an interstate highway to enforce a "zero tolerance" for violators of the 65-mph speed limit. The number of tickets issued for speeding for the 5 days of the zero tolerance enforcement is shown in the table at the right.

Tickets issued, x	Day of the week, y
86	Monday
93	Tuesday
75	Wednesday
86	Thursday
59	Friday

a. On what day were 93 tickets issued?
b. Does this table represent a relation? Why or why not?
c. Does this table represent a function? Why or why not?
a. Tuesday; b. yes; c. no

67. Consider the inequality $y < 3x + 1$.
a. Does this inequality represent a relation? Why or why not?
b. Does this inequality represent a function? Why or why not?
a. Yes; b. no

68. Consider the inequality $y \geq -2x + 3$.
a. Does this inequality represent a relation? Why or why not?
b. Does this inequality represent a function? Why or why not?
a. Yes; b. no

Exploration

69. ***Properties of Graphs of Linear Functions*** Recall that a linear function is one that can be written in the form $f(x) = mx + b$. As the values of m and b change, the graph of the function changes.

a. Graph $f(x) = mx + 2$ for m equal to −3, −2, −1, 1, 2, and 3. From your graphs, complete the following sentences.
  1. When m is negative, y ___________ as x increases.
  2. When m is positive, y ___________ as x increases.
  3. Every line passes through the point whose coordinates are __________.

b. Graph $f(x) = x + b$ for b equal to −3, −2, −1, 0, 1, 2, 3. From your graphs, answer the following questions.
  1. How are the lines similar?
  2. How are the lines different?
  3. How is the value of b related to the y-coordinate of the point at which the line crosses the y-axis?

c. Using inductive reasoning, will the values of y for the graph of $f(x) = -\frac{3}{4}x + 2$ increase or decrease as x increases? Graph this equation and verify your answer.

d. Using inductive reasoning, what are the coordinates of the point at which the graph of $f(x) = 2x - 6$ crosses the y-axis? Graph this equation and verify your answer.

Section 4.2 Equations of the Form $Ax + By = C$

Graphing Equations of the Form $Ax + By = C$

An equation in the form $Ax + By = C$, where A, B, and C are real numbers, is also a linear equation. The graph of the equation is a straight line. Examples of these equations are shown below.

$6x + 5y = 3$ • $A = 6, B = 5$, and $C = 3$

$x - 3y = 7$ • $A = 1, B = -3$, and $C = 7$

$2x + y = 0$ • $A = 2, B = 1$, and $C = 0$

$x = 5$ • $A = 1, B = 0$, and $C = 5$

$y = -2$ • $A = 0, B = 1$, and $C = -2$

One way to graph an equation of the form $Ax + By = C$ involves first solving the equation for y. To solve the equation for y means to rewrite the equation so that y is alone on one side of the equation and the term containing x and the constant are on the other side of the equation.

TAKE NOTE
Being able to solve an equation of the form $Ax + By = C$ for y is important because most graphing utilities require that an equation be in the form $y = mx + b$ when the equation of the line is entered for graphing.

→ Solve $3x + 4y = 12$ for y.

$3x + 4y = 12$

$3x - 3x + 4y = -3x + 12$ • Subtract $3x$ from each side of the equation.

$4y = -3x + 12$ • Simplify.

$\frac{4y}{4} = \frac{-3x + 12}{4}$ • Divide each side of the equation by 4.

$y = \frac{-3x}{4} + \frac{12}{4}$ • Simplify.

$y = -\frac{3}{4}x + 3$

To graph this equation, we can now select values for x and calculate the corresponding values of y. We will use -4, 0, and 4 for x and record the results in an input/output table.

x	-4	0	4
$y = -\frac{3}{4}x + 3$	6	3	0

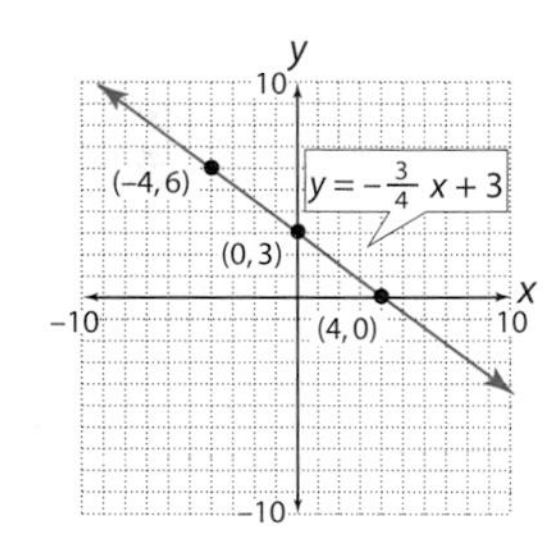

Example 1 Use a graphing calculator to graph $2x - 3y = 10$.

Solution Solve the equation for y.

$2x - 3y = 10$

$2x - 2x - 3y = -2x + 10$

$-3y = -2x + 10$

$\frac{-3y}{-3} = \frac{-2x}{-3} + \frac{10}{-3}$

$y = \frac{2}{3}x - \frac{10}{3}$

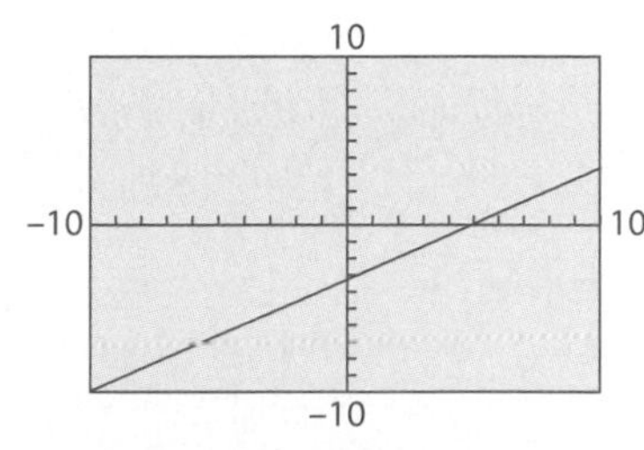

Enter the equation as Y1 = 2X/3 − 10/3. The graph is shown above.

You-Try-It 1 Use a graphing calculator to graph $-x + 2y = 10$.

Solution See page S16.

The graph of an equation in which one of the variables is missing is either a horizontal line or a vertical line.

SUGGESTED ACTIVITY
See the *Instructor's Resource Manual* for an activity that explores vertical and horizontal lines.

Recall that the equation $y = 2$ represents the constant function and could be written as $f(x) = 2$. The graph of this equation is a horizontal line. The equation also could be written in the form $Ax + By = C$ as $0 \cdot x + y = 2$. No matter what value of x is chosen, y is always 2. Some solutions to the equation are (3, 2), (−1, 2), (0, 2), and (−4, 2). The graph is shown at the right.

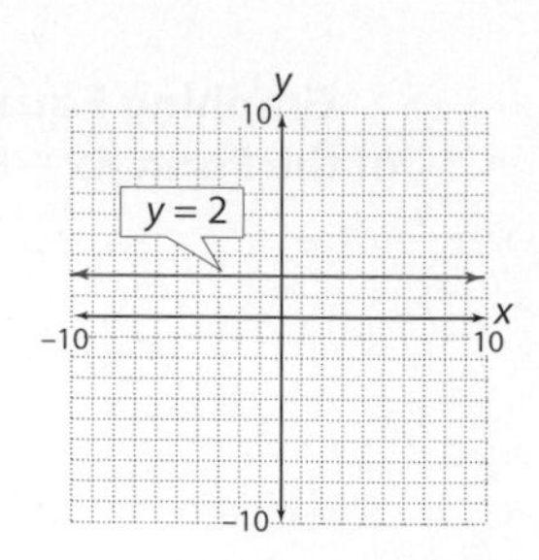

The graph of $y = b$ is a horizontal line passing through the point whose coordinates are $(0, b)$.

INSTRUCTOR NOTE
As an oral exercise, ask students for the equations of the x- and y-axes.

There are two other important points for students: (1) $x = -2$ does not represent a function, which means that not all equations represent functions; and (2) because $x = -2$ is not a function, it cannot be graphed by using the Y= editor.

The equation $x = -2$ could be written as $x + 0 \cdot y = -2$. No matter what value of y is chosen, x is always −2. Some solutions to the equation are (−2, 3), (−2, −2), (−2, 0), and (−2, 2). The graph is shown at the right.

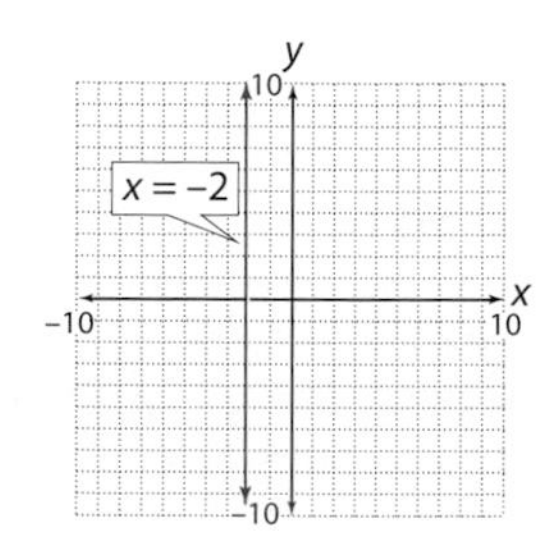

Notice also that there are ordered pairs with the same first coordinate but different second coordinates. Therefore, the graph of $x = -2$ is not the graph of a function.

The graph of $x = a$ is a vertical line passing through the point whose coordinates are $(a, 0)$.

Example 2 Graph: $x = 4$

Solution The graph of $x = 4$ is a vertical line passing through the point (4, 0). The graph is shown at the right.

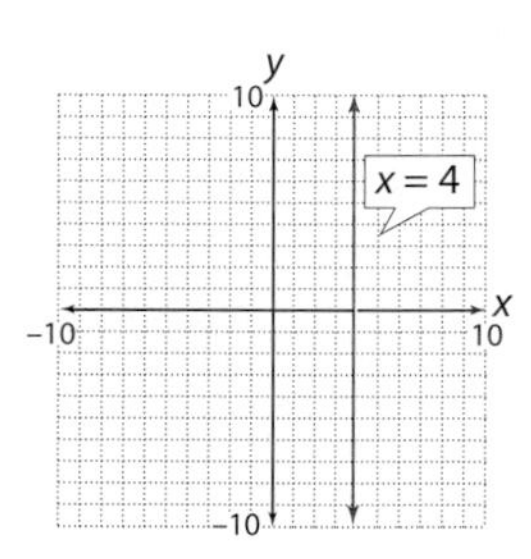

You-Try-It 2 Graph: $y = -3$

Solution See page S16.

x- and y-Intercepts

The graph of the equation $-4x + 3y = 12$ is shown at the right. The graph crosses the x-axis at (−3, 0). This point is called the **x-intercept**. The graph crosses the y-axis at (0, 4). This point is called the **y-intercept**.

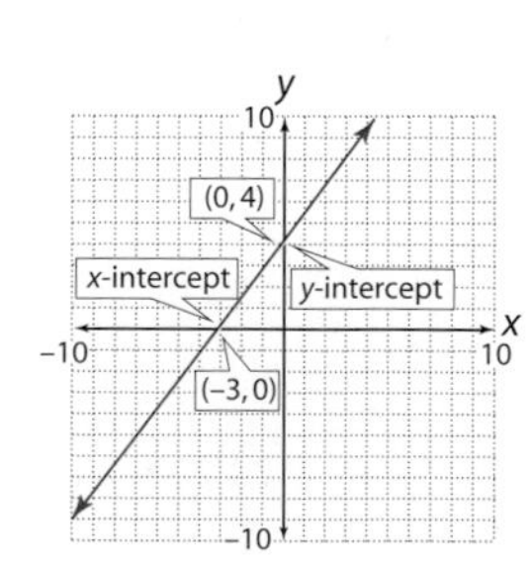

TAKE NOTE
Because the y-coordinate of any point on the x-axis is 0, let $y = 0$ to find the x-intercept.

Because the x-coordinate of any point on the y-axis is 0, let $x = 0$ to find the y-intercept.

➞ Algebraically find the x- and y-intercepts of $-4x + 3y = 12$.

To find the x-intercept, let $y = 0$.

$$\begin{aligned} -4x + 3y &= 12 \\ -4x + 3(0) &= 12 \\ -4x &= 12 \\ x &= -3 \end{aligned}$$

The x-intercept is (−3, 0).

To find the y-intercept, let $x = 0$.

$$\begin{aligned} -4x + 3y &= 12 \\ -4(0) + 3y &= 12 \\ 3y &= 12 \\ y &= 4 \end{aligned}$$

The y-intercept is (0, 4).

Another way to graph some equations of the form $Ax + By = C$ is to find the x- and y-intercepts, plot both intercepts, and then draw a line through the two points.

Example 3 Graph $2x + 5y = 10$ by first finding the x- and y-intercepts.

Solution To find the x-intercept, let $y = 0$.

$$\begin{aligned} 2x + 5y &= 10 \\ 2x + 5(0) &= 10 \\ 2x &= 10 \\ x &= 5 \end{aligned}$$

The x-intercept is $(5, 0)$.

To find the y-intercept, let $x = 0$.

$$\begin{aligned} 2x + 5y &= 10 \\ 2(0) + 5y &= 10 \\ 5y &= 10 \\ y &= 2 \end{aligned}$$

The y-intercept is $(0, 2)$.

To draw the graph, plot the ordered pairs (5, 0) and (0, 2) and then draw a line through the two points.

You-Try-It 3 Graph $3x - 5y = 15$ by first finding the x- and y-intercepts.

Solution See page S17.

Trying to graph an equation of the form $Ax + By = C$ by using intercepts will not work when $C = 0$. For instance, consider the equation $3x + 8y = 0$.

To find the x-intercept, let $y = 0$.

$$\begin{aligned} 3x + 8y &= 0 \\ 3x + 8(0) &= 0 \\ 3x &= 0 \\ x &= 0 \end{aligned}$$

The x-intercept is $(0, 0)$.

To find the y-intercept, let $x = 0$.

$$\begin{aligned} 3x + 8y &= 0 \\ 3(0) + 8y &= 0 \\ 8y &= 0 \\ y &= 0 \end{aligned}$$

The y-intercept is $(0, 0)$.

Note that the x- and y-intercepts are the same. Thus, there are not two different points through which the line can be drawn.

The method for finding the x- and y-intercepts of a linear equation of the form $y = mx + b$ is exactly the same as that for an equation of the form $Ax + By = C$.

Example 4 Graph $y = -2x + 4$ by first finding the x- and y-intercepts.

Solution To find the x-intercept, let $y = 0$.

$$\begin{aligned} y &= -2x + 4 \\ 0 &= -2x + 4 \\ -4 &= -2x \\ 2 &= x \end{aligned}$$

The x-intercept is $(2, 0)$.

To find the y-intercept, let $x = 0$.

$$\begin{aligned} y &= -2x + 4 \\ y &= -2(0) + 4 \\ y &= 4 \end{aligned}$$

The y-intercept is $(0, 4)$.

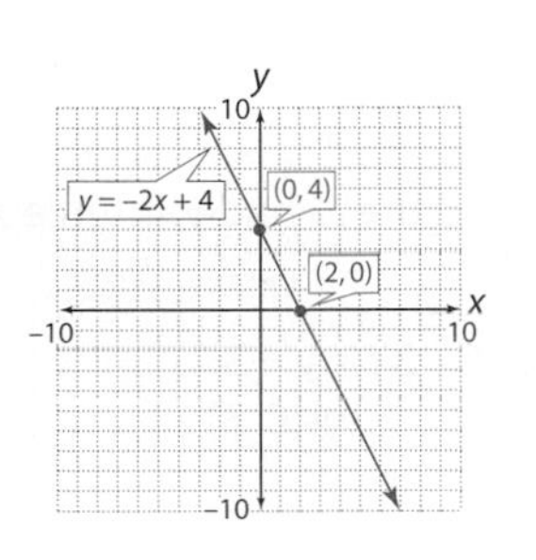

TAKE NOTE
Observe that the y-coordinate of the y-intercept is 4, the constant term of $y = -2x + 4$. This will be discussed more in the next section.

The graph is shown at the right.

You-Try-It 4 Graph $y = \frac{2x}{3} + 2$ by first finding the x- and y-intercepts.

Solution See page S17.

4.2 EXERCISES

Topics for Discussion

1. What is an x-intercept of a graph? What is a y-intercept of a graph?
   A point at which a graph crosses the x-axis. A point at which a graph crosses the y-axis.

2. Explain how to find the x- and y-intercepts of a graph.
   Answers will vary.

3. Is the graph of $Ax + By = C$ always a straight line?
   Yes.

4. Describe the graph of $x = a$ and the graph of $y = b$.
   A line parallel to the y-axis passing through $(a, 0)$. A line parallel to the x-axis passing through $(0, b)$.

Equations of the Form Ax +By = C

Solve the equation for y.

5. $3x + y = 10$
$y = -3x + 10$

6. $2x + y = 5$
$y = -2x + 5$

7. $4x - y = 3$
$y = 4x - 3$

8. $5x - y = 7$
$y = 5x - 7$

9. $2x + 7y = 14$
$y = -\frac{2}{7}x + 2$

10. $3x + 2y = 6$
$y = -\frac{3}{2}x + 3$

11. $2x + 3y = 9$
$y = -\frac{2}{3}x + 3$

12. $2x - 5y = 10$
$y = \frac{2}{5}x - 2$

13. $5x - 2y = 4$
$y = \frac{5}{2}x - 2$

14. $x + 3y = 6$
$y = -\frac{1}{3}x + 2$

15. $x - 4y = 12$
$y = \frac{1}{4}x - 3$

16. $6x - 5y = 10$
$y = \frac{6}{5}x - 2$

Find the x- and y-intercepts.

17. $3x + y = 3$
$(1, 0), (0, 3)$

18. $2x + y = 4$
$(2, 0), (0, 4)$

19. $3x + 2y = 6$
$(2, 0), (0, 3)$

20. $2x - y = 6$
$(3, 0), (0, -6)$

21. $2x - 3y = 12$
$(6, 0), (0, -4)$

22. $3x - 2y = 8$
$\left(\frac{8}{3}, 0\right), (0, -4)$

23. $3x - 2y = 4$
$\left(\frac{4}{3}, 0\right), (0, -2)$

24. $5x - 3y = 15$
$(3, 0), (0, -5)$

25. $y = -4$
No x-intercept, $(0, -4)$

26. $x = -2$
$(0, -2)$, no y-intercept

27. $3x + 4y = 10$
$\left(\frac{10}{3}, 0\right), \left(0, \frac{5}{2}\right)$

28. $y = -5$
No x-intercept, $(0, -5)$

29. $5x - 3y = 14$
$\left(\frac{14}{5}, 0\right), \left(0, -\frac{14}{3}\right)$

30. $y = 5x + 10$
$(-2, 0), (0, 10)$

31. $y = 2x - 6$
$(3, 0), (0, -6)$

32. $5x - 2y = 14$
$\left(\frac{14}{5}, 0\right), (0, -7)$

33. $2x - 3y = 0$
$(0, 0), (0, 0)$

34. $x - 5y = 0$
$(0, 0), (0, 0)$

35. $y = \frac{1}{2}x + 3$
$(-6, 0), (0, 3)$

36. $3x - 4y = 0$
$(0, 0), (0, 0)$

37. $y = -\frac{3}{2}x - 3$
$(-2, 0), (0, -3)$

38. $y = \frac{2}{3}x - 4$
$(6, 0), (0, -4)$

39. $y = 2x$
$(0, 0), (0, 0)$

40. $y = -3x$
$(0, 0), (0, 0)$

Graph by using the x- and y-intercepts.

41. $5x + 2y = 10$

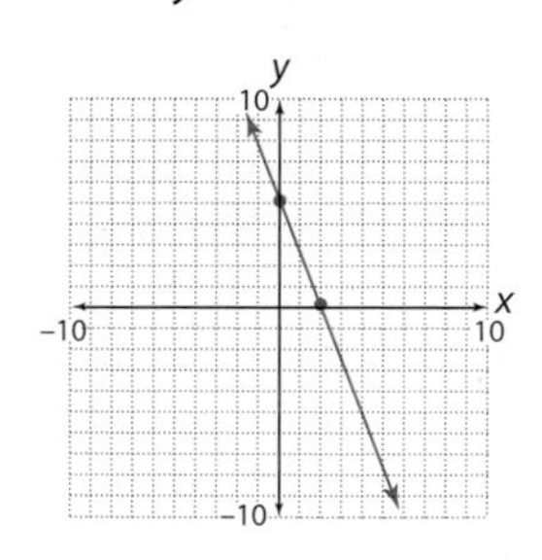

42. $x - 3y = 6$

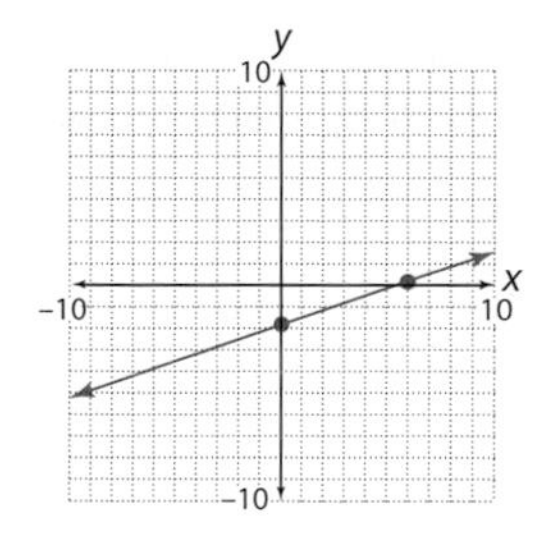

43. $2x + y = 3$

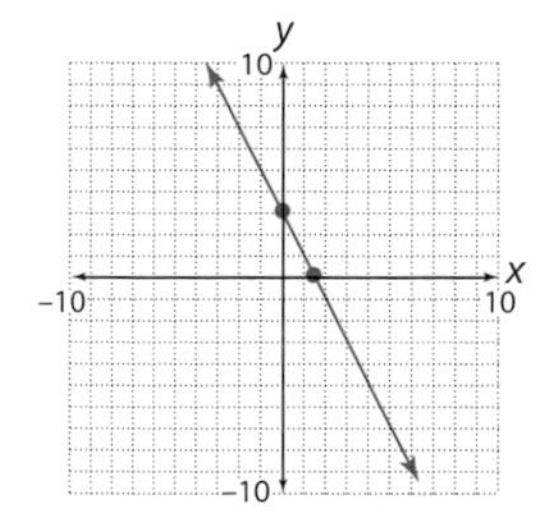

44. $2x + 3y = 6$

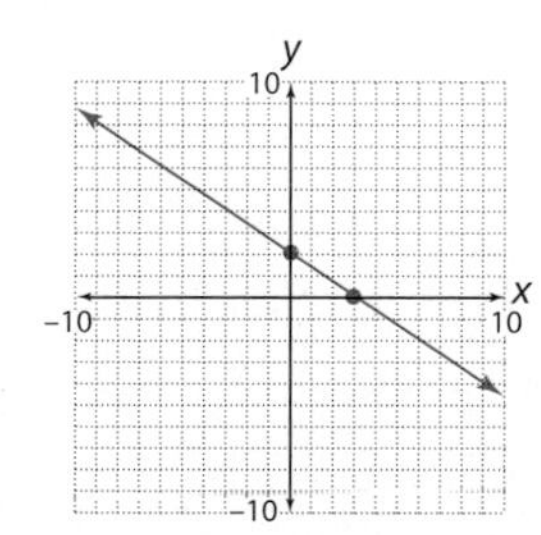

45. $3x - 4y = 12$

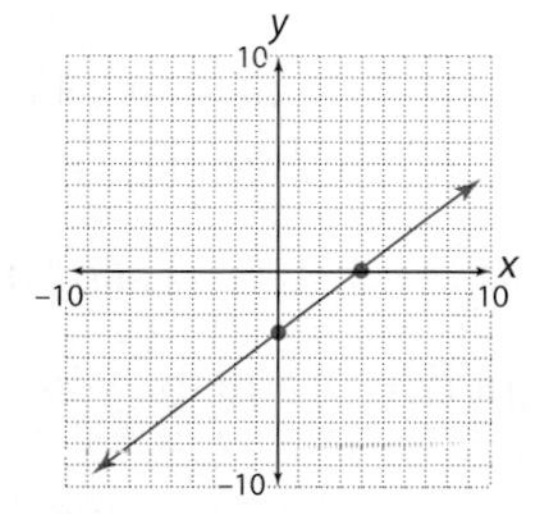

46. $3x + 5y = 15$

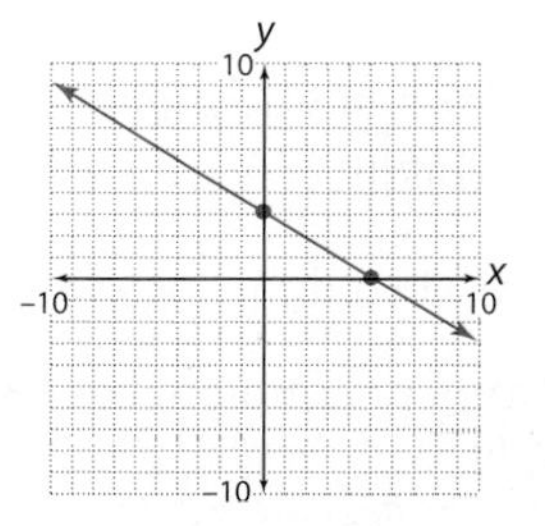

47. $x - 4y = 8$

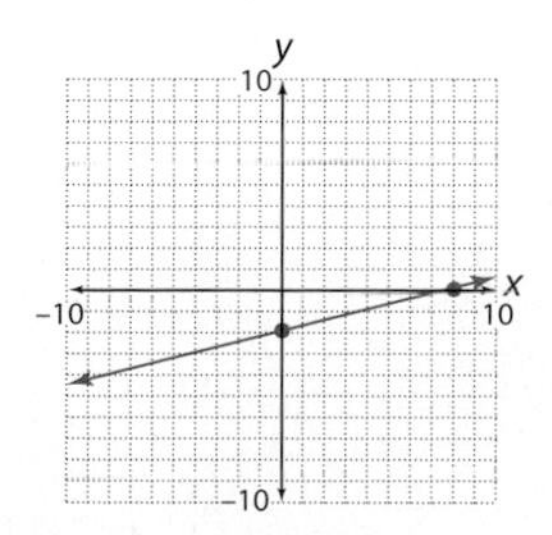

48. $x + 2y = 4$

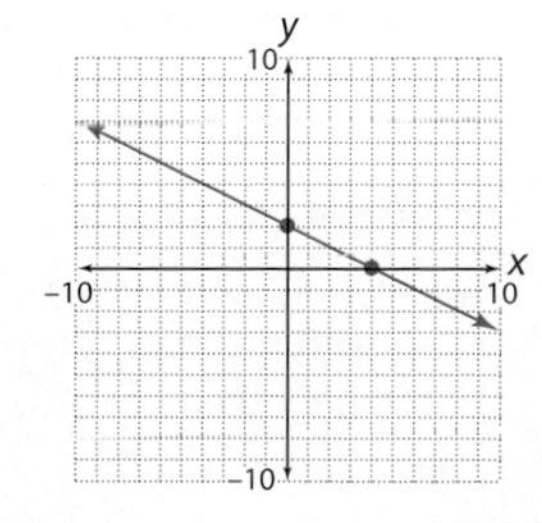

49. $4x + 3y = 12$

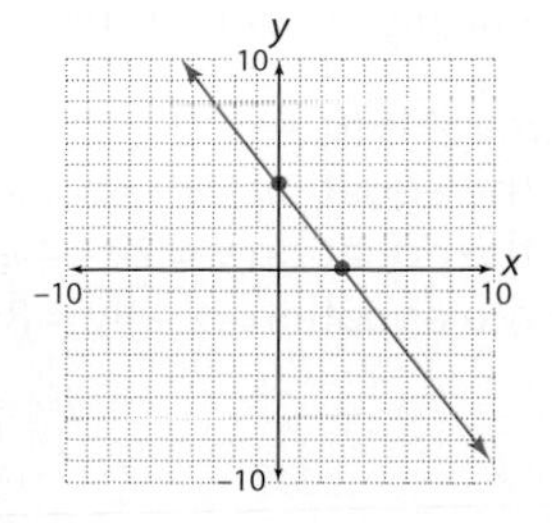

50. $3x + 4y = 12$

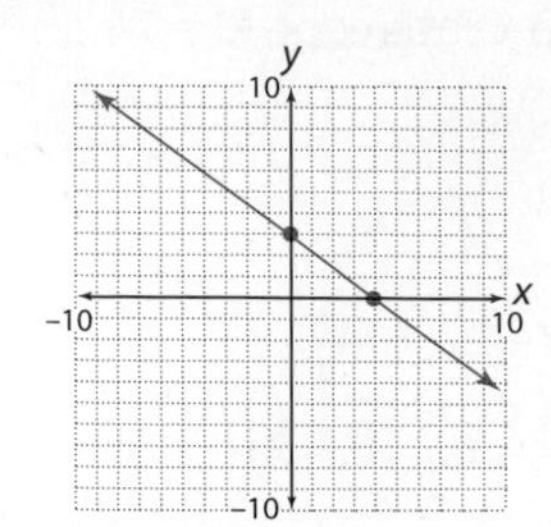

51. $2x - y = 6$

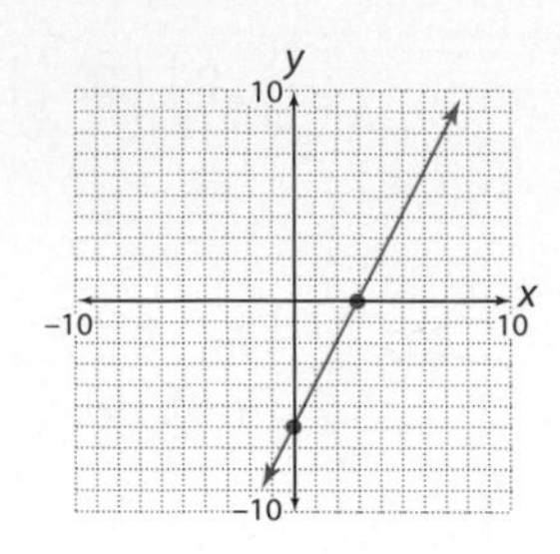

52. $x + 5y = 5$

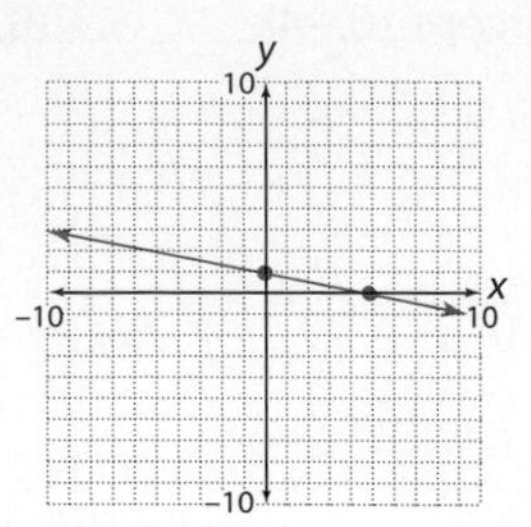

Applying Concepts

Complete the following sentences.

53. The *x*-coordinate of the *y*-intercept of a graph is ___0___.

54. The *y*-coordinate of the *x*-intercept of a graph is ___0___.

55. Assuming that A is not equal to zero, what is the *x*-coordinate of the *x*-intercept of the graph of $Ax + By = C$? Why must we assume that A is not equal to zero?
$\frac{C}{A}$; division by zero is not allowed.

56. Assuming that B is not equal to zero, what is the *y*-coordinate of the *y*-intercept of the graph of $Ax + By = C$? Why must we assume that B is not equal to zero?
$\frac{C}{B}$; division by zero is not allowed.

57. Assuming that $m \neq 0$, what are the coordinates of the intercepts for the graph of $y = mx + b$?
$\left(-\frac{b}{m}, 0\right)$, $(0, b)$

58. Draw a graph that has two *x*-intercepts.
Answers will vary.

Exploration

59. ***Intercept Form of a Straight Line*** We have discussed two equations whose graph is a straight line: $y = mx + b$ and $Ax + By = C$. There are other equations that represent straight lines. One such equation is $\frac{x}{a} + \frac{y}{b} = 1$. This is called the *intercept form* of a straight line because the *x*-intercept is $(a, 0)$ and the *y*-intercept is $(0, b)$.

a. Draw the graphs of $\frac{x}{3} + \frac{y}{4} = 1$ and $\frac{x}{2} - \frac{y}{5} = 1$.

b. Algebraically show that the *x*-intercept of $\frac{x}{3} + \frac{y}{4} = 1$ is $(3, 0)$ and that the *y*-intercept is $(0, 4)$.

c. Write the equation $3x + 5y = 15$ in intercept form.

d. Write the equation $y = 2x - 4$ in intercept form.

e. Write the equation $3x - 4y = 8$ in intercept form.

The graph of the line containing the points whose coordinates are $(-2, 3)$ and $(4, -3)$ is shown at the right. To find the slope of the line, let $P_1 = (-2, 3)$ and $P_2 = (4, -3)$. Then $x_1 = -2$, $y_1 = 3$, $x_2 = 4$, and $y_2 = -3$.

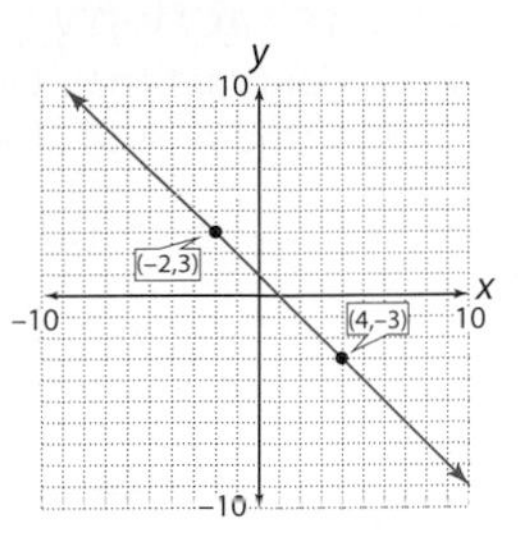

Negative slope

$$m = \frac{y_2 - y_1}{x_2 - x_1} = \frac{-3 - 3}{4 - (-2)} = \frac{-6}{6} = -1$$

The slope is -1.

Here the slope is a negative number. A line that slants downward to the right has **negative slope.**

TAKE NOTE
Negative slope means that the value of y decreases as the value of x increases.

The graph of the line containing the points whose coordinates are $(-5, 6)$ and $(7, 6)$ is shown at the right. To find the slope of the line, let $P_1 = (-5, 6)$ and $P_2 = (7, 6)$. Then $x_1 = -5$, $y_1 = 6$, $x_2 = 7$, and $y_2 = 6$.

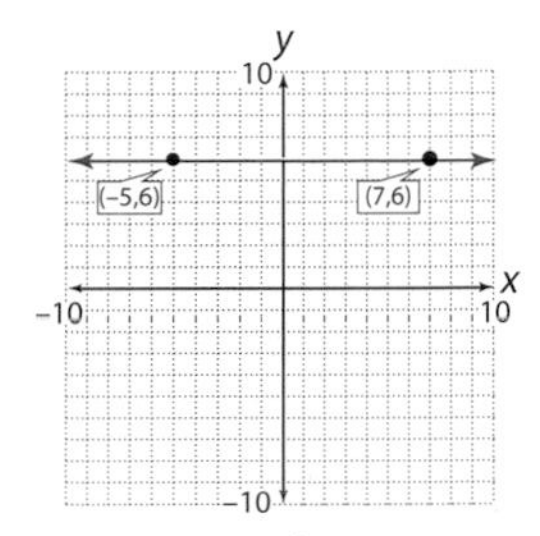

Zero slope

$$m = \frac{y_2 - y_1}{x_2 - x_1} = \frac{6 - 6}{7 - (-5)} = \frac{0}{12} = 0$$

The slope is 0.

A horizontal line has **zero slope**.

SUGGESTED ACTIVITY
After completing the discussion of the different possibilities for slope, draw several lines with different slopes and ask students to identify whether each line has positive slope, negative slope, zero slope, or no slope. When students respond to lines with positive, negative, or zero slope, ask them whether y increases, decreases, or remains the same as x increases.

Another possibility is to draw two lines with different positive slopes and ask for which line y increases more for a given increase in x. For lines with negative slope, ask for which line y decreases more for a given increase in x.

The graph of the line containing the points whose coordinates are $(-4, 7)$ and $(-4, -5)$ is shown at the right. To find the slope of the line, let $P_1 = (-4, 7)$ and $P_2 = (-4, -5)$. Then $x_1 = -4$, $y_1 = 7$, $x_2 = -4$, and $y_2 = -5$.

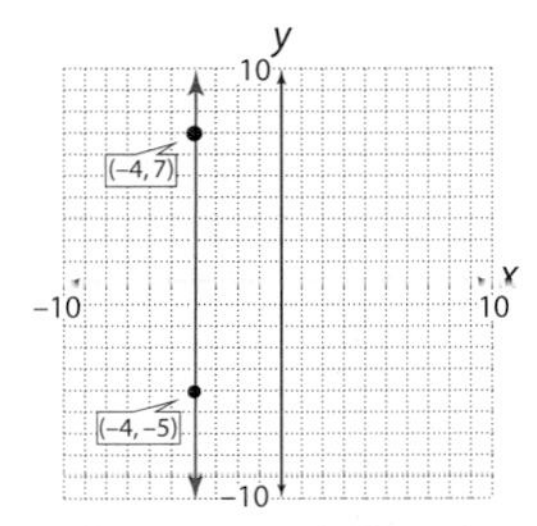

Slope is undefined.

$$m = \frac{y_2 - y_1}{x_2 - x_1} = \frac{-5 - 7}{-4 - (-4)} = \frac{-12}{0}$$

Because division by 0 is not defined, the slope of a vertical line is not defined.

We say that a vertical line has **no slope** or that the slope of a vertical line is **undefined.**

It is important to remember that *zero slope* and *no slope* are different. The graph of a line with zero slope is horizontal; the graph of a line with no slope is vertical.

Example 1 Find the slope of the line containing the points P_1 and P_2.

a. $P_1(-2, -1), P_2(3, 4)$ **b.** $P_1(-1, 4), P_2(-1, 0)$

Solution **a.** $m = \frac{y_2 - y_1}{x_2 - x_1} = \frac{4 - (-1)}{3 - (-2)} = \frac{5}{5} = 1$

The slope is 1.

b. $m = \frac{y_2 - y_1}{x_2 - x_1} = \frac{0 - 4}{-1 - (-1)} = \frac{-4}{0}$

The slope is undefined.

You-Try-It 1 Find the slope of the line containing the points P_1 and P_2.

a. $P_1(-3, 1), P_2(2, -2)$ **b.** $P_1(-1, 2), P_2(4, 2)$

Solution See page S17.

SUGGESTED ACTIVITY
Have students set their calculators to integer mode and then graph $y = \frac{2}{3}x - 2$. Select TRACE; move the cursor to some point and record the ordered pair; move to a different point and record the ordered pair. Students should now calculate the slope. Repeat this to convince students that any two points on the line can be used to find the slope.

It does not matter which two points on a line are used to calculate the slope of the line. The value of the slope will be the same.

$$\text{Slope using } P_1 \text{ and } P_2 = \frac{-4-(-8)}{-3-(-9)} = \frac{4}{6} = \frac{2}{3}$$

$$\text{Slope using } P_2 \text{ and } P_3 = \frac{2-(-4)}{6-(-3)} = \frac{6}{9} = \frac{2}{3}$$

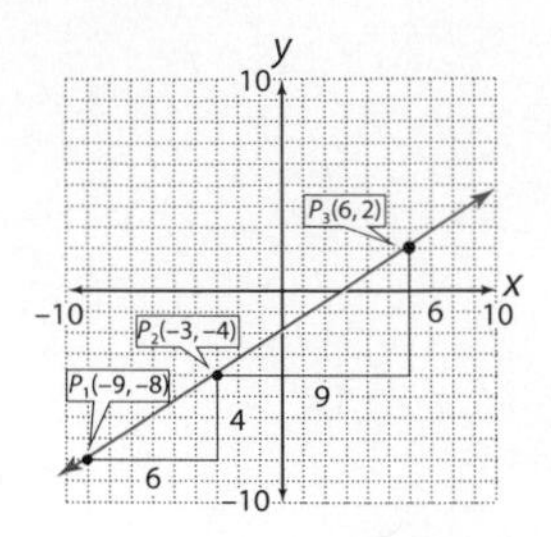

There are many applications of the concept of slope. Here are two possibilities.

In 1988, when Florence Griffith-Joyner set the world record for the 200-meter dash, her average speed was approximately **9.4** meters per second. The graph at the right shows the distance she ran during her record-setting run. From the graph, note that after 5 seconds she had run 47 meters, and after 15 seconds she had run 141 meters. The slope of the line between these two points is

$$m = \frac{141-47}{15-5} = \frac{94}{10} = \mathbf{9.4}$$

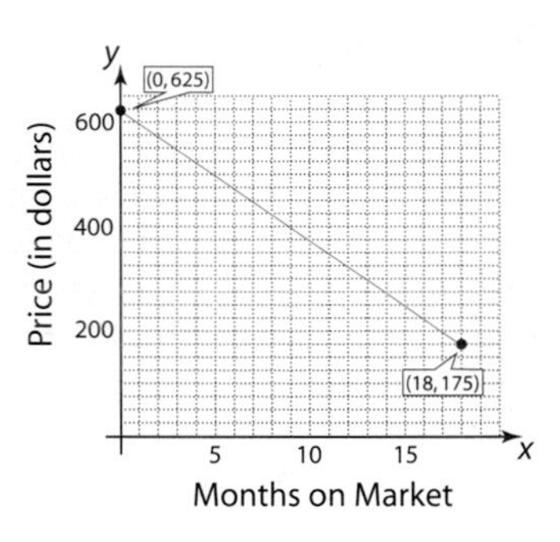

Note that the slope of the line is the same as the rate she was running, **9.4** meters per second. The average speed of an object is related to slope.

In mid-1997, when Intel first introduced its 350-megahertz computer chip, the price was approximately $625. From that time to the end of 1998, the price of the chip dropped an average of $**25** per month. The graph at the right shows the approximate price of the chip over time. From the graph, we learn that the price of the chip when it was released ($x = 0$) was $625 and the approximate price of the chip 18 months later ($x = 18$) was $175. The slope of the line between these two points is

$$m = \frac{175-625}{18-0} = \frac{-450}{18} = \mathbf{-25}$$

Observe that the slope of the line is the same as the rate at which the price was decreasing each month, $25.

Example 2 In 1990, Arie Luyendyk had the record winning time of approximately 2.7 hours in the Indianapolis 500. The graph at the right shows the distance he traveled during the race. Find the slope of the line between the two given points. Write a sentence that states the meaning of the slope.

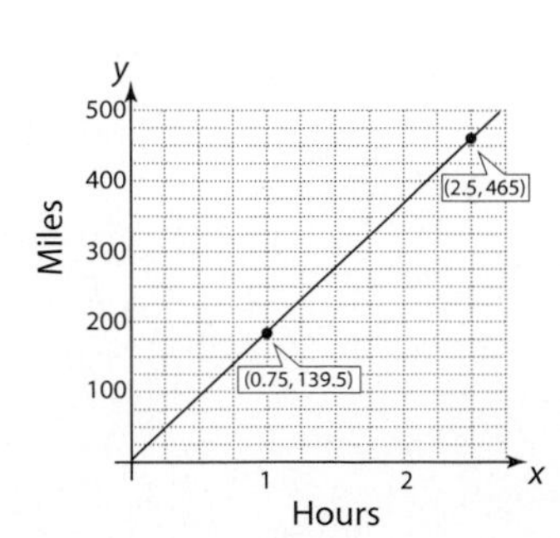

Solution

$$m = \frac{465-139.5}{2.5-0.75} = \frac{325.5}{1.75} = 186$$

A slope of 186 means that Luyendyk's average speed for the race was 186 miles per hour.

You-Try-It 2 In 1988, when Florence Griffith-Joyner set the world record for the 200-meter dash, she also set the world record for the 100-meter dash. The graph at the right shows the distance she ran during her record-setting run. Find the slope of the line using the points shown on the graph. Write a sentence that states the meaning of the slope.

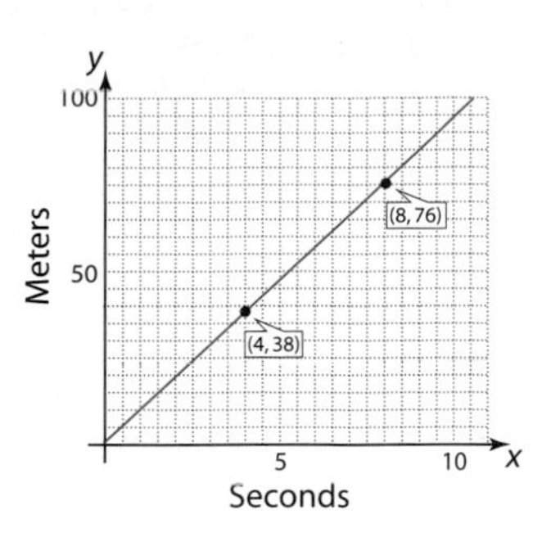

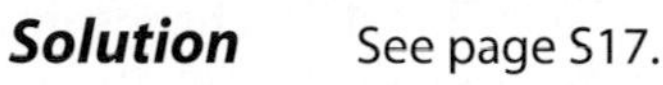

Solution See page S17.

Graph a Line Using the Slope and *y*-intercept

INSTRUCTOR NOTE
See the Instructor's Resource Manual for an alternative method of presenting this section using a discovery approach with a graphing calculator.

Recall that to find the *y*-intercept of the graph of an equation, let $x = 0$. Therefore, to find the *y*-intercept of $y = \frac{3}{4}x + 3$, let $x = 0$.

$$y = \frac{3}{4}x + 3$$

$$y = \frac{3}{4}(0) + 3 \qquad \text{• Let } x = 0.$$

$$y = 3$$

The *y*-intercept is (0, 3).

The constant term of $y = \frac{3}{4}x + 3$ is the *y*-coordinate of the *y*-intercept. In general, for any equation of the form $y = mx + b$, the *y*-intercept is $(0, b)$.

Two points on the graph of $y = \frac{3}{4}x + 3$ are shown at the right. The coordinates of the points are (4, 6) and (–8, –3).

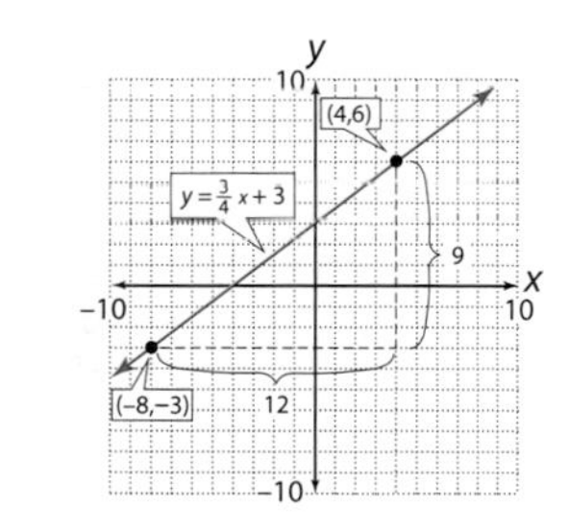

The slope of the line passing through the points is

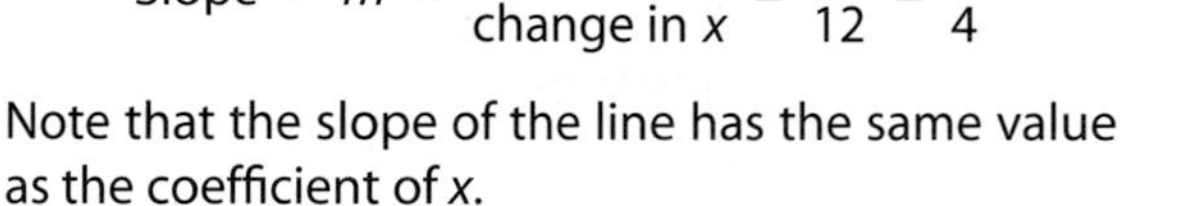

$$\text{Slope} = m = \frac{\text{change in } y}{\text{change in } x} = \frac{9}{12} = \frac{3}{4}$$

Note that the slope of the line has the same value as the coefficient of *x*.

> **Slope-Intercept Form of a Straight Line**
> For any equation of the form $y = mx + b$, the slope of the line is *m*, the coefficient of *x*. The *y*-intercept is $(0, b)$. The equation
> $$y = mx + b$$
> is called the **slope-intercept form of a straight line.**

The slope of the graph of $y = -\frac{2}{3}x + 4$ is $-\frac{2}{3}$, the coefficient of *x*. The *y*-intercept is (0, 4).

$$y = \boxed{m}\,x + \boxed{b}$$

$$y = \boxed{-\frac{2}{3}}\,x + \boxed{4}$$

Slope $= m = -\frac{2}{3}$ ⟶ ↑ ↑ ⟵ *y*-intercept $= (0, b) = (0, 4)$

When the equation of a straight line is in the form $y = mx + b$, the graph can be drawn using the slope and the y-intercept. First locate the y-intercept. Use the slope to find a second point on the line. Then draw a line through the two points.

→ Graph $y = 2x - 3$ by using the slope and y-intercept.

y-intercept $= (0, b) = (0, -3)$

$m = 2 = \frac{2}{1} = \frac{\text{change in } y}{\text{change in } x}$

To graph the line, begin at the y-intercept and move right 1 unit (the change in x) and up 2 units (the change in y). The new point has coordinates $(1, -1)$. Draw a line through $(0, -3)$ and $(1, -1)$.

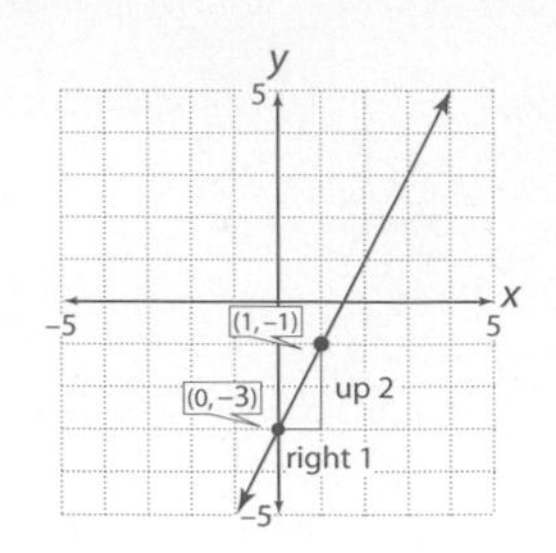

Example 3 Graph $y = -\frac{2}{3}x + 3$ by using the slope and y-intercept.

Solution y-intercept $= (0, b) = (0, 3)$.

$m = -\frac{2}{3} = \frac{-2}{3}$ • When the slope is negative, write the negative sign in the numerator.

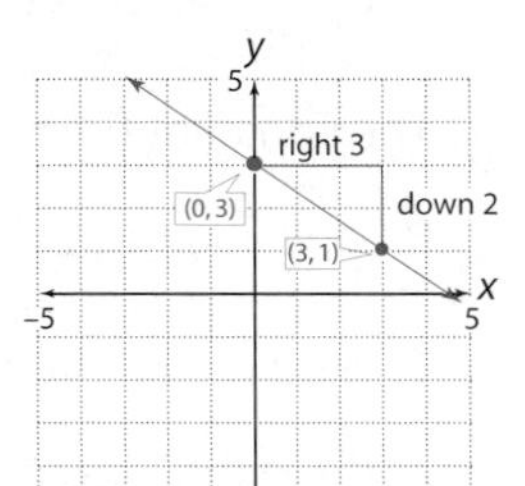

• The change in x is *positive* 3, so from the y-intercept we move 3 units to the *right*. The change in y is *negative* 2, so we move 2 units *down*.

TAKE NOTE
When the slope is negative, we will always write the negative sign in the numerator. Then the change in y is *negative* and we move *down*. The change in x is *positive* and we move *right*. However, it is not wrong to place the negative sign in the denominator. In that case, the change in y is *positive* and we move *up*. The change in x is *negative* and we move *left*.

You-Try-It 3 Graph $y = \frac{1}{4}x - 1$ by using the slope and y-intercept.

Solution See page S17.

When an equation is given in the form $Ax + By = C$, first solve for y.

Example 4 Graph $2x - 3y = 6$ by using the slope and y-intercept.

Solution

$$2x - 3y = 6$$
$$-3y = -2x + 6 \quad \text{• Solve the equation for } y.$$
$$y = \frac{2}{3}x - 2$$

y-intercept $= (0, b) = (0, -2)$.

$m = \frac{2}{3}$

The change in x is *positive*, so from the y-intercept we move 3 units to the *right*. The change in y is *positive*, so we move 2 units *up*.

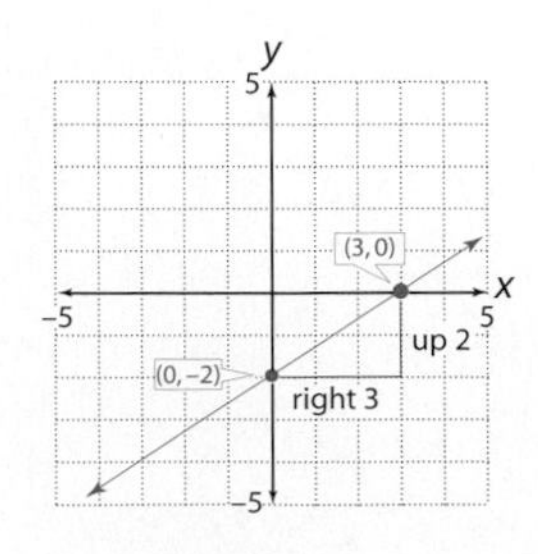

You-Try-It 4 Graph $x - 2y = 4$ by using the slope and y-intercept.

Solution See page S17.

37. $y = \frac{2}{3}x$

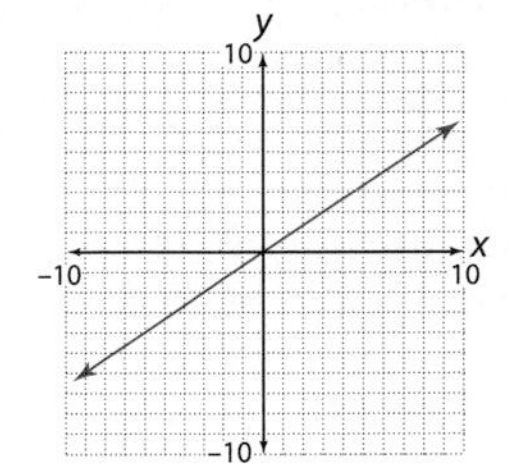

38. $y = \frac{1}{2}x$

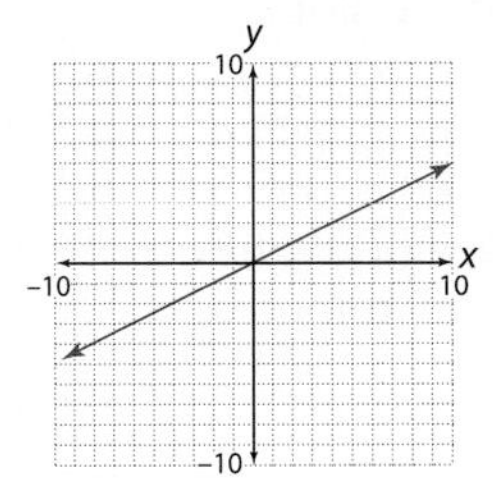

39. $y = -x + 1$

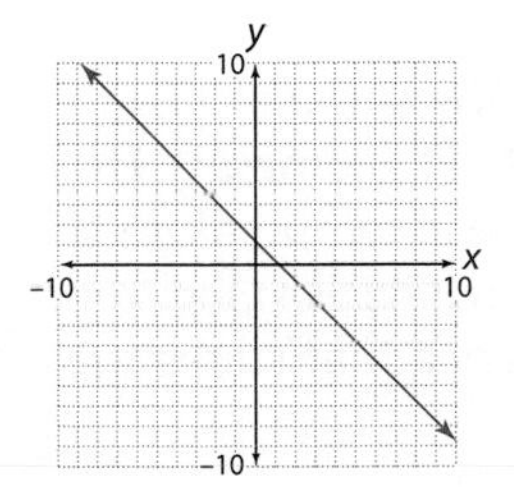

40. $y = -x - 2$

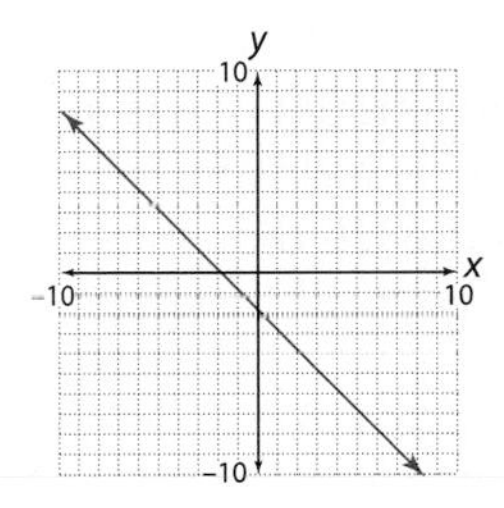

41. $3x - 4y = 12$

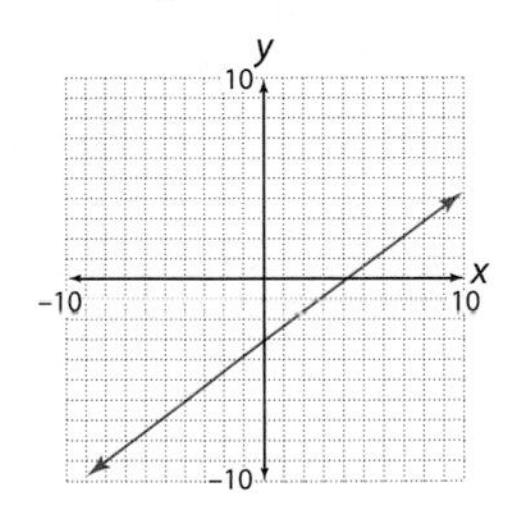

42. $5x - 2y = 10$

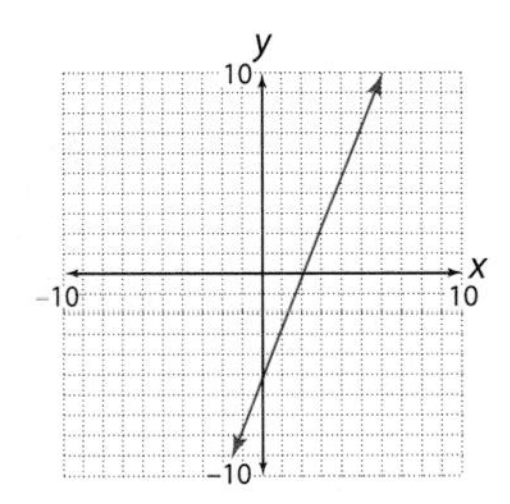

43. $y = -4x + 2$

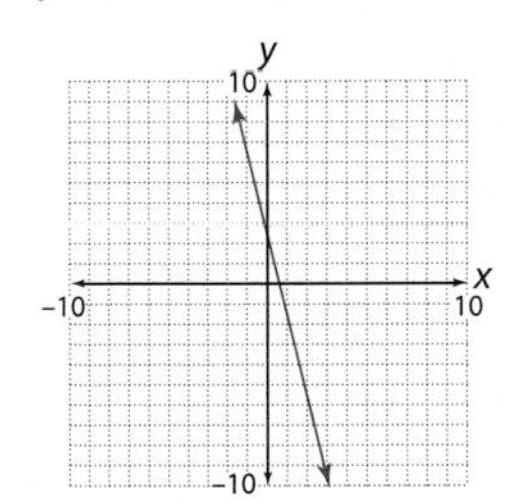

44. $y = 5x - 2$

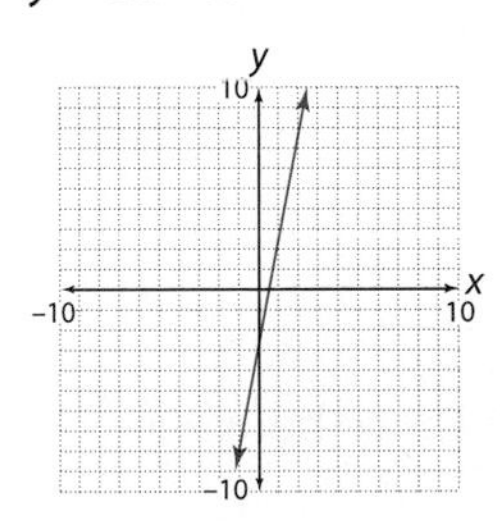

45. $4x - 5y = 20$

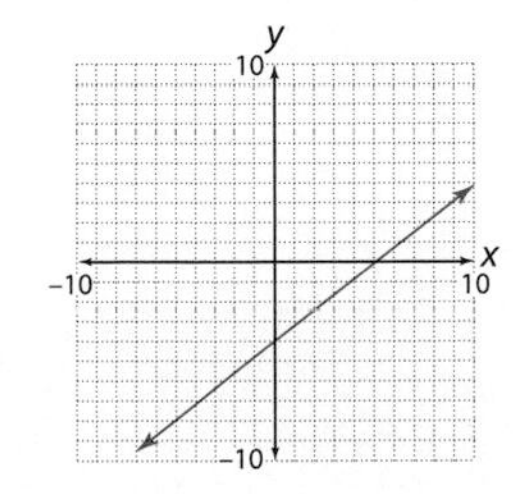

46. $x - 3y = 6$

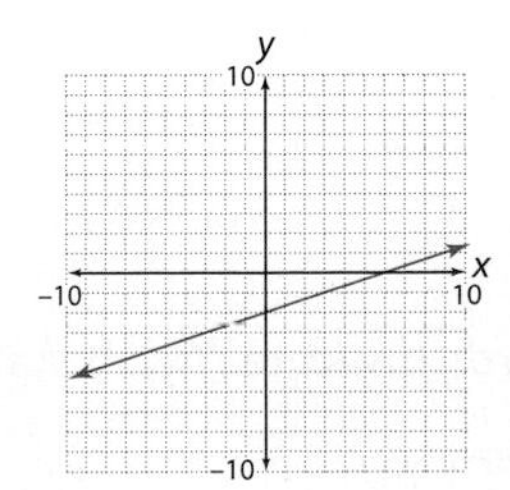

47. The graph below shows the total cost of a cellular phone call. Find the slope of the line. Write a sentence that states the meaning of the slope.

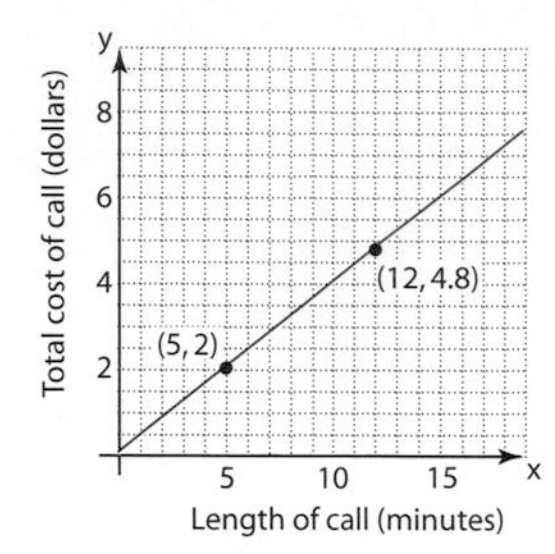

$m = 0.40$
The cellular call costs \$.40 per minute.

48. The graph below shows how the altitude of an airplane above the runway changes after takeoff. Find the slope of the line. Write a sentence that states the meaning of the slope.

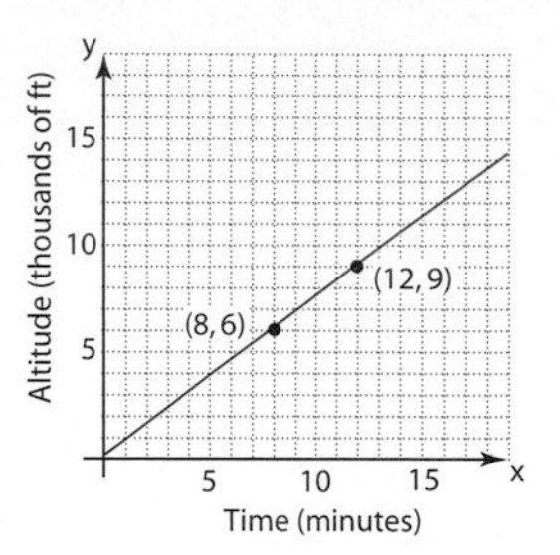

$m = 750$
The altitude of the plane is increasing at 750 feet per minute.

49. The graph below shows how the amount of gas in the tank of a car decreases as the car is driven. Find the slope of the line. Write a sentence that states the meaning of the slope.

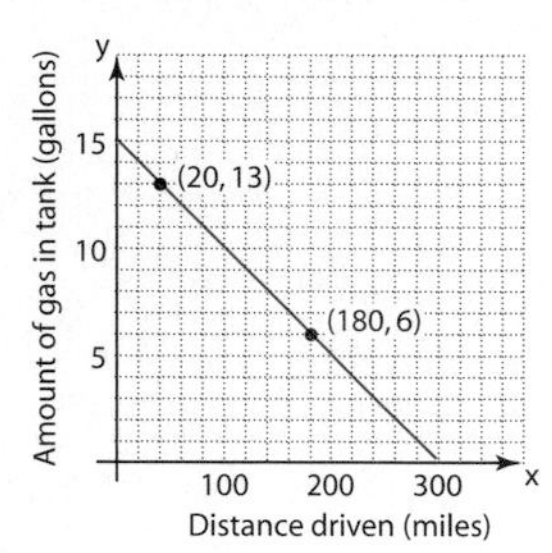

$m = -0.04375$
For each mile the car is driven, approximately 0.04 gallon of fuel is used.

50. The troposphere extends from the surface of Earth to an elevation of approximately 11 kilometers. The graph below shows the decrease in temperature of the troposphere as altitude increases. Find the slope of the line. Write a sentence that states the meaning of the slope.

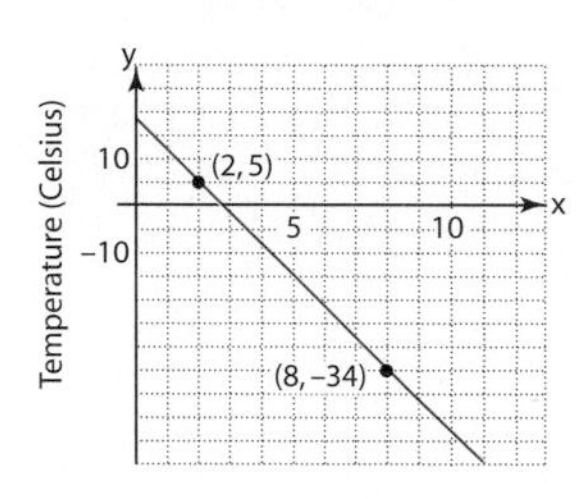

$m = -6.5$
The temperature of the troposphere decreases 6.5°C per kilometer.

51. Lois and Tanya start from the same place on a jogging course. Lois is jogging at 9 kilometers per hour and Tanya is jogging at 6 kilometers per hour. The graph below shows the total distance traveled by each jogger and the total distance between Lois and Tanya. Which lines represent which distances?

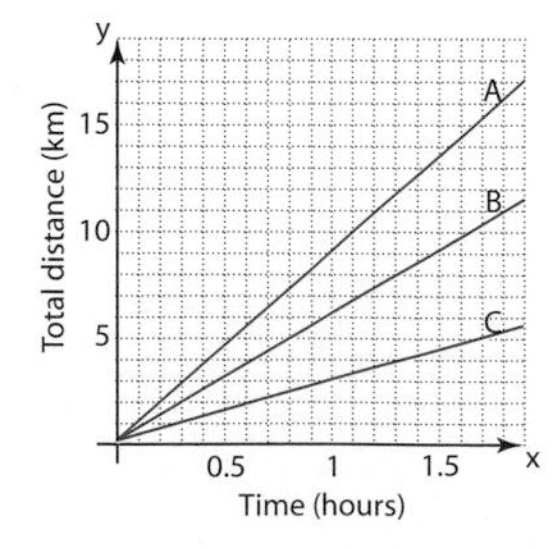

A: Lois
B: Tanya
C: Distance between

52. A chemist is filling two cans from a faucet that releases water at a constant rate. Can 1 has a diameter of 20 millimeters and can 2 has a diameter of 30 millimeters. The depth of the water in each can is measured at 5-second intervals. The graph of the results is shown below. On the graph, which line represents the depth of the water for which can?

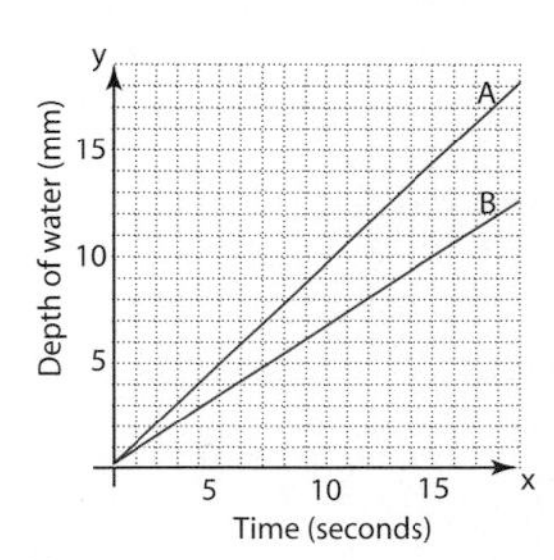

Can 1: A
Can 2: B

Applying Concepts

53. The American National Standards Institute (ANSI) states that the slope for a wheelchair ramp must not exceed $\frac{1}{12}$.

a. Does a ramp that is 6 inches high and 5 feet long meet the requirements of ANSI?

b. Does a ramp that is 12 inches high and 170 inches long meet the requirements of ANSI?

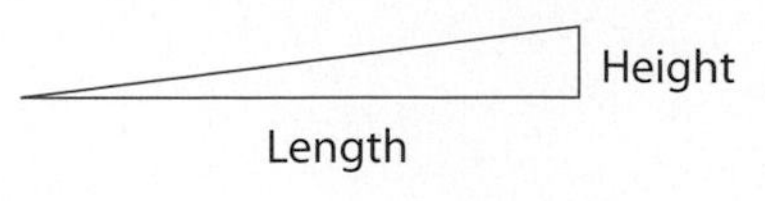

a. No; **b.** yes

54. A ramp for a wheelchair must be 14 inches high. What is the minimum length of this ramp so that it meets the ANSI requirements in the previous problem?
168 inches

55. What does the highway sign shown at the right have to do with slope?
Answers will vary.

56. If (2, 3) are the coordinates of a point on a line that has slope 2, what is the y-coordinate of the point on the line when $x = 4$?
7

57. If (−1, 2) are the coordinates of a point on a line that has slope −3, what is the y-coordinate of the point on the line when $x = 1$?
−4

58. If (1, 4) are the coordinates of a point on a line that has slope $\frac{2}{3}$, what is the y-coordinate of the point on the line when $x = -2$?
2

59. If (−2, −1) are the coordinates of a point on a line that has slope $\frac{3}{2}$, what is the y-coordinate of the point on the line when $x = -6$?
−7

60. If (−1, 3) are the coordinates of a point on a line that has slope $-\frac{3}{4}$, what is the y-coordinate of the point on the line when $x = -5$?
6

61. If (2, −1) are the coordinates of a point on a line that has slope $-\frac{2}{5}$, what is the y-coordinate of the point on the line when $x = -8$?
3

62. What effect does increasing the coefficient of x have on the graph of $y = mx + b$?
Increases the slope

63. What effect does decreasing the coefficient of x have on the graph of $y = mx + b$?
Decreases the slope

64. What effect does increasing the constant term have on the graph of $y = mx + b$?
Increases the y-intercept

65. What effect does decreasing the constant term have on the graph of $y = mx + b$?
Decreases the y-intercept

66. Do the graphs of all straight lines have a y-intercept? If not, give an example of one that does not.
No; for example, $x = 2$.

67. If two lines have the same slope and the same y-intercept, must the graphs of the lines be the same? If not, give an example.
Yes

Exploration

68. ***Designing a Staircase*** When you climb a staircase, the flat part of a stair that you step on is called the *tread* of the stair. The *riser* is the vertical part of the stair. The slope of a staircase is the ratio of the length of the riser to the length of the tread. Because the design of a staircase may affect safety, most cities have building codes that give rules for the design of a staircase.

Tread
Riser

a. The traditional design of a staircase called for a 9-inch tread and an 8.25-inch riser. What is the slope of this staircase?

b. A newer design for a staircase uses an 11-inch tread and a 7-inch riser. What is the slope of this staircase?

c. An architect is designing a house with a staircase that is 8 feet high and 12 feet long. Is the architect using the traditional design given in part a or the newer design given in part b? Explain your answer.

d. Staircases that have a slope between 0.5 and 0.7 are usually considered safer than those with a slope greater than 0.7. Design a safe staircase that goes from the first floor of a house to the basement, which is 9 feet below the first floor.

e. Measure the tread and riser for 3 staircases you encounter. Do these staircases match the traditional design in part a or the newer design in part b?

f. When an escalator step is fully extended, does it match the traditional design in part a or the newer design in part b?

Section 4.4 Equations of Straight Lines

Finding Equations of Lines Using $y = mx + b$

The graph at the right shows the height of a skydiver above the ground after deploying a parachute. From the graph, the skydiver was 1400 feet above the ground when the parachute was opened, and the skydiver landed ($h = 0$) 50 seconds later. By reading the graph, we can approximate the height of the skydiver for various times after the parachute is opened. For instance, the skydiver is approximately 1000 feet above ground after 15 seconds.

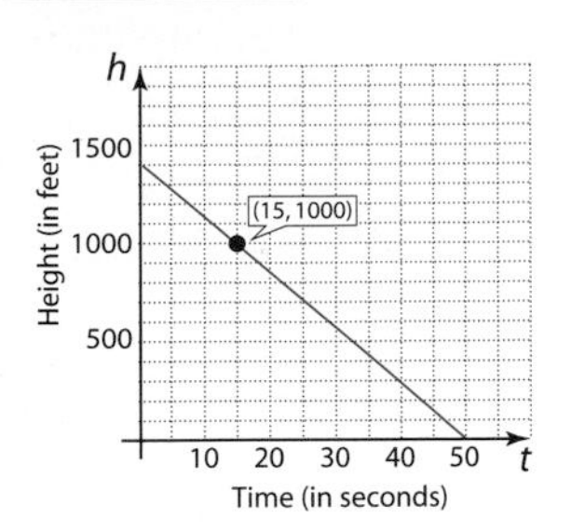

We can only approximate the height of the skydiver from the graph because the graph is not accurate enough to locate a point precisely. However, if we knew the equation of the line, then we could determine the height exactly for any time after the parachute was opened.

TAKE NOTE
The equation $h = -28t + 1400$ is a linear equation in which we are using the variables t and h instead of x and y. The slope of the graph of this line is -28, the coefficient of t. The h-intercept (instead of y-intercept) is $(0, 1400)$.

Using techniques that we will explain in this section, we can determine that the equation of the line is $h = -28t + 1400$, where h is the height of the skydiver t seconds after the parachute is opened. Substituting 15 for t allows us to determine the exact height of the skydiver 15 seconds after the parachute is opened.

$$h = -28t + 1400$$
$$= -28(15) + 1400 \qquad \text{• Replace } t \text{ by 15.}$$
$$= 980$$

The skydiver is 980 feet above the ground.

TAKE NOTE
The exact height 15 seconds after opening the parachute is 980 feet. This corresponds to the estimate of 1000 feet that we obtained from the graph.

There are several methods for finding the equation of a line. Two of these are discussed below.

When the slope of a line and a point on the line are known, the equation of the line can be written using the slope-intercept form, $y = mx + b$. In the first example below, the known point is the y-intercept. In the second example, the known point is a point other than the y-intercept.

⟶ Find the equation of the line that has slope -2 and y-intercept $(0, -1)$.

$y = mx + b$

$y = -2x + b$ • The given slope, -2, is m. Replace m by -2.

$y = -2x - 3$ • The given point, $(0, -3)$, is the y-intercept. Replace b by -3.

The equation of the line with slope -2 and y-intercept -3 is $y = -2x - 3$.

TAKE NOTE
You can and should check that the equation you find is the correct one. Compare the slope of your equation to the required slope and be sure that the given point is a solution of the equation. The slope of the graph of $y = 2x + 5$ is 2, the coefficient of x. This matches the information given in the problem.

Check that $(-2, 1)$ is a solution of the equation.

$y = 2x + 5$	
$1 \mid 2(-2) + 5$	• Replace y by 1 and x by -2.
$1 \mid -4 + 5$	
$1 = 1$	√

⟶ Find the equation of the line that has slope 2 and contains the point whose coordinates are $(-2, 1)$.

$y = mx + b$

$y = 2x + b$ • The given slope, 2, is m. Replace m by 2.

$1 = 2(-2) + b$ • The given point, $(-2, 1)$, is a solution of the equation. Replace x by -2 and y by 1.

$1 = -4 + b$ • Solve for b.

$5 = b$

$y = mx + b$ • Using the slope-intercept equation $y = mx + b$, replace m by 2 and replace b by 5.

$y = 2x + 5$

The equation of the line that has slope 2 and contains the point whose coordinates are $(-2, 1)$ is $y = 2x + 5$.

Example 1 Find the equation of the line that has slope $-\frac{3}{4}$ and that passes through the point whose coordinates are (–4, 1).

Solution

$y = mx + b$

$y = -\frac{3}{4}x + b$ • The given slope, $-\frac{3}{4}$, is m. Replace m by $-\frac{3}{4}$.

$1 = -\frac{3}{4}(-4) + b$ • The given point, (–4, 1), is a solution of the equation. Replace x by –4 and y by 1.

$1 = 3 + b$ • Solve for b.

$-2 = b$

$y = -\frac{3}{4}x - 2$ • Using the slope-intercept equation $y = mx + b$, write the equation of the line by replacing m by $-\frac{3}{4}$ and replacing b by –2.

You-Try-It 1 Find the equation of the line that has slope $\frac{2}{3}$ and that passes through the point whose coordinates are (3, –2).

Solution See page S17.

Finding Equations of Lines Using the Point-Slope Formula

A second method for finding the equation of a line involves use of the *point-slope formula.* This formula is derived from the definition of slope.

Let (x_1, y_1) be the coordinates of a known point on a line and let (x, y) be the coordinates of any other point on the line. See the graph at the left.

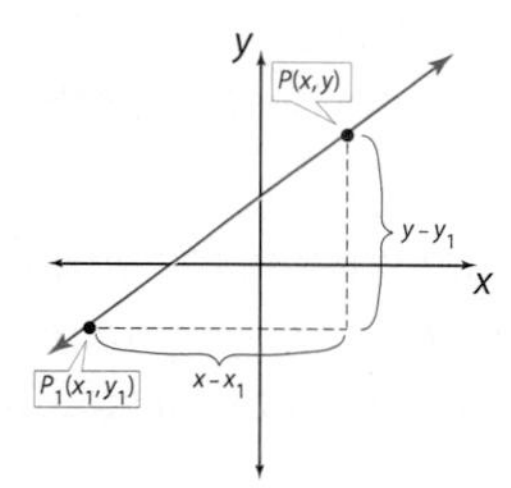

$\frac{y - y_1}{x - x_1} = m$ • Use the formula for slope.

$\frac{y - y_1}{x - x_1}(x - x_1) = m(x - x_1)$ • Multiply each side of the equation by $x - x_1$.

$y - y_1 = m(x - x_1)$ • Simplify.

Point-Slope Formula

The equation of the line that has slope m and passes through the point whose coordinates are (x_1, y_1) can be found by the point-slope formula:

$$y - y_1 = m(x - x_1)$$

TAKE NOTE

A model of the point-slope formula with open parentheses may help you substitute correctly.

$$y - y_1 = m(x - x_1)$$

$$y - (\quad) = (\quad)[x - (\quad)]$$

→ Use the point-slope formula to find the equation of the line that passes through the point whose coordinates are (–3, 1) and that has slope –2.

The given point has coordinates (–3, 1). Therefore, $x_1 = -3$ and $y_1 = 1$. The slope is $m = -2$.

$y - y_1 = m(x - x_1)$ • This is the point-slope formula.

$y - 1 = -2[x - (-3)]$ • Replace m by –2, x_1 by –3, and y_1 by 1.

$y - 1 = -2(x + 3)$ • Simplify until the equation is in the form $y = mx + b$.

$y - 1 = -2x - 6$

$y = -2x - 5$

The equation of the line that passes through the point whose coordinates are (–3, 1) and that has slope –2 is $y = -2x - 5$.

SUGGESTED ACTIVITY

Have students find the equation of the line passing through $P(2, -3)$ with slope 2 using the two methods discussed in this section and have them discuss which approach they prefer. Try this again when the known point is the y-intercept of the line. For instance, ask students to find the equation of the line through (0, 4) with slope $\frac{1}{2}$. Again have students discuss which method they prefer.

Example 2 Use the point-slope formula to find the equation of the line that passes through the point whose coordinates are (2, −3) and that has slope $\frac{3}{4}$.

Solution The given point has coordinates (2, −3). Therefore, $x_1 = 2$ and $y_1 = -3$. The slope is $m = \frac{3}{4}$.

$$y - y_1 = m(x - x_1)$$
• This is the point-slope formula.

$$y - (-3) = \frac{3}{4}(x - 2)$$
• Replace m by $\frac{3}{4}$, x_1 by 2, and y_1 by −3.

$$y + 3 = \frac{3}{4}(x - 2)$$
• Simplify until the equation is in the form $y = mx + b$.

$$y + 3 = \frac{3}{4}x - \frac{3}{2}$$

$$y = \frac{3}{4}x - \frac{9}{2}$$

You-Try-It 2 Use the point-slope formula to find the equation of the line that passes through the point whose coordinates are (1, 4) and that has slope −3.

Solution See page S17.

Example 3 In 1950, there were 13 million adults 65 years old or older in the United States Data from the Census Bureau shows that the population of these adults has been increasing at a constant rate of approximately 0.5 million per year. This rate of increase is expected to continue through the year 2010. Find the equation of the line that approximates the growth of the population of adults 65 years old or older in terms of the year. Use your equation to approximate the population of these adults in 2005.

Solution

State the goal. The goal is to find the equation of the line that approximates the growth of the population of adults 65 years old or older.

Describe a strategy. Let x represent years and y represent the population of these adults in millions. Then (1950, 13) are the coordinates of a given point on this line. The slope of the line is the rate of increase in the population, which is given as 0.5 million per year. Thus, $m = 0.5$. (It is helpful to remember that quantities that represent slope frequently contain the word *per.*) Use the point-slope formula to find the equation.

Solve the problem.

$$y - y_1 = m(x - x_1)$$

$$y - 13 = 0.5(x - 1950)$$
• Replace m by 0.5, x_1 by 1950, and y_1 by 13.

$$y - 13 = 0.5x - 975$$
• Simplify until the equation is in the form $y = mx + b$.

$$y = 0.5x - 962$$

The equation of the line is $y = 0.5x - 962$.

To find the population of these adults in 2005, replace x by 2005 and solve for y.

$$y = 0.5x - 962$$
$$y = 0.5(2005) - 962 = 40.5$$

There will be approximately 40.5 million of these adults in 2005.

Check your work. In 1950, there were 13 million adults 65 years or older. Therefore, the ordered pair (1950, 13) should be a solution of our equation.

$$y = 0.5x - 962$$
$$y = 0.5(1950) - 962 = 975 - 962 = 13 \ \checkmark$$

You-Try-It 3 A small plane flying 8000 feet above an airport begins a gradual descent of 500 feet per minute. Find an equation for the path of the descent. Use your equation to determine the height of the plane above the airport after 2.5 minutes. (Suggestion: Let x represent time and y represent height.)

Solution See pages S17–S18.

Example 4 As of 1997, there was approximately \$37 billion in the Social Security fund. However, projections by the Congressional Budget Office suggest that the amount in the fund will decrease at a constant rate of \$5.6 billion per year through the year 2015.

a. Write an equation for the amount of money in the Social Security fund in terms of the year.
b. Use your equation to predict the amount in the fund in 2010.
c. Write a sentence that explains your answer to part b.

Solution

State the goal. The goal is to find the equation of the line that approximates the decrease in the Social Security fund and then to use the equation to predict the amount in the fund in 2010.

Describe a strategy. **a.** Let x represent years and y represent the amount of money in the fund in billions of dollars. Then (1997, 37) are the coordinates of a given point on this line. The slope of the line is the rate of decrease in the amount of money in the fund, which is given as \$5.6 billion per year. Thus, $m = -5.6$. (The amount is decreasing so the slope is negative.) Use the point-slope formula to find the equation.

Solve the problem.

$$y - y_1 = m(x - x_1)$$

$$y - 37 = -5.6(x - 1997)$$ • Replace m by -5.6, x_1 by 1997, and y_1 by 37.

$$y - 37 = -5.6x + 11{,}183.2$$ • Simplify until the equation is in the form $y = mx + b$.

$$y = -5.6x + 11{,}220.2$$

The equation of the line is $y = -5.6x + 11{,}220.2$.

b. To find the amount of money in the fund in 2010, replace x by 2010 and solve for y.

$$y = -5.6x + 11{,}220.2$$

$$y = -5.6(2010) + 11{,}220.2$$

$$y = -35.8$$

The fund will have –\$35.8 billion in 2010.

c. Because the answer to part b is negative, the Social Security fund will have a deficit of \$35.8 billion in 2010.

Check your work. In 1997, the Social Security fund had \$37 billion. Therefore, the ordered pair (1997, 37) should be a solution of our equation.

$$y = -5.6x + 11{,}220.2$$

$$y = -5.6(1997) + 11{,}220.2 = -11{,}183.2 + 11{,}220.2 = 37 \quad \surd$$

You-Try-It 4 A diver on the surface of the water begins a dive, descending at a constant rate of 20 feet per minute. Find an equation for the depth of the diver below the surface of the water in terms of the time of the dive. (Suggestion: Let t represent the time of the dive and let d represent the depth below the surface. When $t = 0$, $d = 0$. Therefore (0, 0) is an ordered pair solution of the equation.)

Solution See page S18.

4.4 EXERCISES

Topics for Discussion

1. Discuss why the equation $y = mx + b$ is called the slope-intercept form of a straight line.
   The slope is m and the y-intercept is $(0, b)$.

2. Discuss why the equation $y - y_1 = m(x - x_1)$ is called the point-slope formula for a straight line.
   The slope is m and a given point is (x_1, y_1).

3. Why is having the equation of a line helpful?
   Answers will vary.

4. Explain how the point-slope formula is used to find the equation of a line.
   Answers will vary.

5. In this section, we wrote all equations of lines in the form $y = mx + b$. Could we have written the equations using functional notation as $f(x) = mx + b$?
   Yes. $f(x)$ and y can be used interchangeably for functions.

6. Do all equations of lines represent functions?
   No. For instance, $x = 5$ is the equation of a vertical line and does not represent a function.

Finding Equations of Lines

Find the equation of the line using the slope-intercept form.

7. Find the equation of the line that contains the point whose coordinates are $(0, 2)$ and that has slope 2. $y = 2x + 2$

8. Find the equation of the line that contains the point whose coordinates are $(0, -1)$ and that has slope -2. $y = -2x - 1$

9. Find the equation of the line that contains the point whose coordinates are $(-1, 2)$ and that has slope -3. $y = -3x - 1$

10. Find the equation of the line that contains the point whose coordinates are $(2, -3)$ and that has slope 3. $y = 3x - 9$

11. Find the equation of the line that contains the point whose coordinates are $(3, 1)$ and that has slope $\frac{1}{3}$. $y = \frac{1}{3}x$

12. Find the equation of the line that contains the point whose coordinates are $(-2, 3)$ and that has slope $\frac{1}{2}$. $y = \frac{1}{2}x + 4$

13. Find the equation of the line that contains the point whose coordinates are $(4, -2)$ and that has slope $\frac{3}{4}$. $y = \frac{3}{4}x - 5$

14. Find the equation of the line that contains the point whose coordinates are $(2, 3)$ and that has slope $-\frac{1}{2}$. $y = -\frac{1}{2}x + 4$

15. Find the equation of the line that contains the point whose coordinates are $(5, -3)$ and that has slope $-\frac{3}{5}$. $y = -\frac{3}{5}x$

16. Find the equation of the line that contains the point whose coordinates are $(5, -1)$ and that has slope $\frac{1}{5}$. $y = \frac{1}{5}x - 2$

17. Find the equation of the line that contains the point whose coordinates are $(2, 3)$ and that has slope $\frac{1}{4}$. $y = \frac{1}{4}x + \frac{5}{2}$

18. Find the equation of the line that contains the point whose coordinates are $(-1, 2)$ and that has slope $-\frac{1}{2}$. $y = -\frac{1}{2}x + \frac{3}{2}$

19. Find the equation of the line that contains the point whose coordinates are $(-3, -5)$ and that has slope $-\frac{2}{3}$. $y = -\frac{2}{3}x - 7$

20. Find the equation of the line that contains the point whose coordinates are $(-4, 0)$ and that has slope $\frac{5}{2}$. $y = \frac{5}{2}x + 10$

Find the equation of the line using the point-slope formula.

21. Find the equation of the line that passes through the point whose coordinates are $(1, -1)$ and that has slope 2. $y = 2x - 3$

22. Find the equation of the line that passes through the point whose coordinates are $(2, 3)$ and that has slope -1. $y = -x + 5$

23. Find the equation of the line that passes through the point whose coordinates are $(-2, 1)$ and that has slope -2. $y = -2x - 3$

24. Find the equation of the line that passes through the point whose coordinates are $(-1, -3)$ and that has slope -3. $y = -3x - 6$

25. Find the equation of the line that passes through the point whose coordinates are $(0, 0)$ and that has slope $\frac{2}{3}$. $y = \frac{2}{3}x$

26. Find the equation of the line that passes through the point whose coordinates are $(0, 0)$ and that has slope $-\frac{1}{5}$. $y = -\frac{1}{5}x$

27. Find the equation of the line that passes through the point whose coordinates are $(2, 3)$ and that has slope $\frac{1}{2}$. $y = \frac{1}{2}x + 2$

28. Find the equation of the line that passes through the point whose coordinates are (3, −1) and that has slope $\frac{2}{3}$. $y = \frac{2}{3}x - 3$

29. Find the equation of the line that passes through the point whose coordinates are (−4, 1) and that has slope $-\frac{3}{4}$. $y = -\frac{3}{4}x - 2$

30. Find the equation of the line that passes through the point whose coordinates are (−5, 0) and that has slope $-\frac{1}{5}$. $y = -\frac{1}{5}x - 1$

31. Find the equation of the line that passes through the point whose coordinates are (−2, 1) and that has slope $\frac{3}{4}$. $y = \frac{3}{4}x + \frac{5}{2}$

32. Find the equation of the line that passes through the point whose coordinates are (3, −2) and that has slope $\frac{1}{6}$. $y = \frac{1}{6}x - \frac{5}{2}$

33. Find the equation of the line that passes through the point whose coordinates are (−3, −5) and that has slope $-\frac{4}{3}$. $y = -\frac{4}{3}x - 9$

34. Find the equation of the line that passes through the point whose coordinates are (3, −1) and that has slope $\frac{3}{5}$. $y = \frac{3}{5}x - \frac{14}{5}$

35. The pilot of a Boeing 757 jet takes off from Boston's Logan Airport, which is at sea level, and climbs to a cruising altitude of 32,000 feet at a constant rate of 1200 feet per minute. Write a linear equation for the height of the plane in terms of the time after take-off. Use your equation to find the height of the plane 11 minutes after take-off. (Suggestion: Let t represent the time after take-off and let h represent the height above sea level. When $t = 0, h = 0$. Therefore $(0, 0)$ is an ordered pair solution of the equation.)
$h = 1200t$; 13,200 feet

36. A jogger running at 9 miles per hour burns approximately 14 calories per minute. Write a linear equation for the number of calories burned by a jogger in terms of the number of minutes run. Use your equation to find the number of calories burned after 32 minutes. (Suggestion: Let t represent the time running and let C represent the number of calories burned. When $t = 0$, $C = 0$. Therefore $(0, 0)$ is an ordered pair solution of the equation.)
$C = 14t$; 448 calories

37. A cellular phone company offers several different options for using a cellular telephone. One option, for people who plan on using the phone only in emergencies, costs the user $4.95 per month plus $.59 per minute for each minute the phone is used. Write a linear equation for the monthly cost of the phone in terms of the number of minutes the phone is used. Use your equation to find the cost of using the cellular phone for 13 minutes in one month.
$C = 0.59t + 4.95$; $12.62

38. An Airbus 320 plane takes off from Stapleton Airport in Denver, Colorado, which is 5200 feet above sea level, and climbs to 30,000 feet at a constant rate of 1000 feet per minute. Write a linear equation for the height of the plane in terms of the time after take-off. Use your equation to find the height of the plane 8 minutes after take-off. (Suggestion: Let t represent the time after take-off and let h represent the height above sea level. When $t = 0$, $h = 5200$. Therefore $(0, 5200)$ is an ordered pair solution of the equation.)
$h = 1000t + 5200$; 13,200 feet

39. Based on data from the U. S. Department of Commerce, there were 24 million homes with computers in 1991. The average number of computers in homes is expected to increase by 2.4 million per year through 2005. Write a linear equation for the number of computers in homes in terms of the year. Use your equation to find the number of computers that will be in homes in 2001. (Suggestion: Let x represent the year and let y represent the number of homes, in millions, with computers. When $x = 1991$, $y = 24$ million. Therefore $(1991, 24)$ is an ordered pair solution of the equation.)
$y = 2.4x - 4754.4$; 48 million

40. The gas tank of a certain car contains 16 gallons when the driver of the car begins a trip. Each mile driven by the driver decreases the amount of gas in the tank by 0.032 gallon. Write a linear equation for the number of gallons of gas in the tank in terms of the number of miles driven. Use your equation to find the number of gallons in the tank after driving 150 miles. (Suggestion: Let x represent the number of miles driven and let y represent the number of gallons of gas in the tank. When $x = 0$, $y = 16$. Therefore $(0, 16)$ is an ordered pair solution of the equation.)
$y = -0.032x + 16$; 11.2 gallons

Applying Concepts

Is there a linear equation that contains all the given ordered pairs? If there is, find the equation.

41. (5, 1), (4, 2), (0, 6)

Yes; $y = -x + 6$

42. (−2, −4), (0, −3), (4, −1)

Yes; $y = \frac{1}{2}x - 3$

43. (−1, −5), (2, 4), (0, 2)

No

44. (3, −1), (12, −4), (−6, 2)

Yes; $y = -\frac{1}{3}x$

The given ordered pairs are solutions to the same linear equation. Find n.

45. (0, 1), (4, 9), (3, n)
7

46. (2, 2), (−1, 5), (3, n)
1

47. (2, −2), (−2, −4), (4, n)
−1

48. (1, −2), (−2, 4), (4, n)
−8

49. Using the techniques presented in this section, find the equation of the height of the person parachuting given at the beginning of this section.
$h = -28t + 1400$

Exploration

50. ***Equation of a Line Through Two Points*** The equation of a line between two points can be found by using the point-slope formula. First find the slope of the line between the two points. With this value of the slope and either one of the two given points, use the point-slope formula to find the equation of the line.

a. Find the equation of the line through $P_1(-2, -3)$ and $P_2(3, 7)$.

b. Find the equation of the line through $P_1(-1, 6)$ and $P_2(2, 3)$.

c. Verify that it does not matter which point is chosen as the given point in the point-slope formula when two points are known by again finding the equation of the line in part a, this time using the point you did not use the first time.

d. The graph at the right shows the relationship between the Fahrenheit and Celsius temperature scales. Use the graph to find a linear function that gives the Celsius temperature in terms of the Fahrenheit temperature.

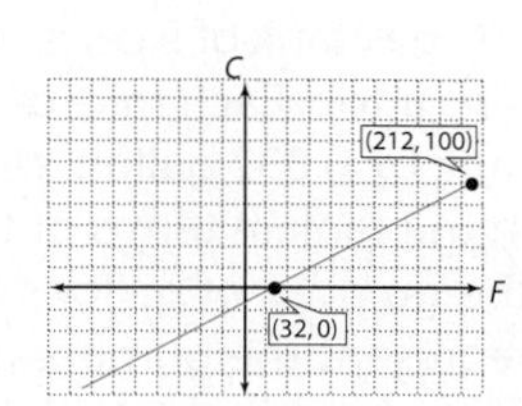

e. The price of a new minivan in 1997 was \$18,500. According to a consumer group, the value of the van will be approximately \$7200 in 6 years. Write a linear function that gives the value of the minivan in terms of its age.

Section 4.5 Solving Systems of Equations by Graphing or Substitution

Solving a System of Equations by Graphing

Suppose there is \$5 in a coin bank that contains only dimes and quarters. How many dimes and quarters are in the bank? With the given information, there are several possible answers to our question. We will list them in the table below.

Dimes, d	0	5	10	15	20	25	30	35	40	45	50
Quarters, q	20	18	16	14	12	10	8	6	4	2	0

Each of these combinations of dimes and quarters can be graphed on a coordinate grid where the number of dimes is represented on the horizontal axis and the number of quarters is represented on the vertical axis. The ordered pairs are solutions of the equation $0.10d + 0.25q = 5$. For instance, to show that the ordered pair (30, 8) is a solution of the equation, replace d by 30 and q by 8.

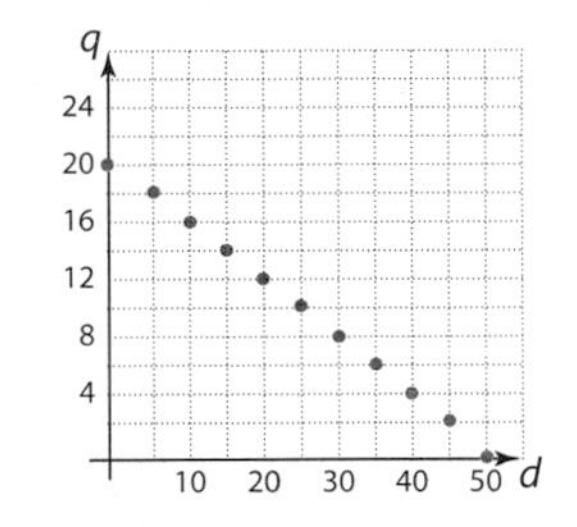

$$0.10d + 0.25q = 5$$
$$0.10(30) + 0.25(8) = 3 + 2 = 5$$

Now suppose that you are given the additional information that the total number of coins in the bank is 35. This is represented by the equation $d + q = 35$. Some of the possible ordered pairs are given in the table below.

Dimes, d	10	15	20	25	30	35
Quarters, q	25	20	15	10	5	0

The ordered pairs of this table are graphed at the right using an **×**. The ordered pairs of the original table are also shown. Note that the ordered pair (25,10) belongs to each graph. This ordered pair occurs in each table and is a solution of both equations. The only combination of coins that satisfies both the condition that there be \$5 in the bank and the condition that the bank contains 35 coins is (25, 10).

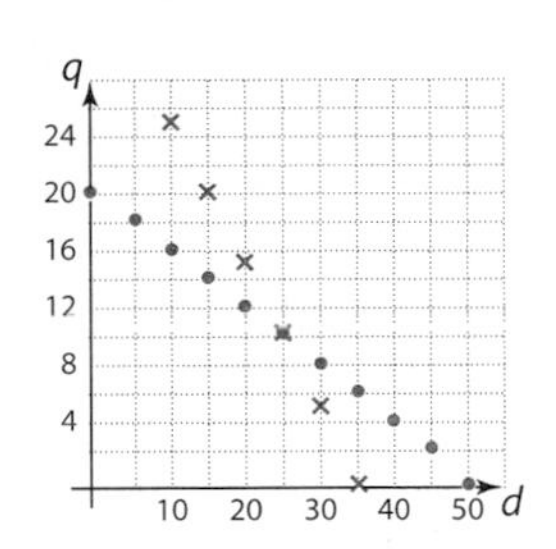

Equations considered together are called a **system of equations.** The two equations for the coin bank problem are an example of a system of equations. The equations are shown at the right.

$$\begin{cases} 0.10d + 0.25q = 5 \\ d + q = 35 \end{cases}$$

TAKE NOTE
The main point of the discussion of a solution of a system of equations is that there are *many* combinations of dimes and quarters that equal \$5 and *many* combinations of dimes and quarters that total 35 coins. However, there is *only one* combination of coins that satisfies both conditions at the same time. That one combination, (25, 10), is the solution of the system of equations.

A **solution of a system of equations in two variables** is an ordered pair that is a solution of each equation. Because, as shown below, the ordered pair (25, 10) is a solution of each equation, it is a solution of the system of equations.

$$\begin{array}{r|l} 0.10d + 0.25q = & 5 \\ 0.10(25) + 0.25(10) & 5 \\ 2.50 + 2.50 & 5 \\ 5 = & 5 \end{array} \qquad \begin{array}{r|l} d + q = & 35 \\ 25 + 10 & 35 \\ 35 = & 35 \end{array}$$

Now consider the ordered pair (10, 16). Substituting this ordered pair into the system of equations, we have

$$\begin{array}{r|l} 0.10d + 0.25q = & 5 \\ 0.10(10) + 0.25(16) & 5 \\ 1 + 4 & 5 \\ 5 = & 5 \end{array} \qquad \begin{array}{r|l} d + q = & 35 \\ 10 + 16 & 35 \\ 26 \neq & 35 \end{array}$$

Because (10, 16) is not a solution of each equation, it is not a solution of the system of equations.

→ Is $(1, -3)$ a solution of the system of equations $\begin{aligned} 3x + 2y &= -3 \\ x - 3y &= 6 \end{aligned}$?

$$\begin{array}{r|l} 3x + 2y = & -3 \\ \hline 3(1) + 2(-3) & -3 \\ 3 + (-6) & -3 \\ -3 = & -3 \end{array} \qquad \begin{array}{r|l} x - 3y = & 6 \\ \hline 1 - 3(-3) & 6 \\ 1 + 9 & 6 \\ 10 \neq & 6 \end{array}$$

Because $(1, -3)$ is not a solution of both equations, it is not a solution of the system of equations.

For linear systems of equations, each equation in the system of equations is a linear equation. Therefore, the graphs of the equations will be straight lines. For a linear system of equations containing two equations, there are three possible conditions. These are shown below.

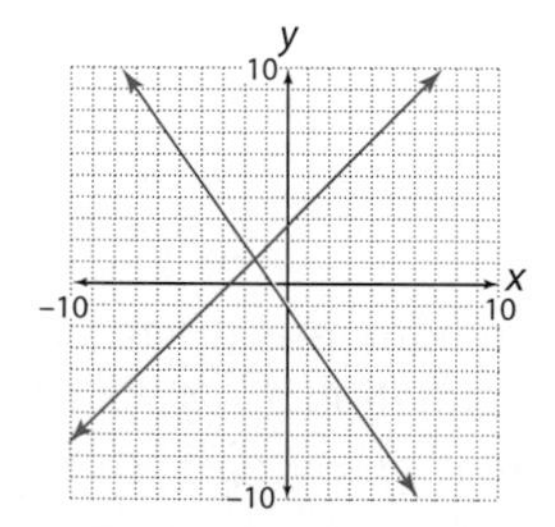

The lines can intersect at one point. The point of intersection of the lines is the ordered pair that is a solution of each equation of the system. It is the solution of the system of equations. The system of equations is **independent.**

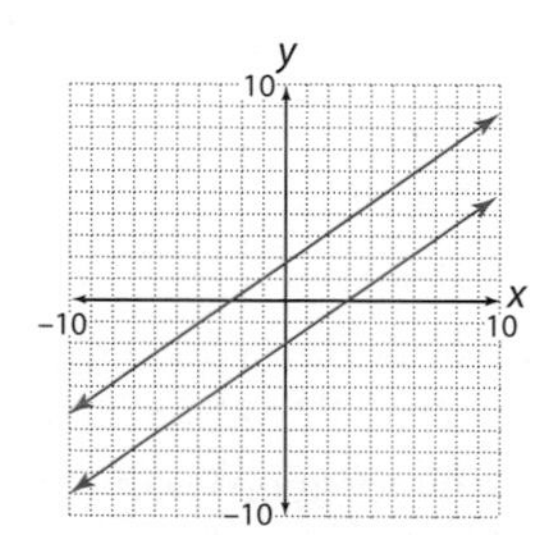

The lines can be parallel and therefore do not intersect. Since they do not intersect, there is no point on both lines and the system of equations has no solution. This is an **inconsistent** system of equations.

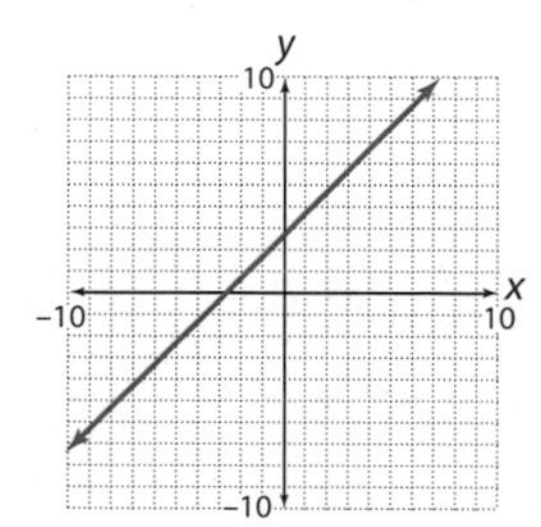

Each equation has the same graph. The lines graph one on top of the other. Therefore, they intersect at infinitely many points. The solutions are the ordered pairs of one of the equations. This system of equations is called **dependent.**

To **solve a system of equations** means to find the ordered pair solutions of the system of equations. There are several methods to do this. The first method we will discuss is a graphical approach.

Example 1 Solve by graphing: $\begin{aligned} 2x + 3y &= 6 \\ 2x + y &= -2 \end{aligned}$

Solution Solve each equation for y.

$$\begin{aligned} 2x + 3y &= 6 \\ 3y &= -2x + 6 \\ y &= -\frac{2}{3}x + 2 \end{aligned} \qquad \begin{aligned} 2x + y &= -2 \\ y &= -2x - 2 \end{aligned}$$

Using a graphing calculator, enter Y1 = −2X/3 + 2 and Y2 = −2X − 2 and then graph the two equations. Then use the INTERSECT feature of the calculator to find the point of intersection of the two lines.

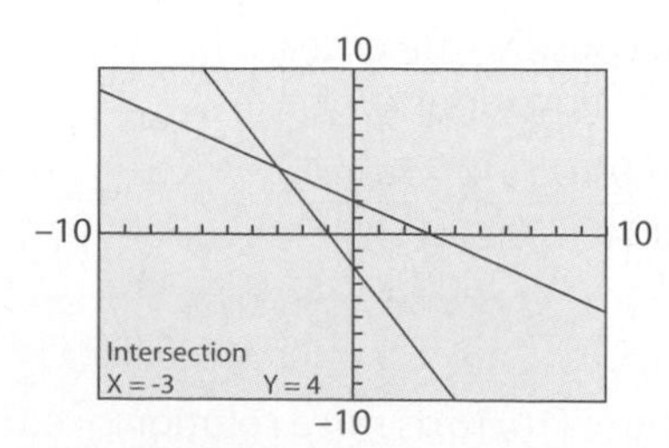

Algebraic check

$$\begin{array}{r|l} 2x + 3y = & 6 \\ \hline 2(-3) + 3(4) & 6 \\ -6 + 12 & 6 \\ 6 = & 6 \end{array} \qquad \begin{array}{r|l} 2x + y = & -2 \\ \hline 2(-3) + 4 & -2 \\ -6 + 4 & -2 \\ -2 = & -2 \end{array}$$

The solution is $(-3, 4)$.

TAKE NOTE
Graphs can be very deceiving. Consider the system of equations $\begin{aligned} y &= x+2 \\ y &= 0.99x \end{aligned}$. Graph these equations with a graphing calculator and the standard viewing window. The graphs will appear parallel but, because the slopes are not equal, they are not parallel and thus intersect. Using the INTERSECT feature of your calculator produces an error message when you attempt to find the point of intersection. However, you can algebraically verify that $(-200, -198)$ is a solution of the system of equations.

You-Try-It 1 Solve by graphing: $\begin{aligned} 3x + 4y &= 2 \\ y &= 2x - 5 \end{aligned}$

Solution See page S18.

When solving a system of equations by graphing, solving each equation for y is helpful in determining whether the system of equations has one solution, no solution, or an infinite number of solutions. For instance, in Example 1, when each equation was solved for y, we could see that the slopes of the graphs of the lines were different (one had slope $-\frac{2}{3}$ and the other had slope -2). Because the slopes were different, the lines were not parallel and therefore had to intersect. The fact that the slopes were not equal is important, as shown in the following.

⟶ Solve the system of equations by graphing: $\begin{aligned} 2x - y &= -2 \\ 6x - 3y &= -15 \end{aligned}$

Solve each equation for y.

$$\begin{aligned} 2x - y &= -2 \\ -y &= -2x - 2 \\ y &= 2x + 2 \end{aligned} \qquad \begin{aligned} 6x - 3y &= -15 \\ -3y &= -6x - 15 \\ y &= 2x + 5 \end{aligned}$$

The graphs are shown at the right. Because the slopes of the lines are equal and the y-intercepts are different, the lines are parallel and never intersect. Therefore, the system of equations has no solution.

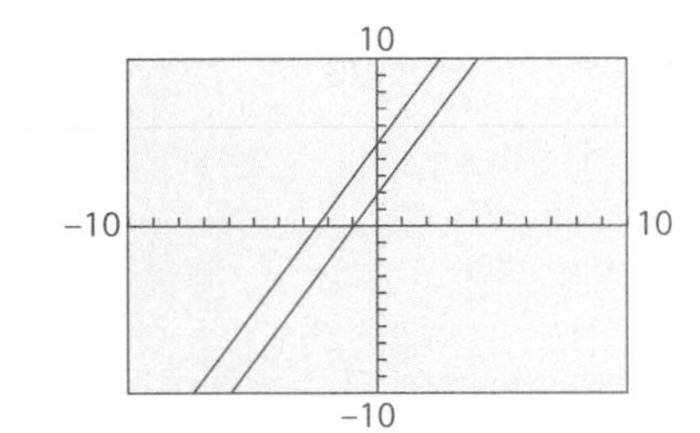

For the previous example, the slopes of the lines were equal but the y-intercepts were different. Therefore, the lines were parallel and the system of equations had no solution. If the slopes of the lines are equal and the y-intercepts are equal, the graphs are the same line. In this case, the solutions are the ordered pairs of one of the equations.

SUGGESTED ACTIVITY
Have students verify that any solution of one equation is a solution of the other equation. After they have done this for selected ordered pairs, have them verify that for each one of their solutions, the relationship between the x- and y-coordinates is $(x, 2x - 3)$. This will help some students understand why we write ordered pair solutions of dependent systems in this manner.

Example 2 Solve by graphing: $\begin{aligned} 4x - 2y &= 6 \\ y &= 2x - 3 \end{aligned}$

Solution Solve $4x - 2y = 6$ for y.

$$\begin{aligned} 4x - 2y &= 6 \\ -2y &= -4x + 6 \\ y &= 2x - 3 \end{aligned}$$

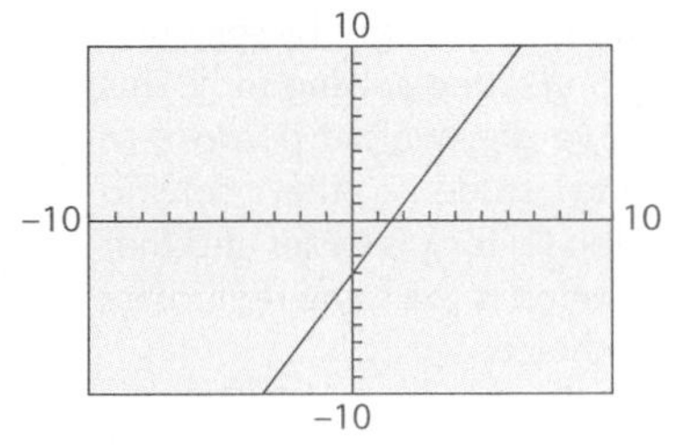

When solved for y, this equation is exactly the same as the second equation of the system of equations. Therefore, the graphs of the two lines are exactly the same and the graphs intersect at an infinite number of points. Any solution of one equation is a solution of the other equation. The solutions are the ordered pairs of $y = 2x - 3$. Because a solution of a system of equations is written as an ordered pair (x, y), we write the solutions of this system of equations as $(x, 2x - 3)$, where we have replaced y by $2x - 3$. The graphs are shown above at the right. Note that because one graph is on top of the other, it appears that there is only one graph.

You-Try-It 2 Solve by graphing:
$$\begin{aligned} 2x + 3y &= 6 \\ 4x + 6y &= 12 \end{aligned}$$

Solution See page S18.

Solving Systems of Equations by Substitution

Besides graphical methods of solving a system of equations, there are algebraic methods that can be used. One algebraic method is the **substitution method**. Because it is necessary to refer to an equation in the system of equations, we will number the equations, as shown in the example below.

In the system of equations at the right, Equation (2) states that $y = \mathbf{3x - 9}$.

$$\begin{aligned} 2x + 5y &= -11 \quad (1) \\ y &= 3x - 9 \quad (2) \end{aligned}$$

Substitute $\mathbf{3x - 9}$ for y in Equation (1).
Solve for x.

$$\begin{aligned} 2x + 5(3x - 9) &= -11 \\ 2x + 15x - 45 &= -11 \\ 17x - 45 &= -11 \\ 17x &= 34 \\ x &= 2 \end{aligned}$$

Substitute the value of x into Equation (2) and solve for y.

$$\begin{aligned} y &= 3x - 9 \\ y &= 3(2) - 9 \\ y &= 6 - 9 \\ y &= -3 \end{aligned}$$

The solution is $(2, -3)$.

The graph of the equations in this system is shown at the right. Note that the lines intersect at the point whose coordinates are $(2, -3)$, which is the algebraic solution we obtained by using the substitution method.

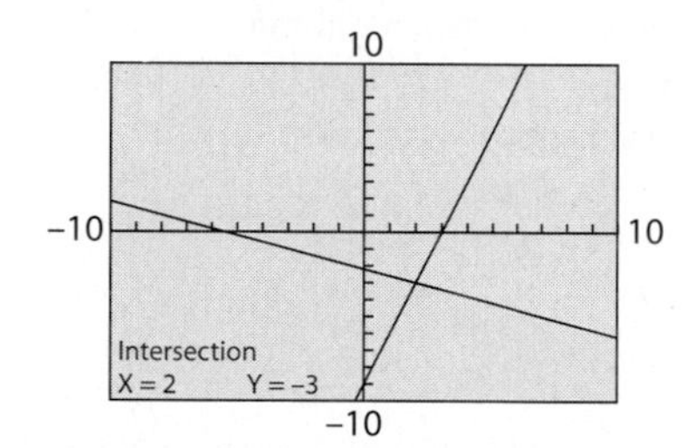

In the example above, Equation (2) was solved for y. To use the substitution method, one of the equations of the system must be solved in terms of the other variable. It does not matter which variable is used.

SUGGESTED ACTIVITY
Have students solve the system of equations at the right by selecting Equation (2) and solving for y. This will serve the twofold purpose of showing students that solving Equation (1) for x is easier and that the solution is the same regardless of the choice.

Example 3 Solve by the substitution method:
$$\begin{aligned} x - 4y &= 9 \quad (1) \\ 2x - 3y &= 11 \quad (2) \end{aligned}$$

Solution Solve Equation (1) for x. (We could choose either equation of the system of equations and solve for either x or y. Solving Equation (1) for x is the easiest choice.)

$$\begin{aligned} x - 4y &= 9 \\ x &= 4y + 9 \quad (3) \end{aligned}$$

$$\begin{aligned} 2x - 3y &= 11 && \text{• This is Equation (2).} \\ 2(4y + 9) - 3y &= 11 && \text{• Replace } x \text{ by } 4y + 9. \\ 8y + 18 - 3y &= 11 && \text{• Solve for } y. \\ 5y + 18 &= 11 \\ 5y &= -7 \\ y &= -\frac{7}{5} \end{aligned}$$

$$\begin{aligned} x &= 4y + 9 && \text{• This is Equation (1).} \\ x &= 4\left(-\frac{7}{5}\right) + 9 && \text{• Replace } y \text{ by } -\tfrac{7}{5}. \\ x &= -\frac{28}{5} + 9 && \text{• Multiply by 4.} \\ x &= -\frac{28}{5} + \frac{45}{5} && \text{• The common denominator is 5. Write 9 as } \tfrac{45}{5}. \\ x &= \frac{17}{5} \end{aligned}$$

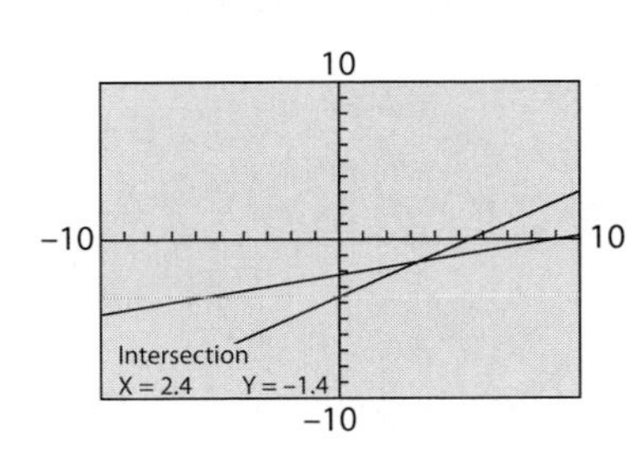

The solution is $\left(\frac{17}{5}, -\frac{7}{5}\right)$. A graphical check using a graphing calculator is shown at the left. Note that the answers are given as decimals.

You-Try-It 3 Solve by the substitution method:

$$\begin{aligned} 3x + 4y &= 18 && (1) \\ 2x - y &= 1 && (2) \end{aligned}$$

Solution See page S18.

Example 4 Solve by the substitution method:

$$\begin{aligned} 6x - 3y &= 3 && (1) \\ y &= 2x + 3 && (2) \end{aligned}$$

Solution

$$\begin{aligned} 6x - 3y &= 3 && \text{• This is Equation (1).} \\ 6x - 3(2x + 3) &= 3 && \text{• Replace } y \text{ by } 2x + 3. \\ 6x - 6x - 9 &= 3 \\ -9 &= 3 \end{aligned}$$

$-9 = 3$ is not a true equation. The system of equations is inconsistent and does not have a solution.

TAKE NOTE
If we solve Equation (1) at the right for y, we obtain

$$\begin{aligned} 6x - 3y &= 3 \\ -3y &= -6x + 3 \\ y &= 2x - 1 \end{aligned}$$

From this we can observe that the slopes are the same but the y-intercepts are different. The lines are parallel and do not intersect.

You-Try-It 4 Solve by the substitution method:

$$\begin{aligned} 4x + 2y &= 6 && (1) \\ 2x + y &= 1 && (2) \end{aligned}$$

Solution See page S18.

Example 5 Solve by the substitution method:

$$\begin{aligned} 4x + 2y &= 6 && (1) \\ y &= -2x + 3 && (2) \end{aligned}$$

Solution

$$\begin{aligned} 4x + 2y &= 6 && \text{• This is Equation (1).} \\ 4x + 2(-2x + 3) &= 6 && \text{• Replace } y \text{ by } -2x + 3. \\ 4x - 4x + 6 &= 6 \\ 6 &= 6 \end{aligned}$$

$6 = 6$ is a true equation. The system of equations is dependent. The solutions are the ordered pair solutions (x, y) that satisfy $y = -2x + 3$. The solutions are written as $(x, -2x + 3)$, where y has been replaced by $-2x + 3$.

TAKE NOTE
If we solve Equation (1) at the right for y, we obtain

$$\begin{aligned} 4x + 2y &= 6 \\ 2y &= -4x + 6 \\ y &= -2x + 3 \end{aligned}$$

From this we can observe that the slopes are the same and the y-intercepts are the same. The graphs are the same lines.

You-Try-It 5 Solve by the substitution method:

$$\begin{aligned} 4x + 2y &= 6 && (1) \\ 2x + y &= 3 && (2) \end{aligned}$$

Solution See page S18.

Applications

The annual simple interest that an investment earns is given by the equation $I = Pr$, where I is the simple interest, P is the principal (the amount invested), and r is the simple interest rate.

→ The annual simple interest rate on a $1250 investment is 5%. Find the annual simple interest earned on the investment.

$I = Pr$ • Use the annual simple interest equation.

$I = 1250(0.05)$ • $P = 1250, r = 5\% = 0.05$.

$I = 62.5$

The annual simple interest is $62.50.

Some annual simple interest problems are solved by using a system of equations.

Example 6 An investor has $10,000 to deposit into two simple interest accounts. On one account, the annual simple interest rate is 7%. On the second account, the annual simple interest rate is 8%. How much should be invested in each account so that the total annual interest earned is $785?

Solution

State the goal. The goal is to find the amount invested in each account.

Describe a strategy. There are two accounts. Let x represent the amount invested in the 7% account and y represent the amount invested in the 8% account. The total of the two investments is $10,000. This information gives one equation of the system of equations.

$x + y = 10{,}000$ • Total amount invested is the sum of the amounts invested in each account

The interest earned from each account can be found from the equation $I = Pr$.

The interest earned on the 7% account is the product of the principal, x, and the interest rate as a decimal, 0.07. The interest earned on the 8% account is the product of the principal, y, and the interest rate as a decimal, 0.08.

Interest from 7% account: $0.07x$
Interest from 8% account: $0.08y$

The total annual interest earned must be $785. This gives the second equation of the system of equations.

$0.07x + 0.08y = 785$ • Total amount of interest earned is the sum of the amounts earned from each account.

The two equations together form the system of equations.

$$x + y = 10{,}000 \quad (1)$$
$$0.07x + 0.08y = 785 \quad (2)$$

Solve the problem. Solve the system of equations by substitution.

$x + y = 10{,}000$ • This is Equation (1).

$y = 10{,}000 - x$ • Solve Equation (1) for y.

$0.07x + 0.08y = 785$ • This is Equation (2).

$0.07x + 0.08(10{,}000 - x) = 785$ • Replace y by $10{,}000 - x$.

$$0.07x + 800 - 0.08x = 785$$
$$-0.01x + 800 = 785$$
$$-0.01x = -15$$
$$x = 1500$$

To find y, replace x by 1500 in Equation (1) and solve for y.

$$x + y = 10{,}000$$
$$1500 + y = 10{,}000$$
$$y = 8500$$

There was $1500 invested at 7% and $8500 invested at 8%.

Check your work. You can check your work by calculating the interest earned from each investment and ensuring that the total is \$785.

Interest earned at 7%: $0.07(1500) = 105$
Interest earned at 8%: $0.08(8500) = 680$

Total interest earned
$105 + 680 = 785$ √

You-Try-It 6 A grocer deposited an amount of money into a high-yield mutual fund that earns 13% annual simple interest. A second deposit, \$2500 more than the first, was placed in a certificate of deposit earning 7% annual simple interest. In one year, the total interest earned on both investments was \$475. How much money was invested in the mutual fund?

Solution See pages S18–S19.

Example 7 The perimeter of a rectangular wall hanging is 150 inches. The length of the wall hanging is 15 inches more than the width. Find the dimensions of the wall hanging.

Solution

State the goal. The goal is to find the length and width of the wall hanging.

Describe a strategy. There are two unknowns, the length and width of the wall hanging. Let L represent the length and W represent the width. A possible diagram of the wall hanging is shown at the right. The perimeter is 150 inches. This information gives one of the equations of the system of equations.

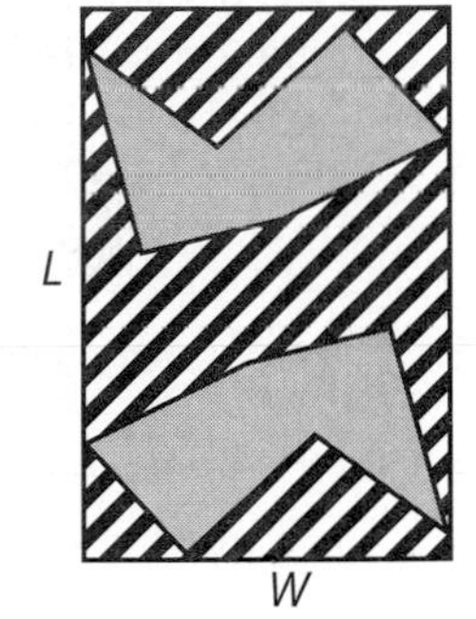

$$2L + 2W = 150$$

We are also given that the length is 15 inches more than the width. This gives the second equation of the system.

$$L = W + 15$$

The two equations together form the system of equations.

$$2L + 2W = 150 \qquad (1)$$
$$L = W + 15 \qquad (2)$$

Solve the problem. Solve the system of equations by substitution.

$$2L + 2W = 150 \qquad \bullet\ \text{This is Equation (1).}$$
$$2(W + 15) + 2W = 150 \qquad \bullet\ \text{Replace } L \text{ by } W + 15.$$
$$2W + 30 + 2W = 150$$
$$4W + 30 = 150$$
$$4W = 120$$
$$W = 30$$

To find L, replace W by 30 in Equation (2) and solve for L.

$$L = W + 15$$
$$= 30 + 15 = 45$$

The width is 30 inches and the length is 45 inches.

Check your work. Check your work by verifying that the perimeter of the wall hanging for these dimensions is 150 inches.

You-Try-It 7 The perimeter of an isosceles triangle is 180 centimeters. The length of each of the two equal sides is 4 times the length of the other side. Find the dimensions of the triangle.

Solution See page S19.

4.5 EXERCISES

Topics for Discussion

1. What is a system of equations?
Two or more equations considered together.

2. What is meant by the terms *independent*, *dependent*, and *inconsistent* when they are applied to a system of linear equations?
Independent systems have exactly one solution; dependent systems have an infinite number of solutions; inconsistent systems have no solution.

3. Explain how the substitution method is used to solve a system of equations.
Answers will vary.

4. When you solve a system of equations by the substitution method, explain how you can tell whether the system of equations has no solution or an infinite number of solutions.
If the result is a false equation, the system has no solution; if the result is a true equation, the system has an infinite number of solutions.

5. Draw the graph of a system of linear equations that has exactly one solution.
Answers will vary.

Solving Systems of Equations

Solve by substitution.

6. $2x + 3y = 7$
$x = 2$
$(2, 1)$

7. $y = 3$
$3x - 2y = 6$
$(4, 3)$

8. $y = x - 3$
$x + y = 5$
$(4, 1)$

9. $y = x + 2$
$x + y = 6$
$(2, 4)$

10. $x = y - 2$
$x + 3y = 2$
$(-1, 1)$

11. $x = y + 1$
$x + 2y = 7$
$(3, 2)$

12. $2x + 3y = 0$
$y = x + 5$
$(-3, 2)$

13. $3x + 2y = 11$
$y = x + 3$
$(1, 4)$

14. $3x - y = 2$
$y = 2x - 1$
$(1, 1)$

15. $y = 4 - 3x$
$3x + y = 5$
Inconsistent

16. $y = 2 - 3x$
$6x + 2y = 7$
Inconsistent

17. $y = 3x + 1$
$6x - 2y = -2$
Dependent, $(x, 3x + 1)$

18. $y = 2 - x$
$3x + 3y = 6$
$(x, -x + 2)$

19. $3x + 5y = -6$
$x = 5y + 3$
$\left(-\frac{3}{4}, -\frac{3}{4}\right)$

20. $x = 4y - 3$
$2x - 3y = 0$
$\left(\frac{9}{5}, \frac{6}{5}\right)$

21. $2x - y = 4$
$3x + 2y = 6$
$(2, 0)$

22. $x + y = 12$
$3x - 2y = 6$
$(6, 6)$

23. $4x - 3y = 5$
$x + 2y = 4$
$(2, 1)$

24. $3x - 5y = 2$
$2x - y = 4$
$\left(\frac{18}{7}, \frac{8}{7}\right)$

25. $7x - y = 4$
$5x + 2y = 1$
$\left(\frac{9}{19}, -\frac{13}{19}\right)$

26. $3x - y = 6$
$x + 3y = 2$
$(2, 0)$

27. $3x + y = 4$
$9x + 3y = 12$
Dependent, $(x, -3x + 4)$

28. $6x - 3y = 6$
$2x - y = 2$
Dependent, $(x, 2x - 2)$

29. $y = 2x + 11$
$y = 5x - 19$
$(10, 31)$

30. $y = 2x - 8$
$y = 3x - 13$
$(5, 2)$

31. $x = 4y - 2$
$x = 6y + 8$
$(-22, -5)$

32. $x = 3y + 7$
$x = 2y - 1$
$(-17, -8)$

33. $x = 3 - 2y$
$x = 5y - 10$
$\left(-\frac{5}{7}, \frac{13}{7}\right)$

34. $y = -4x + 2$
$y = -3x - 1$
$(3, -10)$

35. $x = 2y + 1$
$y = 3x - 13$
$(5, 2)$

36. $y = 3x - 1$
$x = -y - 9$
$(-2, -7)$

37. $3x - y = 11$
$2x + 5y = -4$
$(3, -2)$

38. $-x + 6y = 8$
$2x + 5y = 1$
$(-2, 1)$

Applications

Solve.

39. An engineer invested a portion of $15,000 in a 7% annual simple interest account and the remainder in a 6.5% annual simple interest government bond. The two investments earn $1020 in interest annually. How much was invested in each account?
$9000 at 7%; $6000 at 6.5%

40. An investment club invested part of $20,000 in a preferred stock that earns 8% annual simple interest and the remainder in a municipal bond that earns 7% annual simple interest. The amount of interest earned each year is $1520. How much was invested in each account?
$12,000 at 8%; $8000 at 7%

41. A grocer deposited an amount of money in a high-yield mutual fund that earns 13% annual simple interest. A second deposit, $2500 more than the first, was placed in a certificate of deposit earning 7% annual simple interest. In one year, the total interest earned on both investments was $475. How much money was invested in the mutual fund?
$1500

42. A deposit was made into a 7% annual simple interest account. Another deposit, $1500 less than the first, was placed in a certificate of deposit earning 9% annual simple interest. The total interest earned on both investments for one year was $505. How much money was deposited in the certificate of deposit?
$2500

43. A corporation gave a university $300,000 to support product safety research. The university deposited some of the money in a 10% annual simple interest account and the remainder in an 8.5% annual simple interest account. How much was deposited in each account if the annual interest is $28,500?
$200,000 at 10%; $100,000 at 8.5%

44. A financial consultant invested part of a client's $30,000 in municipal bonds that earn 6.5% annual simple interest and the remainder of the money in 8.5% corporate bonds. How much is invested in each account if the total annual interest earned is $2190?
$18,000 at 6.5%; $12,000 at 8.5%

45. An investment of $2500 is made at an annual simple interest rate of 7%. How much money is invested at an annual simple interest rate of 11% if the total interest earned is 9% of the total investment?
$2500

46. A total of $6000 is invested in two simple interest accounts. The annual simple interest rate on one account is 9%. The annual simple interest rate on the second account is 6%. How much should be invested in each account so that both accounts earn the same amount of interest?
$3600 at 6%; $2400 at 9%

47. A charity deposited a total of $54,000 in two simple interest accounts. The annual simple interest rate on one account is 8%. The annual simple interest rate on the second account is 12%. How much was invested in each account if the total annual interest earned is 9% of the total investment?
$40,500 at 8%; $13,500 at 12%

48. A college sports foundation deposited a total of $24,000 in two simple interest accounts. The annual simple interest rate on one account is 7%. The annual simple interest rate on the second account is 11%. How much is invested in each account if the total annual interest earned is 10% of the total investment?
$6000 at 7%; $18,000 at 11%

49. In a right triangle, the measure of one acute angle is twice the measure of the second acute angle. Find the measure of the two acute angles. (Recall that a right triangle has a right angle whose measure is 90° and that the sum of the measures of the angles of a triangle is 180°.)
$30^\circ, 60^\circ$

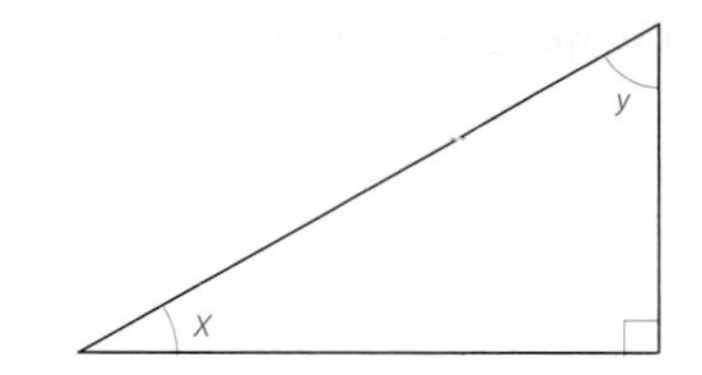

50. Two angles are complementary. The larger angle is 3 degrees more than twice the smaller angle. Find the measure of each angle.
$29^\circ, 61^\circ$

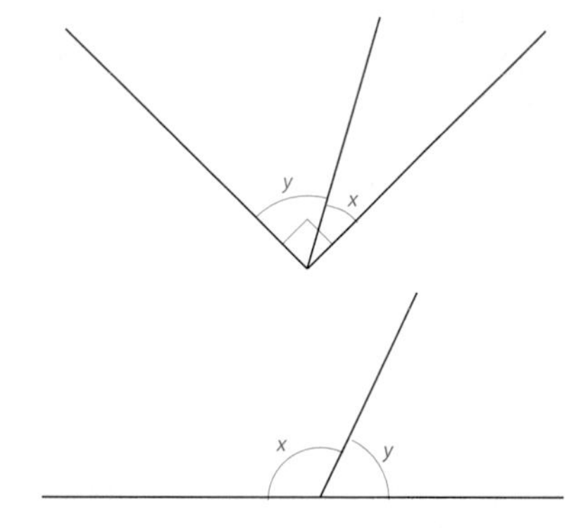

51. Two angles are supplementary. The smaller angle is 6 degrees more than one-half the larger angle. Find the measure of each angle.
$64^\circ, 116^\circ$

52. The perimeter of a rectangular wheat field is 1000 feet. The length of the field is 20 feet less than three times the width. Find the dimensions of the wheat field.
Width: 130 feet; length: 370 feet

53. To raise money to purchase new books for a school library, the Friends of Lincoln School held a bake sale and sold 59 pies and cakes. The income from the sale of the pies and cakes was \$296.25. If the selling price of one pie is \$5.25 and the selling price of one cake is \$4.75, find the number of cakes sold.
27

54. A theater that has only orchestra and loge seats sold 580 tickets for a recent performance of the musical *Stomp*. The total income from the sale of the tickets was \$28,260. If the cost of a loge seat is \$42 and the cost of an orchestra seat is \$55, how many loge seats were sold?
280

Applying Concepts

55. When you solve a system of equations by the substitution method, how do you determine whether the system of equations is inconsistent? How do you determine if it is dependent?
Answers will vary.

56. Write three different systems of equations:
a. one that has $(-3, 5)$ as its only solution,
b. one for which there is no solution, and
c. one that is a dependent system of equations.
Answers will vary. One possible answer is

a. $y = x + 8$, $y = -x + 2$ **b.** $y = 2x - 3$, $y = 2x + 4$ **c.** $y = 3x - 5$, $3x - y = 5$

For what value of k does the system of equations have no solution?

57. $2x - 3y = 7$
$kx - 3y = 4$
2

58. $8x - 4y = 1$
$2x - ky = 3$
1

59. $x = 4y + 4$
$kx - 8y = 4$
2

60. A bank offers a customer a 2-year certificate of deposit (CD) that earns 8% compound annual interest. This means that the interest earned each year is added to the principal before the interest for the next year is calculated. Find the value in 2 years of a nurse's investment of $2500 in this CD.
$2916

61. A bank offers a customer a 3-year certificate of deposit (CD) that earns 8.5% compound annual interest. This means that the interest earned each year is added to the principal before the interest for the next year is calculated. Find the value in 3 years of an accountant's investment of $3000 in this CD.
$3831.87

Exploration

62. ***Break-even Analysis*** Break-even analysis is a method used to determine the sales volume required for a company to break even, or experience neither a profit nor a loss, on the sale of its product. The break-even point represents the number of units that must be made and sold in order for income from sales to equal the cost of the product.

The break-even point can be determined by graphing two equations on the same coordinate grid. The first equation is $R = SN$, where R is the revenue earned, S is the selling price per unit, and N is the number of units sold. The second equation is $T = VN + F$, where T is the total cost, F is the fixed cost, V is the variable cost per unit, and N is the number of units sold. The break-even point is the point where the graphs of the two equations intersect, which is the point where revenue is equal to cost.

a. A company manufactures and sells digital watches. The fixed cost is $20,000, the variable cost per unit is $25, and the selling price per watch is $125. Write two equations for this information.

b. Graph the two equations on the same coordinate grid, using only quadrant I. The horizontal axis is the number of units sold. Use the model shown at the right.

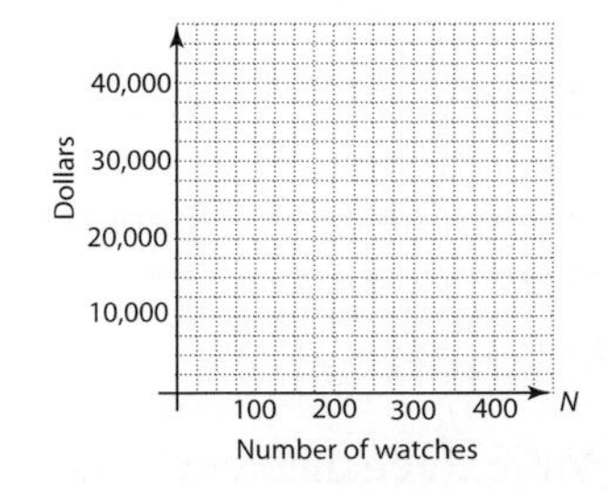

c. How many watches must the company sell in order to break even?

d. What is the significance of the point of intersection of the two graphs?

e. Shade the region between the two lines of the graph to the right of the intersection point. What does this shaded region represent? What does the unshaded region between the two lines to the left of the point of intersection represent?

Section 4.6 Solving Systems of Equations by the Addition Method

Addition Method

Another method of solving a system of equations is called the **addition method**. It is based on the Addition Property of Equations. Recall that this property states that the same number can be added to each side of an equation without changing the solution of the equation.

INSTRUCTOR NOTE

Although somewhat difficult for students, the explanation below of the addition method as it applies to the equations at the right may help some students.

Starting with $3x + 2y = 4$, add 12 to each side of the equation. (12 is the constant term for the second equation.) We now have

$3x + 2y + 12 = 4 + 12$. Now substitute $5x - 2y$ for 12 on the left side of the equation.

$3x + 2y + (5x - 2y) = 4 + 12$

After simplifying, we have $8x = 16$.

In the system of equations at the right, note the effect of adding Equation (2) to Equation (1). Because $2y$ and $-2y$ are opposites, adding the equations results in an equation of one variable. The solution of this equation is the x-coordinate of the ordered pair solution of the system of equations.

$$\begin{aligned} 3x + 2y &= 4 \quad (1) \\ 5x - 2y &= 12 \quad (2) \\ 8x + 0y &= 16 \\ 8x &= 16 \\ x &= 2 \end{aligned}$$

The y-coordinate of the ordered pair is found by substituting the value of x (which is 2) into either Equation (1) or Equation (2) and then solving for y. Equation (1) is used here.

$$\begin{aligned} 3x + 2y &= 4 \quad (1) \\ 3(2) + 2y &= 4 \\ 6 + 2y &= 4 \\ 2y &= -2 \\ y &= -1 \end{aligned}$$

The solution is $(2, -1)$.

The success we had in solving the system of equations above was based on the fact that the coefficients of the y terms were opposites. When neither the x nor y coefficients are opposites, use the Multiplication Property of Equations to rewrite one or both of the equations so that when the equations are added, one of the variable terms is eliminated.

To do this, first choose which variable to eliminate. The coefficients of that variable must be opposites. Multiply each equation by a constant that will produce coefficients that are opposites.

To eliminate x in the system of equations at the right, multiply Equation (1) by -3. This will make the coefficients of x opposites.

$$\begin{aligned} 2x + 3y &= 9 \quad (1) \\ 6x - 5y &= 13 \quad (2) \end{aligned}$$

$-3(2x + 3y) = -3(9)$ • Multiply each side of Equation (1) by -3.

$6x - 5y = 13$ • This is Equation (2).

$-6x - 9y = -27$ • Simplify the product of -3 and Equation (1). The coefficients of x are now opposites.

$6x - 5y = 13$

$0x - 14y = -14$ • Add the equations and then solve for y.

$-14y = -14$

$y = 1$

Substitute the value of y into one of the original equations and solve for x. Equation (1) is used here.

$2x + 3y = 9$ • This is Equation (1).

$2x + 3(1) = 9$ • Replace y by 1.

$2x + 3 = 9$

$2x = 6$

$x = 3$

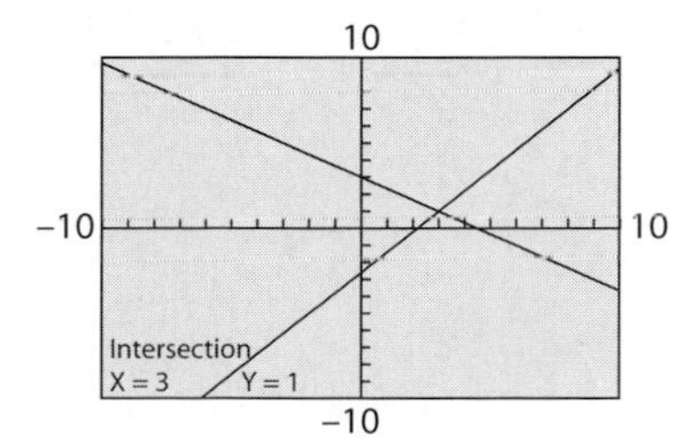

The solution of the system of equations is (3, 1). A graphical check of the solution is shown at the left.

Sometimes it is necessary to multiply each equation of the system of equations by a different number so that the coefficients of one of the terms are opposites.

SUGGESTED ACTIVITY
Have students solve the system of equations at the right by first eliminating y.

To eliminate x in the system of equations at the right, multiply Equation (1) by 2 and Equation (2) by −5.

$$5x + 6y = 3 \quad (1)$$
$$2x - 5y = 16 \quad (2)$$

$2(5x + 6y) = 2 \cdot 3$ • Multiply each side of Equation (1) by 2. Multiply each side of Equation (2) by −5.

$-5(2x - 5y) = -5 \cdot 16$ Note how the constants are selected. The negative sign is used so that the coefficients will be opposites.

$10x + 12y = 6$ • Simplify each product.
$-10x + 25y = -80$ The coefficients of x are now opposites.
$0x + 37y = -74$ • Add the equations and then solve for y.
$37y = -74$
$y = -2$

Substitute the value of y into one of the original equations and solve for x. Equation (1) is used here.

$5x + 6y = 3$ • This is Equation (1).
$5x + 6(-2) = 3$ • Replace y by −2.
$5x - 12 = 3$
$5x = 15$
$x = 3$

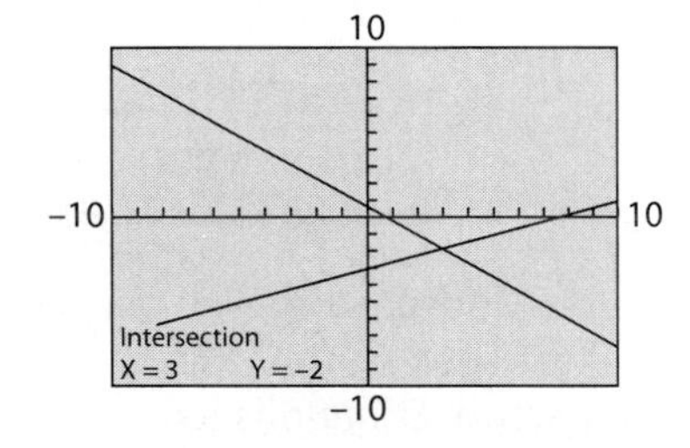

The solution of the system of equations is (3, −2). A graphical check of the solution is shown at the left.

To solve the system of equations at the right, first subtract $2y$ from each side of Equation (1) so that it is in the form $Ax + By = C$.

$$5x = 2y - 7 \quad (1)$$
$$3x + y = -2 \quad (2)$$

$5x - 2y = -7$ • This is the result of subtracting $2y$ from each side of Equation (1).
$3x + y = -2$ • This is Equation (2).

$5x - 2y = -7$
$2(3x + y) = -2 \cdot 2$ • Eliminate y. Multiply Equation (2) by 2.

$5x - 2y = -7$
$6x + 2y = -4$ • Simplify the product of 2 and Equation (2). The coefficients of x are now opposites.
$11x + 0y = -11$ • Add the equations and then solve for x.
$11x = -11$
$x = -1$

Substitute the value of x into one of the original equations and solve for y. Equation (1) is used here.

$5x = 2y - 7$ • This is Equation (1).
$5(-1) = 2y - 7$ • Replace x by −1.
$-5 = 2y - 7$
$2 = 2y$
$1 = y$

The solution of the system of equations is (−1, 1). An algebraic check of the solution is shown at the left.

Check:

$5x = 2y - 7$	
$5(-1)$	$2(1) - 7$
-5	$2 - 7$
$-5 = -5$	

$3x + y = -2$	
$3(-1) + 1$	-2
$-3 + 1$	-2
$-2 = -2$	

The solution checks.

INSTRUCTOR NOTE
Here is an example of a situation that results in an inconsistent system of equations. The perimeter of a rectangle is 100 meters. The sum of the length and width of the rectangle is 40 meters. Find the dimensions of the rectangle. The fact that this system of equations is inconsistent means that no such rectangle exists.

To eliminate y from the system of equations at the right, multiply Equation (1) by –2.

$$2x + y = 2 \quad (1)$$
$$4x + 2y = -5 \quad (2)$$

$$2x + y = 2$$
$$4x + 2y = -5$$

$$-4x - 2y = -4$$ • This is the product of –2 and Equation (1). $-2(2x + y) = -2 \cdot 2 \Rightarrow -4x - 2y = -4$
$$4x + 2y = -5$$
$$0x + 0y = -9$$ • Add the equations.
$$0 = -9$$ • This is not a true equation.

The fact that the last equation is not true means that the system of equations is inconsistent. Therefore, the system of equations has no solution.

Example 1 Solve by the addition method:
$$2x + 4y = 5 \quad (1)$$
$$3x + 5y = 4 \quad (2)$$

Solution

$$3(2x + 4y) = 3 \cdot 5$$ • Eliminate x. Multiply Equation (1) by 3 and multiply Equation (2) by –2.
$$-2(3x + 5y) = -2 \cdot 4$$
$$6x + 12y = 15$$
$$-6x - 10y = -8$$
$$2y = 7$$ • Add the equations.
$$y = \frac{7}{2}$$

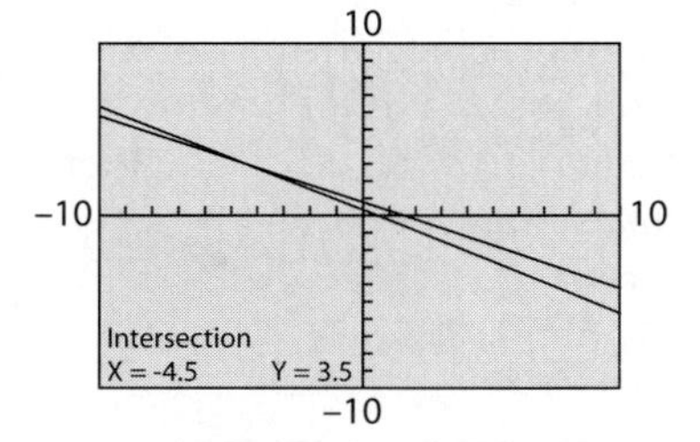

$$2x + 4\left(\frac{7}{2}\right) = 5$$ • Substitute the value of y into Equation (1) and solve for x.
$$2x + 14 = 5$$
$$2x = -9$$
$$x = -\frac{9}{2}$$

The solution is $\left(-\frac{9}{2}, \frac{7}{2}\right)$. A graphical check is shown at the left.

You-Try-It 1 Solve by the addition method:
$$3x - y = 10$$
$$2x + 5y = 1$$

Solution See page S19.

Example 2 Solve by the addition method:
$$6x + 9y = 15 \quad (1)$$
$$4x + 6y = 10 \quad (2)$$

Solution

$$4(6x + 9y) = 4 \cdot 15$$ • Eliminate x. Multiply Equation (1) by 4 and multiply Equation (2) by –6.
$$-6(4x + 6y) = -6 \cdot 10$$
$$24x + 36y = 60$$
$$-24x - 36y = -60$$
$$0 = 0$$ • Add the equations.

$0 = 0$ is a true equation. The system of equations is dependent. The solutions of the system of equations are the ordered pair solutions of either equation. Solve Equation (1) for y. (See the TAKE NOTE.)

$$6x + 9y = 15$$
$$9y = -6x + 15$$
$$y = -\frac{2}{3}x + \frac{5}{3}$$

The solutions are the ordered pairs $\left(x, -\frac{2}{3}x + \frac{5}{3}\right)$.

TAKE NOTE
We could have solved Equation (2) for y. The result would have been the same.

$$4x + 6y = 10$$
$$6y = -4x + 10$$
$$y = -\frac{2}{3}x + \frac{5}{3}$$

You-Try-It 2 Solve by the addition method:

$$\begin{aligned} 2x - 3y &= 4 \quad (1) \\ -4x + 6y &= -8 \quad (2) \end{aligned}$$

Solution See page S19.

Applications

Solving motion problems that involve an object moving with or against a wind or current normally requires two variables. One variable represents the speed of the moving object in calm conditions, and a second variable represents the rate of the wind or current.

A plane flying with the wind will travel a greater distance per hour than it would travel without the wind. The resulting rate of the plane is represented by the sum of the plane's speed and the rate of the wind.

A plane traveling against the wind, on the other hand, will travel a shorter distance per hour than it would travel without the wind. The resulting rate of the plane is represented by the difference between the plane's speed and the rate of the wind.

The same principle is used to describe the rate of a boat traveling with or against a water current.

Example 3 Flying with the wind, a small plane can fly 750 miles in 3 hours. Against the wind, the plane can fly the same distance in 5 hours. Find the rate of the plane in calm air and the rate of the wind.

Solution

State the goal. The goal is to find the rate of the plane in calm air and the rate of the wind.

Describe a strategy. There are two unknowns, the rate of the plane in calm air and the rate of the wind. Let p represent the rate of the plane in calm conditions and let w represent the rate of the wind. Using these variables and the equation $d = rt$, write equations for the distance traveled by the plane with and against the wind.

With the wind, the rate of the plane is $p + w$. The time is 3 hours.

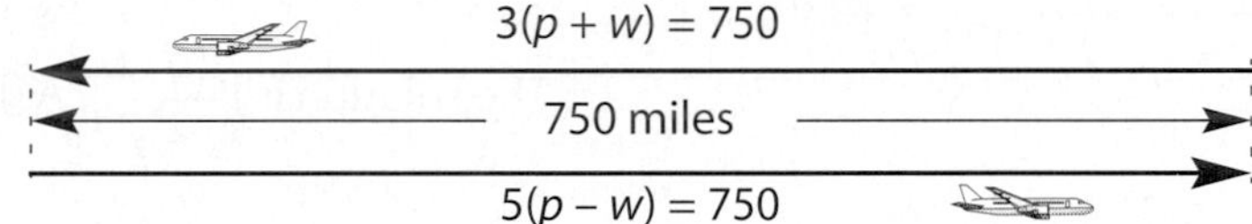

Against the wind, the rate of the plane is $p - w$. The time is 5 hours.

Use the equations in the diagram to write a system of equations.

Solve the problem.

$$3(p + w) = 750$$
$$5(p - w) = 750$$

$$\frac{3(p + w)}{3} = \frac{750}{3} \quad \text{• Simplify Equation (1) by dividing each side by 3.}$$

$$\frac{5(p - w)}{5} = \frac{750}{5} \quad \text{• Simplify Equation (2) by dividing each side by 5.}$$

$$\begin{aligned} p + w &= 250 \\ p - w &= 150 \\ 2p &= 400 \quad \text{• Add the equations.} \\ p &= 200 \end{aligned}$$

Substitute the value of p into $p + w = 250$ and solve for w.

$$\begin{aligned} p + w &= 250 \\ 200 + w &= 250 \\ w &= 50 \end{aligned}$$

The rate of the plane in calm air is 200 miles per hour. The rate of the wind is 50 miles per hour.

Check your work. With $p = 200$ and $w = 50$, the rate of the plane with the wind is 250 miles per hour. Using the equation $d = rt$, check that the distance the plane travels in 3 hours is 750 miles.

$$d = rt$$
$$= 250(3) = 750 \quad \surd$$

You-Try-It 3 A canoeist paddling with the current can travel 24 miles in 3 hours. Against the current, it takes 4 hours to travel the same distance. Find the rate of the current and the rate of the canoeist in calm water.

Solution See pages S19–S20.

The amount of a substance in a solution can be given as a percent of the total solution. For example, in a 5% saltwater solution, 5% of the total solution is salt. The remaining 95% is water.

The solution of a percent mixture problem is based on the equation $Q = Ar$, where Q is the quantity of a substance in the solution, r is the percent of concentration, and A is the amount of solution.

SUGGESTED ACTIVITY
See the Instructor's Resource Manual for an activity that will help prepare students for percent mixture problems. Part 1 can be completed now and Part 2 after the student has completed Example 4.

⟶ A 500-milliliter bottle contains a 3% solution of hydrogen peroxide. Find the amount of hydrogen peroxide in the solution.

$Q = Ar$

$Q = 500(0.03)$ • $A = 500, r = 3\% = 0.03$

$Q = 15$

The bottle contains 15 milliliters of hydrogen peroxide.

Some percent mixture problems can be solved by using a system of equations. The solution of these equations is based on the concept that the sum of the quantities being mixed equals the desired quantity.

Example 4 A chemist wishes to make 3 liters of a 7% acid solution by mixing a 9% solution and a 4% solution. How many liters of each solution should the chemist use?

Solution

State the goal. The goal is to find the number of liters of a 9% acid solution that must be mixed with a 4% acid solution to make 3 liters of a solution that is 7% acid.

Describe a strategy. There are two solutions. Let x represent the number of liters of the 9% solution and y represent the number of liters of the 4% solution. The total of the two solutions is 3 liters. This information gives one equation of the system of equations.

$x + y = 3$ • Total amount of solution is the sum of the number of liters of the 9% solution and the 4% solution.

The quantity of acid in each solution can be found from the equation $Q = Ar$.

The quantity of acid in the 9% solution is the product of the amount of acid, x, and the concentration as a decimal, 0.09. The quantity of acid in the 4% solution is the product of the amount of acid, y, and the concentration as a decimal, 0.04.

Quantity of acid in 9% solution: $0.09x$
Quantity of acid in 4% solution: $0.04y$

The quantity of acid in 3 liters of the final 7% solution is $0.07(3) = 0.21$. This quantity of acid is a result of combining the 9% and 4% solutions. Therefore, it is the sum of the quantities of acid in these two solutions. This information gives the second equation of the system of equations.

$$0.09x + 0.04y = 0.07(3)$$

The two equations together form the system of equations.

$$x + y = 3 \quad (1)$$
$$0.09x + 0.04y = 0.21 \quad (2)$$

Solve the problem.

$$-0.09(x+y) = -0.09 \cdot 3 \qquad \text{• Multiply Equation (1) by } -0.09.$$
$$0.09x + 0.04y = 0.21$$
$$-0.09x - 0.09y = -0.27 \qquad \text{• Simplify the product of } -0.09 \text{ and Equation (1).}$$
$$0.09x + 0.04y = 0.21$$
$$-0.05y = -0.06 \qquad \text{• Add the equations.}$$
$$y = 1.2$$

Substitute the value of y into Equation (1) and solve for x.

$$x + y = 3$$
$$x + 1.2 = 3$$
$$x = 1.8$$

The chemist should mix 1.8 liters of the 9% solution with 1.2 liters of the 4% solution.

Check your work. Replace the values of x and y in Equation (2) to ensure they satisfy that equation.

$$0.09x + 0.04y = 0.21$$
$$0.09(1.8) + 0.04(1.2) = 0.162 + 0.048 = 0.21 \quad \surd$$

You-Try-It 4 A paint that is 21% green dye is mixed with a paint that contains 15% green dye. How many gallons of each must be mixed to make 60 gallons of a paint that is 19% green dye?

Solution See page S20.

Keep in mind when solving application problems that it is not the type of application that requires us to use a system of equations; rather, it is the number of unknown quantities in the application.

Example 5 How many gallons of a 15% salt solution must be mixed with 4 gallons of a 20% salt solution to make a 17% salt solution?

Solution

State the goal. The goal is to find the number of gallons of a 15% salt solution to mix with a 20% salt solution to produce a 17% salt solution.

Describe a strategy. There is only one unknown, so a system of equations is not necessary. Let x represent the number of gallons of the 15% salt solution.

The quantity of salt in each solution can be found from the equation $Q = Ar$.

Salt in 15% solution: $0.15x$ Salt in 20% solution: $0.20(4)$

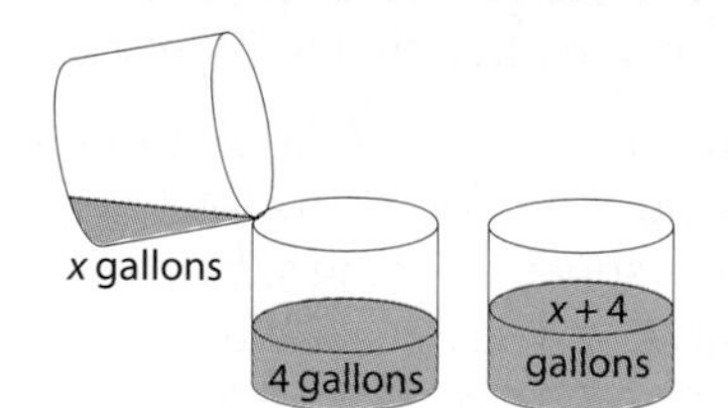

As shown in the diagram at the left, there was originally 4 gallons of a 20% solution to which x gallons of a 15% solution is added. The final mixture contains $(x+4)$ gallons which is to be 17% salt. Thus the salt in the 17% solution is $0.17(x+4)$. Because the salt in the 15% solution is being mixed with the salt in the 20% solution, the amount of salt in the final solution is the sum of the amounts of salt in each solution. This information provides our equation.

Solve the problem.

$$0.15x + 0.20(4) = 0.17(x+4)$$
$$0.15x + 0.8 = 0.17x + 0.68$$
$$-0.02x = -0.12 \qquad \text{• Subtract } 0.17x \text{ from each side of the equation and subtract 0.8 from each side of the equation.}$$
$$x = 6$$

6 gallons of the 15% solution are required.

Check your work. Be sure to check your work.

You-Try-It 5 How many quarts of pure orange juice must be added to 5 quarts of fruit drink that is 10% orange juice to produce an orange drink that is 25% orange juice?

Solution See page S20.

4.6 EXERCISES

Topics for Discussion

1. Explain how to solve a system of equations using the addition method.
   Answers will vary.

2. When you solve a system of equations by the addition method, explain how you can tell whether the system of equations has no solution or an infinite number of solutions.
   If the result is a false equation, the system has no solution; if the result is a true equation, the system has an infinite number of solutions.

3. In percent mixture problems, how are the quantities of a substance before mixing related to the amount of the substance after mixing?
   The amount before mixing is equal to the amount after mixing.

4. In rate-of-wind and rate-of-current problems, how does the wind or current affect the rate of an object?
   Going with the wind or current, the speed of an object is increased; traveling against the wind or current decreases the speed of the object.

Solving Systems of Equations

Solve the system of equations using the addition method.

5. $2x + y = 3$
   $x - y = 3$
   $(2, -1)$

6. $3x + 2y = 1$
   $-3x + y = -4$
   $(1, -1)$

7. $x - 2y = 4$
   $3x + 4y = 2$
   $(2, -1)$

8. $4x + 3y = 15$
   $2x - 5y = 1$
   $(3, 1)$

9. $3x - 7y = 27$
   $6x + 5y = -3$
   $(2, -3)$

10. $2x + 3y = 7$
    $5x + 2y = 1$
    $(-1, 3)$

11. $2x - 3y = 5$
    $4x - 6y = 3$
    Inconsistent

12. $5x + 4y = 9$
    $10x + 8y = 1$
    Inconsistent

13. $5x + 7y = 10$
    $3x - 14y = 6$
    $(2, 0)$

14. $7x - 3y = 6$
    $5x + 2y = -4$
    $(0, -2)$

15. $7x + 8y = 0$
    $5x - 3y = 0$
    $(0, 0)$

16. $x - 9y = 0$
    $6x + 4y = 0$
    $(0, 0)$

17. $3x - 5y = 2$
    $2x - 4y = 1$
    $\left(\frac{3}{2}, \frac{1}{2}\right)$

18. $6x + 9y = 1$
    $5x - 2y = 4$
    $\left(\frac{2}{3}, -\frac{1}{3}\right)$

19. $2x - 5y = 4$
    $x + 5y = 1$
    $\left(\frac{5}{3}, -\frac{2}{15}\right)$

20. $3x - 4y = 1$
$4x + 3y = 1$
$\left(\frac{7}{25}, -\frac{1}{25}\right)$

21. $3x - 5 = -4y + 5$
$4x + 3y = 11$
$(2, 1)$

22. $2y = 4 - 9x$
$3x + 2y = -8$
$(2, -7)$

23. $4x - 8y = 32$
$3x - 6y = 24$
Dependent, $(x, \frac{1}{2}x - 4)$

24. $5x + 2y = 10$
$10x + 4y = 20$
Dependent, $(x, -\frac{5}{2}x + 5)$

25. $2x + 3y = 7 - 2x$
$6x + 2y = 9 - x$
$(1, 1)$

26. $9y = 16 - 2x$
$5x = 1 - 3y$
$(-1, 2)$

27. $2x = 3y + 16$
$4y = -3x + 7$
$(5, -2)$

28. $3x + 2 = 4y + 4$
$3y + 4 = 5x - 4$
$(4, 4)$

Applications

Solve.

29. A plane flying with the jet stream flew from Los Angeles to Chicago, a distance of 2250 miles, in 5 hours. Flying against the jet stream, the plane could fly only 1750 miles in the same amount of time. Find the rate of the plane in calm air and the rate of the wind. Plane, 400 mph; wind, 50 mph

30. A rowing team rowing with the current traveled 40 kilometers in 2 hours. Rowing against the current, the team could travel only 16 kilometers in 2 hours. Find the team's rowing rate in calm water and the rate of the current.
Rowing team, 14 kilometers/hour; current, 6 kilometers/hour

31. A motorboat traveling with the current went 35 miles in 3.5 hours. Traveling against the current, the boat went 12 miles in 3 hours. Find the rate of the boat in calm water and the rate of the current. Boat, 7 mph; current, 3 mph

32. A small plane, flying into a headwind, flew 270 miles in 3 hours. Flying with the wind, the plane traveled 260 miles in 2 hours. Find the rate of the plane in calm air and the rate of the wind. Plane, 110 mph; wind, 20 mph

33. A plane flying with a tailwind flew 300 miles in 2 hours. Against the wind, it took 3 hours to travel the same distance. Find the rate of the plane in calm air and the rate of the wind. Plane, 125 mph; wind, 25 mph

34. A rowing team rowing with the current traveled 18 miles in 2 hours. Against the current, the team rowed a distance of 8 miles in the same amount of time. Find the rate of the rowing team in calm water and the rate of the current.
Rowing team, 6.5 mph; current, 2.5 mph

35. A seaplane flying with the wind flew from an ocean port to a lake, a distance of 240 miles, in 2 hours. Flying against the wind, it made the trip from the lake to the ocean port in 3 hours. Find the rate of the plane in calm air and the rate of the wind.
Plane, 100 mph; wind, 20 mph

36. Rowing with the current, a canoeist paddled 14 miles in 2 hours. Against the current, the canoeist could paddle only 10 miles in the same amount of time. Find the rate of the canoeist in calm water and the rate of the current.
Canoeist, 6 mph; current, 1 mph

37. Flying with the wind, a small plane flew 280 miles in 2 hours. Flying against the wind, the plane flew 160 miles in 2 hours. Find the rate of the plane in calm air and the rate of the wind. Plane, 110 mph; wind, 30 mph

38. With the wind, a quarterback passes a football 140 feet in 2 seconds. Against the wind, the same pass would have traveled 80 feet in 2 seconds. Find the rate of the pass and the rate of the wind. Pass, 55 ft/s; wind, 15 ft/s

39. Flying with the wind, a plane flew 1000 kilometers in 5 hours. Against the wind, the plane could fly only 800 kilometers in the same amount of time. Find the rate of the plane in calm air and the rate of the wind.
Plane, 180 km/h; wind, 20 km/h

40. Traveling with the current, a cruise ship sailed between two islands, a distance of 90 miles, in 3 hours. The return trip against the current required 4 hours and 30 minutes. Find the rate of the cruise ship in calm water and the rate of the current.
Ship, 25 mph; current, 5 mph

41. A motorboat took 2 hours to make a downstream trip with a current of 3 miles per hour. The return trip against the same current took 3 hours. Find the speed of the motorboat in calm water. 15 mph

42. A pilot left an airport for a 3-hour trip with a tailwind of 30 miles per hour to deliver medicines to a rural community. The return trip took 5 hours against the same wind. Find the distance to the rural community. 450 miles

43. A Boeing 767 flies from New York's JFK airport to Charles de Gaulle airport in Paris, France, at an average speed of 550 miles per hour. In 1927, Charles Lindbergh made a similar trip at an average speed of approximately 110 miles per hour and it took him about 26 hours longer than it takes today in the Boeing 767. How long did it take Charles Lindbergh to make the flight from New York to Paris? 32.5 hours

44. A guided tour on the Green River through Red Canyon, Utah, takes about 2 hours by boat. The boat picks up tourists at a dock and returns them to the same dock after the tour. If the boat can travel at a speed of 6 miles per hour and the rate of the current is 2 miles per hour, how far along the river can the tour guide take the tourists?
$\frac{16}{3}$ miles

45. A chemist wants to make 50 milliliters of a 16% acid solution by mixing a 13% acid solution and an 18% acid solution. How many milliliters of each solution should the chemist use? 13% solution, 20 milliliters; 18% solution, 30 milliliters

46. How many pounds of coffee that is 40% java beans must be mixed with 80 pounds of coffee that is 30% java beans to make a coffee blend that is 32% java beans? 20 pounds

47. Thirty ounces of pure silver are added to 50 ounces of a silver alloy that is 20% silver. What is the percent concentration of silver in the resulting alloy?
50%

48. Two hundred liters of punch that contains 35% fruit juice is mixed with 300 liters of a second punch. The resulting fruit punch is 20% fruit juice. Find the percent concentration of fruit juice in the second punch. 10%

49. The manager of a garden shop mixes grass seed that is 60% rye grass with 70 pounds of grass seed that is 80% rye grass to make a mixture that is 74% rye grass. How much of the 60% rye grass is used? 30 pounds

50. Five grams of sugar are added to a 45-gram serving of a breakfast cereal that is 10% sugar. What is the percent concentration of sugar in the resulting mixture? 19%

51. A dermatologist mixes 50 grams of a cream that is 0.5% hydrocortisone with 150 grams of a second hydrocortisone cream. The resulting mixture is 0.68% hydrocortisone. Find the percent concentration of hydrocortisone in the second hydrocortisone cream. 0.74%

52. A carpet manufacturer blends two fibers, one 20% wool and the second 50% wool. How many pounds of each fiber should be woven together to produce 600 pounds of a fabric that is 28% wool?
20% wool, 440 pounds; 50% wool, 160 pounds

53. A hair dye is made by blending a 7% hydrogen peroxide solution and a 4% hydrogen peroxide solution. How many milliliters of each are used to make a 300-milliliter solution that is 5% hydrogen peroxide? 7% solution, 100 milliliters; 4% solution, 200 milliliters

54. How many grams of pure salt must be added to 40 grams of a 20% salt solution to make a solution that is 36% salt? 10 grams

55. How many ounces of an 8% saline solution must be added to 40 ounces of a 15% saline solution to make a saline solution that is 10% salt? 100 ounces

56. A goldsmith mixes 8 ounces of a 30% gold alloy with 12 ounces of a 25% gold alloy. What is the percent concentration of the resulting alloy? 27%

57. A physicist mixes 40 liters of liquid oxygen with 50 liters of liquid air that is 64% liquid oxygen. What is the percent concentration of liquid oxygen in the resulting mixture? 80%

58. How many ounces of pure bran flakes must be added to 50 ounces of cereal that is 40% bran flakes to produce a mixture that is 50% bran flakes? 10 ounces

59. How many milliliters of pure chocolate must be added to 150 milliliters of chocolate topping that is 50% chocolate to make a topping that is 75% chocolate? 150 milliliters

60. A tea that is 20% jasmine is blended with a tea that is 15% jasmine. How many pounds of each tea are used to make 5 pounds of tea that is 18% jasmine? 20% jasmine, 3 pounds; 15% jasmine, 2 pounds

61. A clothing manufacturer has some pure silk thread and some thread that is 85% silk. How many kilograms of each must be woven together to make 75 kilograms of cloth that is 96% silk? Pure silk, 55 kilograms; 85% silk, 20 kilograms

62. How many ounces of dried apricots must be added to 18 ounces of a snack mix that contains 20% dried apricots to make a mixture that is 25% dried apricots? 1.2 ounces

63. A recipe for a rice dish calls for 12 ounces of a rice mixture that is 20% wild rice and 8 ounces of pure wild rice. What is the percent concentration of wild rice in the 20-ounce mixture? 52%

Applying Concepts

64. Find an equation so that the system of equations formed by your equation and $2x - 5y = 9$ will have $(2, -1)$ as a solution.
Answers will vary; $3x + 2y = 4$ is a possibility.

65. The point of intersection of the graphs of the equations $Ax + 2y = 2$ and $2x + By = 10$ is $(2, -2)$. Find A and B. $A = 3, B = -3$

66. The point of intersection of the graphs of the equations $Ax - 4y = 9$ and $4x + By = -1$ is $(-1, -3)$. Find A and B. $A = 3, B = -1$

67. Given that the graphs of the equations $2x - y = 6$, $3x - 4y = 4$, and $Ax - 2y = 0$ all intersect at the same point, find A. $A = 1$

68. Given that the graphs of the equations $3x - 2y = -2$, $2x - y = 0$, and $Ax + y = 8$ all intersect at the same point, find A. $A = 2$

69. For what value of k is the system of equations dependent?

a. $2x + 3y = 7$ 14
$4x + 6y = k$

b. $x = ky - 1$ $\frac{1}{2}$
$y = 2x + 2$

70. For what values of k is the system of equations independent?

a. $2x + ky = 1$ $k \neq 4$
$x + 2y = 2$

b. $x + 2y = 4$ $k \neq \frac{3}{2}$
$kx + 3y = 2$

Explorations

71. ***History*** The Babylonians had a method for solving a system of equations. Here is an adaptation of a problem from an ancient Babylonian text (around 1500 B.C.). See if you can solve the ancient problem. "There are two silver blocks. The sum of $\frac{1}{7}$ of the first block and $\frac{1}{11}$ of the second block is one sheqel (a weight). The first block diminished by $\frac{1}{7}$ of its weight equals the second diminished by $\frac{1}{11}$ of its weight. What are the weights of the two blocks?"

72. ***Exploring the Addition Method for Solving a System of Equations*** Consider the system of equations shown at the right.

$$2x - 5y = -14 \quad (1)$$
$$3x + 2y = 17 \quad (2)$$

a. Graph the equations of this system and verify that the solution is $(3, 4)$.

b. Multiplying Equation (1) by 3 and Equation (2) by -2 gives Equation (3) and Equation (4). Graph these equations and determine the solution of the system of equations. Has the solution changed? Explain how this demonstrates the Multiplication Property of Equations.

$$6x - 15y = -42 \quad (3)$$
$$-6x - 4y = -34 \quad (4)$$

c. Add Equation (3) and Equation (4) and graph the resulting equation without erasing the graphs from part (b). How does the new graph relate to the graphs of Equation (3) and Equation (4)? Explain how this demonstrates the Addition Property of Equations.

Chapter Summary

Definitions

A *relation* is a set of ordered pairs that describes how one quantity is related to another. The *domain* of a relation is the set of first components of the ordered pairs. The *range* of a relation is the set of second components of the ordered pairs. The *graph of a relation* is the graph of the ordered pairs that belong to the relation.

A *function* is a special type of relation in which no two ordered pairs have the same first coordinate and different second coordinates.

In the equation $y = 2x - 1$, the variable y is called the *dependent variable* because its value depends on the value of x. The variable x is called the *independent variable*.

A *linear function* is one that can be represented by the equation $f(x) = mx + b$.

The *constant function* is written as $f(x) = c$, where c is some real number. No matter what value of x is chosen, the value of the constant function is c.

The linear function that approximates the points in a scatter diagram is called a *regression equation*. This equation is used to predict values that are not part of the original data. The *regression line* is the graph of the regression equation.

The point at which a graph crosses the x-axis is called the *x-intercept*. The point at which a graph crosses the y-axis is called the *y-intercept*.

The rate of the change in one quantity to the change in another quantity is called *slope*. A line that slants upward to the right has *positive slope*. A line that slants downward to the right has *negative slope*. A horizontal line has *zero slope*. We say that a vertical line has *no slope* or that the slope of a vertical line is *undefined*.

For any equation of the form $y = mx + b$, the slope of the line is m, the coefficient of x. The y-intercept is $(0, b)$. The equation $y = mx + b$ is called the *slope-intercept form of a straight line*.

The equation of the line that has slope m and passes through the point whose coordinates are (x_1, y_1) can be found by the *point-slope formula*: $y - y_1 = m(x - x_1)$.

Equations considered together are called a *system of equations*.

A *solution of a system of equations in two variables* is an ordered pair that is a solution of each equation.

To *solve a system of equations* means to find the ordered pair solutions of the system of equations.

Procedures

To find the x-intercept, let $y = 0$ and solve for x.

To find the y-intercept, let $x = 0$ and solve for y.

To find the slope of a line containing two points $P_1(x_1, y_1)$ and $P_2(x_2, y_2)$, use the formula for slope: $m = \frac{y_2 - y_1}{x_2 - x_1}, x_1 \neq x_2$.

To find the equation of a line, use the point-slope formula $y - y_1 = m(x - x_1)$. Replace m by the slope of the line and replace (x_1, y_1) by the coordinates of a known point of the line.

For a linear system of equations containing two equations, there are three possible conditions:

The lines can intersect at one point. The point of intersection of the lines is the ordered pair that is a solution of each equation of the system. It is the solution of the system of equations. The system of equations is *independent*.

The lines can be parallel and therefore do not intersect. Since they do not intersect, there is no point on both lines and the system of equations has no solution. This is an *inconsistent* system of equations.

Each equation has the same graph. The lines graph one on top of the other. Therefore, they intersect at infinitely many points. The solutions are the ordered pairs of one of the equations. This system of equations is called *dependent*.

To solve a system of linear equations in two variables by *graphing*, graph each equation on the same coordinate system. If the lines graphed intersect at one point, the point of intersection is the ordered pair that is a solution of the system. If the lines graphed are parallel, the system is inconsistent. If the lines represent the same line, the system is dependent.

To solve a system of equations in two variables by the *substitution method*, one of the equations of the system must be solved in terms of the other variable. It does not matter which variable is used.

To solve a system of linear equations by the *addition method*, use the Multiplication Property of Equations to rewrite one or both of the equations so that the coefficients of one variable are opposites. Then add the two equations and solve for the variables.

The equation $Q = Ar$ is used to solve percent mixture problems. Q is the quantity of a substance in the solution, r is the percent concentration, and A is the amount of solution.

The equation $I = Pr$ is used to solve simple interest problems. I is the amount of interest, P is the principal, and r is the annual simple interest rate.

The equation $d = rt$ is used to solve problems concerning objects moving with or against a wind or current. The distance traveled by the object is d, the rate of the object is r, and the time of travel is t. The speed of an object moving with the wind or current is increased by the wind or current. The speed of an object moving against the wind or current is decreased by the wind or current.

Chapter Review Exercises

1. Find the domain and range of the relation
$\{(-2,-1),(-1,0),(0,1),(1,2),(2,3)\}$
State whether or not the relation is a function.
D: $\{-2,-1,0,1,2\}$; R: $\{-1,0,1,2,3\}$; yes

2. Evaluate the function at the given value of x. Write an ordered pair that is an element of the function.
$f(x) = 4x - 7;\ x = 3$ 5; (3, 5)

3. Evaluate the function at the given value of x. Write an ordered pair that is an element of the function.
$f(x) = x^2 - 4;\ x = -5$ 21; (−5, 21)

4. Graph: $f(x) = x + 5$

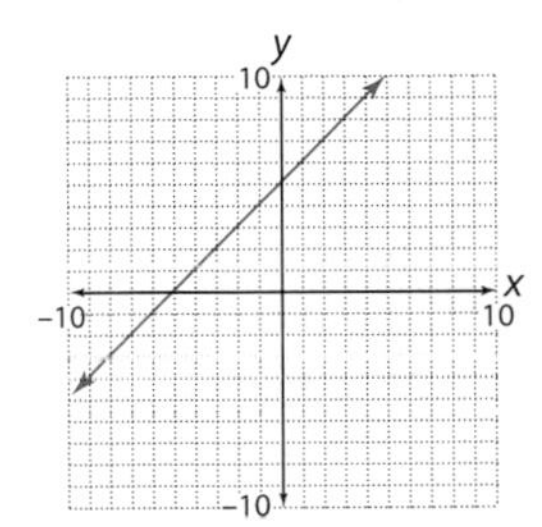

5. Graph: $f(x) = -\frac{3}{4}x - 6$

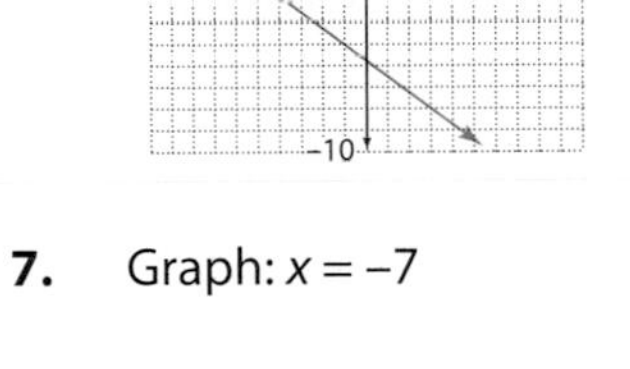

6. Graph by first finding the x- and y- intercepts:
$3x + 2y = 12$

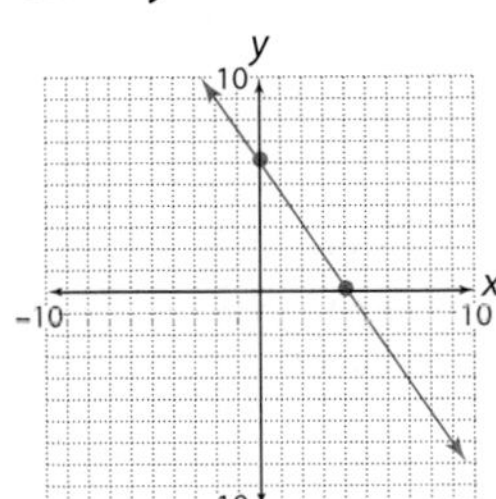

(4, 0), (0, 6)

7. Graph: $x = -7$

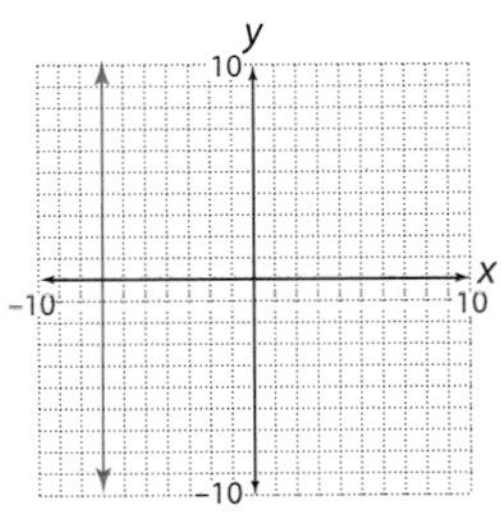

8. Find the slope of the line passing through the given points. $P_1(2,-2)$, $P_2(6,-4)$
$-\frac{1}{2}$

9. Find the slope of the line passing through the given points. $P_1(-2,6)$, $P_2(-3,3)$
3

10. Graph the line whose slope is $-\frac{2}{3}$ and whose y-intercept is (0, 4).

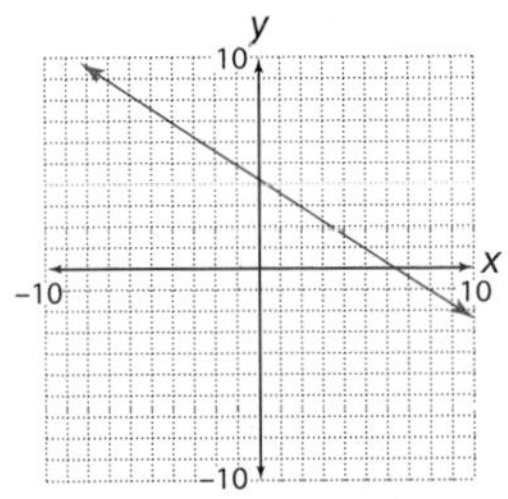

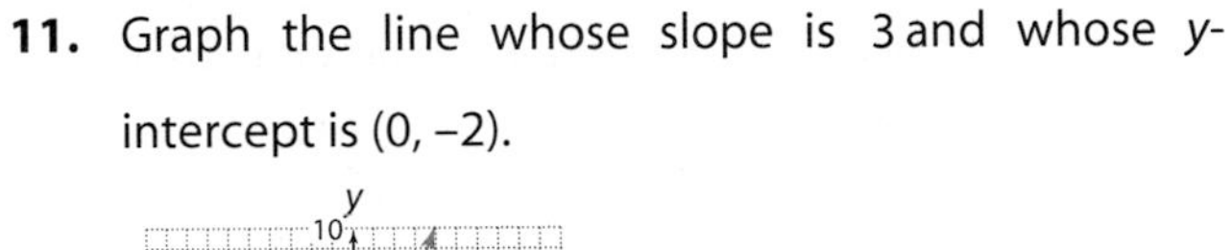

11. Graph the line whose slope is 3 and whose y-intercept is (0, −2).

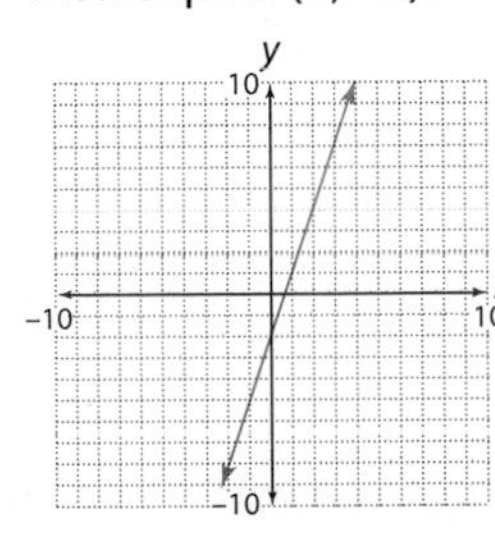

12. Find the equation of the line that passes through the point whose coordinates are $(-3,-5)$ and that has slope $-\frac{4}{3}$. $y = -\frac{4}{3}x - 9$

13. Find the equation of the line that passes through the point whose coordinates are $(0,0)$ and that has slope $\frac{2}{3}$. $y = \frac{2}{3}x$

14. Solve by using a graphing calculator: $\begin{aligned} 5x + 2y &= -14 \\ 3x - 4y &= 2 \end{aligned}$

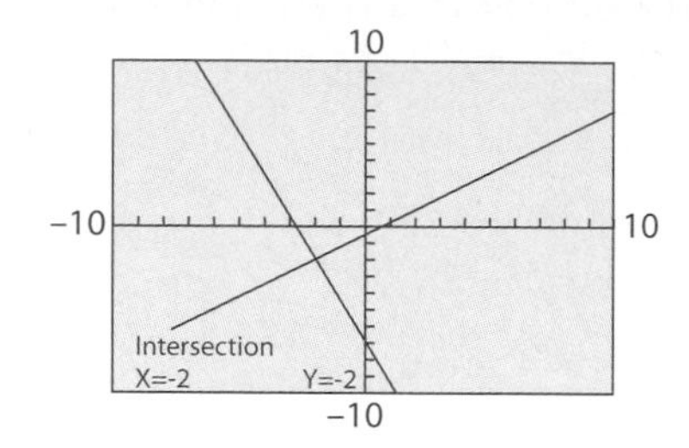

15. Solve by substitution: $\begin{aligned} y &= x + 4 \\ 2x - y &= 6 \end{aligned}$

(10, 14)

16. Solve by substitution: $\begin{aligned} y &= 2 - 3x \\ 6x + 2y &= 7 \end{aligned}$

Inconsistent

17. Solve by the addition method: $\begin{aligned} 2x - 5y &= 19 \\ 3x + 4y &= -6 \end{aligned}$

(2, −3)

18. Solve by the addition method: $\begin{aligned} 5x + 3y &= -3 \\ 3x + 7y &= 19 \end{aligned}$

(−3, 4)

19. A pilot flying with the wind flew 2100 miles from one city to another in 6 hours. The return trip against the wind took 7 hours. Find the rate of the plane in calm air and the rate of the wind. Plane, 325 mph; wind, 25 mph

20. Rowing with the wind, a sculling team went 24 miles in 2 hours. Rowing against the current, the team went 18 miles in 3 hours. Find the rate of the sculling team in calm water and the rate of the current.
Sculling team, 9 mph; current, 3 mph

21. A total of \$7000 is deposited into two simple interest accounts. On one account, the annual simple interest rate is 10%, and on the second account, the annual simple interest rate is 15%. How much should be invested in each account so that the total annual interest earned is \$800?
\$5000 at 10%; \$2000 at 15%

22. A dairy owner wants to make 13 gallons of a dairy mixture that is 14% butterfat by mixing cream that is 30% butterfat with milk that is 4% butterfat. How many gallons of each mixture must be used to make the desired concentration of butterfat? 30% butterfat, 5 gallons; 4% butterfat, 8 gallons

23. The length of a rectangle is 4 times the width and the perimeter is 200 feet. Find the dimensions of the rectangle.
Length: 80 feet, width: 20 feet

Cumulative Review Exercises

1. What is the next term in the sequence: 1, 6, 11, 16, 21, ...? 26

2. Write the negation: "All animals do not have fur." Some animals have fur.

3. Evaluate: $15 + (7 + 3^2)$ 31

4. Evaluate: $90 \div 3^2 - (4 \cdot 3)$ -2

5. If four coins are tossed, what is the probability that none of the coins shows a tail? $\frac{1}{16}$

6. The numbers of rooms occupied in a hotel on 6 consecutive days were 234, 321, 222, 246, 312, and 396. Find the mean and median for the number of rooms occupied. Mean = 288.5, median = 279

7. Translate into a variable expression: Three times a number divided by the sum of twice the number and 7. $\frac{3n}{2n+7}$

8. Evaluate: $-|-5|$ -5

9. Add: $-24 + 7 + (-13)$ -30

10. Multiply: $-10(-5)$ 50

11. What number is 8 more than -7? 1

12. Simplify: $\frac{(-4)^2 + 5}{(3)^3}$ $\frac{7}{9}$

13. Evaluate $\frac{a^2}{a - 2b}$ when $a = -3$ and $b = -4$. $\frac{9}{5}$

14. Draw the figure that is symmetric with respect to the x-axis.

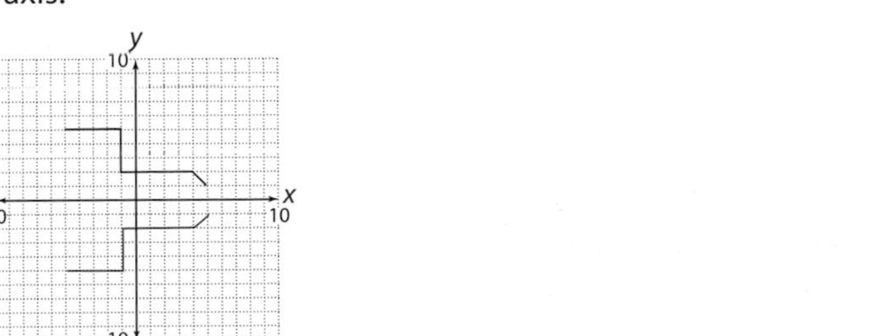

15. What are the coordinates of the point labeled A after a translation of 2 units up and 3 units to the left?

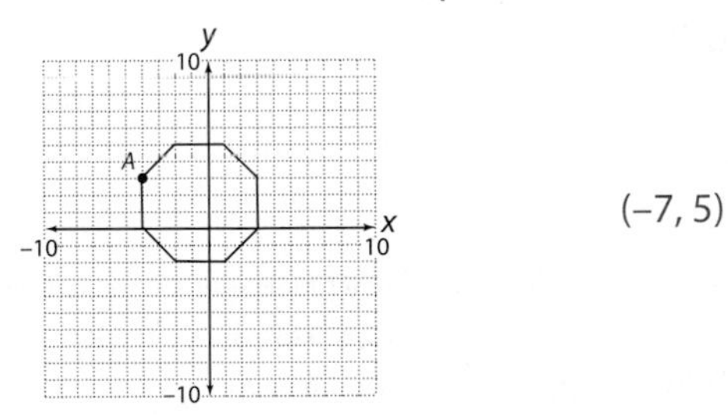

(−7, 5)

16. Simplify: $\left(\frac{5x}{9}\right)\left(\frac{3}{15}\right)$ $\frac{x}{9}$

17. Simplify: $8a - 3[4a - 2(a - 7)]$ $2a - 42$

18. Solve $3x - 8y = 24$ for y. $y = \frac{3}{8}x - 3$

19. Solve and check: $\frac{1}{2} = x + \frac{2}{3}$ $x = -\frac{1}{6}$

20. Identify the property that justifies the statement $4 \cdot 5 = 5 \cdot 4$.
Commutative Property of Multiplication

21. Solve and check: $-3x - 4 = 2x + 6$ -2

22. Find the domain and range of the relation $\{(-5, 1), (-3, 3), (-1, 5)\}$. Is the relation a function? D: $\{-5, -3, -1\}$; R: $\{1, 3, 5\}$; yes

23. Evaluate $f(x) = 2x^2 - x$ when $x = -2$. Write an ordered pair that is an element of the function. 10; $(-2, 10)$

24. Graph: $f(x) = 2x - 3$

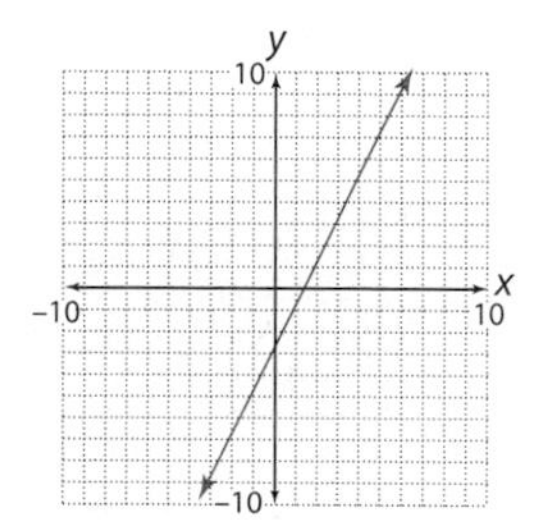

25. Graph the line whose slope is $-\frac{1}{2}$ and whose y-intercept is $(0, -3)$.

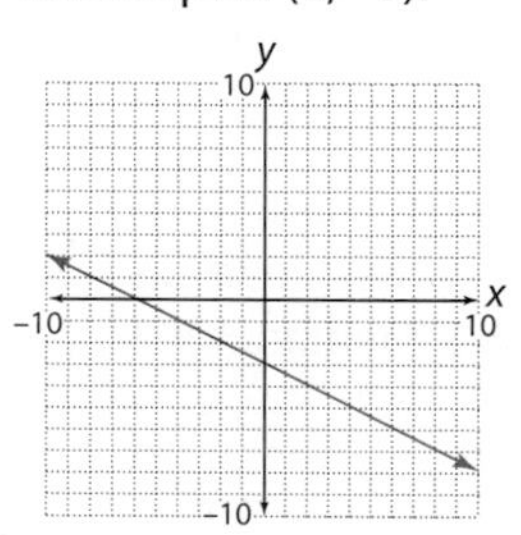

26. Solve by the addition method: $\begin{aligned} 2x + 5y &= -18 \\ 7x - 2y &= 15 \end{aligned}$

$(1, -4)$

27. Solve by substitution: $\begin{aligned} 3x - y &= 6 \\ x + 3y &= 2 \end{aligned}$

$(2, 0)$

28. Find the equation of the line that contains the point whose coordinates are $(0, 7)$ and that has slope $-\frac{2}{5}$. $y = -\frac{2}{5}x + 7$

29. A football stadium increased its 60,000-seat capacity by 15%. How many seats were added to the stadium? 9000 seats

30. How many pounds of lima beans that cost \$.90 per pound must be mixed with 16 pounds of corn that costs \$.50 per pound to make a mixture of vegetables that costs \$.65 per pound? 9.6 pounds

31. A student's grades on five sociology exams were 68, 82, 90, 73, and 95. What is the lowest score this student can receive on the sixth exam and still have earned a total of at least 480 points? 72

32. Two planes start at the same time from the same point and fly in opposite directions. The first plane is flying 100 mph faster than the second plane. In 3 hours, the two planes are 1050 miles apart. Find the rate of each plane.
1st plane, 225 mph; 2nd plane, 125 mph

33. In a triangle, the first angle is 15° more than the second angle. The third angle is three times the second angle. Find the measure of each angle.
48°, 33°, and 99°

Chapter 5

Polynomials

5

$$(x+1)(x+3) = x^2 + 4x + 3$$

$$(x+1)(x+3) = x^2 + 4x + 3$$

Section 5.1 Operations on Monomials and Scientific Notation

Addition and Subtraction of Monomials

The floor plan of a mobile home is shown below. All dimensions given are in feet. Not shown in the diagram is the fact that the height of each room is 8 feet.

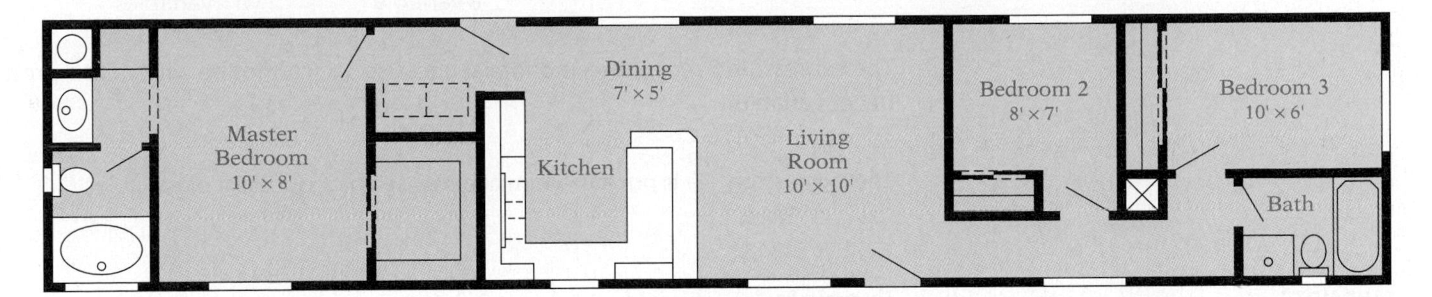

→ What is the combined length of the dining room and the living room?

The length of the dining room is 7 ft.
The length of the living room is 10 ft.

7 ft + 10 ft = 17 ft

The combined length of the dining room and the living room is 17 ft.

→ What is the combined area of the two smaller bedrooms?

The area of bedroom 3 is 10 ft x 6 ft = 60 ft^2.

The area of bedroom 2 is 8 ft x 7 ft = 56 ft^2.

60 ft^2 + 56 ft^2 = 116 ft^2

The combined area of the two smaller bedrooms is 116 ft^2.

→ What is the difference between the areas of the two smaller bedrooms?

60 ft^2 – 56 ft^2 = 4 ft^2

The difference in the areas of the two smaller bedrooms is 4 ft^2.

As shown above, we can find the sum of two areas or the sum of two lengths. However, we cannot find the sum of an area and a length. For example, the area of the dining room is 7 ft x 5 ft = 35 ft^2. The length of the living room is 10 ft.

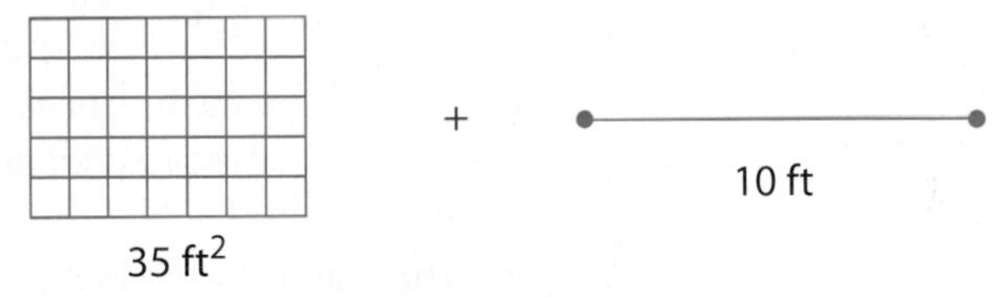

INSTRUCTOR NOTE
Addition and subtraction of like terms has been developed previously. It is included here so that (1) students will see it as addition and subtraction of monomials, (2) for completeness in covering operations on monomials, and (3) to provide students with experience in differentiating addition of monomials from multiplication of monomials. (The exercise set provides mixed practice on the operations.)

The sum 35 ft^2 + 10 ft cannot be simplified.

Just as we cannot add square feet and feet, we cannot add algebraic terms that do not have the same variable part. The same is true for subtraction.

$60x^2 + 56x^2 = 116x^2$

$60x^2 - 56x^2 = 4x^2$

Both $60x^2$ and $56x^2$ have the same variable part: x^2. Add or subtract the coefficients; the variable part stays the same.

$35x^2 + 10x$

$35x^2$ and $10x$ do not have the same variable part. The terms cannot be combined.

When adding and subtracting like terms, we are actually adding and subtracting monomials. A **monomial** is a number, a variable, or a product of a number and variables. For instance,

7	b	$\frac{2}{3}a$	$12xy^2$
A number	A variable	A product of a number and a variable	A product of a number and variables

The expression $3\sqrt{x}$ is not a monomial because $\sqrt{x}$ cannot be written as a product of variables.

The expression $\frac{2x}{y^2}$ is not a monomial because it is a quotient of variables.

TAKE NOTE
Addition of monomials involves using the Distributive Property:
$14x^3y^2z + 8x^3y^2z + 7x^3y^2z$
$= (14 + 8 + 7)x^3y^2z = 29x^3y^2z$

Example 1 Simplify: $14x^3y^2z + 8x^3y^2z + 7x^3y^2z$

Solution $14x^3y^2z + 8x^3y^2z + 7x^3y^2z$
$= 29x^3y^2z$

• The terms have the same variable part. Add the coefficients. The variable part stays the same.

You-Try-It 1 Simplify: $16a^4b^3 + 10a^4b^3 + 5a^4b^3$

Solution See page S21.

Example 2 Simplify: $29c^4d^5 - 6c^4d^5$

Solution $29c^4d^5 - 6c^4d^5$
$= 23c^4d^5$

• The terms have the same variable part. Subtract the coefficients. The variable part stays the same.

You-Try-It 2 Simplify: $37m^3n^2p - 14m^3n^2p$

Solution See page S21.

Multiplication of Monomials

→ What is the volume of air that must be heated in the master bedroom of the mobile home pictured on the previous page? Note that the height of each room is 8 ft.

The volume of air in the master bedroom $= 10\text{ ft} \times 8\text{ ft} \times 8\text{ ft}$
$= (10\text{ ft} \times 8\text{ ft}) \times 8\text{ ft}$
$= 80\text{ ft}^2 \times 8\text{ ft}$
$= 640\text{ ft}^3$

There are 640 ft^3 of air to heat in the master bedroom.

This illustrates that we can multiply square feet by feet. The result is cubic feet, or volume.

Let's look at multiplication with monomials.

POINT OF INTEREST
A billion, which is 10^9, is too large a number for most of us to comprehend. If a computer were to start counting from 1 to 1 billion, writing to the screen one number every second of every day, it would take over 31 years for the computer to complete the task.

And if a billion is a large number, consider a googol. A googol is 1 with 100 zeros after it, or 10^{100}. Edward Kasner is the mathematician credited with thinking up this number, and his nine-year-old nephew is said to have thought up the name. The two then coined the word googolplex, which is 10^{googol}.

SUGGESTED ACTIVITY
You might prefer to have students develop the rule for multiplying exponential expressions. Provide them with several expressions to simplify, for example

x^4x^3
y^2y^6
a^5a^7

Ask them to simplify the expressions by writing each expression in factored form and then writ ing the result with an exponent. After they have completed several examples, ask them to write a rule for x^mx^n. This same approach can be used for the Rule for Simplifying the Power of an Exponential Expression and the Rule for Dividing Exponential Expressions.

Recall that the exponential expression 3^4 means to multiply 3, the base, 4 times. Therefore, $3^4 = 3 \cdot 3 \cdot 3 \cdot 3 = 81$. For the variable exponential expression x^6, x is the base and 6 is the exponent. The exponent indicates the number of times the base occurs as a factor. Therefore,

$$x^6 = \overbrace{x \cdot x \cdot x \cdot x \cdot x \cdot x}^{\text{Multiply } x \text{ 6 times}}$$

The product of exponential expressions with the *same* base can be simplified by writing each expression in factored form and writing the result with an exponent.

$$x^3 \cdot x^2 = \underbrace{\overbrace{(x \cdot x \cdot x)}^{3 \text{ factors}} \cdot \overbrace{(x \cdot x)}^{2 \text{ factors}}}_{5 \text{ factors}}$$
$$= x \cdot x \cdot x \cdot x \cdot x$$
$$= x^5$$

Note that adding the exponents results in the same product.

$$x^3 \cdot x^2 = x^{3+2} = x^5$$

This suggests the following rule for multiplying exponential expressions.

Rule for Multiplying Exponential Expressions

If m and n are positive integers, then $x^m \cdot x^n = x^{m+n}$.

Example 3 Simplify: $a^4 \cdot a^5$

Solution $a^4 \cdot a^5 = a^{4+5} = a^9$ • The bases are the same. Add the exponents.

You-Try-It 3 Simplify: $t^3 \cdot t^8$

Solution See page S21.

Example 4 Simplify: $c^3 \cdot c^4 \cdot c$

Solution $c^3 \cdot c^4 \cdot c = c^{3+4+1}$
$= c^8$ • The bases are the same. Add the exponents. Note that $c = c^1$.

You-Try-It 4 Simplify: $n^6 \cdot n \cdot n^2$

Solution See page S21.

Question: Why can the exponential expression x^5y^3 not be simplified?[1]

1. The bases are not the same. The Rule for Multiplying Exponential Expressions applies only to expressions with the *same* base.

Example 5 Simplify: $(a^3b^2)(a^4)$

Solution $(a^3b^2)(a^4) = a^{3+4}b^2$
$= a^7b^2$

• Multiply variables with the same base by adding the exponents.

You-Try-It 5 Simplify: $c^9(c^5d^8)$

Solution See page S21.

TAKE NOTE
Note on page 326 that
10 ft x 8 ft x 8 ft
= (10 · 8 · 8) · (ft · ft · ft)
= 640 · (ft^{1+1+1}) = 640 ft^3

Example 6 Simplify: $(4x^3)(2x^6)$

Solution $(4x^3)(2x^6) = (4 \cdot 2)(x^3 \cdot x^6)$

• Use the Commutative and Associative Properties of Multiplication to group the coefficients and variables with the same base.

$= 8x^{3+6}$
$= 8x^9$

• Multiply the coefficients. Multiply variables with the same base by adding the exponents.

You-Try-It 6 Simplify: $(5y^4)(3y^2)$

Solution See page S21.

Example 7 Simplify: $(-2v^3z^5)(7v^2z^6)$

Solution $(-2v^3z^5)(7v^2z^6)$
$= [-2(7)](v^{3+2})(z^{5+6})$
$= -14v^5z^{11}$

• Multiply the coefficients of the monomials. Multiply variables with the same base by adding the exponents.

You-Try-It 7 Simplify: $(12p^4q^3)(-3p^5q^2)$

Solution See page S21.

The expression $(x^4)^3$ is an example of a *power of a monomial*; the monomial x^4 is raised to the third (3) power.

The power of a monomial can be simplified by writing the power in factored form and then using the Rule for Multiplying Exponential Expressions.

$$(x^4)^3 = x^4 \cdot x^4 \cdot x^4 = x^{4+4+4} = x^{12}$$

Note that multiplying the exponent inside the parentheses by the exponent outside the parentheses results in the same product.

$$(x^4)^3 = x^{4 \cdot 3} = x^{12}$$

This suggests the following rule for simplifying powers of monomials.

Rule for Simplifying the Power of an Exponential Expression

If m and n are positive integers, then $(x^m)^n = x^{m \cdot n}$.

Question: Which expression is the multiplication of two exponential expressions and which is the power of an exponential expression?[2] **a.** $q^4 \cdot q^{10}$ **b.** $(q^4)^{10}$

Example 8 Simplify: $(z^2)^5$

Solution $(z^2)^5 = z^{2 \cdot 5} = z^{10}$

• z^2 is raised to the power of 5. Simplify the power of an exponential expression by multiplying the exponents.

You-Try-It 8 Simplify: $(t^3)^6$

Solution See page S21.

The expression $(a^2b^3)^2$ is the *power of the product* of two exponential expressions, a^2 and b^3. The power of the product of exponential expressions can be simplified by writing the product in factored form and then using the Rule for Multiplying Exponential Expressions.

Write the exponential expression in factored form. Use the Rule for Multiplying Exponential Expressions.

$$\begin{aligned}(a^2b^3)^2 &= (a^2b^3)(a^2b^3)\\ &= a^{2+2}b^{3+3}\\ &= a^4b^6\end{aligned}$$

Note that multiplying each exponent inside the parentheses by the exponent outside the parentheses results in the same product.

$$\begin{aligned}(a^2b^3)^2 &= a^{2 \cdot 2}b^{3 \cdot 2}\\ &= a^4b^6\end{aligned}$$

Rule for Simplifying Powers of Products

If m, n, and p are positive integers, then $(x^my^n)^p = x^{m \cdot p}y^{n \cdot p}$.

Question: In the expression $(a^8b^6)^5$, what is the product and what is the power?[3]

Example 9 Simplify: $(x^4y)^6$

Solution

$$\begin{aligned}(x^4y)^6 &= x^{4 \cdot 6}y^{1 \cdot 6}\\ &= x^{24}y^6\end{aligned}$$

• Multiply each exponent inside the parentheses by the exponent outside the parentheses. Remember that $y = y^1$.

You-Try-It 9 Simplify: $(bc^7)^8$

Solution See page S21.

---

2. **a.** This is the multiplication of two exponential expressions. q^4 is multiplied times q^{10}.
   b. This is the power of an exponential expression. q^4 is raised to the 10th power.
3. The product is a^8b^6; a^8 is multiplied times b^6. The power is 5; a^8b^6 is raised to the 5th power.

Example 10 Simplify: $(5z^3)^2$

Solution

$$(5z^3)^2 = 5^{1 \cdot 2}z^{3 \cdot 2}$$
$$= 5^2z^6$$

• Multiply each exponent inside the parentheses by the exponent outside the parentheses. Note that $5 = 5^1$.

$$= 25z^6$$

• Evaluate 5^2.

You-Try-It 10 Simplify: $(4y^6)^3$

Solution See page S21.

Example 11 Simplify: $(3m^5p^2)^4$

Solution

$$(3m^5p^2)^4 = 3^{1 \cdot 4}m^{5 \cdot 4}p^{2 \cdot 4}$$
$$= 3^4m^{20}p^8$$

• Multiply each exponent inside the parentheses by the exponent outside the parentheses.

$$= 81m^{20}p^8$$

• Evaluate 3^4.

You-Try-It 11 Simplify: $(2v^6w^9)^5$

Solution See page S21.

Example 12 Simplify: $(-a^5b^8)^6$

Solution

$$(-a^5b^8)^6 = (-1)^{1 \cdot 6}a^{5 \cdot 6}b^{8 \cdot 6}$$
$$= (-1)^6a^{30}b^{48}$$

• Multiply each exponent inside the parentheses by the exponent outside the parentheses. Note that $-a^5b^8 = -1a^5b^8 = (-1)^1a^5b^8$.

$$= a^{30}b^{48}$$

• Evaluate $(-1)^6$. $(-1)^6 = 1$.

You-Try-It 12 Simplify: $(-2x^3y^7)^3$

Solution See page S21.

In some products, it is necessary to use the Rule for Simplifying Powers of Products and the Rule for Multiplying Exponential Expressions.

⟶ Simplify: $(3x^4)^2(4x^3)$

$$(3x^4)^2(4x^3) = (3^{1 \cdot 2}x^{4 \cdot 2})(4x^3)$$
$$= (3^2x^8)(4x^3)$$

• Use the Rule for Simplifying Powers of Products to simplify $(3x^4)^2$.

$$= (9x^8)(4x^3)$$
$$= (9 \cdot 4)(x^8 \cdot x^3)$$
$$= 36x^{8+3}$$

• Use the Rule for Multiplying Exponential Expressions.

$$= 36x^{11}$$

Example 13 Simplify: $(2a^2b)(2a^3b^2)^3$

Solution $(2a^2b)(2a^3b^2)^3$

$= (2a^2b)(2^{1\cdot 3}a^{3\cdot 3}b^{2\cdot 3})$ • Use the Rule for Simplifying Powers of Products.

$= (2a^2b)(2^3a^9b^6)$

$= (2a^2b)(8a^9b^6)$

$= (2\cdot 8)(a^{2+9})(b^{1+6})$ • Use the Rule for Multiplying Exponential Expressions.

$= 16a^{11}b^7$

You-Try-It 13 Simplify: $(-xy^4)(-2x^3y^2)^2$

Solution See page S21.

Division of Monomials

The quotient of two exponential expressions with the *same* base can be simplified by writing each expression in factored form, dividing by the common factors, and then writing the result with an exponent.

$$\frac{x^6}{x^2} = \frac{\overset{1}{\cancel{x}}\cdot\overset{1}{\cancel{x}}\cdot x\cdot x\cdot x\cdot x}{\underset{1}{\cancel{x}}\cdot\underset{1}{\cancel{x}}} = x^4$$

Note that subtracting the exponents results in the same quotient.

$$\frac{x^6}{x^2} = x^{6-2} = x^4$$

This example suggests that to divide monomials with like bases, subtract the exponents.

Rule for Dividing Exponential Expressions

If m and n are positive integers and $x \neq 0$, then $\frac{x^m}{x^n} = x^{m-n}$.

Example 14 Simplify: $\frac{c^8}{c^5}$

Solution $\frac{c^8}{c^5} = c^{8-5}$ • The bases are the same. Subtract exponents.

$= c^3$

You-Try-It 14 Simplify: $\frac{t^{10}}{t^4}$

Solution See page S22.

SUGGESTED ACTIVITY
See the *Instructor's Resource Manual* for an activity that introduces students to integer exponents.

Question: Why can the expression $\frac{x^8}{y^2}$ not be simplified?[4]

Example 15 Simplify: $\frac{x^5y^7}{x^4y^2}$

Solution $\frac{x^5y^7}{x^4y^2} = x^{5-4}y^{7-2}$

$= xy^5$

• Use the Rule for Dividing Exponential Expressions by subtracting the exponents of like bases. Note that $x^{5-4} = x^1$ but the exponent 1 is not written.

You-Try-It 15 Simplify: $\frac{a^7b^6}{ab^3}$

Solution See page S22.

The expression at the right has been simplified in two ways: by dividing by common factors, and by using the Rule for Dividing Exponential Expressions.

$$\frac{x^3}{x^3} = \frac{\overset{1}{\cancel{x}} \cdot \overset{1}{\cancel{x}} \cdot \overset{1}{\cancel{x}}}{\underset{1}{\cancel{x}} \cdot \underset{1}{\cancel{x}} \cdot \underset{1}{\cancel{x}}} = 1$$

Because $\frac{x^3}{x^3} = 1$ and $\frac{x^3}{x^3} = x^0$, 1 must equal x^0. Therefore, the following definition of zero as an exponent is used.

$$\frac{x^3}{x^3} = x^{3-3} = x^0$$

> **Zero as an Exponent**
>
> If $x \neq 0$, then $x^0 = 1$. The expression 0^0 is undefined.

Example 16 Simplify: $(-15y^4)^0, y \neq 0$

Solution $(-15y^4)^0 = 1$

• Any nonzero expression to the zero power is 1.

You-Try-It 16 Simplify: $(-8x^2y^7)^0, x \neq 0, y \neq 0$

Solution See page S22.

---

4. The bases are not the same. The Rule for Dividing Exponential Expressions applies only to expressions with the *same* base.

Example 17 Simplify: $-(6r^3t^2)^0, r \neq 0, t \neq 0$

Solution $-(6r^3t^2)^0 = -1$

• $(6r^3t^2)^0 = 1$. The negative sign in front of the parentheses can be read "the opposite of." The opposite of 1 is -1.

You-Try-It 17 Simplify: $-(9c^7d^4)^0, c \neq 0, d \neq 0$

Solution See page S22.

The expression at the right has been simplified in two ways: by dividing by common factors, and by using the Rule for Dividing Exponential Expressions.

$$\frac{x^3}{x^5} = \frac{\overset{1}{\cancel{x}} \cdot \overset{1}{\cancel{x}} \cdot \overset{1}{\cancel{x}}}{\underset{1}{\cancel{x}} \cdot \underset{1}{\cancel{x}} \cdot \underset{1}{\cancel{x}} \cdot x \cdot x} = \frac{1}{x^2}$$

$$\frac{x^3}{x^5} = x^{3-5} = x^{-2}$$

Because $\frac{x^3}{x^5} = \frac{1}{x^2}$ and $\frac{x^3}{x^5} = x^{-2}$, $\frac{1}{x^2}$ must equal x^{-2}. Therefore the following definition of a negative exponent is used.

Definition of Negative Exponents

If n is a positive integer and $x \neq 0$, then $x^{-n} = \frac{1}{x^n}$ and $\frac{1}{x^{-n}} = x^n$.

An exponential expression is in simplest form when there are no negative exponents in the expression.

POINT OF INTEREST
In the 15th century, the expression $12^{2\bar{m}}$ was used to mean $12x^{-2}$. The use of $\bar{m}$ reflected an Italian influence. In Italy, m was used for minus and p was used for plus. It was understood that $2\bar{m}$ referred to an unnamed variable. Isaac Newton, in the 17th century, advocated the use of a negative exponent.

→ Simplify: y^{-7}

$y^{-7} = \frac{1}{y^7}$ • Use the Definition of Negative Exponents to rewrite the expression with a positive exponent.

→ Simplify: $\frac{1}{c^{-4}}$

$\frac{1}{c^{-4}} = c^4$ • Use the Definition of Negative Exponents to rewrite the expression with a positive exponent.

Question: How are **a.** b^{-8} and **b.** $\frac{1}{w^{-5}}$ rewritten with positive exponents?[5]

---

5. **a.** $b^{-8} = \frac{1}{b^8}$ **b.** $\frac{1}{w^{-5}} = w^5$

Example 18 Simplify: $\frac{3n^{-5}}{4}$

Solution $\frac{3n^{-5}}{4} = \frac{3}{4}n^{-5} = \frac{3}{4} \cdot \frac{1}{n^5}$ • Use the Definition of Negative Exponents to rewrite the expression with a positive exponent.

$= \frac{3}{4n^5}$

You-Try-It 18 Simplify: $\frac{2}{c^{-4}}$

Solution See page S22.

A numerical expression with a negative exponent can be evaluated by first rewriting the expression with a positive exponent.

→ Evaluate: 2^{-3}

TAKE NOTE
Note from the example at the right that 2^{-3} is a *positive* number. A negative exponent does not indicate a negative number.

$2^{-3} = \frac{1}{2^3}$ • Use the Definition of Negative Exponents to rewrite the expression with a positive exponent.

$= \frac{1}{8}$ • Evaluate 2^3.

Sometimes applying the Rule for Dividing Exponential Expressions results in a quotient that contains a negative exponent. If this happens, use the Definition of Negative Exponents to rewrite the expression with a positive exponent.

→ Simplify: $\frac{6x^2}{8x^9}$

$\frac{6x^2}{8x^9} = \frac{3x^2}{4x^9} = \frac{3x^{2-9}}{4}$ • Divide the coefficients by their common factors. Then use the Rule for Dividing Exponential Expressions.

$= \frac{3x^{-7}}{4} = \frac{3}{4} \cdot \frac{x^{-7}}{1} = \frac{3}{4} \cdot \frac{1}{x^7}$ • Rewrite the expression with only positive exponents.

$= \frac{3}{4x^7}$

Example 19 Simplify: $\frac{-35a^6b^{-2}}{25a^{-3}b^5}$

Solution $\frac{-35a^6b^{-2}}{25a^{-3}b^5} = -\frac{7a^6b^{-2}}{5a^{-3}b^5} = -\frac{7a^{6-(-3)}b^{-2-5}}{5} = -\frac{7a^9b^{-7}}{5} = -\frac{7a^9}{5b^7}$

You-Try-It 19 Simplify: $\dfrac{12x^{-8}y}{-16xy^{-3}}$

Solution See page S22.

The rules for simplifying exponential expressions and powers of exponential expressions are true for all integers. These rules are restated here.

Rules of Exponents

If m, n, and p are integers, then:

$x^m \cdot x^n = x^{m+n}$ $\quad (x^m)^n = x^{m \cdot n}$ $\quad (x^m y^n)^p = x^{m \cdot p} y^{n \cdot p}$

$\dfrac{x^m}{x^n} = x^{m-n}, x \neq 0$ $\quad x^0 = 1, x \neq 0$ $\quad x^{-n} = \dfrac{1}{x^n}, \dfrac{1}{x^{-n}} = x^n, x \neq 0$

Simplifying the expressions in Example 20 and You-Try-It 20 requires the rules for multiplying exponential expressions as well as the definition of negative exponents.

Example 20 Simplify: $(-2x)(3x^{-2})^{-3}$

Solution $(-2x)(3x^{-2})^{-3}$

$= (-2x)(3^{-3}x^6)$ • Use the Rule for Simplifying Powers of Products.

$= \dfrac{-2x \cdot x^6}{3^3}$ • Write the expression with positive exponents.

$= -\dfrac{2x^7}{27}$ • Use the Rule for Multiplying Exponential Expressions. Simplify 3^3.

You-Try-It 20 Simplify: $(-3ab)(2a^3b^{-2})^{-3}$

Solution See page S22.

Scientific Notation

Very large and very small numbers are encountered in the fields of science and engineering. For example, the charge of an electron is 0.000000000000000000160 coulomb. These numbers can be written more easily in scientific notation. In scientific notation, a number is expressed as a product of two factors, one a number between 1 and 10 and the other a power of 10.

To change a number written in decimal notation to one written in scientific notation, write it in the form $a \times 10^n$, where $1 \leq a < 10$ and n is an integer.

TAKE NOTE
There are two steps in writing a number in scientific notation: (1) determine the number between 1 and 10, and (2) determine the exponent on 10.

For numbers greater than 10, move the decimal point to the right of the first digit. The exponent n is positive and equal to the number of places the decimal point has been moved.

$240{,}000 = 2.4 \times 10^5$

$93{,}000{,}000 = 9.3 \times 10^7$

For numbers less than 1, move the decimal point to the right of the first nonzero digit. The exponent n is negative. The absolute value of the exponent is equal to the number of places the decimal point has been moved.

$0.00030 = 3.0 \times 10^{-4}$

$0.0000832 = 8.32 \times 10^{-5}$

Look at the last example above: $0.0000832 = 8.32 \times 10^{-5}$. Using the Definition of Negative Exponents,

$$10^{-5} = \frac{1}{10^5} = \frac{1}{100{,}000} = 0.00001$$

Since $10^{-5} = 0.00001$, we can write

$$8.32 \times 10^{-5} = 8.32 \times 0.00001 = 0.0000832$$

which is the number we started with. We have not changed the value of the number; we have just written it in another form.

Example 21 Write the number in scientific notation.

a. 824,300,000,000 **b.** 0.000000961

Solution

a. $824{,}300{,}000{,}000 = 8.243 \times 10^{11}$

- Move the decimal point 11 places to the left. The exponent on 10 is 11.

b. $0.000000961 = 9.61 \times 10^{-7}$

- Move the decimal point 7 places to the right. The exponent on 10 is −7.

You-Try-It 21 Write the number in scientific notation.

a. 57,000,000,000 **b.** 0.000000017

Solution See page S22.

Changing a number written in scientific notation to decimal notation also requires moving the decimal point.

When the exponent on 10 is positive, move the decimal point to the right the same number of places as the exponent.

$3.45 \times 10^9 = 3{,}450{,}000{,}000$

$2.3 \times 10^8 = 230{,}000{,}000$

When the exponent on 10 is negative, move the decimal point to the left the same number of places as the absolute value of the exponent.

$8.1 \times 10^{-3} = 0.0081$

$6.34 \times 10^{-6} = 0.00000634$

Example 22 Write the number in decimal notation.

a. 7.329×10^6 **b.** 6.8×10^{-10}

Solution

a. 7.329×10^6
$= 7{,}329{,}000$

• The exponent on 10 is positive. Move the decimal point 6 places to the right.

b. 6.8×10^{-10}
$= 0.00000000068$

• The exponent on 10 is negative. Move the decimal point 10 places to the left.

You-Try-It 22 Write the number in decimal notation.

a. 5×10^{12} **b.** 4.0162×10^{-9}

Solution See page S22.

Question: Is the expression written in scientific notation?[6]

a. 2.84×10^{-4} **b.** 36.5×10^7 **c.** 0.91×10^{-12}

The rules for multiplying and dividing with numbers in scientific notation are the same as those for operating on algebraic expressions. The power of 10 corresponds to the variable and the number between 1 and 10 corresponds to the coefficient of the variable.

	Algebraic Expressions	Scientific Notation
Multiplication	$(4x^{-3})(2x^5) = 8x^2$	$(4 \times 10^{-3})(2 \times 10^5) = 8 \times 10^2$
Division	$\frac{6x^5}{3x^{-2}} = 2x^{5-(-2)} = 2x^7$	$\frac{6 \times 10^5}{3 \times 10^{-2}} = 2 \times 10^{5-(-2)} = 2 \times 10^7$

Example 23 Multiply or divide.

a. $(3.0 \times 10^5)(1.1 \times 10^{-8})$ **b.** $\frac{7.2 \times 10^{13}}{2.4 \times 10^{-3}}$

Solution

a. $(3.0 \times 10^5)(1.1 \times 10^{-8})$
$= 3.3 \times 10^{-3}$

• Multiply 3.0 times 1.1. Add the exponents on 10.

b. $\frac{7.2 \times 10^{13}}{2.4 \times 10^{-3}}$
$= 3 \times 10^{16}$

• Divide 7.2 by 2.4. Subtract the exponents on 10.

You-Try-It 23 Multiply or divide.

a. $(2.4 \times 10^{-9})(1.6 \times 10^3)$ **b.** $\frac{5.4 \times 10^{-2}}{1.8 \times 10^{-4}}$

Solution See page S22.

---

6. **a.** 2.84×10^{-4} is written in scientific notation. **b.** 36.5×10^7 is not written in scientific notation. 36.5 is not a number between 1 and 10. **c.** 0.91×10^{-12} is not written in scientific notation. 0.91 is not a number between 1 and 10.

5.1 EXERCISES

Topics for Discussion

1. Explain why each of the following is or is not a monomial.

a. $32a^3b$ This is a monomial because it is the product of a number, 32, and variables, a^3b.

b. $\frac{5n^3}{7}$ This is a monomial because it is the product of a number, $\frac{5}{7}$, and a variable, n.

c. $\frac{6c^4}{25d}$ This is not a monomial because there is a variable in the denominator.

2. Explain each of the following. Provide an example of each.

a. The Rule for Multiplying Exponential Expressions
To multiply exponential expressions with the same base, add the exponents.

b. The Rule for Simplifying the Power of an Exponential Expression
The power of an exponential expression is simplified by multiplying the exponents.

c. The Rule for Simplifying Powers of Products
To simplify the power of the product of exponential expressions, multiply each exponent inside the parentheses by the exponent outside the parentheses.

d. The Rule for Dividing Exponential Expressions
To divide two exponential expressions with the same base, subtract the exponents.

3. Explain the error in each of the following. Then correct the error.

a. $6x^{-3} = \frac{1}{6x^3}$ The exponent on 6 is positive one; it should not be moved to the denominator. $6x^{-3} = \frac{6}{x^3}$

b. $xy^{-2} = \frac{1}{xy^2}$ The exponent on x is positive one; it should not be moved to the denominator. $xy^{-2} = \frac{x}{y^2}$

c. $\frac{1}{8a^{-4}} = 8a^4$ The exponent on 8 is positive one; it should not be moved to the numerator. $\frac{1}{8a^{-4}} = \frac{a^4}{8}$

d. $\frac{1}{b^{-5}c} = b^5c$ The exponent on c is positive one; it should not be moved to the numerator. $\frac{1}{b^{-5}c} = \frac{b^5}{c}$

4. In your own words, explain how you know that a number is written in scientific notation.
Answers will vary.

Operations on Monomials

Simplify.

5. $a^4 \cdot a^5$
a^9

6. $y^5 \cdot y^8$
y^{13}

7. $z^3 \cdot z \cdot z^4$
z^8

8. $b \cdot b^2 \cdot b^6$
b^9

9. $(x^3)^5$
x^{15}

10. $(b^2)^4$
b^8

11. $(x^2y^3)^6$
$x^{12}y^{18}$

12. $(m^4n^2)^3$
$m^{12}n^6$

13. $12s^4t^3 + 5s^4t^3$

$17s^4t^3$

14. $8b^6c^5 + 9b^6c^5$

$17b^6c^5$

15. 27^0

1

16. $-(17)^0$

-1

17. $\frac{a^8}{a^2}$

a^6

18. $\frac{c^{12}}{c^5}$

c^7

19. w^{-8}

$\frac{1}{w^8}$

20. m^{-9}

$\frac{1}{m^9}$

21. $(-m^3n)(m^6n^2)$

$-m^9n^3$

22. $(-r^4t^3)(r^2t^9)$

$-r^6t^{12}$

23. $(2x^4)^3$

$8x^{12}$

24. $(3n^3)^3$

$27n^9$

25. $(2x)(3x^2)(4x^4)$

$24x^7$

26. $(5a^2)(4a)(3a^5)$

$60a^8$

27. $(-2a^2)^3$

$-8a^6$

28. $(-3b^3)^2$

$9b^6$

29. $\frac{1}{a^{-5}}$

a^5

30. $\frac{1}{c^{-6}}$

c^6

31. $11p^4q^5 - 7p^4q^5$

$4p^4q^5$

32. $16c^2d^3 - 9c^2d^3$

$7c^2d^3$

33. $(6r^2)(-4r)$

$-24r^3$

34. $(7v^3)(-2v)$

$-14v^4$

35. $(2a^3bc^2)^3$

$8a^9b^3c^6$

36. $(4xy^3z^2)^2$

$16x^2y^6z^4$

37. $\frac{m^4n^7}{m^3n^5}$

mn^2

38. $\frac{a^5b^6}{a^3b^2}$

a^2b^4

39. $(3x)^0$

1

40. $(2a)^0$

1

41. $\frac{6r^4}{4r^2}$

$\frac{3r^2}{2}$

42. $\frac{8x^9}{28x^6}$

$\frac{2x^3}{7}$

43. $\frac{-16a^7}{24a^6}$

$-\frac{2a}{3}$

44. $\frac{18b^5}{-45b^4}$

$-\frac{2b}{5}$

45. $(9mn^4p)(-3mp^2)$

$-27m^2n^4p^3$

46. $(-3v^2wz)(-4vz^4)$

$12v^3wz^5$

47. $(\ xy^5)(3x^2)(5y^3)$

$-15x^3y^8$

48. $(-6m^3n)(-mn^2)(m)$

$6m^5n^3$

49. $(2x)(3x^3)^2$

$18x^7$

50. $(3a)(6a^4)^2$

$108a^9$

51. 4^{-3}

$\frac{1}{64}$

52. 5^{-2}

$\frac{1}{25}$

53. $\dfrac{x^4}{x^9}$

$\dfrac{1}{x^5}$

54. $\dfrac{b}{b^5}$

$\dfrac{1}{b^4}$

55. $(-2n^2)(-3n^4)^3$

$54n^{14}$

56. $(-3m^3n)(-2m^2n^3)^3$

$24m^9n^{10}$

57. $4x^{-7}$

$\dfrac{4}{x^7}$

58. $-6y^{-1}$

$-\dfrac{6}{y}$

59. $\dfrac{1}{3x^{-2}}$

$\dfrac{x^2}{3}$

60. $\dfrac{2}{5c^{-6}}$

$\dfrac{2c^6}{5}$

61. $(3a^3)^2(2a^2)^3$

$72a^{12}$

62. $(2x^2)^4(4x^4)^3$

$1024x^{20}$

63. $\dfrac{3x^4y^5}{6x^4y^8}$

$\dfrac{1}{2y^3}$

64. $\dfrac{7a^3b^6}{21a^5b^6}$

$\dfrac{1}{3a^2}$

65. $\dfrac{14x^4y^6z^2}{16x^3y^9z}$

$\dfrac{7xz}{8y^3}$

66. $\dfrac{25x^4y^7z^2}{20x^5y^9z^{11}}$

$\dfrac{5}{4xy^2z^9}$

67. $(-2x^3y^2)^3(-xy^2)^4$

$-8x^{13}y^{14}$

68. $(-m^4n^2)^5(-2m^3n^3)^3$

$8m^{29}n^{19}$

69. $8x^2y + 5x^2y + 6x^2y$

$19x^2y$

70. $7cd^4 + 9cd^4 + 10cd^4$

$26cd^4$

71. $(3x^{-1}y^{-2})^2$

$\dfrac{9}{x^2y^4}$

72. $(5xy^{-3})^{-2}$

$\dfrac{y^6}{25x^2}$

73. $(2x^{-1})(x^{-3})$

$\dfrac{2}{x^4}$

74. $(-2x^{-5})(x^7)$

$-2x^2$

75. $\dfrac{3x^{-2}y^2}{6xy^2}$

$\dfrac{1}{2x^3}$

76. $\dfrac{2x^{-2}y}{8xy}$

$\dfrac{1}{4x^3}$

77. $\dfrac{2x^{-1}y^{-4}}{4xy^2}$

$\dfrac{1}{2x^2y^6}$

78. $\dfrac{3a^{-2}b}{ab}$

$\dfrac{3}{a^3}$

79. $\dfrac{-16xy^5}{96x^4y^4}$

$-\dfrac{y}{6x^3}$

80. $\dfrac{-8x^2y^4}{44y^2z^5}$

$-\dfrac{2x^2y^2}{11z^5}$

81. Find the length of line segment AC.

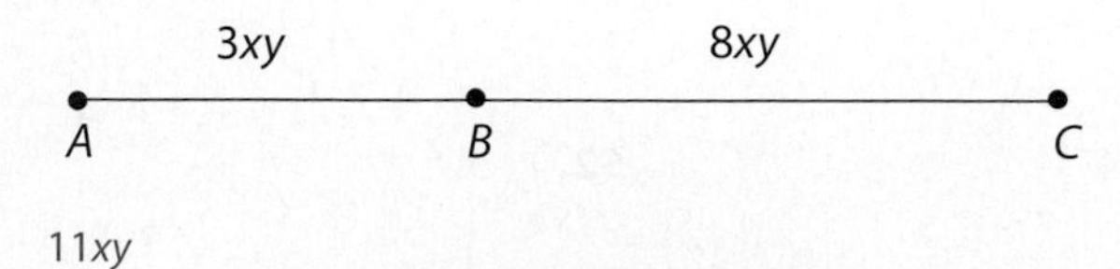

$11xy$

82. Find the length of line segment DF.

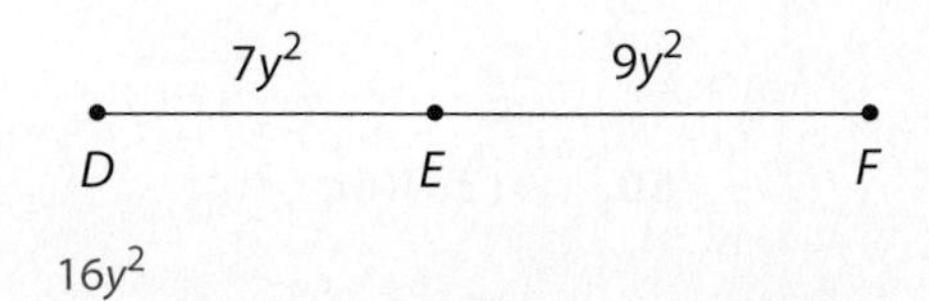

$16y^2$

83. The length of line segment *LN* is $27a^2b$. Find the length of line segment *MN*.

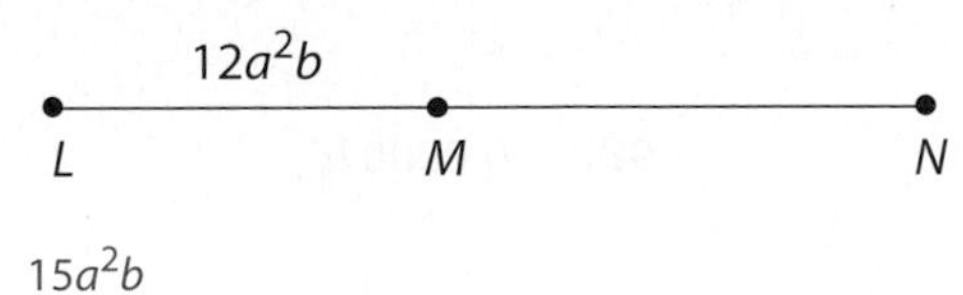

$15a^2b$

84. The length of line segment *QS* is $18c^3$. Find the length of line segment *QR*.

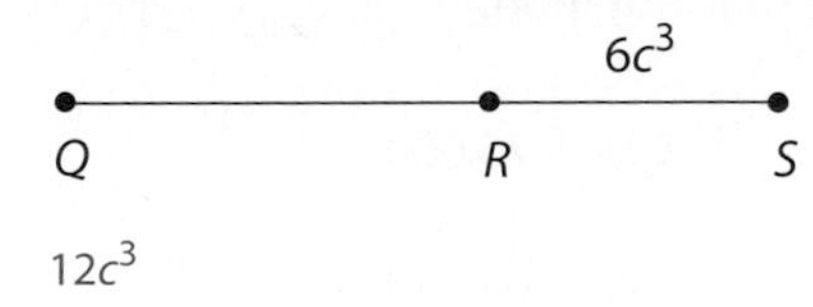

$12c^3$

85. Find the area of the square. The dimension given is in meters.

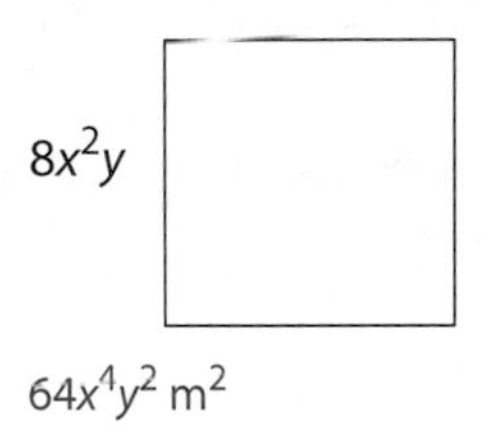

$64x^4y^2$ m^2

86. Find the area of rectangle *ABCD*. The dimensions given are in feet.

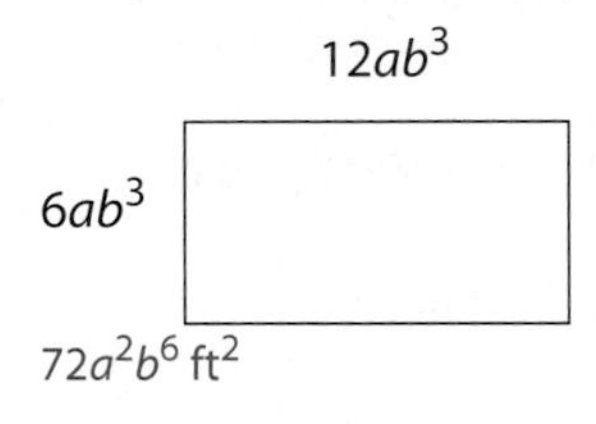

$72a^2b^6$ ft^2

87. Find the perimeter of the rectangle. The dimensions given are in miles.

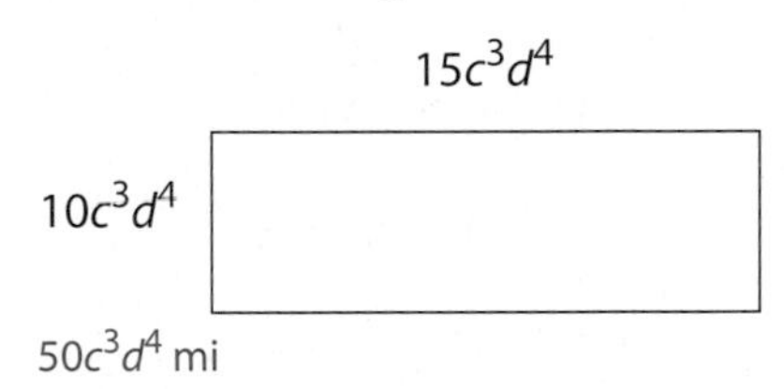

$50c^3d^4$ mi

88. Find the perimeter of the square. The dimension given is in centimeters.

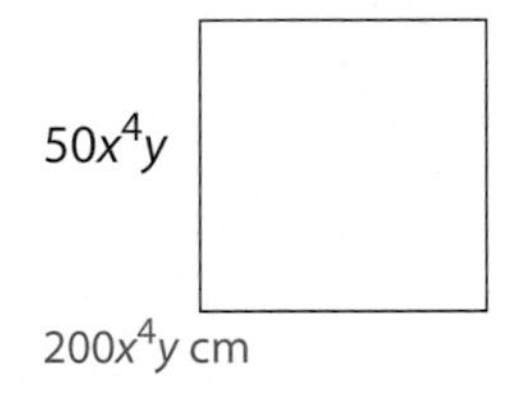

$200x^4y$ cm

89. Find the area of the rectangle. The dimensions given are in kilometers.

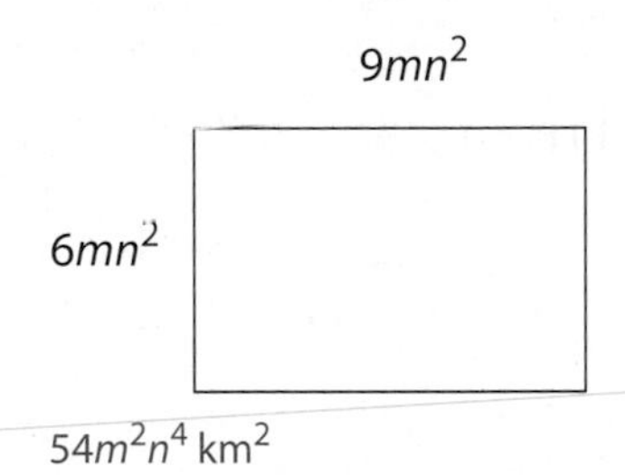

$54m^2n^4$ km^2

90. Find the area of the parallelogram. The dimensions given are in inches.

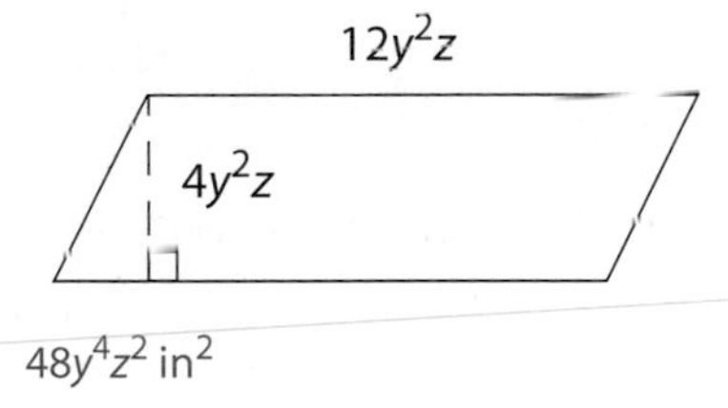

$48y^4z^2$ in^2

91. The area of the rectangle is $24a^3b^5$ square yards. Find the length of the rectangle.

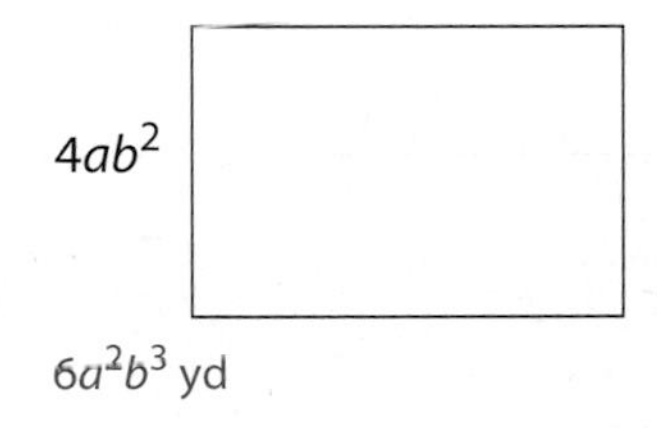

$6a^2b^3$ yd

92. The area of the parallelogram is $56w^4z^6$ square meters. Find the height of the parallelogram.

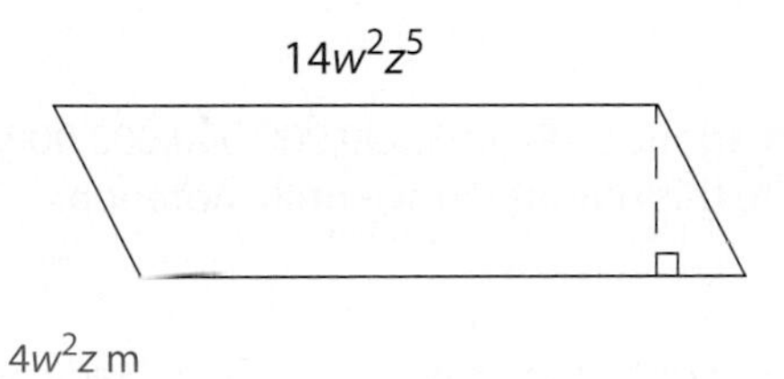

$4w^2z$ m

93. The product of a monomial and $4b$ is $12a^2b$. Find the monomial. $3a^2$

94. The product of a monomial and $8y^2$ is $32x^2y^3$. Find the monomial. $4x^2y$

Scientific Notation

Write the number in scientific notation.

95. 2,370,000

2.37×10^6

96. 75,000

7.5×10^4

97. 0.00045

4.5×10^{-4}

98. 0.000076

7.6×10^{-5}

99. 309,000

3.09×10^5

100. 819,000,000

8.19×10^8

101. 0.000000601

6.01×10^{-7}

102. 0.00000000096

9.6×10^{-10}

103. 57,000,000,000

5.7×10^{10}

104. 934,800,000,000

9.348×10^{11}

105. 0.000000017

1.7×10^{-8}

106. 0.0000009217

9.217×10^{-7}

Write the number in decimal notation.

107. 7.1×10^5

710,000

108. 2.3×10^7

23,000,000

109. 4.3×10^{-5}

0.000043

110. 9.21×10^{-7}

0.000000921

111. 6.71×10^8

671,000,000

112. 5.75×10^9

5,750,000,000

113. 7.13×10^{-6}

0.00000713

114. 3.54×10^{-8}

0.0000000354

115. 5×10^{12}

5,000,000,000,000

116. 1.0987×10^{11}

109,870,000,000

117. 8.01×10^{-3}

0.00801

118. 4.0162×10^{-9}

0.0000000040162

Solve.

119. Light travels approximately 16,000,000,000 miles in one day. Write this number in scientific notation.
1.6×10^{10}

120. Earth's mass is about 5,980,000,000,000,000,000,000,000 kilograms. Write this number in scientific notation.
5.98×10^{24}

121. The graph at the right shows the spending on Medicaid in 1990, 1993, and 1996. Write the dollar amount spent on Medicaid in 1996 in scientific notation.
$\$1.67 \times 10^{11}$

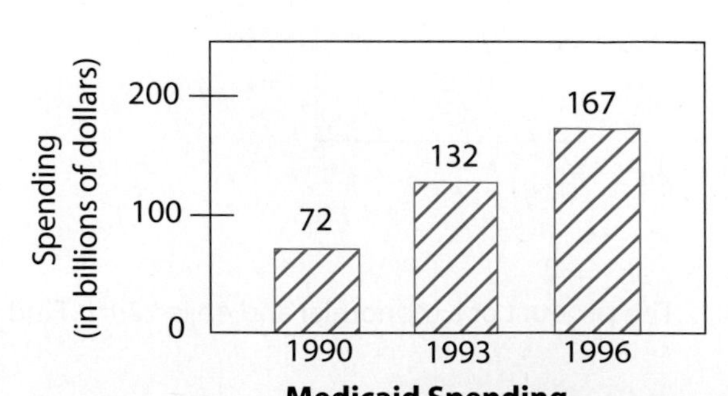

Medicaid Spending
Source: Health Care Financing Administration

122. The length of an infrared light wave is approximately 0.0000037 meter. Write this number in scientific notation.
3.7×10^{-6}

123. The electric charge on an electron is 0.00000000000000000016 coulomb. Write this number in scientific notation.

1.6×10^{-19}

124. A unit used to measure the speed of a computer is the picosecond. One picosecond is 0.000000001 of a second. Write one picosecond in scientific notation.

1×10^{-9}

125. Avogadro's number is used in chemistry. Its value is approximately 602,300,000,000,000,000,000,000. Write this number in scientific notation.

6.023×10^{23}

126. The graph at the right shows the monetary cost of four wars. Write the monetary cost of World War II in scientific notation.

3.1×10^{12}

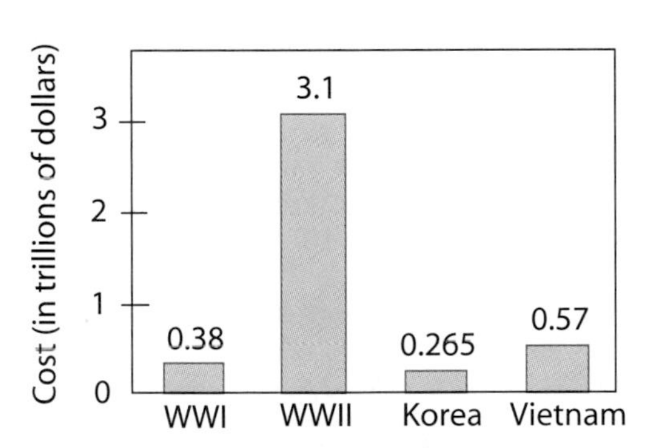

Monetary Cost of War

Source: Congressional Research Service Using Numbers from the *Statistical Abstract of the United States*

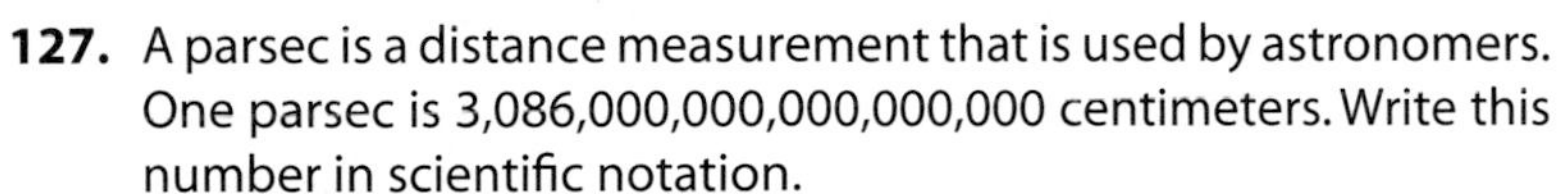

127. A parsec is a distance measurement that is used by astronomers. One parsec is 3,086,000,000,000,000,000 centimeters. Write this number in scientific notation.

3.086×10^{18}

128. One light-year is the distance traveled by light in one year. One light-year is 5,880,000,000,000 miles. Write this number in scientific notation.

5.88×10^{12}

Simplify.

129. $(1.9 \cdot 10^{12})(3.5 \cdot 10^{7})$

6.65×10^{19}

130. $(4.2 \cdot 10^{7})(1.8 \cdot 10^{-5})$

7.56×10^{2}

131. $(2.3 \cdot 10^{-8})(1.4 \cdot 10^{-6})$

3.22×10^{-14}

132. $(3 \cdot 10^{-20})(2.4 \cdot 10^{9})$

7.2×10^{-11}

133. $\dfrac{6.12 \cdot 10^{14}}{1.7 \cdot 10^{9}}$

3.6×10^{5}

134. $\dfrac{6 \cdot 10^{-8}}{2.5 \cdot 10^{-2}}$

2.4×10^{-6}

135. $\dfrac{5.58 \cdot 10^{-7}}{3.1 \cdot 10^{11}}$

1.8×10^{-18}

136. $\dfrac{9.03 \cdot 10^{6}}{4.3 \cdot 10^{-5}}$

2.1×10^{11}

Applying Concepts

137. Evaluate.

a. $8^{-2} + 2^{-5}$ $\quad \dfrac{3}{64}$

b. $9^{-2} + 3^{-3}$ $\quad \dfrac{4}{81}$

138. Determine whether the statement is always true, sometimes true, or never true.

a. The phrase "a power of a monomial" means that a monomial is the base of an exponential expression. Always true

b. To multiply $x^m \cdot x^n$, multiply the exponents. Never true

c. The rules of exponents can be applied to expressions that contain an exponent of zero or contain negative exponents. Always true

d. Like terms have the same coefficient and the same variable part. Sometimes true

e. The expression 3^{-2} represents the reciprocal of 3^2. Always true

139. Evaluate 2^x and 2^{-x} when $x = -2, -1, 0, 1,$ and 2. $\frac{1}{4}, \frac{1}{2}, 1, 2, 4; 4, 2, 1, \frac{1}{2}, \frac{1}{4}$

140. Write in decimal notation.

a. 2^{-4} 0.0625

b. 25^{-2} 0.0016

141. If $m = n + 1$ and $a \neq 0$, then $\dfrac{a^m}{a^n} =$ _____. a

142. Solve: $(-8.5)^x = 1$ 0

Explorations

143. ***Scientific Notation and Order Relations***

a. Place the correct symbol, $<$ or $>$, between the two numbers.

(1) 5.23×10^{18} **?** 5.23×10^{17}
(2) 3.12×10^{11} **?** 3.12×10^{12}
(3) 3.45×10^{-14} **?** 3.45×10^{-15}

b. Write a rule for ordering two numbers written in scientific notation.

144. ***Expressions with Negative Exponents***

a. If x is a nonzero real number, is x^{-2} always positive, always negative, or positive or negative depending on whether x is positive or negative? Explain your answer.

b. If x is a nonzero real number, is x^{-3} always positive, always negative, or positive or negative depending on whether x is positive or negative? Explain your answer.

145. ***Negative Exponents on Fractional Expressions***

a. Simplify each of the following expressions.

(1) $\left(\dfrac{a^2}{b^3}\right)^{-2}$ **(2)** $\left(\dfrac{x^4}{y}\right)^{-3}$ **(3)** $\left(\dfrac{c^5}{d^2}\right)^{-4}$ **(4)** $\left(\dfrac{2^3}{3^4}\right)^{-1}$

b. Write a rule for rewriting with a positive exponent a fraction raised to a negative exponent.

Section 5.2 Operations on Polynomials

Introduction to Polynomials

Some forecasters predict that revenue generated by business on the Internet from 1997 to 2002 can be approximated by the function

$$R(t) = 15.8t^2 - 17.2t + 10.2$$

where R is the annual revenue in billions of dollars and t is the time in years, with $t = 0$ corresponding to the year 1997. Use this function to approximate the annual revenue in the year 2000.

Since $t = 0$ corresponds to 1997, $t = 3$ corresponds to the year 2000. Evaluate the given function for $t = 3$.

$$\begin{aligned} R(t) &= 15.8t^2 - 17.2t + 10.2 \\ R(3) &= 15.8(3)^2 - 17.2(3) + 10.2 \\ &= 15.8(9) - 17.2(3) + 10.2 \\ &= 142.2 - 51.6 + 10.2 \\ &= 100.8 \end{aligned}$$

According to this function, in the year 2000, the revenue generated by business conducted on the Internet will be approximately \$100.8 billion.

In the function $R(t) = 15.8t^2 - 17.2t + 10.2$, the variable expression

$$15.8t^2 - 17.2t + 10.2$$

is a polynomial. A **polynomial** is a variable expression in which the terms are monomials. The polynomial $15.8t^2 - 17.2t + 10.2$ has three terms: $15.8t^2$, $-17.2t$, and 10.2. Note that each of these terms is a monomial.

INSTRUCTOR NOTE
An analogy may help students understand these terms. Dogs, lions, and monkeys are specific types of animals, just as monomials, binomials, and trinomials are specific types of polynomials.

A polynomial of *one* term is a **monomial.** — $-7x^2$ is a monomial.

A polynomial of *two* terms is a **binomial.** — $4y + 3$ is a binomial.

A polynomial of *three* terms is a **trinomial.** — $6b^2 + 5b - 8$ is a trinomial.

Question: Is the expression a polynomial? If it is a polynomial, is it a monomial, a binomial, or a trinomial?[1]

a. $16a^2 - 9b^2$ **b.** $-\frac{2}{3}xy$ **c.** $x^2 + 2xy - 8$ **d.** $\frac{3}{x} - 5$

The terms of a polynomial in one variable are usually arranged so that the exponents of the variable decrease from left to right. This is called **descending order.** The polynomials at the right are written in descending order.

$2x^3 - 3x^2 + 6x - 1$

$5y^4 - 9y^3 + y^2 - 7y + 8$

$t - 4$

1. **a.** It is a polynomial. It is a binomial because it has two terms ($16a^2$ and $9b^2$). **b.** It is a polynomial. It is a monomial because it has one term. **c.** It is a polynomial. It is a trinomial because it has three terms (x^2, $2xy$, and -8). **d.** $\frac{3}{x}$ is not a monomial, so $\frac{3}{x} - 5$ is not a polynomial.

Question: Is the polynomial written in descending order?[2]

a. $3a^2 - 2a^3 + 4a$ **b.** $6d^5 + 4d^3 - 7$

The **degree of a polynomial in one variable** is its largest exponent.

The degree of $t - 4$ is 1. It is a first-degree or **linear** polynomial.
The degree of $6b^2 + 5b - 8$ is 2. It is a second-degree or **quadratic** polynomial.
The degree of $2x^3 - 3x^2 + 6x - 1$ is 3. It is a third-degree or **cubic** polynomial.
The degree of $5y^4 - 9y^3 + y^2 - 7y + 8$ is 4. It is a fourth-degree polynomial.

POINT OF INTEREST
The dimples on a golf ball have a dramatic effect on its flight. A golf ball with a well-designed dimple pattern will travel 2 to 3 times farther than a ball with no dimples.

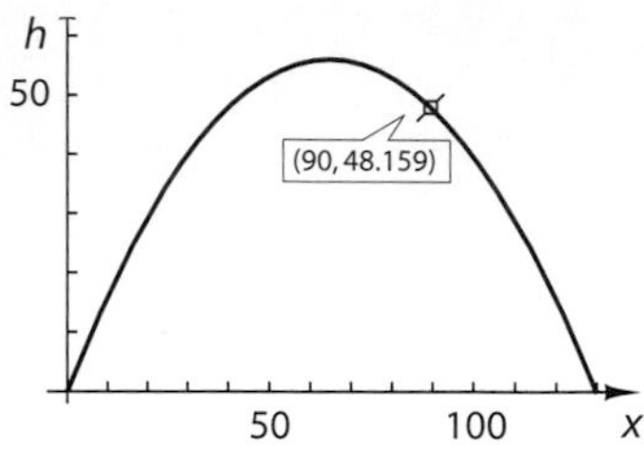

Polynomial functions are used to model many different situations. For instance, ignoring forces other than that of gravity, the height, h, of a golf ball x feet from the point from where it was hit is given by the quadratic polynomial function $h(x) = -0.0133x^2 + 1.7321x$. By evaluating this function, we can determine the height of the ball at various distances from the point at which it was struck. The graph above shows that the ball is 48.159 feet high at a distance of 90 feet from the point at which it was hit. Evaluating the polynomial when $x = 90$, we have

$$\begin{aligned} h(x) &= -0.0133x^2 + 1.7321x \\ h(90) &= -0.0133(90)^2 + 1.7321(90) \quad \text{• Replace } x \text{ by 90.} \\ &= 48.159 \end{aligned}$$

The polynomial function above gave the height of the ball in terms of its distance from where it was hit. That is, the height depended on distance. A different polynomial function can be written that will give the height of the ball in terms of the amount of *time* the ball is in flight. Using this function, you can determine the height of the ball at various *times* during its flight. This is shown in Example 1.

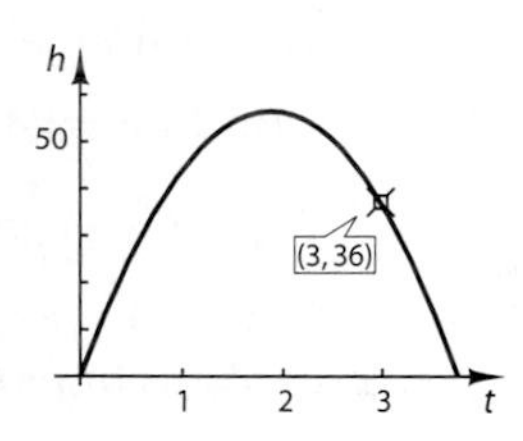

SUGGESTED ACTIVITY
Have students investigate the graphs on this page by answering such questions as:
1. Does the golf ball reach a maximum height?
2. Are there different distances from where the ball was struck that it is the same distance above the ground?
3. Are there different times when the ball is the same distance above the ground?
4. How can you find the time it takes the ball to be 90 feet from where it was struck?

Example 1 The height, h, of a golf ball t seconds after it has been struck is given by $h(t) = -16t^2 + 60t$. Determine the height of the ball 3 seconds after it is hit.

Solution

$$\begin{aligned} h(t) &= -16t^2 + 60t \\ h(3) &= -16(3)^2 + 60(3) \quad \text{• Replace } t \text{ by 3.} \\ &= 36 \end{aligned}$$

The golf ball is 36 feet high 3 seconds after being hit. See the graph at the left.

You-Try-It 1 If \$2000 is deposited into an Individual Retirement Account (IRA), then the value, V, of that investment three years later is given by the cubic polynomial function $V(r) = 2000r^3 + 6000r^2 + 6000r + 2000$, where r is the interest rate (as a decimal) earned on the investment. Determine the value after three years of \$2000 deposited in an IRA that earns an interest rate of 7%.

Solution See page S22.

---

2. **a.** No. The exponents on a (2 and 3) do not decrease from left to right. $-2a^3 + 3a^2 + 4$ is the same polynomial written in descending order. **b.** Yes. The exponents on d (5 and 3) decrease from left to right.

Addition and Subtraction of Polynomials

Polynomials can be added by combining like terms.

Example 2 Add: $(3x^2 - 7x + 4) + (5x^2 + 2x - 8)$

Solution $(3x^2 - 7x + 4) + (5x^2 + 2x - 8)$

$= (3x^2 + 5x^2) + (-7x + 2x) + (4 - 8)$

- Use the Properties of Addition to rearrange and group like terms.

$= 8x^2 - 5x - 4$

- Combine like terms. Write the polynomial in descending order.

You-Try-It 2 Add: $(-4x^2 - 3xy + 2y^2) + (3x^2 - 4y^2)$

Solution See page S22.

In Example 2, you can use a graphing calculator to check your work. Here are some typical graphing calculator screens that you might use to produce the graphs.

TAKE NOTE
Note the equal signs for Y3 and Y4. These indicate that Y3 and Y4 will be graphed but that Y1 and Y2 will not be graphed.

Enter the polynomials to be added.

Enter the sum.

Enter your answer.

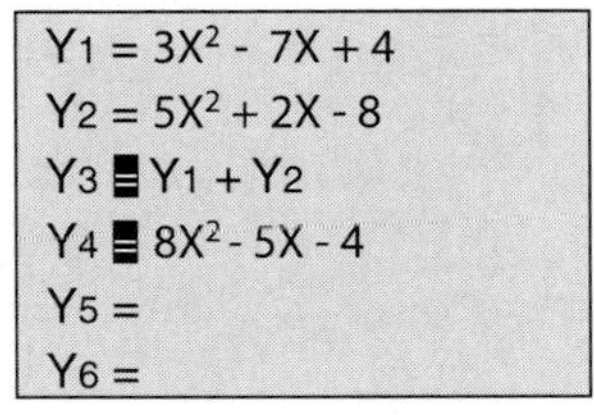

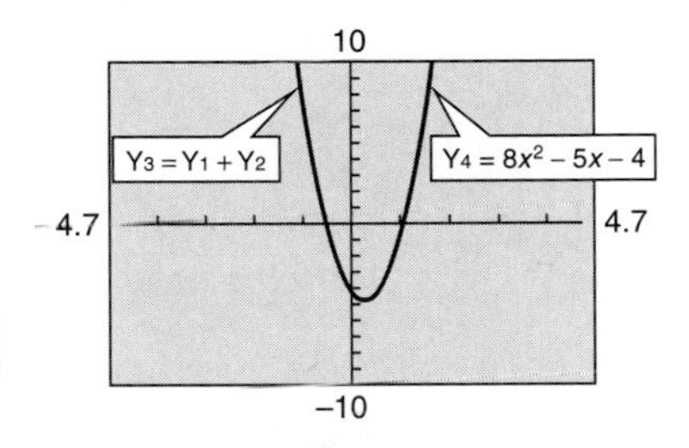

When the graphs above were produced, the graph of Y3 lies on top of Y4. This means that the two graphs are the same and therefore that the addition is correct. If the two graphs do not match exactly, the addition is incorrect. It is not always possible to do a graphical check of your work. In You-Try-It 2, the polynomials have two variables. These polynomials cannot be graphed with a graphing calculator.

Recall that the definition of subtraction is addition of the opposite.

$$a - b = a + (-b)$$

This definition holds true for polynomials. Polynomials can be subtracted by adding the opposite of the second polynomial to the first. The **opposite** of a polynomial is the polynomial with the sign of every term changed.

The opposite of the polynomial $x^2 - 2x + 3$ is $-x^2 + 2x - 3$.

The opposite of the polynomial $-4y^3 + 5y - 8$ is $4y^3 - 5y + 8$.

Question: What is the opposite of the polynomial $5d^4 - 6d^2 + 9$?[3]

---

3. The opposite of $5d^4 - 6d^2 + 9$ is $-5d^4 + 6d^2 - 9$.

Example 3 Subtract: $(-3a^2 - 7) - (-8a^2 + a - 4)$

Solution $(-3a^2 - 7) - (-8a^2 + a - 4)$

$= (-3a^2 - 7) + (8a^2 - a + 4)$ • Rewrite subtraction as addition of the opposite.

$= (-3a^2 + 8a^2) - a + (-7 + 4)$ • Use the Properties of Addition to rearrange and group like terms.

$= 5a^2 - a - 3$ • Combine like terms. Write the polynomial in descending order.

You-Try-It 3 Subtract: $(5x^2 - 3x + 4) - (-6x^3 - 2x + 8)$

Solution See page S22.

A company's **revenue** is the money the company earns by selling its products. A company's **cost** is the money it spends to manufacture and sell its products. A company's **profit** is the difference between its revenue and its cost. This relationship is expressed by the formula $P = R - C$, where P is the profit, R is the revenue, and C is the cost. This formula is used in Example 4 and You-Try-It 4.

Example 4 A company manufactures and sells wood stoves. The total monthly cost, in dollars, to produce n wood stoves is $30n + 2000$. The company's revenue, in dollars, obtained from selling all n wood stoves is $-0.4n^2 + 150n$. Express in terms of n the company's monthly profit.

Solution

State the goal. Our goal is to write a variable expression for the company's profit from manufacturing and selling n wood stoves.

Describe a strategy. Use the formula $P = R - C$. Substitute the given polynomials for R and C. Then subtract the polynomials.

Solve the problem.

$P = R - C$

$P = (-0.4n^2 + 150n) - (30n + 2000)$ • $R = -0.4n^2 + 150n, C = 30n + 2000$

$P = (-0.4n^2 + 150n) + (-30n - 2000)$ • Rewrite subtraction as addition of the opposite.

$P = -0.4n^2 + (150n - 30n) - 2000$

$P = -0.4n^2 + 120n - 2000$

The company's monthly profit, in dollars, is $-0.4n^2 + 120n - 2000$.

Check your work. √

You-Try-It 4 A company's total monthly cost, in dollars, for manufacturing and selling n videotapes per month is $35n + 2000$. The company's revenue, in dollars, from selling all n videotapes is $-0.2n^2 + 175n$. Express in terms of n the company's monthly profit.

Solution See page S23.

Multiplication of Polynomials

To multiply a polynomial by a monomial, use the Distributive Property and the Rule for Multiplying Exponential Expressions.

The monomial $-2x$ is multiplied by the trinomial $x^2 - 4x - 3$ as follows:

$-2x(x^2 - 4x - 3)$

$= -2x(x^2) - (-2x)(4x) - (-2x)(3)$ • Use the Distributive Property.

$= -2(x^{1+2}) - (-2 \cdot 4)(x^{1+1}) - (-2 \cdot 3)x$ • Use the Rule for Multiplying Exponential Expressions.

$= -2x^3 + 8x^2 + 6x$

Example 5 Multiply. **a.** $(5y + 4)(-2y)$ **b.** $x^3(2x^2 - 3x + 2)$

Solution **a.** $(5y + 4)(-2y) = 5y(-2y) + 4(-2y)$

$= -10y^2 - 8y$

b. $x^3(2x^2 - 3x + 2) = x^3(2x^2) - x^3(3x) + x^3(2)$

$= 2x^5 - 3x^4 + 2x^3$

You-Try-It 5 Multiply. **a.** $(-2d + 3)(-4d)$ **b.** $-a^3(3a^2 + 2a - 7)$

Solution See page S23.

Multiplication of two polynomials requires the repeated application of the Distributive Property.

Shown below is the binomial $y - 2$ multiplied by the trinomial $y^2 + 3y + 1$.

$(y - 2)(y^2 + 3y + 1)$

$= (y - 2)(y^2) + (y - 2)(3y) + (y - 2)(1)$ • Use the Distributive Property to multiply y – 2 times each term of the trinomial.

$= y^3 - 2y^2 + 3y^2 - 6y + y - 2$ • Use the Distributive Property to multiply each term of the trinomial times y – 2.

$= y^3 + y^2 - 5y - 2$ • Combine like terms.

INSTRUCTOR NOTE
Before doing an example similar to the one at the right, provide an illustration of multiplying two whole numbers, for example, 473 x 28. Compare the procedure for multiplying two polynomials to this example.

Two polynomials can also be multiplied using a vertical format similar to that used for multiplication of whole numbers.

$$\begin{array}{r} y^2 + 3y + 1 \\ \underline{y - 2} \\ -2y^2 - 6y - 2 \\ \underline{y^3 + 3y^2 + y \quad\;\;} \\ y^3 + y^2 - 5y - 2 \end{array}$$

• Multiply each term in the trinomial by –2.

• Multiply each term in the trinomial by y. Like terms must be written in the same column.

• Add the terms in each column.

Example 6 Multiply: $(2b^3 - b + 1)(2b + 3)$

Solution

$$\begin{array}{rl} 2b^3 - b + 1 & \\ 2b + 3 & \\ \hline 6b^3 \quad\quad - 3b + 3 & = 3(2b^3 - b + 1) \\ 4b^4 \quad\quad - 2b^2 + 2b \quad\quad & = 2b(2b^3 - b + 1) \\ \hline & \text{Like terms are in the same column.} \\ 4b^4 + 6b^3 - 2b^2 - b + 3 & \text{Add the terms in each column.} \end{array}$$

You-Try-It 6 Multiply: $(3c^3 - 2c^2 + c - 3)(2c + 5)$

Solution See page S23.

The product of two binomials can be found by using a method called **FOIL**, which is based on the Distributive Property. The letters of FOIL stand for **F**irst, **O**uter, **I**nner, and **L**ast.

TAKE NOTE
The FOIL method is not really a different way of multiplying two polynomials. It is based on the Distributive Property.
$(2x + 3)(x + 5)$
$= (2x + 3)x + (2x + 3)5$
$= 2x^2 + 3x + 10x + 15$
$= 2x^2 + 13x + 15$

Note that the terms
$2x^2 + 3x + 10x + 15$
are the same products found using the FOIL method.

To multiply $(2x + 3)(x + 5)$:

Multiply the First terms.	$(\mathbf{2x} + 3)(\mathbf{x} + 5)$	$2x \cdot x = 2x^2$
Multiply the Outer terms.	$(\mathbf{2x} + 3)(x + \mathbf{5})$	$2x \cdot 5 = 10x$
Multiply the Inner terms.	$(2x + \mathbf{3})(\mathbf{x} + 5)$	$3 \cdot x = 3x$
Multiply the Last terms.	$(2x + \mathbf{3})(x + \mathbf{5})$	$3 \cdot 5 = 15$
		F O I L
Add the products.	$(2x + 3)(x + 5)$	$= 2x^2 + 10x + 3x + 15$
Combine like terms.		$= 2x^2 + 13x + 15$

Example 7 Multiply. **a.** $(4x - 3)(3x - 2)$ **b.** $(3x - 2y)(x + 4y)$

Solution **a.** This is the product of two binomials. Use the FOIL method.

$(4x - 3)(3x - 2)$
$= 4x(3x) + (4x)(-2) + (-3)(3x) + (-3)(-2)$
$= 12x^2 - 8x - 9x + 6$
$= 12x^2 - 17x + 6$

b. This is the product of two binomials. Use the FOIL method.

$(3x - 2y)(x + 4y)$
$= 3x(x) + (3x)(4y) + (-2y)(x) + (-2y)(4y)$
$= 3x^2 + 12xy - 2xy - 8y^2$
$= 3x^2 + 10xy - 8y^2$

You-Try-It 7 Multiply. **a.** $(4y - 5)(3y - 3)$ **b.** $(3a + 2b)(3a - 5b)$

Solution See page S23.

SUGGESTED ACTIVITY
See the *Instructor's Resource Manual* for an activity involving squaring a binomial.

The expression $(a + b)^2$ is the square of a binomial. We are squaring $(a + b)$, which means that we are multiplying it times itself.

$$(a + b)^2 = (a + b)(a + b)$$

$(a + b)(a + b)$ is the product of two binomials. Use the FOIL method to multiply.

$$(a + b)^2 = (a + b)(a + b) = a^2 + ab + ab + b^2 = a^2 + 2ab + b^2$$

Example 8 Multiply: $(3x - 2)^2$

Solution

$(3x - 2)^2$ • This is the square of a binomial. Multiply $(3x - 2)$ times itself.
$= (3x - 2)(3x - 2)$
$= 9x^2 - 6x - 6x + 4$ • Use the FOIL method.
$= 9x^2 - 12x + 4$ • Combine like terms.

You-Try-It 8 Multiply: $(3x + 2y)^2$

Solution See page S23.

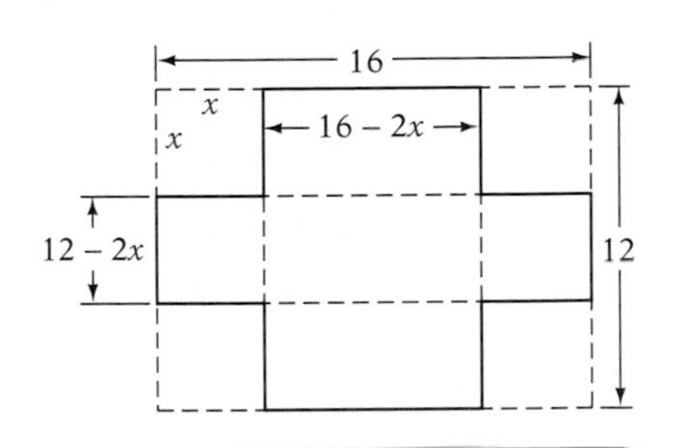

Example 9 A rectangular piece of cardboard measures 12 inches by 16 inches. An open box is formed by cutting four squares that measure x inches on a side from the corners of the cardboard and then folding up the sides as shown in the figure at the left. Determine the volume of the box in terms of x. Use your equation to find the volume of the box when x is 2 inches.

Solution

State the goal. The goal is to determine the volume of the box in terms of x, and then to find the volume when $x = 2$.

Describe a strategy. To determine the volume of the box in terms of x, use the formula for the volume of a box, $V = LWH$. Substitute variable expressions for L, W, and H. Then multiply. To determine the volume when $x = 2$, substitute 2 for x in the equation. Then simplify the numerical expression.

Solve the problem.

$V = LWH$
$= (16 - 2x)(12 - 2x)x$ • $L = 16 - 2x, W = 12 - 2x, H = x$.
$= (4x^2 - 56x + 192)x$ • Multiply $(16 - 2x)(12 - 2x)$.
$= 4x^3 - 56x^2 + 192x$ • Multiply by x.

The volume of the box in terms of x is $(4x^3 - 56x^2 + 192x)$ cubic inches.

$V = 4x^3 - 56x^2 + 192x$
$= 4(2)^3 - 56(2)^2 + 192(2)$ • Replace x by 2.
$= 192$

When x is 2 inches, the volume of the box is 192 cubic inches.

Check your work. √

You-Try-It 9 The radius of a circle is $(x - 4)$ feet. Find the area of the circle in terms of the variable x. Leave the answer in terms of π.

Solution See page S23.

Division of Polynomials

As shown below, $\frac{8+4}{2}$ can be simplified by first adding the terms in the numerator and then dividing the result by the denominator. It can also be simplified by first dividing each term in the numerator by the denominator and then adding the results.

$$\frac{8+4}{2} = \frac{12}{2} = 6 \qquad \frac{8+4}{2} = \frac{8}{2} + \frac{4}{2} = 4 + 2 = 6$$

It is this second method that is used to divide a polynomial by a monomial: divide each term in the numerator by the denominator, and then write the sum of the quotients.

TAKE NOTE
Recall that the fraction bar can be read "divided by."

To divide $\frac{6x^2+4x}{2x}$, divide each term of the polynomial $6x^2 + 4x$ by the monomial $2x$. Then simplify each quotient.

$$\frac{6x^2+4x}{2x} = \frac{6x^2}{2x} + \frac{4x}{2x}$$
$$= 3x + 2$$

Example 10 Divide: $\frac{6x^3 - 3x^2 + 9x}{3x}$

Solution $\frac{6x^3 - 3x^2 + 9x}{3x}$

$= \frac{6x^3}{3x} - \frac{3x^2}{3x} + \frac{9x}{3x}$ • Divide each term in the numerator by the denominator.

$= 2x^2 - x + 3$ • Simplify each quotient.

You-Try-It 10 Divide: $\frac{4x^3y + 8x^2y^2 - 4xy^3}{2xy}$

Solution See page S23.

The method illustrated above is appropriate only when the divisor is a monomial. To divide two polynomials in which the divisor is not a monomial, a method similar to that used for division of whole numbers is used.

INSTRUCTOR NOTE
It may help students if you start with the division algorithm for whole numbers and show them that a similar procedure is used to divide polynomials.

To divide $(x^2 - 5x + 8) \div (x - 3)$:

Step 1

$$\begin{array}{r} x \phantom{{}-5x+8} \\ x-3\,\overline{)\,x^2 - 5x + 8} \\ \underline{x^2 - 3x} \phantom{{}+8} \\ -2x + 8 \end{array}$$

Think: $x\overline{)x^2} = \frac{x^2}{x} = x$

Multiply: $x(x - 3) = x^2 - 3x$
Subtract: $(x^2 - 5x) - (x^2 - 3x)$
Bring down the 8.

Step 2

$$\begin{array}{r} x - 2 \\ x-3\,\overline{)\,x^2 - 5x + 8} \\ \underline{x^2 - 3x} \phantom{{}+8} \\ -2x + 8 \\ \underline{-2x + 6} \\ 2 \end{array}$$

Think: $x\overline{)-2x} = \frac{-2x}{x} = -2$

Multiply: $-2(x - 3) = -2x + 6$
Subtract: $(-2x + 8) - (2x + 6) = 2$
The remainder is 2.

The same equation used to check division of whole numbers is used to check polynomial division.

(Quotient x Divisor) + Remainder = Dividend

Check: $(x - 2)(x - 3) + 2 = x^2 - 3x - 2x + 6 + 2 = x^2 - 5x + 8$

$$(x^2 - 5x + 8) \div (x - 3) = x - 2 + \frac{2}{x - 3}$$

Note that the remainder is used to write a fraction with the remainder over the divisor. This is similar to arithmetic, in which the answer to $15 \div 4$ is written $3\frac{3}{4}$.

Example 11 Divide: $(6x + 2x^3 + 26) \div (x + 2)$

Solution Arrange the terms of the dividend in descending order. There is no term of x^2 in $2x^3 + 6x + 26$. Insert a $0x^2$ for the missing term so that like terms will be in columns.

$$\begin{array}{r} 2x^2 - 4x + 14 \\ x+2\,\overline{)\,2x^3 + 0x^2 + 6x + 26} \\ \underline{2x^3 + 4x^2} \phantom{{}+6x+26} \\ -4x^2 + 6x \phantom{{}+26} \\ \underline{-4x^2 - 8x} \phantom{{}+26} \\ 14x + 26 \\ \underline{14x + 28} \\ -2 \end{array}$$

$$(6x + 2x^3 + 26) \div (x + 2) = 2x^2 - 4x + 14 - \frac{2}{x + 2}$$

You-Try-It 11 Divide: $(x^3 - 7 - 2x) \div (x - 2)$

Solution See page S24.

TAKE NOTE
Recall that a factor of a number divides that number evenly. A factor of a polynomial divides that polynomial evenly. The quotient is another factor of the polynomial.

Example 12 Given that $x + 1$ is a factor of $x^3 + x^2 + 4x + 4$, find another factor of $x^3 + x^2 + 4x + 4$.

Solution

$$\begin{array}{r} x^2 + 4 \\ x + 1 \overline{)\, x^3 + x^2 + 4x + 4} \\ \underline{x^3 + x^2} \\ 0 + 4x + 4 \\ \underline{4x + 4} \\ 0 \end{array}$$

Another factor of $x^3 + x^2 + 4x + 4$ is $x^2 + 4$.

You-Try-It 12 Given that $x + 5$ is a factor of $x^4 + 5x^3 + 2x + 10$, find another factor of $x^4 + 5x^3 + 2x + 10$.

Solution See page S24.

Example 13 The area of a rectangle is $(2x^2 - 3x - 9)$ square meters. The width of the rectangle is $(x - 3)$ meters. Find the length of the rectangle in terms of the variable x. Use the formula $L = \frac{A}{W}$, where L is the length, A is the area, and W is the width of a rectangle.

$x - 3$

Solution

State the goal. The goal is to write a variable expression for the length of a rectangle that has an area of $(2x^2 - 3x - 9)$ square meters and a width of $(x - 3)$ meters.

Describe a strategy. Use the formula $L = \frac{A}{W}$. Substitute the given polynomials for A and W. Then divide.

Solve the problem.

$$L = \frac{A}{W} = \frac{2x^2 - 3x - 9}{x - 3}$$

$$\begin{array}{r} 2x + 3 \\ x - 3 \overline{)\, 2x^2 - 3x - 9} \\ \underline{2x^2 - 6x} \\ 3x - 9 \\ \underline{3x - 9} \\ 0 \end{array}$$

The length of the rectangle is $(2x + 3)$ meters.

Check your work. √

You-Try-It 13 The area of a parallelogram is $(3x^2 + 2x - 8)$ square feet. The length of the base is $(x + 2)$ feet. Find the height of the parallelogram in terms of the variable x. Use the formula $h = \frac{A}{b}$, where h is the height, A is the area, and b is the length of the base of a parallelogram.

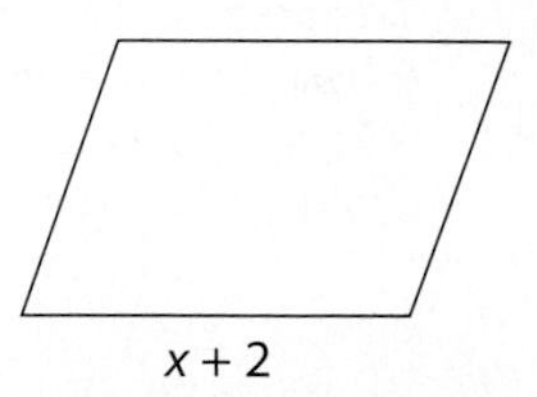

Solution See page S24.

5.2 EXERCISES

Topics for Discussion

1. State whether the polynomial is a monomial, a binomial, or a trinomial. Explain your answer.

a. $8x^4 - 6x^2$ This is a binomial. It contains two terms, $8x^4$ and $-6x^2$.

b. $4a^2b^2 + 9ab + 10$ This is a trinomial. It contains three terms, $4a^2b^2$, $9ab$, and 10.

c. $7x^3y^4$ This is a monomial. It is one term, $7x^3y^4$. (Note: It is a product of a number and variables. There is no addition or subtraction operation in the expression.)

2. Explain each of the following terms. Give an example of each.

a. polynomial A polynomial is a variable expression in which the terms are monomials. Examples will vary.

b. monomial A monomial is a polynomial of one term. Examples will vary.

c. binomial A binomial is a polynomial of two terms. Examples will vary.

d. trinomial A trinomial is a polynomial of three terms. Examples will vary.

3. State whether or not the expression is a polynomial. Explain your answer.

a. $\frac{1}{5}x^3 + \frac{1}{2}x$ Yes. Both $\frac{1}{5}x^3$ and $\frac{1}{2}x$ are monomials. (Note: The coefficients of variables can be fractions.)

b. $\frac{1}{5x^2} + \frac{1}{2x}$ No. A polynomial does not have a variable in the denominator of a fraction.

c. $x + \sqrt{5}$ Yes. Both x and $\sqrt{5}$ are monomials. (Note: The variable is not under a radical sign.)

4. Determine whether the statement is always true, sometimes true, or never true.

a. The terms of a polynomial are monomials. Always true.

b. Subtraction is addition of the opposite. Always true.

c. The FOIL method is used to multiply two binomials. Sometimes true.

d. Using the FOIL method, the terms $3x$ and 5 are the "First" terms in $(3x + 5)(2x + 7)$. Never true.

e. To square a binomial means to multiply it times itself. Always true.

Introduction to Polynomials

5. As sand that is very fine is poured into a pile, the volume, V, of the cone-shaped pile is given by $V(h) = \frac{8}{3}\pi h^3$, where h is the height of the cone. Find the volume of a sand pile that is 2 feet high. Round to the nearest hundredth.

67.02 ft^3

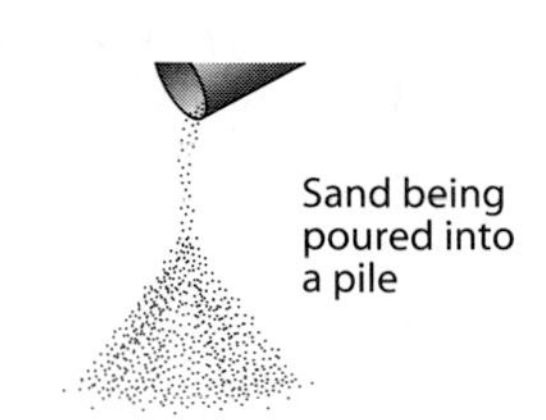

6. The area, A, of a rectangle with a perimeter of 100 meters is given by $A(w) = 50w - w^2$, where w is the width of the rectangle. What is the area of this rectangle when the width is 10 meters?

400 m^2

7. The wavelength, L (in meters), of a deep-water wave can be approximated by the function $L(v) = 0.6411v^2$, where v is the wave speed in meters per second. Find the length of a deep-water wave that has a speed of 30 meters per second.

576.99 m

8. The height, h, of a cliff diver t seconds after beginning a dive can be modeled by the function $h(t) = -16t^2 + 5t + 50$. How high is the cliff from which the diver is jumping? (*Hint*: Determine the value of t before the diver starts the dive.)
50 ft

9. The probability, P, that all three security lights in a garage will fail is given by the function $P(x) = 1 - 3x + 3x^2 - x^3$, where x is the probability that one light will not fail. Find the probability that all three lights will fail if the probability that one light will fail is 0.01. Write the answer to the nearest tenth of a percent.
97.0%

10. In the diagram at the right, the total number of circles, T, when there are n rows is given by $T = 0.5n^2 + 0.5n$. Verify the formula for the four figures shown. What is the total number of circles when there are 10 rows?
For $n = 1, T = 1$. For $n = 2, T = 3$. For $n = 3, T = 6$. For $n = 4, T = 10$. For $n = 10, T = 55$.

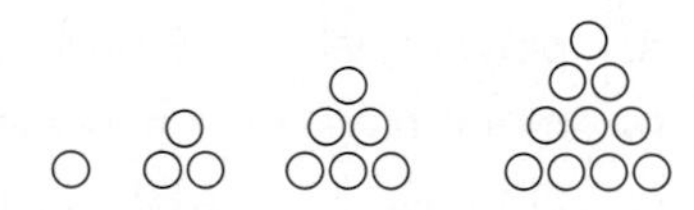

Addition and Subtraction of Polynomials

Add or subtract.

11. $(x^2 + 7x) + (-3x^2 - 4x)$
$-2x^2 + 3x$

12. $(3y^2 - 2y) + (5y^2 + 6y)$
$8y^2 + 4y$

13. $(x^2 - 6x) - (x^2 - 10x)$
$4x$

14. $(y^2 + 4y) - (y^2 + 10y)$
$-6y$

15. $(4x^2 - 5xy) + (3x^2 + 6xy - 4y^2)$
$7x^2 + xy - 4y^2$

16. $(2x^2 - 4y^2) + (6x^2 - 2xy + 4y^2)$
$8x^2 - 2xy$

17. $(2y^2 - 4y) - (-y^2 + 2)$
$3y^2 - 4y - 2$

18. $(-3a^2 - 2a) - (4a^2 - 4)$
$-7a^2 - 2a + 4$

19. $(2a^2 - 7a + 10) + (a^2 + 4a + 7)$
$3a^2 - 3a + 17$

20. $(-6x^2 + 7x + 3) + (3x^2 + x + 3)$
$-3x^2 + 8x + 6$

21. $(x^2 - 2x + 1) - (x^2 + 5x + 8)$
$-7x - 7$

22. $(3x^2 + 2x - 2) - (5x^2 - 5x + 6)$
$-2x^2 + 7x - 8$

23. $(-2x^3 + x - 1) - (-x^2 + x - 3)$
$-2x^3 + x^2 + 2$

24. $(2x^2 + 5x - 3) - (3x^3 + 2x - 5)$
$-3x^3 + 2x^2 + 3x + 2$

25. $(x^3 - 7x + 4) + (2x^2 + x - 10)$
$x^3 + 2x^2 - 6x - 6$

26. $(3y^3 + y^2 + 1) + (-4y^3 - 6y - 3)$
$-y^3 + y^2 - 6y - 2$

27. $(5x^3 + 7x - 7) + (10x^2 - 8x + 3)$
$5x^3 + 10x^2 - x - 4$

28. $(3y^3 + 4y + 9) + (2y^2 + 4y - 21)$
$3y^3 + 2y^2 + 8y - 12$

29. $(2y^3 + 6y - 2) - (y^3 + y^2 + 4)$
$y^3 - y^2 + 6y - 6$

30. $(-2x^2 - x + 4) - (-x^3 + 3x - 2)$
$x^3 - 2x^2 - 4x + 6$

31. $(4y^3 - y - 1) - (2y^2 - 3y + 3)$
$4y^3 - 2y^2 + 2y - 4$

32. $(3x^2 - 2x - 3) - (2x^3 - 2x^2 + 4)$
$-2x^3 + 5x^2 - 2x - 7$

33. Find the length of line segment AC.

$3x^2 - 4x + 5$ $8x^2 + 6x - 1$

A   B   C

$11x^2 + 2x + 4$

34. Find the length of line segment DF.

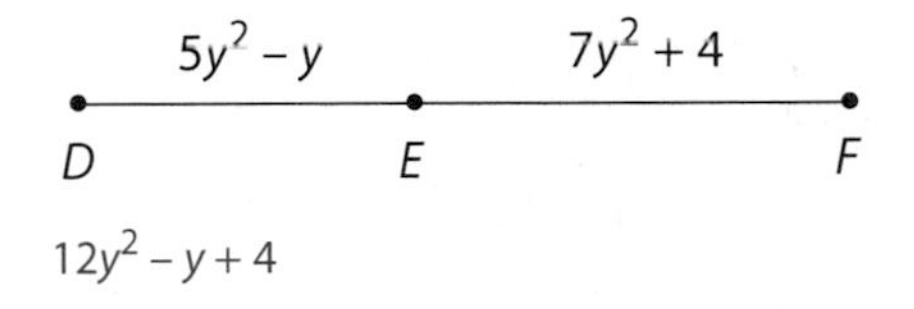

$12y^2 - y + 4$

35. The length of line segment LN is $7a^2 + 4a - 3$. Find the length of line segment MN.

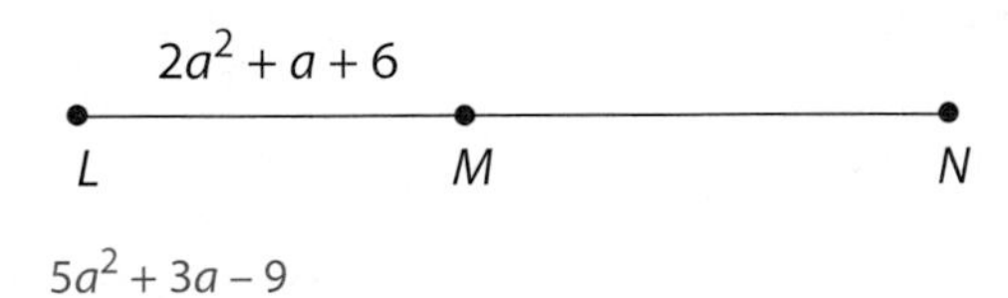

$5a^2 + 3a - 9$

36. The length of line segment QS is $12c^3 + 4c^2 - 6$. Find the length of line segment QR.

$5c^3 - 3c + 9$

Q   R   S

$7c^3 + 4c^2 + 3c - 15$

37. Find the perimeter of the rectangle. The dimensions given are in kilometers.

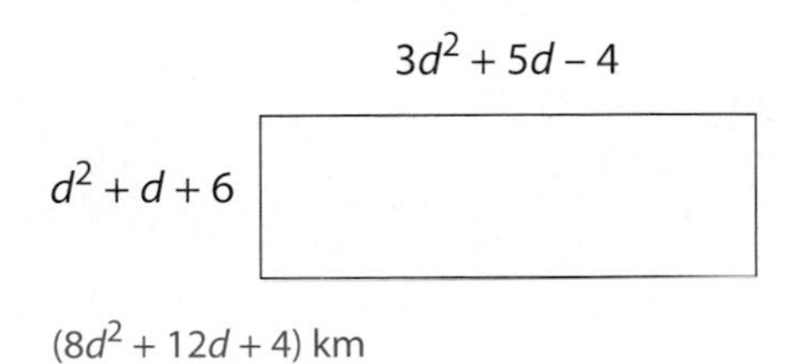

$(8d^2 + 12d + 4)$ km

38. Find the perimeter of the rectangle. The dimensions given are in meters.

$4b^2 - 8b + 6$

$b^2 + 5b - 1$

$(10b^2 - 6b + 10)$ m

39. The total monthly cost, in dollars, for a company to produce and sell n guitars per month is $240n + 1200$. The company's revenue, in dollars, from selling all n guitars is $-2n^2 + 400n$. Express in terms of n the company's monthly profit. Use the formula $P = R - C$.
$(-2n^2 + 160n - 1200)$ dollars

40. A company's total monthly cost, in dollars, for manufacturing and selling n cameras per month is $40n + 1800$. The company's revenue, in dollars, from selling all n cameras is $-n^2 + 250n$. Express in terms of n the company's monthly profit. Use the formula $P = R - C$.
$(-n^2 + 210n - 1800)$ dollars

41. What polynomial must be added to $3x^2 - 4x - 2$ so that the sum is $-x^2 + 2x + 1$?
$-4x^2 + 6x + 3$

42. What polynomial must be added to $-2x^3 + 4x - 7$ so that the sum is $x^2 - x - 1$?
$2x^3 + x^2 - 5x + 6$

43. What polynomial must be subtracted from $6x^2 - 4x - 2$ so that the difference is $2x^2 + 2x - 5$?
$4x^2 - 6x + 3$

44. What polynomial must be subtracted from $2x^3 - x^2 + 4x - 2$ so that the difference is $x^3 + 2x - 8$?
$x^3 - x^2 + 2x + 6$

Multiplication and Division of Polynomials

Multiply or divide.

45. $(x^2 + 3x + 2)(x + 1)$
$x^3 + 4x^2 + 5x + 2$

46. $(x^2 - 2x + 7)(x - 2)$
$x^3 - 4x^2 + 11x - 14$

47. $(a - 3)(a^2 - 3a + 4)$
$a^3 - 6a^2 + 13a - 12$

48. $(2x - 3)(x^2 - 3x + 5)$
$2x^3 - 9x^2 + 19x - 15$

49. $(-2b^2 - 3b + 4)(b - 5)$
$-2b^3 + 7b^2 + 19b - 20$

50. $(-a^2 + 3a - 2)(2a - 1)$
$-2a^3 + 7a^2 - 7a + 2$

51. $(x^3 - 3x + 2)(x - 4)$
$x^4 - 4x^3 - 3x^2 + 14x - 8$

52. $(y^3 + 4y^2 - 8)(2y - 1)$
$2y^4 + 7y^3 - 4y^2 - 16y + 8$

53. $(y + 2)(y^3 + 2y^2 - 3y + 1)$
$y^4 + 4y^3 + y^2 - 5y + 2$

54. $(2a - 3)(2a^3 - 3a^2 + 2a - 1)$
$4a^4 - 12a^3 + 13a^2 - 8a + 3$

55. $(x + 1)(x + 3)$
$x^2 + 4x + 3$

56. $(y + 2)(y + 5)$
$y^2 + 7y + 10$

57. $(a - 3)(a + 4)$
$a^2 + a - 12$

58. $(b - 6)(b + 3)$
$b^2 - 3b - 18$

59. $(y - 7)(y - 3)$
$y^2 - 10y + 21$

60. $(a - 8)(a - 9)$
$a^2 - 17a + 72$

61. $(2x + 1)(x + 7)$
$2x^2 + 15x + 7$

62. $(y + 2)(5y + 1)$
$5y^2 + 11y + 2$

63. $(3x - 1)(x + 4)$
$3x^2 + 11x - 4$

64. $(7x - 2)(x + 4)$
$7x^2 + 26x - 8$

65. $(4x - 3)(x - 7)$
$4x^2 - 31x + 21$

66. $(2x - 3)(4x - 7)$
$8x^2 - 26x + 21$

67. $(3y - 8)(y + 2)$
$3y^2 - 2y - 16$

68. $(5y - 9)(y + 5)$
$5y^2 + 16y - 45$

69. $(7a - 16)(3a - 5)$
$21a^2 - 83a + 80$

70. $(5a - 12)(3a - 7)$
$15a^2 - 71a + 84$

71. $(3b + 13)(5b - 6)$
$15b^2 + 47b - 78$

72. $(x + y)(2x + y)$
$2x^2 + 3xy + y^2$

73. $(2a + b)(a + 3b)$
$2a^2 + 7ab + 3b^2$

74. $(3x - 4y)(x - 2y)$
$3x^2 - 10xy + 8y^2$

75. $(2a - b)(3a + 2b)$
$6a^2 + ab - 2b^2$

76. $(5a - 3b)(2a + 4b)$
$10a^2 + 14ab - 12b^2$

77. $(d - 6)(d + 6)$
$d^2 - 36$

78. $(y - 5)(y + 5)$
$y^2 - 25$

79. $(2x + 3)(2x - 3)$
$4x^2 - 9$

80. $(4x - 7)(4x + 7)$
$16x^2 - 49$

81. $(x + 1)^2$
$x^2 + 2x + 1$

82. $(y - 3)^2$
$y^2 - 6y + 9$

83. $(3a - 5)^2$
$9a^2 - 30a + 25$

84. $(6x - 5)^2$
$36x^2 - 60x + 25$

85. $\frac{2x + 2}{2}$
$x + 1$

86. $\frac{5y + 5}{5}$
$y + 1$

87. $\frac{10a - 25}{5}$
$2a - 5$

88. $\frac{16b - 40}{8}$
$2b - 5$

89. $\frac{3a^2 + 2a}{a}$
$3a + 2$

90. $\frac{6y^2 + 4y}{y}$
$6y + 4$

91. $\frac{4b^3 - 3b}{b}$
$4b^2 - 3$

92. $\frac{12x^2 - 7x}{x}$
$12x - 7$

93. $\frac{3x^2 - 6x}{3x}$
$x - 2$

94. $\frac{10y^2 - 6y}{2y}$
$5y - 3$

95. $\frac{5x^2 - 10x}{-5x}$
$-x + 2$

96. $\frac{3y^2 - 27y}{-3y}$
$-y + 9$

97. $\frac{x^3 + 3x^2 - 5x}{x}$
$x^2 + 3x - 5$

98. $\frac{a^3 - 5a^2 + 7a}{a}$
$a^2 - 5a + 7$

99. $\frac{x^6 - 3x^4 - x^2}{x^2}$
$x^4 - 3x^2 - 1$

100. $\frac{a^8 - 5a^5 - 3a^3}{a^2}$
$a^6 - 5a^3 - 3a$

101. $\frac{5x^2y^2 + 10xy}{5xy}$
$xy + 2$

102. $\frac{8x^2y^2 - 24xy}{8xy}$
$xy - 3$

103. $(b^2 - 14b + 49) \div (b - 7)$
$b - 7$

104. $(x^2 - x - 6) \div (x - 3)$
$x + 2$

105. $(2x^2 + 5x + 2) \div (x + 2)$
$2x + 1$

106. $(2y^2 - 13y + 21) \div (y - 3)$
$2y - 7$

107. $(x^2 + 1) \div (x - 1)$
$x + 1 + \frac{2}{x - 1}$

108. $(x^2 + 4) \div (x + 2)$
$x - 2 + \frac{8}{x + 2}$

109. $(6x^2 - 7x) \div (3x - 2)$
$2x - 1 - \frac{2}{3x - 2}$

110. $(6y^2 + 2y) \div (2y + 4)$
$3y - 5 + \frac{20}{2y + 4}$

111. $(a^2 + 5a + 10) \div (a + 2)$
$a + 3 + \frac{4}{a + 2}$

112. $(b^2 - 8b - 9) \div (b - 3)$
$b - 5 - \frac{24}{b - 3}$

113. $(2y^2 - 9y + 8) \div (2y + 3)$
$y - 6 + \frac{26}{2y + 3}$

114. $(3x^2 + 5x - 4) \div (x - 4)$
$3x + 17 + \frac{64}{x - 4}$

115. $(8x + 3 + 4x^2) \div (2x - 1)$
$2x + 5 + \frac{8}{2x - 1}$

116. $(10 + 21y + 10y^2) \div (2y + 3)$
$5y + 3 + \frac{1}{2y + 3}$

117. $(x^3 + 3x^2 + 5x + 3) \div (x + 1)$
$x^2 + 2x + 3$

118. $(x^3 - 6x^2 + 7x - 2) \div (x - 1)$
$x^2 - 5x + 2$

119. $(x^4 - x^2 - 6) \div (x^2 + 2)$
$x^2 - 3$

120. $(x^4 + 3x^2 - 10) \div (x^2 - 2)$
$x^2 + 5$

121. Find the area of the square.
The dimension given is in meters.

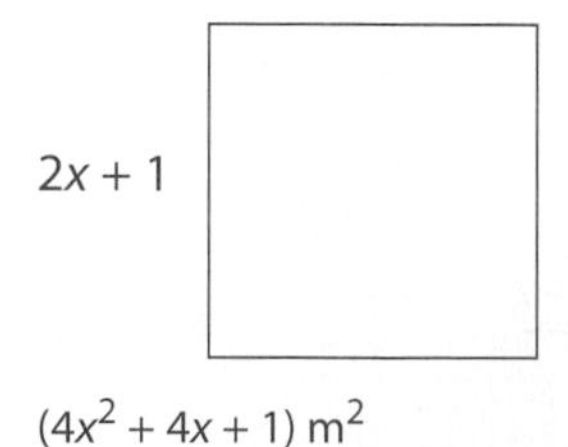

$(4x^2 + 4x + 1)$ m^2

122. Find the area of the square.
The dimension given is in yards.

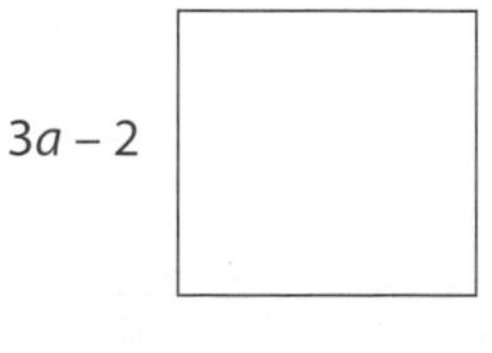

$(9a^2 - 12a + 4)$ yd^2

123. Find the area of the rectangle.
The dimensions given are in miles.

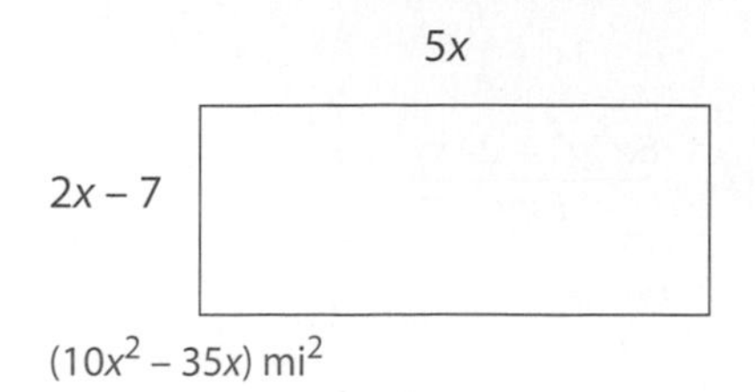

$(10x^2 - 35x)$ mi^2

124. Find the area of the rectangle.
The dimensions given are in feet.

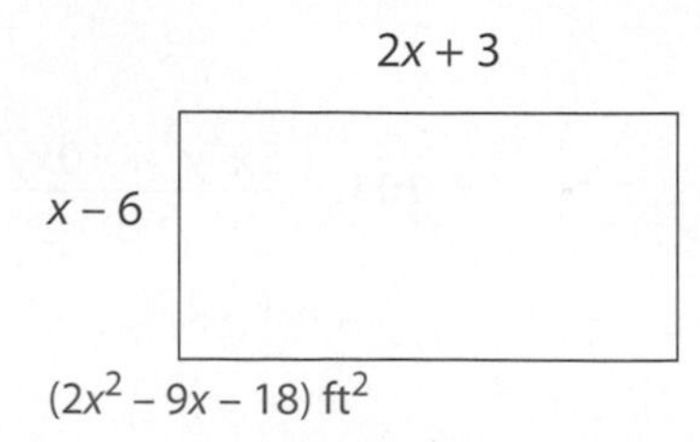

$(2x^2 - 9x - 18)$ ft^2

125. The radius of a circle is $(x + 4)$ inches. Find the area of the circle in terms of the variable x. Leave the answer in terms of π.

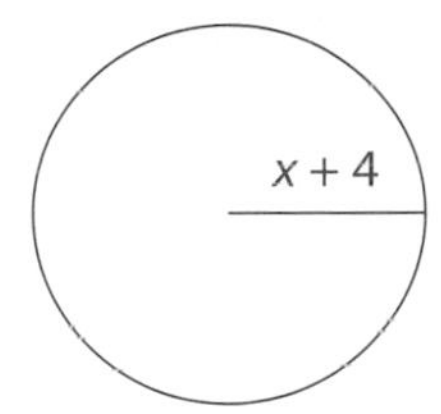

$(\pi x^2 + 8\pi x + 16\pi)$ in^2

126. The radius of a circle is $(x - 3)$ centimeters. Find the area of the circle in terms of the variable x. Leave the answer in terms of π.

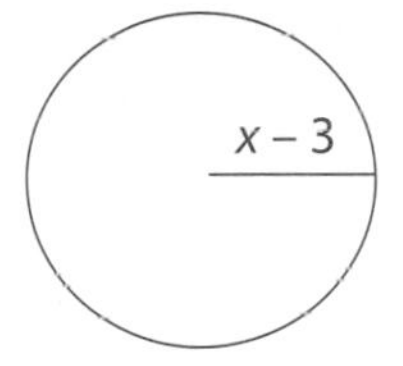

$(\pi x^2 - 6\pi x + 9\pi)$ cm^2

127. The length of a side of a cube is $(4x + 1)$ inches. Find the volume of the cube in terms of the variable x.
$(64x^3 + 48x^2 + 12x + 1)$ in^3

128. A box has a length of $(5x + 3)$ centimeters, a width of $(2x - 1)$ centimeters, and a height of $4x$ centimeters. Find the volume of the box in terms of the variable x.
$(40x^3 + 4x^2 - 12x)$ cm^3

129. The base of a triangle is $4x$ meters and the height is $(2x + 5)$ meters. Find the area of the triangle in terms of the variable x.
$(4x^2 + 10x)$ m^2

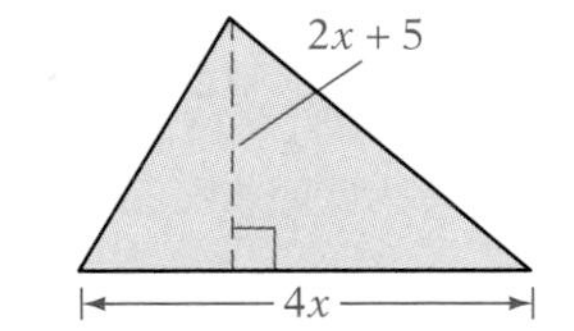

130. The base of a triangle is $(2x + 6)$ inches and the height is $(x - 8)$ inches. Find the area of the triangle in terms of the variable x.
$(x^2 - 5x - 24)$ in^2

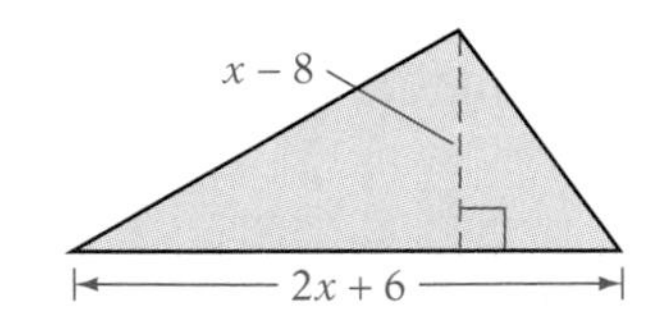

131. The width of a rectangle is $(3x + 1)$ inches. The length of the rectangle is twice the width. Find the area of the rectangle in terms of the variable x.
$(18x^2 + 12x + 2)$ in^2

132. The width of a rectangle is $(4x - 3)$ centimeters. The length of the rectangle is twice the width. Find the area of the rectangle in terms of the variable x.
$(32x^2 - 48x + 18)$ cm^2

133. The area of a rectangle is $(3x^2 - 22x - 16)$ feet. The width of the rectangle is $(x - 8)$ feet. Find the length of the rectangle in terms of the variable x. Use the formula $L = \frac{A}{W}$, where L is the length, A is the area, and W is the width of a rectangle.
$(3x + 2)$ ft

134. The area of a rectangle is $(10x^2 + 7x - 12)$ square meters. The length of the rectangle is $(5x - 4)$ meters. Find the width of the rectangle in terms of the variable x. Use the formula $W = \frac{A}{L}$, where W is the width, A is the area, and L is the length of a rectangle.

$(2x + 3)$ m

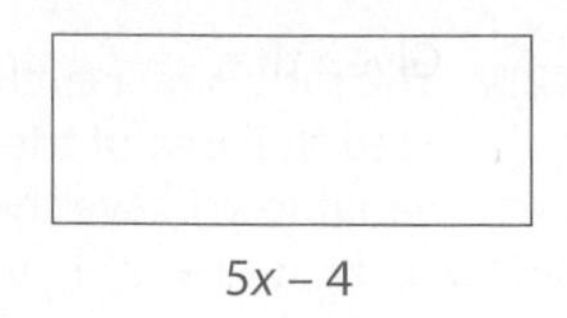

135. The area of a parallelogram is $(2x^3 - 9x^2 - 6x + 5)$ square inches. The height is $(x - 5)$ inches. Find the length of the base of the parallelogram in terms of the variable x. Use the formula $b = \frac{A}{h}$, where b is the length of the base, A is the area, and h is the height of a parallelogram.

$(2x^2 + x - 1)$ in.

136. The area of a parallelogram is $(2x^3 + 6x^2 - 4x - 12)$ square meters. The length of the base is $(x + 3)$ meters. Find the height of the parallelogram in terms of the variable x. Use the formula $h = \frac{A}{b}$, where h is the height, A is the area, and b is the length of the base of a parallelogram.

$(2x^2 - 4)$ m

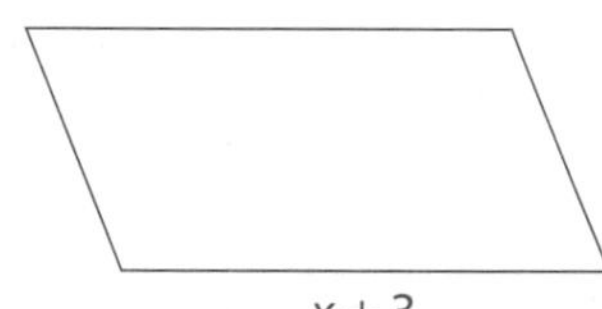

137. A softball diamond has dimensions 45 feet by 45 feet. A base path border x feet wide lies on both the first–base side and the third–base side of the diamond. Express the total area of the softball diamond and the base path in terms of the variable x.

$(90x + 2025)$ ft^2

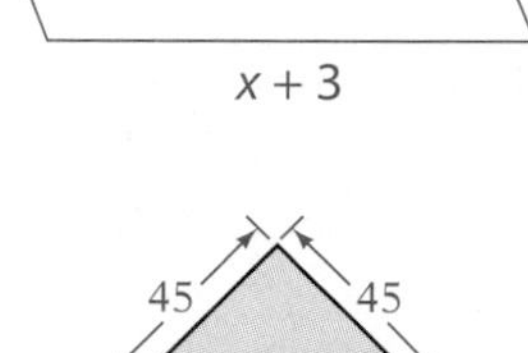

138. An athletic field has dimensions 30 yards by 100 yards. An end zone that is w yards wide borders each end of the field. Express the total area of the field and the end zones in terms of the variable w.

$(60w + 3000)$ yd^2

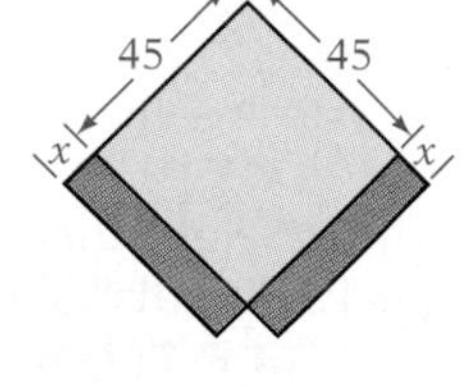

139. An open box is made from a square piece of cardboard that measures 40 inches on each side. To construct the box, squares that measure x inches on a side are cut from each corner. Express the volume of the box in terms of x. What is the volume of the box when x is 3 inches?

$(4x^3 - 160x^2 + 1600x)$ in^3; 3468 in^3

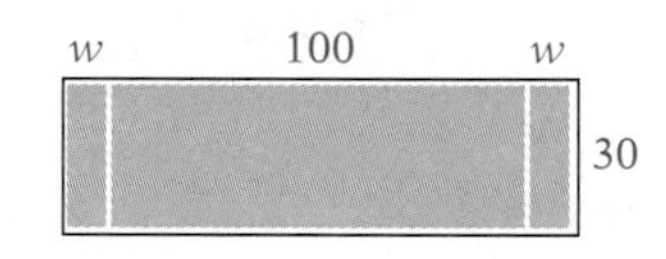

140. A sheet of tin 50 centimeters wide and 200 centimeters long is made into a trough by bending up two sides, each of length x, until they are perpendicular to the bottom. Express the volume of the trough in terms of x. What is the volume of the trough when x is 10 centimeters?

$(10{,}000x - 400x^2)$ cm^3; 60,000 cm^3

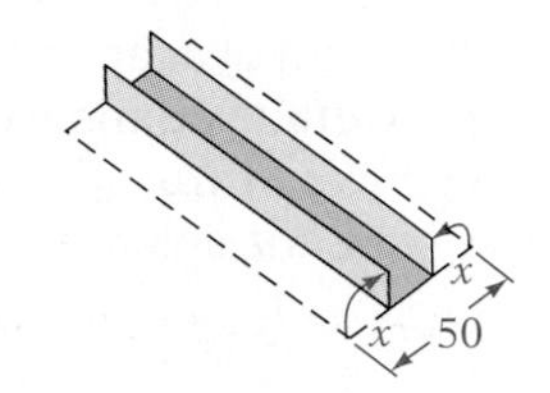

141. What polynomial has quotient $x^2 + 2x - 1$ when divided by $x + 3$?

$x^3 + 5x^2 + 5x - 3$

142. What polynomial has quotient $3x - 4$ when divided by $4x + 5$?

$12x^2 - x - 20$

143. Given that $x + 5$ is a factor of $x^3 + 12x^2 + 36x + 5$, find another factor of $x^3 + 12x^2 + 36x + 5$.
$x^2 + 7x + 1$

144. Given that $x - 2$ is a factor of $x^3 + 2x^2 - 9x + 2$, find another factor of $x^3 + 2x^2 - 9x + 2$.
$x^2 + 4x - 1$

145. Subtract $4x^2 - x - 5$ from the product of $x^2 + x + 3$ and $x - 4$. $x^3 - 7x^2 - 7$

146. Add $x^2 + 2x - 3$ to the product of $2x - 5$ and $3x + 1$. $7x^2 - 11x - 8$

147. The quotient of a polynomial and $2x + 1$ is $2x - 4 + \frac{7}{2x + 1}$. Find the polynomial. $4x^2 - 6x + 3$

148. The quotient of a polynomial and $x - 3$ is $x^2 - x + 8 + \frac{22}{x - 3}$. Find the polynomial. $x^3 - 4x^2 + 11x - 2$

Applying Concepts

149. Determine whether the statement is always true, sometimes true, or never true.

a. The opposite of the polynomial $ax^3 - bx^2 + cx - d$ is $-ax^3 + bx^2 - cx + d$. Always true

b. A binomial is a polynomial of degree 2. Sometimes true

c. To multiply two polynomials, multiply each term of one polynomial by the other polynomial. Always true

d. The square of a binomial is a trinomial. Always true

150. Is it possible to add two polynomials, each of degree 3, and have the sum be a polynomial of degree 2? If so, give an example. If not, explain why not.
Yes. For example, $(3x^3 - 2x^2 + 3x - 4) + (-3x^3 + 4x^2 - 6x + 5) = 2x^2 - 3x + 1$

151. If a polynomial of degree 3 is multiplied by a polynomial of degree 2, what is the degree of the resulting polynomial?
5

152. Is it possible to multiply a polynomial of degree 2 by a polynomial of degree 2 and have the product be a polynomial of degree 3? If so, give an example. If not, explain why not.
No. Two polynomials of degree 2 will have terms ax^2 and bx^2, $a \neq 0$, $b \neq 0$. Multiplying these terms yields abx^4, where $ab \neq 0$. Therefore, the product will have an x^4 term and will be of degree 4.

153. An open box is made from a square piece of cardboard that measures 20 inches on each side. To construct the box, squares that measure x inches on a side are cut from each corner. Express the volume of the box in terms of x. Can x be 10 inches? Explain your answer.
$(400x - 80x^2 + 4x^3)$ in^3; no; explanations will vary.

154. Add or subtract. Then use a graphing calculator to check the sum or difference.

a. $(4x^2 + 2x) + (x^2 + 6x)$ $5x^2 + 8x$

b. $(x^2 + x - 1) + (3x^2 - 10x + 4)$ $4x^2 - 9x + 3$

c. $(3x^2 + x - 3) - (x^2 + 4x - 2)$ $2x^2 - 3x - 1$

Explorations

155. ***Patterns in Products of Polynomials***

a. Multiply: $(x + 1)(x - 1)$

b. Multiply: $(x + 1)(-x^2 + x - 1)$

c. Multiply: $(x + 1)(x^3 - x^2 + x - 1)$

d. Multiply: $(x + 1)(-x^4 + x^3 - x^2 + x - 1)$

e. Use the pattern of the answers to parts a to d to multiply $(x + 1)(x^5 - x^4 + x^3 - x^2 + x - 1)$.

f. Use the pattern of the answers to parts a to e to multiply $(x + 1)(-x^6 + x^5 - x^4 + x^3 - x^2 + x - 1)$.

156. ***Pascal's Triangle***

Simplifying the power of a binomial is called *expanding the binomial.* The expansion of the first three powers of a binomial is shown below.

$$(a + b)^1 = a + b$$

$$(a + b)^2 = (a + b)(a + b) = a^2 + 2ab + b^2$$

$$(a + b)^3 = (a + b)^2(a + b) = (a^2 + 2ab + b^2)(a + b) = a^3 + 3a^2b + 3ab^2 + b^3$$

a. Find $(a + b)^4$. [*Hint*: $(a + b)^4 = (a + b)^3(a + b)$]

b. Find $(a + b)^5$. [*Hint*: $(a + b)^5 = (a + b)^4(a + b)$]

If we continue in this way, the results for $(a + b)^6$ would be:

$$(a + b)^6 = a^6 + 6a^5b + 15a^4b^2 + 20a^3b^3 + 15a^2b^4 + 6ab^5 + b^6$$

c. Now expand $(a + b)^8$. Before you begin, see if you can find a pattern that will assist in writing the expansion of $(a + b)^8$ without having to multiply it out. Here are some hints.

1. Write out the variable terms of each binomial expansion from $(a + b)^1$ through $(a + b)^6$. Observe how the exponents on the variables change.
2. Write out the coefficients of all the terms without the variable parts. It will be helpful to make a triangular arrangement as shown below. Note that each row begins and ends with a 1. Also note in the two shaded regions that any number in a row is the sum of the two closest numbers above it. For instance, $1 + 5 = 6$ and $6 + 4 = 10$.

```
            1   1
          1   2   1
        1   3   3   1
      1   4   6   4   1
    1   5  10  10   5   1
  1   6  15  20  15   6   1
```

d. The triangle of numbers shown above is called Pascal's Triangle. To find the expansion of $(a + b)^7$, you will need to find the seventh row of Pascal's Triangle. Use the patterns you have observed to write the expansion $(a + b)^7$.

e. Find the eighth row of Pascal's Triangle.

f. Find the expansion of $(a + b)^8$.

g. Pascal's Triangle has been the subject of extensive analysis, and many patterns have been found. See if you can find some of them.

Section 5.3 Factoring Polynomials

Common Monomial Factors

In the last section, we multiplied polynomials. For example,

$$2x(x^2 + 3x - 5) = 2x^3 + 6x^2 - 10x$$

Here the Distributive Property is used to multiply a monomial times a trinomial.

$$(x + 4)(x - 7) = x^2 - 7x + 4x - 28 = x^2 - 3x - 28$$

Here the FOIL method is used to multiply two binomials.

In this section we will write polynomials in factored form. A polynomial is in **factored form** when it is written as a product of other polynomials. It can be thought of as the reverse of multiplication. In the examples above, the factored form is on the left and the polynomial is on the right. In this section, we will be given polynomials to write in factored form.

Polynomial		**Factored Form**
$2x^3 + 6x^2 - 10x$	=	$2x(x^2 + 3x - 5)$
$x^2 - 3x - 28$	=	$(x + 4)(x - 7)$

Factoring is an important problem-solving technique in many applications in science, engineering, and advanced mathematics. Factoring enables us to write a more complicated polynomial as the product of simpler polynomials.

Question: Is the expression written in factored form? In other words, is it written as a product of polynomials?[1]

a. $a^3(4b + 9)$ **b.** $2y^2 - y + 1$ **c.** $(5c + 6)(c - 8)$

To factor out a common monomial from the terms of a polynomial, first find the greatest common factor of the terms.

TAKE NOTE
12 is the GCF of 24 and 60 because 12 is the largest integer that divides evenly into both 24 and 60.

The **greatest common factor (GCF)** of two or more integers is the greatest integer that is a factor of all the integers.

$24 = \mathbf{2} \cdot \mathbf{2} \cdot 2 \cdot \mathbf{3}$
$60 = \mathbf{2} \cdot \mathbf{2} \cdot \mathbf{3} \cdot 5$
$\text{GCF} = \mathbf{2} \cdot \mathbf{2} \cdot \mathbf{3} = 12$

The GCF of two or more monomials is the product of the GCF of the coefficients and the common variable factors.

$6x^3y = \mathbf{2} \cdot 3 \cdot \mathbf{x} \cdot \mathbf{x} \cdot x \cdot \mathbf{y}$
$8x^2y^2 = \mathbf{2} \cdot 2 \cdot 2 \cdot \mathbf{x} \cdot \mathbf{x} \cdot \mathbf{y} \cdot y$
$\text{GCF} = \mathbf{2} \cdot \mathbf{x} \cdot \mathbf{x} \cdot \mathbf{y} = 2x^2y$

Note that the exponent of each variable in the GCF is the same as the smallest exponent of that variable in any of the monomials.

The GCF of $6x^3y$ and $8x^2y^2$ is $2x^2y$.

1. **a.** Yes, this is in factored form. It is a monomial (a^3) times a binomial ($4b + 9$). **b.** No, this is not in factored form. **c.** Yes, this is in factored form. It is a binomial ($5c + 6$) times a binomial ($c - 8$).

→ Factor: $5x^3 - 35x^2 + 10x$

$5x^3 = 5 \cdot x^3$
$35x^2 = 5 \cdot 7 \cdot x^2$
$10x = 2 \cdot 5 \cdot x$

• List the factors of each term of the polynomial.

The GCF is $5x$.

• The GCF of 5, 35, and 10 is 5. The smallest exponent on x^3, x^2, and x is 1. The GCF of the terms is $5x$.

$\frac{5x^3 - 35x^2 + 10x}{5x}$
$= x^2 - 7x + 2$

• Divide the polynomial by the GCF.

$5x^3 - 35x^2 + 10x$
$= 5x(x^2 - 7x + 2)$

• Write the polynomial as the product of the GCF and the quotient found above.

$5x(x^2 - 7x + 2)$
$= 5x^3 - 35x^2 + 10x$

• Check the factorization by multiplying.

Note that to factor $5x^3 - 35x^2 + 10x$ as $5x(x^2 - 7x + 2)$, we are using the Distributive Property in reverse.

Recall that the Distributive Property is: $a(b + c) = ab + ac$

The Distributive Property in reverse is: $ab + ac = a(b + c)$

Example 1 Factor: $16x^4y^5 + 8x^4y^2 - 12x^3y$

Solution

$16x^4y^5 = \mathbf{2} \cdot \mathbf{2} \cdot 2 \cdot 2 \cdot x^4 \cdot y^5$
$8x^4y^2 = \mathbf{2} \cdot \mathbf{2} \cdot 2 \cdot x^4 \cdot y^2$
$12x^3y = \mathbf{2} \cdot \mathbf{2} \cdot 3 \cdot \mathbf{x^3} \cdot \mathbf{y}$

• List the factors of each term of the polynomial.

The GCF is $4x^3y$.

• Find the GCF of the terms.

$\frac{16x^4y^5 + 8x^4y^2 - 12x^3y}{4x^3y}$
$= 4xy^4 + 2xy - 3$

• Divide the polynomial by the GCF.

$16x^4y^5 + 8x^4y^2 - 12x^3y$
$= 4x^3y(4xy^4 + 2xy - 3)$

• Write the polynomial as a product of the GCF and the quotient found above.

$4x^3y(4xy^4 + 2xy - 3)$
$= 16x^4y^5 + 8x^4y^2 - 12x^3y$

• Check by using the Distributive Property.

You-Try-It 1 Factor: $6x^4y^2 - 9x^3y^2 + 12x^2y^4$

Solution See page S24.

Factoring by Grouping

In the example below, the common monomial factor a is factored out of the binomial.

$$2xa + 5ya = a(2x + 5y)$$

If we replace the a with $(a + b)$ in $2xa + 5ya$, the result is

$$2x(a + b) + 5y(a + b)$$

Now instead of a common monomial factor, there is a **common binomial factor**. The Distributive Property is used to factor a common binomial factor from an expression.

$$2x(a + b) + 5y(a + b) = (a + b)(2x + 5y)$$

Example 2 Factor: $y(x + 2) + 3(x + 2)$

Solution

$$y(x + 2) + 3(x + 2)$$
$$= (x + 2)(y + 3)$$

• The common binomial factor is $x + 2$.

You-Try-It 2 Factor: $a(b - 7) + b(b - 7)$

Solution See page S24.

Some polynomials can be factored by grouping the terms so that a common binomial factor is found.

→ Factor: $2x^3 - 3x^2 + 4x - 6$

$$2x^3 - 3x^2 + 4x - 6$$
$$= (2x^3 - 3x^2) + (4x - 6)$$
$$= x^2(2x - 3) + 2(2x - 3)$$
$$= (2x - 3)(x^2 + 2)$$

• Group the first two terms and the last two terms. (Put them in parentheses.)
• Factor out the GCF from each group.
• Write the expression as a product of factors.

$$(2x - 3)(x^2 + 2)$$
$$= 2x^3 + 4x - 3x^2 - 6$$
$$= 2x^3 - 3x^2 + 4x - 6$$

• Check the factorization by multiplying the binomials. Use the FOIL method.

Example 3 Factor: $3y^3 - 4y^2 - 6y + 8$

Solution

$$3y^3 - 4y^2 - 6y + 8$$
$$= (3y^3 - 4y^2) - (6y - 8)$$

• Group the first two terms and the last two terms. Note that $-6y + 8 = -(6y - 8)$.

$$= y^2(3y - 4) - 2(3y - 4)$$

• Factor out the GCF from each group.

$$= (3y - 4)(y^2 - 2)$$

• Write the expression as a product of two binomials.

$$(3y - 4)(y^2 - 2)$$
$$= 3y^3 - 6y - 4y^2 + 8$$

• Check by using the FOIL method.

TAKE NOTE
By the Commutative Property of Addition, $3y^3 - 6y - 4y^2 + 8 = 3y^3 - 4y^2 - 6y + 8$. The factorization checks.

You-Try-It 3 Factor: $y^5 - 5y^3 + 4y^2 - 20$

Solution See page S24.

Factoring Trinomials of the Form $ax^2 + bx + c$

Trinomials of the form $ax^2 + bx + c$ are shown below.

$$\begin{array}{ll} x^2 + 9x + 14 & a = 1, b = 9, c = 14 \\ x^2 - 2x - 15 & a = 1, b = -2, c = -15 \\ 3x^2 - x + 4 & a = 3, b = -1, c = 4 \\ 4x^2 - 16 & a = 4, b = 0, c = -16 \end{array}$$

Question: What is the value of a, b, and c in $4x^2 + 5x - 8$?[2]

SUGGESTED ACTIVITY
You may prefer to have students create models of factoring, as shown on page 380, before you present this material on factoring trinomials. The activity may help them develop a better understanding of the concept of factoring before performing the calculations.

There are various methods of factoring trinomials. The method presented here is based on factoring by grouping.

In Example 3 and You-Try-It 3, factoring by grouping was used to factor a polynomial containing four terms. To use factoring by grouping to factor a trinomial, we need to rewrite the trinomial so that it has four terms. Here is an example.

To factor $x^2 - 2x - 24$:

Create a table that has four regions in it. Place the first term of the trinomial in the first region and the last term of the trinomial in the last region. Find the product of these two terms.

$-24x^2$

x^2			-24

Find two factors of the product $-24x^2$ whose sum is the middle term of the trinomial, $-2x$. It may be helpful to systematically list the possible factors.

$$\begin{array}{ll} (-1x)(+24x) & (+1x)(-24x) \\ (-2x)(+12x) & (+2x)(-12x) \\ (-3x)(+8x) & (+3x)(-8x) \\ (-4x)(+6x) & (+4x)(-6x) \end{array}$$

Each of the pairs of factors listed has a product of $-24x^2$, but only the factors $+4x$ and $-6x$ have the sum $-2x$. Write these two factors in the two middle regions of the table.

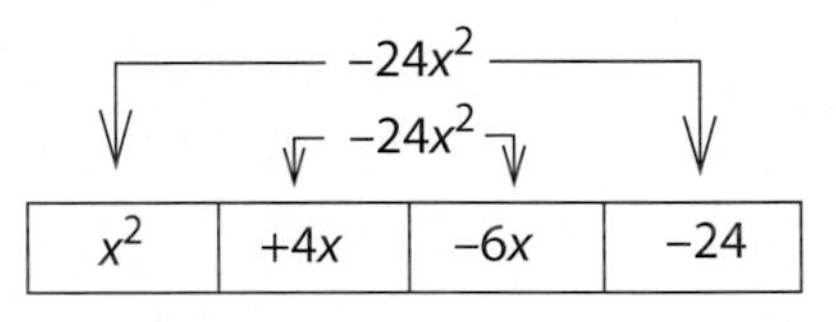

TAKE NOTE
$x^2 + 4x - 6x - 24 = x^2 - 2x - 24$
We have not changed the polynomial.

Factor this polynomial by grouping,

$$\begin{aligned} &x^2 + 4x - 6x - 24 \\ &= (x^2 + 4x) - (6x + 24) \\ &= x(x + 4) - 6(x + 4) \\ &= (x + 4)(x - 6) \end{aligned}$$

TAKE NOTE
The graph of $y = x^2 - 2x - 24$ is shown below.

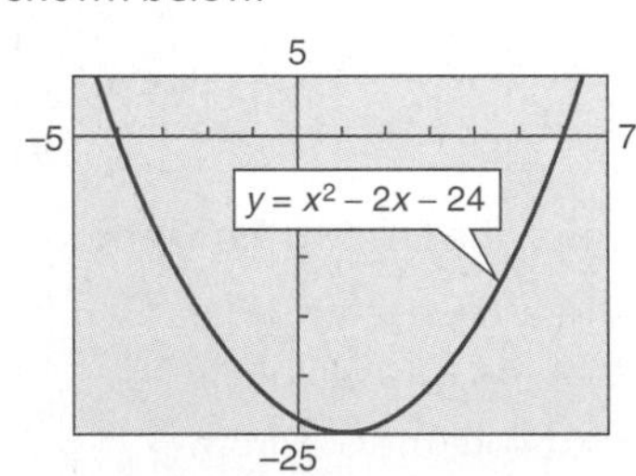

Note that the x-coordinates of the x-intercepts of the graph can be used to determine the factorization of the polynomial.

$$\begin{aligned} [x-(-4)](x-6) &= (x+4)(x-6) \\ &= x^2 - 2x - 24 \end{aligned}$$

Check.

$$\begin{aligned} &(x + 4)(x - 6) \\ &= x^2 - 6x + 4x - 24 \\ &= x^2 - 2x - 24 \end{aligned}$$

If the two factors $+4x$ and $-6x$ are put in the table in the reverse order, the factorization is the same.

$-24x^2$

$-24x^2$

x^2	$-6x$	$+4x$	-24

$$\begin{aligned} &x^2 - 6x + 4x - 24 \\ &= (x^2 - 6x) + (4x - 24) \\ &= x(x - 6) + 4(x - 6) \\ &= (x - 6)(x + 4) \end{aligned}$$

Note that by the Commutative Property of Multiplication $(x + 4)(x - 6) = (x - 6)(x + 4)$.

2. In $4x^2 + 5x - 8$, the value of a is 4, the value of b is 5, and the value of c is -8.

To factor $2x^2 + 13x + 15$:

Create a table that has four regions in it. Place the first term of the trinomial in the first region and the last term of the trinomial in the last region. Find the product of these two terms.

$30x^2$

$2x^2$			$+15$

Find two factors of the product $30x^2$ whose sum is the middle term of the trinomial, $+13x$. It may be helpful to systematically list the possible factors.

$(+1x)(+30x)$
$(+2x)(+15x)$
$(+3x)(+10x)$
$(+5x)(+6x)$

Only the factors $3x$ and $10x$ have the sum $13x$. Write these two factors in the two middle regions of the table.

$30x^2$
$30x^2$

$2x^2$	$+3x$	$+10x$	$+15$

TAKE NOTE
$2x^2 + 3x + 10x + 15 =$
$2x^2 + 13x + 15$
We have not changed the original polynomial.

Factor this polynomial by grouping.

$$\begin{aligned} &2x^2 + 3x + 10x + 15 \\ &\quad = (2x^2 + 3x) + (10x + 15) \\ &\quad = x(2x + 3) + 5(2x + 3) \\ &\quad = (2x + 3)(x + 5) \end{aligned}$$

Check.

$$\begin{aligned} &(2x + 3)(x + 5) \\ &\quad = 2x^2 + 10x + 3x + 15 \\ &\quad = 2x^2 + 13x + 15 \end{aligned}$$

Example 4 Factor: $x^2 - 6x - 16$

Solution $x^2 - 6x - 16$

$-16x^2$

x^2			-16

$(-1x)(+16x)$ $(+1x)(-16x)$
$(-2x)(+8x)$ $(+2x)(-8x)$ ← These factors have a sum of $-6x$.
$(-4x)(+4x)$

x^2	$+2x$	$-8x$	-16

$$\begin{aligned} x^2 + 2x - 8x - 16 &= (x^2 + 2x) - (8x + 16) \\ &= x(x + 2) - 8(x + 2) \\ &= (x + 2)(x - 8) \end{aligned}$$

Check: $(x + 2)(x - 8) = x^2 - 8x + 2x - 16$
$= x^2 - 6x - 16$

You-Try-It 4 Factor: $x^2 + 3x - 18$

Solution See page S24.

SUGGESTED ACTIVITY
Present students with the following problem: The area of a rectangular dance floor is $(2x^2 + 12x + 18)$ square feet, and its perimeter is $(6x + 18)$ feet. What are possible dimensions of the dance floor.
[Answer: $(x + 3)$ ft by $2(x + 3)$ ft]

Example 5 Factor: $6x^2 - 13x + 6$

Solution $6x^2 - 13x + 6$

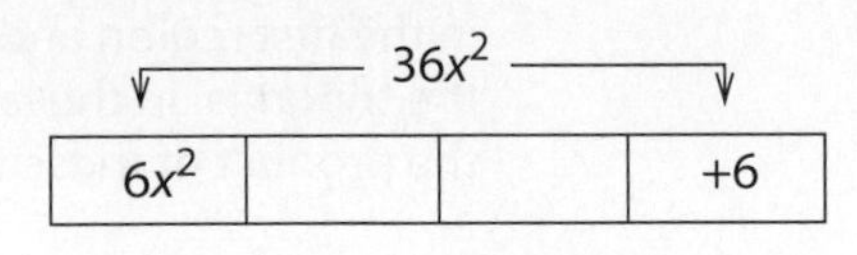

$(-1x)(-36x)$	• The middle term, $-13x$, is negative. Find two negative factors of $36x^2$ whose sum is $-13x$.
$(-2x)(-18x)$	
$(-3x)(-12x)$	
$(-4x)(-9x)$ ←	These two factors have a sum of $-13x$.
$(-6x)(-6x)$	

$6x^2$	$-4x$	$-9x$	$+6$

$$\begin{aligned} 6x^2 - 4x - 9x + 6 &= (6x^2 - 4x) - (9x - 6) \\ &= 2x(3x - 2) - 3(3x - 2) \\ &= (3x - 2)(2x - 3) \end{aligned}$$

$$\begin{aligned} \text{Check: } (3x - 2)(2x - 3) &= 6x^2 - 9x - 4x + 6 \\ &= 6x^2 - 13x + 6 \end{aligned}$$

You-Try-It 5 Factor: $6x^2 - 11x + 5$

Solution See page S25.

In the examples above, all the possible factors of the product of the first and last terms were listed. However, once the factors whose sum is the middle term have been found, no other factors need to be checked.

Also, in every example above, the given trinomial was factorable. This is not always the case. For example, consider the trinomial $x^2 - 6x - 8$.

The product of the first and last terms is $-8x^2$.

x^2			-8

(arrows from x^2 and -8 labelled $-8x^2$)

The possible factors of -8 are listed at the right. None of the pairs of factors has a sum of -6.

$(-1x)(+8x)$	$(+1x)(-8x)$
$(-2x)(+4x)$	$(+2x)(-4x)$

The trinomial $x^2 - 6x - 8$ is said to be **nonfactorable over the integers**. It is also called a **prime polynomial**.

TAKE NOTE
The graph of $y = x^2 - 6x - 8$ is shown below. By using the x-intercepts of the x-coordinates, it is possible to find an approximate factorization of the polynomial $x^2 - 6x - 8$.

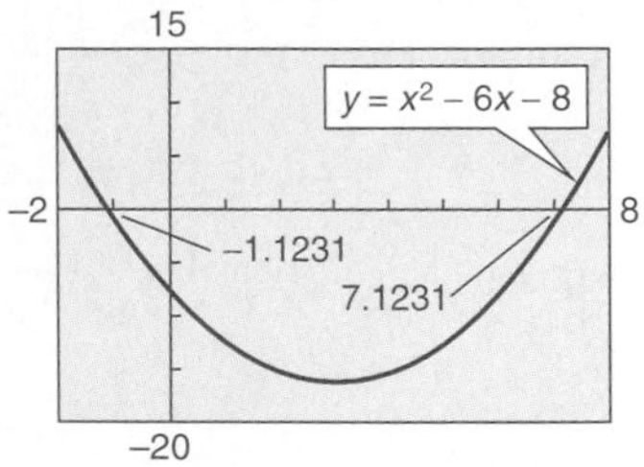

$$\begin{aligned} &[x - (-1.1231)](x - 7.1231) \\ &\quad = (x + 1.1231)(x - 7.1231) \\ &\quad \approx x^2 - 6x - 8 \end{aligned}$$

Thus, although the polynomial does not factor using integers, it is possible to find an approximation to the factorization.

The product of $(a + b)(a - b)$ can be found using the FOIL method.

$$\begin{aligned} (a + b)(a - b) &= a^2 - ab + ab - b^2 \\ &= a^2 - b^2 \end{aligned}$$

The expression $(a + b)(a - b)$ is called the **sum and difference of two terms.** The result of multiplying the sum and difference of two terms is the binomial $a^2 - b^2$, which is called the **difference of two squares.**

We can factor the difference of two squares using the same procedure used to factor trinomials.

Example 6 Factor: $4x^2 - 9$

Solution $4x^2 - 9$ • Recall that $b = 0$. The middle term is $0x$.

$-36x^2$

$4x^2$			-9

$(-1x)(+36x)$ • Find two factors of $-36x^2$ whose sum is $0x$.
$(-2x)(+18x)$
$(-3x)(+12x)$
$(-4x)(+9x)$
$(-6x)(+6x)$ ← These two factors have a sum of $0x$.

$4x^2$	$-6x$	$+6x$	-9

$$\begin{aligned} 4x^2 - 6x + 6x - 9 &= (4x^2 - 6x) + (6x - 9) \\ &= 2x(2x - 3) + 3(2x - 3) \\ &= (2x - 3)(2x + 3) \end{aligned}$$

$$\begin{aligned} \text{Check: } (2x - 3)(2x + 3) &= 4x^2 + 6x - 6x - 9 \\ &= 4x^2 - 9 \end{aligned}$$

You-Try-It 6 Factor: $25x^2 - 1$

Solution See page S25.

Example 7 The area of a rectangle is $(x^2 + 18x + 32)$ square meters. Find the dimensions of the rectangle in terms of the variable x.

Area = $(x^2 + 18x + 32)$ m^2

Solution

State the goal. The goal is to find the length and the width of a rectangle that has an area of $(x^2 + 18x + 32)$ square meters. We are looking for two expressions that when multiplied equal $x^2 + 18x + 32$.

Describe a strategy. Factor the trinomial $x^2 + 18x + 32$.

Solve the problem.

$32x^2$

x^2			$+32$

Two factors of $32x^2$ whose sum is $18x$ are $2x$ and $16x$.

x^2	$+2x$	$+16x$	$+32$

$$\begin{aligned} x^2 + 2x + 16x + 32 &= (x^2 + 2x) + (16x + 32) \\ &= x(x + 2) + 16(x + 2) \\ &= (x + 2)(x + 16) \end{aligned}$$

The dimensions of the rectangle are $(x + 2)$ meters and $(x + 16)$ meters.

Check your work. √

You-Try-It 7 The area of a parallelogram is $(4x^2 + 8x + 3)$ square feet. Find the dimensions of the parallelogram in terms of the variable x.

Area = $(4x^2 + 8x + 3)$ ft^2

Solution See page S25.

Factoring Completely

A polynomial is factored completely when it is written as a product of factors that are nonfactorable over the integers.

The first step in *any* factoring problem is to determine whether the terms of the polynomial have a common factor. If they do, factor it out first.

SUGGESTED ACTIVITY
Present students with the following problem: A rectangular swimming pool has an area of $(x^2 + x - 2)$ square meters. A deck, x meters wide, is added around the pool. What is the area of the deck?
[Answer: $(8x^2 + 2x)$ square meters]

Example 8 Factor: $6x^3 - 46x^2 + 28x$

Solution

$6x^3 - 46x^2 + 28x$ • The terms have a GCF of $2x$.

$= 2x(3x^2 - 23x + 14)$ • Factor out the GCF.

$42x^2$

$3x^2$			$+14$

• Factor $3x^2 - 23x + 14$.

$(-1x)(-42x)$
$(-2x)(-21x)$
$(-3x)(-14x)$
$(-6x)(-7x)$

• The middle term, $-23x$, is negative. Find two negative factors of $42x^2$ whose sum is $-23x$. The factors $-2x$ and $-21x$ have a sum of $-23x$.

x^2	$-2x$	$-21x$	$+14$

$$3x^2 - 2x - 21x + 14 = (3x^2 - 2x) - (21x - 14)$$
$$= x(3x - 2) - 7(3x - 2)$$
$$= (3x - 2)(x - 7)$$

$$6x^3 - 46x^2 + 28x = 2x(3x^2 - 23x + 14)$$
$$= 2x(3x - 2)(x - 7)$$

Check: $2x(3x - 2)(x - 7) = (6x^2 - 4x)(x - 7)$
$$= 6x^3 - 42x^2 - 4x^2 + 28x$$
$$= 6x^3 - 46x^2 + 28x$$

You-Try-It 8 Factor: $4a^2b - 30ab - 14b$

Solution See page S25.

Question: Is the expression factored completely? That is, are all the factors nonfactorable over the integers?[3]

a. $(3x + 6)(4x + 1)$ **b.** $5y(x^2 - 4)$ **c.** $(2x + 7)(x^2 - 2x + 9)$

---

3. **a.** No. The binomial $3x + 6$ has a common factor of 3 and can be factored as $3(x + 2)$.
   b. No. The binomial $x^2 - 4$ is the difference of two squares and can be factored as $(x + 2)(x - 2)$.
   c. Yes. Both $2x + 7$ and $x^2 - 2x + 9$ are nonfactorable over the integers.

5.3 EXERCISES

Topics for Discussion

1. Explain the meaning of "a factor" and the meaning of "to factor."
A factor is a number or expression in a multiplication. To factor means to write a polynomial as a product of other polynomials.

2. **a.** Provide an example of how the Distributive Property is used to multiply a monomial times a binomial. Examples will vary.
b. Provide an example of how the Distributive Property is used to factor a binomial. Examples will vary.

3. Determine whether the statement is always true, sometimes true, or never true.
a. To factor a polynomial means to write it as a multiplication. Always true
b. The greatest common factor of two numbers is the largest number that divides evenly into both numbers. Always true
c. To factor a trinomial of the form $ax^2 + bx + c$ means to rewrite the polynomial as a product of two binomials. Sometimes true
d. The first step in factoring a trinomial is to determine whether the terms have a common factor. Always true
e. A binomial is factorable. Sometimes true
f. The expression $x^2 - 12$ is an example of the difference of two squares. Never true
g. The expression $(y + 8)(y - 8)$ is the product of the sum and difference of two terms. The two terms are y and 8. Always true

4. Explain why the statement is true.
a. The terms of the binomial $3x - 9$ have a common factor. The common factor is 3.
b. The expression $3x^2 + 15$ is not in factored form. $3x^2 + 15$ is a sum; it is not a product.
c. $2x - 1$ is a factor of $x(2x - 1)$. In $x(2x - 1)$, x is multiplied times $2x - 1$. Both x and $2x - 1$ are factors.
d. The trinomial $x^2 + 3x + 5$ is a prime polynomial. It cannot be expressed as the product of two binomials.
e. The factored form of $2x^2 - 7x - 15$ is $(2x + 3)(x - 5)$. $(2x + 3)(x - 5)$ is a product, both $2x + 3$ and $x - 5$ are nonfactorable, and when the two binomials are multiplied, the result is the trinomial.

Factoring Polynomials

Find the greatest common factor.

5. x^2y^4, xy^6
xy^4

6. a^5b^3, a^3b^8
a^3b^3

7. $14a^3, 49a^7$
$7a^3$

8. $12y^2, 27y^4$
$3y^2$

9. $9a^2b^4, 24a^4b^2$
$3a^2b^2$

10. $15a^4b^2, 9ab^5$
$3ab^2$

11. $2x^2y, 4xy, 8x$
$2x$

12. $16x^2, 8x^4y^2, 12xy$
$4x$

13. $3x^2y^2, 6x, 9x^3y^3$
$3x$

Complete the table by finding two integers whose product is given in the column headed ab and whose sum is given in the column headed $a + b$. Assume $a \le b$.

	ab	$a + b$	a	b	
14.	100	20			10, 10
15.	40	13			5, 8
16.	−42	−11			−14, 3
17.	−72	−1			−9, 8
18.	75	−20			−15, −5
19.	44	−15			−11, −4

Factor.

20. $8x + 12$
$4(2x + 3)$

21. $16a - 24$
$8(2a - 3)$

22. $7x^2 - 3x$
$x(7x - 3)$

23. $12y^2 - 5y$
$y(12y - 5)$

24. $10x^4 - 12x^2$
$2x^2(5x^2 - 6)$

25. $12a^5 - 32a^2$
$4a^2(3a^3 - 8)$

26. $6a^2bc + 4ab^2c$
$2abc(3a + 2b)$

27. $10x^2yz^2 + 15xy^3z$
$5xyz(2xz + 3y^2)$

28. $x^3 - 3x^2 - x$
$x(x^2 - 3x - 1)$

29. $a^3 + 4a^2 + 8a$
$a(a^2 + 4a + 8)$

30. $2x^2 + 8x - 12$
$2(x^2 + 4x - 6)$

31. $5x^2 - 15x + 35$
$5(x^2 - 3x + 7)$

32. $3x^3 + 6x^2 + 9x$
$3x(x^2 + 2x + 3)$

33. $5y^3 - 20y^2 + 10y$
$5y(y^2 - 4y + 2)$

34. $2x^4 - 4x^3 + 6x^2$
$2x^2(x^2 - 2x + 3)$

35. $3y^4 - 9y^3 - 6y^2$
$3y^2(y^2 - 3y - 2)$

36. $x^3y - 3x^2y^2 + 7xy^3$
$xy(x^2 - 3xy + 7y^2)$

37. $x^4y^4 - 3x^3y^3 + 6x^2y^2$
$x^2y^2(x^2y^2 - 3xy + 6)$

38. $4x^5y^5 - 8x^4y^4 + x^3y^3$
$x^3y^3(4x^2y^2 - 8xy + 1)$

39. $16x^2y - 8x^3y^4 - 48x^2y^2$
$8x^2y(2 - xy^3 - 6y)$

40. $x^3 + 4x^2 + 3x + 12$
$(x + 4)(x^2 + 3)$

41. $x^3 - 4x^2 - 3x + 12$
$(x - 4)(x^2 - 3)$

42. $2y^3 + 4y^2 + 3y + 6$
$(y + 2)(2y^2 + 3)$

43. $3y^3 - 12y^2 + y - 4$
$(y - 4)(3y^2 + 1)$

44. $ab + 3b - 2a - 6$
$(a + 3)(b - 2)$

45. $yz + 6z - 3y - 18$
$(y + 6)(z - 3)$

46. $x^2a - 2x^2 - 3a + 6$
$(a - 2)(x^2 - 3)$

47. $x^2y + 4x^2 + 3y + 12$
$(y + 4)(x^2 + 3)$

48. $x^2 - 3x + 4ax - 12a$
$(x - 3)(x + 4a)$

49. $t^2 + 4t - st - 4s$
$(t + 4)(t - s)$

50. $10xy^2 - 15xy + 6y - 9$
$(2y - 3)(5xy + 3)$

51. $10a^2b - 15ab - 4a + 6$
$(2a - 3)(5ab - 2)$

52. $x^2 + 5x + 6$
$(x + 2)(x + 3)$

53. $x^2 + x - 2$
$(x - 1)(x + 2)$

54. $x^2 + x - 6$
$(x + 3)(x - 2)$

55. $a^2 + a - 12$
$(a + 4)(a - 3)$

56. $a^2 - 2a - 35$
$(a + 5)(a - 7)$

57. $a^2 - 3a + 2$
$(a - 1)(a - 2)$

58. $a^2 - 5a + 4$
$(a - 1)(a - 4)$

59. $b^2 + 7b - 8$
$(b + 8)(b - 1)$

60. $y^2 + 6y - 55$
$(y + 11)(y - 5)$

61. $z^2 - 4z - 45$
$(z + 5)(z - 9)$

62. $y^2 - 8y + 15$
$(y - 3)(y - 5)$

63. $z^2 - 14z + 45$
$(z - 5)(z - 9)$

64. $p^2 + 12p + 27$
$(p + 3)(p + 9)$

65. $b^2 + 9b + 20$
$(b + 4)(b + 5)$

66. $y^2 - 8y + 32$
Nonfactorable

67. $y^2 - 9y + 81$
Nonfactorable

68. $p^2 + 24p + 63$
$(p + 3)(p + 21)$

69. $x^2 - 15x + 56$
$(x - 7)(x - 8)$

70. $5x^2 + 6x + 1$
$(x + 1)(5x + 1)$

71. $2y^2 + 7y + 3$
$(y + 3)(2y + 1)$

72. $2a^2 - 3a + 1$
$(a - 1)(2a - 1)$

73. $3a^2 - 4a + 1$
$(a - 1)(3a - 1)$

74. $4x^2 - 3x - 1$
$(x - 1)(4x + 1)$

75. $2x^2 - 5x - 3$
$(x - 3)(2x + 1)$

76. $6t^2 - 11t + 4$
$(2t - 1)(3t - 4)$

77. $10t^2 + 11t + 3$
$(2t + 1)(5t + 3)$

78. $8x^2 + 33x + 4$
$(x + 4)(8x + 1)$

79. $10z^2 + 3z - 4$
$(2z - 1)(5z + 4)$

80. $3x^2 + 14x - 5$
$(x + 5)(3x - 1)$

81. $3z^2 + 95z + 10$
Nonfactorable

82. $8z^2 - 36z + 1$
Nonfactorable

83. $2t^2 - t - 10$
$(t + 2)(2t - 5)$

84. $2t^2 + 5t - 12$
$(t + 4)(2t - 3)$

85. $12y^2 + 19y + 5$
$(3y + 1)(4y + 5)$

86. $5y^2 - 22y + 8$
$(y - 4)(5y - 2)$

87. $11a^2 - 54a - 5$
$(a - 5)(11a + 1)$

88. $4z^2 + 11z + 6$
$(z + 2)(4z + 3)$

89. $6b^2 - 13b + 6$
$(2b - 3)(3b - 2)$

90. $6x^2 + 35x - 6$
$(x + 6)(6x - 1)$

91. $a^2 - 81$
$(a + 9)(a - 9)$

92. $a^2 - 49$
$(a + 7)(a - 7)$

93. $4x^2 - 1$
$(2x + 1)(2x - 1)$

94. $9x^2 - 1$
$(3x + 1)(3x - 1)$

95. $1 - 49x^2$
$(1 + 7x)(1 - 7x)$

96. $1 - 64x^2$
$(1 + 8x)(1 - 8x)$

97. $t^2 + 36$
Nonfactorable

98. $x^2 + 64$
Nonfactorable

99. $3x^2 + 15x + 18$
$3(x+ 2)(x + 3)$

100. $3a^2 + 3a - 18$
$3(a + 3)(a - 2)$

101. $4x^2 - 4x - 8$
$4(x + 1)(x - 2)$

102. $ab^2 + 2ab - 15a$
$a(b + 5)(b - 3)$

103. $ab^2 + 7ab - 8a$
$a(b + 8)(b - 1)$

104. $2a^3 + 6a^2 + 4a$
$2a(a + 1)(a + 2)$

105. $3y^3 - 15y^2 + 18y$
$3y(y - 2)(y - 3)$

106. $2y^4 - 28y^3 - 64y^2$
$2y^2(y + 2)(y - 16)$

107. $3y^4 + 54y^3 + 135y^2$
$3y^2(y + 3)(y + 15)$

108. $4x^2 + 6x + 2$
$2(x + 1)(2x + 1)$

109. $12x^2 + 33x - 9$
$3(x + 3)(4x - 1)$

110. $2x^3 - 11x^2 + 5x$
$x(x - 5)(2x - 1)$

111. $2x^3 + 3x^2 - 5x$
$x(x - 1)(2x + 5)$

112. $10t^2 - 5t - 50$
$5(t + 2)(2t - 5)$

113. $16t^2 + 40t - 96$
$8(t + 4)(2t - 3)$

114. $3p^3 - 16p^2 + 5p$
$p(p - 5)(3p - 1)$

115. $6p^3 + 5p^2 + p$
$p(2p + 1)(3p + 1)$

116. $2x^2 - 18$
$2(x + 3)(x - 3)$

117. $50 - 2x^2$
$2(5 + x)(5 - x)$

118. $72 - 2x^2$
$2(6 + x)(6 - x)$

119. $b^4 - a^2b^2$
$b^2(b + a)(b - a)$

120. $2x^4y^2 - 2x^2y^2$
$2x^2y^2(x + 1)(x - 1)$

121. The area of a rectangle is $(2x^2 + 9x + 4)$ square inches. Find the dimensions of the rectangle in terms of the variable x.

$A = 2x^2 + 9x + 4$

$(2x + 1)$ in. by $(x + 4)$ in.

122. The area of a rectangle is $(3x^2 + 17x + 10)$ square miles. Find the dimensions of the rectangle in terms of the variable x.

$A = 3x^2 + 17x + 10$

$(3x + 2)$ mi by $(x + 5)$ mi

123. The area of a square is $(4x^2 + 12x + 9)$ square centimeters. Find the length of a side of the square in terms of the variable x.

$A =$
$4x^2 + 12x + 9$

$(2x + 3)$ cm

124. The area of a square is $(9x^2 + 6x + 1)$ square meters. Find the length of a side of the square in terms of the variable x.

$A =$
$9x^2 + 6x + 1$

$(3x + 1)$ m

125. The area of a parallelogram is $(30x^2 + 23x + 3)$ square yards. Find the dimensions of the parallelogram in terms of the variable x.

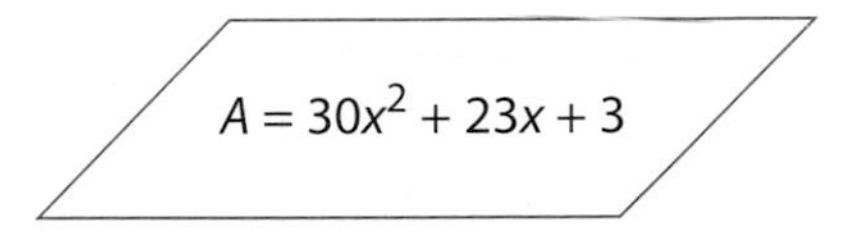

$(6x + 1)$ yd by $(5x + 3)$ yd

126. The area of a parallelogram is $(4x^2 + 23x + 15)$ square feet. Find the dimensions of the parallelogram in terms of the variable x.

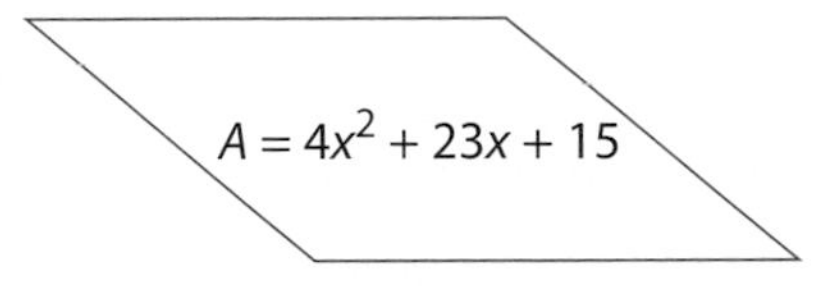

$(4x + 3)$ ft by $(x + 5)$ ft

127. The volume of a box is $(3xy^2 + 21xy + 18x)$ square centimeters. Find the dimensions of the box in terms of the variables x and y.

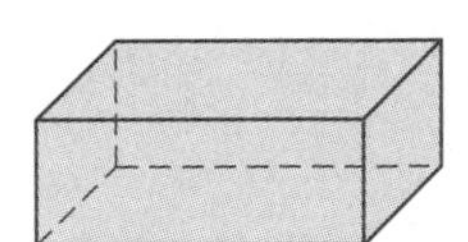

$3x$ cm by $(y + 1)$ cm by $(y + 6)$ cm

128. The volume of a box is $(4x^2y + 32xy + 60y)$ square inches. Find the dimensions of the box in terms of the variables x and y.

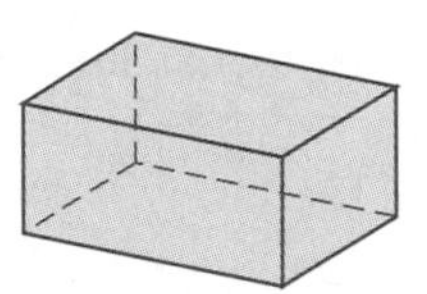

$4y$ in. by $(x + 5)$ in. by $(x + 3)$ in.

129. Write an expression in factored form for the shaded portion of the diagram.

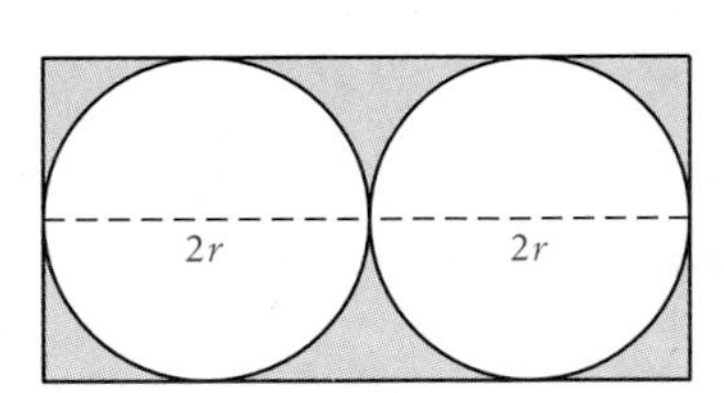

$2r^2(4 - \pi)$

130. Write an expression in factored form for the shaded portion of the diagram.

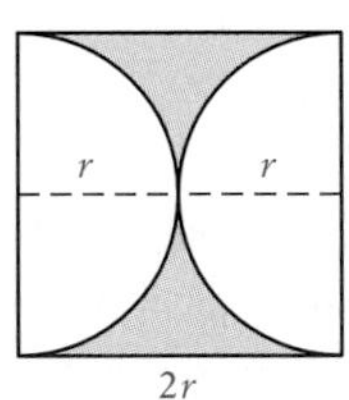

$r^2(4 - \pi)$

131. Write an expression in factored form for the shaded portion of the diagram.

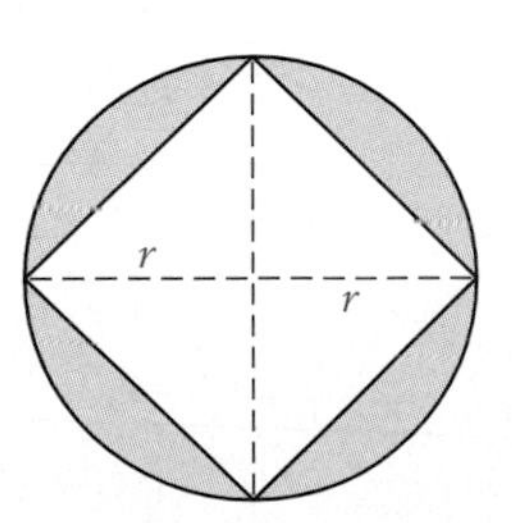

$r^2(\pi - 2)$

132. Find the dimensions of a rectangle that has the same area as the shaded region in the diagram below. Write the dimensions in terms of the variable a.

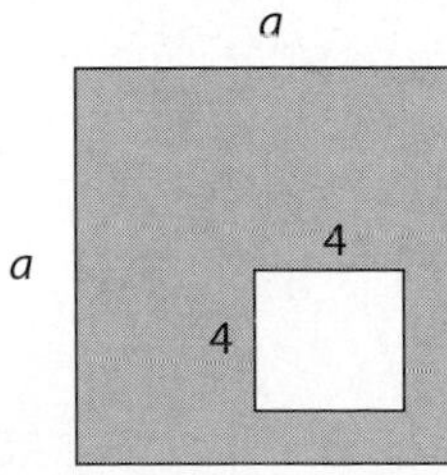

$(a + 4)$ by $(a - 4)$

133. Find the dimensions of a rectangle that has the same area as the shaded region in the diagram below. Write the dimensions in terms of the variable x.

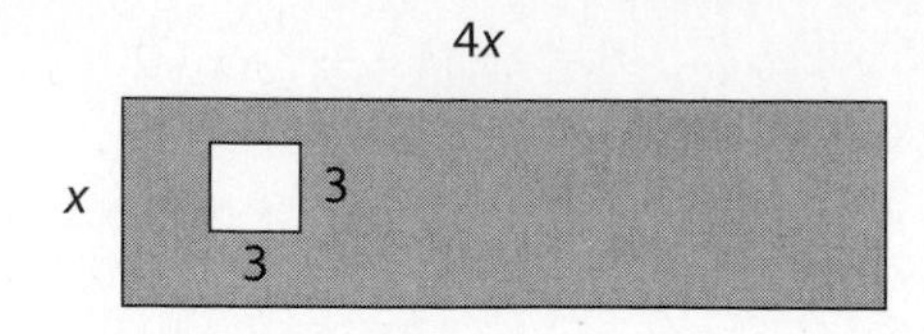

$(2x - 3)$ by $(2x + 3)$

134. Find the dimensions of a rectangle that has the same area as the shaded region in the diagram below. Write the dimensions in terms of the variable a.

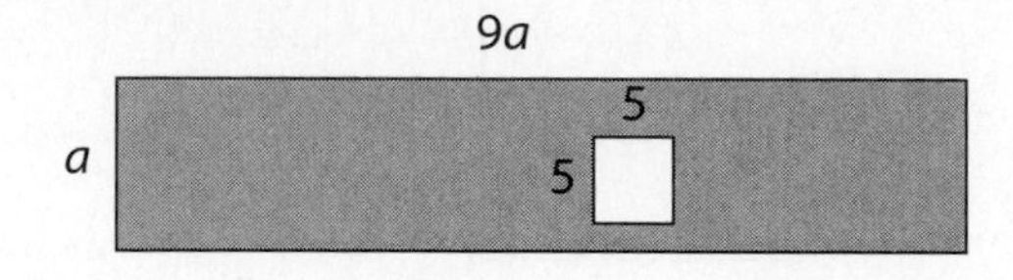

$(3a + 5)$ by $(3a - 5)$

Applying Concepts

135. Find all integers k such that the trinomial can be factored over the integers.

a. $x^2 + kx + 35$ $-36, 36, -12, 12$

b. $x^2 + kx + 18$ $-19, 19, -11, 11, -9, 9$

c. $x^2 - kx + 21$ $-22, 22, -10, 10$

d. $x^2 - kx + 14$ $-15, 15, -9, 9$

136. Determine the positive integer values of k for which the following polynomials are factorable over the integers.

a. $y^2 + 4y + k$ 3, 4

b. $z^2 + 7z + k$ 6, 10, 12

c. $a^2 - 6a + k$ 5, 8, 9

d. $c^2 - 7c + k$ 6, 10, 12

e. $x^2 - 3x + k$ 2

f. $y^2 + 5y + k$ 4, 6

137. Exercise 136 included the requirement that $k > 0$. If k is allowed to be any integer, how many different values of k are possible for each polynomial? Explain your answer.
An infinite number of different values of k are possible. Explanations will vary.

138. Find all integers k such that the trinomial can be factored over the integers.

a. $2x^2 + kx + 3$ $-7, 7, -5, 5$

b. $2x^2 + kx - 3$ $-5, 5, -1, 1$

c. $3x^2 + kx + 2$ $-7, 7, -5, 5$

d. $3x^2 + kx - 2$ $-5, 5, -1, 1$

e. $2x^2 + kx + 5$ $-11, 11, -7, 7$

f. $2x^2 + kx - 5$ $-9, 9, -3, 3$

139. Given that $x + 2$ is a factor of $x^3 - 2x^2 - 5x + 6$, factor $x^3 - 2x^2 - 5x + 6$ completely.
$(x + 2)(x - 3)(x - 1)$

140. In the expression $P = 2L + 2W$, what is the effect on P when the quantity $L + W$ doubles?
P doubles

141. The area of a rectangle is $(3x^2 + x - 2)$ square feet. Find the dimensions of the rectangle in terms of the variable x. Given that $x > 0$, specify the dimension that is the length and the dimension that is the width. Can $x < 0$? Can $x = 0$?

$A = 3x^2 + x - 2$

The dimensions are $3x - 2$ and $x + 1$. If $x = 1.5$, then the rectangle is a square. If $x < 1.5$, the length is $(x + 1)$ feet and the width is $(3x - 2)$ feet. If $x > 1.5$, then the width is $(x + 1)$ feet and the length is $(3x - 2)$ feet. If $x < 0$, then $3x - 2$ is negative, which is not possible. Therefore, x cannot be less than 0. If $x = 0$, $3x^2 + x - 2 < 0$, which is not possible. Therefore, x cannot be equal to 0.

142. The area of a square is $(16x^2 + 24x + 9)$ square feet. Find the dimensions of the square in terms of the variable x. Can $x = 0$? What are the possible values of x?

$(4x + 3)$ ft by $(4x + 3)$ ft; yes; $x > -\frac{3}{4}$

143. Select any odd integer greater than 1, square it, and then subtract 1. Is the result evenly divisible by 8? Prove that this procedure always produces a number divisible by 8. (*Suggestion*: Any odd integer greater than 1 can be expressed as $2n + 1$, where n is a natural number.)

$(2n + 1)^2 - 1 = (2n + 1)(2n + 1) - 1 = (4n^2 + 4n + 1) - 1 = 4n^2 + 4n = 4n(n + 1)$.
Since n or $n + 1$ is an even number, $4n(n + 1)$ is divisible by 8.

144. **a.** The graph of $y = x^2 + 3x - 2$ is shown at the right. Use a graphing calculator and the x-intercepts of the x-coordinates of $y = x^2 + 3x - 2$ to determine the approximate factorization of the polynomial $x^2 + 3x - 2$. Round to the nearest ten-thousandth.

b. Use a graphing calculator to determine the approximate factorization of the polynomial $x^2 + x - 5$. Round to the nearest ten-thousandth.

a. $(x - 0.5616)(x + 3.5616)$ **b.** $(x - 1.7913)(x + 2.7913)$

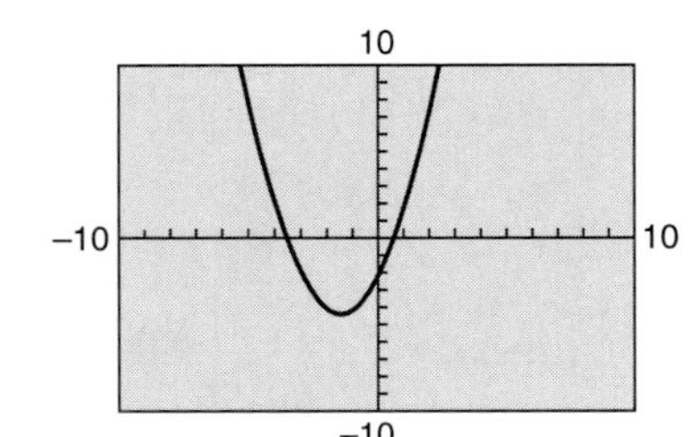

Explorations

145. ***Graphs of Second-Degree Polynomials and Their Factors***

The product of the binomials $(x + 2)(x - 4)$ is $x^2 - 2x - 8$. Use a graphing calculator and the standard window to graph $y = x^2 - 2x - 8$. The result is the graph shown at the right. Now graph $y = (x + 2)(x - 4)$ on the graphing calculator using the standard window. Explain the result.

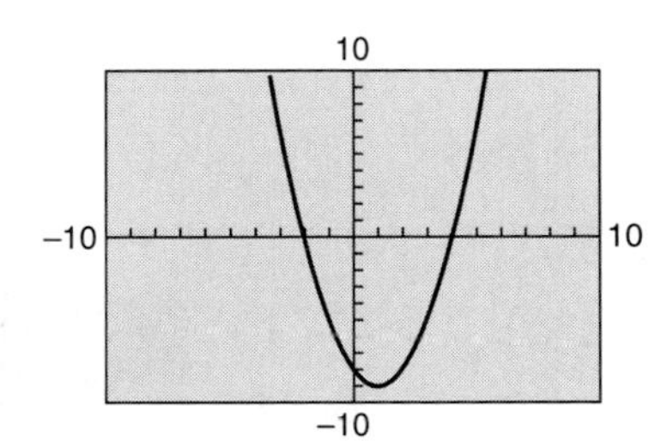

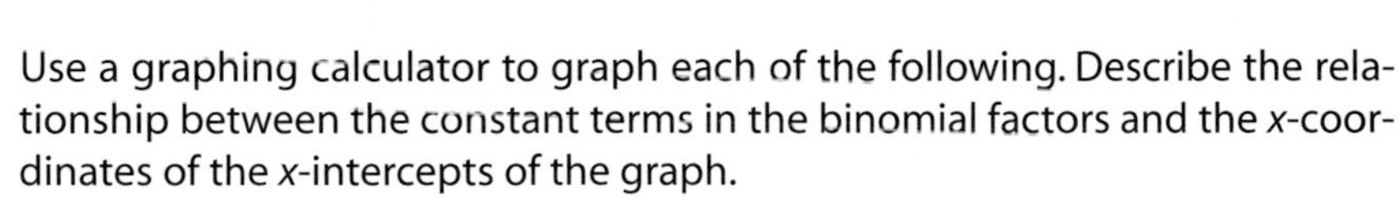
Use a graphing calculator to graph each of the following. Describe the relationship between the constant terms in the binomial factors and the x-coordinates of the x-intercepts of the graph.

a. $y = x^2 - 2x - 3$; $y = (x - 3)(x + 1)$

b. $y = x^2 + 3x - 10$; $y = (x - 2)(x + 5)$

c. $y = x^2 - 2x - 15$; $y = (x + 3)(x - 5)$

d. $y = x^2 + 3x - 4$; $y = (x - 1)(x + 4)$

146. ***Models of Factoring***

Geometric figures and a mat can be used to model factoring. The figures shown below represent 1, x, and x^2.

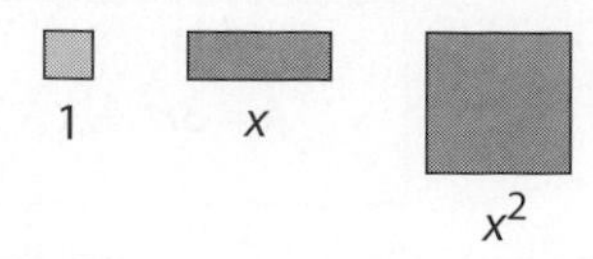

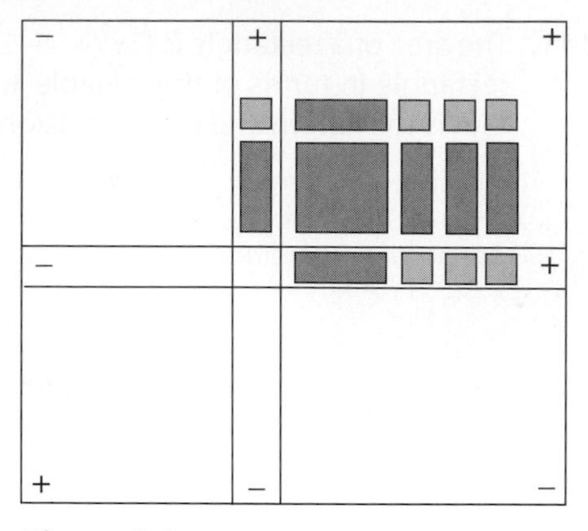

Figure 5-1

The mat in Figure 5-1 at the right is modeling $(x + 1)(x + 3) = x^2 + 4x + 3$. Note that the factor $x + 1$ is represented down the positive vertical axis. The factor $x + 3$ is represented along the positive horizontal axis. The trinomial $x^2 + 4x + 3$ is pictured in Quadrant 1.

a. What factors are pictured in Figure 5-2 at the right? What trinomial is represented in Quadrant I?

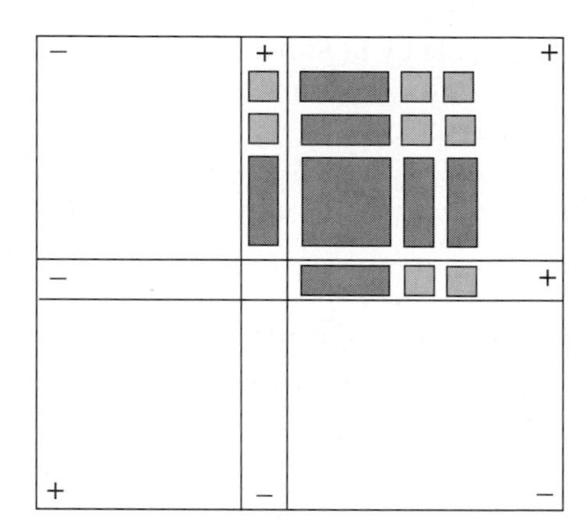

Figure 5-2

b. What factors are pictured in Figure 5-3 at the right below? What trinomial is represented in the quadrants? (*Hint*: You will need to "combine like terms.")

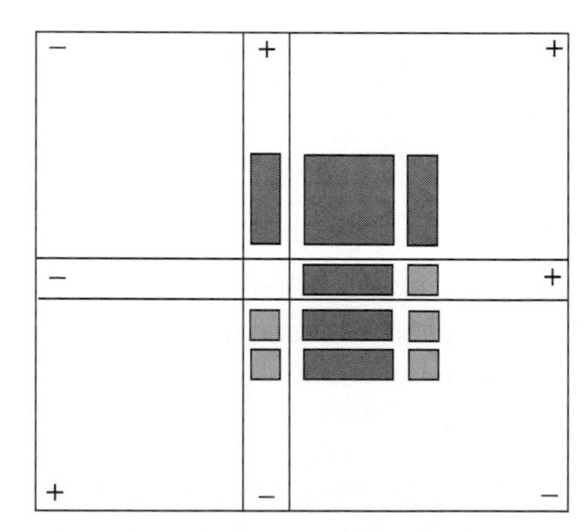

Figure 5-3

c. In the mat below, the binomial factors are shown. Use geometric figures to represent the trinomial that results from multiplying the two binomials.

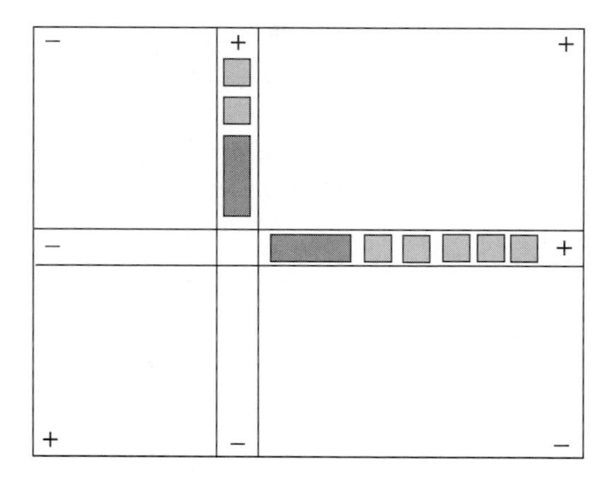

d. In the mat below, a trinomial is pictured. Use geometric figures to represent the binomial factors of the trinomial.

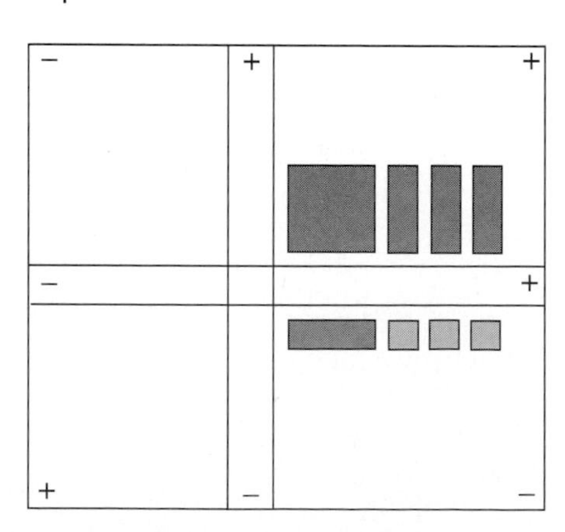

e. Use geometric figures and a mat to represent two binomial factors and their product.

Chapter Summary

Definitions

A *monomial* is a number, a variable, or a product of a number and variables.

A *polynomial* is a variable expression in which the terms are monomials. A polynomial of one term is a *monomial.* A polynomial of two terms is a *binomial.* A polynomial of three terms is a *trinomial.*

The terms of a polynomial in one variable are usually arranged so that the exponents of the variable decrease from left to right. This is called *descending order.*

The *degree of a polynomial in one variable* is its largest exponent.

The *greatest common factor (GCF)* of two or more integers is the greatest integer that is a factor of all the integers.

To *factor a polynomial* means to write the polynomial as a product of other polynomials. To *factor a trinomial of the form* $ax^2 + bx + c$ means to express the trinomial as the product of two binomials.

A polynomial that does not factor using only integers is *nonfactorable over the integers.* A polynomial is *factored completely* when it is written as a product of factors that are nonfactorable over the integers.

The expression $(a + b)(a - b)$ is called the *sum and difference of two terms.* The product of the sum and difference of two terms is $a^2 - b^2$, called the *difference of two squares.*

Procedures

Addition and Subtraction of Monomials — To add or subtract monomials with the same variable part, add or subtract the coefficients. The variable part stays the same.

Rule for Multiplying Exponential Expressions — $x^m \cdot x^n = x^{m+n}$

Rule for Simplifying the Power of an Exponential Expression — $(x^m)^n = x^{m \cdot n}$

Rule for Simplifying Powers of Products — $(x^m y^n)^p = x^{m \cdot p} y^{n \cdot p}$

Rule for Dividing Exponential Expressions — For $x \neq 0$, $\dfrac{x^m}{x^n} = x^{m-n}$

Zero as an Exponent — For $x \neq 0$, $x^0 = 1$

Definition of Negative Exponents — For $x \neq 0$, $x^{-n} = \dfrac{1}{x^n}$ and $\dfrac{1}{x^{-n}} = x^n$

Addition of Polynomials — To add polynomials, add the coefficients of the like terms.

Subtraction of Polynomials — To subtract two polynomials, add the opposite of the second polynomial to the first.

Multiplication of Polynomials	To multiply two polynomials, multiply each term of one polynomial times each term of the other polynomial.
The FOIL Method	To multiply two binomials, add the products of the **F**irst terms, the **O**uter terms, the **I**nner terms, and the **L**ast terms.
Division of Polynomials	To divide a polynomial by a monomial, divide each term of the polynomial by the monomial. If the divisor is not a monomial, use the long division method, which is similar to that used for division of whole numbers.
Common Monomial Factor	To factor a monomial from a polynomial, use the Distributive Property in reverse: $ab + ac = a(b + c)$.
Factor by Grouping	To factor a polynomial with four terms, group the first two terms and the last two terms (put them in parentheses). Factor out the GCF from each group. Write the expression as a product of factors by factoring out the common binomial factor.
Factor a Trinomial by Grouping	To factor a trinomial by grouping, find the product of the first and last terms. Find two factors of this product whose sum is the middle term of the trinomial. Replace the middle term of the trinomial with these two factors. The resulting polynomial has four terms. Factor it by grouping.
First Step in Factoring a Polynomial	The first step in any factoring problem is to determine whether the terms of the polynomial have a common factor. If they do, factor it out first.

Scientific Notation

To express a number in scientific notation, write it in the form $a \times 10^n$, where a is a number between 1 and 10 and n is an integer. If the number is greater than 10, the exponent on 10 will be positive. If the number is less than 1, the exponent on 10 will be negative.	$367{,}000{,}000 = 3.67 \times 10^8$ $0.0000059 = 5.9 \times 10^{-6}$
To change a number written in scientific notation to decimal notation, move the decimal point to the right if the exponent on 10 is positive and to the left if the exponent on 10 is negative. Move the decimal point the same number of places as the absolute value of the exponent on 10.	$2.418 \times 10^7 = 24{,}180{,}000$ $9.06 \times 10^{-5} = 0.0000906$

Chapter Review Exercises

1. Subtract: $47a^2b^3c - 23a^2b^3c$
$24a^2b^3c$

2. Add: $(3x^3 - 2x^2 - 4) + (8x^2 - 8x + 7)$
$3x^3 + 6x^2 - 8x + 3$

3. Multiply: $(5xy^2)(-4x^2y^3)$
$-20x^3y^5$

4. Simplify: $\dfrac{12x^2}{-3x^{-4}}$
$-4x^6$

5. Factor: $4a^2 - 12a + 9$
$(2a - 3)(2a - 3)$

6. Factor: $5x^2 - 45x - 15$
$5(x^2 - 9x - 3)$

7. Simplify: $(2ab^{-3})(3a^{-2}b^4)$
$\dfrac{6b}{a}$

8. Divide: $\dfrac{16x^5 - 8x^3 + 20x}{4x}$
$4x^4 - 2x^2 + 5$

9. Factor: $a^2 - 19a + 48$
$(a - 3)(a - 16)$

10. Factor: $x^3 + 2x^2 - 15x$
$x(x + 5)(x - 3)$

11. Multiply: $-3y^2(-2y^2 + 3y - 6)$
$6y^4 - 9y^3 + 18y^2$

12. Simplify: $(2x - 5)^2$
$4x^2 - 20x + 25$

13. Factor: $6x^2 + 19x + 8$
$(2x + 1)(3x + 8)$

14. Factor: $2b^2 - 32$
$2(b + 4)(b - 4)$

15. Simplify: $(a^2b^{-3})^2$
$\dfrac{a^4}{b^6}$

16. Write 0.000029 in scientific notation.
2.9×10^{-5}

17. Factor: $8x^2 + 20x - 48$
$4(x + 4)(2x - 3)$

18. Factor: $ab + 6a - 3b - 18$
$(b + 6)(a - 3)$

19. Multiply: $(4y - 3)(4y + 3)$
$16y^2 - 9$

20. Subtract: $(3y^2 - 5y + 8) - (-2y^2 + 5y + 8)$
$5y^2 - 10y$

21. Simplify: $(-3a^3b^2)^2$
$9a^6b^4$

22. Multiply: $(2a - 7)(5a^2 - 2a + 3)$
$10a^3 - 39a^2 + 20a - 21$

23. Factor: $2x^2 + 4x - 5$
Nonfactorable

24. Factor: $2y^4 - 14y^3 - 16y^2$
$2y^2(y + 1)(y - 8)$

25. Simplify: $\dfrac{-2a^2b^3}{8a^4b^8}$

$-\dfrac{1}{4a^2b^5}$

26. Divide: $(8x^2 + 4x - 3) \div (2x - 3)$

$4x + 8 + \dfrac{21}{2x - 3}$

27. Write 3.5×10^{-8} in decimal notation.
0.000000035

28. Factor: $p^2 + 5p + 6$
$(p + 2)(p + 3)$

29. The length of the side of a square is $(2x + 3)$ meters. Find the area of the square in terms of the variable x.
$(4x^2 + 12x + 9)$ m^2

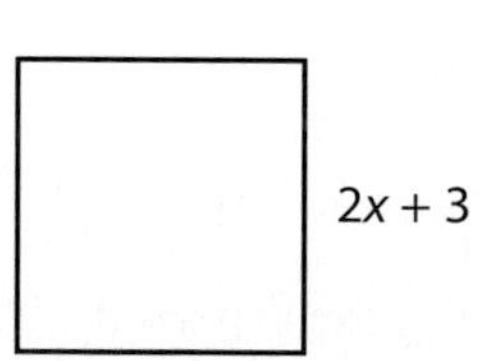

30. The area of a rectangle is $(3x^2 + 17x + 10)$ square inches. Find the dimensions of the rectangle in terms of the variable x.
$(3x + 2)$ in. by $(x + 5)$ in.

$A = 3x^2 + 17x + 10$

Cumulative Review Exercises

1. What are the next two letters of the sequence A, B, F, G, K, L, . . .?
P, Q

2. Given that ΔΔΔΔ = ◊◊◊◊◊ and ◊◊◊◊◊ = ÔÔ, then ÔÔÔÔ equal how many Δ?
8 Δ

3. Determine which of the following are true.
a. $65 \geq 49$ **b.** $37 \leq 37$ **c.** $16 < 24 > 8$
a and b are true.

4. Evaluate: $4^3 \cdot 3^2$
576

5. Evaluate: $\frac{12+4}{8} + 5(6-3)^2$
47

6. Evaluate the variable expression $5ab^3 - 4a^2b$ when $a = 3$ and $b = 2$.
48

7. Express the fact that 14 is between 8 and 25 using inequality symbols.
$8 < 14 < 25$

8. Simplify $-|42|$ and $|-81|$.
-42 and 81

9. Add: $-28 + (-61)$
-89

10. Subtract: $34 - (-53)$
87

11. Evaluate $-4ab + 6b^2$ when $a = 3$ and $b = -1$.
18

12. Write the negation of the sentence "No taxpayers file their returns electronically."
Some taxpayers file their returns electronically.

13. Find a counterexample to the statement "If two numbers are divided, the quotient is smaller than either the divisor or the dividend."
Answers will vary. For example, $3 \div \frac{1}{2} = 6$.

14. Subtract: $-\frac{3}{8} - \left(-\frac{5}{12}\right)$
$\frac{1}{24}$

15. Divide: $-\frac{5}{6} \div \frac{10}{9}$
$-\frac{3}{4}$

16. Simplify: $-8 + 5(3y - 4)$
$15y - 28$

17. Convert 4.5 gallons to quarts.
18 qt

18. Solve: $-8 = -5 + t$
-3

19. Solve: $-\frac{3d}{5} = 9$
-15

20. Solve: $7 - 2(4x - 1) = 5(6 - x)$
-7

21. 18 is 24% of what number?
75

22. Find the slope of the line containing the points (–1, 1) and (2, 3).

$\frac{2}{3}$

23. Solve the system of equations:
$3x + 8y = -1$
$x - 2y = -5$
(–3, 1)

24. Subtract: $(3y^2 - 5y + 7) - (4y^2 + 6y - 9)$
$-y^2 - 11y + 16$

25. Simplify: $(-3a^4b^2)^3$
$-27a^{12}b^6$

26. Multiply: $(x + 2)(x^2 - 5x + 4)$
$x^3 - 3x^2 - 6x + 8$

27. Divide: $(a^3 + a^2 + 18) \div (a + 3)$
$a^2 - 2a + 6$

28. Factor: $6a^4 + 22a^3 + 12a^2$
$2a^2(3a + 2)(a + 3)$

29. Graph $y = 3x - 2$ for $x = -2, -1, 0, 1, 2$.

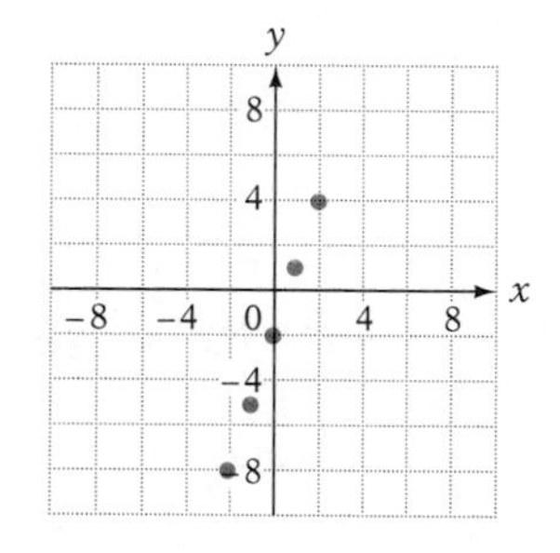

30. Translate the rectangle 1 unit to the left and 4 units down. What are the coordinates of the new rectangle?
(–5, 2), (1, 2), (1, –6), (–5, –6)

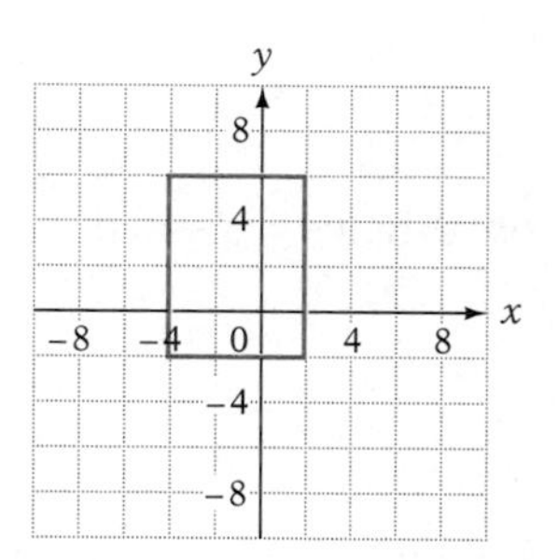

31. Complete the input/output table.

Input, n	–4	–3	–2	–1	0	1	2
Output, $(-6) + n$							

–10, –9, –8, –7, –6, –5, –4

32. The Phoenix Mercury in the Western Conference of the Women's National Basketball Association played 29 games during the 1997 season. Their scores in these games were: 76, 55, 68, 57, 77, 69, 69, 51, 57, 84, 70, 69, 83, 64, 67, 67, 70, 83, 78, 70, 59, 77, 61, 71, 66, 69, 78, 73, 41. Draw a stem-and-leaf plot for these data.

Stem	Leaves
8	4, 3, 3
7	6, 7, 0, 0, 8, 0, 7, 1, 8, 3
6	8, 9, 9, 9, 4, 7, 7, 1, 6, 9
5	5, 7, 1, 7, 9
4	1

33. The melting point of radon is –71°C. The melting point of oxygen is three times the melting point of radon. Find the melting point of oxygen.
–213°C

You-Try-It 3 The weight of an object is related to its distance above the surface of the earth. A formula for this relationship is $d = 4000\sqrt{\frac{E}{S}} - 4000$, where E is the object's weight on the surface of the earth and S is the object's weight at a distance of d miles above the earth's surface. An astronaut who weighs 184 pounds on the surface of the earth weighs 46 pounds in space. How far above the earth's surface is the astronaut?

Solution See page S26.

Thus far in this section, the radicand in each radical expression has been a perfect square. Since the square root of a perfect square is an integer, the exact value of each radical expression could be found.

If the radicand is not a perfect square, the square root can only be approximated. For example, the radicand in the radical expression $\sqrt{2}$ is 2, and 2 is not a perfect square. The square root of 2 can be approximated to any desired place value. Each of the approximations shown below was found using a calculator.

To the nearest tenth:	$\sqrt{2} \approx 1.4$	$(1.4)^2 = 1.96$
To the nearest hundredth:	$\sqrt{2} \approx 1.41$	$(1.41)^2 = 1.9881$
To the nearest thousandth:	$\sqrt{2} \approx 1.414$	$(1.414)^2 = 1.999396$
To the nearest ten-thousandth:	$\sqrt{2} \approx 1.4142$	$(1.4142)^2 = 1.99996164$

The square of the approximation gets closer and closer to 2 as the number of place values in the decimal approximation increases. But no matter how many place values are used to approximate $\sqrt{2}$, the digits never terminate or repeat. $\sqrt{2}$ is not a rational number.

In general, the square root of any number that is not a perfect square is not a rational number, and the square root can only be approximated. These numbers are *irrational numbers*. Recall that an **irrational number** is a number whose decimal representation never terminates or repeats.

Question: Which of the following represent irrational numbers?[2]

a. $\sqrt{15}$ **b.** $\sqrt{81}$ **c.** $\sqrt{97}$

→ Approximate $\sqrt{19}$ to the nearest ten-thousandth.

19 is not a perfect square. Use a calculator to approximate $\sqrt{19}$.

$\sqrt{19} \approx 4.3589$

---

2. **a.** 15 is not a perfect square. $\sqrt{15}$ is an irrational number. **b.** 81 is a perfect square. $\sqrt{81} = 9$, and it is not an irrational number. **c.** 97 is not a perfect square. $\sqrt{97}$ is an irrational number.

Example 4 Approximate $\sqrt{72.6}$ to the nearest ten-thousandth.

Solution $\sqrt{72.6} \approx 8.5206$ • Use a calculator to find the approximation.

You-Try-It 4 Approximate $\sqrt{113.4}$ to the nearest ten-thousandth.

Solution See page S26.

→ Approximate $3\sqrt{19}$ to the nearest ten-thousandth.

$3\sqrt{19}$ means 3 times $\sqrt{19}$. Use a calculator to find the approximation.

$3\sqrt{19} \approx 13.0767$

Example 5 Approximate $8\sqrt{21}$ to the nearest ten-thousandth.

Solution $8\sqrt{21} \approx 36.6606$ • Use a calculator to find the approximation.

You-Try-It 5 Approximate $7\sqrt{34}$ to the nearest ten-thousandth.

Solution See page S26.

Example 6 If an object is dropped from a plane, how long will it take for the object to fall 150 feet? Use the formula $t = \frac{\sqrt{d}}{4}$, where t is the time in seconds that the object falls and d is the distance in feet that the object falls. Round to the nearest hundredth.

State the goal. The goal is to find the time it takes for an object dropped from a plane to fall 150 feet.

Describe a strategy. Replace the variable d in the given formula by 150. Evaluate the variable expression.

Solve the problem.

$t = \frac{\sqrt{d}}{4}$

$t = \frac{\sqrt{150}}{4}$ • Replace d by 150.

$t \approx 3.06$ • First calculate $\sqrt{150}$. Divide the result by 4.

It takes 3.06 seconds for an object dropped from a plane to fall 150 feet.

Check your work. √

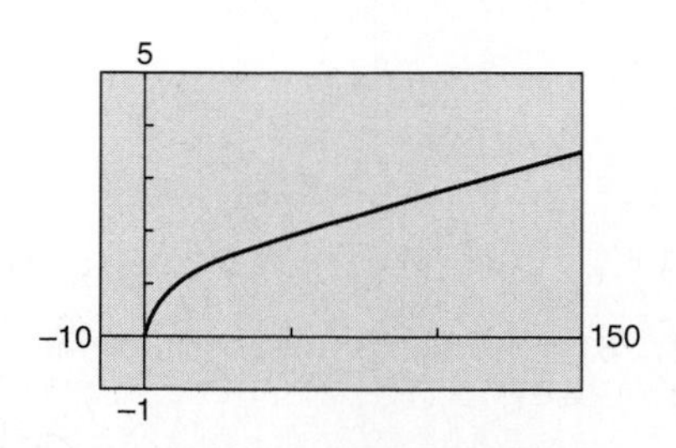

The graph of $t = \frac{\sqrt{d}}{4}$ is shown at the left. Note that the domain is $\{d \mid d \geq 0\}$, and the range is $\{t \mid t \geq 0\}$. This is reasonable in the context of the application: Neither the distance the object falls nor the time of the fall can be negative.

Note that when X = 150, Y ≈ 3. This is the answer we calculated in Example 6.

You-Try-It 6 Weather satellites are able to measure the diameter of a storm. The duration of the storm can then be determined by using the formula $t = \sqrt{\frac{d^3}{216}}$, where t is the duration of the storm in hours and d is the diameter of the storm in miles. Find the duration of a storm that has a diameter of 10 miles. Round to the nearest tenth.

Solution See page S26.

Sometimes we are not interested in an approximation of the square root of a number, but rather want the exact value in simplest form.

A radical expression is in simplest form when the radicand contains no factor, other than 1, that is a perfect square. The Product Property of Square Roots is used to simplify radical expressions.

Product Property of Square Roots

If a and b are positive numbers, then $\sqrt{a \cdot b} = \sqrt{a} \cdot \sqrt{b}$.

The Product Property of Square Roots states that the square root of a product is equal to the product of the square roots. For example:

$$\sqrt{4 \cdot 9} = \sqrt{4} \cdot \sqrt{9}$$

Note that $\sqrt{4 \cdot 9} = \sqrt{36} = 6$ and $\sqrt{4} \cdot \sqrt{9} = 2 \cdot 3 = 6$.

→ Simplify: $\sqrt{50}$

Think: What perfect square is a factor of 50?
Begin with a perfect square that is larger than 50.
Then test each successively smaller perfect square.
$8^2 = 64$; 64 is too big.
$7^2 = 49$; 49 is not a factor of 50.
$6^2 = 36$; 36 is not a factor of 50.
$5^2 = 25$; 25 is a factor of 50. $(50 = 25 \cdot 2)$

Write $\sqrt{50}$ as $\sqrt{25 \cdot 2}$. $\quad \sqrt{50} = \sqrt{25 \cdot 2}$

Use the Product Property of Square Roots. $\quad = \sqrt{25} \cdot \sqrt{2}$

Simplify $\sqrt{25}$. $\quad = 5 \cdot \sqrt{2}$

The radicand 2 contains no factor other than 1 that is a perfect square. The radical expression $5\sqrt{2}$ is in simplest form. $\quad = 5\sqrt{2}$

This last example shows that $\sqrt{50} = 5\sqrt{2}$. The two expressions are different representations of the same number. Using a calculator to evaluate each expression, we find that $\sqrt{50} \approx 7.07106781$ and $5\sqrt{2} \approx 7.07106781$.

Example 7 Simplify: $\sqrt{32}$

SUGGESTED ACTIVITY
See the *Instructor's Resource Manual* for an activity involving simplifying a square root.

Solution $6^2 = 36$; 36 is too big.
$5^2 = 25$; 25 is not a factor of 32.
$4^2 = 16$; 16 is a factor of 32. ($32 = 16 \cdot 2$)

$$\begin{aligned}\sqrt{32} &= \sqrt{16 \cdot 2} \\ &= \sqrt{16} \cdot \sqrt{2} \\ &= 4 \cdot \sqrt{2} \\ &= 4\sqrt{2}\end{aligned}$$

You-Try-It 7 Simplify: $\sqrt{80}$

Solution See page S26.

Question: In Example 7, the expression $\sqrt{32}$ was simplified as $4\sqrt{2}$. Why is the following simplification of $\sqrt{32}$ incorrect?[3]

$$\sqrt{32} = \sqrt{4 \cdot 8} = \sqrt{4} \cdot \sqrt{8} = 2 \cdot \sqrt{8} = 2\sqrt{8}$$

Example 8 Simplify: $5\sqrt{90}$

Solution $10^2 = 100$; 100 is too big.
Not factors of 90: $9^2 = 81; 8^2 = 64; 7^2 = 49; 6^2 = 36; 5^2 = 25; 4^2 = 16$
$3^2 = 9$; 9 is a factor of 90. ($90 = 9 \cdot 10$)

$$\begin{aligned}5\sqrt{90} &= 5\sqrt{9 \cdot 10} \\ &= 5\sqrt{9} \cdot \sqrt{10} \\ &= 5 \cdot 3 \cdot \sqrt{10} \\ &= 15\sqrt{10}\end{aligned}$$

You-Try-It 8 Simplify: $6\sqrt{200}$

Solution See page S27.

Simplify Variable Radical Expressions

As stated on page 390, the square root of a negative number is not a real number.

$\sqrt{-9}$ is not a real number.

A radical expression in which the radicand is a variable does not always represent a real number.

The expression $\sqrt{x}$ does not represent a real number if $x < 0$.

---

3. The answer $2\sqrt{8}$ is not in simplest form because the radicand (8) contains a factor that is a perfect square: 4. $\sqrt{8}$ can be simplified as $\sqrt{8} = \sqrt{4 \cdot 2} = \sqrt{4} \cdot \sqrt{2} = 2 \cdot \sqrt{2} = 2\sqrt{2}$.

The expression $\sqrt{x^3}$ does not represent a real number when $x < 0$. For example, if $x = -4$, then

$$\sqrt{(-4)^3} = \sqrt{-64}\text{, which is not a real number.}$$

Now consider the expression $\sqrt{x^2}$ and evaluate the expression for $x = 2$ and $x = -2$.

$$\sqrt{2^2} = \sqrt{4} = 2 = |2| \qquad \sqrt{(-2)^2} = \sqrt{4} = 2 = |-2|$$

This suggests the following:

For any real number x, $\sqrt{x^2} = |x|$.

If $x \geq 0$, then $\sqrt{x^2} = x$.

In order to avoid variable expressions that do not represent real numbers, and so that absolute value signs are not needed for certain expressions, the variables in this chapter will represent positive numbers unless otherwise stated.

A variable or a product of variables written in exponential form is a **perfect square** when each exponent is an even number.

y^6 is a perfect square because it is the square of y^3: $(y^3)^2 = y^6$.
Note that the exponent, 6, is an even number.

b^{14} is a perfect square because it is the square of b^7: $(b^7)^2 = b^{14}$.
Note that the exponent, 14, is an even number.

Question: Which of the following represent perfect squares?[4]

a. d^8 **b.** n^{11} **c.** a^{24}

To find the square root of a perfect square, remove the radical sign and divide the exponent by 2.

TAKE NOTE

$\sqrt{c^{10}} = c^5$ because $(c^5)^2 = c^{10}$.

Example 9 Simplify: $\sqrt{c^{10}}$

Solution $\sqrt{c^{10}} = c^5$ • c^{10} is a perfect square. Remove the radical sign and divide the exponent by 2.

You-Try-It 9 Simplify: $\sqrt{t^{16}}$

Solution See page S27.

---

4. d^8 and a^{24} are perfect squares because the exponents on the variables are even numbers. n^{11} is not a perfect square because the exponent, 11, is not an even number.

A variable radical expression is in simplest form when the radicand contains no factor greater than 1 that is a perfect square.

→ Simplify: $\sqrt{x^7}$

Write x^7 as the product of a perfect square and x. $\sqrt{x^7} = \sqrt{x^6 \cdot x}$

Use the Product Property of Square Roots. $= \sqrt{x^6} \cdot \sqrt{x}$

Simplify the square root of the perfect square. $= x^3\sqrt{x}$

Example 10 Simplify: $\sqrt{y^{15}}$

Solution

$$\sqrt{y^{15}} = \sqrt{y^{14} \cdot y}$$
$$= \sqrt{y^{14}} \cdot \sqrt{y} = y^7\sqrt{y}$$

You-Try-It 10 Simplify: $\sqrt{z^9}$

Solution See page S27.

→ Simplify: $3x\sqrt{12x^3y^8}$

Write the radicand as the product of a perfect square and factors that do not contain a perfect square. $3x\sqrt{12x^3y^8} = 3x\sqrt{4x^2y^8 \cdot 3x}$

Use the Product Property of Square Roots. $= 3x\sqrt{4x^2y^8} \cdot \sqrt{3x}$

Simplify the square root of the perfect square. $= 3x \cdot 2xy^4 \cdot \sqrt{3x}$

Multiply $3x$ and $2xy^4$. $= 6x^2y^4\sqrt{3x}$

Example 11 Simplify. **a.** $\sqrt{24y^5}$ **b.** $2a\sqrt{18ab^{14}}$

Solution

a.
$$\sqrt{24y^5} = \sqrt{4y^4 \cdot 6y}$$
$$= \sqrt{4y^4} \cdot \sqrt{6y}$$
$$= 2y^2\sqrt{6y}$$

b.
$$2a\sqrt{18ab^{14}} = 2a\sqrt{9b^{14} \cdot 2a}$$
$$= 2a\sqrt{9b^{14}} \cdot \sqrt{2a}$$
$$= 2a \cdot 3b^7\sqrt{2a}$$
$$= 6ab^7\sqrt{2a}$$

You-Try-It 11 Simplify. **a.** $\sqrt{45x^{11}}$ **b.** $5a\sqrt{28a^4b^{13}}$

Solution See page S27.

6.1 EXERCISES

Topics for Discussion

1. Fill in the blanks.

a. The square of an integer is called a ________. perfect square

b. In the expression $\sqrt{41}$, 41 is called the __________. radicand

c. Because $11^2 = 121$, 11 is a __________ of 121. square root

d. Since $(4.5)(4.5) = 20.25$, __________ is a square root of __________. 4.5; 20.25

e. A square has an area of 100 cm^2. The length of a side of the square is __________. 10 cm

2. Use the roster method to list the whole numbers between $\sqrt{6}$ and $\sqrt{95}$. {3, 4, 5, 6, 7, 8, 9}

3. Explain why $2\sqrt{3}$ is in simplest form and $\sqrt{12}$ is not in simplest form.

The radicand in $\sqrt{12}$ contains a factor that is a perfect square (4).

4. Describe how to simplify a square root in which:

a. the radicand is a number that is a perfect square. Answers will vary.

b. the radicand is a number that is not a perfect square. Answers will vary.

c. the radicand is a variable raised to an even exponent. Answers will vary.

d. the radicand is a variable raised to an odd exponent. Answers will vary.

5. Explain what is meant by the fact that a radical symbol is a grouping symbol.

All operations under the radical symbol must be performed as Step 1 in the Order of Operations Agreement (or as the first step in simplifying a numerical expression).

6. Explain why the square root of a negative number is not a real number.

There is no real number that, when squared, is equal to a negative number.

7. The table at the right shows nine numbers and the square of each of those numbers. Use the table to answer each of the following questions. Do not use a calculator.

N	*N* Squared
16.1	259.21
16.2	262.44
16.3	265.69
16.4	268.96
16.5	272.25
16.6	275.56
16.7	278.89
16.8	282.24
16.9	285.61

a. What is the square root of 275.56?

b. Find $\sqrt{265.69}$.

c. $\sqrt{281.5}$ is between what two numbers?

d. $\sqrt{293}$ is greater than what number?

e. Approximate the square root of 261.5.

f. Approximate $\sqrt{273.8}$.

a. 16.6; **b.** 16.3; **c.** 16.7 and 16.8; **d.** 16.9; **e.** answers will vary (between 16.1 and 16.2); **f.** answers will vary (between 16.5 and 16.6).

Simplifying Numerical Radical Expressions

Simplify.

8. $\sqrt{36}$
6

9. $\sqrt{1}$
1

10. $-\sqrt{9}$
−3

11. $-\sqrt{1}$
−1

12. $\sqrt{169}$
13

13. $\sqrt{81}$
9

14. $-\sqrt{25}$
–5

15. $-\sqrt{64}$
-8

16. $\sqrt{-9}$
Not a real number

17. $\sqrt{-49}$
Not a real number

18. $\sqrt{196}$
14

19. $\sqrt{225}$
15

20. $-\sqrt{100}$
–10

21. $-\sqrt{4}$
–2

22. $\sqrt{8 + 17}$
5

23. $\sqrt{40 + 24}$
8

24. $\sqrt{49} + \sqrt{9}$
10

25. $\sqrt{100} + \sqrt{16}$
14

26. $\sqrt{121} - \sqrt{4}$
9

27. $\sqrt{144} - \sqrt{25}$
7

28. $3\sqrt{81}$
27

29. $8\sqrt{36}$
48

30. $-2\sqrt{49}$
–14

31. $-6\sqrt{121}$
-66

32. $5\sqrt{16} - 4$
16

33. $7\sqrt{64} + 9$
65

34. $3 + 10\sqrt{1}$
13

35. $14 - 3\sqrt{144}$
-22

36. $\sqrt{4} - 2\sqrt{16}$
-6

37. $\sqrt{144} + 3\sqrt{9}$
21

38. $5\sqrt{25} + \sqrt{49}$
32

39. $20\sqrt{1} - \sqrt{36}$
14

Evaluate the expression for the given values of the variables.

40. $-4\sqrt{xy}$, where $x = 3$ and $y = 12$
–24

41. $-3\sqrt{xy}$, where $x = 20$ and $y = 5$
–30

42. $8\sqrt{x + y}$, where $x = 19$ and $y = 6$
40

43. $7\sqrt{x + y}$, where $x = 34$ and $y = 15$
49

44. $5 + 2\sqrt{ab}$, where $a = 27$ and $b = 3$
23

45. $6\sqrt{ab} - 9$, where $a = 2$ and $b = 32$
39

46. $\sqrt{a^2 + b^2}$, where $a = 3$ and $b = 4$
5

47. $\sqrt{c^2 - a^2}$, where $a = 6$ and $c = 10$
8

48. $\sqrt{c^2 - b^2}$, where $b = 12$ and $c = 13$
5

49. $\sqrt{b^2 - 4ac}$, where $a = 1$, $b = -4$, and $c = -5$
6

Solve.

50. What is the sum of five and the square root of nine?
8

51. Find eight more than the square root of four.
10

52. Find the difference between six and the square root of twenty-five.
1

53. What is seven decreased by the square root of sixteen?
3

54. What is negative four times the square root of eighty-one?
−36

55. Find the product of negative three and the square root of forty-nine.
−21

Approximate to the nearest ten-thousandth.

56. $\sqrt{3}$
1.7321

57. $\sqrt{7}$
2.6458

58. $\sqrt{10}$
3.1623

59. $\sqrt{19}$
4.3589

60. $\sqrt{300}$
17.3205

61. $\sqrt{245}$
15.6525

62. $\sqrt{52.8}$
7.2664

63. $\sqrt{73.4}$
8.5674

64. $2\sqrt{6}$
4.8990

65. $10\sqrt{21}$
45.8258

66. $3\sqrt{14}$
11.2250

67. $6\sqrt{15}$
23.2379

68. $-4\sqrt{2}$
−5.6569

69. $-5\sqrt{13}$
−18.0278

70. $-8\sqrt{30}$
−43.8178

71. $-12\sqrt{53}$
−87.3613

Solve.

72. In pinhole photography, there is no camera lens. A tiny hole replaces the lens. Light passes through the hole, and an image is formed in the camera. The size of the pinhole affects the quality of the photograph. One formula for the optimum size of the pinhole is $d = 0.0073\sqrt{D}$, where d is the diameter of the pinhole and D is the distance, in inches, from the pinhole to the film.

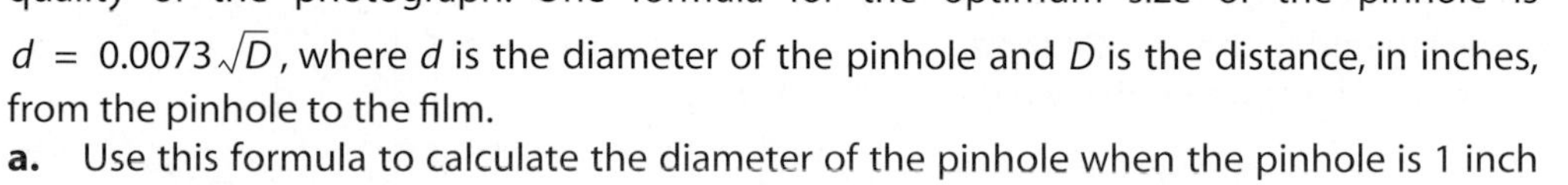

a. Use this formula to calculate the diameter of the pinhole when the pinhole is 1 inch from the film.
b. Use this formula to calculate the diameter of the pinhole when the pinhole is 4 inches from the film.
c. According to the formula, should the diameter of the pinhole increase or decrease as the distance between the pinhole and the film increases?

a. 0.0073 in.; **b.** 0.0146 in.; **c.** increase

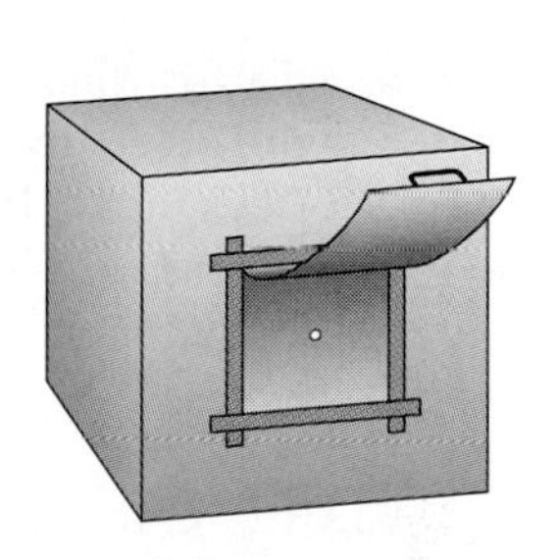

73. The formula $v = \sqrt{12r}$ can be used to calculate the speed at which a person riding a merry-go-round is traveling. In this formula, v is the speed in feet per second, and r is the distance in feet from the center of the merry-go-round to the rider.

a. A child is sitting 3 feet from the center of a merry-go-round. How fast is the child traveling when the merry-go-round is operating?

b. A person is sitting 12 feet from the center of a moving merry-go-round. How fast is the person traveling?

c. Does the speed of the rider increase or decrease as the distance between the rider and the center of the merry-go-round increases?

a. 6 ft/s; **b.** 12 ft/s; **c.** increase

74. The formula $d = \sqrt{12h}$ is used to approximate the distance d, in kilometers, to the horizon from a point h meters above the earth's surface.

a. On a clear day, how far can you see from the top of the Sears Tower in Chicago? The height of the Sears Tower is 443 meters. Round to the nearest hundredth.

b. On a clear day, how much farther can you see from the top of the World Trade Center in New York City than from the top of the Empire State Building? The height of the World Trade Center is 419 meters, and the height of the Empire State Building is 381 meters. Round to the nearest hundredth.

a. 72.91 km **b.** 3.29 km

75. Traffic accident investigators can use the formula $v = 2\sqrt{5L}$ to determine the speed of a car that leaves skid marks after the driver hits the brake pedal to stop the car. In this formula, the speed, v, of the car is measured in miles per hour, and the length, L, of the skid mark is measured in feet.

a. Find the speed of a car that left a skid mark that was 150 feet long. Round to the nearest tenth.

b. Two cars were involved in an accident on a two-lane highway. The length of the skid mark left by the sports car was 198 feet. The length of the skid mark left by the minivan was 172 feet. Which car was traveling at the faster speed? How much faster was that car traveling? Round to the nearest tenth.

a. 54.8 mph; **b.** the sports car, 4.3 mph

76. The time it takes for an object to fall a distance of d feet on the moon is given by the formula $t = \sqrt{\dfrac{d}{2.75}}$, where t is the time in seconds. The time it takes for an object to fall a distance of d feet on Earth is given by the formula $t = \sqrt{\dfrac{d}{16}}$.

a. If an astronaut drops an object on the moon, how long will it take for the object to fall 25 feet? Round to the nearest tenth.

b. Does it take longer for an object to drop 25 feet on Earth or on the moon? How much longer? Round to the nearest tenth.

c. Use your answers to parts a and b to determine which graph shown at the right is a graph of $t = \sqrt{\dfrac{d}{2.75}}$ and which is a graph of $t = \sqrt{\dfrac{d}{16}}$.

a. 3.0 s; **b.** the moon, 1.8 s;

c. Figure A is the graph of $t = \sqrt{\dfrac{d}{2.75}}$; Figure B is the graph of $t = \sqrt{\dfrac{d}{16}}$.

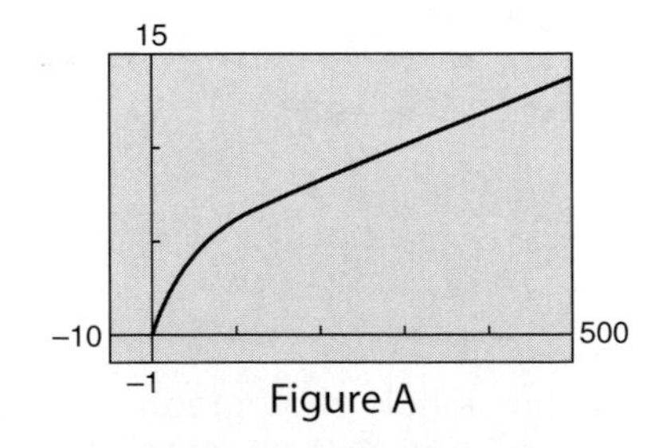

Figure A

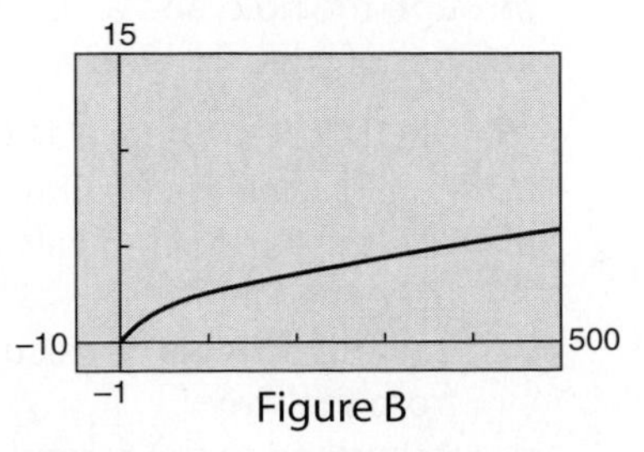

Figure B

77. The time it takes for an object to fall a distance of d feet on Jupiter is given by the formula $t = \sqrt{\dfrac{d}{43.4}}$, where t is the time in seconds. The time it takes for an object to fall a distance of d feet on Earth is given by the formula $t = \sqrt{\dfrac{d}{16}}$.

a. How long does it take for an object to fall 200 feet on Jupiter? Round to the nearest tenth.

b. Does it take longer for an object to drop 200 feet on Jupiter or on Earth? How much longer? Round to the nearest tenth.

c. Use your answers to parts a and b to determine which graph shown at the right is a graph of $t = \sqrt{\dfrac{d}{43.4}}$ and which is a graph of $t = \sqrt{\dfrac{d}{16}}$.

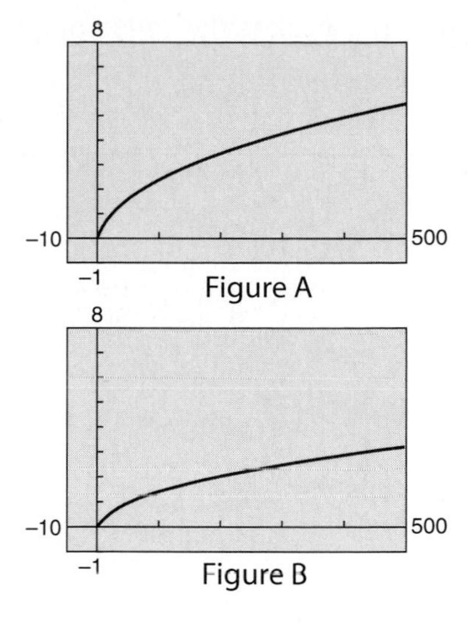

Figure A

Figure B

a. 2.1 s; **b.** Earth, 1.4 s;

c. Figure A is the graph of $t = \sqrt{\dfrac{d}{16}}$; Figure B is the graph of $t = \sqrt{\dfrac{d}{43.4}}$.

78. A pendulum is an object suspended from a fixed support so that it swings freely back and forth under the force of gravity. The formula used to determine the time it takes for one full swing of a pendulum is $T = 2\pi\sqrt{\dfrac{L}{32}}$, where T is the time in seconds and L is the length of the pendulum in feet. If the length of the pendulum is given in meters, then the formula $T = 2\pi\sqrt{\dfrac{L}{9.8}}$ is used.

a. How long does it take for a pendulum that is 2 feet long to complete one full swing? Round to the nearest tenth.

b. Does it take longer for a pendulum that is 2 feet long or one that is 2 meters long to complete one full swing? How much longer? Round to the nearest tenth.

c. Use your answers to parts a and b to determine which graph shown at the right is a graph of $T = 2\pi\sqrt{\dfrac{L}{32}}$ and which is a graph of $T = 2\pi\sqrt{\dfrac{L}{9.8}}$.

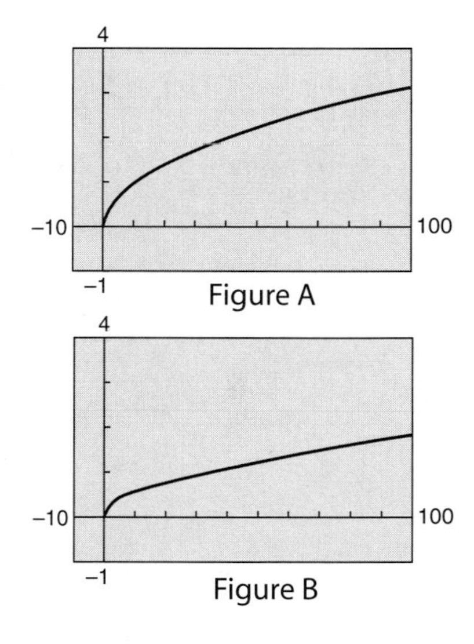

Figure A

Figure B

a. 1.6 s; **b.** 2 meters long, 1.3 s;

c. Figure A is the graph of $T = \sqrt{\dfrac{L}{9.8}}$; Figure B is the graph of $T = \sqrt{\dfrac{L}{32}}$.

79. The speed of sound in different air temperatures is calculated using the formula $v = \dfrac{1087\sqrt{t + 273}}{16.52}$, where v is the speed in feet per second and t is the temperature in degrees Celsius.

a. Find the speed of sound when the temperature is 10°C. Round to the nearest tenth.

b. As the temperature increases, does the speed of sound increase or decrease?

a. 1106.9 ft/s; **b.** increase

80. To calculate the speed of a satellite in orbit above the surface of Earth, use the formula $v = \sqrt{\dfrac{4 \times 10^{14}}{h + (6.4 \times 10^{6})}}$, where v is the speed in meters per second and h is the height in meters of the satellite above Earth's surface.

a. Find the speed of an orbiting satellite that is 700,000 meters from Earth's surface. Round to the nearest whole number.

b. As an orbiting satellite gets farther from Earth, does the speed increase or decrease?

a. 7506 m; **b.** decrease

81. The formula given below is used to calculate the wind chill factor in degrees Fahrenheit.

$$c = 0.0817(3.71\sqrt{s} + 5.81 - 0.25s)(t - 91.4) + 91.4$$

In this formula, c is the wind chill factor in degrees Fahrenheit, s is the wind speed in miles per hour, and t is the temperature in degrees Fahrenheit.

a. Find the wind chill factor when the temperature is 10°F and the wind is blowing at 15 miles per hour. Round to the nearest whole number.

b. Use the same wind speed given in part a, but lower the temperature by 10°F. Recalculate the wind chill factor. Round to the nearest whole number.

c. Use the same temperature given in part a, but lower the wind speed by 10 miles per hour. Recalculate the wind chill factor. Round to the nearest whole number.

d. Based on your answers to parts a, b, and c, which resulted in a greater difference in the wind chill factor, a change in the temperature or a change in the wind speed?

e. An important note about the formula for calculating the wind chill factor: If $c > t$, then $c = t$. Find values of c and t for which $c > t$. Explain why $c = t$ for these values.

a. −18°F; **b.** −31°F; **c.** 6°F; **d.** a change in the wind speed; **e.** answers will vary.

Simplify.

82. $\sqrt{8}$ $2\sqrt{2}$

83. $\sqrt{12}$ $2\sqrt{3}$

84. $\sqrt{45}$ $3\sqrt{5}$

85. $\sqrt{18}$ $3\sqrt{2}$

86. $\sqrt{20}$ $2\sqrt{5}$

87. $\sqrt{44}$ $2\sqrt{11}$

88. $\sqrt{27}$ $3\sqrt{3}$

89. $\sqrt{56}$ $2\sqrt{14}$

90. $\sqrt{48}$ $4\sqrt{3}$

91. $\sqrt{28}$ $2\sqrt{7}$

92. $\sqrt{75}$ $5\sqrt{3}$

93. $\sqrt{96}$ $4\sqrt{6}$

94. $\sqrt{63}$ $3\sqrt{7}$

95. $\sqrt{72}$ $6\sqrt{2}$

96. $\sqrt{98}$ $7\sqrt{2}$

97. $\sqrt{108}$ $6\sqrt{3}$

98. $\sqrt{200}$ $10\sqrt{2}$

99. $\sqrt{175}$ $5\sqrt{7}$

100. $\sqrt{21}$ $\sqrt{21}$

101. $\sqrt{55}$ $\sqrt{55}$

102. $5\sqrt{40}$
$10\sqrt{10}$

103. $11\sqrt{80}$
$44\sqrt{5}$

104. $-2\sqrt{300}$
$-20\sqrt{3}$

105. $-5\sqrt{180}$
$-30\sqrt{5}$

106. $4\sqrt{250}$
$20\sqrt{10}$

107. $3\sqrt{120}$
$6\sqrt{30}$

108. $-2\sqrt{160}$
$-8\sqrt{10}$

109. $-6\sqrt{128}$
$-48\sqrt{2}$

Simplify Variable Radical Expressions

Simplify.

110. $\sqrt{x^6}$
x^3

111. $\sqrt{x^{12}}$
x^6

112. $\sqrt{y^{15}}$
$y^7\sqrt{y}$

113. $\sqrt{c^{11}}$
$c^5\sqrt{c}$

114. $\sqrt{a^{20}}$
a^{10}

115. $\sqrt{a^{16}}$
a^8

116. $\sqrt{x^4y^4}$
x^2y^2

117. $\sqrt{x^{12}y^8}$
x^6y^4

118. $\sqrt{4x^4}$
$2x^2$

119. $\sqrt{25y^8}$
$5y^4$

120. $\sqrt{24x^2}$
$2x\sqrt{6}$

121. $\sqrt{x^3y^{15}}$
$xy^7\sqrt{xy}$

122. $\sqrt{x^3y^7}$
$xy^3\sqrt{xy}$

123. $\sqrt{a^{15}b^5}$
$a^7b^2\sqrt{ab}$

124. $\sqrt{a^3b^{11}}$
$ab^5\sqrt{ab}$

125. $\sqrt{48y^7}$
$4y^3\sqrt{3y}$

126. $\sqrt{60x^5}$
$2x^2\sqrt{15x}$

127. $\sqrt{72y^7}$
$6y^3\sqrt{2y}$

128. $\sqrt{49a^4b^8}$
$7a^2b^4$

129. $\sqrt{144x^2y^8}$
$12xy^4$

130. $\sqrt{18x^5y^7}$
$3x^2y^3\sqrt{2xy}$

131. $\sqrt{32a^5b^{15}}$
$4a^2b^7\sqrt{2ab}$

132. $\sqrt{40x^{11}y^7}$
$2x^5y^3\sqrt{10xy}$

133. $\sqrt{72x^9y^3}$
$6x^4y\sqrt{2xy}$

134. $\sqrt{80a^9b^{10}}$
$4a^4b^5\sqrt{5a}$

135. $\sqrt{96a^5b^7}$
$4a^2b^3\sqrt{6ab}$

136. $2\sqrt{16a^2b^3}$
$8ab\sqrt{b}$

137. $5\sqrt{25a^4b^7}$
$25a^2b^3\sqrt{b}$

138. $x\sqrt{x^4y^2}$
x^3y

139. $y\sqrt{x^3y^6}$
$xy^4\sqrt{x}$

140. $4\sqrt{20a^4b^7}$
$8a^2b^3\sqrt{5b}$

141. $5\sqrt{12a^3b^5}$
$10ab^2\sqrt{3ab}$

142. $3x\sqrt{12x^2y^7}$
$6x^2y^3\sqrt{3y}$

143. $4y\sqrt{18x^5y^4}$
$12x^2y^3\sqrt{2x}$

144. $2x^2\sqrt{8x^2y^3}$
$4x^3y\sqrt{2y}$

145. $3y^2\sqrt{27x^4y^3}$
$9x^2y^3\sqrt{3y}$

Applying Concepts

146. Determine whether the statement is always true, sometimes true, or never true.

a. The square root of a positive number is a positive number, and the square root of a negative number is a negative number. Never true

b. Every positive number has two square roots, one of which is the opposite of the other. Always true

c. The square root of a number that is not a perfect square is an irrational number. Always true

d. If the radicand of a radical expression is evenly divisible by a perfect square greater than 1, then the radical expression is not in simplest form. Always true

For any real numbers a and b, $\sqrt{ab} = \sqrt{a} \cdot \sqrt{b}$. Sometimes true

When an expression that is a perfect square is written in exponential form, the exponents are multiples of 2. Always true

147. Simplify.

a. $\sqrt{0.64}$ **b.** $-\sqrt{0.81}$ **c.** $\sqrt{6\frac{1}{4}}$ **d.** $-\sqrt{5\frac{4}{9}}$

a. 0.8; **b.** −0.9; **c.** $2\frac{1}{2}$; **d.** $-2\frac{1}{3}$

148. List the expressions in order from smallest to largest: $\sqrt{\frac{1}{4}+\frac{1}{8}}$, $\sqrt{\frac{1}{3}+\frac{1}{9}}$, $\sqrt{\frac{1}{5}+\frac{1}{6}}$

$\sqrt{\frac{1}{5}+\frac{1}{6}}$, $\sqrt{\frac{1}{4}+\frac{1}{8}}$, $\sqrt{\frac{1}{3}+\frac{1}{9}}$

149. Find a number n such that $\sqrt{n}$ is greater than 9 and less than 10.

Answers will vary. $81 < n < 100$

150. a. Use the expressions $\sqrt{9+16}$ and $\sqrt{9}+\sqrt{16}$ to show that $\sqrt{9+16} \neq \sqrt{9}+\sqrt{16}$.

b. Use the expressions $\sqrt{16-9}$ and $\sqrt{16}-\sqrt{9}$ to show that $\sqrt{16-9} \neq \sqrt{16}-\sqrt{9}$.

c. How do these examples relate to the Order of Operations Agreement?

a. $\sqrt{9+16} = \sqrt{25} = 5 \neq \sqrt{9}+\sqrt{16} = 3+4 = 7$. **b.** $\sqrt{16-9} = \sqrt{7} \neq \sqrt{16}-\sqrt{9} = 4-3 = 1$.

c. The correct simplifications are $\sqrt{9+16} = \sqrt{25} = 5$ and $\sqrt{16-9} = \sqrt{7}$ because the radical symbol is a grouping symbol.

151. Given $f(x) = \sqrt{2x-1}$, find:

a. $f(1)$ **b.** $f(5)$ **c.** $f(13)$ **d.** $f(14)$ **e.** $f(32)$

a. 1; **b.** 3; **c.** 5; **d.** $3\sqrt{3}$; **e.** $3\sqrt{7}$

152. For what values of x is the radical expression a real number? Write the answer in set builder notation.

a. $\sqrt{x}$ $\{x \mid x \geq 0\}$

b. $\sqrt{5x}$ $\{x \mid x \geq 0\}$

c. $\sqrt{x-3}$ $\{x \mid x \geq 3\}$

d. $\sqrt{x+6}$ $\{x \mid x \geq -6\}$

Explorations

153. ***Finding Square Roots of Expressions in Factored Form***

The square root of a number is one of two equal factors of the number. The square root of a number x is a number that, when squared, equals x.

$\sqrt{4} = 2$ because $2^2 = 4$.

In this section, we found the square roots of radical expressions by finding perfect-square factors of the radicand. For example,

$$\sqrt{50} = \sqrt{25 \cdot 2} = \sqrt{25} \cdot \sqrt{2} = 5 \cdot \sqrt{2} = 5\sqrt{2}$$

An alternative approach is to write the prime factorization of the radicand in factored form. Then find pairs of identical factors. For example:

$\sqrt{50} = \sqrt{5 \cdot 5 \cdot 2}$ • Write the prime factorization of the radicand in factored form.

$= \sqrt{(5 \cdot 5) \cdot 2}$ • Find pairs of identical factors.

$= 5\sqrt{2}$ • Each pair of identical factors is removed from under the radical symbol; one number of the pair is written in front of the radical symbol.

Note that this means that $\sqrt{5 \cdot 5} = 5$.

To simplify $\sqrt{72}$:

$\sqrt{72} = \sqrt{2 \cdot 2 \cdot 2 \cdot 3 \cdot 3}$ • Write the prime factorization of 72 in factored form.

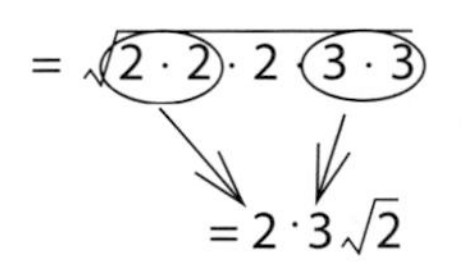

$= \sqrt{(2 \cdot 2) \cdot 2 (3 \cdot 3)}$ • Find pairs of identical factors.

$= 2 \cdot 3\sqrt{2}$ • Each pair of identical factors is removed from under the radical symbol; one number of the pair is written in front of the radical symbol.

Note that this means that $\sqrt{2 \cdot 2} = 2$ and $\sqrt{3 \cdot 3} = 3$.

$= 6\sqrt{2}$ • The numbers written in front of the radical symbol are multiplied.

Use this method to simplify each of the following.

a. $\sqrt{100}$

b. $\sqrt{90}$

c. $\sqrt{48}$

d. $\sqrt{75}$

e. $2\sqrt{45}$

f. $3\sqrt{98}$

The same approach can be used to simplify radical expressions in which the radicand is a variable expression.

To simplify $\sqrt{x^6y^3}$:

$\sqrt{x^6y^3} = \sqrt{x \cdot x \cdot x \cdot x \cdot x \cdot x \cdot y \cdot y \cdot y}$ • Write the prime factorization of the radicand in factored form.

$= \sqrt{(x \cdot x)(x \cdot x)(x \cdot x)(y \cdot y)\, y}$ • Find pairs of identical factors.

$= x \cdot x \cdot x \cdot y\sqrt{y}$ • Each pair of identical factors is removed from under the radical symbol; one variable of each pair is written in front of the radical symbol. Note that this means that $\sqrt{x \cdot x} = x$ and $\sqrt{y \cdot y} = y$.

$= x^3y\sqrt{y}$

Use this method to simplify each of the following.

g. $\sqrt{a^2b^8}$

h. $\sqrt{x^4y^{10}}$

i. $\sqrt{x^5y^6}$

j. $\sqrt{a^9b}$

k. $\sqrt{20a^7b^2}$

l. $\sqrt{72x^3y^{12}}$

154. ***Expressway On-Ramp Curves***

Highway engineers design expressway on-ramps with both efficiency and safety in mind. An on-ramp cannot be built with too sharp a curve or the speed at which drivers take the curve will cause them to skid off the road. However, an on-ramp built with too wide a curve requires more building material, takes longer to travel, and requires more land usage. The following formula states the relationship between the maximum speed at which a car can travel around a curve without skidding and the radius of the curve: $v = \sqrt{2.5r}$. In this formula, v is speed in miles per hour, and r is the radius, in feet, of an unbanked curve.

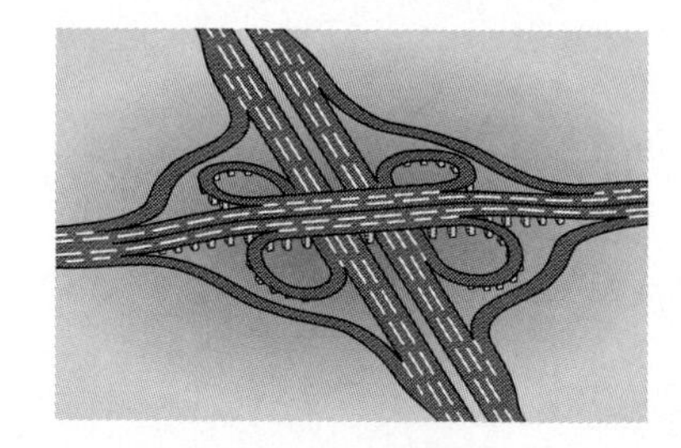

a. Use a graphing calculator to graph this equation. Use a window of Xmin = 0, Xmax = 2000, Ymin = 0, Ymax = 100.

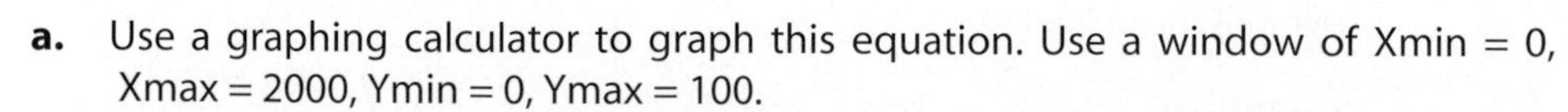

b. What do the values used for Xmin, Xmax, Ymin, and Ymax represent?

c. As the radius of the curve increases, does the maximum safe speed increase or decrease? Explain your answer based on the graph of the equation.

d. Use the graph to approximate the maximum safe speed when the radius of the curve is 100 feet. Round to the nearest whole number.

e. Use the graph to approximate the radius of the curve for which the maximum safe speed is 40 mph. Round to the nearest whole number.

f. Arie Luyendyk won the Indianapolis 500 in 1997 driving at an average speed of 145.827 mph. At this speed, should the radius of an unbanked curve be more or less than one mile?

Section 6.2 Operations on Radical Expressions

Addition and Subtraction of Radical Expressions

Triangle *ABC* in the rectangular coordinate system at the right has vertices at (0, 3), (3, 0), and (−4, −1).

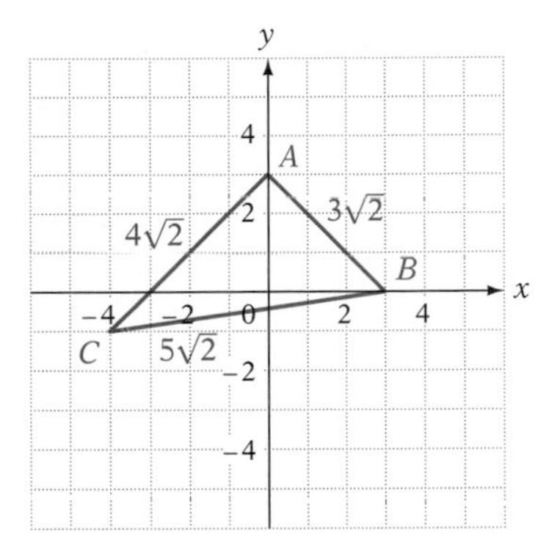

TAKE NOTE
In the next section of this chapter, you will learn how to determine the lengths of line segments *AB*, *BC*, and *AC* given only the vertices of the triangle.

The length of line segment *AB* is $3\sqrt{2}$ units, the length of line segment *BC* is $5\sqrt{2}$ units, and the length of line segment *AC* is $4\sqrt{2}$ units. The perimeter of triangle *ABC* is:

$$3\sqrt{2} + 5\sqrt{2} + 4\sqrt{2} = (3 + 5 + 4)\sqrt{2}$$
$$= 12\sqrt{2}$$

The perimeter of triangle *ABC* is $12\sqrt{2}$ units.

Note from this example that the Distributive Property is used to simplify the sum of radical expressions with the same radicand. It is also used to simplify the difference of radical expressions with the same radicand.

$$7\sqrt{3} + 8\sqrt{3} = (7 + 8)\sqrt{3} = 15\sqrt{3}$$

$$6\sqrt{2x} - 4\sqrt{2x} = (6 - 4)\sqrt{2x} = 2\sqrt{2x}$$

Radical expressions that are in simplest form and have different radicands cannot be simplified by the Distributive Property.

$2\sqrt{3} + 4\sqrt{2}$ cannot be simplified by the Distributive Property.

Note that adding and subtracting radical expressions is similar to combining like terms in a variable expression.

$4\sqrt{x} + 3\sqrt{x} = 7\sqrt{x}$	$4x + 3x = 7x$
$8\sqrt{x} - 6\sqrt{x} = 2\sqrt{x}$	$8x - 6x = 2x$
$5\sqrt{x} + 9\sqrt{y}$ cannot be simplified.	$5x + 9y$ cannot be simplified.

Question: Which of the following are like radical expressions?[1]

a. $6\sqrt{5}$ and $9\sqrt{5}$ **b.** $3\sqrt{2}$ and $2\sqrt{3}$ **c.** $11\sqrt{7x}$ and $9\sqrt{7x}$

1. a and c are examples of like radical expressions. b is not an example of like radical expressions because $\sqrt{2} \neq \sqrt{3}$.

SUGGESTED ACTIVITY
The length of each side of an octagon is $2\sqrt{2}$ inches. What is the perimeter of the octagon?
[Answer: $16\sqrt{2}$ inches]

Example 1 Simplify: $5\sqrt{2} - 3\sqrt{2} + 12\sqrt{2}$

Solution

$5\sqrt{2} - 3\sqrt{2} + 12\sqrt{2}$

$= (5 - 3 + 12)\sqrt{2}$ • Use the Distributive Property.

$= 14\sqrt{2}$

You-Try-It 1 Simplify: $9\sqrt{3} + 3\sqrt{3} - 18\sqrt{3}$

Solution See page S27.

In the examples below, the radical expressions must be simplified before they can be added or subtracted.

Example 2 Simplify: $3\sqrt{12} - 5\sqrt{27}$

Solution

$3\sqrt{12} - 5\sqrt{27}$

$= 3\sqrt{4 \cdot 3} - 5\sqrt{9 \cdot 3}$ • Simplify each radical expression.

$= 3\sqrt{4} \cdot \sqrt{3} - 5\sqrt{9} \cdot \sqrt{3}$

$= 3 \cdot 2\sqrt{3} - 5 \cdot 3\sqrt{3}$

$= 6\sqrt{3} - 15\sqrt{3}$

$= (6 - 15)\sqrt{3}$ • Use the Distributive Property to subtract the expressions.

$= -9\sqrt{3}$

You-Try-It 2 Simplify: $2\sqrt{50} - 5\sqrt{32}$

Solution See page S27.

Example 3 Simplify: $8x\sqrt{18x} - 2x\sqrt{32x}$

Solution

$8x\sqrt{18x} - 2x\sqrt{32x}$

$= 8x\sqrt{9 \cdot 2x} - 2x\sqrt{16 \cdot 2x}$ • Simplify each radical expression.

$= 8x\sqrt{9} \cdot \sqrt{2x} - 2x\sqrt{16} \cdot \sqrt{2x}$

$= 8x \cdot 3\sqrt{2x} - 2x \cdot 4\sqrt{2x}$

$= 24x\sqrt{2x} - 8x\sqrt{2x}$

$= (24x - 8x)\sqrt{2x}$ • Use the Distributive Property to subtract the expressions.

$= 16x\sqrt{2x}$

You-Try-It 3 Simplify: $y\sqrt{28y} + 7y\sqrt{63y}$

Solution See page S27.

Multiplication of Radical Expressions

Rectangle *ABCD* in the rectangular coordinate system at the right has a length of $\sqrt{50}$ units and a width of $\sqrt{32}$ units. The area of the rectangle is:

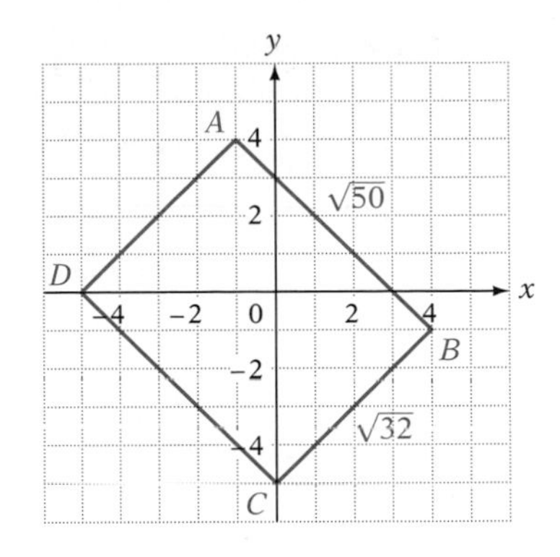

$$A = LW$$
$$A = (\sqrt{50})(\sqrt{32})$$
$$A = \sqrt{50 \cdot 32}$$
$$A = \sqrt{1600}$$
$$A = 40$$

The area of the rectangle is 40 square units.

TAKE NOTE
Verify that the area of the rectangle is 40 square units by counting the squares. Note that each fractional part of a square within the rectangle is one-half of a whole square.

The product $(\sqrt{32})(\sqrt{50})$ was simplified by using the Product Property of Square Roots, which was used in Section 6.1 to simplify radical expressions.

Product Property of Square Roots

If a and b are positive numbers, then $\sqrt{a} \cdot \sqrt{b} = \sqrt{a \cdot b}$.

→ Multiply: $\sqrt{2x}\,\sqrt{18x}$

Use the Product Property of Square Roots to multiply the radicands. $\quad \sqrt{2x}\,\sqrt{18x} = \sqrt{2x \cdot 18x}$

$= \sqrt{36x^2}$

Simplify. $\quad = 6x$

Example 4 Multiply: $\sqrt{3x^3}\,\sqrt{10x}\,\sqrt{6x^2}$

Solution

$$\sqrt{3x^3}\,\sqrt{10x}\,\sqrt{6x^2} = \sqrt{3x^3 \cdot 10x \cdot 6x^2}$$
$$= \sqrt{180x^6}$$
$$= \sqrt{36x^6 \cdot 5}$$
$$= \sqrt{36x^6} \cdot \sqrt{5}$$
$$= 6x^3\sqrt{5}$$

You-Try-It 4 Multiply: $\sqrt{5a}\,\sqrt{15a^3}\,\sqrt{3a}$

Solution See page S27.

Note that when the expression $\sqrt{x}$ is squared, the result is x.

$$(\sqrt{x})^2 = \sqrt{x}\,\sqrt{x} = \sqrt{x \cdot x} = \sqrt{x^2} = x$$

The Square of a Square Root

For $x > 0$, $(\sqrt{x})^2 = x$.

Example 5 Simplify: $(\sqrt{5a})^2$

Solution $(\sqrt{5a})^2 = 5a$ • The square of a square root equals the radicand.

You-Try-It 5 Simplify: $(\sqrt{7y})^2$

Solution See page S27.

In Example 6, the Distributive Property is used to remove the parentheses.

Example 6 Multiply: $\sqrt{2x}(x + \sqrt{2x})$

Solution

$$\begin{aligned}\sqrt{2x}(x + \sqrt{2x}) &= \sqrt{2x}(x) + \sqrt{2x}\sqrt{2x} \\ &= x\sqrt{2x} + (\sqrt{2x})^2 \\ &= x\sqrt{2x} + 2x\end{aligned}$$

You-Try-It 6 Multiply: $\sqrt{5x}(\sqrt{5x} - \sqrt{25x})$

Solution See page S27.

In Example 7, the FOIL method is used to multiply the two radical expressions.

Example 7 Multiply: $(\sqrt{2} - 3x)(\sqrt{2} + x)$

Solution

$$\begin{aligned}(\sqrt{2} - 3x)(\sqrt{2} + x) &= (\sqrt{2})^2 + x\sqrt{2} - 3x\sqrt{2} - 3x^2 \\ &= 2 + (x - 3x)\sqrt{2} - 3x^2 \\ &= 2 - 2x\sqrt{2} - 3x^2\end{aligned}$$

You-Try-It 7 Multiply: $(3\sqrt{x} - y)(5\sqrt{x} - y)$

Solution See page S28.

The expressions $a + b$ and $a - b$, which are the sum and difference of two terms, are called **conjugates** of each other.

The conjugate of $\sqrt{x} - 7$ is $\sqrt{x} + 7$.

The conjugate of $\sqrt{y} + \sqrt{3}$ is $\sqrt{y} - \sqrt{3}$.

Question: What is the conjugate of the expression?[2]

a. $\sqrt{d}+5$ **b.** $6-\sqrt{3}$ **c.** $\sqrt{y}-\sqrt{2}$

The product of conjugates is the difference of two squares.

$$(a+b)(a-b)=a^2-b^2$$

$$(2+\sqrt{7})(2-\sqrt{7})=2^2-(\sqrt{7})^2=4-7=-3$$

$$(\sqrt{3}+z)(\sqrt{3}-z)=(\sqrt{3})^2-z^2=3-z^2$$

Example 8 Multiply: $(\sqrt{a}-\sqrt{b})(\sqrt{a}+\sqrt{b})$

Solution $(\sqrt{a}-\sqrt{b})(\sqrt{a}+\sqrt{b})$

- $\sqrt{a}-\sqrt{b}$ and $\sqrt{a}+\sqrt{b}$ are conjugates.

$=(\sqrt{a})^2-(\sqrt{b})^2$

- The product is the square of the first term, $\sqrt{a}$, minus the square of the second term, $\sqrt{b}$.

$=a-b$

You-Try-It 8 Multiply: $(\sqrt{x}+7)(\sqrt{x}-7)$

Solution See page S28.

Division of Radical Expressions

Thus far in this section, we have added, subtracted, and multiplied radical expressions. We are now going to simplify expressions in which radical expressions are divided. In doing so, we will use the Quotient Property of Square Roots, which states that the square root of a quotient is equal to the quotient of the square roots.

Quotient Property of Square Roots

If a and b are positive numbers, then $\sqrt{\frac{a}{b}}=\frac{\sqrt{a}}{\sqrt{b}}$.

For example, $\sqrt{\frac{16}{4}}$ is equal to $\frac{\sqrt{16}}{\sqrt{4}}$.

$$\sqrt{\frac{16}{4}}=\sqrt{4}=2 \quad \text{and} \quad \frac{\sqrt{16}}{\sqrt{4}}=\frac{4}{2}=2$$

2. **a.** $\sqrt{d}-5$; **b.** $6+\sqrt{3}$; **c.** $\sqrt{y}+\sqrt{2}$

TAKE NOTE

By the definition of square root, $\sqrt{\frac{4}{25}} = \frac{2}{5}$ because $\left(\frac{2}{5}\right)^2 = \frac{4}{25}$.

Example 9 Simplify: $\sqrt{\frac{4}{25}}$

Solution

$\sqrt{\frac{4}{25}} = \frac{\sqrt{4}}{\sqrt{25}}$ • Use the Quotient Property of Square Roots.

$= \frac{2}{5}$ • Simplify the numerator and the denominator.

Note: You can use a calculator to verify this. Convert the fraction $\frac{4}{25}$ to the decimal 0.16. Use a calculator to find $\sqrt{0.16} = 0.4$. The fraction $\frac{2}{5} = 0.4$. The expressions $\sqrt{\frac{4}{25}}$ and $\frac{2}{5}$ are equal.

You-Try-It 9 Simplify: $\sqrt{\frac{9}{16}}$

Solution See page S28.

Example 10 Simplify: $\sqrt{\frac{4x^2}{z^6}}$

Solution

$\sqrt{\frac{4x^2}{z^6}} = \frac{\sqrt{4x^2}}{\sqrt{z^6}}$ • Use the Quotient Property of Square Roots.

$= \frac{2x}{z^3}$ • Simplify the numerator and the denominator.

You-Try-It 10 Simplify: $\sqrt{\frac{16y^2}{x^4}}$

Solution See page S28.

A radical expression is not in simplest form if a radical remains in the denominator. The procedure used to remove a radical from the denominator is called **rationalizing the denominator.**

→ Simplify: $\frac{2}{\sqrt{3}}$

The expression has a radical in the denominator. Multiply the expression by $\frac{\sqrt{3}}{\sqrt{3}}$, which equals 1. Then simplify.

$$\frac{2}{\sqrt{3}} = \frac{2}{\sqrt{3}} \cdot \frac{\sqrt{3}}{\sqrt{3}}$$

$$= \frac{2\sqrt{3}}{(\sqrt{3})^2}$$

$$= \frac{2\sqrt{3}}{3}$$

Thus $\frac{2}{\sqrt{3}} = \frac{2\sqrt{3}}{3}$, but $\frac{2}{\sqrt{3}}$ is not in simplest form. $\frac{2\sqrt{3}}{3}$ is in simplest form because no radical remains in the denominator and the radicand in the numerator contains no perfect-square factors other than 1.

Use a calculator to verify that, to the limits of the calculator, $\frac{2}{\sqrt{3}} = \frac{2\sqrt{3}}{3}$.

SUGGESTED ACTIVITY
Have the students simplify each of the following expressions:

$\frac{3}{\sqrt{3}}, \frac{5}{\sqrt{5}}, \frac{7}{\sqrt{7}}, \frac{11}{\sqrt{11}}$, and $\frac{13}{\sqrt{13}}$.

Then have them write a rule for simplifying $\frac{a}{\sqrt{a}}$.

[Answers: $\sqrt{3}, \sqrt{5}, \sqrt{7}, \sqrt{11}, \sqrt{13}$. The rule is $\frac{a}{\sqrt{a}} = \sqrt{a}$.]

Example 11 Simplify: $\frac{6}{\sqrt{5}}$

Solution

$\frac{6}{\sqrt{5}}$ • There is a radical in the denominator.

$= \frac{6}{\sqrt{5}} \cdot \frac{\sqrt{5}}{\sqrt{5}}$ • Multiply by $\frac{\sqrt{5}}{\sqrt{5}}$, which equals 1.

$= \frac{6\sqrt{5}}{(\sqrt{5})^2}$

$= \frac{6\sqrt{5}}{5}$

You-Try-It 11 Simplify: $\frac{x}{\sqrt{2}}$

Solution See page S28.

The following list summarizes the form of a radical expression in simplest form.

Radical Expressions in Simplest Form

A radical expression is in simplest form if:

1. The radicand contains no factor greater than 1 that is a perfect square.
2. There is no fraction under the radical sign.
3. There is no radical in the denominator of a fraction.

Question: Is the expression in simplest form?[3]

a. $\sqrt{\frac{2}{5}}$ **b.** $\frac{3}{\sqrt{7}}$ **c.** $\frac{\sqrt{9}}{11}$ **d.** $\frac{\sqrt{x}}{13}$

---

3. **a.** $\sqrt{\frac{2}{5}}$ is not in simplest form because there is a fraction under the radical sign. **b.** $\frac{3}{\sqrt{7}}$ is not in simplest form because there is a radical in the denominator of the fraction. **c.** $\frac{\sqrt{9}}{11}$ is not in simplest form because the radicand 9 contains a factor greater than 1 that is a perfect square: $\sqrt{9} = 3$. **d.** $\frac{\sqrt{x}}{13}$ is in simplest form.

INSTRUCTOR NOTE

Some students will want to simplify $\frac{\sqrt{21}}{3}$ by dividing the denominator into 21. These students need to be reminded that $\sqrt{21} \neq 21$.

Example 12 Simplify: $\sqrt{\frac{7}{3}}$

Solution

$\sqrt{\frac{7}{3}}$ • There is a fraction under the radical sign.

$= \frac{\sqrt{7}}{\sqrt{3}}$ • Use the Quotient Property of Square Roots.

$= \frac{\sqrt{7}}{\sqrt{3}} \cdot \frac{\sqrt{3}}{\sqrt{3}}$ • There is a radical in the denominator. Multiply by $\frac{\sqrt{3}}{\sqrt{3}}$, which equals 1.

$= \frac{\sqrt{7 \cdot 3}}{(\sqrt{3})^2}$ • Use the Product Property of Square Roots to multiply the numerators.

$= \frac{\sqrt{21}}{3}$ • The expression is now in simplest form.

You-Try-It 12 Simplify: $\sqrt{\frac{5}{11}}$

Solution See page S28.

Example 13 Simplify: $\frac{x\sqrt{8}}{2\sqrt{50}}$

Solution

$\frac{x\sqrt{8}}{2\sqrt{50}}$

$= \frac{x\sqrt{4 \cdot 2}}{2\sqrt{25 \cdot 2}}$ • Simplify the radical expressions in the numerator and denominator.

$= \frac{x\sqrt{4}\sqrt{2}}{2\sqrt{25}\sqrt{2}}$

$= \frac{x \cdot 2\sqrt{2}}{2 \cdot 5\sqrt{2}}$

$= \frac{x \cdot \overset{1}{\cancel{2}}\overset{1}{\cancel{\sqrt{2}}}}{\underset{1}{\cancel{2}} \cdot 5\underset{1}{\cancel{\sqrt{2}}}}$ • Divide the numerator and denominator by the common factors.

$= \frac{x}{5}$ • Write the fraction in simplest form.

You-Try-It 13 Simplify: $\frac{y\sqrt{27}}{3\sqrt{75}}$

Solution See page S28.

6.2 EXERCISES

Topics for Discussion

1. Fill in the blanks.
   a. The sum or difference of like radical expressions is simplified by using the ________ Property. Distributive
   b. The expressions $\sqrt{a}+\sqrt{b}$ and $\sqrt{a}-\sqrt{b}$ are called __________. conjugates
   c. To multiply $\sqrt{x}$ times $\sqrt{x}+\sqrt{2}$, use the __________. Distributive Property
   d. To multiply $\sqrt{a}+\sqrt{2}$ times $\sqrt{a}-\sqrt{3}$, use the __________. FOIL method
   e. The procedure used to rewrite a radical expression so that it does not have a radical in the denominator is called __________. rationalizing the denominator

2. Explain how to add or subtract two radical expressions.
   Answers will vary.

3. Explain the meaning of the sentence: "The square root of a product is equal to the product of the square roots."
   Answers will vary.

4. Explain why $\frac{\sqrt{5}}{5}$ is in simplest form and $\frac{1}{\sqrt{5}}$ is not in simplest form.

   The expression $\frac{1}{\sqrt{5}}$ has a radical in the denominator.

5. Why can we multiply $\frac{1}{\sqrt{3}}$ by $\frac{\sqrt{3}}{\sqrt{3}}$ without changing the value of the fraction?

   By the Multiplication Property of One, we can multiply an expression by 1 without changing its value. $\frac{\sqrt{3}}{\sqrt{3}}=1$.

6. Describe a radical expression that is in simplest form.
   Answers will vary. See the description on page 415.

Operations on Radical Expressions

Add, subtract, or multiply.

7. $2\sqrt{2}+\sqrt{2}$
   $3\sqrt{2}$

8. $-3\sqrt{7}+2\sqrt{7}$
   $-\sqrt{7}$

9. $4\sqrt{5}-10\sqrt{5}$
   $-6\sqrt{5}$

10. $-3\sqrt{3}-5\sqrt{3}$
    $-8\sqrt{3}$

11. $2\sqrt{x}+8\sqrt{x}$
    $10\sqrt{x}$

12. $8\sqrt{y}-10\sqrt{y}$
    $-2\sqrt{y}$

13. $-5\sqrt{2a}+2\sqrt{2a}$
    $-3\sqrt{2a}$

14. $-2\sqrt{3b}-9\sqrt{3b}$
    $-11\sqrt{3b}$

15. $2y\sqrt{3}-9y\sqrt{3}$
    $-7y\sqrt{3}$

16. $-4\sqrt{xy} + 6\sqrt{xy}$
$2\sqrt{xy}$

17. $4\sqrt{2} - 5\sqrt{2} + 8\sqrt{2}$
$7\sqrt{2}$

18. $3\sqrt{3} + 8\sqrt{3} - 16\sqrt{3}$
$-5\sqrt{3}$

19. $5\sqrt{x} - 8\sqrt{x} + 9\sqrt{x}$
$6\sqrt{x}$

20. $\sqrt{x} - 7\sqrt{x} + 6\sqrt{x}$
0

21. $\sqrt{45} + \sqrt{125}$
$8\sqrt{5}$

22. $\sqrt{32} - \sqrt{98}$
$-3\sqrt{2}$

23. $2\sqrt{2} + 3\sqrt{8}$
$8\sqrt{2}$

24. $4\sqrt{128} - 3\sqrt{32}$
$20\sqrt{2}$

25. $5\sqrt{18} - 2\sqrt{75}$
$15\sqrt{2} - 10\sqrt{3}$

26. $5\sqrt{75} - 2\sqrt{18}$
$25\sqrt{3} - 6\sqrt{2}$

27. $5\sqrt{4x} - 3\sqrt{9x}$
$\sqrt{x}$

28. $-3\sqrt{25y} + 8\sqrt{49y}$
$41\sqrt{y}$

29. $3\sqrt{3x^2} - 5\sqrt{27x^2}$
$-12x\sqrt{3}$

30. $-2\sqrt{8y^2} + 5\sqrt{32y^2}$
$16y\sqrt{2}$

31. $2x\sqrt{xy^2} - 3y\sqrt{x^2y}$
$2xy\sqrt{x} - 3xy\sqrt{y}$

32. $4a\sqrt{b^2a} - 3b\sqrt{a^2b}$
$4ab\sqrt{a} - 3ab\sqrt{b}$

33. $8\sqrt{8} - 4\sqrt{32} - 9\sqrt{50}$
$-45\sqrt{2}$

34. $2\sqrt{12} - 4\sqrt{27} + \sqrt{75}$
$-3\sqrt{3}$

35. $-2\sqrt{3} + 5\sqrt{27} - 4\sqrt{45}$
$13\sqrt{3} - 12\sqrt{5}$

36. $-2\sqrt{8} - 3\sqrt{27} + 3\sqrt{50}$
$11\sqrt{2} - 9\sqrt{3}$

37. $4\sqrt{75} + 3\sqrt{48} - \sqrt{99}$
$32\sqrt{3} - 3\sqrt{11}$

38. $2\sqrt{75} - 5\sqrt{20} + 2\sqrt{45}$
$10\sqrt{3} - 4\sqrt{5}$

39. $\sqrt{25x} - \sqrt{9x} + \sqrt{16x}$
$6\sqrt{x}$

40. $\sqrt{4x} - \sqrt{100x} - \sqrt{49x}$
$-15\sqrt{x}$

41. $3\sqrt{3x} + \sqrt{27x} - 8\sqrt{75x}$
$-34\sqrt{3x}$

42. $5\sqrt{5x} + 2\sqrt{45x} - 3\sqrt{80x}$
$-\sqrt{5x}$

43. $2a\sqrt{75b} - a\sqrt{20b} + 4a\sqrt{45b}$
$10a\sqrt{3b} + 10a\sqrt{5b}$

44. $2b\sqrt{75a} - 5b\sqrt{27a} + 2b\sqrt{20a}$
$-5b\sqrt{3a} + 4b\sqrt{5a}$

45. $\sqrt{5}\sqrt{5}$
5

46. $\sqrt{11}\sqrt{11}$
11

47. $\sqrt{3}\sqrt{12}$
6

48. $\sqrt{2}\sqrt{8}$
4

49. $\sqrt{x}\sqrt{x}$
x

50. $\sqrt{y}\sqrt{y}$
y

51. $\sqrt{xy^3}\sqrt{x^5y}$
x^3y^2

52. $\sqrt{a^3b^5}\sqrt{ab^5}$
a^2b^5

53. $\sqrt{3a^2b^5}\sqrt{6ab^7}$
$3ab^6\sqrt{2a}$

54. $\sqrt{5x^3y}\sqrt{10x^2y}$
$5x^2y\sqrt{2x}$

55. $\sqrt{6a^3b^2}\sqrt{24a^5b}$
$12a^4b\sqrt{b}$

56. $\sqrt{8ab^5}\sqrt{12a^7b}$
$4a^4b^3\sqrt{6}$

57. $\sqrt{2}(\sqrt{2} - \sqrt{3})$
$2 - \sqrt{6}$

58. $3(\sqrt{12} - \sqrt{3})$
$3\sqrt{3}$

59. $\sqrt{x}(\sqrt{x} - \sqrt{y})$
$x - \sqrt{xy}$

60. $\sqrt{b}(\sqrt{a} - \sqrt{b})$
$\sqrt{ab} - b$

61. $\sqrt{5}(\sqrt{10} - \sqrt{x})$
$5\sqrt{2} - \sqrt{5x}$

62. $\sqrt{6}(\sqrt{y} - \sqrt{18})$
$\sqrt{6y} - 6\sqrt{3}$

63. $\sqrt{8}(\sqrt{2} - \sqrt{5})$
$4 - 2\sqrt{10}$

64. $\sqrt{10}(\sqrt{20} - \sqrt{a})$
$10\sqrt{2} - \sqrt{10a}$

65. $\sqrt{2ac}\sqrt{5ab}\sqrt{10cb}$
$10abc$

66. $\sqrt{3xy}\sqrt{6x^3y}\sqrt{2y^2}$
$6x^2y^2$

67. $\sqrt{3a}(\sqrt{3a} - \sqrt{3b})$
$3a - 3\sqrt{ab}$

68. $\sqrt{5x}(\sqrt{10x} - \sqrt{x})$
$5x\sqrt{2} - x\sqrt{5}$

69. $(3 - \sqrt{y})(5 - \sqrt{y})$
$15 - 8\sqrt{y} + y$

70. $(\sqrt{x} - \sqrt{y})(\sqrt{x} + \sqrt{y})$
$x - y$

71. $(\sqrt{x} - 3)(\sqrt{x} - 3)$
$x - 6\sqrt{x} + 9$

72. $(2\sqrt{a} - y)(2\sqrt{a} - y)$
$4a - 4y\sqrt{a} + y^2$

73. $(\sqrt{3x} + y)(\sqrt{3x} - y)$
$3x - y^2$

74. $(5 + 2\sqrt{y})(3 - \sqrt{y})$
$15 + \sqrt{y} - 2y$

75. $(2\sqrt{x} + 1)(5\sqrt{x} + 4)$
$10x + 13\sqrt{x} + 4$

76. $(5\sqrt{x} - 2)(5\sqrt{x} + 2)$
$25x - 4$

Solve.

77. Find **a.** the perimeter and **b.** the area of the square shown below.

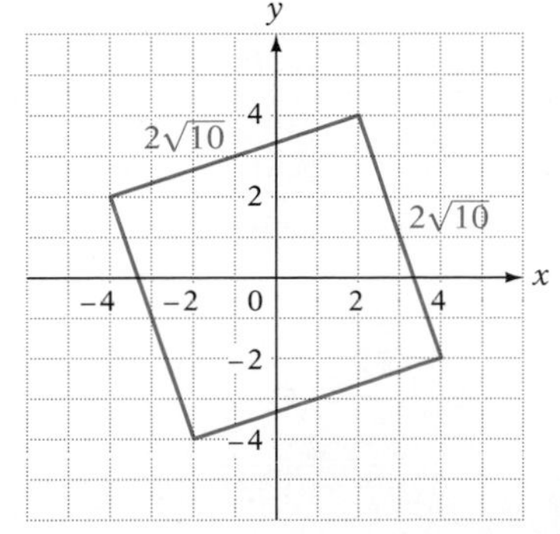

a. $8\sqrt{10}$
b. 40

78. Find **a.** the perimeter and **b.** the area of the rectangle shown below.

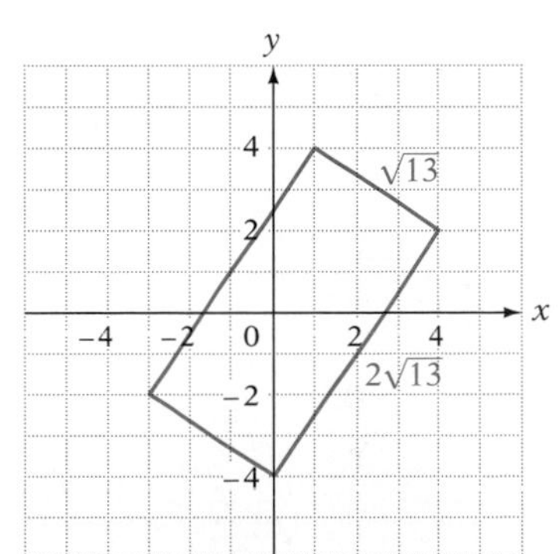

a. $6\sqrt{13}$
b. 26

79. Find **a.** the perimeter and **b.** the area of the triangle shown below.

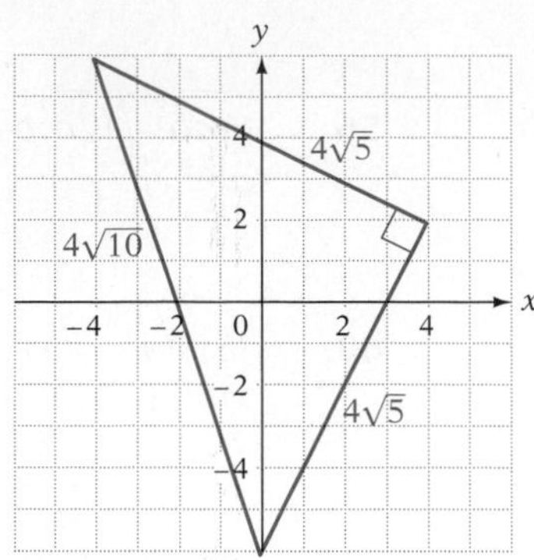

a. $8\sqrt{5} + 4\sqrt{10}$
b. 40

80. Find **a.** the perimeter and **b.** the area of the triangle shown below.

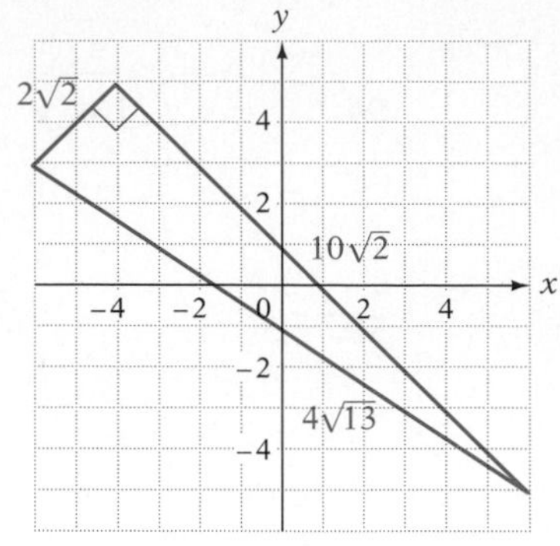

a. $12\sqrt{2} + 4\sqrt{13}$
b. 20

81. The lengths of the sides of a triangle are $4\sqrt{3}$ cm, $2\sqrt{3}$ cm, and $2\sqrt{15}$ cm. Find the perimeter of the triangle.

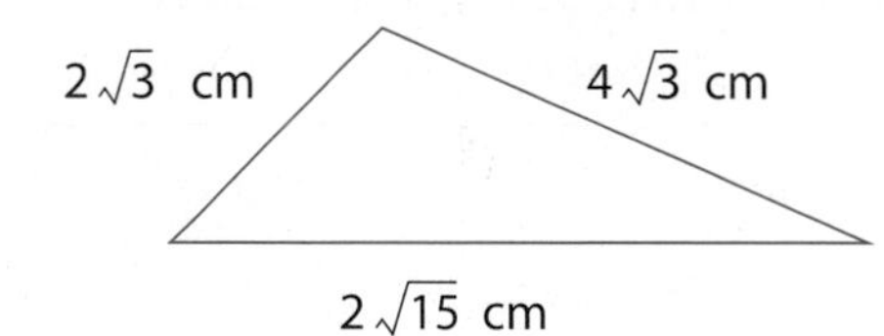

$(6\sqrt{3} + 2\sqrt{15})$ cm

82. The length of a rectangle is $3\sqrt{2}$ cm. The width is $\sqrt{2}$ cm. Find the perimeter of the rectangle.

$3\sqrt{2}$ cm

$\sqrt{2}$ cm

$8\sqrt{2}$ cm

83. The length of a rectangle is $5\sqrt{5}$ cm. The width is $\sqrt{5}$ cm. Find the perimeter of the rectangle. Round to the nearest tenth.
26.8 cm

84. The lengths of the sides of a triangle are $5\sqrt{3}$ m, $3\sqrt{3}$ m, and 6 m. Find the perimeter of the triangle. Round to the nearest tenth.
19.9 cm

Simplify.

85. $\sqrt{\dfrac{81}{100}}$

$\dfrac{9}{10}$

86. $\sqrt{\dfrac{25}{49}}$

$\dfrac{5}{7}$

87. $\sqrt{\dfrac{16a^2}{b^8}}$

$\dfrac{4a}{b^4}$

88. $\sqrt{\dfrac{a^6}{36y^4}}$

$\dfrac{a^3}{6y^2}$

89. $\dfrac{5}{\sqrt{2}}$

$\dfrac{5\sqrt{2}}{2}$

90. $\dfrac{8}{\sqrt{7}}$

$\dfrac{8\sqrt{7}}{7}$

91. $\dfrac{x}{\sqrt{3}}$

$\dfrac{x\sqrt{3}}{3}$

92. $\dfrac{b}{\sqrt{5}}$

$\dfrac{b\sqrt{5}}{5}$

93. $\sqrt{\frac{3}{4}}$

$\frac{\sqrt{3}}{2}$

94. $\sqrt{\frac{5}{9}}$

$\frac{\sqrt{5}}{3}$

95. $\sqrt{\frac{2}{3}}$

$\frac{2\sqrt{3}}{3}$

96. $\sqrt{\frac{6}{7}}$

$\frac{6\sqrt{7}}{7}$

97. $\frac{1}{\sqrt{3}}$

$\frac{\sqrt{3}}{3}$

98. $\frac{1}{\sqrt{8}}$

$\frac{\sqrt{2}}{4}$

99. $\frac{\sqrt{32}}{\sqrt{2}}$

4

100. $\frac{\sqrt{45}}{\sqrt{5}}$

3

101. $\frac{3}{\sqrt{x}}$

$\frac{3\sqrt{x}}{x}$

102. $\frac{4}{\sqrt{2x}}$

$\frac{2\sqrt{2x}}{x}$

103. $\frac{\sqrt{98}}{\sqrt{2}}$

7

104. $\frac{\sqrt{48}}{\sqrt{3}}$

4

105. $\frac{\sqrt{27a}}{\sqrt{3a}}$

3

106. $\frac{\sqrt{72x}}{\sqrt{2x}}$

6

107. $\frac{14}{\sqrt{7y}}$

$\frac{2\sqrt{7y}}{y}$

108. $\frac{12}{\sqrt{6x}}$

$\frac{2\sqrt{6x}}{x}$

109. $\frac{5\sqrt{8}}{4\sqrt{50}}$

$\frac{1}{2}$

110. $\frac{2\sqrt{18}}{5\sqrt{32}}$

$\frac{3}{10}$

111. $\frac{y\sqrt{18}}{\sqrt{27}}$

$\frac{y\sqrt{6}}{3}$

112. $\frac{x\sqrt{12}}{\sqrt{75}}$

$\frac{2x}{5}$

Solve.

113. The area of the rectangle shown below is 40 square units. The length is $4\sqrt{5}$ units. Find the width of the rectangle.

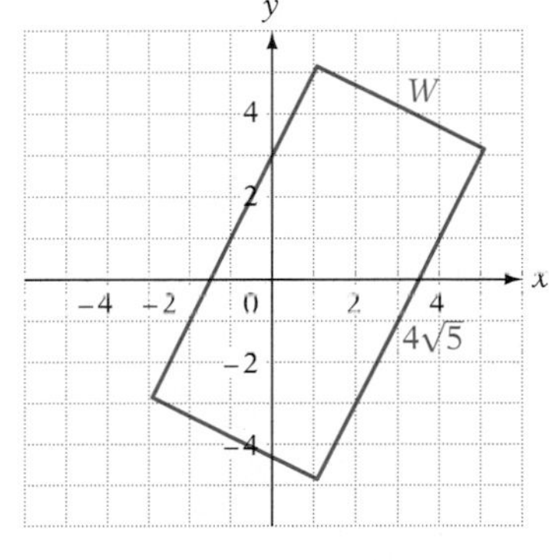

$2\sqrt{5}$

114. The area of the rectangle shown below is 48 square units. The width is $4\sqrt{2}$ units. Find the length of the rectangle.

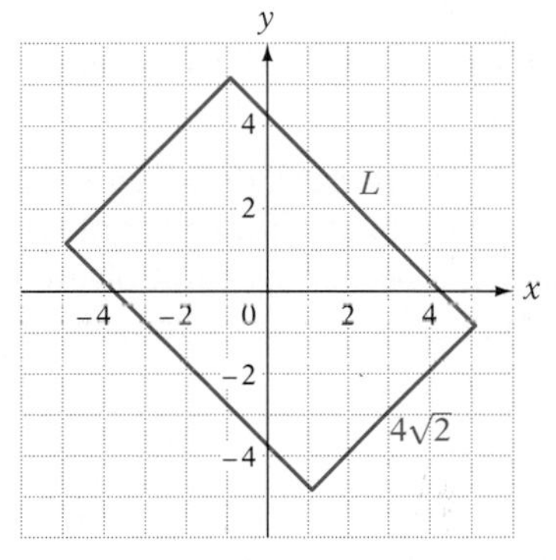

$6\sqrt{2}$

Applying Concepts

115. Given $f(x) = \sqrt{x+5} + \sqrt{5x+3}$, write $f(3)$ in simplest form.

$5\sqrt{2}$

116. **a.** Show that $1 + \sqrt{6}$ and $1 - \sqrt{6}$ are solutions of the equation $x^2 - 2x - 5 = 0$.

b. Show that $-3 + 3\sqrt{2}$ and $-3 - 3\sqrt{2}$ are solutions of the equation $x^2 + 6x - 9 = 0$.

c. Show that $\frac{2 + \sqrt{3}}{2}$ and $\frac{2 - \sqrt{3}}{2}$ are solutions of the equation $4x^2 - 8x + 1 = 0$.

The complete solution is in the Solutions Manual.

117. Determine whether the statement is always true, sometimes true, or never true.

a. The procedure for rationalizing the denominator of a fraction is used when a fraction has a radical expression in the denominator. Always true

b. The square root of a fraction is equal to the square root of the numerator over the square root of the denominator. Always true

c. $\sqrt{a^2 + b^2} = \sqrt{a} + \sqrt{b}$ Sometimes true

d. A radical expression is in simplest form if the radicand contains no factor other than 1 that is a perfect square. Sometimes true

118. Determine whether the statement is true or false. If it is false, correct the right side of the equation.

a. $(\sqrt{y})^4 = y^2$

b. $(2\sqrt{x})^3 = 8x^3$

c. $(\sqrt{x} + 1)^2 = x + 1$

a. True; **b.** $(2\sqrt{x})^3 = 8x\sqrt{x}$; **c.** $(\sqrt{x} + 1)^2 = x + 2\sqrt{x} + 1$

Explorations

119. ***Comparing Radical Expressions with Polynomial Expressions***

a. Write a paragraph that compares simplifying $5x + 9x$ to simplifying $5\sqrt{x} + 9\sqrt{x}$.

b. Write a paragraph that compares simplifying $7x + 10y$ to simplifying $7\sqrt{x} + 10\sqrt{y}$.

c. Write a paragraph that compares simplifying $(3x)(6x)$ to simplifying $(3\sqrt{x})(6\sqrt{x})$.

d. Write a paragraph that compares simplifying $(4x)(2y)$ to simplifying $(4\sqrt{x})(2\sqrt{y})$.

120. ***Rationalizing Denominators with Two Terms***

When the denominator of a fraction contains a radical expression with two terms, simplify the radical expression by multiplying the numerator and denominator by the conjugate of the denominator. Here is an example.

Simplify: $\frac{\sqrt{y}}{\sqrt{y} + 3}$

$$\frac{\sqrt{y}}{\sqrt{y} + 3} = \frac{\sqrt{y}}{\sqrt{y} + 3} \cdot \frac{\sqrt{y} - 3}{\sqrt{y} - 3}$$

• Multiply the numerator and denominator by $\sqrt{y} - 3$, the conjugate of the denominator.

$$= \frac{(\sqrt{y})^2 - 3\sqrt{y}}{(\sqrt{y})^2 - 3^2} = \frac{y - 3\sqrt{y}}{y - 9}$$

Simplify each of the following.

a. $\frac{3}{5 + \sqrt{5}}$ **b.** $\frac{\sqrt{2}}{\sqrt{3} - \sqrt{2}}$ **c.** $\frac{\sqrt{x}}{\sqrt{x} - \sqrt{y}}$ **d.** $\frac{\sqrt{xy}}{\sqrt{x} + \sqrt{y}}$

Section 6.3 Radical Equations

Solving Equations Containing Radical Expressions

In Section 6.1, we used the formula $R = 1.4\sqrt{h}$ to determine the distance a person looking through a submarine periscope can see. Recall that in this formula, R is the distance in miles a person can see and h is the height in feet of the periscope above the surface of the water. Given the height of the periscope, we found the distance the person could see. In the example below, the distance we want to be able to see is given, and we are asked to find what height the periscope must be above the surface of the water.

→ How far would a submarine periscope have to be above the surface of the water in order for the lookout to see a ship 3 miles away? Use the formula $R = 1.4\sqrt{h}$, where R is the distance in miles and h is the height in feet of the periscope above the surface of the water. Round to the nearest tenth.

State the goal. The goal is to find the height of the periscope above the surface of the water when the lookout can see a distance of 3 miles.

Describe a strategy. Replace the variable R in the given formula by 3. Then solve the equation for h.

Solve the problem.

$R = 1.4\sqrt{h}$

$3 = 1.4\sqrt{h}$ • Replace R by 3.

$\frac{3}{1.4} = \sqrt{h}$ • Solve the equation for $\sqrt{h}$. Divide each side by 1.4.

$\left(\frac{3}{1.4}\right)^2 = (\sqrt{h})^2$ • We want to solve the equation for h. Since the square of $\sqrt{h} = h$, square each side of the equation.

$4.6 \approx h$

To see a ship 3 miles away, the periscope must be 4.6 feet above the surface of the water.

Check your work. √

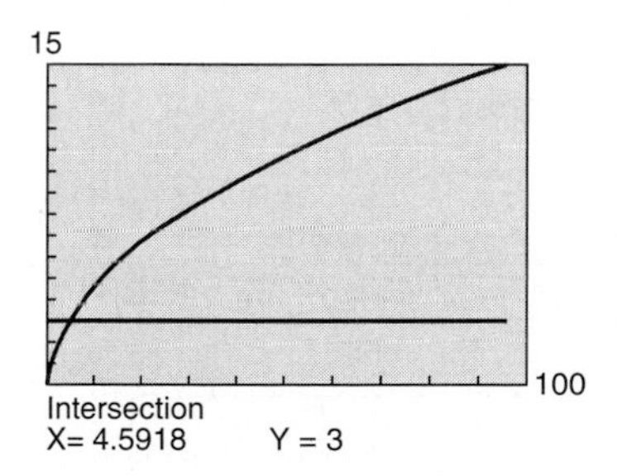

The graph of $Y1 = 1.4\sqrt{X}$ is shown at the left along with the graph of $Y2 = 3$ (the distance the lookout wants to see). The window is Xmin = 0, Xmax = 100, Ymin = 0, Ymax = 15. Note that the coordinates of the point of intersection of the two graphs verify that a lookout can see 3 miles when the periscope is approximately 4.6 feet above the surface of the water.

SUGGESTED ACTIVITY

Have students graph $Y1 = 1.4\sqrt{X}$ and find, to the nearest tenth, how far a submarine periscope has to be above the surface of the water in order to see a ship 4 miles away. [Answer: 8.2 feet]

In solving the equation $\frac{3}{1.4} = \sqrt{h}$, we used the Property of Squaring Both Sides of an Equation. This property states that if two numbers are equal, then the squares of the numbers are equal. This property is used to solve radical equations.

Property of Squaring Both Sides of an Equation

If a and b are real numbers and $a = b$, then $a^2 = b^2$.

The Property of Squaring Both Sides of an Equation states that if a and b are real numbers and $a = b$, then $a^2 = b^2$. It is not true that if a and b are real numbers and $a^2 = b^2$, then $a = b$. For example, let $a = 2$ and $b = -2$. Then $(2)^2 = (-2)^2$, but $2 \neq -2$.

Here are two examples of using the Property of Squaring Both Sides of an Equation to solve equations. It is important to note that **whenever this property is used, it is necessary to check the solutions of the equation because the resulting equation may have a solution that is not a solution of the original equation.**

Example 1 Solve: $\sqrt{3x} + 2 = 5$

Solution

$\sqrt{3x} + 2 = 5$

$\sqrt{3x} = 3$ • Rewrite the equation so that the radical is alone on one side of the equation.

$(\sqrt{3x})^2 = 3^2$ • Square each side of the equation.

$3x = 9$ • Solve for x.

$x = 3$

Algebraic check: • Each side of the equation was squared. The proposed solution must be checked.

$\sqrt{3x} + 2$	5
$\sqrt{3 \cdot 3} + 2$	5
$\sqrt{9} + 2$	5
$3 + 2$	5

• Replace x by 3.

$5 = 5$ • This is a true equation. The solution checks.

Graphical check:

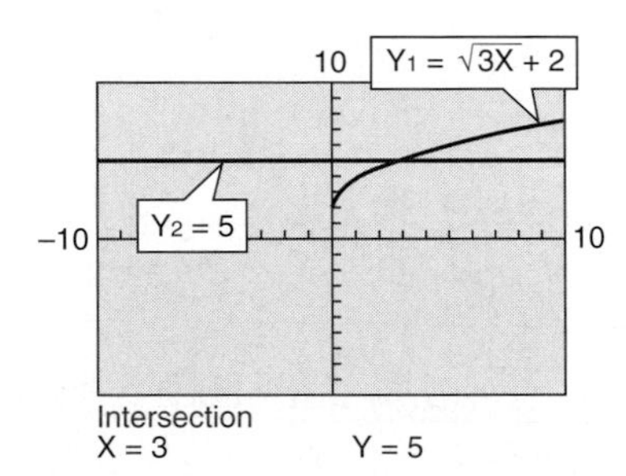

The solution of $\sqrt{3x} + 2 = 5$ is the x-coordinate of the intersection of the graph of Y1 = $\sqrt{3X} + 2$ and the graph of Y2 = 5.

You-Try-It 1 Solve: $\sqrt{4x} + 3 = 7$

Solution See page S28.

Example 2 Solve: $\sqrt{2x-5} + 3 = 0$

Solution

$\sqrt{2x-5} + 3 = 0$

$\sqrt{2x-5} = -3$ • Rewrite the equation so that the radical is alone on one side of the equation.

$(\sqrt{2x-5})^2 = (-3)^2$ • Square each side of the equation.

$2x - 5 = 9$ • Solve for x.

$2x = 14$

$x = 7$

Algebraic check: • Each side of the equation was squared. The proposed solution must be checked.

$\sqrt{2x-5} + 3$	0
$\sqrt{2(7)-5} + 3$	0
$\sqrt{14-5} + 3$	0
$\sqrt{9} + 3$	0
$3 + 3$	0

• Replace x by 7.

$6 \neq 0$ • This is a not a true equation. The solution does not check.

Graphical check:

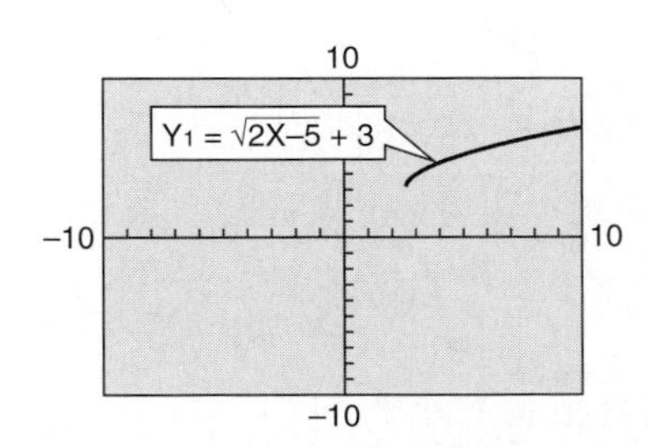

The graph of Y1 = $\sqrt{2X-5} + 3$ is shown above. Note that the graph does not cross the x-axis. Therefore, there is no value of x for which $\sqrt{2x-5} + 3 = 0$.

You-Try-It 2 Solve: $\sqrt{3x-2} - 5 = 0$

Solution See page S28.

In Examples 1 and 2, each equation contained only one radical. Example 3 below illustrates the procedure for solving a radical equation containing two radical expressions. Note that the process of squaring both sides of the equation is performed twice.

Example 3 Solve: $\sqrt{x+5} + \sqrt{x} = 5$

Solution

$\sqrt{x+5} + \sqrt{x} = 5$

$\sqrt{x+5} = 5 - \sqrt{x}$ • Solve for one of the radical expressions. Subtract $\sqrt{x}$ from each side of the equation.

$(\sqrt{x+5})^2 = (5 - \sqrt{x})^2$ • Square each side of the equation.

$x + 5 = 25 - 10\sqrt{x} + x$ • $(5 - \sqrt{x})^2 = (5 - \sqrt{x})(5 - \sqrt{x})$

$-20 = -10\sqrt{x}$ • Subtract x and 25 from each side of the equation.

$2 = \sqrt{x}$ • Divide each side by -10. This is still a radical equation.

$2^2 = (\sqrt{x})^2$ • Square each side of the equation.

$4 = x$

Algebraic check: • Each side of the equation was squared. The proposed solution must be checked.

$\sqrt{x+5} + \sqrt{x} = 5$	
$\sqrt{4+5} + \sqrt{4}$	5
$\sqrt{9} + \sqrt{4}$	5
$3 + 2$	5

$5 = 5$ • This is a true equation. The solution checks.

Graphical check:

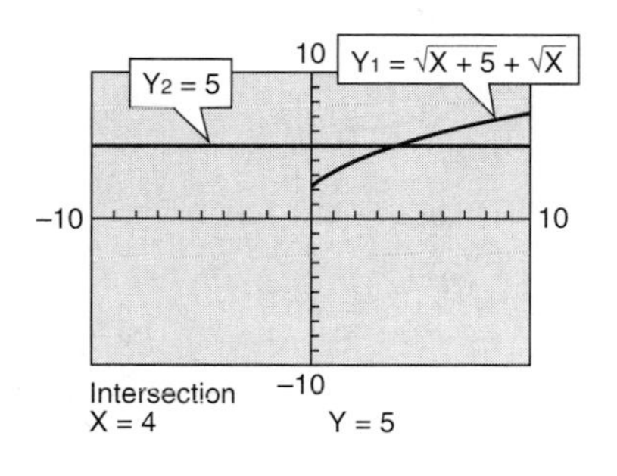

The graphs of $Y1 = \sqrt{X+5} + \sqrt{X}$ and $Y2 = 5$ are shown above. The x-coordinate of the point of intersection is the solution of the equation.

You-Try-It 3 Solve: $\sqrt{x+5} = \sqrt{x} + 1$

Solution See page S28.

Question: The graph of the equation $Y1 = \sqrt{X+5} + \sqrt{X}$ is shown above in the graphical check of the equation $\sqrt{x+5} + \sqrt{x} = 5$. **a.** Why is there no point on the graph of the equation for which $x < 0$? **b.** What is the domain?[1]

---

1. **a.** If $x < 0$, then x is a negative number. If x is a negative number, then $\sqrt{x}$ is not a real number.
b. The domain is $x \geq 0$.

Example 4 The formula $v = 2\sqrt{5L}$ is used to determine the speed of a car that leaves skid marks after the driver hits the brake pedal to stop the car. In this formula, the speed, v, of the car is measured in miles per hour, and the length, L, of the skid mark is measured in feet. How long a skid mark will be left by a car that was traveling at 50 mph? Round to the nearest whole number.

State the goal. The goal is to find the length of a skid mark that will be left by a car that was traveling at 50 mph.

Describe a strategy. Replace the variable v in the given formula by 50. Then solve the equation for L.

Solve the problem.

$v = 2\sqrt{5L}$

$50 = 2\sqrt{5L}$ • Replace v by 50.

$25 = \sqrt{5L}$ • Solve the equation for the radical. Divide each side by 2.

$25^2 = (\sqrt{5L})^2$ • Square each side of the equation.

$625 = 5L$

$125 = L$ • Solve for L.

A car traveling at 50 mph will leave a skid mark 125 feet long.

Check your work.

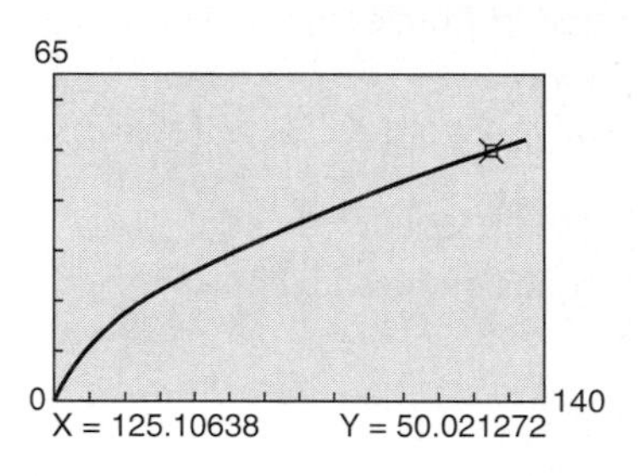

The graph of $Y1 = 2\sqrt{5X}$ is shown at the left. The point X = 125.10638, Y = 50.021272 is on the graph.

Question: Use the graph to answer the question: As the speed of the vehicle increases, does the length of the skid mark increase or decrease? Explain.[2]

You-Try-It 4 A formula for the optimum size of the pinhole in a pinhole camera is $d = 0.0073\sqrt{D}$, where d is the diameter of the pinhole and D is the distance, in inches, from the pinhole to the film. Use this formula to find the distance the film should be placed from the pinhole when the diameter of the pinhole is 0.008 inch. Round to the nearest tenth.

Verify the solution by graphing the equation $Y1 = 0.0073\sqrt{X}$ on a graphing calculator. Note: The number 0.0073 is very small, so the graph will not be visible using a standard viewing window. You might start with a window of Xmin = 0, Xmax = 3, Ymin = 0, Ymax = 0.025.

Solution See page S28.

2. As the speed increases, the length of the skid mark increases. This is evidenced by the fact that as y increases, x increases.

6.3 EXERCISES

Topics for Discussion

1. An equation that contains a variable expression in a radicand is a radical equation. Which of the following equations are radical equations?

a. $8 = \sqrt{5} + x$

b. $\sqrt{x - 7} = 9$

c. $\sqrt{x} + 4 = 6$

d. $12 = \sqrt{3}\,x$

b and c are radical equations.

2. What does the Property of Squaring Both Sides of an Equation state?
Answers will vary.

3. When the Property of Squaring Both Sides of an Equation is used to solve a radical equation, why is it necessary to check the solutions?
The resulting equation may have a solution that is not a solution of the original equation.

4. Describe two methods by which the solution to a radical equation can be checked.
Answers will vary. (Solutions can be checked by using an algebraic or a graphical method.)

5. Explain how to solve an equation containing two radical expressions.
Answers will vary.

6. Example 4 on page 426 used the formula $v = 2\sqrt{5L}$, which determines the speed of a car that leaves skid marks after the driver hits the brake pedal to stop the car. In this formula, v is the speed of the car measured in miles per hour and L is the length of the skid mark measured in feet. Suppose you graph the equation $Y1 = 2\sqrt{5L}$ on a graphing calculator and locate the point (45, 30) on the graph. What is the meaning of this ordered pair in the context of the given formula?
A car traveling at 45 mph will leave a skid mark 30 feet long.

Solving Equations Containing Radical Expressions

Solve the equation algebraically. Check the solution algebraically.

7. $\sqrt{x} = 5$
25

8. $\sqrt{y} = 7$
49

9. $\sqrt{a} = 12$
144

10. $\sqrt{a} = 9$
81

11. $\sqrt{5x} = 5$
5

12. $\sqrt{3x} = 4$
$\frac{16}{3}$

13. $\sqrt{4x} = 8$
16

14. $\sqrt{6x} = 3$
$\frac{3}{2}$

15. $\sqrt{2x} - 4 = 0$
8

16. $3 - \sqrt{5x} = 0$
$\frac{9}{5}$

17. $\sqrt{4x} + 5 = 2$
No solution

18. $\sqrt{3x} + 9 = 4$
No solution

Solve the equation algebraically. Check the solution graphically.

19. $\sqrt{3x-2} = 4$
6

20. $\sqrt{5x+6} = 1$
-1

21. $\sqrt{2x+1} = 7$
24

22. $\sqrt{5x+4} = 3$
1

23. $0 = 2 - \sqrt{3-x}$
-1

24. $0 = 5 - \sqrt{10+x}$
15

25. $\sqrt{5x+2} = 0$
$-\frac{2}{5}$

26. $\sqrt{3x-7} = 0$
$\frac{7}{3}$

27. $\sqrt{3x} - 6 = -4$
$\frac{4}{3}$

28. $\sqrt{5x} + 8 = 23$
45

29. $0 = \sqrt{3x-9} - 6$
15

30. $0 = \sqrt{2x+7} - 3$
1

31. $\sqrt{5x-1} = \sqrt{3x+9}$
5

32. $\sqrt{3x+4} = \sqrt{12x-14}$
2

33. $\sqrt{5x-3} = \sqrt{4x-2}$
1

34. $\sqrt{5x-9} = \sqrt{2x-3}$
2

35. $\sqrt{x^2-5x+6} = \sqrt{x^2-8x+9}$
1

36. $\sqrt{x^2-2x+4} = \sqrt{x^2+5x-12}$
$\frac{16}{7}$

37. $\sqrt{x} = \sqrt{x+3} - 1$
1

38. $\sqrt{x+5} = \sqrt{x} + 1$
4

39. $\sqrt{2x+5} = 5 - \sqrt{2x}$
2

40. $\sqrt{2x} + \sqrt{2x+9} = 9$
8

41. $\sqrt{3x} - \sqrt{3x+7} = 1$
No solution

42. $\sqrt{x} - \sqrt{x+9} = 1$
No solution

Solve.

43. The perimeter of the rectangle shown below is 24 meters. Find the value of x.

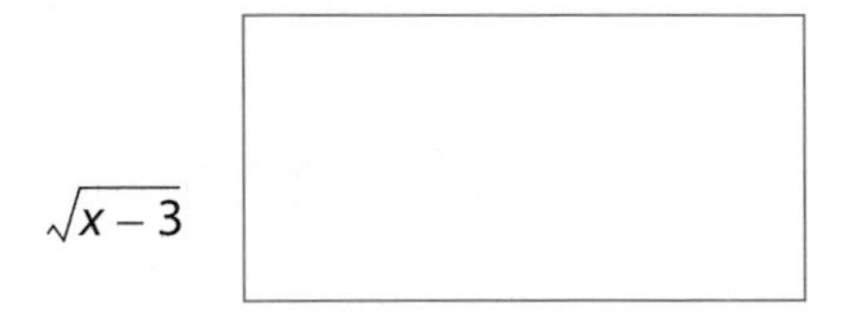

8

19 m

44. The perimeter of the square shown below is 32 inches. Find the value of x.

57 in.

45. How far from the center of a merry-go-round is a child sitting when the child is traveling at a speed of 9 feet per second? Use the formula $v = \sqrt{12r}$, where v is the speed in feet per second and r is the distance in feet from the center of the merry-go-round to the rider.
6.75 ft

46. How high a hill must you climb in order to be able to see a distance of 30 kilometers? Use the formula $d = \sqrt{12h}$, where d is the distance in kilometers to the horizon from a point h meters above the earth's surface.
A 75 meter hill

47. A tsunami is a great sea wave produced by underwater earthquakes or volcanic eruption. Use the formula $v = 3\sqrt{d}$, where v is the velocity in feet per second of a tsunami as it approaches land and d is the depth in feet of the water, to find the depth in feet of the water when the velocity of a tsunami reaches 60 feet per second.
400 ft

48. If an object is dropped from an airplane, how far will it fall in 10 seconds? Use the formula $t = \frac{\sqrt{d}}{4}$, where t is the time in seconds that the object falls and d is the distance in feet that the object falls.
1600 ft

49. The time it takes for an object to fall a distance of d feet on the moon is given by the formula $t = \sqrt{\frac{d}{2.75}}$, where t is the time in seconds. If an astronaut drops an object on the moon, how far will it fall in 5 seconds?
68.75 ft

50. A stone is dropped into a mine shaft and hits the bottom 3.5 seconds later. How deep is the mine shaft? The equation for the distance an object falls in T seconds is $T = \sqrt{\frac{d}{16}}$, where d is the distance in feet.
196 ft

51. The weight of an object is related to its distance above the surface of the earth. A formula for this relationship is $d = 4000\sqrt{\frac{E}{S}} - 4000$, where E is the object's weight on the surface of the earth and S is the object's weight at a distance of d miles above the earth's surface. An astronaut weighs 25 pounds when she is 6000 miles above the earth's surface. How much does the astronaut weigh on the earth's surface?
156.25 lb

52. Find the length of a pendulum that makes one swing in 3 seconds. The equation for the time of one swing is $T = 2\pi\sqrt{\frac{L}{32}}$, where T is the time in seconds and L is the length in feet. Round to the nearest hundredth.
7.30 ft

53. The speed of sound in different air temperatures is calculated using the formula $v = \frac{1087\sqrt{t+273}}{16.52}$, where v is the speed in feet per second and t is the temperature in degrees Celsius. What must the temperature be in order for sound to travel at a speed of 1200 feet per second? Round to the nearest tenth.

59.6°C

Applying Concepts

54. Explain how solving the equation $3\sqrt{x-8} - 5 = 7$ is similar to solving the equation $3x - 5 = 7$.

Answers will vary.

55. Determine whether the statement is always true, sometimes true, or never true.

a. We can square both sides of an equation without changing the solutions of the equation. Sometimes true

b. The Property of Squaring Both Sides of an Equation is used to eliminate a radical expression from an equation. Always true

c. For any real numbers a and b, $a^2 = b^2$ means $a = b$. Sometimes true

d. The first step in solving a radical equation is to square both sides of the equation. Sometimes true

56. Solve.

a. $\sqrt{\frac{5y+2}{3}} = 3$ **b.** $\sqrt{\frac{3y}{5}} - 1 = 2$ **c.** $\sqrt{9x^2+49} + 1 = 3x + 2$

a. 5; **b.** 15; **c.** 8

Exploration

57. ***Graphing Radical Functions***

a. Graph each of the following equations: $Y1 = \sqrt{X}$, $Y2 = \sqrt{X} + 3$, $Y3 = \sqrt{X} - 2$
Explain how adding a number to or subtracting a number from $\sqrt{X}$ affects the graph.

b. Graph each of the following equations: $Y1 = \sqrt{X}$, $Y2 = \sqrt{X+3}$, $Y3 = \sqrt{X-2}$
Explain how adding a number to or subtracting a number from the radicand X affects the graph.

c. The graph of $y = \sqrt{x}$ is translated 5 units up. Find the equation of the graph. Check your answer by graphing the equation.

d. The graph of $y = \sqrt{x}$ is translated 6 units left. Find the equation of the graph. Check your answer by graphing the equation.

e. The graph of $y = \sqrt{x}$ is translated 4 units down. Find the equation of the graph. Check your answer by graphing the equation.

f. The graph of $y = \sqrt{x}$ is translated 7 units right. Find the equation of the graph. Check your answer by graphing the equation.

Section 6.4 The Pythagorean Theorem

The Pythagorean Theorem

A **right triangle** contains one right angle. The side opposite the right angle is called the **hypotenuse**. The other two sides are called **legs**.

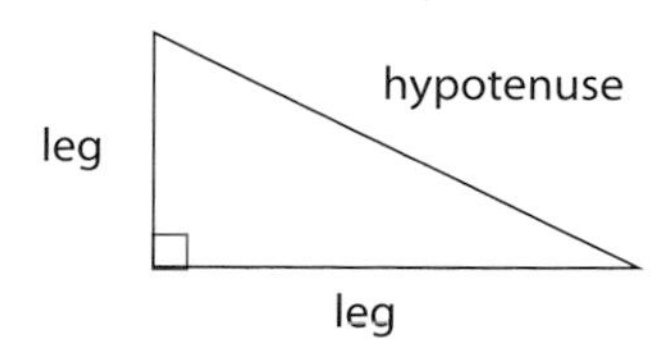

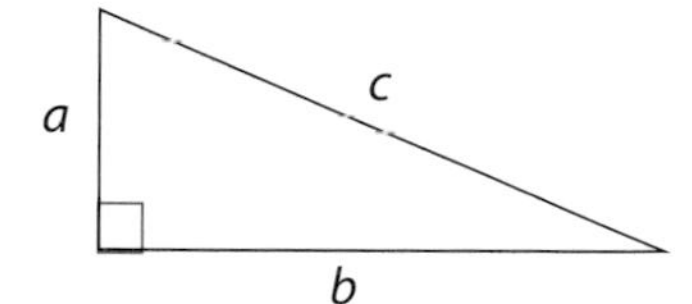

Question: For the triangle at the left, which side is the hypotenuse?[1]

The angles in a right triangle are usually labeled with the capital letters *A*, *B*, and *C*, with *C* reserved for the right angle. The side opposite angle *A* is side *a*, the side opposite angle *B* is side *b*, and *c* is the hypotenuse.

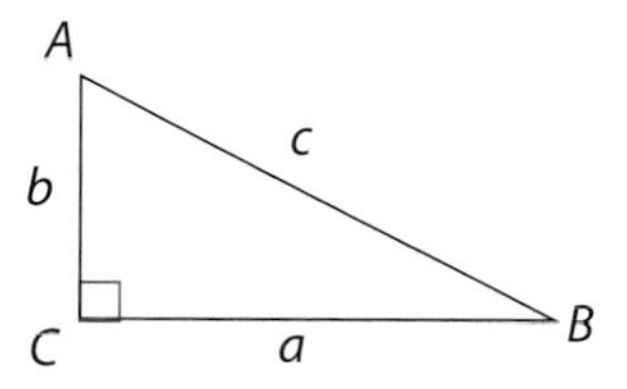

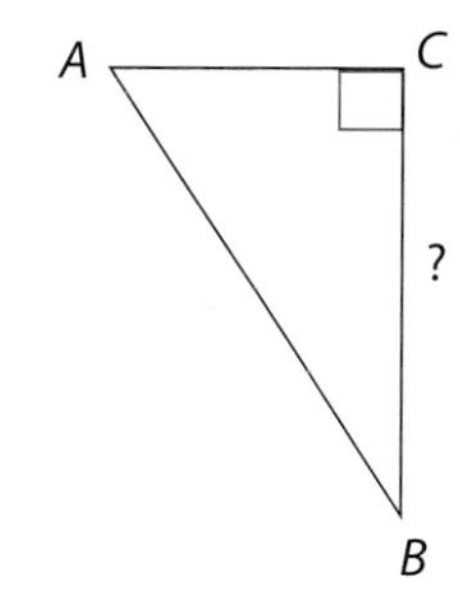

Question: Using the convention stated in the above paragraph, what letter should replace the question mark in the triangle at the left?[2]

The Pythagorean Theorem states an important relationship that exists among the sides of a right triangle. The theorem is named after Pythagoras, a Greek mathematician and philosopher who lived from about 572 to 501 B.C. Although Pythagoras is generally credited with the discovery, there is evidence that it was known years before his birth. There is reference to this theorem in Chinese writings from about 1100 B.C., and it is believed that the Egyptians and Babylonians knew of it as early as 2000 B.C.

POINT OF INTEREST
Historians believe that surveyors in ancient Egypt used stretched ropes with 12 equally spaced knots to measure right angles when laying out land boundaries. The surveyors knew that a triangle with sides measuring 3, 4, and 5 units was a right triangle.

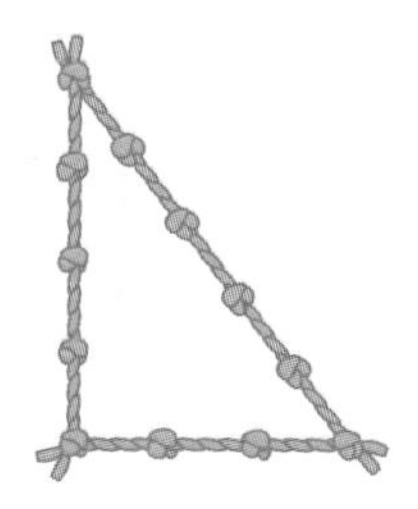

The **Pythagorean Theorem** states that the square of the hypotenuse of a right triangle is equal to the sum of the squares of the two legs.

The figure at the right is a right triangle with legs measuring 3 units and 4 units and a hypotenuse measuring 5 units. Each side of the triangle is also the side of a square. The number of square units in the area of the largest square is equal to the sum of the areas of the smaller squares.

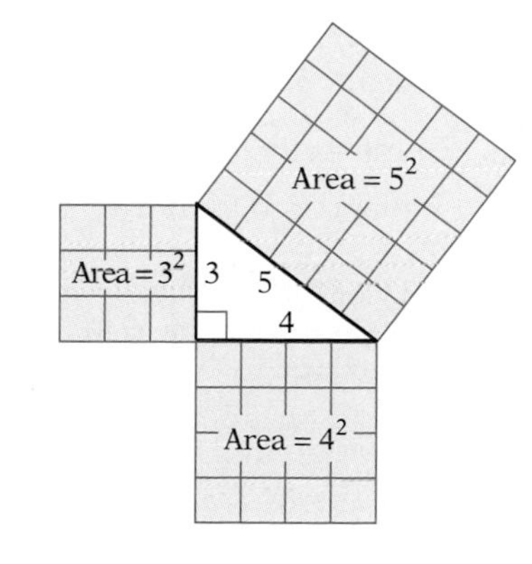

Square of the hypotenuse	=	sum of the squares of the two legs

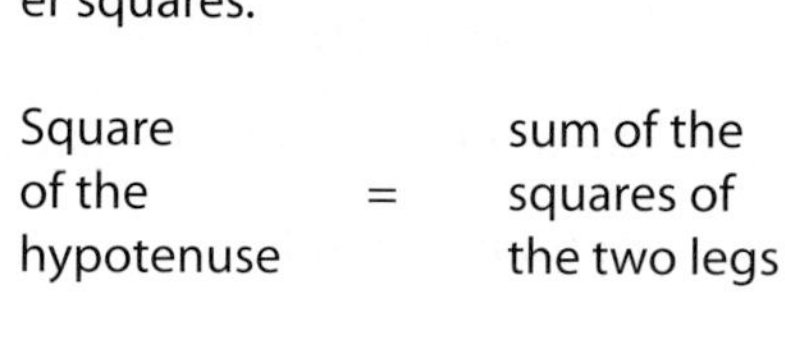

$$5^2 = 3^2 + 4^2$$
$$25 = 9 + 16$$
$$25 = 25$$

1. Side *c* is the hypotenuse because it is the side opposite the right angle.
2. The letter *a* (lowercase *a*) should replace the question mark because it is the side opposite angle *A*.

Pythagorean Theorem

If a and b are the lengths of the legs of a right triangle and c is the length of the hypotenuse, then $c^2 = a^2 + b^2$.

If the lengths of two sides of a right triangle are known, the Pythagorean Theorem can be used to find the length of the third side.

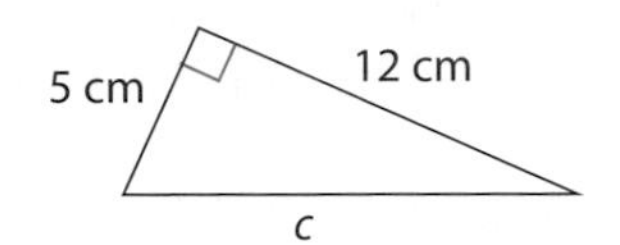

Consider a right triangle with legs that measure 5 cm and 12 cm. Use the Pythagorean Theorem, with $a = 5$ and $b = 12$, to find the length of the hypotenuse. (If you let $a = 12$ and $b = 5$, the result is the same.)

$$c^2 = a^2 + b^2$$
$$c^2 = 5^2 + 12^2$$
$$c^2 = 25 + 144$$
$$c^2 = 169$$

This equation states that the square of c is 169. Since $13^2 = 169$, $c = 13$, and the length of the hypotenuse is 13 cm. We can find c by taking the square root of 169: $\sqrt{169} = 13$. This suggests the following property.

The Principal Square Root Property

If $r^2 = s$, then $r = \sqrt{s}$, and r is called the square root of s.

The Principal Square Root Property and its application are illustrated as follows:

Because $4^2 = 16, 4 = \sqrt{16}$. Therefore, if $c^2 = 16, c = \sqrt{16} = 4$.

Because $8^2 = 64, 8 = \sqrt{64}$. Therefore, if $c^2 = 64, c = \sqrt{64} = 8$.

Question: If $c^2 = 81$, what is the value of c?[3]

→ The length of one leg of a right triangle is 9 inches. The hypotenuse is 12 inches. Find the length of the other leg. Round to the nearest tenth.

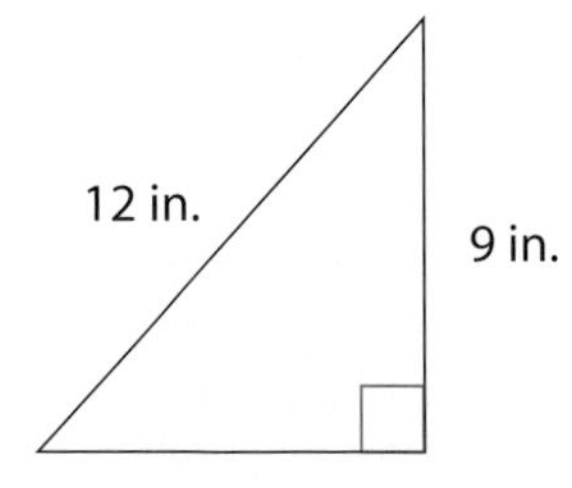

Use the Pythagorean Theorem.	$a^2 + b^2 = c^2$
$a = 9$ and $c = 12$.	$9^2 + b^2 = 12^2$
Square 9 and square 12.	$81 + b^2 = 144$
We want to solve for b^2. Subtract 81 from each side of the equation.	$81 - 81 + b^2 = 144 - 81$
	$b^2 = 63$
Use the Principal Square Root Property.	$b = \sqrt{63}$
Use a calculator to approximate $\sqrt{63}$.	$b \approx 7.9$

The length of the other leg is approximately 7.9 inches.

TAKE NOTE
In this example, if you let $b = 9$ and solve for a^2, the result is the same.

---

3. $c = \sqrt{81} = 9$

Example 1 The two legs of a right triangle measure 5 meters and 13 meters. Find the hypotenuse of the right triangle. Round to the nearest hundredth.

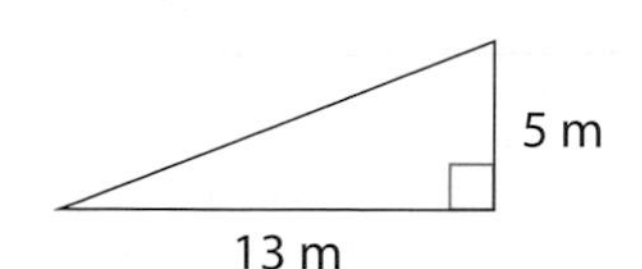

Solution

$c^2 = a^2 + b^2$ • Use the Pythagorean Theorem.

$c^2 = 5^2 + 13^2$ • $a = 5$ and $b = 13$.

$c^2 = 25 + 169$ • Square 5 and square 13.

$c^2 = 194$

$c = \sqrt{194}$ • Use the Principal Square Root Property.

$c \approx 13.93$

The length of the hypotenuse is 13.93 meters.

You-Try-It 1 The two legs of a right triangle measure 7 meters and 14 meters. Find the hypotenuse of the right triangle. Round to the nearest hundredth.

Solution See page S29.

Example 2 The hypotenuse of a right triangle measures 14 feet, and one leg measures 11 feet. Find the measure of the other leg. Round to the nearest tenth.

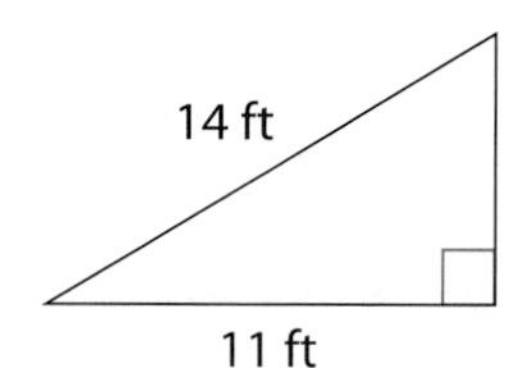

Solution

$a^2 + b^2 = c^2$ • Use the Pythagorean Theorem.

$11^2 + b^2 = 14^2$ • $a = 11$ and $c = 14$.

$121 + b^2 = 196$ • Square 11 and square 14.

$121 - 121 + b^2 = 196 - 121$ • Solve for b^2. Subtract 121 from each side of the equation.

$b^2 = 75$

$b = \sqrt{75}$ • Use the Principal Square Root Property.

$b \approx 8.7$

The measure of the other leg is approximately 8.7 feet.

You-Try-It 2 The hypotenuse of a right triangle measures 16 feet, and one leg measures 5 feet. Find the measure of the other leg. Round to the nearest hundredth.

Solution See page S29.

SUGGESTED ACTIVITY
See the *Instructor's Resource Manual* for an activity involving the Pythagorean Theorem.

The hypotenuse is always the longest side of a right triangle. This fact can be used as a quick check when you are solving a right triangle for the length of one of the sides. For example, in Example 2, we found that the length of a leg was 8.7 feet. Since $8.7 < 14$ (the hypotenuse), we know that the answer is "in the right ballpark."

Question: Two legs of a right triangle measure 3 meters and 9 meters. Is the length of the hypotenuse less than 3 meters, between 3 and 9 meters, or greater than 9 meters?[4]

---

4. The length of the hypotenuse is longer than either leg, so the length is greater than 9 meters.

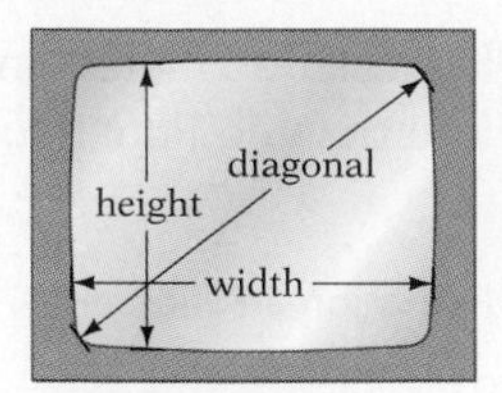

Example 3 High Definition Television (HDTV) gives consumers a wider viewing area, more like a film seen in a movie theater. A regular television with a 27-inch diagonal measurement has a screen 16.2 inches tall. An HDTV screen with the same 16.2-inch height would have a diagonal measuring 33 inches. How many inches wider is the HDTV screen? Round to the nearest tenth.

State the goal.

The goal is to find the difference between the widths of an HDTV screen and a regular television screen given that both have a height of 16.2 inches, the diagonal of the HDTV screen is 33 inches, and the diagonal of the regular TV is 27 inches.

Describe a strategy.

- Use the Pythagorean Theorem to find the width of an HDTV screen that has a 16.2-inch height and a 33-inch diagonal.
- Use the Pythagorean Theorem to find the width of a regular TV screen that has a 16.2-inch height and a 27-inch diagonal.
- Subtract the width of the regular television screen from the width of the HDTV screen.

Solve the problem.

$a^2 + b^2 = c^2$

$(16.2)^2 + b^2 = 33^2$ • For the HDTV, $a = 16.2$ and $c = 33$.

$262.44 + b^2 = 1089$

$b^2 = 826.56$ • Solve for b^2. Subtract 262.44 from each side of the equation.

$b = \sqrt{826.56}$ • Use the Principal Square Root Property.

$b \approx 28.7$ • The width of the HDTV screen is 28.7 inches.

$a^2 + b^2 = c^2$

$(16.2)^2 + b^2 = 27^2$ • For the regular TV, $a = 16.2$ and $c = 27$.

$262.44 + b^2 = 729$

$b^2 = 466.56$ • Solve for b^2. Subtract 262.44 from each side of the equation.

$b = \sqrt{466.56}$ • Use the Principal Square Root Property.

$b = 21.6$ • The width of the regular TV screen is 21.6 inches.

$28.7 - 21.6 = 7.1$ • Find the difference between the two widths.

The HDTV screen is 7.1 inches wider than the regular TV screen.

Check your work.

√

SUGGESTED ACTIVITY
See the *Instructor's Resource Manual* for an activity that requires the Pythagorean Theorem.

You-Try-It 3 Dave Marshall needs to clean the gutters of his home. The gutters are 28 feet above the ground. For safety, the distance a ladder reaches up a wall should be four times the distance from the bottom of the ladder to the base of the side of the house. Therefore, the ladder must be 7 feet from the base of the house. Will a 30-foot ladder be long enough to reach the gutters?

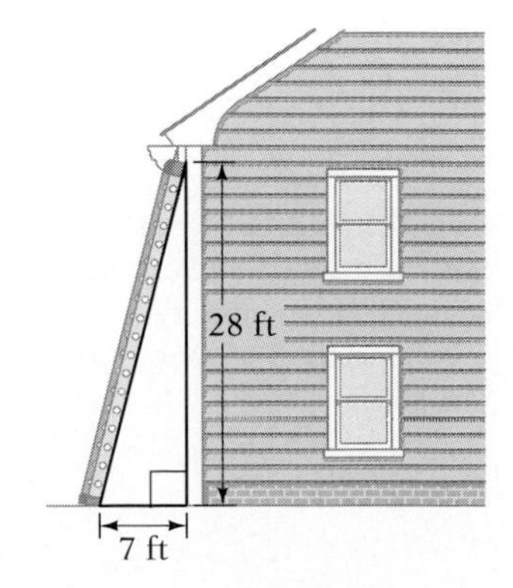

Solution See page S29.

The Distance Formula

In the rectangular coordinate system at the right, points are graphed at (1, 3) and (5, 3). A line segment is drawn between the two points. We can determine the length of the line segment by counting the number of units between the points; it is 4 units. We can also determine the length by finding the absolute value of the difference between the x-coordinates of the two points.

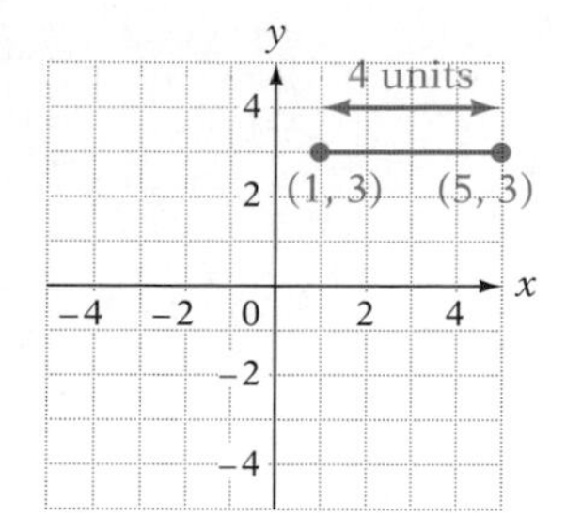

$$|5 - 1| = |4| = 4$$

It does not matter in what order the two numbers are subtracted; the result is the same.

$$|1 - 5| = |-4| = 4$$

In the rectangular coordinate grid at the right, points are graphed at (3, 0) and (3, 4). A line segment is drawn between the two points. We can determine the length of the line segment by counting the number of units between the points; it is 4 units. We can also determine the length by finding the absolute value of the difference between the y-coordinates of the two points.

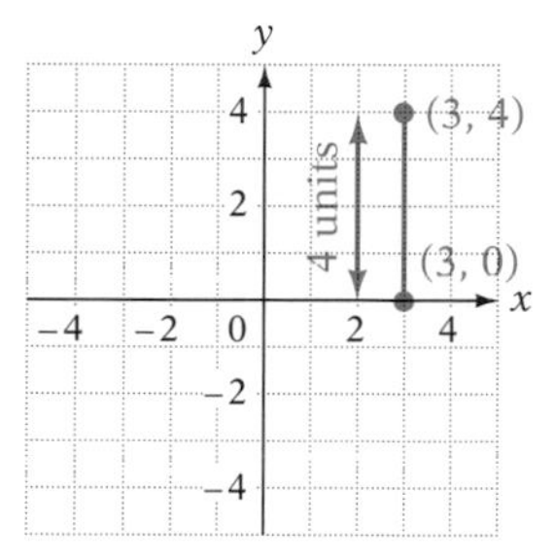

$$|4 - 0| = |4| = 4$$

It does not matter in what order the two numbers are subtracted; the result is the same.

$$|0 - 4| = |-4| = 4$$

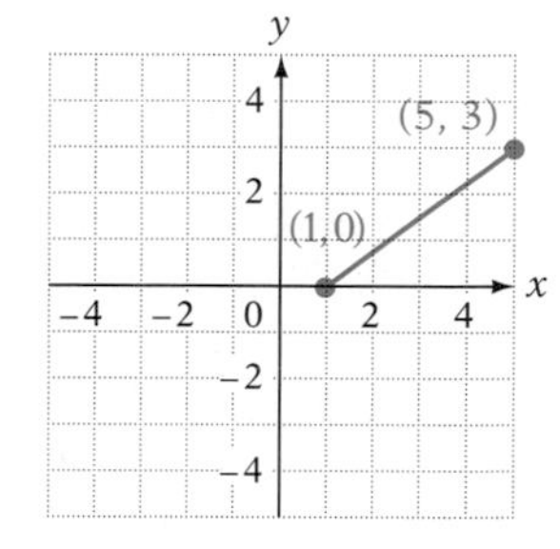

In the coordinate grid at the left, points are graphed at (1, 0) and (5, 3). A line segment is drawn between the two points. Because the line segment is neither horizontal nor vertical, we cannot determine its length by counting the number of units between the points. We can, however, draw a right triangle (as shown below) in which the line segment drawn is the hypotenuse. The vertex of the right angle is at the point (5, 0). One leg of the right triangle is a horizontal line segment, and the other leg is a vertical line segment. We can determine the length of each leg and then use the Pythagorean Theorem to calculate the length of the hypotenuse.

Find the length of each leg.

$$a = |5 - 1| = |4| = 4$$

$$b = |3 - 0| = |3| = 3$$

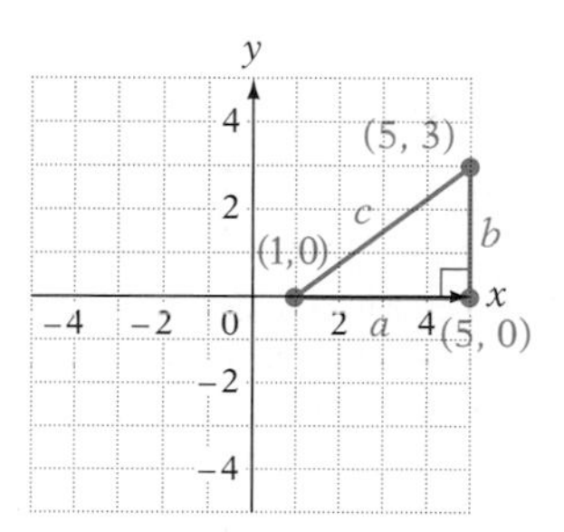

Use the Pythagorean Theorem.

$$c^2 = a^2 + b^2$$
$$c^2 = 4^2 + 3^2$$
$$c^2 = 16 + 9$$
$$c^2 = 25$$
$$c = \sqrt{25}$$
$$c = 5$$

The length of the line segment between the points (1, 0) and (5, 3) is 5 units.

We can apply this method of using the Pythagorean Theorem to find the distance between two points in the plane to the points (x_1, y_1) and (x_2, y_2).

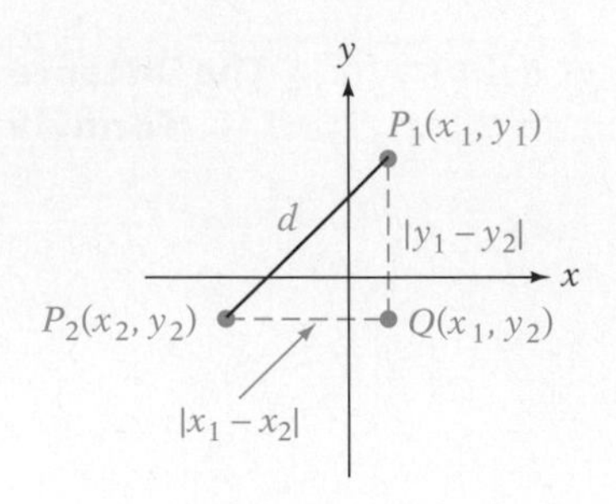

The vertical distance between (x_1, y_1) and (x_2, y_2) is $|y_1 - y_2|$. The horizontal distance between (x_1, y_1) and (x_2, y_2) is $|x_1 - x_2|$.

Apply the Pythagorean Theorem to the right triangle.

$$d^2 = |x_1 - x_2|^2 + |y_1 - y_2|^2$$

Because the square of a number cannot be negative, the absolute value signs are not necessary.

$$d^2 = (x_1 - x_2)^2 + (y_1 - y_2)^2$$

Use the Principal Square Root Property to solve the equation for d. This equation is known as the distance formula.

$$d = \sqrt{(x_1 - x_2)^2 + (y_1 - y_2)^2}$$

The Distance Formula

If (x_1, y_1) and (x_2, y_2) are two points in the plane, then the distance d between the two points is given by

$$d = \sqrt{(x_1 - x_2)^2 + (y_1 - y_2)^2}$$

Example 4 Find the distance between the points (–3, 2) and (4, –1). Round to the nearest tenth.

Solution

$$d = \sqrt{(x_1 - x_2)^2 + (y_1 - y_2)^2}$$ • Use the distance formula.

$$d = \sqrt{(-3 - 4)^2 + [2 - (-1)]^2}$$ • Let $(x_1, y_1) = (-3, 2)$ and $(x_2, y_2) = (4, -1)$.

$$d = \sqrt{(-7)^2 + (3)^2}$$

$$d = \sqrt{49 + 9}$$

$$d = \sqrt{58}$$

$$d \approx 7.6$$

The distance between the points is approximately 7.6 units.

TAKE NOTE
If $(x_1, y_1) = (-3, 2)$, then $x_1 = -3$ and $y_1 = 2$. If $(x_2, y_2) = (4, -1)$, then $x_2 = 4$ and $y_2 = -1$.

SUGGESTED ACTIVITY
Have the students solve Example 4 using $(x_1, y_1) = (4, -1)$ and $(x_2, y_2) = (-3, 2)$ to show that the answer is the same.

You-Try-It 4 Find the distance between the points (5, –2) and (–4, 3). Round to the nearest tenth.

Solution See page S29.

6.4 EXERCISES

Topics for Discussion

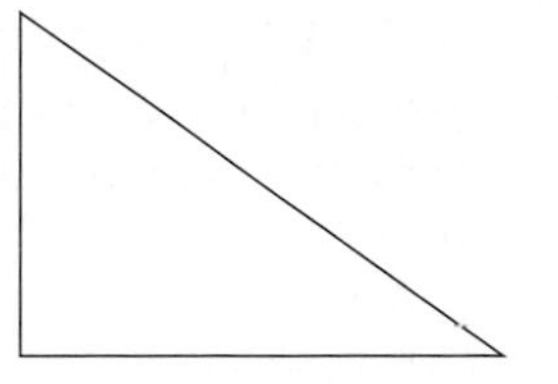

1. Label the right triangle shown at the right. Include the right angle symbol, the three angles, and the three sides.
   The right angle symbol must be at the 90° angle. The right angle must be labeled *C* and the hypotenuse labeled *c*. One acute angle should be labeled *A*, with the side opposite it labeled *a*. The other acute angle should be labeled *B*, with the side opposite it labeled *b*.

2. What does the Pythagorean Theorem state?
   The square of the hypotenuse of a right triangle is equal to the sum of the squares of the two legs.

3. Can the Pythagorean Theorem be used to find the length of side *c* of the triangle at the right? If so, determine *c*. If not, explain why the theorem cannot be used.
   No. The triangle is not a right triangle.

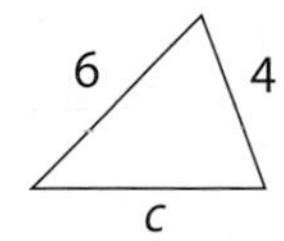

4. What does the Principal Square Root Property state?
   If $r^2 = s$, then $r = \sqrt{s}$, and r is called the square root of s.

5. What does the distance formula state?
   If (x_1, y_1) and (x_2, y_2) are two points in the plane, then the distance between the two points is given by $d = \sqrt{(x_1 - x_2)^2 + (y_1 - y_2)^2}$.

The Pythagorean Theorem

Determine whether the triangle is a right triangle.

6. 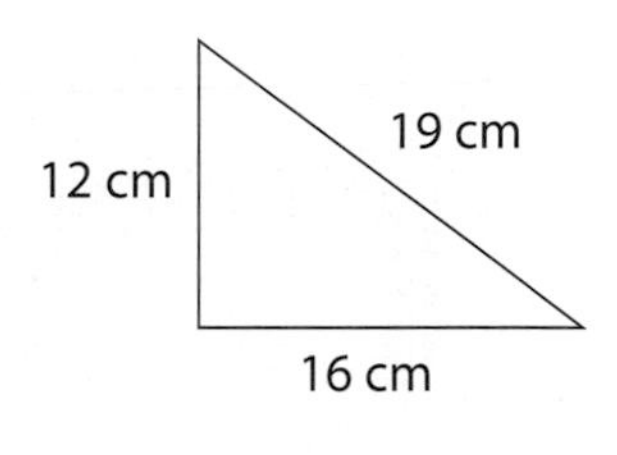

   No

7. 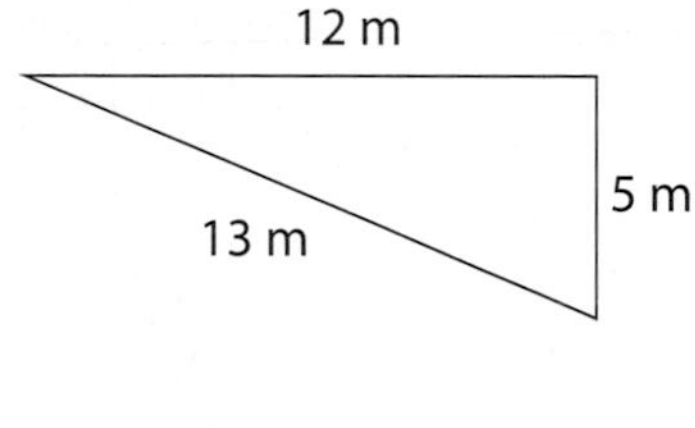

   Yes

8. 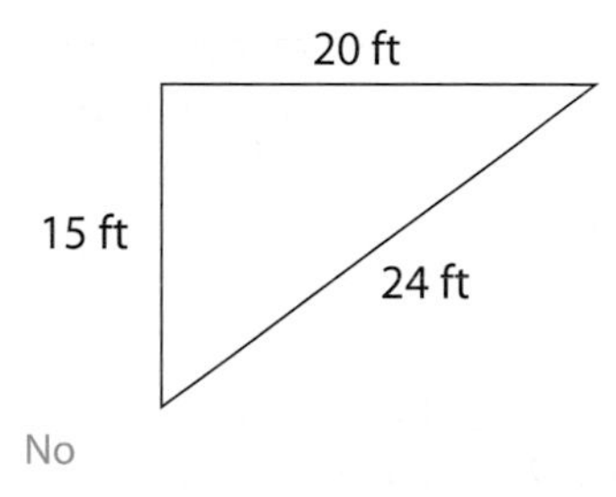

   No

9. The two legs of a right triangle measure 5 centimeters and 9 centimeters. Find the length of the hypotenuse. Round to the nearest tenth.
   10.3 cm

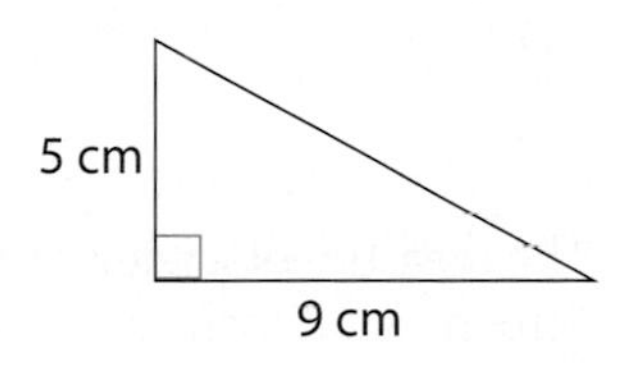

10. The two legs of a right triangle measure 8 inches and 4 inches. Find the length of the hypotenuse. Round to the nearest tenth.
    8.9 in.

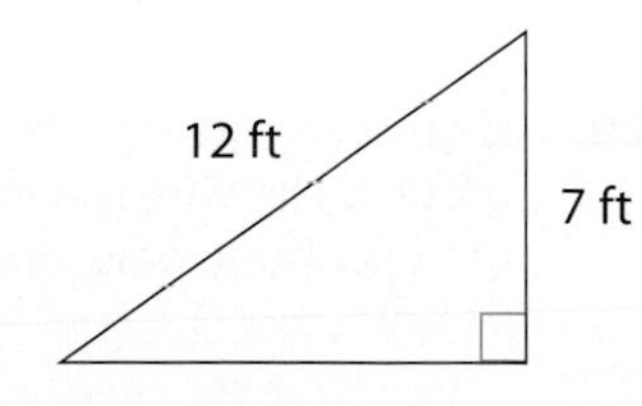

11. The hypotenuse of a right triangle measures 12 feet. One leg of the triangle measures 7 feet. Find the length of the other leg of the triangle. Round to the nearest hundredth.
    9.75 ft

12. The hypotenuse of a right triangle measures 20 centimeters. One leg of the triangle measures 16 centimeters. Find the length of the other leg of the triangle.
12 cm

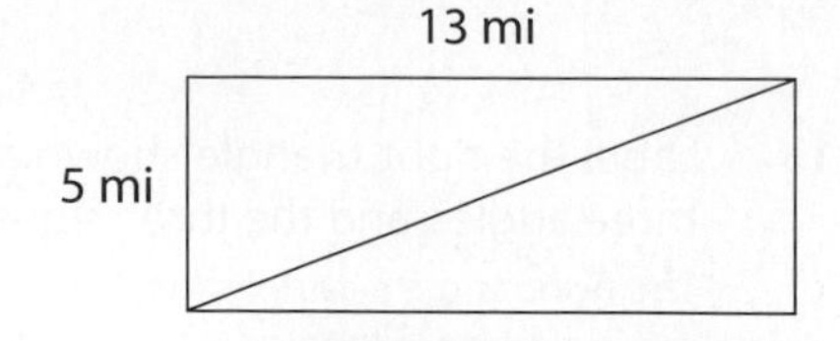

13. The diagonal of a rectangle is a line drawn from one vertex to the opposite vertex. Find the length of the diagonal in the rectangle shown at the right. Round to the nearest tenth.
13.9 mi

14. The infield of a baseball diamond is a square. The distance between successive bases is 90 feet. The pitcher's mound is on the diagonal between home plate and second base at a distance of 60.5 feet from home plate. Is the pitcher's mound more or less than halfway between home plate and second base?
Less than halfway

15. The infield of a softball diamond is a square. The distance between successive bases is 60 feet. The pitcher's mound is on the diagonal between home plate and second base at a distance of 46 feet from home plate. Is the pitcher's mound more or less than halfway between home plate and second base?
More than halfway

16. An L–shaped sidewalk from the parking lot to a memorial is shown in the figure at the right. The distance directly across the grass to the memorial is 650 feet. The distance to the corner is 600 feet. Find the distance from the corner to the memorial.
250 ft

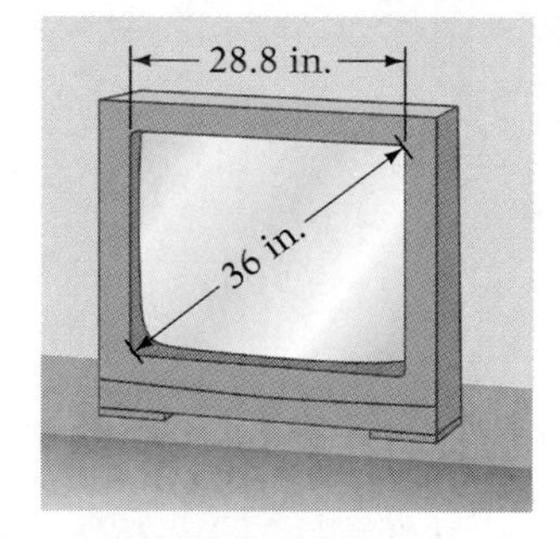

17. A commuter plane leaves an airport traveling due south at 400 mph. Another plane leaving at the same time travels due east at 300 mph. Find the distance between the two planes after 2 hours.
1000 mi

18. The measure of a television screen is given by the length of a diagonal across the screen. A big-screen 36-inch television has a width of 28.8 inches. Find the height of the screen. Round to the nearest tenth.
21.6 in.

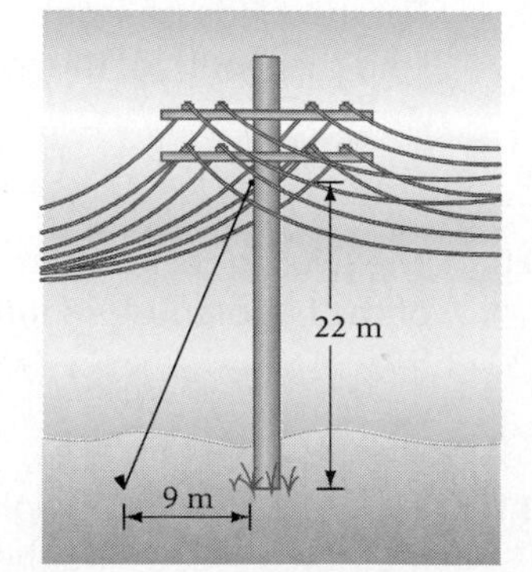

19. The measure of a television screen is given by the length of a diagonal across the screen. A 33-inch television has a width of 26.4 inches. Find the height of the screen. Round to the nearest tenth.
19.8 in.

20. A guy wire is to be attached to a telephone pole at a point 22 meters above the ground. The wire is anchored to the ground at a point 9 meters from the base of the telephone pole. How long a guy wire is required? Round to the nearest tenth.
23.8 m

21. Melissa leaves a dock in her sailboat and sails 4 miles due east. She then tacks and sails 2.5 miles due south. The walkie-talkie Melissa has on board has a range of 5 miles. Will she be able to call a friend on the dock from her location using the walkie-talkie?
Yes

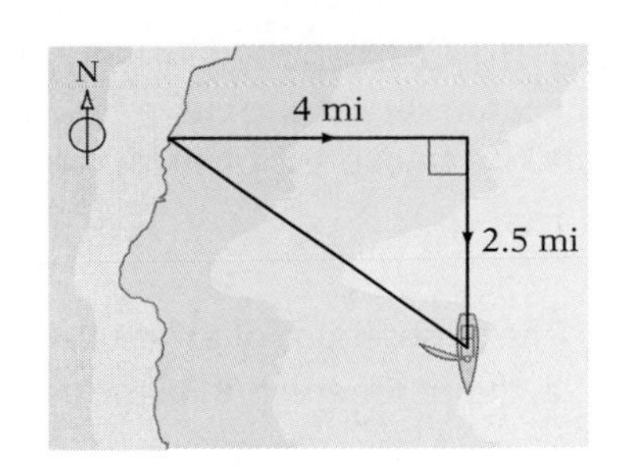

The Distance Formula

Find the distance between the two points. Give the exact value.

22. (3, 5) and (5, 1)
$2\sqrt{5}$

23. (–2, 3) and (4, –1)
$2\sqrt{13}$

24. (0, 3) and (–2, 4)
$\sqrt{5}$

Find the distance between the two points. Round to the nearest tenth.

25. (6, –1) and (–3, –2)
9.1

26. (–3, –5) and (2, –4)
5.1

27. (–7, –5) and (–2, –1)
6.4

28. (5, –2) and (–2, 5)
9.9

29. (3, –6) and (6, 0)
6.7

30. (–5, 5) and (2, –5)
12.2

31. A triangle has vertices at (3, –2), (–3, –2), and (–3, 2). Find **a.** the perimeter and **b.** the area of the triangle. Round to the nearest tenth.
a. 17.2 units; **b.** 12 square units

32. A triangle has vertices at (1, 3), (1, –3), and (–6, –3). Find **a.** the perimeter and **b.** the area of the triangle. Round to the nearest tenth.
a. 22.2 units; **b.** 21 square units

33. A parallelogram has vertices at (–4, 1), (2, 1), (4, –3), and (–2, –3). Find **a.** the perimeter and **b.** the area of the parallelogram. Round to the nearest tenth.
a. 20.9 units; **b.** 24 square units

34. A parallelogram has vertices at (4, 6), (–1, 6), (–3, –1), and (2, –1). Find **a.** the perimeter and **b.** the area of the parallelogram. Round to the nearest tenth.
a. 24.6 units; **b.** 35 square units

35. A trapezoid has vertices at (4, 1), (–6, 1), (–2, –5), and (2, –5). Find **a.** the perimeter and **b.** the area of the trapezoid. Round to the nearest tenth.
a. 27.5 units; **b.** 42 square units

36. A trapezoid has vertices at (–2, 2), (6, 5), (6, –5), and (–2, –4). Find **a.** the perimeter and **b.** the area of the trapezoid. Round to the nearest tenth.
a. 32.6 units; **b.** 64 square units

Applying Concepts

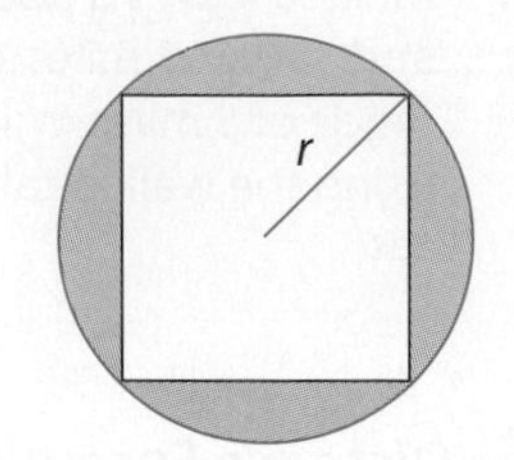

37. Write an expression in factored form for the shaded region in the diagram at the right.
$r^2(\pi - 2)$

38. The hypotenuse of a right triangle is $5\sqrt{2}$ centimeters, and one leg is $3\sqrt{2}$ centimeters. Find **a.** the perimeter and **b.** the area of the triangle.
a. $12\sqrt{2}$ cm; **b.** 12 cm^2

39. Explain why, when using the distance formula, it makes no difference which point you designate as (x_1, y_1) and which you designate as (x_2, y_2).
Answers will vary.

40. A triangle has vertices at (4, 5), (−3, 9), and (1, 2). Is the triangle scalene, isosceles, or equilateral?
Isosceles

41. A triangle has vertices at (4, 5), (−3, 9), and (1, 3). Is the triangle a right triangle?
Yes

42. For what positive value of y will the distance between the points (5, −2) and (−3, y) be 10 units?
4

43. Determine whether the statement is always true, sometimes true, or never true.

a. $(x_1 - x_2) = |x_1 - x_2|$ Sometimes true

b. $(x_1 - x_2) = (x_2 - x_1)$ Sometimes true

c. $|x_1 - x_2| = |x_2 - x_1|$ Always true

d. $|x_1 - x_2|^2 = (x_1 - x_2)^2$ Always true

e. $(x_1 - x_2)^2 = (x_2 - x_1)^2$ Always true

Explorations

44. ***Pythagorean Triples***
The Pythagorean Theorem states that if a and b are the legs of a right triangle and c is the hypotenuse, then $a^2 + b^2 = c^2$. For instance, the triangle with legs 3 and 4 and hypotenuse 5 is a right triangle because $3^2 + 4^2 = 5^2$. The numbers 3, 4, and 5 are called a **Pythagorean triple** because they are natural numbers that satisfy the equation of the Pythagorean Theorem.

a. Determine whether the numbers are a Pythagorean triple.
1. 5, 7, and 9 **2.** 8, 15, and 17 **3.** 1, 60, and 61 **4.** 8, 45, and 53

b. Determine whether multiples of Pythagorean triples are also Pythagorean triples. Do fractional multiples of Pythagorean triples satisfy the Pythagorean Theorem? Explain how you arrived at your conclusions.

45. ***Pythagorean Relationships***
Complete the statement using the symbol $<$, $=$, or $>$. Explain how you determined which symbol to use.

a. For an acute triangle with side c the longest side, $a^2 + b^2 \square c^2$.

b. For a right triangle with side c the longest side, $a^2 + b^2 \square c^2$.

c. For an obtuse triangle with side c the longest side, $a^2 + b^2 \square c^2$.

Section 6.5 Introduction to Trigonometry

Geometry of a Right Triangle

Using the Pythagorean Theorem, given two of the sides of the triangle, it is possible to determine the length of the third side. In some situations, however, it may not be practical or possible to know two of the sides of a right triangle.

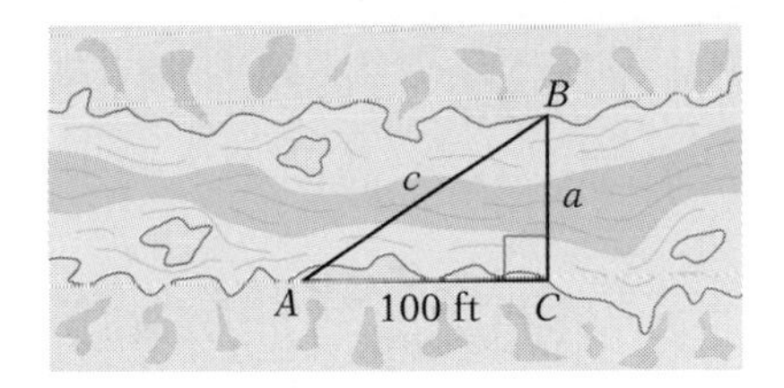

Consider, for example, the problem of an engineer trying to determine the distance across a ravine so that a bridge can be built connecting the two sides. Look at the triangle drawn in the diagram at the left. It would be fairly easy to measure the length of the side of the triangle that is on the land (100 feet), but measuring the lengths of sides a and c cannot be accomplished easily because of the ravine.

To solve problems similar to this one, relationships between the sides and angles of a triangle were explored. This study became known as *trigonometry,* which comes from two Greek words meaning "triangle measurement."

TAKE NOTE
An isosceles triangle is a triangle with two equal sides and two equal angles.

An isosceles *right* triangle is a special kind of isosceles triangle.

We begin this study with a special triangle called an **isosceles right triangle**. In this triangle, the two legs are equal and the angles opposite the two legs are equal. The measure of the equal angles is 45°. For this reason, an isosceles right triangle is also called a **45°–45°–90° triangle**.

We can use the Pythagorean Theorem to find the hypotenuse of a 45°–45°–90° triangle in terms of the legs.

In an isosceles right triangle, $a = b$. $\quad c^2 = a^2 + b^2$

Substitute a for b in the Pythagorean Theorem. $\quad c^2 = a^2 + a^2$

Solve for c. $\quad c^2 = 2a^2$

Use the Principal Square Root Property. $\quad c = \sqrt{2a^2}$

$$c = a\sqrt{2}$$

Therefore, the hypotenuse (c) is equal to $\sqrt{2}$ times the length of a leg (a).

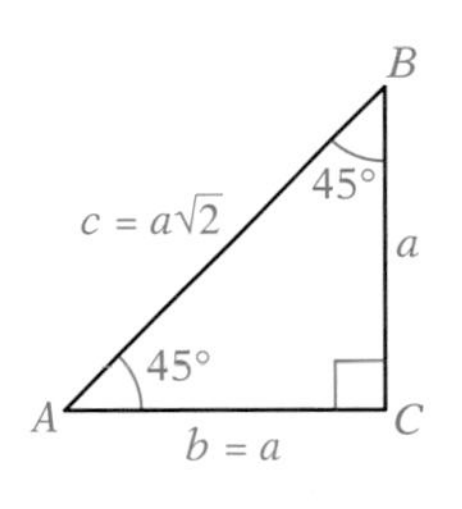

Relationship Among the Sides of an Isosceles Right Triangle

For any 45°–45°–90° triangle, the hypotenuse is $\sqrt{2}$ times the length of a leg.

Example 1 The hypotenuse of a 45°–45°–90° triangle is 6 centimeters. Find the length of a side opposite one of the 45° angles.

Solution

$a\sqrt{2} = c$ • The hypotenuse is $\sqrt{2}$ times the length of a leg.

$a\sqrt{2} = 6$ • The hypotenuse is 6 centimeters.

$a = \frac{6}{\sqrt{2}}$ • Solve for a.

$a = \frac{6\sqrt{2}}{(\sqrt{2})^2}$ • Multiply the fraction by $\frac{\sqrt{2}}{\sqrt{2}}$.

$a = 3\sqrt{2}$ • Simplify.

The length of a side opposite one of the 45° angles is $3\sqrt{2}$ centimeters.

SUGGESTED ACTIVITY
See the *Instructor's Resource Manual* for an activity involving 45°–45°–90° triangles.

You-Try-It 1 The hypotenuse of a 45°–45°–90° triangle is 8 inches. Find the length of a side opposite one of the 45° angles.

Solution See page S29.

Example 2 In an isosceles right triangle, the length of a side opposite a 45° angle is $5\sqrt{2}$ meters. Find the length of the hypotenuse.

Solution

$a\sqrt{2} = c$ • The hypotenuse is $\sqrt{2}$ times the length of a leg.

$(5\sqrt{2})\sqrt{2} = c$ • The length of a leg is $5\sqrt{2}$.

$5(\sqrt{2})^2 = c$ • Simplify the left side of the equation.

$5(2) = c$

$10 = c$

The hypotenuse is 10 meters.

You-Try-It 2 The length of a side opposite a 45° triangle in a right triangle is $4\sqrt{2}$ inches. Find the length of the hypotenuse.

Solution See page S29.

Example 3 A rope attached to the top of a flagpole 50 feet high is anchored to the ground. The rope makes a 45° angle with the ground. How long is the rope? Round to the nearest tenth.

State the goal.

The goal is to find the length of a rope that stretches from the top of a 50-foot flagpole to the ground and makes a 45° angle with the ground.

Describe a strategy.

This is a 45°–45°–90° triangle. The height of the flagpole is the length of a leg opposite the 45° angle. The length of the rope is the hypotenuse. Use the relationship among the sides of an isosceles right triangle: the hypotenuse is $\sqrt{2}$ times the length of a leg.

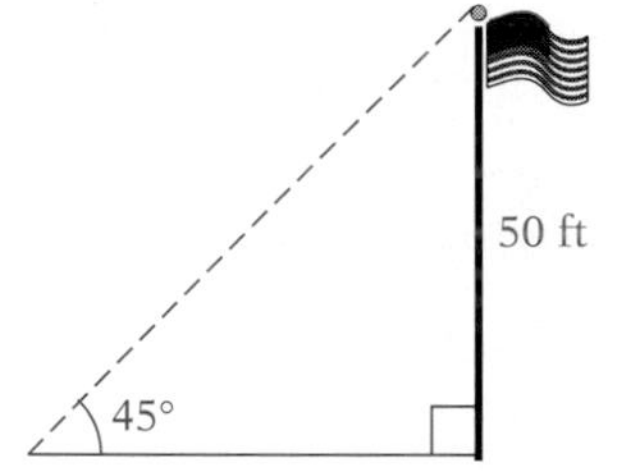

Solve the problem.

$a\sqrt{2} = c$ • The hypotenuse is $\sqrt{2}$ times the length of a leg.

$50\sqrt{2} = c$ • The length of a leg is 50 feet.

$70.7 \approx c$

The rope is 70.7 feet long.

Check your work.

√

You-Try-It 3 The Bedford village green is in the shape of a square that measures 65 meters on each side. Along the diagonal of the square is a path through the park. What is the length of the path through the park? Round to the nearest tenth.

Solution See page S30.

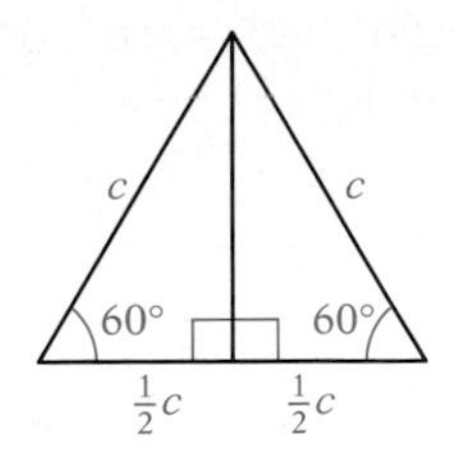

A right triangle in which the two acute angles measure 30° and 60° is called a **30°–60°–90° triangle.** If two 30°–60°–90° triangles, each with a hypotenuse c, are positioned so that the longer leg of each triangle lies on the same line segment, then an equilateral triangle is formed, and the shorter leg of each triangle is $\frac{1}{2}c$. See the diagram at the left.

We can use the Pythagorean Theorem to find the length of the longer leg.

Let b = the length of the longer leg.	$a^2 + b^2 = c^2$
Since $a = \frac{1}{2}c$, $2a = c$. Substitute $2a$ for c.	$a^2 + b^2 = (2a)^2$
	$a^2 + b^2 = 4a^2$
Solve for b.	$b^2 = 3a^2$
Use the Principal Square Root Property.	$b = \sqrt{3a^2}$
	$b = a\sqrt{3}$

Therefore, the longer leg (b) is equal to $\sqrt{3}$ times the length of the shorter leg (a).

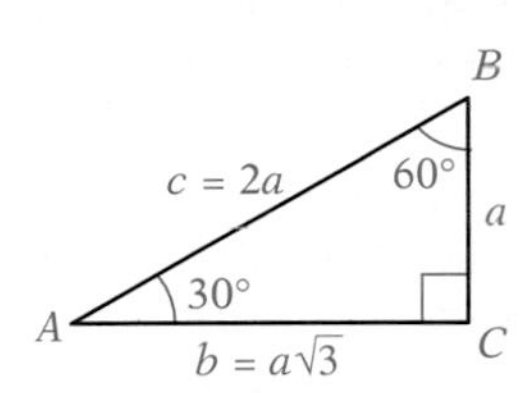

Relationships Among the Sides of a 30°–60°–90° Triangle

For any 30°–60°–90° triangle, the hypotenuse is twice the shorter leg, or the leg opposite the 30° angle. The longer leg, or the leg opposite the 60° angle, is $\sqrt{3}$ times the length of the shorter leg.

Example 4 The hypotenuse in a 30°–60°–90° triangle is 6 centimeters. Find the length of the leg opposite the 30° angle.

Solution The shorter leg of a 30°–60°–90° triangle is opposite the 30° angle.

$2a = c$	• Let a = the length of the shorter leg.
$2a = 6$	• The hypotenuse is twice the shorter leg.
$a = 3$	• Solve for a.

The length of the leg opposite the 30° angle is 3 centimeters.

SUGGESTED ACTIVITY
See the *Instructor's Resource Manual* for an activity involving 30°–60°–90° triangles.

You-Try-It 4 The leg opposite the 30° angle in a 30°–60°–90° triangle measures 5 inches. Find the lengths of the other two sides of the triangle.

Solution See page S30.

Trigonometric Functions of an Acute Angle

Given triangles ABC and $A'B'C'$, in which $\angle A = \angle A'$, $\angle B = \angle B'$, and $\angle C = \angle C'$, then triangle ABC is similar to $A'B'C'$.

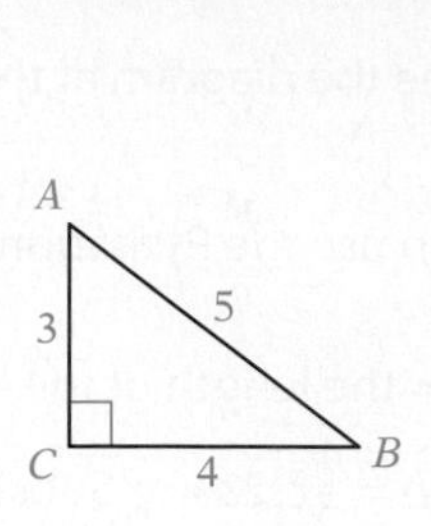

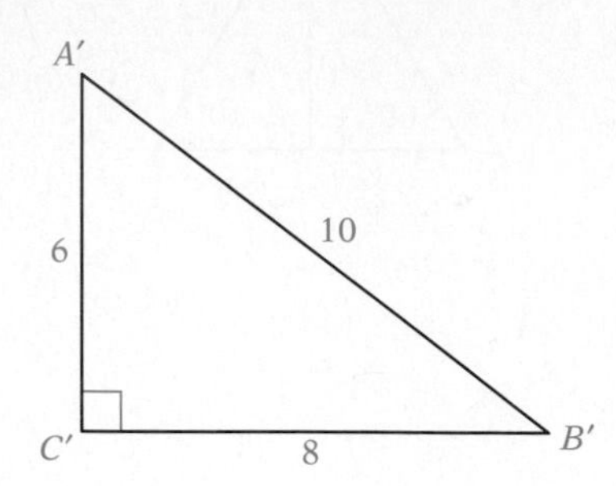

The lengths of the corresponding sides of these triangles are different; however, the *ratios* of the lengths of the corresponding sides are equal.

For example,

$$\frac{a}{c} = \frac{4}{5} \text{ and } \frac{a'}{c'} = \frac{8}{10} = \frac{4}{5}. \quad \text{Therefore, } \frac{a}{c} = \frac{a'}{c'}.$$

$$\frac{b}{a} = \frac{3}{4} \text{ and } \frac{b'}{a'} = \frac{6}{8} = \frac{3}{4}. \quad \text{Therefore, } \frac{b}{a} = \frac{b'}{a'}.$$

These ratios remain constant regardless of the lengths of the sides of the triangles. They depend only on the sizes of the acute angles in any two similar triangles.

Using the three sides, a, b, and c, of a right triangle, three ratios can be written. These ratios are used to define three trigonometric functions: sine, cosine, and tangent (abbreviated sin, cos, and tan, respectively).

In defining these trigonometric functions, "opposite" is used to mean the length of the side opposite the given angle, and "adjacent" is used to mean the length of the side adjacent to (next to) the given angle.

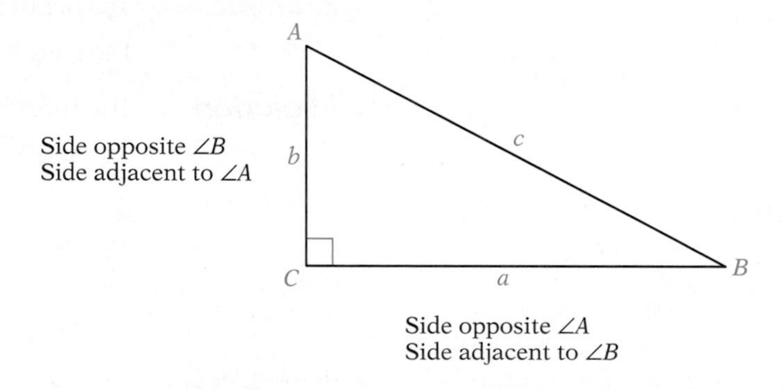

TAKE NOTE
The hypotenuse, c, is never referred to as the side opposite angle C, and it is never referred to as the side adjacent to angle A or angle B. It is referred to only as the hypotenuse.

The Trigonometric Functions

$$\sin A = \frac{\text{opposite}}{\text{hypotenuse}} = \frac{a}{c} \qquad \sin B = \frac{\text{opposite}}{\text{hypotenuse}} = \frac{b}{c}$$

$$\cos A = \frac{\text{adjacent}}{\text{hypotenuse}} = \frac{b}{c} \qquad \cos B = \frac{\text{adjacent}}{\text{hypotenuse}} = \frac{a}{c}$$

$$\tan A = \frac{\text{opposite}}{\text{adjacent}} = \frac{a}{b} \qquad \tan B = \frac{\text{opposite}}{\text{adjacent}} = \frac{b}{a}$$

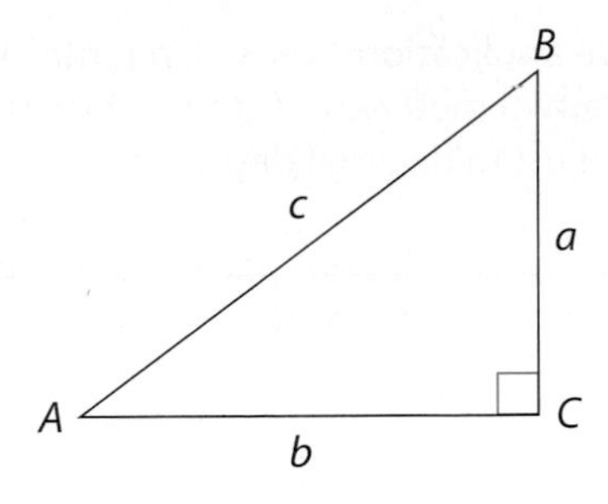

Question: Which trigonometric function is defined by each of the following ratios?[1]

a. $\frac{\text{adjacent}}{\text{hypotenuse}}$ **b.** $\frac{\text{opposite}}{\text{adjacent}}$ **c.** $\frac{\text{opposite}}{\text{hypotenuse}}$

For the right triangle shown at the left, which side is:

d. adjacent to $\angle A$ **e.** opposite $\angle A$

f. adjacent to $\angle B$ **g.** opposite $\angle B$

TAKE NOTE
For a trigonometric function, the input is an angle and the output is a number. Recall that the output of a function is called the value of the function.

For each angle, θ, of an acute triangle, there is one and only one number associated with $\sin\theta$. The same is true for $\cos\theta$ and $\tan\theta$. The values of these functions for 30°, 45°, and 60° angles can be found from the relationships we found earlier for these triangles.

Use the relationships among the sides of a 45°–45°–90° triangle to find the values of the three trigonometric functions of a 45° angle.

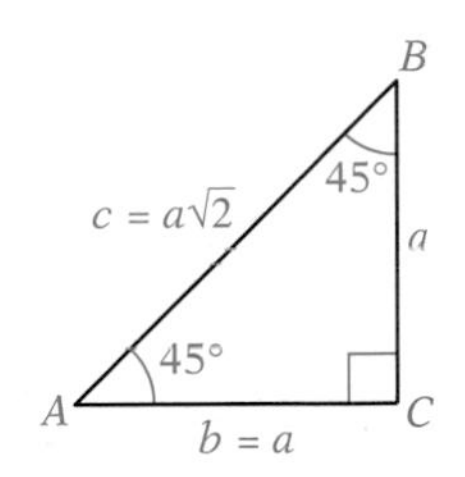

$$\sin 45^\circ = \frac{\text{opposite}}{\text{hypotenuse}} = \frac{a}{a\sqrt{2}} = \frac{1}{\sqrt{2}} = \frac{\sqrt{2}}{2} \approx 0.70710678$$

$$\cos 45^\circ = \frac{\text{adjacent}}{\text{hypotenuse}} = \frac{a}{a\sqrt{2}} = \frac{1}{\sqrt{2}} = \frac{\sqrt{2}}{2} \approx 0.70710678$$

$$\tan 45^\circ = \frac{\text{opposite}}{\text{adjacent}} = \frac{a}{b} = \frac{a}{a} = 1$$

Use the relationships among the sides of a 30°–60°–90° triangle to find the values of the three trigonometric functions of a 30° and a 60° angle.

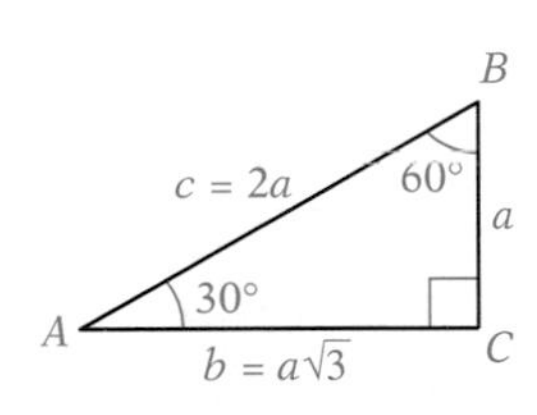

$$\sin 30^\circ = \frac{a}{2a} = \frac{1}{2} = 0.5 \qquad \sin 60^\circ = \frac{a\sqrt{3}}{2a} = \frac{\sqrt{3}}{2} \approx 0.8660254$$

$$\cos 30^\circ = \frac{a\sqrt{3}}{2a} = \frac{\sqrt{3}}{2} \approx 0.8660254 \qquad \cos 60^\circ = \frac{a}{2a} = \frac{1}{2} = 0.5$$

$$\tan 30^\circ = \frac{a}{a\sqrt{3}} = \frac{1}{\sqrt{3}} = \frac{\sqrt{3}}{3} \approx 0.57735027 \qquad \tan 60^\circ = \frac{a\sqrt{3}}{a} = \sqrt{3} \approx 1.7320508$$

SUGGESTED ACTIVITY
Have students work in groups to discover these trigonometric functions of 45°, 30°, and 60° angles rather than presenting them with the values.

The values of trigonometric functions of most angles between 0° and 90° cannot be found using geometric methods. To find the values of the trigonometric functions of these angles, a calculator is used.

→ Use a calculator to find cos 16°.

Put the calculator in the degree mode.
Press the cos key and enter 16.
Press ENTER. cos 16° = 0.9612617

Despite the fact that the values of these trigonometric functions are approximate, it is customary to use the equals sign rather than the approximately equal sign when writing these functions.

1. **a.** Cosine (abbreviated cos); **b.** tangent (abbreviated tan); **c.** sine (abbreviated sin); **d.** *b*; **e.** *a*; **f.** *a*; **g.** *b*

Although 1° is a fairly small angle, there are applications for which it may be necessary to have parts of a degree. There are two methods of doing this: using the Degree–Minutes–Seconds system (DMS) or using decimal degrees.

In the DMS system, a degree is subdivided into 60 equal parts called **minutes**. One minute is subdivided into 60 equal parts called **seconds**.

$$1 \text{ degree } (1^\circ) = 60 \text{ minutes } (60')$$

$$1 \text{ minute } (1') = 60 \text{ seconds } (60'')$$

The angle measure 62 degrees, 53 minutes, 14 seconds is written $62^\circ 53' 14''$.

Using decimal degrees, a degree is subdivided into smaller units using decimals. For example, 28.53° equals 28° plus $\frac{53}{100}$ of a degree.

Question: Is the angle measure written in the DMS system or in decimal degrees?[2]

a. 36.78° **b.** $41^\circ 29' 50''$

The examples which follow illustrate how to convert measurement between the DMS system and decimal degrees.

CALCULATOR NOTE
The TI-83 graphing calculator can be used to convert from the DMS system to decimal degrees. To convert $36^\circ 28' 45''$, enter:

36 [2nd] [ANGLE] 1

28 [2nd] [ANGLE] 2

45 [2nd] [ALPHA] ″

[ENTER]

(The ″ is above the + key.) The display will read 36.47916667.

→ Convert $20^\circ 14'$ to decimal degrees. Round to the nearest thousandth of a degree.

Since $1' = \frac{1^\circ}{60}$, $14' = \frac{14^\circ}{60}$.

$$20^\circ 14' = 20^\circ + \frac{14^\circ}{60}$$
$$\approx 20^\circ + 0.233^\circ$$
$$\approx 20.233^\circ$$

Example 5 Convert $43^\circ 28'$ to decimal degrees. Round to the nearest thousandth of a degree.

Solution

$$43^\circ 28' = 43^\circ + \frac{28^\circ}{60}$$
$$\approx 43^\circ + 0.467^\circ$$
$$\approx 43.467^\circ$$

• Since $1' = \frac{1^\circ}{60}$, $28' = \frac{28^\circ}{60}$.

You-Try-It 5 Convert $78^\circ 52'$ to decimal degrees. Round to the nearest thousandth of a degree.

Solution See page S30.

---

2. **a.** 36.78° is written in decimal degrees. **b.** $41^\circ 29' 50''$ is written in the DMS system.

CALCULATOR NOTE
The TI-83 graphing calculator can be used to convert from decimal degrees to the DMS system. To convert 55.215°, enter:

55.215 [2nd] [ANGLE] 1

[2nd] [ANGLE] 4 [ENTER]

The display will read 55°12′54″.

→ Convert 32.82° to degrees, minutes, and seconds.

Let m = the number of minutes in 0.82°.

$$\frac{m}{60} = 0.82$$
$$m = 49.2$$
$$32.82° = 32°49.2'$$

Let s = the number of seconds in 0.2′.

$$\frac{s}{60} = 0.2$$
$$s = 12$$
$$32.82° = 32°49'12''$$

Example 6 Convert 96.43° to degrees, minutes, and seconds.

Solution

$\frac{m}{60} = 0.43$ • Let m = the number of minutes in 0.43°.

$m = 25.8$ • 96.43° = 96°25.8′

$\frac{s}{60} = 0.8$ • Let s = the number of seconds in 0.8′.

$s = 48$

$96.43° = 96°25'48''$

You-Try-It 6 Convert 135.64° to degrees, minutes, and seconds.

Solution See page S30.

TAKE NOTE
Remember to put the calculator in the degree mode.

Example 7 Use a calculator to find sin 43.8° to the nearest ten-thousandth.

Solution $\sin 43.8° = 0.6921$ • The calculator must be in the degree mode. Press the sin key. Enter 43.8.

You-Try-It 7 Use a calculator to find tan 37.1° to the nearest ten-thousandth.

Solution See page S30.

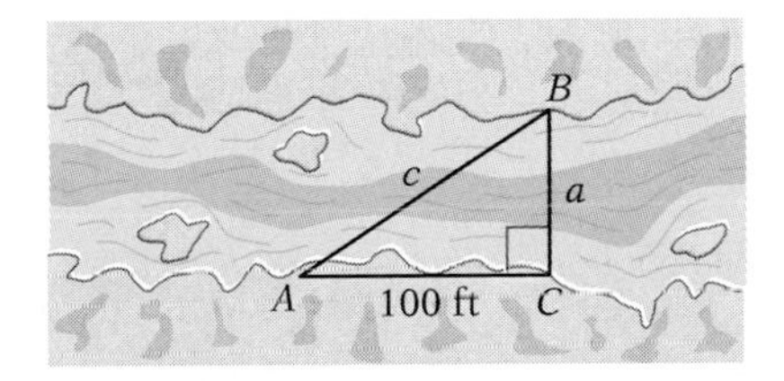

Using trigonometry, the engineer mentioned at the beginning of this section could determine the distance across the ravine after determining the measure of $\angle A$. Suppose the engineer measures the angle as 33.8°. Now the engineer would ask, "Which trigonometric function involves the side opposite an angle and the side adjacent to that angle?" Knowing that the tangent function is the required function, the engineer could write and solve the equation $\tan 33.8° = \frac{a}{100}$.

$$\tan 33.8° = \frac{a}{100}$$

$100(\tan 33.8°) = a$ • Multiply each side of the equation by 100.

$66.9 \approx a$ • Use a calculator to find tan 33.8°. Multiply the result in the display by 100.

The length of the bridge must be approximately 66.9 feet.

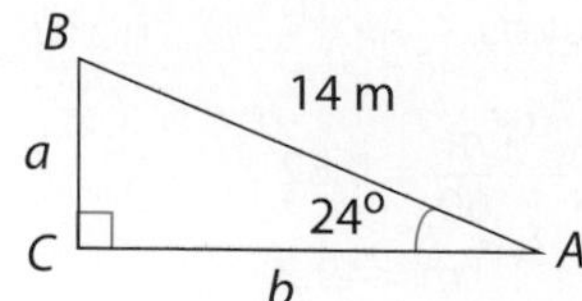

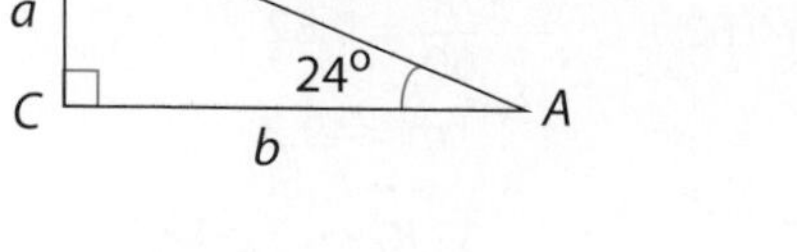

Example 8 For the right triangle shown at the left, find the length of side a. Round to the nearest hundredth.

Solution We are given the measure of $\angle A$ and the hypotenuse. We want to find the length of side a. Side a is opposite $\angle A$. The trigonometric function that involves the side opposite an angle and the hypotenuse is the sine function.

$$\sin A = \frac{\text{side opposite } A}{\text{hypotenuse}}$$

$$\sin 24^\circ = \frac{a}{14}$$

$$14(\sin 24^\circ) = a$$ • Multiply each side by 14.

$$5.69 \approx a$$ • Use a calculator to find sin 24°. Multiply the result in the display by 14.

The length of side a is approximately 5.69 meters.

A
12 ft
b
48°
C
B
a

You-Try-It 8 For the right triangle shown at the left, find the length of side a. Round to the nearest hundredth.

Solution See page S30.

Inverse Trigonometric Functions

Sometimes it is necessary to find one of the acute angles in a right triangle. For instance, suppose it is necessary to find $\angle A$ in the figure at the right. Because the side adjacent to $\angle A$ is known and the hypotenuse is known, we can write:

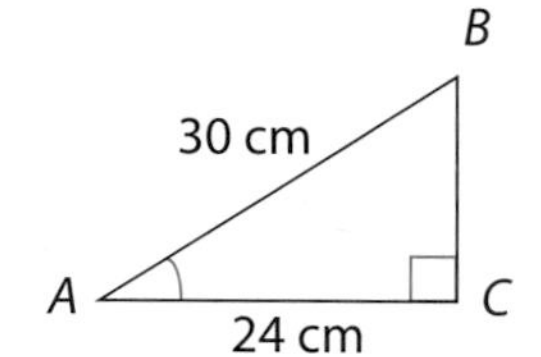

$$\cos A = \frac{\text{side adjacent to } A}{\text{hypotenuse}}$$

$$\cos A = \frac{24}{30}$$

$$\cos A = 0.8$$

The solution of this equation is the angle whose cosine is 0.8. This angle can be found by using the $\cos^{-1}$ key on a calculator.

$$\cos^{-1}(0.8) = 36.86989765$$

To the nearest tenth, $\angle A$ is 36.9°.

The function $\cos^{-1}$ is called the *inverse cosine function.*

Definition of the Inverse Sine, Cosine, and Tangent Functions

For $0^\circ < x < 90^\circ$:

$y = \sin^{-1} x$ can be read "y is the angle whose sine is x."

$y = \cos^{-1} x$ can be read "y is the angle whose cosine is x."

$y = \tan^{-1} x$ can be read "y is the angle whose tangent is x."

Note that $\sin^{-1} x$ is used to denote the inverse of the sine function. It is not the reciprocal of $\sin x$ but the notation used for its inverse. The same is true for $\cos^{-1}$ and $\tan^{-1}$.

The expression $y = \sin^{-1} x$ is sometimes written $y = \arcsin x$. The two expressions are equivalent. The expressions $y = \cos^{-1} x$ and $y = \arccos x$ are equivalent, as are $y = \tan^{-1} x$ and $y = \arctan x$.

To find an inverse function on a calculator, usually the INV or 2nd key is pressed prior to pushing the function key. Some calculators have $\sin^{-1}$, $\cos^{-1}$, and $\tan^{-1}$ keys. Consult the manual of instruction for your calculator.

Example 9 Use a calculator to find $\sin^{-1} 0.9171$. Round to the nearest tenth.

Solution $\sin^{-1} 0.9171 = 66.5°$

• The calculator must be in the degree mode. Press the keys for the inverse sine function. Enter 0.9171.

You-Try-It 9 Use a calculator to find $\tan^{-1} 0.3165$. Round to the nearest tenth.

Solution See page S30.

→ Find $\tan^{-1} 1$.

The value of $\tan^{-1} 1$ is an angle θ such that $\tan \theta = 1$. Note on page 445 that $\tan 45° = 1$.

$\tan^{-1} 1 = 45°$

Example 10 Find $\sin^{-1} \frac{\sqrt{2}}{2}$.

Solution $\sin^{-1} \frac{\sqrt{2}}{2} = 45°$

• Find an angle θ such that $\sin \theta = \frac{\sqrt{2}}{2}$.

Note on page 445 that $\sin 45° = \frac{\sqrt{2}}{2}$.

You-Try-It 10 Find $\cos^{-1} \frac{\sqrt{3}}{2}$.

Solution See page S30.

→ Given $\sin \theta = 0.7239$, find θ. Use a calculator. Round to the nearest tenth.

The calculator must be in the degree mode.
This is equivalent to finding $\sin^{-1} 0.7239$.

$\theta = 46.4°$

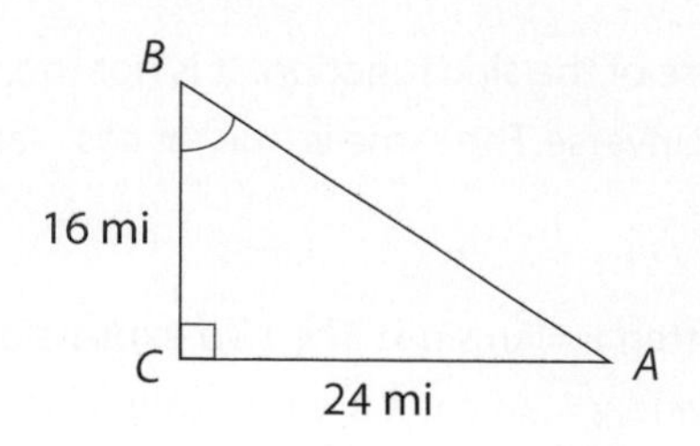

Example 11 For the right triangle shown at the left, find $\angle B$. Round to the nearest tenth.

Solution We want to find the measure of $\angle B$, and we are given the length of the side opposite $\angle B$ and the side adjacent to $\angle B$. The trigonometric function that involves the side opposite an angle and the side adjacent to that angle is the tangent function.

$$\tan B = \frac{\text{side opposite } B}{\text{side adjacent to } B}$$

$$\tan B = \frac{24}{16}$$

$$\tan B = 1.5$$

$$B \approx 56.3^0$$ • Use the $\tan^{-1}$ key on a calculator.

The measure of $\angle B$ is approximately 56.3°.

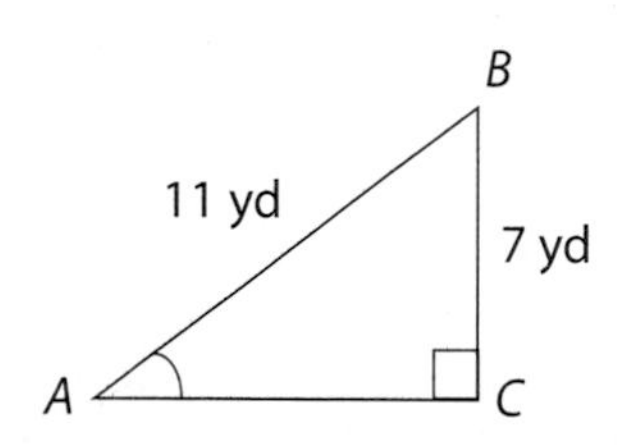

You-Try-It 11 For the right triangle shown at the left, find $\angle A$. Round to the nearest tenth.

Solution See page S30.

Solving Right Triangles

To solve a right triangle means to use the information given about it to find the unknown sides and angles. A right triangle can be solved when two sides are known or when one acute angle and one side of the triangle are known. In either case, begin by drawing a diagram of the right triangle and labeling the given parts.

In Example 12 and You-Try-It 12, two sides of the triangle are given. In Example 13 and You-Try-It 13, one acute angle and one side are given.

Example 12 Solve right triangle *ABC* given $b = 12$ centimeters and $c = 15$ centimeters. Round to the nearest whole number.

Solution

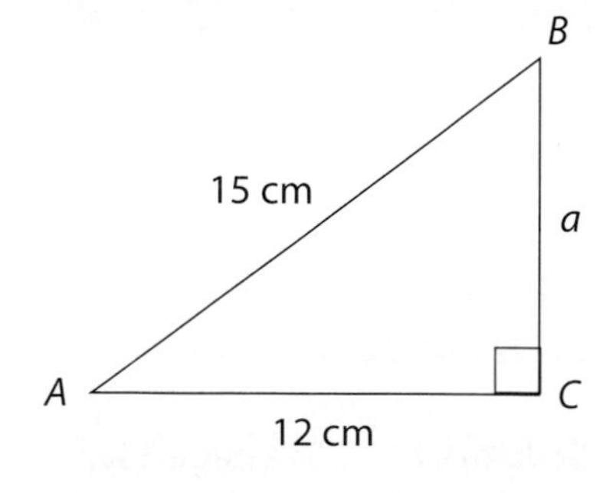

• Draw a diagram, labeling the given parts. Remember that $\angle C$ is always the right angle.

$$a^2 + b^2 = c^2$$
$$a^2 + 12^2 = 15^2$$
$$a^2 + 144 = 225$$
$$a^2 = 81$$
$$a = 9$$

• Use the Pythagorean Theorem to find the third side.

$\sin B = \dfrac{\text{side opposite } B}{\text{hypotenuse}}$ • Write a trigonometric function that relates the two given sides and one unknown angle. Solve for the unknown angle by finding $\sin^{-1} 0.8$.

$\sin B = \dfrac{12}{15}$

$\sin B = 0.8$

$B \approx 53^\circ$

$\angle A = 90^\circ - 53^\circ$ • Angle A and angle B are complementary angles.

$= 37^\circ$

Side $a = 9$ centimeters, $\angle A = 37^\circ$, and $\angle B = 53^\circ$.

You-Try-It 12 Solve right triangle ABC given $a = 8$ centimeters and $b = 10$ centimeters. Round to the nearest whole number.

Solution See page S31.

Example 13 Solve right triangle ABC given $c = 24$ meters and $\angle A = 35^\circ$. Round to the nearest whole number.

Solution

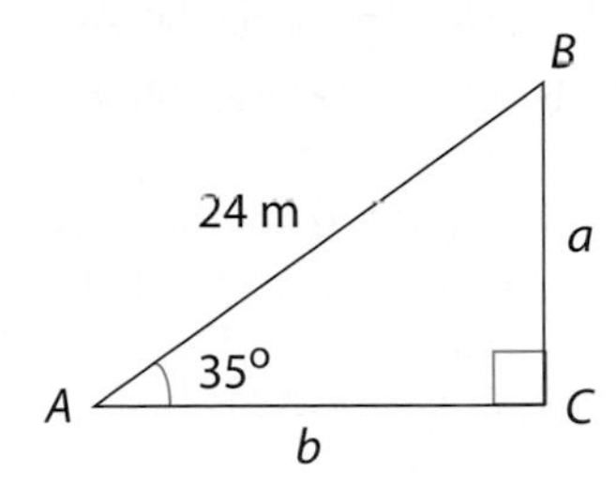

• Draw a diagram, labeling the given parts. Remember that $\angle C$ is always the right angle.

$\angle B = 90^\circ - 35^\circ$ • Find $\angle B$, the complement of $\angle A$.

$= 55^\circ$

$\sin A = \dfrac{\text{side opposite } A}{\text{hypotenuse}}$ • Write a trigonometric function that relates the given angle and the given side to one of the unknown sides.

$\sin 35^\circ = \dfrac{a}{24}$

$24(\sin 35^\circ) = a$ Solve the equation for the unknown side.

$14 \approx a$

$\cos A = \dfrac{\text{side adjacent to } A}{\text{hypotenuse}}$ • Write a trigonometric function that relates the given angle and the given side to the other unknown side.

$\cos 35^\circ = \dfrac{b}{24}$

$24(\cos 35^\circ) = b$ Solve the equation for the unknown side.

$20 \approx b$

$\angle B = 55^\circ$, side $a = 14$ meters, and side $b = 20$ meters.

You-Try-It 13 Solve right triangle ABC given $b = 18$ centimeters and $\angle B = 25^{\circ}$. Round to the nearest whole number.

Solution See page S31.

Angles of Elevation and Depression

Solving right triangles is necessary in a variety of situations. One application, called **line of sight problems**, concerns an observer looking at an object.

Angles of elevation and depression are measured with respect to a horizontal line. If the object being sighted is above the observer, the acute angle formed by the line of sight and the horizontal line is an **angle of elevation**. If the object being sighted is below the observer, the acute angle formed by the line of sight and the horizontal line is an **angle of depression**.

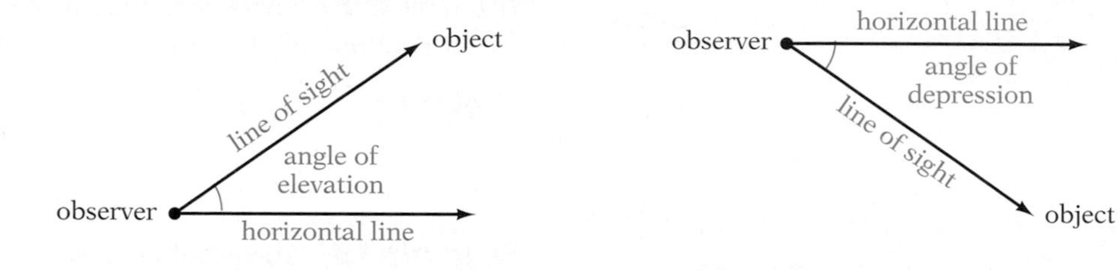

Example 14 The angle of elevation of the top of a flagpole 62 feet away is 34°. Find the height of the flagpole. Round to the nearest tenth.

State the goal.

The goal is to find the height of the flagpole given the horizontal distance to the flagpole and the angle from the horizontal to the top of the flagpole.

Describe a strategy.

Draw a diagram. To find the height, write a trigonometric function that relates the given information and the unknown side of the triangle.

Solve the problem.

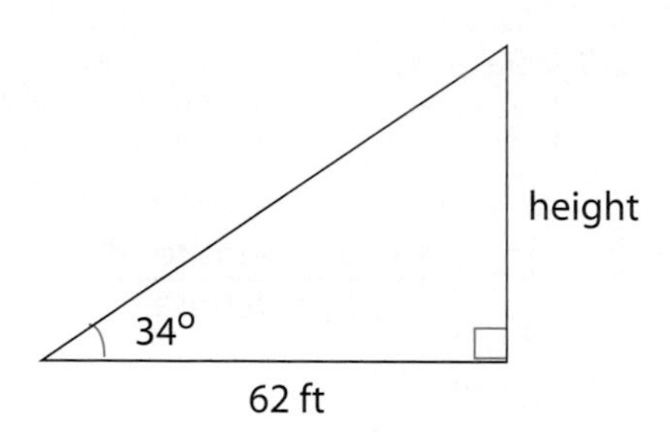

$$\tan 34^{\circ} = \frac{h}{62}$$

$$62(\tan 34^{\circ}) = h$$

$$41.8 \approx h$$

The height of the flagpole is 41.8 feet.

Check your work.

√

You-Try-It 14 From the top of a lighthouse that is 20 meters high, the angle of depression of a boat on the water is 25°. How far is the boat from the lighthouse? Round to the nearest tenth.

Solution See page S31.

6.5 EXERCISES

Topics for Discussion

1. Explain how to derive the formula for the length of the hypotenuse in relation to the length of a leg in a 45°–45°–90° triangle.
Use the Pythagorean Theorem, substituting a for b and solving for c.

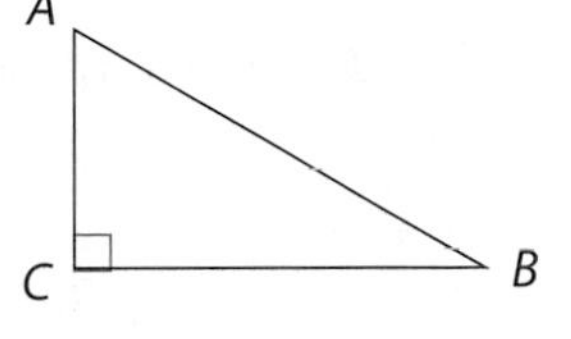

2. Name the sides of the 30°–60°–90° triangle shown at the right in order from longest to shortest.
side c, side a, side b

3. Explain the difference between angle measure written in the DMS system and decimal degrees.
Answers will vary.

4. For the right triangle at the right, use the sides a, b, and c to:

a. name the ratio for sin A.
b. name the ratio for sin B.
c. name the ratio for cos A.
d. name the ratio for cos B.
e. name the ratio for tan A.
f. name the ratio for tan B.

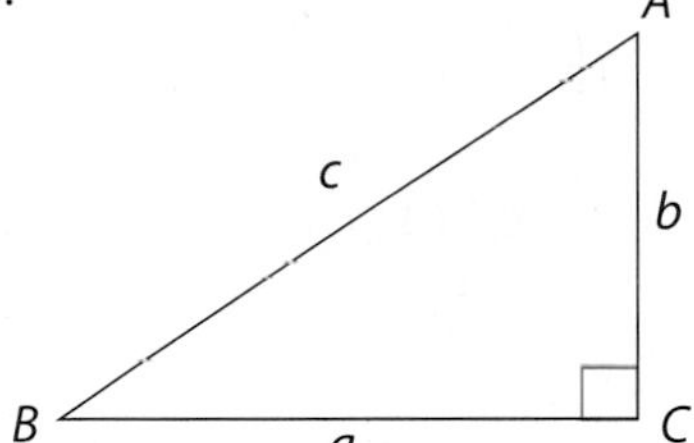

a. $\frac{a}{c}$ **b.** $\frac{b}{c}$ **c.** $\frac{b}{c}$ **d.** $\frac{a}{c}$ **e.** $\frac{a}{b}$ **f.** $\frac{b}{a}$

5. Explain the meaning of the notation $y = \sin^{-1} x$, $y = \cos^{-1} x$, and $y = \tan^{-1} x$.
Answers will vary.

Geometry of a Right Triangle

6. The hypotenuse of an isosceles right triangle is 30 centimeters. Find the length of a side opposite one of the 45° angles.
$15\sqrt{2}$ cm

7. The hypotenuse of a 45°–45°–90° triangle is 5 feet. Find the length of a side opposite one of the 45° angles.
$\frac{5\sqrt{2}}{2}$ ft

8. Find the hypotenuse of a 45°–45°–90° triangle if the length of a side opposite one of the 45° angles is 6 inches.
$6\sqrt{2}$ in.

9. In a 45°–45°–90° triangle, the length of a side opposite one of the 45° angles is 12 meters. Find the hypotenuse.
$12\sqrt{2}$ m

10. Find the hypotenuse of a 30°–60°–90° triangle if the side opposite the 30° angle is 4 feet.
8 ft

11. In a 30°–60°–90° triangle, the side opposite the 30° angle is 7 inches. Find the hypotenuse.
14 in.

12. In a 30°–60°–90° triangle, the side opposite the 60° angle is 20 centimeters. Find the hypotenuse.
$\frac{40\sqrt{3}}{3}$ cm

13. Find the hypotenuse of a 30°–60°–90° triangle if the side opposite the 60° angle is $5\sqrt{3}$ yards.
10 yd

14. The hypotenuse in a 30°–60°–90° triangle is 16 meters. Find the lengths of the other two sides of the triangle.
8 m and $8\sqrt{3}$ m

15. The hypotenuse in a 30°–60°–90° triangle is 8 inches. Find the lengths of the other two sides of the triangle.
4 in. and $4\sqrt{3}$ in.

16. An 18-foot ladder resting against a house makes a 45° angle with the ground. The ladder just reaches a window in the second story of the house. How high is the window from the ground? Round to the nearest tenth.
12.7 ft

17. A rope attached to the top of a tent that has a height of 6 feet is anchored to the ground. The rope makes a 45° angle with the ground. How long is the rope? Round to the nearest tenth.
8.5 ft

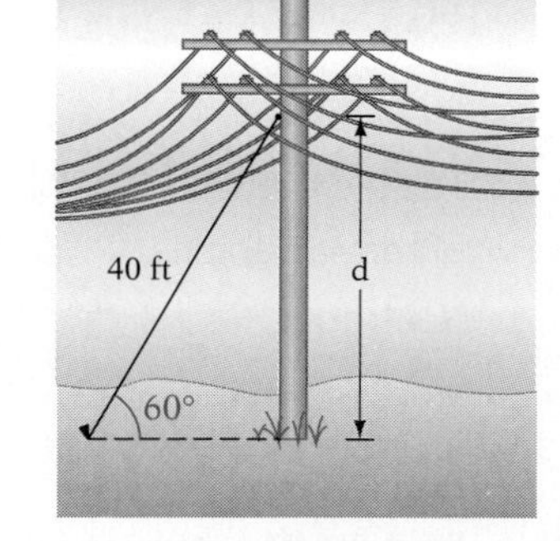

18. A 40-foot guy wire is attached to a telephone pole and makes an angle of 60° with the ground. Find the distance from the base of the pole to the point on the pole where the guy wire is attached. Round to the nearest tenth.
34.6 ft

19. The Wilmington recreational park is in the shape of a square. Along the diagonal of the square is a path through the park. The path is 80 feet long. Find the area of the park.
3200 ft^2

20. The distance from the first floor of a home to the second floor is 10 feet. A stairway from the first to the second floor makes a 30° angle with the floor. Find the length of the stairway.
20 ft

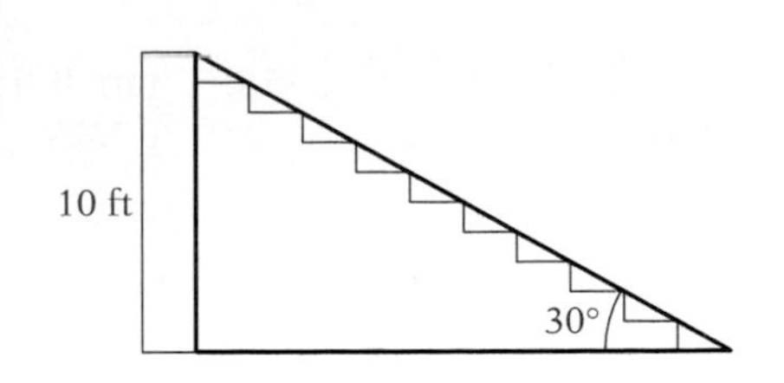

Trigonometric Functions of an Acute Angle

Convert to decimal degrees. Round to the nearest thousandth of a degree.

21. $47^\circ52'$
47.867°

22. $68^\circ30'$
68.5°

23. $76^\circ19'$
76.317°

24. $50^\circ28'$
50.467°

25. $8^\circ35'$
8.583°

26. $21^\circ8'$
21.133°

27. $4^\circ43'$
4.717°

28. $86^\circ53'$
86.883°

29. $38^\circ17'$
38.283°

30. $16^\circ24'$
16.4°

31. $70^\circ36'$
70.6°

32. $25^\circ40'$
25.667°

Convert to degrees, minutes, and seconds.

33. 33.76°
$33^\circ45'36''$

34. 54.52°
$54^\circ31'12''$

35. 76.08°
$76^\circ4'48''$

36. 9.31°
$9^\circ18'36''$

37. 40.3°
$40^\circ18'$

38. 76.25°
$76^\circ15'$

39. 52.46°
$52^\circ27'36''$

40. 17.89°
$17^\circ53'24''$

41. 82.345°
$82^\circ20'42''$

42. 28.215°
$28^\circ12'54''$

43. 43.335°
$43^\circ20'6''$

44. 66.425°
$66^\circ25'30''$

Use your calculator to find the value of each of the following. Round to the nearest ten-thousandth.

45. $\cos 47^\circ$
0.6820

46. $\sin 62^\circ$
0.8829

47. $\tan 55^\circ$
1.4281

48. $\cos 11^\circ$
0.9816

49. $\sin 85.6^\circ$
0.9971

50. $\cos 21.9^\circ$
0.9278

51. $\tan 63.4^\circ$
1.9970

52. $\sin 7.8^\circ$
0.1357

53. $\tan 41.6^\circ$
0.8878

54. $\cos 73^\circ$
0.2924

55. $\sin 57.7^\circ$
0.8453

56. $\tan 39.2^\circ$
0.8156

57. $\sin 58.3^\circ$
0.8508

58. $\tan 35.1^\circ$
0.7028

59. $\cos 46.9^\circ$
0.6833

60. $\sin 50^\circ$
0.766

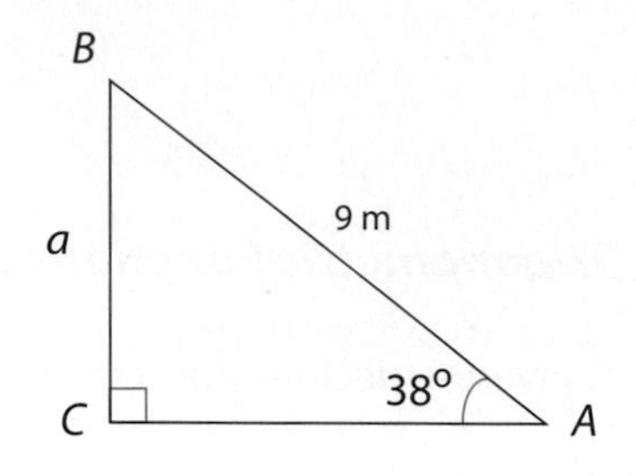

61. For right triangle *ABC*, the hypotenuse is 9 meters and $\angle A = 38^\circ$ as shown at the right. Find the length of side *a*. Round to the nearest hundredth.
5.54 m

62. For right triangle *ABC*, the hypotenuse is 15 centimeters and $\angle B = 52^\circ$. Find the length of side *a*. Round to the nearest hundredth.
9.23 cm

63. For right triangle *ABC*, the hypotenuse is 6 inches and $\angle A = 26^\circ$. Find the length of side *b*. Round to the nearest hundredth.
5.39 in.

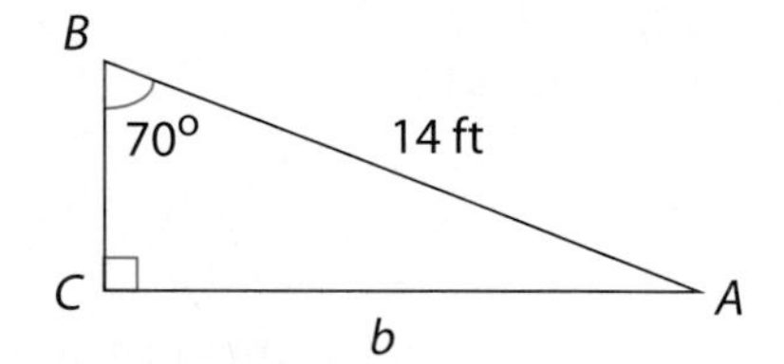

64. For right triangle *ABC*, the hypotenuse is 14 feet and $\angle B = 70^\circ$, as shown at the right. Find the length of side *b*. Round to the nearest hundredth.
13.16 ft

65. For right triangle *ABC*, side *a* is 20 centimeters and $\angle B = 41^\circ$. Find the length of the hypotenuse. Round to the nearest hundredth.
26.50 cm

66. For right triangle *ABC*, side *a* is 8 yards and $\angle A = 63^\circ$. Find the length of the hypotenuse. Round to the nearest hundredth.
8.98 yd

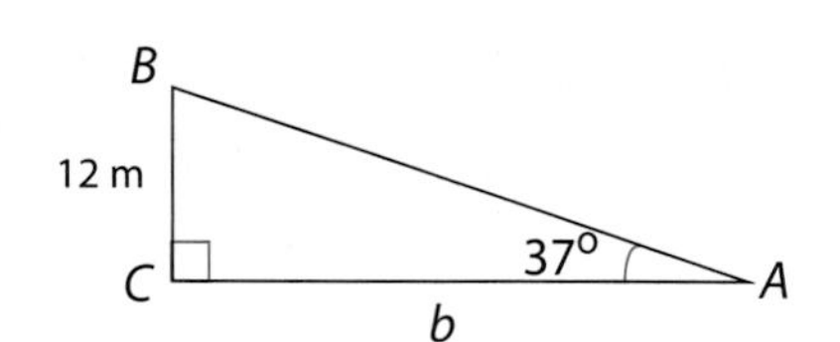

67. For right triangle *ABC*, side *a* is 12 meters and $\angle A = 37^\circ$, as shown at the right. Find the length of side *b*. Round to the nearest hundredth.
15.92 m

68. For right triangle *ABC*, side *b* is 5 miles and $\angle A = 24^\circ$. Find the length of side *a*. Round to the nearest hundredth.
2.23 mi

Inverse Trigonometric Functions

Use a calculator. Round to the nearest tenth.

69. Given $\sin \theta = 0.6239$, find θ.
38.6°

70. Given $\cos \phi = 0.9516$, find ϕ.
17.9°

71. Find $\cos^{-1} 0.7536$.
41.1°

72. Find $\sin^{-1} 0.4478$.
26.6°

73. Given $\tan \phi = 0.3899$, find ϕ.
21.3°

74. Given $\sin \theta = 0.7349$, find θ.
47.3°

75. Find $\tan^{-1} 0.7815$.
38.0°

76. Find $\cos^{-1} 0.6032$.
52.9°

77. Given $\cos \theta = 0.3007$, find θ.
72.5°

78. Given $\tan \phi = 1.588$, find ϕ.
57.8°

79. Find $\sin^{-1} 0.0105$.
0.6°

80. Find $\tan^{-1} 0.2438$.
13.7°

81. Given $\sin \beta = 0.9143$, find β.
66.1°

82. Given $\cos \theta = 0.4756$, find θ.
61.6°

83. Find $\cos^{-1} 0.8704$.
29.5°

84. Find $\sin^{-1} 0.2198$.
12.7°

85. Given $\sin \beta = \frac{\sqrt{3}}{2}$, find β.
60°

86. Given $\cos \theta = \frac{\sqrt{2}}{2}$, find θ.
45°

87. Find $\tan^{-1} \frac{\sqrt{3}}{3}$.
30°

88. Find $\tan^{-1} \sqrt{3}$.
60°

89. In right triangle *ABC*, side *a* is 16 meters and side *b* is 18 meters. Find $\angle A$. Round to the nearest tenth.
41.6°

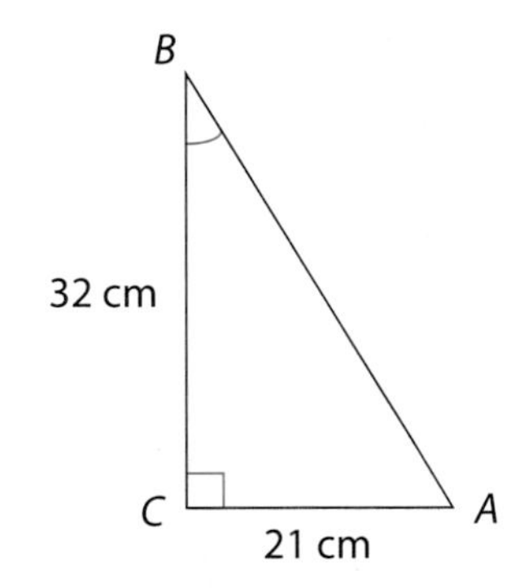

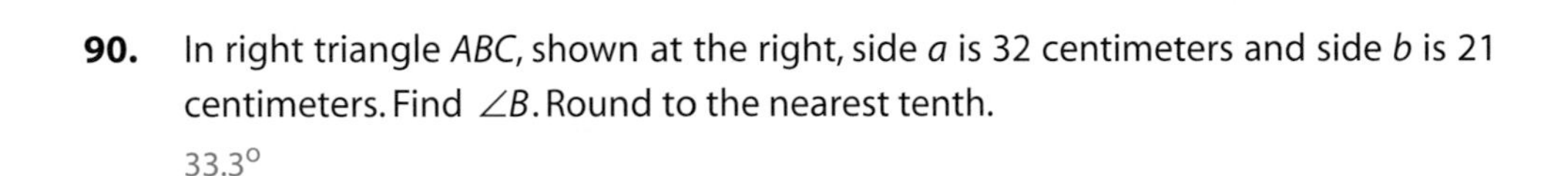

90. In right triangle *ABC*, shown at the right, side *a* is 32 centimeters and side *b* is 21 centimeters. Find $\angle B$. Round to the nearest tenth.
33.3°

91. In right triangle *ABC*, side *a* is 10 inches and the hypotenuse is 14 inches. Find $\angle B$. Round to the nearest tenth.
44.4°

92. In right triangle *ABC*, side *b* is 25 yards and the hypotenuse is 40 yards. Find $\angle A$. Round to the nearest tenth.
51.3°

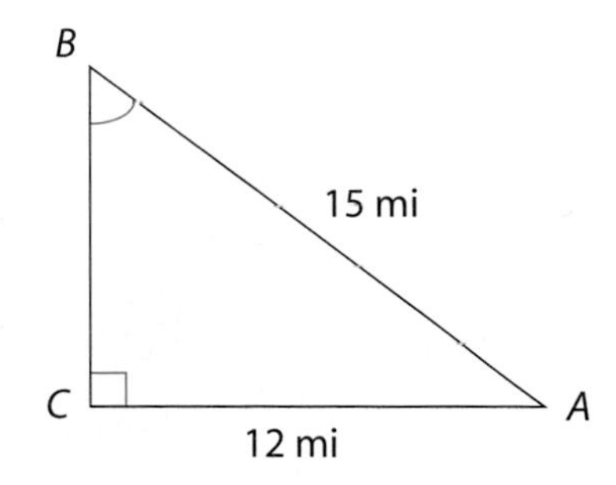

93. In right triangle *ABC*, shown at the right, side *b* is 12 miles and the hypotenuse is 15 miles. Find $\angle B$. Round to the nearest tenth.
53.1°

94. In right triangle ABC, side a is 7 feet and the hypotenuse is 9 feet. Find $\angle A$. Round to the nearest tenth.
51.1°

Solving Right Triangles

Solve right triangle ABC. Round to the nearest whole number.

95. $a = 15$ centimeters, $\angle A = 30^{\circ}$
$\angle B = 60^{\circ}$, $b = 26$ cm, $c = 30$ cm

96. $b = 6$ feet, $\angle A = 60^{\circ}$
$\angle B = 30^{\circ}$, $a = 10$ ft, $c = 12$ ft

97. $a = 5$ meters, $\angle B = 45^{\circ}$
$\angle A = 45^{\circ}$, $b = 5$ m, $c = 7$ m

98. $b = 20$ inches, $\angle B = 45^{\circ}$
$\angle A = 45^{\circ}$, $a = 20$ in., $c = 28$ in.

99. $a = 8$ centimeters, $b = 12$ centimeters
$\angle A = 34^{\circ}$, $\angle B = 56^{\circ}$, $c = 14$ cm

100. $a = 20$ meters, $b = 30$ meters
$\angle A = 34^{\circ}$, $\angle B = 56^{\circ}$, $c = 36$ m

101. $a = 10$ feet, $\angle A = 42^{\circ}$
$\angle B = 48^{\circ}$, $b = 11$ ft, $c = 15$ ft

102. $b = 20$ inches, $\angle B = 76^{\circ}$
$\angle A = 14^{\circ}$, $a = 5$ in., $c = 21$ in.

103. $a = 5$ meters, $c = 9$ meters
$\angle A = 34^{\circ}$, $\angle B = 56^{\circ}$, $b = 7$ m

104. $b = 14$ feet, $c = 15$ feet
$\angle A = 21^{\circ}$, $\angle B = 69^{\circ}$, $a = 5$ ft

105. $a = 18$ inches, $\angle B = 72^{\circ}$
$\angle A = 18^{\circ}$, $b = 55$ in., $c = 58$ in.

106. $b = 6$ centimeters, $\angle A = 23^{\circ}$
$\angle B = 67^{\circ}$, $a = 3$ cm, $c = 7$ cm

107. $a = 12$ feet, $c = 30$ feet
$\angle A = 24^{\circ}$, $\angle B = 66^{\circ}$, $b = 27$ ft

108. $b = 25$ inches, $c = 40$ inches
$\angle A = 51^{\circ}$, $\angle B = 39^{\circ}$, $a = 31$ in.

109. $a = 4$ centimeters, $\angle B = 40^{\circ}$
$\angle A = 50^{\circ}$, $b = 3$ cm, $c = 5$ cm

110. $b = 10$ meters, $\angle A = 50^{\circ}$
$\angle B = 40^{\circ}$, $a = 12$ m, $c = 16$ m

111. The distance between the two floors in a department store is 22 feet. An escalator from the first floor to the second floor makes an angle of 34° with the floor. How long is the escalator? Round to the nearest tenth.
39.3 ft

112. A straight road up a hill makes an angle of 6° with the horizontal. The length of the road from the bottom of the hill to the top of the hill is 2.25 miles. Find the height of the hill. Round to the nearest hundredth.
0.24 mi

Angles of Elevation and Depression

Solve. Round to the nearest tenth.

113. The angle of elevation of the top of a telephone pole 5 meters away is 70°. Find the height of the telephone pole.
13.7 m

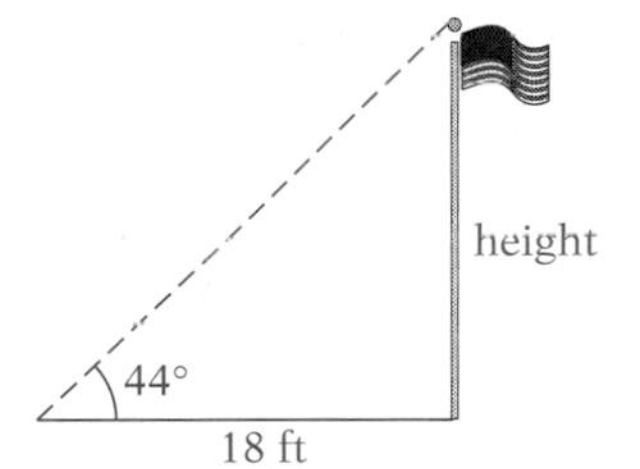

114. A flagpole casts a shadow of 18 feet. Find the height of the flagpole if the angle of elevation of the top of the pole from the tip of the shadow is 44°.
17.4 ft

115. From the top of a lighthouse that is 150 feet high, the angle of depression of a boat on the water is 38°. How far is the boat from the base of the lighthouse?
192.0 ft

116. From a helicopter at an altitude of 1000 feet, the angle of depression of the landing site is 25°. Find the direct distance from the helicopter to the landing site.
2366.2 ft

117. Find the height of an office building that casts an 80-foot shadow when the angle of elevation of the sun is 55°.
114.3 ft

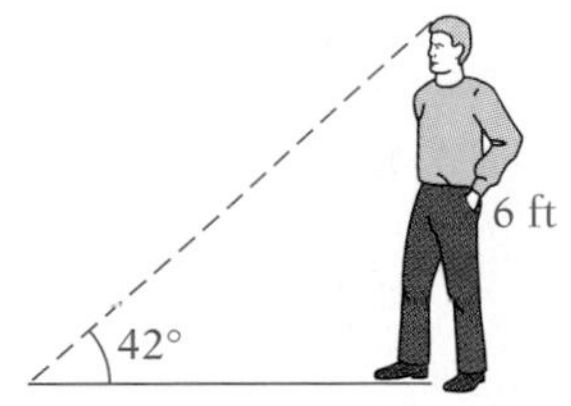

118. Find the length of a shadow cast by a person who is 6 feet tall when the angle of elevation of the sun is 42°.
6.7 ft

119. From the lookout post on a ship in the water, the angle of depression to the shoreline is 38°. If the lookout post is 115 meters above sea level, how far is the ship from the shore?
147.2 m

120. From a glass elevator on the outside of a building, the angle of depression to a landmark is 40°. If the landmark is 100 feet from the base of the building, what is the height of the elevator?
83.9 ft

Applying Concepts

121. Use your calculator to find the value of each of the following. Round to the nearest ten-thousandth.

a. $\sin 48^\circ 10'$ **b.** $\cos 44^\circ 30'$ **c.** $\tan 12^\circ 20'$

a. 0.7451; **b.** 0.7133; **c.** 0.2186

122. Explain why, given that $\angle A$ is an acute angle in right triangle *ABC*, $\sin A < 1$ and $\cos A < 1$.

Explanations will vary. For example, sin *A* is the ratio of the side opposite angle *A* to the hypotenuse. Since the hypotenuse is always longer than either of the two legs, the denominator of the ratio is larger than the numerator and the ratio is less than 1. Also, cos *A* is the ratio of the side adjacent to angle *A* to the hypotenuse. By the same reasoning, the ratio is less than 1.

123. Explain why, given that $\angle A$ is an acute angle in right triangle *ABC*, tan *A* can be less than 1, greater than 1, or equal to 1.

Explanations will vary. For example, tan *A* is the ratio of the side opposite angle *A* to the side adjacent to angle *A*. The length of the side opposite angle *A* can be greater than, less than, or equal to the length of the side adjacent to angle *A*. Therefore, in the fraction formed, the numerator can be either greater than, less than, or equal to the denominator.

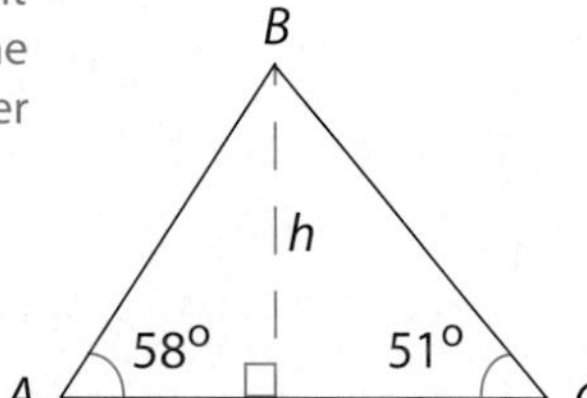

124. In the triangle at the right, the base $\overline{AC}$ is 20 meters. Find *h*. Round to the nearest tenth.

13.9 m

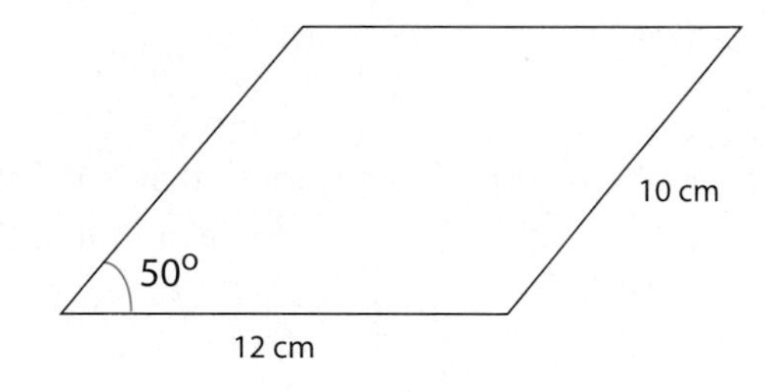

125. Find the area of the parallelogram shown at the right. Round to the nearest tenth.

91.9 cm^2

Explorations

126. ***Relationships Among Trigonometric Functions***

Determine whether the statement is true or false. Explain how you arrived at your conclusion.

a. The tangents of the two acute angles in a right triangle are reciprocals.

b. In a right triangle *ABC*, $\sin A = \cos B$ and $\cos A = \sin B$.

c. $\tan \theta = \dfrac{\sin \theta}{\cos \theta}$

d. $\sin(90^\circ - \theta) = \cos \theta$

127. ***Slope and the Tangent Function***

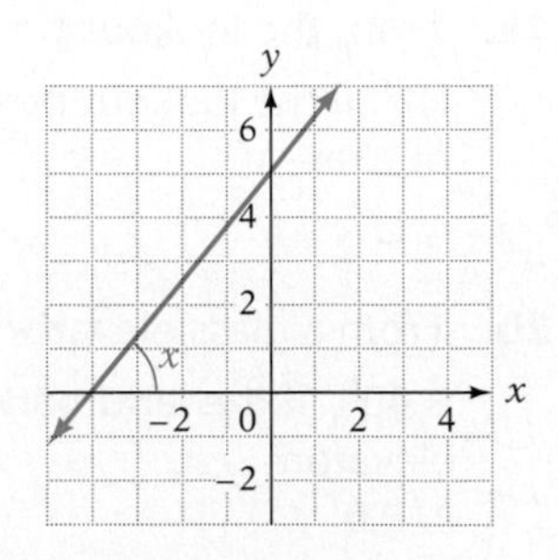

a. The measure of $\angle x$ shown at the right is 51.34019°. Find tan *x*. Round to the nearest hundredth.

b. Find the slope of the line.

c. Write the slope as a decimal.

d. What do you notice about tan *x* and the slope of the line?

e. A line passes through points (0, 6) and (−3, 0). Find the tangent of the positive acute angle that the line makes with the *x*-axis. Round to the nearest tenth.

Chapter Summary

Definitions

The *square root* of a number is one of two equal factors of the number; the square root of a number x is a number that, when squared, equals x. The symbol for square root, called a *radical sign*, is $\sqrt{}$. The number under the radical sign is called the *radicand.*

Every positive number has two square roots, one a positive number and one a negative number. The symbol $\sqrt{}$ is used to indicate the positive square root of a number, called the *principal square root.* When the negative square root of a number is to be found, a negative sign is placed in front of the radical sign. The square root of a negative number is not a real number.

The square of an integer is called a *perfect square*. A variable or a product of variables written in exponential form is a perfect square when each exponent is an even number. A radical expression is in simplest form when the radicand contains no factor, other than 1, that is a perfect square.

The expressions $a + b$ and $a - b$, which are the sum and difference of two terms, are called *conjugates* of each other. For example, the conjugate of $\sqrt{y} - 8$ is $\sqrt{y} + 8$, and the conjugate of $\sqrt{2} - \sqrt{a}$ is $\sqrt{2} + \sqrt{a}$.

The procedure used to remove a radical from the denominator of a fraction is called *rationalizing the denominator.*

In the *DMS* system, a degree is subdivided into 60 equal parts called *minutes*, and one minute is subdivided into 60 equal parts called *seconds*. The angle measure 47 degrees, 28 minutes, 36 seconds is written $47°28'36''$. Alternatively, a degree is subdivided into smaller units using decimals. For example, 59.17° equals 59° plus 17/100 of a degree. An angle measure involving decimals is written in *decimal degrees.*

To *solve a right triangle* means to use the information given about it to find the unknown sides and angles.

Line of sight problems involve an observer looking at an object. If the object being sighted is above the observer, the acute angle formed by the line of sight and a horizontal line is an *angle of elevation.* If the object being sighted is below the observer, the acute angle formed by the line of sight and a horizontal line is an *angle of depression.*

Procedures

Product Property of Square Roots

If a and b are positive numbers, then $\sqrt{a \cdot b} = \sqrt{a} \cdot \sqrt{b}$.
The Product Property of Square Roots is used to simplify radical expressions. For example: $\sqrt{12} = \sqrt{4 \cdot 3} = \sqrt{4} \cdot \sqrt{3} = 2\sqrt{3}$

Adding and Subtracting Radical Expressions

The Distributive Property is used to add and subtract like radical expressions. For example: $16\sqrt{3x} - 4\sqrt{3x} = (16 - 4)\sqrt{3x} = 12\sqrt{3x}$

Multiplying Radical Expressions

The Product Property of Square Roots is used to multiply radical expressions. For example: $\sqrt{5x}\sqrt{7y} = \sqrt{5x \cdot 7y} = \sqrt{35xy}$
Use the Distributive Property to remove parentheses. For example: $\sqrt{y}(2 + \sqrt{3x}) = 2\sqrt{y} + \sqrt{3xy}$

Use the FOIL method to multiply two radical expressions that each have two terms. For example: $(5 - \sqrt{x})(11 + \sqrt{x}) = 55 - 6\sqrt{x} - x$

The Square of a Square Root

For $x > 0$, $(\sqrt{x})^2 = x$.

The Product of Conjugates

$(a + b)(a - b) = a^2 - b^2$
For example: $(\sqrt{x} + 4)(\sqrt{x} - 4) = (\sqrt{x})^2 - 4^2 = x - 16$

Quotient Property of Square Roots

If a and b are positive numbers, then

$\sqrt{\frac{a}{b}} = \frac{\sqrt{a}}{\sqrt{b}}$. For example: $\sqrt{\frac{9x^2}{y^8}} = \frac{\sqrt{9x^2}}{\sqrt{y^8}} = \frac{3x}{y^4}$

Property of Squaring Both Sides of an Equation

If a and b are real numbers and $a = b$, then $a^2 = b^2$.

Pythagorean Theorem

If a and b are the lengths of the legs of a right triangle and c is the hypotenuse, then $c^2 = a^2 + b^2$.

The Principal Square Root Property

If $r^2 = s$, then $r = \sqrt{s}$, and r is called the square root of s.

The Distance Formula

If (x_1, y_1) and (x_2, y_2) are two points in the plane, then the distance d between the two points is given by $d = \sqrt{(x_1 - x_2)^2 + (y_1 - y_2)^2}$.

Relationship Among the Sides of an Isosceles Right Triangle

For any 45°–45°–90° triangle, the hypotenuse is $\sqrt{2}$ times the length of a leg.

Relationships Among the Sides of a 30°–60°–90° Triangle

For any 30°–60°–90° triangle, the hypotenuse is twice the shorter leg, or the leg opposite the 30° angle. The longer leg, or the leg opposite the 60° angle, is $\sqrt{3}$ times the length of the shorter leg.

Radical Expressions in Simplest Form A radical expression is in simplest form if:
1. The radicand contains no factor greater than 1 that is a perfect square.
2. There is no fraction under the radical sign.
3. There is no radical in the denominator of a fraction.

Trigonometric Functions

$\sin A = \frac{\text{opposite}}{\text{hypotenuse}} = \frac{a}{c}$ $\qquad$ $\sin B = \frac{\text{opposite}}{\text{hypotenuse}} = \frac{b}{c}$

$\cos A = \frac{\text{adjacent}}{\text{hypotenuse}} = \frac{b}{c}$ $\qquad$ $\cos B = \frac{\text{adjacent}}{\text{hypotenuse}} = \frac{a}{c}$

$\tan A = \frac{\text{opposite}}{\text{adjacent}} = \frac{a}{b}$ $\qquad$ $\tan B = \frac{\text{opposite}}{\text{adjacent}} = \frac{b}{a}$

Definition of the Inverse Sine, Cosine, and Tangent Functions

For $0^\circ < x < 90^\circ$:
$y = \sin^{-1} x$ can be read "y is the angle whose sine is x."
$y = \cos^{-1} x$ can be read "y is the angle whose cosine is x."
$y = \tan^{-1} x$ can be read "y is the angle whose tangent is x."

Chapter Review Exercises

1. Simplify: $\sqrt{25}$
5

2. Simplify: $-\sqrt{121}$
-11

3. Simplify: $4\sqrt{36}$
24

4. Simplify: $\sqrt{100} - 2\sqrt{49}$
-4

5. Simplify: $5\sqrt{48}$

$20\sqrt{3}$

6. Simplify: $-3\sqrt{120}$

$-6\sqrt{30}$

7. Simplify: $\sqrt{c^{18}}$

c^9

8. Simplify: $\sqrt{36x^4y^5}$

$6x^2y^2\sqrt{y}$

9. Simplify: $y\sqrt{24y^6}$

$2y^4\sqrt{6}$

10. Simplify: $3\sqrt{18x^4y^5}$

$9x^2y^2\sqrt{2y}$

11. Evaluate $-3\sqrt{x+y}$ when $x = 18$ and $y = 31$.
-21

12. Approximate $3\sqrt{47}$ to the nearest ten-thousandth.
20.5670

13. Add: $6\sqrt{7} + \sqrt{7}$
$7\sqrt{7}$

14. Subtract: $9x\sqrt{5} - 5x\sqrt{5}$
$4x\sqrt{5}$

15. Add: $3\sqrt{12x} + 5\sqrt{48x}$
$26\sqrt{3x}$

16. Subtract: $\sqrt{20a^5} - 2a\sqrt{45a^3}$
$-4a^2\sqrt{5a}$

17. Multiply: $\sqrt{2}\sqrt{50}$
10

18. Multiply: $\sqrt{a^3b^4c}\ \sqrt{a^7b^2c^3}$
$a^5b^3c^2$

19. Multiply: $\sqrt{3}(\sqrt{12} - \sqrt{3})$
3

20. Multiply: $\sqrt{6a}(\sqrt{3a} + \sqrt{2a})$
$3a\sqrt{2} + 2a\sqrt{3}$

21. Multiply: $(\sqrt{a} + 5)(2\sqrt{a} + 3)$
$2a + 13\sqrt{a} + 15$

22. Multiply: $(4\sqrt{y} - \sqrt{5})(2\sqrt{y} + \sqrt{5})$
$8y + 2\sqrt{5y} - 5$

23. Multiply: $(\sqrt{5ab} - \sqrt{7})(\sqrt{5ab} + \sqrt{7})$
$5ab - 7$

24. Simplify: $\dfrac{16}{\sqrt{a}}$
$\dfrac{16\sqrt{a}}{a}$

25. Simplify: $\dfrac{\sqrt{250}}{\sqrt{10}}$
5

26. Simplify: $\dfrac{\sqrt{54a^3}}{\sqrt{6a}}$
$3a$

27. Simplify: $\dfrac{8}{\sqrt{2x}}$
$\dfrac{4\sqrt{2x}}{x}$

28. Solve: $\sqrt{5x} = 10$
20

29. Solve: $3 - \sqrt{7x} = 5$
No solution

30. Solve: $\sqrt{10x + 4} - 8 = 0$
6

31. Solve: $\sqrt{2x + 9} = \sqrt{8x - 9}$
3

32. Solve: $\sqrt{x + 1} - \sqrt{x - 2} = 1$
3

33. Convert $83^\circ 45'$ to decimal degrees.
83.75°

34. Convert 72.48° to degrees, minutes, and seconds.
$72^\circ 28' 48''$

35. Find cos 39°.
0.7771

36. Find $\sin^{-1} 0.5736$. Round to the nearest tenth.
35.0°

37. Traffic accident investigators can use the formula $v = 2\sqrt{5L}$ to determine the speed of a car that leaves skid marks after the driver hits the brake pedal to stop the car. In this formula, the speed, v, of the car is measured in miles per hour, and the length, L, of the skid mark is measured in feet. Find the speed of a car that left a skid mark that was 80 feet long.
40 mph

38. A bicycle will overturn if it rounds a corner too sharply or too quickly. The equation for the maximum velocity at which a cyclist can turn a corner without tipping over is given by the equation $v = 4\sqrt{r}$, where v is the velocity of the bicycle in miles per hour and r is the radius of the corner in feet. Find the radius of the sharpest corner that a cyclist can safely turn when riding at a speed of 20 mph.
25 ft

39. The time it takes for an object to fall a distance of d feet on Mars is given by the formula $t = \sqrt{\frac{d}{6.45}}$, where t is the time in seconds. The time it takes for an object to fall a distance of d feet on Earth is given by the formula $t = \sqrt{\frac{d}{16}}$.

a. How long does it take for an object to fall 150 feet on Mars? Round to the nearest tenth.

b. Does it take longer for an object to drop 150 feet on Mars or on Earth? How much longer? Round to the nearest tenth.

c. Use your answers to parts a and b to determine which graph shown at the right is a graph of $t = \sqrt{\frac{d}{6.45}}$ and which is a graph of $t = \sqrt{\frac{d}{16}}$.

a. 4.8 s; **b.** Mars, 1.8 s;

c. Figure A is the graph of $t = \sqrt{\frac{d}{6.45}}$; Figure B is the graph of $t = \sqrt{\frac{d}{16}}$.

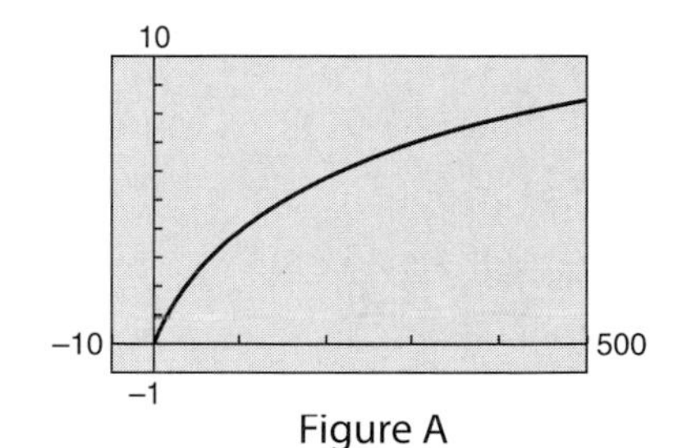

Figure A

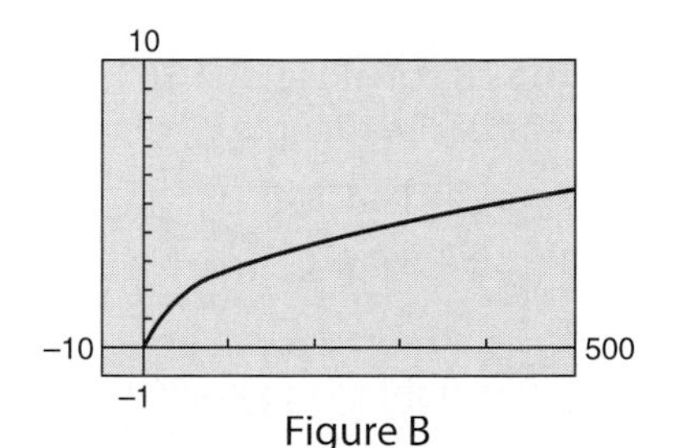

Figure B

40. A tsunami is a great sea wave produced by underwater earthquakes or volcanic eruption. Use the formula $v = 3\sqrt{d}$, where v is the velocity in feet per second of a tsunami as it approaches land and d is the depth in feet of the water, to find the depth in feet of the water when the velocity of a tsunami reaches 30 feet per second.
100 ft

41. Find **a.** the perimeter and **b.** the area of the rectangle shown at the right.
a. $14\sqrt{5}$ units; **b.** 60 square units

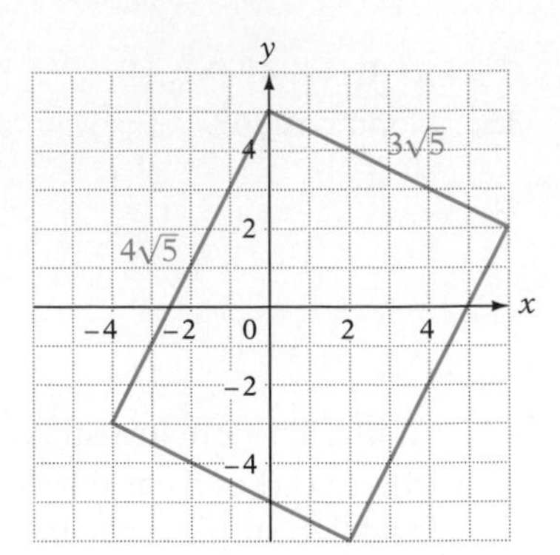

42. The hypotenuse of a right triangle measures 18 centimeters. One leg of the triangle measures 11 centimeters. Find the length of the other leg of the triangle. Round to the nearest hundredth.
14.25 cm

43. A guy wire is attached to a point 25 feet above the ground on a telephone pole that is perpendicular to the ground. The wire is anchored to the ground at a point 8 feet from the base of the pole. Find the length of the guy wire. Round to the nearest hundredth.
26.25 ft

44. Find the distance between the points (−2, 4) and (3, 5). Round to the nearest tenth.
5.1 units

45. A triangle has vertices at (−2, 3), (−2, −3), and (2, 3). Find **a.** the perimeter and **b.** the area of the triangle. Round to the nearest tenth.
a. 17.2 units **b.** 12 square units

46. The hypotenuse of a 45°–45°–90° triangle is 8 meters. Find the length of a side opposite one of the 45° angles.
$4\sqrt{2}$ m

47. The hypotenuse of a 30°–60°–90° triangle is 12 centimeters. Find the lengths of the two legs of the triangle.
6 cm and $6\sqrt{3}$ cm

48. Solve right triangle *ABC* given $a = 6$ meters and $c = 10$ meters. Round to the nearest whole number.
$b = 8$ m, $\angle A = 37^\circ$, $\angle B = 53^\circ$

49. Solve right triangle *ABC* given $b = 4$ inches and $\angle A = 25^\circ$. Round to the nearest whole number.
$\angle B = 65^\circ$, $a = 2$ in., $c = 4$ in.

50. The angle of elevation of the top of a flagpole 50 feet away is 28°. Find the height of the flagpole. Round to the nearest tenth.
26.6 ft

Cumulative Review Exercises

1. Given $A = \{-6, -3, 0, 3, 6\}$ and $B = \{-4, -2, 0, 2, 4\}$, find:
a. $A \cup B$ **b.** $A \cap B$
a. $\{-6, -4, -3, -2, 0, 2, 3, 4, 6\}$; **b.** $\{0\}$

2. Find a counterexample: "All numbers that end in 6 are divisible by 3."
Answers will vary. For example, 16.

3. Evaluate: $24 - 8 \div 2^3$
23

4. Translate into a variable expression: The sum of 3 times the square of b and the difference between b and 9.
$3b^2 + (b - 9)$

5. Evaluate: $-|-9|$
-9

6. Subtract: $-12 - 23 - (-7)$
-28

7. What is the quotient of -56 and -8?
7

8. Write $\frac{4}{9}$ as a repeating decimal.
$0.\overline{4}$

9. Simplify: $-\frac{2}{5}\left(\frac{5}{8}\right) \div \left(-\frac{1}{2}\right)$
$\frac{1}{2}$

10. Convert 8 quarts to gallons.
2 gal

11. Simplify: $\left(-\frac{2}{3}a\right)3$
$-2a$

12. Simplify: $5(2x - 3y) + 4(-6x - y)$
$-14x - 19y$

13. What percent of 112 is 14?
12.5%

14. Solve: $5(3x - 20) + x = 10(x - 4)$
10

15. Find the slope of the line containing the points $(-3, 4)$ and $(2, -1)$.
-1

16. Solve the system of equations:
$6x + 5y = 7$
$3x - 7y = 13$
$(2, -1)$

17. Simplify:
$(6x^2 - 2x - 1) + (4x^2 + 3x + 5) - (x^2 + x + 1)$
$9x^2 + 3$

18. Multiply: $(-2a^2)(3ab^3)(a^4b)$
$-6a^7b^4$

19. Simplify: $(2x^{-5}y^3)^4$
$\frac{16y^{12}}{x^{20}}$

20. Multiply: $(x + 4)(3x^2 - 5x - 1)$
$3x^3 + 7x^2 - 21x - 4$

21. Factor: $2x^2 + 3x - 20$

$(2x - 5)(x + 4)$

22. Factor: $2b^2 - 18$

$2(b + 3)(b - 3)$

23. Simplify: $\sqrt{32a^5b^{11}}$

$4a^2b^5\sqrt{2ab}$

24. Subtract: $5\sqrt{8} - 3\sqrt{50}$

$-5\sqrt{2}$

25. Multiply: $(\sqrt{y} + 3)(\sqrt{y} + 5)$

$y + 8\sqrt{y} + 15$

26. Simplify: $\dfrac{4}{\sqrt{8}}$

$\sqrt{2}$

27. Solve: $6 - \sqrt{2y} = 2$

8

28. Find $\cos^{-1} 0.5446$. Round to the nearest tenth.

57.0°

29. Graph $y = -3x + 1$ for $x = -2, -1, 0, 1, 2$.

Points at $(-2, 7), (-1, 4), (0, 1), (1, -2), (2, -5)$

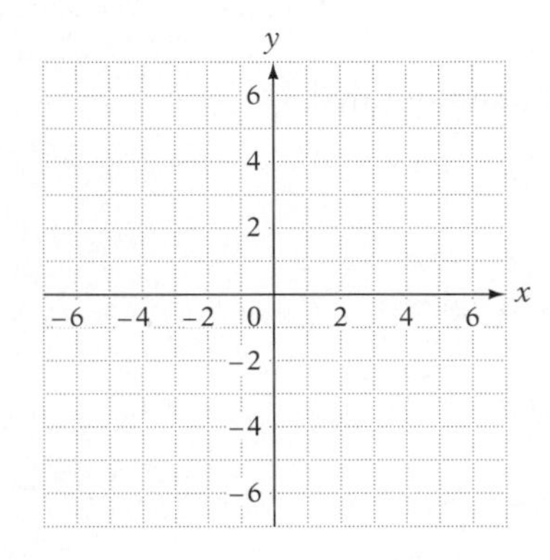

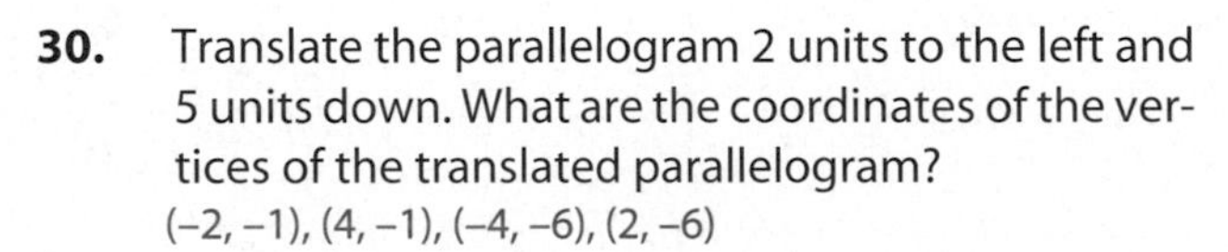

30. Translate the parallelogram 2 units to the left and 5 units down. What are the coordinates of the vertices of the translated parallelogram?

$(-2, -1), (4, -1), (-4, -6), (2, -6)$

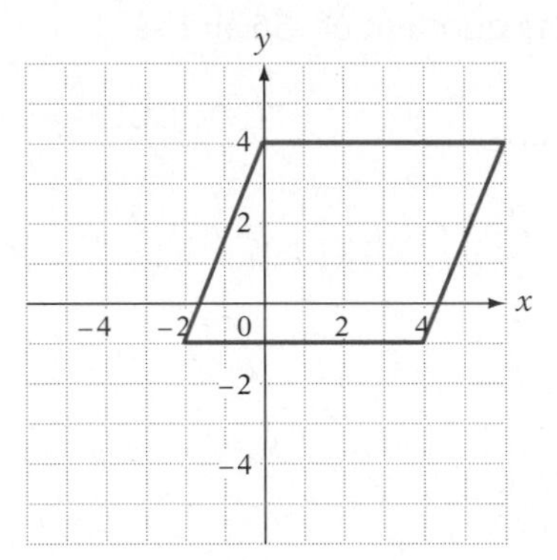

31. Chris, Pat, Leslie, and Dana are investors. Each has invested in a different stock (utility, computer, oil, or automotive). From the following statements, determine which stock each has invested in.

a. The stock owned by Chris and the utility company stock have increased in price per share during the past month, while the stock owned by Leslie and the oil company stock have decreased in price per share.

b. Chris and the automotive stock investor purchased their shares on NASDAQ, while Pat and the utility company investor purchased their shares on the New York Stock Exchange.

Dana, utility; Leslie, automotive; Chris, computer; Pat, oil

32. Use a graphing calculator to produce a graph of $y = -0.5x^3 + 1$ using the decimal viewing window. Find the value of x, to the nearest tenth, for which the output is -1.

1.6

33. Translate "eight less than twice a number equals four times the sum of the number and six" into an equation and solve.

$2x - 8 = 4(x + 6); -16$

TAKE NOTE
You should always check your solutions by substituting the proposed solutions back into the *original* equation. There are several ways to do this using a graphics calculator. Check the appendix Guidelines for Using Graphing Calculators for suggestions on how to use your calculator to check solutions to an equation without graphing.

Example 1 Solve by factoring: $3x^2 = x + 4$

Solution

$$3x^2 = x + 4$$

Write the equation in standard form.

$$3x^2 - x - 4 = 0$$

Factor the polynomial.

$$(3x - 4)(x + 1) = 0$$

Use the Principle of Zero Products.

$$3x - 4 = 0 \quad \text{or} \quad x + 1 = 0$$
$$3x = 4 \qquad x = -1$$
$$x = \frac{4}{3}$$

Graphical check:
Graph the standard form of the equation with 0 replaced by *y*.

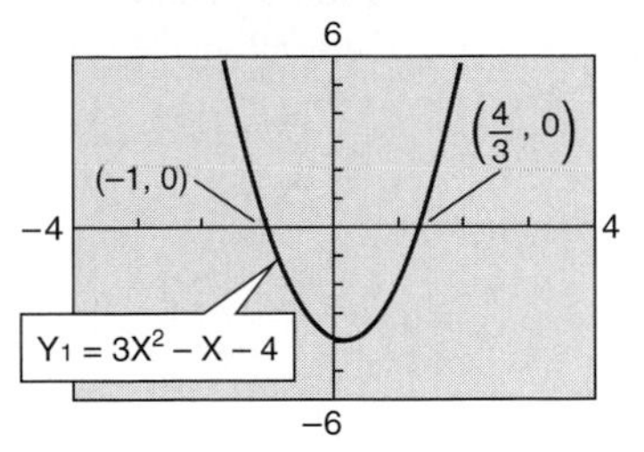

The solutions are −1 and $\frac{4}{3}$.

You-Try-It 1 Solve by factoring: $2x^2 - x = 1$

Solution See page S32.

SUGGESTED ACTIVITY
The *Instructor's Resource Manual* contains an activity that will help convince students that the quadratic equations with a double root have a graph that is tangent to the *x*-axis.

Some quadratic equations have *repeated* solutions. Consider the equation $x^2 + 6x + 9 = 0$. By factoring and using the Principle of Zero Products, we have

$$x^2 + 6x + 9 = 0$$
$$(x + 3)(x + 3) = 0$$
$$x + 3 = 0 \quad \text{or} \quad x + 3 = 0$$
$$x = -3 \qquad x = -3$$

−3 is called a **double root** of the equation.

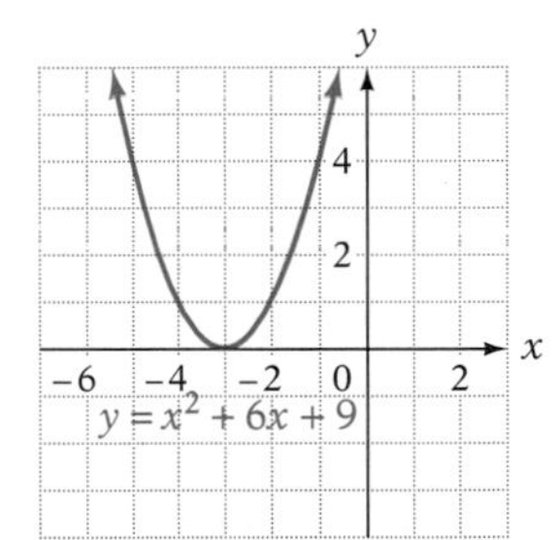

Note that the graph of $y = x^2 + 6x + 9$ just touches the *x*-axis at (−3, 0). We say that the parabola is *tangent* to the *x*-axis at (−3, 0). It is a characteristic of quadratic equations that have double roots to have a graph that is tangent to the *x*-axis at the *x*-coordinate that is the double root of the equation.

Solve Quadratic Equations by Taking Square Roots

Consider the quadratic equation $x^2 = 16$. The graph $Y_1 = x^2$ (the left side of $x^2 = 16$) and $Y_2 = 16$ (the right side of the equation) are shown at the right.

Using the INTERSECT feature of a graphing calculator, we can determine that the graphs intersect at (−4, 16) and (4, 16).

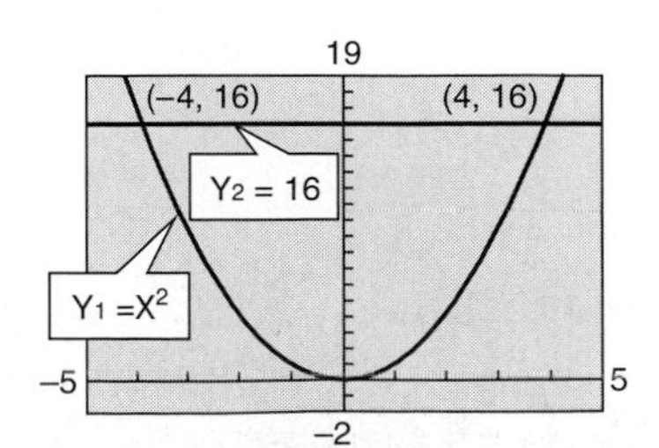

Because the graphs intersect at (4, 16) and (−4, 16), the solution of $x^2 = 16$ is 4 and −4.

Solutions that are plus or minus the same number (such as 4 and −4 above) are frequently written: "The solutions are ±4." The ±4 is read "plus or minus 4."

TAKE NOTE
Recall that if x is any real number, then $\sqrt{x^2} = |x|$. Because of this, for the equation at the right, we have $|x| = 4$. The numbers –4 and 4 are solutions of this equation.

The equation $x^2 = 16$ can also be solved by taking the square root of each side of the equation.

$x^2 = 16$

$\sqrt{x^2} = \sqrt{16}$ • Take the square root of each side of the equation.

$|x| = 4$

$x = \pm 4$

The solutions are 4 and –4.

> **Principle of Taking the Square Root of Each Side of an Equation**
>
> If $x^2 = a$, then $x = \pm\sqrt{a}$.

Example 2 Solve by taking square roots: $3x^2 - 36 = 0$

Solution

$3x^2 - 36 = 0$

$3x^2 = 36$ • Solve for x^2.

$x^2 = 12$

$\sqrt{x^2} = \sqrt{12}$ • Take the square root of each side of the equation.

$x = \pm 2\sqrt{3}$ • Write the answer in simplest form.

Graphical check:

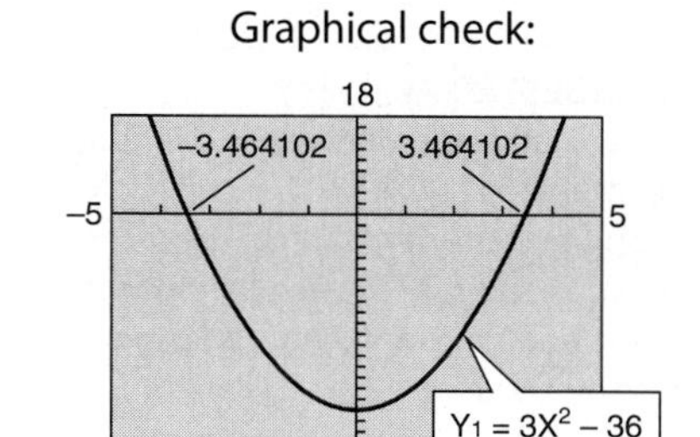

The *exact* solutions are $-2\sqrt{3}$ and $2\sqrt{3}$. The *approximate* solutions are –3.464102 and 3.464102. The graphical check shows the approximate solutions.

You-Try-It 2 Solve by taking square roots: $4t^2 - 234 = 0$

Solution See page S32.

The graph of $y = x^2 + 16$ is shown at the right. Note that the graph does not pass through the x-axis and therefore has no x-intercepts. This means that there are no real numbers for which $x^2 + 16 = 0$. Consequently, $x^2 + 16 = 0$ has no real number solutions.

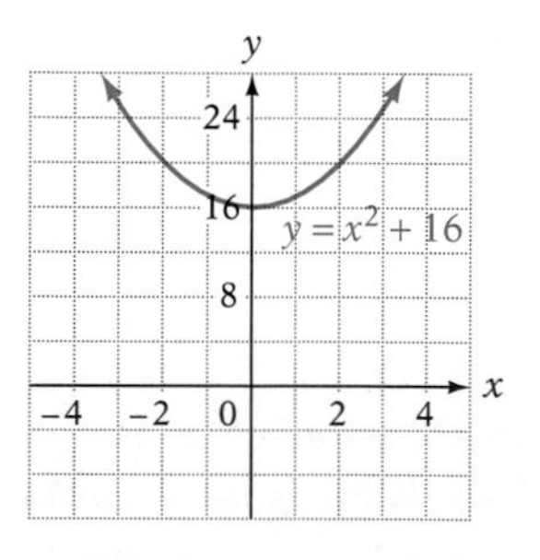

The graph shows that there is no value of x for which y is 0. Therefore, $x^2 + 16 = 0$ has no real number solutions.

INSTRUCTOR NOTE
Another way to algebraically show that $x^2 + 16 = 0$ has no real number solutions is to have students recall that the sum of squares does not factor using real numbers. Consequently, $x^2 + 16 = 0$ has no real number solutions.

An attempt to solve $x^2 + 16 = 0$ by taking square roots leads to the following.

$x^2 + 16 = 0$

$x^2 = -16$

$\sqrt{x^2} = \sqrt{-16}$ • $\sqrt{-16}$ is not a real number.

Because $\sqrt{-16}$ is not a real number, the equation has no real number solutions.

Some equations that contain the square of a binomial can be solved by taking square roots.

→ Solve: $(x-3)^2 = 25$

$(x-3)^2 = 25$

$\sqrt{(x-3)^2} = \sqrt{25}$ • Take the square root of each side of the equation.

$x - 3 = \pm 5$ • Simplify and solve for x.

$x - 3 = 5 \qquad x - 3 = -5$

$x = 8 \qquad x = -2$

The solutions are 8 and −2.

Example 3 Solve by taking square roots: $2(x-1)^2 - 36 = 0$

Solution

$2(x-1)^2 - 36 = 0$

$2(x-1)^2 = 36$ • Solve for $(x-1)^2$.

$(x-1)^2 = 18$ • Take the square root of each side of the equation.

$\sqrt{(x-1)^2} = \sqrt{18}$

$x - 1 = \pm 3\sqrt{2}$ • Solve for x. Write the answer in simplest form.

$x - 1 = 3\sqrt{2} \qquad x - 1 = -3\sqrt{2}$

$x = 1 + 3\sqrt{2} \qquad x = 1 - 3\sqrt{2}$

The solutions are $1 + 3\sqrt{2} \approx 5.2426407$ and $1 - 3\sqrt{2} \approx -3.2426407$.

Graphical check:

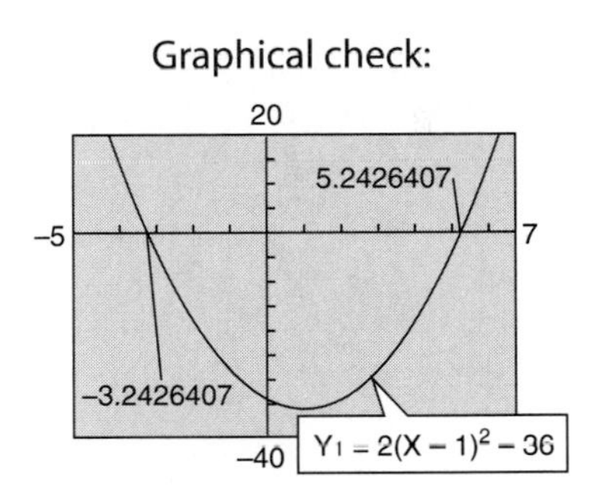

You-Try-It 3 Solve by taking square roots: $3(x+2)^2 + 4 = 64$

Solution See page S32.

Applications

Consider the situation of a punter for a football team. Sometimes the goal of the punter is not to punt the ball as far as possible but to try to kick the ball in such a way that it lands as close as possible to the goal line without going past it.

A model of the height, $h(x)$, of a football x feet from the punter can be approximated by $h(x) = -0.01782x^2 + 2.47509x + 5$. This is an example of a quadratic function. A **quadratic function** is one of the form $f(x) = ax^2 + bx + c$, where $a \neq 0$.

Suppose a punter is 45 yards from the goal line and punts a football according to the model above. Will the football cross the goal line before hitting the ground?

There are several ways to answer this question. By tracing along the graph we find that when $x = 135$ (45 yards), $y \approx 14.4$. Thus, the answer to the question is yes. When the football is 45 yards from the punter, it is still approximately 14 feet high and will land over the goal line.

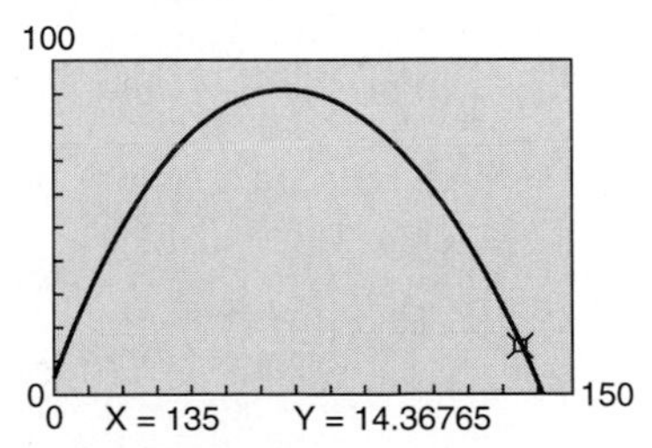

TAKE NOTE
In the description of the function, x was given in feet. To evaluate the function, the input value must be feet. Since 1 yard = 3 feet, the unit conversion rate is $\frac{3\text{ feet}}{1\text{ yard}}$. Thus,

$$45\text{ yards} = \frac{45\text{ yards}}{1} \cdot \frac{3\text{ feet}}{1\text{ yard}} = 135\text{ feet}$$

We could have also answered the question by evaluating the function for $x = 135$.

$$\begin{aligned} h(x) &= -0.01782x^2 + 2.47509x + 5 \\ h(135) &= -0.01782(135)^2 + 2.47509(135) + 5 \\ &\approx 14.36765 \end{aligned}$$

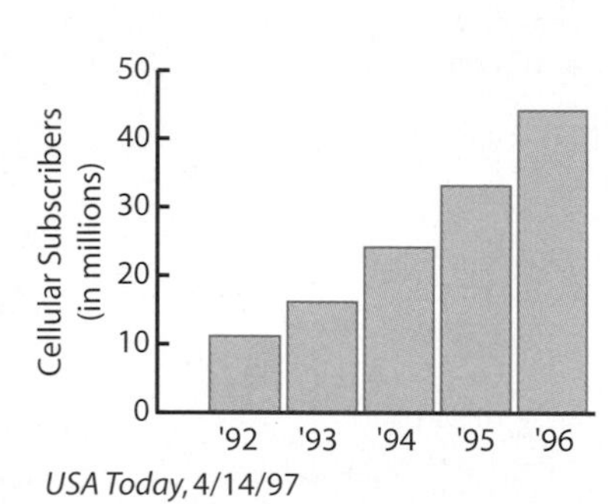

USA Today, 4/14/97

Example 4 The bar graph at the left shows the increase in the number of cellular phone subscribers. The number of subscribers, N (in millions), can be modeled by the equation

$$N = 0.93(t + 1.47)^2 + 5.73$$

where $t = 1$ corresponds to 1992.

a. In 1996, the actual number of subscribers was 44,000,000. How many subscribers are predicted by the model?

b. Assuming that the model is valid through the year 2000, in what year will the number of cellular subscribers first exceed 60 million?

Solution

a. Because $t = 1$ corresponds to 1992, $t = 5$ for 1996. Evaluate the expression for N when $t = 5$.

$$\begin{aligned} N &= 0.93(t + 1.47)^2 + 5.73 \\ &= 0.93(5 + 1.47)^2 + 5.73 && \text{• Replace } t \text{ by 5.} \\ &= 0.93(6.47)^2 + 5.73 \\ &= 44.660637 \end{aligned}$$

The model predicts approximately 44.7 million subscribers in 1996.

b. To find the year in which the number of cellular subscribers first exceeded 60 million, replace N by 60 and solve for t.

$$\begin{aligned} N &= 0.93(t + 1.47)^2 + 5.73 \\ 60 &= 0.93(t + 1.47)^2 + 5.73 && \text{• Replace } N \text{ by 60.} \\ 60 - 5.73 &= 0.93(t + 1.47)^2 + 5.73 - 5.73 && \text{• Solve for } (t + 1.47)^2. \\ 54.27 &= 0.93(t + 1.47)^2 \\ 58.35 &\approx (t + 1.47)^2 && \text{• Divide each side by 0.93.} \\ \sqrt{58.35} &\approx \sqrt{(t + 1.47)^2} && \text{• Take the square root of each side of the equation.} \\ \pm 7.64 &\approx t + 1.47 \end{aligned}$$

$$7.64 = t + 1.47 \qquad -7.64 = t + 1.47$$

$$6.17 = t \qquad -9.11 = t \qquad \text{• } -9.11 \text{ is not possible.}$$

The year corresponding to $t = 6$ (the smallest whole number less than 6.17) is 1997. Therefore, in 1997 the number of cellular subscribers first exceeded 60 million.

You-Try-It 4 The length of a rectangle is 3 times the width and the area is 19 square meters. Find the length and width of the rectangle. Round to the nearest hundredth.

$3x$

x

Solution See page S32.

7.1 EXERCISES

Topics for Discussion

1. How does a quadratic equation differ from a linear equation?
A quadratic equation has a second-degree term.

2. What is the standard form of a quadratic equation?
$ax^2 + bx + c = 0$

3. What is the Principle of Zero Products?
If $ab = 0$, then $a = 0$ or $b = 0$.

4. Explain why the equation $x(x + 4) = 12$ cannot be solved by saying $x = 12$ or $x + 4 = 12$.
Answers will vary.

5. What is a double root of a quadratic equation?
One of the two equal roots of a quadratic equation.

Quadratic Equations

For the given quadratic equation, find the values of *a*, *b*, and *c*.

6. $3x^2 - 4x + 1 = 0$
$a = 3, b = -4, c = 1$

7. $x^2 + 2x - 5 = 0$
$a = 1, b = 2, c = -5$

8. $2x^2 - 5 = 0$
$a = 2, b = 0, c = -5$

9. $4x^2 + 1 = 0$
$a = 4, b = 0, c = 1$

10. $6x^2 - 3x = 0$
$a = 6, b = -3, c = 0$

11. $-x^2 + 7x = 0$
$a = -1, b = 7, c = 0$

Write the quadratic equation in standard form.

12. $x^2 - 8 = 3x$
$x^2 - 3x - 8 = 0$

13. $2x^2 = 4x - 1$
$2x^2 - 4x + 1 = 0$

14. $x^2 = 16$
$x^2 - 16 = 0$

15. $x + 5 = x(x - 3)$
$x^2 - 4x - 5 = 0$

16. $2(x + 3)^2 = 5$
$2x^2 + 12x + 13 = 0$

17. $4(x - 1)^2 = 3$
$4x^2 - 8x + 1 = 0$

Solve by factoring.

18. $x^2 + 2x - 15 = 0$
$-5, 3$

19. $p^2 + 3p + 2 = 0$
$-1, -2$

20. $6x^2 - 9x = 0$
$0, \frac{3}{2}$

21. $3v^2 - 5v + 2 = 0$
$\frac{2}{3}, 1$

22. $6r^2 = 12 - r$
$-\frac{3}{2}, \frac{4}{3}$

23. $9s^2 - 6s + 1 = 0$
$\frac{1}{3}$

24. $x^2 + 16x + 64 = 0$

-8

25. $x^2 + x = 12$

$-4, 3$

26. $4y^2 - 4y = -1$

$\frac{1}{2}$

27. $9z^2 - 4 = 0$

$-\frac{2}{3}, \frac{2}{3}$

28. $\frac{2x^2}{9} + x = 2$

$\frac{3}{2}, -6$

29. $p + 18 = p(p - 2)$

$-3, 6$

30. $r^2 - r - 2 = (2r - 1)(r - 3)$

$5, 1$

31. $s^2 + 5s - 4 = (2s + 1)(s - 4)$

$0, 12$

32. $x^2 + x + 5 = (3x + 2)(x - 4)$

$-1, \frac{13}{2}$

Solve by taking square roots.

33. $x^2 = 36$

± 6

34. $4x^2 - 49 = 0$

$\pm\frac{7}{2}$

35. $16v^2 - 9 = 0$

$\pm\frac{3}{4}$

36. $w^2 - 24 = 0$

$\pm 2\sqrt{6}$

37. $2(x + 5)^2 = 8$

$-3, -7$

38. $4(x + 5)^2 = 64$

$-1, -9$

39. $12(x + 3)^2 = 27$

$-\frac{3}{2}, -\frac{9}{2}$

40. $49(v + 1)^2 - 25 = 0$

$-\frac{2}{7}, -\frac{12}{7}$

41. $(x + 1)^2 + 36 = 0$

No real number solution

42. $4\left(x - \frac{2}{3}\right)^2 = 16$

$\frac{8}{3}, -\frac{4}{3}$

43. The value P of an initial investment of A dollars after 2 years is given by $P = A(1 + r)^2$, where r is the annual percentage rate earned by the investment. If an initial investment of \$1500 grew to a value of \$1782.15 in 2 years, what was the annual percentage rate?
9%

44. The kinetic energy of a moving body is given by $E = \frac{1}{2}mv^2$, where E is the kinetic energy, m is the mass, and v is the velocity. What is the velocity of a moving body whose mass is 5 kilograms and whose kinetic energy is 250 newton-meters?
10 m/s

45. On a certain type of street surface, the equation $d = 0.0074v^2$ can be used to approximate the distance, d, a car traveling v miles per hour will slide when its brakes are applied. After applying the brakes, the owner of a car involved in an accident skidded 40 feet. Did the traffic officer investigating the accident issue the car owner a ticket for speeding if the speed limit was 65 mph?
Yes ($v \approx 73.5$ mph)

46. According to Torricelli's principle, the velocity, v (in feet per second), of water that pours through a hole in a bucket depends on the height h (in feet) of the water level above the hole and is given by $64h = v^2$. Find the velocity of water pouring through a hole in a bucket when the water level is 9 inches above the hole. Round to the nearest hundredth.
6.93 feet per second

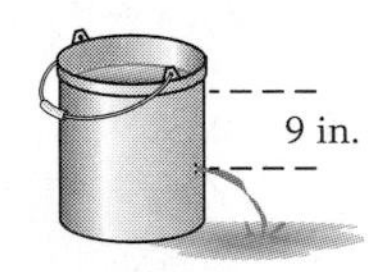

47. A circular swimming pool has a radius of 12 feet. A decking of constant width is built around the pool so that the pool and decking have a total area of 900 square feet. Find the width of the decking to the nearest hundredth of a foot.
4.93 feet

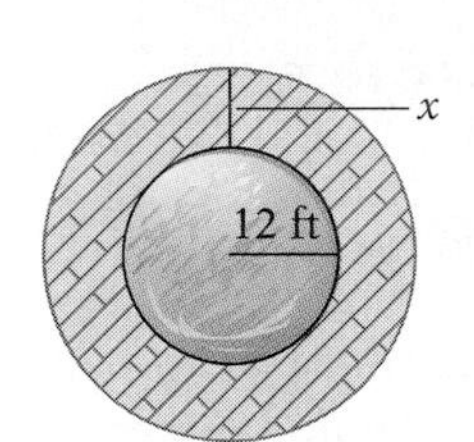

48. The height of a triangle is 1 meter less than the base. Find the dimensions of the triangle if its area is 3 square meters.
base: 3 meters; height: 2 meters

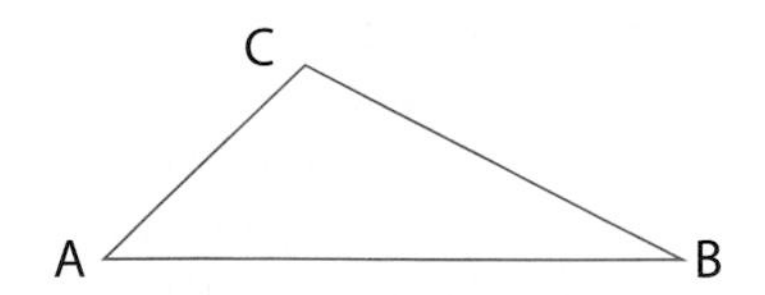

Applying Concepts

Solve for x.

49. $(x^2 - 1)^2 = 9$
± 2

50. $(x^2 + 3)^2 = 25$
$\pm\sqrt{2}$

51. Find the real number solutions of $x^4 = 25$ by applying the Principle of Taking the Square Root of Each Side of an Equation twice.
$\pm\sqrt{5}$

52. Solve for x.

a. $ax^2 - bx = 0, a \neq 0$

$0, \frac{b}{a}$

b. $ax^2 - b = 0, a > 0, b > 0$

$\pm\frac{\sqrt{ab}}{a}$

53. Use the Principle of Zero Products to find a quadratic equation for which the given numbers are solutions. Write the equation in standard form with integer coefficients.

a. 2, 4

$x^2 - 6x + 8 = 0$

b. −3, 2

$x^2 + x - 6 = 0$

c. $\frac{1}{2}, \frac{3}{4}$

$8x^2 - 10x + 3 = 0$

d. $-\sqrt{2}, \sqrt{2}$

$x^2 - 2 = 0$

54. The following was offered as the solution of the equation $(x - 2)(x + 3) = 4$.

$$(x - 2)(x + 3) = 4$$

$$x - 2 = 4 \quad \text{or} \quad x + 3 = 4$$

$$x = 6 \qquad\qquad x = 1$$

Explain why the solution is incorrect and provide the correct solution.
Answers will vary.

Exploration

55. ***Inductive Reasoning*** By using inductive reasoning, formulas for certain sums of natural numbers can be found.

a. Using the procedure for finding the sum of the first 10,000 natural numbers that was used in Chapter 1, find a formula for the sum of the first n natural numbers.

b. Find each of the following: $1 + 3$; $1 + 3 + 5$; $1 + 3 + 5 + 7$; $1 + 3 + 5 + 7 + 9$. On the basis of your findings, make a conjecture as to the sum of the first n odd natural numbers.

c. If n is a natural number, then $2n - 1$ is always an odd number. The sum of the first n odd natural numbers is $1 + 3 + 5 + \cdots + 2n - 1$. Use the method of part **a.** to verify the conjecture you made in part **b.**

d. Find each of the following: $2 + 4$; $2 + 4 + 6$; $2 + 4 + 6 + 8$; $2 + 4 + 6 + 8 + 10$. On the basis of your findings, make a conjecture as to the sum of the first n even natural numbers.

e. If n is a natural number, then $2n$ is always an even number. The sum of the first n even natural numbers is $2 + 4 + 6 + \cdots + 2n$. Use the method of part **a.** to verify the conjecture you made in part **d.**

Section 7.2 Solving Quadratic Equations by Completing the Square

Completing the Square

Not all quadratic equations can be solved easily by factoring. There is a method, however, called *completing the square* that can be used to solve all quadratic equations.

Recall that a perfect-square trinomial is the square of a binomial.

Perfect-square trinomial		*Square of a binomial*
$x^2 + 6x + 9$	=	$(x + 3)^2$
$x^2 - 12x + 36$	=	$(x - 6)^2$
$x^2 + 10x + 25$	=	$(x + 5)^2$

INSTRUCTOR NOTE
Completing the square is very difficult for students. Writing a procedure on the board similar to the one at the top of the next page and allowing students to use it will help many of them. Then erase the procedure and ask groups of students to create their own. Have each group present its procedure to the class for critique. The result of the discussions should be that each group ends up with a workable plan.

For each perfect-square trinomial, the square of $\frac{1}{2}$ of the coefficient of x equals the constant term.

$$\left(\frac{1}{2} \cdot \text{coefficient of } x\right)^2 = \text{constant term}$$

$x^2 + 6x + 9$ $\qquad \left(\frac{1}{2} \cdot 6\right)^2 = 3^2 = 9$

$x^2 - 12x + 36$ $\qquad \left[\frac{1}{2}(-12)\right]^2 = (-6)^2 = 36$

$x^2 + 10x + 25$ $\qquad \left(\frac{1}{2} \cdot 10\right)^2 = 5^2 = 25$

Adding a constant term to a binomial of the form $x^2 + bx$ that creates a perfect-square trinomial is called **completing the square.**

Example 1 Complete the square of $z^2 + 3z$. Write the resulting perfect-square trinomial as the square of a binomial.

Solution

$z^2 + 3z$

$\left[\frac{1}{2}(3)\right]^2 = \left(\frac{3}{2}\right)^2 = \frac{9}{4}$ • Find the constant term.

$z^2 + 3z + \frac{9}{4}$ • Complete the square of $z^2 + 3z$ by adding the constant term.

$z^2 + 3z + \frac{9}{4} = \left(x + \frac{3}{2}\right)^2$ • Write the resulting perfect-square trinomial as the square of a binomial.

You-Try-It 1 Complete the square of $x^2 - 8x$. Write the resulting perfect-square trinomial as the square of a binomial.

Solution See page S32.

POINT OF INTEREST

Early mathematicians solved quadratic equations by literally completing the square. For these mathematicians, all equations had geometric interpretations. They found that a quadratic equation could be solved by making certain figures into squares. See the Exploration at the end of this section for an idea of how this was done.

A quadratic equation that cannot be solved easily by factoring can be solved by completing the square. The purpose of completing the square is to rewrite a quadratic equation in a form that can be solved by taking the square root of each side of the equation. Here is the procedure.

1. Write the equation in the form $x^2 + bx = c$.
2. Add to each side of the equation the term that completes the square on $x^2 + bx$.
3. Factor the perfect-square trinomial. Write it as the square of a binomial.
4. Take the square root of each side of the equation.
5. Solve for x.

Example 2 Solve by completing the square: $x^2 = 6x + 3$

Solution

$x^2 = 6x + 3$

$x^2 - 6x = 3$ • Write the equation in the form $x^2 + bx = c$.

$x^2 - 6x + 9 = 3 + 9$ • Complete the square. Add $\left[\frac{1}{2}(-6)\right]^2 = 9$ to each side of the equation.

$(x-3)^2 = 12$ • Factor the trinomial.

$\sqrt{(x-3)^2} = \sqrt{12}$ • Take the square root of each side of the equation.

$x - 3 = \pm 2\sqrt{3}$ • Simplify the radical.

$x - 3 = 2\sqrt{3}$ $\quad$ $x - 3 = -2\sqrt{3}$ • Solve for x.

$x = 3 + 2\sqrt{3}$ $\quad$ $x = 3 - 2\sqrt{3}$

Graphical check:

For the graphical check, we have graphed the left and right sides of the original equation and then used the INTERSECT feature of the calculator to find the points of intersection.

The exact solutions are $3 + 2\sqrt{3}$ and $3 - 2\sqrt{3}$. The approximate solutions are 6.464102 and −0.464102.

You-Try-It 2 Solve by completing the square: $y^2 + 4y - 6 = 0$

Solution See page S32.

SUGGESTED ACTIVITY

The *Instructor's Resource Manual* contains an activity that lets students investigate the relationship between the x-intercepts of the graph of $y = ax^2 + bx + c$ and the roots of $ax^2 + bx + c = 0$.

It is important to note that completing the square does not change the equation. For instance, in Example 2,

$$x^2 = 6x + 3 \qquad \text{became} \qquad (x-3)^2 = 12$$

If we simplify $(x-3)^2 = 12$, we have

$(x-3)^2 = 12$

$x^2 - 6x + 9 = 12$ • Square the binomial.

$x^2 = 6x + 3$ • Add $6x$ to each side of the equation and subtract 9 from each side.

This is the original equation. This shows that the two equations are the same.

A graphical illustration also shows that the equations are the same. Write $x^2 = 6x + 3$ as $x^2 - 6x - 3 = 0$ and $(x-3)^2 = 12$ as $(x-3)^2 - 12 = 0$. Now replace 0 by y and then graph each equation. The graphs are shown on the next page.

TAKE NOTE
We have shown the graphs on separate screens. Try drawing the graphs on the *same* screen. You should notice that the two graphs appear one on top of the other.

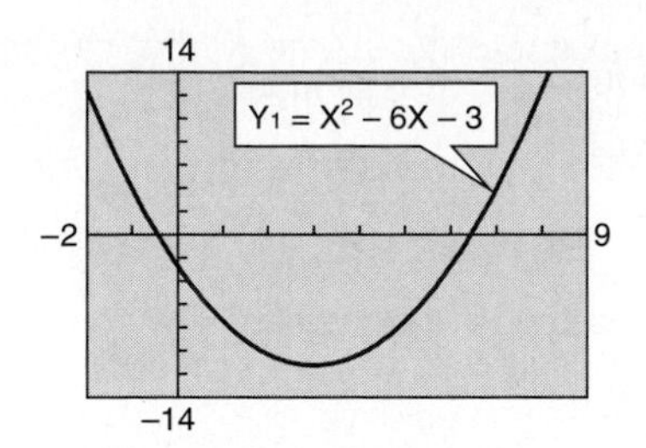

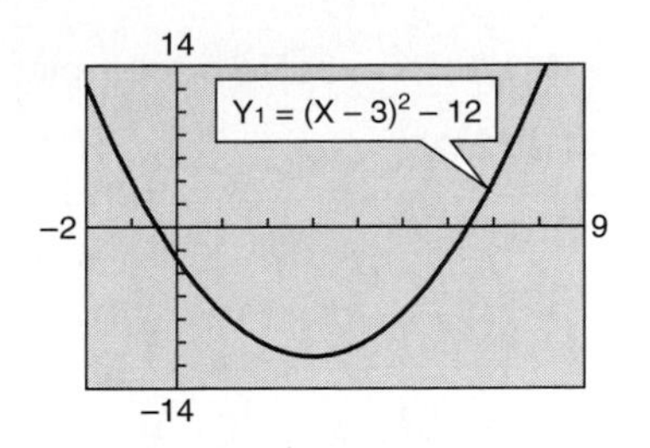

For the completing the square method to work, a, the coefficient of x^2, must be 1. If it is not 1, first multiply each side of the equation by $\frac{1}{a}$.

Example 3 Solve by completing the square: $3x^2 - 4x - 9 = 0$

Solution

Algebraic solution

$$3x^2 - 4x - 9 = 0$$

$$\frac{1}{3}(3x^2 - 4x - 9) = \frac{1}{3}(0)$$ • Multiply each side of the equation by $\frac{1}{3}$.

$$x^2 - \frac{4}{3}x - 3 = 0$$

$$x^2 - \frac{4}{3}x = 3$$ • Write the equation in the form $x^2 + bx = c$.

$$x^2 - \frac{4}{3}x + \frac{4}{9} = 3 + \frac{4}{9}$$ • Complete the square. To each side of the equation, add $\left[\frac{1}{2}\left(-\frac{4}{3}\right)\right]^2 = \left(-\frac{2}{3}\right)^2 = \frac{4}{9}$.

$$\left(x - \frac{2}{3}\right)^2 = \frac{31}{9}$$ • Factor the trinomial. Simplify the right side.

$$\sqrt{\left(x - \frac{2}{3}\right)^2} = \sqrt{\frac{31}{9}}$$ • Take the square root of each side of the equation.

$$x - \frac{2}{3} = \pm\frac{\sqrt{31}}{3}$$ • Simplify the radical.

$$x - \frac{2}{3} = \frac{\sqrt{31}}{3} \qquad x - \frac{2}{3} = -\frac{\sqrt{31}}{3}$$ • Solve for x.

$$x = \frac{2}{3} + \frac{\sqrt{31}}{3} \qquad x = \frac{2}{3} - \frac{\sqrt{31}}{3}$$

$$x = \frac{2 + \sqrt{31}}{3} \qquad x = \frac{2 - \sqrt{31}}{3}$$

Graphical check:

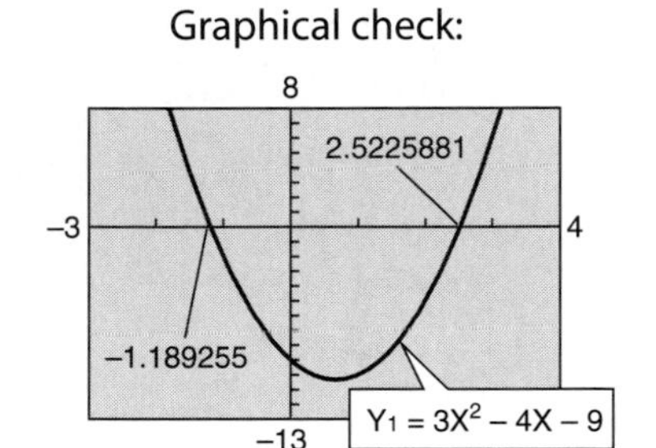

The exact solutions are $\frac{2 + \sqrt{31}}{3}$ and $\frac{2 - \sqrt{31}}{3}$. The approximate solutions are 2.5225881 and −1.1892548.

You-Try-It 3 Solve by completing the square: $2x^2 + 8x = 3$

Solution See page S33.

Question: Referring to Example 3, how are the x-intercepts for the graph of $Y_1 = 3x^2 - 4x - 9$ related to the solutions of the equation?[1]

1. The x-coordinates of the x-intercepts are the solutions of the equation.

Example 4 Solve by completing the square: $2x^2 - 4x - 1 = 0$

Solution

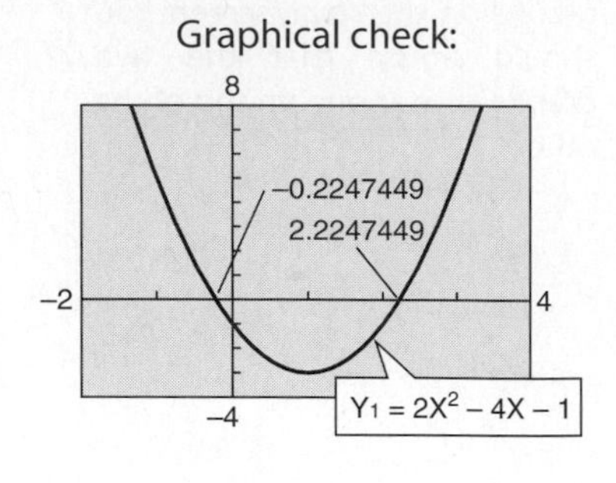

$2x^2 - 4x - 1 = 0$

$\frac{1}{2}(2x^2 - 4x - 1) = \frac{1}{2}(0)$ • Multiply each side of the equation by $\frac{1}{2}$.

$x^2 - 2x - \frac{1}{2} = 0$

$x^2 - 2x = \frac{1}{2}$ • Write the equation in the form $x^2 + bx = c$.

$x^2 - 2x + 1 = \frac{1}{2} + 1$ • Complete the square. To each side of the equation, add $\left[\frac{1}{2}(-2)\right]^2 = (-1)^2 = 1$.

$(x-1)^2 = \frac{3}{2}$ • Factor the trinomial. Simplify the right side.

$\sqrt{(x-1)^2} = \sqrt{\frac{3}{2}}$ • Take the square root of each side of the equation.

$x - 1 = \pm\frac{\sqrt{3}}{\sqrt{2}}$ • Simplify the radical.

$x - 1 = \pm\frac{\sqrt{6}}{2}$ • Rationalize the denominator. $\frac{\sqrt{3}}{\sqrt{2}} = \frac{\sqrt{3}}{\sqrt{2}} \cdot \frac{\sqrt{2}}{\sqrt{2}} = \frac{\sqrt{6}}{2}$

$x - 1 = \frac{\sqrt{6}}{2}$ $\qquad$ $x - 1 = -\frac{\sqrt{6}}{2}$ • Solve for x.

$x = 1 + \frac{\sqrt{6}}{2}$ $\qquad$ $x = 1 - \frac{\sqrt{6}}{2}$

$= \frac{2}{2} + \frac{\sqrt{6}}{2} = \frac{2 + \sqrt{6}}{2}$ $\qquad$ $= \frac{2}{2} - \frac{\sqrt{6}}{2} = \frac{2 - \sqrt{6}}{2}$

The exact solutions are $\frac{2 + \sqrt{6}}{2}$ and $\frac{2 - \sqrt{6}}{2}$. The approximate solutions are 2.2247449 and −0.2247449.

You-Try-It 4 Solve by completing the square: $3x^2 - 6x - 2 = 0$

Solution See page S33.

Not all quadratic equations that are solved by completing the square result in solutions that are real numbers. Consider the equation $x^2 - 6x + 13 = 0$.

$x^2 - 6x + 13 = 0$

$x^2 - 6x = -13$ • Write the equation in the form $x^2 + bx = c$.

$x^2 - 6x + 9 = -13 + 9$ • Complete the square. Add $\left[\frac{1}{2}(-6)\right]^2 = 9$ to each side of the equation.

$(x-3)^2 = -4$ • Factor the trinomial.

$\sqrt{(x-3)^2} = \sqrt{-4}$ • Take the square root of each side of the equation.

Since $\sqrt{-4}$ is not a real number, the equation has no real number solutions.

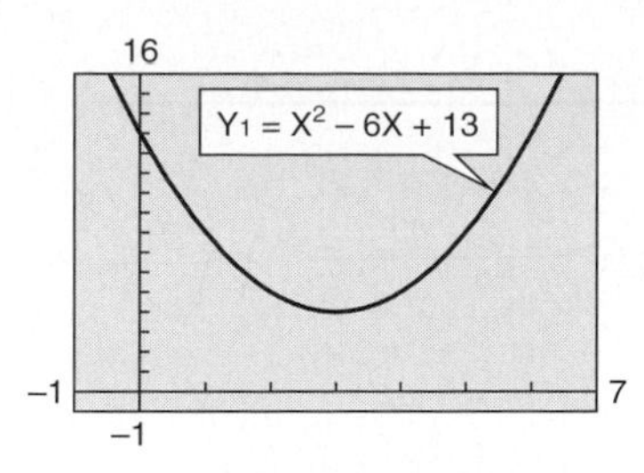

The graph of $y = x^2 - 6x + 13$ is shown at the left. Note that the graph does not pass through the x-axis. Thus there are no real numbers x for which $y = 0$. Therefore, there are no real numbers x for which $x^2 - 6x + 13 = 0$.

Question: The graph at the left does not have any x-intercepts. Does it have a y-intercept and, if so, what is the y-intercept?[2]

Applications

Example 5 The owners of a family campground have decided to enclose a rectangular area for a children's playground, using a stream running through the campground as one of the sides. The length of the rectangle is 50 meters more than the width, and the area of the rectangle is to be 600 square meters. What are the length and width of the rectangle?

Solution

State the goal. We must find the length and width of the playground.

Describe a strategy. The length of the rectangle is 50 meters more than the width. Letting W represent the width and L represent the length, we have $L = W + 50$.

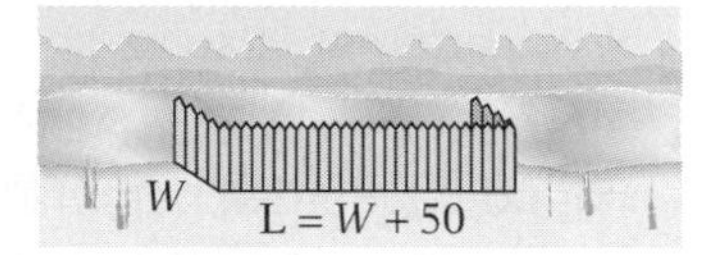

Because we are given the area ($A = 600$ m^2) of the rectangle, we use the formula for the area of a rectangle, $A = LW$.

Solve the problem.

$$A = LW$$

$$600 = (W+50)W \qquad \bullet\ L = W + 50. \text{ Replace } L \text{ by } W + 50.$$

$$600 = W^2 + 50W \qquad \bullet \text{ Simplify.}$$

$$600 + 625 = W^2 + 50W + 625 \qquad \bullet \text{ Complete the square. } \left(\frac{1}{2}\cdot 50\right)^2 = 25^2 = 625$$

$$1225 = (W+25)^2$$

$$\sqrt{1225} = \sqrt{(W+25)^2}$$

$$\pm 35 = W + 25 \qquad \bullet \text{ Solve for } W.$$

$$35 = W + 25 \qquad -35 = W + 25$$

$$10 = W \qquad -60 = W$$

A width of −60 meters would not make sense. Thus the width is 10 meters.

$$L = W + 50 \qquad \bullet \text{ The length is 50 meters more than the width.}$$

$$= 10 + 50 = 60 \qquad \bullet\ W = 10. \text{ Replace } W \text{ by } 10.$$

The length is 60 meters.

Check your work. Check that the dimensions produce a rectangle with an area of 600 m^2.

$$A = LW$$

$$= 60(10) = 600 \qquad \bullet\ L = 60, W = 10$$

The solution checks.

You-Try-It 5 A rock is tossed upward from the top of a cliff that is 100 feet above the ocean. The height, $h(t)$, of the rock above the ocean t seconds after it has been released is given by $h(t) = -16t^2 + 64t + 100$. How many seconds after the rock is released is it 36 feet above the ocean? Round to the nearest thousandth.

Solution See page S33.

---

2. Yes. The y-intercept is (0, 13). This can be determined from the equation by replacing x by 0 in the equation $y = x^2 - 6x + 13$ and solving for y.

7.2 EXERCISES

Topics for Discussion

1. Explain how to complete the square of a binomial.
   Answers will vary.

2. What are the advantages and disadvantages of using completing the square to solve a quadratic equation?
   Answers will vary.

3. What is the next step in solving the quadratic equation $2x^2 - 4x = 7$ by completing the square?
   Divide both sides of the equation by 2, the coefficient of x^2.

4. If you attempt to solve $x^2 - 6x + 9 = 0$ by completing the square, the result after simplifying is $(x-3)^2 = 0$. Does this mean that the original equation has no solution? If not, what are the solutions of the equation?
   No. The equation has a double root, 3.

Quadratic Equations

Complete the square. Write the resulting perfect-square trinomial as the square of a binomial.

5. $x^2 + 12x$
   $x^2 + 12x + 36;\ (x+6)^2$

6. $x^2 - 4x$
   $x^2 - 4x + 4;\ (x-2)^2$

7. $x^2 + 10x$
   $x^2 + 10x + 25;\ (x+5)^2$

8. $x^2 + 3x$
   $x^2 + 3x + \frac{9}{4};\ \left(x+\frac{3}{2}\right)^2$

9. $x^2 - x$
   $x^2 - x + \frac{1}{4};\ \left(x-\frac{1}{2}\right)^2$

10. $x^2 + 5x$
    $x^2 + 5x + \frac{25}{4};\ \left(x+\frac{5}{2}\right)^2$

Solve by completing the square.

11. $x^2 + 2x - 3 = 0$
    $-3, 1$

12. $y^2 + 4y - 5 = 0$
    $-5, 1$

13. $v^2 + 4v + 1 = 0$
    $-2 \pm \sqrt{3}$

14. $y^2 - 2y - 5 = 0$
    $1 \pm \sqrt{6}$

15. $v^2 - 6v + 13 = 0$
    No real number solution

16. $x^2 + 4x + 13 = 0$
    No real number solution

17. $x^2 + 6x = 5$
    $-3 \pm \sqrt{14}$

18. $w^2 - 8w = 3$
    $4 \pm \sqrt{19}$

19. $x^2 = 4x - 4$
    2

20. $z^2 = 8z - 16$
    4

21. $z^2 = 2z + 1$
    $1 \pm \sqrt{2}$

22. $y^2 = 10y - 20$
    $5 \pm \sqrt{5}$

Solve. First try to solve the equation by factoring. If you are unable to solve the equation by factoring, solve the equation by completing the square.

23. $p^2 + 3p = 1$
$\frac{-3 \pm \sqrt{13}}{2}$

24. $r^2 + 5r = 2$
$\frac{-5 \pm \sqrt{33}}{2}$

25. $w^2 + 7w = 8$
$-8, 1$

26. $y^2 + 5y = -4$
$-1, -4$

27. $x^2 + 6x + 4 = 0$
$-3 \pm \sqrt{5}$

28. $y^2 - 8y - 1 = 0$
$4 \pm \sqrt{17}$

29. $r^2 - 8r = -2$
$4 \pm \sqrt{14}$

30. $s^2 + 6s = 5$
$-3 \pm \sqrt{14}$

31. $t^2 - 3t = -2$
$1, 2$

32. $y^2 = 4y + 12$
$6, -2$

33. $w^2 = 3w + 5$
$\frac{3 \pm \sqrt{29}}{2}$

34. $x^2 = 1 - 3x$
$\frac{-3 \pm \sqrt{13}}{2}$

35. $x^2 - x - 1 = 0$
$\frac{1 \pm \sqrt{5}}{2}$

36. $x^2 - 7x = -3$
$\frac{7 \pm \sqrt{37}}{2}$

37. $y^2 - 5y + 3 = 0$
$\frac{5 \pm \sqrt{13}}{2}$

38. $z^2 - 5z = -2$
$\frac{5 \pm \sqrt{17}}{2}$

39. $v^2 + v - 3 = 0$
$\frac{-1 \pm \sqrt{13}}{2}$

40. $x^2 - x = 1$
$\frac{1 \pm \sqrt{5}}{2}$

41. $y^2 = 7 - 10y$
$-5 \pm 4\sqrt{2}$

42. $v^2 = 14 + 16v$
$8 \pm \sqrt{78}$

43. $s^2 + 3s = -1$
$\frac{-3 \pm \sqrt{5}}{2}$

44. $r^2 - 3r = 5$
$\frac{3 \pm \sqrt{29}}{2}$

45. $t^2 - t = 4$
$\frac{1 \pm \sqrt{17}}{2}$

46. $y^2 + y - 4 = 0$
$\frac{-1 \pm \sqrt{17}}{2}$

47. $x^2 - 3x + 5 = 0$
No real number solution

48. $z^2 + 5z + 7 = 0$
No real number solution

49. $2t^2 - 3t + 1 = 0$
$1, \frac{1}{2}$

50. $2x^2 - 7x + 3 = 0$
$3, \frac{1}{2}$

51. $2r^2 + 5r = 3$
$-3, \frac{1}{2}$

52. $2y^2 - 3y = 9$
$3, -\frac{3}{2}$

53. $4v^2 - 4v - 1 = 0$
$\frac{1 \pm \sqrt{2}}{2}$

54. $2s^2 - 4s - 1 = 0$
$\frac{2 \pm \sqrt{6}}{2}$

55. $4z^2 - 8z = 1$
$\frac{2 \pm \sqrt{5}}{2}$

56. $3r^2 - 2r = 2$
$\frac{1 \pm \sqrt{7}}{3}$

57. $3y - 5 = (y - 1)(y - 2)$
$3 \pm \sqrt{2}$

58. $4p + 2 = (p - 1)(p + 3)$
$1 \pm \sqrt{6}$

59. $\frac{x^2}{4} - \frac{x}{2} = 3$
$1 \pm \sqrt{13}$

60. $\frac{x^2}{6} - \frac{x}{3} = 1$
$1 \pm \sqrt{7}$

61. $\frac{2x^2}{3} = 2x + 3$
$\frac{3 \pm 3\sqrt{3}}{2}$

62. $\frac{3x^2}{2} = 3x + 2$
$\frac{3 \pm \sqrt{21}}{3}$

63. $\frac{x}{3} + \frac{3}{x} = \frac{8}{3}$
$4 \pm \sqrt{7}$

64. $\frac{x}{4} - \frac{2}{x} = \frac{3}{4}$
$\frac{3 \pm \sqrt{41}}{2}$

65. A rectangular corral is constructed in a pasture. The length of the rectangle is 8 feet longer than twice the width. The area of the rectangle is to be 640 square feet. What are the length and width of the rectangle?
Length: 40 feet; width: 16 feet.

66. In Germany, there are no speed limits on some portions of the autobahn (highway). Other portions have a speed limit of 180 kilometers per hour (approximately 112 miles per hour). The distance, d (in meters), required to stop a car traveling at v kilometers per hour is $d = 0.0056v^2 + 0.14v$. Approximate, to the nearest tenth, the maximum speed a driver can be going and still be able to stop within 150 meters.
151.6 km/h

67. A penalty kick in soccer is made from a penalty mark that is 36 feet from a goal which is 8 feet high. A possible equation for the flight of a penalty kick is $h - -0.002x^2 + 0.36x$, where h is the height (in feet) of the ball x feet from the penalty mark. Assuming that the flight of the kick is toward the goal and that it is not touched by the goalie, will the ball land in the net?
No. The ball is too high by about 2.4 feet.

68. The Water Arc is a fountain that shoots water across the Chicago River from a water cannon. The path of the water can be approximated by the equation $h = -0.006x^2 + 1.2x + 10$, where x is the horizontal distance from the cannon and h is the height of the water above the river. On one particular day, some people were walking along the opposite side of the river from the Water Arc when a pulse of water was shot in their direction. If the distance from the Water Arc to the people was 220 feet, did they get wet from the cannon's water?
No. The water traveled only about 208 feet.

Applying Concepts

69. Solve: $\sqrt{2x+7} - 4 = x$
-3

70. Solve: $\sqrt{x+5} = \sqrt{x} + 1$
4

Find b so that the trinomial is a perfect square.

71. $x^2 + bx + 25$
$-10, 10$

72. $x^2 + bx + 100$
$-20, 20$

73. $x^2 + bx + 12$
$-4\sqrt{3}, 4\sqrt{3}$

74. $x^2 + bx + 18$
$-6\sqrt{2}, 6\sqrt{2}$

75. $4x^2 + bx + 49$
$-28, 28$

76. $x^2 + bx + c^2$
$-2c, 2c$

Recall that division by zero is not allowed. For the problems below, what values of x are not possible for each of the fractions?

77. $\dfrac{1}{(x+2)(x-3)}$

$-2, 3$

78. $\dfrac{1}{(2x+1)(x-1)}$

$-\frac{1}{2}, 1$

79. $\dfrac{1}{x^2+2x-8}$

$-4, 2$

80. $\dfrac{1}{x^2+4x-5}$

$-5, 1$

81. $\dfrac{1}{x^2+2x-5}$

$-1+\sqrt{6}, -1-\sqrt{6}$

82. $\dfrac{1}{x^2-3x-6}$

$\dfrac{3+\sqrt{33}}{2}, \dfrac{3-\sqrt{33}}{2}$

Exploration

83. ***Completing the Square*** The method of completing the square has been known since at least A.D. 825, when al-Khwarizmi used it in a textbook. The method is very geometrical and literally completes a square. In the figure below are three figures: a square whose area is x^2, a rectangle whose area is x, and a smaller square whose area is 1.

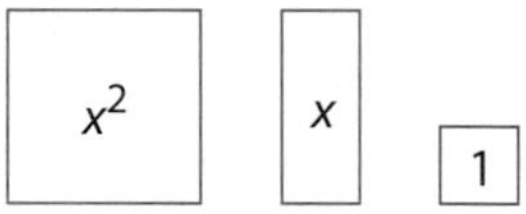

x^2 x 1

Now consider the expression $x^2 + 6x$. To complete the square of this, we would add $\left(\frac{1}{2}\cdot 6\right)^2 = 3^2 = 9$ to the expression. Here is the geometric construction that al-Khwarizmi used.

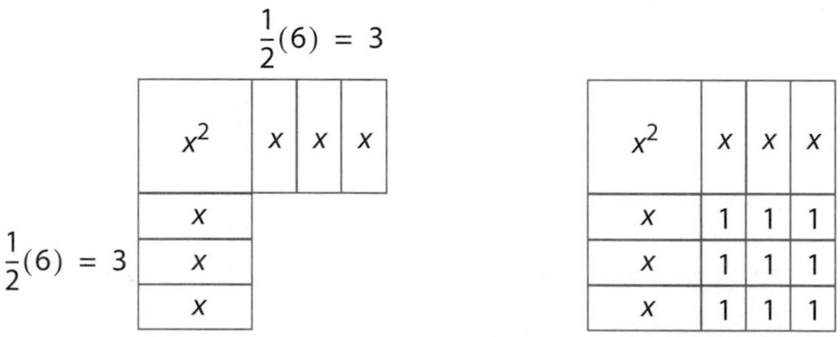

$\frac{1}{2}(6) = 3$

x^2	x	x	x
x			
x			
x			

$\frac{1}{2}(6) = 3$

x^2	x	x	x
x	1	1	1
x	1	1	1
x	1	1	1

9 squares were added

Note that it is necessary to add 9 squares to the figure to "complete the square." One of the difficulties in using a geometric approach such as this is that it cannot easily be extended to $x^2 - 6x$. There is no way to draw $-6x$! That did not bother al-Khwarizmi because negative numbers were not a significant part of mathematics until well into the 13th century.

a. Show how al-Khwarizmi would have completed the square for $x^2 + 4x$.

b. Show how al-Khwarizmi would have completed the square for $x^2 + 10x$.

Section 7.3 Solving Quadratic Equations Using the Quadratic Formula

Quadratic Formula

INSTRUCTOR NOTE
Students may wonder why we discuss the different methods for solving quadratic equations. Explain to students that factoring, when it is easy to determine the factors, is the quickest method. Completing the square is a method that is used in other situations and is the method we used to derive the quadratic formula. Completing the square is rarely used to solve an equation. The quadratic formula is used for those equations for which it is not easy to determine the factors. Of course, it is also possible to find a graphical solution. A possible class discussion topic is the weaknesses and strengths of each method.

Any quadratic equation can be solved by completing the square. Applying this method to the standard form of a quadratic equation produces a formula, called the *quadratic formula,* that can be used to solve a quadratic equation.

To solve $ax^2 + bx + c = 0, a \neq 0$, by completing the square, subtract c from each side of the equation.

$$ax^2 + bx + c = 0$$
$$ax^2 + bx + c - c = 0 - c$$
$$ax^2 + bx = -c$$

Multiply each side of the equation by $\frac{1}{a}$, the reciprocal of a.

$$\frac{1}{a}(ax^2 + bx) = \frac{1}{a}(-c)$$
$$x^2 + \frac{b}{a}x = -\frac{c}{a}$$

Complete the square by adding $\left(\frac{1}{2} \cdot \frac{b}{a}\right)^2 = \frac{b^2}{4a^2}$ to each side of the equation.

$$x^2 + \frac{b}{a}x + \frac{b^2}{4a^2} = \frac{b^2}{4a^2} - \frac{c}{a}$$

Simplify the right side of the equation.

$$x^2 + \frac{b}{a}x + \frac{b^2}{4a^2} = \frac{b^2}{4a^2} - \frac{c}{a}\left(\frac{4a}{4a}\right)$$
$$x^2 + \frac{b}{a}x + \frac{b^2}{4a^2} = \frac{b^2}{4a^2} - \frac{4ac}{4a^2}$$
$$x^2 + \frac{b}{a}x + \frac{b^2}{4a^2} = \frac{b^2 - 4ac}{4a^2}$$

Factor the left side of the equation.

$$\left(x + \frac{b}{2a}\right)^2 = \frac{b^2 - 4ac}{4a^2}$$

Take the square root of each side of the equation.

$$\sqrt{\left(x + \frac{b}{2a}\right)^2} = \sqrt{\frac{b^2 - 4ac}{4a^2}}$$
$$x + \frac{b}{2a} = \pm\frac{\sqrt{b^2 - 4ac}}{2a}$$

Solve for x.

$$x + \frac{b}{2a} = \frac{\sqrt{b^2 - 4ac}}{2a} \qquad x + \frac{b}{2a} = -\frac{\sqrt{b^2 - 4ac}}{2a}$$
$$x = -\frac{b}{2a} + \frac{\sqrt{b^2 - 4ac}}{2a} \qquad x = -\frac{b}{2a} - \frac{\sqrt{b^2 - 4ac}}{2a}$$
$$x = \frac{-b + \sqrt{b^2 - 4ac}}{2a} \qquad x = \frac{-b - \sqrt{b^2 - 4ac}}{2a}$$

The Quadratic Formula

The solution of the equation $ax^2 + bx + c = 0, a \neq 0$, is

$$x = \frac{-b + \sqrt{b^2 - 4ac}}{2a} \text{ or } x = \frac{-b - \sqrt{b^2 - 4ac}}{2a}$$

The quadratic formula is frequently written in the form

$$x = \frac{-b \pm \sqrt{b^2 - 4ac}}{2a}$$

Example 1 Solve using the quadratic formula: $x^2 - 4x - 12 = 0$

Solution

$x^2 - 4x - 12 = 0$ • The equation is in standard form. $a = 1, b = -4$, and $c = -12$.

$$x = \frac{-b \pm \sqrt{b^2 - 4ac}}{2a}$$

$$= \frac{-(-4) \pm \sqrt{(-4)^2 - 4(1)(-12)}}{2(1)}$$ • Replace a, b, and c in the quadratic formula by these values.

$$= \frac{4 \pm \sqrt{16 + 48}}{2}$$ • Simplify.

$$= \frac{4 \pm \sqrt{64}}{2} = \frac{4 \pm 8}{2}$$

$$x = \frac{4+8}{2} = \frac{12}{2} = 6 \qquad x = \frac{4-8}{2} = \frac{-4}{2} = -2$$

The solutions are −2 and 6.

Graphical check:

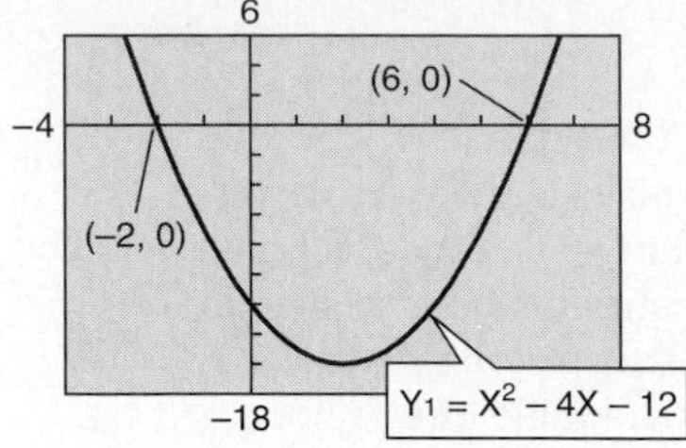

The solutions of $x^2 - 4x - 12 = 0$ are the x-coordinates of the x-intercepts of the graph of $y = x^2 - 4x - 12$. Since the x-intercepts are (−2, 0) and (6, 0), the solutions are −2 and 6.

You-Try-It 1 Solve by using the quadratic formula: $x^2 + 3x + 2 = 0$

Solution See page S33.

The quadratic equation in Example 1 could have been solved by factoring instead of using the quadratic formula. Here is the solution. In those cases where it is fairly easy to find the factors of the quadratic polynomial, factoring is the quickest method of solving the quadratic equation. If the factors are not readily apparent, use the quadratic formula.

$$x^2 - 4x - 12 = 0$$
$$(x+2)(x-6) = 0$$
$$x + 2 = 0 \qquad x - 6 = 0$$
$$x = -2 \qquad x = 6$$

Example 2 Solve by using the quadratic formula: $2x^2 = 4x + 3$

Solution

$$2x^2 = 4x + 3$$

$2x^2 - 4x - 3 = 0$ • Write the equation in standard form. $a = 2, b = -4$, and $c = -3$

$$x = \frac{-b \pm \sqrt{b^2 - 4ac}}{2a}$$

$$= \frac{-(-4) \pm \sqrt{(-4)^2 - 4(2)(-3)}}{2(2)}$$ • Replace a, b, and c in the quadratic formula by these values.

$$= \frac{4 \pm \sqrt{16 + 24}}{4}$$ • Simplify.

$$= \frac{4 \pm \sqrt{40}}{4} = \frac{4 \pm 2\sqrt{10}}{4}$$

$$= \frac{2(2 \pm \sqrt{10})}{2 \cdot 2} = \frac{2 \pm \sqrt{10}}{2}$$

Graphical check:

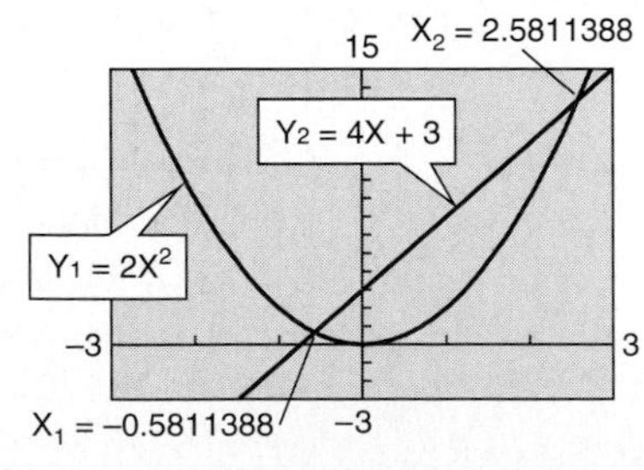

Let $Y_1 = 2x^2$ and $Y_2 = 4x + 3$. Then $2x^2 = 4x + 3$ when $Y_1 = Y_2$. By graphing each equation, $Y_1 = Y_2$ when $x \approx -0.5811388$ and $x \approx 2.5811388$.

The exact solutions are $\frac{2 + \sqrt{10}}{2}$ and $\frac{2 - \sqrt{10}}{2}$. The approximate solutions are 2.5811388 and −0.5811388.

You-Try-It 2 Solve by using the quadratic formula: $x^2 - 1 = 3x$

Solution See page S34.

TAKE NOTE
Note that for the solution at the right, $a = -2, b = 4, c = 3$. In Example 2, $a = 2, b = -4$, and $c = -3$. The solution of the equation, however, is the same.

In Example 2, we wrote the equation $2x^2 = 4x + 3$ in standard form by subtracting $4x$ and subtracting 3 from each side of the equation. We could have just subtracted $2x^2$ from each side of the equation. The solutions would have been the same.

$$2x^2 = 4x + 3$$

$$0 = -2x^2 + 4x + 3 \quad \text{• Subtract } 2x^2 \text{ from each side of the equation.}$$

$$x = \frac{-4 \pm \sqrt{4^2 - 4(-2)(3)}}{2(-2)} \quad \text{• } a = -2, b = 4, \text{ and } c = 3.$$

$$= \frac{-4 \pm \sqrt{16 + 24}}{-4}$$

$$= \frac{-4 \pm \sqrt{40}}{-4}$$

$$= \frac{-4 \pm 2\sqrt{10}}{-4}$$

$$= \frac{-2(2 \pm \sqrt{10})}{-2(2)}$$

$$= \frac{2 \pm \sqrt{10}}{2}$$

SUGGESTED ACTIVITY
The *Instructor's Resource Manual* contains an activity that lets students investigate the relationship between a quadratic equation with no real number solutions and its graph.

This suggests that you can write a quadratic equation as either $ax^2 + bx + c = 0$ or as $0 = ax^2 + bx + c$ before applying the quadratic formula.

Applying the quadratic formula to some quadratic equations will show that the equation has no real number solutions.

Example 3 Solve by using the quadratic formula: $-2x^2 + x - 4 = 0$

Solution

$$-2x^2 + x - 4 = 0 \quad \text{• The equation is in standard form.}$$

$$x = \frac{-b \pm \sqrt{b^2 - 4ac}}{2a}$$

$$= \frac{-1 \pm \sqrt{1^2 - 4(-2)(-4)}}{2(-2)} \quad \text{• } a = -2, b = 1, \text{ and } c = -4.$$

$$= \frac{-1 \pm \sqrt{1 - 32}}{-4}$$

$$= \frac{-1 \pm \sqrt{-31}}{-4} \quad \text{• } \sqrt{-31} \text{ is not a real number.}$$

Graphical check:

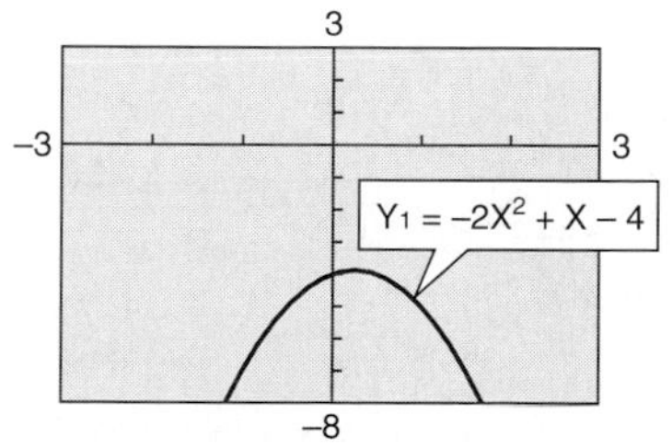

Because the graph does not cross the x-axis, there are no x-intercepts and therefore no real number solutions of the equation $-2x^2 + x - 4 = 0$.

Because $\sqrt{-31}$ is not a real number, the equation has no real number solutions.

You-Try-It 3 Solve by using the quadratic formula: $x^2 = 3x - 6$

Solution See page S34.

Applications

Example 4 In a slow-pitch softball game, the height of the ball thrown by a pitcher can be modeled by the equation $h = -16t^2 + 24t + 4$, where h is the height of the ball and t is the time since it was released by the pitcher. If the batter hits the ball when it is 2 feet off the ground, how many seconds has the ball been in the air? Round to the nearest hundredth of a second.

Solution

State the goal. We must find the number of seconds the ball has been in the air.

Describe a strategy. The batter hits the ball when it is 2 feet off the ground. Therefore, $h = 2$ ft. Substitute this value of h into the equation $h = -16t^2 + 24t + 4$ and solve for t.

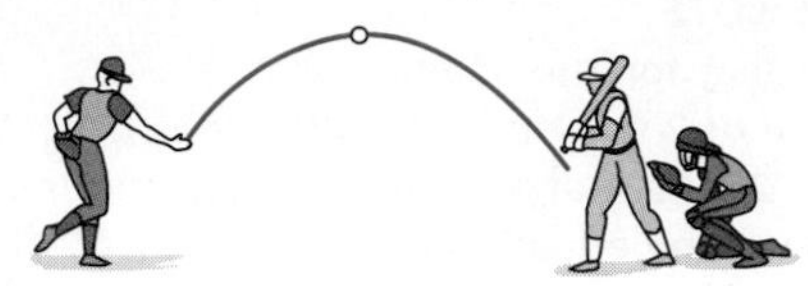

Solve the problem.

$$h = -16t^2 + 24t + 4$$

$$2 = -16t^2 + 24t + 4$$ • Replace h by 2.

$$0 = -16t^2 + 24t + 2$$ • Write the equation in standard form.

$$t = \frac{-24 \pm \sqrt{24^2 - 4(-16)(2)}}{2(-16)}$$ • Solve for t.

$$= \frac{-24 \pm \sqrt{704}}{-32}$$

$$t = \frac{-24 + \sqrt{704}}{-32} \approx -0.0791562 \qquad t = \frac{-24 - \sqrt{704}}{-32} \approx 1.5791562$$

A time of −0.079 second would not make sense in the context of this problem. Therefore, the time the ball is in the air is approximately 1.58 seconds.

Check your work. One way to check your work is to replace t by 1.58 seconds in the equation $h = -16t^2 + 24t + 4$. The value of h should be approximately 2 feet.

$$h = -16t^2 + 24t + 4$$

$$= -16(1.58)^2 + 24(1.58) + 4$$ • Replace t by 1.58.

$$= 1.9776$$

Since $h \approx 2$ when $t = 1.58$, it appears that our answer is correct.

You-Try-It 4 The path of water from a hose on a fire tugboat can be modeled by the equation $y = -0.005x^2 + 1.2x + 10$, where y is the height of the water above the ocean when it is x feet from the tugboat. At what distance from the tugboat is the water from the hose 5 feet above the ocean? Round to the nearest hundredth of a foot.

Solution See page S34.

7.3 EXERCISES

Topics for Discussion

1. What is the quadratic formula?
   A formula that is used to solve a quadratic equation and given by $x = \dfrac{-b \pm \sqrt{b^2 - 4ac}}{2a}$.

2. Are there quadratic equations that cannot be solved by using the quadratic formula? If so, give an example of one.
   No.

3. Discuss the pros and cons of using factoring, graphing, completing the square, and the quadratic formula as a method of solving a quadratic equation.
   Answers will vary.

Quadratic Equations

Solve by using the quadratic formula.

4. $z^2 + 6z - 7 = 0$
   $-7, 1$

5. $s^2 + 3s - 10 = 0$
   $-5, 2$

6. $w^2 = 3w + 18$
   $6, -3$

7. $r^2 = 5 - 4r$
   $-5, 1$

8. $t^2 - 2t = 5$
   $1 \pm \sqrt{6}$

9. $y^2 - 4y = 6$
   $2 \pm \sqrt{10}$

10. $t^2 + 6t - 1 = 0$
    $-3 \pm \sqrt{10}$

11. $z^2 + 4z + 1 = 0$
    $-2 \pm \sqrt{3}$

12. $w^2 + 3w + 5 = 0$
    No real number solution

13. $x^2 - 2x + 6 = 0$
    No real number solution

14. $w^2 = 4w + 9$
    $2 \pm \sqrt{13}$

15. $y^2 = 8y + 3$
    $4 \pm \sqrt{19}$

Solve. First try to solve the equation by factoring. If you are unable to solve the equation by factoring, solve the equation by using the quadratic formula.

16. $p^2 - p = 0$
    $0, 1$

17. $2v^2 + v = 0$
    $0, -\frac{1}{2}$

18. $4t^2 - 4t - 1 = 0$
    $\dfrac{1 \pm \sqrt{2}}{2}$

19. $4x^2 - 8x - 1 = 0$
    $\dfrac{2 \pm \sqrt{5}}{2}$

20. $4t^2 - 9 = 0$
    $\pm\frac{3}{2}$

21. $4s^2 - 25 = 0$
    $\pm\frac{5}{2}$

22. $3x^2 - 6x + 2 = 0$
    $\dfrac{3 \pm \sqrt{3}}{3}$

23. $5x^2 - 6x = 3$
    $\dfrac{3 \pm 2\sqrt{6}}{5}$

24. $3t^2 = 2t + 3$
    $\dfrac{1 \pm \sqrt{10}}{3}$

25. $4n^2 = 7n - 2$

$\frac{7 \pm \sqrt{17}}{8}$

26. $2x^2 + x + 1 = 0$

No real number solution

27. $3r^2 - r + 2 = 0$

No real number solution

28. $2y^2 + 3 = 8y$

$\frac{4 \pm \sqrt{10}}{2}$

29. $5x^2 - 1 = x$

$\frac{1 \pm \sqrt{21}}{10}$

30. $3t^2 = 7t + 6$

$-\frac{2}{3}, 3$

31. $3x^2 = 10x + 8$

$-\frac{2}{3}, 4$

32. $3y^2 - 4 = 5y$

$\frac{5 \pm \sqrt{73}}{6}$

33. $6x^2 - 5 = 3x$

$\frac{3 \pm \sqrt{129}}{12}$

34. $3x^2 = x + 3$

$\frac{1 \pm \sqrt{37}}{6}$

35. $2n^2 = 7 - 3n$

$\frac{-3 \pm \sqrt{65}}{4}$

36. $5d^2 - 2d - 8 = 0$

$\frac{1 \pm \sqrt{41}}{5}$

37. $x^2 - 7x - 10 = 0$

$\frac{7 \pm \sqrt{89}}{2}$

38. $5z^2 + 11z = 12$

$\frac{4}{5}, -3$

39. $4v^2 = v + 3$

$1, -\frac{3}{4}$

40. $v^2 + 6v + 1 = 0$

$-3 \pm 2\sqrt{2}$

41. $s^2 + 4s - 8 = 0$

$-2 \pm 2\sqrt{3}$

42. $4t^2 - 12t - 15 = 0$

$\frac{3 \pm 2\sqrt{6}}{2}$

43. $4w^2 - 20w + 5 = 0$

$\frac{5 \pm 2\sqrt{5}}{2}$

44. $2x^2 = 4x - 5$

No real number solution

45. $3r^2 = 5r - 6$

No real number solution

46. $9y^2 + 6y - 1 = 0$

$\frac{-1 \pm \sqrt{2}}{3}$

47. $9s^2 - 6s - 2 = 0$

$\frac{1 \pm \sqrt{3}}{3}$

48. $6s^2 - s - 2 = 0$

$-\frac{1}{2}, \frac{2}{3}$

49. $6y^2 + 5y - 4 = 0$

$\frac{1}{2}, -\frac{4}{3}$

50. $4p^2 + 16p = -11$

$\frac{-4 \pm \sqrt{5}}{2}$

51. $4y^2 - 12y = -1$

$\frac{3 \pm 2\sqrt{2}}{2}$

52. $4x^2 = 4x + 11$

$\frac{1 \pm 2\sqrt{3}}{2}$

53. $4s^2 + 12s = 3$

$\frac{-3 \pm 2\sqrt{3}}{2}$

54. $4p^2 = -12p - 9$

$-\frac{3}{2}$

55. $3y^2 + 6y = -3$

-1

56. $9v^2 = -30v - 23$

$\frac{-5 \pm \sqrt{2}}{3}$

57. $9t^2 = 30t + 17$

$\frac{5 \pm \sqrt{42}}{3}$

58. $\frac{x^2}{2} - \frac{x}{3} = 1$

$\frac{1 \pm \sqrt{19}}{3}$

59. $\frac{x^2}{4} - \frac{x}{2} = 5$

$1 \pm \sqrt{21}$

60. $\frac{2x^2}{5} = x + 1$

$\frac{5 \pm \sqrt{65}}{4}$

61. $\frac{3x^2}{2} + 2x = 1$

$\frac{-2 \pm \sqrt{10}}{3}$

62. $\frac{x}{5} + \frac{5}{x} = \frac{12}{5}$

$6 \pm \sqrt{11}$

63. $\frac{x}{4} + \frac{3}{x} = \frac{5}{2}$

$5 \pm \sqrt{13}$

64. A baseball player hits a ball. The height h of the ball above the ground at time t can be approximated by the equation $h = -16t^2 + 76t + 5$. When will the ball hit the ground? Round to the nearest hundredth. (*Hint*: The ball strikes the ground when $h = 0$ feet.)
4.81 s

65. A basketball player shoots at a basket 25 feet away. The height h of the ball above the ground at time t is given by the equation $h = -16t^2 + 32t + 6.5$. How many seconds after the ball is released does it hit the basket? Round to the nearest hundredth. (*Hint*: When it hits the basket, $h = 10$ feet.)
1.88 s

66. A lob shot in tennis results in the tennis ball following an arc that can be approximated by the equation $h = -0.16x^2 + 4.16x + 3$, where h is the height of the ball x feet from where it is hit. What is the closest distance from the place where the lob was hit that a tennis player with a maximum reach of 8 feet (including the racket) can be in order to return the ball as it descends before it hits the ground? What is the farthest distance the player can be and still return the shot before it hits the ground? Round to the nearest hundredth. 24.74 feet

67. A pizza restaurant offers two 9-inch-diameter personal pizzas for $9.00. What must be the diameter of one pizza so that the total area of the one pizza is equal to the area of the two 9-inch pizzas? Round to the nearest tenth.
12.7 inches

Applying Concepts

68. Solve: $\sqrt{x^2 + 2x + 1} = x - 1$
No solution

69. Consider the expression $b^2 - 4ac$ that is part of the quadratic formula.
a. If $b^2 - 4ac = 0$, what is true about the solutions of the quadratic equation?
b. If $b^2 - 4ac < 0$, what is true about the solutions of the quadratic equation?
c. If $b^2 - 4ac > 0$, what is true about the solutions of the quadratic equation?
a. The solutions are equal; **b.** there are no real solutions; **c.** there are two real, unequal solutions.

70. Consider the two rectangles shown at the right. Each rectangle has the same perimeter, but the areas of the two rectangles are different.

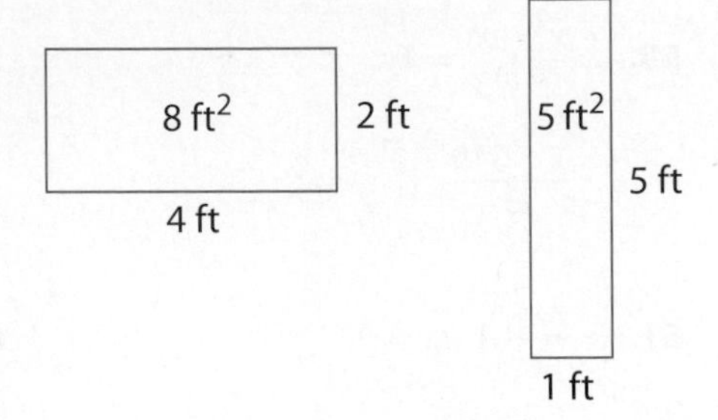

a. Write a formula for a rectangle whose perimeter is 12 feet using L for the length and W for the width.

b. Write a formula for the area of a rectangle using L and W.

c. Solve the formula in part a for L and use that expression to substitute for L in the formula in part b. Then simplify.

d. The formula you created in part c gives the area of the rectangle in terms of the width of one of the sides. Experiment with this formula until you find the dimensions of the rectangle that has the largest area.

a. $2L + 2W = 12$ or $L + W = 6$; **b.** $A = LW$; **c.** $A = -W^2 + 6W$;
d. length = 3 feet; width = 3 feet

Exploration

71. ***Gravity*** There is a story that Galileo (1564–1642) dropped two rocks of different weights from the Leaning Tower of Pisa to show that both rocks reached the ground at the same time, thereby disputing the theory that heavier objects fall faster (with no air resistance) than lighter ones. This means, for example, that on the moon (where there is no air), a feather and a rock dropped from the same height would reach the ground at the same time. The only force acting on the feather and on the rock is gravity. Modern historians question whether Galileo actually performed this experiment. However, he did investigate gravity and show that the force of gravity was the same for all objects. In this Exploration, you will try to duplicate some of Galileo's work. Get two meter sticks and tape them together to form a small wedge that will be used as a track. Place one end of the meter sticks (the end with 0) on a stack of books so that the angle of the meter sticks and a table top is approximately 10°.

a. Place a large marble or other solid ball on the track at the 10-centimeter mark and let it roll to the bottom. Using a stop watch, measure the time it takes to get to the bottom. Repeat this 5 times and calculate the mean of the 5 time readings. Record the ordered pair (time, distance) in a table.

b. Repeat part a by choosing the 15-centimeter position. Continue to repeat this experiment for the 20-, 25-, 30-, 35-, 40-, 45-, and 50-centimeter positions. Make a scatter plot of this data using the horizontal axis as time and the vertical axis as distance.

c. Using the STAT feature of your calculator, enter the time values in L1 and the distance values in L2. Press the STAT key, and select CALC and then QuadReg. This will give you a quadratic equation that approximates the points on your graph. Use this equation to approximate the time it would take the ball to roll to the bottom from the 70-centimeter mark. Now let the ball roll from that mark and compare the two results.

d. Now repeat this experiment for another solid ball of different weight. Are the results approximately the same?

e. From your data, what effect does doubling the time have on the distance a ball rolls?

f. Repeat parts a through e using an angle of 20°. Are your answers to parts c and d for a 20° angle different from your answers when the angle was 10°?

g. Do this experiment again for a 30° ramp. If these experiments were done accurately, you will have found that heavier objects do not travel to the end of the ramp in a shorter time than lighter ones. How did Galileo conclude from this that balls of different weights dropped from the Leaning Tower of Pisa would reach the ground in the same time?

Section 7.4 Properties of Quadratic Functions

Vertex and Axis of Symmetry for a Parabola

TAKE NOTE
It does not matter whether we write $y = x^2 + 2x - 3$ or $f(x) = -2x^2 + 4x + 5$. In each case, the equation represents a function. Remember that, for functions, y and $f(x)$ are different symbols for the *same* number.

Recall that a quadratic function is one of the form $f(x) = ax^2 + bx + c, a \neq 0$. The graph of this function is a **parabola**. The graph is "cup-shaped" and opens either up or down depending on the value of *a*. Examples of two parabolas are shown below.

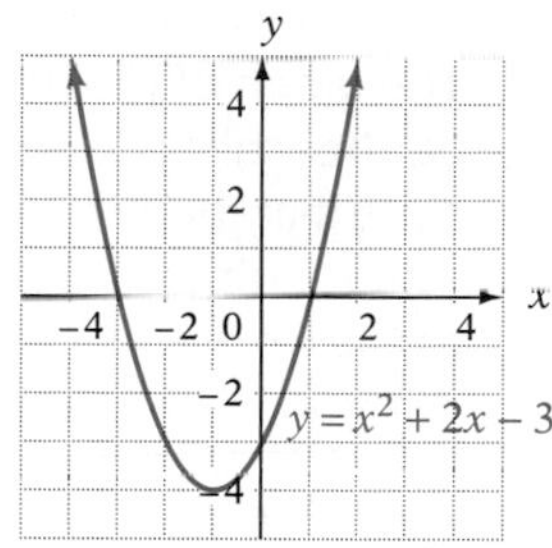

Because $a > 0$ $(a = 1)$, the graph opens up.

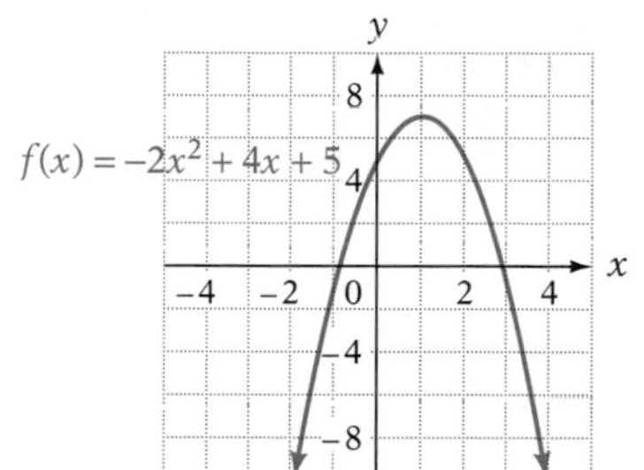

Because $a < 0$ $(a = -2)$, the graph opens down.

Question: Does the graph of $y = \frac{1}{2}x^2 - 4x - 10$ open up or down?[1]

POINT OF INTEREST
The mirror in the telescope at the Palomar Mountain Observatory is 2 feet thick at the ends and weighs 14.75 tons. The mirror has been ground to a true paraboloid (the three-dimensional version of a parabola) to within 0.0000015 inch.

Every parabola has an *axis of symmetry* and a *vertex*. The **vertex** is the lowest point on the graph of a parabola when the parabola opens up (as shown at the right) or the highest point on the parabola when it opens down.

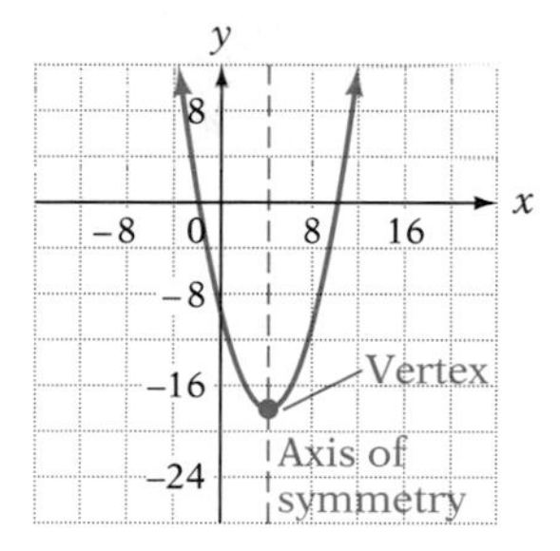

The **axis of symmetry** is a line through the vertex of the parabola such that if the parabola were folded along the axis of symmetry, the two halves of the parabola would match.

Completing the square can be used to find the vertex of a parabola.

TAKE NOTE
It is important to recognize that $y = x^2 - 6x + 4$ and $y = (x - 3)^2 - 5$ are exactly the same.

$$y = (x-3)^2 - 5$$
$$= x^2 - 6x + 9 - 5$$
$$= x^2 - 6x + 4$$

→ Find the vertex and axis of symmetry for $y = x^2 - 6x + 4$.

Complete the square of $x^2 - 6x$ by adding $\left[\frac{1}{2}(-6)\right]^2 = 9$ to the expression. Because we are adding 9 to the expression, we must also subtract 9 so that the value of the expression does not change.

$$y = x^2 - 6x + 4$$
$$= (x^2 - 6x + 9) + (4 - 9)$$

Add and subtract 9.

Factor the trinomial and simplify $(4 - 9)$.

$$y = (x-3)^2 - 5$$

Because the square of any term is greater than or equal to zero, $(x-3)^2 \geq 0$. Subtracting 5 from each side of this inequality and then replacing $(x-3)^2 - 5$ by *y*, we have $y \geq -5$.

$$(x-3)^2 \geq 0$$
$$(x-3)^2 - 5 \geq 0 - 5$$
$$(x-3)^2 - 5 \geq -5$$
$$y \geq -5$$

1. $a = \frac{1}{2}$, which is greater than 0. The graph opens up.

SUGGESTED ACTIVITY
The *Instructor's Resource Manual* contains an activity that lets students investigate the concept of maximum and minimum for a quadratic function.

Because the coefficient of x^2 is 1, the graph of the parabola opens up and the vertex is the lowest point on the graph. From the inequality $y \geq -5$, the lowest y value is –5. This occurs when $(x - 3)^2 = 0$, which is when $x = 3$, as shown at the right.

$$\begin{aligned} y &= (x-3)^2 - 5 \\ &= (3-3)^2 - 5 \\ &= -5 \end{aligned}$$

The graph of the parabola is shown at the right. The vertex is (3, –5). Note that the axis of symmetry passes through the vertex and is parallel to the y-axis. Therefore the equation of the axis of symmetry is $x = 3$.

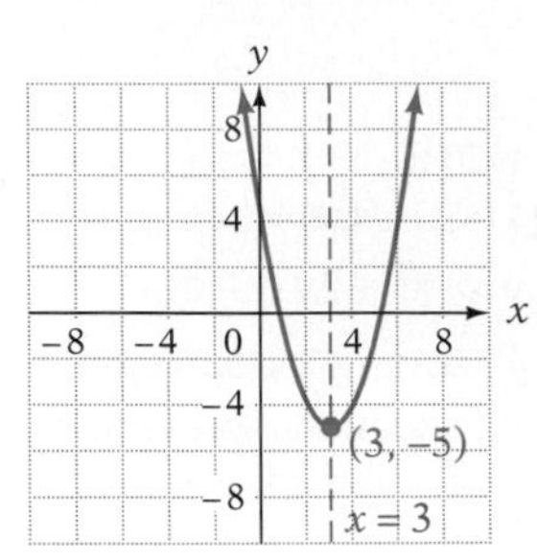

TAKE NOTE
The formula at the right says that to find the coordinates of the vertex, first find the x-coordinate, which is $-\frac{b}{2a}$. Now evaluate the function at $-\frac{b}{2a}$ to obtain the y-coordinate. Mathematically, this is $f\left(-\frac{b}{2a}\right)$.

A formula for the vertex can be found in a manner similar to the one we used to derive the quadratic formula. We complete the square of $ax^2 + bx$ for $y = ax^2 + bx + c$. The result is the following formula.

Vertex and Axis of Symmetry for a Parabola

The coordinates of the vertex of the parabola given by $f(x) = ax^2 + bx + c$ are $\left[-\frac{b}{2a}, f\left(-\frac{b}{2a}\right)\right]$. The axis of symmetry is $x = -\frac{b}{2a}$.

Example 1 For the graph of $f(x) = 2x^2 - 3x + 1$, find the vertex and the axis of symmetry.

Solution From $f(x) = 2x^2 - 3x + 1$, we have $a = 2$ and $b = -3$.

x-coordinate of the vertex: $-\frac{b}{2a} = -\frac{(-3)}{2(2)} = \frac{3}{4}$ • Replace a by 2 and b by –3.

To find the y-coordinate of the vertex, evaluate the function for $x = \frac{3}{4}$.

$$\begin{aligned} f(x) &= 2x^2 - 3x + 1 \\ f\left(\frac{3}{4}\right) &= 2\left(\frac{3}{4}\right)^2 - 3\left(\frac{3}{4}\right) + 1 \\ &= \frac{9}{8} - \frac{9}{4} + 1 \\ &= -\frac{1}{8} \end{aligned}$$

• Replace x by $\frac{3}{4}$.

The coordinates of the vertex are $\left(\frac{3}{4}, -\frac{1}{8}\right)$. The axis of symmetry is $x = \frac{3}{4}$.

The graph of the parabola is shown at the left.

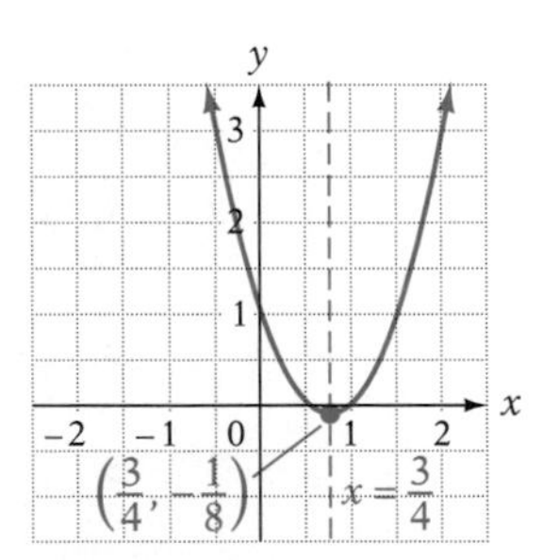

You-Try-It 1 For the graph of $f(x) = x^2 + 4x - 2$, find the vertex and the axis of symmetry.

Solution See page S34.

7.4 EXERCISES

Topics for Discussion

1. What is a quadratic function? Give some examples of quadratic functions.
A function given by $f(x) = ax^2 + bx + c, a \neq 0$.

2. Explain the axis of symmetry of a parabola.
Answers will vary.

3. When defining a quadratic function as $f(x) = ax^2 + bx + c$, there is a requirement placed on a that $a \neq 0$. Explain why that requirement is made.
If $a = 0$, the function would be a linear function instead of a quadratic one.

4. What is the vertex of a parabola?
The lowest point on a parabola that opens up or the highest point on a parabola that opens down.

Determine whether the parabola opens up or down.

5. $y = x^2 - 3$ Up

6. $y = -x^2 + 4$ Down

7. $y = \frac{1}{2}x^2 - 2$ Up

8. $y = -\frac{1}{3}x^2 + 5$ Down

9. $y = x^2 - 2x + 3$ Up

10. $y = -x^2 + 4x - 1$ Down

11. The x-coordinate of the vertex of a parabola is 2. If the equation of the parabola is $f(x) = x^2 - 4x + 1$, determine the y-coordinate of the vertex.
-3

12. The x-coordinate of the vertex of a parabola is -3. If the equation of the parabola is $f(x) = 2x^2 + 12x - 1$, determine the y-coordinate of the vertex.
-19

13. The x-coordinate of the vertex of a parabola is $\frac{1}{2}$. If the equation of the parabola is $f(x) = 2x^2 - 2x + 1$, determine the y-coordinate of the vertex.

1

14. The x-coordinate of the vertex of a parabola is $-\frac{3}{2}$. If the equation of the parabola is $f(x) = 2x^2 + 6x - 1$, determine the y-coordinate of the vertex.

$-\frac{11}{2}$

Determine the vertex and axis of symmetry of the parabola whose equation is given below.

15. $y = x^2$ $(0, 0), x = 0$

16. $y = -x^2$ $(0, 0), x = 0$

17. $y = -x^2 + 1$ $(0, 1), x = 0$

18. $y = x^2 - 1$ $(0, -1), x = 0$

19. $y = 2x^2$ $(0, 0), x = 0$

20. $y = \frac{1}{2}x^2$ $(0, 0), x = 0$

21. $y = -\frac{1}{2}x^2 + 1$ $(0, 1), x = 0$

22. $y = 2x^2 - 1$ $(0, -1), x = 0$

23. $y = x^2 - 6x + 2$ $(3, -7), x = 3$

24. $y = x^2 + 4x - 3$
$(-2, -7), x = -2$

25. $y = x^2 + 4x$
$(-2, -4), x = -2$

26. $y = x^2 - 4x$
$(2, -4), x = 2$

27. $y = -2x^2 - 6x + 1$
$(-\frac{3}{2}, \frac{11}{2}), x = -\frac{3}{2}$

28. $y = 2x^2 - 4x + 3$
$(1, 1), x = 1$

29. $y = 2x^2 - 5x + 2$
$\left(\frac{5}{4}, -\frac{9}{8}\right), x = \frac{5}{4}$

30. $y = -2x^2 + 3x + 3$
$\left(\frac{3}{4}, \frac{33}{8}\right), x = \frac{3}{4}$

31. $y = 4x^2 + 2x + 1$
$\left(-\frac{1}{4}, \frac{3}{4}\right), x = -\frac{1}{4}$

32. $y = -2x^2 + 3x - 5$
$\left(\frac{3}{4}, -\frac{31}{8}\right), x = \frac{3}{4}$

Applying Concepts

33. Given that $x = 3$ is the axis of symmetry of a parabola and that $P(4, 9)$ and $Q(1, 12)$ are points on the graph of the parabola, name the coordinates of the two points that are symmetric with the axis of symmetry to the given points.
$(2, 9); (5, 12)$

34. Given that $x = -4$ is the axis of symmetry of a parabola and that $P(0, 7)$ and $Q(-2, 1)$ are points on the graph of the parabola, name the coordinates of the two points that are symmetric with the axis of symmetry to the given points.
$(-8, 7); (-6, 1)$

Exploration

35. ***Standard Form of the Equation of a Parabola*** The standard form of the equation of a parabola is given by $f(x) = a(x - h)^2 + k$, where (h, k) are the coordinates of the vertex of the parabola.

a. In which direction does the graph of the parabola whose equation is $f(x) = -2(x - 3)^2 + 4$ open?

b. What are the coordinates of the vertex of a parabola whose graph is given by $f(x) = -(x + 4)^2 - 1$?

c. What is the equation of the axis of symmetry of the parabola whose graph is given by $f(x) = -3(x - 1)^2 + 5$?

Chapter Summary

Definitions

An equation of the form $ax^2 + bx + c = 0, a \neq 0$, is a *quadratic equation*. A quadratic equation is also called a second-degree equation.

A quadratic equation is in *standard form* when the polynomial is in descending order and equal to zero.

Adding to a binomial the constant term that makes it a perfect-square trinomial is called *completing the square*.

The graph of an equation of the form $y = ax^2 + bx + c, a \neq 0$, is a parabola. The *vertex* is the lowest point on a parabola that opens up or the highest point on a parabola that opens down.

Procedures

Solving a Quadratic Equation by Factoring Write the equation in standard form, factor the left side of the equation, apply the Principle of Zero Products, solve for the variable.

Principle of Taking the Square Root of Each Side of an Equation If $x^2 = a$, then $x = \pm\sqrt{a}$. This principle is used to solve quadratic equations by taking square roots.

Solving a Quadratic Equation by Completing the Square When the quadratic equation is in the form $ax^2 + bx = c$, multiply each side of the equation by $\frac{1}{a}$ and then add to each side of the equation the term that completes the square on $x^2 + bx$. Factor the perfect-square trinomial, and write it as the square of a binomial. Take the square root of each side of the equation and solve for x.

The Quadratic Formula The solutions of $ax^2 + bx + c = 0, a \neq 0$, are

$$x = \frac{-b \pm \sqrt{b^2 - 4ac}}{2a}$$

The Axis of Symmetry of a Parabola The axis of symmetry for a graph of $f(x) = ax^2 + bx + c, a \neq 0$, is the line $x = -\frac{b}{2a}$.

The Vertex of a Parabola The vertex for a graph of $f(x) = ax^2 + bx + c, a \neq 0$, is the point whose coordinates are $\left[-\frac{b}{2a}, f\left(-\frac{b}{2a}\right)\right]$.

Chapter Review Exercises

1. Solve: $b^2 - 16 = 0$
$4, -4$

2. Solve: $x^2 - x - 3 = 0$
$\frac{1 \pm \sqrt{13}}{2}$

3. Solve: $x^2 - 3x - 5 = 0$
$\frac{3 \pm \sqrt{29}}{2}$

4. Solve: $49x^2 = 25$
$\pm\frac{5}{7}$

5. Solve: $x^2 = 10x - 2$
$5 \pm \sqrt{23}$

6. Solve: $6x(x + 1) = x - 1$
$-\frac{1}{2}, -\frac{1}{3}$

7. Solve: $4y^2 + 9 = 0$
No real number solution

8. Solve: $5x^2 + 20x + 12 = 0$
$\frac{-10 \pm 2\sqrt{10}}{5}$

9. Solve: $x^2 - 4x + 1 = 0$
$2 \pm \sqrt{3}$

10. Solve: $x^2 - x = 30$
$6, -5$

11. Solve: $6x^2 + 13x - 28 = 0$
$\frac{4}{3}, -\frac{7}{2}$

12. Solve: $x^2 = 40$
$\pm 2\sqrt{10}$

13. Solve: $3x^2 - 4x = 1$
$\frac{2 \pm \sqrt{7}}{3}$

14. Solve: $x^2 - 2x - 10 = 0$
$1 \pm \sqrt{11}$

15. Solve: $x^2 - 12x + 27 = 0$
3, 9

16. Solve: $(x - 7)^2 = 81$
16, −2

17. Find the vertex and axis of symmetry for $y = x^2 + 1$.
Vertex: (0, 1); axis of symmetry: $x = 0$

18. Find the vertex and axis of symmetry for $y = x^2 - 4x + 3$.
Vertex: (2, −1); axis of symmetry: $x = 2$

19. Solve: $(y + 4)^2 - 25 = 0$
1, −9

20. Solve: $4x^2 + 16x = 7$
$\frac{-4 \pm \sqrt{23}}{2}$

21. Solve: $24x^2 + 34x + 5 = 0$
$-\frac{1}{6}, -\frac{5}{4}$

22. Solve: $x^2 = 4x - 8$
No real number solution

23. Solve: $x^2 - 5 = 8x$
$4 \pm \sqrt{21}$

24. Solve: $25(2x^2 - 2x + 1) = (x + 3)^2$
$\frac{4}{7}$

25. Solve: $x(x - 3) = 2x - 6$
3, 2

26. Solve: $2x(x + 2) = x + 2$
$-2, \frac{1}{2}$

27. Solve: $4(x - 3)^2 = 20$
$3 \pm \sqrt{5}$

28. Solve: $x^2 + 8x - 3 = 0$
$-4 \pm \sqrt{19}$

29. Find the vertex and axis of symmetry for $y = 2x^2 + 4x - 1$.
Vertex: (−1, −3); axis of symmetry: $x = -1$

30. Find the vertex and axis of symmetry for $y = -3x^2 - 9x + 3$.
Vertex: $\left(-\frac{3}{2}, \frac{39}{4}\right)$; axis of symmetry: $x = -\frac{3}{2}$

31. If (2, 3) are the coordinates of a point on the graph of a parabola whose axis of symmetry is $x = 5$, give the coordinates of another point on the graph.
(8, 3)

32. If (–2, 4) are the coordinates of a point on the graph of a parabola whose axis of symmetry is $x = -1$, give the coordinates of another point on the graph.
(0, 4)

33. Solve $x^2 + 6x - 2 = 0$ and then check the solution by graphing.

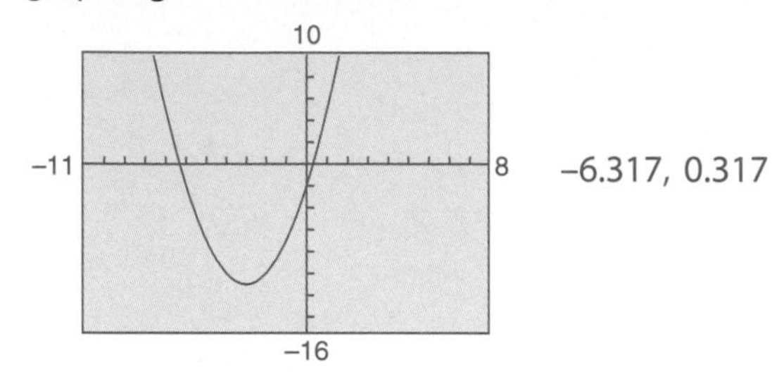

–6.317, 0.317

34. Solve $x^2 - x + 3 = 0$ and then check the solution by graphing.

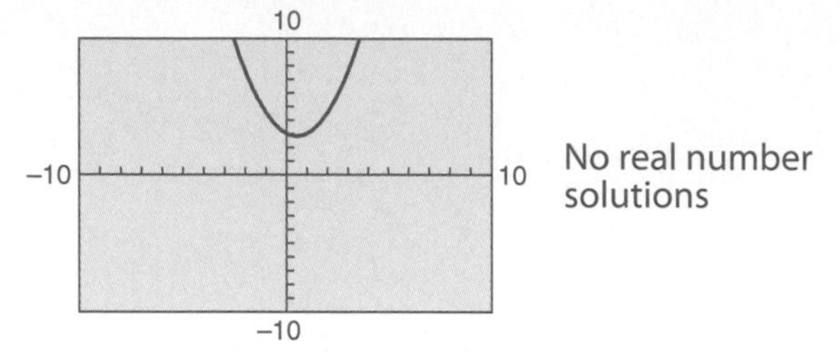

No real number solutions

35. In a soccer league, each team plays the other teams in the league 3 times. The number, N, of games that are necessary to play in a league that has n teams is given by $N = \frac{3}{2}n^2 - \frac{3}{2}n$. Find the number of games that are played in a league that has 9 teams. 108

36. According to Torricelli's principle, the velocity, v (in feet per second), of water that pours through a hole in a bucket depends on the height h (in feet) of the water level above the hole and is given by $64h = v^2$. Find the velocity of water pouring through a hole in a bucket when the water level is 11 inches above the hole. Round to the nearest hundredth.
7.66 feet per second.

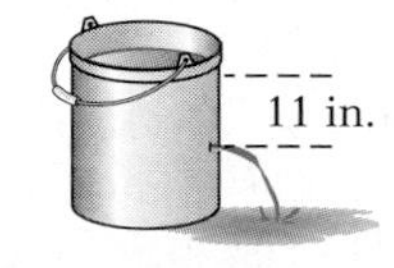

37. A circular swimming pool has a radius of 15 feet. A decking of constant width is built around the pool so that the pool and decking have a total area of 1000 square feet. Find the width of the decking to the nearest hundredth of a foot.
2.84 feet

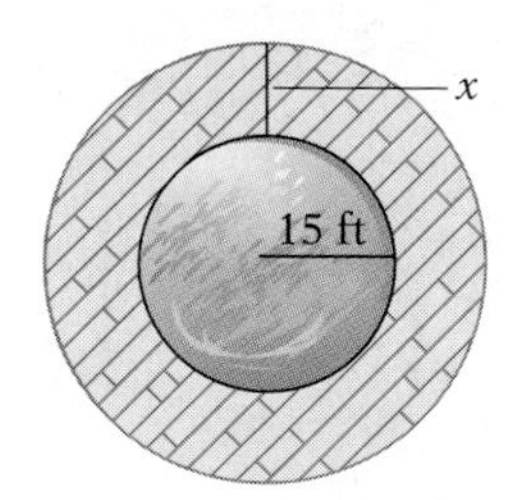

38. A model rocket is launched with an initial velocity of 200 feet per second. The height, h, of the rocket t seconds after launch is given by $h = -16t^2 + 200t$. How many seconds after the launch will the rocket be 300 feet above the ground? Round to the nearest hundredth of a second.
1.74 seconds, 10.76 seconds

39. The graph at the right is the graph of $y = x^2 + 6x + 12$. Explain, by using the graph, why the equation $x^2 + 6x + 12 = 0$ has no real number solutions.
The graph does not cross the x-axis. Therefore, there are no real numbers x for which y is zero.

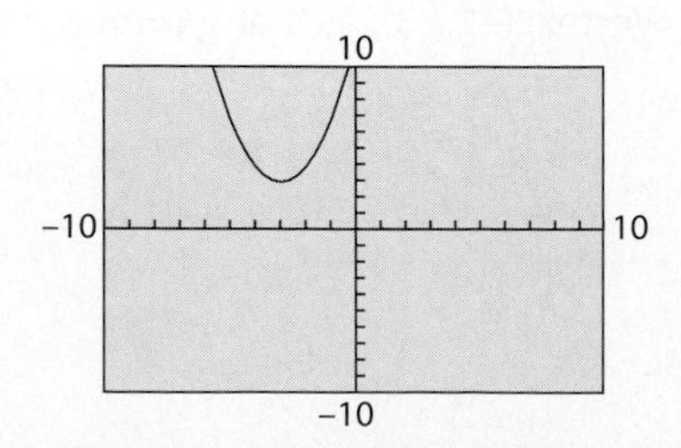

Cumulative Review Exercises

1. What is the next term in the sequence 0, 3, 8, 15, 24, ...? 35

2. Find a counterexample: "All right triangles have two equal angles."
Answers will vary.

3. Simplify: $2y - 3[2y - 4(3 - y)] + 4$

$-16y + 40$

4. Solve: $-\frac{3x}{5} = -\frac{9}{10}$ $\frac{3}{2}$

5. Find the x- and y-intercepts of the graph of $4x - 3y = 12$.
(3, 0) and (0, –4)

6. If a card is randomly chosen from a regular deck of playing cards, what is the probability that the card is a 4 or a spade?
$\frac{4}{13}$

7. The bar graph at the right shows the average number of gallons of ice cream consumed per person for various countries.

a. Of the countries shown, which countries consumed the same average number of gallons per person?

b. Of the countries shown, which country consumed the greatest average number of gallons per person?

c. Of the countries shown, which country consumed the least average number of gallons per person?

a. Canada and Italy; b. U.S.A.; c. France

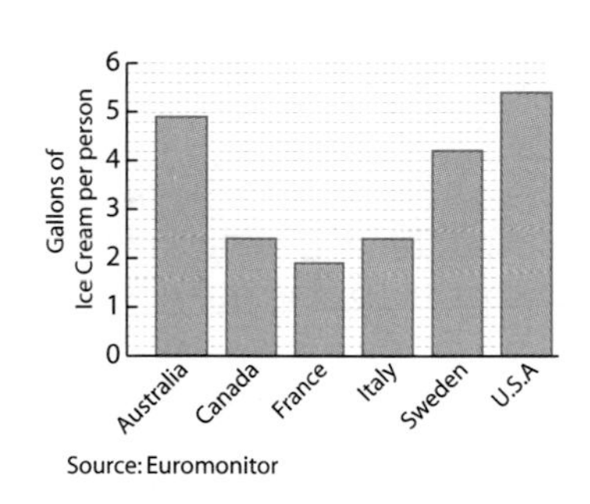

Source: Euromonitor

8. Find the equation of the line that contains the point whose coordinates are $(-3, 2)$ and has a slope of $-\frac{4}{3}$. $y = -\frac{4}{3}x - 2$

9. Evaluate $f(x) = 2x^2 - 3x - 1$ at $x = -3$. 26

10. Find the domain and range of the relation {(0, 1), (–1, 4), (3, 1), (4, –1), (5, 3)}. Is the relation a function? Domain: {–1, 0, 3, 4, 5}; range: {–1, 1, 3, 4}; yes

11. Solve: $2x^2 - 7x = -3$

$\frac{1}{2}, 3$

12. Solve: $3x^2 - 4x - 5 = 0$

$\frac{2 \pm \sqrt{19}}{3}$

13. Simplify: $\dfrac{(2a^{-2}b)^2}{-3a^{-5}b^4}$

$-\dfrac{4a}{3b^2}$

14. Solve by the addition method: $\begin{aligned} 3x + 2y &= 2 \\ 5x - 2y &= 14 \end{aligned}$

$(2, -2)$

15. Multiply: $(\sqrt{a} - \sqrt{2})(\sqrt{a} + \sqrt{2})$

$a - 2$

16. Solve: $3 = 8 - \sqrt{5x}$

5

17. Solve: $3(x - 2)^2 = 36$

$2 \pm 2\sqrt{3}$

18. Solve: $3x - 2 \geq 7$

$\{x \mid x \geq 3\}$

19. Complete the input/output table and graph the ordered pairs.

Input, x	−3	−2	0	1	3
Output, $y = x^2 - 4$	5	0	−4	−3	5

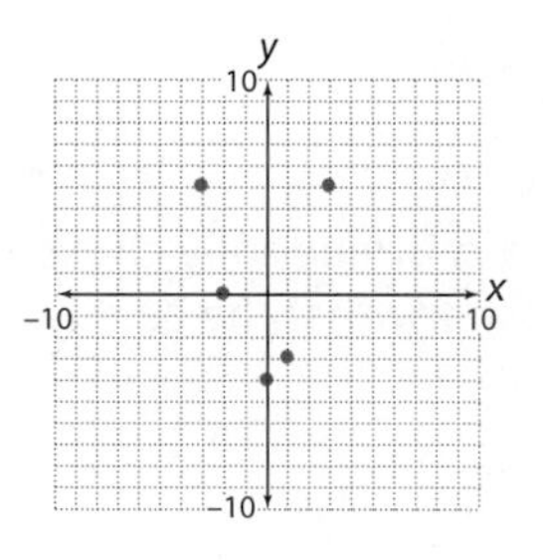

20. What are the coordinates of points A and B after the figure at the right has been translated 3 units to the right and 2 units down?

(−1, 1); (5, −3)

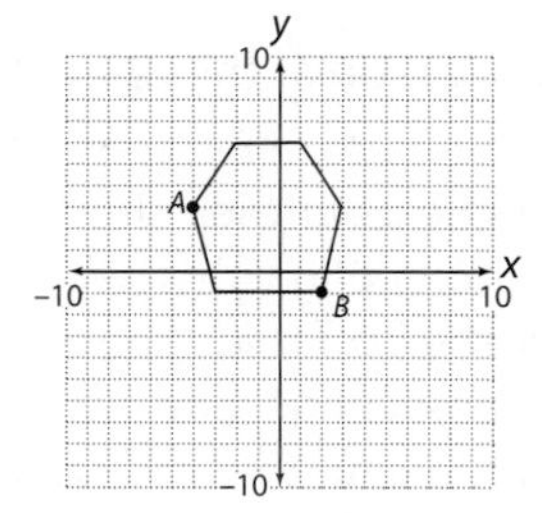

21. Find the cost per pound of a mixture made from 20 pounds of cashews that cost \$3.50 per pound and 50 pounds of peanuts that cost \$1.75 per pound.

\$2.25

22. The angle of depression from the top of the Transamerica building in San Francisco to the top of the Bank of America building 4500 feet away is approximately 1°. How much taller is the Transamerica building than the Bank of America building? Round to the nearest foot. 79 feet

23. A 720-mile trip from one city to another takes 3 hours when a plane is flying with the wind. The return trip, against the wind, takes 4.5 hours. Find the rate of the plane in calm air and the rate of the wind.

Plane, 200 mph; wind, 40 mph

24. A guy wire is attached to a point 30 meters above the ground on a telephone pole that is perpendicular to the ground. The wire is anchored to the ground at a point 10 meters from the base of the pole. Find the length of the guy wire. Round to the nearest hundredth. 31.62 meters

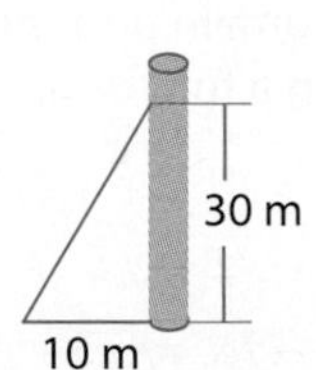

25. The measurement of heights (in inches) of 5 tomato plants 4 weeks after being planted were: 4.5, 4.2, 5.1, 4.8, 4.4. Find the mean and standard deviation of the heights. Round to the nearest hundredth. Mean, 4.6; standard deviation, 0.32

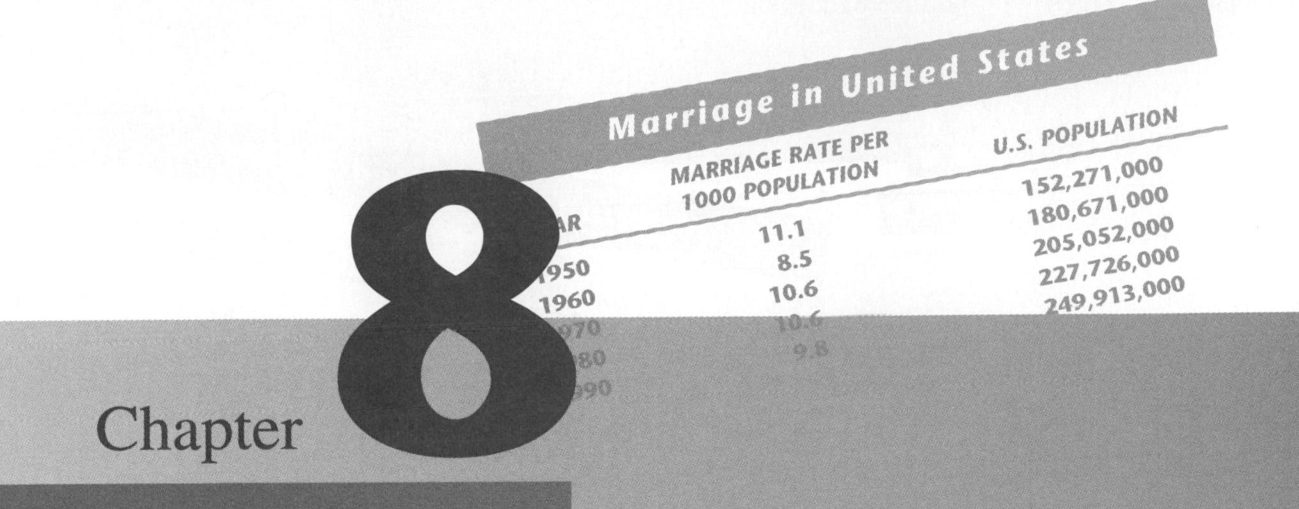

Chapter 8

Rational Expressions

Marriage in United States

YEAR	MARRIAGE RATE PER 1000 POPULATION	U.S. POPULATION
1950	11.1	152,271,000
1960	8.5	180,671,000
1970	10.6	205,052,000
1980	10.6	227,726,000
1990	9.8	249,913,000

Section 8.1 Multiplication and Division of Rational Expressions

Simplify Rational Expressions

A tour boat used for river excursions takes passengers from a loading dock to an island 5 miles up the river. The boat can travel at a speed of 10 mph. However, there is a current, and when the boat is traveling up the river, the river's current is flowing in the direction opposite from that in which the boat is traveling, and it slows the boat down.

The time it takes to complete the trip can be represented by the expression $\frac{5}{10-x}$, where x is the rate of the current. For example, if the rate of the current is 5 mph, then the time it takes to travel from the dock to the island is calculated as:

$$\frac{5}{10-x} = \frac{5}{10-5} = \frac{5}{5} = 1$$

The trip takes one hour when the rate of the river's current is 5 mph.

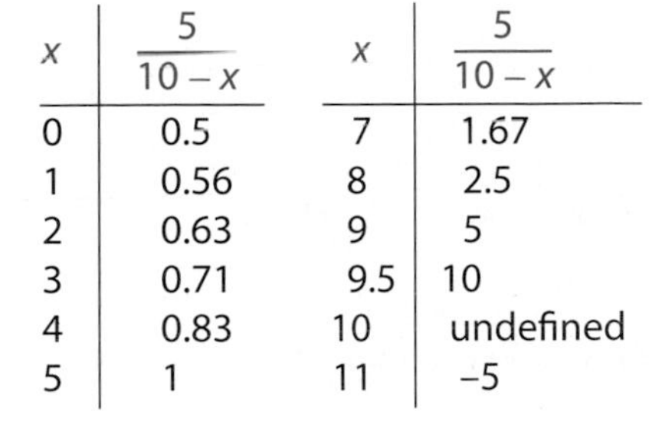

x	$\frac{5}{10-x}$	x	$\frac{5}{10-x}$
0	0.5	7	1.67
1	0.56	8	2.5
2	0.63	9	5
3	0.71	9.5	10
4	0.83	10	undefined
5	1	11	–5

A table of values for different values of x is shown at the right. Note that the faster the rate of the river's current, the longer the trip to the island. Also note that when the rate of the current is 10 mph, the expression $\frac{5}{10-x}$ is undefined because the denominator is zero. If the rate is 10 mph, the boat will never get to the island! If the rate of the current is greater than 10 mph, the time is a negative number, which is not possible.

The graph of $Y1 = \frac{5}{10-X}$ is shown at the left below as it would be graphed on a graphing calculator that is in the CONNECTED mode. The vertical line at $x = 10$ appears to be a part of the graph, but it is not. The calculator is "connecting" the plotted points to the left of $x = 10$ with the plotted points to the right of $x = 10$.

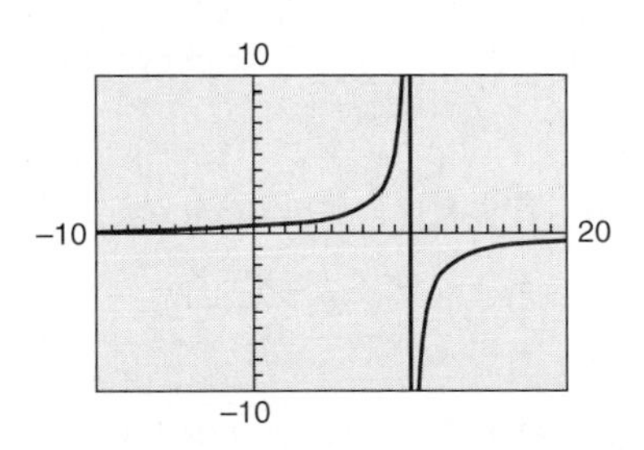

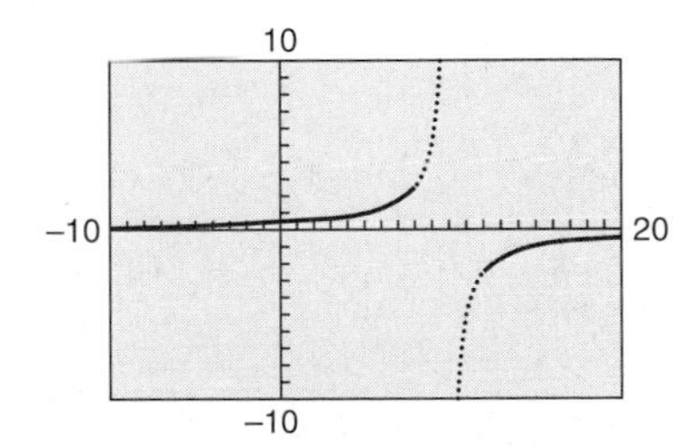

We can avoid this problem by putting the graphing calculator in the DOT mode. The graph will then appear as shown at the right above. There is no point on the graph at $x = 10$.

The expression $\frac{5}{10-x}$ is a *rational expression.* Fractions in which the numerator and denominator are polynomials are **rational expressions.** Further examples of rational expressions are shown below.

$$\frac{3}{x} \qquad \frac{y+2}{y^2-5y+1} \qquad \frac{ab-7}{a^3b^2+ab^2}$$

POINT OF INTEREST
There are many instances of rational expressions in application problems. For example, the expression $\frac{wtk}{1000}$ represents the cost to operate an electrical appliance. In this expression, w is the number of watts, t is the number of hours, and k is the cost per kilowatt-hour. What is the cost to operate a 1000-watt microwave for 8 hours if the cost per kilowatt-hour is \$.00125?

INSTRUCTOR NOTE
The answer to the question posed in the Point of Interest above is \$.01.

Question: Which of the following are rational expressions?[1]

a. $\frac{4}{\sqrt{x}}$ **b.** $\frac{b^2-9}{b}$ **c.** $2y^{-2}+3y^{-1}+4$

When simplifying or operating on rational expressions, we must be sure that, when the variables are replaced with numbers, the denominator is not zero.

For the rational expression $\frac{x+6}{2x-4}$, x cannot be 2 because the denominator would then be zero.

$$\frac{x+6}{2x-4}$$

$$\frac{2+6}{2(2)-4}=\frac{8}{0}$$ Not a real number.

Example 1 For what value of the variable is the expression undefined?

a. $\frac{7x}{2x+8}$ **b.** $\frac{x-5}{x^2+x-6}$

Solution **a.** $2x+8=0$
$2x=-8$
$x=-4$

• We want to know what value of x makes the denominator zero, so we set the denominator equal to zero and then solve the equation for x.

When $x=-4$, the denominator is zero.
The expression is undefined when $x=-4$.

b. $x^2+x-6=0$
$(x+3)(x-2)=0$
$x+3=0 \qquad x-2=0$
$x=-3 \qquad x=2$

• This is a quadratic equation. Solve by factoring.

When $x=-3$ or 2, the denominator is zero.
The expression is undefined when $x=-3$ or 2.

You-Try-It 1 For what value of the variable is the expression undefined?

a. $\frac{x+8}{3x-12}$ **b.** $\frac{x-6}{x^2+3x-4}$

Solution See page S35.

A rational expression is in simplest form when the numerator and denominator have no common factors. The Multiplication Property of One is used to write a rational expression in simplest form.

---

1. **a.** This is not a rational expression because $\sqrt{x}$ is not a polynomial. **b.** This is a rational expression because b^2-9 and b are polynomials. **c.** This is not a rational expression because $2y^{-2}+3y^{-1}+4$ is not a polynomial.

INSTRUCTOR NOTE
Simplifying a rational expression is closely related to simplifying a rational number—the common factors are removed. Making this connection will help some students.

→ Simplify: $\dfrac{x^2-4}{x^2+x-2}$

$$\frac{x^2-4}{x^2+x-2} = \frac{(x-2)(x+2)}{(x-1)(x+2)} \quad \text{• Factor the numerator and denominator.}$$

$$= \frac{x-2}{x-1}\cdot\frac{x+2}{x+2}$$

$$= \frac{x-2}{x-1}\cdot 1 \quad \text{• } \frac{x+2}{x+2} = 1$$

$$= \frac{x-2}{x-1},\ x \neq -2, 1$$

This simplification is usually performed with slashes through the common factors, as shown below.

$$\frac{x^2-4}{x^2+x-2} = \frac{(x-2)\overset{1}{\cancel{(x+2)}}}{(x-1)\underset{1}{\cancel{(x+2)}}} = \frac{x-2}{x-1},\ x \neq -2, 1$$

The restrictions $x \neq -2, 1$ are necessary to prevent division by zero. Note that we are listing the values of x that result in a denominator of zero in the *original* expression. The denominator $x^2 + x - 2 = (x - 1)(x + 2)$, and $(x - 1)(x + 2) = 0$ when $x = 1$ or $x = -2$.

One method of checking the simplification performed above is to substitute the same number for x into the original expression and the simplified expression. The number 3 is used here.

$$\text{When } x = 3,\ \frac{x^2-4}{x^2+x-2} = \frac{3^2-4}{3^2+3-2} = \frac{9-4}{9+3-2} = \frac{5}{10} = \frac{1}{2}.$$

$$\text{When } x = 3,\ \frac{x-2}{x-1} = \frac{3-2}{3-1} = \frac{1}{2}.$$

Both expressions are equal to $\frac{1}{2}$ when $x = 3$. Note that any number can be substituted for x except -2 and 1, since they result in division by zero.

Question: When $x = 4$, what is the value of $\dfrac{x^2-4}{x^2+x-2}$ and $\dfrac{x-2}{x-1}$?[2]

For the remainder of this chapter, we will omit the restrictions on the variables that prevent division by zero and assume that the values of the variables are such that division by zero is not possible.

---

2. When $x = 4$, $\dfrac{x^2-4}{x^2+x-2} = \dfrac{4^2-4}{4^2+4-2} = \dfrac{16-4}{16+4-2} = \dfrac{12}{18} = \dfrac{2}{3}$ and $\dfrac{x-2}{x-1} = \dfrac{4-2}{4-1} = \dfrac{2}{3}$.

SUGGESTED ACTIVITY
See the *Instructor's Resource Manual* for an activity involving simplifying rational expressions.

INSTRUCTOR NOTE
You may want to explain to your students that it is possible to incorrectly simplify a rational expression, substitute a number into the original and simplified expressions, and still have the two expressions equal. However, if this "check" does not produce equal answers, then the simplification is not correct.

Example 2 Simplify.

a. $\dfrac{4x^3y^4}{6x^4y}$ **b.** $\dfrac{x^2-9}{x^2+x-12}$

Solution

a. $\dfrac{4x^3y^4}{6x^4y} = \dfrac{2x^3y^4}{3x^4y}$ • Simplify the numerical coefficients.

$= \dfrac{2y^3}{3x}$ • Simplify using the rules of exponents.

Check: We chose to substitute 2 for x and 3 for y in the original expression and the simplified expression.

$$\frac{4x^3y^4}{6x^4y} = \frac{4(2^3)(3^4)}{6(2^4)(3)} = \frac{4\cdot 8\cdot 81}{6\cdot 16\cdot 3} = 9$$

$$\frac{2y^3}{3x} = \frac{2(3^3)}{3(2)} = \frac{2\cdot 27}{3\cdot 2} = 9$$

b. $\dfrac{x^2-9}{x^2+x-12} = \dfrac{(x+3)\overset{1}{\cancel{(x-3)}}}{(x+4)\underset{1}{\cancel{(x-3)}}}$ • Factor the numerator and denominator. Divide by the common factors.

$= \dfrac{x+3}{x+4}$

Check: We chose to substitute 5 for x in the original expression and the simplified expression.

$$\frac{x^2-9}{x^2+x-12} = \frac{5^2-9}{5^2+5-12} = \frac{25-9}{25+5-12} = \frac{16}{18} = \frac{8}{9}$$

$$\frac{x+3}{x+4} = \frac{5+3}{5+4} = \frac{8}{9}$$

You-Try-It 2 Simplify.

a. $\dfrac{6a^5b}{12a^2b^3}$ **b.** $\dfrac{x^2+2x-24}{x^2-16}$

Solution See page S35.

Multiply Rational Expressions

The product of two fractions is a fraction whose numerator is the product of the numerators of the two fractions and whose denominator is the product of the denominators of the two fractions.

$$\frac{a}{b}\cdot\frac{c}{d} = \frac{ac}{bd} \qquad \frac{2}{3}\cdot\frac{4}{5} = \frac{8}{15}$$

$$\frac{3x}{y}\cdot\frac{2}{z} = \frac{6x}{yz} \qquad \frac{x+2}{x}\cdot\frac{3}{x-4} = \frac{3x+6}{x^2-4x}$$

INSTRUCTOR NOTE
You may want to instruct your students to use the same method of checking their work when multiplying and dividing rational expressions as was used in simplifying rational expressions: substitute the same number into the original expression and the product or quotient; then check that the results are equal.

→ Multiply: $\frac{x^2+3x}{x^2-3x-4} \cdot \frac{x^2-5x+4}{x^2+2x-3}$

$$\frac{x^2+3x}{x^2-3x-4} \cdot \frac{x^2-5x+4}{x^2+2x-3}$$

Factor the numerator and denominator of each fraction.

$$= \frac{x(x+3)}{(x-4)(x+1)} \cdot \frac{(x-4)(x-1)}{(x+3)(x-1)}$$

Multiply the fractions and divide by the common factors.

$$= \frac{x\overset{1}{\cancel{(x+3)}}\overset{1}{\cancel{(x-4)}}\overset{1}{\cancel{(x-1)}}}{\underset{1}{\cancel{(x-4)}}(x+1)\underset{1}{\cancel{(x+3)}}\underset{1}{\cancel{(x-1)}}}$$

Write the answer in simplest form.

$$= \frac{x}{x+1}$$

Example 3 Multiply: $\frac{x^2+x-6}{x^2+7x+12} \cdot \frac{x^2+3x-4}{x^2-4}$

Solution

$$\frac{x^2+x-6}{x^2+7x+12} \cdot \frac{x^2+3x-4}{x^2-4}$$

$$= \frac{(x+3)(x-2)}{(x+3)(x+4)} \cdot \frac{(x+4)(x-1)}{(x+2)(x-2)}$$ • Factor.

$$= \frac{\overset{1}{\cancel{(x+3)}}\overset{1}{\cancel{(x-2)}}\overset{1}{\cancel{(x+4)}}(x-1)}{\underset{1}{\cancel{(x+3)}}\underset{1}{\cancel{(x+4)}}(x+2)\underset{1}{\cancel{(x-2)}}}$$ • Multiply. Divide by the common factors.

$$= \frac{x-1}{x+2}$$

You-Try-It 3 Multiply: $\frac{x^2+2x-15}{x^2-9} \cdot \frac{x^2-3x-18}{x^2-7x+6}$

Solution See page S35.

Divide Rational Expressions

The **reciprocal** of a fraction is the fraction with the numerator and denominator interchanged.

Note that $x^2 = \frac{x^2}{1}$.

Fraction	Reciprocal
$\frac{a}{b}$	$\frac{b}{a}$
x^2	$\frac{1}{x^2}$
$\frac{x+2}{x}$	$\frac{x}{x+2}$

To divide two fractions, multiply by the reciprocal of the divisor.

$$\frac{a}{b} \div \frac{c}{d} = \frac{a}{b} \cdot \frac{d}{c} = \frac{ad}{bc}$$

$$\frac{4}{x} \div \frac{y}{5} = \frac{4}{x} \cdot \frac{5}{y} = \frac{20}{xy}$$

$$\frac{x+3}{x} \div \frac{x-2}{4} = \frac{x+3}{x} \cdot \frac{4}{x-2} = \frac{4x+12}{x^2-2x}$$

SUGGESTED ACTIVITY
The product of a rational number and its reciprocal is 1. Show that this is true for the rational expressions $\frac{5}{7}$ and $\frac{2x^2-4x+2}{x(x-1)}$.

The basis for the division rule is shown below.

$$\frac{a}{b} \div \frac{c}{d} = \frac{\frac{a}{b}}{\frac{c}{d}} = \frac{\frac{a}{b}}{\frac{c}{d}} \cdot \frac{\frac{d}{c}}{\frac{d}{c}} = \frac{\frac{a}{b} \cdot \frac{d}{c}}{\frac{c}{d} \cdot \frac{d}{c}} = \frac{\frac{a}{b} \cdot \frac{d}{c}}{1} = \frac{a}{b} \cdot \frac{d}{c}$$

$$\frac{a}{b} \div \frac{c}{d} = \frac{a}{b} \cdot \frac{d}{c}$$

Example 4 Divide: $\frac{3x^2y - xy^2}{z^2} \div \frac{6x^2 - 2xy}{z^3}$

Solution

$$\frac{3x^2y - xy^2}{z^2} \div \frac{6x^2 - 2xy}{z^3}$$

$$= \frac{3x^2y - xy^2}{z^2} \cdot \frac{z^3}{6x^2 - 2xy}$$

• Rewrite division as multiplication by the reciprocal.

$$= \frac{xy\overset{1}{\cancel{(3x-y)}} \cdot z^3}{z^2 \cdot 2x\underset{1}{\cancel{(3x-y)}}}$$

$$= \frac{yz}{2}$$

You-Try-It 4 Divide: $\frac{a^2}{4bc^3 - 2b^2c^2} \div \frac{a}{6bc - 3b^2}$

Solution See page S35.

8.1 EXERCISES

Topics for Discussion

1. Determine whether the statement is always true, sometimes true, or never true.

a. A fraction is in simplest form when the only factor common to both the numerator and the denominator is 1. Always true

b. Before multiplying two rational expressions, we must write the fractions in terms of a common denominator. Never true

c. The expression $\frac{x}{x+2}$ is not a real number if $x = 0$. Never true

d. The procedure for multiplying rational expressions is the same as that for multiplying arithmetic fractions. Always true

e. To divide two rational expressions, multiply the reciprocal of the first fraction by the second fraction. Never true

f. When a rational expression is written in simplest form, its value is less than it was before it was rewritten in simplest form. Never true

g. If a rational expression has the same variable in the numerator and the denominator, then it can be simplified. Sometimes true

2. Explain how to:

a. simplify a rational expression. Answers will vary.

b. multiply a rational expression. Answers will vary.

c. divide a rational expression. Answers will vary.

3. Write the reciprocal of the expression.

a. $\frac{x^2 - x - 30}{x^2 - 6x - 7}$ $\qquad \frac{x^2 - 6x - 7}{x^2 - x - 30}$

b. $\frac{3y^2z - 12yz}{8yz - 32z}$ $\qquad \frac{8yz - 32z}{3y^2z - 12yz}$

c. $45ab^2$ $\qquad \frac{1}{45ab^2}$

4. For each expression, replace the variables with numbers and simplify. Provide enough examples of each to determine whether the expression can always, sometimes, or never be simplified. Explain your conclusions.

a. $\frac{x+1}{x-1}$ **b.** $\frac{x}{x+4}$ **c.** $\frac{xy}{x}$ **d.** $\frac{x+y}{x}$ **e.** $\frac{x^2+2x}{x}$

a. Sometimes; **b.** sometimes; **c.** always; **d.** sometimes; **e.** always. Explanations will vary.

5. The graph of $Y1 = \frac{1}{X+2}$ is shown at the right as graphed on a graphing calculator in the CONNECTED mode. Provide an explanation for the vertical line at $x = -2$ and explain what it means.

Answers will vary. For example, the calculator is "connecting" the plotted points to the left of $x = -2$ with the plotted points to the right of $x = -2$. When $x = -2$, the value of this expression is not a real number because the denominator equals zero.

Simplifying Rational Expressions

For what value of the variable is the expression undefined?

6. $\frac{8}{3x}$ 0

7. $\frac{x^2 + 20x + 24}{x}$ 0

8. $\frac{x}{5x + 15}$ −3

9. $\frac{7x}{12x - 18}$ $\frac{3}{2}$

10. $\frac{x}{(x + 6)(x - 1)}$ −6, 1

11. $\frac{x}{(x - 2)(x + 5)}$ 2, −5

12. $\frac{6}{x^2 - 1}$ 1, −1

13. $\frac{9}{x^2 - 16}$ 4, −4

14. $\frac{x - 4}{x^2 - 2x + 6}$ 3, −2

15. $\frac{3x}{x^2 + 6x + 9}$ −3

16. $\frac{4x + 8}{2x^2 - 3x - 5}$ $\frac{5}{2}, -1$

17. $\frac{3x - 6}{3x^2 - 10x - 8}$ $-\frac{2}{3}, 4$

Simplify.

18. $\frac{9x^3}{12x^4}$ $\frac{3}{4x}$

19. $\frac{16x^2y}{24xy^3}$ $\frac{2x}{3y^2}$

20. $\frac{6y(y + 2)}{9y^2(y + 2)}$ $\frac{2}{3y}$

21. $\frac{12x^3(x - 3)}{18x(x - 3)}$ $\frac{2x^2}{3}$

22. $\frac{a^2 + 4a}{ab + 4b}$ $\frac{a}{b}$

23. $\frac{x^2 - 3x}{2x - 6}$ $\frac{x}{2}$

24. $\frac{6x - 4}{3x^2 - 2x}$ $\frac{2}{x}$

25. $\frac{5xy - 3y}{15x - 9}$ $\frac{y}{3}$

26. $\frac{y^2 - 3y + 2}{y^2 - 4y + 3}$ $\frac{y - 2}{y - 3}$

27. $\frac{x^2 + 5x + 6}{x^2 + 8x + 15}$ $\frac{x + 2}{x + 5}$

28. $\frac{x^2 + 3x - 10}{x^2 + 2x - 8}$ $\frac{x + 5}{x + 4}$

29. $\frac{a^2 + 7a - 8}{a^2 + 6a - 7}$ $\frac{a + 8}{a + 7}$

30. $\frac{x^2 + x - 12}{x^2 - 6x + 9}$ $\frac{x + 4}{x - 3}$

31. $\frac{x^2 + 8x + 16}{x^2 - 2x - 24}$ $\frac{x + 4}{x - 6}$

32. $\frac{x^2 - 3x - 10}{x^2 - 25}$ $\frac{x + 2}{x + 5}$

33. $\frac{y^2 - 4}{y^2 - 3y - 10}$ $\frac{y - 2}{y - 5}$

34. **a.** Complete the tables by evaluating the rational expressions for the given values of x.

x	$\frac{x}{x^2+3x}$
−5	−0.5
−4	−1
−3	undefined
−2	1
−1	0.5
0	undefined
1	0.25
2	0.2
3	$0.1\overline{6}$
4	0.142857...
5	0.125

x	$\frac{1}{x+3}$
−5	−0.5
−4	−1
−3	undefined
−2	1
−1	0.5
0	$0.\overline{3}$
1	0.25
2	0.2
3	$0.1\overline{6}$
4	0.142857...
5	0.125

b. How do the values of the rational expressions differ? Explain why they differ in this regard.

For $x = 0$, the expression $\frac{x}{x^2+3x}$ is undefined, whereas the expression $\frac{1}{x+3}$ equals $0.\overline{3}$. Explanations will vary; for example: when $x = 0$, the denominator of the first expression is 0; however, when the expression is in simplest form, the denominator is 3 when $x = 0$.

35. **a.** Complete the tables by evaluating the rational expressions for the given values of x.

x	$\frac{x-4}{x^2-3x-4}$
−5	−0.25
−4	$-0.\overline{3}$
−3	−0.5
−2	−1
−1	undefined
0	1
1	0.5
2	$0.\overline{3}$
3	0.25
4	undefined
5	$0.1\overline{6}$

x	$\frac{1}{x+1}$
−5	−0.25
−4	$-0.\overline{3}$
−3	−0.5
−2	−1
−1	undefined
0	1
1	0.5
2	$0.\overline{3}$
3	0.25
4	0.2
5	$0.1\overline{6}$

b. How do the values of the rational expressions differ? Explain why they differ in this regard.

For $x = 4$, the expression $\frac{x-4}{x^2-3x-4}$ is undefined, whereas the expression $\frac{1}{x+1}$ equals 0.2. Explanations will vary; for example: when $x = 4$, the denominator of the first expression is 0; however, when the expression is in simplest form, the denominator is 5 when $x = 4$.

36. Write a rational expression that is not in simplest form. Evaluate it for several values of the variable. Write the expression in simplest form and then evaluate it for the same values of the variable. Describe how the values differ. Explain why they differ in this regard.
Answers will vary.

Multiply and Divide Rational Expressions

Multiply or divide.

37. $\frac{8x^2}{9y^3} \cdot \frac{3y^2}{4x^3}$

$\frac{2}{3xy}$

38. $\frac{4a^2b^3}{15x^5y^2} \cdot \frac{25x^2y}{16ab}$

$\frac{5ab^2}{12x^3y}$

39. $\frac{12x^3y^4}{7a^2b^3} \cdot \frac{14a^3b^4}{9x^2y^2}$

$\frac{8xy^2ab}{3}$

40. $\frac{18a^4b^2}{25x^2y^3} \cdot \frac{50x^5y^6}{27a^6b^3}$

$\frac{4x^3y^3}{3a^2b}$

41. $\frac{3x-6}{5x-20} \cdot \frac{10x-40}{27x-54}$

$\frac{2}{9}$

42. $\frac{8x-12}{14x+7} \cdot \frac{42x+21}{32x-48}$

$\frac{3}{4}$

43. $\frac{3x^2+2x}{2xy-3y} \cdot \frac{2xy^3-3y^3}{3x^3+2x^2}$

$\frac{y^2}{x}$

44. $\frac{4a^2x-3a^2}{2by+5b} \cdot \frac{2b^3y+5b^3}{4ax-3a}$

ab^2

45. $\frac{x^2+5x+4}{x^3y^2} \cdot \frac{x^2y^3}{x^2+2x+1}$

$\frac{y(x+4)}{x(x+1)}$

46. $\frac{x^2+x-2}{xy^2} \cdot \frac{x^3y}{x^2+5x+6}$

$\frac{x^2(x-1)}{y(x+3)}$

47. $\frac{x^4y^2}{x^2+3x-28} \cdot \frac{x^2-49}{xy^4}$

$\frac{x^3(x-7)}{y^2(x-4)}$

48. $\frac{x^2-2x-24}{x^2-5x-6} \cdot \frac{x^2+5x+6}{x^2+6x+8}$

$\frac{x+3}{x+1}$

49. $\frac{x^2-8x+7}{x^2+3x-4} \cdot \frac{x^2+3x-10}{x^2-9x+14}$

$\frac{x+5}{x+4}$

50. $\frac{y^2+y-20}{y^2+2y-15} \cdot \frac{y^2+4y-21}{y^2+3y-28}$

1

51. $\frac{2x^2+5x+2}{2x^2+7x+3} \cdot \frac{x^2-7x-30}{x^2-6x-40}$

$\frac{x+2}{x+4}$

52. $\frac{x^2-4x-32}{x^2-8x-48} \cdot \frac{3x^2+17x+10}{3x^2-22x-16}$

$\frac{x+5}{x-12}$

53. $\frac{4x^2y^3}{15a^2b^3} \div \frac{6xy}{5a^3b^5}$

$\frac{2xy^2ab^2}{9}$

54. $\frac{9x^3y^4}{16a^4b^3} \div \frac{45x^4y^2}{14a^7b}$

$\frac{7a^3y^2}{40b^2x}$

55. $\frac{6x-12}{8x+32} \div \frac{18x-36}{10x+40}$

$\frac{5}{12}$

56. $\frac{28x+14}{45x-30} \div \frac{14x+7}{30x-20}$

$\frac{4}{3}$

57. $\frac{6x^3+7x^2}{12x-3} \div \frac{6x^2+7x}{36x-9}$

$3x$

58. $\frac{5a^2y+3a^2}{2x^3+5x^2} \div \frac{10ay+6a}{6x^3+15x^2}$

$\frac{3a}{2}$

59. $\frac{x^2+4x+3}{x^2y} \div \frac{x^2+2x+1}{xy^2}$

$\frac{y(x+3)}{x(x+1)}$

60. $\frac{x^2-49}{x^4y^3} \div \frac{x^2-14x+49}{x^4y^3}$

$\frac{x+7}{x-7}$

61. $\dfrac{x^2y^5}{x^2-11x+30} \div \dfrac{xy^6}{x^2-7x+10}$

$\dfrac{x(x-2)}{y(x-6)}$

62. $\dfrac{x^2-5x+6}{x^2-9x+18} \div \dfrac{x^2-6x+8}{x^2-9x+20}$

$\dfrac{x-5}{x-6}$

63. $\dfrac{x^2+3x-40}{x^2+2x-35} \div \dfrac{x^2+2x-48}{x^2+3x-18}$

$\dfrac{(x+6)(x-3)}{(x+7)(x-6)}$

64. $\dfrac{y^2-y-56}{y^2+8y+7} \div \dfrac{y^2-13y+40}{y^2-4y-5}$

1

65. $\dfrac{2x^2-3x-20}{2x^2-7x-30} \div \dfrac{2x^2-5x-12}{4x^2+12x+9}$

$\dfrac{2x+3}{x-6}$

66. $\dfrac{6n^2+13n+6}{4n^2-9} \div \dfrac{6n^2+n-2}{4n^2-1}$

$\dfrac{2n+1}{2n-3}$

Applying Concepts

Simplify.

67. $\dfrac{y^2}{x} \cdot \dfrac{x}{2} \div \dfrac{y}{x}$

$\dfrac{xy}{2}$

68. $\dfrac{ab}{3} \cdot \dfrac{a}{b^2} \div \dfrac{a}{4}$

$\dfrac{4a}{3b}$

69. $\left(\dfrac{2x}{y}\right)^3 \div \left(\dfrac{x}{3y}\right)^2$

$\dfrac{72x}{y}$

70. $\left(\dfrac{c}{3}\right)^2 \div \left(\dfrac{c}{2} \cdot \dfrac{c}{4}\right)$

$\dfrac{8}{9}$

71. $\left(\dfrac{a-3}{b}\right)^2 \left(\dfrac{b}{a-3}\right)^3$

$\dfrac{b}{a-3}$

72. $\left(\dfrac{x-4}{y^2}\right)^3 \left(\dfrac{y}{x-4}\right)^2$

$\dfrac{x-4}{y^4}$

73. $\dfrac{x^2+3x-40}{x^2+2x-35} \div \dfrac{x^2+2x-48}{x^2+3x-18} \cdot \dfrac{x^2-36}{x^2-9}$

$\dfrac{(x+6)(x+6)}{(x+7)(x+3)}$

74. $\dfrac{x^2+x-6}{x^2+7x+12} \cdot \dfrac{x^2+3x-4}{x^2+x-2} \div \dfrac{x^2-16}{x^2-4}$

$\dfrac{(x-2)(x-2)}{(x-4)(x+4)}$

75. Find two different pairs of rational expressions whose product is $\dfrac{3x^2-13x+4}{2x^2+7x+3}$.

$\dfrac{x-4}{x+3}$ and $\dfrac{3x-1}{2x+1}$, or $\dfrac{x-4}{2x+1}$ and $\dfrac{3x-1}{x+3}$

76. Write a rational expression for which the following values of the variable must be excluded: –2 and 5.
Answers will vary.

77. If $\dfrac{(x+2)^m}{(x+2)^n} = x^2 + 4x + 4$, what is the relationship between m and n?
$m = n + 2$

Explorations

78. ***Experimenting with Values of Rational Expressions***

a. Given the expression $\dfrac{9}{x^2+1}$, choose some values of x and evaluate the expression for those values. Is it possible to choose a value of x for which the value of the expression is greater than 10? If so, what is that value of x? If not, explain why it is not possible.

b. Given the expression $\dfrac{1}{y-3}$, choose some values of y and evaluate the expression for those values. Is it possible to choose a value of y for which the value of the expression is greater than 10,000,000? If so, what is that value of y? If not, explain why it is not possible.

79. ***Graphs of Rational Expressions***

a. The graph of $Y1 = \dfrac{2}{X-3}$ is shown at the right. Use a graphing calculator to find the value of $\dfrac{2}{X-3}$ for several values of x between 1 and 3. Describe the value of $\dfrac{2}{X-3}$ as x gets closer and closer to 3. Find the value of $\dfrac{2}{X-3}$ for several values of x between 3 and 5. Describe the value of $\dfrac{2}{X-3}$ as x gets closer and closer to 3.

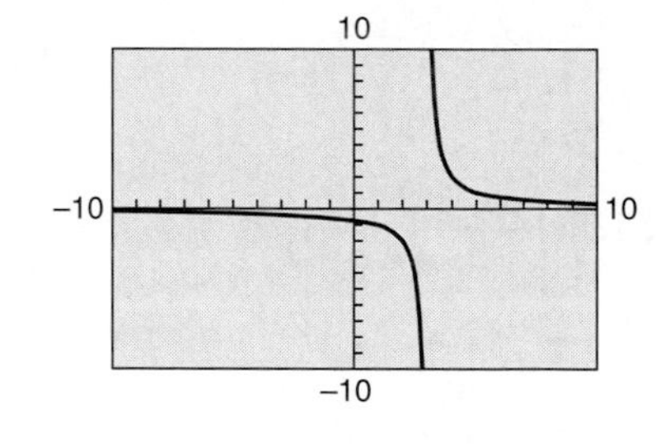

b. Use a graphing calculator to graph $Y1 = \dfrac{3}{X+4}$. Find the value of $\dfrac{3}{X+4}$ for several values of x between –6 and –4. Describe the value of $\dfrac{3}{X+4}$ as x gets closer and closer to –4. Find the value of $\dfrac{3}{X+4}$ for several values of x between –4 and –2. Describe the value of $\dfrac{3}{X+4}$ as x gets closer to –4.

c. Based on your answers to parts a and b, describe the value of $\dfrac{5}{x-2}$ as x gets closer and closer to 2 from 1 and gets closer and closer to 2 from 3.

Section 8.2 Addition and Subtraction of Rational Expressions

Addition and Subtraction of Rational Expressions with the Same Denominator

If a painter can paint a room in 4 hours, then in 1 hour, the painter can paint $\frac{1}{4}$ of the room. The painter's rate of work is $\frac{1}{4}$ of the room each hour. The **rate of work** is that part of a task that is completed in 1 unit of time.

If you can type a research paper in 5 hours, then you can type $\frac{1}{5}$ of the paper each hour. Your rate of work is $\frac{1}{5}$ of the paper each hour.

Question: A computer can load a national basketball team's official Web site in 3 minutes. What is the computer's rate of work for loading the Web site?[1]

A pipe can fill a tank in 30 minutes. This pipe's rate of work is $\frac{1}{30}$ of the tank each minute. A second pipe can fill the tank in x minutes. The rate of work for the second pipe is $\frac{1}{x}$ of the tank each minute.

Question: A computer can download a program in y minutes. What is the computer's rate of work for downloading the program?[2]

Let's return to the example of typing the research paper. You can type the entire paper in 5 hours, so your rate of work is $\frac{1}{5}$ of the paper each hour. Therefore,

in 1 hour, you have completed $\frac{1}{5}$ of the paper,

in 2 hours, you have completed $\frac{2}{5}$ of the paper,

in 3 hours, you have completed $\frac{3}{5}$ of the paper,

in 4 hours you have completed $\frac{4}{5}$ of the paper,

and in 5 hours you have completed $\frac{5}{5}$ of the paper, or the whole paper.

We can calculate the part of a task completed in a given amount of time by using the formula:

Rate of work · Time worked = Part of task completed

---

1. The computer's rate of work is $\frac{1}{3}$ of the job each minute.

2. The computer's rate of work is $\frac{1}{y}$ of the job each minute.

For example, an open faucet can fill a sink in 6 minutes. The faucet's rate of work is $\frac{1}{6}$ of the sink each minute. We can calculate the fraction of the sink that the faucet will fill in 5 minutes.

$$\text{Rate of work} \cdot \text{Time worked} = \text{Part of task completed}$$
$$\frac{1}{6} \cdot 5 = \frac{5}{6}$$

The faucet will fill $\frac{5}{6}$ of the sink in 5 minutes.

Question: An open drain can empty a sink in 8 minutes. What fraction of the sink will the drain empty in 3 minutes?[3]

Suppose an open pipe can fill a tank in x hours. We can use the same formula to express the fraction of the tank that the pipe will fill in 4 hours.

$$\text{Rate of work} \cdot \text{Time worked} = \text{Part of task completed}$$
$$\frac{1}{x} \cdot 4 = \frac{4}{x}$$

The pipe will fill $\frac{4}{x}$ of the tank in 4 hours.

Question: An open drain can empty a pool in y hours. What fraction of the pool will the drain empty in 2 hours?[4]

Suppose it takes you x hours to write a program for your computer class. You work on the program for 2 hours on Monday and 6 hours on Tuesday. What fraction of the program have you completed?

You completed $\frac{1}{x} \cdot 2 = \frac{2}{x}$ of the program on Monday.

You completed $\frac{1}{x} \cdot 6 = \frac{6}{x}$ of the program on Tuesday.

You have completed $\frac{2}{x} + \frac{6}{x} = \frac{2+6}{x} = \frac{8}{x}$ of the program.

In this example, we have added two rational expressions. Both expressions have the same denominator (x). We add the two expressions by adding the numerators. The denominator of the sum is the common denominator.

$$\frac{a}{b} + \frac{c}{b} = \frac{a+c}{b}$$

$$\frac{5x}{18} + \frac{7x}{18} = \frac{5x+7x}{18} = \frac{12x}{18} = \frac{2x}{3}$$

TAKE NOTE
After adding the rational expressions, the sum is written in simplest form.

$$\frac{x}{x^2-1} + \frac{1}{x^2-1} = \frac{x+1}{x^2-1} = \frac{\overset{1}{\cancel{x+1}}}{\underset{1}{\cancel{(x+1)}}(x-1)} = \frac{1}{x-1}$$

3. Rate of work · Time worked = Part of task completed; $\frac{1}{8} \cdot 3 = \frac{3}{8}$; the drain will empty $\frac{3}{8}$ of the sink in 3 minutes.

4. Rate of work · Time worked = Part of task completed; $\frac{1}{y} \cdot 2 = \frac{2}{y}$; the drain will empty $\frac{2}{y}$ of the pool in 2 hours.

Example 1 Add: $\frac{7}{b^2} + \frac{9}{b^2}$

Solution $\frac{7}{b^2} + \frac{9}{b^2} = \frac{7+9}{b^2}$

$= \frac{16}{b^2}$

• The denominators are the same. Add the numerators. The denominator of the sum is the common denominator.

You-Try-It 1 Add: $\frac{3}{xy} + \frac{12}{xy}$

Solution See page S35.

Example 2 A mason can build a fireplace in x hours. If the mason worked on the fireplace for 4 hours on Thursday and 6 hours on Friday, what fraction of the fireplace was completed on Thursday and Friday?

Solution The mason completed $\frac{1}{x} \cdot 4 = \frac{4}{x}$ of the fireplace on Thursday.

The mason completed $\frac{1}{x} \cdot 6 = \frac{6}{x}$ of the fireplace on Friday.

$\frac{4}{x} + \frac{6}{x} = \frac{4+6}{x} = \frac{10}{x}$

$\frac{10}{x}$ of the fireplace was completed on Thursday and Friday.

You-Try-It 2 An open drain in a pool can empty the water in y hours. If the drain was left open for 2 hours on Saturday and 3 hours on Sunday, what fraction of the pool was drained on Saturday and Sunday?

Solution See page S35.

SUGGESTED ACTIVITY
As shown at the right, $\frac{2x}{x-2} - \frac{4}{x-2}$ is equal to 2. Have students evaluate the expression for several values of x. They should note that, with the exception of $x = 2$, the calculation always yields 2.

When subtracting rational expressions with the same denominator, subtract the numerators. The denominator of the difference is the common denominator. Write the answer in simplest form.

$$\frac{2x}{x-2} - \frac{4}{x-2} = \frac{2x-4}{x-2} = \frac{2\overset{1}{\cancel{(x-2)}}}{\underset{1}{\cancel{x-2}}} = 2$$

TAKE NOTE
Be careful with signs when subtracting rational expressions. In the example at the right, we must subtract the *entire* numerator $2x + 3$.

$$\frac{3x-1}{x^2-5x+4} - \frac{2x+3}{x^2-5x+4} = \frac{(3x-1)-(2x+3)}{x^2-5x+4}$$

$$= \frac{3x-1-2x-3}{x^2-5x+4}$$

$$= \frac{x-4}{x^2-5x+4} = \frac{\overset{1}{\cancel{x-4}}}{\underset{1}{\cancel{(x-4)}}(x-1)} = \frac{1}{x-1}$$

Example 3 Subtract: $\frac{3x^2}{x^2-1} - \frac{x+4}{x^2-1}$

Solution

$\frac{3x^2}{x^2-1} - \frac{x+4}{x^2-1}$ • The denominators are the same.

$= \frac{3x^2-(x+4)}{x^2-1}$ • Subtract the numerators.

$= \frac{3x^2-x-4}{x^2-1}$

$= \frac{(3x-4)\overset{1}{\cancel{(x+1)}}}{\underset{1}{\cancel{(x+1)}}(x-1)}$

$= \frac{3x-4}{x-1}$

You-Try-It 3 Subtract: $\frac{2x^2}{x^2-x-12} - \frac{7x+4}{x^2-x-12}$

Solution See page S36.

Example 4 It takes a total of x hours to plant your vegetable garden. You spent 3 hours on the garden on Saturday, and your spouse worked for 5 hours. How much more of the entire job did your spouse complete?

Solution You completed $\frac{1}{x}\cdot 3 = \frac{3}{x}$ of the planting.

Your spouse completed $\frac{1}{x}\cdot 5 = \frac{5}{x}$ of the planting.

$\frac{5}{x} - \frac{3}{x} = \frac{5-3}{x} = \frac{2}{x}$

Your spouse completed $\frac{2}{x}$ more of the planting than you did.

You-Try-It 4 It takes a total of h hours to wallpaper the living room. You spent 2 hours wallpapering on Sunday, and your roommate worked on it for 3 hours. How much more of the entire job did your roommate complete?

Solution See page S36.

Addition and Subtraction of Rational Expressions with Different Denominators

Thus far in this section, each of the examples involved adding or subtracting rational expressions in which the denominators were the same.

Question: What is the sum of $\frac{1}{2}$ and $\frac{1}{3}$?[5]

---

5. $\frac{1}{2}+\frac{1}{3}=\frac{3}{6}+\frac{2}{6}=\frac{5}{6}$

In the sum $\frac{1}{2} + \frac{1}{3}$, the denominators are not the same. Therefore, the fractions $\frac{1}{2}$ and $\frac{1}{3}$ had to be rewritten as fractions with a common denominator before they could be added. The same is true for adding or subtracting rational expressions; if the denominators are not the same, they must first be rewritten as fractions with a common denominator. The common denominator is generally the least common multiple (LCM) of the denominators.

The least common multiple of two or more numbers is the smallest number that contains the prime factorization of each number.

The LCM of 12 and 18 is 36. 36 contains the prime factors of 12 and the prime factors of 18. 36 is the smallest number that both 12 and 18 divide evenly into.

$$12 = 2 \cdot 2 \cdot 3$$
$$18 = 2 \cdot 3 \cdot 3$$
$$\text{LCM} = 36 = \overbrace{2 \cdot \underbrace{2 \cdot 3}_{}}^{\text{Factors of 12}} \underbrace{\cdot\, 3}_{\text{Factors of 18}}$$

The least common multiple of two or more polynomials is the polynomial of least degree that contains the factors of each polynomial.

To find the LCM of two or more polynomials, first factor each polynomial completely. The LCM is the product of each factor the greatest number of times it occurs in any one factorization.

The LCM of $8x^2y$ and $4xy^2$ is the product of the LCM of the numerical coefficients and each variable factor the greatest number of times it occurs in either factorization.

$$8x^2y = 2 \cdot 2 \cdot 2 \cdot x \cdot x \cdot y$$
$$4xy^2 = 2 \cdot 2 \cdot x \cdot y \cdot y$$
$$\text{LCM} = \overbrace{2 \cdot 2 \cdot 2 \cdot x \cdot x \cdot y}^{\text{Factors of } 8x^2y} \cdot y = 8x^2y^2 \quad (\text{Factors of } 4xy^2: 2 \cdot 2 \cdot x \cdot y \cdot y)$$

Example 5 Find the LCM of $4a^3b$ and $6ab^2$.

Solution

$4a^3b = 2 \cdot 2 \cdot a \cdot a \cdot a \cdot b$

$6ab^2 = 2 \cdot 3 \cdot a \cdot b \cdot b$

- Factor each polynomial completely.

$\text{LCM} = 2 \cdot 2 \cdot 3 \cdot a \cdot a \cdot a \cdot b \cdot b$

$= 12a^3b^2$

- Write the product of the LCM of the numerical coefficients and each variable factor the greatest number of times it occurs in either factorization.

You-Try-It 5 Find the LCM of $8uv^4$ and $12uv$.

Solution See page S36.

Note that in each case above, the exponent on a variable in the LCM is always the highest exponent on that variable in either of the given monomials. For instance, in Example 5, in $4a^3b$ and $6ab^2$, the highest exponent on a is 3 and the highest exponent on b is 2, and the LCM contains a^3 and b^2.

→ Add: $\dfrac{2}{15a} + \dfrac{7}{5a^2}$

Find the LCM of $15a$ and $5a^2$.

The LCM of the numerical coefficients 15 and 5 is 15.

The highest exponent on the variable a is a^2.

The LCM of the denominators is $15a^2$.

Write each fraction with a denominator of $15a^2$.

Note that the denominator $15a$ must be multiplied by a to equal $15a^2$, and the denominator $5a^2$ must be multiplied by 3 to equal $15a^2$.

Therefore, multiply $\dfrac{2}{15a}$ by $\dfrac{a}{a}$ and multiply $\dfrac{7}{5a^2}$ by $\dfrac{3}{3}$.

TAKE NOTE
Once the fractions have been rewritten with the same denominator, then we can add the fractions by adding the numerators and placing the sum over the common denominator.

$$\frac{2}{15a} + \frac{7}{5a^2} = \frac{2}{15a} \cdot \frac{a}{a} + \frac{7}{5a^2} \cdot \frac{3}{3}$$
$$= \frac{2a}{15a^2} + \frac{21}{15a^2}$$
$$= \frac{2a + 21}{15a^2}$$

Example 6 Add or subtract. **a.** $\dfrac{y}{x} - \dfrac{4y}{3x} + \dfrac{3y}{4x}$ **b.** $x - \dfrac{3}{5x}$

Solution **a.** The LCM of x, $3x$, and $4x$ is $12x$.
Rewrite each fraction with a denominator of $12x$.

$$\frac{y}{x} - \frac{4y}{3x} + \frac{3y}{4x} = \frac{y}{x} \cdot \frac{12}{12} - \frac{4y}{3x} \cdot \frac{4}{4} + \frac{3y}{4x} \cdot \frac{3}{3}$$
$$= \frac{12y}{12x} - \frac{16y}{12x} + \frac{9y}{12x}$$
$$= \frac{12y - 16y + 9y}{12x}$$
$$= \frac{5y}{12x}$$

b. The denominator of x is 1 ($x = \frac{x}{1}$).

The LCM of the denominators 1 and $5x$ is $5x$.
Rewrite x as a fraction with a denominator of $5x$.

$$x - \frac{3}{5x} = \frac{x}{1} - \frac{3}{5x}$$
$$= \frac{x}{1} \cdot \frac{5x}{5x} - \frac{3}{5x}$$
$$= \frac{5x^2}{5x} - \frac{3}{5x}$$
$$= \frac{5x^2 - 3}{5x}$$

You-Try-It 6 Add or subtract. **a.** $\dfrac{z}{8y} - \dfrac{4z}{3y} + \dfrac{5z}{4y}$ **b.** $y + \dfrac{5}{7y}$

Solution See page S36.

SUGGESTED ACTIVITY
Ask students to complete the following equation:

$$\frac{b^2-1}{b^2+b} + \underline{\qquad} = \frac{2b+1}{b}$$

[Answer: $\frac{b+2}{b}$]

Example 7 Subtract: $\frac{2x}{x-3} - \frac{1}{x+1}$

Solution The denominator $x - 3$ is prime; it cannot be factored.
The denominator $x + 1$ is prime; it cannot be factored.
The LCM of $x - 3$ and $x + 1$ is $(x - 3)(x + 1)$.
Rewrite each fraction with a denominator of $(x - 3)(x + 1)$.

$$\frac{2x}{x-3} - \frac{1}{x+1} = \frac{2x}{x-3}\cdot\frac{x+1}{x+1} - \frac{1}{x+1}\cdot\frac{x-3}{x-3}$$

$$= \frac{2x^2+2x}{(x-3)(x+1)} - \frac{x-3}{(x-3)(x+1)}$$

$$= \frac{2x^2+2x-(x-3)}{(x-3)(x+1)}$$

$$= \frac{2x^2+2x-x+3}{(x-3)(x+1)}$$

$$= \frac{2x^2+x+3}{(x-3)(x+1)}$$

You-Try-It 7 Subtract: $\frac{4x}{3x-1} - \frac{9}{x+4}$

Solution See page S36.

Example 8 A painter can paint a garage in x hours. The painter's assistant can paint the same garage in $2x$ hours. What fraction of the job has been completed after they both work for 3 hours?

Solution The painter completes $\frac{1}{x}\cdot 3 = \frac{3}{x}$ of the job.

The assistant completes $\frac{1}{2x}\cdot 3 = \frac{3}{2x}$ of the job.

We need to add the fractions $\frac{3}{x}$ and $\frac{3}{2x}$.

The LCM of the denominators x and $2x$ is $2x$.

Rewrite the fraction $\frac{3}{x}$ with a denominator of $2x$ by multiplying it by $\frac{2}{2}$.

$$\frac{3}{x} + \frac{3}{2x} = \frac{3}{x}\cdot\frac{2}{2} + \frac{3}{2x}$$

$$= \frac{6}{2x} + \frac{3}{2x} = \frac{9}{2x}$$

After 3 hours, they have completed $\frac{9}{2x}$ of the job.

You-Try-It 8 A large pipe can fill a tank in y hours. A smaller pipe can fill the same tank in $3y$ hours. What fraction of the tank has been filled after both pipes have been open for two hours?

Solution See page S36.

8.2 EXERCISES

Topics for Discussion

1. Determine whether the statement is always true, sometimes true, or never true.
   a. To add two fractions, add the numerators and add the denominators. Never true
   b. The procedure for subtracting two rational expressions containing variables is the same as that for subtracting arithmetic fractions. Always true
   c. To add two rational expressions with different denominators, first multiply both fractions by the LCM of their denominators. Never true
   d. Rewriting two fractions in terms of the LCM of their denominators is the reverse of simplifying the fractions. Always true

2. Explain how to:
   a. add two rational expressions with the same denominator. Answers will vary.
   b. subtract two rational expressions with the same denominator when the numerators are binomials. Answers will vary.
   c. add two rational expressions with different denominators. Answers will vary.

3. If the statement is correct, write "Correct." If the statement is not correct, explain the error.
   a. We can rewrite $\frac{x}{y}$ as $\frac{4x}{4y}$ by using the Multiplication Property of One. Correct.
   b. The LCM of x^2, x^5, and x^8 is x^2. The LCM is x^8.
   c. If $x \neq -2$ and $x \neq 0$, then $\frac{x}{x+2} + \frac{3}{x+2} = \frac{x+3}{x+2} = \frac{3}{2}$. $\frac{x+3}{x+2}$ is in simplest form.
   d. If $y \neq 8$, then $\frac{y}{y-8} - \frac{8}{y-8} = \frac{y-8}{y-8} = 1$. Correct.
   e. If $x \neq 0$, then $\frac{4x-3}{x} - \frac{3x+1}{x} = \frac{4x-3-3x+1}{x} = \frac{x-2}{x}$. $\frac{4x-3}{x} - \frac{3x+1}{x} = \frac{4x-3-(3x+1)}{x}$

Addition and Subtraction of Rational Expressions

Find the LCM of the expressions.

4. $8x^2$, $4x$ — $8x^2$

5. $6y^2$, $4y$ — $12y^2$

6. $8x^3y$, $12xy^2$ — $24x^3y^2$

7. $6a^2$, $18ab^3$ — $18a^2b^3$

8. $10x^4y^2$, $15x^3y$ — $30x^4y^2$

9. $12a^2b$, $18ab^3$ — $36a^2b^3$

10. $x-1$, $x-2$ — $(x-1)(x-2)$

11. $x+4$, $x-3$ — $(x+4)(x-3)$

Add or subtract.

12. $\frac{3}{y^2} + \frac{8}{y^2}$

$\frac{11}{y^2}$

13. $\frac{6}{ab} - \frac{2}{ab}$

$\frac{4}{ab}$

14. $\frac{3}{x+4} - \frac{10}{x+4}$

$-\frac{7}{x+4}$

15. $\frac{x}{x+6} - \frac{2}{x+6}$

$\frac{x-2}{x+6}$

16. $\frac{3x}{2x+3} + \frac{5x}{2x+3}$

$\frac{8x}{2x+3}$

17. $\frac{6y}{4y+1} - \frac{11y}{4y+1}$

$-\frac{5y}{4y+1}$

18. $\frac{2x+1}{x-3} + \frac{3x+6}{x-3}$

$\frac{5x+7}{x-3}$

19. $\frac{4x+3}{2x-7} + \frac{3x-8}{2x-7}$

$\frac{7x-5}{2x-7}$

20. $\frac{5x-1}{x+9} - \frac{3x+4}{x+9}$

$\frac{2x-5}{x+9}$

21. $\frac{6x-5}{x-10} - \frac{3x-4}{x-10}$

$\frac{3x-1}{x-10}$

22. $\frac{4x-3}{2x+7} - \frac{x-7}{2x+7}$

$\frac{3x+4}{2x+7}$

23. $\frac{2n}{3n+4} - \frac{5n-3}{3n+4}$

$\frac{-3n+3}{3n+4}$

24. $\frac{x}{x^2+2x-15} - \frac{3}{x^2+2x-15}$

$\frac{1}{x+5}$

25. $\frac{3x}{x^2+3x-10} - \frac{6}{x^2+3x-10}$

$\frac{3}{x+5}$

26. $\frac{2x+3}{x^2-x-30} - \frac{x-2}{x^2-x-30}$

$\frac{1}{x-6}$

27. $\frac{3x-1}{x^2+5x-6} - \frac{2x-7}{x^2+5x-6}$

$\frac{1}{x-1}$

28. $\frac{4y+7}{2y^2+7y-4} - \frac{y-5}{2y^2+7y-4}$

$\frac{3}{2y-1}$

29. $\frac{x+1}{2x^2-5x-12} + \frac{x+2}{2x^2-5x-12}$

$\frac{1}{x-4}$

30. $\frac{4}{x} + \frac{5}{y}$

$\frac{4y+5x}{xy}$

31. $\frac{7}{a} + \frac{5}{b}$

$\frac{7b+5a}{ab}$

32. $\frac{12}{x} - \frac{5}{2x}$

$\frac{19}{2x}$

33. $\frac{5}{3a} - \frac{3}{4a}$

$\frac{11}{12a}$

34. $\frac{1}{2x} - \frac{5}{4x} + \frac{7}{6x}$

$\frac{5}{12x}$

35. $\frac{7}{4y} + \frac{11}{6y} - \frac{8}{3y}$

$\frac{11}{12y}$

36. $\frac{5}{3x} - \frac{2}{x^2} + \frac{3}{2x}$

$\frac{19x-12}{6x^2}$

37. $\frac{6}{y^2} + \frac{3}{4y} - \frac{2}{5y}$

$\frac{7y+120}{20y^2}$

38. $\frac{2x+1}{3x} + \frac{x-1}{5x}$

$\frac{13x+2}{15x}$

39. $\dfrac{4x-3}{6x}+\dfrac{2x+3}{4x}$

$\dfrac{14x+3}{12x}$

40. $\dfrac{x-3}{6x}+\dfrac{x+4}{8x}$

$\dfrac{7}{24}$

41. $\dfrac{2x-3}{2x}+\dfrac{x+3}{3x}$

$\dfrac{8x-3}{6x}$

42. $\dfrac{2x+9}{9x}-\dfrac{x-5}{5x}$

$\dfrac{x+90}{45x}$

43. $\dfrac{3y-2}{12y}-\dfrac{y-3}{18y}$

$\dfrac{7}{36}$

44. $y+\dfrac{8}{3y}$

$\dfrac{3y^2+8}{3y}$

45. $\dfrac{7}{2n}-n$

$\dfrac{7-2n^2}{2n}$

46. $\dfrac{4}{x+4}+x$

$\dfrac{(x+2)(x+2)}{x+4}$

47. $x+\dfrac{3}{x+2}$

$\dfrac{x^2+2x+3}{x+2}$

48. $5-\dfrac{x-2}{x+1}$

$\dfrac{4x+7}{x+1}$

49. $3+\dfrac{x-1}{x+1}$

$\dfrac{4x+2}{x+1}$

50. $\dfrac{4}{x-2}+\dfrac{5}{x+3}$

$\dfrac{9x+2}{(x-2)(x+3)}$

51. $\dfrac{2}{x-3}+\dfrac{5}{x-4}$

$\dfrac{7x-23}{(x-3)(x-4)}$

52. $\dfrac{6}{x-7}-\dfrac{4}{x+3}$

$\dfrac{2(x+23)}{(x-7)(x+3)}$

53. $\dfrac{3}{y+6}-\dfrac{4}{y-3}$

$\dfrac{-y-33}{(y+6)(y-3)}$

54. $\dfrac{2x}{x+1}+\dfrac{1}{x-3}$

$\dfrac{2x^2-5x+1}{(x+1)(x-3)}$

55. $\dfrac{3x}{x-4}+\dfrac{2}{x+6}$

$\dfrac{3x^2+20x-8}{(x-4)(x+6)}$

56. $\dfrac{4x}{2x-1}-\dfrac{5}{x-6}$

$\dfrac{4x^2-34x+5}{(2x-1)(x-6)}$

57. A mason can build a retaining wall in x hours. If the mason worked on the retaining wall for 3 hours on Monday and 5 hours on Tuesday, what fraction of the wall was completed on those two days?

$\frac{8}{x}$ of the wall

58. An oil pipeline can fill a tank in m hours. If the pipeline is open for 4 hours on Thursday and 7 hours on Friday, what fraction of the tank was filled by the pipeline on Thursday and Friday?

$\frac{11}{m}$ of the tank

59. A printing press can print the first edition of a book in h hours. During one day, the press ran for 6 hours during the first shift and 2 hours during the second shift. How much more of the entire job was completed during the first shift than during the second shift?

$\frac{4}{h}$ of the job

60. It takes the students in a dormitory a total of y hours to build an ice sculpture. The students in the east wing spent 11 hours on the sculpture, and the students in the west wing spent 8 hours on it. How much more of the entire job was completed by the students in the east wing?

$\frac{3}{y}$ of the job

61. A company has a contract to manufacture 10,000 aluminum cans. The company has a new machine that can make the cans in c minutes. An older machine requires $4c$ minutes to do the same job. What fraction of the job has been completed after both machines have been operating for 45 minutes?

$\frac{225}{4c}$ of the job

62. Two welders are welding the girders of a building. The first welder could complete the entire job in w hours, while the second would require $2w$ hours to complete the job. What fraction of the job has been completed after both welders have been working on it for 9 hours?

$\frac{27}{2w}$ of the job

63. One side of a triangle measures $\frac{9}{2x}$ inches. The other two sides measure

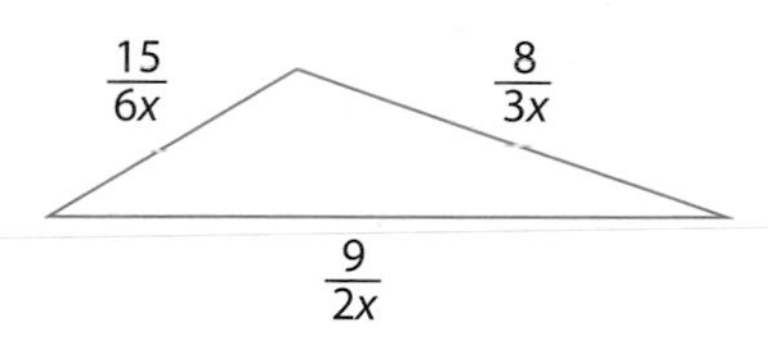

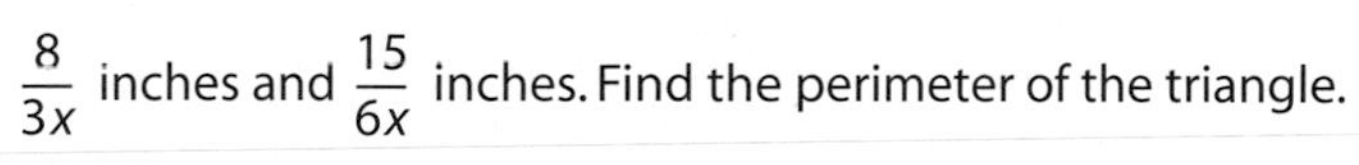

$\frac{8}{3x}$ inches and $\frac{15}{6x}$ inches. Find the perimeter of the triangle.

$\frac{29}{3x}$ inches

64. In a gubernatorial election, $\frac{1}{3x}$ of the voters cast their ballots for the Republican candidate, $\frac{5}{12x}$ voted for the Democratic candidate, and $\frac{1}{6x}$ voted for the Independent candidate. The remainder of the voters chose a write-in candidate. What fraction of the voters did not vote for a write-in candidate?

$\frac{11}{12x}$

65. A river ferry requires $\frac{10}{r-2}$ hours to travel upstream and $\frac{10}{r+2}$ hours to make the return trip down the river. Write an expression for the total time of the round trip.

$\frac{20r}{(r+2)(r-2)}$ h

66. Flying with the wind, a small plane flew the distance between two airports in $\frac{250}{w+10}$ hours. On the return trip, flying against the wind, the plane flew the distance in $\frac{250}{w-10}$ hours. Write an expression for the total time of the round trip.

$\frac{500w}{(w+10)(w-10)}$ h

Applying Concepts

67. When is the LCM of two expressions equal to their product?

When the two expressions have no common factors.

Simplify.

68. $b - 3 + \frac{5}{b+4}$

$\frac{b^2 + b - 7}{b + 4}$

69. $2y - 1 + \frac{6}{y+5}$

$\frac{2y^2 + 9y + 1}{y + 5}$

Rewrite the expression as the sum of two fractions in simplest form.

70. $\frac{5b + 4a}{ab}$

$\frac{5}{a} + \frac{4}{b}$

71. $\frac{6x + 7y}{xy}$

$\frac{6}{y} + \frac{7}{x}$

72. $\frac{3x^2 + 4xy}{x^2y^2}$

$\frac{3}{y^2} + \frac{4}{xy}$

73. $\frac{2mn^2 + 8m^2n}{m^3n^3}$

$\frac{2}{m^2n} + \frac{8}{mn^2}$

Exploration

74. ***Patterns in Mathematics***

a. Find the sum of each of the following:

$\frac{1}{1 \cdot 2} + \frac{1}{2 \cdot 3}$

$\frac{1}{1 \cdot 2} + \frac{1}{2 \cdot 3} + \frac{1}{3 \cdot 4}$

$\frac{1}{1 \cdot 2} + \frac{1}{2 \cdot 3} + \frac{1}{3 \cdot 4} + \frac{1}{4 \cdot 5}$

Note the pattern in these sums.
Then find the sum of 50 terms, of 100 terms, and of 1000 terms.

b. Find the sum of each of the following:

$\frac{1}{6} + \frac{1}{9} + \frac{1}{18}$

$\frac{1}{8} + \frac{1}{12} + \frac{1}{24}$

$\frac{1}{10} + \frac{1}{15} + \frac{1}{30}$

$\frac{1}{12} + \frac{1}{18} + \frac{1}{36}$

Continue the pattern for the next three sums.
Using a variable, write a pattern for these sums. Then find the sum of the expression you wrote.

Section 8.3 Complex Fractions

Simplifying Complex Fractions

The line graphed at the right goes through the points $(\frac{3}{a}, \frac{11}{a})$ and $(-\frac{3}{2a}, -\frac{5}{2a})$. We can find the slope of the line by using the formula for slope. Let $(\frac{3}{a}, \frac{11}{a})$ be (x_1, y_1) and $(-\frac{3}{2a}, -\frac{5}{2a})$ be (x_2, y_2). Then

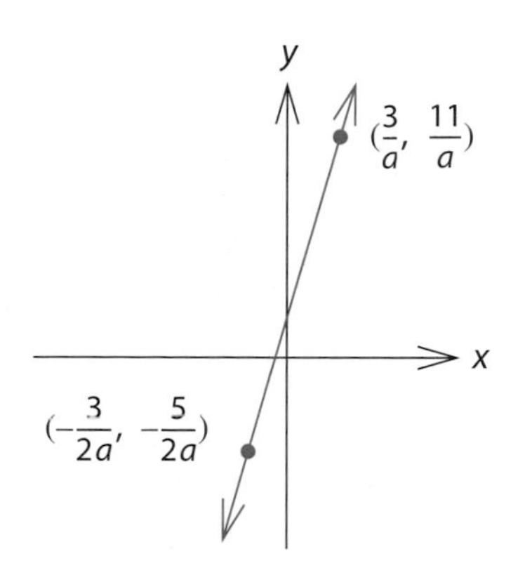

$$m = \frac{y_2 - y_1}{x_2 - x_1} = \frac{\frac{5}{2a} - \frac{11}{a}}{-\frac{3}{2a} - \frac{3}{a}}$$

The expression $\dfrac{\frac{5}{2a} - \frac{11}{a}}{-\frac{3}{2a} - \frac{3}{a}}$ is a complex fraction. A **complex fraction** is a fraction whose numerator or denominator contains one or more fractions.

To simplify a complex fraction, first find the LCM of the denominators of the fractions in the numerator and denominator. Multiply the numerator and denominator of the complex fraction by the LCM. Then simplify.

The LCM of the denominators of $\dfrac{-\frac{5}{2a} - \frac{11}{a}}{-\frac{3}{2a} - \frac{3}{a}}$ is $2a$. Multiply the numerator and denominator of the expression by $2a$.

TAKE NOTE
We are multiplying the expression by $\frac{2a}{2a}$, which equals 1. Multiplying an expression by 1 does not change the value of the expression.

$$\frac{-\frac{5}{2a} - \frac{11}{a}}{-\frac{3}{2a} - \frac{3}{a}} = \frac{-\frac{5}{2a} - \frac{11}{a}}{-\frac{3}{2a} - \frac{3}{a}} \cdot \frac{2a}{2a}$$

$$= \frac{\left(-\frac{5}{2a} - \frac{11}{a}\right)2a}{\left(-\frac{3}{2a} - \frac{3}{a}\right)2a}$$

$$= \frac{-\frac{5}{2a} \cdot 2a - \frac{11}{a} \cdot 2a}{-\frac{3}{2a} \cdot 2a - \frac{3}{a} \cdot 2a}$$

$$= \frac{-5 - 22}{-3 - 6}$$

$$= \frac{-27}{-9} = 3$$

The slope of the line through the points $(\frac{3}{a}, \frac{11}{a})$ and $(-\frac{3}{2a}, -\frac{5}{2a})$ is 3.

Here are two more examples of simplifying complex fractions.

Example 1 Simplify. **a.** $\dfrac{\dfrac{1}{x}+\dfrac{1}{2}}{\dfrac{1}{x^2}-\dfrac{1}{4}}$ **b.** $\dfrac{1-\dfrac{2}{x}-\dfrac{15}{x^2}}{1-\dfrac{11}{x}+\dfrac{30}{x^2}}$

Solution **a.** The LCM of $x, 2, x^2$, and 4 is $4x^2$.

$$\frac{\frac{1}{x}+\frac{1}{2}}{\frac{1}{x^2}-\frac{1}{4}} = \frac{\frac{1}{x}+\frac{1}{2}}{\frac{1}{x^2}-\frac{1}{4}} \cdot \frac{4x^2}{4x^2}$$

$$= \frac{\frac{1}{x}\cdot 4x^2 + \frac{1}{2}\cdot 4x^2}{\frac{1}{x^2}\cdot 4x^2 - \frac{1}{4}\cdot 4x^2}$$

$$= \frac{4x+2x^2}{4-x^2}$$

$$= \frac{2x\overset{1}{\cancel{(2+x)}}}{(2-x)\underset{1}{\cancel{(2+x)}}}$$

$$= \frac{2x}{2-x}$$

TAKE NOTE
We are using the Distributive Property to multiply $4x^2$ times each term in the numerator and each term in the denominator.

SUGGESTED ACTIVITY
By Newton's second law, the acceleration, a, of an object is equal to the force, F, in newtons divided by the mass, m, in kilograms. This is expressed in the formula $a = \frac{F}{m}$, where acceleration is measured in meters per second squared. The force on an object is $\frac{x^2-25}{x-2}$ newtons and the mass is $(6x-30)$ kilograms. Find the acceleration of the object.

[Answer: $\frac{x+5}{6(x-2)}$ m/s^2]

b. The LCM of x and x^2 is x^2.

$$\frac{1-\frac{2}{x}-\frac{15}{x^2}}{1-\frac{11}{x}+\frac{30}{x^2}} = \frac{1-\frac{2}{x}-\frac{15}{x^2}}{1-\frac{11}{x}+\frac{30}{x^2}} \cdot \frac{x^2}{x^2}$$

$$= \frac{1\cdot x^2 - \frac{2}{x}\cdot x^2 - \frac{15}{x^2}\cdot x^2}{1\cdot x^2 - \frac{11}{x}\cdot x^2 + \frac{30}{x^2}\cdot x^2}$$

$$= \frac{x^2-2x-15}{x^2-11x+30}$$

$$= \frac{\overset{1}{\cancel{(x-5)}}(x+3)}{\underset{1}{\cancel{(x-5)}}(x-6)}$$

$$= \frac{x+3}{x-6}$$

You-Try-It 1 Simplify. **a.** $\dfrac{\dfrac{1}{3}-\dfrac{1}{x}}{\dfrac{1}{9}-\dfrac{1}{x^2}}$ **b.** $\dfrac{1+\dfrac{4}{x}+\dfrac{3}{x^2}}{1+\dfrac{10}{x}+\dfrac{21}{x^2}}$

Solution See page S37.

8.3 EXERCISES

Topics for Discussion

1. What is a complex fraction? Provide an example of a complex fraction.
A fraction whose numerator or denominator contains one or more fractions. Examples will vary.

2. Explain how to simplify a complex fraction.
Answers will vary.

3. Determine whether the statement is always true, sometimes true, or never true.

a. A complex fraction is a fraction that has fractions in the numerator and denominator. Sometimes true

b. To simplify a complex fraction, multiply the complex fraction by the LCM of the denominators of the fractions in the numerator and denominator of the complex fraction. Never true

c. Our goal in simplifying a complex fraction is to rewrite it so that there are no fractions in the numerator or in the denominator. We then write the fraction in simplest form. Always true

Simplifying Complex Fractions

Simplify.

4. $\dfrac{1+\dfrac{3}{x}}{1-\dfrac{9}{x^2}}$ $\quad \dfrac{x}{x-3}$

5. $\dfrac{1+\dfrac{4}{x}}{1-\dfrac{16}{x^2}}$ $\quad \dfrac{x}{x-4}$

6. $\dfrac{2-\dfrac{8}{x+4}}{3-\dfrac{12}{x+4}}$ $\quad \dfrac{2}{3}$

7. $\dfrac{5-\dfrac{25}{x+5}}{1-\dfrac{3}{x+5}}$ $\quad \dfrac{5x}{x+2}$

8. $\dfrac{1+\dfrac{5}{y-2}}{1-\dfrac{2}{y-2}}$ $\quad \dfrac{y+3}{y-4}$

9. $\dfrac{2-\dfrac{11}{2x-1}}{3-\dfrac{17}{2x-1}}$ $\quad \dfrac{4x-13}{2(3x-10)}$

10. $\dfrac{4-\dfrac{2}{x+7}}{5+\dfrac{1}{x+7}}$ $\quad \dfrac{2(2x+13)}{5x+36}$

11. $\dfrac{5+\dfrac{3}{x-8}}{2-\dfrac{1}{x-8}}$ $\quad \dfrac{5x-37}{2x-17}$

12. $\dfrac{\dfrac{3}{x-2}+3}{\dfrac{4}{x-2}+4}$ $\quad \dfrac{3}{4}$

13. $\dfrac{\dfrac{3}{2x+1}-3}{2-\dfrac{4x}{2x+1}}$ $\quad -3x$

14. $\dfrac{2-\dfrac{3}{x}-\dfrac{2}{x^2}}{2+\dfrac{5}{x}+\dfrac{2}{x^2}}$ $\quad \dfrac{x-2}{x+2}$

15. $\dfrac{2+\dfrac{5}{x}-\dfrac{12}{x^2}}{4-\dfrac{4}{x}-\dfrac{3}{x^2}}$ $\quad \dfrac{x+4}{2x+1}$

16. $\dfrac{1-\dfrac{1}{x}-\dfrac{6}{x^2}}{1-\dfrac{9}{x^2}}$ $\quad \dfrac{x+2}{x+3}$

17. $\dfrac{1+\dfrac{4}{x}+\dfrac{4}{x^2}}{1-\dfrac{2}{x}-\dfrac{8}{x^2}}$ $\quad \dfrac{x+2}{x-4}$

18. $\dfrac{1-\dfrac{5}{x}-\dfrac{6}{x^2}}{1+\dfrac{6}{x}+\dfrac{5}{x^2}}$ $\quad \dfrac{x-6}{x+5}$

19. $\dfrac{1-\dfrac{7}{a}+\dfrac{12}{a^2}}{1+\dfrac{1}{a}-\dfrac{20}{a^2}}$ $\quad \dfrac{a-3}{a+5}$

20. $\dfrac{1 - \frac{6}{x} + \frac{8}{x^2}}{1 - \frac{3}{x} - \frac{4}{x^2}}$

$\dfrac{x-2}{x+1}$

21. $\dfrac{1 + \frac{3}{x} - \frac{18}{x^2}}{1 + \frac{4}{x} - \frac{21}{x^2}}$

$\dfrac{x+6}{x+7}$

22. $\dfrac{x - \frac{4}{x+3}}{1 + \frac{1}{x+3}}$

$x - 1$

23. $\dfrac{y + \frac{1}{y-2}}{1 + \frac{1}{y-2}}$

$y - 1$

24. $\dfrac{1 - \frac{x}{2x+1}}{x - \frac{1}{2x+1}}$

$\dfrac{1}{2x-1}$

25. $\dfrac{1 - \frac{2x-2}{3x-1}}{x - \frac{4}{3x-1}}$

$\dfrac{1}{3x-4}$

26. $\dfrac{x - 5 + \frac{14}{x+4}}{x + 3 - \frac{2}{x+4}}$

$\dfrac{x-3}{x+5}$

27. $\dfrac{a + 4 + \frac{5}{a-2}}{a + 6 + \frac{15}{a-2}}$

$\dfrac{a-1}{a+1}$

28. $\dfrac{x + 3 - \frac{10}{x-6}}{x + 2 - \frac{20}{x-6}}$

$\dfrac{x-7}{x-8}$

29. $\dfrac{x - 7 + \frac{5}{x-1}}{x - 3 + \frac{1}{x-1}}$

$\dfrac{x-6}{x-2}$

30. $\dfrac{y - 6 + \frac{22}{2y+3}}{y - 5 + \frac{11}{2y+3}}$

$\dfrac{2y-1}{2y+1}$

31. $\dfrac{x + 2 - \frac{12}{2x-1}}{x + 1 - \frac{9}{2x-1}}$

$\dfrac{2x+7}{2x+5}$

32. Find the slope of the line that passes through the points $(\frac{5}{a}, -\frac{2}{a})$ and $(-\frac{1}{3a}, -\frac{4}{3a})$.

$-\frac{1}{8}$

33. Find the slope of the line that passes through the points $(-\frac{2}{3a}, \frac{4}{3a})$ and $(\frac{7}{2a}, -\frac{9}{2a})$.

$-\frac{7}{5}$

Applying Concepts

Simplify.

34. $1 - \dfrac{1}{1 - \frac{1}{2}}$

-1

35. $1 - \dfrac{1}{1 - \frac{1}{y+1}}$

$-\frac{1}{y}$

36. $\dfrac{x^{-2} - y^{-2}}{x^{-2}y^{-2}}$

$y^2 - x^2$

37. $\left(\frac{y}{4} - \frac{4}{y}\right) \div \left(\frac{4}{y} - 3 + \frac{y}{2}\right)$

$\dfrac{y+4}{2(y-2)}$

38. Explain why we multiply the numerator and denominator of a complex fraction by the LCM of the denominators of the fractions in the numerator and denominator.
Answers will vary.

Exploration

39. ***An Alternative Method of Simplifying a Complex Fraction***
A different approach to simplifying a complex fraction is to rewrite the numerator and the denominator of the complex fraction as single fractions. Then divide the numerator by the denominator. For example,

$$\frac{x + \frac{1}{x}}{2 - \frac{3}{x}} = \frac{\frac{x^2}{x} + \frac{1}{x}}{\frac{2x}{x} - \frac{3}{x}} = \frac{\frac{x^2+1}{x}}{\frac{2x-3}{x}} = \frac{x^2+1}{x} \div \frac{2x-3}{x} = \frac{x^2+1}{x} \cdot \frac{x}{2x-3} = \frac{(x^2+1)x}{x(2x-3)} = \frac{x^2+1}{2x-3}$$

Use this method to simplify Exercises 4 to 15. Which method do you prefer? Why?

Section 8.4 Applications of Equations Containing Fractions

Equations Containing Fractions

In this section, we will be solving two types of application problems: work problems and uniform motion problems. Each of these types of problems involves solving equations containing fractions, so we will first look at solving these types of equations.

In Chapter 3, equations containing fractions were solved by the method of clearing denominators. To **clear denominators**, we multiply each side of the equation by the LCM of the denominators. The result is an equation that contains no fractions.

In this section, we will again solve equations containing fractions by clearing denominators. The difference between this section and Section 3.3 is that the fractions in these equations contain variables in the denominator. Here is an example.

→ Solve: $\frac{3x-1}{4x} + \frac{2}{3x} = \frac{7}{6x}$

$$\frac{3x-1}{4x} + \frac{2}{3x} = \frac{7}{6x}$$

• The LCM of the denominators $4x$, $3x$, and $6x$ is $12x$.

$$12x\left(\frac{3x-1}{4x} + \frac{2}{3x}\right) = 12x\left(\frac{7}{6x}\right)$$

• Multiply each side of the equation by $12x$.

$$12x\left(\frac{3x-1}{4x}\right) + 12x\left(\frac{2}{3x}\right) = 12x\left(\frac{7}{6x}\right)$$

• Use the Distributive Property on the left side of the equation.

$$3(3x-1) + 4(2) = 2(7)$$
$$9x - 3 + 8 = 14$$
$$9x + 5 = 14$$
$$9x = 9$$
$$x = 1$$

Algebraic check:

$$\frac{3x-1}{4x} + \frac{2}{3x} = \frac{7}{6x}$$

$\frac{3(1)-1}{4(1)} + \frac{2}{3(1)}$	$\frac{7}{6(1)}$
$\frac{2}{4} + \frac{2}{3}$	$\frac{7}{6}$
$\frac{1}{2} + \frac{2}{3}$	$\frac{7}{6}$
$\frac{3}{6} + \frac{4}{6}$	$\frac{7}{6}$

$$\frac{7}{6} = \frac{7}{6}$$

The solution $x = 1$ checks.

Graphical check:

The solution to this equation can also be checked graphically by using a graphing calculator and the same method used earlier in the text. Let $Y1 = \frac{3x-1}{4x} + \frac{2}{3x}$ and $Y2 = \frac{7}{6x}$. Note how the expressions are entered into the calculator:

Y1 = (3X − 1)/(4X) + 2/(3X)

Y2 = 7/(6X)

The denominator of each fraction is within parentheses. For example, $\frac{7}{6x}$ is entered with parentheses around the 6X. If 7/6X is entered instead, the calculator interprets the expression as $\frac{7}{6}x$, which is $\frac{7x}{6}$, with x in the numerator, not the denominator.

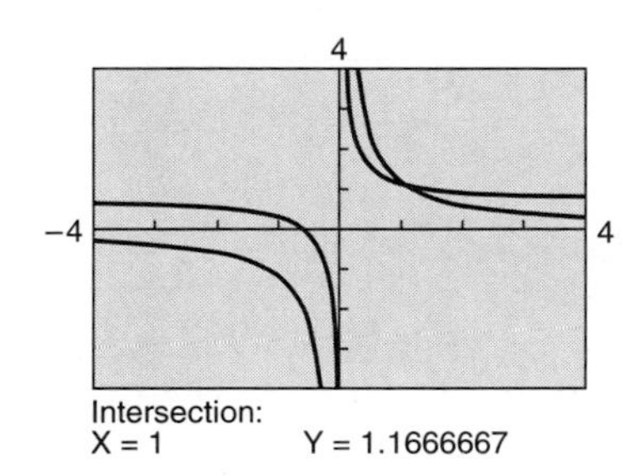

Use the INTERSECT feature of the calculator to determine the point of intersection.

The x-coordinate of the point of intersection is the solution of the equation.

The solution $x = 1$ checks.

Example 1 Solve $\frac{4}{x} - \frac{x}{2} = \frac{7}{2}$ and check the solutions graphically and algebraically.

Solution

$$\frac{4}{x} - \frac{x}{2} = \frac{7}{2}$$ • The LCM of x and 2 is $2x$.

$$2x\left(\frac{4}{x} - \frac{x}{2}\right) = 2x\left(\frac{7}{2}\right)$$ • Multiply each side by $2x$.

$$2x\left(\frac{4}{x}\right) - 2x\left(\frac{x}{2}\right) = 2x\left(\frac{7}{2}\right)$$

$$8 - x^2 = 7x$$ • This is a quadratic equation.

$$0 = x^2 + 7x - 8$$

$$0 = (x+8)(x-1)$$

$$x + 8 = 0 \qquad x - 1 = 0$$

$$x = -8 \qquad x = 1$$

Graphical check:

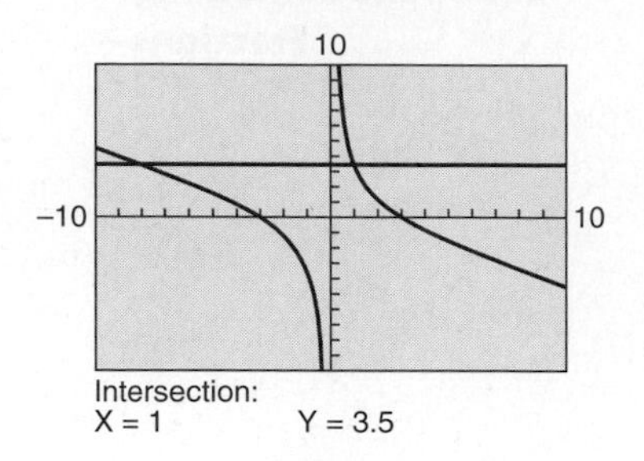

Use the INTERSECT feature a second time to check the solution $x = -8$. The solutions check.

Algebraic check:

$$\frac{4}{x} - \frac{x}{2} = \frac{7}{2}$$

$\frac{4}{-8} - \frac{-8}{2}$	$\frac{7}{2}$
$-\frac{1}{2} + \frac{8}{2}$	$\frac{7}{2}$

$$\frac{7}{2} = \frac{7}{2}$$

$$\frac{4}{x} - \frac{x}{2} = \frac{7}{2}$$

$\frac{4}{1} - \frac{1}{2}$	$\frac{7}{2}$
$\frac{8}{2} - \frac{1}{2}$	$\frac{7}{2}$

$$\frac{7}{2} = \frac{7}{2}$$

The solutions check.
The solutions are −8 and 1.

You-Try-It 1 Solve $x + \frac{1}{3} = \frac{4}{3x}$ and check the solutions graphically and algebraically.

Solution See page S37.

Occasionally, a value of a variable in a fractional equation makes one of the denominators zero. In that case, that value of the variable is not a solution of the equation. This is illustrated in Example 2.

Example 2 Solve $\frac{2x}{x-2} = 1 + \frac{4}{x-2}$ and check the solution algebraically.

Solution

$$\frac{2x}{x-2} = 1 + \frac{4}{x-2}$$

$$(x-2)\left(\frac{2x}{x-2}\right) = (x-2)\left(1 + \frac{4}{x-2}\right)$$

$$(x-2)\left(\frac{2x}{x-2}\right) = (x-2)(1) + (x-2)\left(\frac{4}{x-2}\right)$$

$$2x = x - 2 + 4$$

$$x = 2$$

Algebraic check:

$$\frac{2x}{x-2} = 1 + \frac{4}{x-2}$$

$\frac{2(2)}{2-2}$	$1 + \frac{4}{2-2}$
$\frac{4}{0}$	$1 + \frac{4}{0}$

2 does not check as a solution.
The equation has no solution.

27. A small air conditioner will cool a room 2° in 15 minutes. A larger air conditioner will cool the room 2° in 10 minutes. How long would it take to cool the room 2° with both air conditioners operating?
6 min

28. One printing press can print the first edition of a book in 55 minutes. A second printing press requires 66 minutes to print the same number of copies. How long would it take to print the first edition of the book with both presses operating?
30 min

29. Two welders working together can complete a job in 6 hours. One of the welders, working alone, can complete the task in 10 hours. How long would it take the second welder, working alone, to complete the task?
15 h

30. Two pipelines can fill a small tank in 30 minutes. Working alone, the larger pipeline can fill the tank in 45 minutes. How long would it take the smaller pipeline, working alone, to fill the tank?
90 min

31. Working together, two dock workers can load a crate in 6 minutes. One of the dock workers, working alone, can load the crate in 15 minutes. How long would it take the other dock worker, working alone, to load the crate?
10 min

32. With two harvesters operating, a plot of land can be harvested in 1 hour. If only the newer harvester is used, the land can be harvested in 1.5 hours. How long would it take to harvest the field using only the older harvester?
3 h

33. A cement mason can build a barbecue in 8 hours. A second mason requires 12 hours to do the same task. After working alone for 4 hours, the first mason quits. How long will it take the second mason to complete the job?
6 h

34. A mechanic requires 2 hours to repair a transmission. An apprentice requires 6 hours to make the same repairs. If the mechanic works alone on a transmission for 1 hour and then stops, how long will it take the apprentice to complete the repairs?
3 h

35. One computer technician can wire a modem in 4 hours. A second technician requires 6 hours to do the same job. After working alone for 2 hours, the first technician quits. How long will it take the second technician to complete the wiring?
3 h

36. A wallpaper hanger requires 2 hours to hang the wallpaper on one wall of a room. A second wallpaper hanger requires 4 hours to hang the same amount of wallpaper. The first wallpaper hanger works alone for 1 hour and then quits. How long will it take the second hanger, working alone, to finish papering the wall?
2 h

37. Two machines fill cereal boxes at the same rate. After the two machines work together for 7 hours, one machine breaks down. The second machine requires 14 more hours to finish filling the boxes. How long would it have taken one of the machines, working alone, to fill the boxes?
28 h

38. Two welders who work at the same rate are riveting the girders of a building. After they work together for 10 hours, one of the welders quits. The second welder requires 20 more hours to complete the welds. Find the time it would have taken one of the welders, working alone, to complete the welds.
40 h

39. A camper drove 90 miles to a recreational area and then hiked 5 miles into the woods. The rate of the camper while driving was nine times the rate while hiking. The time spent hiking and driving was 3 hours. Find the rate at which the camper hiked.
5 mph

40. The president of a company traveled 1800 miles by jet and 300 miles on a prop plane. The rate of the jet was four times the rate of the prop plane. The entire trip took 5 hours. Find the rate of the jet plane.
600 mph

41. To assess the damage done by a fire, a forest ranger traveled 1080 miles by jet and then an additional 180 miles by helicopter. The rate of the jet was four times the rate of the helicopter. The entire trip took 5 hours. Find the rate of the jet.
360 mph

42. An engineer traveled 165 miles by car and then an additional 660 miles by plane. The rate of the plane was four times the rate of the car. The total trip took 6 hours. Find the rate of the car.
55 mph

43. After sailing 15 miles, a sailor changed direction and increased the boat's speed by 2 mph. An additional 19 miles was sailed at the increased speed. The total sailing time was 4 hours. Find the rate of the boat for the first 15 miles.
7.5 mph

44. On a recent trip, a trucker traveled 330 miles at a constant rate. Because of road conditions, the trucker then reduced the speed by 25 mph. An additional 30 miles was traveled at the reduced rate. The entire trip took 7 hours. Find the rate of the trucker for the first 330 miles.
55 mph

45. Commuting from work to home, a lab technician traveled 10 miles at a constant rate through congested traffic. Upon reaching the expressway, the technician increased the speed by 20 mph. An additional 20 miles was traveled at the increased speed. The total time for the trip was 1 hour. How fast did the technician travel through the congested traffic?
20 mph

10 mi 20 mi
r $r + 20$

46. As part of a conditioning program, a jogger ran 8 miles in the same amount of time as a cyclist rode 20 miles. The rate of the cyclist was 12 mph faster than the rate of the jogger. Find the rate of the jogger and the rate of the cyclist.
Jogger: 8 mph; cyclist: 20 mph

47. An express train traveled 600 miles in the same amount of time as it took a freight train to travel 360 miles. The rate of the express train was 20 mph greater than the rate of the freight train. Find the rate of each train.
Freight train: 30 mph; express train: 50 mph

48. A twin-engine plane flies 800 miles in the same amount of time as it takes a single-engine plane to fly 600 miles. The rate of the twin-engine plane is 50 mph greater than the rate of the single-engine plane. Find the rate of the twin-engine plane.
200 mph

49. A car is traveling at a rate that is 36 mph faster than the rate of a cyclist. The car travels 384 miles in the same time it takes the cyclist to travel 96 miles. Find the rate of the car.
48 mph

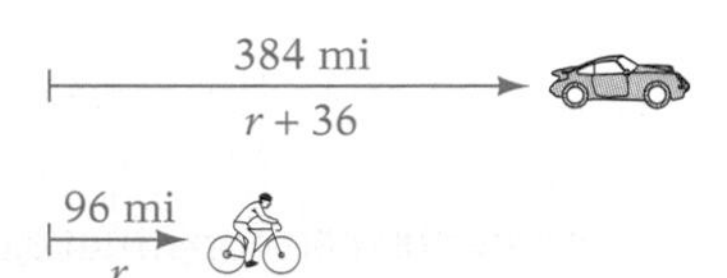

50. A small motor on a fishing boat can move the boat at a rate of 6 mph in calm water. Traveling with the current, the boat can travel 24 miles in the same amount of time as it takes to travel 12 miles against the current. Find the rate of the current.
2 mph

51. A commercial jet can fly 550 mph in calm air. Traveling with the jet stream, the plane flew 2400 miles in the same amount of time as it took to fly 2000 miles against the jet stream. Find the rate of the jet stream.
50 mph

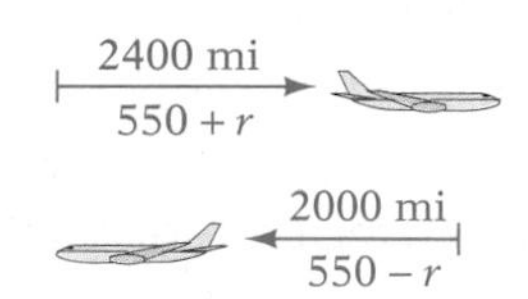

52. A cruise ship can sail 28 mph in calm water. Sailing with the gulf stream, the ship can sail 170 miles in the same amount of time as it takes to sail 110 miles against the gulf stream. Find the rate of the gulf stream.
6 mph

53. Rowing with the current of a river, a rowing team can row 25 miles in the same amount of time as it takes to row 15 miles against the current. The rate of the rowing team in calm water is 20 mph. Find the rate of the current.
5 mph

54. A plane can fly 180 mph in calm air. Flying with the wind, the plane can fly 600 miles in the same amount of time as it takes to fly 480 miles against the wind. Find the rate of the wind.
20 mph

Applying Concepts

55. The sum of a number and its reciprocal is $\frac{25}{12}$. Find the number.

$\frac{3}{4}$ or $\frac{4}{3}$

56. The sum of the multiplicative inverses of two consecutive integers is $\frac{11}{30}$. Find the integers.

5 and 6

57. The numerator of a fraction is 3 less than the denominator. If both the numerator and denominator of the fraction are increased by 5, the new fraction is $\frac{4}{5}$. Find the original fraction.

$\frac{7}{10}$

58. One press can complete the weekly edition of a newspaper in 12 hours, a second press can complete the job in 8 hours, and a third press can complete the job in 6 hours. How long would it take to print the newspaper with all three presses operating?

$2\frac{2}{3}$ h

59. A carpenter can construct a curio cabinet in 4 weeks. It would take the carpenter's more experienced apprentice 6 weeks. How long would it take the carpenter's less experienced apprentice to build the curio cabinet if, working together, all three can do the job in 2 weeks?

12 weeks

60. By increasing your speed by 5 mph, you can drive the 165-mile trip to your hometown in 15 minutes less time than it usually takes you to drive the trip. How fast do you usually drive?

55 mph

61. Explain the procedure for solving an equation containing fractions. Include in your discussion an explanation of how the LCM of the denominators is used to eliminate fractions in the equation.

Explanations will vary.

Explorations

62. ***Writing Rational Equations***

a. Write an original rational equation that has 0 as its only solution.

b. Write an original rational equation that has no solution.

c. Write an original rational equation that has one solution that does not check and a second solution that does check.

63. ***Solutions of Equations***

Are the solutions of the equations shown below the same? Explain your answer. Use both algebraic methods and graphical methods to justify your answer.

$$2x = 9x$$

$$\frac{1}{2x} = \frac{1}{9x}$$

Section 8.5 Literal Equations

Solve a Literal Equation for One of the Variables

A **literal equation** is an equation that contains more than one variable. Examples of literal equations are shown below.

$$3x + 5y = 7$$

$$4a - 6b + c = 0$$

Formulas are used to express a relationship among physical quantities. A **formula** is a literal equation that states rules about measurements. Examples of formulas are shown below.

$$\frac{1}{R_1} + \frac{1}{R_2} = \frac{1}{R} \qquad \text{(Physics)}$$

$$s = a + (n - 1)d \qquad \text{(Mathematics)}$$

$$A = P + Prt \qquad \text{(Business)}$$

POINT OF INTEREST
Johannes Kepler, a German astronomer, discovered in the early 17th century that the orbits of the planets in our solar system were not circular but elliptical. He also discovered the relationship between the distance of a planet from the sun and the rate at which the planet orbits the sun: $\frac{d^3}{t^2}$, in which d is the mean distance of the planet from the sun and t is the orbital period, a constant. This knowledge is vital to understanding the orbits of planets, moons, and satellites that are launched from Earth.

In Chapter 4 you learned that equations of the form $y = mx + b$ and $Ax + By = C$ are linear equations. You also learned that an equation of the form $Ax + By = C$ can be written in the form $y = mx + b$ by solving the equation for y. One reason this is an important skill is that an equation must be in the form $y = mx + b$ in order to be entered into a graphing calculator.

An equation of the form $Ax + By = C$, such as $2x + 5y = 10$, is a literal equation because it has two variables, x and y. Solving an equation of this form for y is solving a literal equation for one of the variables. This is the focus of this lesson.

The Addition and Multiplication Properties of Equations can be used to solve a literal equation for one of the variables. The goal is to rewrite the equation so that the letter being solved for is alone on one side of the equation and all numbers and other letters are on the other side.

→ Solve $A = P(1 + i)$ for i.

The goal is to rewrite the equation so that i is on one side of the equation and all other variables are on the other side.

$A = P(1 + i)$

$A = P + Pi$ • Use the Distributive Property to remove parentheses.

$A - P = P - P + Pi$ • Subtract P from each side of the equation. Note that we are subtracting the term that does not contain the variable being solved for.

$A - P = Pi$

$\frac{A - P}{P} = \frac{Pi}{P}$ • Divide each side of the equation by P.

$\frac{A - P}{P} = i$ • i is alone on one side. All other variables are on the other side.

INSTRUCTOR NOTE
Some students will want to simplify the expression $\frac{A - P}{P}$. These students need to be reminded that we can divide by common **factors** only.

Question: What variable has the equation $V = \frac{1}{3}\pi r^2 h$ been solved for if it is rewritten in the form $h = \frac{3V}{\pi r^2}$?[1]

SUGGESTED ACTIVITY
The formula for the volume, V, of a right circular cone is $V = \frac{1}{3}\pi r^2 h$, where r is the radius of the base and h is the height. What is the radius of the base of a cone that has a volume of π^2 square inches and a height of 4 inches?
[Answer: $\frac{\sqrt{3\pi}}{2}$ inches]

Example 1 **a.** Solve $I = \frac{E}{R+r}$ for R.

b. Solve $L = a(1 + ct)$ for c.

c. Solve $S = C - rC$ for C.

Solution **a.**

$$I = \frac{E}{R+r}$$
• We are solving for R.

$$(R+r)I = (R+r)\frac{E}{R+r}$$
• Multiply each side by the LCM of the denominators.

$$RI + rI = E$$
$$RI + rI - rI = E - rI$$
$$RI = E - rI$$
• Subtract rI (the term that does not contain R) from each side of the equation.

$$\frac{RI}{I} = \frac{E - rI}{I}$$
• Divide each side by I.

$$R = \frac{E - rI}{I}$$

b.

$$L = a(1 + ct)$$
• We are solving for c.

$$L = a + act$$
• Use the Distributive Property.

$$L - a = a - a + act$$
$$L - a = act$$
• Subtract a (the term that does not contain c) from each side.

$$\frac{L-a}{at} = \frac{act}{at}$$
• Divide each side by at.

$$\frac{L-a}{at} = c$$

c.

$$S = C - rC$$
• We are solving for C.

$$S = C(1 - r)$$
• Factor C from $C - rC$.

$$\frac{S}{1-r} = \frac{C(1-r)}{1-r}$$
• Divide each side by $1 - r$.

$$\frac{S}{1-r} = C$$

INSTRUCTOR NOTE
Example 1c will be difficult for students. Remind them that when solving $2x + 3x = 10$, they are using the Distributive Property to combine $2x$ and $3x$.

$2x + 3x = (2 + 3)x = 5x$

Then each side of the equation is divided by 5.

For $ax + bx = 5$, the procedure is exactly the same except that $a + b$ does not simplify further.

Here is an equation that can be used as a class exercise: Solve $cx - y = bx + 5$ for x.

You-Try-It 1 **a.** Solve $s = \frac{A+L}{2}$ for L.

b. Solve $S = a + (n - 1)d$ for n.

c. Solve $S = C + rC$ for C.

Solution See page S39.

---

1. In the equation $h = \frac{3V}{\pi r^2}$, h is alone on one side of the equation and all numbers and other variables are on the other side. Therefore, the equation was solved for h.

8.5 EXERCISES

Topics for Discussion

1. Which of the following are literal equations? Why?

a. $\frac{1}{a}+\frac{1}{b}=\frac{1}{c}$ **b.** $\frac{1}{5}+\frac{1}{2}=\frac{1}{x}$ **c.** $\frac{RR_2}{R_2-R}$ **d.** $S=vt-16t^2$

a and d are literal equations because they are equations that contain more than one variable. b is not a literal equation because it contains only one variable. c is not an equation.

2. Write two literal equations. Explain why they are literal equations.
Answers will vary.

3. Determine whether the statement is always true, sometimes true, or never true.

a. A literal equation contains three variables. Sometimes true

b. A literal equation is solved for one of its variables by using the same properties used to solve equations in one variable. Always true

c. In solving a literal equation, the goal is to get the variable being solved for alone on one side of the equation and all numbers and other variables on the other side of the equation. Always true

Literal Equations

4. The equation $P = C + M$ is used to find the price of a product, P, given the cost, C, and the markup, M. Explain how to solve the equation $P = C + M$ for C. Then solve the equation for C.
Subtract M from each side of the equation; $C = P - M$.

5. The equation $A = P - D$ expresses the amount financed, A, on an installment purchase in terms of the price, P, of the product and the down payment, D. Explain how to solve the equation $A = P - D$ for P. Then solve the equation for P.
Add D to each side of the equation; $P = A + D$.

6. The equation $A = MN$ is used to find the total amount paid in paying off a loan, A, given the monthly payment, M, and the number of payments, N. Explain how to solve the equation $A = MN$ for N. Then solve the equation for N.
Divide each side of the equation by M; $N = \frac{A}{M}$.

7. The equation $S = \frac{A}{N}$ expresses an employee's salary per pay period, S, in terms of the employee's annual salary, A, and the number of pay periods per year, N. Explain how to solve the equation $S = \frac{A}{N}$ for A. Then solve the equation for A.
Multiply each side of the equation by N; $A = SN$.

Solve the formula for the given variable.

8. $d = rt$; t (Physics)

$t = \frac{d}{r}$

9. $E = IR$; R (Physics)

$R = \frac{E}{I}$

10. $A = bh; h$ (Geometry)

$h = \frac{A}{b}$

11. $PV = nRT; T$ (Chemistry)

$T = \frac{PV}{nR}$

12. $P = a + b + c; b$ (Geometry)

$b = P - a - c$

13. $P = R - C; C$ (Business)

$C = R - P$

14. $A = \frac{1}{2}bh; h$ (Geometry)

$h = \frac{2A}{b}$

15. $V = \frac{1}{3}Ah; h$ (Geometry)

$h = \frac{3V}{A}$

16. $P = 2L + 2W; L$ (Geometry)

$L = \frac{P - 2W}{2}$

17. $P = 2L + 2W; W$ (Geometry)

$W = \frac{P - 2L}{2}$

18. $F = \frac{9}{5}C + 32; C$ (Temperature Conversion)

$C = \frac{5F - 160}{9}$

19. $C = \frac{5}{9}(F - 32); F$ (Temperature Conversion)

$F = \frac{9}{5}C + 32$

20. $R = \frac{C - S}{t}; S$ (Business)

$S = C - Rt$

21. $P = \frac{R - C}{n}; R$ (Business)

$R = Pn + C$

22. $A = P + Prt; P$ (Business)

$P = \frac{A}{1 + rt}$

23. $T = fm - gm; m$ (Engineering)

$m = \frac{T}{f - g}$

24. $A = Sw + w; w$ (Physics)

$w = \frac{A}{S + 1}$

25. $a = S - Sr; S$ (Mathematics)

$S = \frac{a}{1 - r}$

26. The surface area of a right circular cylinder is given by the formula $S = 2\pi rh + 2\pi r^2$, where r is the radius of the base and h is the height of the cylinder.

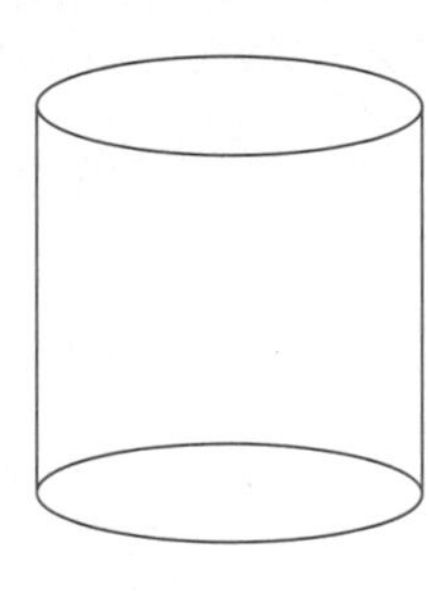

a. Solve the formula $S = 2\pi rh + 2\pi r^2$ for h.

b. Use your answer to part a to find the height of a right circular cylinder when the surface area is 12π square inches and the radius is 1 inch.

c. Use your answer to part a to find the height of a right circular cylinder when the surface area is 24π square inches and the radius is 2 inches.

a. $h = \frac{S - 2\pi r^2}{2\pi r}$ **b.** 5 in. **c.** 4 in.

27. Break-even analysis is a method used to determine the sales volume required for a company to break even, or experience neither a profit nor a loss, on the sale of its product. The break-even point represents the number of units that must be made and sold for income from sales to equal the cost of producing the product. The break-even point can be calculated using the formula $B = \frac{F}{S - V}$, where F is the fixed costs, S is the selling price per unit, and V is the variable costs per unit.

a. Solve the formula $B = \frac{F}{S - V}$ for S.

b. Use your answer to part a to find the selling price per snowboard required for a company to break even. The fixed costs are \$20,000, the variable costs per snowboard are \$80, and the company plans to make and sell 200 snowboards.

c. Use your answer to part a to find the selling price per portable CD player required for a company to break even. The fixed costs are \$15,000, the variable costs per CD player are \$50, and the company plans to make and sell 600 portable CD players.

a. $S = \frac{F + BV}{B}$ **b.** \$180 **c.** \$75

28. When markup is based on selling price, the selling price of a product is given by the formula $S = \frac{C}{1-r}$, where C is the cost of the product and r is the markup rate.

a. Solve the formula $S = \frac{C}{1-r}$ for r.

b. Use your answer to part a to find the markup rate on a tennis racket when the cost is \$112 and the selling price is \$140.

c. Use your answer to part a to find the markup rate on a video game when the cost is \$50.40 and the selling price is \$72.

a. $r = \frac{S-C}{S}$ **b.** 20% **c.** 30%

29. Resistors are used to control the flow of current. The total resistance of two resistors in a circuit can be given by the formula $R = \frac{1}{\frac{1}{R_1} + \frac{1}{R_2}}$, where R_1 and R_2 are the resistances of the two resistors in the circuit. Resistance is measured in ohms.

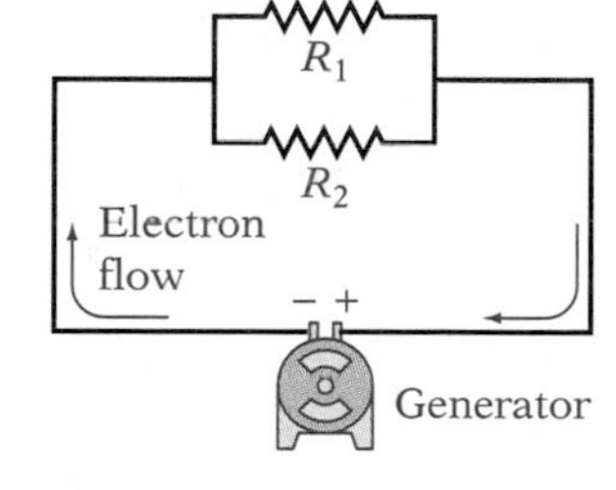

a. Solve the formula $R = \frac{1}{\frac{1}{R_1} + \frac{1}{R_2}}$ for R_1.

b. Use your answer to part a to find the resistance in R_1 if the resistance in R_2 is 30 ohms and the total resistance is 12 ohms.

c. Use your answer to part a to find the resistance in R_1 if the resistance in R_2 is 15 ohms and the total resistance is 6 ohms.

a. $R_1 = \frac{RR_2}{R_2 - R}$ **b.** 20 ohms **c.** 10 ohms

Applying Concepts

The equation $\frac{x^2}{16} + \frac{y^2}{9} = 1$ cannot be entered into a graphing calculator to be graphed. However, we can solve this equation for y, as shown below.

$$\frac{y^2}{9} = 1 - \frac{x^2}{16}$$ • Subtract $\frac{x^2}{16}$ from each side of the equation.

$$y^2 = 9\left(1 - \frac{x^2}{16}\right)$$ • Multiply each side of the equation by 9.

$$y = \pm 3\sqrt{1 - \frac{x^2}{16}}$$ • Take the square root of each side of the equation.

There are two solutions for y: $y = 3\sqrt{1 - \frac{x^2}{16}}$ and $y = -3\sqrt{1 - \frac{x^2}{16}}$. Enter both of these equations into a graphing calculator and graph both equations. The result is the ellipse shown at the right.

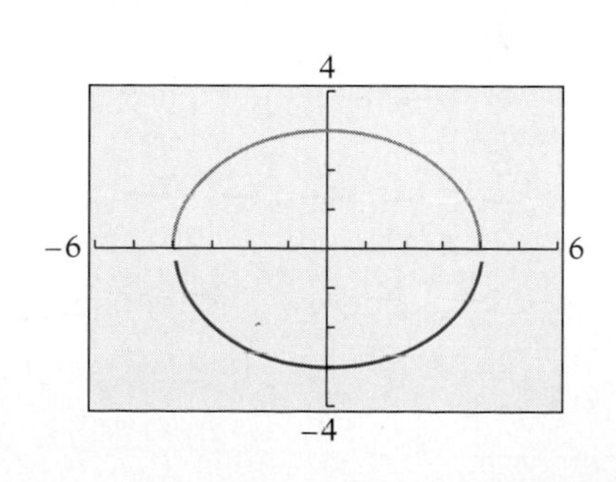

Solve each equation for y. Then graph using a graphing calculator.

30. $\frac{x^2}{25} + \frac{y^2}{49} = 1$

An ellipse with center at the origin, through (5, 0), (–5, 0), (0, 7), and (0, –7)

31. $\frac{x^2}{4} + \frac{y^2}{64} = 1$

An ellipse with center at the origin, through (2, 0), (–2, 0), (0, 8), and (0, –8)

32. $\frac{x^2}{36} + \frac{y^2}{36} = 1$

A circle with center at the origin and a radius of 6 units

33. $\frac{x^2}{81} + \frac{y^2}{81} = 1$

A circle with center at the origin and a radius of 9 units

Exploration

34. ***Dimensional Analysis***

In solving application problems, scientists, engineers, and other professionals find it useful to include the units as they work through the solutions to problems so that the answers are in the proper units. Using units to organize and check the correctness of an application is called **dimensional analysis.** Applying dimensional analysis to application problems requires converting units as well as multiplying and dividing units.

Equivalent measures are used to form conversion factors to change one unit of measurement to another. For example, the equivalent measures **1 mi** and **5280 ft** are used to form the following conversion factors:

$$\frac{1 \text{ mi}}{5280 \text{ ft}} \qquad \frac{5280 \text{ ft}}{1 \text{ mi}}$$

Because **1 mi = 5280 ft**, both of the conversion factors $\frac{1 \text{ mi}}{5280 \text{ ft}}$ and $\frac{5280 \text{ ft}}{1 \text{ mi}}$ are equal to 1.

To convert 4 miles to feet, multiply 4 miles by the conversion factor $\frac{5280 \text{ ft}}{1 \text{ mi}}$.

$$4 \text{ mi} = 4 \text{ mi} \cdot 1 = \frac{4 \text{ mi}}{1} \cdot \frac{5280 \text{ ft}}{1 \text{ mi}} = \frac{4 \cancel{\text{mi}} \cdot 5280 \text{ ft}}{1 \cancel{\text{mi}}} = 4 \cdot 5280 \text{ ft} = 21{,}120 \text{ ft}$$

Note in the above example: First, you can think of dividing the numerator and denominator by the common unit "mile" just as you would divide the numerator and denominator of a rational expression by a common factor. Second, the conversion factor $\frac{5280 \text{ ft}}{1 \text{ mi}}$ is equal to 1, and multiplying an expression by 1 does not change the value of the expression. Third, we had the choice of two conversion factors, $\frac{1 \text{ mi}}{5280 \text{ ft}}$ or $\frac{5280 \text{ ft}}{1 \text{ mi}}$. We chose $\frac{5280 \text{ ft}}{1 \text{ mi}}$ because the unit in the numerator is the same as the unit desired in the answer (ft), and the unit in the denominator is the same as the unit in the given measurement (mi).

Use dimensional analysis to convert each of the following. Round to the nearest hundredth.

a. 44,880 feet to miles

b. 800 yards to miles

c. 32 cups to gallons

d. 88 feet/second to miles per hour

e. 66 mph to feet per second

f. 90 kilometers/hour to meters per second

g. 8 miles/second to kilometers per hour

h. 100 yards in 9.6 seconds to miles per hour

Section 8.6 Proportions and Similar Triangles

Solving Proportions

The table below shows the automobile fatality rate per 100,000 people in the United States for the years 1990 through 1996. Also shown is the U.S. population for those years.

Year	Fatality Rate per 100,000 Population	U.S. Population
1990	17.88	250,000,000
1991	16.46	253,000,000
1992	15.39	255,000,000
1993	15.57	258,000,000
1994	15.64	260,000,000
1995	15.91	263,000,000
1996	15.80	265,000,000

Sources: National Highway Traffic Safety Administration;
U. S. Bureau of the Census

What was the number of automobile fatalities in the United States in 1991? Was this more or less than the number in 1995? We can answer these questions by writing and solving proportions. But first we need to define a rate and a ratio.

A **rate** is the quotient of two quantities that have different units. The 1990 automobile fatality rate of 17.88 per 100,000 population can be written:

$$\frac{17.88 \text{ deaths}}{100{,}000 \text{ people}}$$

POINT OF INTEREST
If we solve the literal equation $C = \pi d$ for π, the result is $\pi = \frac{C}{d}$. This means that π is equal to the ratio of the circumference of a circle to the diameter.

A **ratio** is the quotient of two quantities that have the same unit. The ratio of the number of unemployed persons in the United States in September of 1997 to the number of unemployed persons in the United States in September of 1996 is:

$$\frac{6{,}752{,}000 \text{ unemployed persons}}{7{,}043{,}000 \text{ unemployed persons}} = \frac{6{,}752}{7{,}043}$$

Note that units are written as part of a rate, while units are not written as part of a ratio.

A **proportion** is the equality of two rates or ratios. The following are examples of proportions.

$$\frac{200 \text{ miles}}{4 \text{ hours}} = \frac{50 \text{ miles}}{1 \text{ hour}}$$

Note that the units in the numerators (miles) are the same and the units in the denominators (hours) are the same.

$$\frac{7}{14} = \frac{1}{2}$$

The definition of a proportion can be stated as follows:

If $\frac{a}{b}$ and $\frac{c}{d}$ are equal ratios or rates, then $\frac{a}{b} = \frac{c}{d}$ is a proportion.

Each of the four members in a proportion is called a **term.** Each term is numbered as shown below.

$$\begin{matrix}\text{first term} \longrightarrow & a & & c & \longleftarrow \text{third term}\\ & - & = & - & \\ \text{second term} \longrightarrow & b & & d & \longleftarrow \text{fourth term}\end{matrix}$$

SUGGESTED ACTIVITY
Here is a challenging problem involving rates.

For the first 5 miles of a 10-mile race, Ray's rate was 10 mph. For the last 5 miles, his rate slowed to 8 mph. How long did it take Ray to complete the race? What was his average rate for the race?

[Answer: 1 1/8 hours; 8 8/9 mph]

The second and third terms of the proportion are called the **means,** and the first and fourth terms are called the **extremes.**

If we multiply both sides of the proportion by the LCM of the denominators, *bd*, we obtain the following result:

$$\frac{a}{b} = \frac{c}{d}$$

$$bd\left(\frac{a}{b}\right) = bd\left(\frac{c}{d}\right)$$

$$ad = bc$$

In a proportion, **the product of the means equals the product of the extremes.** This is sometimes phrased "the cross products are equal."

For example, in the proportion $\frac{2}{3} = \frac{8}{12}$, the cross products are equal.

The product of the means = The product of the extremes

$$3 \cdot 8 = 2 \cdot 12$$

$$24 = 24$$

Question: For the proportion $\frac{3}{8} = \frac{9}{24}$: **a.** Which are the first and third terms? **b.** Which terms are the means? **c.** What is the product of the extremes?[1]

When one of the numbers in a proportion is unknown, we can solve the proportion for the unknown number.

Example 1 Solve: $\frac{8}{3} = \frac{12}{x}$

Solution

$$\frac{8}{3} = \frac{12}{x}$$

$$3 \cdot 12 = 8 \cdot x$$ • The product of the means equals the product of the extremes.

$$36 = 8x$$

$$\frac{36}{8} = \frac{8x}{8}$$ • Divide both sides of the equation by the coefficient of *x*.

$$4.5 = x$$

1. **a.** The first term is 3. The third term is 9. **b.** The 8 and the 9 are the means. **c.** The product of the extremes is $3 \cdot 24 = 72$.

Check: $\frac{8}{3} = \frac{12}{4.5}$

$$3(12) = 8(4.5)$$
$$36 = 36$$

The solution is 4.5.

You-Try-It 1 Solve: $\frac{n}{5} = \frac{12}{25}$

Solution See page S39.

Now let's return to the questions related to automobile fatalities. The table of rates and populations is repeated below.

Year	Fatality Rate per 100,000 Population	U.S. Population
1990	17.88	250,000,000
1991	16.46	253,000,000
1992	15.39	255,000,000
1993	15.57	258,000,000
1994	15.64	260,000,000
1995	15.91	263,000,000
1996	15.80	265,000,000

Source: National Highway Safety Administration; U. S. Bureau of the Census

To approximate the number of automobile fatalities in the United States in 1991, we can write a proportion using any variable to represent the number of fatalities. The variable N is used in the proportion below.

TAKE NOTE
The fatality rate per 100,000 people is equal to the rate of the total number of fatalities to the total population.

$$\frac{16.46 \text{ deaths}}{100{,}000 \text{ people}} = \frac{N \text{ deaths}}{253{,}000{,}000 \text{ people}}$$
$$100{,}000 \cdot N = 16.46(253{,}000{,}000)$$
$$100{,}000N = 4{,}164{,}380{,}000$$
$$N \approx 41{,}644$$

There were approximately 41,644 automobile fatalities in the United States in 1991.

To answer the question, "Was this more or less than the number in 1995?" we need to solve another proportion.

TAKE NOTE
The same unit is in both numerators (deaths) and the same unit is in both denominators (people).

$$\frac{15.91 \text{ deaths}}{100{,}000 \text{ people}} = \frac{N \text{ deaths}}{263{,}000{,}000 \text{ people}}$$
$$100{,}000 \cdot N = 15.91(263{,}000{,}000)$$
$$100{,}000N = 4{,}184{,}330{,}000$$
$$N \approx 41{,}843$$

$$41{,}644 < 41{,}843$$

The number of automobile fatalities in 1991 was less than the number in 1995.

It is important to remember, in setting up a proportion, to keep the same units in the numerators and the same units in the denominators. In the previous example, "deaths" is the unit in the numerators and "people" is the unit in the denominators. The following proportion, with the unit "people" in the numerators and "deaths" in the denominators, could also be used to solve the problem.

$$\frac{100{,}000 \text{ people}}{15.91 \text{ deaths}} = \frac{263{,}000{,}000 \text{ people}}{N \text{ deaths}}$$

Question: Write two proportions that can be used to determine the number of automobile fatalities in 1996. Use the data in the table on the previous page.[2]

Example 2 The table below shows the marriage rate per 1000 people aged 15 years and over in the United States during selected years. The U.S. population for those years is also given. Use this data to find the difference between the number of marriages in 1950 and the number of marriages in 1990. Round to the nearest hundred thousand.

Year	Marriage Rate per 1000 Population	U.S. Population
1950	11.1	152,271,000
1960	8.5	180,671,000
1970	10.6	205,052,000
1980	10.6	227,726,000
1990	9.8	249,913,000

Sources: National Center for Health Statistics; U.S. Bureau of the Census

State the goal.

The goal is to find the difference between the number of marriages in 1950 and the number of marriages in 1990.

Describe a strategy.

- Write and solve a proportion to find the number of marriages in 1950.
- Write and solve a proportion to find the number of marriages in 1990.
- Subtract to find the difference.

Solve the problem.

$$\frac{11.1 \text{ marriages}}{1000 \text{ people}} = \frac{M \text{ marriages}}{152{,}271{,}000 \text{ people}}$$

$$1000 \cdot M = 11.1(152{,}271{,}000)$$
$$1000M = 1{,}690{,}208{,}100$$
$$M \approx 1{,}690{,}208$$

• Find the number of marriages in 1950.

$$\frac{9.8 \text{ marriages}}{1000 \text{ people}} = \frac{M \text{ marriages}}{249{,}913{,}000 \text{ people}}$$

$$1000 \cdot M = 9.8(249{,}913{,}000)$$
$$1000M = 2{,}449{,}147{,}400$$
$$M \approx 2{,}449{,}147$$

• Find the number of marriages in 1990.

---

2. Two possibilities are the proportion $\frac{15.80 \text{ deaths}}{100{,}000 \text{ people}} = \frac{N \text{ deaths}}{265{,}000{,}000 \text{ people}}$, which has the unit "deaths" in the numerators, and the proportion $\frac{100{,}000 \text{ people}}{15.80 \text{ deaths}} = \frac{265{,}000{,}000 \text{ people}}{N \text{ deaths}}$, which has the unit "deaths" in the denominators.

$2{,}449{,}147 - 1{,}690{,}208 = 758{,}939 \approx 800{,}000$ • Subtract to find the difference.

There were approximately 800,000 more marriages in the United States in1990 than in 1950.

Check your work.

Be sure to check the solution.

You-Try-It 2 A major concern in regard to the Social Security system is the fact that fewer and fewer workers are paying into the system to support more and more beneficiaries.

a. The ratio of workers to beneficiaries in 1960 was 5 to 1. If the number of beneficiaries in 1960 was 14 million, approximately how many workers were there in 1960?

b. The ratio of the number of workers to beneficiaries in 2030 is expected to be 2 to 1. How many Social Security beneficiaries are there expected to be in 2030 if the number of workers in 2030 is expected to be 167 million?

Sources: Social Security Administration; U.S. Bureau of the Census

Solution See page S39.

Similar Triangles

Proportions have applications to the field of geometry. We will present here their application to similar triangles.

Similar objects have the same shape but not necessarily the same size. A tennis ball is similar to a basketball. A model airplane is similar to an actual airplane.

Similar objects have corresponding parts. For example, the wing on a model airplane corresponds to the wing on the actual airplane. The relationship between the sizes of each of the corresponding parts can be written as a ratio, and each ratio will be the same. If the wing on the model airplane is $\frac{1}{100}$ the size of the wing on the actual airplane, then the model fuselage is $\frac{1}{100}$ the size of the actual fuselage, the wheels of the landing gear on the model are $\frac{1}{100}$ the size of the wheels on the actual airplane, and so on.

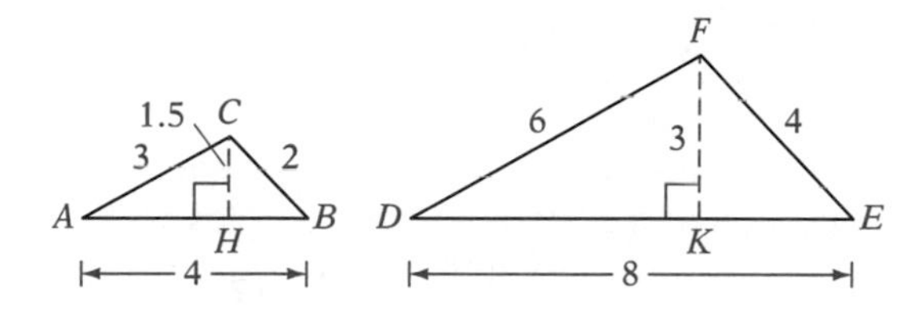

The two triangles *ABC* and *DEF* shown at the left are similar. Side *AB* corresponds to side *DE*, side *AC* corresponds to side *DF*, and side *BC* corresponds to side *EF*. The height *CH* corresponds to height *FK*. The ratios of corresponding parts are equal.

$$\frac{AB}{DE} = \frac{4}{8} = \frac{1}{2}, \qquad \frac{AC}{DF} = \frac{3}{6} = \frac{1}{2}, \qquad \frac{BC}{EF} = \frac{2}{4} = \frac{1}{2}, \qquad \frac{CH}{FK} = \frac{1.5}{3} = \frac{1}{2}$$

Since the ratios of corresponding parts are equal, three proportions can be formed using the sides of the triangles.

$$\frac{AB}{DE} = \frac{AC}{DF}, \qquad \frac{AB}{DE} = \frac{BC}{EF}, \quad \text{and} \quad \frac{AC}{DF} = \frac{BC}{EF}$$

Three proportions can also be formed by using the sides and heights of the triangles.

$$\frac{AB}{DE} = \frac{CH}{FK}, \qquad \frac{AC}{DF} = \frac{CH}{FK}, \quad \text{and} \quad \frac{BC}{EF} = \frac{CH}{FK}$$

The corresponding angles in similar triangles are equal. Therefore,

$$\angle A = \angle D, \quad \angle B = \angle E, \quad \text{and} \quad \angle C = \angle F$$

SUGGESTED ACTIVITY
See the *Instructor's Resource Manual* for an activity involving similar figures.

→ Triangles *ABC* and *DEF* shown below are similar. Find the area of triangle *ABC*.

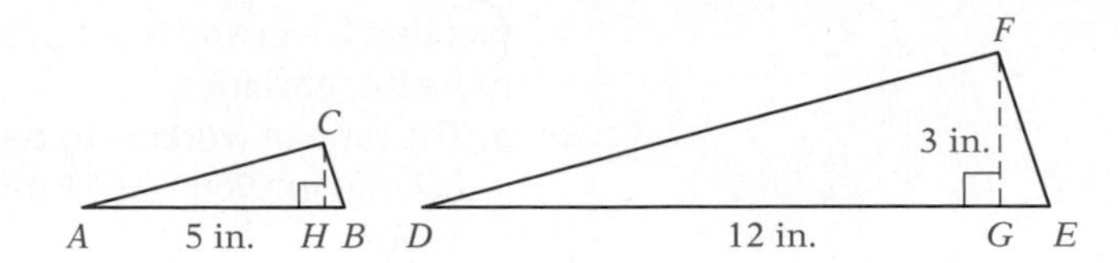

State the goal. The goal is to find the area of triangle *ABC*.

Describe a strategy.

- Write and solve a proportion to find *CH*, the height of triangle *ABC*.
- Use the formula for the area of a triangle: $A = \frac{1}{2}bh$.

Solve the problem.

$$\frac{AB}{DE} = \frac{CH}{FG}$$

$$\frac{5}{12} = \frac{CH}{3}$$

$$12(CH) = 5(3)$$

$$12(CH) = 15$$

$$CH = 1.25$$

$$A = \frac{1}{2}bh$$

$$A = \frac{1}{2}(5)(1.25) = 3.125$$

The area of triangle *ABC* is 3.125 in^2.

Check your work. Be sure to check the solution.

It is also true that if the three angles of one triangle are equal respectively to the three angles of another triangle, then the two triangles are similar.

In the triangle at the right, line segment *DE* is drawn parallel to the base *AB*. $\angle x = \angle m$ and $\angle y = \angle n$ because corresponding angles are equal, and $\angle C = \angle C$. Therefore, the three angles of triangle *DEC* are equal respectively to the three angles of triangle *ABC*. Triangle *DEC* is similar to triangle *ABC*.

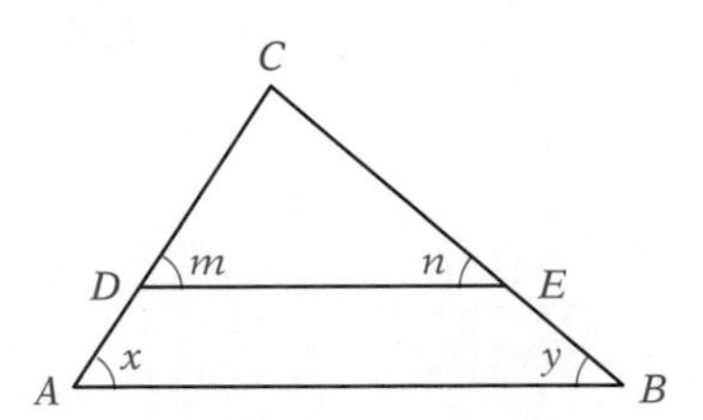

The sum of the three angles of a triangle is 180°. If two angles of one triangle are equal to two angles of another triangle, then the third angles must be equal. Thus we can say that if two angles of one triangle are equal to two angles of another triangle, then the two triangles are similar.

→ Line segments AB and CD intersect at point O in the figure at the right. Angles C and D are right angles. Find the length of DO.

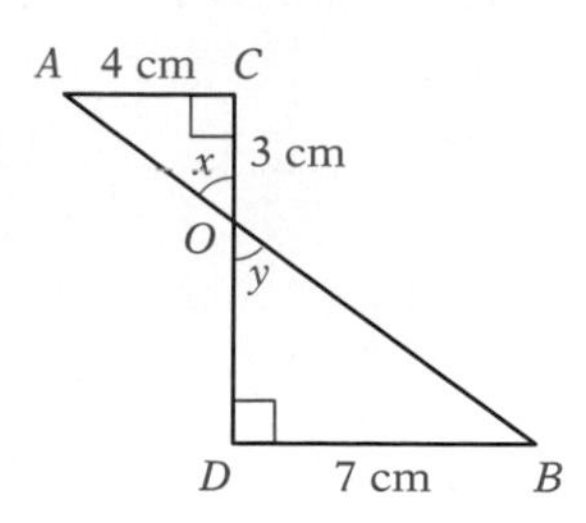

State the goal. The goal is to find the length of DO.

Describe a strategy.

- First determine if triangle AOC is similar to triangle BOD:
  $\angle C = \angle D$ because they are right angles. $\angle x = \angle y$ because they are vertical angles. Therefore, triangle AOC is similar to triangle BOD because two angles of one triangle are equal to two angles of the other triangle.
- Use a proportion to find the length of DO.

Solve the problem.

$$\frac{AC}{DB} = \frac{CO}{DO}$$

$$\frac{4}{7} = \frac{3}{DO}$$

$$7(3) = 4(DO)$$

$$21 = 4(DO)$$

$$5.25 = DO$$

The length of DO is 5.25 centimeters.

Check your work. Be sure to check the solution.

Example 3 In the figure at the right, AB is parallel to DC and angles B and D are right angles. $AB = 12$ meters, $DC = 4$ meters, and $AC = 18$ meters. Find the length of CO.

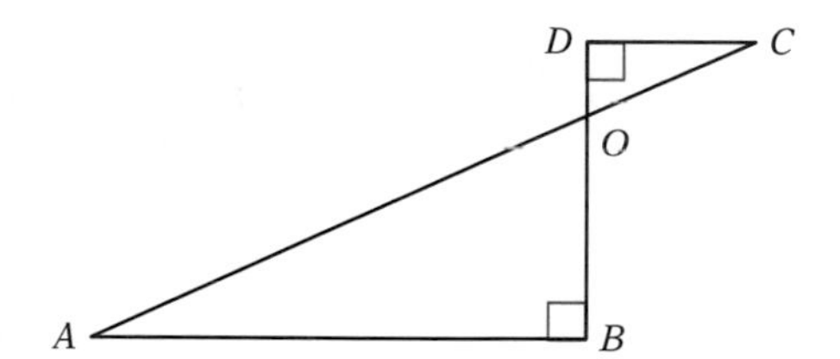

State the goal. The goal is to find the length of CO.

Describe a strategy. Triangle AOB is similar to triangle COD. Solve a proportion to find the length of CO. Let x represent the length of CO. Then $18 - x$ represents the length of AO.

Solve the problem.

$$\frac{DC}{AB} = \frac{CO}{AO}$$

$$\frac{4}{12} = \frac{x}{18 - x}$$

$$12(x) = 4(18 - x)$$

$$12x = 72 - 4x$$

$$16x = 72$$

$$x = 4.5$$

The length of CO is 4.5 meters.

Check your work. Be sure to check the solution.

You-Try-It 3 In the figure at the right, AB is parallel to DC and angles A and D are right angles. $AB = 10$ centimeters, $CD = 4$ centimeters, and $DO = 3$ centimeters. Find the area of triangle AOB.

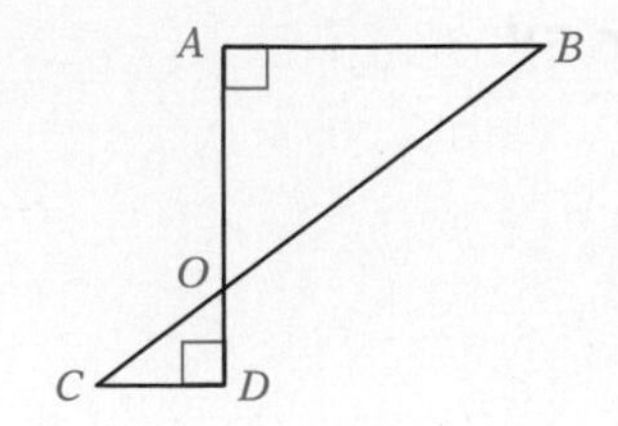

Solution See page S40.

Surveyors use similar triangles to find the measure of distances that cannot be measured directly. This is illustrated in Example 4 and You-Try-It 4.

Example 4 The diagram at the right represents a river of width CD. Triangles AOB and DOC are similar. The distances AB, BO, and OC were measured and found to have the lengths given in the diagram. Find the width of the river.

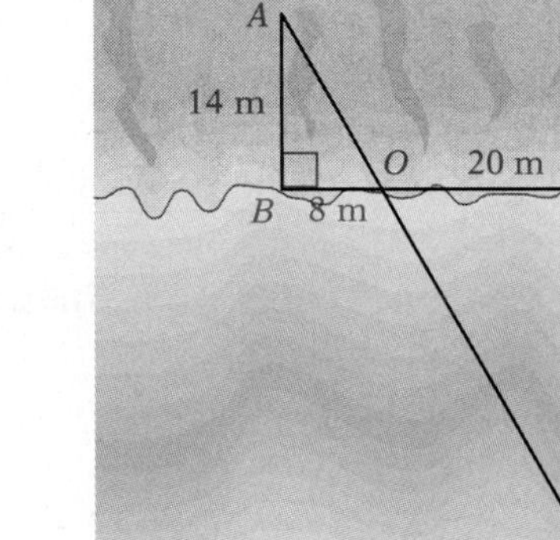

State the goal. The goal is to find CD, the width of the river.

Describe a strategy. Write and solve a proportion to find the length of side CD.

Solve the problem.

$$\frac{AB}{CD} = \frac{BO}{OC}$$

$$\frac{14}{CD} = \frac{8}{20}$$

$$CD(8) = 14(20)$$

$$CD(8) = 280$$

$$CD = 35$$

The width of the river is 35 meters.

Check your work. Be sure to check the solution.

You-Try-It 4 The diagram below shows how surveyors laid out similar triangles along the Winnepaugo River. Find the width, w, of the river.

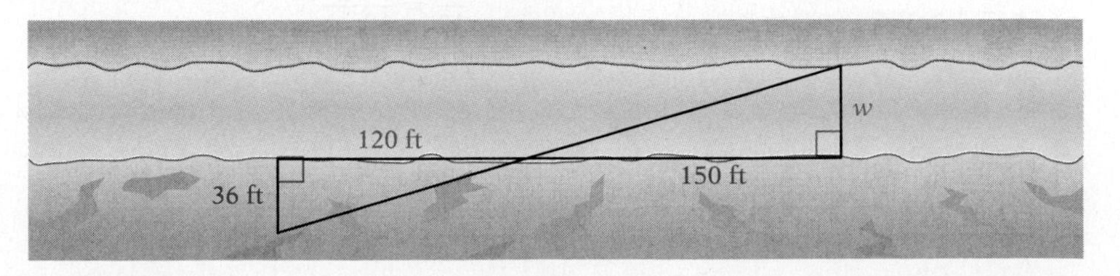

Solution See page S40.

8.6 EXERCISES

Topics for Discussion

1. **a.** What is a rate? **b.** What is a ratio? **c.** What is a proportion?
Answers will vary. For example: **a.** A rate is the quotient of two quantities that have different units. **b.** A ratio is the quotient of two quantities that have the same units. **c.** A proportion is the equality of two rates or ratios.

2. Provide two examples of proportions, one involving rates and one involving ratios.
Examples will vary.

3. Explain what the means and the extremes in a proportion are.
Answers will vary. For example, the second and third terms are the means; the first and fourth terms are the extremes.

4. Explain why the product of the means in a proportion is equal to the product of the extremes.
Answers will vary. For example, it is equivalent to multiplying both sides of the proportion by the LCM of the denominators.

5. Determine whether the statement is always true, sometimes true, or never true.

a. If an acute angle of a right triangle is equal to an acute angle of another right triangle, then the triangles are similar. Always true

b. Two isosceles triangles are similar triangles. Sometimes true

c. Two equilateral triangles are similar triangles. Always true

d. Two squares are similar. Always true

e. Two rectangles are similar. Sometimes true

6. Explain how proportions relate to similar triangles.
Answers will vary.

Solving Proportions

Solve.

7. $\frac{4}{5} = \frac{12}{x}$
15

8. $\frac{6}{x} = \frac{2}{3}$
9

9. $\frac{20}{9} = \frac{64}{x}$
28.8

10. $\frac{n}{27} = \frac{5}{8}$
16.875

11. $\frac{8}{n+3} = \frac{4}{n}$
3

12. $\frac{3}{x-2} = \frac{4}{x}$
8

13. $\frac{6}{x+4} = \frac{12}{5x-13}$
7

14. $\frac{2}{3x-1} = \frac{3}{4x+1}$
5

15. $\frac{2}{x+3} = \frac{6}{5x+5}$
2

16. $\frac{5}{n+3} = \frac{3}{n-1}$
7

17. $\frac{5}{2x-3} = \frac{10}{x+3}$
3

18. $\frac{4}{5y-1} = \frac{2}{2y-1}$
-1

19. $\frac{x}{x-1} = \frac{8}{x+2}$
2, 4

20. $\frac{x}{x+12} = \frac{1}{x+5}$
$-6, 2$

21. $\frac{2x}{x+4} = \frac{3}{x-1}$
$-\frac{3}{2}, 4$

22. $\frac{5}{3y-8} = \frac{y}{y+2}$
$-\frac{2}{3}, 5$

23. A cord is a quantity of cut wood, to be used for fuel, that is equal to 128 cubic feet in a stack measuring 4 feet by 4 feet by 8 feet. Cutting 8 cords of wood produces 1 cord of sawdust. At this rate, how much sawdust is produced by cutting 14 cords of wood?
1.75 cords

24. In a city of 25,000 homes, a survey was taken to determine the number with cable television. Of the 300 homes surveyed, 210 had cable television. Estimate the number of homes in the city that have cable television.
17,500 homes

25. The lighting for some billboards is provided by using solar energy. If 3 small solar energy panels can generate 10 watts of power, how many panels are necessary to provide 600 watts of power?
180 panels

26. The sales tax on a car that sells for \$12,000 is \$780. At this rate, what is the sales tax on a car that sells for \$18,500?
\$1202.50

27. To conserve energy and still allow for as much natural lighting as possible, an architect suggests that the ratio of the area of window to the area of the total wall surface be 5 to 12. Using this ratio, determine the recommended area of a window to be installed in a wall that measures 8 feet by 12 feet.
40 ft^2

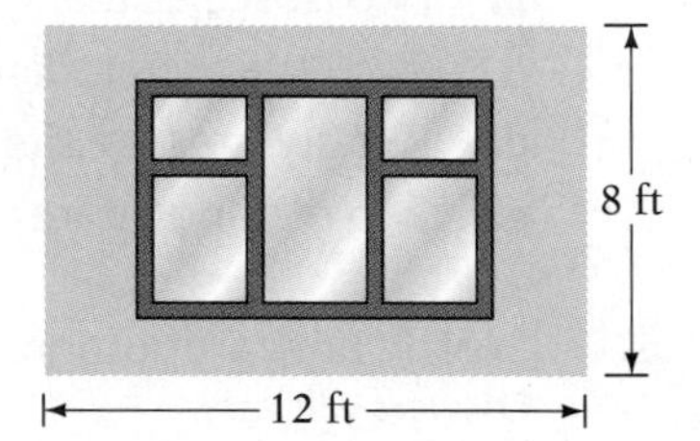

28. As part of a conservation effort for a lake, 40 fish are caught, tagged, and then released. Later 80 fish are caught. Four of the 80 fish are found to have tags. Estimate the number of fish in the lake.
800 fish

29. The table below shows the incarceration rates for those states with the highest rates in the nation. Also shown is the population of each state.

State	Incarceration Rate per 100,000	Population
Texas	659	18,724,000
Louisiana	573	4,342,000
Oklahoma	536	3,278,000
South Carolina	510	3,673,000
Arizona	473	4,218,000

Sources: *Time*, January 15, 1996; U.S. Bureau of the Census

a. Find the number of inmates in the state of Oklahoma.
b. Which has more inmates, South Carolina or Arizona?
c. Find the difference between the number of prisoners in Texas and the number in Louisiana.

a. 17,570 inmates **b.** Arizona **c.** 98,511 prisoners

30. The table below shows the vehicle theft rate per 100,000 residents in the United States for the years 1960, 1970, 1980, and 1990. Also shown is the U.S. population for those years.

Year	Vehicle Theft Rate per 100,000 Residents	Population
1960	183.0	179,323,175
1970	456.8	203,302,031
1980	502.2	226,545,805
1990	657.8	248,709,873

Sources: FBI; National Insurance Crime Bureau; U. S. Bureau of the Census

a. Find the number of motor vehicle thefts in 1960.

b. How many more vehicle thefts were there in 1990 than in 1980?

c. Find the percent increase in the number of car thefts from 1980 to 1990. Round to the nearest percent.

d. Find the percent increase in the vehicle theft rate per 100,000 residents from 1980 to 1990. Round to the nearest percent.

e. Why can you determine without performing any calculations that there were more vehicle thefts in 1980 than in 1970?

f. Provide an explanation for providing statistics on the rate of theft per 100,000 residents rather than on the number of thefts.

a. 328,161 motor vehicle thefts; **b.** 498,301 more thefts; **c.** 44%; **d.** 31%; **e.** answers will vary; for example, both the rate and the population increased from 1970 to 1980; **f.** answers will vary.

31. The table below shows the states with the lowest death rates per 100,000 people in the United States in 1996. Also shown is the population of each state.

State	Death Rate per 100,000	Population
Hawaii	389.1	1,187,000
Utah	407.1	1,951,000
Minnesota	411.9	4,610,000
North Dakota	419.7	691,000

Sources: Centers for Disease Control and Prevention; U.S. Bureau of the Census

a. Find the number of deaths in Hawaii in 1996.

b. Which state had more deaths in 1996, Minnesota or North Dakota?

c. The death rate in the United States in 1996 was 493.6 per 100,000 people. What percent of the deaths in the United States in 1996 occurred in Utah? Use a U.S. population of 265,000,000. Round to the nearest tenth of a percent.

d. The District of Columbia had the highest death rate in the country in 1996: 771.7 deaths per 100,000 people. The land area of Washington, D.C., is less than 0.1% of the land area of North Dakota. Does this mean that the number of deaths in Washington, D.C., was less than the number in North Dakota?

e. The population of Washington, D.C., is 567,000. Find the difference between the number of deaths in Washington, D.C., in 1996 and the number of deaths in North Dakota.

f. What factors might contribute to a state's having a low death rate?

a. 4619 deaths; **b.** Minnesota; **c.** 0.6%; **d.** no; answers will vary; for example, it is the population, not the land size, that influences the number of deaths given the death rate; **e.** 1476 deaths; **f.** answers will vary.

32. The graph below shows the motor vehicle crash rate per 100,000 drivers in different age groups in the United States.

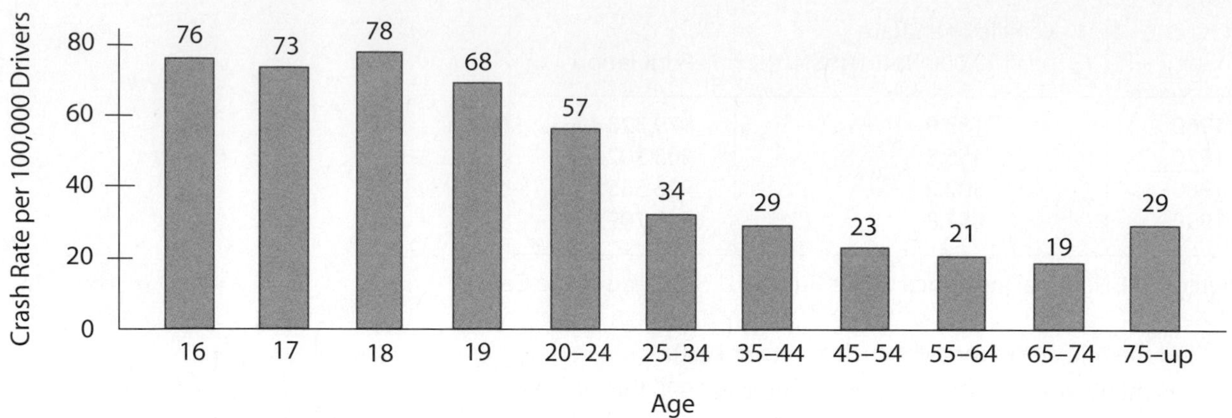

Source: National Safety Council

a. The U.S. population of 25- to 34-year-olds is 38,979,000. Find the number of motor vehicle crashes this age group was involved in.

b. The U.S. population of 20- to 24-year-olds is 17,633,000. The U.S. population of 35- to 44-year-olds is 44,353,000. Which age group was involved in more motor vehicle crashes? How many more?

c. If ages 16–19 were combined, rather than shown separately, would the combined rate for this group of drivers be greater than or less than the individual rates shown in the graph?

d. How are statistics such as these used by automobile insurance companies to determine premiums for drivers of different ages?

a. 13,253 crashes; **b.** 35- to 44-year-olds; 2811 more; **c.** greater than the individual rates; **d.** answers will vary.

Similar Triangles

Triangles *ABC* and *DEF* in Exercises 33 to 40 are similar. Round answers to the nearest tenth.

33. Find side *AC*.

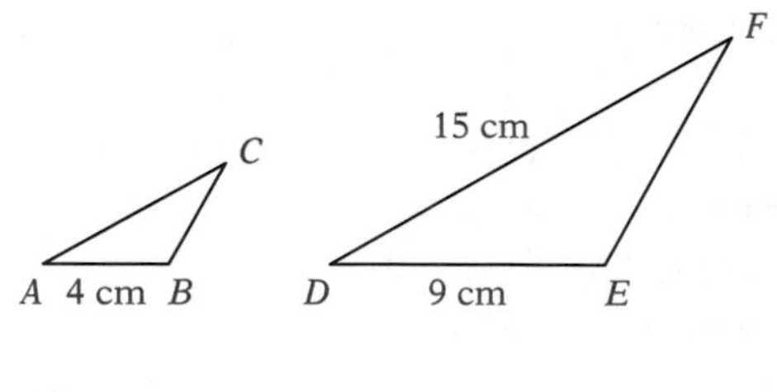

6.7 cm

34. Find side *DE*.

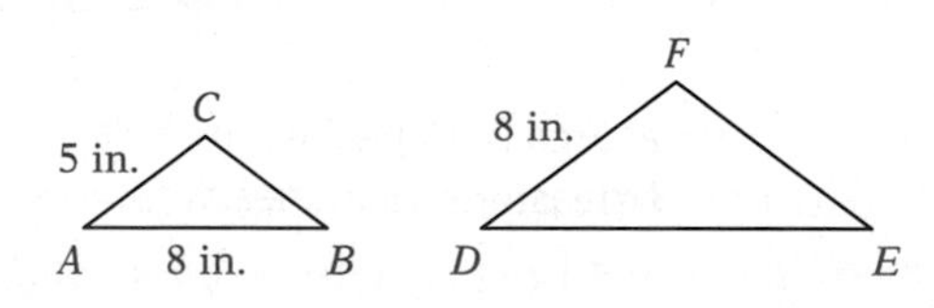

12.8 in.

35. Find the height of triangle *ABC*.

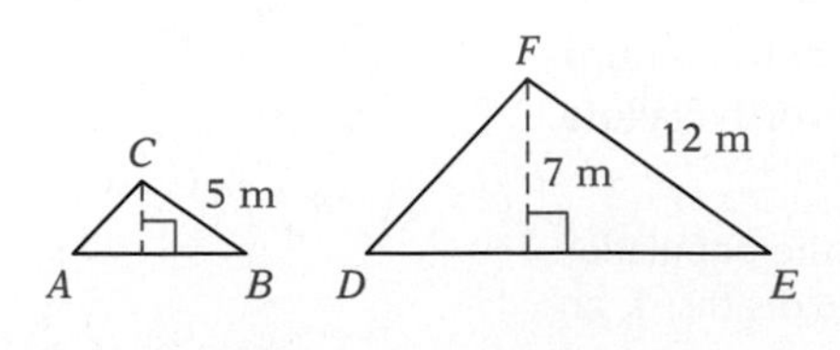

2.9 m

36. Find the height of triangle *DEF*.

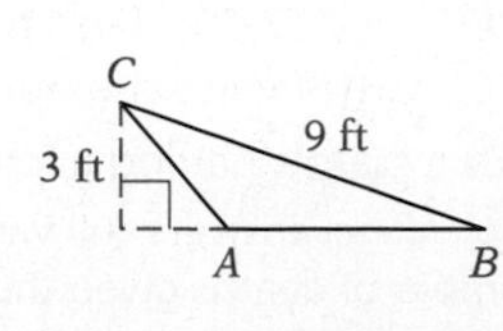

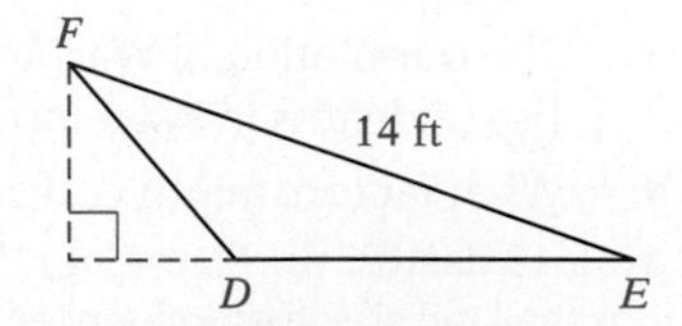

4.7 ft

37. Find the perimeter of triangle *DEF*.

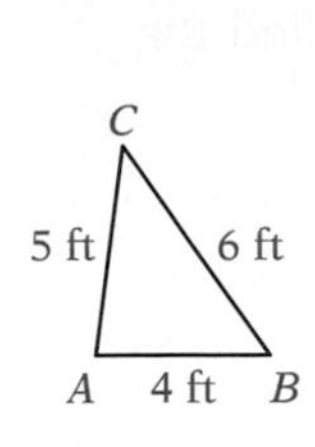

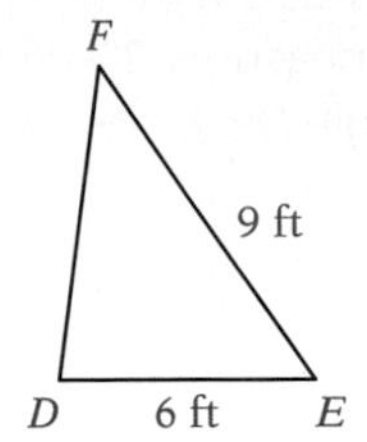

22.5 ft

38. Find the perimeter of triangle *ABC*.

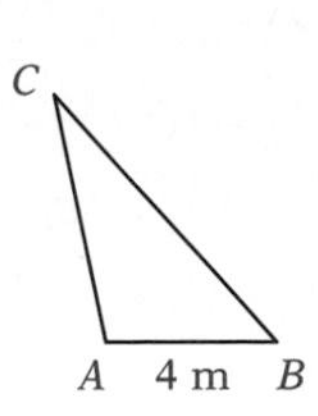

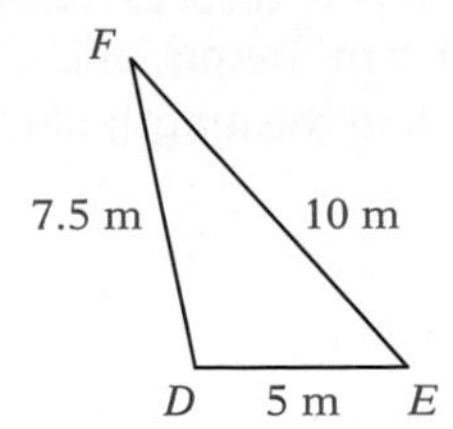

18 m

39. Find the area of triangle *ABC*.

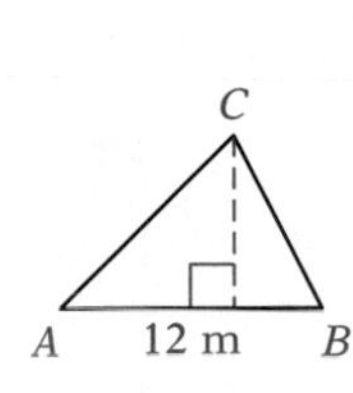

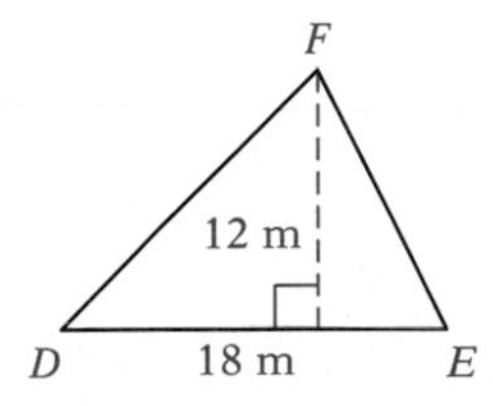

48 m^2

40. Find the area of triangle *ABC*.

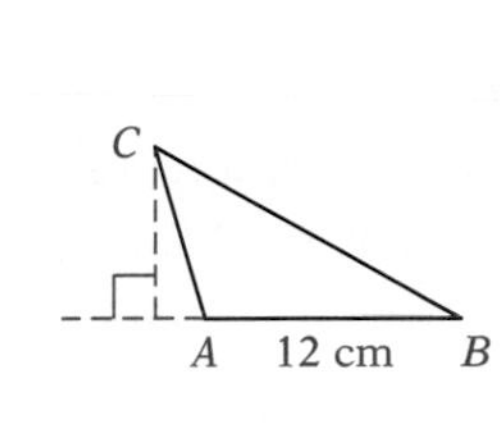

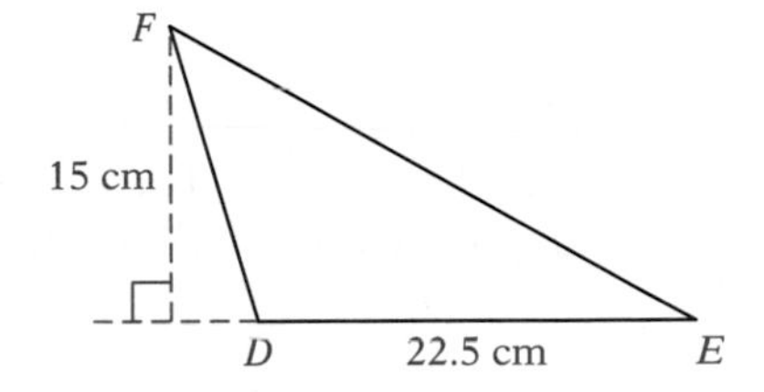

48 cm^2

41. Given *BD* || *AE*, *BD* measures 5 centimeters, *AE* measures 8 centimeters, and *AC* measures 10 centimeters, find the length of *BC*.

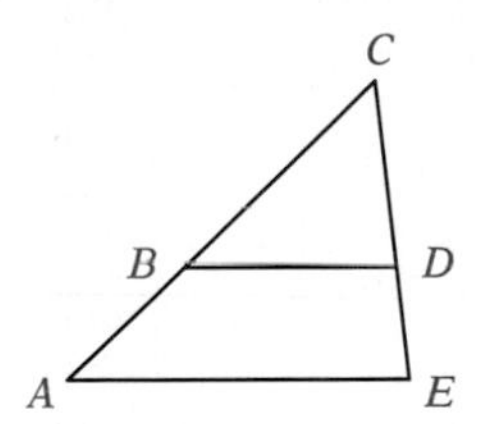

6.25 cm

42. Given *AC* || *DE*, *BD* measures 8 meters, *AD* measures 12 meters, and *BE* measures 6 meters, find the length of *BC*.

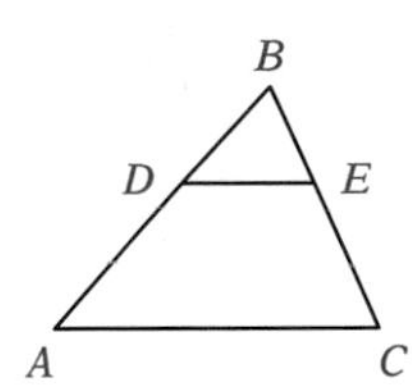

15 m

43. Given *DE* || *AC*, *DE* measures 6 inches, *AC* measures 10 inches, and *AB* measures 15 inches, find the length of *DA*.

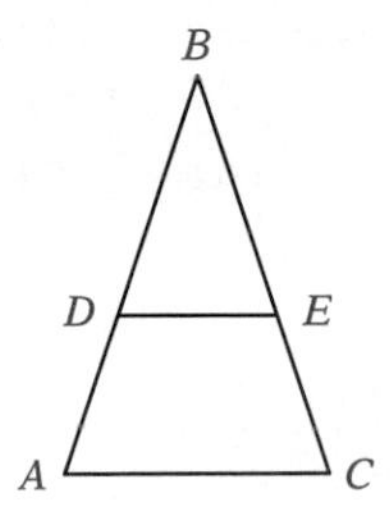

6 in.

44. Given *MP* and *NQ* intersect at *O*, *NO* measures 25 feet, *MO* measures 20 feet, and *PO* measures 8 feet, find the length of *QO*.

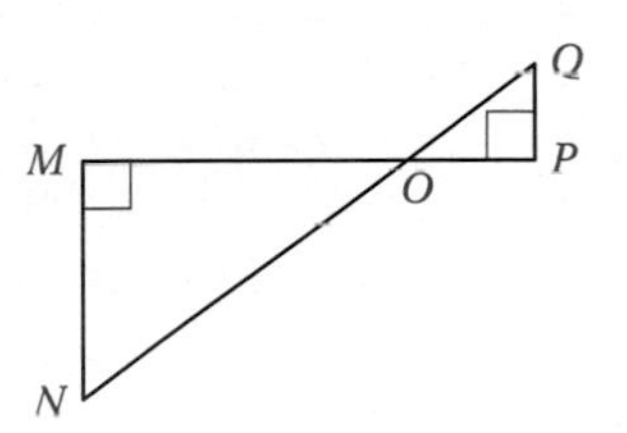

10 ft

45. Given MP and NQ intersect at O, NO measures 24 centimeters, MN measures 10 centimeters, MP measures 39 centimeters, and QO measures 12 centimeters, find the length of OP.

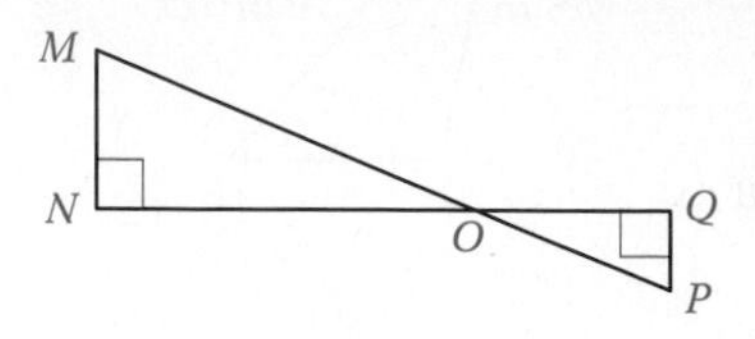

13 cm

46. Given MQ and NP intersect at O, NO measures 12 meters, MN measures 9 meters, PQ measures 3 meters, and MQ measures 20 meters, find the perimeter of triangle OPQ.

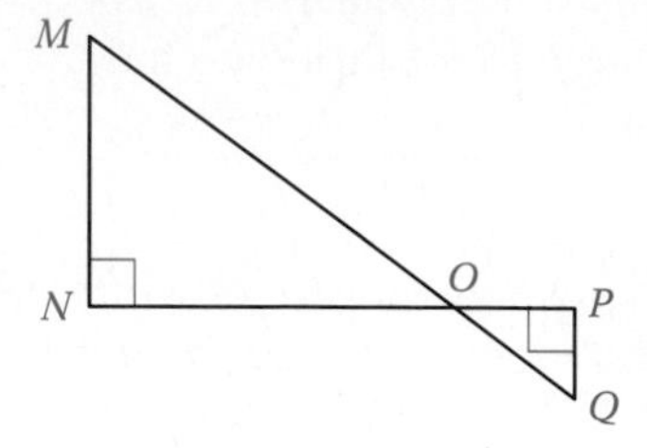

12 m

47. The sun's rays cast a shadow as shown in the diagram at the right. Find the height of the flagpole. Write the answer in terms of feet.
14.375 ft

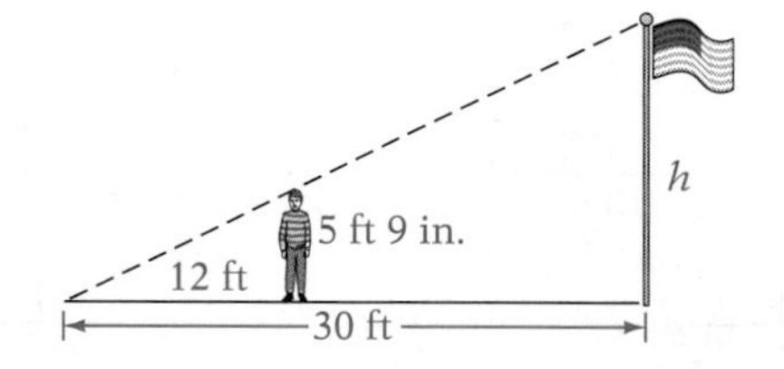

48. The diagram at the right represents a river of width CD. The distances AB, BO, and OC were measured and found to have the lengths given in the diagram. Find the width of the river.
36 m

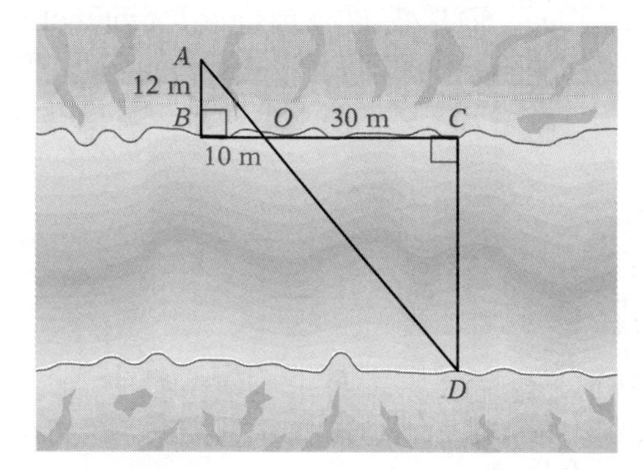

49. The diagram at the right shows how surveyors laid out similar triangles along a ravine. Find the width, w, of the ravine.
82.5 ft

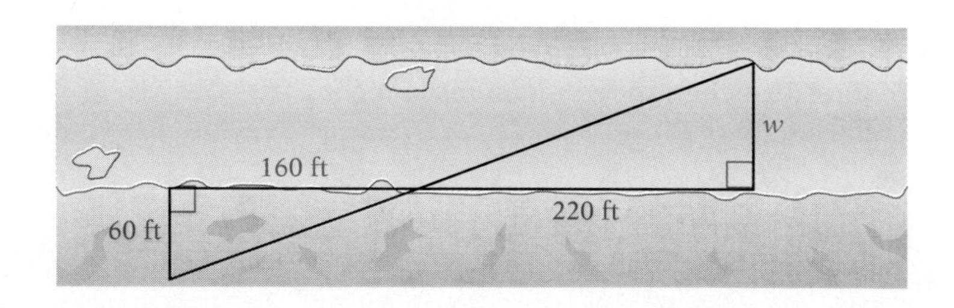

Applying Concepts

50. The table on page 559 shows that in 1991 the automobile fatality rate was 16.46 and the number of fatalities was 41,644. In 1995, the automobile fatality rate was 15.91 and the number of fatalities was 41,843. Discuss the fact that in 1991 the rate was higher than in 1995 but the number of fatalities was lower.
Answers will vary.

51. A basketball player has made 5 out of every 6 foul shots attempted. If 48 foul shots were missed in the player's career, how many foul shots were made in the player's career?
240 foul shots

52. The "sitting fee" for school pictures is \$8. If 10 photos cost \$20, including the sitting fee, what would 24 photos cost, including the sitting fee?
\$36.80

53. Three people put their money together to buy lottery tickets. The first person put in \$20, the second person put in \$25, and the third person put in \$30. One of their tickets was a winning ticket. If they won \$15 million, what was the first person's share of the winnings?
\$4 million

54. After the skiing-related deaths of William Kennedy and Sonny Bono, the following statistics were published.

Ski Season	1996–97	1995–96	1994–95	1993–94	1992–93	1991–92	1990–91
Fatalities	36	45	49	41	42	35	38
Rate per Million Skier Days	0.69	0.65	0.93	0.75	0.78	0.69	0.60

Source: National Ski Areas Association

a. How many skier days were there during the 1996–97 ski season?
b. Based on the data in the table and your answer to part a, what do you think the definition of a "skier day" is?
c. Based on the data, how dangerous do you consider the sport of skiing to be?
a. 52,173,913 skier days; **b.** answers will vary; **c.** answers will vary.

55. The height of a right triangle is drawn from the right angle to the hypotenuse. Explain why the two triangles formed are similar to the original triangle and similar to each other.
Explanations will vary.

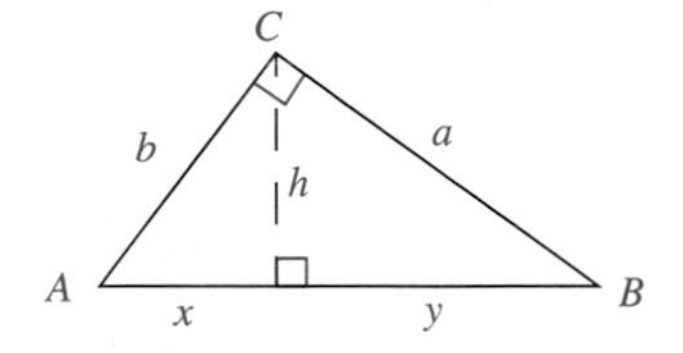

Exploration

56. ***Topology***
In this section, we discussed similar figures, that is, figures with the same shape. The branch of geometry called **topology** is interested in even more basic properties of figures than their size and shape. For example, look at the figures below. We could take a rubber band and stretch it into any one of these shapes.

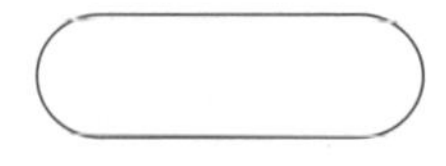
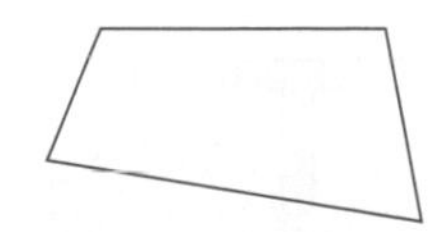
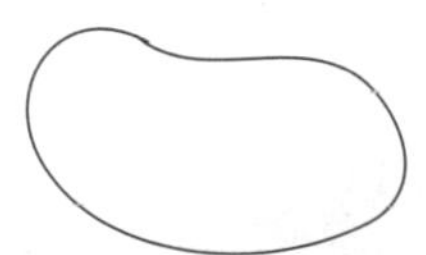

All three of these figures are different shapes, but each can be turned into one of the others by stretching the rubber band.

In topology, figures that can be stretched, molded, or bent into the same shape without puncturing or cutting belong to the same family. They are called topologically equivalent.

Rectangles, triangles, and circles are topologically equivalent.

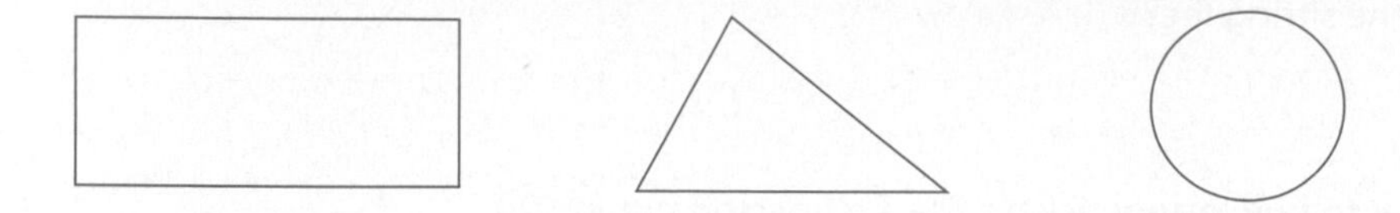

Line segments and wavy curves are topologically equivalent.

Note that the figures formed from a rubber band and those formed from a line segment are not topologically equivalent: to form a line segment from a rubber band, we would have to cut the rubber band.

In the following plane figures, the lines are joined where they cross. They are topologically equivalent. They are not topologically equivalent to any of the figures shown above.

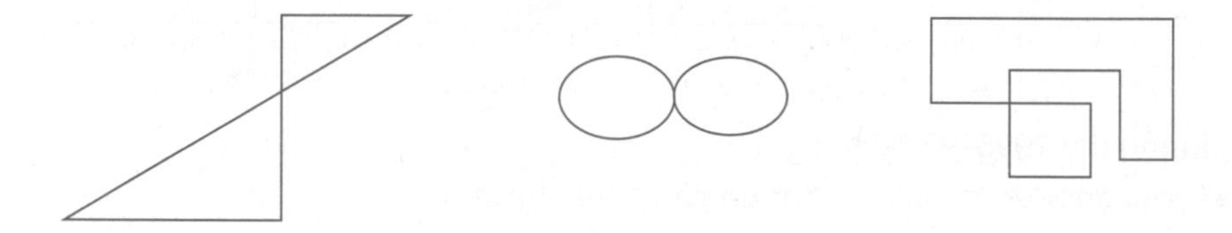

A topologist (a person who studies topology) is interested in identifying and describing different families of equivalent figures. This applies to solids as well as to plane figures. For example, a topologist considers a brick, a potato, and a cue ball to be equivalent to each other. Think of using modeling clay to form each of these shapes.

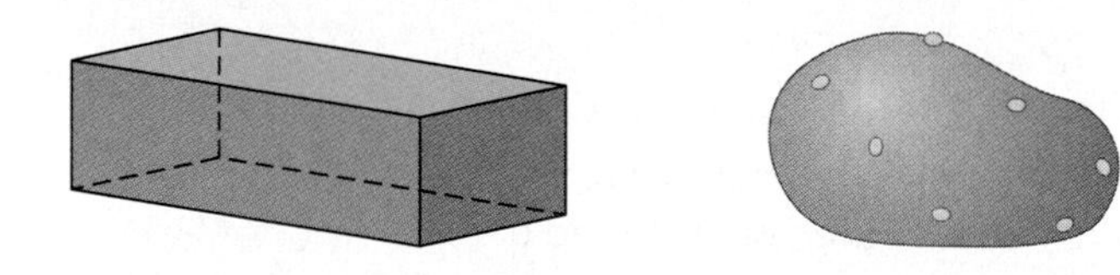

For parts a, b, and c below, which of the figures listed is not topologically equivalent to the others?

a. parallelogram square ray trapezoid

b. wedding ring doughnut fork sewing needle

c. A D O P T

d. Make a list of three objects that are topologically equivalent.

Section 8.7 Probability of Compound Events

Probability of Compound Events

In this section, we will continue the study of probability to include calculating probabilities of the occurrence of two or more events. Recall that the probability of an event E is the ratio of the number of elements in the event to the number of elements in the sample space.

For instance, consider the experiment of rolling a single die once. The sample space is $S = \{1, 2, 3, 4, 5, 6\}$. To calculate the probability that a number greater than 4 is rolled, first find the number of elements in the event. The event is $E = \{5, 6\}$. Therefore, $n(E) = 2$. The probability of the event is

$$P(E) = \frac{n(E)}{n(S)} = \frac{2}{6} = \frac{1}{3}$$

The probability of rolling a number greater than 4 is $\frac{1}{3}$.

TAKE NOTE
A regular deck of playing cards consists of 52 cards in 4 suits, spades, hearts, diamonds, and clubs. Each suit contains the 13 cards 2 through 10, jack, queen, king, and ace. Spades and clubs are normally black, and hearts and diamonds are normally red.

Earlier in the text we calculated the probabilities of what are called *simple* events. This did not mean that it was easy to calculate the probability, it meant that the event was a single event. In this section, we will calculate the probability of *compound* events. These events are characterized by using the words *or* and *and*.

Suppose a single card is selected from a regular deck of playing cards. Two examples of a compound event are:

- A two *or* a king is selected.
- A five *and* a spade is selected.

Question: Two dice are rolled. Give an example of a compound event.[1]

TAKE NOTE
Recall that $A \cup B$ contains the elements of set A *or* set B. $A \cap B$ contains the elements that belong to set A *and* set B.

Calculating the probability of a compound event depends on the concepts of *union* and *intersection* of sets, which were discussed earlier in the text.

Calculating Probabilities Involving *Or* or *And*

If A and B are two events in a sample space S, then

$$P(A \text{ or } B) = \frac{n(A \cup B)}{n(S)}$$

$$P(A \text{ and } B) = \frac{n(A \cap B)}{n(S)}$$

Example 1 A dodecahedral die (a 12-sided die with the numbers 1 through 12 on the faces) is rolled once. Find the probability that

a. the number rolled is an even number or divisible by 3.
b. the number rolled is an even number and divisible by 3.

Solution The sample space is $S = \{1, 2, 3, 4, 5, 6, 7, 8, 9, 10, 11, 12\}$. Let A be the event that an even number is rolled and B be the event that a number divisible by 3 is rolled. Then $A = \{2, 4, 6, 8, 10, 12\}$ and $B = \{3, 6, 9, 12\}$.

a. To determine the probability of A or B, first find

$A \cup B = \{2, 3, 4, 6, 8, 9, 10, 12\}$. Then

$$P(A \text{ or } B) = \frac{n(A \cup B)}{n(S)} = \frac{8}{12} = \frac{2}{3}$$

The probability of an even number or a number divisible by 3 is $\frac{2}{3}$.

1. Here are two examples: The sum of the pips is 7 *or* 11. One die shows a 3 *and* one die shows a 4.

b. To find the probability of A and B, first determine $A \cap B = \{6, 12\}$. Then

$$P(A \text{ and } B) = \frac{n(A \cap B)}{n(S)} = \frac{2}{12}$$
$$= \frac{1}{6}$$

The probability of an even number and a number divisible by 3 is $\frac{1}{6}$.

You-Try-It 1 One number is randomly chosen from
$S = \{1, 2, 3, 4, 5, 6, 7, 8, 9, 10, 11, 12, 13, 14, 15, 16, 17, 18, 19, 20\}$.
Find the probability that
a. the number selected is an odd number or divisible by 5.
b. the number selected is an odd number and divisible by 5.

Solution See page S40.

Suppose we draw a single card from a deck of playing cards. The sample space S is the 52 cards. Therefore, $n(S) = 52$. Now consider the events

E_1 = A four is drawn = {♠4, ♥4, ♦4, ♣4}

E_2 = A spade is drawn = {♠A, ♠2, ♠3, ♠4, ♠5, ♠6, ♠7, ♠8, ♠9, ♠10, ♠J, ♠Q, ♠K}

It is possible, on one draw, to satisfy the conditions of both events: The ♠4 could be drawn. This card is an element of both E_1 and E_2.

Now consider the events

E_3 = A five is drawn = {♠5, ♥5, ♦5, ♣5}

E_4 = A king is drawn = {♠K, ♥K, ♦K, ♣K}

In this case, it is not possible to draw one card that satisfies the conditions of both events. There are no elements common to both sets.

Two events that cannot both occur at the same time are called *mutually exclusive events*. The events E_3 and E_4 are mutually exclusive events. E_1 and E_2 are not mutually exclusive events.

Mutually Exclusive Events

Two events A and B are mutually exclusive if they cannot occur at the same time. That is, A and B are mutually exclusive when $A \cap B = \varnothing$.

Question: A die is rolled once. Let E be the event that an even number of pips is on the upward face, and let O be the event that an odd number of pips is on the upward face. Are the events E and O mutually exclusive?[2]

An important probability formula is used to calculate the probabilities of mutually exclusive events.

Addition Formula for the Probability of Mutually Exclusive Events

If A and B are two mutually exclusive events, then

$$P(A \text{ or } B) = P(A) + P(B)$$

2. Yes. It is not possible to roll an even number and an odd number on a single roll of the die.

Example 2 Suppose a single card is drawn from a regular deck of playing cards. Find the probability of drawing a five or a king.

Solution Let $A = \{♠5, ♥5, ♦5, ♣5\}$ and $B = \{♠K, ♥K, ♦K, ♣K\}$. There are 52 cards in a regular deck of playing cards; thus $n(S) = 52$. Because the events are mutually exclusive, we can use the Addition Formula for the Probability of Mutually Exclusive Events.

$$P(A \text{ or } B) = P(A) + P(B)$$

• Addition Formula for the Probability of Mutually Exclusive events

$$= \frac{1}{13} + \frac{1}{13} = \frac{2}{13}$$

• $P(A) = \frac{4}{52} = \frac{1}{13}, P(B) = \frac{4}{52} = \frac{1}{13}$

The probability of drawing a five or a king is $\frac{2}{13}$.

You-Try-It 2 Two fair dice are tossed once. What is the probability of rolling a 7 or an 11? For the sample space for this experiment, see page 70.

Solution See page S40.

Consider the experiment of rolling two dice. Let A be the event rolling a sum of 8, and let B be the event rolling a double (the same number on both dice).

A = {⚁⚅ ⚅⚁ ⚂⚄ ⚄⚂ ⚃⚃}

B = {⚀⚀ ⚁⚁ ⚂⚂ ⚃⚃ ⚄⚄ ⚅⚅}

These events are *not* mutually exclusive because it is possible to satisfy the conditions of each event on one toss of the dice—a ⚃⚃ could be rolled. Therefore, $P(A \text{ or } B)$, the probability of a sum of 8 or a double, cannot be calculated using the Addition Formula given on the preceding page. However, a modification of that formula can be used.

Addition Formula for the Probability of Any Two Events

Let A and B be two events. Then

$$P(A \text{ or } B) = P(A) + P(B) - P(A \text{ and } B)$$

Using this formula, with

A = {⚁⚅ ⚅⚁ ⚂⚄ ⚄⚂ ⚃⚃}

B = {⚀⚀ ⚁⚁ ⚂⚂ ⚃⚃ ⚄⚄ ⚅⚅}

A and B = {⚃⚃}

the probability of A or B can be calculated.

$$P(A \text{ or } B) = P(A) + P(B) - P(A \text{ and } B)$$

$$= \frac{5}{36} + \frac{6}{36} - \frac{1}{36} = \frac{10}{36} = \frac{5}{18}$$

On a single roll of two dice, the probability of a sum of 8 or a double is $\frac{5}{18}$.

	F	*No F*	*Total*
V	21	97	118
No V	48	128	176
Total	69	225	294

V: Vaccinated
F: Contracted the flu

Example 3 The table at the left shows data from an experiment to test the effectiveness of a flu vaccine. If one person is selected from this population, what is the probability that the person was vaccinated or contracted the flu?

Solution Let $V = \{\text{people vaccinated}\}$ and $F = \{\text{people who contracted the flu}\}$. These events are not mutually exclusive because there are 21 people who were vaccinated and who contracted the flu. The sample space S consists of the 294 people who participated in the experiment. From the table, $n(V) = 118$, $n(F) = 69$, and $n(V \text{ and } F) = 21$.

$$\begin{aligned} P(V \text{ or } F) &= P(V) + P(F) - P(V \text{ and } F) \\ &= \frac{118}{294} + \frac{69}{294} - \frac{21}{294} \\ &= \frac{166}{294} \approx 0.56 \end{aligned}$$

The probability of selecting a person who was vaccinated or who contracted the flu is approximately 0.56.

SUGGESTED ACTIVITY
A new family moved next door to the Nelsons. The Nelsons' son Jim became friends with their oldest child, William. If the new neighbors have four children, what is the probability that they have exactly one other boy?

[Answer: $\frac{3}{8}$]

You-Try-It 3 The data in the table below shows the starting salaries for college graduates for selected degrees. If one person is chosen from this population, what is the probability that that person has a degree in business or has a starting salary between $20,000 and $24,999?

	Engineering	Business	Chemistry	Psychology
Less than 20,000	0	4	1	12
20,000–24,999	4	16	3	16
25,000–29,999	7	21	5	15
30,000–34,999	12	35	5	7
35,000 or more	12	22	4	5

Solution See page S40.

Conditional Probability

Suppose a die is rolled once and you are presented the following two situations:

1. What is the probability of rolling a ⚁?
2. What is the probability of rolling a ⚁ if you are also told that the upward face contains an even number of pips?

In the first case, if we let $B = \{⚁\}$ and $S = \{⚀\ ⚁\ ⚂\ ⚃\ ⚄\ ⚅\}$, then $P(B) = \frac{1}{6}$. In the second case, however, we have some additional information. We are told that the result of the rolled die was an even number of pips on the upward face. Thus, the only possibilities are $\{⚁\ ⚃\ ⚅\}$. The probability that there are two pips on the upward face is therefore $\frac{1}{3}$.

The probability of an event B based on knowing that some other event A has occurred is called a *conditional* probability.

Conditional Probability Formula

Let A and B be two events in a sample space S. Then the conditional probability of B given that A has occurred is given by

$$P(B|A) = \frac{P(A \text{ and } B)}{P(A)}$$

The symbol $P(B|A)$ is read "the probability of B given A."

Returning to situation 2 on the preceding page, let $B = \{⚁\}$ (a two is rolled), and $A = \{⚁\ ⚃\ ⚅\}$ (an even number is rolled). Then A and $B = \{⚁\}$. Using the Conditional Probability Formula, the probability of B given A is

$$P(B|A) = \frac{P(A \text{ and } B)}{P(A)} = \frac{\frac{1}{6}}{\frac{3}{6}} = \frac{1}{3}$$

Example 4 The data in the table below shows the results of a survey to determine the number of adults who have had financial help from their parents for certain purchases.

Age	College Tuition	Buy a Car	Buy a House	Total
18–28	405	253	261	919
29–39	389	219	392	1000
40–49	291	146	245	682
50–59	150	71	112	333
60+	62	15	98	175
Total	1297	704	1108	3109

If one person is selected from this survey, what is the probability that the person received financial help for purchasing a home given that the person is between the ages of 29 and 39?

Solution Let B = {adults receiving financial help for home purchase} and A = {adults between 29 and 39}. From the table, $n(A \text{ and } B) = 392$, $n(A) = 1000$, and $n(S) = 3109$. Then

$$P(A \text{ and } B) = \frac{n(A \text{ and } B)}{n(S)} = \frac{392}{3109} \quad \text{and} \quad P(A) = \frac{n(A)}{n(S)} = \frac{1000}{3109}$$

Using the Conditional Probability Formula, we have

$$P(B|A) = \frac{P(A \text{ and } B)}{P(A)} = \frac{\frac{392}{3109}}{\frac{1000}{3109}} = \frac{392}{1000} = 0.392$$

The conditional probability of B given A has occurred is 0.392.

You-Try-It 4 A pair of dice is tossed once. What is the probability that the result is a sum of six given that the result is not a sum of seven?

Solution See page S40.

Probability of Events Occurring in Succession

Consider a situation in which a box contains five balls, three white and two black. If one and then a second ball are randomly selected from this box without replacement,[3] then the sample space of this experiment is {BB, BW, WB, WW}.

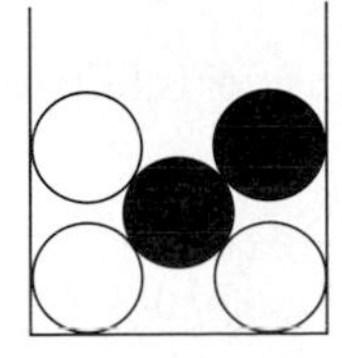

This diagram shows the results of choosing a black and then a white ball, BW.

3. Without replacement means that one ball is drawn and then the second ball is drawn *without* replacing the first ball. With replacement would mean that the first ball is returned to the box before a second ball is drawn.

TAKE NOTE

The idea here is that $\frac{2}{5}$ of the time a black ball is drawn as the first ball, and in $\frac{3}{4}$ of the $\frac{2}{5}$ cases, a white ball is drawn as the second ball. Translating *of* in the phrase $\frac{3}{4}$ ***of*** $\frac{2}{5}$ into an expression gives $\frac{3}{4} \cdot \frac{2}{5} = \frac{3}{10}$. That is, the probabilities are multiplied.

Consider the event {BW}. Originally, there are 5 balls from which to choose. The probability of choosing a black ball on the first draw is $\frac{2}{5}$. On the second draw, there are only 4 balls remaining, of which 3 are white. Therefore, the probability of a white ball on the second draw is $\frac{3}{4}$. Thus a black ball is drawn $\frac{2}{5}$ of the time on the first draw, and then in $\frac{3}{4}$ of those cases, a white ball is drawn on the second draw. That is, $\frac{3}{10}$ of the time you draw a black and then a white ball.

Multiplication Rule for Probabilities

The probability of two or more events occurring in succession is the product of the probabilities of each of the events.

Example 5 Two cards are randomly drawn in succession, without replacement, from a regular deck of playing cards. What is the probability that the two cards are aces?

SUGGESTED ACTIVITY

The letters A, L, G, E, B, R, A are written one to a card, and the cards are placed in a box. If the cards are drawn one at a time, without replacement, what is the probability that they are drawn in the order ALGEBRA?

[Answer: $\frac{1}{2520}$]

Solution In a regular deck of playing cards, there are 52 cards, of which 4 are aces. The probability of an ace on the first draw is therefore $\frac{4}{52} = \frac{1}{13}$. On the second draw, there are 51 cards remaining in the deck (this is where we use the *without replacement* condition), of which 3 are aces. (An ace was drawn on the first draw.) Thus the probability of an ace on the second draw is $\frac{3}{51} = \frac{1}{17}$. By the Multiplication Rule for Probabilities, the probability of two aces in a row is $\frac{1}{13} \cdot \frac{1}{17} = \frac{1}{221}$.

You-Try-It 5 A fair die is rolled and then a fair coin is tossed. What is the probability of rolling a ⚃ and then tossing a tail?

Solution See page S40.

In Example 5 we stressed that the drawing of the cards was *without replacement*. If the experiment is *with replacement*, the probability of two aces in succession changes.

First draw: There are 52 cards, 4 of which are aces. Therefore, $P(\text{Ace}) = \frac{4}{52} = \frac{1}{13}$.

Second draw: There are 52 cards, 4 of which are aces. Therefore, $P(\text{Ace}) = \frac{1}{13}$.

The probability of drawing two aces in succession with replacement is $\frac{1}{13} \cdot \frac{1}{13} = \frac{1}{169}$. This answer is different from the one calculated in Example 5.

For experiments that are conducted *with* replacement, a selected item is replaced before the next item is selected. This means that the total number of items from which to choose is the same for each selection. For experiments that are conducted *without* replacement, a selected item is *not* replaced before the next item is selected. This means that the total number of items from which to choose decreases by one each time a selection is made.

8.7 EXERCISES

Topics for Discussion

1. What are compound events?
Events characterized by using the words *or* and *and.*

2. What are mutually exclusive events?
Events that cannot occur at the same time.

3. How are probabilities of events combined with the words *and* and *or* related to operations with sets?
Answers will vary.

4. What is a conditional probability?
A probability calculated knowing that some other event has occurred.

Probability of Compound Events

Solve.

5. Suppose a number is chosen from the positive integers that are less than or equal to 20.
a. If *A* is the event that the number is divisible by 3 and *B* is the event that the number is divisible by 4, are *A* and *B* mutually exclusive events?
b. If *A* is the event that the number is less than 9 and *B* is the event that the number is greater than 12, are *A* and *B* mutually exclusive events?
a. No; **b.** yes

6. Suppose a family is selected at random from all families in the United States.
a. If *A* is the event that the family's annual income is less than $30,000 and *B* is the event that the family consists of 4 people, are the events *A* and *B* mutually exclusive?
b. If *A* is the event that one person in the family has red hair and *B* is the event that one person in the family has blue eyes, are the events *A* and *B* mutually exclusive?
a. No; **b.** no

7. One card is drawn from a regular deck of playing cards. Find the probability of each event.
a. The card is a heart or a diamond.
b. The card is a club or a nine.
a. $\frac{1}{2}$; **b.** $\frac{4}{13}$

8. A dodecahedral die is tossed once. Find the probability of each event.
a. The number is even or divisible by 5.
b. The number is greater than 4 and less than 9.
a. $\frac{7}{12}$; **b.** $\frac{1}{3}$

9. Two dice are rolled once. Find the probability of each event.
a. The sum of the pips on the upward faces is 7 or 11.
b. The sum of the pips on the upward faces is 2, 3, or 12.
a. $\frac{2}{9}$; **b.** $\frac{1}{9}$

10. A box contains 2000 envelopes, of which 50 have a $100 bill, 75 have a $50 bill, 100 have a $20 bill, and 125 have a $10 bill. If you select one envelope from this box, what is the probability that the envelope contains more than $20?
$\frac{1}{16}$

11. If a coin is flipped 5 times, is it more likely that the result will be HHHHH or HTHHT?
The probabilities are the same.

12. Six children, three boys and three girls, are randomly seated in six chairs that are placed in a row. Find the probability of each event.
a. The boys and girls are in alternating seats, beginning with a girl.
b. The boys and girls are in alternating seats, beginning with a boy.
c. The boys and girls are in alternating seats.
a. $\frac{1}{20}$ **b.** $\frac{1}{20}$ **c.** $\frac{1}{10}$

13. Two cards are drawn, without replacement, from a regular deck of playing cards. Find the probability of each event. Write the answer as a decimal rounded to the nearest thousandth.
a. The first card is an ace and the second card is a 10, jack, queen, or king.
b. The first card is a 10, jack, queen, or king and the second card is an ace.
c. An ace is on one card and a 10, jack, queen, or king is on the other card.
a. 0.024; **b.** 0.024; **c.** 0.048

14. A fair coin is tossed three times. What is the probability of a head on either the first toss, second toss, or third toss?
$\frac{7}{8}$

15. A missile radar detection system consists of two radar screens. The probability that any one of the radar screens will detect an incoming missile is 0.98. If radar detections are assumed to be independent events, what is the probability that a missile that enters the detection space of the radar will be detected by both screens? Write the answer as a decimal.
0.9604

16. An oil drilling venture involves drilling four wells in different parts of the country. For each well, the probability that it will be profitable is 0.25, and the probability that it will be unprofitable is 0.75. If these events are independent, what is the probability of one or more unprofitable wells? Write the answer as a decimal rounded to the nearest one-thousandth.
0.996

17. One pair of dice is rolled once. What is the probability that the sum of the pips on the upward faces is five given that it is not seven?
$\frac{2}{15}$

18. One card is drawn at random from a regular deck of playing cards. What is the probability that the card is a diamond given that it is a red card?

$\frac{1}{2}$

19. The table below gives the ages and political party affiliation of a random sample of 1000 students at a college.

Ages	Republican	Democrat	Independent	TOTAL
18 to 23	83	140	71	294
24 to 29	78	125	40	243
30 to 35	113	128	37	278
36 or greater	71	90	24	185
TOTAL	345	483	172	

Suppose one student is selected at random from this sample.

a. What is the probability that the student is a Republican?
b. What is the probability that the student's age is between 24 and 29?
c. What is the probability that the student is a Republican given that the student's age is between 24 and 29? Round to the nearest thousandth.
d. What is the probability that the student is between 30 and 35 given that the student is an independent? Round to the nearest thousandth.
e. What is the probability that the student is a Republican and the student's age is between 24 and 29?

a. 0.345; **b.** 0.243; **c.** 0.321; **d.** 0.215; **e.** 0.078

20. The table below shows the results of a survey that asked people if they owned stock in a company listed on the New York Stock Exchange.

Annual Income	Yes	No
Less than \$20,000	78	257
\$20,000 to \$49,999	95	203
\$50,000 to \$99,999	245	98
\$100,000 or greater	78	25
TOTAL	496	583

Suppose one person is selected at random from this sample. Calculate the following probabilities. Round to the nearest thousandth.

a. The person has an income that is \$50,000 or more.
b. The person does not own a stock listed on the New York Stock Exchange.
c. The person has an income between \$20,000 and \$49,000 given that the person does not own a stock listed on the New York Stock Exchange.
d. The person owns a stock listed on the New York Stock Exchange given that the person has an income of \$100,000 or greater.

a. 0.413; **b.** 0.540; **c.** 0.348; **d.** 0.757

Applying Concepts

21. A fair coin is tossed until a head shows on the upward face. What is the probability that four tosses are required before the first head occurs?

$\frac{1}{16}$

22. If events *A* and *B* are mutually exclusive, what is the probability of *A* given *B*?

0

23. Show by example that the probability of A given B is not always the same as the probability of B given A. (*Suggestion:* Try using a deck of cards and define two events that are not mutually exclusive.)

Answers will vary. For instance, choose one card from a deck of playing cards. Let A be the event that the card is an ace and B be the event that the card is a heart. Then $P(A|B) = \frac{1}{13}$ and $P(B|A) = \frac{1}{4}$.

Exploration

24. ***Expectation*** *Expectation*, E, is a number used to determine the fairness of a gambling game. It is defined as the probability, p, of winning a bet times the total amount, A, that can be won. For instance, suppose you and a friend bet on being able to guess the outcome of the flip of a coin. Each of you bet \$1, so the total amount to win is \$2 (your bet and your friend's bet). The probability of guessing the outcome of a toss of a coin is $\frac{1}{2}$. Thus the expectation is

$$E = pA = \frac{1}{2}(2) = 1$$

The expectation is \$1, the amount of your bet. Therefore, the game is fair because the amount of your bet equals your expectation.

When a game is unfair, it benefits one of the players. For instance, suppose you bet \$1 and a friend bets \$2 on the flip of a coin. Your expectation is

$$E = pA = \frac{1}{2}(3) = 1.50$$

Your expectation is \$1.50. This is \$.50 more than your bet, so the game is an advantage to you. Your friend's expectation is also \$1.50. However, this is \$.50 less than the bet of your friend. This is a disadvantage to your friend. The amount of advantage or disadvantage affects how much you win or lose after playing the game for a long time.

$$\text{Amount won} = \text{Advantage} \cdot \text{Number of games played}$$
$$\text{Amount lost} = \text{Disadvantage} \cdot \text{Number of games played}$$

This means that if your advantage is \$.50 and you play the game 200 times, the amount you will win is approximately \$100 $(0.50 \cdot 200)$.

a. Your expectation is less than your bet for all gambling games in a casino and all state lotteries. Explain why this is true.

b. In the California lottery, a person can bet \$1 and choose 6 numbers from 51. The amounts you can win and the probabilities for these are shown in the table below. Calculate your expectation for winning each of the prizes.

	Probability	Amt. won (\$)	Expectation
3 matching numbers	0.015758385	5	
4 matching numbers	0.000824567	80	
5 matching numbers	0.000014992	2000	
6 matching numbers	0.000000056	4,000,000	

c. The *total expectation* for a game is the sum of the expectations of winning each prize. What is the total expectation for the lottery game in part b?

d. If the amounts available to win for matching 3, 4, and 5 numbers remain the same, how large must the prize for matching all 6 numbers be if your expectation equals your bet?

Chapter Summary

Definitions

Fractions in which the numerator and denominator are polynomials are *rational expressions.* A rational expression is in simplest form when the numerator and denominator have no common factors other than 1.

The *reciprocal* of a fraction is the fraction with the numerator and denominator interchanged.

The *least common multiple* (LCM) of two or more polynomials is the simplest polynomial that contains the prime factorization of each polynomial.

A *complex fraction* is a fraction whose numerator or denominator contains one or more fractions.

A *literal equation* is an equation that contains more than one variable.

$$\frac{\text{first term}}{\text{second term}} = \frac{\text{third term}}{\text{fourth term}}$$

A *rate* is the quotient of two quantities that have different units. A *ratio* is the quotient of two quantities that have the same unit. A *proportion* is the equality of two rates or ratios. Each of the four members in a proportion is called a *term*. The second and third terms of a proportion are called the *means*, and the first and fourth terms are called the *extremes*. In a proportion, the product of the means equals the product of the extremes.

Similar objects have the same shape but not necessarily the same size. For similar triangles, the ratios of corresponding sides are equal. The ratio of corresponding heights is equal to the ratio of corresponding sides. The corresponding angles in similar triangles are equal.

In the field of probability, *compound* events are events characterized by using the words *or* and *and*.

Procedures

Simplifying Rational Expressions

Factor the numerator and denominator. Divide by the common factors. For example:

$$\frac{x^2 - 3x - 10}{x^2 - 25} = \frac{(x+2)\cancel{(x-5)}}{(x+5)\cancel{(x-5)}} = \frac{x+2}{x+5}$$

Multiplying Rational Expressions

$$\frac{a}{b} \cdot \frac{c}{d} = \frac{ac}{bd}$$

Multiply the numerators. Multiply the denominators. Write the answer in simplest form. For example:

$$\frac{x^2 - 3x}{x^2 + x} \cdot \frac{x^2 + 5x + 4}{x^2 - 4x + 3} = \frac{x(x-3)}{x(x+1)} \cdot \frac{(x+4)(x+1)}{(x-3)(x-1)}$$

$$= \frac{\cancel{x}\cancel{(x-3)}(x+4)\cancel{(x+1)}}{\cancel{x}\cancel{(x+1)}\cancel{(x-3)}(x-1)} = \frac{x+4}{x-1}$$

Dividing Rational Expressions

$$\frac{a}{b} \div \frac{c}{d} = \frac{a}{b} \cdot \frac{d}{c} = \frac{ad}{bc}$$

To divide two fractions, multiply the dividend by the reciprocal of the divisor. For example:

$$\frac{3x^2}{8a^4} \div \frac{6x^4}{4a^7} = \frac{3x^2}{8a^4} \cdot \frac{4a^7}{6x^4} = \frac{3x^2 \cdot 4a^7}{8a^4 \cdot 6x^4} = \frac{a^3}{4x^2}$$

Simplifying Complex Fractions

Multiply the numerator and denominator of the complex fraction by the LCM of the denominators of the fractions in the numerator and the denominator. For example:

$$\frac{\frac{1}{x} + \frac{1}{y}}{\frac{1}{x} - \frac{1}{y}} = \frac{\frac{1}{x} + \frac{1}{y}}{\frac{1}{x} - \frac{1}{y}} \cdot \frac{xy}{xy} = \frac{\frac{1}{x} \cdot xy + \frac{1}{y} \cdot xy}{\frac{1}{x} \cdot xy - \frac{1}{y} \cdot xy} = \frac{y+x}{y-x}$$

Adding and Subtracting Rational Expressions

1. Find the LCM of the denominators.
2. Write each fraction with the LCM as the denominator.
3. Add or subtract the numerators. The denominator of the sum or difference is the common denominator.
4. Write the answer in simplest form.

For example:

$$\frac{x}{2y} - \frac{3x}{4y} + \frac{5x}{6y} = \frac{x}{2y}\cdot\frac{6}{6} - \frac{3x}{4y}\cdot\frac{3}{3} + \frac{5x}{6y}\cdot\frac{2}{2}$$

$$= \frac{6x}{12y} - \frac{9x}{12y} + \frac{10x}{12y} = \frac{6x - 9x + 10x}{12y} = \frac{7x}{12y}$$

Solving Equations Containing Fractions

Clear denominators by multiplying each side of the equation by the LCM of the denominators. Then solve for the variable. For example:

$$\frac{1}{2a} = \frac{2}{a} - \frac{3}{8}$$

$$8a\left(\frac{1}{2a}\right) = 8a\left(\frac{2}{a} - \frac{3}{8}\right)$$

$$4 = 16 - 3a$$

$$-12 = -3a$$

$$4 = a$$

Work Equation

Rate of work · Time worked = Part of task completed

Uniform Motion Problems

$$\frac{\text{Distance}}{\text{Rate}} = \text{Time}$$

Solving a Literal Equation

Rewrite the equation so that the letter being solved for is alone on one side of the equation and all numbers and other letters are on the other side.

Probabilities Involving *Or* or *And*

If A and B are two events in a sample space S, then $P(A \text{ or } B) = \dfrac{n(A \cup B)}{n(S)}$ and $P(A \text{ and } B) = \dfrac{n(A \cap B)}{n(S)}$.

Probability of Mutually Exclusive Events

Two events A and B are mutually exclusive if they cannot occur at the same time. That is, A and B are mutually exclusive when $A \cap B = \varnothing$.

Addition Formula for the Probability of Mutually Exclusive Events

If A and B are two mutually exclusive events, then $P(A \text{ or } B) = P(A) + P(B)$.

Addition Formula for the Probability of Any Two Events

Let A and B be two events. Then $P(A \text{ or } B) = P(A) + P(B) - P(A \text{ and } B)$.

Conditional Probability Formula

Let A and B be two events in a sample space S. Then the conditional probability of B given that A has occurred is given by $P(B|A) = \dfrac{P(A \text{ and } B)}{P(A)}$.

Multiplication Rule for Probabilities

The probability of two or more events occurring in succession is the product of the probabilities of each of the events.

Chapter Review Exercises

1. Simplify: $\frac{2x^2-13x-45}{2x^2-x-15}$

$\frac{x-9}{x-3}$

2. Multiply: $\frac{3x^2+4x-15}{x^2-11x+28} \cdot \frac{x^2-5x-14}{3x^2+x-10}$

$\frac{x+3}{x-4}$

3. Divide: $\frac{x^2-5x-14}{x^2-3x-10} \div \frac{x^2-4x-21}{x^2-9x+20}$

$\frac{x-4}{x+3}$

4. Add: $\frac{3}{4ab} + \frac{5}{4ab}$

$\frac{2}{ab}$

5. Subtract: $\frac{5x+3}{2x^2+5x-3} - \frac{3x+4}{2x^2+5x-3}$

$\frac{1}{x+3}$

6. Simplify: $\frac{5x-1}{x^2-9} + \frac{4x-3}{x^2-9} - \frac{8x-1}{x^2-9}$

$\frac{1}{x+3}$

7. Add: $\frac{6}{a} + \frac{9}{b}$

$\frac{6b+9a}{ab}$

8. Add: $\frac{x+7}{15x} + \frac{x-2}{20x}$

$\frac{7x+22}{60x}$

9. Subtract: $\frac{2x+9}{3x} - \frac{5}{2x}$

$\frac{4x+3}{6x}$

10. Simplify: $\frac{1+\frac{1}{x}-\frac{12}{x^2}}{1+\frac{2}{x}-\frac{8}{x^2}}$

$\frac{x-3}{x-2}$

11. Solve: $\frac{x+8}{x+4} = 1 + \frac{5}{x+4}$

No solution

12. Solve: $\frac{20}{2x+3} = \frac{17x}{2x+3} - 5$

5

13. Solve $i = \frac{100m}{c}$ for c.

$c = \frac{100m}{i}$

14. Solve $f = v + at$ for t.

$t = \frac{f-v}{a}$

15. Solve: $\frac{x}{5} = \frac{x+12}{9}$

15

16. Solve: $\frac{3}{x+4} = \frac{5}{x+6}$

−1

17. For what value of the variable is the expression undefined?

a. $\frac{x}{4x+12}$ **b.** $\frac{x+1}{x^2+5x-6}$

a. −3; **b.** −6, 1

18. Find the slope of the line that passes through the points $(-\frac{3}{2a}, \frac{5}{2a})$ and $(\frac{1}{a}, -\frac{5}{a})$.

−3

19. Given MP and NQ intersect at O, NQ measures 25 centimeters, MO measures 6 centimeters, and PO measures 9 centimeters, find the length of QO.
15 cm

20. Given $AE \parallel BD$, AB measures 5 feet, ED measures 8 feet, and BC measures 3 feet, find the length of CE.
12.8 ft

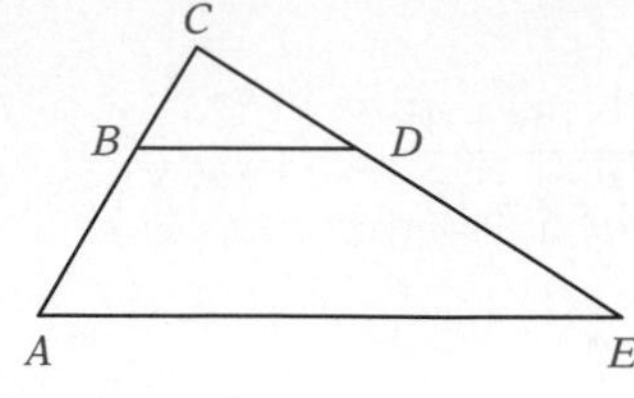

21. Two brick masons are constructing a walkway. The first mason could complete the entire job in x hours. The second mason would require $2x$ hours to complete the job. What fraction of the job has been completed after both masons have been working on it for 3 hours?

$\frac{9}{2x}$ of the job

22. A brick mason can construct a patio in 3 hours. If the mason works with an apprentice, they can construct the patio in 2 hours. How long would it take the apprentice, working alone, to construct the patio?
6 h

23. The rate of a jet is 400 mph in calm air. Flying with the wind, the jet can fly 2100 miles in the same amount of time as it takes to fly 1900 miles against the wind. Find the rate of the wind.
20 mph

24. A car travels 315 miles in the same amount of time as it takes a bus to travel 245 miles. The rate of the car is 10 mph faster than that of the bus. Find the rate of the car.
45 mph

25. A weight of 21 pounds stretches a spring 14 inches. At the same rate, how far would a weight of 12 pounds stretch the spring?
8 in.

26. The table at the right shows the states with the highest rates of tuberculosis per 100,000 people in 1995. Also shown is the population of each state in 1995. Which state had more cases of TB in 1995, Alaska or New York? How many more?

State	TB Rate per 100,000	Population
Hawaii	16.9	1,187,000
Alaska	15.8	604,000
New York	14.2	18,136,000

Sources: Centers for Disease Control and Prevention; U.S. Bureau of the Census

New York; 2480 more cases

27. A dodecahedral die is tossed once. Find the probability that **a.** the number is odd or divisible by 4, and **b.** the number is greater than 3 and less than 8.

a. $\frac{3}{4}$; **b.** $\frac{1}{3}$

28. One pair of dice is rolled once. What is the probability that the sum of the pips on the upward faces is eleven given that it is not nine?

$\frac{1}{16}$

29. Two dice are rolled and then a coin is tossed. What is the probability of rolling a three and then tossing a head?

$\frac{1}{36}$

Cumulative Review Exercises

1. A charity raffle sells 1500 raffle tickets for a big-screen television set. If you purchase 5 tickets, what is the probability that you will win the television?

$\frac{1}{300}$

2. Translate into a variable expression: The difference between 3 times t and the sum of t and 6.

$3t - (t + 6)$

3. Evaluate: $24 + 16 \div 2^2$

28

4. Subtract: $-14 - (-23) - 8$

1

5. What number is 50 more than -66?

-16

6. Simplify: $-9a^2b + 6a^2b$

$-3a^2b$

7. Solve: $-12 = -\frac{3}{4}x$

16

8. Solve: $18 - 3x = 2(3 - 2x)$

-12

9. Find the x- and y-intercepts of $3x - 2y = 24$.

$(8, 0)$ and $(0, -12)$

10. Solve the system of equations:

$$6x - y = 1$$
$$y = 3x + 1$$

$(\frac{2}{3}, 3)$

11. Multiply: $(3xy^4)(-2x^3y)$

$-6x^4y^5$

12. Simplify: $\frac{a^2b^{-5}}{a^{-1}b^{-3}}$

$\frac{a^3}{b^2}$

13. Factor: $y^2 - 7y + 6$

$(y - 6)(y - 1)$

14. Factor: $12x^2 - x - 1$

$(4x + 1)(3x - 1)$

15. Write 8.921×10^{-6} in decimal notation.

0.000008921

16. Simplify: $4\sqrt{250}$

$20\sqrt{10}$

17. Simplify: $\sqrt{121x^8y^2}$

$11x^4y$

18. Evaluate $4 + 3\sqrt{ab}$ when $a = 2$ and $b = 8$.

16

19. Multiply: $\sqrt{3}(\sqrt{6} - \sqrt{x})$

$3\sqrt{2} - \sqrt{3x}$

20. Add: $2\sqrt{27a} + 8\sqrt{48a}$

$38\sqrt{3a}$

21. Solve: $\sqrt{2x-3} - 5 = 0$

14

22. Solve: $6x(x+1) = x - 1$

$-\frac{1}{2}, -\frac{1}{3}$

23. Solve: $x^2 = 4x - 1$

$2 \pm \sqrt{3}$

24. Divide: $\dfrac{x^2 - 3x - 10}{x^2 - 4x - 12} \div \dfrac{x^2 - x - 20}{x^2 - 2x - 24}$

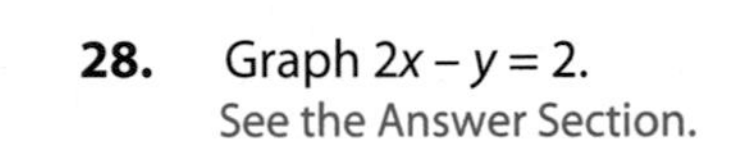

1

25. Subtract: $\dfrac{3x}{x^2 + 5x - 24} - \dfrac{9}{x^2 + 5x - 24}$

$\dfrac{3}{x+8}$

26. Solve: $\dfrac{x}{x+6} = \dfrac{3}{x}$

$-3, 6$

27. Graph the line that has slope $-\frac{2}{3}$ and y-intercept (0, 4).

See the Answer Section.

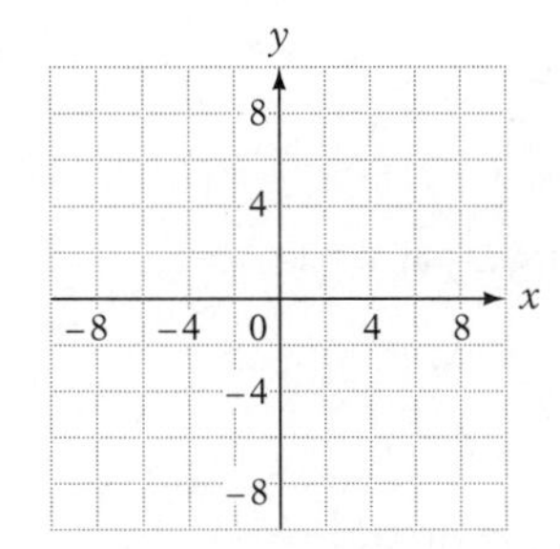

28. Graph $2x - y = 2$.

See the Answer Section.

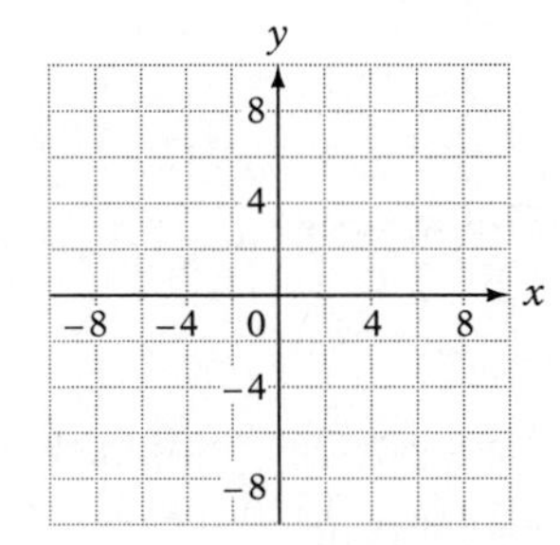

29. The cholesterol levels of 14 adults are recorded in the table below. Draw a box-and-whiskers plot of the data.

Cholesterol Levels for 14 Adults

345	195	257	221	181	246	278
290	216	173	239	204	192	215

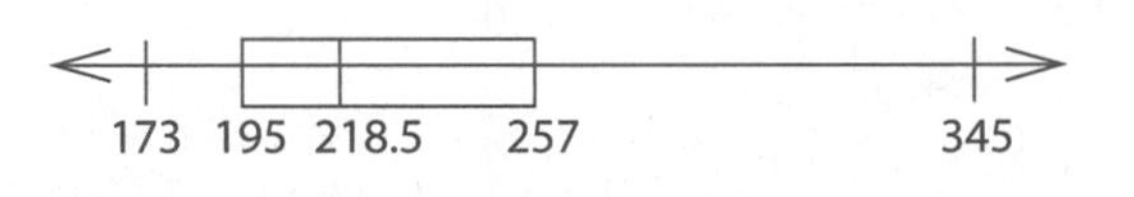

30. Use a graphing calculator to draw a graph of $y = -x^3 + 4x - 2$. Find the value of x, to the nearest tenth, for which the output is 4.

−2.5

31. How many pounds of cashews that cost \$3.50 per pound must be combined with 50 pounds of peanuts that cost \$1.75 per pound to make a mixture that costs \$2.25 per pound?

20 lb

32. The length of a side of a square is $(3x + 1)$ meters. Find the area of the square in terms of the variable x.

$(9x^2 + 6x + 1)\text{ m}^2$

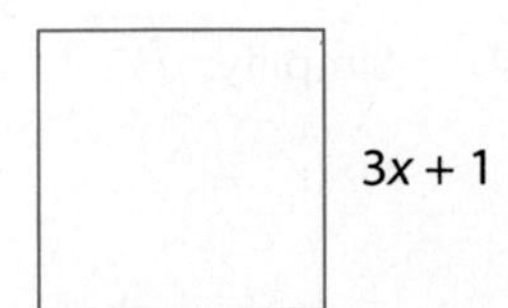

33. Solve right triangle ABC given $a = 11$ meters and $b = 8$ meters. Round to the nearest whole number.

$\angle A = 54^\circ, \angle B = 36^\circ, c = 14\text{ m}$

Appendix: Guidelines for Using Graphing Calculators

TEXAS INSTRUMENTS TI-83

To evaluate an expression:

a. Press the [Y=] key. A menu showing \Y1 = through \Y7 = will be displayed vertically with a blinking cursor to the right of \Y1 =. Press [CLEAR], if necessary, to delete an unwanted expression.

b. Input the expression to be evaluated. For example, to input the expression $-3a^2b - 4c$, use the following keystrokes:

[(−)] 3 [ALPHA] A [∧] 2 [ALPHA] B [−] 4 [ALPHA] C [2nd] QUIT

Note the difference between the keys for a *negative* sign [(−)] and a *minus* sign [−].

c. Store the value of each variable that will be used in the expression. For example, to evaluate the expression above when $a = 3$, $b = -2$, and $c = -4$, use the following keystrokes:

3 [STO▷] [ALPHA] A [ENTER] [(−)] 2 [STO▷] [ALPHA] B [ENTER] [(−)] 4 [STO▷] [ALPHA] C [ENTER]

These steps store the value of each variable.

d. Press [VARS] [▷] [1] [1] [ENTER]. The value for the expression, Y1, for the given values is displayed; in this case, Y1 = 70.

To graph a function:

a. Press the [Y=] key. A menu showing \Y1 = through \Y7 = will be displayed vertically with a blinking cursor to the right of \Y1 =. Press [CLEAR], if necessary, to delete an unwanted expression.

b. Input the expression for each function that is to be graphed. Press [X,T,θ,n] to input x. For example, to input $y = x^3 + 2x^2 - 5x - 6$, use the following keystrokes:

[X,T,θ,n] [∧] 3 [+] 2 [X,T,θ,n] [∧] 2 [−] 5 [X,T,θ,n] [−] 6

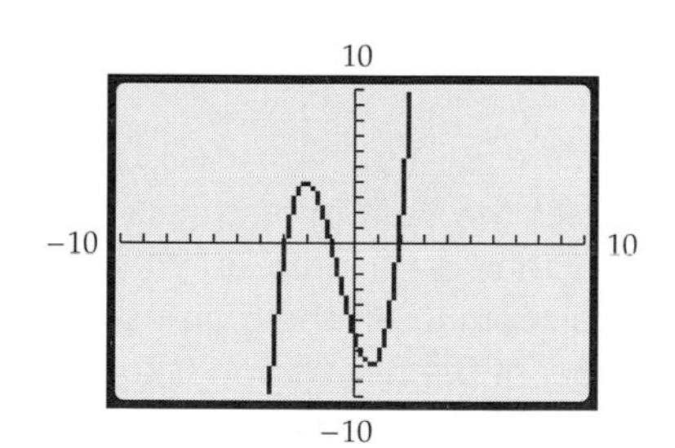

c. Set the domain and range by pressing [WINDOW]. Enter the values for the minimum x-value (Xmin), the maximum x-value (Xmax), the distance between tick marks on the x-axis (Xscl), the minimum y-value (Ymin), the maximum y-value (Ymax), and the distance between tick marks on the y-axis (Yscl). Now press [GRAPH]. For the graph shown at the left, Xmin = −10, Xmax = 10, Xscl = 1, Ymin = −10, Ymax = 10, and Yscl = 1. This is called the standard viewing rectangle. Pressing [ZOOM] [6] is a quick way to set the calculator to the standard viewing rectangle. *Note:* This will also immediately graph the function in that window.

d. Press the [Y=] key. The equal sign has a black rectangle around it. This indicates that the function is active and will be graphed when the [GRAPH] key is pressed. A function is deactivated by using the arrow keys. Move the cursor over the equal sign and press [ENTER]. When the cursor is moved to the right, the black rectangle will not be present and that equation will not be active.

e. Graphing some radical equations requires special care. To graph the function $y = \sqrt{2x + 3}$, enter the following keystrokes:

[Y=] [2nd] √‾ 2 [X,T,θ,n] [+] 3 [)]

The graph is shown below.

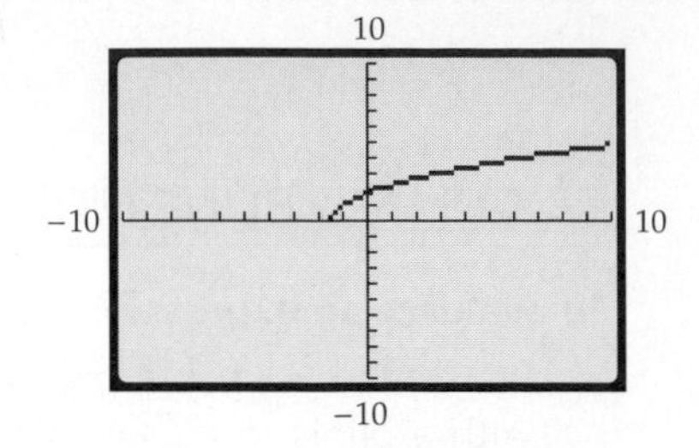

To display the *x*-coordinates of rectangular coordinates as integers:

a. Set the viewing window as follows: Xmin = −47, Xmax = 47, Xscl = 10, Ymin = −31, Ymax = 31, Yscl = 10. You can also press [ZOOM] 8 [ENTER].

b. Graph the function. Press [TRACE] and then move the cursor with the [◁] and [▷] keys. The values of x and $y = f(x)$ displayed on the bottom of the screen are the coordinates of a point on the graph.

To display the *x*-coordinates of rectangular coordinates in tenths:

a. Set the viewing window as follows: [ZOOM] [4]

b. Graph the function. Press [TRACE] and then move the cursor with the [◁] and [▷] keys. The values of x and $y = f(x)$ displayed on the bottom of the screen are the coordinates of a point on the graph.

To evaluate a function for a given value of *x*, or to produce ordered pairs of a function:

a. Input the equation; for example, input $y = 2x^3 - 3x + 2$.

b. Press [2nd] QUIT.

c. Input a value for x; for example, to input 3 press 3 [STO▷] [X,T,θ,*n*] [ENTER].

d. Press [VARS] [▷] [1] [1] [ENTER]. The value for the expression, Y_1, for the given x-value is displayed, in this case, $Y_1 = 47$. An ordered pair of the function is (3, 47).

e. Repeat steps **c.** and **d.** to produce as many pairs as desired. The TABLE feature of the *TI-83* can also be used to determine ordered pairs.

Zoom Features

To zoom in or out on a graph:

a. Here are two methods of using ZOOM. The first method uses the built-in features of the calculator. Move the cursor to a point on the graph that is of interest. Press [ZOOM]. The ZOOM menu will appear. Press [2] [ENTER] to zoom in on the graph by the amount shown under the SET FACTORS menu. The center of the new graph is the location at which you placed the cursor. Press [ZOOM] [3] [ENTER] to zoom out on the graph by the amount under the SET FACTORS menu. (The SET FACTORS menu is accessed by pressing [ZOOM] [▷] [4].)

b. The second method uses the ZBOX option under the ZOOM menu. To use this method, press [ZOOM] [1]. A cursor will appear on the graph. Use the arrow keys to move the cursor to a portion of the graph that is of interest. Press [ENTER]. Now use the arrow keys to draw a box around the portion of the graph you wish to see. Press [ENTER]. The portion of the graph defined by the box will be drawn.

c. Pressing [ZOOM] [6] resets the window to the standard 10×10 viewing window.

Solving Equations

This discussion is based on the fact that the solution of an equation can be related to the x-intercepts of a graph. For instance, the real solutions of the equation $x^2 = x + 1$ are the x-

intercepts of the graph of $f(x) = x^2 - x - 1$, which are the zeros of f.

To solve $x^2 = x + 1$, rewrite the equation with all terms on one side. The equation is now $x^2 - x - 1 = 0$. Think of this equation as $Y_1 = x^2 - x - 1$. The x-intercepts of the graph of Y_1 are the solutions of the equation $x^2 = x + 1$.

a. Enter $x^2 - x - 1$ into Y_1.

b. Graph the equation. You may need to adjust the viewing window so that the x-intercepts are visible.

c. Press [2nd] CALC [2].

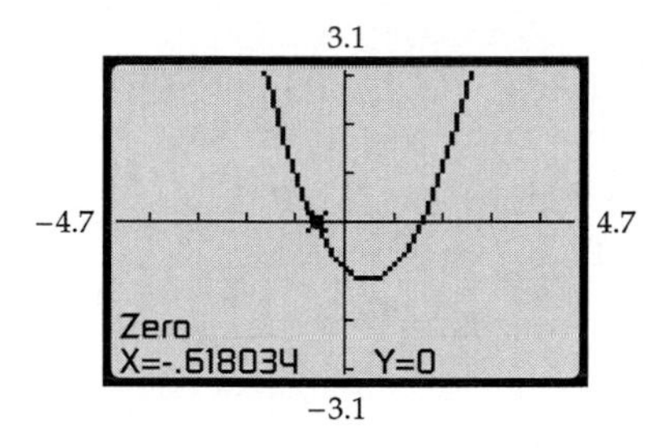

d. Move the cursor to a point on the curve that is to the left of an x-intercept. Press [ENTER].

e. Move the cursor to a point on the curve that is to the right of the same x-intercept. Press [ENTER].

f. Move the cursor to the approximate x-intercept. Press [ENTER].

g. The root is shown as the x-coordinate on the bottom of the screen; in this case, the root is approximately −0.618034. To find the next intercept, repeat steps **c.** through **f.** The SOLVER feature under the MATH menu can also be used to find solutions of equations.

Solving Systems of Equations in Two Variables

To solve a system of equations:

To solve $\begin{array}{l} y = x^2 - 1 \\ \frac{1}{2}x + y = 1 \end{array}$,

a. Solve each equation for y.

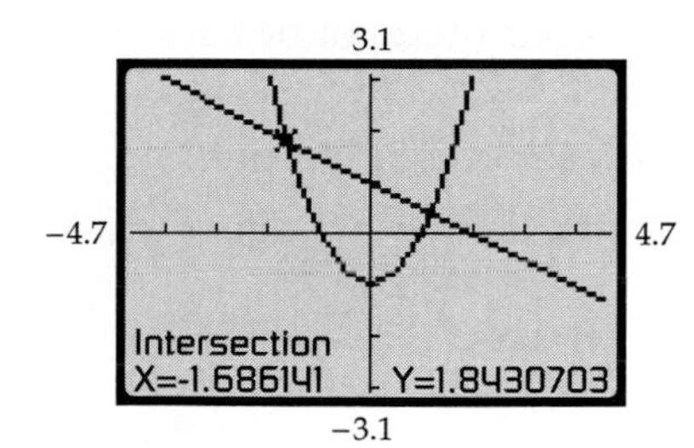

b. Enter the first equation as Y_1. For instance, $Y_1 = x^2 - 1$.

c. Enter the second equation as Y_2. For instance, $Y_2 = 1 - \frac{1}{2}x$.

d. Graph both equations. (*Note:* The point of intersection must appear on the screen. It may be necessary to adjust the viewing window so that the points of intersection are displayed.)

e. Press [2nd] CALC [5].

f. Move the cursor to the left of the first point of intersection. Press [ENTER].

g. Move the cursor to the right of the first point of intersection. Press [ENTER].

h. Move the cursor to the approximate point of intersection. Press [ENTER].

i. The first point of intersection is (−1.686141, 1.8430703).

j. Repeat steps **e.** through **h.** for each point of intersection.

Finding Minimum or Maximum Values of a Function

a. Enter the function into Y_1. The equation $y = x^2 - x - 1$ is used here.

b. Graph the equation. You may need to adjust the viewing window so that the maximum or minimum points are visible.

c. Press [2nd] CALC [3] to determine a minimum value or press [2nd] CALC [4] to determine a maximum value.

d. Move the cursor to a point on the curve that is to the left of the minimum (maximum). Press [ENTER].

e. Move the cursor to a point on the curve that is to the right of the minimum (maximum). Press [ENTER].

f. Move the cursor to the approximate minimum (maximum). Press [ENTER].

g. The minimum (maximum) is shown as the *y*-coordinate on the bottom of the screen; in this case the minimum value is –1.25.

Logic Operators

To evaluate a logical expression:

a. The logical operators *and, or, xor* (exclusive or), and *not* are accessed using the following keystrokes: [2nd] TEST [▷]. The relational operators $=$, $\neq$, $>$, $<$, $\geq$, and $\leq$ are accessed by pressing [2nd] TEST.

b. To evaluate the logical expression ($x < -4$) or ($x \geq 2$) when $x = 3$, enter the following keystrokes:

3 [STO▷] [X,T,θ,*n*] [ENTER] [(] [X,T,θ,*n*] [2nd] TEST 5 [(–)] 4 [)]

[2nd] TEST [▷] 2 [(] [X,T,θ,*n*] [2nd] TEST 4 2 [)] [ENTER]

After pressing ENTER, the value on the screen should be 1, indicating that the expression is true.

c. You can modify the above keystrokes to evaluate other logical expressions.

Using Tables

To use a table:

a. Press [2nd] TBLSET to activate the table setup menu.

b. TblStart is the beginning number for the table; ΔTbl is the difference between any two *x*-values in the table.

c. The portion of the table that appears as Indpnt: **Auto** Ask / Depend: **Auto** Ask allows you to choose between automatically having the calculator produce the results (Auto) or by having the calculator ask you for values of *x*. You can choose Ask by using the arrow keys.

d. Once a table has been set up, enter an expression for Y1. Now select TABLE by pressing [2nd] TABLE. A table showing values of the expression will be displayed on the screen.

Statistics

To calculate various statistical measures:

a. Press [STAT] to access the statistics menu. Press 1 to Edit or enter a new list of data. To delete data already in L1, press the up arrow key to highlight L1. Then press [CLEAR] and [ENTER]. Now enter each data value, pressing [ENTER] after each value.

b. When all the data has been entered, press [STAT] [▷] 1 [ENTER]. The values of the mean, standard deviation, median, first quartile (Q_1), third quartile (Q_3), minimum data value, and maximum data value will be calculated.

SHARP EL-9600

To evaluate an expression:

a. The SOLVER mode of the calculator is used to evaluate expressions. To enter SOLVER mode, press [2ndF] SOLVER [CL]. The expression $-3a^2b - 4c$ must be entered as the equation $-3a^2b - 4c = t$. The letter t can be any letter other than one used in the expression. Use the following keystrokes to input $-3a^2b - 4c = t$:

[(–)] 3 [ALPHA] A [a^b] 2 [▷] [ALPHA] B [–] 4 [ALPHA] C [ALPHA] [=] [ALPHA] T [ENTER]

Note the difference between the keys for a *negative* sign (–) and a *minus* sign.

b. After you press [ENTER], variables used in the equation will be displayed on the screen. To evaluate the expression for $a = 3$, $b = -2$, and $c = -4$, input each value, pressing [ENTER] after each number. When the cursor moves to T, press [2ndF] EXE. T = 70 will appear on the screen. This is the value of the expression. To evaluate the expression again for different values of a, b, and c, press [2ndF] QUIT and then [2ndF] SOLVER.

c. Press [+ − × ÷] to return to normal operation.

To graph a function:

a. Press the [Y=] key. The screen will show Y_1 through Y_8.

b. Input the expression for a function that is to be graphed. Press [X/θ/T/n] to enter an x.

For example, to input $y = \frac{1}{2}x - 3$, use the following keystrokes:

c. Set the viewing window by pressing [WINDOW]. Enter the values for the minimum x-value (Xmin), the maximum x-value (Xmax), the distance between tick marks on the x-axis (Xscl), the minimum y-value (Ymin), the maximum y-value (Ymax), and the distance between tick marks on the y-axis (Yscl). Press [ENTER] after each entry. Press [GRAPH]. For the graph shown at the left, enter Xmin = −10, Xmax = 10, Xscl = 1, Ymin = −10, Ymax = 10, Yscl = 1. Press [GRAPH].

d. Press [Y=] to return to the equation. The equal sign has a black rectangle around it. This indicates that the function is active and will be graphed when the [GRAPH] key is pressed. A function is deactivated by using the arrow keys. Move the cursor over the equal sign and press [ENTER]. When the cursor is moved to the right, the black rectangle will not be present and that equation will not be active.

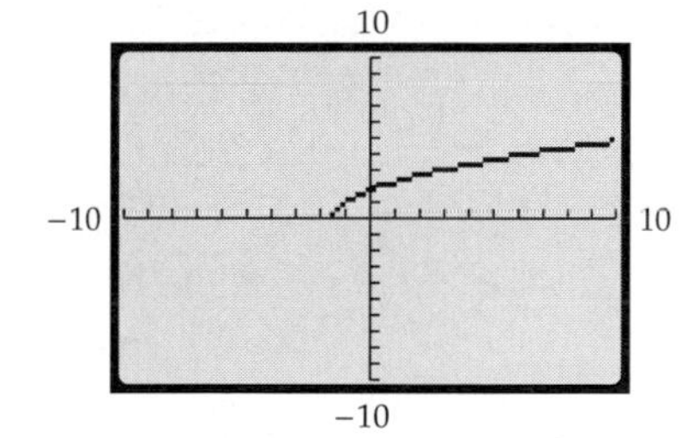

e. Graphing some radical equations requires special care. To graph the function $y = \sqrt{2x + 3}$, enter the following keystrokes:

[Y=] [CL] [2ndF] √ 2 [X/θ/T/n] [+] 3 [▷] [GRAPH]

The graph is shown at the left.

To display the *xy*-coordinates as integers:

a. Press [ZOOM] [▷] 8.

b. Graph the function. Press [TRACE]. Use the left and right arrow keys to trace along the graph of the function. The x- and y-coordinates of the function are shown on the bottom of the screen.

To display the *xy*-coordinates in tenths:

a. Press [ZOOM] [▷] 7.

b. Graph the function. Press [TRACE]. Use the left and right arrow keys to trace along the graph of the function. The x- and y-coordinates of the function are shown on the bottom of the screen.

To evaluate a function for a given value of *x*, or to produce ordered pairs of the function:

a. Press [Y=]. Input the expression. For instance, input

[Y=] [CL] 2 [X/θ/T/n] [∧] 3 [−] 3 [X/θ/T/n] [+] 2. Press [ENTER].

b. Press [+ − × ÷]. Store the x-coordinate of the ordered pair you want in [X/θ/T/n].

For instance, enter 3 [STO] [X/θ/T/n] [ENTER].

c. Press [VARS] [ENTER] 1 [ENTER]. The value of y, 47, will be displayed on the screen. The ordered pair is (3, 47). The TABLE feature of the calculator can also be used to find many ordered pairs for a function.

Zoom Features

To zoom in or out on a graph:

a. Here are two methods of using ZOOM. The first method uses the built-in features of the calculator. Move the cursor to a point on the graph that is of interest. Press [ZOOM]. The ZOOM menu will appear. Press 3 to zoom in on the graph by the amount shown by FACTOR. The center of the new graph is the location at which you placed the cursor. Press [ZOOM] 4 to zoom out on the graph by the amount shown in FACTOR.

b. The second method uses the BOX option under the ZOOM menu. To use this method, press [ZOOM] 2. A cursor will appear on the screen. Use the arrow keys to move the cursor to a portion of the graph that is of interest. Press [ENTER]. Use the arrow keys to draw a box around the portion of the graph you wish to see. Press [ENTER].

Solving Equations or Systems of Equations in Two Variables

This discussion is based on the fact that the real solutions of an equation can be related to the x-intercepts of a graph. For instance, the real solutions of $x^2 = x + 1$ are the x-intercepts of the graph of $f(x) = x^2 - x - 1$, which are the zeros of f.

To solve $x^2 = x + 1$, rewrite the equation with all terms on one side of the equation. The equation is now $x^2 - x - 1 = 0$. Think of this equation as $Y_1 = x^2 - x - 1$. The x-intercepts of the graph of Y_1 are the solutions of the equation $x^2 = x + 1$.

a. Enter $x^2 - x - 1$ into Y_1.

b. Graph the equation. You may need to adjust the viewing window so that the x-intercepts are visible.

c. Press [2ndF] CALC 5.

d. A solution is shown as the x-coordinate at the bottom of the screen. To find the next intercept, move the cursor to the right of the first x-intercept. Then press [2ndF] CALC 5.

Solving Systems of Equations

To solve a system of equations:

a. Solve each equation for y.

b. Press [Y=] and then enter both equations.

c. Graph the equations. You may need to adjust the viewing window so that the points of intersection are visible.

d. Press [2ndF] CALC 2 to find a point of intersection. The x- and y-coordinates at the bottom of the screen are the coordinates for the point of intersection.

e. Pressing [2ndF] CALC 2 again will find the next point of intersection.

Finding Maximum and Minimum Values of a Function

a. Press [Y=] and then enter the function.

b. Graph the equation. You may need to adjust the viewing window so that the maximum (minimum) are visible.

c. Press [2ndF] CALC 3 for the minimum value of the function or [2ndF] CALC 4 for the maximum value of the function.

d. The y-coordinate at the bottom of the screen is the maximum (minimum).

Logic Operators

To evaluate a logical expression:

a. The logical operators *and, or, xor* (exclusive or), and *not* are accessed using the following keystrokes: [MATH] G. The relational operators $=, \neq, >, <, \geq$, and $\leq$ are accessed by pressing [MATH] F.

b. To evaluate the logical expression $(x < -4)$ or $(x \geq 2)$ when $x = 3$, enter the following keystrokes:

[X/θ/T/n] [STO] 3 [(] [X/θ/T/n] [MATH] F 5 [(−)] 4 [)]

[MATH] G 2 [(] [X/θ/T/n] [MATH] F 4 2 [)] [ENTER]

After pressing ENTER, the value on the screen should be 1, indicating that the expression is true.

c. You can modify the above keystrokes to evaluate other logical expressions.

Using Tables

To use a table:

a. Press [2nd] TBLSET to activate the table setup menu.

b. The portion of the table that appears as Input: **Auto** User allows you to choose between automatically having the calculator produce the results (Auto) or having the calculator ask you for values of x. You can choose User by using the arrow keys.

c. TBLStart is the beginning number for the table; TBLStep is the difference between any two x-values in the table.

d. Once a table has been set up, enter an expression for Y1. Now select TABLE by pressing [TABLE]. A table showing values of the expression will display on the screen.

Statistics

To calculate various statistical measures:

a. Press [STAT] to access the statistics menu. Press A and [ENTER] to Edit or enter a new list of data. To delete data already in L1, press the up arrow key to highlight L1. Then press [DEL], and then [ENTER]. Now enter each data value, pressing [ENTER] after each value.

b. When all the data has been entered, press [+ − × ÷] [STAT] C 1. The values of the mean, standard deviation, median, first quartile (Q_1), third quartile (Q_3), minimum data value, and maximum data value will be calculated.

CASIO *CFX-9850G*

To evaluate an expression:

a. Press [MENU] [5]. Use the arrow keys to highlight Y1.

b. Input the expression to be evaluated. For example, to input the expression $-3A^2B - 4C$, use the following keystrokes:

[(−)] 3 [ALPHA] A [x^2] [ALPHA] B [−] 4 [ALPHA] C [EXE]

Note the difference between the keys for a *negative* sign [(−)] and a *minus* sign [−].

c. Press [MENU] 1. Store the value of each variable that will be used in the expression. For example, to evaluate the expression above when $A = 3$, $B = -2$, and $C = -4$, use the following keystrokes:

3 [→] [ALPHA] A [EXE] [(−)] 2 [→] [ALPHA] B [EXE] [(−)] 4 [→] [ALPHA] C [EXE]

These steps store the value of each variable.

d. Press [VARS] [F4] [F1] 1 [EXE].

The value of the expression, Y1, for the given values is displayed; in this case, Y1 = 70.

To graph a function:

a. Press Menu [5] to obtain the GRAPH FUNCTION Menu.

b. Input the function that you desire to graph. Press [X,θ,T] to input the variable x. For example, to input $y = x^3 + 2x^2 - 5x - 6$, use the following keystrokes:

[X,θ,T] [∧] 3 [+] 2 [X,θ,T] [x^2] [−] 5 [X,θ,T] [−] 6 [EXE]

c. Set the viewing window by pressing [SHIFT] [F3] and the Range Parameter Menu will appear. Enter the values for the minimum x-value (Xmin), maximum x-value (Xmax), units between tick marks on the x-axis (Xscl), minimum y-value (Ymin), maximum y-value (Ymax), and the units between tick marks on the y-axis (Yscl). Press [EXE] after each of the 6 entries above. Press [EXIT], or [SHIFT] [QUIT], to leave the Range Parameter Menu.

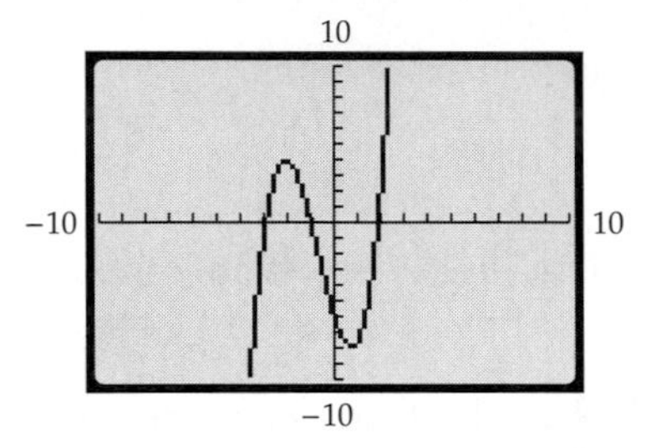

d. Press [F6] to draw the graph. For the graph shown at the left, Xmin = −10, Xmax = 10, Xscl = 1, Ymin = −10, Ymax = 10, Yscl = 1.

e. In the equation for Y_1, there is a rectangle around the equal sign. This indicates that this function is *active* and will be graphed when the [F6] key is pressed. A function is deactivated by using the [F1] key. After using this key once, the rectangle around the equal sign will not be present and that function will not be graphed.

To display the x-coordinates of rectangular coordinates as integers:

a. Set the Range as follows: For example, set Xmin = −63, Xmax = 63, Xscl = 10, Ymin = −32, Ymax = 32, Yscl = 10.

b. Graph a function and use the Trace feature. Press [F1] and then move the cursor with the [◁] and the [▷] keys. The values of x and $y = f(x)$ displayed on the bottom of the screen are the coordinates of a point on the graph. Observe that the x-value is given as an integer.

To display the x-coordinates of rectangular coordinates in tenths:

a. Set the Range as follows: For example, set Xmin = −6.3, Xmax = 6.3. A quick way to choose these range parameter settings is to press [F1] from the V-Window Menu.

b. Graph a function and use the Trace feature. Press [F1] and then move the cursor with the [◁] and the [▷] keys. The values of x and $y = f(x)$ displayed on the bottom of the screen are the coordinates of a point on the graph. Observe that the x-value is given as a decimal that terminates in the first decimal place (tenths).

To evaluate a function for a given value of x, or to produce ordered pairs of the function:

a. Press [MENU] [5].

b. Input the function to be evaluated. For example, input $2x^3 - 3x + 2$ into Y_1.

c. Press [MENU] 1.

d. Input a value for x; for example, to input 3 press

3 [→] [X,θ,T] [EXE]

e. Press [VARS] [F4] [F1] 1 [EXE].

The value of Y_1 for the given value $x = 3$ is displayed. In this case, $Y_1 = 47$.

Zoom Features

To zoom in or out on a graph:

a. After drawing a graph, press [SHIFT] Zoom to display the Zoom/Auto Range menu. To zoom in on a graph by a factor of 2 on the x-axis and a factor of 1.5 on the y-axis:

Press [F2] to display the Factor Input Screen. Input the zoom factors for each axis: 2 [EXE] 1 [·] 5 [EXE] [EXIT]. Press [F3] to redraw the graph according to the factors specified above. To specify the center point of the enlarged (reduced) display after pressing SHIFT Zoom use the arrow keys to move the pointer to the position you wish to become the center of the next display. You can repeat the zoom procedures as needed. If you wish to see the original graph, press F6 F1. This procedure resets the range parameters to their original values and redraws the graph.

b. A second method of zooming makes use of the Box Zoom Function. To use this method, first draw a graph. Then press SHIFT Zoom F1. Now use the arrow (cursor) keys to move the pointer. Once the pointer is located at a portion of the graph that is of interest, press EXE. Now use the arrow keys to draw a box around the portion of the graph you wish to see. Press EXE. The portion of the graph defined by the box will be drawn.

Solving Equations

This discussion is based on the fact that the real solutions of an equation can be related to the x-intercepts of a graph. For instance, the real solutions of $x^2 = x + 1$ are the x-intercepts of the graph of $f(x) = x^2 - x - 1$, which are the zeros of f.

To solve $x^2 = x + 1$, rewrite the equation with all terms on one side. The equation is now $x^2 - x - 1 = 0$. Think of this equation as $Y_1 = x^2 - x - 1$. The x-intercepts of the graph of Y_1 are the solutions of the equation $x^2 = x + 1$.

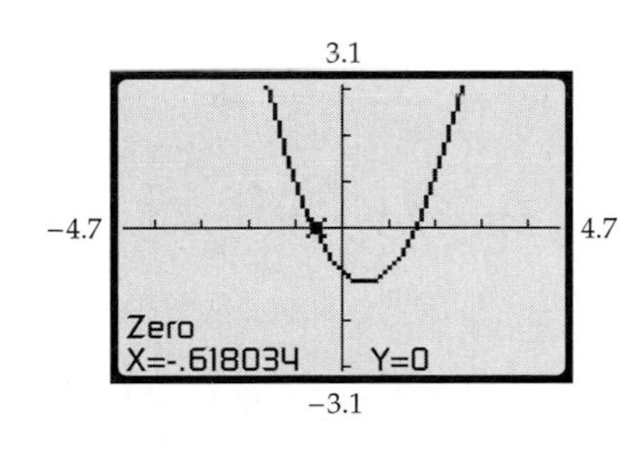

a. Enter $x^2 - x - 1$ into Y_1.

b. Graph the equation. You may need to adjust the viewing window so that the x-intercepts are visible.

c. Press SHIFT G-SOLV F1.

d. The root is shown as the x-coordinate on the bottom of the screen; in this case, the root is approximately −0.618034. To find the next x-intercept, press the right arrow key.

The EQUA Mode (Press MENU ALPHA A) can also be used to find solutions of linear, quadratic, and cubic equations.

Solving Systems of Two Equations in Two Variables

The following discussion is based on the concept that the solutions of a system of two equations are represented by the points of intersection of the graphs.

The system of equations $\begin{array}{c} y = x^2 - 1 \\ \frac{1}{2}x + y = 1 \end{array}$ will be solved.

a. Solve each equation for y.

b. Enter the first equation in the Graph Menu as Y_1. For instance, let $Y_1 = x^2 - 1$.

c. Enter the second equation as Y_2. For instance, let $Y_2 = 1 - \frac{1}{2}x$.

d. Graph both equations. (*Note:* The point of intersection must appear on the screen. It may be necessary to adjust the viewing window so that the point of intersection that is of interest is the only intersection point that is displayed.)

e. Press SHIFT G-SOL F5 EXE.

f. The display will show that the graphs intersect at (−1.686141, 1.8430703). To find the next intersect, repeat step **e.**

Finding Minimum or Maximum Values of a Function

a. Enter the function into the graphing menu. For this example we have used $y = x^2 - x - 1$.

b. Graph the function. Adjust the viewing window so that the maximum or minimum is visible.

c. Press [SHIFT] G-SOL [F2] [EXE] for a maximum and [F3] [EXE] for a minimum.

d. The local maximum (minimum) is shown as the y-coordinate on the bottom of the screen; in this case, the minimum value is −1.25.

Logic Operators

To evaluate a logical expression:

a. The logical operators *and, or, xor* (exclusive or), and *not* are accessed using the following keystrokes: [MENU] 1 [OPTN] [F6] [F6] [F4]. The relational operators $=, \neq, >, <, \geq$, and $\leq$ are accessed by pressing [SHIFT] PRGM [F6] [F3].

b. To evaluate the logical expression ($x < -4$) or ($x \geq 2$) when $x = 3$, enter the following keystrokes:

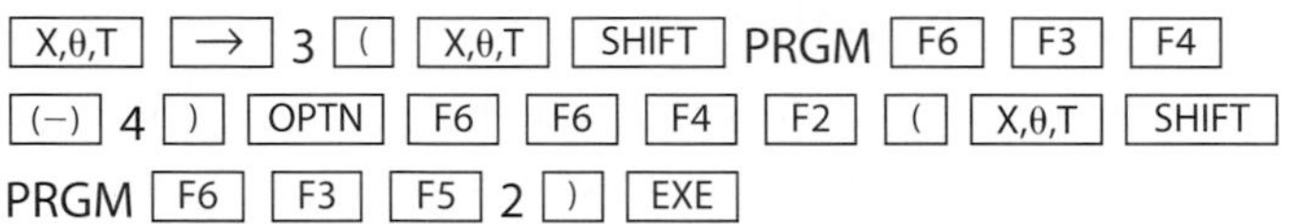

After pressing EXE, the value on the screen should be 1, indicating that the expression is true.

c. You can modify the above keystrokes to evaluate other logical expressions.

Using Tables

To use a table:

a. Highlight TABLE on the MAIN MENU screen. Press [EXE] 7 to activate the table setup menu.

b. Enter the expression for which you wish to create a table of values into Y1.

c. Press [F5] and enter the starting x-value for the table, the ending x-value, and the pitch. The pitch is the difference between successive x-values in the table. Press [EXIT] to return to the expression editing window.

d. Press [F6] to create the values in the table.

Statistics

To calculate various statistical measures:

a. Press [MENU] 2 to activate the statistics highlight DEL-A (you may need to press [F6] first), press [F4], and then [F1]. Now enter each data value, pressing [EXE] after each value.

b. When all the data has been entered, make sure that CALC is on the screen (you may need to press [F6] first). Press [F2] [F1]. The values of the mean, standard deviation, median, first quartile (Q_1), third quartile (Q_3), minimum data value, and maximum data value will be calculated.

Solutions to Chapter 1 You-Try-Its

SECTION 1.1

You-Try-It 1

Goal: To find the units digit of 5^{92}.
Strategy: Try to find a pattern by evaluating some powers of 5.
$5^1 = 5; 5^2 = 25; 5^3 = 125; 5^4 = 625$
Solve: It appears that every power of 5 ends with a 5. Therefore, the units digit of 5^{92} is 5.
Check.

You-Try-It 2

It appears that the middle letter is one of the vowels a, e, i, o, or u. Therefore, the next term is *bun*.

You-Try-It 3

A pattern is the sum of the number and the number times ten.

$11 \cdot 23 = 23 + 230 = 253$

$11 \cdot 36 = 36 + 360 = 396$

Therefore, $11 \cdot 78 = 78 + 780 = 858$.

You-Try-It 4

We are given that 🍎🍎🍎 = ♣♣ and ♣♣♣♣ = ♦♦, then ♦♦♦♦ = ♣♣♣♣♣♣♣♣. Therefore,
♣♣♣♣♣♣♣♣ = 🍎🍎🍎🍎🍎🍎🍎🍎🍎🍎🍎🍎.
Four ♦'s equal 12 🍎's.

You-Try-It 5

a. A general conclusion is reached from specific facts. Inductive reasoning is being used.
b. A conclusion is reached based on facts. Deductive reasoning is being used.

SECTION 1.2

You-Try-It 1

The statement is combined with *or*; therefore the statement is true when at least one of the inequalities is true. Since $(x < 6)$ is true when $x = 3$, the statement is true.

You-Try-It 2

$7 \le 3x + 1$	$3x + 1 \le 10$
$7 \le 3(3) + 1$	$3(3) + 1 \le 10$
$7 \le 9 + 1$	$9 + 1 \le 10$
$7 \le 10$ True	$10 \le 10$ True

Statements combined with *and* are true when each statement is true. Because both statements are true, the result is true.

You-Try-It 3

$A \cup B = \{1, 2, 3, 4, 5, 6, 7, 8\}$

$A \cap B = \{3, 6, 7\}$

You-Try-It 4

a. Smog is healthy.
b. Saturn does have rings around it.

You-Try-It 5

Some animals cannot fly.

You-Try-It 6

No houses are old.

You-Try-It 7

The number 2 is a counterexample to the statement.

SECTION 1.3

You-Try-It 1

a. Because the bar for women is higher than the bar for men, more women were willing to wait up to 3 minutes.
b. From the graph, more men were willing to wait 1 minute or less.
c. The bar does not quite reach the 15 mark. Therefore, there were less than 15 men who were willing to wait up to 10 minutes.

You-Try-It 2

a. From the graph, between1960 and 1970 life expectancy for men increased the least.
b. From the graph, between1940 and 1950 life expectancy for women increased the most.
c. Life expectancy for men in 1940: 62.8
Life expectancy for men in 1990: 72.9
$72.9 - 62.8 = 10.1$
The change in life expectancy was 10.1 years.

You-Try-It 3

Count the number of salaries in each of the salary ranges.

Class (Salary, in millions of $)	Frequency (Number of players with a salary in that class)
0.2 – 0.69	16
0.7 – 1.19	10
1.2 – 1.69	9
1.7 – 2.19	4
2.2 – 2.69	6
2.7 – 3.19	2
3.2 – 3.69	6
3.7 – 4.19	5

You-Try-It 4

From the graph, find the average income for each of the following.

25,595 – 28,595: 8

28,595 – 31,595: 10

31,595 – 34,595: 10

$8 + 10 + 10 = 28$

In 28 states, the average income is less than $34,595.

You-Try-It 5

10	4 7 4 4
11	5 3 9 7 5
12	9 2 7 6 5 6 8 5 0
13	5 4
14	2
15	6 4 2

SECTION 1.4

You-Try-It 1

a. $2^6 = 2 \cdot 2 \cdot 2 \cdot 2 \cdot 2 \cdot 2 = 64$

b. $3^3 \cdot 2^2 = (3 \cdot 3 \cdot 3) \cdot (2 \cdot 2)$

$= 27 \cdot 4 = 108$

You-Try-It 2

a. $48 \div 2^3 - 2 \cdot 3 = 48 \div 8 - 2 \cdot 3$

$= 6 - 6$

$= 0$

b. $82 - 8(4 + 2 \cdot 3) + (4 - 1)^3 = 82 - 8(10) + (3)^3$

$= 82 - 8(10) + 27$

$= 82 - 80 + 27$

$= 2 + 27$

$= 29$

You-Try-It 3

$1.30g$

$1.30(9.7) = 12.61$

The cost of 9.7 gallons of gas is $12.61.

You-Try-It 4

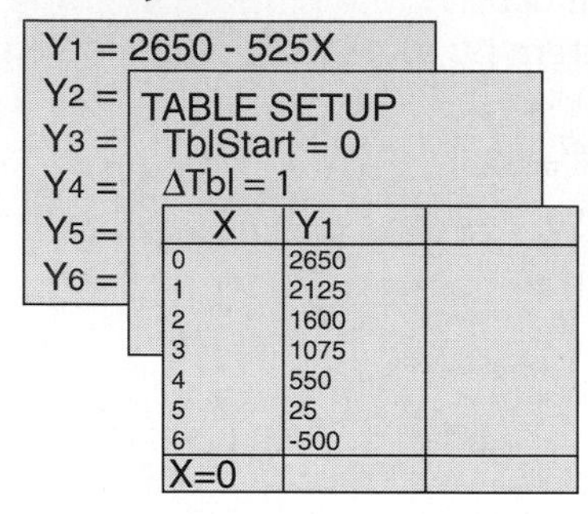

You-Try-It 5

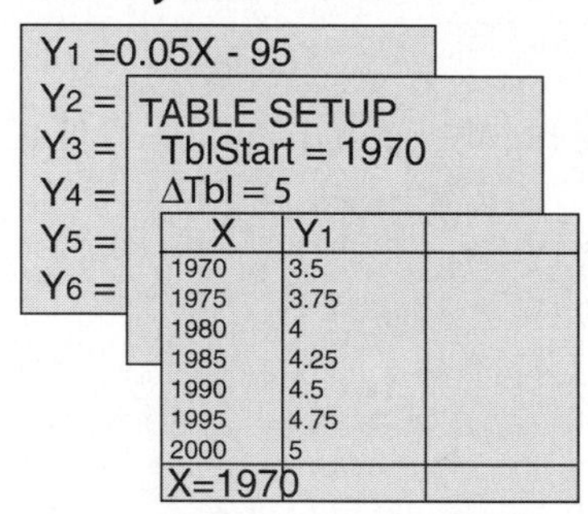

a. From the input/output table, 4.5 pounds of garbage were generated per person per day in 1990.

b. From the input/output table, there were 4.75 pounds of garbage generated per person per day in 1995.

You-Try-It 6

$3xy^2 - 3x^2y$

$3(2)(5)^2 - 3(2)^2(5) = 3(2)(25) - 3(4)(5)$

$= 150 - 60 = 90$

SECTION 1.5

You-Try-It 1

a. The unknown number: n

7 more than the product of a number and 12

$12n + 7$

b. The unknown number: n

The total of 18 and the quotient of a number and 9

$18 + \frac{n}{9}$

c. The unknown number: n

The square of a number subtracted from the number increased by 9

$(n - 9) - n^2$

You-Try-It 2
Let n represent the amount of rye flour.
Wheat flour is 4 times rye flour: $4n$

You-Try-It 3
Let h represent the number of overtime hours worked by the chef.
The weekly earnings for the chef in terms of the number of overtime hours is $640 + 32h$.

You-Try-It 4

Input	0	1	2	3	4	5	6	n
Output	1	3	5	7	9	11	**13**	$\mathbf{2n+1}$

$2n + 1$

$2(9) + 1 = 18 + 1 = 19$

You-Try-It 5
The number of dots is the square of the pattern number. Therefore the output is n^2 when the input is n.

n^2

$7^2 = 49$

There will be 49 dots when the pattern number is 7.

SECTION 1.6

You-Try-It 1

$$\bar{x} = \frac{19 + 16 + 18 + 14 + 18 + 15 + 17 + 19}{8}$$

$$= \frac{136}{8} = 17$$

The mean miles per gallon is 17.

You-Try-It 2
Arrange the data from smallest to largest.
8 15 17 18 18 19 21 21 22 22 23 24 25 25 28 28
Because there is an even number of data, the median is the mean of the middle two numbers.

$$\text{Median} = \frac{21 + 22}{2} = 21.5$$

The median number of hours worked is 21.5.

You-Try-It 3
The modal response is the category that has the largest number. Therefore, the modal response is record stores.

You-Try-It 4

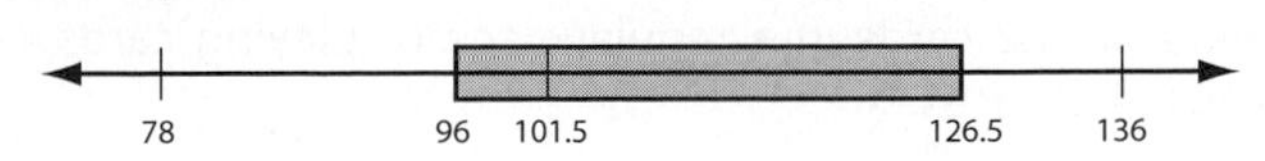

The interquartile range is $126.5 - 96 = 30.5$.

You-Try-It 5
The values of 25% of the data in a data set lie above Q_3. Therefore, $0.25(583{,}000) = 145{,}750$ physicians earned over \$240,000.

You-Try-It 6
Arrange the numbers from smallest to largest.
12 12 13 13 15 15 16 16 17 17 17 21

The median equals $\dfrac{15 + 16}{2} = 15.5$.

Q_1 is the median of 12 12 13 13 15 15, which is 13.
Q_3 is the median of 16 16 17 17 17 21, which is 17.
Interquartile range $= 17 - 13 = 4$; range $= 21 - 12 = 9$.

SECTION 1.7

You-Try-It 1
There are six possible outcomes. The sample space is $S = \{1, 2, 3, 4, 5, 6\}$.

You-Try-It 2
There are 5 digits from which to choose, so there are 5 choices for the first digit, 5 choices for the second digit, and 5 choices for the third digit. By the Counting Principle, there are $5 \cdot 5 \cdot 5 = 125$ different three-digit numbers.

You-Try-It 3
There are 10 choices for the first number, 26 choices for the next letter, 26 choices for the next letter, 26 choices for the next letter, 10 choices for the next number, 10 choices for the next number, and 10 choices for the last number. By the Counting Principle, there are

$$10 \cdot 26 \cdot 26 \cdot 26 \cdot 10 \cdot 10 \cdot 10 = 175{,}760{,}000$$

possible license plates.

You-Try-It 4
From the table of outcomes of two dice, there are 6 outcomes that have a sum of 7. Therefore,

$$P(\text{sum of } 7) = \frac{6}{36} = \frac{1}{6}$$

The probability that the sum of 7 is $\frac{1}{6}$.

You-Try-It 5

There are 52 cards in a regular deck of playing cards, of which 4 are aces.

$$P(\text{ace}) = \frac{4}{52} = \frac{1}{13}$$

The probability of an ace is $\frac{1}{13}$.

You-Try-It 6

Because a dodecahedral die has 12 sides, there are 12 possible outcomes from one roll of the die. Of the possible outcomes, 4, 8, and 12 are the only numbers divisible by 4.

$$P(\text{divisible by } 4) = \frac{3}{12} = \frac{1}{4}$$

The probability of a number divisible by 4 is $\frac{1}{4}$.

You-Try-It 7

From the table of outcomes of two dice, there are 5 outcomes for which the sum is 6.

$$P(\text{sum of } 6) = \frac{5}{36}$$

The probability of a sum of 6 is $\frac{5}{36}$. Therefore, the probability that Erin will land on Boardwalk is $\frac{5}{36}$.

You-Try-It 8

$$P(E) = \frac{\text{number of observations of } E}{\text{total number of observations}} = \frac{336}{5892} \approx 0.057$$

You-Try-It 9

From the table of outcomes from the roll of two dice, there are 2 outcomes whose sum is 11. There are 34 outcomes for a sum not equal to 11.

The odds in favor of 11 $= \frac{2}{34} = \frac{1}{17}$. The odds in favor of a sum of 11 are 1 TO 17.

You-Try-It 10

The probability of winning $= \frac{1}{1+9} = \frac{1}{10}$

The probability is $\frac{1}{10}$.

Solutions to Chapter 2 You-Try-Its

SECTION 2.1

You-Try-It 1

a. $|-42| = 42$

$-|-31| = -31$

b. $-(|-23| - |-17|) = -(23 - 17) = -6$

You-Try-It 2

a. $|-28| = 28, |71| = 71$

$71 - 28 = 43$

$(-28) + 71 = 43$

b. Recall that *total* means to add.

$|-42| = 42, |28| = 28$

$42 - 28 = 14$

$(-42) + (28) = -14$

You-Try-It 3

Evaluate the expression $x + (-6)$ for each of the input values.

$x + (-6)$	$x + (-6)$	$x + (-6)$
$(-3) + (-6) = -9$	$(-2) + (-6) = -8$	$(-1) + (-6) = -7$

$x + (-6)$	$x + (-6)$	$x + (-6)$	$x + (-6)$
$0 + (-6) = -6$	$1 + (-6) = -5$	$2 + (-6) = -4$	$3 + (-6) = -3$

Input	−3	−2	−1	0	1	2	3
Output	−9	−8	−7	−6	−5	−4	−3

You-Try-It 4

a. $46 - 72 = 46 + (-72) = -26$

b. $-15 - 12 - 9 - (-36) = -15 + (-12) + (-9) + (36)$

$= 0$

You-Try-It 5

Evaluate the expression $z - 6$ for each of the input values.

$z - 6$	$z - 6$	$z - 6$
$(-3) - 6 = -9$	$(-2) - 6 = -8$	$(-1) - 6 = -7$

$z - 6$	$z - 6$	$z - 6$	$z - 6$
$0 - 6 = -6$	$1 - 6 = -5$	$2 - 6 = -4$	$3 - 6 = -3$

Input	−3	−2	−1	0	1	2	3
Output	−9	−8	−7	−6	−5	−4	−3

You-Try-It 6

$d(A, B) = |-8 - (-1)| = |-7| = 7$

You-Try-It 7

a. To find the difference, subtract the temperature on Monday from the temperature on Wednesday. Temperature on Monday: −12. Temperature on Wednesday: −10.

$-10 - (-12) = -10 + 12 = 2$

The difference is 2 degrees.

b. To find the difference, subtract the temperature on Thursday from the temperature on Friday.

Temperature on Thursday: −7. Temperature on Friday: 8.

$8 - (-7) = 8 + 7 = 15$

The difference is 15 degrees.

SECTION 2.2

You-Try-It 1

a. Recall that *times* means to multiply.

$-12(-32) = 384$

b. $4(-3)(-7)(-6) = -12(-7)(-6)$

$= 84(-6) = -504$

You-Try-It 2

Evaluate the expression $5x$ for each of the input values.

$5x$	$5x$	$5x$	$5x$
$5(-3) = -15$	$5(-2) = -10$	$5(-1) = -5$	$5(0) = 0$

$5x$	$5x$	$5x$
$5(1) = 5$	$5(2) = 10$	$5(3) = 15$

Input	−3	−2	−1	0	1	2	3
Output	−15	−10	−5	0	5	10	15

You-Try-It 3

$b^2 - 4ac$

$(-2)^2 - 4(5)(-4) = 4 + 80 = 84$

You-Try-It 4

a. $\frac{96}{-8} = -12$

b. $\frac{-12^2}{-8} = \frac{-144}{-8} = 18$

You-Try-It 5

Evaluate the expression $\frac{4}{x}$ for each of the input values.

$\frac{4}{x}$	$\frac{4}{x}$	$\frac{4}{x}$
$\frac{4}{-4} = -1$	$\frac{4}{-2} = -2$	$\frac{4}{-1} = -4$

$\frac{4}{x}$	$\frac{4}{x}$	$\frac{4}{x}$	$\frac{4}{x}$
$\frac{4}{1} = 4$	$\frac{4}{2} = 2$	$\frac{4}{4} = 1$	$\frac{4}{8} = \frac{1}{2}$

Input	−4	−2	−1	1	2	4	8
Output	−1	−2	−4	4	2	1	$\frac{1}{2}$

You-Try-It 6

$$\frac{-2a}{b^3}$$

$$\frac{-2(4)}{(-2)^3} = \frac{-8}{-8} = 1$$

You-Try-It 7

Goal You must determine the student's score.

Strategy To find the student's score, multiply the value of each answer on the test by the number of answers in that category. Then add the results.

Solution $2(48) = 96, -4(14) = -56, -2(8) = -16$

$96 + (-56) + (-16) = 40 + (-16) = 24$

The student's score was 24.

Check √

SECTION 2.3

You-Try-It 1

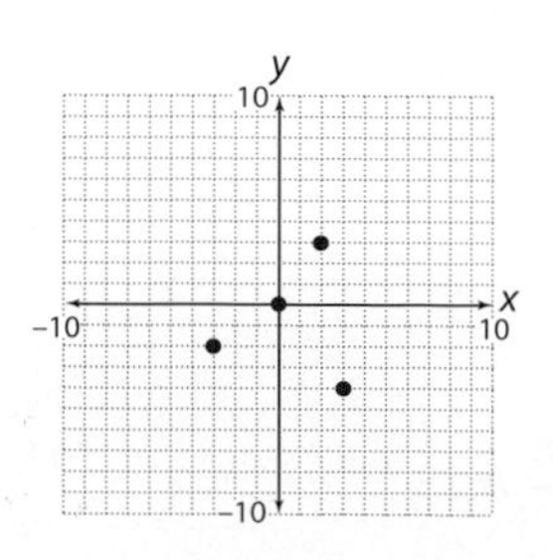

You-Try-It 2

$A(0, -2), B(-2, 4), C(-4, -3), D(1, 0)$

You-Try-It 3

The horizontal axis is the amount of sugar and the vertical axis is the amount of fiber. Graph each of the ordered pairs (4, 3), (3, 0), (5, 3), (6, 2), (3, 1), and (7, 5).

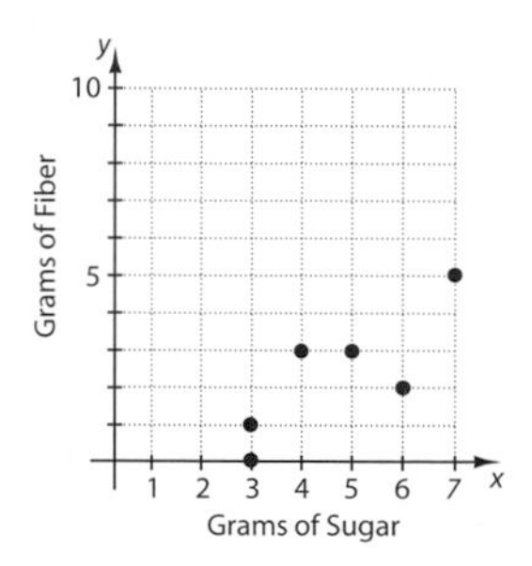

You-Try-It 4

First complete the table. Evaluate the expression $1 - 2x$ for each of the input values.

$1 - 2x$	$1 - 2x$	$1 - 2x$
$1 - 2(-3) = 7$	$1 - 2(-2) = 5$	$1 - 2(-1) = 3$

$1 - 2x$	$1 - 2x$	$1 - 2x$	$1 - 2x$
$1 - 2(0) = 1$	$1 - 2(1) = -1$	$1 - 2(2) = -3$	$1 - 2(3) = -5$

Input	−3	−2	−1	0	1	2	3
Output	7	5	3	1	−1	−3	−5

We will graph the output on an xy–coordinate system. Therefore, the output, $1 - 2x$, will be designated as y. The ordered pairs to graph are (−3, 7), (−2, 5), (−1, 3), (0, 1), (1, −1), (2, −3), and (3, −5).

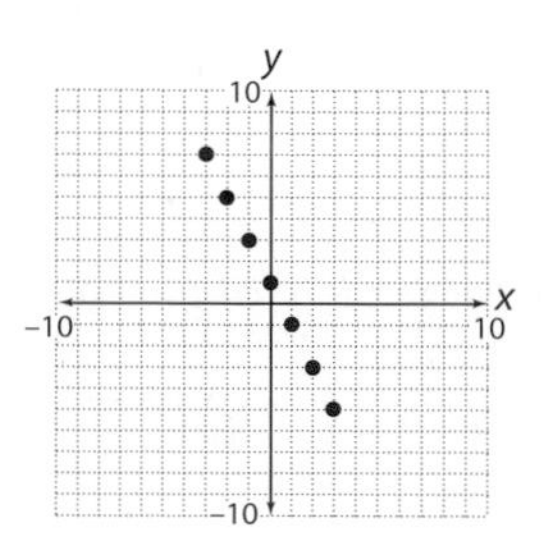

You-Try-It 5

Create an input/output table for the equation where the inputs are the values of x and the outputs are the values of y, which, in this case, are $2x$. Then graph the ordered pairs. We will show the input/output table vertically rather than horizontally.

Input, x	Output, $2x$
−4	−8
−3	−6
−2	−4
0	0
1	2
3	6
5	10

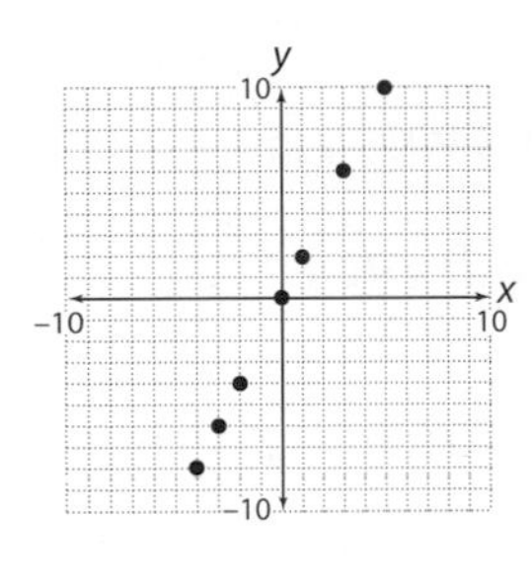

You-Try-It 6

Create an input/output table for the equation where the inputs are the values of t and the outputs are the values of the car, which, in this case, are $20{,}000 - 2000t$. Then graph the ordered pairs. We will show the input/output table vertically rather than horizontally. On the graph, the horizontal axis is t and the vertical axis is V.

Input, t	Output, $20{,}000 - 2000t$
0	20,000
1	18,000
2	16,000
3	14,000
4	12,000
5	10,000

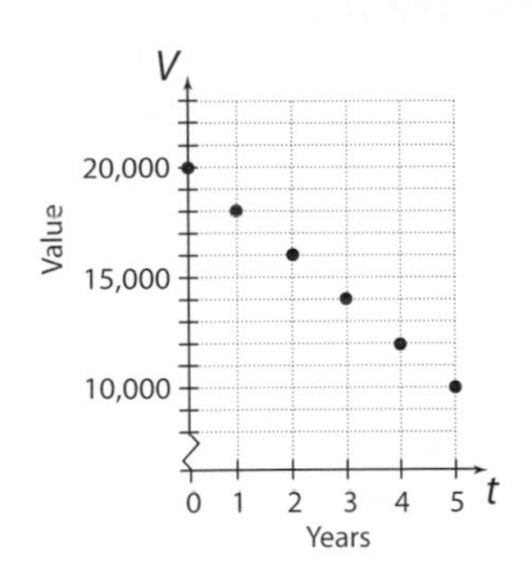

SECTION 2.4

You-Try-It 1

Goal To determine which of the regions in the graph, adult contemporary or news/talk radio, is larger.

Strategy We could look at the graph to see which section appears larger. In this case, however, the two regions are so close to the same size that a different method of solving the problem is necessary. One possibility is to compare the fraction $\frac{3}{16}$ to the fraction $\frac{3}{20}$.

Solution To compare fractions, rewrite the fractions as equivalent fractions with a common denominator. Use the least common multiple (LCM) of the denominators as the common denominator. The LCM of 16 and 20 is 80.

$$\frac{3}{16} = \frac{15}{80} \qquad \frac{3}{20} = \frac{12}{80}$$

Because $15 > 12$, $\frac{15}{80} > \frac{12}{80}$. Therefore, more people listen to adult contemporary radio stations than to news/talk radio stations.

Check √

You-Try-It 2

Find a common denominator. The LCM of 9 and 6 is 18.

$$-\frac{8}{9} + \frac{5}{6} = -\frac{16}{18} + \frac{15}{18}$$
$$= \frac{-16 + 15}{18} = \frac{-1}{18}$$
$$= -\frac{1}{18}$$

You-Try-It 3

Replace a, b, and c with their values and then simplify.
$-a + b - c$

$$-\left(-\frac{5}{6}\right) + \frac{1}{2} - \left(-\frac{2}{3}\right) = \frac{5}{6} + \frac{1}{2} + \frac{2}{3}$$
$$= \frac{5}{6} + \frac{3}{6} + \frac{4}{6}$$
$$= \frac{5 + 3 + 4}{6} = \frac{12}{6} = 2$$

You-Try-It 4

$-5.1 - (-10.39) = -5.1 + 10.39$
$|-5.1| = 5.1$
$|10.39| = 10.39$
$10.39 - 5.1 = 5.29$
$-5.1 - (-10.39) = 5.29$

You-Try-It 5

$$\left(-\frac{16}{9}\right)\left(-\frac{15}{28}\right) = \frac{16 \cdot 15}{9 \cdot 28} = \frac{20}{21}$$

You-Try-It 6

Because $\frac{3}{4}$ is the scale factor chosen, that means we want to find $\frac{3}{4}$ of the length of the line, or $\frac{3}{4}$ of $2\frac{2}{3}$. Remember that *of* means to multiply.

$$\frac{3}{4} \cdot 2\frac{2}{3} = \frac{3}{4} \cdot \frac{8}{3} = \frac{24}{12} = 2$$

The length of the new line is 2 inches.

You-Try-It 7

$$\frac{1}{8} \div \left(-\frac{5}{12}\right) = -\left(\frac{1}{8} \div \frac{5}{12}\right)$$
$$= -\left(\frac{1}{8} \cdot \frac{12}{5}\right)$$
$$= -\frac{12}{40} = -\frac{3}{10}$$

You-Try-It 8

To convert from one unit to another, it is necessary to use the appropriate conversion factor. If you do not know the factor, you may need to look in a reference book. For feet and yards, the conversion factors are $\frac{1 \text{ yard}}{3 \text{ feet}}$ and $\frac{3 \text{ feet}}{1 \text{ yard}}$. Because we are converting from feet to yards, use the conversion factor that contains yards in the numerator. We use the abbreviation ft for feet and yd for yards.

$$29 \text{ ft} = 29 \text{ ft} \cdot 1 = \frac{29 \text{ ft}}{1} \cdot \frac{1 \text{ yd}}{3 \text{ ft}} = \frac{29}{3} \text{ yd} = 9.\overline{66} \text{ yards}$$

You-Try-It 9

Goal To find the number of $\frac{2}{3}$-cup servings in $7\frac{1}{2}$ gallons of iced tea.

Strategy To find the number of $\frac{2}{3}$-cup servings, first convert $7\frac{1}{2}$ gallons to cups. Then divide the result by $\frac{2}{3}$.

Solution

$$7\frac{1}{2} \text{ gal} = \frac{15}{2} \text{ gal} \cdot \frac{16 \text{ cups}}{1 \text{ gal}} = \frac{240}{2} \text{ cups} = 120 \text{ cups}$$

$$120 \div \frac{2}{3} = \frac{120}{1} \cdot \frac{3}{2} = \frac{360}{2} = 180$$

There are 180 $\frac{2}{3}$-cup servings in $7\frac{1}{2}$ gallons of iced tea.

Check √

SECTION 2.5

You-Try-It 1

Use the formula for the area of a triangle, $A = \frac{1}{2}bh$. The base is the horizontal distance between Q and R.

$b = |4 - (-2)| = |4 + 2| = |6| = 6$

The height is the vertical distance between P and Q or between P and R.

$h = |4 - (-2)| = |6| = 6$

Find the area.

$A = \frac{1}{2}bh = \frac{1}{2}(6)(6) = 3(6) = 18$

The area is 18 square units.

You-Try-It 2

We will use a prime (′) to indicate the vertices of the triangle that is formed by subtracting 3 from the y-coordinates of the vertices of the original triangle.

$A(-4,-1) \rightarrow A'(-4,-4)$
$B(1,3) \rightarrow B'(1,0)$
$C(-1,5) \rightarrow C'(-1,2)$

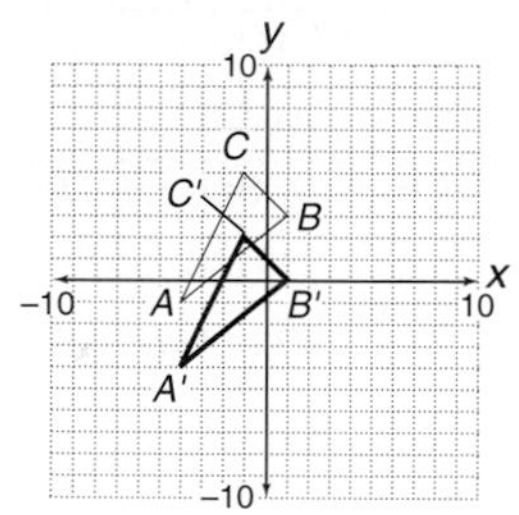

You-Try-It 3

A translation of 4 units to the right means that 4 is added to each x-coordinate and a translation of 3 units down means that 3 is subtracted from each y-coordinate.

$A(-5,2) \rightarrow A'(-1,-1)$
$B(-2,5) \rightarrow B'(2,2)$
$C(1,2) \rightarrow C'(5,-1)$
$D(-2,-1) \rightarrow D'(2,-4)$

You-Try-It 4

We will use a prime (′) to indicate the vertices of the trapezoid that is formed by multiplying the x-coordinates of the original trapezoid by $\frac{1}{2}$.

$A(-6,-3) \rightarrow A'(-3,-3)$
$B(-4,4) \rightarrow B'(-2,4)$
$C(2,4) \rightarrow C'(1,4)$
$D(4,-3) \rightarrow D'(2,-3)$

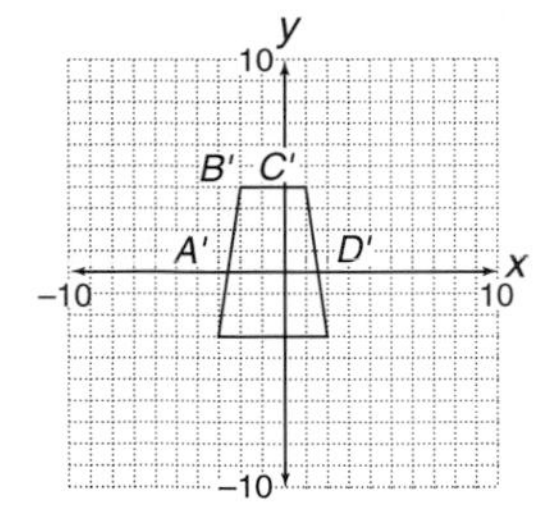

You-Try-It 5

a. If the coordinate grid were folded along the y-axis, Figure A would lie on top of Figure B. Thus Figure A and Figure B are symmetric with respect to the y-axis.

b. Because Figure C is closer to the x-axis than Figure A, if the coordinate grid were folded along the x-axis, Figure A would not lie on top of Figure C. Thus Figure A and Figure C are not symmetric with respect to the x-axis.

You-Try-It 6

a. For each of the given coordinate pairs, multiply the x-coordinate by -1. Graph the new ordered pairs and connect them so the resultant graph is symmetric with respect to the y-axis. See graph at left below.

b. For each of the given coordinate pairs, multiply the y-coordinate by -1. Graph the new ordered pairs and connect them so the resultant graph is symmetric with respect to the x-axis. See graph at right below.

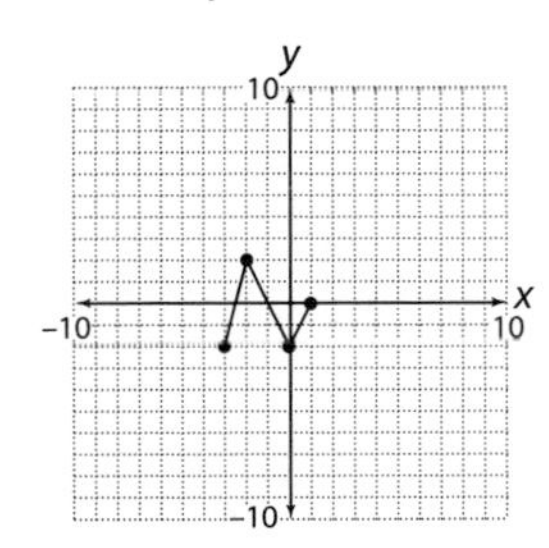

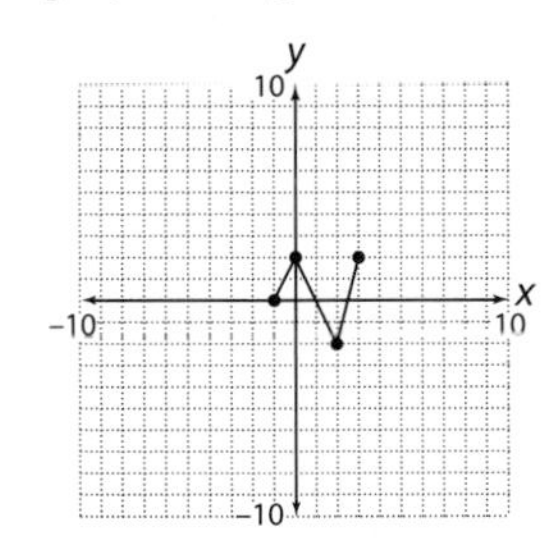

SECTION 2.6

You-Try-It 1

Place your pencil at 12 and then move up to the curve. Approximately 20% of the smokers were not smoking after 12 months.

You-Try-It 2

Place your pencil at 1972 and then move up to the curve. Approximately \$98 million was spent on presidential elections in 1972.

You-Try-It 3

a. Since $y = \frac{2}{3}x + 1$, the output value of $\frac{2}{3}x + 1$ when the input is 2.1 will be the y-coordinate on the graph of $y = \frac{2}{3}x + 1$ when $x = 2.1$.

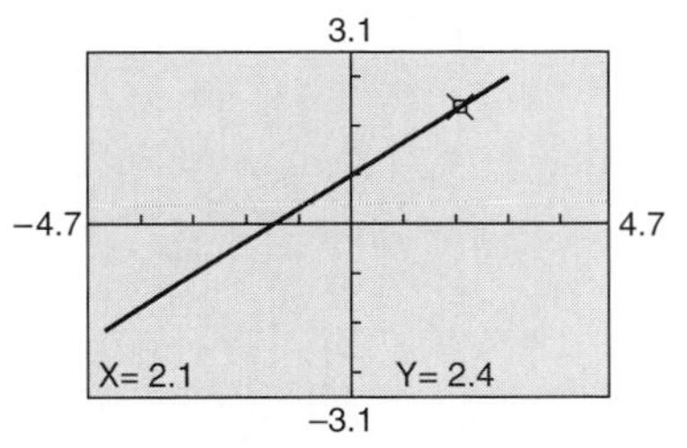

b. To find the input value for a given output, −0.8, trace along the graph until the given output value, Y, is shown at the bottom of the screen.

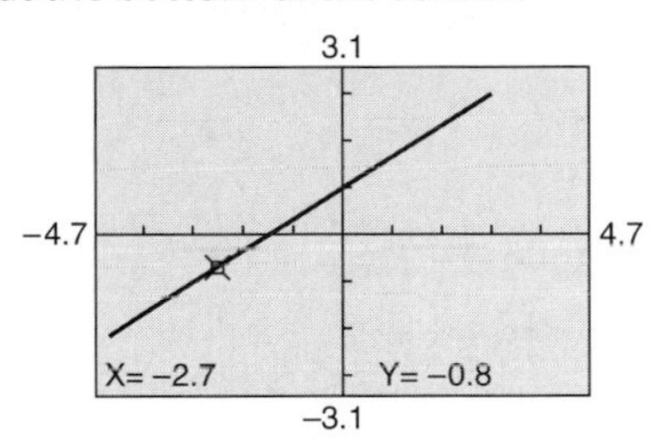

You-Try-It 4

To find the input, x, for which the output of $x^3 + 2$ is -3, graph Y1 = X^3 + 2 and Y2 = −3. Use a viewing window with Xmin = −3, Xmax = 1, Ymin = −5, Ymax = 5.

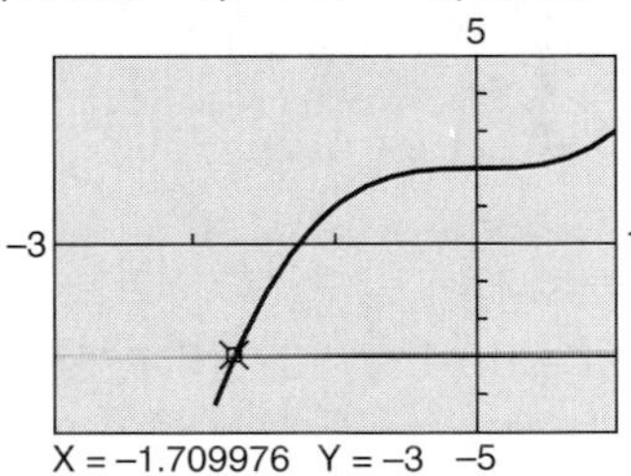

When $x = -1.709976$, the output is -3.

You-Try-It 5

We must find the time it takes the ferry to be 15 miles from Port Raven. That is, we are given the output, 15 miles, and must determine the input. Draw the graph of Y1 = 46 − 12X and Y2 = 15. Then use the INTERSECT feature to determine the coordinates of the point where the curves intersect.

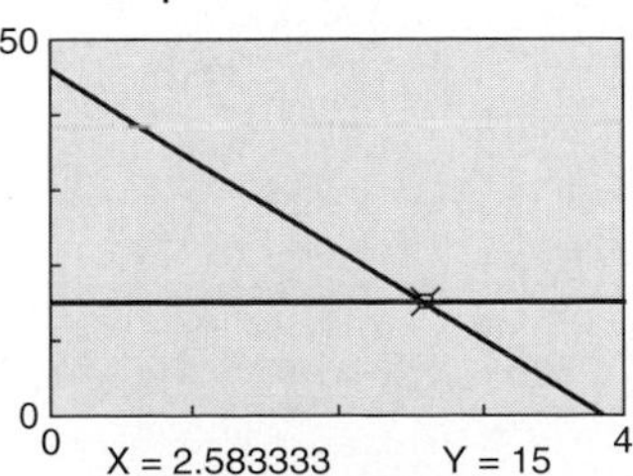

To the nearest tenth of an hour, it took 2.6 hours before the ferry was 15 miles from Port Raven.

You-Try-It 6

To find the number of grams of $AgNO_3$, draw the graphs of Y1 = 100X/(10 + X) and Y2 = 5. Then use the INTERSECT feature to determine the coordinates of the point where the curves intersect.

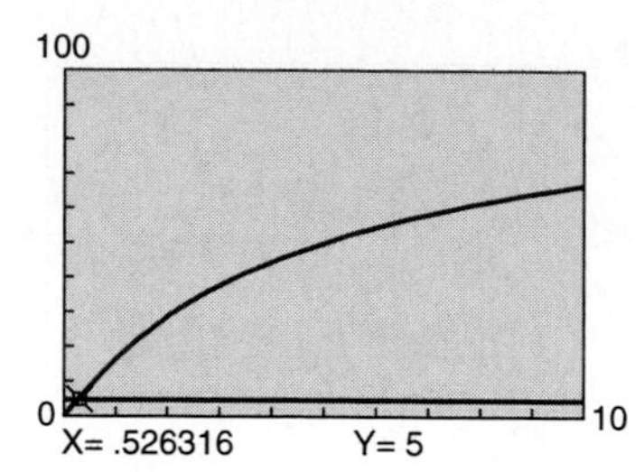

To the nearest hundredth of a gram, 0.53 gram of $AgNO_3$ is required to make a solution that has an $AgNO_3$ concentration of 5%.

Solutions to Chapter 3 You-Try-Its

SECTION 3.1

You-Try-It 1

When three numbers are multiplied together, the numbers can be grouped in any order and the product will be the same. Therefore, $4(3x) = (4 \cdot 3)x$.

You-Try-It 2

The sum of a number and its opposite is zero. Therefore, $12 + (-12) = 0$.

You-Try-It 3

a. $$\begin{aligned} -5(-3a) &= -5(-3 \cdot a) \\ &= [(-5) \cdot (-3)]a \\ &= 15a \end{aligned}$$

b. $$\begin{aligned} \left(-\frac{1}{2}c\right)2 &= \left(-\frac{1}{2} \cdot 2\right)c \\ &= -1 \cdot c \\ &= -c \end{aligned}$$

You-Try-It 4

a. $$\begin{aligned} 3a - 2b + 5a &= 3a + 5a - 2b \\ &= (3a + 5a) - 2b \\ &= 8a - 2b \end{aligned}$$

b. $$\begin{aligned} 2z^2 - 5z - 3z^2 + 6z &= 2z^2 - 3z^2 - 5z + 6z \\ &= (2z^2 - 3z^2) + (-5z + 6z) \\ &= -z^2 + z \end{aligned}$$

You-Try-It 5

For **a** through **d**, use the Distributive Property.

a. $-3(5y - 2) = -15y + 6$

b. $-(6c + 5) = -6c - 5$

c. $(3p - 7)(-3) = -9p + 21$

d. $-2(4x + 2y - 6z) = -8x - 4y + 12z$

You-Try-It 6

Use the Distributive Property, then combine like terms.

$$\begin{aligned} 7(-3x - 4y) - 3(3x + y) &= -21x - 28y - 9x - 3y \\ &= -30x - 31y \end{aligned}$$

You-Try-It 7

Use the Distributive Property first on the parentheses, combine like terms, then use the Distributive Property on the brackets, and again combine like terms.

$$\begin{aligned} 2v - 3[5 - 3(3 + 2v)] &= 2v - 3[5 - 9 - 6v] \\ &= 2v - 3[-4 - 6v] \\ &= 2v + 12 + 18v \\ &= 20v + 12 \end{aligned}$$

SECTION 3.2

You-Try-It 1

$3x$	$= x^2 - 4$
$3(4)$	$(4)^2 - 4$
12	$16 - 4$
12 =	12

Yes, 4 is a solution of the equation.

You-Try-It 2

$$\begin{aligned} y - 7 &= -10 \\ y - 7 + 7 &= -10 + 7 \\ y + 0 &= -3 \\ y &= -3 \end{aligned}$$

The solution is -3.

You-Try-It 3

$$\begin{aligned} -7 &= -3 + v \\ -7 + 3 &= -3 + 3 + v \\ -4 &= 0 + v \\ -4 &= v \end{aligned}$$

The solution is -4.

You-Try-It 4

$$\begin{aligned} 15 &= \frac{3y}{5} \\ 15 &= \frac{3}{5}y \\ \frac{5}{3}(15) &= \frac{5}{3}\left(\frac{3}{5}y\right) \\ 25 &= 1 \cdot y \\ 25 &= y \end{aligned}$$

The solution is 25.

You-Try-It 5

$$\begin{aligned} 9 &= 27q \\ \frac{9}{27} &= \frac{27q}{27} \\ \frac{1}{3} &= q \end{aligned}$$

The solution is $\frac{1}{3}$.

You-Try-It 6

$$\begin{aligned} 18 &= 2n + 6n \\ 18 &= 8n \\ \frac{18}{8} &= \frac{8n}{8} \\ \frac{9}{4} &= n \end{aligned}$$

The solution is $\frac{9}{4}$.

You-Try-It 7

$$\begin{aligned} PB &= A \\ P(60) &= 27 \\ 60P &= 27 \\ \frac{60P}{60} &= \frac{27}{60} \\ P &= 0.45 \end{aligned}$$

27 is 45% of 60.

You-Try-It 8

The goal is to find how many workers took a lunch break that was 15 minutes or less.

$$\begin{aligned} PB &= A \\ (0.55)(1260) &= A \\ 693 &= A \end{aligned}$$

693 workers took a lunch break that was 15 minutes or less.

You-Try-It 9

The goal is to find the percent decrease in sight distance. The amount of decrease is 1000 − 550 = 450. The base is 1000, the sight distance before the fog. The amount is 450, the decrease in sight distance.

$$\begin{aligned} PB &= A \\ 1000P &= 450 \\ \frac{1000P}{1000} &= \frac{450}{1000} \\ P &= 0.45 \end{aligned}$$

There was a 45% decrease in sight distance.

SECTION 3.3

You-Try-It 1

$$\begin{aligned} 8 &= 4x - 6 \\ 8 + 6 &= 4x - 6 + 6 \\ 14 &= 4x \\ \frac{14}{4} &= \frac{4x}{4} \\ \frac{7}{2} &= x \end{aligned}$$

The solution is $\frac{7}{2}$.

You-Try-It 2

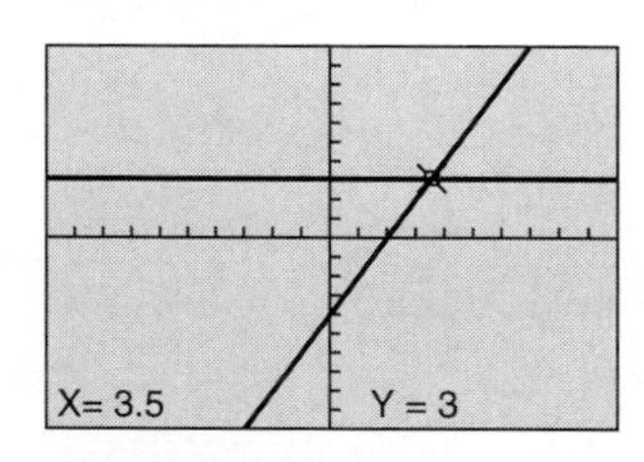

$3 = 2x - 4$	
3	$2(3.5) - 4$
3	$7 - 4$
3	= 3

The solution is 3.5.

You-Try-It 3

$$\begin{aligned} 2z - 5 &= 7z + 8 \\ 2z - 7z - 5 &= 7z - 7z + 8 \\ -5z - 5 &= 8 \\ -5z - 5 + 5 &= 8 + 5 \\ -5z &= 13 \\ \frac{-5z}{-5} &= \frac{13}{-5} \\ z &= -\frac{13}{5} \end{aligned}$$

The solution is $-\frac{13}{5}$.

You-Try-It 4

$$\begin{aligned} 4x + 3(2x - 1) &= 2(4x - 3) \\ 4x + 6x - 3 &= 8x - 6 \\ 10x - 3 &= 8x - 6 \\ 10x - 8x - 3 &= 8x - 8x - 6 \\ 2x - 3 &= -6 \\ 2x - 3 + 3 &= -6 + 3 \\ 2x &= -3 \\ \frac{2x}{2} &= \frac{-3}{2} \\ x &= -\frac{3}{2} \end{aligned}$$

The solution is $-\frac{3}{2}$.

You-Try-It 5

Find the LCM of the denominators. The LCM is 12.

$$\begin{aligned}
\frac{2x}{3} - \frac{1}{4} &= \frac{3x}{4} + \frac{1}{3} \\
12\left(\frac{2x}{3} - \frac{1}{4}\right) &= 12\left(\frac{3x}{4} + \frac{1}{3}\right) \\
12\left(\frac{2x}{3}\right) - 12\left(\frac{1}{4}\right) &= 12\left(\frac{3x}{4}\right) + 12\left(\frac{1}{3}\right) \\
8x - 3 &= 9x + 4 \\
8x - 9x - 3 &= 9x - 9x + 4 \\
-x - 3 &= 4 \\
-x &= 7 \\
x &= -7
\end{aligned}$$

Graphical check:

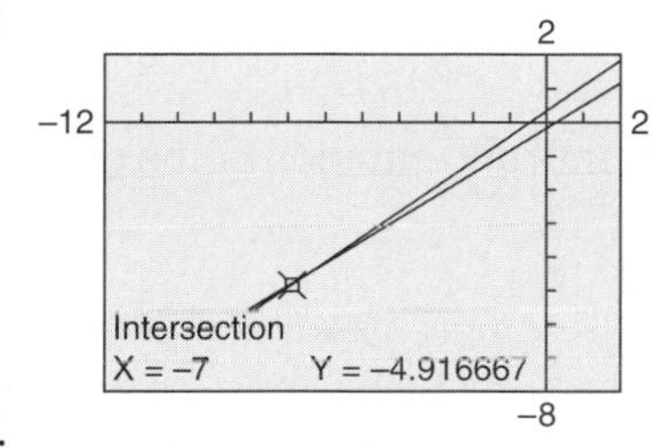

The solution is −7.

You-Try-It 6 The unknown number: n

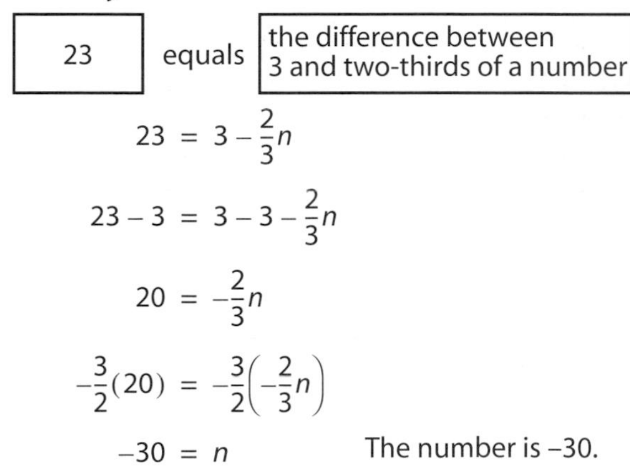

23	equals	the difference between 3 and two-thirds of a number

$$\begin{aligned}
23 &= 3 - \frac{2}{3}n \\
23 - 3 &= 3 - 3 - \frac{2}{3}n \\
20 &= -\frac{2}{3}n \\
-\frac{3}{2}(20) &= -\frac{3}{2}\left(-\frac{2}{3}n\right) \\
-30 &= n
\end{aligned}$$

The number is −30.

You-Try-It 7 The unknown number: n

9 less than twice a number	is	5 times the sum of the number and 12

$$\begin{aligned}
2n - 9 &= 5(n + 12) \\
2n - 9 &= 5n + 60 \\
2n - 5n - 9 &= 5n - 5n + 60 \\
-3n - 9 &= 60 \\
-3n - 9 + 9 &= 60 + 9 \\
-3n &= 69 \\
\frac{-3n}{-3} &= \frac{69}{-3} \\
n &= -23
\end{aligned}$$

The number is −23.

You-Try-It 8

Goal: To find the original value of the car.

Strategy: Use the basic percent equation, $pB = A$. The percent is 55%, the base is unknown, and the amount is \$12,237.50.

Solution:

$$\begin{aligned}
PB &= A \\
0.55B &= 12{,}237.50 \\
\frac{0.55B}{0.55} &= \frac{12{,}237.50}{0.55} \\
B &= 22{,}250
\end{aligned}$$

The original value of the car was \$22,250.

Be sure to check your work.

SECTION 3.4

You-Try-It 1

Let x represent the measure of one angle. Because the angles are supplements of each other, the measure of the supplementary angle is $180° - x$. Note that $x + (180° - x) = 180°$.

One angle	is	3 degrees more than twice its supplement

$$\begin{aligned}
x &= 2(180 - x) + 3 \\
x &= 360 - 2x + 3 \\
x &= 363 - 2x \\
3x &= 363 \\
x &= 121 \\
180 - x &= 180 - 121 = 59
\end{aligned}$$

The angles are 121° and 59°.

You-Try-It 2

Note that the angles are adjacent angles of intersecting lines; therefore the angles are supplementary angles.

$$\begin{aligned}
2x + 3x &= 180 \\
5x &= 180 \\
\frac{5x}{5} &= \frac{180}{5} \\
x &= 36 \\
3(36) &= 108 \\
2(36) &= 72 \\
108 + 72 &= 180
\end{aligned}$$

The larger angle is 108°.

You-Try-It 3

Alternate interior angles are equal.

$$\begin{aligned}
3x &= x + 40 \\
2x &= 40 \\
x &= 20
\end{aligned}$$

The value of x is 20°.

You-Try-It 4

The goal is to find $m\angle d$. Redraw the figure and label the figure as shown.

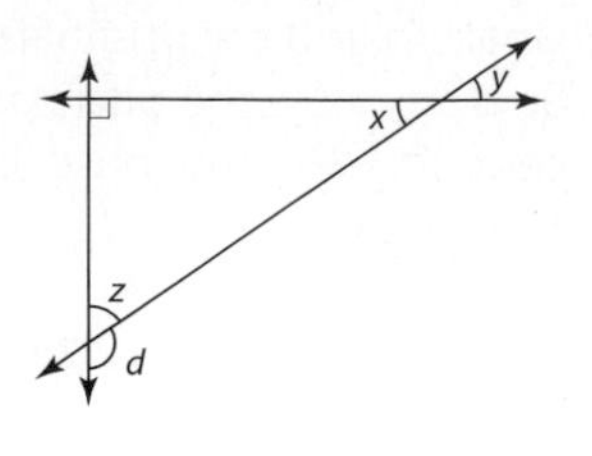

$\angle x$ and $\angle y$ are vertical angles. Thus, $m\angle x = 55°$. Because the triangle is a right triangle, $\angle x$ and $\angle z$ are complementary angles.

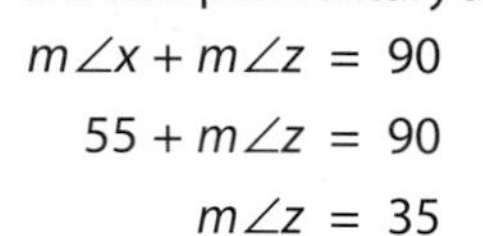

$$\begin{aligned} m\angle x + m\angle z &= 90 \\ 55 + m\angle z &= 90 \\ m\angle z &= 35 \end{aligned}$$

$\angle z$ and $\angle d$ are supplementary angles. Therefore,

$$\begin{aligned} m\angle d + m\angle z &= 180 \\ m\angle d + 35 &= 180 \\ m\angle d &= 145 \end{aligned}$$

The measure of angle d is 145°.

SECTION 3.5

You-Try-It 1

The goal is to find the rate of each train. Let r be the rate of the slower train. Then the rate of the faster train is $2r$. (It is traveling twice as fast.)
Use the equation $d = rt$.
Distance traveled by the slower train in 3 hours: $3r$
Distance traveled by the faster train in 3 hours: $3(2r)$
In 3 hours, the sum of the distances is 306 miles.

$$\begin{aligned} 3r + 3(2r) &= 306 \\ 3r + 6r &= 306 \\ 9r &= 306 \\ r &= 34 \end{aligned}$$

The slower train is going 34 miles per hour; the faster train is going 2(34) = 68 miles per hour.

You-Try-It 2

The goal is to find the distance to the countryside.
Let t represent the time traveling to the countryside. The time returning is 7 minus the time going, or $7 - t$.
Use the equation $d = rt$.
Distance traveled to the countryside: $16t$
Distance traveled returning: $12(7 - t)$
The distance to the countryside equals the distance returning.

$$\begin{aligned} 16t &= 12(7 - t) \\ 16t &= 84 - 12t \\ 28t &= 84 \\ t &= 3 \end{aligned}$$

The time traveling to the countryside is 3 hours. To find the distance to the countryside, use $d = rt$.
Distance = 16(3) = 48 miles
The distance to the countryside is 48 miles.

SECTION 3.6

You-Try-It 1

The goal is to find the percent concentration of the mixture.
Let r represent the percent concentration of the mixture.
Quantity of acid in 9% solution: 1.8(0.09)
Quantity of acid in 4% solution: 1.2(0.04)
There is 1.8 + 1.2 = 3 total liters of solution.
Quantity of acid in the mixture: $3r$
The quantity of acid in the mixture is the sum of the quantities of acid in each solution.

$$\begin{aligned} 1.8(0.09) + 1.2(0.04) &= 3r \\ 0.162 + 0.048 &= 3r \\ 0.21 &= 3r \\ \frac{0.21}{3} &= \frac{3r}{3} \\ 0.07 &= r \end{aligned}$$

The percent concentration of the mixture is 7%.

You-Try-It 2

The goal is to find how many pounds of a mixture that is 30% mocha java beans must be added to 80 pounds of a mixture that is 40% mocha java beans to produce a new mixture that is 32% mocha java beans.
Let x represent the number of pounds of the 30% mixture of mocha java beans.
Quantity of mocha java in 30% mixture: $0.30x$
Quantity of mocha java in 40% mixture: 0.40(80)
There is $x + 80$ total pounds of mixture.
Quantity of mocha java in the final mixture: $0.32(x + 80)$
The quantity of mocha java in the final mixture is the sum of the quantities of mocha java beans in each mixture.

$$\begin{aligned} 0.30x + 0.40(80) &= 0.32(x + 80) \\ 0.30x + 32 &= 0.32x + 25.6 \\ -0.02x &= -6.4 \\ x &= 320 \end{aligned}$$

320 pounds of the 40% mixture must be added.

You-Try-It 3

The goal is to find the number of ounces of a 25% gold alloy that must be mixed with 8 ounces of a 30% gold alloy to produce a new alloy that is 27% gold.
Let x represent the number of ounces of the 25% gold alloy.
Quantity of gold in 25% alloy: $0.25x$
Quantity of gold in 30% alloy: 0.30(8)
There is $x + 8$ total ounces of alloy.
Quantity of gold in the final alloy: $0.27(x + 8)$
The quantity of gold in the final alloy is the sum of the quantities of gold in each mixture.

$$\begin{aligned} 0.25x + 0.30(8) &= 0.27(x + 8) \\ 0.25x + 2.4 &= 0.27x + 2.16 \\ -0.02x &= -0.24 \\ x &= 12 \end{aligned}$$

The jeweler must use 12 ounces of the 25% alloy.

SECTION 3.7

You-Try-It 1

$\{x \mid x \le 5, x \in \text{integers}\}$

You-Try-It 2

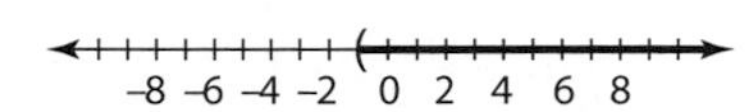

You-Try-It 3

$$x - 4 \le 1$$
$$x - 4 + 4 \le 1 + 4$$
$$x \le 5$$

–8 –6 –4 –2 0 2 4 6 8

The solution set is $\{x \mid x \le 5\}$.

You-Try-It 4

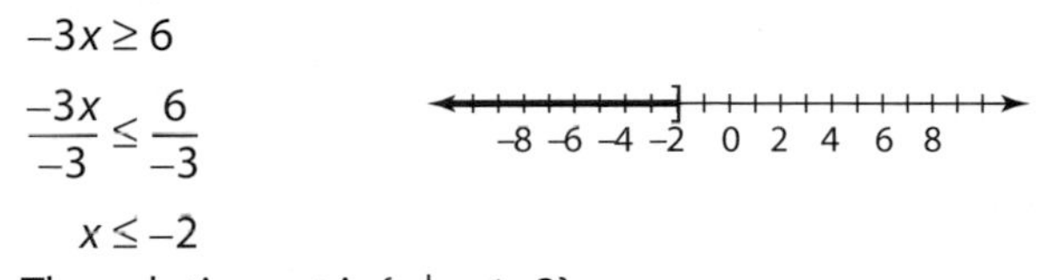

$$-3x \ge 6$$
$$\frac{-3x}{-3} \le \frac{6}{-3}$$
$$x \le -2$$

The solution set is $\{x \mid x \le -2\}$.

You-Try-It 5

$$3x - 5 > x + 7$$
$$3x - x - 5 > x - x + 7$$
$$2x - 5 > 7$$
$$2x - 5 + 5 > 7 + 5$$
$$2x > 12$$
$$\frac{2x}{2} > \frac{12}{2}$$
$$x > 6$$

The solution set is $\{x \mid x > 6\}$.

You-Try-It 6

$$3(3 - 2x) \ge -5x - 2(3 - x)$$
$$9 - 6x \ge -5x - 6 + 2x$$
$$9 - 6x \ge -3x - 6$$
$$9 - 6x + 3x \ge -3x + 3x - 6$$
$$9 - 3x \ge -6$$
$$9 - 9 - 3x \ge -6 - 9$$
$$-3x \ge -15$$
$$\frac{-3x}{-3} \le \frac{-15}{-3}$$
$$x \le 5$$

The solution set is $\{x \mid x \le 5\}$.

You-Try-It 7

The goal is to find the maximum height as an integer so that the area is less than 112 square inches.

The base, b, of the triangle is given as 8 inches and the height is given as $(3x - 5)$ inches. The area of a triangle is given by $A = \frac{1}{2}bh$, where b is the base and h is the height.

$$\frac{1}{2}bh < A$$
$$\frac{1}{2}(8)(3x - 5) < 112$$
$$4(3x - 5) < 112$$
$$12x - 20 < 112$$
$$12x - 20 + 20 < 112 + 20$$
$$12x < 132$$
$$\frac{12x}{12} < \frac{132}{12}$$
$$x < 11$$

Because $x < 11$,

$3x - 5 < 3(11) - 5 = 28$

The height must be an integer less than 28. Therefore, the maximum height of the triangle is 27 inches.

Solutions to Chapter 4 You-Try-Its

SECTION 4.1

You-Try-It 1

Domain: {−3, −2, −1, 0, 1, 2, 3}
Range: {0, 1, 2, 3}

Because no two ordered pairs have the same first coordinate, the relation is a function.

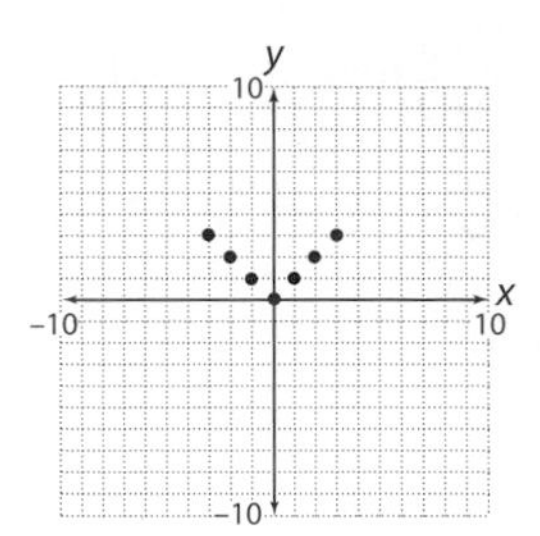

You-Try-It 2

$V(x) = x^3$

$V(2.5) = (2.5)^3$

$V(2.5) = 15.625$

The volume is 15.625 cubic meters.

You-Try-It 3

$f(x) = 3x - 2$

$f(-2) = 3(-2) - 2 = -6 - 2 = -8$

$f(-1) = 3(-1) - 2 = -3 - 2 = -5$

$f(0) = 3(0) - 2 = 0 - 2 = -2$

$f(1) = 3(1) - 2 = 3 - 2 = 1$

$f(2) = 3(2) - 2 = 6 - 2 = 4$

When the domain is {−2, −1, 0, 1, 2}, the range of $f(x) = 3x - 2$ is {−8, −5, −2, 1, 4}.

You-Try-It 4

x	$y = f(x) = 2 - \frac{x}{2}$	y	(x, y)
−6	$f(-6) = 2 - \frac{-6}{2}$	5	(−6, 5)
0	$f(0) = 2 - \frac{0}{2}$	2	(0, 2)
6	$f(6) = 2 - \frac{6}{2}$	−1	(6, −1)

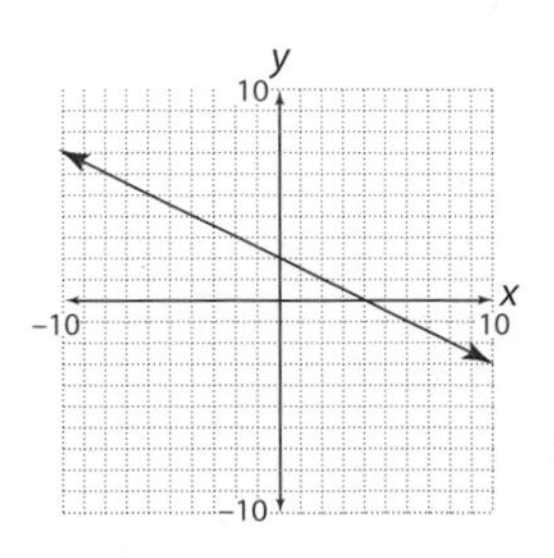

You-Try-It 5

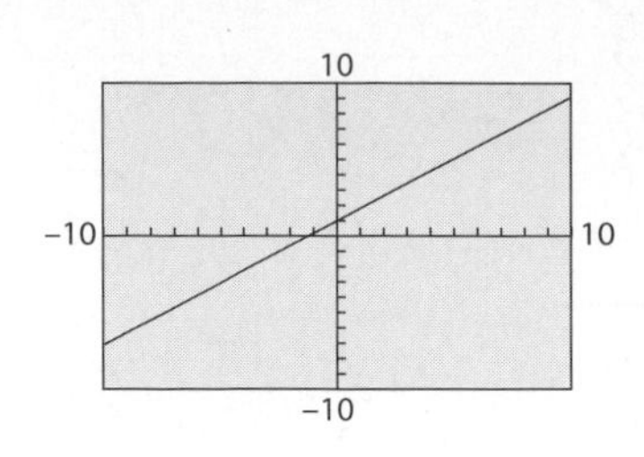

$$g(x) = \frac{4}{5}x + 1$$

$$g(5) = \frac{4(5)}{5} + 1 = 5$$

$$g(0) = \frac{4(0)}{5} + 1 = 1$$

$$g(-10) = \frac{4(-10)}{5} + 1 = -7$$

You-Try-It 6

$$\begin{aligned} f(x) &= 2x + 5 \\ 1 &= 2x + 5 \\ -4 &= 2x \\ -2 &= x \end{aligned}$$

The value of x is −2.

You-Try-It 7

Evaluate the regression equation when $x = 31$.

$$\begin{aligned} y &= 2x + 18.7 \\ &= 2(31) + 18.7 \\ &= 62 + 18.7 \\ &= 80.7 \end{aligned}$$

When the student's score on the placement test is 31, the student's course average in physics is predicted to be 81.

SECTION 4.2

You-Try-It 1

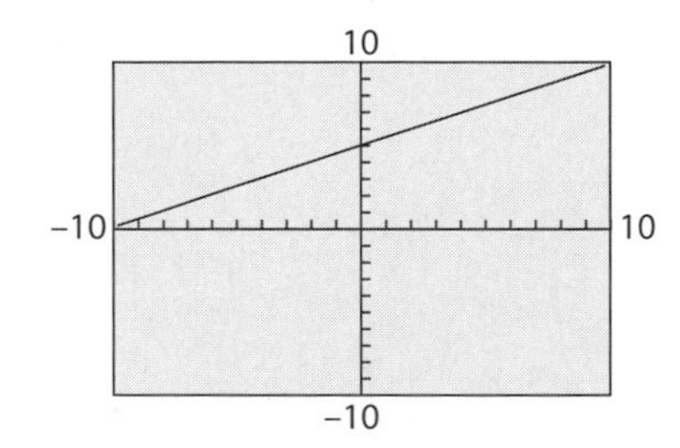

You-Try-It 2

The graph of $y = -3$ is a horizontal line passing through the point (0, −3).

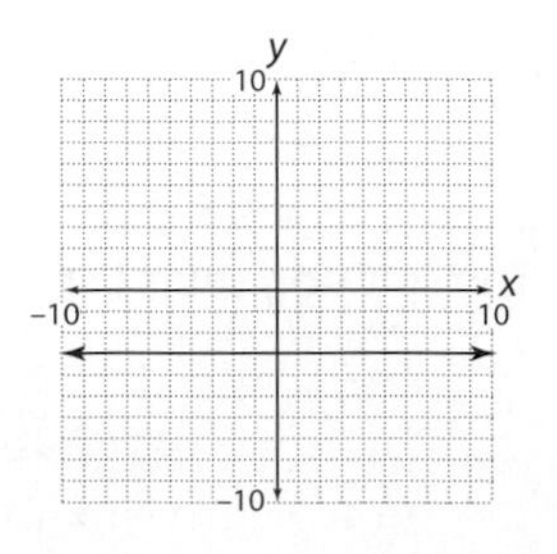

You-Try-It 3

To find the x-intercept, let $y = 0$.

$$\begin{aligned} 3x - 5y &= 15 \\ 3x - 5(0) &= 15 \\ 3x &= 15 \\ x &= 5 \end{aligned}$$

The x-intercept is $(5, 0)$.

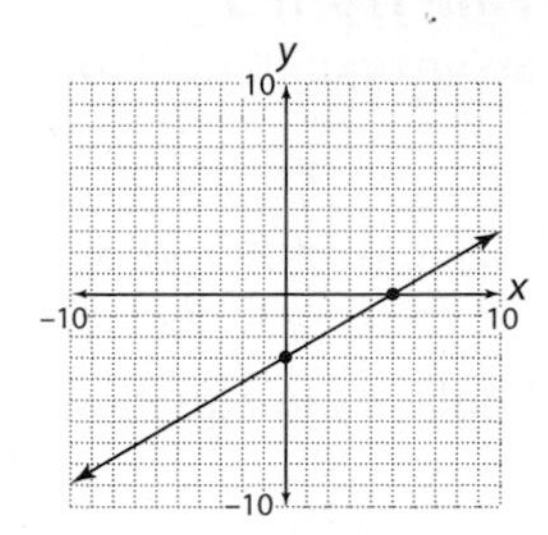

To find the y-intercept, let $x = 0$.

$$\begin{aligned} 3x - 5y &= 15 \\ 3(0) - 5y &= 15 \\ -5y &= 15 \\ y &= -3 \end{aligned}$$

The y-intercept is $(0, -3)$.

You-Try-It 4

To find the x-intercept, let $y = 0$.

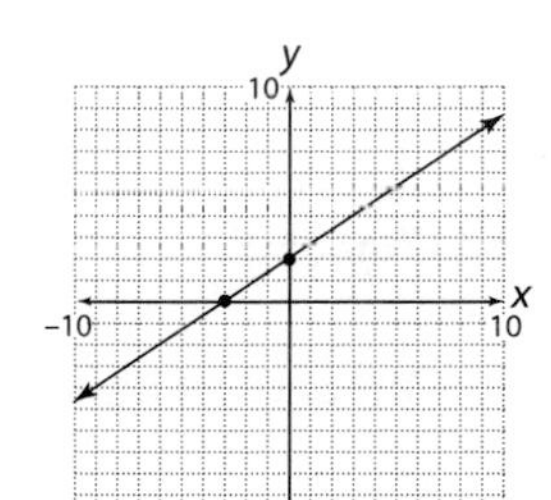

$$\begin{aligned} y &= \frac{2}{3}x + 2 \\ 0 &= \frac{2}{3}x + 2 \\ -2 &= \frac{2}{3}x \\ -6 &= 2x \\ -3 &= x \end{aligned}$$

The x-intercept is $(-3, 0)$.

To find the y-intercept, let $x = 0$.

$$\begin{aligned} y &= \frac{2}{3}x + 2 \\ y &= 0 + 2 \\ y &= 2 \end{aligned}$$

The y-intercept is $(0, 2)$.

SECTION 4.3

You-Try-It 1

a. $m = \dfrac{y_2 - y_1}{x_2 - x_1} = \dfrac{-2 - 1}{2 - (-3)} = \dfrac{-3}{5}$. The slope is $-\dfrac{3}{5}$.

b. $m = \dfrac{y_2 - y_1}{x_2 - x_1} = \dfrac{2 - 2}{4 - (-1)} = \dfrac{0}{5}$. The slope is 0.

You-Try-It 2

$$m = \frac{76 - 38}{8 - 4} = \frac{38}{4} = \frac{19}{2} = 9.5$$

A slope of 9.5 means that Griffith-Joyner's average speed for her run was 9.5 meters per second.

You-Try-It 3

$y = \dfrac{1}{4}x - 1$

y-intercept $= (0, b) = (0, -1)$

$m = \dfrac{1}{4}$

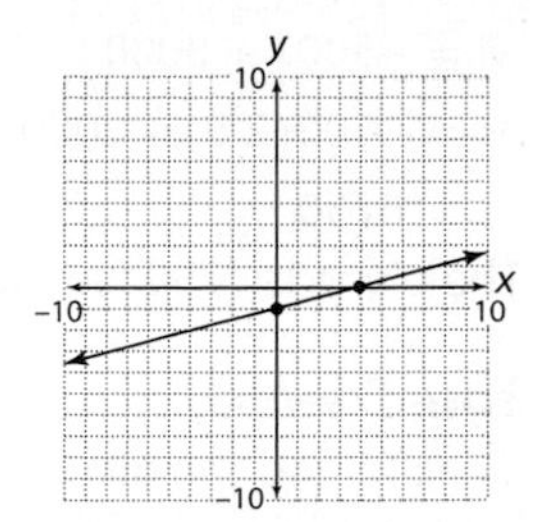

You-Try-It 4

$$\begin{aligned} x - 2y &= 4 \\ -2y &= -x + 4 \\ y &= \frac{1}{2}x - 2 \end{aligned}$$

y-intercept $= (0, b) = (0, -2)$.

$m = \dfrac{1}{2}$

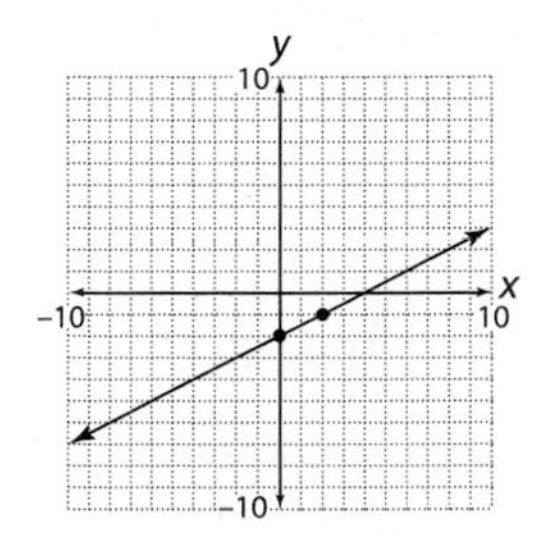

SECTION 4.4

You-Try-It 1

$$\begin{aligned} y &= mx + b \\ y &= \frac{2}{3}x + b && \bullet\ m = \frac{2}{3}. \\ -2 &= \frac{2}{3}(3) + b && \bullet\ x_1 = 3, y_1 = -2. \\ -2 &= 2 + b \\ -4 &= b \\ y &= \frac{2}{3}x - 4 \end{aligned}$$

You-Try-It 2

Given point coordinates $(1, 4)$, $x_1 = 1$ and $y_1 = 4$. The slope is $m = -3$.

$$\begin{aligned} y - y_1 &= m(x - x_1) \\ y - 4 &= -3(x - 1) \\ y - 4 &= -3x + 3 \\ y &= -3x + 7 \end{aligned}$$

You-Try-It 3

Goal: To find the plane's path of descent.

Strategy: Use the point-slope formula to find the equation.

$m = -500$; $(0, 8000)$.

$$\begin{aligned} y - y_1 &= m(x - x_1) \\ y - 8000 &= -500(x - 0) \\ y - 8000 &= -500x \\ y &= -500x + 8000 \end{aligned}$$

The equation of the path is $y = -500x + 8000$.

$$y = -500x + 8000$$
$$y = -500(2.5) + 8000$$
$$= 6750$$

The plane is 6750 feet above the airport after 2.5 minutes.

You-Try-It 4

Goal: To find the diver's depth below the surface.
Strategy: Use the point-slope formula to find the equation.
$(0, 0)$; $m = -20$.

$$y - y_1 = m(x - x_1)$$
$$y - 0 = -20(x - 0)$$
$$y = -20x$$

SECTION 4.5

You-Try-It 1

Solve the equation $3x + 4y = 2$ for y.

$$3x + 4y = 2$$
$$4y = -3x + 2$$
$$y = -\frac{3}{4}x + \frac{1}{2}$$

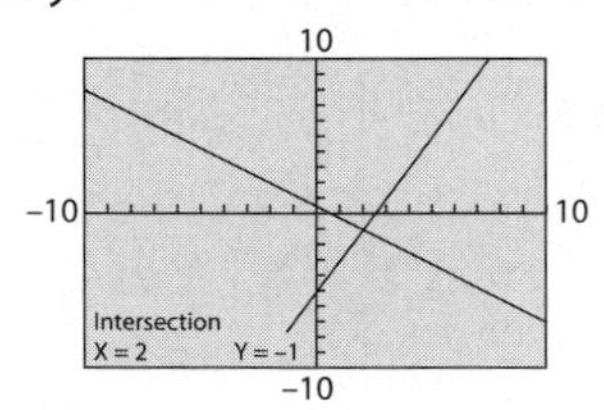

Graph $y = 2x - 5$ and
$y = -\frac{3}{4}x + \frac{1}{2}$.
The solution is $(2, -1)$.

You-Try-It 2

Solve each equation for y.

$$2x + 3y = 6 \qquad 4x + 6y = 12$$
$$3y = -2x + 6 \qquad 6y = -4x + 12$$
$$y = -\frac{2}{3}x + 2 \qquad y = -\frac{2}{3}x + 2$$

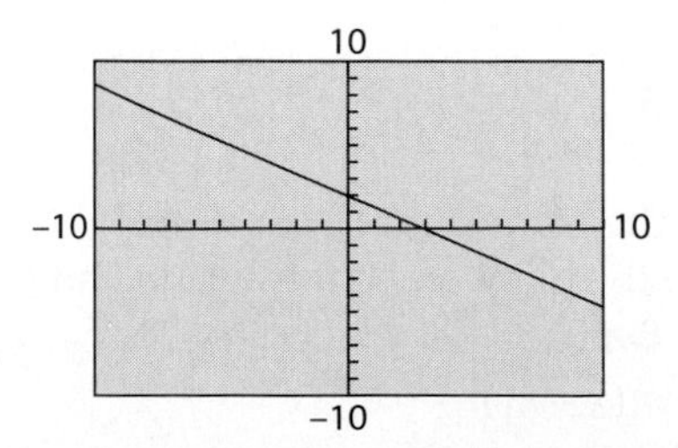

The graphs of the two lines are exactly the same. Therefore the solutions of this system of equations are $(x, -\frac{2}{3}x + 2)$.

You-Try-It 3

Solve equation (2) for y.

$$2x - y = 1$$
$$-y = -2x + 1$$
$$y = 2x - 1$$

Replace y in equation (1) and solve for x.

$$3x + 4(2x - 1) = 18$$
$$3x + 8x - 4 = 18$$
$$11x - 4 = 18$$
$$11x = 22$$
$$x = 2$$

$$2(2) - y = 1$$
$$4 - y = 1$$
$$y = 3$$

The solution is $(2, 3)$.

You-Try-It 4

Solve the second equation for y.

$$2x + y = 1$$
$$y = -2x + 1$$

Substitute into the first equation and solve for x.

$$4x + 2(-2x + 1) = 6$$
$$4x - 4x + 2 = 6$$
$$2 = 6 \qquad \text{Not a true equation.}$$

This is not a true equation. The system of equations is inconsistent and does not have a solution.

You-Try-It 5

Solve Equation (2) for y.
$y = -2x + 3$
Replace y by $-2x + 3$ in the first equation.

$$4x + 2(-2x + 3) = 6$$
$$4x - 4x + 6 = 6$$
$$6 = 6$$

The system of equations is dependent. The solutions are the ordered pairs $(x, -2x + 3)$.

You-Try-It 6

Goal: To find the amount invested in the mutual fund.
Strategy: Let x represent the amount in the mutual fund which is invested at 13% and y represent the amount invested at 7%.

The amount invested at 7% is \$2500 more than the amount invested at 13%.

$y = x + 2500$

Interest earned at 13%: $0.13x$

Interest earned at 7%: $0.07y$

The sum of the interest earned is \$475.

$0.13x + 0.07y = 475$

Solve:

$$\begin{aligned} y &= x + 2500 \\ 0.13x + 0.07y &= 475 \end{aligned}$$

$$\begin{aligned} 0.13x + 0.07(x + 2500) &= 475 \\ 0.13x + 0.07x + 175 &= 475 \\ 0.20x &= 300 \\ x &= 1500 \end{aligned}$$

There was \$1500 invested in the mutual fund.

Check your work.

You-Try-It 7

Goal: To find the dimensions of a triangle.

Strategy: Let x represent the length of the side of the triangle that is not equal to the other two. Let y represent the length of one of the two equal sides.

The length of one of the equal sides is 4 times the length of the other side.

$y = 4x$

The perimeter is 180 centimeters.

$x + 2y = 180$

Solve:

$$\begin{aligned} y &= 4x \quad (1) \\ x + 2y &= 180 \quad (2) \end{aligned}$$

$$\begin{aligned} x + 2(4x) &= 180 \\ x + 8x &= 180 \\ 9x &= 180 \\ x &= 20 \end{aligned}$$

One side is 20 centimeters.

To find the length of one of the equal sides, replace x by 20 in Equation (1).

$$\begin{aligned} y &= 4x \\ &= 4(20) \\ &= 80 \end{aligned}$$

The length of one of the equal sides is 80 centimeters.

Be sure to check your work.

SECTION 4.6

You-Try-It 1

$$\begin{aligned} 3x - y &= 10 \\ 2x + 5y &= 1 \end{aligned}$$

$$\begin{aligned} 5(3x - y) &= 5(10) \\ 2x + 5y &= 1 \end{aligned}$$

• Eliminate y. Multiply Equation (1) by 5.

$$\begin{aligned} 15x - 5y &= 50 \\ 2x + 5y &= 1 \\ 17x &= 51 \\ x &= 3 \end{aligned}$$

$$\begin{aligned} 3x - y &= 10 \\ 3(3) - y &= 10 \\ 9 - y &= 10 \\ -y &= 1 \\ y &= -1 \end{aligned}$$

• This is Equation (1).

• Replace x by 3.

The solution is $(3, -1)$.

You-Try-It 2

$$\begin{aligned} 2x - 3y &= 4 \\ -4x + 6y &= -8 \end{aligned}$$

$$\begin{aligned} 2(2x - 3y) &= 2(4) \\ -4x + 6y &= -8 \end{aligned}$$

• Eliminate x. Multiply Equation (2) by 2.

$$\begin{aligned} 4x - 6y &= 8 \\ -4x + 6y &= -8 \\ 0 &= 0 \end{aligned}$$

• This is a true equation.

The system of equations is dependent. Solve Equation (1) for y.

$$\begin{aligned} 2x - 3y &= 4 \\ -3y &= -2x + 4 \\ y &= \frac{2}{3}x - \frac{4}{3} \end{aligned}$$

The ordered pair solutions are $\left(x, \frac{2}{3}x - \frac{4}{3}\right)$.

You-Try-It 3

Goal: To find the rate of the current and the rate of the canoe in calm water.

Strategy: Let r represent the rate of the canoe and let c represent the rate of the current. With the current, the rate of the canoe is $r + c$. Against the current, the rate of the canoe is $r-c$. Use the equation $d = rt$ to write equations for the distance traveled by the canoe with and against the current.

Solution:

$$\begin{aligned} 3(r+c) &= 24 \\ 4(r-c) &= 24 \\ r+c &= 8 && \text{• Divide by 3.} \\ r-c &= 6 && \text{• Divide by 4.} \\ 2r &= 14 && \text{• Add the equations.} \\ r &= 7 \end{aligned}$$

The rate of the canoe in calm water is 7 miles per hour.

Replace r by 7 in the equation $r + c = 8$ and solve for c.

$$\begin{aligned} r+c &= 8 \\ 7+c &= 8 \\ c &= 1 \end{aligned}$$

The rate of the current is 1 mile per hour.

You-Try-It 4

Goal: To find the number of gallons of each paint required to make the desired mixture.

Strategy: Let x represent the number of gallons of the paint that is 21% green dye and let y represent the number of gallons of the paint that is 15% green dye. The total of the two mixtures is 60 gallons.

$$x + y = 60$$

Quantity of green dye in 21% paint: $0.21x$

Quantity of green dye in 15% paint: $0.15y$

Quantity of green dye in 19% paint: $0.19(60)$

The sum of the quantities of green dye in the two mixtures equals the quantity of green dye in the final mixture.

$$0.21x + 0.15y = 11.4$$

Solution:

$$\begin{aligned} x+y &= 60 && (1) \\ 0.21x + 0.15y &= 11.4 && (2) \end{aligned}$$

$$\begin{aligned} -0.21(x+y) &= -0.21(60) \\ 0.21x + 0.15y &= 11.4 \end{aligned}$$

$$\begin{aligned} -0.21x - 0.21y &= -12.6 \\ 0.21x + 0.15y &= 11.4 \\ -0.06y &= -1.2 \\ y &= 20 \end{aligned}$$

Replace y in Equation (1) by 20 and solve for x.

$$\begin{aligned} x+y &= 60 \\ x+20 &= 60 \\ x &= 40 \end{aligned}$$

40 gallons of the 21% green dye paint must be mixed with 20 gallons of the 15% green dye paint.

You-Try-It 5

Goal: To find how many quarts of pure orange juice are required to make the desired mixture.

Strategy: There is only one unknown. Let x represent the number of quarts of pure orange juice to add. Because this orange juice is being added to 5 quarts of an existing mixture, the final mixture will contain $(x + 5)$ quarts.

Quantity of orange juice in pure juice: $1x$

Quantity of orange juice in 10% juice: $0.10(5)$

Quantity of orange juice in 25% juice: $0.25(x + 5)$

The sum of the quantities of orange juice in the two mixtures equals the quantity of orange juice in the final mixture.

$$x + 0.10(5) = 0.25(x + 5)$$

Solution:

$$\begin{aligned} x + 0.10(5) &= 0.25(x+5) \\ x + 0.5 &= 0.25x + 1.25 \\ 0.75x &= 0.75 \\ x &= 1 \end{aligned}$$

One quart of orange juice must be added.

Solutions to Chapter 5 You-Try-Its

SECTION 5.1

You-Try-It 1

$$16a^4b^3 + 10a^4b^3 + 5a^4b^3 = (16 + 10 + 5)a^4b^3$$
$$= 31a^4b^3$$

You-Try-It 2

$$37m^3n^2p - 14m^3n^2p = (37 - 14)m^3n^2p$$
$$= 23m^3n^2p$$

You-Try-It 3

$$t^3 \cdot t^8 = t^{3+8} = t^{11}$$

You-Try-It 4

$$n^6 \cdot n \cdot n^2 = n^{6+1+2} = n^9$$

You-Try-It 5

$$c^9(c^5d^8) = c^{9+5}d^8 = c^{14}d^8$$

You-Try-It 6

$$(5y^4)(3y^2) = (5 \cdot 3)(y^4 \cdot y^2) = 15y^{4+2} = 15y^6$$

You-Try-It 7

$$(12p^4q^3)(-3p^5q^2) = [12(-3)](p^{4+5})(q^{3+2})$$
$$= -36p^9q^5$$

You-Try-It 8

$$(t^3)^6 = t^{3 \cdot 6} = t^{18}$$

You-Try-It 9

$$(bc^7)^8 = b^{1 \cdot 8}c^{7 \cdot 8} = b^8c^{56}$$

You-Try-It 10

$$(4y^6)^3 = 4^{1 \cdot 3}y^{6 \cdot 3}$$
$$= 4^3y^{18}$$
$$= 64y^{18}$$

You-Try-It 11

$$(2v^6w^9)^5 = 2^{1 \cdot 5}v^{6 \cdot 5}w^{9 \cdot 5}$$
$$= 2^5v^{30}w^{45}$$
$$= 32v^{30}w^{45}$$

You-Try-It 12

$$(-2x^3y^7)^3 = (-2)^{1 \cdot 3}x^{3 \cdot 3}y^{7 \cdot 3}$$
$$= (-2)^3x^9y^{21}$$
$$= -8x^9y^{21}$$

You-Try-It 13

$$(-xy^4)(-2x^3y^2)^2$$
$$= (-xy^4)[(-2)^{1 \cdot 2}x^{3 \cdot 2}y^{2 \cdot 2}]$$
$$= (-xy^4)[(-2)^2x^6y^4]$$
$$= (-xy^4)(4x^6y^4)$$
$$= (-1 \cdot 4)(x^{1+6})(y^{4+4})$$
$$= -4x^7y^8$$

You-Try-It 14

$$\frac{t^{10}}{t^4} = t^{10-4} = t^6$$

You-Try-It 15

$$\frac{a^7b^6}{ab^3} = a^{7-1}b^{6-3} = a^6b^3$$

You-Try-It 16

$$(-8x^2y^7)^0 = 1$$

You-Try-It 17

$$-(9c^7d^4)^0 = -1$$

You-Try-It 18

$$\frac{2}{c^{-4}} = 2 \cdot \frac{1}{c^{-4}} = 2 \cdot c^4 = 2c^4$$

You-Try-It 19

$$\frac{12x^{-8}y}{-16xy^{-3}} = -\frac{3x^{-8}y}{4xy^{-3}} = -\frac{3x^{-8-1}y^{1-(-3)}}{4} = -\frac{3x^{-9}y^4}{4} = -\frac{3y^4}{4x^9}$$

You-Try-It 20

$$\begin{aligned}(-3ab)(2a^3b^{-2})^{-3} &= (-3ab)(2^{-3}a^{-9}b^6)\\ &= \frac{-3ab \cdot b^6}{2^3a^9}\\ &= -\frac{3b^7}{8a^8}\end{aligned}$$

You-Try-It 21

a. $57{,}000{,}000{,}000 = 5.7 \times 10^{10}$
b. $0.000000017 = 1.7 \times 10^{-8}$

You-Try-It 22

a. $5 \times 10^{12} = 5{,}000{,}000{,}000{,}000$
b. $4.0162 \times 10^{-9} = 0.0000000040162$

You-Try-It 23

a. $(2.4 \times 10^{-9})(1.6 \times 10^3) = 3.84 \times 10^{-6}$

b. $\dfrac{5.4 \times 10^{-2}}{1.8 \times 10^{-4}} = 3 \times 10^2$

SECTION 5.2

You-Try-It 1

$$\begin{aligned}V(r) &= 2000r^3 + 6000r^2 + 6000r + 2000\\ V(0.07) &= 2000(0.07)^3 + 6000(0.07)^2 + 6000(0.07) + 2000\\ &= 0.686 + 29.4 + 420 + 2000\\ &= 2450.086\end{aligned}$$

The value after three years of \$2000 deposited in an IRA earning an interest rate of 7% is \$2450.09.

You-Try-It 2

$$\begin{aligned}&(-4x^2 - 3xy + 2y^2) + (3x^2 - 4y^2)\\ &= (-4x^2 + 3x^2) - 3xy + (2y^2 - 4y^2)\\ &= -x^2 - 3xy - 2y^2\end{aligned}$$

You-Try-It 3

$$\begin{aligned}&(5x^2 - 3x + 4) - (-6x^3 - 2x + 8)\\ &= (5x^2 - 3x + 4) + (6x^3 + 2x - 8)\\ &= 6x^3 + 5x^2 + (-3x + 2x) + (4 - 8)\\ &= 6x^3 + 5x^2 - x - 4\end{aligned}$$

You-Try-It 4

Goal

The goal is to write a variable expression for the company's monthly profit from selling n videotapes.

Strategy

Use the formula $P = R - C$. Substitute the given polynomials for R and C. Then subtract the polynomials.

Solution

$P = R - C$

$P = (-0.2n^2 + 175n) - (35n + 2000)$

$P = (-0.2n^2 + 175n) + (-35n - 2000)$

$P = -0.2n^2 + (175n - 35n) - 2000$

$P = -0.2n^2 + 140n - 2000$

The company's monthly profit, in dollars, is $-0.2n^2 + 140n - 2000$.

Check 

You-Try-It 5

a. $(-2d + 3)(-4d) = (-2d)(-4d) + (3)(-4d)$

$= 8d^2 - 12d$

b. $(-a^3)(3a^2 + 2a - 7)$

$= (-a^3)(3a^2) + (-a^3)(2a) + (-a^3)(-7)$

$= -3a^5 - 2a^4 + 7a^3$

You-Try-It 6

$$
\begin{array}{rrrrr}
 & 3c^3 & -\,2c^2 & +\,c & -\,3 \\
 & & & 2c & +\,5 \\
\hline
 & 15c^3 & -\,10c^2 & +\,5c & -\,15 \\
6c^4 & -\,4c^3 & +\,2c^2 & -\,6c & \\
\hline
6c^4 & +\,11c^3 & -\,8c^2 & -\,c & -\,15
\end{array}
$$

You-Try-It 7

a. $(4y - 5)(3y - 3)$

$= (4y)(3y) + (4y)(-3) + (-5)(3y) + (-5)(-3)$

$= 12y^2 - 12y - 15y + 15$

$= 12y^2 - 27y + 15$

b. $(3a + 2b)(3a - 5b)$

$= (3a)(3a) + (3a)(-5b) + (2b)(3a) + (2b)(-5b)$

$= 9a^2 - 15ab + 6ab - 10b^2$

$= 9a^2 - 9ab - 10b^2$

You-Try-It 8

$(3x + 2y)^2 = (3x + 2y)(3x + 2y)$

$= 9x^2 + 6xy + 6xy + 4y^2$

$= 9x^2 + 12xy + 4y^2$

You-Try-It 9

Goal

The goal is to write a variable expression for the area of a circle that has a radius of $(x - 4)$ feet.

Strategy

Use the formula $A = \pi r^2$. Substitute the given binomial for r. Then multiply.

Solution

$A = \pi r^2$

$A = \pi(x - 4)^2$

$A = \pi(x - 4)(x - 4)$

$A = \pi(x^2 - 4x - 4x + 16)$

$A = \pi(x^2 - 8x + 16)$

$A = \pi x^2 - 8\pi x + 16\pi$

The area of the circle is $(\pi x^2 - 8\pi x + 16\pi)$ square feet.

Check √

You-Try-It 10

$$\frac{4x^3y + 8x^2y^2 - 4xy^3}{2xy} = \frac{4x^3y}{2xy} + \frac{8x^2y^2}{2xy} - \frac{4xy^3}{2xy}$$

$$= 2x^2 + 4xy - 2y^2$$

You-Try-It 11

$x^3 - 7 - 2x = x^3 + 0x^2 - 2x - 7$

$$\begin{array}{r}
x^2 + 2x + 2 \\
x - 2\,\overline{)\,x^3 + 0x^2 - 2x - 7} \\
\underline{x^3 - 2x^2} \qquad\qquad \\
2x^2 - 2x \qquad \\
\underline{2x^2 - 4x} \qquad \\
2x - 7 \\
\underline{2x - 4} \\
-3
\end{array}$$

$(x^3 - 2x - 7) \div (x - 2) = x^2 + 2x + 2 - \dfrac{3}{x - 2}$

You-Try-It 12

$$\begin{array}{r}
x^3 + 2 \qquad\qquad\qquad \\
x + 5\,\overline{)\,x^4 + 5x^3 + 0x^2 + 2x + 10} \\
\underline{x^4 + 5x^3} \qquad\qquad\qquad\quad \\
0 \qquad\quad + 2x + 10 \\
\underline{2x + 10} \\
0
\end{array}$$

Another factor of $x^4 + 5x^2 + 2x + 10$ is $x^3 + 2$.

You-Try-It 13

Goal

The goal is to write a variable expression for the height of a parallelogram that has an area of $(3x^2 + 2x - 8)$ feet and a base of $(x + 2)$ feet.

Strategy

Use the formula $h = \dfrac{A}{b}$. Substitute the given polynomials for A and b. Then divide.

Solution

$h = \dfrac{A}{b} = \dfrac{3x^2 + 2x - 8}{x + 2}$

$$\begin{array}{r}
3x - 4 \\
x + 2\,\overline{)\,3x^2 + 2x - 8} \\
\underline{3x^2 + 6x} \quad \\
-4x - 8 \\
\underline{-4x - 8} \\
0
\end{array}$$

The height of the parallelogram is $(3x - 4)$ feet.

Check √

SECTION 5.3

You-Try-It 1

$6x^4y^2 - 9x^3y^2 + 12x^2y^4$

$6x^4y^2 = 2 \cdot 3 \cdot x^4 \cdot y^2$
$9x^3y^2 = 3 \cdot 3 \cdot x^3 \cdot y^2$
$12x^2y^4 = 2 \cdot 2 \cdot 3 \cdot x^2 \cdot y^4$

The GCF is $3x^2y^2$.

$\dfrac{6x^4y^2 - 9x^3y^2 + 12x^2y^4}{3x^2y^2} = 2x^2 - 3x + 4y^2$

$6x^4y^2 - 9x^3y^2 + 12x^2y^4 = 3x^2y^2(2x^2 - 3x + 4y^2)$

You-Try-It 2

$a(b - 7) + b(b - 7) = (b - 7)(a + b)$

You-Try-It 3

$y^5 - 5y^3 + 4y^2 - 20$
$= (y^5 - 5y^3) + (4y^2 - 20)$
$= y^3(y^2 - 5) + 4(y^2 - 5)$
$= (y^2 - 5)(y^3 + 4)$

You-Try-It 4

$x^2 + 3x - 18$

$-18x^2$ (product of first and last boxes)

x^2			-18

$(-1x)(+18x)$ $(+1x)(-18x)$
$(-2x)(+9x)$ $(+2x)(-9x)$
$(-3x)(+6x)$ $(+3x)(-6x)$

The factors $-3x$ and $+6x$ have a sum of $+3x$.

x^2	$-3x$	$+6x$	-18

$x^2 - 3x + 6x - 18$
$= (x^2 - 3x) + (6x - 18)$
$= x(x - 3) + 6(x - 3)$
$= (x - 3)(x + 6)$

You-Try-It 5

$6x^2 - 11x + 5$

$\downarrow$ ——— $30x^2$ ——— $\downarrow$

$6x^2$			$+5$

$(-1x)(-30x)$ • The middle term is $-11x$. Find two negative factors of $30x^2$ whose sum is $-11x$.
$(-2x)(-15x)$
$(-3x)(-10x)$
$(-5x)(-6x)$ ← These factors have a sum of $-11x$.

$6x^2$	$-5x$	$-6x$	$+5$

$6x^2 - 5x - 6x + 5$
$= (6x^2 - 5x) - (6x - 5)$ • $6x$ and -5 have no common factor other than 1.
$= x(6x - 5) - 1(6x - 5)$
$= (6x - 5)(x - 1)$

You-Try-It 6

$25x^2 - 1$ • The middle term is $0x$.

$\downarrow$ ——— $-25x^2$ ——— $\downarrow$

$25x^2$			-1

$(-1x)(+25x)$
$(+1x)(-25x)$
$(-5x)(+5x)$ ← These two factors have a sum of $0x$.

$25x^2$	$-5x$	$+5x$	-1

$25x^2 - 5x + 5x - 1$
$= (25x^2 - 5x) + (5x - 1)$
$= 5x(5x - 1) + 1(5x - 1)$
$= (5x - 1)(5x + 1)$

You-Try-It 7

Goal

The goal is to find the base and height of a parallelogram that has an area of $(4x^2 + 8x + 3)$ square feet. We are looking for two expressions that when multiplied equal $4x^2 + 8x + 3$.

Strategy

Factor the trinomial $4x^2 + 8x + 3$.

Solution

$\downarrow$ ——— $12x^2$ ——— $\downarrow$

$4x^2$			$+3$

$(+1x)(+12x)$
$(+2x)(+6x)$ ← These factors have a sum of $+8x$.
$(+3x)(+4x)$

$4x^2$	$+2x$	$+6x$	$+3$

$4x^2 + 2x + 6x + 3$
$= (4x^2 + 2x) + (6x + 3)$
$= 2x(2x + 1) + 3(2x + 1)$
$= (2x + 1)(2x + 3)$

The dimensions of the parallelogram are $(2x + 1)$ feet and $(2x + 3)$ feet.

Check √

You-Try-It 8

$4a^2b - 30ab + 14b$
$= 2b(2a^2 - 15a + 7)$

$\downarrow$ ——— $14a^2$ ——— $\downarrow$

$2a^2$			$+7$

$(-1a)(-14a)$ ← These factors have a sum of $-15a$.
$(-2a)(-7a)$

$2a^2$	$-1a$	$-14a$	$+7$

$2a^2 - 1a - 14a + 7$
$= (2a^2 - 1a) - (14a - 7)$
$= a(2a - 1) - 7(2a - 1)$
$= (2a - 1)(a - 7)$

$4a^2b - 30ab + 14b = 2b(2a^2 - 15a + 7)$
$= 2b(2a - 1)(a - 7)$

Solutions to Chapter 6 You-Try-Its

SECTION 6.1

You-Try-It 1

$$9\sqrt{49} - \sqrt{4} = 9 \cdot 7 - 2$$
$$= 63 - 2$$
$$= 61$$

You-Try-It 2

$$6\sqrt{a+b}$$
$$6\sqrt{15+21} = 6\sqrt{36}$$
$$= 6 \cdot 6$$
$$= 36$$

You-Try-It 3

Goal
The goal is to find the astronaut's distance above the surface of the earth.

Strategy
In the given formula, replace the variable E by 184 and the variable S by 46. Evaluate the variable expression.

Solution

$$d = 4000\sqrt{\frac{E}{S}} - 4000$$
$$d = 4000\sqrt{\frac{184}{46}} - 4000$$
$$d = 4000\sqrt{4} - 4000$$
$$d = 4000 \cdot 2 - 4000$$
$$d = 8000 - 4000$$
$$d = 4000$$

The astronaut is 4000 miles above the earth's surface.

Check √

You-Try-It 4

$$\sqrt{113.4} \approx 10.6489$$

You-Try-It 5

$$7\sqrt{34} \approx 40.8167$$

You-Try-It 6

Goal
The goal is to determine how long a storm that has a diameter of 10 miles will last.

Strategy
Replace the variable d in the given formula by 10. Evaluate the variable expression.

Solution

$$t = \sqrt{\frac{d^3}{216}}$$
$$t = \sqrt{\frac{10^3}{216}}$$
$$t = \sqrt{\frac{1000}{216}}$$
$$t \approx 2.2$$

A storm that has a diameter of 10 miles will last approximately 2.2 hours.

Check √

You-Try-It 7

$9^2 = 81$; 81 is too big.

$8^2 = 64$; 64 is not a factor of 80.

$7^2 = 49$; 49 is not a factor of 80.

$6^2 = 36$; 36 is not a factor of 80.

$5^2 = 25$; 25 is not a factor of 80.

$4^2 = 16$; 16 is a factor of 80. $(16 \cdot 5 = 80)$

$$\sqrt{80} = \sqrt{16 \cdot 5}$$
$$= \sqrt{16} \cdot \sqrt{5}$$
$$= 4 \cdot \sqrt{5}$$
$$= 4\sqrt{5}$$

You-Try-It 8

$15^2 = 225$; 225 is too big.

Not factors of 200: $14^2 = 196$; $13^2 = 169$; $12^2 = 144$; $11^2 = 121$

$10^2 = 100$; 100 is a factor of 200. $(200 = 100 \cdot 2)$

$$\begin{aligned} 6\sqrt{200} &= 6\sqrt{100 \cdot 2} \\ &= 6\sqrt{100} \cdot \sqrt{2} \\ &= 6 \cdot 10\sqrt{2} \\ &= 60\sqrt{2} \end{aligned}$$

You-Try-It 9

$$\sqrt{t^{16}} = t^8$$

You-Try-It 10

$$\begin{aligned} \sqrt{z^9} &= \sqrt{z^8 \cdot z} \\ &= \sqrt{z^8} \cdot \sqrt{z} \\ &= z^4\sqrt{z} \end{aligned}$$

You-Try-It 11

a. $$\begin{aligned} \sqrt{45x^{11}} &= \sqrt{9x^{10} \cdot 5x} \\ &= \sqrt{9x^{10}} \cdot \sqrt{5x} \\ &= 3x^5\sqrt{5x} \end{aligned}$$

b. $$\begin{aligned} 5a\sqrt{28a^4b^{13}} &= 5a\sqrt{4a^4b^{12} \cdot 7b} \\ &= 5a\sqrt{4a^4b^{12}} \cdot \sqrt{7b} \\ &= 5a \cdot 2a^2b^6\sqrt{7b} \\ &= 10a^3b^6\sqrt{7b} \end{aligned}$$

SECTION 6.2

You-Try-It 1

$$9\sqrt{3} + 3\sqrt{3} - 18\sqrt{3} = (9 + 3 - 18)\sqrt{3} = -6\sqrt{3}$$

You-Try-It 2

$$\begin{aligned} 2\sqrt{50} - 5\sqrt{32} &= 2\sqrt{25 \cdot 2} - 5\sqrt{16 \cdot 2} \\ &= 2\sqrt{25} \cdot \sqrt{2} - 5\sqrt{16} \cdot \sqrt{2} \\ &= 2 \cdot 5\sqrt{2} - 5 \cdot 4\sqrt{2} \\ &= 10\sqrt{2} - 20\sqrt{2} \\ &= (10 - 20)\sqrt{2} \\ &= -10\sqrt{2} \end{aligned}$$

You-Try-It 3

$$\begin{aligned} y\sqrt{28y} + 7y\sqrt{63y} &= y\sqrt{4 \cdot 7y} + 7y\sqrt{9 \cdot 7y} \\ &= y\sqrt{4} \cdot \sqrt{7y} + 7y\sqrt{9} \cdot \sqrt{7y} \\ &= y \cdot 2\sqrt{7y} + 7y \cdot 3\sqrt{7y} \\ &= 2y\sqrt{7y} + 21y\sqrt{7y} \\ &= (2y + 21y)\sqrt{7y} \\ &= 23y\sqrt{7y} \end{aligned}$$

You-Try-It 4

$$\begin{aligned} \sqrt{5a}\sqrt{15a^3}\sqrt{3a} &= \sqrt{5a \cdot 15a^3 \cdot 3a} \\ &= \sqrt{225a^5} \\ &= \sqrt{225a^4 \cdot a} \\ &= \sqrt{225a^4} \cdot \sqrt{a} \\ &= 15a^2\sqrt{a} \end{aligned}$$

You-Try-It 5

$$(\sqrt{7y})^2 = 7y$$

You-Try-It 6

$$\begin{aligned} \sqrt{5x}(\sqrt{5x} - \sqrt{25x}) &= \sqrt{5x}\sqrt{5x} - \sqrt{5x}\sqrt{25x} \\ &= (\sqrt{5x})^2 - \sqrt{125x^2} \\ &= 5x - \sqrt{25x^2 \cdot 5} \\ &= 5x - \sqrt{25x^2} \cdot \sqrt{5} \\ &= 5x - 5x\sqrt{5} \end{aligned}$$

You-Try-It 7

$(3\sqrt{x} - y)(5\sqrt{x} - y)$

$= 15(\sqrt{x})^2 - 3y\sqrt{x} - 5y\sqrt{x} + y^2$

$= 15x - 8y\sqrt{x} + y^2$

You-Try-It 8

$(\sqrt{x} + 7)(\sqrt{x} - 7) = (\sqrt{x})^2 - 7^2 = x - 49$

You-Try-It 9

$$\sqrt{\frac{9}{16}} = \frac{\sqrt{9}}{\sqrt{16}} = \frac{3}{4}$$

You-Try-It 10

$$\sqrt{\frac{16y^2}{x^4}} = \frac{\sqrt{16y^2}}{\sqrt{x^4}} = \frac{4y}{x^2}$$

You-Try-It 11

$$\frac{x}{\sqrt{2}} = \frac{x}{\sqrt{2}} \cdot \frac{\sqrt{2}}{\sqrt{2}} = \frac{x\sqrt{2}}{(\sqrt{2})^2} = \frac{x\sqrt{2}}{2}$$

You-Try-It 12

$$\sqrt{\frac{5}{11}} = \frac{\sqrt{5}}{\sqrt{11}} = \frac{\sqrt{5}}{\sqrt{11}} \cdot \frac{\sqrt{11}}{\sqrt{11}} = \frac{\sqrt{5 \cdot 11}}{(\sqrt{11})^2} = \frac{\sqrt{55}}{11}$$

You-Try-It 13

$$\frac{y\sqrt{27}}{3\sqrt{75}} = \frac{y\sqrt{9 \cdot 3}}{3\sqrt{25 \cdot 3}} = \frac{y\sqrt{9}\sqrt{3}}{3\sqrt{25}\sqrt{3}} = \frac{y \cdot \cancel{3}\cancel{\sqrt{3}}}{\cancel{3} \cdot 5\cancel{\sqrt{3}}} = \frac{y}{5}$$

SECTION 6.3

You-Try-It 1

$\sqrt{4x} + 3 = 7$

$\sqrt{4x} = 4$

$(\sqrt{4x})^2 = 4^2$

$4x = 16$

$x = 4$

The solution checks.
The solution is 4.

You-Try-It 2

$\sqrt{3x - 2} - 5 = 0$

$\sqrt{3x - 2} = 5$

$(\sqrt{3x - 2})^2 = 5^2$

$3x - 2 = 25$

$3x = 27$

$x = 9$

The solution checks.
The solution is 9.

You-Try-It 3

$\sqrt{x + 5} = \sqrt{x} + 1$

$(\sqrt{x + 5})^2 = (\sqrt{x} + 1)^2$

$x + 5 = x + 2\sqrt{x} + 1$

$4 = 2\sqrt{x}$

$2 = \sqrt{x}$

$2^2 = (\sqrt{x})^2$

$4 = x$

The solution checks.
The solution is 4.

You-Try-It 4

Goal

The goal is to find the distance the film should be placed from the pinhole when the diameter of the pinhole is 0.008 inch.

Strategy

Replace the variable d in the given formula by 0.008, then solve the equation for D.

Solution

$d = 0.0073\sqrt{D}$

$0.008 = 0.0073\sqrt{D}$

$\frac{0.008}{0.0073} = \sqrt{D}$

$\left(\frac{0.008}{0.0073}\right)^2 = (\sqrt{D})^2$

$1.2 \approx D$

The film should be placed 1.2 inches from the pinhole.

Check √

You-Try-It 12

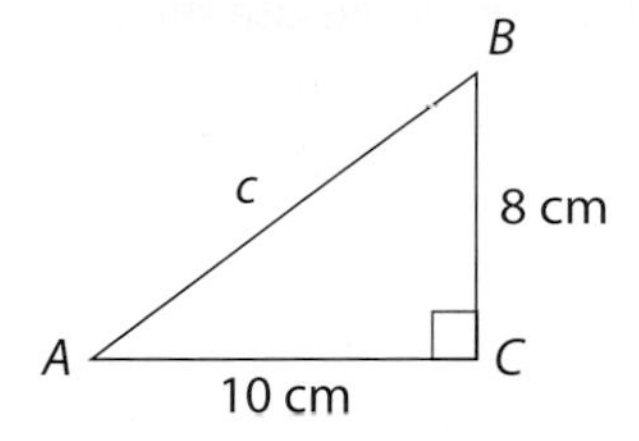

$c^2 = a^2 + b^2$
$c^2 = 8^2 + 10^2$
$c^2 = 64 + 100$
$c^2 = 164$
$c = \sqrt{164}$
$c \approx 13$

$\tan A = \dfrac{\text{side opposite } A}{\text{side adjacent to } A}$

$\tan A = \dfrac{8}{10}$

$\tan A = \dfrac{4}{5}$

$\tan A = 0.8$

$A \approx 39^\circ$

$\angle B = 90^\circ - 39^\circ = 51^\circ$

Side $c = 13$ centimeters, $\angle A = 39^\circ$, and $\angle B = 51^\circ$.

You-Try-It 13

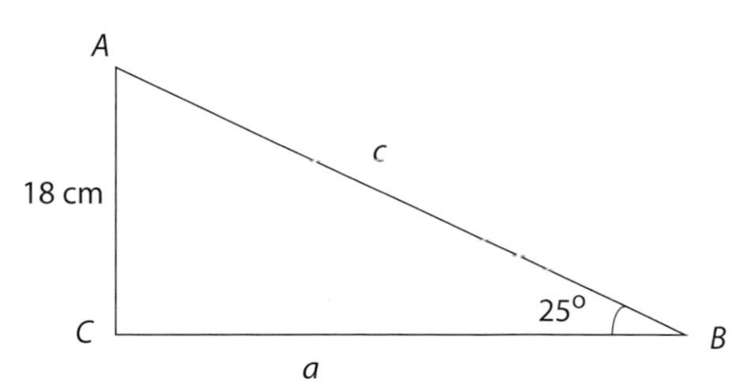

$\angle A = 90^\circ - 25^0 = 65^\circ$

$\sin B = \dfrac{\text{side opposite } B}{\text{hypotenuse}}$

$\sin 25^0 = \dfrac{18}{c}$

$c = \dfrac{18}{\sin 25^\circ}$

$c \approx 43$

$\tan B = \dfrac{\text{side opposite } B}{\text{side adjacent to } B}$

$\tan 25^\circ = \dfrac{18}{a}$

$a = \dfrac{18}{\tan 25^\circ}$

$a \approx 39$

$\angle A = 65^\circ$, side $c = 43$ centimeters, and side $a = 39$ centimeters.

You-Try-It 14

Goal

The goal is to find how far the boat is from the lighthouse given that the lighthouse is 20 meters high and the angle of depression from the top of the lighthouse is 25°.

Strategy

- Draw a diagram. Label the unknown distance *d*.
- Find the complement of the angle of depression.
- Write a trigonometric function that relates the given information and side *d* of the triangle.

Solution

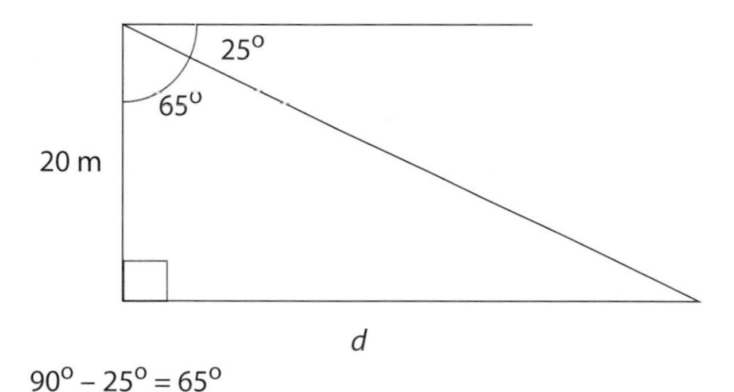

$90^\circ - 25^\circ = 65^\circ$

$\tan 65^\circ = \dfrac{d}{20}$

$20(\tan 65^\circ) = d$

$42.9 \approx d$

The boat is 42.9 meters from the lighthouse.

Check √

Solutions to Chapter 7 You-Try-Its

SECTION 7.1

You-Try-It 1

$$2x^2 - x = 1$$

Write the equation in standard form.

$$2x^2 - x - 1 = 0$$

Factor the polynomial.

$$(2x + 1)(x - 1) = 0$$

Use the Principle of Zero Products.

$$2x + 1 = 0 \quad \text{or} \quad x - 1 = 0$$

$$2x = -1 \qquad x = 1$$

$$x = -\frac{1}{2}$$

The solutions are $-\frac{1}{2}$ and 1.

You-Try-It 2

$$4t^2 - 234 = 0$$

$4t^2 = 234$ • Solve for t^2.

$$t^2 = 58.5$$

$\sqrt{t^2} = \sqrt{58.5}$ • Take the square root of each side.

$$t = \pm\sqrt{58.5}$$

The *exact* solutions are $-\sqrt{58.5}$ and $\sqrt{58.5}$.
The *approximate* solutions are −7.6485 and 7.6485.

You-Try-It 3

$$3(x + 2)^2 + 4 = 64$$

$3(x + 2)^2 = 60$ • Solve for $(x + 2)^2$.

$$(x + 2)^2 = 20$$

$\sqrt{(x + 2)^2} = \sqrt{20}$ • Take the square root of each side.

$x + 2 = \pm 2\sqrt{5}$ • Solve for x.

$$x + 2 = -2\sqrt{5} \qquad x + 2 = 2\sqrt{5}$$

$$x = -2 - 2\sqrt{5} \qquad x = -2 + 2\sqrt{5}$$

The solutions are $-2 - 2\sqrt{5} \approx -6.472136$ and $-2 + 2\sqrt{5} \approx 2.472136$.

You-Try-It 4

To find the length and width of the rectangle, use the equation $A = LW$.
Let the width $= x$.

$$A = LW$$

$19 = LW$ • Replace A with 19 m^2.

$19 = 3x \cdot x$ • Replace W with x and L with $3x$.

$$19 = 3x^2$$

$\frac{19}{3} = x^2$ • Solve for x.

$\sqrt{\frac{19}{3}} = \sqrt{x^2}$ • Take the square root of each side.

$$x \approx \pm 2.5166115$$

Since the width of a rectangle cannot be negative, the solution −2.52 is not possible.

$$3x \approx 3(2.52) = 7.56$$

The width of the rectangle is approximately 2.52 meters.
The length is approximately 7.56 meters.

SECTION 7.2

You-Try-It 1

$$x^2 - 8x$$

$\left[\frac{1}{2}(-8)\right]^2 = (-4)^2 = 16$ • Find the constant term.

$x^2 - 8x + 16$ • Complete the square.

$x^2 - 8x + 16 = (x - 4)^2$ • Write the resulting perfect-square trinomial as the square of a binomial.

You-Try-It 2

$$y^2 + 4y - 6 = 0$$

$$y^2 + 4y = 6$$

$$y^2 + 4y + 4 = 6 + 4$$

$$(y + 2)^2 = 10$$

$$\sqrt{(y + 2)^2} = \sqrt{10}$$

$$y + 2 = \pm\sqrt{10}$$

$$y + 2 = \sqrt{10} \qquad y + 2 = -\sqrt{10}$$

$$y = -2 + \sqrt{10} \qquad y = -2 - \sqrt{10}$$

The *exact* solutions are $-2 + \sqrt{10}$ and $-2 - \sqrt{10}$.
The *approximate* solutions are 1.1622777 and −5.1622777.

You-Try-It 3

$$2x^2 + 8x = 3$$

$$\frac{1}{2}(2x^2 + 8x) = \frac{1}{2}(3)$$

$$x^2 + 4x = \frac{3}{2}$$

$$x^2 + 4x + 4 = \frac{3}{2} + 4$$

$$(x+2)^2 = \frac{11}{2}$$

$$\sqrt{(x+2)^2} = \sqrt{\frac{11}{2}} = \frac{\sqrt{22}}{2}$$

$$x + 2 = \pm\frac{\sqrt{22}}{2}$$

$$x + 2 = \frac{\sqrt{22}}{2} \qquad x + 2 = -\frac{\sqrt{22}}{2}$$

$$x = -2 + \frac{\sqrt{22}}{2} \qquad x = -2 - \frac{\sqrt{22}}{2}$$

The *exact* solutions are $\frac{-4+\sqrt{22}}{2}$ and $\frac{-4-\sqrt{22}}{2}$.

The *approximate* solutions are 0.3452079 and −4.3452079.

You-Try-It 4

$$3x^2 - 6x - 2 = 0$$

$$3x^2 - 6x = 2$$

$$\frac{1}{3}(3x^2 - 6x) = \frac{1}{3}(2)$$

$$x^2 - 2x = \frac{2}{3}$$

$$x^2 - 2x + 1 = \frac{2}{3} + 1$$

$$(x-1)^2 = \frac{5}{3}$$

$$\sqrt{(x-1)^2} = \sqrt{\frac{5}{3}}$$

$$x - 1 = \pm\sqrt{\frac{5}{3}}$$

$$x - 1 = \pm\frac{\sqrt{15}}{3}$$

$$x - 1 = \frac{\sqrt{15}}{3} \qquad x - 1 = -\frac{\sqrt{15}}{3}$$

$$x = 1 + \frac{\sqrt{15}}{3} \qquad x = 1 - \frac{\sqrt{15}}{3}$$

$$x = \frac{3}{3} + \frac{\sqrt{15}}{3} \qquad x = \frac{3}{3} - \frac{\sqrt{15}}{3}$$

$$= \frac{3+\sqrt{15}}{3} \qquad = \frac{3-\sqrt{15}}{3}$$

The *exact* solutions are $\frac{3+\sqrt{15}}{3}$ and $\frac{3-\sqrt{15}}{3}$.

The *approximate* solutions are 2.2909944 and −0.2909944.

You-Try-It 5

Goal

We must find the time after the rock is released when it is 36 feet above the ocean.

Strategy

Solve the equation $h(t) = -16t^2 + 64t + 100$ for t when $h(t) = 36$.

Solution

$$h(t) = -16t^2 + 64t + 100$$

$$36 = -16t^2 + 64t + 100$$

$$16t^2 - 64t = 64$$

$$\frac{1}{16}(16t^2 - 64t) = \frac{1}{16}(64)$$

$$t^2 - 4t = 4$$

$$t^2 - 4t + 4 = 4 + 4$$

$$(t-2)^2 = 8$$

$$\sqrt{(t-2)^2} = \sqrt{8}$$

$$t - 2 = \pm 2\sqrt{2}$$

$$t - 2 = 2\sqrt{2} \qquad t - 2 = -2\sqrt{2}$$

$$t = 2 + 2\sqrt{2} \qquad t = 2 - 2\sqrt{2}$$

$$\approx 4.828 \qquad \approx -0.828$$

A time of −0.828 does not make sense.

The rock is 36 ft above the ocean after 4.828 seconds.

Check √

SECTION 7.3

You-Try-It 1

$$x^2 + 3x + 2 = 0$$

$$x = \frac{-b \pm \sqrt{b^2 - 4ac}}{2a} = \frac{-(3) \pm \sqrt{(3)^2 - 4(1)(2)}}{2(1)}$$

$$= \frac{-3 \pm \sqrt{9-8}}{2} = \frac{-3 \pm \sqrt{1}}{2} = \frac{-3 \pm 1}{2}$$

$$x = \frac{-3+1}{2} = \frac{-2}{2} = -1 \qquad x = \frac{-3-1}{2} = \frac{-4}{2} = -2$$

The solutions are −1 and −2.

You-Try-It 2

$$x^2 - 1 = 3x$$
$$x^2 - 3x - 1 = 0$$
$$x = \frac{-b \pm \sqrt{b^2 - 4ac}}{2a}$$
$$= \frac{-(-3) \pm \sqrt{(-3)^2 - 4(1)(-1)}}{2(1)}$$
$$= \frac{3 \pm \sqrt{9-(-4)}}{2}$$
$$= \frac{3 \pm \sqrt{13}}{2}$$

The *exact* solutions are $\frac{3+\sqrt{13}}{2}$ and $\frac{3-\sqrt{13}}{2}$.

The *approximate* solutions are 3.303 and –0.303.

You-Try-It 3

$$x^2 = 3x - 6$$
$$x^2 - 3x + 6 = 0$$
$$x = \frac{-b \pm \sqrt{b^2 - 4ac}}{2a}$$
$$= \frac{-(-3) \pm \sqrt{(-3)^2 - 4(1)(6)}}{2(1)}$$
$$= \frac{3 \pm \sqrt{9-24}}{2}$$
$$= \frac{3 \pm \sqrt{-15}}{2}$$

Because $\sqrt{-15}$ is not a real number, the equation has no real number solutions.

You-Try-It 4

Goal

We must find the distance the water is from the tugboat when it is 5 feet above the ocean.

Strategy

When the water is 5 feet above the ocean, $y = 5$ feet. Substitute this value of y into the equation $y = -0.005x^2 + 1.2x + 10$ and then solve for x.

Solution

$$y = -0.005x^2 + 1.2x + 10$$
$$5 = -0.005x^2 + 1.2x + 10$$
$$0 = -0.005x^2 + 1.2x + 5$$
$$x = \frac{-b \pm \sqrt{b^2 - 4ac}}{2a}$$
$$= \frac{-(1.2) \pm \sqrt{(1.2)^2 - 4(-0.005)(5)}}{2(-0.005)}$$
$$= \frac{(-1.2) \pm \sqrt{1.44-(-0.1)}}{-0.01}$$
$$= \frac{(-1.2) \pm \sqrt{1.54}}{-0.01}$$
$$x = \frac{(-1.2) - \sqrt{1.54}}{-0.01} \approx 244.1$$
$$x = \frac{(-1.2) + \sqrt{1.54}}{-0.01} \approx -4.1$$

A distance of –4.10 does not make sense. Therefore, the water is approximately 244.10 feet from the tugboat.

Check √

SECTION 7.4

You-Try-It 1

$$f(x) = x^2 + 4x - 2$$
$$-\frac{b}{2a} = -\frac{4}{2(1)} = -2$$
$$f\left(-\frac{b}{2a}\right) = f(-2) = (-2)^2 + 4(-2) - 2$$
$$= -6$$

Vertex: (–2, –6)

Axis of symmetry: $x = -2$

Solutions to Chapter 8 You-Try-Its

SECTION 8.1

You-Try-It 1

a. $\dfrac{x+8}{3x-12}$

$$3x - 12 = 0$$
$$3x = 12$$
$$x = 4$$

The expression is undefined when $x = 4$.

b. $\dfrac{x-6}{x^2+3x-4}$

$$x^2 + 3x - 4 = 0$$
$$(x+4)(x-1) = 0$$
$$x + 4 = 0 \qquad x - 1 = 0$$
$$x = -4 \qquad x = 1$$

The expression is undefined when $x = -4$ or 1.

You-Try-It 2

a. $\dfrac{6a^5b}{12a^2b^3} = \dfrac{a^5b}{2a^2b^3}$

$$= \frac{a^3}{2b^2}$$

Check: We chose to substitute 2 for a and 3 for b.

$$\frac{6a^5b}{12a^2b^3} = \frac{6 \cdot 2^5 \cdot 3}{12 \cdot 2^2 \cdot 3^3} = \frac{6 \cdot 32 \cdot 3}{12 \cdot 4 \cdot 27} = \frac{4}{9}$$

$$\frac{a^3}{2b^2} = \frac{2^3}{2 \cdot 3^2} = \frac{8}{2 \cdot 9} = \frac{4}{9}$$

b. $\dfrac{x^2+2x-24}{x^2-16} = \dfrac{(x+6)\cancel{(x-4)}}{(x+4)\cancel{(x-4)}}$

$$= \frac{x+6}{x+4}$$

Check: We chose to substitute 5 for x.

$$\frac{x^2+2x-24}{x^2-16} = \frac{(5)^2+2(5)-24}{(5)^2-16} = \frac{25+10-24}{25-16} = \frac{11}{9}$$

$$\frac{x+6}{x+4} = \frac{5+6}{5+4} = \frac{11}{9}$$

You-Try-It 3

$$\frac{x^2+2x-15}{x^2-9} \cdot \frac{x^2-3x-18}{x^2-7x+6}$$
$$= \frac{(x+5)(x-3)}{(x+3)(x-3)} \cdot \frac{(x-6)(x+3)}{(x-6)(x-1)}$$
$$= \frac{(x+5)\cancel{(x-3)}\cancel{(x-6)}\cancel{(x+3)}}{\cancel{(x+3)}\cancel{(x-3)}\cancel{(x-6)}(x-1)}$$
$$= \frac{x+5}{x-1}$$

You-Try-It 4

$$\frac{a^2}{4bc^3-2b^2c^2} \div \frac{a}{6bc-3b^2}$$
$$= \frac{a^2}{4bc^3-2b^2c^2} \cdot \frac{6bc-3b^2}{a}$$
$$= \frac{a^2 \cdot 3b\cancel{(2c-b)}}{2bc^2\cancel{(2c-b)} \cdot a}$$
$$= \frac{3a}{2c^2}$$

SECTION 8.2

You-Try-It 1

$$\frac{3}{xy} + \frac{12}{xy} = \frac{3+12}{xy} = \frac{15}{xy}$$

You-Try-It 2

$\dfrac{1}{y} \cdot 2 = \dfrac{2}{y}$ of the pool was drained on Saturday.

$\dfrac{1}{y} \cdot 3 = \dfrac{3}{y}$ of the pool was drained on Sunday.

$$\frac{2}{y} + \frac{3}{y} = \frac{2+3}{y} = \frac{5}{y}$$

$\dfrac{5}{y}$ of the pool was drained on Saturday and Sunday.

You-Try-It 3

$$\frac{2x^2}{x^2-x-12}-\frac{7x+4}{x^2-x-12}$$
$$=\frac{2x^2-(7x+4)}{x^2-x-12}$$
$$=\frac{2x^2-7x-4}{x^2-x-12}$$
$$=\frac{(2x+1)\cancel{(x-4)}}{(x+3)\cancel{(x-4)}}$$
$$=\frac{2x+1}{x+3}$$

You-Try-It 4

You completed $\frac{1}{h}\cdot 2=\frac{2}{h}$ of the wallpapering.

Your roommate completed $\frac{1}{h}\cdot 3=\frac{3}{h}$ of the wallpapering.

$$\frac{3}{h}-\frac{2}{h}=\frac{3-2}{h}=\frac{1}{h}$$

Your roommate completed $\frac{1}{h}$ more of the wallpapering than you did.

You-Try-It 5

Find the LCM of $8uv^4$ and $12uv$.

$$8uv^4=2\cdot 2\cdot 2\cdot u\cdot v\cdot v\cdot v\cdot v$$
$$12uv=2\cdot 2\cdot 3\cdot u\cdot v$$
$$\text{LCM}=2\cdot 2\cdot 2\cdot 3\cdot u\cdot v\cdot v\cdot v\cdot v$$
$$=24uv^4$$

You-Try-It 6

a. The LCM of $8y$, $3y$, and $4y$ is $24y$.
Rewrite each fraction with a denominator of $24y$.

$$\frac{z}{8y}-\frac{4z}{3y}+\frac{5z}{4y}=\frac{z}{8y}\cdot\frac{3}{3}-\frac{4z}{3y}\cdot\frac{8}{8}+\frac{5z}{4y}\cdot\frac{6}{6}$$
$$=\frac{3z}{24y}-\frac{32z}{24y}+\frac{30z}{24y}$$
$$=\frac{3z-32z+30z}{24y}=\frac{z}{24y}$$

b. The denominator of y is 1.
The LCM of the denominators 1 and $7y$ is $7y$.
Rewrite y as a fraction with a denominator of $7y$.

$$y+\frac{5}{7y}=\frac{y}{1}+\frac{5}{7y}$$
$$=\frac{y}{1}\cdot\frac{7y}{7y}+\frac{5}{7y}$$
$$=\frac{7y^2}{7y}+\frac{5}{7y}$$
$$=\frac{7y^2+5}{7y}$$

You-Try-It 7

$$\frac{4x}{3x-1}-\frac{9}{x+4}=\frac{4x}{3x-1}\cdot\frac{x+4}{x+4}-\frac{9}{x+4}\cdot\frac{3x-1}{3x-1}$$
$$=\frac{4x^2+16x}{(3x-1)(x+4)}-\frac{27x-9}{(3x-1)(x+4)}$$
$$=\frac{4x^2+16x-(27x-9)}{(3x-1)(x+4)}$$
$$=\frac{4x^2+16x-27x+9}{(3x-1)(x+4)}$$
$$=\frac{4x^2-11x+9}{(3x-1)(x+4)}$$

You-Try-It 8

The large pipe fills $\frac{1}{y}\cdot 2=\frac{2}{y}$ of the tank.

The small pipe fills $\frac{1}{3y}\cdot 2=\frac{2}{3y}$ of the tank.

$$\frac{2}{y}+\frac{2}{3y}=\frac{2}{y}\cdot\frac{3}{3}+\frac{2}{3y}$$
$$=\frac{6}{3y}+\frac{2}{3y}=\frac{8}{3y}$$

After 2 hours, the pipes have filled $\frac{8}{3y}$ of the tank.

SECTION 8.3

You-Try-It 1

a. The LCM of $3, x, 9$, and x^2 is $9x^2$.

$$\frac{\frac{1}{3}-\frac{1}{x}}{\frac{1}{9}-\frac{1}{x^2}} = \frac{\frac{1}{3}-\frac{1}{x}}{\frac{1}{9}-\frac{1}{x^2}} \cdot \frac{9x^2}{9x^2}$$

$$= \frac{\frac{1}{3}\cdot 9x^2 - \frac{1}{x}\cdot 9x^2}{\frac{1}{9}\cdot 9x^2 - \frac{1}{x^2}\cdot 9x^2}$$

$$= \frac{3x^2-9x}{x^2-9}$$

$$= \frac{3x(x-3)}{(x+3)(x-3)} = \frac{3x}{x+3}$$

b. The LCM of x and x^2 is x^2.

$$\frac{1+\frac{4}{x}+\frac{3}{x^2}}{1+\frac{10}{x}+\frac{21}{x^2}} = \frac{1+\frac{4}{x}+\frac{3}{x^2}}{1+\frac{10}{x}+\frac{21}{x^2}} \cdot \frac{x^2}{x^2}$$

$$= \frac{1\cdot x^2 + \frac{4}{x}\cdot x^2 + \frac{3}{x^2}\cdot x^2}{1\cdot x^2 + \frac{10}{x}\cdot x^2 + \frac{21}{x^2}\cdot x^2}$$

$$= \frac{x^2+4x+3}{x^2+10x+21}$$

$$= \frac{(x+3)(x+1)}{(x+3)(x+7)} = \frac{x+1}{x+7}$$

SECTION 8.4

You-Try-It 1

$$x+\frac{1}{3} = \frac{4}{3x}$$

$$3x\left(x+\frac{1}{3}\right) = 3x\left(\frac{4}{3x}\right)$$

$$3x(x)+3x\left(\frac{1}{3}\right) = 3x\left(\frac{4}{3x}\right)$$

$$3x^2+x = 4$$

$$3x^2+x-4 = 0$$

$$(3x+4)(x-1) = 0$$

$$3x+4=0 \qquad x-1=0$$

$$3x=-4 \qquad x=1$$

$$x=-\frac{4}{3}$$

Graphical check:

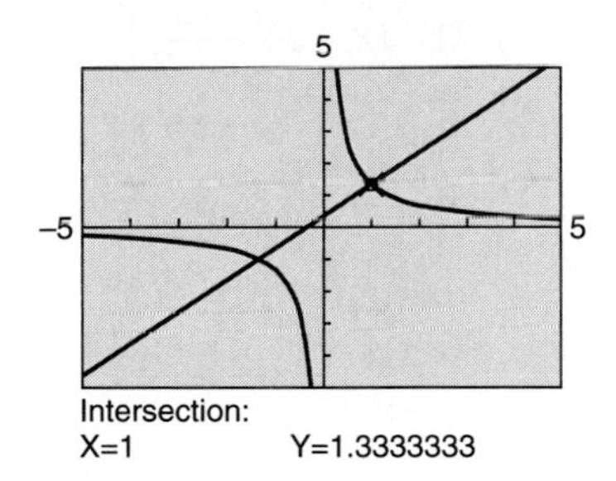

The solutions check.

Algebraic check:

$$x+\frac{1}{3} = \frac{4}{3x}$$

$$-\frac{4}{3}+\frac{1}{3} \;\Big|\; \frac{4}{3\left(-\frac{4}{3}\right)}$$

$$-\frac{3}{3} \;\Big|\; \frac{4}{-4}$$

$$-1=-1$$

$$x+\frac{1}{3} = \frac{4}{3x}$$

$$1+\frac{1}{3} \;\Big|\; \frac{4}{3\cdot 1}$$

$$\frac{4}{3}=\frac{4}{3}$$

The solutions check.

The solutions are $-\frac{4}{3}$ and 1.

You-Try-It 2

$$\frac{5x}{x+2} = 3 - \frac{10}{x+2}$$
$$(x+2)\left(\frac{5x}{x+2}\right) = (x+2)\left(3 - \frac{10}{x+2}\right)$$
$$(x+2)\left(\frac{5x}{x+2}\right) = (x+2)(3) - (x+2)\left(\frac{10}{x+2}\right)$$
$$5x = 3x + 6 - 10$$
$$5x = 3x - 4$$
$$2x = -4$$
$$x = -2$$

Algebraic check:

$\frac{5x}{x+2}$	$= 3 - \frac{10}{x+2}$
$\frac{5(-2)}{-2+2}$	$3 - \frac{10}{-2+2}$
$\frac{-10}{0}$	$3 - \frac{10}{0}$

−2 does not check as a solution.
The equation has no solution.

You-Try-It 3

Goal
The goal is to find the time it would take one printer, working alone, to print the company's payroll checks.

Strategy
- Using t to represent the amount of time it would take one printer to complete the job, find the fraction of the total job performed by each printer.
- Write an equation using the fact that the sum of the parts completed by each printer equals 1, the complete task.

Solution
Part completed by first printer $= \frac{1}{t} \cdot 3 = \frac{3}{t}$

Part completed by second printer $= \frac{1}{t} \cdot (3+2) = \frac{5}{t}$

$$\frac{3}{t} + \frac{5}{t} = 1$$
$$t\left(\frac{3}{t} + \frac{5}{t}\right) = t(1)$$
$$t \cdot \frac{3}{t} + t \cdot \frac{5}{t} = t$$
$$3 + 5 = t$$
$$8 = t$$

It would take one printer 8 hours to print the checks.

Check √

You-Try-It 4

Goal
The goal is to find the rate of the current.

Strategy
- Using r to represent the rate of the river's current, write one expression for the time it took to canoe with the current and another expression for the time it took to canoe against the current. The speed of the canoe is increased by r when traveling with the current and decreased by r when traveling against the current.
- Write an equation using the fact that the time it took for the canoe to travel up the river equals the time it took for the canoe to travel down the river.

Solution

Time paddling with the current: $\frac{30}{8+r}$

Time paddling against the current: $\frac{18}{8-r}$

Time with the current = Time against the current

$$\frac{30}{8+r} = \frac{18}{8-r}$$
$$(8+r)(8-r)\frac{30}{8+r} = (8+r)(8-r)\frac{18}{8-r}$$
$$(8-r)30 = (8+r)18$$
$$240 - 30r = 144 + 18r$$
$$240 = 144 + 48r$$
$$96 = 48r$$
$$2 = r$$

The river's current is 2 mph.

Check √

You-Try-It 5

Goal

The goal is to find the rate of the sailboat sailing across the lake.

Strategy

- Using r to represent the rate of the sailboat going across the lake, write one expression for the time it took to sail across the lake and one expression for the time it took to sail back.
- Write an equation using the fact that the round trip took a total of 3 hours.

Solution

Time across the lake: $\frac{6}{r}$

Time for the return trip: $\frac{6}{2r}$

Time sailing across + Time sailing back = 3 hours

$$\frac{6}{r} + \frac{6}{2r} = 3$$

$$2r\left(\frac{6}{r} + \frac{6}{2r}\right) = 2r(3)$$

$$2r\left(\frac{6}{r}\right) + 2r\left(\frac{6}{2r}\right) = 2r(3)$$

$$12 + 6 = 6r$$

$$18 = 6r$$

$$3 = r$$

The rate going across the lake was 3 mph.

Check √

SECTION 8.5

You-Try-It 1

a.

$$s = \frac{A+L}{2}$$

$$(2)s = (2)\left(\frac{A+L}{2}\right)$$

$$2s = A + L$$

$$2s - A = L$$

b.

$$S = a + (n-1)d$$

$$S = a + nd - d$$

$$S - a + d = nd$$

$$\frac{S-a+d}{d} = \frac{nd}{d}$$

$$\frac{S-a+d}{d} = n$$

c.

$$S = C + rC$$

$$S = C(1+r)$$

$$\frac{S}{1+r} = \frac{C(1+r)}{1+r}$$

$$\frac{S}{1+r} = C$$

SECTION 8.6

You-Try-It 1

$$\frac{n}{5} = \frac{12}{25}$$

$$5(12) = n(25)$$

$$60 = 25n$$

$$2.4 = n$$

Check: $\frac{2.4}{5} = \frac{12}{25}$

$$5(12) = 2.4(25)$$

$$60 = 60$$

The solution is 2.4.

You-Try-It 2

Goal

a. The goal is to find the number of workers in 1960.
b. The goal is to find the expected number of beneficiaries in 2030.

Strategy

a. Write and solve a proportion to find the number of workers in 1960.
b. Write and solve a proportion to find the number of beneficiaries in 2030.

Solution

a. $\frac{5 \text{ workers}}{1 \text{ beneficiary}} = \frac{W \text{ workers}}{14{,}000{,}000 \text{ beneficiaries}}$

$$1 \cdot W = 5(14{,}000{,}000)$$

$$W = 70{,}000{,}000$$

There were 70,000,000 workers in 1960.

b. $\frac{2 \text{ workers}}{1 \text{ beneficiary}} = \frac{167{,}000{,}000 \text{ workers}}{B \text{ beneficiaries}}$

$$1 \cdot 167{,}000{,}000 = 2 \cdot B$$

$$167{,}000{,}000 = 2B$$

$$83{,}500{,}000 = B$$

There will be 83,500,000 beneficiaries in 2030.

Check √

You-Try-It 3

Goal

The goal is to find the area of triangle *AOB*.

Strategy

- Write and solve a proportion to find *AO*, the height of triangle *AOB*.
- Use the equation for the area of a triangle, $A = \frac{1}{2}bh$.

Solution

$$\frac{AB}{CD} = \frac{AO}{DO}$$

$$\frac{10}{4} = \frac{AO}{3}$$

$$4(AO) = 10(3)$$

$$4(AO) = 30$$

$$AO = 7.5$$

$$A = \frac{1}{2}bh$$

$$A = \frac{1}{2}(10)(7.5) = 5(7.5) = 37.5$$

The area of the triangle is 37.5 square centimeters.

Check √

You-Try-It 4

Goal

The goal is to find *w*, the width of the river.

Strategy

Write and solve a proportion to find the length of *w*.

Solution

$$\frac{120}{150} = \frac{36}{w}$$

$$150(36) = 120(w)$$

$$5400 = 120w$$

$$45 = w$$

The width of the river is 45 feet.

Check √

SECTION 8.7

You-Try-It 1

Let A be the event that the number is odd and B be the event that the number is divisible by 5. Then,
$A = \{1, 3, 5, 7, 9, 11, 13, 15, 17, 19\}$ and
$B = \{5, 10, 15, 20\}$.

a. $A \cup B = \{1, 3, 5, 7, 9, 10, 11, 13, 15, 17, 19, 20\}$

$$P(A \text{ or } B) = \frac{n(A \cup B)}{n(S)} = \frac{12}{20} = \frac{3}{5}$$

The probability that the number is odd or divisible by 5 is $\frac{3}{5}$.

b. $A \cap B = \{5, 15\}$

$$P(A \text{ and } B) = \frac{n(A \cap B)}{n(S)} = \frac{2}{20} = \frac{1}{10}$$

The probability that the number is odd and divisible by 5 is $\frac{1}{10}$.

You-Try-It 2

$A = \{(1,6), (6,1), (2,5), (5,2), (3,4), (4,3)\}$
$B = \{(5,6), (6,5)\}$

$P(A) = \frac{1}{6}$ and $P(B) = \frac{1}{18}$.

$$P(A \text{ or } B) = P(A) + P(B) = \frac{1}{6} + \frac{1}{18} = \frac{3}{18} + \frac{1}{18} = \frac{4}{18} = \frac{2}{9}$$

The probability of rolling a 7 or 11 is $\frac{2}{9}$.

You-Try-It 3

Let A = {degree in business}.
Let B = {starting salary between \$20,000 and \$24,999}.
$n(A) = 98$, $n(B) = 39$, and $n(A \cap B) = 16$

$$P(A \text{ or } B) = P(A) + P(B) - P(A \text{ and } B)$$

$$= \frac{98}{206} + \frac{39}{206} - \frac{16}{206} = \frac{121}{206}$$

The probability is $\frac{121}{206}$.

You-Try-It 4

Let A be the event that the sum is not seven:
$A = \{(1,1), (1,2), (1,3), (1,4), (1,5), (2,1), (2,2), (2,3), (2,4), (2,6), (3,1), (3,2), (3,3), (3,5), (3,6), (4,1), (4,2), (4,4), (4,5), (4,6), (5,1), (5,3), (5,4), (5,5), (5,6), (6,2), (6,3), (6,4), (6,5), (6,6)\}$.
Let B be the event that the sum is six:
$B = \{(1,5), (5,1), (2,4), (4,2), (3,3)\}$
$n(A \cap B) = 5$

$$P(B|A) = \frac{P(A \text{ and } B)}{P(A)} = \frac{5/36}{30/36} = \frac{5}{30} = \frac{1}{6}$$

The probability is $\frac{1}{6}$.

You-Try-It 5

Let A be the event of rolling a ⚃ and then tossing a tail.

$P(⚃) = \frac{1}{6}$; $P(\text{Tail}) = \frac{1}{2}$. The probability of A is the product of the two probabilities.

$$P(A) = \frac{1}{6} \cdot \frac{1}{2} = \frac{1}{12}$$

The probability is $\frac{1}{12}$.

Answers to Chapter 1 Exercises

1.1 EXERCISES

1. Use Polya's suggestions as a guideline. **3.** Drawing conclusions based on known facts **5.** Once **7.** 7 **9.** 1:59 P.M. **11.** Complete solution in solutions manual. However, to start, place four coins on each pan of the balance scale. **13.** 12 **15.** 36 **17.** 21 **19.** d **21.** septagon **23.** ||||••• **25.** The difference is always 3087. **27.** 2 **29.** 4 **31.** 1 **33.** 3 **35.** Anna—golf; Kay—sailing; Megan—tennis; Nicole—horseback riding **37.** Lomax—compact car; Parish—sports car; Thorpe—station wagon; Wong—sedan **39.** 4 **41.** No

1.2 EXERCISES

1. *And, or,* and *not* are the ones discussed here. **3.** The truth value of a true statement is true, and the truth value of a false statement is false. **5.** Answers will vary. **7.** Yes **9.** No **11.** No **13.** Yes **15.** True **17.** False **19.** True **21.** True **23.** False **25.** True **27.** True **29.** True **31. a.** True; **b.** false; **c.** true; **d.** false **33.** $\{2, 4\}$ **35.** $\varnothing$ **37.** $\{1, 2, 3, 5, 7, 9, 11\}$ **39.** A square has four sides. **41.** Water does not freeze at zero degrees Celsius. **43.** $41 \neq 63$ **45.** No prime numbers are odd numbers. **47.** Some even numbers are odd numbers. **49.** No apples are green. **51.** All printers print in color. **53.** Some music does not sound nice. **55.** Answers will vary. **57.** Answers will vary. **59.** Answers will vary. **61. a.** Sometimes true; **b.** sometimes true; **c.** sometimes true **63. a.** A; **b.** A; **c.** A; **d.** $\varnothing$ **65. a.** True; **b.** true; **c.** true; **d.** false **67.** No

1.3 EXERCISES

1. Answers will vary. **3.** A frequency table shows the actual number in each class; a relative frequency table shows the percent of the total in each class. **5.** A frequency table shows only the number of data in a class. A stem-and-leaf plot shows the value of every number in the data set. **7. a.** $15 million; **b.** $10 million; **c.** less than; **d.** yes **9. a.** Michigan; **b.** Ohio and Illinois; **c.** 36,000; **d.** 20,000 **11. a.** 50–59; **b.** 20–29; **c.** 18%; **d.** 40–59 **13. a.** 2nd; **b.** 4th; **c.** 1st quarter of 1995; **d.** 1996 **15. a.** Decreasing; **b.** no; **c.** 1997; **d.** increasing **17.** 2; 2; 12; 8; 6; 1 **19.** 1; 4; 5; 6; 13; 15; 7 **21.** 6.5%, 6.5%, 38.7%, 25.8%. 19.4%, 3.2% **23.** 2.0%, 7.8%, 9.8%, 11.8%, 25.5%, 31.4%, 13.7% **25. a.** 4; **b.** 8; **c.** no **27. a.** 4; **b.** 80 and 100; **c.** no

29.

Stem	Leaves
1	0, 7, 9
2	3, 8
3	8, 9, 8
4	1, 6, 7, 8
5	3, 8, 9
6	0, 1, 1, 8, 4, 3, 9, 2
7	5, 7
8	2, 2
9	0, 2, 1

31.

Stem	Leaves
6	8, 5, 2
7	9, 9, 8, 8, 8, 8, 7, 7, 7, 7, 6, 6, 6, 5, 5, 5, 4, 4, 4, 4, 3, 3, 3, 3, 3, 3, 3, 0, 0
8	7, 6, 5, 5, 5, 4, 4, 3, 3, 3, 2, 2, 2, 1, 0, 0
9	4, 1, 1

33. a. Hawaii and New Jersey; **b.** $741.60; **c.** $597.40

1.4 EXERCISES

1. To ensure that everyone performs the order of mathematical operations in the same way. **3.** The first means the number substituted for the variable; the second means the result. **5.** 81 **7.** 1 **9.** 196 **11.** 3 **13.** 5 **15.** 8 **17.** 37 **19.** 10 **21.** 197 **23.** 16 **25.** 72 **27.** 72 **29.** 13 **31.** 14 **33.** 28 **35.** 34 **37.** 31 **39.** 48 **41.** 220 centimeters **43.** 2800 square centimeters **45.** 3 square meters

47.

Input, x	0	1	2	3	4	5
Output, $y = 3x + 2$	2	5	8	11	14	17

a. $y = 11$ **b.** $x = 1$ **49.**

Input, x	1	1.5	2	2.5	3	3.5
Output, $y = x^2$	1	2.25	4	6.25	9	12.25

a. $y = 4$ **b.** $x = 3.5$ **51.**

Input, x	1.5	1.75	2	2.25	2.5	2.75
Output, $y = 2x^2 + 1$	5.5	7.125	9	11.125	13.5	16.125

a. $y = 7.125$ **b.** $x = 2.75$

53.

Input, x	0	1	2	3	4	5
Output, $y = x^2 + 3x + 1$	1	5	11	19	29	41

a. $y = 19$ **b.** $x = 4$ **55. a.** The temperature is 91° F. **b.** The elevation is 2450 feet. **57. a.** The next eruption will be in 75.4 minutes. **b.** The length of the last eruption was 2.5 minutes. **59. a.** The height is 69 feet. **b.** The ball was 65 feet above the ground at 1.5 and 2.5 seconds.

1.5 EXERCISES

1. Answers will vary. **3.** Answers will vary. **5.** $\frac{4}{p-6}$ **7.** $\frac{d+8}{d}$ **9.** $\frac{3}{8}(t+15)$ **11.** $13-n$ **13.** $\frac{3}{7}n$ **15.** $\frac{2n}{5}$ **17.** $15n-8$

19. $(n^3+n)-6$ **21.** $\frac{3}{4+n}$ **23.** $11+\frac{1}{2}n$ **25.** $n(2n-9)$ **27.** $(9-n)n$ **29.** $\frac{2n+7}{n}$ **31.** $n-(n^3-10)$ **33.** $7n^2-4$

35. $80-13n$ **37.** $2s$ **39.** $4c$ **41.** $s+25$ **43.** $\frac{1}{2}L-3$ **45.** $0.32+0.23(w-1)$ **47.** S and $12-S$ **49.** x and $20-x$

51. $640+24h$ **53.** $5n+10d$ **55.** $5x$ **57.** $\frac{1}{4}x$

1.6 EXERCISES

1. Answers will vary. **3.** Smallest data value, Q_1, median, Q_3, largest data value **5.** Answers will vary. **7.** mean = 69.9; median = 70; mode = 70 **9.** mean = 97.2; median = 98; mode = 98, 101 **11.** 14.3 **13.** mean = 49.64; median = 49.605 **15.** modal response = Download types **17.** modal response = Paid more than the minimum on all cards **19.** $Q_1 = 189$; $Q_3 = 234$ **21.** 75% **23.** 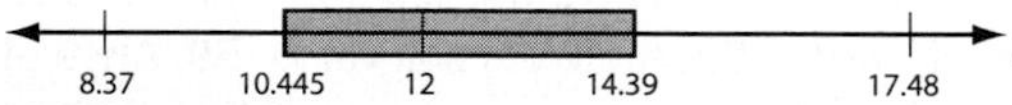

; interquartile range = 3.945

25. 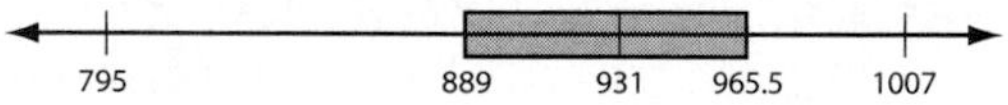

; interquartile range = 76.5 **27.** Yes

1.7 EXERCISES

1. Zero and one. **3.** Probabilities calculated from gathered data. **5.** Answers will vary. **7.** {HH, HT, TH, TT} **9.** {22, 23, 25, 32, 33, 35, 52, 53, 55} **11.** {*xyz, xzy, yxz, yzx, zxy, zyx*} **13.** Yes **15.** {king of hearts, king of spades, king of clubs, king of diamonds}

17. {Monday, Tuesday, Wednesday, Thursday, Friday} **19.** {(5, 6), (6, 5), (6, 6)} **21.** 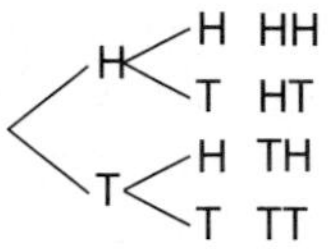

23. 256 **25.** 46,656

27. 144 **29.** $\frac{1}{4}$ **31.** $\frac{1}{4}$ **33.** $\frac{1}{8}$ **35.** $\frac{1}{6}$ **37.** $\frac{1}{3125}$ **39.** $\frac{1}{19}$ **41.** 1 TO 5 **43.** 2 TO 1 **45.** $\frac{1}{16}$ **47.** $\frac{2}{7}$ **49.** $\frac{13}{20}$
51. a. mode; **b.** the probability of being greater than the median.

CHAPTER REVIEW EXERCISES

1. a. {1, 2, 3, 4, 5, 6, 7, 9}; **b.** {4, 6}; **c.** ∅ **2.** 720 **3.** Some animals have feathers. **4.** Answers will vary. **5.** True **6.** $13.50

7. 18 **8.** 1 **9.** 19 **10.** 64 **11.** $\frac{4}{7}n$ **12.** $5n-4$ **13.** $\frac{2n}{3n-7}$ **14.** $4d^2-(d+8)$ **15.** 2 **16.** $3r$ **17.** $\frac{1}{8}$ **18.** 1 TO 3 **19.** 1024; 2^n = number of folds **20.**

4	5
5	9 8
6	4 1 7
7	8 2 3 4 8 4 9
8	7 9 3 2 4 3
9	4 2

21. a. \$.7 billion; **b.** 1996 and 1997; **c.** more than **22.** Steve, soccer; Mario, football; Ray, basketball; Bill, hockey **23.** mean = 249; median = 252.5; mode = none **24.** Very unsatisfied **25.** $Q_1 = 453$; $Q_3 = 533$ **26.** 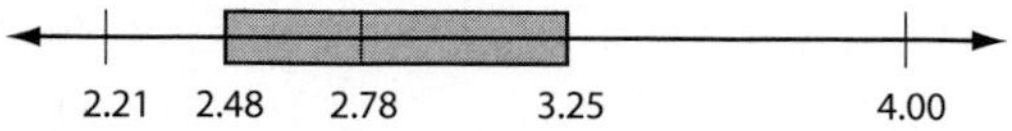

interquartile range = 0.77 **27.** {*abc, acb, bac, bca, cab, cba*}.

Answers to Chapter 2 Exercises

2.1 EXERCISES

1. No **3.** No **5.** Zero **7.** −25 **9.** 34 **11.** 0 **13.** −12 **15.** 16 **17.** −49 **19.** 16 **21.** 32 **23.** −86 **25.** −54 **27.** 42 **29.** −25 **31.** −24 **33.** −95 **35.** 9 **37.** −40 **39.** 43 **41.** 37 **43.** −10 **45.** 3 **47.** −20 **49.** −10 **51.** −5, −4, −3, −2, −1, 0, 1 **53.** 2, 3, 4, 5, 6, 7, 8 **55.** −14 **57.** −28 **59.** 6 **61.** 0 **63.** 12 **65.** −138 **67.** −7 **69.** 6 **71.** 48 **73.** −7, −6, −5, −4, −3, −2, −1 **75.** 4, 3, 2, 1, 0, −1, −2 **77.** The difference is 399°C. **79.** The temperature rose 49°F. **81. a.** 6 **b.** 11 **c.** 6 **83. a.** The difference is 20 m. **b.** The submarine rose 75 m. **c.** The submarine is more than 60 m below sea level. **d.** The submarine rose more during the last 60 seconds. **85. a.** 1992 **b.** The difference is 8 million dollars. **c.** Between 1992 and 1993. **87.** −11 **89.** 11 **91.** 3 **93.** 2, $4+(-n)$ **95.** ≤ **97. a.** 5°F with a 20-mph wind feels colder. **b.** −15°F with a 20-mph wind feels colder. **99. a.** 25 **b.** 10, 1, 3, 0, 3, 1, 8, 8, 3, 1, 0, 0, 2, 1, 5 **c.** $3.0\overline{6}$ **d.** 3.6 **e.** No **f.** All data has the same value.

2.2 EXERCISES

1. $x = 0$ **3.** Sometimes **5.** The product of any number and zero is zero; it is not possible to find the other number. **7.** −84 **9.** 132 **11.** 0 **13.** −160 **15.** 90 **17.** −160 **19.** −64 **21.** −625 **23.** −136 **25.** −162 **27.** 90 **29.** 12, 8, 4, 0, −2, −6, −10 **31.** 9, 4, 1, 0, 1, 4, 9 **33.** −15 **35.** −8 **37.** 5 **39.** −28 **41.** −9 **43.** 6 **45.** −1 **47.** 2 **49.** 3 **51.** −6 **53.** −3 **55.** 14 **57.** 3 **59.** 3, 2, 1, 0, −1, −2, −3 **61.** 2, 4, 8, −8, −4, −2 **63.** 96 **65.** −24 **67.** 4 **69.** −31 **71.** The average temperature was −1°. **73.** The person's score is 214. **75.** = **77.** > **79.** Opposite signs **81.** Yes, when $x = 0$. **83. a.**

Triangles	1	2	3	4
Toothpicks	3	5	7	9

b. 21 **c.** $2n + 1$ = number of toothpicks required **85. a.**

Pentagons	1	2	3	4
Toothpicks	5	9	13	17

b. 41 **c.** $4n + 1$ = number of toothpicks required

2.3 EXERCISES

1. Answers will vary. **3.** 0 **5. a.** x and y both positive; **b.** x negative, y positive; **c.** x and y both negative; **d.** x positive, y negative. **7.** **9.** **11.**

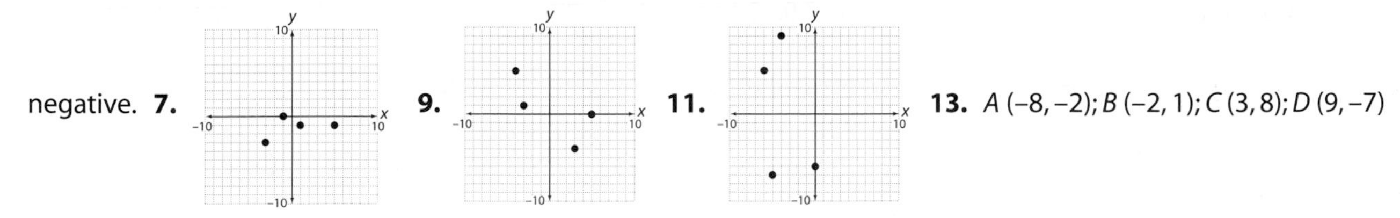

13. $A(-8,-2)$; $B(-2,1)$; $C(3,8)$; $D(9,-7)$

15. $A(0,-7)$; $B(0,0)$; $C(-4,0)$; $D(6,9)$ **17.** Abscissa: A, 0; B, −8; ordinate: C, −4; D, −2 **19.**

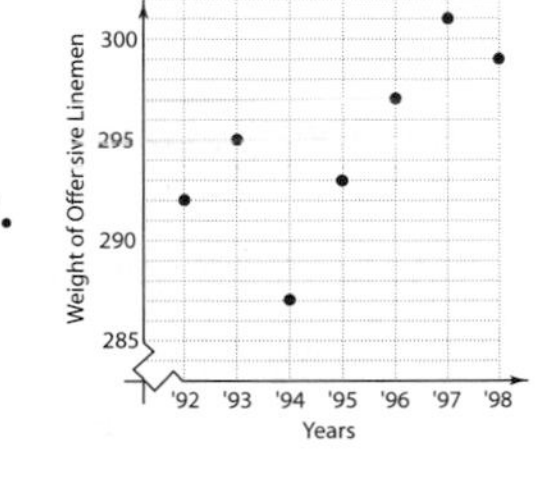

21.

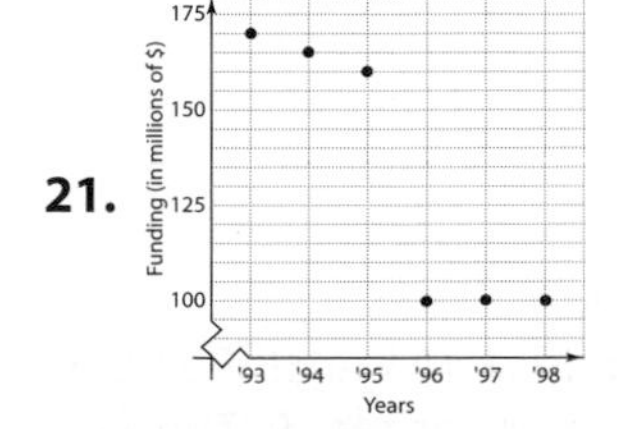

23.

Year	1992	1993	1994	1995	1996	1997
% who repay	31	32	33	34	36	39

25.

Year	1992	1993	1994	1995	1996	1997
Average charge	18	19	21	23	24	26

27. **29.** **31.** **33.** **35.** **37.**

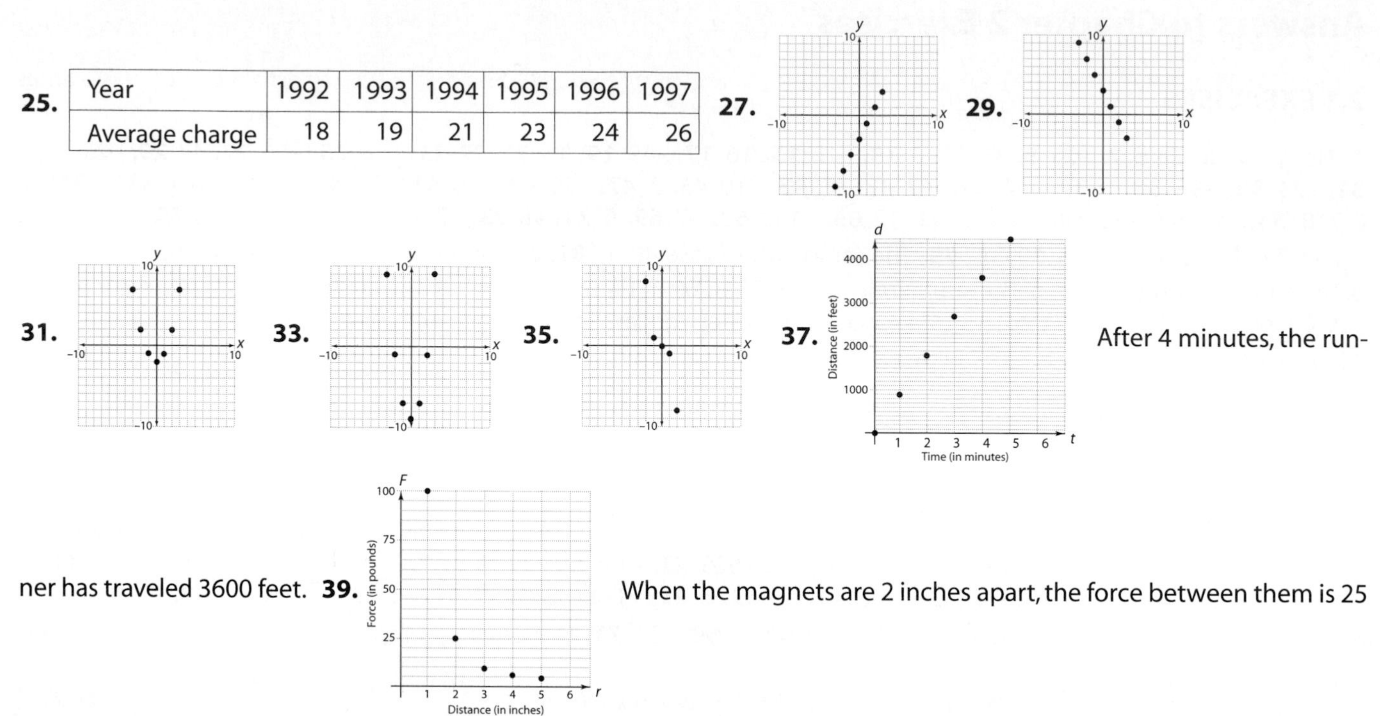

After 4 minutes, the runner has traveled 3600 feet. **39.** When the magnets are 2 inches apart, the force between them is 25 pounds. **41.** The line is vertical passing through the point (3, 0). **43.** The line is a diagonal that cuts quadrants I and III equally. **45.** The figure is a square. **47. a.** Yes; **b.** yes; **c.** no **49.** Computer coordinate system: $A(0, 0)$; $B(100, 150)$; $C(550, 100)$; $D(350, 250)$; $E(150, 400)$; $F(600, 450)$; Rectangular coordinate system: $A(-350, 250)$; $B(-250, 100)$; $C(200, 150)$; $D(0, 0)$; $E(-200, -150)$; $F(250, -200)$.

2.4 EXERCISES

1. No **3.** Rational and irrational numbers together **5.** Answers will vary. **7.** Integer, rational, real **9.** Rational, real **11.** Rational, real **13.** $0.\overline{6}$ **15.** -0.8 **17.** $1.\overline{571428}$ **19.** $0.4\overline{123}$ **21.** $\frac{4}{5}$ **23.** $\frac{1}{9}$ **25.** $\frac{2}{3}$ **27.** $\frac{7}{9}$ **29.** $\frac{3}{16}$ **31.** $\frac{1}{6}$ **33.** $=$ **35.** $<$ **37.** $<$ **39.** $<$ **41.** More Americans eat out because of the atmosphere. **43.** $\frac{1}{6}$ **45.** $\frac{35}{24}$ **47.** $\frac{1}{24}$ **49.** $-\frac{2}{3}$ **51.** $\frac{11}{24}$ **53.** $\frac{17}{24}$ **55.** $-\frac{4}{15}$ **57.** $-\frac{1}{6}$ **59.** $-\frac{2}{5}$ **61.** $\frac{35}{24}$ **63.** $-\frac{3}{8}$ **65.** $\frac{1}{10}$ **67.** $\frac{15}{64}$ **69.** $\frac{3}{2}$ **71.** $-\frac{8}{9}$ **73.** $\frac{2}{3}$ **75.** $-\frac{1}{2}$ **77.** $\frac{8}{9}$ **79.** $\frac{21}{10}$ **81.** 16 feet X $18\frac{2}{3}$ feet **83.** $2\frac{1}{3}$ feet **85.** 12 quarts **87.** 54 inches **89.** $1\frac{3}{4}$ pounds **91.** $\frac{1}{2}$ pint **93.** 30 miles per hour **95.** $\frac{1}{20}$ gallon per second **97.** The profit on each package is \$.46. **99.** The recycler will pay the charity \$234.38. **101.** 11 studs are required. **103.** The new fraction is greater than $\frac{4}{7}$. **105.** Yes. Add the 2 numbers and then divide by 2. **107.**

x	$-\frac{3}{2}$	$-\frac{1}{2}$	$\frac{1}{2}$	$\frac{3}{2}$
y	$-\frac{9}{2}$	$-\frac{3}{2}$	$\frac{3}{2}$	$\frac{9}{2}$

109. a. 85.0875, 86.1875, 87.55; **b.** yes; **c.** no; **d.** fluctuates less.

2.5 EXERCISES

1. A movement that does not change the shape of the figure or rotate the figure. **3.** Draw the mirror image of the figure using the x-axis as the mirror. The x-coordinates of the two figures are equal. The y-coordinates are opposites. **5.** The x-coordinates of the two figures are equal. The y-coordinates are opposites. **7.** 48 square units **9.** 54 square units **11.** 15 square units **13.** 45 square units **15.** 99 square units **17.** (5, 6), (5, −2), (−1, −2), (−1, 6) **19.** (−4, −2), (5, −2), (−7, −8), (−2, −8) **21.** (−1, 5), (5, 5), (10, 0) **23.** (−9, −7), (7, 3), (0, −7); Area = 45 square units **25.** (−4, 0), (0, 9), (8, 9), (10, 0); Area = 99 square units

27. (4, 6), (4, −3), (−6, −3), (−6, 6) **29.** (4, 9), (2, −6), (−4, 6), (−2, 9) **31.** $(12, \frac{3}{4})$, $(18, -\frac{3}{4})$, $(-12, -\frac{3}{4})$ **33.** The area of the transformed figure is 6 times the original area. The product of the factors is 6. **35.** The area of the transformed figure is 2 times the original area. The product of the factors is 2. **37.**

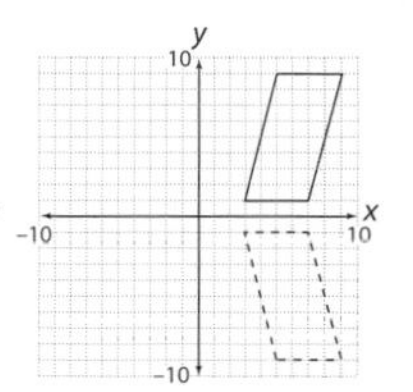

39.

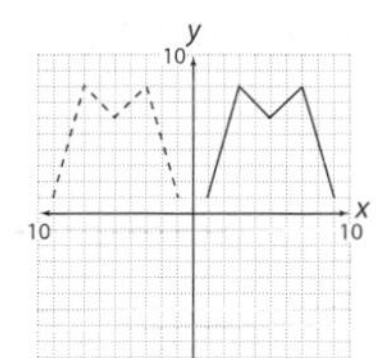

41.

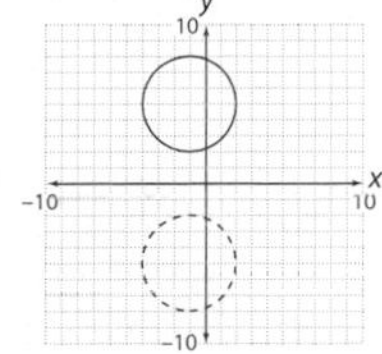

43.

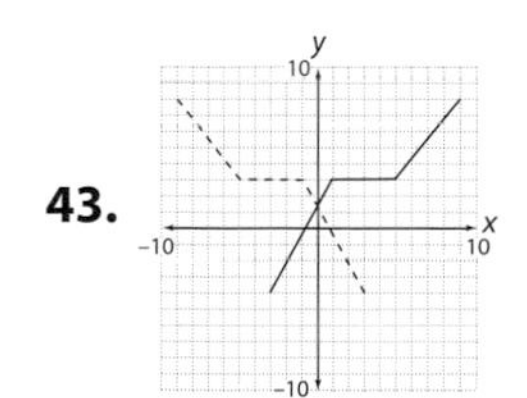

45. It is a line segment on the *y*-axis. **47.**

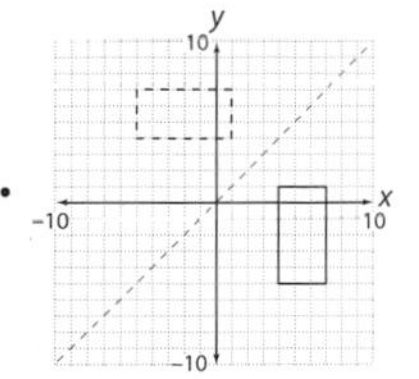

49. a.

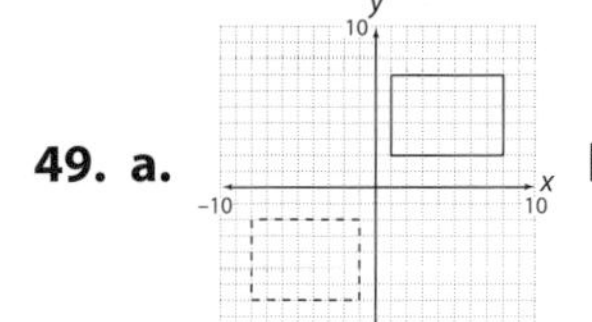

b.

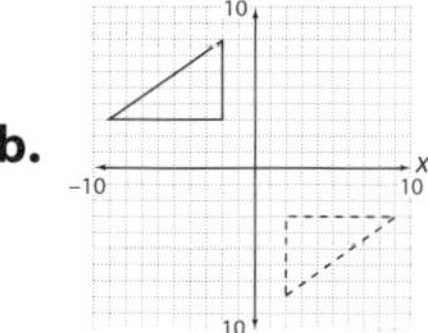

2.6 EXERCISES

1. Input is *x*-axis, output is *y*-axis. **3.** Answers will vary. **5.** The *x*- and *y*-coordinates are equal. **7. a.** 20 teams; **b.** 1994; **c.** 23 teams; **d.** 1991 and 1992 **9. a.** 80 grams; **b.** 70 grams; **c.** the amount increases; **d.** 105 grams **11. a.** 3 millimeters; **b.** yes; **c.** between 50 and 100; **d.** between 250 and 300 **13. a.** 3; **b.** 1 **15. a.** 1.44; **b.** 1.6 or −1.6 **17. a.** 7.759; **b.** −1.7 **19. a.** 42 miles; **b.** 0.6 minute **21. a.** 22,500 feet; **b.** 26.7 minutes **23.** 13.5 ounces **25.** 450 grams **27. a.** No; **b.** 34.2 million households; **c.** home PCs; **d.** the average increase was 3.167 million; **e.** 53.934 million households **29.** 1.3 billion light-years **31. a.** (75, 50); **b.** Answers will vary.

CHAPTER REVIEW EXERCISES

1. 14 **2.** −3 **3.** 18 **4.** 0 **5.** 74 **6.** −44 **7.** −276 **8.** 51 **9.** 9 **10.** −8 **11.** −5 **12.** −7 **13.** 56 **14.** −27 **15.** $0.\overline{45}$ **16.** $\frac{1}{12}$

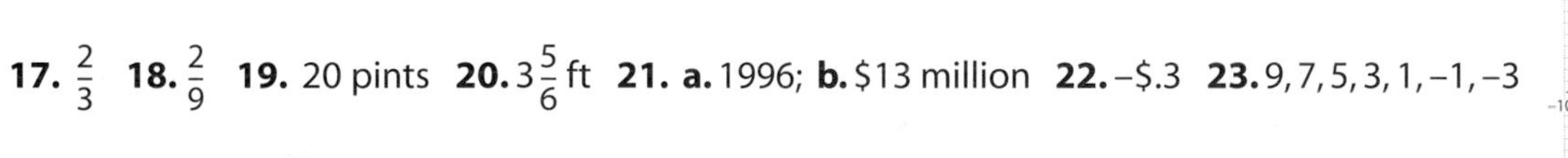

17. $\frac{2}{3}$ **18.** $\frac{2}{9}$ **19.** 20 pints **20.** $3\frac{5}{6}$ ft **21. a.** 1996; **b.** \$13 million **22.** −\$.3 **23.** 9, 7, 5, 3, 1, −1, −3

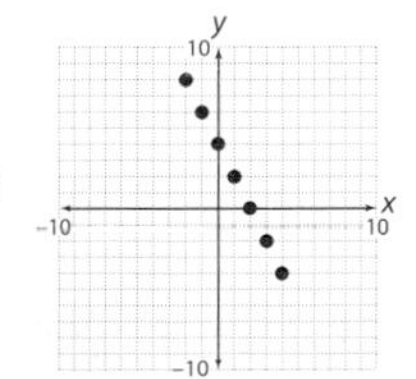

24.

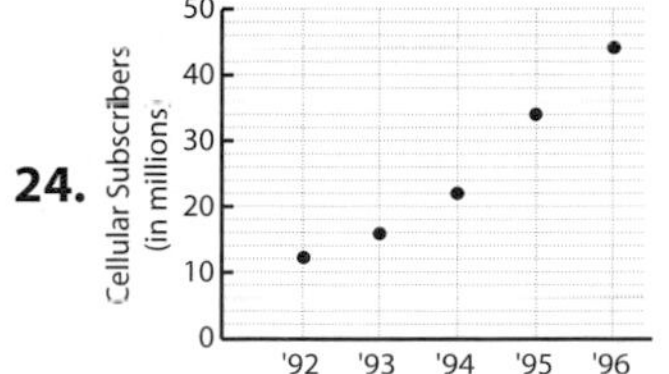

25. 54 square units **26.** (0, 4), (5, −6), (9, −2), (7, 2) **27.**

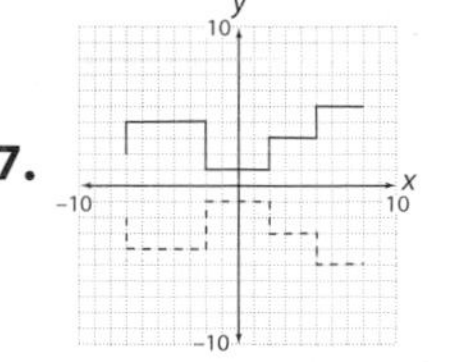

28.

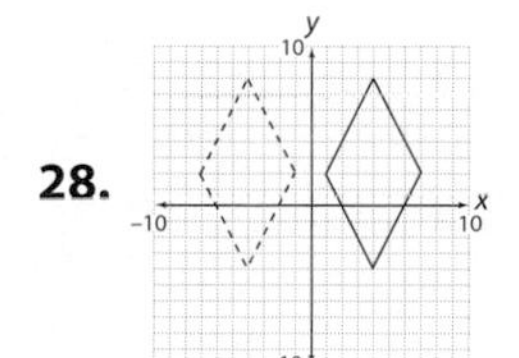

29. 3.42 **30.** 6.2 minutes **31.** 6 ounces

CUMULATIVE REVIEW EXERCISES

1. True **2.** −96 **3.** Mean = 64.48; median = 65 **4.** No planes have jet engines. **5.** $\frac{6}{11}$ **6.**

9	1, 3, 0
8	2, 9, 0, 2, 1, 5
7	0, 8, 7, 3, 9, 0, 1, 8, 6, 2
6	3, 9, 7, 2, 2, 2

7. True **8.** $x^2 + 2x + 4$ **9.** 19, $3n + 1$; 28 **10.** (−6, 0), (−4, 4), (6, 2), (4, −2) **11.** 50 square units **12.**

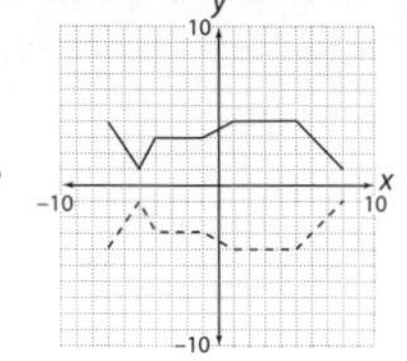

13. (−2, 3), (8, 2), (0, −4), (7, −3) **14.** 8, 6, 4, 2, 0, −2, −4; **15.** −2.09 **16.** 3.68 hours **17.** 5.6 grams

Answers to Chapter 3 Exercises

3.1 EXERCISES

1. Answers will vary. **3.** Answers will vary. **5.** No; $-x = -1 \cdot x$. If x represents a negative number, then $-x$ is a positive number; if x represents a positive number, then $-x$ is a negative number. **7.** 0 **9.** 6 **11.** −4 **13.** −8 **15.** 1 **17.** The Associative Property of Multiplication **19.** The Inverse Property of Addition **21.** The Distributive Property **23.** The Distributive Property **25.** The Commutative Property of Addition **27.** $21y$ **29.** $-6y$ **31.** Simplest form **33.** $5xy$ **35.** $-14x^2$ **37.** $-3x-8y$ **39.** $-2x$ **41.** $22y^2$ **43.** 0 **45.** $-\frac{7}{20}y$ **47.** $12x$ **49.** $-6a$ **51.** x **53.** $2x$ **55.** $3y$ **57.** $-2a-14$ **59.** $15x^2+6x$ **61.** $-6y^2+21$ **63.** $-12a^2-20a+28$ **65.** $a-7$ **67.** $18y-51$ **69.** $4x-4$ **71.** 0 **73.** $-a+b$ **75. a.** False; $8 \div 4 \neq 4 \div 8$. **b.** False; $(8 \div 4) \div 2 \neq 8 \div (4 \div 2)$. **c.** False; $7-(5-2) \neq (7-5)-2$. **d.** False; $7-4 \neq 4-7$. **e.** True.

3.2 EXERCISES

1. Answers will vary. **3.** Answers will vary. **5.** An equation has an equal sign; an expression does not have an equal sign. **7.** No **9.** No **11.** No **13.** 6 **15.** −2 **17.** 2 **19.** 14 **21.** $-\frac{3}{4}$ **23.** $-\frac{1}{3}$ **25.** −7 **27.** −8 **29.** 12 **31.** 6 **33.** 9 **35.** 3 **37.** 400 **39.** 200% **41.** 400 **43.** 400 **45.** 200 **47.** 30 **49.** 13.9% **51.** 22.9% **53.** 1032 wiretaps **55.** 67 votes **57.** −15 **59.** 5 **61.** 6 **63.** 41,493 **65.** Answers will vary. $x+9-7$ is a possibility. **67.** 2

3.3 EXERCISES

1. Answers will vary. **3.** Answers will vary. **5.** Answers will vary. **7.** −1 **9.** 3 **11.** 4 **13.** $\frac{3}{4}$ **15.** $-\frac{1}{6}$ **17.** $\frac{2}{5}$ **19.** −16 **21.** −16 **23.** 2 **25.** 3.95 **27.** 3 **29.** −3 **31.** −7 **33.** −2 **35.** 10 **37.** $-\frac{3}{4}$ **39.** 3 **41.** 2 **43.** $-\frac{1}{3}$ **45.** $n-15=7; n=22$ **47.** $7n=-21; n=-3$ **49.** $3n-4=5; n=3$ **51.** $4(2n+3)=12; n=0$ **53.** $12=6(n-3); n=5$ **55.** $22=6n-2; n=4$ **57.** $4n+7=2n+3; n=-2$ **59.** $5n-8=8n+4; n=-4$ **61.** $2(n-25)=3n; n=-50$ **63.** $3n=2(20-n)$; 8 and 12 **65.** $3n+2(18-n)=44$; 8 and 10 **67.** \$16,000 **69.** 45 mips **71.** 15 pounds **73.** \$117.75 **75.** 5 hours **77.** \$2000 and \$3000 **79.** Length: 80 feet; width 50 feet **81.** 8 feet **83.** 3 feet **85.** $\frac{1}{3}$ **87.** 15 minutes **89.** 75 coins

3.4 EXERCISES

1. The opposite angles of intersecting lines. The lines are perpendicular. **3.** They are not parallel. **5.** 40°, acute **7.** 108°, obtuse **9.** 90°, right **11.** 15° **13.** 37° **15.** 28°, 62° **17.** 48°, 132° **19.** 15° **21.** 17° **23.** 40° **25.** 20° **27.** 126° **29.** 11° **31.** $m\angle a = 38°, m\angle b = 142°$ **33.** $m\angle a = 47°, m\angle b = 133°$ **35.** 20° **37.** 47° **39.** $m\angle x = 145°$; $m\angle y = 60°$ **41.** $m\angle a = 45°; m\angle b = 135°$ **43.** 22° and 68° **45.** 62.83 feet **47.** 30° **49. a.** Always true; **b.** always true; **c.** sometimes true **d.** sometimes true.

3.5 EXERCISES

1. Answers will vary. For instance, $d = rt$; $t = \frac{d}{r}$; and $r = \frac{d}{t}$ **3.** $(t-1)$ hours **5.** 50 feet **7.** $d = 45t$;

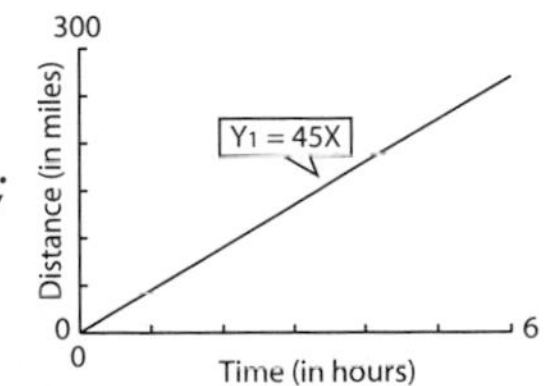

9. a. Davadene; **b.** Yes. Her graph intersects Jacob's graph. **11.** 105 miles per hour; 130 miles per hour **13.** 11 A.M. **15.** 2 hours **17.** 120 miles **19.** 68 miles per hour **21.** 300 miles **23.** 2.5 hours **25.** 1.5 hours **27.** 180 miles **29.** Michele; her graph is above Sarakit's graph. **31.** 15 miles per hour

3.6 EXERCISES

1. a. Decreases; **b.** increases; **c.** remains the same **3. a.** Increases; **b.** increases **5.** 24% **7.** 20 gallons **9.** 30 pounds **11.** 6.25 gallons **13.** 19% **15.** 20 pounds **17.** 7% solution: 100 milliliters; 4% solution: 200 milliliters **19.** 150 ounces **21.** Pure silk: 55 kilograms; 85% silk: 20 kilograms **23.** 3.75 gallons **25.** Answers will vary.

3.7 EXERCISES

1. No **3.** Answers will vary. **5.** $\{x \mid x < 2\}$; 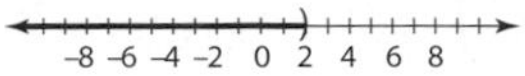 **7.** $\{x \mid x \ge -1\}$; 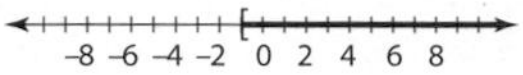 **9.** $\{x \mid x \le -3\}$; 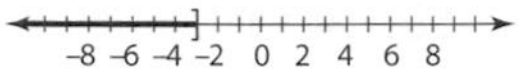 **11.** $\{x \mid x > 0\}$; 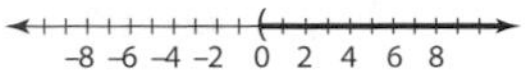 **13.** $\{y \mid y \ge -9\}$ **15.** $\{x \mid x \ge 5\}$ **17.** $\{x \mid x \le 10\}$

19. $\{x \mid x > 2\}$ **21.** $\left\{x \middle| x \ge -\frac{31}{24}\right\}$ **23.** $\{x \mid x \le 0.70\}$ **25.** $\left\{x \middle| x < \frac{5}{3}\right\}$ **27.** $\left\{x \middle| x > -\frac{5}{2}\right\}$ **29.** $\{x \mid x < -18\}$ **31.** $\{y \mid y \ge 6\}$

33. $\{b \mid b \le 33\}$ **35.** $\left\{y \middle| y \le \frac{5}{6}\right\}$ **37.** $\left\{y \middle| y > -\frac{3}{2}\right\}$ **39.** $\{y \mid y \ge -0.8\}$ **41.** $\{x \mid x \le 5\}$ **43.** $\{x \mid x < 4\}$ **45.** $\{x \mid x \ge 1\}$

47. $\{x \mid x < 0\}$ **49.** $\{x \mid x > 500\}$ **51.** $\{x \mid x \le -5\}$ **53.** $\left\{x \middle| x < \frac{25}{11}\right\}$ **55.** $\left\{n \middle| n \le \frac{11}{18}\right\}$ **57.** $\left\{x \middle| x \le \frac{2}{5}\right\}$ **59.** $\left\{n \middle| n > \frac{7}{10}\right\}$

61. 2 **63.** More than 440 lb **65.** More than $5714 **67.** More than 60 minutes **69.** More than 76 miles **71.** {1, 2} **73.** {1, 2, 3} **75.** {3, 4, 5} **77.** {10, 11, 12, 13}

CHAPTER REVIEW EXERCISES

1. 4 **2.** Associative Property of Multiplication **3.** $-12xy$ **4.** x **5.** $6x^2 + 3xy - 9y^2$ **6.** $-x + 50$ **7.** 7 **8.** 1 **9.** −10 **10.** 25 **11.** 20 **12.** 5 **13.** $\frac{9}{7}$ **14.** $\frac{8}{3}$ **15.** 2 **16.** 0 **17.** $\frac{2}{3}$ **18.** 4 **19.** 1 **20.** −4 **21.** 22° **22.** 57° **23. a.** 51,000 people; **b.** 20,000 people **24.** 15°, 75° **25.** 20.5% **26.** 250 tickets **27.** 50 mph, 65 mph **28.** 17.25 pounds **29.** Less than 60,000 gallons

CUMULATIVE REVIEW EXERCISES

1. 16 **2.** For example, 16 **3.** 82 **4.** −72 **5.** $\frac{3n}{4n+3}$ **6.** $\frac{1}{16}$ **7. a.** 1995; **b.** 1991; **c.** $8.7 billion **8.** Mean = 99.1°; median = 98° **9.** $Q_1 = 453$; $Q_3 = 533$ **10.** 15 **11.** −17 **12.** 38 **13.** 25 **14.** 64 **15.** −72 **16.** −26 **17.** $-\frac{1}{15}$

18. 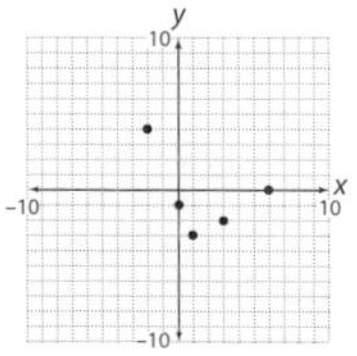

19. Associate Property of Multiplication **20.** −5 **21.** −1 **22.** $\{x \mid x \ge -1\}$ **23.** $\{x \mid x > -3\}$ **24.** $7\frac{1}{7}\%$

25. 43° **26.** 75°, 60°, and 45° **27.** 180 miles **28.** 75 liters or less **29.** $A'(-5, -6)$, $B'(-1, 3)$, $C'(5, -9)$

Answers to Chapter 4 Exercises

4.1 EXERCISES

1. Answers will vary. **3.** It is the value of the function at x. **5.** Answers will vary. **7.** D: {0, 2, 4, 6}; R: {0}; yes **9.** D: {2}; R: {2, 4, 6, 8}; no **11.** D: {0, 1, 2, 3}; R: {0, 1, 2, 3}; yes **13.** D: {−3, −2, −1, 1, 2, 3}; R: {−3, 2, 3, 4}; yes **15.** 40; (10, 40) **17.** −11; (−6, −11) **19.** 12; (−2, 12) **21.** $\frac{7}{2}$; $\left(\frac{1}{2}, \frac{7}{2}\right)$ **23.** 15; (3, 15) **25.** $\frac{1}{3}$; $\left(3, \frac{1}{3}\right)$ **27.** R: {−19, −13, −7, −1, 5}; (−5, −19), (−3, −13), (−1, −7), (1, −1), (3, 5) **29.** R: {1, 2, 3, 4, 5}; (−4, 1), (−2, 2), (0, 3), (2, 4), (4, 5) **31.** R: {6, 7, 15}; (−3, 15), (−1, 7), (0, 6), (1, 7), (3, 15)

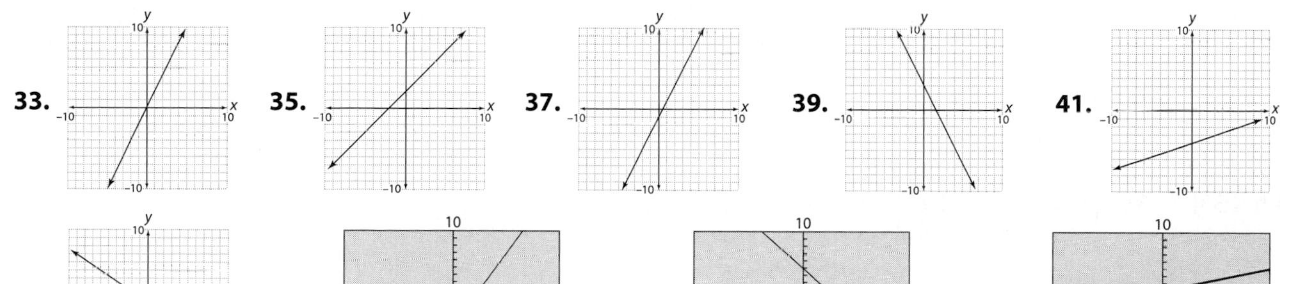

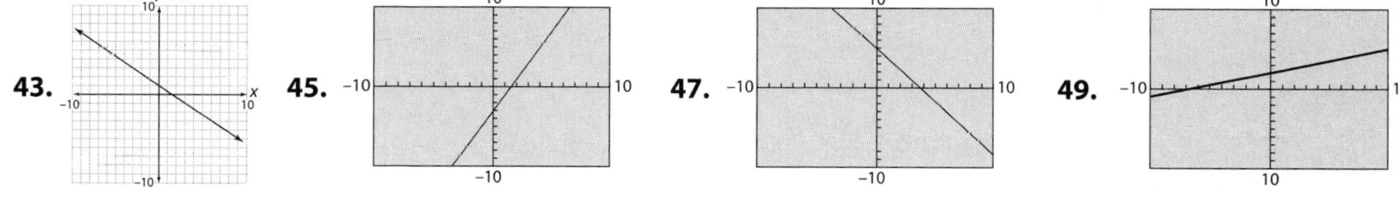

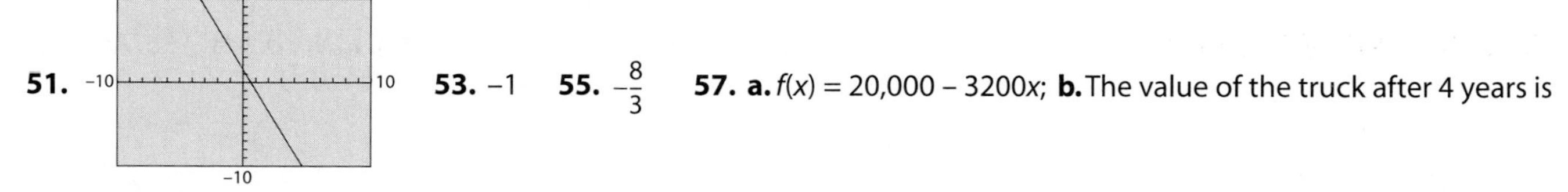

53. −1 **55.** $-\frac{8}{3}$ **57. a.** $f(x) = 20{,}000 - 3200x$; **b.** The value of the truck after 4 years is $7200. **59. a.** Yes; **b.** 27; **c.** 10; **d.** 44; **e.** 8; **f.** 65 **61. a.** Yes; **b.** 1.2 grams; **c.** 86%; **d.** 95% **63.** 6 **65. a.** Yes; **b.** $20.88; **c.** $20.52; **d.** $227.04; **e.** $324.56 **67. a.** Yes; **b.** no

4.2 EXERCISES

1. A point at which a graph crosses the x-axis. A point at which a graph crosses the y-axis. **3.** Yes **5.** $y = -3x + 10$ **7.** $y = 4x - 3$ **9.** $y = -\frac{2}{7}x + 2$ **11.** $y = -\frac{2}{3}x + 3$ **13.** $y = \frac{5}{2}x - 2$ **15.** $y = \frac{1}{4}x - 3$ **17.** (1, 0), (0, 3) **19.** (2, 0), (0, 3) **21.** (6, 0), (0, −4) **23.** $\left(\frac{4}{3}, 0\right)$, (0, −2) **25.** No x-intercept, (0, −4) **27.** $\left(\frac{10}{3}, 0\right)$, $\left(0, \frac{5}{2}\right)$ **29.** $\left(\frac{14}{5}, 0\right)$, $\left(0, -\frac{14}{3}\right)$

31. (3, 0), (0, −6) **33.** (0, 0), (0, 0) **35.** (−6, 0), (0, 3) **37.** (−2, 0), (0, −3) **39.** (0, 0), (0, 0)

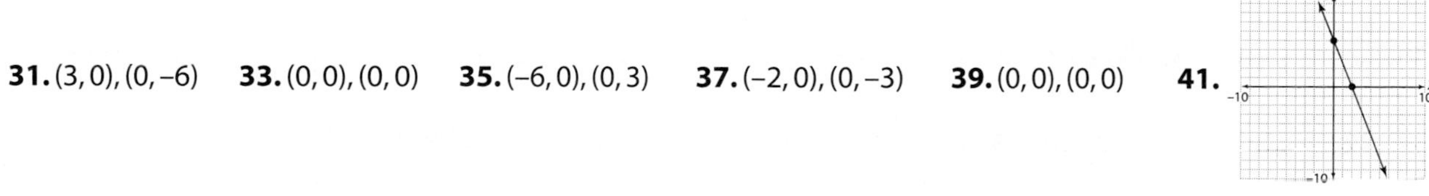

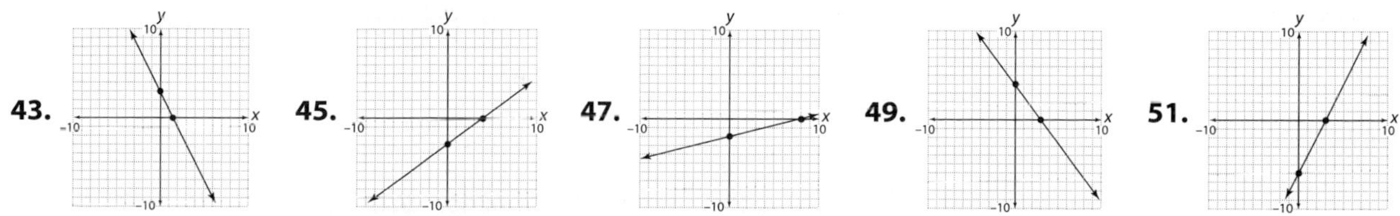

53. 0 **55.** $\frac{C}{A}$; division by 0 is not allowed. **57.** $\left(-\frac{b}{m}, 0\right)$, $(0, b)$

4.3 EXERCISES

1. The graph of a line with zero slope is horizontal; the graph of a line with no slope is vertical. **3.** If $x_1 = x_2$, the denominator would equal zero, and division by zero is not allowed. **5.** -2 **7.** $-\frac{5}{2}$ **9.** $\frac{1}{4}$ **11.** $-\frac{1}{2}$ **13.** -1 **15.** Undefined **17.** 0

19. $-\frac{1}{3}$ **21.** 1 **23.** -5 **25.** 0 **27.** Undefined **29.**

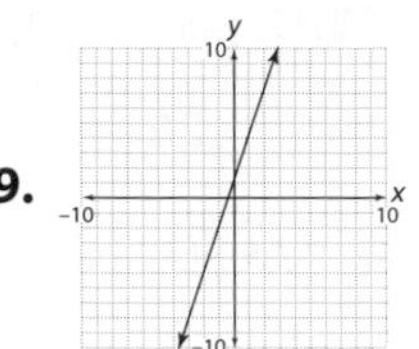

31.

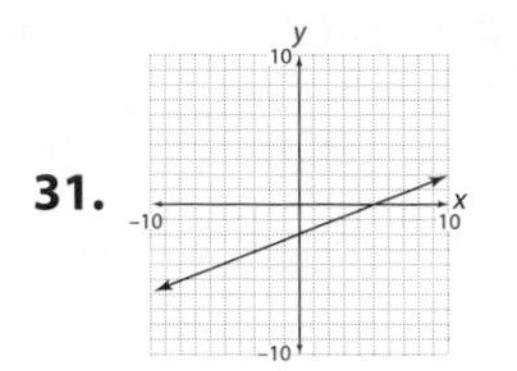

33.

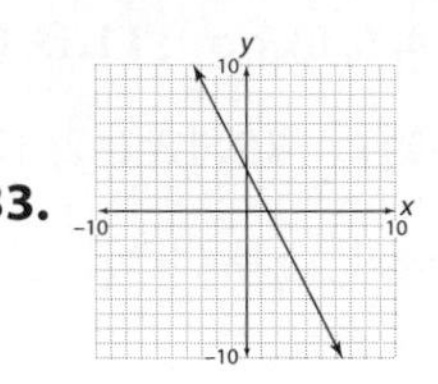

35.

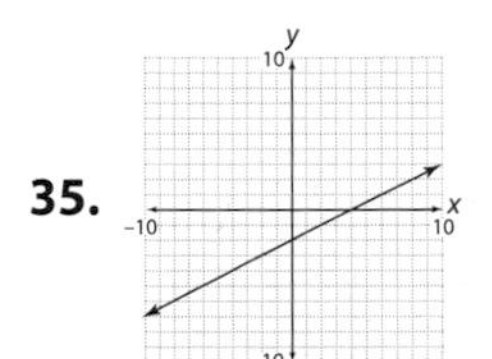

37.

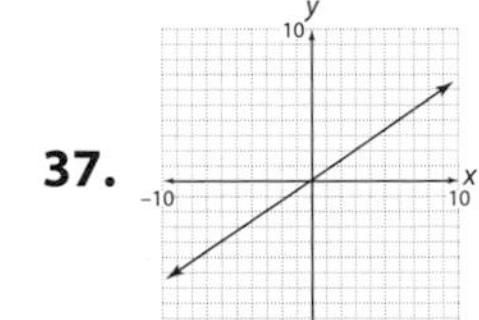

39.

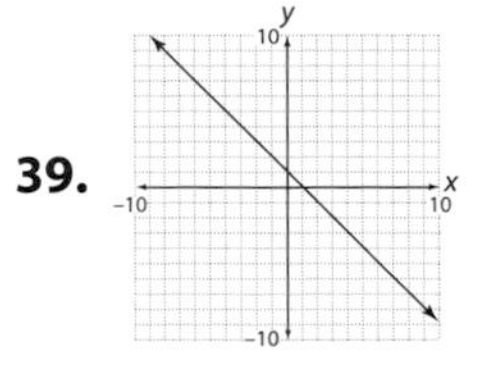

41.

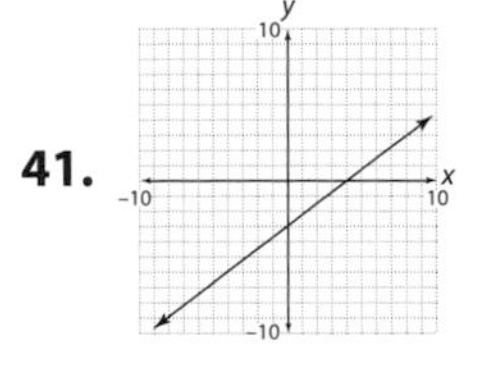

43.

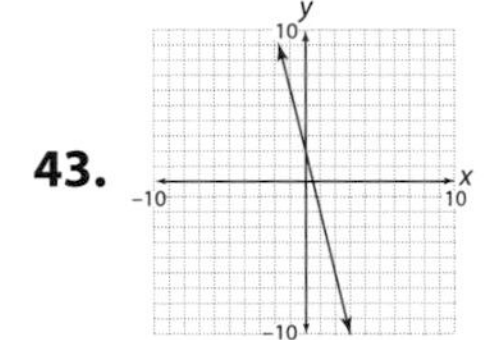

45.

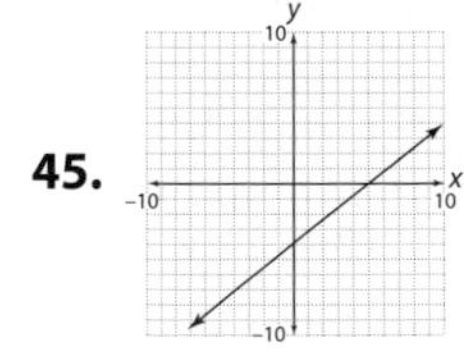

47. $m = 0.40$. The cellular call costs \$.40 per minute. **49.** $m = -0.04375$. For each mile the car is driven, approximately 0.04 gallon of fuel is used. **51.** A: Lois, B: Tanya, C: distance between them **53. a.** No; **b.** yes **55.** Answers will vary. **57.** -4 **59.** -7 **61.** 3 **63.** Decreases the slope **65.** Decreases the *y*-intercept **67.** Yes

4.4 EXERCISES

1. The slope is *m* and the *y*-intercept is $(0, b)$. **3.** Answers will vary. **5.** Yes. $f(x)$ and *y* can be used interchangeably for functions. **7.** $y = 2x + 2$ **9.** $y = -3x - 1$ **11.** $y = \frac{1}{3}x$ **13.** $y = \frac{3}{4}x - 5$ **15.** $y = -\frac{3}{5}x$ **17.** $y = \frac{1}{4}x + \frac{5}{2}$ **19.** $y = -\frac{2}{3}x - 7$ **21.** $y = 2x - 3$ **23.** $y = -2x - 3$ **25.** $y = \frac{2}{3}x$ **27.** $y = \frac{1}{2}x + 2$ **29.** $y = -\frac{3}{4}x - 2$ **31.** $y = \frac{3}{4}x + \frac{5}{2}$ **33.** $y = -\frac{4}{3}x - 9$ **35.** $h = 1200t$; 13,200 feet **37.** $C = 0.59t + 4.95$; \$12.62 **39.** $y = 2.4x - 4754.4$; 48 million **41.** Yes; $y = -x + 6$ **43.** No **45.** 7 **47.** -1 **49.** $h = -28t + 1400$

4.5 EXERCISES

1. Two or more equations considered together **3.** Answers will vary. **5.** Answers will vary. **7.** (4, 3) **9.** (2, 4) **11.** (3, 2) **13.** (1, 4) **15.** Inconsistent **17.** Dependent, $(x, 3x + 1)$ **19.** $\left(-\frac{3}{4}, -\frac{3}{4}\right)$ **21.** (2, 0) **23.** (2, 1) **25.** $\left(\frac{9}{19}, -\frac{13}{19}\right)$ **27.** Dependent, $(x, -3x + 4)$ **29.** (10, 31) **31.** $(-22, -5)$ **33.** $\left(-\frac{5}{7}, \frac{13}{7}\right)$ **35.** (5, 2) **37.** $(3, -2)$ **39.** \$9000 at 7%; \$6000 at 6.5% **41.** \$1500 **43.** \$200,000 at 10%; \$100,000 at 8.5% **45.** \$2500 **47.** \$40,500 at 8%; \$13,500 at 12% **49.** 30°, 60° **51.** 64°, 116° **53.** 27 **55.** Answers will vary. **57.** 2 **59.** 2 **61.** \$3831.87

4.6 EXERCISES

1. Answers will vary. **3.** The amount before mixing is equal to the amount after mixing. **5.** $(2, -1)$ **7.** $(2, -1)$ **9.** $(2, -3)$ **11.** Inconsistent **13.** (2, 0) **15.** (0, 0) **17.** $\left(\frac{3}{2}, \frac{1}{2}\right)$ **19.** $\left(\frac{5}{3}, -\frac{2}{15}\right)$ **21.** (2, 1) **23.** Dependent, $(x, \frac{1}{2}x - 4)$ **25.** (1, 1) **27.** $(5, -2)$ **29.** Plane, 400 mph; wind, 50 mph **31.** Boat, 7 mph; current, 3 mph **33.** Plane, 125 mph; wind, 25 mph **35.** Plane, 100 mph; wind, 20 mph **37.** Plane, 110 mph; wind, 30 mph **39.** Plane, 180 km/h; wind, 20 km/h **41.** 15 mph **43.** 32.5 hours **45.** 13% solution, 20 milliliters; 18% solution, 30 milliliters **47.** 50%

49. 30 pounds **51.** 0.74% **53.** 7% solution, 100 milliliters; 4% solution, 200 milliliters **55.** 100 ounces **57.** 80% **59.** 150 milliliters **61.** Pure silk, 55 kilograms; 85% silk, 20 kilograms **63.** 52% **65.** $A = 3, B = -3$ **67.** $A = 1$

69. a. 14; **b.** $\frac{1}{2}$

CHAPTER REVIEW EXERCISES

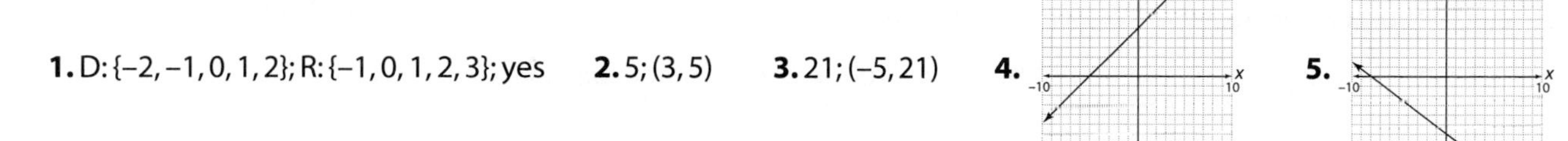

1. D: {−2, −1, 0, 1, 2}; R: {−1, 0, 1, 2, 3}; yes **2.** 5; (3, 5) **3.** 21; (−5, 21) **4.** [graph] **5.** [graph]

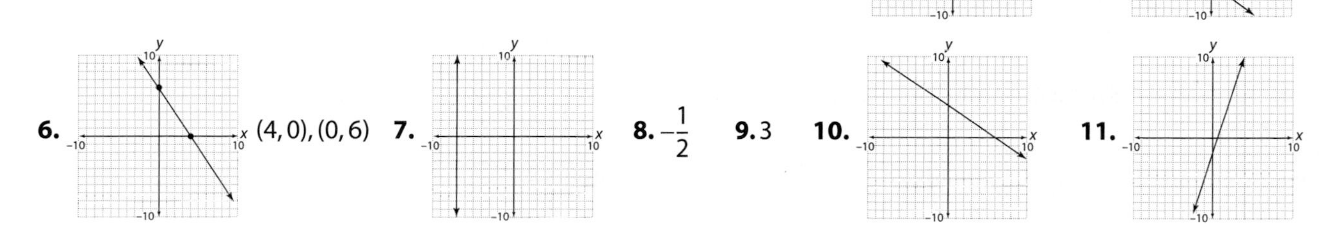

6. [graph] (4, 0), (0, 6) **7.** [graph] **8.** $-\frac{1}{2}$ **9.** 3 **10.** [graph] **11.** [graph]

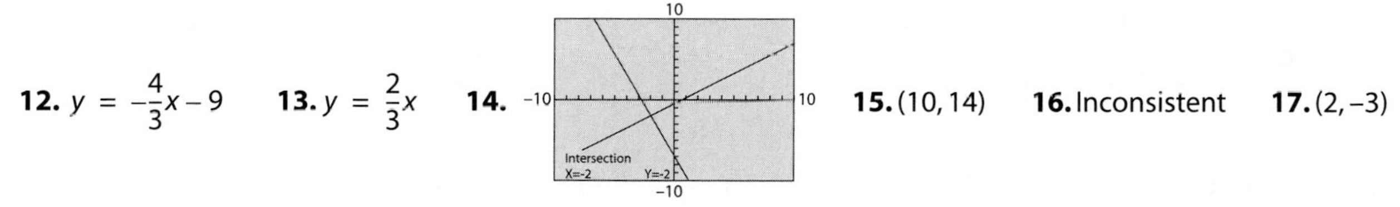

12. $y = -\frac{4}{3}x - 9$ **13.** $y = \frac{2}{3}x$ **14.** [graph] **15.** (10, 14) **16.** Inconsistent **17.** (2, −3)

18. (−3, 4) **19.** Plane, 325 mph; wind, 25 mph **20.** Sculling team, 9 mph; current, 3 mph **21.** \$5000 at 10%; \$2000 at 15% **22.** 30% butterfat, 5 gallons; 4% butterfat, 8 gallons **23.** Length: 80 feet, width: 20 feet

CUMULATIVE REVIEW EXERCISES

1. 26 **2.** Some animals have fur. **3.** 31 **4.** −2 **5.** $\frac{1}{16}$ **6.** Mean = 288.5; median = 279 **7.** $\frac{3n}{2n+7}$ **8.** −5 **9.** −30

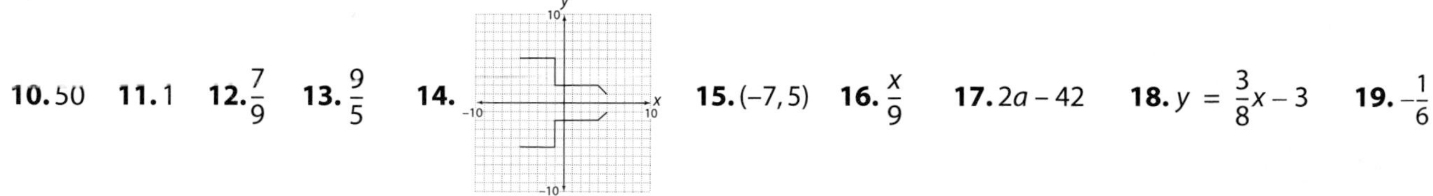

10. 50 **11.** 1 **12.** $\frac{7}{9}$ **13.** $\frac{9}{5}$ **14.** [graph] **15.** (−7, 5) **16.** $\frac{x}{9}$ **17.** $2a - 42$ **18.** $y = \frac{3}{8}x - 3$ **19.** $-\frac{1}{6}$

20. Commutative Property of Multiplication **21.** −2 **22.** D: {−5, −3, −1}; R: {1, 3, 5}; yes **23.** 10; (−2, 10)

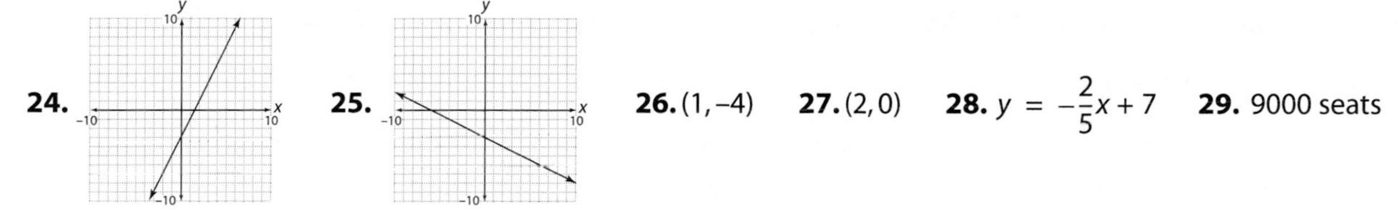

24. [graph] **25.** [graph] **26.** (1, −4) **27.** (2, 0) **28.** $y = -\frac{2}{5}x + 7$ **29.** 9000 seats

30. 9.6 pounds **31.** 72 **32.** 1st plane, 225 mph; 2nd plane, 125 mph **33.** 48°, 33°, and 99°

Answers to Chapter 5 Exercises

5.1 EXERCISES

1a. This is a monomial because it is the product of a number, 32, and variables, a^3b.

b. This is a monomial because it is the product of a number, $\frac{5}{7}$, and a variable, n.

c. This is not a monomial because there is a variable in the denominator.

3a. The exponent on 6 is positive; it should not be moved to the denominator. $6x^{-3} = \frac{6}{x^3}$

b. The exponent on x is positive; it should not be moved to the denominator. $xy^{-2} = \frac{x}{y^2}$

c. The exponent on 8 is positive; it should not be moved to the numerator. $\frac{1}{8a^{-4}} = \frac{a^4}{8}$

d. The exponent on c is positive; it should not be moved to the numerator. $\frac{1}{b^{-5}c} = \frac{b^5}{c}$

5. a^9 **7.** z^8 **9.** x^{15} **11.** $x^{12}y^{18}$ **13.** $17s^4t^3$ **15.** 1 **17.** a^6 **19.** $\frac{1}{w^8}$ **21.** $-m^9n^3$ **23.** $8x^{12}$

25. $24x^7$ **27.** $-8a^6$ **29.** a^5 **31.** $4p^4q^5$ **33.** $-24r^3$ **35.** $8a^9b^3c^6$ **37.** mn^2 **39.** 1 **41.** $\frac{3r^2}{2}$

43. $-\frac{2a}{3}$ **45.** $-27m^2n^4p^3$ **47.** $-15x^3y^8$ **49.** $18x^7$ **51.** $\frac{1}{64}$ **53.** $\frac{1}{x^5}$ **55.** $54n^{14}$ **57.** $\frac{4}{x^7}$ **59.** $\frac{x^2}{3}$

61. $72a^{12}$ **63.** $\frac{1}{2y^3}$ **65.** $\frac{7xz}{8y^3}$ **67.** $-8x^{13}y^{14}$ **69.** $19x^2y$ **71.** $\frac{9}{x^2y^4}$ **73.** $\frac{2}{x^4}$ **75.** $\frac{1}{2x^3}$ **77.** $\frac{1}{2x^2y^6}$

79. $-\frac{y}{6x^3}$ **81.** $11xy$ **83.** $15a^2b$ **85.** $64x^4y^2$ m^2 **87.** $50c^3d^4$ mi **89.** $54m^2n^4$ km^2 **91.** $6a^2b^3$ yd

93. $3a^2$ **95.** 2.37×10^6 **97.** 4.5×10^{-4} **99.** 3.09×10^5 **101.** 6.01×10^{-7} **103.** 5.7×10^{10} **105.** 1.7×10^{-8}

107. 710,000 **109.** 0.000043 **111.** 671,000,000 **113.** 0.00000713 **115.** 5,000,000,000,000 **117.** 0.00801
119. 1.6×10^{10} **121.** $\$1.67 \times 10^{11}$ **123.** 1.6×10^{-19} **125.** 6.023×10^{23} **127.** 3.086×10^{18} **129.** 6.65×10^{19}

131. 3.22×10^{-14} **133.** 3.6×10^5 **135.** 1.8×10^{-18} **137 a.** $\frac{3}{64}$ **b.** $\frac{4}{81}$ **139.** $\frac{1}{4}, \frac{1}{2}, 1, 2, 4; 4, 2, 1, \frac{1}{2}, \frac{1}{4}$ **141.** a

5.2 EXERCISES

1. a. This is a binomial. It contains two terms, $8x^4$ and $-6x^2$. **b.** This is a trinomial. It contains three terms, $4a^2b^2$, $9ab$, and 10. **c.** This is a monomial. It is one term, $7x^3y^4$. (Note: It is a product of a number and variables. There is no addition or subtraction operation in the expression.) **3. a.** Yes. Both $\frac{1}{5}x^3$ and $\frac{1}{2}x$ are monomials. (Note: The coefficients of variables can be fractions.) **b.** No. A polynomial does not have a variable in the denominator of a fraction. **c.** Yes. Both x and $\sqrt{5}$ are monomials. (Note: The variable is not under a radical sign.) **5.** 67.02 ft^3 **7.** 576.99 m **9.** 97.0%
11. $-2x^2 + 3x$ **13.** $4x$ **15.** $7x^2 + xy - 4y^2$ **17.** $3y^2 - 4y - 2$ **19.** $3a^2 - 3a + 17$ **21.** $-7x - 7$
23. $-2x^3 + x^2 + 2$ **25.** $x^3 + 2x^2 - 6x - 6$ **27.** $5x^3 + 10x^2 - x - 4$ **29.** $y^3 - y^2 + 6y - 6$ **31.** $4y^3 - 2y^2 + 2y - 4$
33. $11x^2 + 2x + 4$ **35.** $5a^2 + 3a - 9$ **37.** $(8d^2 + 12d + 4)$ km **39.** $(-2n^2 + 160n - 1200)$ dollars **41.** $-4x^2 + 6x + 3$
43. $4x^2 - 6x + 3$ **45.** $x^3 + 4x^2 + 5x + 2$ **47.** $a^3 - 6a^2 + 13a - 12$ **49.** $-2b^3 + 7b^2 + 19b - 20$
51. $x^4 - 4x^3 - 3x^2 + 14x - 8$ **53.** $y^4 + 4y^3 + y^2 - 5y + 2$ **55.** $x^2 + 4x + 3$ **57.** $a^2 + a - 12$ **59.** $y^2 - 10y + 21$
61. $2x^2 + 15x + 7$ **63.** $3x^2 + 11x - 4$ **65.** $4x^2 - 31x + 21$ **67.** $3y^2 - 2y - 16$ **69.** $21a^2 - 83a + 80$
71. $15b^2 + 47b - 78$ **73.** $2a^2 + 7ab + 3b^2$ **75.** $6a^2 + ab - 2b^2$ **77.** $d^2 - 36$ **79.** $4x^2 - 9$ **81.** $x^2 + 2x + 1$

83. $9a^2 - 30a + 25$ **85.** $x + 1$ **87.** $2a - 5$ **89.** $3a + 2$ **91.** $4b^2 - 3$ **93.** $x - 2$ **95.** $-x + 2$ **97.** $x^2 + 3x - 5$

99. $x^4 - 3x^2 - 1$ **101.** $xy + 2$ **103.** $b - 7$ **105.** $2x + 1$ **107.** $x + 1 + \frac{2}{x-1}$ **109.** $2x - 1 - \frac{2}{3x-2}$

111. $a + 3 + \frac{4}{a+2}$ **113.** $y - 6 + \frac{26}{2y+3}$ **115.** $2x + 5 + \frac{8}{2x-1}$ **117.** $x^2 + 2x + 3$ **119.** $x^2 - 3$

121. $(4x^2 + 4x + 1)$ m^2 **123.** $(10x^2 - 35x)$ mi^2 **125.** $(\pi x^2 + 8\pi x + 16\pi)$ in^2 **127.** $(64x^3 + 48x^2 + 12x + 1)$ in^3
129. $(4x^2 + 10x)$ m^2 **131.** $(18x^2 + 12x + 2)$ in^2 **133.** $(3x + 2)$ ft **135.** $(2x^2 + x - 1)$ in. **137.** $(90x + 2025)$ ft^2
139. $(4x^3 - 160x^2 + 1600x)$ in^3; 3468 in^3 **141.** $x^3 + 5x^2 + 5x - 3$ **143.** $x^2 + 7x + 1$ **145.** $x^3 - 7x^2 - 7$
147. $4x^2 - 6x + 3$ **149 a.** Always true; **b.** sometimes true; **c.** always true; **d.** always true; **151.** 5
153. $(400x - 80x^2 + 4x^3)$ in^3; no; explanations will vary.

5.3 EXERCISES

1. A factor is a number or expression in a multiplication. To factor means to write a polynomial as a product of other polynomials. **3.a.** Always true; **b.** always true; **c.** sometimes true; **d.** always true; **e.** sometimes true; **f.** never true; **g.** always true **5.** xy^4 **7.** $7a^3$ **9.** $3a^2b^2$ **11.** $2x$ **13.** $3x$ **15.** $a = 5, b = 8$
17. $a = -9, b = 8$ **19.** $a = -11, b = -4$ **21.** $8(2a - 3)$ **23.** $y(12y - 5)$ **25.** $4a^2(3a^3 - 8)$
27. $5xyz(2xz + 3y^2)$ **29.** $a(a^2 + 4a + 8)$ **31.** $5(x^2 - 3x + 7)$ **33.** $5y(y^2 - 4y + 2)$ **35.** $3y^2(y^2 - 3y - 2)$
37. $x^2y^2(x^2y^2 - 3xy + 6)$ **39.** $8x^2y(2 - xy^3 - 6y)$ **41.** $(x - 4)(x^2 - 3)$ **43.** $(y - 4)(3y^2 + 1)$ **45.** $(y + 6)(z - 3)$
47. $(y + 4)(x^2 + 3)$ **49.** $(t + 4)(t - s)$ **51.** $(2a - 3)(5ab - 2)$ **53.** $(x - 1)(x + 2)$ **55.** $(a + 4)(a - 3)$
57. $(a - 1)(a - 2)$ **59.** $(b + 8)(b - 1)$ **61.** $(z + 5)(z - 9)$ **63.** $(z - 5)(z - 9)$ **65.** $(b + 4)(b + 5)$
67. Nonfactorable over the integers **69.** $(x - 7)(x - 8)$ **71.** $(y + 3)(2y + 1)$ **73.** $(a - 1)(3a - 1)$
75. $(x - 3)(2x + 1)$ **77.** $(2t + 1)(5t + 3)$ **79.** $(2z - 1)(5z + 4)$ **81.** Nonfactorable over the integers
83. $(t + 2)(2t - 5)$ **85.** $(3y + 1)(4y + 5)$ **87.** $(a - 5)(11a + 1)$ **89.** $(2b - 3)(3b - 2)$ **91.** $(a + 9)(a - 9)$
93. $(2x + 1)(2x - 1)$ **95.** $(1 + 7x)(1 - 7x)$ **97.** Nonfactorable over the integers **99.** $3(x + 2)(x + 3)$
101. $4(x + 1)(x - 2)$ **103.** $a(b + 8)(b - 1)$ **105.** $3y(y - 2)(y - 3)$ **107.** $3y^2(y + 3)(y + 15)$
109. $3(x + 3)(4x - 1)$ **111.** $x(x - 1)(2x + 5)$ **113.** $8(t + 4)(2t - 3)$ **115.** $p(2p + 1)(3p + 1)$
117. $2(5 + x)(5 - x)$ **119.** $b^2(b + a)(b - a)$ **121.** $(2x + 1)$ in. by $(x + 4)$ in. **123.** $(2x + 3)$ cm
125. $(6x + 1)$ yd by $(5x + 3)$ yd **127.** $3x$ cm by $(y + 1)$ cm by $(y + 6)$ cm
129. $2r^2(4 - \pi)$ **131.** $r^2(\pi - 2)$ **133.** $(2x - 3)$ by $(2x + 3)$ **135.a.** $-36, 36, -12, 12$; **b.** $-19, 19, -11, 11, -9, 9$; **c.** $-22, 22, -10, 10$; **d.** $-15, 15, -9, 9$ **137.** An infinite number of different values of k are possible. Explanations will vary.
139. $(x + 2)(x - 3)(x - 1)$ **141.** The dimensions are $3x - 2$ and $x + 1$. If $x = 1.5$, then the rectangle is a square. If $x < 1.5$, the length is $(x + 1)$ feet and the width is $(3x - 2)$ feet. If $x > 1.5$, then the width is $(x + 1)$ feet and the length is $(3x - 2)$ feet. If $x < 0$, then $3x - 2$ is negative, which is not possible. Therefore, x cannot be less than 0. If $x = 0$, $3x^2 + x - 2 < 0$, which is not possible. Therefore, x cannot be equal to 0. **143.** Answers will vary.

CHAPTER REVIEW EXERCISES

1. $24a^2b^3c$ **2.** $3x^3 + 6x^2 - 8x + 3$ **3.** $-20x^3y^5$ **4.** $-4x^6$ **5.** $(2a - 3)(2a - 3)$ **6.** $5(x^2 - 9x - 3)$

7. $\frac{6b}{a}$ **8.** $4x^4 - 2x^2 + 5$ **9.** $(a - 3)(a - 16)$ **10.** $x(x + 5)(x - 3)$ **11.** $6y^4 - 9y^3 + 18y^2$ **12.** $4x^2 - 20x + 25$

13. $(2x + 1)(3x + 8)$ **14.** $2(b + 4)(b - 4)$ **15.** $\frac{a^4}{b^6}$ **16.** 2.9×10^{-5} **17.** $4(x + 4)(2x - 3)$ **18.** $(b + 6)(a - 3)$

19. $16y^2 - 9$ **20.** $5y^2 - 10y$ **21.** $9a^6b^4$ **22.** $10a^3 - 39a^2 + 20a - 21$ **23.** Nonfactorable over the integers

24. $2y^2(y + 1)(y - 8)$ **25.** $-\frac{1}{4a^2b^5}$ **26.** $4x + 8 + \frac{21}{2x-3}$ **27.** 0.000000035 **28.** $(p + 2)(p + 3)$

29. $(4x^2 + 12x + 9)$ m^2 **30.** $(3x + 2)$ in. by $(x + 5)$ in.

CUMULATIVE REVIEW EXERCISES

1. P, Q **2.** 8 Δ **3.** a and b are true. **4.** 576 **5.** 47 **6.** 48 **7.** $8 < 14 < 25$ **8.** -42 and 81
9. -89 **10.** 87 **11.** 18 **12.** Some taxpayers file their returns electronically.

13. Answers will vary. For example, $3 \div \frac{1}{2} = 6$. **14.** $\frac{1}{24}$ **15.** $-\frac{3}{4}$ **16.** $15y - 28$ **17.** 18 qt

18. −3 **19.** −15 **20.** −7 **21.** 75 **22.** $\frac{2}{3}$ **23.** (−3, 1) **24.** $-y^2 - 11y + 16$ **25.** $-27a^{12}b^6$

26. $x^3 - 3x^2 - 6x + 8$ **27.** $a^2 - 2a + 6$ **28.** $2a^2(3a + 2)(a + 3)$ **29.**

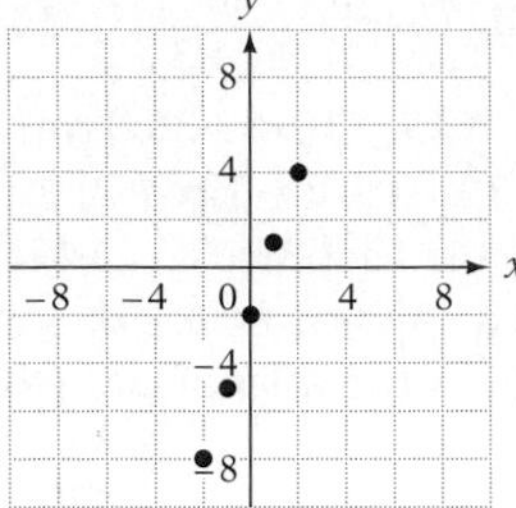

30. The coordinates of the vertices of the new rectangle are (−5, 2), (1, 2), (1, −6), (−5, −6).

31. −10, −9, −8, −7, −6, −5, −4 **32.**

Stem	Leaves
8	4, 3, 3
7	6, 7, 0, 0, 8, 0, 7, 1, 8, 3
6	8, 9, 9, 9, 4, 7, 7, 1, 6, 9
5	5, 7, 1, 7, 9
4	1

33. 213°C

Answers to Chapter 6 Exercises

6.1 EXERCISES

1. a. Perfect square; **b.** radicand; **c.** square root; **d.** 4.5; 20.25; **e.** 10 cm **3.** The radicand in $\sqrt{12}$ contains a factor that is a perfect square (4). **5.** All operations under the radical symbol must be performed as Step 1 in the Order of Operations Agreement (or as the first step in simplifying a numerical expression). **7. a.** 16.6; **b.** 16.3; **c.** 16.7 and 16.8; **d.** 16.9; **e.** answers will vary (between 16.1 and 16.2); **f.** answers will vary (between 16.5 and 16.6). **9.** 1 **11.** −1 **13.** 9 **15.** −8 **17.** $\sqrt{-49}$ is not a real number. **19.** 15 **21.** −2 **23.** 8 **25.** 14 **27.** 7 **29.** 48 **31.** −66 **33.** 65 **35.** −22 **37.** 21 **39.** 14 **41.** −30 **43.** 49 **45.** 39 **47.** 8 **49.** 6 **51.** 10 **53.** 3 **55.** −21 **57.** 2.6458 **59.** 4.3589 **61.** 15.6525 **63.** 8.5674 **65.** 45.8258 **67.** 23.2379 **69.** −18.0278 **71.** −87.3613 **73. a.** 6 ft/s; **b.** 12 ft/s; **c.** increase **75. a.** 54.8 mph; **b.** the sports car, 4.3 mph **77. a.** 2.1 s; **b.** Earth, 1.4 s; **c.** Figure A is the graph of $t = \sqrt{\frac{d}{16}}$; Figure B is the graph of $t = \sqrt{\frac{d}{43.4}}$. **79. a.** 1106.9 ft/s; **b.** increase **81. a.** −18°F; **b.** −31°F; **c.** 6°F; **d.** a change in the wind speed; **e.** answers will vary. **83.** $2\sqrt{3}$ **85.** $3\sqrt{2}$ **87.** $2\sqrt{11}$ **89.** $2\sqrt{14}$ **91.** $2\sqrt{7}$ **93.** $4\sqrt{6}$ **95.** $6\sqrt{2}$ **97.** $6\sqrt{3}$ **99.** $5\sqrt{7}$ **101.** $\sqrt{55}$ **103.** $44\sqrt{5}$ **105.** $-30\sqrt{5}$ **107.** $6\sqrt{30}$ **109.** $-48\sqrt{2}$ **111.** x^6 **113.** $c^5\sqrt{c}$ **115.** a^8 **117.** x^6y^4 **119.** $5y^4$ **121.** $xy^7\sqrt{xy}$ **123.** $a^7b^2\sqrt{ab}$ **125.** $4y^3\sqrt{3y}$ **127.** $6y^3\sqrt{2y}$ **129.** $12xy^4$ **131.** $4a^2b^7\sqrt{2ab}$ **133.** $6x^4y\sqrt{2xy}$ **135.** $4a^2b^3\sqrt{6ab}$ **137.** $25a^2b^3\sqrt{b}$ **139.** $xy^4\sqrt{x}$ **141.** $10ab^2\sqrt{3ab}$ **143.** $12x^2y^3\sqrt{2x}$ **145.** $9x^2y^3\sqrt{3y}$ **147. a.** 0.8; **b.** −0.9; **c.** $2\frac{1}{2}$; **d.** $-2\frac{1}{3}$ **149.** Answers will vary. $81 < n < 100$ **151. a.** 1; **b.** 3; **c.** 5; **d.** $3\sqrt{3}$; **e.** $3\sqrt{7}$

6.2 EXERCISES

1. a. Distributive; **b.** conjugates; **c.** Distributive Property; **d.** FOIL method; **e.** rationalizing the denominator. **3.** Answers will vary. **5.** By the Multiplication Property of One, we can multiply an expression by 1 without changing its value. $\frac{\sqrt{3}}{\sqrt{3}} = 1$. **7.** $3\sqrt{2}$ **9.** $-6\sqrt{5}$ **11.** $10\sqrt{x}$ **13.** $-3\sqrt{2a}$ **15.** $-7y\sqrt{3}$ **17.** $7\sqrt{2}$ **19.** $6\sqrt{x}$ **21.** $8\sqrt{5}$ **23.** $8\sqrt{2}$ **25.** $15\sqrt{2} - 10\sqrt{3}$ **27.** $\sqrt{x}$ **29.** $-12x\sqrt{3}$ **31.** $2xy\sqrt{x} - 3xy\sqrt{y}$ **33.** $-45\sqrt{2}$ **35.** $13\sqrt{3} - 12\sqrt{5}$ **37.** $32\sqrt{3} - 3\sqrt{11}$ **39.** $6\sqrt{x}$ **41.** $-34\sqrt{3x}$ **43.** $10a\sqrt{3b} + 10a\sqrt{5b}$ **45.** 5 **47.** 6 **49.** x **51.** x^3y^2 **53.** $3ab^6\sqrt{2a}$ **55.** $12a^4b\sqrt{b}$ **57.** $2 - \sqrt{6}$ **59.** $x - \sqrt{xy}$ **61.** $5\sqrt{2} - \sqrt{5x}$ **63.** $4 - 2\sqrt{10}$ **65.** $10abc$ **67.** $3a - 3\sqrt{ab}$ **69.** $15 - 8\sqrt{y} + y$ **71.** $x - 6\sqrt{x} + 9$ **73.** $3x - y^2$ **75.** $10x + 13\sqrt{x} + 4$ **77. a.** $8\sqrt{10}$ units; **b.** 40 square units **79. a.** $8\sqrt{5} + 4\sqrt{10}$ units; **b.** 40 square units **81.** $(6\sqrt{3} + 2\sqrt{15})$ cm **83.** 26.8 cm **85.** $\frac{9}{10}$ **87.** $\frac{4a}{b^4}$ **89.** $\frac{5\sqrt{2}}{2}$ **91.** $\frac{x\sqrt{3}}{3}$ **93.** $\frac{\sqrt{3}}{2}$ **95.** $\frac{\sqrt{6}}{3}$ **97.** $\frac{\sqrt{3}}{3}$ **99.** 4 **101.** $\frac{3\sqrt{x}}{x}$ **103.** 7 **105.** 3 **107.** $\frac{2\sqrt{7y}}{y}$ **109.** $\frac{1}{2}$ **111.** $\frac{y\sqrt{6}}{3}$ **113.** $2\sqrt{5}$ units **115.** $5\sqrt{2}$ **117. a.** Always true; **b.** always true; **c.** sometimes true; **d.** sometimes true

6.3 EXERCISES

1. b and c are radical equations. **3.** The resulting equation may have a solution that is not a solution of the original equation. **5.** Answers will vary. **7.** 25 **9.** 144 **11.** 5 **13.** 16 **15.** 8 **17.** The equation has no solution.

19. 6 **21.** 24 **23.** −1 **25.** $-\frac{2}{5}$ **27.** $\frac{4}{3}$ **29.** 15 **31.** 5 **33.** 1 **35.** 1 **37.** 1 **39.** 2
41. The equation has no solution. **43.** 19 m **45.** 6.75 ft **47.** 400 ft **49.** 68.75 ft **51.** 156.25 lb
53. 59.6°C **55. a.** Sometimes true; **b.** always true; **c.** sometimes true; **d.** sometimes true

6.4 EXERCISES

1. The right angle symbol must be at the 90° angle. The right angle must be labeled *C* and the hypotenuse labeled *c*. One acute angle should be labeled *A*, with the side opposite it labeled *a*. The other acute angle should be labeled *B*, with the side opposite it labeled *b*. **3.** No. The triangle is not a right triangle. **5.** If (x_1, y_1) and (x_2, y_2) are two points in the plane, then the distance between the two points is given by $d = \sqrt{(x_1 - x_2)^2 + (y_1 - y_2)^2}$. **7.** Yes **9.** 10.3 cm
11. 9.75 ft **13.** 13.9 mi **15.** The pitcher's mound is more than halfway between home plate and second base.
17. 1000 mi **19.** 19.8 in. **21.** Yes **23.** $2\sqrt{13}$ **25.** 9.1 **27.** 6.4 **29.** 6.7 **31. a.** 17.2 units;
b. 12 square units **33. a.** 20.9 units; **b.** 24 square units **35. a.** 27.5 units; **b.** 42 square units **37.** $r^2(\pi - 2)$
39. Answers will vary. **41.** Yes **43. a.** Sometimes true; **b.** sometimes true; **c.** always true; **d.** always true;
e. always true

6.5 EXERCISES

1. Use the Pythagorean Theorem, substituting *a* for *b* and solving for *c*. **3.** Answers will vary. **5.** Answers will vary.
7. $\frac{5\sqrt{2}}{2}$ ft **9.** $12\sqrt{2}$ m **11.** 14 in. **13.** 10 yd **15.** 4 in. and $4\sqrt{3}$ in. **17.** 8.5 ft **19.** 3200 ft^2 **21.** 47.867°
23. 76.317° **25.** 8.583° **27.** 4.717° **29.** 38.283° **31.** 70.6° **33.** 33°45′36″ **35.** 76°4′48″
37. 40°18′ **39.** 52°27′36″ **41.** 82°20′42″ **43.** 43°20′6″ **45.** 0.6820 **47.** 1.4281 **49.** 0.9971
51. 1.9970 **53.** 0.8878 **55.** 0.8453 **57.** 0.8508 **59.** 0.6833 **61.** 5.54 m **63.** 5.39 in. **65.** 26.50 cm
67. 15.92 m **69.** 38.6° **71.** 41.1° **73.** 21.3° **75.** 38.0° **77.** 72.5° **79.** 0.6° **81.** 66.1° **83.** 29.5°
85. 60° **87.** 30° **89.** 41.6° **91.** 44.4° **93.** 53.1° **95.** $\angle B = 60°, b = 26$ cm, $c = 30$ cm
97. $\angle A = 45°, b = 5$ m, $c = 7$ m **99.** $\angle A = 34°, \angle B = 56°, c = 14$ cm **101.** $\angle B = 48°, b = 11$ ft, $c = 15$ ft
103. $\angle A = 34°, \angle B = 56°, b = 7$ m **105.** $\angle A = 18°, b = 55$ in., $c = 58$ in. **107.** $\angle A = 24°, \angle B = 66°$, $b = 27$ ft **109.** $\angle A = 50°, b = 3$ cm, $c = 5$ cm **111.** 39.3 ft **113.** 13.7 m **115.** 192.0 ft **117.** 114.3 ft
119. 147.2 m **121. a.** 0.7451; **b.** 0.7133; **c.** 0.2186 **123.** Explanations will vary. **125.** 91.9 cm^2

CHAPTER REVIEW EXERCISES

1. 5 **2.** −11 **3.** 24 **4.** −4 **5.** $20\sqrt{3}$ **6.** $-6\sqrt{30}$ **7.** c^9 **8.** $6x^2y^2\sqrt{y}$ **9.** $2y^4\sqrt{6}$
10. $9x^2y^2\sqrt{2y}$ **11.** −21 **12.** 20.5670 **13.** $7\sqrt{7}$ **14.** $4x\sqrt{5}$ **15.** $26\sqrt{3x}$ **16.** $-4a^2\sqrt{5a}$ **17.** 10
18. $a^5b^3c^2$ **19.** 3 **20.** $3a\sqrt{2} + 2a\sqrt{3}$ **21.** $2a + 13\sqrt{a} + 15$ **22.** $8y + 2\sqrt{5y} - 5$ **23.** $5ab - 7$
24. $\frac{16\sqrt{a}}{a}$ **25.** 5 **26.** $3a$ **27.** $\frac{4\sqrt{2x}}{x}$ **28.** 20 **29.** The equation has no solution. **30.** 6
31. 3 **32.** 3 **33.** 83.75° **34.** 72°28′48″ **35.** 0.7771 **36.** 35.0° **37.** 40 mph **38.** 25 ft
39. a. 4.8 s; **b.** Mars, 1.8 s; **c.** Figure A is the graph of $t = \sqrt{\frac{d}{6.45}}$; Figure B is the graph of $t = \sqrt{\frac{d}{16}}$. **40.** 100 ft
41. a. $14\sqrt{5}$ units; **b.** 60 square units **42.** 14.25 cm **43.** 26.25 ft **44.** 5.1 units **45. a.** 17.2 units;
b. 12 square units **46.** $4\sqrt{2}$ m **47.** 6 cm and $6\sqrt{3}$ cm **48.** $b = 8$ m, $\angle A = 37°, \angle B = 53°$
49. $\angle B = 65°, a = 2$ in., $c = 4$ in. **50.** 26.6 ft

CUMULATIVE REVIEW EXERCISES

1. a. $\{-6, -4, -3, -2, 0, 2, 3, 4, 6\}$; **b.** $\{0\}$ **2.** Answers will vary. For example, 16. **3.** 23 **4.** $3b^2 + (b - 9)$ **5.** -9 **6.** -28 **7.** 7 **8.** $0.\overline{4}$ **9.** $\frac{1}{2}$ **10.** 2 gallons **11.** $-2a$ **12.** $-14x - 19y$ **13.** 12.5% **14.** 10 **15.** -1 **16.** $(2, -1)$ **17.** $9x^2 + 3$ **18.** $-6a^7b^4$ **19.** $\frac{16y^{12}}{x^{20}}$ **20.** $3x^3 + 7x^2 - 21x - 4$ **21.** $(2x - 5)(x + 4)$ **22.** $2(b + 3)(b - 3)$ **23.** $4a^2b^5\sqrt{2ab}$ **24.** $-5\sqrt{2}$ **25.** $y + 8\sqrt{y} + 15$ **26.** $\sqrt{2}$ **27.** 8 **28.** 57.0° **29.** Points at $(-2, 7), (-1, 4), (0, 1), (1, -2), (2, -5)$ **30.** $(-2, -1), (4, -1), (2, -6), (-4, -6)$ **31.** Dana, utility; Leslie, automotive; Chris, computer; Pat, oil **32.** 1.6 **33.** $2x - 8 = 4(x + 6); -16$

Answers to Chapter 7 Exercises

7.1 EXERCISES

1. A quadratic equation has a second-degree term. **3.** If $ab = 0$, then $a = 0$ or $b = 0$. **5.** One of the two equal roots of a quadratic equation. **7.** $a = 1, b = 2, c = -5$ **9.** $a = 4, b = 0, c = 1$ **11.** $a = -1, b = 7, c = 0$ **13.** $2x^2 - 4x + 1 = 0$ **15.** $x^2 - 4x - 5 = 0$ **17.** $4x^2 - 8x + 1 = 0$ **19.** $-1, -2$ **21.** $\frac{2}{3}, 1$ **23.** $\frac{1}{3}$ **25.** $-4, 3$ **27.** $-\frac{2}{3}, \frac{2}{3}$ **29.** $-3, 6$ **31.** $0, 12$ **33.** ± 6 **35.** $\pm\frac{3}{4}$ **37.** $-3, -7$ **39.** $-\frac{3}{2}, -\frac{9}{2}$ **41.** No real number solution **43.** 9% **45.** Yes ($v \approx 73.5$ mph) **47.** 4.93 feet **49.** ± 2 **51.** $\pm\sqrt{5}$ **53. a.** $x^2 - 6x + 8 = 0$; **b.** $x^2 + x - 6 = 0$; **c.** $8x^2 - 10x + 3 = 0$; **d.** $x^2 - 2 = 0$

7.2 EXERCISES

1. Answers will vary. **3.** Divide both sides of the equation by 2, the coefficient of x^2. **5.** $x^2 + 12x + 36$; $(x + 6)^2$ **7.** $x^2 + 10x + 25$; $(x + 5)^2$ **9.** $x^2 - x + \frac{1}{4}$; $\left(x - \frac{1}{2}\right)^2$ **11.** $-3, 1$ **13.** $-2 \pm \sqrt{3}$ **15.** No real number solution **17.** $-3 \pm \sqrt{14}$ **19.** 2 **21.** $1 \pm \sqrt{2}$ **23.** $\frac{-3 \pm \sqrt{13}}{2}$ **25.** $-8, 1$ **27.** $-3 \pm \sqrt{5}$ **29.** $4 \pm \sqrt{14}$ **31.** $1, 2$ **33.** $\frac{3 \pm \sqrt{29}}{2}$ **35.** $\frac{1 \pm \sqrt{5}}{2}$ **37.** $\frac{5 \pm \sqrt{13}}{2}$ **39.** $\frac{-1 \pm \sqrt{13}}{2}$ **41.** $-5 \pm 4\sqrt{2}$ **43.** $\frac{-3 \pm \sqrt{5}}{2}$ **45.** $\frac{1 \pm \sqrt{17}}{2}$ **47.** No real number solution **49.** $1, \frac{1}{2}$ **51.** $-3, \frac{1}{2}$ **53.** $\frac{1 \pm \sqrt{2}}{2}$ **55.** $\frac{2 \pm \sqrt{5}}{2}$ **57.** $3 \pm \sqrt{2}$ **59.** $1 \pm \sqrt{13}$ **61.** $\frac{3 \pm 3\sqrt{3}}{2}$ **63.** $4 \pm \sqrt{7}$ **65.** Length: 40 feet; width: 16 feet **67.** No. The ball is too high by about 2.4 feet. **69.** -3 **71.** $-10, 10$ **73.** $-4\sqrt{3}, 4\sqrt{3}$ **75.** $-28, 28$ **77.** $-2, 3$ **79.** $-4, 2$ **81.** $-1 + \sqrt{6}, -1 - \sqrt{6}$

7.3 EXERCISES

1. A formula that is used to solve a quadratic equation and given by $x = \frac{-b \pm \sqrt{b^2 - 4ac}}{2a}$. **3.** Answers will vary. **5.** $-5, 2$ **7.** $-5, 1$ **9.** $2 \pm \sqrt{10}$ **11.** $-2 \pm \sqrt{3}$ **13.** No real number solution **15.** $4 \pm \sqrt{19}$ **17.** $0, -\frac{1}{2}$ **19.** $\frac{2 \pm \sqrt{5}}{2}$ **21.** $\pm\frac{5}{2}$ **23.** $\frac{3 \pm 2\sqrt{6}}{5}$ **25.** $\frac{7 \pm \sqrt{17}}{8}$ **27.** No real number solution **29.** $\frac{1 \pm \sqrt{21}}{10}$ **31.** $-\frac{2}{3}, 4$ **33.** $\frac{3 \pm \sqrt{129}}{12}$ **35.** $\frac{-3 \pm \sqrt{65}}{4}$ **37.** $\frac{7 \pm \sqrt{89}}{2}$ **39.** $1, -\frac{3}{4}$ **41.** $-2 \pm 2\sqrt{3}$ **43.** $\frac{5 \pm 2\sqrt{5}}{2}$ **45.** No real number solution **47.** $\frac{1 \pm \sqrt{3}}{3}$ **49.** $\frac{1}{2}, -\frac{4}{3}$ **51.** $\frac{3 \pm 2\sqrt{2}}{2}$ **53.** $\frac{-3 \pm 2\sqrt{3}}{2}$ **55.** -1 **57.** $\frac{5 \pm \sqrt{42}}{3}$ **59.** $1 \pm \sqrt{21}$ **61.** $\frac{-2 \pm \sqrt{10}}{3}$ **63.** $5 \pm \sqrt{13}$ **65.** 1.88 seconds **67.** 12.7 inches **69. a.** The solutions are equal; **b.** there are no real solutions; **c.** there are two real, unequal solutions.

7.4 EXERCISES

1. A function given by $f(x) = ax^2 + bx + c, a \neq 0$. **3.** If $a = 0$, the function would be a linear function instead of a quadratic one. **5.** Up **7.** Up **9.** Up **11.** -3 **13.** 1 **15.** $(0, 0), x = 0$ **17.** $(0, 1), x = 0$ **19.** $(0, 0), x = 0$ **21.** $(0, 1), x = 0$ **23.** $(3, -7), x = 3$ **25.** $(-2, -4), x = -2$ **27.** $(-\frac{3}{2}, \frac{11}{2}), x = -\frac{3}{2}$ **29.** $\left(\frac{5}{4}, -\frac{9}{8}\right), x = \frac{5}{4}$ **31.** $\left(-\frac{1}{4}, \frac{3}{4}\right), x = -\frac{1}{4}$ **33.** $(2, 9)$; $(5, 12)$

CHAPTER REVIEW EXERCISES

1. 4, −4 **2.** $\frac{1 \pm \sqrt{13}}{2}$ **3.** $\frac{3 \pm \sqrt{29}}{2}$ **4.** $\pm\frac{5}{7}$ **5.** $5 \pm \sqrt{23}$ **6.** $-\frac{1}{2}, -\frac{1}{3}$ **7.** No real number solution **8.** $\frac{-10 \pm 2\sqrt{10}}{5}$

9. $2 \pm \sqrt{3}$ **10.** 6, −5 **11.** $\frac{4}{3}, -\frac{7}{2}$ **12.** $\pm 2\sqrt{10}$ **13.** $\frac{2 \pm \sqrt{7}}{3}$ **14.** $1 \pm \sqrt{11}$ **15.** 3, 9 **16.** 16, −2

17. Vertex: (0, 1); axis of symmetry: $x = 0$ **18.** Vertex: (2, −1); axis of symmetry: $x = 2$ **19.** 1, −9 **20.** $\frac{-4 \pm \sqrt{23}}{2}$

21. $-\frac{1}{6}, -\frac{5}{4}$ **22.** No real number solution **23.** $4 \pm \sqrt{21}$ **24.** $\frac{4}{7}$ **25.** 3, 2 **26.** $-2, \frac{1}{2}$ **27.** $3 \pm \sqrt{5}$

28. $-4 \pm \sqrt{19}$ **29.** Vertex: (−1, −3); axis of symmetry: $x = -1$

30. Vertex: $\left(-\frac{3}{2}, \frac{39}{4}\right)$; axis of symmetry: $x = -\frac{3}{2}$ **31.** (8, 3) **32.** (0, 4)

33. 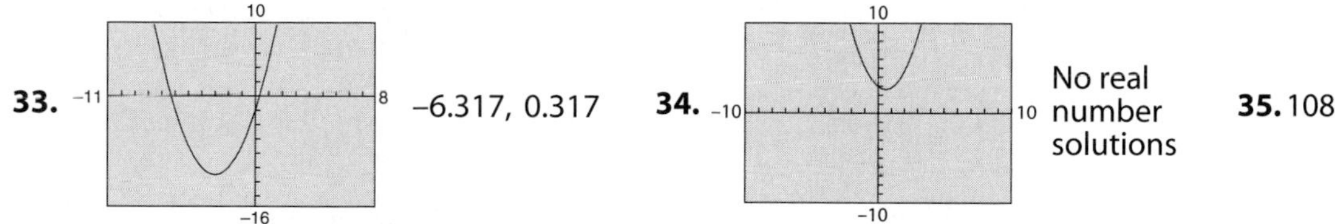

−6.317, 0.317 **34.** No real number solutions **35.** 108

36. 7.66 feet per second **37.** 2.84 feet **38.** 1.74 seconds, 10.76 seconds **39.** The graph does not cross the x-axis. Therefore, there are no real numbers x for which y is zero.

CUMULATIVE REVIEW EXERCISES

1. 35 **2.** Answers will vary. **3.** $-16y + 40$ **4.** $\frac{3}{2}$ **5.** (3, 0) and (0, −4) **6.** $\frac{4}{13}$ **7. a.** Canada and Italy; **b.** U.S.A.; **c.** France **8.** $y = -\frac{4}{3}x - 2$ **9.** 26 **10.** Domain: {−1, 0, 3, 4, 5}; range: {−1, 1, 3, 4}; yes **11.** $\frac{1}{2}, 3$ **12.** $\frac{2 \pm \sqrt{19}}{3}$

13. $-\frac{4a}{3b^2}$ **14.** (2, −2) **15.** $a - 2$ **16.** 5 **17.** $2 \pm 2\sqrt{3}$ **18.** $\{x \mid x \geq 3\}$ **19.** 5, 0, −4, −3, 5;

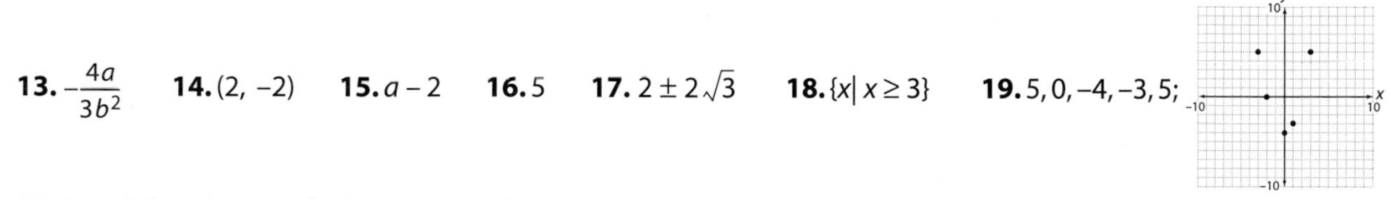

20. (−1, 1); (5, −3) **21.** \$2.25 **22.** 79 feet **23.** Plane, 200 mph; wind, 40 mph **24.** 31.62 meters
25. Mean: 4.6; standard deviation: 0.32

Answers to Chapter 8 Exercises

8.1 EXERCISES

1. a. Always true; **b.** never true; **c.** never true; **d.** always true; **e.** never true; **f.** never true; **g.** sometimes true

3. a. $\frac{x^2-6x-7}{x^2-x-30}$ **b.** $\frac{8yz-32z}{3y^2z-12yz}$ **c.** $\frac{1}{45ab^2}$ **5.** Answers will vary. **7.** 0 **9.** $\frac{3}{2}$ **11.** 2, –5

13. 4, –4 **15.** –3 **17.** $-\frac{2}{3}, 4$ **19.** $\frac{2x}{3y^2}$ **21.** $\frac{2x^2}{3}$ **23.** $\frac{x}{2}$ **25.** $\frac{y}{3}$ **27.** $\frac{x+2}{x+5}$ **29.** $\frac{a+8}{a+7}$

31. $\frac{x+4}{x-6}$ **33.** $\frac{y-2}{y-5}$ **35. a.** Column 1: $-0.25, -0.\overline{3}, -0.5, -1$, undefined, $1, 0.5, 0.\overline{3}, 0.25$, undefined, $0.1\overline{6}$;

Column 2: $-0.25, -0.\overline{3}, -0.5, -1$, undefined, $1, 0.5, 0.\overline{3}, 0.25, 0.2, 0.1\overline{6}$; **b.** explanations will vary. **37.** $\frac{2}{3xy}$

39. $\frac{8xy^2ab}{3}$ **41.** $\frac{2}{9}$ **43.** $\frac{y^2}{x}$ **45.** $\frac{y(x+4)}{x(x+1)}$ **47.** $\frac{x^3(x-7)}{y^2(x-4)}$ **49.** $\frac{x+5}{x+4}$ **51.** $\frac{x+2}{x+4}$ **53.** $\frac{2xy^2ab^2}{9}$

55. $\frac{5}{12}$ **57.** $3x$ **59.** $\frac{y(x+3)}{x(x+1)}$ **61.** $\frac{x(x-2)}{y(x-6)}$ **63.** $\frac{(x+6)(x-3)}{(x+7)(x-6)}$ **65.** $\frac{2x+3}{x-6}$ **67.** $\frac{xy}{2}$ **69.** $\frac{72x}{y}$

71. $\frac{b}{a-3}$ **73.** $\frac{(x+6)(x+6)}{(x+7)(x+3)}$ **75.** $\frac{x-4}{x+3}$ and $\frac{3x-1}{2x+1}$, or $\frac{x-4}{2x+1}$ and $\frac{3x-1}{x+3}$ **77.** $m = n + 2$

8.2 EXERCISES

1. a. Never true; **b.** always true; **c.** never true; **d.** always true **3. a.** Correct; **b.** the LCM is x^8;
c. $\frac{x+3}{x+2}$ is in simplest form; **d.** correct; **e.** $\frac{4x-3}{x} - \frac{3x+1}{x} = \frac{4x-3-(3x+1)}{x}$ **5.** $12y^2$ **7.** $18a^2b^3$

9. $36a^2b^3$ **11.** $(x+4)(x-3)$ **13.** $\frac{4}{ab}$ **15.** $\frac{x-2}{x+6}$ **17.** $-\frac{5y}{4y+1}$ **19.** $\frac{7x-5}{2x-7}$ **21.** $\frac{3x-1}{x-10}$ **23.** $\frac{-3n+3}{3n+4}$

25. $\frac{3}{x+5}$ **27.** $\frac{1}{x-1}$ **29.** $\frac{1}{x-4}$ **31.** $\frac{7b+5a}{ab}$ **33.** $\frac{11}{12a}$ **35.** $\frac{11}{12y}$ **37.** $\frac{7y+120}{20y^2}$ **39.** $\frac{14x+3}{12x}$

41. $\frac{8x-3}{6x}$ **43.** $\frac{7}{36}$ **45.** $\frac{7-2n^2}{2n}$ **47.** $\frac{x^2+2x+3}{x+2}$ **49.** $\frac{4x+2}{x+1}$ **51.** $\frac{7x-23}{(x-3)(x-4)}$ **53.** $\frac{-y-33}{(y+6)(y-3)}$

55. $\frac{3x^2+20x-8}{(x-4)(x+6)}$ **57.** $\frac{8}{x}$ of the wall **59.** $\frac{4}{h}$ of the job **61.** $\frac{225}{4c}$ of the job **63.** $\frac{29}{3x}$ in.

65. $\frac{20r}{(r+2)(r-2)}$ h **67.** The LCM is equal to their product when the two expressions have no common factors.

69. $\frac{2y^2+9y+1}{y+5}$ **71.** $\frac{6}{y} + \frac{7}{x}$ **73.** $\frac{2}{m^2n} + \frac{8}{mn^2}$

8.3 EXERCISES

1. A fraction whose numerator or denominator contains one or more fractions. Examples will vary.

3. a. Sometimes true; **b.** never true; **c.** always true **5.** $\frac{x}{x-4}$ **7.** $\frac{5x}{x+2}$ **9.** $\frac{4x-13}{2(3x-10)}$ **11.** $\frac{5x-37}{2x-17}$

13. $-3x$ **15.** $\frac{x+4}{2x+1}$ **17.** $\frac{x+2}{x-4}$ **19.** $\frac{a-3}{a+5}$ **21.** $\frac{x+6}{x+7}$ **23.** $y-1$ **25.** $\frac{1}{3x-4}$ **27.** $\frac{a-1}{a+1}$

29. $\frac{x-6}{x-2}$ **31.** $\frac{2x+7}{2x+5}$ **33.** $-\frac{7}{5}$ **35.** $-\frac{1}{y}$ **37.** $\frac{y+4}{2(y-2)}$

8.4 EXERCISES

1. a. Always true; **b.** always true; **c.** sometimes true; **d.** never true; **e.** never true **3.** $\frac{x}{5}$ of the job

5. $\frac{t}{30} + \frac{t}{20} = 1$ **7. a.** 2; **b.** Figure 1. Explanations will vary. **9.** 4 **11.** −3 **13.** $\frac{3}{4}$ **15.** −1
17. No solution **19.** −1, 6 **21.** $\frac{2}{3}$, 4 **23.** 12 min **25.** 2 h **27.** 6 min **29.** 15 h **31.** 10 min
33. 6 h **35.** 3 h **37.** 28 h **39.** 5 mph **41.** 360 mph **43.** 7.5 mph **45.** 20 mph
47. Freight train: 30 mph; express train: 50 mph **49.** 48 mph **51.** 50 mph **53.** 5 mph **55.** $\frac{3}{4}$ or $\frac{4}{3}$
57. $\frac{7}{10}$ **59.** 12 weeks **61.** Explanations will vary.

8.5 EXERCISES

1. a and d are literal equations because they are equations that contain more than one variable. b is not a literal equation because it contains only one variable. c is not an equation. **3. a.** Sometimes true;
b. always true; **c.** always true **5.** Add D to each side of the equation; $P = A + D$ **7.** Multiply each side of the equation by N; $A = SN$ **9.** $R = \frac{E}{I}$ **11.** $T = \frac{PV}{nR}$ **13.** $C = R - P$ **15.** $h = \frac{3V}{A}$ **17.** $W = \frac{P - 2L}{2}$
19. $F = \frac{9}{5}C + 32$ **21.** $R = Pn + C$ **23.** $m = \frac{T}{f - g}$ **25.** $S = \frac{a}{1 - r}$ **27. a.** $S = \frac{F + BV}{B}$; **b.** \$180; **c.** \$75
29. a. $R_1 = \frac{RR_2}{R_2 - R}$; **b.** 20 ohms; **c.** 10 ohms **31.** This is the graph of an ellipse with center at the origin, through (2, 0), (−2, 0), (0, 8), and (0, −8). **33.** This is the graph of a circle with center at the origin and a radius of 9 units.

8.6 EXERCISES

1. a. Answers will vary; **b.** answers will vary; **c.** answers will vary. **3.** Answers will vary.
5. a. Always true; **b.** sometimes true; **c.** always true; **d.** always true; **e.** sometimes true **7.** 15
9. 28.8 **11.** 3 **13.** 7 **15.** 2 **17.** 3 **19.** 2, 4 **21.** $-\frac{3}{2}$, 4 **23.** 1.75 cords **25.** 180 panels
27. 40 ft^2 **29. a.** 17,570 inmates; **b.** Arizona; **c.** 98,511 prisoners **31. a.** 4619 deaths; **b.** Minnesota;
c. 0.6%; **d.** no; answers will vary; **e.** 1476 deaths; **f.** answers will vary. **33.** 6.7 cm **35.** 2.9 m
37. 22.5 ft **39.** 48 m^2 **41.** 6.25 cm **43.** 6 in. **45.** 13 cm **47.** 14.375 ft **49.** 82.5 ft
51. 240 foul shots **53.** \$4 million **55.** Explanations will vary.

8.7 EXERCISES

1. Compound events are events characterized by using the words *or* and *and*. **3.** Answers will vary. **5. a.** No; **b.** yes
7. a. $\frac{1}{2}$; **b.** $\frac{4}{13}$ **9. a.** $\frac{2}{9}$; **b.** $\frac{1}{9}$ **11.** The probabilities are the same. **13. a.** 0.024; **b.** 0.024; **c.** 0.048
15. 0.9604 **17.** $\frac{2}{15}$ **19. a.** 0.345; **b.** 0.243; **c.** 0.321; **d.** 0.215; **e.** 0.078 **21.** $\frac{1}{16}$ **23.** Answers will vary.

CHAPTER REVIEW EXERCISES

1. $\frac{x - 9}{x - 3}$ **2.** $\frac{x + 3}{x - 4}$ **3.** $\frac{x - 4}{x + 3}$ **4.** $\frac{2}{ab}$ **5.** $\frac{1}{x + 3}$ **6.** $\frac{1}{x + 3}$ **7.** $\frac{6b + 9a}{ab}$ **8.** $\frac{7x + 22}{60x}$

9. $\frac{4x + 3}{6x}$ **10.** $\frac{x - 3}{x - 2}$ **11.** No solution **12.** 5 **13.** $c = \frac{100m}{i}$ **14.** $t = \frac{f - v}{a}$ **15.** 15 **16.** −1

17. a. -3; **b.** $-6, 1$ **18.** -3 **19.** 15 cm **20.** 12.8 ft **21.** $\frac{9}{2x}$ of the job **22.** 6 h **23.** 20 mph **24.** 45 mph **25.** 8 in. **26.** New York; 2480 more cases **27. a.** $\frac{3}{4}$; **b.** $\frac{1}{3}$ **28.** $\frac{1}{16}$ **29.** $\frac{1}{36}$

CUMULATIVE REVIEW EXERCISES

1. $\frac{1}{300}$ **2.** $3t-(t+6)$ **3.** 28 **4.** 1 **5.** -16 **6.** $-3a^2b$ **7.** 16 **8.** -12 **9.** $(8, 0)$ and $(0, -12)$ **10.** $(\frac{2}{3}, 3)$ **11.** $-6x^4y^5$ **12.** $\frac{a^3}{b^2}$ **13.** $(y-6)(y-1)$ **14.** $(4x+1)(3x-1)$ **15.** 0.000008921 **16.** $20\sqrt{10}$ **17.** $11x^4y$ **18.** 16 **19.** $3\sqrt{2}-\sqrt{3x}$ **20.** $38\sqrt{3a}$ **21.** 14 **22.** $-\frac{1}{2}, -\frac{1}{3}$ **23.** $2+\sqrt{3}$ and $2-\sqrt{3}$ **24.** 1 **25.** $\frac{3}{x+8}$ **26.** $-3, 6$

27.

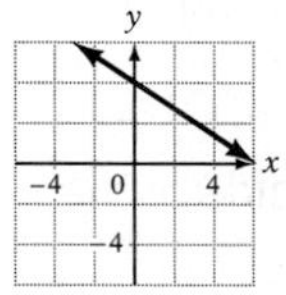

28.

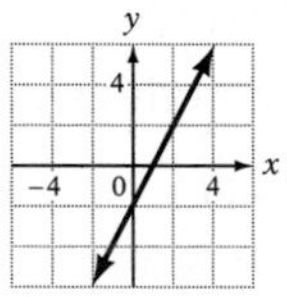

29.

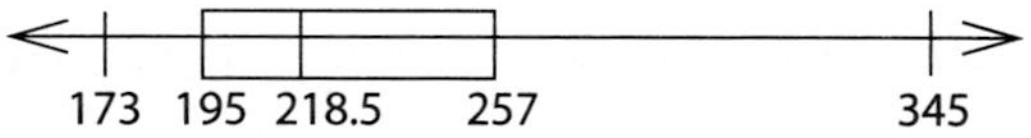

30. -2.5 **31.** 20 lb **32.** $(9x^2+6x+1)\text{ m}^2$ **33.** $\angle A = 54^\circ$, $\angle B = 36^\circ$, $c = 14$ m

Index